MATHEMATICAL MODELLING

of
MATERIALS PROCESSING OPERATIONS

# of MATERIALS PROCESSING OPERATIONS

Edited by
J. Szekely, L. B. Hales, H. Henein,
N. Jarrett, K. Rajamani,
and I. Samarsekera

Proceedings of a symposium sponsored by the
Metallurgical Society, Inc. held in Palm Springs,
California, USA November 29 through December
2, 1987 at the fifth Extractive and Process
Metallurgical Fall Meeting.

A Publication of  The Metallurgical Society, Inc.

A Publication of The Metallurgical Society, Inc.
420 Commonwealth Drive
Warrendale, Pennsylvania 15086
(412) 776-9000

Printed in the United States of America.
Library of Congress Catalogue Number 87-42884
ISBN NUMBER 0-87339-071-7

PREFACE

The papers collected in this volume were presented at the fifth Extractive and Process Metallurgy Fall Meeting sponsored by The Metallurgical Society, Inc. at Palm Springs, California, USA November 29 through December 2, 1987.

Despite recent declines in the more developed industrial nations, the processing of materials to produce commercial chemicals, metals, and fabricated products is, and will continue to be, an important element of their economy. Nearly all of the classic processing techniques in use, inspite of a few recent improvements, were basically developed many years ago through empirical correlations resulting from trial and error operations research well before high-speed computation much basic thermodynamic and physical properties data were readily available. Today, economi optimization, as well as the development of novel process technology and waste recycling techniques, is necessary for industry survival in highly competitive world markets where some of the participants already enjoy significant advantages in ore concentration and labor or energy costs. Necessary developments can only be implemented if processes are sufficiently well understood to be expressed as accurate mathematical models, relating the controllable process variables to efficiency and product properties. Futher advances, permitting flexible computer integrated manufacturing (CIM), automation and close process control, are only possible with a quantitative appreciation of fundamental physical relationships.

In an effort accelerate future development in mathematical modeling, this conference proceedings publication takes its place with other recent Metallurgical Society publications on modern modeling techniques such as Computer Modeling of Phase Diagrams, edited by L.H. Bennett and Modeling of Casting and Welding Processes edited by S. Kou and R. Mehrabian as milestone contributions to the metallurgical literature in this field.

Rather than concentrating narrowly on one subject, this symposia attempts to give a broad overview of the state-of-the-art of modeling of metals processing operations ranging from mineral processing and extractive metallurgy to solid forming and shaping of final products and includes the necessary development of sensors and instrumentation for process control as well as process economics.

Mathematical modeling is important to all aspects of material processing as evidenced by its rapid growth and in this symposia we have been fortunate to have many of the most important areas represented.

The editors and organizing committee wish to thank the plenary speakers, authors and luncheon and dinner speakers who all contributed valuable technical information presented at the meeting and contained in this book. We also wish to express our sincere appreciation to the very competent staff of The Metallurgical Society who assisted us and alleviated the many chores inherent in organizing such a symposium.

Finally, thanks are due to the organizing committee and session chairman without which such a symposium would be impossible.

Julian Szekely
Massachusetts Institute of Technology
Cambridge, Massachusetts

Hani Henein
Carnegie-Mellon University
Pittsburgh, Pennsylvania

Noel Jarrett
Alcoa Laboratories
Alcoa Center, Pennsylvania

November 1988

GENERAL MEETING CHAIRMAN

Julian Szekely
Massachusetts Institute of Technology

ORGANIZING COMMITTEE

Hani Henein
Carnegie Mellon University

Noel Jarrett
Alcoa Laboratories

I.E. Samesekera
University of British Columbia

Lynn B. Hales
Kennecott

K. Rajamani
University of Utah

TABLE OF CONTENTS

FERROUS

NONFERROUS

CASTING

REHEAT AND DEFORMATION

ECONOMICS

PLENARY

MATERIALS-BASED DEFORMATION PROCESS MODELS

O. Richmond

Alcoa Laboratories
Alcoa Center, PA   15069

Abstract

Deformation processes like rolling, forging, extrusion and sheet
forming alter not only the shape of a material, but also the internal and
surface structures and their associated properties. Sometimes the material
is changed so drastically that failure occurs during the process. Accurate
mathematical models of such processes therefore require constitutive equations which describe not only the dependence of mechanical response on
current material parameters, but also the evolution of these material parameters with thermomechanical history. In addition, they require failure
criteria which define the limits of deformation. Examples of constitutive
equations and failure criteria are described which include the evolution of
material parameters like hardness, porosity and crystallographic texture.
In addition, examples of complete process models are described where the
constitutive equations are combined with the classical conservation laws,
and with appropriate boundary conditions. The boundary conditions must
accurately describe transfer of heat and force at free surfaces and at
tool/workpiece interfaces in order for the models to be valid. When this
is done the models are capable of simulation of the evolution of the inhomogeneous strains and structures in actual processes. This requires
sophisticated computational methods like finite elements and large scale
computers. Finally, some remarks are made concerning the needs for further
research.

3

Introduction

The properties of material products depend upon many aspects of material structure, such as the configurations of phases, crystals, pores, precipitates, microcracks and dislocations, all of which can be altered by the thermomechanical histories that are imposed during processing. Thus, in addition to determining the geometry of products, processes also determine the material structure. Optimum process design involves the selection of the heating, cooling and mechanical force histories which optimize product properties, through control of the internal surface structure, while achieving a desired shape.

Until recently process design has been mainly empirical, relying on experience and on trial and error. Even when mathematical models were used, they were based upon material models which ignored the effects of thermomechanical history on properties, and they were limited to two-dimensional approximations like plane stress, plane strain and axial symmetry. But computers now have changed all that. It has become possible to use much more realistic models which include the effects on properties of thermomechanical history, and supercomputers are bringing full three-dimensional modeling within reach. In other words, it has become possible to construct realistic, materials-based mathematical models for use on the quantitative design, control and analysis of processes. The status and expectations of this development are the subjects of this presentation.

The scope will be limited to deformation processes although the general concepts apply to other processes as well, like casting, consolidation and heat treatment. It also will be limited primarily to aspects in which I have had first-hand knowledge or involvement. The intent is not to be comprehensive, but rather to illustrate the essential content, status and expectations of the subject.

General Aspects of Deformation Process Models

In general, a product undergoing deformation processing consists of two types of representative material elements, one in the interior and one at the surface. During processing these elements experience local thermomechanical histories which differ for different elements. Mathematical equations which describe the response of typical elements to arbitrary thermomechanical histories are called constitutive equations. These are either algebraic equations or ordinary differential equations expressing changes only with time because they describe the evolution of strain and structure under spatially uniform thermomechanical histories. The special variation of strain and structure is obtained by combining the constitutive equations with the laws of mass, momentum and energy conservation to form a set of partial differential equations, and then to subject these to boundary conditions which describe the geometry and the heat and force transfer at the free surfaces and tool/workpiece interfaces of a specific process.

· The accurate mathematical description of the heat transfer at free surfaces and of the heat transfer and friction at tool/workpiece interfaces is an essential ingredient of the total model. The development of such descriptions has generally not kept pace with the development of constitutive equations. Thus, this is a key area for current research.

The constitutive equations, classical physical laws and boundary conditions give sufficient information for a complete mathematical model. To be complete from a physical viewpoint, however, the resulting model must also be examined to ensure that the material would not fail during the process. Thus, appropriate failure criteria are also an essential part of a complete process model.

Constitutive Equations

Typical constitutive models have two aspects: state equations and structure evolution equations. State equations describe the relationship of the current flow stress, σ, to the current strain rate, $\dot{\varepsilon}$, and temperature, T, as well as to the current values of a number of internal variables, s_1, s_2, representing the current material structure. Thus,

$$\sigma = f \ (\dot{\varepsilon}, T, s_1, s_2, \ldots). \tag{1}$$

In general, the stress, strain rate and some of the internal variables will be tensor quantities. Here, however, for simplicity, we shall consider only uniaxial stress states. Thus, σ, represents the non-zero stress component and, $\dot{\varepsilon}$, represents the corresponding component of strain rate. Also, in this discussion elastic effects will be ignored so that no distinction is made between total strain rate and its inelastic part.

In addition to the state equation (1), a complete constitutive model of material behavior must include evolution equations describing the changes in the structure with thermomechanical history. Thus,

$$\left.
\begin{aligned}
\dot{s}_1 &= g_1 \ (\dot{\varepsilon}, T, s_1, s_2, \ldots) \\[2mm]
\dot{s}_2 &= g_2 \ (\dot{\varepsilon}, T, s_1, s_2, \ldots) \\[2mm]
&\ldots\ldots\ldots
\end{aligned}
\right\} \tag{2}$$

Here the superposed dot represents time differentiation and s_1, s_2, etc. represent structural parameters like matrix hardness (dislocation structure) and microporosity.

More primitive constitutive equations, which have been used in earlier process models, do not include the evolution of structure. For example, the perfectly-plastic constitutive equation used for cold forming assumes that the hardness is constant. Thus, equation (1) becomes,

$$\sigma = Y, \text{ a material constant.} \tag{3}$$

Also, the simple nonlinear viscous (or creep) constitutive equation used for hot forming processes assumes that the hardness is constant, but that the stress is dependent on strain rate and temperature. Thus, equation (1) becomes,

$$\sigma = f \ (\dot{\varepsilon}, T). \tag{4}$$

The simplest constitutive model which includes structure evolution, though it is not commonly described in this manner, is the strain hardening model for cold forming. In this model the hardness is assumed to depend on prior strain. Thus, equations (1) and (2) give,

$$\left.\begin{array}{l} \sigma = f\ (s_1) \\[2em] \dot{s}_1 = g_1\ (s_1)\ \dot{\varepsilon} \end{array}\right\} \tag{5}$$

The analogous model for hot forming gives:

$$\left.\begin{array}{l} \sigma = f\ (\dot{\varepsilon}, T, s_1) \\[2em] \dot{s}_1 = g_1\ (\dot{\varepsilon}, T, s_1) \end{array}\right\} \tag{6}$$

A recent example of this type of model with considerable experimental validation on commercial purity aluminum has been given by Sample and Lalli [1]. It is remarkable how well such a model can describe fairly complex loading histories. The inference seems to be that the entire effect of the complex dislocation structure on flow resistance can be represented by a single scalar parameter.

An example of a constitutive model with two internal variables is that recently proposed by Richmond and Smelser and compared with data on porous iron in Spitzig, et al. [2]. This model has the form given by equations (1) and (2) where s_1 represents matrix hardness and s_2 represents porosity, both functions of strain history.

An example of a constitutive model with anisotropic behavior is the classical Taylor-Bishop-Hill model of polycrystal deformation. This model and variations of it are now being used to predict anisotropic yield behavior from measured crystallite orientation distribution functions, c.f. Barlat and Richmond [3], and to predict the evolution of deformation texture, c.f. Asaro and Needleman [4].

A Hierarchy of Material Structure

It may be difficult to include all the pertinent aspects of microstructure in the manner indicated by equations (1) and (2). However, another possibility is to treat them in a hierarchical manner. For example, the effect of solute concentration on slip-system resistance might be determined from a dislocation model of single crystal behavior. The results of this model might then be homogenized into a continuum crystal model with latent hardening. This continuum crystal model might then provide the unit crystal behavior for a Taylor-Bishop-Hill polycrystal. The results of this polycrystal model could in turn be homogenized to form an anisotropic continuum plasticity model. The overall result would be a capability to model the effect of solute concentration on polycrystal hardness and anisotropy.

6

Tribology of Tool/Workpiece Interfaces

Primitive process models traditionally assume that the friction at tool interfaces obeys Coulomb's law, or perhaps that it is a constant fraction of the yield stress of the deforming workpiece. With the growing development of constitutive models for interior elements, there is also a need for more focus on interface behavior. There is a need for accurate measurements of the distribution of friction, heat transfer, surface roughness and film thickness within tool/workpiece contacts. There also is a growing need to develop micromechanical models of these phenomena. Applely, et al. [5] have used measured interface velocities determined from deformed grids together with finite element analysis to compute the friction distribution during strip drawing. Wilson [6] has developed micromechanical models for friction at metal working tool interfaces. Various results show strong indication that the Coulomb model is quite inadequate.

Failure Criteria

Barlat and Jalinier [7] have developed failure criteria for sheet stretching based upon the effect of residual porosity on shear localization (see also [3]). Bourcier, et al. [8] have suggested a related criterion for failure due to the presence of porosity in bulk deformation. These works suggest that failure depends not only on mean porosity but also on the spatial distribution of defects. Magnusen, et al. [9] give clear experimental evidence of this effect.

Global Process Models

Global process models are beginning to appear in which the inhomogeneous distribution of structure is modeled. Smelser, Richmond and Thompson [10] modeled hardness distribution resulting from hot extrusion using a constitutive model of the form given by equations (6). Dawson, et al. [11] recently modeled distribution of texture evolution during cold rolling. Such models are the harbingers of a new age in process modeling.

Closing Remarks

Deformation processing involves large plastic deformations. Large plastic deformations involve changes in structure. The theory of large plastic deformation including structure evolution is a complex nonlinear theory requiring sophisticated computational methods and large computers to obtain numerical results. The development of deformation process modeling has paralleled the development of large strain plasticity theory, finite element methods and computer hardware until today these fields have progressed to the point where they are enabling a revolution in process design. They are enabling accurate three-dimensional simulation of complex processes including the capability to predict material structure and damage, as well as deformation.

The vision is clear but there is much to be done. There are needs for treatment of more aspects of microstructure evolution; tool interface models that accurately reflect the variation in friction, surface roughness and damage, and improved models of surface and internal damage. Finally, there is still a need for more efficient computational methods and computer hardware, and a need to create user friendly computer-aided design tools.

References

1. Sample, V. M. and Lalli, L. A., "Effects of Thermomechanical History on Hardness of Aluminum," Materials Science and Technology 3(1), (1987), pp. 28-35.

2. Spitzig, W. A., Smelser, R. E. and Richmond, O., "The Evolution of Damage and Fracture in Iron Compacts with Various Initial Porosities," to be published, Acta Metall.

3. Barlat F. and Richmond, O., "Prediction of Tricomponent Plane Stress Yield Surfaces and Associated Flow and Failure Behavior of Strongly Textured FCC Sheet," to be published, Matls. Sci. & Engrng.

4. Asaro, R. J. and Needleman, A., "Texture Development and Strain Hardening in Rate Dependent Polycrystals," Acta Metall 33(6), (1985), p. 923.

5. Applely, E. J., Lu, C. Y., Rao, R. S., Devenpeck, M. L., Wright, P. K. and Richmond, O., "Strip Drawing: A Theoretical-Experimental Comparison," Int. J. Mech. Sci. 26(5), (1984), pp. 351-362.

6. Wilson, W. R. D., "Friction and Lubrication in Bulk Metal-Forming Processes," J. Applied Metalworking 1(1), (1979), pp. 7-19.

7. Barlat, F. and Jalinier, J. M., "Formability of Sheet Metal with Heterogeneous Damage," J. Matls. Sci. 20(9), (1985), p. 3385.

8. Bourcier, R. J., Koss, D. A., Smelser, R. E. and Richmond, O., "The Influence of Porosity on the Deformation and Fracture of Alloys," Acta Metall 34(12), (1986), pp. 2443-2453.

9. Magnusen, P. E., Koss, D. A. and Dubensky, E. M., "Ductile Fracture and Random vs. Regular Hole/Void Arrays," Office of Naval Research, Contract No. N00014-86-K-0381.

10. Smelser, R. E., Richmond, O. and Thompson, E. G., "A Numerical Study of the Effects of Die Profile on Extrusion," Proc. NUMIFORM '86, (1986), pp. 305-312.

11. Dawson, P. R. and Mathur K. K., unpublished research.

8

MATHEMATICAL MODELING: EXPECTATIONS, FAILURES AND SUCCESSES

J. W. Evans

Department of Materials Science and Mineral Engineering
University of California
Berkeley, CA 94720

Abstract

During the past two decades, mathematical models of metals/materials processing operations have become more common and undergone considerable evolution. In part these changes arise from technological developments such as more powerful computers, but more important has been a change in the expectation of what benefits will be derived from mathematical modeling. In many cases early expectations have not been fulfilled; in a more important sense, the achievement of knowledge, modeling can be described as very successful. The paper is intended to chart this progress using examples of models that have appeared in the literature.

Introduction

It is now twenty years since I personally became involved in mathematical modeling of metallurgical processes. Although I cannot claim to have acquired the wisdom of maturity, I discern changes in these two decades that I believe are worth discussing. The title of this paper has been chosen so that I can comment on the extent to which expectations of twenty years ago have met with success, and on the role that successes and failures have played in shaping our present expectations. In discussing failures I have chosen, for obvious reasons, to write of my own work, although I am bold enough to think that my failures are no worse than those of others! In writing of success, I have attempted to bring in the many admirable models of others but have only partially achieved this goal; those I have omitted will pardon me, recognizing that this paper is not intended as a comprehensive review.

The Era of the Spherical Chicken

In the late 1960s mathematical modeling had not won acceptance, even by the technical members of the metallurgical community. The swamp lands of empiricism and practicality were separated from the trackless jungles of mathematical modeling (containing such threatening beasts as partial differential equations) by unscaleable mountains of poor communication. One of the natives of the former territory, a middle level manager in an engineering company that is a household name (and coincidentally, perhaps, now much diminished from its former size) declared to me that he had never been able to understand calculus and it had never done him any harm! Such as he were able to dismiss mathematical modeling with stale jokes about spherical chickens, which grated on the nerves of modelers because they contained the germ of truth. I refer to the fact that models of that era tended to invoke the grossest simplifications of physical reality while striving for the utmost precision in mathematics. My own first paper (1) is an example of this kind of modeling. It concerned the radiant heat loss from the surface of steel in a ladle and Fig. 1 shows the computed heat loss as a function of time with the ratio of freeboard to diameter as parameter. The ladle's refractory lining heats up, diminishing the heat loss, but no allowance was made in the model for the cooling of the steel surface, even though that cooling would be a few hundred degrees over the time period of the calculations (as the reader will be able to demonstrate by elementary calculations). These results and others in the paper stemmed from mathematics that I can still claim was elegant, coupled with computer programming that I thought sophisticated at the time. I recall that I was not in the least disturbed by the fact that my calculations on heat loss from the steel took no cognizance of its cooling! Neither, apparently, was my co-author, although I believe that we are now both more sensitive to reality.

The Present

I would like to contrast that work on heat loss with recent work on the mathematical modeling of the Hall-Heroult cell that is used in the production of most of the world's aluminum. My own group has done work on this topic but it is minor compared to the efforts (largely unpublished) mounted by the aluminum companies. A manager in one such company recently told me that the very future of the company as an integrated aluminum producer was dependent on cell improvements guided by mathematical modeling.

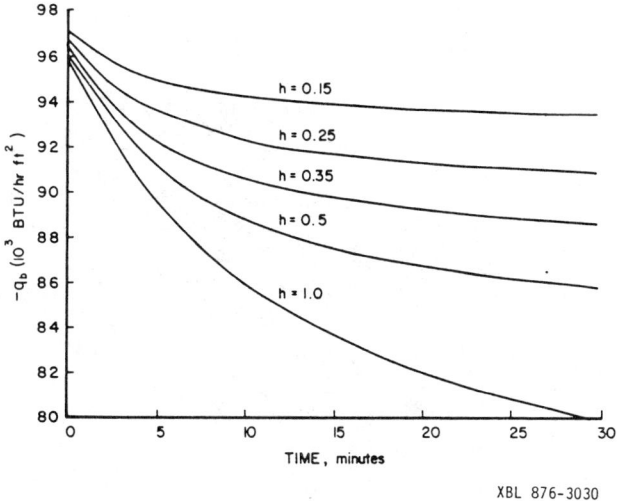

XBL 876-3030

Figure 1 - Calculated rate of heat loss from the surface of molten steel held in a refractory ladle; the effect of freeboard (1).

Key indices of cell performance are current efficiency and electrical energy consumption per unit mass of metal produced. For reasons which there is insufficient space to present, these indices are greatly affected by the flow of the molten salt electrolyte and molten aluminum in the cell. That flow is turbulent, driven by both bubbles and electromagnetic forces, three dimensional, frequently unsteady and within a changing geometry of poor predictability. Nevertheless, at considerable cost in terms of computer time and manpower, mathematical models for Hall cells have been developed that incorporate the important aspects of real cells. A paper from ALCOA is to be presented on this topic at this meeting and Fig. 2 shows the computed electrolyte-metal interface shape for a particular cell from an earlier paper (2) by the author of that paper. Notwithstanding their *realism* the Hall cell models are mathematically *imprecise* e.g. in their treatment of turbulence. Hall cell models are proving extremely useful in the design of new cells (e.g. to provide guidance on the placement of busbars around cells in order to minimize detrimental electromagnetic effects) and in providing insight into the behaviour and improvement of existing cells.

The radiative heat loss model exemplified mathematical modeling of the late sixties; mathematical precision was achieved at the expense of realism with consequent impact on the utility of the model. Present day Hall cell models exemplify the best current practice; mathematical precision is sought after but is definitely secondary to realism and utility.

Knowledge, Not Numbers

This shift from precision to realism and utility has gone hand-in-hand with a change in expectations concerning modeling. Twenty years ago there seems to have been the belief that before long models would be sufficiently good, and our computers sufficiently swift, that models could be used in grand schemes to control and optimize processes or even whole

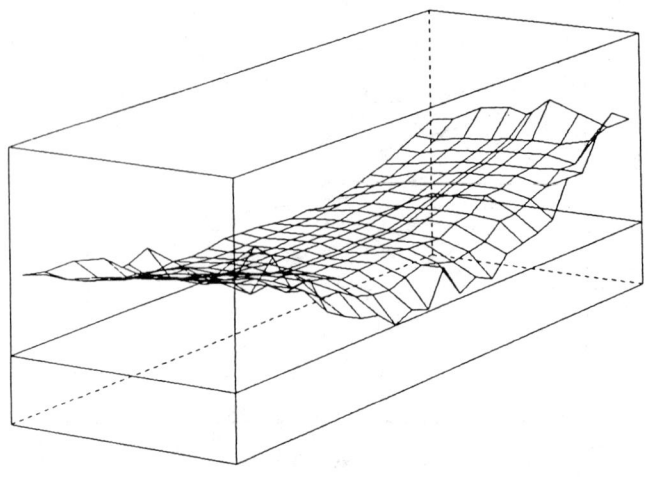

XBL 876-3033

Figure 2 - Computed interface shape between the molten salt electrolyte and the aluminum in an ALCOA Hall cell (2).

plants. In part this expectation was fueled by perceived success in modeling, control and optimization in the chemical industry. Unfortunately very few metallurgical operations are as simple as, say, distillation; the physico-chemical phenomena occurring in a distillation column are transparently simple compared, for example, to those occurring in the iron-blast furnace. That expectation has never been fulfilled.

Some would suggest that mathematical models will shortly be exploited in CAD/CAM. In the sense of mathematical models as they have appeared in the literature I believe this is stretching the meaning of CAD/CAM beyond reason and I suggest that the principal expectation, at present, is that mathematical models of metallurgical processes will provide knowledge (or at least insight).

I illustrate this by an example from our own work on fluidized bed electrodes. In the early days of investigating these electrodes we built a mathematical model which was really an adaptation of a model by Newman and Tobias (3) for porous electrodes. That model treated the bed and the electrolyte permeating it as two overlapping continua through which current passed by electronic and ionic conduction respectively, electrochemical reaction being the diversion of current from one continuum to the other (at the same location). The effective conductivity of the electronic continuum (the bed of particles) was shown by the model to be important in governing the behaviour of the electrode, particularly the way the reactions were distributed across the bed (4). This led to an experimental program to measure that effective conductivity, to the result that it was highly sensitive to bed expansion and finally to bed designs where operating problems (bed defluidization) could be avoided. It is doubtful that the model ever gave us *numbers* that we found useful but the *knowledge* we gained was invaluable in subsequent years of experimentation.

As a second example I cite a recently published paper on the excellent modeling of flow in tundishes carried out by He and Sahai (5). Fig. 3 presents just two calculated flows from that paper (which also describes physical modeling yielding results in good agreement with the mathematical

12

model).   This figure gives the velocity components in a vertical plane
positioned along the long direction of the tundish not far from its center.   Metal enters at the top left and flows out of the bottom right.   The
essential difference between the two parts of the figure is that the upper
part is for a tundish which is a mere rectangular trough while   the   lower
figure is for a tundish with its walls inclined 4° to the vertical.   Both
the magnitudes (indicated by arrow lengths) and directions of the flow are
changed by this inclination.   Real tundishes   are   more   complicated   than
those   of   this   mathematical model.   Consequently it is unlikely that the
numerical values of the computed velocity components   will   be   of   direct
use.   However, the *conclusion*, that inclination of the tundish walls, even
by a few degrees, can radically alter the flow, is an invaluable one.

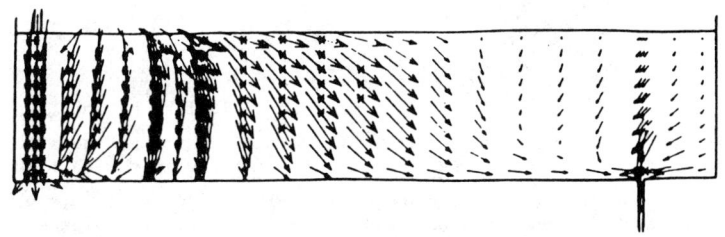

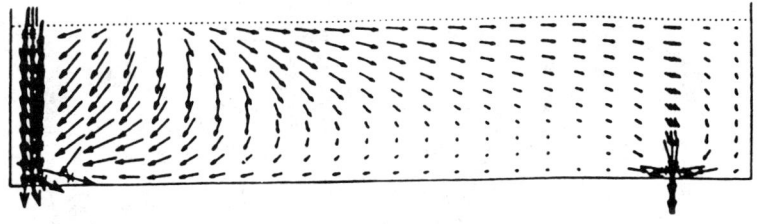

XBL 876-3032

Figure 3 - Velocity fields for a tundish predicted by He and Sahai (5).
The plane shown is a vertical one near the long axis of the tundish.
Upper figure for tundish with vertical walls.   Lower figure for tundish
with inclined walls.

A Failure

As   another example I point to one of our failures.   This was a failure in the sense of the research not leading to the predictive mathematical model that we had promised in proposing the   investigation   although   I
would count it a success in terms of furthering our knowledge of the phenomena   involved.   This investigation concerned the removal of non-metallic
inclusions from steel melts by coalescence, accompanied by   the   inclusion
floating to the top surface (or being transferred to the walls of the containing vessel).   These phenomena undoubtedly take place; for example Fig.
4   shows the results of Quantamet measurements on samples drawn at various

13

times from an inductively stirred laboratory scale melt at Berkeley (6). Coalescence is a topic which has frequently been modeled using the population balance equation:

$$\frac{\partial n_v}{\partial t}(v,t) = \int_0^{v/2} \beta\ (v-\tilde{v},\tilde{v})\ n_v\ (v-\tilde{v},t)\ n_v\ (\tilde{v},t)\ d\tilde{v}$$

$$- n_v(v,t) \int_0^{\infty} \beta\ (v,\tilde{v})\ n_v\ (\tilde{v},t)\ d\tilde{v}$$

$$- s(n_v,v) \tag{1}$$

In this equation $n_v(v,t)dv$ is the number of inclusions per unit volume of melt in the size range v to $v + dv$. The first integral on the right (the "births integral") represents the formation of inclusions within this size range by coalescence of inclusions of volume $\tilde{v}$ and $v-\tilde{v}$. $\beta(v-\tilde{v},v)$ is a rate "constant" for such coalescence events, the rate of coalescence being dependent on β and the number of the corresponding inclusions per unit

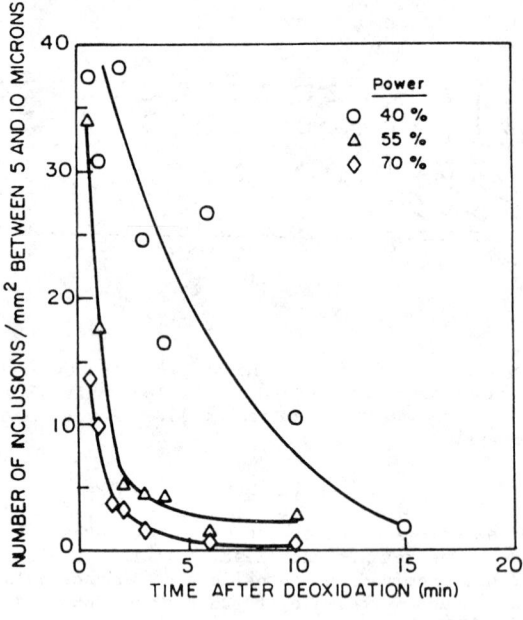

Figure 4 - Variation in the number of oxide inclusions (in the size range 5-10 μm) with time in laboratory experiments carried out in induction furnaces operated at various power levels (6).

14

volume of melt. The second ("deaths") integral represents removal of inclusions from the size range v to v+dv by coalescence to produce larger
inclusions. The third ("sedimentation") term allows for removal of inclusions by floating to the surface or transport to the walls of the vessel
containing the melt. Solution of equation (1), starting with an initial
size distribution, then gives a mathematical prediction of the size distribution (and thence by integration the total volume of inclusion) at any
later times.

Analytical solutions of equation (1) are available for a very limited
set of conditions, namely for a limited number of initial size distributions and a limited number of functional forms of β (i.e. dependence of
coalescence parameter on size). These solutions have been summarized by
Gelbard and Seinfeld (7); they are of little use in the present context
except to test numerical solutions to equation (1). Such numerical solutions face an immediate difficulty. Each integral must be "discretized"
into a sum over a finite range of inclusion size (a finite domain of $\tilde{v}$)
and since the deaths integral has an infinite upper limit a "finite domain
error" is automatically introduced. This error is lessened if the range
of inclusion sizes is extended by moving its upper limit upward (to a few
orders of magnitude greater than the size of a typical inclusion, say),
but this "coarsens" the representation of the size distribution and thereby introduces a different kind of error. In any event, as coalescence
proceeds, inclusions grow outside the size range and are "lost" to the
calculations. The magnitude of this loss can be determined from the
analytical solutions mentioned above and Table I gives results determined
in this way by Gelbard and Seinfeld. The "cases" represent various conditions on β and initial size distribution and τ a dimensionless time; the
reader is referred to the original paper for details. M_1 is the mass
fraction remaining in the calculations as a function of time and M_o is the
number fraction remaining. These cases pertain to situations where there
is no sedimentation and therefore a correct calculation would have M_o and
M_1 remaining at unity. The artificial loss of number and mass revealed in
Table I is indicative of a serious error. It should be noted that this
error is merely a consequence of the need to replace the upper limit of
the deaths integral by a finite number and is independent of whatever
sophistication is employed otherwise in the numerical calculations.

Another difficulty is the evaluation of the rate "constant" β. The
most likely collision mechanism for inclusions is known as "gradient collision" and is the overtaking of one inclusion by another in liquids where
there are velocity gradients. An equation yielding β in turbulent flows
is then

$$\beta(v,\tilde{v}) = \gamma C \sqrt{\epsilon/\nu} \left[\frac{d_v + d_{\tilde{v}}}{2}\right]^3 \qquad (2)$$

where d_v is the diameter of inclusion size v etc.

ϵ is the turbulence kinetic energy dissipation rate
ν is the kinematic viscosity
γ is the fraction of collisions that result in coalescence
and C is a constant.

15

A difficulty is that there is no unanimity concerning the value of C. Values ranging from 1.29 to 12π have been suggested, the extremes being suggestions of Saffman and Turner (8) and of Levich (9). This difficulty is minor compared to the difficulty of estimating γ, which is sometimes known as the "coalescence efficiency". γ must surely depend on the "stickiness" of the inclusions i.e. on morphology and perhaps on whether the inclusions are solid, liquid or two phase. Might not γ therefore depend on inclusion composition and temperature, as well as shape or even size? Two final difficulties are those of predicting ϵ reliably in practical metallurgical systems and the prediction of the sedimentation term S (which is likely to involve its own coalescence efficiency representing the fraction of inclusions that are trapped on reaching the bounding surfaces).

A reasonable approach to the above challenges might be to conduct a series of laboratory experiments wherein inclusion size distributions would be measured as a function of time in melts where the fluid motion is well characterized (and therefore ϵ readily estimated). This was attempted at Berkeley and Fig. 4 is typical of the results; the amount of inclusion in the melt declined with time in a reaonably reproducible manner that was dependent on the degree of agitation. Unfortunately there was almost no shift in the inclusion size distribution with time (e.g. the mean size stayed roughly constant). The explanation was that laboratory melts have a high ratio of surface to volume and therefore inclusion removal dominated over inclusion coalescence. Consequently the necessary information on coalescence efficiencies etc. could *not* be obtained in the laboratory.

It might have been possible to obtain the needed information from an industrial scale experiment but I hope that I have convinced the reader that no coalescence efficiency obtained in this way could be applied to a melt of different temperature or chemistry. Industrial scale experiments for a large number of melts of different temperatures deoxidized/ desulfurized using different reagents seemed the only hope but we were now past the point of diminishing returns. Certainly the sponsors of our research were uninterested in such an ambitious and expensive venture!

There is much to be learned from this mistake; I have embodied these in the following dicta:

(a) Even in this day of supercomputers there are still problems which require numerical solutions of unreasonable cost. Such problems should be identified early in any mathematical modeling effort so that the cost can be faced, alternative approaches sought or the project abandoned.

(b) Mathematical modeling of processes where the physico-chemical phenomena are opaque is unwise. By "opaque" I mean that the phenomena are not understood in a quantitative sense or are not amenable to straightforward experimental measurement. A model for a process that is opaque in this way will inevitably contain a surfeit of adjustable parameters and will be most unlikely to lead to either knowledge or believable numbers.

<u>Some Successes</u>

Despite my arguing that the principal virtue of mathematical modeling is that it leads to knowledge and that the numbers are secondary, I should like to give some examples of models achieving precision along with knowledge and utility.

Fig. 5, from the work of Mazumdar and Guthrie (10) on gas stirred ladles, shows measured vertical velocities (as a function of radial position) at various vertical positions in a water model of an argon stirred ladle. The computed velocities are the lines in the figure and the agreement is more than satisfactory.

Fig. 6 is taken from a review paper by Brimacombe (11) describing research done at the University of British Columbia on the continuous casting of steel. The figure shows good agreement between computed shell thickness or surface temperature and measurements on an actual caster.

Szekely's group at MIT has long been involved in mathematical modeling, most recently with an emphasis on fluid flow, particularly electromagnetically driven flow. Fig. 7, taken from a paper by Szekely and

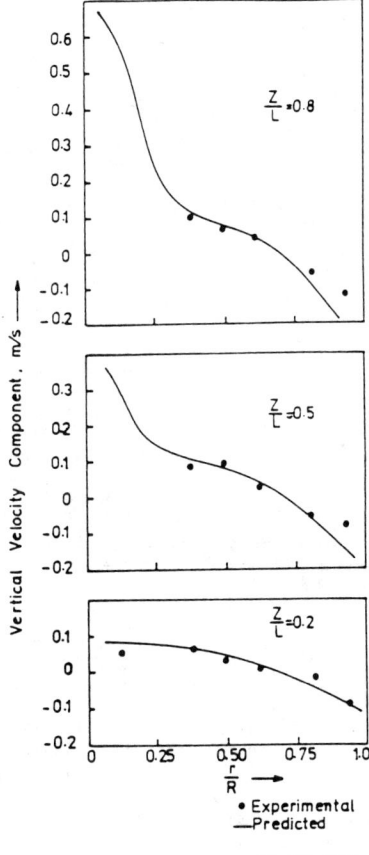

XBL 876-3035

Figure 5 - Comparison of experimentally measured and theoretically predicted vertical velocity component at different depths in the water model of an argon stirred ladle of Mazumdar and Guthrie (10).

El-Kaddah (12), shows a comparison between the measurements of Vives and Ricou (13) on an inductively stirred pool of mercury and their model's predictions for this flow. Evans and Lympany (14) also were able to obtain a good match between their model for an inductively stirred melt and the measurements of Moore and Hunt (15). The successful prediction of velocities in inductively stirred melts now extends to electromagnetic casters, as will be seen from the paper by Sakane and Evans on those casters that appears in these proceedings.

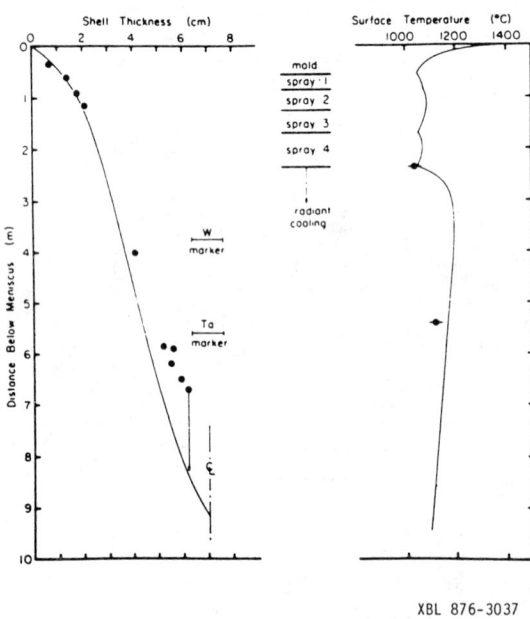

XBL 876-3037

Figure 6 - Comparison of model-predicted shell thickness and surface temperature for a continuous steel caster with values measured from an actual billet caster (11).

Modeling of reactions between gases and solids has been pursued by several research groups, in recent years with considerable vigor by Sohn at the University of Utah. A difficulty here is that chemical reaction rates are not predictable from first principles and therefore mathematical modeling becomes dependent on a great deal of experimental work. Fig. 8 shows a comparison between predicted and measured extent of reduction for a non-isothermal packed bed of taconite pellets, taken from the work of Ranade and Evans (16). The kinetic information input into this model was obtained by isothermal experiments on single taconite pellets.

These are but a few examples of many models where both the numbers and the knowledge that emerge from the modeling have credibility.

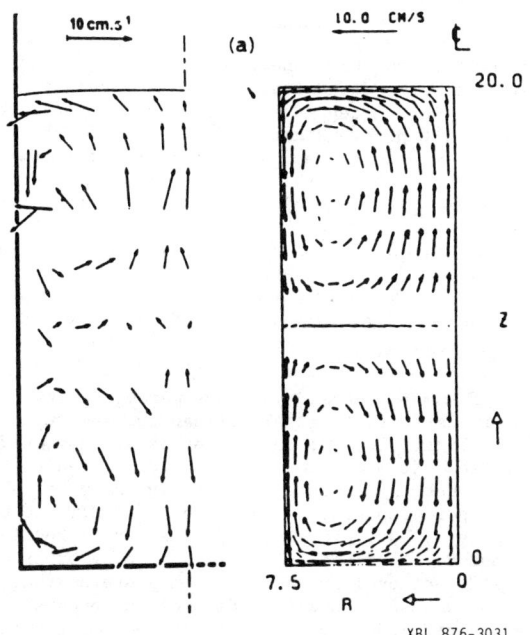

XBL 876-3031

Figure 7 - A comparison of the experimental measurements of Vives and Ricou (13) and the predictions of Szekely and El-Kaddah (12) for an inductively stirred melt.

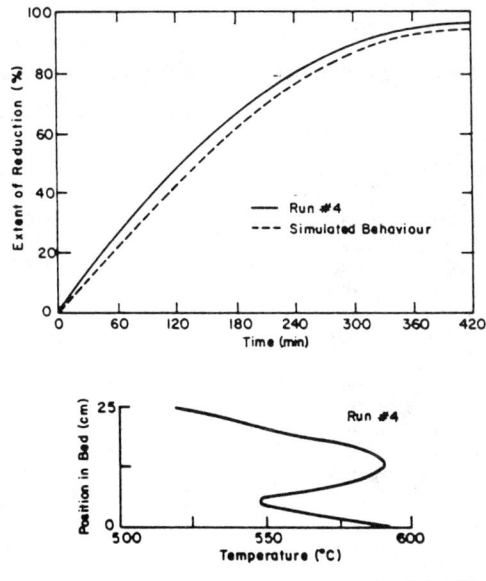

XBL 876-3036

Figure 8 - A comparison between measured and simulated reduction of a bed of taconite pellets from the work of Ranade and Evans (16). The bed temperature profile is indicated in the lower part of the figure.

19

Concluding Remarks

This paper attempts to chart the growth of mathematical modeling from a state of immaturity, two decades ago, to one of acceptance in our technical society today. That acceptance is made clear by the choice of mathematical modeling as a topic for this Fall meeting of TMS. The maturing of the field has been accompanied by an improvement in the communication of its practitioners. I believe that there is now much less of a tendency for authors of modeling papers to bludgeon their readers with mathematics. In a crude attempt to demonstrate this I have counted equations in my own papers in refereed journals, omitting short papers, review articles and a few on which I could not readily lay my hands. Table II is the result; although this limited data (representing forty or so papers) does not prove my point, I believe that a similar survey of others working in the field would show the same results.

What of the future? These days, many metallurgists have become (or are becoming) materials scientists. I believe the same trend will be seen in mathematical modeling. Indeed, this trend is already evident although there is competition from mathematical modelers already in the field of advanced materials, such as Jensen of the University of Minnesota who models chemical vapor deposition, or Brown at MIT who models crystal growth. In this paper I have expressed my view that knowledge is more important than numbers. This view is shared by other modelers and I expect that we shall see modeling papers in the future that de-emphasize the mathematics and emphasize understanding of the process being modeled. This emphasis has helped modeling to win acceptance by the technical community. I look forward to the many interesting and valuable results that my colleagues will generate.

Table I. Results of Gelbard and Seinfeld: Finite Domain Errors

Case	τ	M_o	M_1	Case	τ	M_o	M_1
1	1.0	1.00	1.00	4	1.0	0.998	0.959
	3.0	1.00	1.00		2.0	0.990	0.856
	5.0	1.00	0.996		3.0	0.986	0.806
	7.0	0.998	0.983		5.0	0.983	0.780
	10.0	0.989	0.939		7.0	0.983	0.776
	20.0	0.914	0.703		10.0	0.983	0.776
2	5.0	1.00	1.00	5	1.0	0.999	0.990
	10.0	1.00	1.00		1.5	0.968	0.858
	20.0	0.997	0.980		2.0	0.839	0.545
	30.0	0.982	0.908		2.5	0.627	0.259
	50.0	0.915	0.705		3.0	0.416	0.102
3	0.5	0.999	0.988				
	1.0	0.983	0.776				
	1.5	0.950	0.483				
	2.0	0.921	0.283				
	3.0	0.887	0.0979				

Table II. Equations per paper of the author as a function of time

Before 1971	30.0
1971-74	19.7
1975-78	13.6
1979-82	9.2
1983-86	9.7

References

1. J. Szekely and J. W. Evans, "Radiative Heat Loss from the Surface of Molten Steel Held in a Ladle", Trans. AIME, 245 (1969) 1149-1159.
2. W. E. Wahnsiedler, "Hydrodynamic Modeling of Commercial Hall-Heroult Cells" (Paper presented 116th Mtg. TMS, Denver, Feb. 24-26, 1987) 269.
3. J. S. Newman and C. W. Tobias, "Theoretical Analysis of Current Distribution in Porous Electrodes", J. Electrochem. Soc., 109 (1962) 1183-1191.
4. B. J. Sabacky and J. W. Evans, "Electrodeposition of Metals in Fluidized Bed Electrodes: Part I, Mathematical Models", J. Electrochem. Soc., 126 (1979) 1176-1180.
5. Y. He and Y. Sahai, "The Effect of Tundish Wall Inclination on Fluid Flow and Mixing: A Modeling Study", Met. Trans., 18B (1987) 81-92.
6. B. M. Tracy and J. W. Evans, "Removal of Non-Metallic Inclusions from Steel Melts" (Paper presented SCANINJECT III, 3d Int. Conf. Refining Iron and Steel by Powder Injection, Lulea, Sweden, June 1983) 20:1.
7. F. Gelbard and J. H. Seinfeld, "Numerical Solution of Dynamic Equation for Particulate Systems", J. Comp. Physics, 28 (1978) 357-375.
8. P. G. Saffman and J. S. Turner, "On the Collision of Drops in Turbulent Clouds", J. Fluid Mech., 1 (1956) 16-30.
9. V. Levich, Physicochemical Hydrodynamics (Englewood Cliffs, NJ: Prentice Hall, 1962), 216.
10. D. Mazumdar and R. I. L. Guthrie, "Hydrodyamic Modeling of Some Gas Injection Procedures in Ladle Metallurgy Operations", Met. Trans., 16B (1985) 83-90.
11. J. K. Brimacombe, "Design of Continuous Casting Machines Based on a Heat-Flow Analysis: State-of-the-Art Review", Can. Met. Quart., 15 (1976) 163-175.
12. J. Szekely and N. El-Kaddah, "Turbulent Recirculating Flows in Metals Processing" (Paper presented Int. Sem. Refining and Alloying of Liquid Aluminum and Ferro-Alloys, Norwegian Inst. Tech., August 1985), 249.
13. C. Vives and R. Ricou, "Experimental Study of Continuous Electromagnetic Casting of Aluminum Alloys", Met. Trans., 16B (1985) 377-405.
14. J. W. Evans and S. D. Lympany, "An Improved Mathematical Model for Melt Flow in Induction Furnaces and Comparison with Experimental Data", Met Trans. 14B (1983) 306-308.
15. D. J. Moore and J. C. R. Hunt, (Paper presented 3d Bat-Sheva Symp. MHD Flows and Turbulence, Beer-Sheva, Israel, 1981).
16. M. G. Ranade and J. W. Evans, "The Reaction Between a Gas and a Solid in a Non-Isothermal Packed Bed: Simulation and Experiments", I. and E. C. Proc. Design and Dev., 19 (1980) 118-123.

DEVELOPMENTS IN THE USE OF MODELING AND SIMULATION TECHNIQUES

FOR THE IMPROVEMENT OF MINERAL PROCESSING OPERATIONS

John A. Herbst

Control International, Inc.
2319 Foothill Drive, Suite 200
Salt Lake City, Utah 84109

Abstract

There has been a technical revolution in the use of mathematical modeling in the minerals industry in the last decade. Process models have moved from academic curiosities to important industrial tools. The paper discusses the reasons for this revolution and gives a series of example applications in scale-up design, flowsheet optimization and automatic control which demonstrate the power of the tools of modeling and simulation for industrial problem solving.

Introduction

The 1980's have been difficult years for the mineral industry. Declining head grades and lower selling prices for many metals have caused several mine/mill closures. The surviving operations have been forced to pursue methods of improving plant productivity without significant capital investment. As the chemical and petroleum industries before it, the mineral industry has turned, at least in part, to the methods of process analysis to achieve the necessary improvements. Mathematical modeling and computer simulation have played an important role in these efforts.

The principal steps in process analysis have been outlined by Himmelblau and Bischoff (1) as shown in Figure 1. Basically this procedure involves recognizing a problem (e.g. a bottleneck in throughput, a limitation to produce quality, etc.), formulating a model to describe the effect of important variables on process performance, estimating the parameters of the variables which can be manipulated and confirming the model.

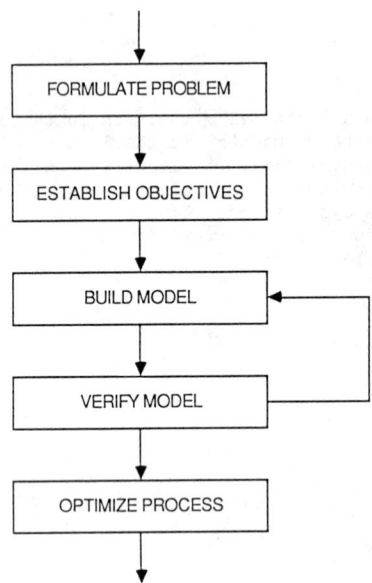

Figure 1 - Steps in Process Analysis

The areas of process analysis which have been most important for the mineral industry in the last decade have been optimization of existing operations, both at steady-state and dynamically. In order to accomplish this optimization a variety of steady-state and dynamic models had to be developed for individual unit operations and overall plant performance. In turn these models had to be incorporated into computer packages which were user friendly enough to be useful to mineral processing engineers.

This paper discusses briefly the type and structure of the models which have provided the most accurate description of the steady state and dynamic behavior of mineral processing unit operations and reviews the simulation packages that are currently available. A series of example applications are presented based on software developed at the University of Utah and by Control International, Inc. with which the author is most familiar.

Modeling and Simulation

Attempts to write equations to predict the behavior of mineral processing operations have been going on for over one hundred years. These efforts have been, until very recently, viewed by the mineral industry as academic exercises with little or no practical importance.

Early modeling efforts were largely empirical or quasi theoretical (2). In recent years there has been much more emphasis placed on the underlying physics and chemistry of the process being modeled. The phenomenological model (which derives its form from theory but whose constants are determined experimentally) has turned out to be the most accurate and practical type of model for mineral processing (3). Typically these models have the following general conservation equation form for any mineral species characterized by its contents, size, density, etc.

FLOW IN - FLOW OUT + GENERATION = ACCUMULATION

where the generation term usually involves kinetic rate constants which are determined experimentally. For dynamic systems the application of this equation results in an integral differential equation or a series of differential equations for each of the species of interest in the unit being modeled. For steady state systems the accumulation term is equal to zero and a set of algebraic equations arise. Table I gives a summary (non exhaustive) of the models which have been developed in this way for all types of mineral processing unit operations. This Table (and Table II) were originally presented in reference (4) and are updated to 1987 in the present paper.

Table I. Examples of Mineral Processing Unit Operations
and Mathematical Models

Unit Operation	References
Comminution	Austin and Gardner (5), Austin and Klimpel (6,16), Austin, et al. (7), Austin et al. (8) Hatch and Mular (9), Herbst, et al. (10), Herbst and Fuerstenau (11), Herbst and Rajamani (12,), Herbst, et al. (13,14) Spring et al. (15), Rajamani and Herbst (17) Hodouin, et al. (18), Lynch (19), Meloy and Bergstrom (20), Whiten (21), Herbst and Oblad (22).
Screening and Classification	Karra (23), Plitt et al. (24), Lynch and Rao (25), Ferrara and Peretti (26,27).
Flotation	Arbiter and Harris (28), Bascur and Herbst (29), Kapur and Mehrotra (30), King (31,32), Lynch, et al. (33), Mehrotra and Saxena (34), Mika and Fuerstenau (35), Moys and King (36), Niemi (37), Niemi, et al. (38), Sastry and Fuerstenau (39), Zaragoza and Herbst (40).
Agglomeration	Sastry (41), Wellstead, et al. (42), Young, et al. (43).
Thickening	Fitch (44), Kos (45), Kynch (46), Tarrer, et al. (47), Bascur and Herbst (48), Concha and Bascur (49).
Filtration	Tiller (50), Wakeman (51,52).

Computer simulation of the performance of a single unit operation or series of unit operations involves the programming of the model equations represented by the general conservation equation to obtain accurate solutions for important variables. The best simulators are those which are not only accurate but also user friendly, which entails that the simulator is flexible and yet data input is easy and output is in an easy to use form. Table II gives a sampling of the major simulation packages available in 1987.

Table II. Examples of Mineral Processing
Circuit/Plant Simulators

Type	Name	Reference
Steady State	FLOTE	King (31)
	ESTIMILL	Herbst, et al. (10)
	SIMPLANT	Ruebush, et al. (53)
	SPOC	CANMET (54)
	ESTILIB	Peterson, Herbst (55)
	MODSIM	King (56)
	UTAH-MODSIM	Fu, Herbst (57)
	GRINDSIM.S	Pate, Herbst (58)
DYNAMIC	DYNAMILL	Rajamani, Herbst (17)
	UCMINPRO	Adel, Sastry (59)
	FLOATSIM.D	Zaragoza, Herbst (40)
	GRINDSIM.D	Pate, Herbst (60)

Steady State Simulators and Their Use

Steady state simulators can and have been used in virtually all aspects of plant analysis to achieve improvement in steady state performance. General categories of such applications involve:

1) Scale-up design, in which the models must contain information of the relation between model parameters and equipment sizes.

2) Flowsheet selection, in which the models must be "connectable" in a variety of configurations.

3) Circuit optimization, in which the dependence of performance on all critical operating variables must be reflected in the models.

In the sections which follow, an example application from each of these areas is described.

Scale-Up Design

A procedure which has been proposed for commercial mill design using a population balance modeling approach and specific power information is shown in Figure 2 (11). Here batch grinding data (size distributions for various energy inputs) obtained in a small batch mill are input to an estimation program which determines the "best" set of breakage rate parameters from the data provided. In turn, these breakage rate parameters are input to a scale-up program along with traditional inputs such as: a) desired circuit throughput and product size, b) circuit operating conditions including fraction of critical speed and fractional ball load and ball size, size distribution and percent solids in feed and circulating load (if closed circuit). In addition, since the population balance approach relies on more detailed models of the grinding circuit subprocesses than the classical design methods do, it is necessary to provide input to the scale-up program on the nature of material transport through the mill (e.g. residence time distribution information) and classifier performance characteristics (e.g. selectivity functions). The scale-up program will make the design calculations including the selection of mill dimensions and drive requirements, will predict product size distributions throughout the commercial circuit and will provide information on the sensitivity of the design to errors in the design assumptions.

**Industrial Mill Design
from Laboratory Grinding Data**

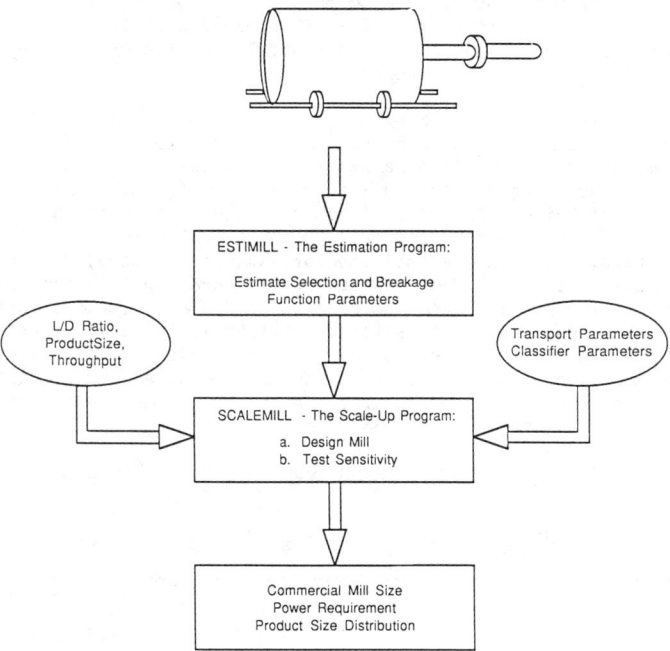

Figure 2 - Procedure for Scale-up Design of Ball Mills Using Population Balance Models

In order to implement such a scale-up scheme it is necessary to have a computer program for parameter estimation and a program for scale-up prediction. A program which is suitable for parameter estimation, ESTIMILL, has been developed and tested on a wide variety of ores at the University of Utah. This program has been in use in academia and industry since 1977 (10). The development of a separate program, SCALEMILL, which is suitable for scale-up prediction has been the subject of a recent paper (12).

SCALEMILL is a FORTRAN program developed for the purpose of selecting industrial-scale grinding circuit equipment based on a known set of grinding model parameters and cyclone model parameters. Design calculations can either be done for open circuit, standard closed circuit (postclassification) or reversed closed circuit (preclassification). For closed circuit, it is assumed that the larger scale milling circuit uses a bank of hydrocyclones operating in parallel for classification. SCALEMILL makes scale-up calculations in the context of a linear-kinetic grinding model and a regression model of the hydrocyclone to arrive at the specific energy input for a given task (energy that must be expended in the mill per ton of ore feed to produce the desired product) from which mill dimensions are calculated. The program also determines the number of hydrocyclones needed to achieve the desired separation and throughput. An example application of the selection of an industrial grinding circuit for copper ore grinding is given briefly below.

Problem Statement: Select a mill to grind 425 tph of copper ore from a feed size of 80% passing 2360 m to a product of 80% passing 210 m. The mill is to be loaded to 37% of its volume with 76.2 mm (3 inch) topsize balls and rotated at a speed of 72% of its critical speed. The L/D ratio for the mill is to be 1.2.

SCALEMILL Solution: Batch experiments were conducted in a 380 x 290 mm (15 x 11.5 inch) mill with a BLH torque sensor for power determination. The ball size distribution approximated that of an "equilibrium charge" with top size of 76.2 mm (3 inch). The mill speed was kept at 72% of its critical speed. Size distributions of feed and product and power data were input to the ESTIMILL to estimate breakage parameters. In turn the estimated breakage parameters, RTD correlation data (N=3.5) plus the design requirements were input to the computer program. The results from the program of the power, the diameter and the length required for this task are shown in Table III. As a confirmation of these predictions the actual plant values obtained are shown for these same conditions. The agreement between predictions and actual performance is very close. In this case the prediction error for the power required is only 3.3%.

Table III. Results for Scale-up Design Example

	Power (kw)	Diameter (m)	Length (m)
SCALEMILL	1784	4.72 (15.5 ft.)	5.64 (18.5 ft.)
PLANT DATA	1845	4.88 (16 ft.)	5.79 (19 ft.)

Additionally, the computer simulation approach provides information on the behavior of the grinding circuit that it is not possible to get from the traditional energy size reduction approach to scale-up. The influence of changes in size distribution of the feed, breakage characteristics of the feed and performance of the classifier can all be readily investigated with the computer simulators. SCALEMILL provides the details of the size distribution of important streams in the grinding circuit. Figure 3 shows a comparison of computer predictions of product size distributions and those actually observed in the full scale plant.

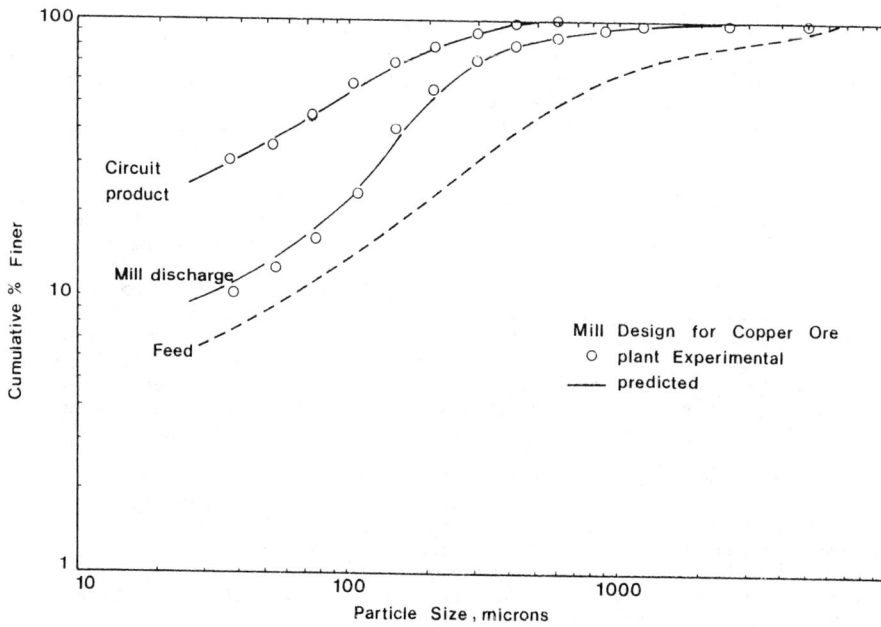

Figure 3 - Comparison of Predicted and Experimental Size
Distributions for Commercial Mill

As a final note on this subject, this scale-up procedure has been
tested for a wide range of material types and mill sizes. Table IV
summarizes the results of typical evaluations. Comparisons of SCALEMILL
prediction and actual plant data in each case illustrate the utility of
this approach for scale-up and mill design.

Table IV. Comparison of Mill Size Prediction Made with
Scalemill to the Actual Mill Size for Four Operations

	Copper	Copper	Iron	Copper
Classification	Pre	Pre	Open	Post
Feed Rate, tph	225	425	132	458
SCALEMILL				
Net Power, kW	1860	1784	2257	2844
Diameter, m	4.72	4.72	4.42	5.49
Length, m	5.64	5.64	9.75	6.12
Plant Data				
Net Power, kW	1840	1845	2212	2975
Diameter, m	4.72	4.88	4.42	5.49
Length, m	5.79	5.79	9.75	6.40
$d_{80,F}$ m	9530	2360	400	7500
$d_{80,P}$ m	180	210	55	205
Error in Power, %	1.1	-3.3	2.0	4.4

29

Flowsheet Design

Steady-state simulators are valuable in the design of flowsheets since the equipment, size of equipment, arrangement, operating conditions and other design variables can be rapidly analyzed and the optimum circuit configuration reached. In addition, the cost of the equipment may be computed for performing economic analysis.

An example of a powerful general purpose program is MODSIM (56) and its counterpart UTAH-MODSIM (57) which contains subroutines especially written for simulating comminution circuits. Flowsheets are created graphically on a computer terminal in any configuration with unit operations such as crushers, screens, tumbling mills, hydrocyclones and flotation cells available for flowsheet construction. The outputs of the program are the ore and water mass balances and the sizes and costs of equipment. Models for each process unit are supplied with the program as FORTRAN language subroutines and relate the output of a unit to its input. User subroutines containing the code for alternative models can easily be substituted since program variables are standardized and follow a certain format.

Another example of a simulator for flowsheet design is SIMPLANT (53,61,62). SIMPLANT simulates integrated grinding and flotation plants. For grinding there are 3 circuit configuration possibilities - open, preclassification and postclassification. Classification can be performed with either single or two stage hydrocyclones. The flotation simulator is the one reported by King (31) and allows any configuration for rougher, cleaner, scavenger and recleaner cells.

The population balance model for the grinding mill uses a linear size discretized model for batch grinding coupled with an n-mizers-in-series transport model for continuous systems. The grinding model accounts for the presence of two mineralogical components, designated A and B. During grinding the breakage of free A particles produces free A particles, breakage of free B particles produces free B particles and breakage of locked AB particles produces free A, free B and locked AB particles (of any system dependent proportion).

The hydrocyclone model equations reported (25) are used with modifications to account for the effect of a mineral's specific gravity on its cut size (d_{50}) in the cyclone. The corrected efficiency curve is the Rosin-Rammler equation as reported in (24).

A recent paper (55) gives an example of the use of SIMPLANT to evaluate the advantages of two stage classification for copper ore. SIMPLANT was used to simulate a ball mill circuit with postclassification followed by a flotation circuit composed of rougher, scavenger cleaner and recleaner cells. Simulations were run for both single stage and two stage sequential classification. Some of the results of interest from these simulations are given in Table V.

	One-Stage	Two-Stage
Grinding		
Ore Feed Rate, tph	225	270
Product Size, % -270 m	68.32	62.54
Chalcopyrite Liberation, % Product	86.4	84.6
Chalcopyrite Cut Size, m		
Cyclone 1	35.1	35.2
Cyclone 2	---	35.6
Flotation		
Feed Grade Chalcopyrite, %	2.39	2.39
Concentrate Grade Chalcopyrite, %	82.10	85.53
Chalcopyrite Recovery	86.98	84.97

In this case, the steady state simulator has been able to provide, in a cost effective manner, predictions of the impact of a change in circuit configuration. Based on these simulations, management is in a position to make a decision concerning changes to the plant and their possible improvement in performance.

Circuit Optimization

In this final example concerned with optimization, simulation was used to find the optimum operating conditions for a 210 tph ball mill circuit grinding phosphate rock (14). The circuit included a 5.33 by 5.79 m. ball mill with preclassification by either hydrocyclones or fine screens. Variables manipulated in this optimization were ball size, mill percent solids and type of classification. The optimization methodology was as follows.

Steady state size distributions from around the grinding circuit were obtained for different operating conditions and in three configurations: open circuit, closed circuit with hydrocyclones and closed circuit with screens. The residence time distribution of the mill was obtained by tracer tests with lithium chloride. The plant tests were performed in order to verify the scale-up predictions from the laboratory tests described below.

Laboratory tests were performed with a 0.38 m. batch ball mill to investigate the effect of percent solids and top ball size on mill performance. The program ESTIMILL was used to estimate the selection function of the ore under the various conditions. The predictions of the size distribution using ESTIMILL compared very well with those from plant tests.

Having verified the predictive capabilities of ESTIMILL, the effect of ball size, mill feed percent solids and configuration on the capacity of the circuit were simulated (Figures 4 and 5). Based on these simulations the recommendations shown in Table VI were made and implemented in the plant. Because of downstream limitations (high percent solids desirable for product acidulation), the change in percent solids was limited to only 2%, however. The predicted increase in capacity for the plant with the recommended changes was 23.3%. The actual results are shown in Table VI with a 25.2% increase in capacity. The closeness of the simulation prediction to the actual plant performance demonstrates the usefulness and power of the simulation approach.

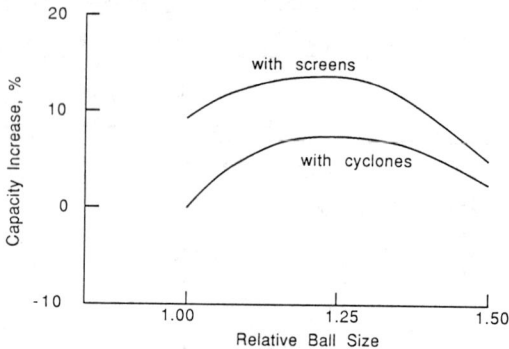

Figure 4 - Computer Predictions of the Influence of Ball Size on Mill Capacity

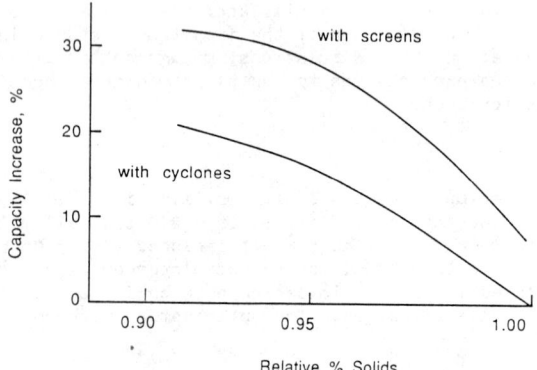

Figure 5 - Computer Predictions of the Influence of % Solids on Mill Capacity

Table VI. Recommendations and Results for Optimization Example

	Before	After	Predicted
Feed Rate	210	263	259
Product Size, % -20 Mesh	98.9	98.9	99.0

	Before	Recommended
Relative Ball Top Size, mm/mm Base Case	1.0	1.25
Relative Mill Feed, % Solids/% Solids		
Base Case	1.0	.97
Classification	hydrocyclones	screens

32

Dynamic Simulators and Their Use

Dynamic simulators can be used in the analysis of the dynamic characteristics of a plant. There are two broad classes of application, off-line and on-line. The off-line class includes control strategy selection, off-line controller tuning and operator training. The on-line application usually involves using a model in real time control system decision making.

Off-line

The principal off-line application for dynamic models and simulators is in the area of control system design. Once a dynamic simulator has been developed for a given unit operation it can provide a very cost effective method for choosing between alternative control strategies and, if the models are accurate enough, it can be used for control loop tuning (17).

In this section, a very simple flotation cell simulator, DYNAFLOAT II, is used to 1) evaluate the dynamic response of a flotation cell to an aeration rate change and 2) compare control strategies in which product grade is controlled by manipulating pulp level and aeration rate.

DYNAFLOAT II Description. DYNAFLOAT II is a program written in BASIC for the dynamic simulation of a flotation cell using a phenomenological model (29). In the development of this model, an attempt was made to include all the geometrical, manipulated and controlled variables. The flotation process involved the interaction of three phases: solid, liquid and gas. For the purpose of model development, an abstraction of the process was made. The flotation cell was divided into two volumes: the pulp volume, in which intimate particle/bubble contact was induced by the turbulent action of the impeller; and the froth volume, which acted as a separating medium to segregate and to remove the valuable minerals.

In summary, the general flotation model is based on a population balance model and the hydraulic characteristics of the three-phase contacting device. The model represents the behavior of each mineralogical species selected and any number of particle sizes. Each of the particle types can exist in one of two states (in the pulp or in the froth). Mechanisms of interphase transfer, represented in the kinetic equations, include attachment/detachment and entrainment/drainage. In each case the influence of important manipulated variables of the flotation process, such as aeration rate, frother addition, agitation, pulp level, froth level or interphase transfer are included in the model equations.

The two example applications of DYNAFLOAT II are given here in problem/solution form as outlined (65).

Problem Statement (Process Matrix Development): Consider a one-cubic-foot pilot-plant flotation cell for coal cleaning. The coal contains 23% ash and is all -30 mesh material. The reagent schedule is 0.3 kg/ton (0.6 lb/ton) of frother MIBC and 0.76 kg/ton (1.5 lb/ton) of kerosene at pH of 6.5. The coal slurry feed is at 7.5 percent solids by weight. The flotation cell operates at 900 rpm, the aeration rate is 40 lt/min and the frother addition rate is 50 ml/min.

Determine the dynamic response of the system to step change to each of the manipulated variables with constant pulp level using a variable-speed pump.

DYNAFLOAT II Solution: The first step in a control system analysis is
to classify manipulated variables according to the magnitude and speed
of their effect on controlled variables. This analysis produces the
system process matrix. Model parameter values were extracted from pilot
scale data involving Illinois No. 6 coal (67).

    To obtain the process matrix with the simulator, start by tuning
the pulp-level controller. Make individual step changes to aeration
rate, impeller speed and frother addition rate using the disturbance
module and the pulp-level controller. Each time the program will evolve
to the next steady state. The evolution of the recovery of clean coal,
clean coal grade, froth depth and pulp level are recorded. Figure 6
shows the approach to the new steady state of the recovery, grade and
froth height after a step change in aeration rate. The process matrix
obtained for this system is shown in Table VII. This table reveals the
strong interactions between the controlled and manipulated variables for
this flotation system.

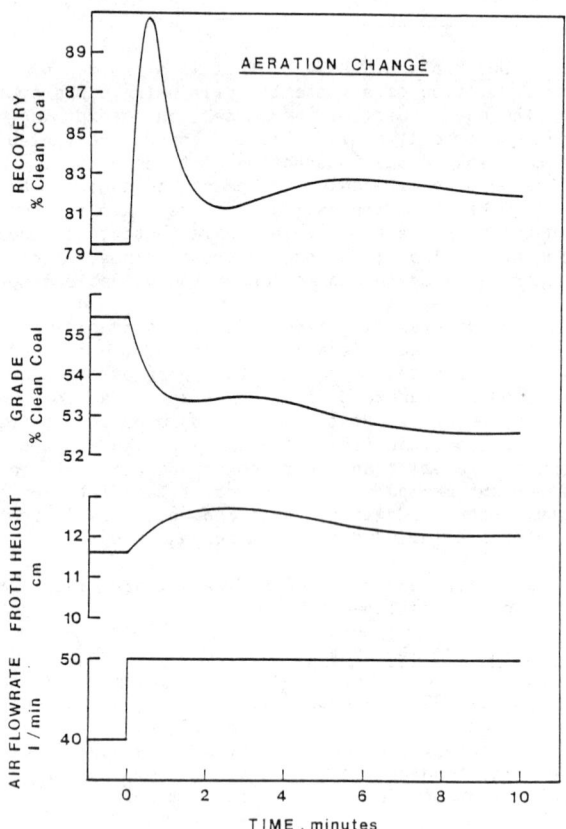

Figure 6 - Simulated Response of Flotation System to
a Step Change in Air Flow Rate

Manipulated Variables	Controlled Variables		
	grade	recovery	froth depth
aeration rate	- fast	+ fast	+ fast
impeller speed	0 - slow	+ - fast	+ slow
pulp level (tailings flowrate)	- slow	+ fast	- slow
frother addition rate	- fast	+ fast	+ slow

Problem Statement (Comparison of control strategies): Configure and
compare the performance of a grade/level controller and a grade/air
controller with constant pulp levels. Examine the response of the
system to a setpoint change in the grade of valuable in the froth
product from 55.5% clean coal to 53% clean coal.

DYNAFLOAT II Solution: In the grade/level, the setpoint for the pulp
level in the flotation cell is cascaded from the grade-controller loop
that defines the pulp level necessary to obtain a desired grade.

For evaluation and tuning of feedback control strategies, a systematic
search minimizing a performance index is recommended. The integral of the
square error between the actual and desired setpoint can be evaluated and
the controller indices such as rise time, settling time and offset error can
be used to decide which controller performs better.

Figure 7 shows the responses of the coal system to a grade setpoint
change from 55.5% to 53% clean coal, with air and level as the manipulated
variables. The grade/air controller response gives a very desirable
response with a short settling time. The grade/level controller, due to its
cascaded nature, is somewhat more oscillatory and displays a longer settling
time. Notice that both controllers produce a very fast response in recovery
with the grade/air controller producing a stable response more rapidly.
From the analysis of these two alternatives, it can be concluded that
grade/air with constant level produces a better control response.

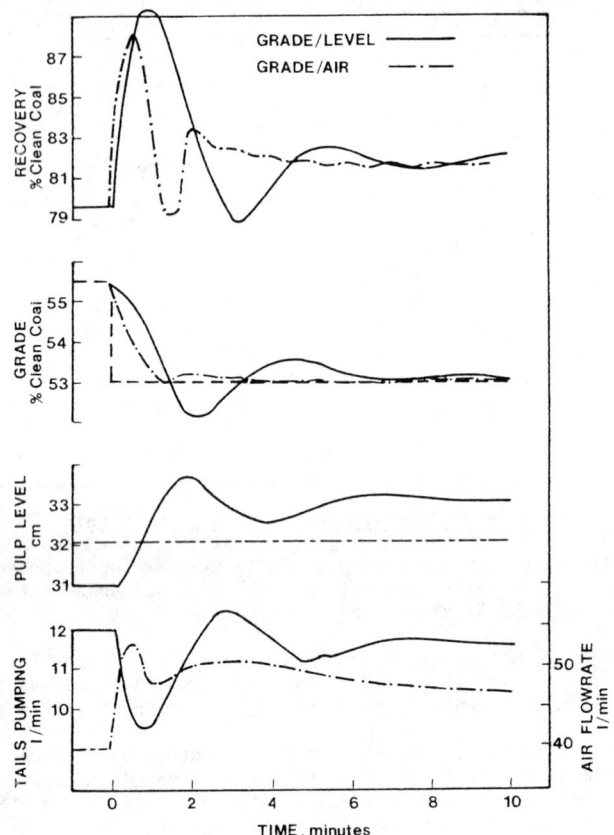

Figure 7 - Comparison of Two Grade Control
Strategies by Simulation

A recent paper (66) has shown that if the predominant disturbance
characteristic of a plant is known, this type of simulation approach to
control strategy development is capable of predicting quantitatively the
expected improvement from one strategy compared with another.

On-line

Limitations in classical control strategies for mineral processing
systems are due to a lack of information about the magnitude of controlled
variable responses to manipulated variable changes and the nature of the
interactions between variables (63,64). The problem is further aggravated
by the fact that important disturbances, such as mineralogical composition
and hardness changes, cannot be directly measured at the present time.
Feedback control methods assume the direction of change in a manipulated
variable for corrective action in a control variable is known and that
controller gains can be found which are suitable for all circumstances. But
often the mineral processing system responses are too complicated to be
characterized by such simple descriptors. An obvious solution to such a

problem is to build a model which contains the missing information about the process into the strategy. By building in such a model, "well informed" responses to disturbances can be made and, ultimately truly "optimal" control performance can be achieved.

The nature of a model based control strategy is revealed in Figure 8. The essential features are: 1) process model which is simple enough to be used for rapid on-line calculations but detailed enough to faithfully reproduce the essential dynamic characteristic of the process, 2) an estimator which combines measurements within the process and model information to determine the state of the system at any instant in time and 3) an optimizer which uses the current state of the system and the model to select the path for manipulated variables that will achieve the process objectives in an optimal way. In such a scheme the optimizer supervises the setpoints of standard regulatory control loops by providing the optimal path to the controller(s). Because the calculations required for such a control strategy are inherently more complex than those for classical strategies, a digital computer is required to implement the model based concept.

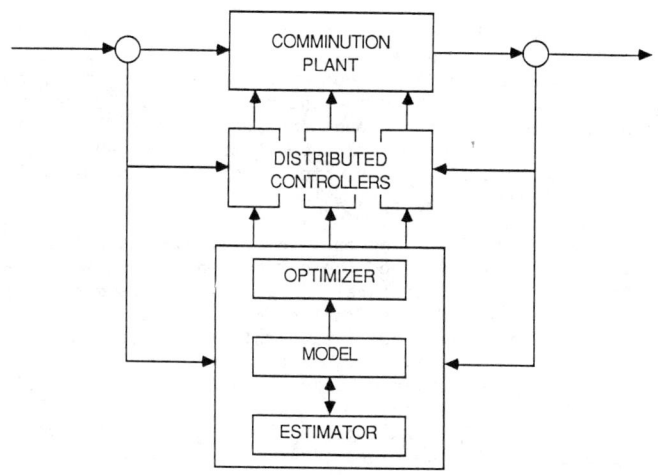

Figure 8 - The Structure of a Model Based Control Strategy

In this section an example of a closed loop supervisory strategy for a 40,000 tpd rod/ball mill grinding circuit at a copper ore concentrator is given. An optimal control law that minimizes the cumulative control error is employed. The controlled variables were overflow particle size (% +100 mesh) and slurry volume in the sump. The manipulated variables were fresh ore feed rate and sump water addition. The on-line model of the circuit consisted of 6 differential equations describing two size fractions in the rod mill, ball mills and sump. The performance of the optimal control strategy is compared to that using conventional PI control loops.

The grinding circuit to be controlled consists of an open-circuit rod mill followed by two closed-circuit ball mills in parallel. All three mills are overflow discharge types. The two hydrocyclone banks that feed the ball mills each contain four hydrocyclones.

The inputs that can be adjusted are rod mill ore feed rate, rod mill water addition and the sump water addition rate. The percent solids in the rod mill are maintained constant by a fixed ratio of sump water to feed rate. The cyclone feed pumps are fixed speed.

The available measurements for this plant are listed in Table VIII.

Table VIII. Grinding Circuit Measurements

Mass flowrate solids, feed
Mass flowrate water, feed
Percent solids, cyclone feed
Volume flowrate, cyclone feed
Percent solids, cyclone overflow
Particle size, cyclone overflow
Sump slurry level
Ball mill power

Each of the grinding circuits in the concentrator are under the control of a Fisher Provox^{TM} system, which is based partly in its microprocessor card controllers and in an HP 1000F minicomputer that performs the graphics, logging and user programs such as the optimal control. The optimal value of feed is downloaded to a microprocessor card responsible for maintaining the feed rate at setpoint. Since no measurement of sump water addition is available the optimal value of sump water is converted to percent value opening and then downloaded to a positioner loop.

On-Line Model Equations. The on-line model must be simple enough to permit the on-line calculations of the estimation and control equations but should retain some essential features of a full off-line model (17). Under certain assumptions a full N-size fraction model can be reduced to only two size fractions, +100 mesh and -100 mesh. This simplification leads to two equations for each of the dynamic elements in the circuit plus the Rao Lynch equations for the two hydrocyclone banks. The model can be simplified further by assuming that both ball mill lines are identical, although this is in general not true. By assuming they are identical, the model value reflects an average of the two lines (68).

The fractional rate of breakage of +100 mesh particles, k hr^{-1}, is assumed to be proportional to the specific power draw of the mill (69). This rate is a function of ore type so to maintain model correctness it must change as the ore does. A way to do this is to estimate the rate and other variable model parameters on-line. This is done by defining them as states, thus allowing the Kalman Filter to update the parameter values at each measurement sample time.

Optimal Control Law. The objective of this control system was to maintain the controlled variables at setpoint. This can be accomplished by finding the sequence of optimal control actions that minimize a quadratic performance index involving deviations from setpoint and control effort.

For a system modeled with linear dynamic equations the performance index can be minimized by applying Pontryagin's maximum principle. The solution is a general optimal feedback control law which is relatively easy to implement with mine computers or distributed control systems.

38

Plant Verification. The optimal control strategy was programmed in the FORTRAN language on the HP 1000F minicomputer at the plant. A flowsheet of the program is shown in Figure 9. The optimal values of feed rate and sump dilution water were transmitted from the host minicomputer to the distributed controller.

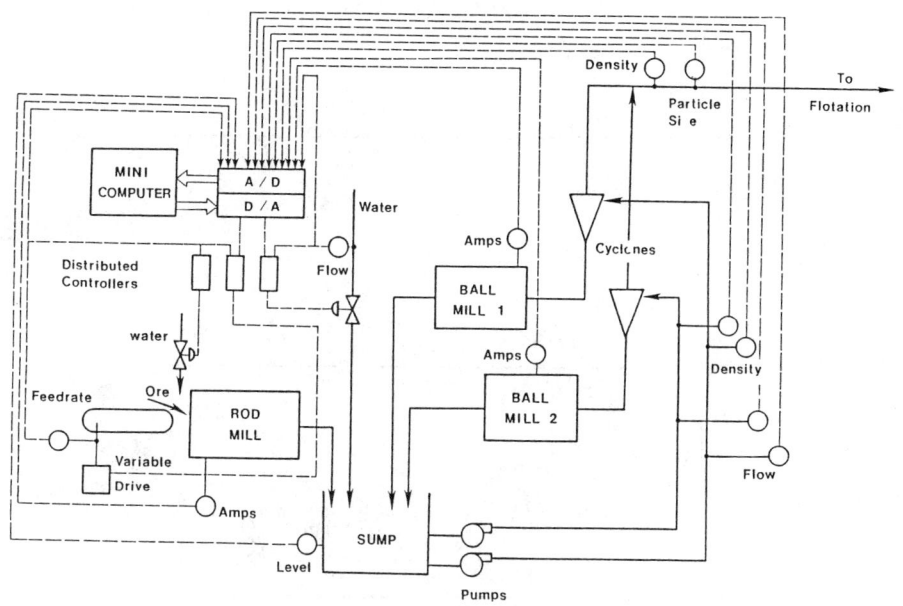

Figure 9 - Schematic of Control Strategy for
Rod/Ball Mill Grinding Circuit

Programming of the computer and calibration of instrumentation required about three months. Because of the pending shutdown of the plant some of the expected instrumentation was not installed. The actual test of the strategy occurred in the two day period prior to shutdown. Unfortunately, the plant performance was rather erratic during this period. As a result there were only a few hours during which the plant and strategy were operating satisfactorily. The data from this period are presented in Figure 10, which shows some of the Kalman Filter estimates. The figure shows two important results concerning the on-line model: 1) the on-line model tracks the performance of the measured variables very well, 2) the model allows the determination of the magnitude and frequency of unmeasured disturbances in ore hardness. Figure 11 shows a comparison between the optimal control strategy and another grinding circuit under conventional PI control. Here the feedrate, product size and sump are plotted for the grinding line optimal control (dotted line) and another grinding line under conventional PI control (solid lines). A quantitative comparison of the performance of the two strategies shows that the model based strategy resulted in an increased throughput of 15% and a reduced product size of 5% +100 mesh when compared with the conventional PI approach.

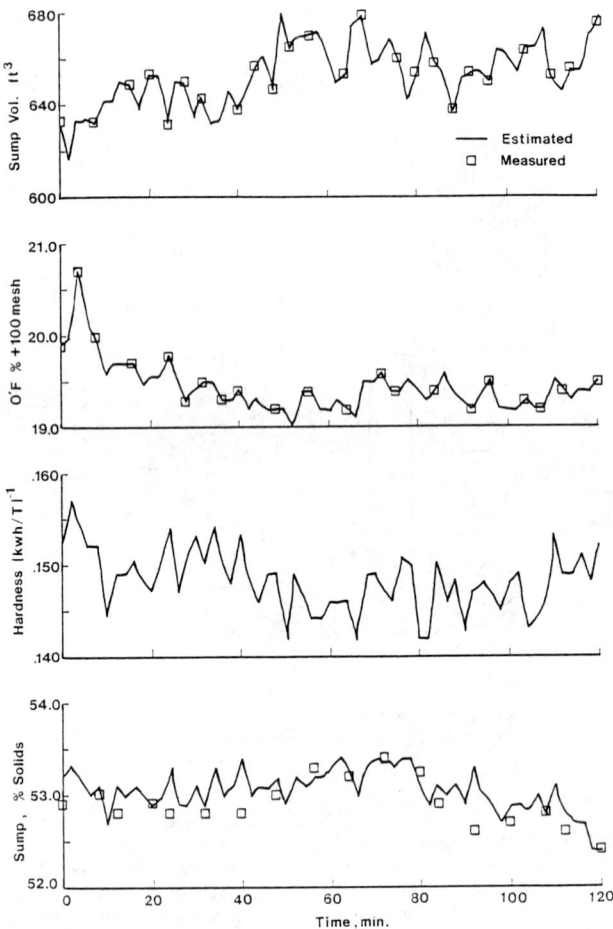

Figure 10 – On-Line Kalman Filter Estimates of Measured and Unmeasured Variables in Grinding Circuit

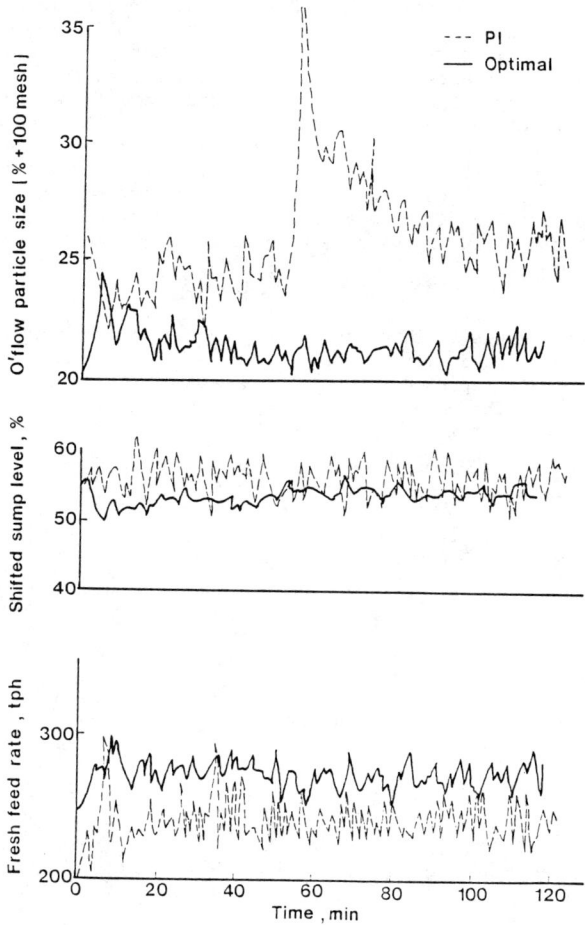

Figure 11 - Comparison of Conventional PI and Model Based Optimal Control System Performance for Grinding Plant

Conclusions

The purpose of this paper was to demonstrate that mathematical modeling and computer simulation have become very useful tools for mineral processing plant improvement. This was shown through a series of example applications taken from plants around the world. Steady state process models and the associated simulators were shown to provide the basis for reducing design risk for tumbling mills from $\pm$ 20% to less than $\pm$ 5% and to provide the basis for mineral processing plant circuit optimization producing increases in throughput of up to 25%. Dynamic models and simulators were shown to minimize the effort required for control system design through off-line evaluations. Finally, on-line models for control were shown to provide improvements in capacity of up to 15% with additional increases in product quality. In the future there will be a larger number of model and simulation applications in mineral processing plants and perhaps as the level of sophistication increases even greater improvements can be expected.

Acknowledgements

The author wishes to thank the many graduate students at the University of Utah who have worked with him to develop verification methodologies and to collect the data reported here. Also a debt of gratitude is owed to his colleagues both at the University of Utah and Control International for their input.

References

1. D.M. Himmelbau and K.B. Bishoff, <u>Process Analysis and Simulation:</u> <u>Deterministic Systems</u> (New York, NY: John Wiley & Sons, 1968), 3.

2. A.L. Mular and J.A. Herbst, "Digital Simulation: An Aid for Mineral Processing Plant Design," <u>Mineral Processing Plant Design,</u> ed. A.L. Mular and R.B. Bhappu (New York, NY: SME-AIME, 1978), 306-338.

3. J.A. Herbst, "Rate Processes in Multiparticle Metallurgical Systems," <u>Rate Processes in Extractive Metallurgy</u>, ed. H.Y. Sohn and M.E. Wadsworth (New York, NY: Plenum Press, 1979), 53-112.

4. K.V.S. Sastry and G.T. Adel, "A Survey of Computer Simulation Software for Mineral Processing Systems," <u>Control '84 Mineral/Metallurgical Processing</u>, ed. J.A. Herbst (New York, NY: SME-AIME, 1984), 121-130.

5 L.G. Austin and R.P. Gardner, "A Chemical Engineering Treatment of Batch Grinding," Symposium Zerkleinern, <u>The 1st European Symposium on Comminution</u>, ed. H. Rumpf and D. Behrens (Weinheim, Germany: Verlag Chemie, 1962), 217-248.

6. L.G. Austin and R.R. Klimpel, "The Theory of Grinding Operations," <u>Industrial Engineering Chemistry</u>, 56 (11) (1964), 18-29.

7. L.G. Austin, et al., "Simulation of Grinding Circuits for Design," <u>Design and Installation of Comminution Circuits</u>, ed. A.L. Mular and G.V. Jergensen (New York, NY: SME-AIME, 1982), 301-324.

8. L.G. Austin, R.R. Klimpel and P.T. Luckie, <u>Process Engineering of Size Reduction: Ball Milling</u>, (New York, NY: SME-AIME, 1984).

9. C.C. Hatch and A.L. Mular, "Simulation of the Brenda Mines Ltd. Secondary Crushing Plant," <u>Transactions of the SME-AIME</u>, 272 (1982), 1354-1362.

10. J.A. Herbst, K. Rajamani and D.J. Kinneberg, "ESTIMILL - A Program for Grinding Simulation and Parameter Estimation with Linear Models" (User's Manual, Dept. of Metallurgy, University of Utah, Salt Lake City, Utah, 1977).

11. J.A. Herbst and D.W. Fuerstenau, "Scale-up Procedure for Continuous Grinding Mill Design Using Population Balance Models," <u>Inter. J. of Mineral Processing</u>, 7 (1980), 1-31.

12. J.A. Herbst and K. Rajamani, "Developing a Simulator for Ball Mill Scale-Up - A Case Study," <u>Design and Installation of Comminution Circuits</u>, ed. A.L. Mular and G.V. Jergensen (New York, NY: SME-AIME, 1982), 325-342.

13. J.A. Herbst, Y.C. Lo and K. Rajamani, "Population Balance Model Predictions of Large Diameter Mills," <u>Minerals and Mineral Processing</u>, May (1985), 114-120.

14. J.A. Herbst, Y.C. Lo and J.E. Bohrer, "Increasing the Capacity of a Phosphate Grinding Circuit with the Aid of Computer Simulation," accepted for publication in <u>Transactions of the SME-AIME</u>, (1987).

15. R. Spring, et al., "Industrial Ball Mill Model Validation," Control '84 Mineral/Metallurgical Processing, ed, J.A. Herbst (New York, NY: SME-AIME, 1984), 71-76.

16. L.G. Austin and R.R. Klimpel, "Modeling for Scale-Up of Tumbling Ball Mills," Control '84 Mineral/Metallurgical Processing, ed. J.A. Herbst (New York, NY: SME-AIME, 1984), 167-184.

17. K. Rajamani and J.A. Herbst, "A Dynamic Simulator for the Evaluation of Grinding Circuit Control Strategies," Preprints - European Symposium on Particle Technology 1980, Volume A, ed. K. Schonert (Amsterdam: Dechema, 1980), 64-81.

18. D. Houdin, F. Flament and M.D. Everell, "Practical Benefits Obtained from Process Analysis, Simulation and Optimization Applied to Industrial Comminution and Classification Units," Preprints - XIV International Mineral Processing Congress (Toronto: 1982), paper III-3.

19. A.J. Lynch, Mineral Crushing and Grinding Circuits, (Amsterdam: Elsevier, 1977).

20. T.P. Meloy and B.H. Bergstrom, "Matrix Simulation of Ball Mill Circuits Considering Impact and Attrition Grinding," Proceedings - VII International Mineral Processing Congress (New York: 1965), 19-31.

21. W.J. Whiten, "Models and Control Techniques for Crushing Plants," Control '84 Mineral/Metallurgical Processing (New York, NY: SME-AIME, 1984), 217-224.

22. J.A. Herbst and A.E. Oblad, "Modern Control Theory Applied to Crushing, Part I," IFAC Symposium on Automation for Mineral Resource Development (Brisbane, Australia: 1985), 301-306.

23. V.K. Karra, "Development of a Model for Predicting the Screening Performance of a Vibrating Screen," The Canadian Mining and Metallurgical Bulletin, 72 (804) (1979), 167-176.

24. L.R. Plitt, J.A. Finch and B.C. Flintoff, "Modelling the Hydrocyclone Classifier," Preprints - European Symposium on Particle Technology 1980, Volume A, ed. K. Schonert (Amsterdam: Dechema, 1980), 790-804.

25. A.G. Lynch and T.C. Rao, "Modeling and Scale-Up of Hydrocyclone Classifiers," Proceedings - XI International Mineral Processing Congress (Cagliari, Italy, 1975), 245-269.

26. G. Ferrara and U. Preti, "A Contribution to Screening Kinetics," Proceedings - XI International Mineral Processing Congress (Cagliari, Italy, 1975).

27. G. Ferrara, U. Preti and G.D. Schena, "Modeling of Screening Operations," accepted for publication in International Journal of Mineral Processing, (1987).

28. N. Arbiter and C.C. Harris, "Flotation Kinetics," Froth Flotation 50th Anniversary Volume, ed. D.W. Fuerstenau, (New York, NY: SME-AIME, 1962), 215-246.

29. O. Bascur and J.A. Herbst, "Dynamic Modeling of a Flotation Cell with a View Toward Automatic Control," Preprints - XIV International Mineral Processing Congress (Toronto, 1982), paper III - 11.

30. P.C. Kapur and S.P. Mehrotra, "Phenomenological Model for Flotation Kinetics," Transactions of the Institute of Mining and Metallurgy, Section C, 82 (1973), C229-234.

31. R.P. King, "A Computer Program for the Simulation of the Performance of a Flotation Plant," (Report No. 1436, National Institute for Metallurgy, Johannesburg, South Africa, 1973).

32. R.P. King, "The Use of Simulation in the Design and Modification of Flotation Plants," Flotation A.M. Gaudin Memorial Volume, Vol. 2 (New York, NY: Society of Mining Engineers, 1976), 937-962.

33. A.G. Lynch, et al. Mineral and Coal Flotation Circuits, Their Simulation and Control (Amsterdam: Elsevier, 1981).

34. S.P. Mehrotra and A.K. Saxena, "Effect of Process Variables on the Residence Time Distribution of a Solid in a Continuously Operated Flotation Cell," International Journal of Mineral Processing, 10 (1983), 255-277.

35. T.S. Mika and D.W. Fuerstenau, "A Microscopic Model of the Flotation Process," Proceedings - VIII International Mineral Processing Congress, Vol. 2 (Leningrad: 1969), 246-269.

36. M.H. Moys and R.P. King, "A Computer Programme for the Estimation of Parameters in Flotation," (Report No. 1568, National Institute of Metallurgy, Johannesburg, South Africa, 1973).

37. A.G. Niemi, "A Study of Dynamic and Control Properties of Industrial Flotation Processes," Acta Polytechnica Scandinavica, Chemical Metallurgy Series No. 48 (1966), 1-111.

38. A.G. Niemi, S. Kurronen and H. Kuopanportti, "Unit Operations, Modeling and Computer Aided Design of Flotation Plants," Preprints - XIV International Mineral Processing Congress (Toronto: 1982), paper III-3.

39. K.V.S. Sastry and D.W. Fuerstenau, "Theoretical Analysis of a Counter Current Flotation Column," Transactions of Society of Mining Engineers, (247) 1970, 46-52.

40. J.A. Herbst, L.B. Hales and R. Zaragoza, "Strategies for the Control of Flotation Plants,"

41. K.V.S. Sastry, "Mathematical Modeling of Pellet Growth Processes and Computer Simulation of Pelletizing Circuits," Proceedings of the 2nd International Symposium on Agglomeration (Nurenberg, West Germany: 1981), 1-15.

42. P.E. Wellsted, N. Munro and M. Cross, "Modeling, Stability and Control of an Iron Ore Balling Drum Circuit," Transactions of the Institute of Measurement and Control, 2 (2) (1980), 86-99.

43. R.M. Young, M. Cross and R.D. Gibson, "Mathematical Model of Grate-Kiln-Cooler Process Used for Induration of Iron Ore Pellets," Ironmaking and Steelmaking, 6 (1) (1979), 1-13.

44. B. Fitch, "Current Theory and Thickener Design," Industrial and Engineering Chemistry, 58 (10), 1966, 18-28.

45. P. Kos, "Fundamentals of Gravity Thickening," Chemical Engineering Progress, 73 (11) (1977), 99-105.

46. G.J. Kynch, "A Theory of Sedimentation," Transactions of the Faraday Society, 48 (1952), 166-176.

47. A.R. Taver, D.L. Vives and D.M. Kennedy, "A Model for Continuous Thickening," AICHE Symposium Series, 74 (174) (1978), 67-74.

48. O.A. Bascur and J.A. Herbst, "Improved Thickener Performance Through the Use of an Extended Kalman Filter," Concentration and Dewatering Symposium, ed. A.L. Mular (New York, NY: SME-AIME, 1986).

49. F. Concha and O.A. Bascur, "Phenomenological Model of Sedimentation," XII International Mineral Processing Congress, vol. 1 (Sao Paulo, Brazil: 1977), 29-46.

50. F.M. Tiller, "Solid-Liquid Separation," Theory and Practice of Solid-Liquid Separation (Houston, TX: University of Houston, 1975).

51. R.J. Wakeman, "Filtration Post-treatment Processes (London: Elsevier, 1975).

52. R.J. Wakeman, "The Prediction and Separation of Cake Dewatering Characteristics," Filtration and Separation, November-December (1979), 655.

53. J.C. Ruebush, J.A. Herbst and K. Rajamani, "SIMPLANT - A Program for the Simulation of Individual Mineral Behavior in an Integrated Grinding and Flotation Circuit" (User's Manual, Dept. of Metallurgy, University of Utah, Salt Lake City, Utah, 1980).

54. CANMET - Canadian Centre for Mineral and Energy Technology, Simulated Processing of Ore and Coal Manual (Ontario, Canada: CANMET, 1981).

55. R.D. Peterson and J.A. Herbst, "Estimation of Kinetic Parameters of a Grinding - Liberation Model," International Journal of Mineral Processing, 14 (1985), 111-126.

56. R.P. King, "MODSIM, Modular Method for Design, Balancing and Simulation of Ore Dressing Plant Flowsheets," (Report No. G9, Dept. of Metallurgy, University of Witwatersrand, Johannesburg, South Africa, 1983).

57. J.A. Herbt, G.D. Schena and L.S. Fu, "Computerized Design of Comminution Circuits," (Paper presented at the SME Annual Meeting, New Orleans, LA, 206, March 1986) 8.

58. W.T. Pate and J.A. Herbst, "GRINDSIM.S - A Fortran Simulator for Steady State Grinding Circuits," (manuscript in preparation, 1987).

59. G.T. Adel and K.V.S. Sastry, "Design Aspects of a Mineral Processing Simulation Package," Proceedings of the 17th Annual APCOM Symposium (New York, NY: SME-AIME, 1982) 681-692.

60. W.T. Pate and J.A. Herbst, "GRINDSIM.D - A Fortran Simulator for Steady State Grinding Circuits," (manuscript in preparation, 1987).

61. R.D. Peterson, "Estimation of Parameters for and Verification of an Integrated Model for Grinding and Flotation Circuit Simulation," (M.S. Thesis, University of Utah, Salt Lake City, 1983).

62. J.C. Ruebush, "Simulation for Individual Mineral Behavior in an Integrated Grinding and Flotation Plant" (M.S. Thesis, University of Utah, 1982).

63. J.A. Herbst and K. Rajamani, "The Application of Modern Control Theory to Mineral Processing Operations," Proceeding, 12th CMMI Congress, vol. 2, ed. H.W. Glenn (Johannesburg, South Africa: South African Institute of Mining and Metallurgy, 1982), 779-792.

64. O.A. Bascur and J.A. Herbst, "Mineral Processing Control in the 1980's," Control '84 Mineral/Metallurgical Processing, ed. J.A. Herbst (New York, NY: SME-AIME, 1984), 197-215.

65. O.A. Bascur and J.A. Herbst, "Dynamic Simulators for Training Personnel in the Control of Grinding/Flotation Systems," IFAC Symposium on Automation for Mineral Resource Development, (Brisbane, Australia: 1985), 307-316.

66. J.A. Herbst, L.B. Hales and W.T. Pate, "Documentation of Improvements Resulting from the Automation of Mineral Processing Operations" (Paper presented at Tecnomin, Lima, Peru, 1986.

67. O.A. Bascur, "Modeling and Computer Control of a Flotation Cell" (Ph.D. dissertation, University of Utah, Salt Lake City, 1982).

68. J.A. Herbst and W.T. Pate, "The Power of Model Based Control" (Paper presented at the IFAC Symposium on Automation in Mining, Mineral and Metal Processing, Tokyo, 24-29 Feb., 1986).

69. J.A. Herbst and D.W. Fuerstenau, "Mathematical Simulation of Dry Ball Milling Using Specific Power Information," Transactions of SME-AIME, 254 (1973), 343-348.

GUEST LUNCHEON SPEAKER

COMPUTERS IN MANUFACTURING

--A FEW HARD-WON LESSONS

David W. Fradin
Manager, R&D Project Office
Alcoa Laboratories
Alcoa Center, Pa   15069

Abstract

    Certain aspects of the application of computers in manufacturing are
discussed within the context of a simplified, 4-part view of a total
manufacturing enterprise. This view is used both to serve as a reference
point for discussing the development of factory software and to organize
a vision of computer-aided-manufacturing (CIM) of the future. The
perspective presented, while not intended to be comprehensive, suggests
some limitations and pitfalls of software development for the factory and
suggests a more organized view of CIM in the distant future than may be
currently understood by the non-expert.

I. Introduction

The concept of Computer-Integrated-Manufacturing (CIM) has descended on American manufacturing in a cloud of confusion, inconsistency, and unbounded promise. Countless articles in trade magazines, in business journals, and in journals specifically devoted to CIM describe aspects and forms of CIM without truly clarifying the concept or providing a definitive set of pointers for the unschooled. Technical descriptions of large-scale CIM projects by arospace firms and by large government manufacturing systems project a logical and monotonic path to CIM salvation and project benefits that should convince the greatest skeptics and most cash-starved firms that CIM is right for American industry.

In fact, CIM is a vaguely defined concept that speaks to several trends and forces that are affecting manufacturing businesses. These trends and forces include:

the recognition that managing manufacturing enterprises successfully requires managing information intelligently;

the growing confidence in large-scale computer use within the factory, a confidence that builds on a gradually increasing base of good experiences with computing systems (amid many painful experiences);

a strong American bent for integrated systems and "top-down" logical consistency;

and continuing pressures to find technology-driven responses to the growing perceived inefficiencies of American manufacturing relative to Japan.

The move to CIM is complicated by the facts that (1) manufacturing companies--even large aerospace firms--are notoriously inefficient in software development and (2) companies differ significantly in their environment, detailed needs and specific opportunities for application of computing systems. Further, where there are consistent lessons to be learned in terms of success factors and indicators of failure, the practitioners of CIM, who have fought the wars and licked the wounds, have an almost unspoken conspiracy to talk about the system in its final, unblemished form rather than to talk about the process of trying to develop a CIM environment, a process with its agonies, waste, occasional successes-- and valuable insights.

The discussion that follows attempts to convey some practical insights about CIM by describing some of the contextual issues for computers in manufacturing and by offering a few lessons based on experiences that are probably typical for many CIM practitioners. The unschooled hoping to apply CIM magic to their manufacturing enterprises will not, unfortunately, find the long-sought recipe from this discussion nor will the CIM professionals (whoever that may be) find the systematic framework for their CIM efforts. The recipe and the full systematic framework wait for the future, providing a healthy challenge and opportunity for the consultant and the academic.

Section II summarizes a particular model of CIM that is useful for the discussions that follow and that is consistent with the basic view of CIM held by the Air Force and the aerospace industry. This model is picked up again in Section III which builds to a CIM vision. Both sections are interrupted repeatedly by lessons that attempt to drive the vision back to reality. Some of the lessons and thoughts about computers in manufacturing are repeated in the final section.

II. Context

For the current discussion, the domain of CIM is taken to be the technical threads of a manufacturing business that impact directly on the factory--more specifically, the design engineering, quality, manufacturing engineering, and production functions of the business. While this view tends to ignore the full logistics cycle so important in aerospace, the generalization to include the full life-cycle from systems concept to field support will not add to the admittedly selective discussions that follow.

This view of the CIM domain leads to an idealized 4-part model for CIM that is shown in the Figure 1. The first part deals with ENGINEERING DRAWINGS, the information tool that represents most of the documented "jewels" of those businesses that deal with finished, high-value-added products. The rest of the documented "jewels" are contained in process instructions and specifications. For material-supply businesses such as the conventional aluminum industry, the documented "jewels" are represented by process instructions and specifications alone. The first part, then, strictly applies only to finished-product businesses.

The 4-Part View....

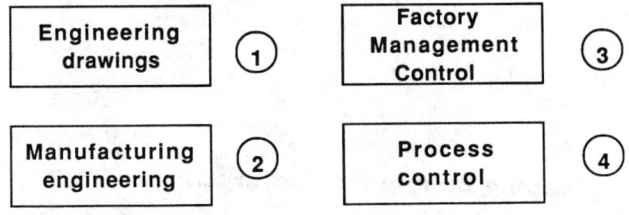

FIGURE 1

MANUFACTURING ENGINEERING is the second part of this idealized CIM domain. It contains process planning and other support functions that are needed to define the production process. Process planning is part of this area, as is numerical control (NC) programming for metal working. FACTORY MANAGEMENT CONTROL refers to the transaction control activities that are imbedded in MRP-II (Manufacturing Resource Planning) and production floor scheduling. Finally, PROCESS CONTROL refers to the systems and techniques that ensure accuracy of the process itself. Process control may involve control of dimensions and pressures and temperature in a rolling process, of pressure and temperature in a polymer curing process, or dimensional control and tool condition monitoring in metal working.

Implicit in this simple model is the existence of interactions that connect these blocks in pairs or in groups. ENGINEERING DRAWINGS, for example, can interact strongly with PROCESS CONTROL. This interaction has been made explicit in the work on quality control methods pioneered by Genichi Taguchi, which seeks to desensitize the manufacturing process to manufacturing variables (such as temperature variations in an autoclave) by careful product design (as, in this example, by the choice of material as specified in the engineering drawings). At a less rigorous level, a number of companies are implementing methods for moving constraints and guidelines from manufacturing upstream into design as design rules. MANUFACTURING ENGINEERING can also be seen to interact with FACTORY MANAGEMENT CONTROL and PROCESS CONTROL in many situations.

Lesson #1--"Farming" Can Yield a Big Harvest

Good manufacturing does not begin with computers. As quality businesses in the US and Japan have shown, there is a significant harvest to reap in terms of efficiency and effectiveness by paying careful attention to such factors as:

> downtime control for manufacturing resources
> factory organization and layout
> product design
> tooling design
> management of the "people" dimension

As suggested by Figure 2, most of the opportunity for benefits comes from intelligent "farming," not from the use of computers. The ability of a product business to respond to changing market conditions with new products or new product designs--an ability that might be called "product agility"-- also depends strongly on "farming," as suggested by Figure 3. Computers, then, do not replace good management in manufacturing enterprise. Instead, they can help good managers do a better job.

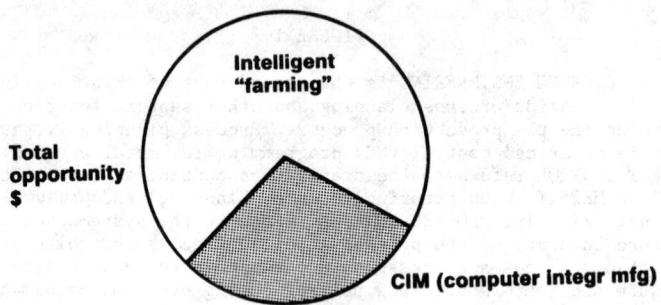

Reaping Benefits in Factory ...

Intelligent "farming"

Total opportunity $

CIM (computer integr mfg)

... CIM Ain't the Whole Bag

FIGURE 2

54

... Even with Leadtime or Product Agility

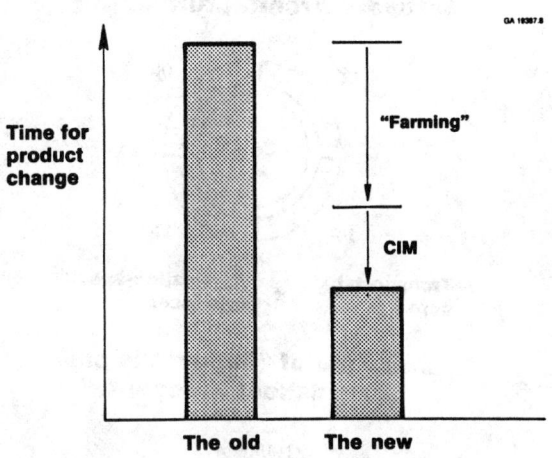

FIGURE 3

<u>Lesson #2--Managing Software Projects is Equivalent to Smoking 3 Packs of
Cigarettes Per day</u>

Few frustrations in the world of CIM are greater than those associated
with developing and implementing large-scale software systems. An actual
experience with a real-time shop-floor scheduling system, known as the
Dynamic Scheduler, illustrates some of the difficulties and lessons.
Although this specific case study involves the FACTORY MANAGEMENT CONTROL
portion of the 4-part model, its implications apply broadly to the full
domain of CIM.

The Dynamic Scheduler was being developed by Company A under a
government contract. It was conceived of as having a core controller to
schedule resources within an aerospace factory that produces discrete
parts on a batch basis. Added to the core, in building-block fashion,
were interfaces to connect the core to the various data sources within
the factory and to users, as suggested by Figure 4. The kinds of decisions
made by the Scheduler were to include, for example, the decision to
schedule a specific part to a specific horizontal lathe. The number of
decisions and thus the complexity of the core increases with the number, N,
of manufacturing stations and resources in the system, scaling roughly as
N! for large systems. Since the factory of Company A for which the
scheduler was to be used was fairly large, the core software development
was a significant task.

Software Architecture ...

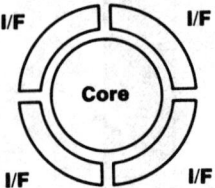

Transportable core + **Application-specific interfaces**

... A Kind of "Super" Visicalc/ Spreadsheet Concept

FIGURE 4

As the original government contract was being proposed, Company B, another aerospace firm, saw an opportunity to use the Dynamic Scheduler in several of its factories. Company B became a major part of the contract, with the role of developing interfaces specific to its factories and products. Its projected effort involved about 15 manyears. A project team was organized for Company B; a detailed project plan was developed; and effective working relationships were set up with Company A. Because of the inherent technical logic of the effort and the careful organization of both Company A's and Company B's efforts, the Dynamic Scheduler was clearly a success on its way to happen.

But then the clouds began to gather. By 8 months into the 2 1/2 year program, significant cost overruns were projected and a 1 year schedule slippage was projected for the delivery of the core software from Company A. In addition, the software functionality in the core was projected to be deficient compared to original expectations. Within Company B, the plant in which the scheduler was to be initially applied and the internal information systems group working a portion of the Company B subcontract sought to withdraw from the program. Despite excellent technical work, the effort at both companies was coming apart before the contract was half completed.

The experiences of this program suggested that the actual progress on the program could not be clearly understood from looking at the technical dimension alone. The technical people were success-oriented and appeared to be capable of moving past any hurdle encountered. In fact, the problems on the program were not strictly technical. In order to work effectively, the Dynamic Scheduler had to satisfy a number of end-users within both companies who often had divergent requirements and biases. The needs of this end-user set were not adequately addressed by the "force-fit" of a standard core scheduler. Compounding the problem of dealing effectively with the end-user set, the software developers were in charge, not the end-users who were the actual customers. It was not always apparent whether technology was trying to fix the end-user or whether the problems perceived by the end-user were being addressed by technology. Another serious problem was that the interdependence of the two companies was too great. Two thousand miles separated the two development groups, and the barrier of two different corporate structures made it difficult to achieve common goals and perceptions.

Most large-scale software projects (for firms outside the computing industry) are "3X" projects. The resources needed to complete them are at least three times the planned resources. Superbly managed software projects tend to be "2X" project. The Dynamic Scheduler was a "3X" project-- but it did not have to be. To make software development and implementation, certain rules-of-thumb are appropriate:

1. Make the end-user part of the project very early in the cycle. Treat him as the customer and the boss throughout, with his expectations clearly documented and updated.

2. Look for the problems that can occur, and identify early indicators of problems. Full contingency planning is probably wasteful, but some is required. Temper the success orientation of the technical personnel with a realistic margin for problems.

3. Document acceptance specs very early in the cycle, and make certain that the early scoping and requirements phases are thorough and adequately documented.

Large-scale software projects that appear to have no risk, that are developed off-line by a crack technical team with little end-user involvement until the end, that seem to ignore the complexities created by existing computing software systems, and that have minimal documentation are suspect. Managing software involves much more than monitoring milestone charts.

III. Building a Vision

A number of vendors offer integrated computer systems for the factory floor. Most of these systems emphasize the FACTORY MANAGEMENT CONTROL portion of the 4-part view and are appropriate only to comparatively small production operations. Commercial systems for such functions as Manufacturing Resource Planning (MRP-II), capacity planning, and batch scheduling have been developed for medium-scale manufacturing enterprises, but these systems are not truly integrated data systems even for FACTORY MANAGEMENT CONTROL. Large-scale enterprises tend to develop major systems pieces--not fully integrated systems--using their own resources or in partnership with software systems houses.

The "vision" for CIM lies beyond these current systems. There are at least three frontiers for CIM that build towards a total CIM vision: Product Definition Systems (PDS) or total electronic drawings, application software for MANUFACTURING ENGINEERING, and PROCESS CONTROL using intelligent systems. Each of these is briefly discussed.

1. Product Definition Systems

This new area of work, which is getting increasing attention within the aerospace community, involves ENGINEERING DRAWINGS. In this area, the industry, led by a number of Air Force sponsored efforts, is seeking to develop systems for computer-recognizable drawings. Such systems are called "Product Definition Systems (PDS)."

In its simplest form, Product Definition seeks to capture product geometry in a manner that can be unambiguously recognized by a computer. Associated with the geometry is a set of non-geometry information involving tolerances and notes. The requirements for such complete electronic drawings lead naturally to technology developments in feature-based data models and solid modeling. In a more complete form, however, Product Definition is much more than data representation technologies. As suggested by Figure 5, the Product Definition System (PDS) involves interfaces, workstations that contain application programs for process planning and other MANUFACTURING ENGINEERING functions, and configuration management.

Product Definition Support System
Data Structures + Interface System

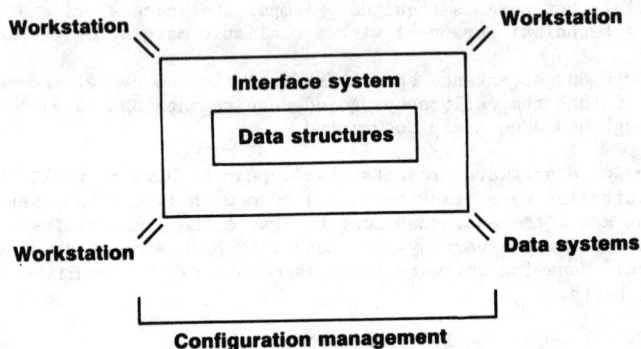

FIGURE 5

The systems issues are particularly significant for PDS because of the complexity of the parts geometry and the sheer volume of data. In the two major aircraft engine suppliers, for example, the engineering drawings for compressor casings are so complex that they may need up to 2 months for interpretation and analysis by an experienced manufacturing engineer. The number of new drawings and major revisions encountered in a single year by such companies approaches 100,000 with several million drawings on file. Building a system that can handle such enormous volumes of data efficiently while interfacing with several thousand frequent and occasional users both within and outside the company (eg, vendors) is a serious challenge that requires a massive commitment of resources.

In a fundamental sense, PDS is the "glue" that can integrate the entire CIM system, at least in product businesses. The principal output of the upstream functions of design and analysis is the set of drawings that define the product characteristics. It is from these drawings--to be automated into PDS data systems--that MANUFACTURING ENGINEERING develops process definition. If this critical "glue" is automated as PDS, then there is the potential for total integration of the product/production cycle, and the environment is created for efficient communication of data between the upstream engineering functions and the downstream manufacturing functions.

2. Application Software

It is well appreciated that current techniques of automation have tended to greatly reduce the levels of direct labor in the plant but at the expense of increased indirect labor. A major component of indirect labor within the technical threads of a manufacturing enterprise is MANUFACTURING ENGINEERING. Automating the functions within this component--process planning, NC programming, quality assurance, inspection-plan generation for inspection systems--requires the development of application software that can be driven off a PDS database.

There are a number of commercially available systems for several MANUFACTURING ENGINEERING functions. Most of these systems are limited by scale, however, and by the fact that they are not capable of being driven off a true PDS database (except for systems that involve only strongly focused product designs). Application software that can interpret geometry and drawing notes without human intervention tends to require sophisticated rules, such as in an expert or artificial intelligence system, or it tends to be significantly limited in scope and functionality.

One example of application software, which is commercially available, is process planning. The role of conventional manual process planning consists of using data from both engineering drawings and a manufacturing database to produce routings, machine settings, tooling plans, and other information associated with production. Computer-aided Process Planning (CAPP) software represents the state-of-the-art in automating this functionality, but, as suggested by Figure 6, CAPP systems are still limited in their ability for automated decision-making as well as in their ability to couple to a true PDS database. CAPP systems of today will evolve into systems that incorporate progressively higher levels of decision-making, including the ability to interpret geometry features and non-geometry attributes of complex product forms.

A Broad Spectrum of Computer Enhancements Possible ...

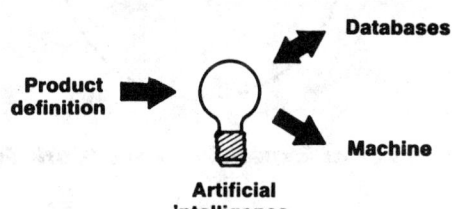

Increasing computerization

- Manual operation with computer query
- Computer-assisted process planning (CAPP)
- Automated decision making

Product definition ➡ Databases

Machine

Artificial intelligence

... But Even Capp Isn't Really Here Today

FIGURE 6

3. Intelligent Processing

At the PROCESS CONTROL level, prototype systems are now under development for intelligent control using sophisticated control architectures and artificial intelligence (AI) techniques. In such systems, there are conceptually two levels of control--on line and off-line or non-real-time. The non-real-time systems may contain physics models of the process (which could involve, for example, curing of polymer composites or conventional maching of metals), a model of the machine or processing equipment, the interface to PDS, and processing logic that is organized with AI techniques. Within the on-line system are the sensors, "reduced" processing models that are driven by the sensor data, and algorithm-based logic for real-time control. Data transfer between the two domains occurs for status and control of exceptions.

A few programs in intelligent control of processing are currently being sponsored through government contracts. These efforts go considerably beyond statistical process control and are capable of being integrated into a full CIM system of the future. Intelligent processing, as well as CAPP, will hang off the PDS "hub" in the future.

Lesson #3--Beware the Demo Trap

Advanced systems for each of the domains of the 4-part view are under development. Most begin their lives with concept designs and demonstration systems or prototypes. And most suffer from the tendency of the technologist to represent these demos as completed technologies ready for the factory. In fact, as suggested by Figure 7, the successful operation of a prototype may involve only a small portion of the resources needed to field a practical system.

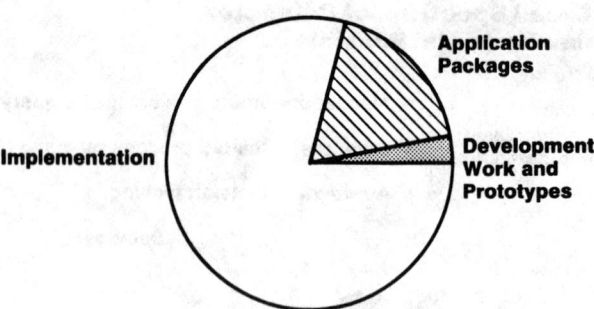

Demos Have the Flash ...

Application Packages

Development Work and Prototypes

Implementation

... But a Lot of Expensive Hard Work Follows

FIGURE 7

The limitations of the demo must be recognized early. Insufficient documentation, "canned" shows, insufficient involvement by the end-user all suggest systems that are early in their development cycle. Though the technologist may believe that the job is "90% completed," the remaining 10% may contain substantial risks as well as substantial resource needs. The practical issues associated with factory hardening or factory acceptance may turn a successful demo into a wasteful failure.

Another example of a demo trap is the prototype CIM system that seeks to create a fully integrated product flow--design through finished part. Such systems are being described in the literature. Demos of such systems may have all the apparent functionality of a CIM "vision", but the scope and interfaces to the demo system may be so limited as to have little practical value except as a research tool.

IV. Conclusions

Computers are firmly entrenched in manufacturing today in applications such as:

- business information systems
- local databases as aids to inventory, time & attendance, and manufacturing engineering functions
- batch scheduler for the factory floor and simulators
- process control subsystems

Stepping up to the large-scale and advanced systems embodied in a full CIM vision, however, is a difficult challenge that is beyond our current reach. Manufacturing enterprises have had considerable difficulty managing software developments, and factory software has been surprisingly resistent to transportability from one factory to another. In addition, many of the key technologies needed for true CIM--PDS, application software for manufacturing, intelligent processing--are still under comparatively early development.

There are no fail-safe rules for developing software, but is clear that certain commonsense guidelines associated with the involvement and buy-in by the end-user, with early and proper documentation of expectations and specs, and with realism associated with demos and limitations of commercial software for a CIM environment probably is one that incorporates skepticism and patience.

Computing systems have a growing place in manufacturing enterprises. The CIM vision suggests a level of automated information flow and control that will completely change the face of manufacturing. One wonders if, within that future New World, will the human role be simpler or more complex? Will those systems that are produced lead to more rapid change in the manufacturing world or will they reduce the pace of change? Will they reduce costs or will they just hide costs more effectively? Where will the man sit among the computers?

GRINDING CIRCUIT SIMULATION

Session Chairmen
K. Rajamani, University of Utah
D. Kinneberg, Leach and Garner

This session explores new mathematical concepts for convenient computer simulation and for real-time process control of dynamic behavior for industrial grinding circuits. Modeling techniques are used to design and simulate filtering systems and to optimize the pressure oxidation of gold ores, highlighting actual plant performance comparisons.

COMPUTER SIMULATION AND NUMERICAL ANALYSIS

OF PARTICLE BREAKAGE DURING METAL GRINDING

K. V. S. Sastry and Sureshan Moothedath[*]
Department of Material Science and Mineral Engineering
University of California, Berkeley
Berkeley, California 94720

Abstract

A mathematical model for grinding of metal powders is presented for the specific case of size reduction by binary breakage. A numerical solution procedure and a simulation algorithm have been developed and used to ascertain the functional dependence of the breakage parameters. It was found that the size distribution of metal powders does not become self-preserving and that the breakage rate function can be represented by a constant value and the breakage distribution by a simple random function. Computer simulation predictions for the metal particle-size distribution have been found to be in excellent agreement with the experimental data.

[*]Current Address: National Council for Cement and Building Research, Ballabgarh, Haryana 121 004, INDIA

Introduction

Grinding of metal particles by vibration milling is an important process by which lustrous, flaky metal powders are manufactured. One of the largest single uses of metal powders is in the paint and pigment industry where the luster and flaky nature of the particles is highly desirable (1). Even though the milling process is quite time consuming it is popular in the industry. Vibration mills are reported to yield better results than ball mills especially for low throughput applications (2).

An extensive experimental investigation of vibration milling of metal particles was conducted and the qualitative analysis of the results was reported elsewhere (3,4). The mechanisms of grinding of metal particles could be represented as a combination of flattening, attrition, and breakage (4,5). During the initial stages of milling, where bulk shaped particles are fed to the mill, flattening was reported to be the rate governing mechanism for aluminum, but for copper flattening and attrition were found to be more predominant. However for the rest of the milling process, it was concluded that the size reduction takes place by the mechanism of binary breakage of particles for both aluminum and copper powders.

A mechanistic model for the vibration milling of metal powders was developed earlier by making use of the population balance approach (5). Special cases of the model were studied where only one of the mechanisms of size change is rate controlling. The study included analytical solution for pure flattening and moment methods for flattening and breakage. This paper provides a formal assessment of the breakage mechanism and breakage parameters during vibration milling by making use of computer simulation and numerical analysis. Also a comparison of experimental results on particle size distribution is made with those predicted by computer simulation.

Mathematical Model for Particle Breakage

The general population balance model for grinding of metal particles in a batch system given by Sastry and Moothedath (5) can be modified for the case of binary breakage to give:

$$\frac{\partial n_{\alpha,\beta}(\alpha,\beta,t)}{\partial t} = - k(\alpha,\beta)\, n_{\alpha,\beta}(\alpha,\beta,t) + \int_{\alpha}^{\infty} 2\, b(\alpha,\alpha';\beta)\, k(\alpha',\beta)\, n_{\alpha,\beta}(\alpha',\beta,t)\, d\alpha$$

$$(1)$$

where

$n_{\alpha,\beta}(\alpha,\beta,t)\, d\alpha\, d\beta$ = number of particles with projected area diameter of α to $\alpha+d\alpha$ and mean thickness of β to $\beta+d\beta$, $n_{\alpha,\beta}(\alpha,\beta,t)$ is also known as joint-density function with reference to α and β;

$k(\alpha,\beta)$ = rate of binary breakage of particles of diameter α and $\alpha+d\alpha$ which is also known as breakage rate function; and

$b(\alpha,\alpha';\beta)\, d\alpha$ = number fraction of particles of size α to $\alpha+d\alpha$ and β produced due to breakage of particles of sizes α and β, and $b(\alpha,\alpha';\beta)$ is also known as breakage distribution function.

Sastry and Moothedath (5) also proposed the following functional relationships for the breakage rate and distribution functions after making use of a number of mathematical and physical reasons:

$$k(\alpha,\beta) = -k_B\,\alpha^{2u}\,\beta^{-v} \qquad (2)$$

where u and v are arbitrary and positive, k_B is the rate parameter dependent on material properties, mill dimensions and operating conditions; and

$$b(\alpha,\alpha';\beta) = b(\alpha,\alpha') = [2g\,(\alpha/\alpha') - 12(g-1)(\alpha/\alpha')^3\{1 - (\alpha/\alpha')^2\}]\,/\,\alpha' \qquad (3)$$

where g is an arbitrary constant bounded by 0 and 1.

It is further reasonable to expect that by the time the particle size reduction is taking place purely by binary breakage, all the particles must have been flattened to a constant critical thickness, of say β_α. Mathematically,

$$n_{\alpha,\beta}(\alpha,\beta,t) = I(\beta - \beta_{cr})\,n_\alpha(\alpha,t) \qquad (4)$$

where

$I(\beta - \beta_{cr})$ = the impulse function for β at β_{cr} and

$n_\alpha(\alpha,t)$ = number of particles with size α to $\alpha+d\alpha$ at time t; $n_\alpha(\alpha,t)$ is also known as marginal density function for α.

Transforming the equation from number-size distribution to mass distribution would be most convenient for the purposes of computation and comparison of model predictions with experimental results. This is accomplished by making use of the following relationship between number and mass-density function:

$$w(\alpha,t) = \int_0^\infty \pi\alpha^2\beta\rho\,n_{\alpha,\beta}(\alpha,\beta,t)\,d\beta\,/\,4m_T \qquad (5)$$

where $w(\alpha,t)d\alpha$ is the weight fraction of particles with size α, ρ is the density of the particles and m_T is the total mass of all particles, which is constant in the case of batch mills.

Multiplication of Eq. 1 by $(\pi\alpha^2\beta\rho/4)$ and making use of Eqs. 2 through 5 yields

$$\frac{\partial w(\alpha,t)}{\partial t} = -k_B^*\,w(\alpha,t) + \int_\alpha^\infty 2\,b(\alpha,\alpha')\,k_B^*\,\alpha^2(\alpha')^{2u-2}w(\alpha',t)\,d\alpha \qquad (6)$$

where $b(\alpha,\alpha')$ is given by Eq. 3 and

$$k_B^* = k_B\,\beta_{cr}^{-v}. \qquad (7)$$

We also find it useful to introduce the moments of the mass-size distribution which are defined by

$$\mu_i = \int_0^\infty \alpha^i\,w(\alpha,t)\,d\alpha \qquad (8)$$

67

where μ_i is known as i-th moment, and one can easily verify that the zero-th moment is unity and the first moment corresponds to the weight-mean diameter (actually weight-mean projected area diameter) of the particles.

Substitution of Eq. 8 into Eq. 6 gives, after a few manipulations (6):

$$\frac{d\mu_i}{dt} = k_B^* \, a_i \, \mu_{2u+i} \tag{9}$$

where

$$a = \frac{4\,g}{i+4} - \frac{48(g-1)(i+2)}{(i+6)(i+8)} - 1. \tag{10}$$

Analytical solution of Eq. 6 does not seem possible. However, there exist definite mathematical similarities between the present model for binary breakage of metal flakes and the model for breakage of fibers or linear polymers (7). Having seen these similarities, it was anticipated that the size distribution of metal particles should become self-preserving as it was shown by Goren (7) for the case of fibers. If this were to be the situation the size distribution of metal particles should become self-preserving; that is, reach an asymptotic form when properly normalized. To verify the validity of the self-preserving nature of the size distribution we introduce the following variables and functions

$$\delta = \alpha \, / \, \mu_1 \tag{11}$$

$$\emptyset(\delta,t)d\delta = w(\alpha,t)d\alpha \tag{12}$$

$$\sigma_i(t) = \int_0^\infty \delta^i \, \emptyset(\delta,t) \, d\delta \tag{13}$$

Here δ, $\emptyset$ and σ_i are respectively the size, mass-density and i-th moment normalized with reference to μ_1, the weight mean diameter. Combining Eqs 8, 11, 12 and 13 we also obtain

$$\sigma_i(t) = \mu_i(t) \, / \, \mu_1^i(t) \tag{14}$$

Also we have

$$W(\alpha,t) = \int_0^\infty w(\alpha,t) \, d\alpha = \Phi(\delta,t) = \int_0^\infty \emptyset(\delta,t) \, d\delta \tag{15}$$

where $W(\alpha,t)$ and $\Phi(\delta,t)$ correspond to cumulative and normalized cumulative weight distributions.

Now, we recognize that if the self-preserving hypothesis is valid then the functions $\emptyset$, Φ, and σ should become independent of t for large values of t. This also implies that these functions should be independent of initial conditions. Actually, it is convenient to analyze the simulated and experimental results by using $\mu_1(t)$ to represent the time variable.

It can be seen that the mathematical model for grinding of metal powders has only three parameters g, u and k_B^*. Clearly, numerical values of these parameters cannot be determined a priori and it is possible to calculate them only by matching model predictions with experimental data.

Numerical Solution for Computer Simulation

As mentioned earlier, analytical solution to the model equation given by Eq. 6 is not expected to be readily obtainable. Also, rather than attempting to normalize Eq. 6 by using Eqs. 11 through 14, it was decided to obtain a numerical solution. Actually, the numerical solution would also be convenient and more versatile for computer simulation. A numerical solution algorithm was developed by making use of the discretization procedure developed by Sastry and Gaschignard (8). The discretized equations were subsequently solved by the fourth-order Adams-Moulton prediction corrector procedure, coupled with the Runge-Kutta method. Numerical solutions were found to be quite accurate and rapid when discretizations were made into 20 size classes. The numerical results thus form the basis to assess - (i) the validity of self-preserving size distributions, (ii) the assessment of model parameters and finally (iii) the simulation of particle size distributions and other data for metal grinding.

Experimental Data

For the purposes of model parameter assessment and validation of model predictions, we make use of the data that are available from Sureshan's work (3). Briefly, the data were obtained in a laboratory vibrating mill by grinding aluminum, brass and copper particles with feed size in the range of 2 to 4 mm. Experiments were conducted under different operating conditions (including, for example, material loading, grinding aid dosage) to different time periods. The particle-size distributions were determined by sieve analysis. In this paper, we used only the data on grinding of aluminum powders. Unless specified, all the data plotted in the figures are for the operating conditions of 3% by weight addition of stearic acid (grinding aid), 70% by weight of mill loaded with grinding media and 0.0875 weight ratio of feed material to ball charge.

Analysis of Computer Simulation Results

Recognizing that the value of the parameter g is bounded by 0 and 1 and that when g=1 the expression for the breakage distribution function becomes the simplest, we decided to begin our computer simulations with that case. Then for the parameter u ($\geq$0) we chose to try the values 0, 0.5, 1 and 2. Finally for the initial particle size distribution, we assumed that the particles were uniformly distributed between 200 and 400 μm (that is they have a weight-mean diameter of 300 μm).

Figure 1 presents a plot of the second normalized moment as a function of the weight-mean diameter for the four different values of u. The results clearly indicate that σ_2 reaches an asymptotic value whenver u is not zero. Also, it is seen from this figure that higher the value of u the faster the second moment reaches its asymptotic value. Subsequently we tested for different initial conditions and also found that the second normalized moment reaches its asymptotic value only when u is not equal to zero. An illustration of this observation is shown in Figures 2 and 3 where the results are presented for the cases of u=1 and 0 respectively.

In order to see the relevance of the numerical findings to the experimental data, σ_2 is plotted as a function of μ_1 in Figure 4. This figure clearly shows that the second normalized moments of the experimental data do not reach an asymptotic value. Also, it is seen that the experimental results resemble most closely to the simulated results for the case of u=0.

Both the computed and experimental size distributions are normalized as per Eqs. 11 through 15 and presented in Figure 5. It is again seen from this figure that the self-preserving distributions are valid only for the cases of u=1 and u=2. Further numerical analysis with different initial feed size distributions, indicated that these self-preserving distributions are reached for sufficiently long grind times and only whenever the value of u is not zero. In the case of u=0, it was observed that the normalized distributions fall within a window which is determined both by the initial conditions and the time of grind. A typical window of these distributions is shown in Figure 5. We also notice that the experimental distributions lie within this window of distributions. Thus, the experimental data on size distributions confirm our earlier observation with the moment data that the value of u=0 in the functional form of the breakage rate function.

From these computer simulation runs we have discovered that the self-preserving distribution are possible only for non-zero values of u. Further, we found that the experimental size distributions do not exhibit self-preserving nature. Thus it appears reasonable to conclude that the value of u in Eq. 2 can be set to zero and consequently the breakage rate function becomes equal to a constant value of k_B^*.

Simplified Model for Computer Simulation

The observation that the breakage rate parameter corresponds to a constant value simplifies the model equations significantly. For example, with u=0, Eq. 9 becomes

$$\frac{d\mu_i}{dt} = k_B^* a_i \mu_i \qquad (16)$$

Whose solution is simply

$$\mu_i(t) = \mu_i(0) \exp [-k_B a_i t] \qquad (17)$$

Eq. 17 indicates that a semilogarithmic plot $\mu_i(t)$ versus t should yield a straight line with a slope of $(-k_B^* a_i)$. Figure 6 provides a plot for the variation of the weight-mean diameter with time t for a number of experiments. It might be observed from this figure that during the initial stages of the grinding experiment, the weight-mean diameter increases as expected for the flattening mechanism. Then of course, we find that for later stages of grinding the straight line relationship is found true. This is a further proof of the validity of a constant breakage rate parameter.

A more detailed analysis of the moment equations and experimental data (6) revealed, with correlational coefficients higher than 95 percent, that the value of g=1 in the breakage distribution function describes the vibration milling of metal powders. This again was a remarkable and a major simplifying experimental observation for the breakage distribution function resulting in the functional form

$$b(\alpha,\alpha') = \alpha / (\alpha')^2 \qquad (18)$$

which physically means that the metal particles undergo a random split in their projected area.

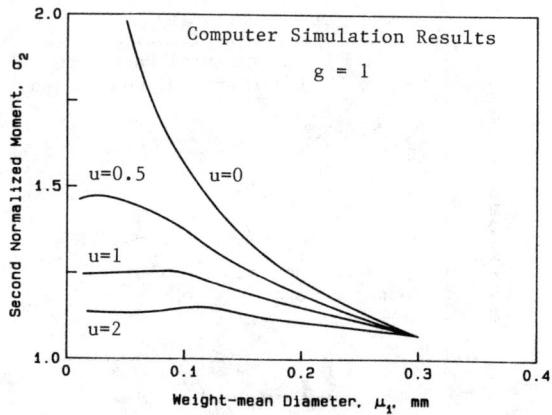

Figure 1. Dependence of second normalized moment on the
weight-mean diameter as predicted by computer simulation.

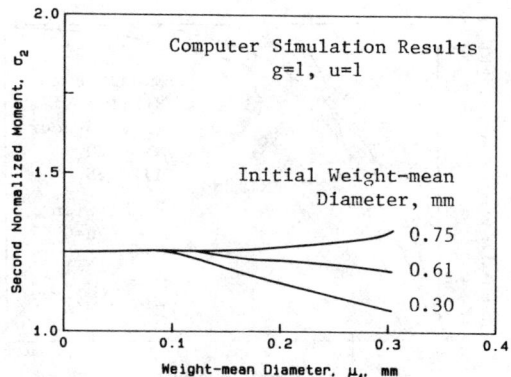

Figure 2. Computer results of the second normalized moment
for different initial conditions for the case of u=1.

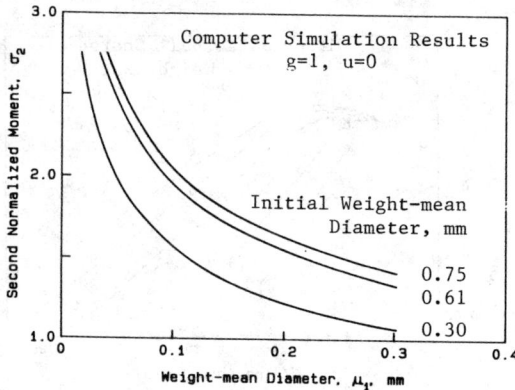

Figure 3. Computer results of the second normalized moment
for different initial conditions for the case of u=0.

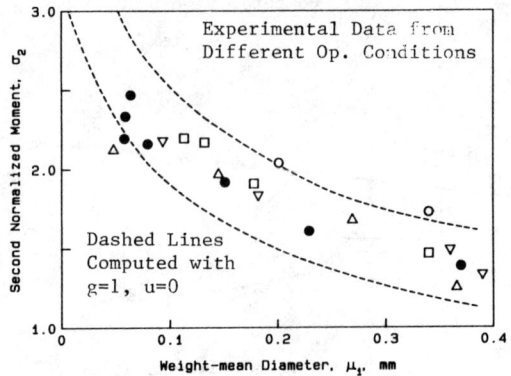

Figure 4. Experimental results for the second normalized moment of aluminum powders.

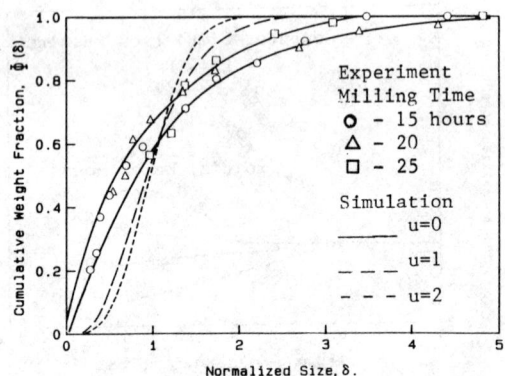

Figure 5. Experimental results and simulation predictions for the normalized size distributions.

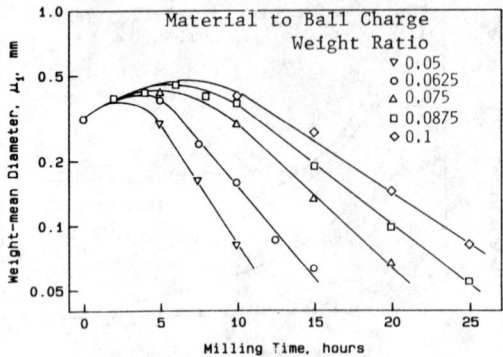

Figure 6. Experimental data on the variation of weight-mean diameter of aluminum powders with milling time.

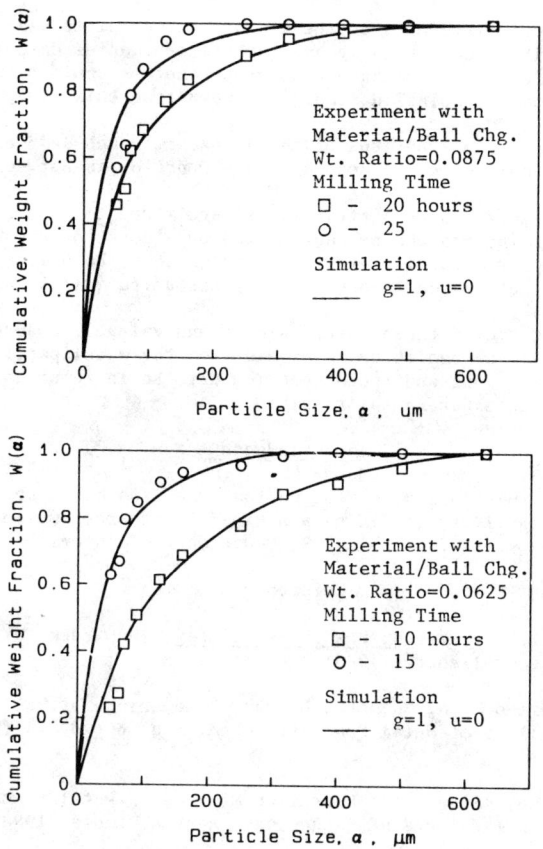

Figure 7. Comparison of computer simulation results and experimental data for the size distribution of aluminum powders produced by vibration milling.

Prediction of Particle Size Distributions

The two major simplifying assumptions for the breakage parameters were incorporated along with the other experimental data observations to make predictions of particle size distributions of metal powders under two different experimental conditions. The results are presented in Figure 7. It can be seen that the size distributions generated by computer simulations are in excellent agreement with the experimental results.

The computer program in its present form allows the prediction of size distribution and moment data under different operating conditions and mill design parameters for batch milling of metal powders.

Summary and Conclusions

This paper presented highlights of computer simulation for metal grinding by binary breakage. The simulation was based on a population balance model for which a solution was obtained by a novel discretization procedure. Numerical results on the moments and complete size distributions were compared with the previously available experimental data in order that the functional forms for breakage process parameters could be determined. Several significant conclusions could be drawn from this study.

1. The size distributions of metal powders reach self-preserving forms only when the breakage rate function is not constant.

2. Experimental size distributions were found to be non-self-preserving and can be represented by a constant breakage rate parameter. Earlier studies revealed that the breakage distribution can be described by the uniform distribution.

3. Making use of these rather simple and valuable observations, simulation predictions were made for the metal particle size distributions and these were found to be in remarkable agreement with the experimental results.

Acknowledgement

Computer simulations were made on the PDP-11/34 computer facilities of the California Institute of Mining and Mineral Resources which is established through a grant from the U. S. Department of Interior.

References

1. C. G. Goetzel, Treatise on Powder Metallurgy, (New York, NY: Inter/Science Publishers, 1963).

2. R. Vedaraman and R. M. Chandrasehkaran, "Comparison of Ball Milling and Vibration Milling of Metal Powders," Chem. Eng. World, 14 (7) (1979), 55-60.

3. M. K. Sureshan, "Studies in Vibration Milling of Metal Powders," (Ph.D Thesis, Indian Institute of Technology, Madras, India, 1982).

4. M. K. Sureshan, R. Vedaraman, and M. Ramanujam, "Influence of Operating Variables on the Vibration Milling of Metal Powders," Particulate Sci. and Tech., 1 (1983), 55-65.

5. K. V. S. Sastry and S. K. Moothedath, " A Mathematical Model for Vibration Milling of Metal Powders," Submitted to Powder Tech., 1987.

6. S. K. Moothedath, K. V. S. Sastry and M. Ramanujam, "Modeling and Experimental Verification of Vibration Milling of Metal Powders," Paper presented at the 1985 Annual Meeting of the Fine Particle Society, Miami, Florida.

7. S. L. Goren, "Distribution of Lengths in the Breakage of Fibers or Linear Polymers," Can. J. Chem. Eng., 46 (1968), 185-188.

8. K. V. S. Sastry and P. Gaschignard, "Discretization Procedure for the Coalescence Equation of Particulate Processes," Ind. Eng. Chem. Fundamentals, 20 (1981), 355-361.

EMPIRICAL MODELS FOR MULTIVARIATE FILTERING

OF CLOSED GRINDING CIRCUITS REAL TIME DATA

R. Lanthier and D. Hodouin

Département de mines et métallurgie
Université Laval
Québec, Canada G1K 7P4

Abstract

The phenomenological approach to dynamic modelling of grinding circuits leads to high-order non-linear models which are convenient for simulation but are not for real-time filtering of the signals delivered by the sensors. Empirical discrete models are more suitable for direct application of the linear systems methods. Low order discrete transfer functions are identified using data generated by a dynamic stochastic simulator calibrated on a two-stage industrial closed grinding circuit. Various model identification techniques are used and their ability to describe grinding dynamics are evaluated. The models are then used for the design of a filter involving circulating load and product fineness as observed variables. The robustness of the filter is tested by simulation and its usefulness discussed.

Introduction

Dynamic modelling of grinding circuits is useful for a wide variety of applications. It is the basic approach to the understanding of the process behavior when it is submitted to disturbances inherent to the industrial environment. Dynamic models incorporated in simulators of complex units are used for optimal adjustment of operating conditions and for designing classical regulators or model-based control strategies. In the latter case the model is explicitly incorporated in the equations of the regulator and of the filters used for an optimal utilization of the information given by the sensors. When building a model, two different approaches, phenomenological or empirical, may be taken, leading to different model characteristics which can be seen as advantages or drawbacks depending on the planned application.

Phenomenological models are based on the application of physical and chemical laws to the studied processes. They are generally robust and can be used over a wide range of operating conditions. This endows them with advantages for simulation applications as process analysis, equipment/flowsheet modifications, tuning of operating conditions, controler design (1,2), etc. However they are generally non-linear and involve complex equations which requires fairly large computational efforts. They are thus unsuitable for real-time control without discretisation and simplification stages which in turn, decrease their accuracy and generality.

On the contrary, empirical models are usually based on linear time-discrete input-output relationships. Since it is difficult to attach physical meaning to their parameters they are not good tools for simulation, and extrapolation out of the calibration zone is not allowed. Nevertheless, because of the small number of parameters involved in their development and because of their implementation facility they are powerful tools for dynamic analysis and modern control techniques as illustrated for grinding mill mixing analysis (3), flotation dynamics analysis (4) and grinding control (5).

In this work, both type of models are used for applications to which they are the most suitable. A phenomenological non-linear model of a two-stage closed grinding industrial circuit is implemented in a dynamic stochastic simulator. This model, although convenient for simulation, is inadequate for real-time filtering and control. Thus, low-order multiple-input-multiple-output (MIMO) empirical models are calibrated on data generated by the simulator, using three different off-line identification methods. The results are discussed with respect to their ability to represent the dynamic variations of the process variables around mean operating conditions. Finally, these models are used to design a multivariate filter and illustrate its potential utility.

Phenomenological Simulation

Phenomenological relationships are implemented in a stochastic non-linear dynamic simulator calibrated with industrial data. This simulator which is representative of industrial operation will be considered as the process in the next sections.

The grinding mill model is based on chemical reaction engineering concepts such as perfect mixers in series, interacting tanks with residual volume for transport dynamics and a first order grinding reaction in each mixer characterized by a rate factor S_i^k and a fragments distribution b_{ij}^k for each size class j and grindability class k. Other elements are described by simpler models such as single mixers for the sumps, pure delays for the pipes and zero order classifications models based on semi-empirical concepts for the hydrocyclones. The detailed mathematical formulation of these models is given in reference (2) and (7).

The structure of the dynamic simulator (6) is depicted in Figure 1. The flowsheet is entered as a coded description of the calculation path between pieces of equipment. When the necessary information (grinding parameters, volume of the sumps, classifier watersplit and cut-size parameters, ...) is defined for simulating a given flowsheet, steady-state simulation is performed. Then, the dynamic simulation based on a fourth-order Runge-Kutta method, is activated with input variables defined as deterministic functions (step, pulses, cyclic variation) or as random functions.

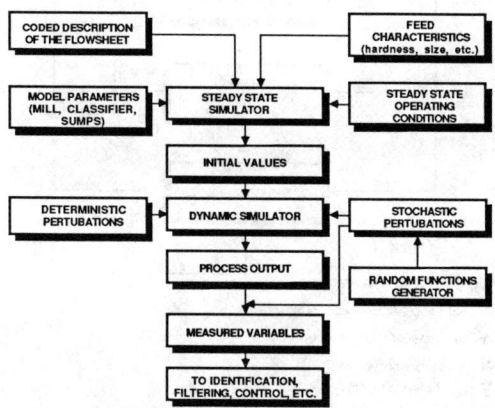

Figure 1 - Structure of the grinding circuit dynamic simulation.

Stochastic functions such as the random input variables, the actuators noises, the process noises and the sensors noises are automatically programmed by a noise generator preprocessor. Random variables are generated by an empirical ARMA equation driven by a centered gaussian white noise ξ_t:

$$x_t + \alpha_1 x_{t-1} + \ldots + \alpha_p x_{t-p} = \xi_t + \beta_1 \xi_{t-1} + \ldots + \beta_q \xi_{t-q} \tag{1}$$

The coefficients (α_j, β_j) are selected by the simulator user or estimated by identification from a typical plant record in order to generate a random signal which exhibits the desired statistical properties of variance, auto-correlation function or power spectrum (8).

Figure 2 defines the methodology and the terminology which are used in this study. The simulator is fed with deterministic variations of the manipulated variables u. These variations are corrupted with noises such that the true input variables to the simulator are stochastic. The output variables are calculated by the simulator, and includes the noises added by the process itself. The variables $(y_s(t))$ will be called in the following stochastic process output. Measurement noises added to them yield to the process measured output $(y(t))$.

In order to generate data for the subsequent empirical model construction, the two-stage grinding circuit depicted in Figure 3 was simulated. Figure 4 gives the equivalent mathematical diagram as implemented in the simulator program. Industrial data (9,10) were used for the calibration of the classifier and grinding mills models. The rod-mill is simulated as a series of ball mills. The variances of the white sequences which drive the

ARMA filters and generate the actuators, process and sensors noises are selected in such a way to be representative of typical signals monitored in the plant.

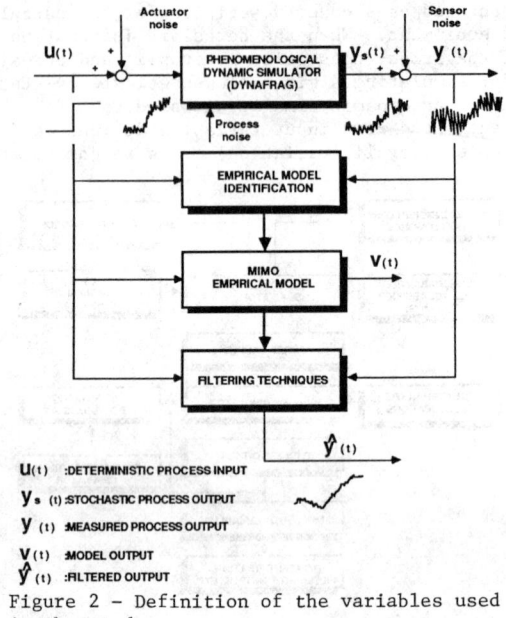

U(t) :DETERMINISTIC PROCESS INPUT

Y$_s$ (t) :STOCHASTIC PROCESS OUTPUT

Y (t) :MEASURED PROCESS OUTPUT

V (t) :MODEL OUTPUT

$\hat{Y}$ (t) :FILTERED OUTPUT

Figure 2 – Definition of the variables used in the study

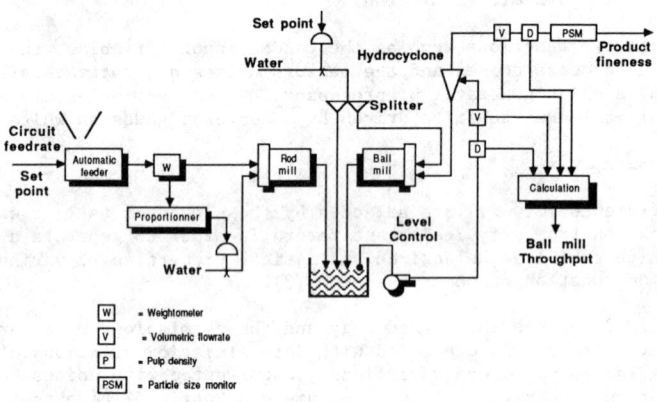

W = Weightometer

V = Volumetric flowrate

P = Pulp density

PSM = Particle size monitor

Figure 3 – Flowsheet of the simulated grinding circuit.

Model Identification

To build the MIMO empirical model of the grinding circuit of Figure 3 two input variables were considered: the ore fresh feed rate (u1) and the water addition rate to the sumps (u2). Since the actuators (the automatic ore feeder and water valve) are incorporated in the system to be modelled, u1 and u2 can be considered as the set-points of the actuators. The two output variables selected as typical variables to be controlled, are the

78

product fineness, measured by the percentage of particles finer than 74 μm, and the mill solids throughput in tons per hour.

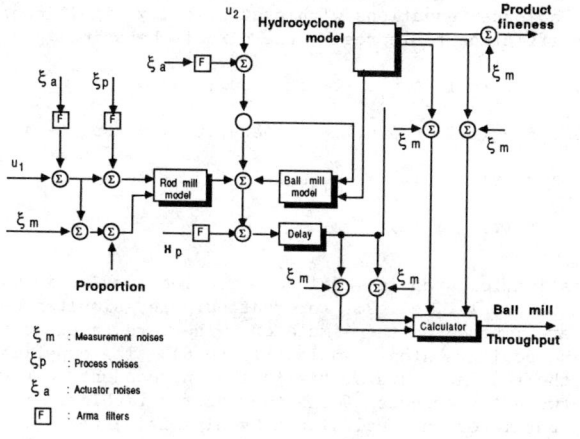

Figure 4 – Computation diagram for the simulation of the grinding circuit of Figure 3.

For the model identification procedure, only the measured process output y (Figure 2) delivered by the simulator will be used since they are the only available signals in a real industrial circuit. However, the advantage of the simulator is to provide also the stochastic process output (Figure 2), a signal free of measurement noise, as well as the deterministic output variables when all random contributions are set to zero. This information, normally not available in a plant, will be used to compare various identification schemes and to assess the validity of the selected model and of the filters which are based on it.

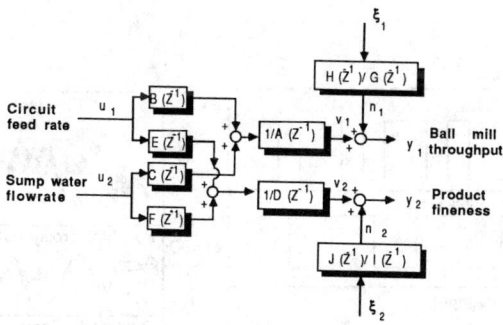

Figure 5 – Block-diagram of the grinding circuit empirical model.

There are three stages in the identification process for empirical model: first, the definition of an appropriate mathematical structure; second, the selection of the order of the model and the pure delays; and third, the estimation of the model parameters. As depicted in Figure 5, a time discrete structure, based on polynomials of the backshift operator z^{-1}, is selected. The random parts of the simulator variables are lumped in additive autoregressive output noises characterized by the polynomials

G and H. The determination of a suitable order is made by trials using various discrimination techniques (11) which always involve some compromise between quality of fitting and model complexity (number of parameters). The following low order equations with statistically significant parameters have been found efficient to represent the circuit behavior:

$$v_1(t) + a_1 v_1(t-1) = b_1 u_1(t-3) + c_1 u_2(t-4)$$

$$v_2(t) + d_1 v_2(t-1) = e_1 u_1(t-3) + e_2 u_1(t-4) + f_1 u_2(t-3) + f_2 u_2(t-4) \qquad (2)$$

$$y_1(t) = v_1(t) + \xi_1(t) + g_1 \xi_1(t-1) + g_2 \xi_1(t-2)$$

$$y_2(t) = v_2(t) + \xi_2(t)$$

The estimation step in model construction consist in estimating the parameters a1,b1,c1,...g1,g2. For that purpose, simultaneous variations of u1 and u2 are imposed to the simulator. They are taken as a sequence of negative and positive step signals (Figure 6). The experimental design is such that the two input are statistically independent to favor the discrimination between the effects of the two input variables. For each output variables, parameters are calculated by minimizing the sum of the squared values of the white sequences $\xi 1$ and $\xi 2$ (Figure 5):

$$J_1(a1,b1,c1,g1,g2) = \Sigma \ \xi_1^2$$

$$J_2(d1,e1,e2,f1,f2) = \Sigma \ \xi_2^2 \qquad (3)$$

When $\xi 1$ and $\xi 2$ are gaussian, the solution is that of the maximum likelihood method (ML) (12). The derivatives of the criteria J1 and J2 are not linear with respect to the parameters; as a consequence, the method requires a non-linear programming algorithm. The Powell (13) and the Marquardt algorithms (14) were used. They both give satisfying convergence behaviors.

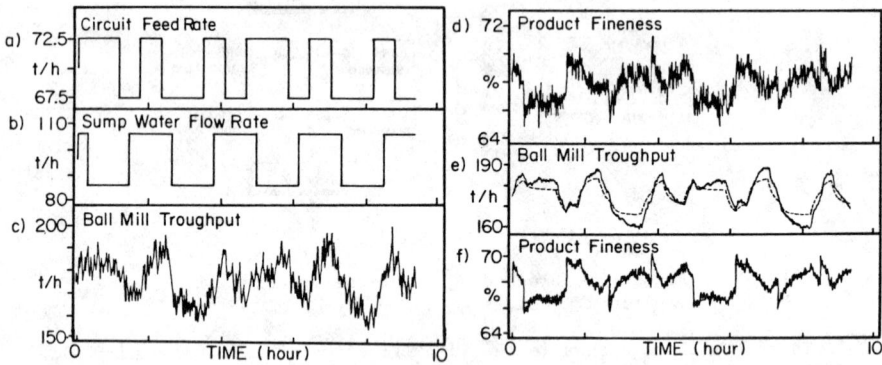

Figure 6 – Signals used for the identification of the empirical model. Input variables u (a,b), output variables: measured y (c,d), stochastic y_S and modelled v(e,f)

80

An alternative method, which leads to the simple least-squares solution (LS), consists in minimizing the sum of the squared equation errors e1 and e2 (Figure 7) :

$$J_1^{*}(a1,b1,c1) = \Sigma \ e_1^2$$

$$J_2^{*}(d1,\ldots,f1) = \Sigma \ e_2^2 \qquad (4)$$

The derivatives of the criteria are now linear with respect to the parameters and the solution is straightforward. However, since the sequences e_1 and e_2

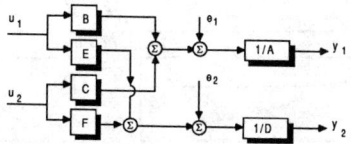

Figure 7 - Modified empirical model for least-squares parameters estimation.

are not white, the parameters estimates are biased. The generalized least squares method (GLS) uses a filter to whiten the equation errors e_1 and e_2. It is an intermediate method which transforms the non-linear ML problem in an iteration of LS linear problems in an attempt to provide unbiased parameters. Results are given in Table 1 for the parameters values, in Figure 6 for the comparison of the ML model output (v) to the simulator stochastic output (y_s) and in Figures 8 and 9 for the comparison of the GLS and ML models output (v) to the simulator deterministic output. It is clear from Figure 8 that the GLS parameters are less accurate than the ML estimates. As a consequence the ML model will be adopted in the next section.

Table I : Results of the identification of the grinding circuit model

PARAMETERS	MAXIMUM LIKELIHOOD	LEAST SQUARES	GENERALIZED LEAST-SQUARES
a_1	- 0,963	- 0,896	- 0,885
b_1	0,146	0,261	0,279
c_1	0,017	0,040	0,043
d_1	- 0,977	- 0,686	- 0,897
e_1	0,120	0,005	0,039
e_2	- 0,129	- 0,016	- 0,050
f_1	0,0851	0,032	0,073
f_2	- 0,0849	- 0,018	- 0,070
g_1	0,899	-	-
g_2	- 0,171	-	-

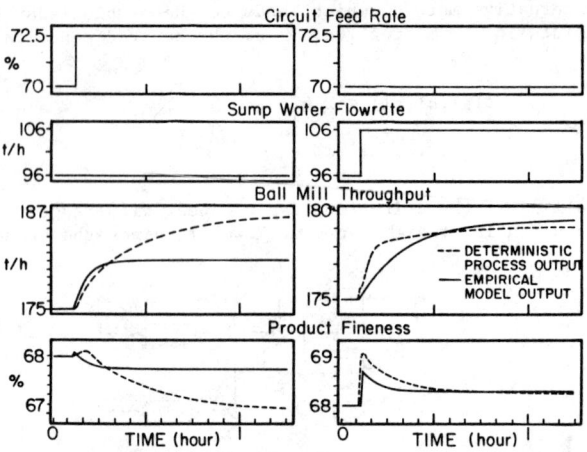

Figure 8 - Step responses of the model calibrated
by the GLS estimation method.

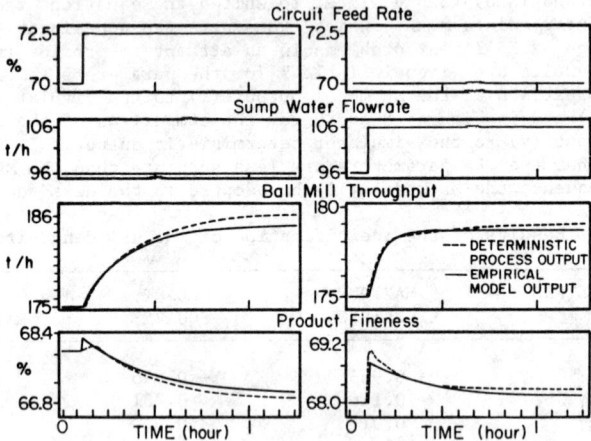

Figure 9 - Step responses of the model calibrated
by the ML estimation method.

Multivariate Filtering

Multivariate filters based on grinding process models are used in a
wide variety of industrial applications. At a first level, they improve
the information given by the sensors by providing measured variables
estimates (15) where the effects of the noises are attenuated. Since they
are designed using a model of the process, they allow a coherent interpretation of the information simultaneously delivered by the various sensors
of a process. Furthermore, these filters can be used for the indirect
estimation of unmeasured variables (17,16) and as a consequence they increase the utility of the sensors. When used in conjunction with control
systems, filters improve feedback loops performances by eliminating

undesired disturbances in controlled variables, thus providing smoother control and allowing higher gains. Finally, they can be used for on-line model identification (17,18).

This section will show that the use of simple low order empirical models in the Kalman filter scheme provides a sufficient and effective way of filtering close grinding circuits realtime data.

The first step in filter design is the rewriting of Equations (2) into a controlable and observable state space formulation (19,20). The block-diagram of Figure 10 corresponds to one of the possible interpretations of the model developped in the previous section. In this scheme the definition of the state variables (x_1 to x_8) is straightforward. This system is equivalent to the following state space stochastic model:

$$x(k+1) = A\,x(k) + B\,u(k) + w(k)$$
$$y(k) = C\,x(k) + v(k) \tag{5}$$

with:

$$x = [\ x_1, x_2, \ldots\ x_8\ ]^T$$
$$u = [\ u_1, u_2\ ]^T$$
$$y = [\ y_1, y_2\ ]^T$$

$$B = \begin{bmatrix} 0 & 0 \\ 0 & 0 \\ 0 & b_1 \\ b_2 & -a_1 b_2 \\ 0 & 0 \\ 0 & 0 \\ d_1 & c_1 \\ d_2 - d_1 c_1 & e_2 - e_1 c_1 \end{bmatrix}$$

$$A = \begin{bmatrix} 0 & 1 & 0 & 0 & & & & \\ 0 & 0 & 1 & 0 & & & & \\ 0 & 0 & 0 & 1 & & 0 & & \\ 0 & 0 & 0 & -a_1 & & & & \\ & & & & 0 & 1 & 0 & 0 \\ & & 0 & & 0 & 0 & 1 & 0 \\ & & & & 0 & 0 & 0 & 1 \\ & & & & 0 & 0 & 0 & -c_1 \end{bmatrix}$$

$$C = \begin{bmatrix} 1 & 0 & 0 & 0 & 0 & 0 & 0 & 0 \\ 0 & 0 & 0 & 0 & 1 & 0 & 0 & 0 \end{bmatrix}$$

The state noise sequence $w(k)$, which accounts for all the process disturbances, and $v(k)$, the measurement noise sequence, are white zero-mean gaussian noises, independent of each other (uncorrelated). They verify the following statistical properties:

$$E\{w(k)w^T(j)\} = 0\ ,\ E\{v(k)v^T(j)\} = 0 \quad \text{for all } k \neq j$$

and: $E\{w(k)w^T(k)\} = Q\ ,\ E\{v(k)v^T(k)\} = R$

where $E\{\ \}$ stands for the mathematical expectation.

A recursive filter is an algorithm which calculate the best estimate of $x(k)$ at the time $t=kT$, denoted $\hat{x}(k)$, using the previous estimates $\hat{x}(k-1)$, the new measurement value $y(k)$, the input value $u(k)$, and the model i.e. the matrices A, B and C and the variances Q and R.

When the statistical properties of the Gaussian noises $w(k)$, $y(k)$ and the initial conditions $x(0)$ (assumed to be gaussian and of variance P_0) are correctly known, then the optimal estimates of $x(k)$ are obtained by the well-known Kalman-filter (21). The equations of the filter are developed in two steps. The prediction step estimates the states at time

83

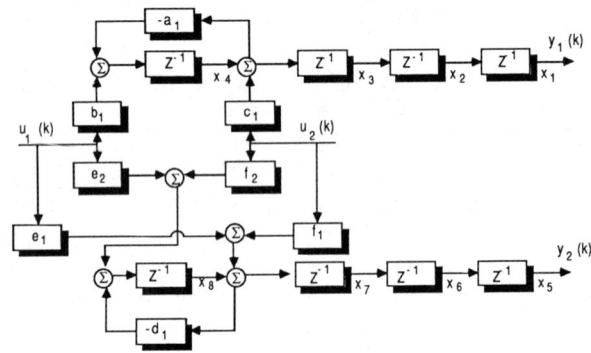

Figure 10 - Block-diagram for the state space
representation of the grinding circuit
empirical model.

k, $\hat{x}(k/k-1)$, from the previous estimates and observations. Then, the
correction step uses the kth observation to update the state estimate
$\hat{x}(k/k)$ and the estimation covariance $P(k/k)$. The recursive equations,
between observations, are:

$$\hat{x}(k/k-1) = A \ \hat{x}(k-1/k-1) + B \ u(k-1)$$

$$P(k/k-1) = A \ P(k-1/k-1) \ A^T + Q(k)$$

(6)

where $P(k/k-1)$ is the state prediction error covariance matrix, and, at
observations, are:

$$\hat{x}(k/k) = \hat{x}(k/k-1) + L(k) \ [\ y(k) - C \ \hat{x}(k/k-1) \ ]$$

$$P(k/k) = P(k/k-1) - L(k) \ C \ P(k/k-1)$$

(7)

$$L(k) = P(k/k-1) \ C^T \ [\ C \ P(k/k-1) \ C^T - R(k) \ ]^{-1}$$

where $L(k)$ is the kalman gain and $P(k/k)$ the state estimation error covariance matrix.

   The filter output $\hat{x}(k/k)$ is essentially a weighted sum of the values
predicted by the model and the new measurement values available. The
relative weights assigned to the model output and the measurement are
determined by the matrix $L(k)$ which is dependent upon the variances Q and
R of the model output and the measurement. When the process disturbance
level is low and the model reliable compared to the measurement errors, L
is low. Reciprocally, for reliable sensors and poor model, L is high.
Generally R can be evaluated from the performances of the sensors; Q is
however more difficult to determine since it embeds the contributions of
the process disturbances as well as the model imperfections, due to the
linear approximation of a non-linear system and to the identification in
a noisy environment. For the present purpose the matrix Q was calculated
such as to minimize the variance of the estimation error for a given
stochastic simulation of the grinding circuit. This was done empirically
by a linear programming method applied to the Kalman filter implemented
on the simulator. For an industrial circuit, the variance adjustment can
be made by empirical tuning of Q. Obviously the filter cannot be truly
optimal, however its structure gives a practical method to design an

84

efficient asymptotic estimator.

To illustrate the utility of this type of filter based on an empirical model, it was implemented in the simulator and used under different situations. First, when the two output variables, the circulating load and the product size are available, it is used as a real-time linear filter which gives the best estimates of the two variables. Occasionally, one of the two measured variables can be missing, due for instance to a sensor failure. Under these circumstances the filter can be used to give an optimal estimate of the missing variable based on the model and the past information. The same equations are used but the gain matrix L(k) is computed using an infinite variance for the missing measurement.

Figures 11 and 12 shows respectively the results for the mill throughput and the product fineness. The filtered output variables satisfactorily follow the y_S output of the simulator (stochastic process variable without measurement noises) as confirmed in Table II. When the flowmeter signal (Figure 11c) or the particle size monitor signal (Figure 12c) is not available the kalman filter is able to follow the changes in the output caused by sudden changes in the input, without loosing its capability of filtering. The estimates are less accurate than they were in Figures 11b and 12b when both sensors were on (see Table II), because they rely only upon the process model; however they are good enough to be used by an operator or in a control loop.

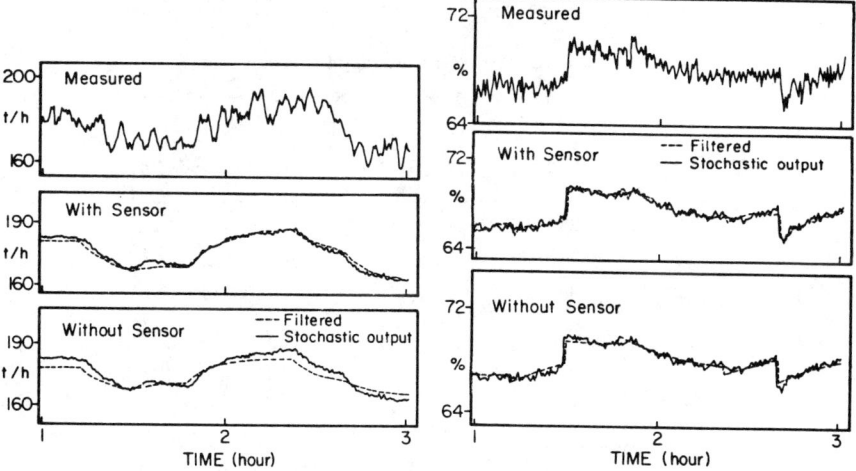

Figure 11 - Results of ball mill throughput filtering.

Figure 12 - Results of product fineness filtering.

So far, the filter has been used with stationary variables i.e. with deviations around constant mean values. Figure 13 illustrates what happens when a drift of the stationary mean, due to a grindability change in this case, occurs. Since the output variation is not caused by a permanent variation of the input variables, which are maintained at their reference values, the model output part of the filter remains unchanged. Only the output error part of the filter can allow the state variables estimates to vary. When the stationary gain L is small (for a good filtering behavior) the filter is unable to follow the output level variations and a bias in the estimates results (Figure 13).

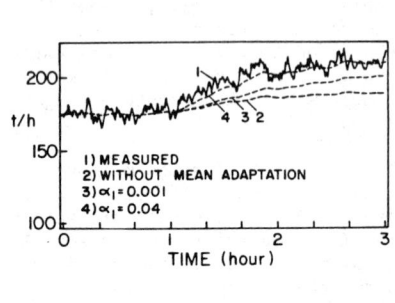

Figure 13 - Effect of a grindability
step on the filtering performances
with and without mean adaptation,
when both inputs are constant.

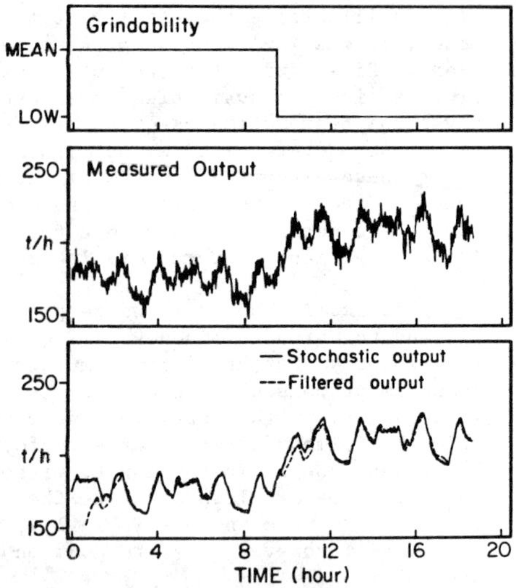

Figure 14 - Response of the augmented
filter to a grindability step and input variations as in Figure 6a,b.

Table II: Mean squared values

	FILTERING	
	WITH SENSOR	WITHOUT SENSOR
BALL MILL THROUGHPUT		
$(\hat{y}-y)^2$	17,4	32,1
$(\hat{y}-y_s)^2$	2,0	12,0
PRODUCT FINENESS		
$(\hat{y}-y)^2$	0,20	0,35
$(\hat{y}-y_s)^2$	0,07	0,10

To overcome this problem, it is possible to use an augmented state
vector which includes the mean values of the outputs (20). Initially, the
system was described in Equations (5) by the deviation y:

$$y(k) = Y(k) - \bar{Y}(k) \qquad (8)$$

where Y(k) is the measured value of the output variables and $\bar{Y}(k)$, its mean
value. Including Equation (8) in the system of Equations (5), it can re-
written as:

$$x(k+1) = Ax(k) + Bu(k) + w(k) \qquad (9)$$
$$Y(k) = Cx(k) + \bar{Y}(k) + v(k)$$

or in an augmented state form :

$$\begin{bmatrix} x(k+1) \\ \bar{Y}(k+1) \end{bmatrix} = \begin{bmatrix} A & 0 \\ 0 & I \end{bmatrix} \begin{bmatrix} x(k) \\ \bar{Y}(k) \end{bmatrix} + \begin{bmatrix} B \\ 0 \end{bmatrix} u(k) + \begin{bmatrix} w(k) \\ e(k) \end{bmatrix}$$

$$Y(k) = \begin{bmatrix} C & I \end{bmatrix} \begin{bmatrix} x(k) \\ \bar{Y}(k) \end{bmatrix} + v(k)$$

with the assumption that $\bar{Y}(k)$ is approximately constant over one sampling period.

It is now possible to apply the Kalman filter equations to this aug-mented system with a new covariance matrix Q defined as

$$Q = \begin{bmatrix} E\{w(k)w^T(k)\} & 0 \\ 0 & E\{e(k)e^T(k)\} \end{bmatrix}$$

where $E\{e(k)e^T(k)\}$ is a diagonal matrix (α_1, α_2) determining the convergence speed of the mean adaptation part of the algorithm. Figure 13 illustrates the adaptation, as a function of α_1, for the mill throughput. It shows that the modified algorithm is capable of following level changes due to experimental perturbation without loosing its filtering capability.

As shown above the grindability variations significantly changes the average behavior of the circuit. As a consequence, the mean adaptation is a necessary aspect in the design of a filter to compensate for the non-zero mean disturbances or the permanent changes in the grinding kinetics. Fur-thermore the dynamic behaviour of the circuit around the stationary means should depends upon the grindability level. To test this aspect, variations of the input variables were applied simultaneously to a step variation of grindability. Results drawn in Figure 14 for the mill throughput show that the Kalman filter still properly follows the dynamic behaviour of the vari-able after the grindability change and when the mean adaptation is achieved. This is an indication that the dynamic model is not very sensitive to the ore grindability. To confirm this result, the identification process based on the ML method was resumed for the low grindability level with the same input sequence and the same noise generators. Table III compares the new parameters to the previously estimated values. It is clear that the dif-ference is not significant and does not justify the design of an adaptation scheme for the model parameters.

Table III: Parameters variations as a function
of ore grindability

PARAMETERS	MEAN GRINDABILITY	LOW GRINDABILITY
a_1	-0,963	-0,973
b_1	0,146	0,143
c_1	0,017	0,015
d_1	-0,977	-0,977
e_1	0,120	0,113
e_2	-0,129	-0,122
f_1	0,0851	0,0873
f_2	-0,0849	-0,0872

87

Conclusion

This paper shows that low order linear MIMO empirical models are appropriate to represent the non-linear dynamics of closed grinding circuits. The time-discrete model which has been developped has the advantages of being simple to calibrate by standard identification techniques and easily usable in filtering applications. Identification techniques based on output errors, as the maximum likelihood method, give better parameters estimates than prediction error methods, as the least squares or generalized least squares methods.

A Kalman filter, based on the MIMO empirical model, is able to produce efficient estimates of the output variables by eliminating the noises introduced by the process, the sensors and the model imperfections. It is shown that the filter still provides good estimates of output variables in case of sensor failures. To compensate the effects of non-stationary disturbances or changes in the ground ore properties, it is essential to design an augmented filter for tracking the variations of the variables means. The model parameters are not very sensitive to the ore grindability and there is no need for adaptive filtering methods. However when the model is unknown, suitable on-line identification methods should be implemented, a study which is now in progress.

Acknowledgments

The Centre de Recherches minérales of the Ministère de l'Energie et des Ressources of the Québec Government and the Natural Sciences and Engineering Research Council of Canada are thanked for the financial support of this work.

References

1. K. Rajamani, Optimal Control of Ball Mill Grinding, Ph.D. Thesis, university of Utah, 1979.

2. Y. Dubé, R. Lanthier, and D. Hodouin "Computer-Aided Dynamic Analysis and Control Design for Grinding Circuits", Can. Inst. Min. Met. Bulletin, (in press), 1987.

3. F. Flament, D. Hodouin, C. Bazin: A Time-discrete Approach to the Modelling of Grinding Mill Mixing Properties. Particulate Science and Technology, Vol. 2(2), 1984, p. 167.

4. C. Bazin, D. Hodouin: Off-line and On-line Identification of Empirical Dynamic Models for a Flotation Process. To be presented at XVIth Int. mineral Processing Cong. Stockholm. Sweden. June 1988.

5. D. Hodouin, C. Bazin: An Empirical Approach to the Modelling and Control of Ball Mill Grinding. Paper 87-110, SME/AIME Annual Meeting, Denver, Colorado, Feb. 1987.

6. D. Hodouin, Y. Dubé, and R. Lanthier, "Stochastic Simulation of Filtering and Control Strategies for Grinding Circuits", Int. J. Miner. Proc., (in press), 1987.

7. Y. Dubé, and D. Hodouin, "Adaptive Filtering for a Pilot Grinding Circuit" Proceedings of the 5th IFAC Symposium, Automation in Mining, Mineral and Metal Processing, Tokyo, Japan, 1986.

8. G.E.P. Box, G.M. Jenkins, "Time Series Analysis; Forecasting and Control", Holden-Day, San Francisco, 1976.

9. D. Hodouin, J. McMullen, and M.D. Everell, "Mathematical Simulation of the Operation of a Three-stages Grinding Circuit for a Fine Grained Zn/Pb/Cu Ore", Proceedings of European Symposium on Particle Technology, Amsterdam, 1980.

10. D. Hodouin, M.D. Everell, D. Laguitton, and L.L. Sirois, "Computer Simulation Approach to Improve Comminution Strategies for Fine-Grained Complex Sulphide Ores From New Brunswick, Canada", Proceedings of Int. Conf. on Complex Sulphide Ores, Pergamon, 1980.

11. L. Ljung, "System Identification; Theory for the User", Prentice-Hall Englewood Cliffs, New Jersey, 1987.

12. G.C. Goodwin, and R.L. Payne, "Dynamic System Identification Experiment Design and Data Analysis", Academic Press New York, 1977.

13. M.J.D. Powell, "An efficient Method For Finding the Minimum of a Function of Several Variables Without Calculating Derivatives", Computer Journal, 1964.

14. D.W. Marquardt, "An Algorithm For Least Squares Estimation of Non-linear Parameters", Siam J. Appl. Math., 1963.

15. G.G. Wyatt-Mair, K.C. Garner and R.P. King. Real Time Digital Computer State Estimate for a Hard Rock Milling Circuit. Proc. 3rd IFAC Symposium on Automation in Mining, Mineral and Metal Processing, Pergamon Press, 1980.

16. W.T. Pate and J.A. Herbst. Dynamic Estimation of Unmeasured Variables in an Industrial Grinding Circuit Using a Kalman Filter. AIME Annual Meeting, Atlanta, Georgia, 1983.

17. J.O. Olsen and T. Lenhartzen. Grinding Circuit Parameter Estimation Using the Maximum Likelihood Method. Int. J. Miner. Proc., 1979.

18. J.A. Herbst, K. Rajamani, W.T. Pate: Identification of Ore Hardness Disturbances in a Grinding Circuit Using a Kalman Filter. Proceedings of the 3rd IFAC Symp. on Automation in Mining, Mineral and Metal Processing, Montréal, Pergamon Press, 1980.

19. J.C. Gille, et M. Clique, "Systèmes linéaires; équations d'état", Eyrolles, Paris, 1984.

20. C. Foulard, S. Gentil, et J.P. Sandraz, "Commande et Régulation par Calculateur numérique: de la théorie aux applications", 5ième éd. Eyrolles, Paris, 1987.

21. A.H. Jazwinski, "Stochastic Processes and Filtering Theory", Academic Press, New York, 1970.

DYNAMIC BEHAVIOR IN AN INDUSTRIAL GRINDING CIRCUIT

Weiping Xiong, Guoxiang Zhang, Songren Li, and Zhen Su
Mineral Engineering Department
Central South University of Technology
Changsha, Hunan
P.R. China

and Weibai Hu
Department of Metallurgy and Metallurgical Engineering
University of Utah
Salt Lake City, Utah 84112
U.S.A.

Abstract

The dynamic behavior of grinding circuits is an important consideration in the development of process automation. The influences of the feed rate, sump water addition, and circulating load on some important process variables, such as product particle size, pulp flow rate, and density, have been investigated with a great amount of dynamic data from an industrial closed grinding circuit collected by a microcomputer sampling system. The transfer functions of the circuit were also established, in which parameters were estimated recursively by the instrumental variable technique. The results showed that the fresh feed rate was the main manipulating variable governing the dynamic behavior. The control of circulating load is one of the feasible approaches in the process optimization of closed grinding circuits. Some manipulating and control strategies are also discussed.

91

Introduction

In general, the dynamic behavior of grinding circuits is characterized by certain individual responses of process variables and by interactions among controllable variables. It is necessary to understand the dynamic behavior of a particular grinding circuit before the manipulation of operating variables and the design of automation systems. In many cases, poor performances of some grinding circuits can be traced to the inadequate understanding of the process dynamic behavior. Over the past ten years, several groups have investigated the responses of some process variables to the changes of operating variables in ball mill grinding circuits with different methods (1-4). Some of their results have been applied in simulation and control systems of several grinding circuits. However, in industrial closed grinding circuits, the circulating load is also an important process variable that is related directly to the grinding efficiency and therefore affects the following process variables, especially the mill product size.

Based on a great amount of dynamic data collected from an industrial closed grinding circuit by the microcomputer system, dynamic responses of pulp flow rate, pulp density, and the overflow particle size to changes of fresh feed rate, sump water addition, as well as circulating load were investigated. The relationships among these variables and reasonable control strategies were also discussed. The stochastic difference equations with a delay factor were used to describe the dynamic responses in this paper, in which the parameter is estimated on-line by the instrumental variable technique.

Grinding Circuit and Sampling System

The principle of a closed loop consisting of a ball mill 16.5 feet in diameter by 21 feet long and a set of 26-inch hydrocyclones is employed in the grinding circuit. The throughput of the circuit is designed to be 250 tonnes per hour, and the required fineness of the cyclone overflow is 65% solids passing 74 μm. The slurry level in the sump is kept constant by a classical PI controller through a variable-speed pump.

Measurements available are fresh feed solids, mill water, mill power, sump water addition, cyclone feed volumetric flow rate and pulp density, overflow pulp density, volumetric flow rate, and particle size. All outputs of the instruments are transmitted on-line to the microcomputer sampling system (5). Thus, dynamic data are sampled, filtered, and logged at sampling intervals of approximately 30 seconds during a period of nearly 3 hours. A schematic of the grinding circuit and instrumentation is shown in Fig. 1.

Description of Dynamics and Parameter Estimation

Much of the previous research work showed that the industrial grinding circuit was a slowly varying, nonlinear system with large inerta, in which the dynamic responses of all process variables exhibited, more or less, pure lag or capacity lag. Considering the characteristics, the dynamic responses are described in the form of stochastic difference equations. The process variable, $y(k)$, is written as a linear combination of the previous process variable, $y(k-1)$, the operating variable, $u(k-d)$, and the process noise, $e(k)$, i.e.,

$$y(k) = -a_1 y(k-1) - a_2 y(k-2) - \ldots - a_n y(k-n) + b_0 u(k-d)$$

$$+ b_1 u(k-d-1) + \ldots + b_n u(k-d-n) + e(k) \qquad (1)$$

92

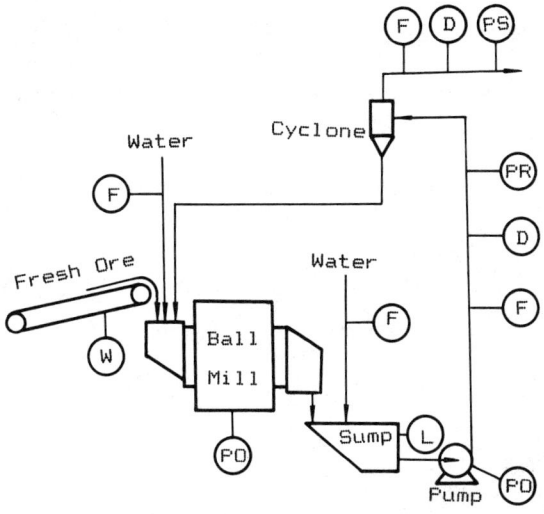

W   feed rate measurement
F   volumetric flow rate measurement
L   pulp level measurement
D   pulp density measurement
PD power measurement
PR inlet pressure measurement
PB particle size measurement

Figure 1.   Schematic diagram of the grinding circuit.

where $k = t/T = 0, 1, 2, \ldots$ is the discrete time, T is the sampling interval time, t is the discrete dead time, and n is the model order. The discrete process time delay d is a nonnegative integer.

Equation 1 can be written in compact form as follows:

$$y(k) = \underline{x}^T(K) \; \underline{Q} + e(k) \qquad (2)$$

with the data vector:

$$\underline{x}\,(k) = [-y(k-1), \ldots, -y(k-n), u(k-d), \ldots, u(k-d-n)]^T$$

and the parameter vector:

$$\underline{Q} = [a_1, \ldots, a_n, b_o, b_1, \ldots, b_n]^T$$

Using the Z transform, the transfer function of Equation 1, $H(z)$, can be expressed as

$$H(z) = \frac{Y(z)}{U(z)} = z^{-d} \frac{b_0 + b_1 z^{-1} + \ldots + b_n z^{-n}}{1 + a_1 z^{-1} + \ldots + a_n z^{-n}} \tag{3}$$

There are a number of well-known parameter estimation techniques to be able to estimate the $2n+1$ parameters Q_i of Q, in which the least squares is a common one.

Because we have available a string of dynamic data $[y(k), u(k)]$ for $k = 1, \ldots, (N + n)$, we can set a system of N equations as (assume $N \gg n$)

$$\underline{y} = \underline{\underline{X}}\underline{Q} + \underline{e} \tag{4}$$

where $\underline{y} = [y(n+1), y(n+2), \ldots, y(n+N)]^T$

$\underline{e} = [e(n+1), e(n+2), \ldots, e(n+N)]^T$

$$\underline{\underline{X}} = \begin{vmatrix} \underline{x}^T(n+1) \\ \underline{x}^T(n+2) \\ \vdots \\ \underline{x}^T(n+N) \end{vmatrix} = \begin{vmatrix} -y(n), \ldots, & -y(1), & u(n+1), \ldots, u(1) \\ -y(n+1), \ldots, & -y(2), & u(n+1), \ldots, u(2) \\ \vdots & & \vdots \\ -y(n+N-1), \ldots, & -y(N), & u(n+N), \ldots, u(N) \end{vmatrix}$$

According to the least-squares theory, we immediatley obtain the least-squares estimate $\hat{Q}$ by

$$\hat{\underline{Q}} = (\underline{\underline{X}}^T\underline{\underline{X}})^{-1}\underline{\underline{X}}^T\underline{y} \tag{5}$$

This solution exists if $\underline{\underline{X}}^T\underline{\underline{X}}$ is nonsingular.

However, the process noise generally is not a white noise for industrial grinding circuits, and its autocorrelation function Ree $\neq 0$. In this case, the parameter estimate $\hat{Q}$ obtained by the least squares is biased and is not a consistent estimate of the true parameters Q. Here, another parameter estimation (6), the instrumental variable technique, was used to overcome the failure of the least squares. The instrumental variable estimate $\hat{\underline{Q}}_{iv}$ can be solved by

$$\hat{\underline{Q}}_{iv} = (\underline{\underline{Z}}^T\underline{\underline{X}})^{-1}\underline{\underline{Z}}^T\underline{y} \tag{6}$$

where the square matrix $\underline{\underline{X}}$ is the same as in Equation 5; the matrix $\underline{\underline{Z}}$ is called the instrumental variable matrix. In this paper, the elements in $\underline{\underline{Z}}$, Zij, are taken as follows:

$$
\underset{=}{Z} = \begin{vmatrix} \underline{z}^T(n+1) \\ \underline{z}^T(n+2) \\ \vdots \\ \underline{z}^T(n+N) \end{vmatrix} = \begin{vmatrix} -\hat{y}(n), \ldots, & -\hat{y}(1) & : & u(n+1), \ldots, u(1) \\ -\hat{y}(n+1), \ldots, & -\hat{y}(2) & : & u(n+1), \ldots, u(2) \\ \vdots & & : & \vdots \\ -\hat{y}(n+N-1), \ldots, & -\hat{y}(N) & : & u(n+N), \ldots, u(N) \end{vmatrix}
$$

where $\hat{y}(k)$ for $k = 1, 2, \ldots, n+N-1$ is obtained by

$$
\underline{\hat{y}} = \underset{=}{X}\underline{\hat{Q}} \tag{7}
$$

where $\underline{\hat{Q}}$ is the least-square estimate from Equation 5.

Based on Equation 6, the following set of recursive equations is used to estimate $\underline{\hat{Q}}_{iv}$:

$$
\underline{\hat{Q}}_{k+1} = \underline{\hat{Q}}_k + \underset{=}{P}_k \underline{z}_{n+k} (1 + \underline{x}_{n+k}^T \underset{=}{P}_k \underline{x}_{n+k})^{-1} [y(n + k - 1) - \underline{x}_{n+k}^T \underline{\hat{Q}}_k]
$$

$$
\underset{=}{P}_{k+1} = \underset{=}{P}_k - \underset{=}{P}_k \underline{z}_{n+k} (1 + \underline{x}_{n+k}^T \underset{=}{P}_k \underline{z}_{n+k})^{-1} \underline{x}_{n+1}^T \underset{=}{P}_k
$$

The order determination of the model (1) is made according to Akaike's information criterion. All computations are completed by a computer with a FORTRAN program package.

Dynamic Responses of Process Variables

Response to a Change in the Feed Rate

The properties of the fresh ore feed, such as specific gravity, size distribution and hardness, and the feed rate are important factors governing the dynamic behavior of the grinding circuit. Any change of feed rate will make many process variables change. Typical step-test responses are shown in Fig. 2 where the sump water addition was kept constant.

The process transfer functions obtained individually by the instrumental variable techniques are as follows:

$$
H(z) = z^{-1} \frac{0.041 - 0.025 \, z^{-1}}{1 - 0.996 \, z^{-1}} \quad \text{for percent solids, cyclone feed, \%;}
$$

$$
H(z) = z^{-1} \frac{0.108 - 0.074 \, z^{-1}}{1 - 0.994 \, z^{-1}} \quad \text{for cyclone feed rate of solids, t/h;}
$$

$$
H(z) = z^{-1} \frac{0.072 - 0.043 \, z^{-1}}{1 - 0.938 \, z^{-1}} \quad \text{for percent solids, cyclone overflow, \%;}
$$

$$
H(z) = z^{-1} \frac{-0.147 - 0.114 \, z^{-1}}{1 - 1.005 \, z^{-1}} \quad \text{for fraction -75 µm solids, cyclone overflow;}
$$

The predicted results from the above models are plotted in Fig. 2 and are very close to the measured results. It indicates that the responses of these proccess variables to the feed rate can be approximately considered as first-order responses with a dead-time lag. The dead times are of the range of 1 to 2 minutes. Thus it can be seen that the responses toa chnge in fresh ore rate are simple as well as fast. Therefore, the feed rate must be carefully controlled to ensure the maximum throughput at a required product size.

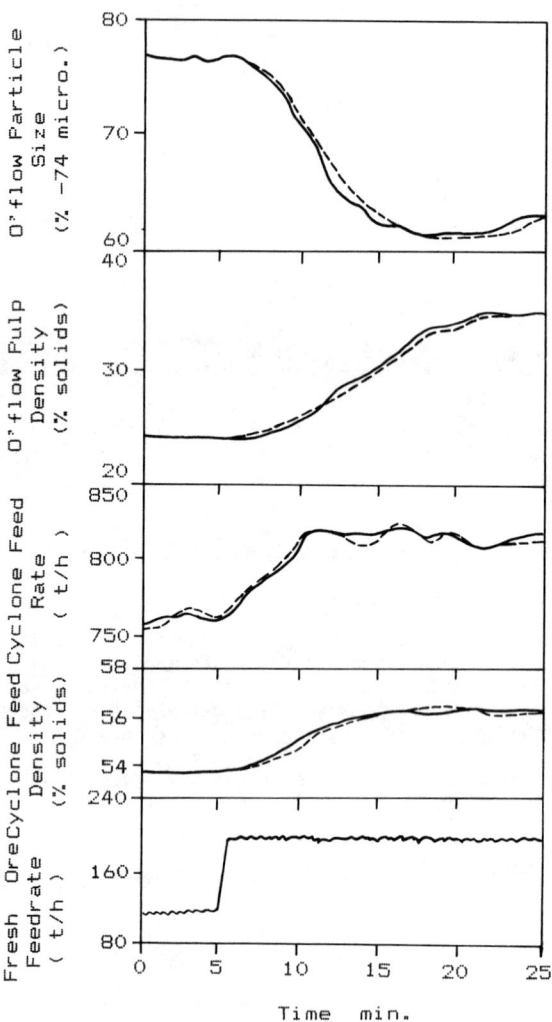

Figure 2.   Measured (——) and predicted (- - -) responses
to step change in the feed rate of fresh ore.

Responses to a Change in Sump Water Addition

The sump water addition in grinding circuits plays an important role in adjusting the performance of cyclones. In other words, any change in the sump water addition will have a fast and direct effect on the way in which the cyclone classifies the solids. The responses of process variables to a change in sump water addition are shown in Fig. 3, where the feed rate of fresh solids was in steady state.

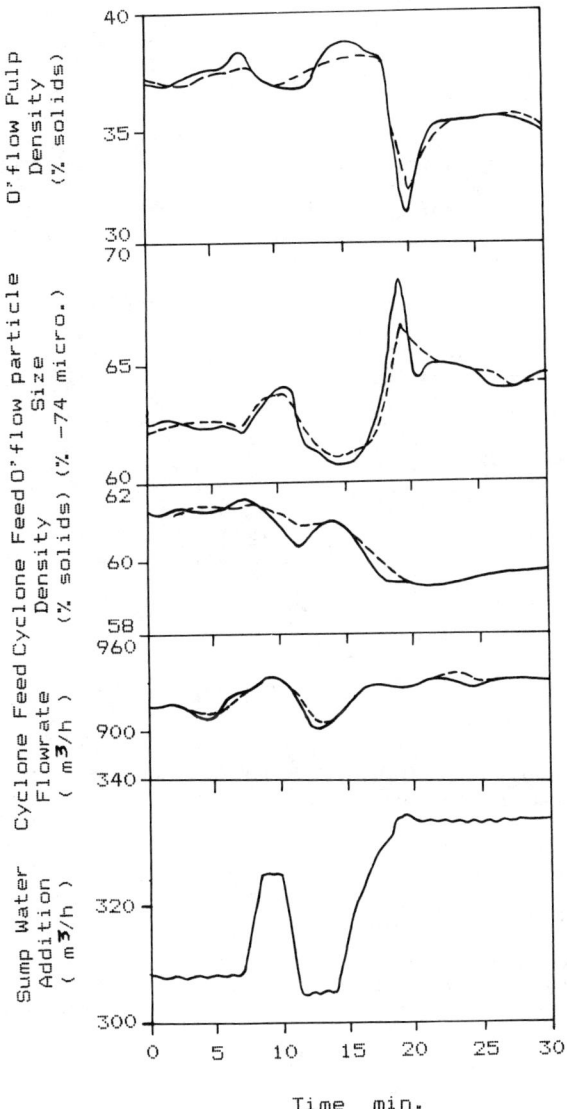

Figure 3. Measured (——) and predicted (---) responses to change in sump water addition.

It has been found from Fig. 3. that all responses of the process variables in cyclone units were very sensitive to the change of the water addition, but the magnitude of effect on the cyclone performance was limited and the effective action time was relatively short, i.e., about 5 minutes. It can be illustrated as follows: Upon a sudden increase in sump water additin, the cyclone feed slurry becomes dilute and also the volumetric flow rate to the cyclone increases as a result of sump level control. The cyclone cut size decreases, thus producing a finer product. Simultaneously, a large portion of coarse solids, i.e., the cyclone underflow, returned to the mill increases. Due to an increase in the solids hold-up of the mill, the mill discharge gets coarser. In this way, the coarse discharge begins to appear in the cyclone feed, resulting in a coarse product. The total time of the cyclic reaction is only a few minutes.

Unfortunately, the control strategies of the sump water addition in many mineral processing plants are still aimed at keeping the constant slurry level or the solids content of the overflow required by the following flotation. In our opinion, the correct control strategy of the water addition should first be aimed at keeping an optimal classification of the cyclone together with the control of the fresh feed rate.

The process transfer functions are as follows:

$$H(z) = \frac{0.671 - 0.530\ z^{-1}}{1 - 0.952\ z^{-1}}$$ for cyclone feed flow rate, m^3/h;

$$H(z) = z^{-1}\ \frac{-0.251 - 0.257\ z^{-1}}{1 - 0.997\ z^{-1}}$$ for percent solids, cyclone feed, %;

$$H(z) = \frac{0.828 - 0.328\ z^{-1} - 0.381\ z^{-2}}{1 - 0.996\ z^{-1} - 0.214\ z^{-2}}$$ for fraction $-75\mu m$ solids, cyclone overflow;

$$H(z) = \frac{-0.213 - 0.353\ z^{-1} - 0.592\ z^{-2}}{1 - 0.874\ z^{-1} - 0.104\ z^{-2}}$$ for percent solids, cyclone overflow, %;

Comparison of the predicted results with the measured results shown in Fig. 3 shows good fit. It has indicated that the responses of flow rate and density of the cyclone feed exhibit approximately the first order while the responses of the density and particle size of the overflow seem to be the second order without a dead time.

Effect of Circulating Load on Process Variables

The circulating load, consisting of the underflow of cyclones, has obvious differences in grindability and size distribution with the fresh ore feed to the mill. Its effect on the mill performance is located between the balls and ores. Therefore, the circulating load is one of the main variables governing the grinding efficiency because it will be conducive to the strength of the grinding process in a reasonable range.

However, the circulating load is still a process variable and subject to the influences of other operating variables, especially the feed rate of fresh solids shown as Fig. 4.

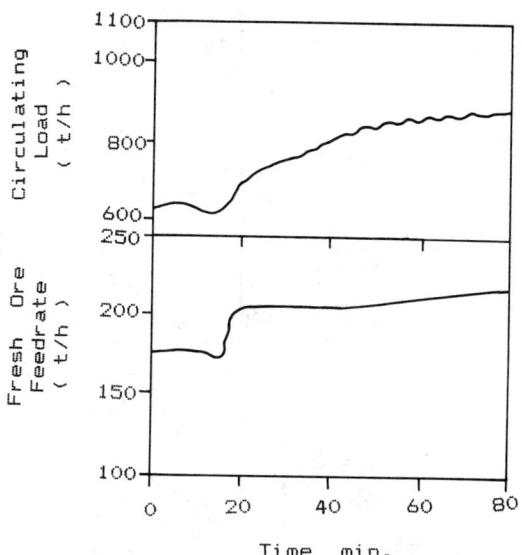

Figure 4.   Response of the circulating load to
change in the feed rate of fresh ore.

Figure 5 gives the response curves of some process variables as circulating load changes where the fresh feed rate was kept approximately at 170 tonnes per hour and the sump water addition at 310 cubic meters per hour. Figure 5 showed that the overflow throughput at required particle size and density increased as the circulating load increased. The fact implicates to us that there is a wide range in which the optimal grinding performance can be achieved by increasing the circulating load. Of course, an increase of the circulating load has an upper limit, and beyond it the grinding performance will sharply become poor even that the mill fills up.

Conclusions

The major conclusions obtained from this study were:

(1) The feed rate of fresh solids is the most important operating variable governing the dynamic behavior in grinding circuits when the properties of the solids have a little fluctuation. Hence it is possible to optimize the performance of the grinding circuit by controlling the feed rate.
(2) The effect of the sump water addition on the particle size of the cyclone overflow, where the slurry level is controlled at a set point, is temporal because this effect will immediately affect, in turn, the mill performance through the change of the cyclone underflow.
(3) The circulating load in closed grinding circuits directly affects the grinding efficiency of mills. It is one of the approaches to optimize closed grinding circuits that the circulating load is controlled based on some type of control strategies.

99

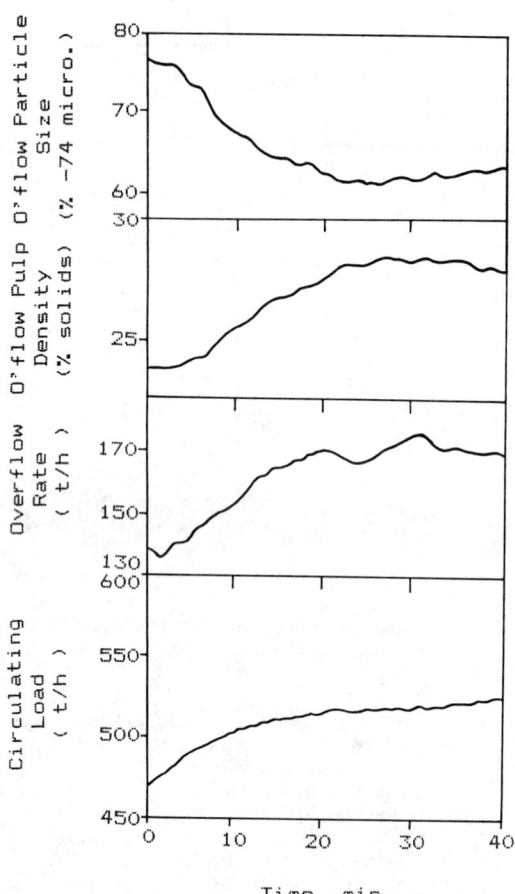

Figure 5.   Responses of process variables to
change in the circulating load.

References

1. M. J. Lees and A. J. Lynch, "Dynamic Behavior of a High Capacity Multi-
    Stage Grinding Circuit," Institution of Mining and Metallurgy Trans., Sec.
    C, 81 (1972), 227-235.

2. R. K. Jaspan et al., "An Analysis of Closed Circuit Wet Grinding Mill
    Control Characteristics by Simulation," The 4th European Symposium on
    Comminution, Nurnberg (1975), 539-557.

3. R. M. Finlayson and D. G. Hulbert, "The Simulation of the Behavior of Individual Minerals in a Closed Grinding Circuit," Proceedings of the 3rd IFAC Symposium, Montreal, Canada (1980), 323-332.

4. I. J. Barker and D. G. Hulbert, "Dynamic Behavior in the Control of the Milling Circuits," Proceedings of the 4th IFAC Symposium, Helsinki, Finland (1983), 139-152.

5. Weiping Xiong et al., "Application of a Microcomputer System in Detection and Sampling of the Dynamic Behavior in an Industrial Grinding Circuit," Nonferrous Metal, Mineral Processing Section, 2 (1987), 33-39 (in Chinese).

6. T. C. Hsia, System Identification (Lexington Books, D. C. Heath and Company, Massachusetts and Toronto, 1977).

EXPERT SYSTEM CONTROL OF A GRINDING CIRCUIT

Lynn B. Hales and Don G. Wardell

Control International, Inc.
2319 Foothill Drive, Suite 200
Salt Lake City, Utah 84109

Abstract

A grinding circuit simulator was used to evaluate a grinding control strategy that has been written using a real-time expert system shell. The strategy was implemented using linguistic rules. The expert strategy was compared to a classical proportional-integral strategy to control product size with sump water and sump level with fresh feed rate. The expert control strategy increased the average tonnage by 5 percent and decreased the average product size by one percentage point on 100 mesh.

Introduction

Expert systems form a class of software that has developed as a result of the large research effort to develop computers and software that think intelligently. By definition, an expert system is a computer program that contains the knowledge component of an expert skill, can offer intelligent advice and can justify its line of reasoning. Expert systems can advise people on what actions to take and when to take them, assist in the diagnosis of the physical condition of a human patient or piece of equipment, or even take actions without human guidance, as in process control applications.

Expert systems can also be used as advisors to newcomers in any field in which mastery is attained by many years of experience. They often can provide expertise more uniformly and rapidly than human experts; they can provide a means of documenting expert knowledge and procedures; and they provide a common repository for a changing knowledge base. The expert system can represent the best expertise available, thought out in the absence of pressure situations. Each of these attributes or characteristics should be helpful in improving typical concentrator control systems.

Implementing computer supervisory control strategies to control grinding plants has been and continues to be a very significant tool capable of increasing plant production and profitability. Implementation has been difficult, however, due to several factors, some of which deal with maintaining FORTRAN based programs. These problems include:

1) a high turnover rate of plant personnel;
2) minimum programming experience of those responsible for writing and maintaining the control software;
3) minimum educational background in computers and computer programming of those responsible for writing and maintaining the control software;
4) poor programming style and non-structured language;
5) a natural difficulty in understanding FORTRAN code written by someone else;
6) a natural difficulty in making program changes and debugging FORTRAN code.

Expert systems offer a potential solution to these problems since they, in theory, can be used to monitor and control using linguistic terms rather than the traditional programmatic approach associated with the FORTRAN programming language.

Expert Systems Concepts and Real-Time Control

There has been significant progress in the past 15 years to implement modern control systems within the mineral's industry. In many cases this has been in the form of distributed control systems. When used properly, distributed control systems can greatly improve the performance and profitability of the plant in which they are applied. This occurs through centralizing the control function, improving plant wide reporting and documenting the decisions of the operators in context with the many operating disturbances typically experienced within a normal day. One problem, perhaps the most significant, still remains however. It is the fact that the operators still have to monitor the process and make decisions to change setpoints of the control loops that have been implemented through the distributed control system. This is where supervisory computer control comes in. Supervisory computer control programs run in real-time, monitor the process and then make setpoint changes as necessary, i.e. as dictated by

104

some control criteria. Historically these programs have been written in FORTRAN or BASIC and are customized for each application.

Why Include Real-Time Expert Systems in Grinding Control Systems?

There are many reasons to consider the use of a real-time expert system as part of a grinding control system. One of the most obvious is to achieve a competitive advantage. Unfortunately, operators are not as effective at optimizing plant performance at three o'clock in the morning as some might hope. With the cutbacks in the number of operators in plants, coupled with a general decrease in their skill levels, control practices have to change. The following quote from the AI Expert Magazine, (1) establishes the main reason so much attention is being given to expert systems.

> The reason so many organizations are investing in AI is the tremendous potential impact to the bottom line. AI can improve the bottom line of corporations by increasing productivity, improving products or services, reducing training costs, decreasing labor costs, ensuring consistent decision making and preserving corporate knowledge.

Observing an operator running a grinding circuit really provides evidence for the need to use expert systems within concentrators. During the course of a shift the operators make, or should make, countless setpoint changes to the process to compensate for variations in feed conditions and equipment conditions. For this reason there is a significant opportunity to use expert systems in concentrator control systems to monitor the process and then supervise the changing of setpoints.

Monitoring the process can be accomplished by one of two methods or combinations of the two. The first is to have the operator enter critical parameter values as they change; the second is to have real-time data from the plant control system be continuously fed to the expert system. The best solution will, of course, use both methods as figure 1 below shows.

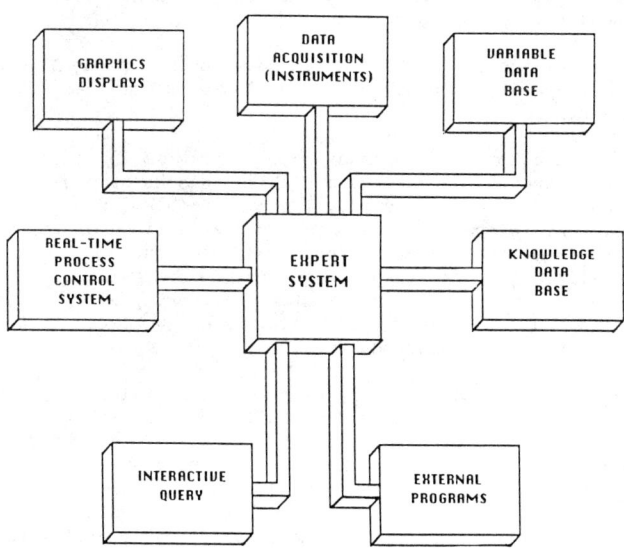

Figure 1: Combined method for monitoring the process

105

Expert Systems Vs. Standard FORTRAN Programs

Expert systems focus on representation of knowledge and on an inference mechanism to perform reasoning on the knowledge base to provide advice to the user. In contrast, conventional computer programs focus on data and procedures to process data. This type of program can only provide answers to problems for which it has been specifically programmed. In other words,

a knowledge-based approach to systems design represents an evolutionary change with revolutionary consequences; for it replaces the software tradition of:

DATA + ALGORITHM = PROGRAM

with a new architecture centered around a knowledge base and an inference engine, so that now

KNOWLEDGE + INFERENCE = SYSTEM.

This second formula is clearly similar, but different enough to have profound consequences (2).

Artificial Intelligence, as the name implies, really does enable a computer to think. By simplifying the way programs are put together, AI imitates the basic human learning process by which new information is absorbed and made available for future reference. The human mind can incorporate new knowledge without changing the way the mind works or disturbing all the other facts that have already been stored away. AI programs based on expert system shells work the same way in that they are far simpler to implement and change or add to than are traditional programs.

If a traditional program needs to be modified in order to accommodate new information, the entire program will have to be scanned to determine the correct place to insert the new code. This procedure is not only time consuming, but often times involves subsequent editing to correct the errors that occur naturally by the process (3).

The following quote from Levine (3) suggests why using AI programming techniques may result in improvement of real-time concentrator process control systems.

AI techniques allow the construction of a program in which each piece of the program represents a highly independent and identifiable step toward the solution of a problem or set of problems. Each piece of the program is like a piece of information in a person's mind. If that information is disputed, the mind can automatically adjust its thinking to accommodate a new set of facts. One doesn't have to go about reconsidering every piece of information one has ever learned, only those few pieces that are relevant to the particular change.

A standard program can do everything an artificial intelligence program can do, but it cannot be programmed as easily nor as quickly. In both types of programs all pieces are interdependent on the way they carry out their designed function. But an AI program possesses a notable characteristic which is equivalent to a vital characteristic of human intelligence. Each minute piece can be modified without affecting the structure of the entire program. This flexibility provides greater programming efficiency and understandability--in a word, intelligence.

Ball Mill Circuit Simulator

To evaluate the concept of controlling a grinding circuit with an expert system, a dynamic simulator of the ball mill grinding circuit, shown in Figure 2, was set up using a simulator package called GRINDSIM.D [TM]. GRINDSIM.D [TM] is based on the dynamic ball mill simulator DYNAMILL developed at the University of Utah (4). The basic stabilizing control loops used were:

1. feedrate
2. water addition to the ball mill (ratioed to new feed rate)
3. sump level (controlled by new feedrate)
4. cyclone overflow particle size (controlled by sump water addition)

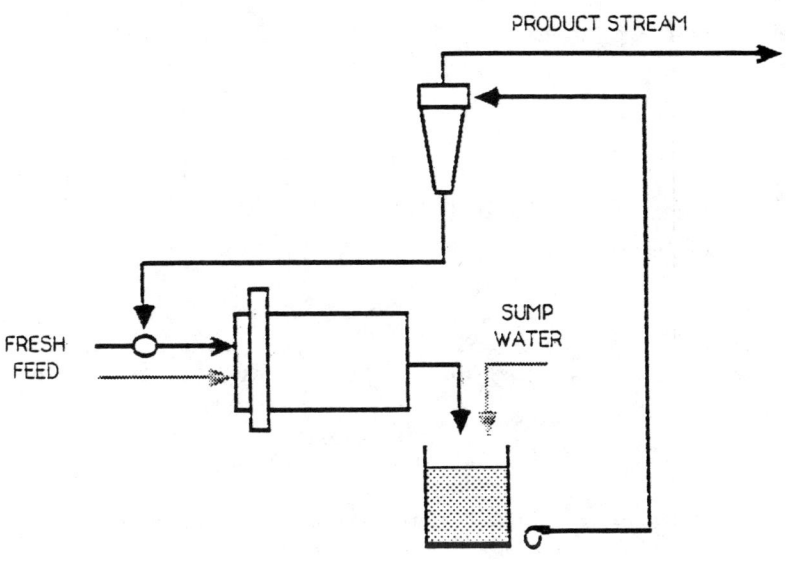

PRODUCT STREAM

SUMP
WATER

FRESH
FEED

Figure 2: Ball Mill grinding circuit flow sheet

For the purpose of evaluating the expert system concept, a 30 hour simulation period was chosen where the ore hardness was varied randomly as shown in Figure 3. In addition to the expert system the control system has been further enhanced by incorporating an on-line process model system that continuously calculates the ore hardness and the percent solids in the ball mill.

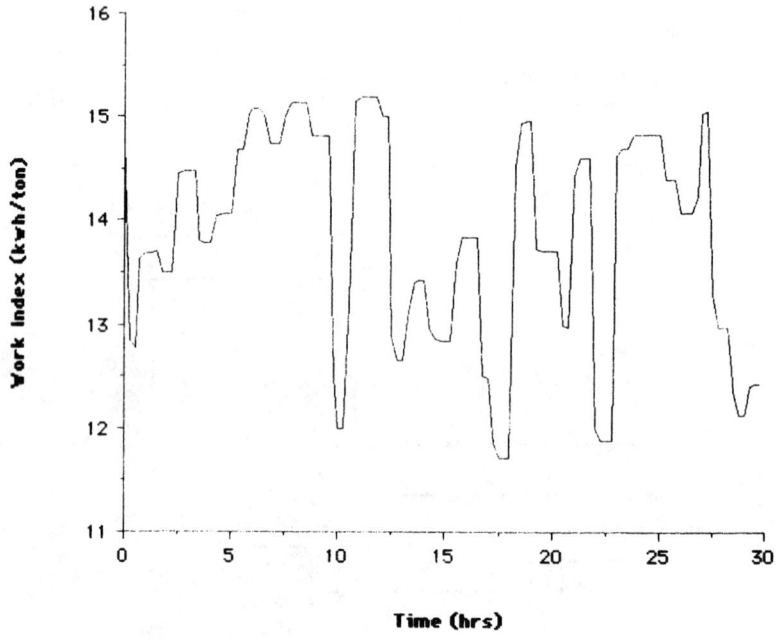

Figure 3: Hardness disturbance for the simulation

Results of the Control Study

Base Case - Stabilizing Control Loops Only

The base case is just the stabilizing control loops by themselves. As the ore hardness variations are experienced the particle size setpoint is maintained constant at 25 percent plus 100 mesh and the feedrate varies to maintain the sump level constant at its setpoint level. The results of the simulation are shown in Figures 4 and 5. The average particle size during the 30 hour simulation period was 25.0 with a standard deviation of .06. The average throughput rate was 190.7 tph with a standard deviation of 11.7 tph.

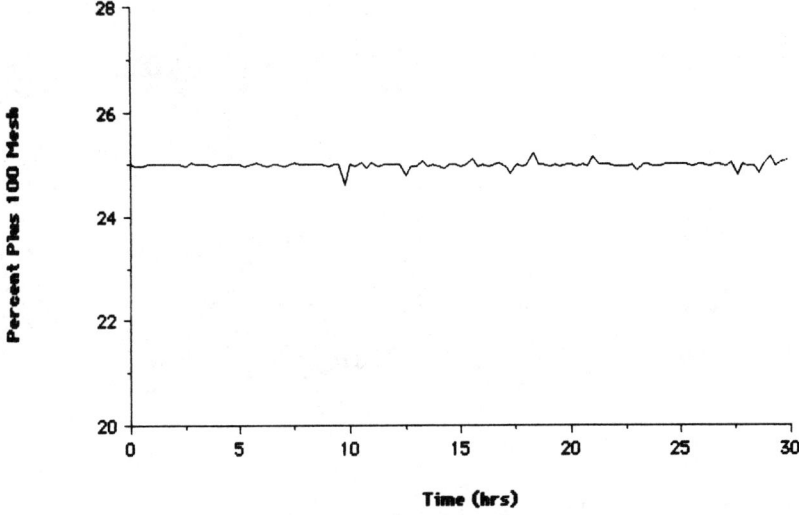

(a) Product Size with PI Control Only

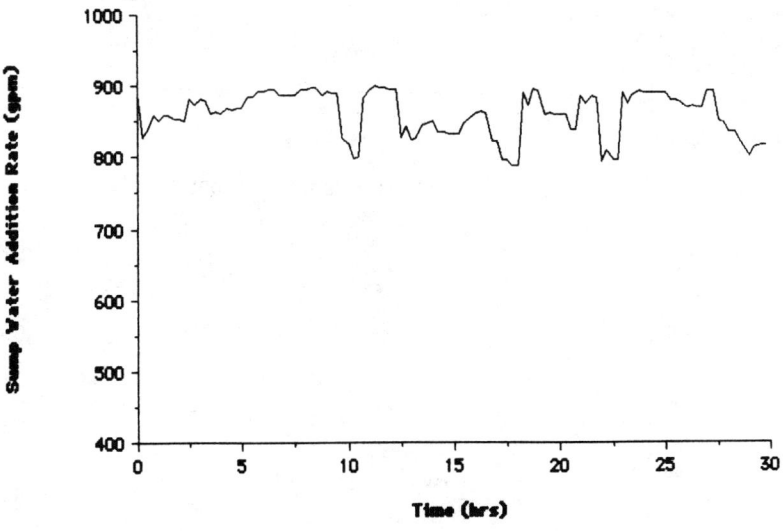

(b) Sump Water Addition with PI Control Only

Figure 4: Size and sump water addition with PI control

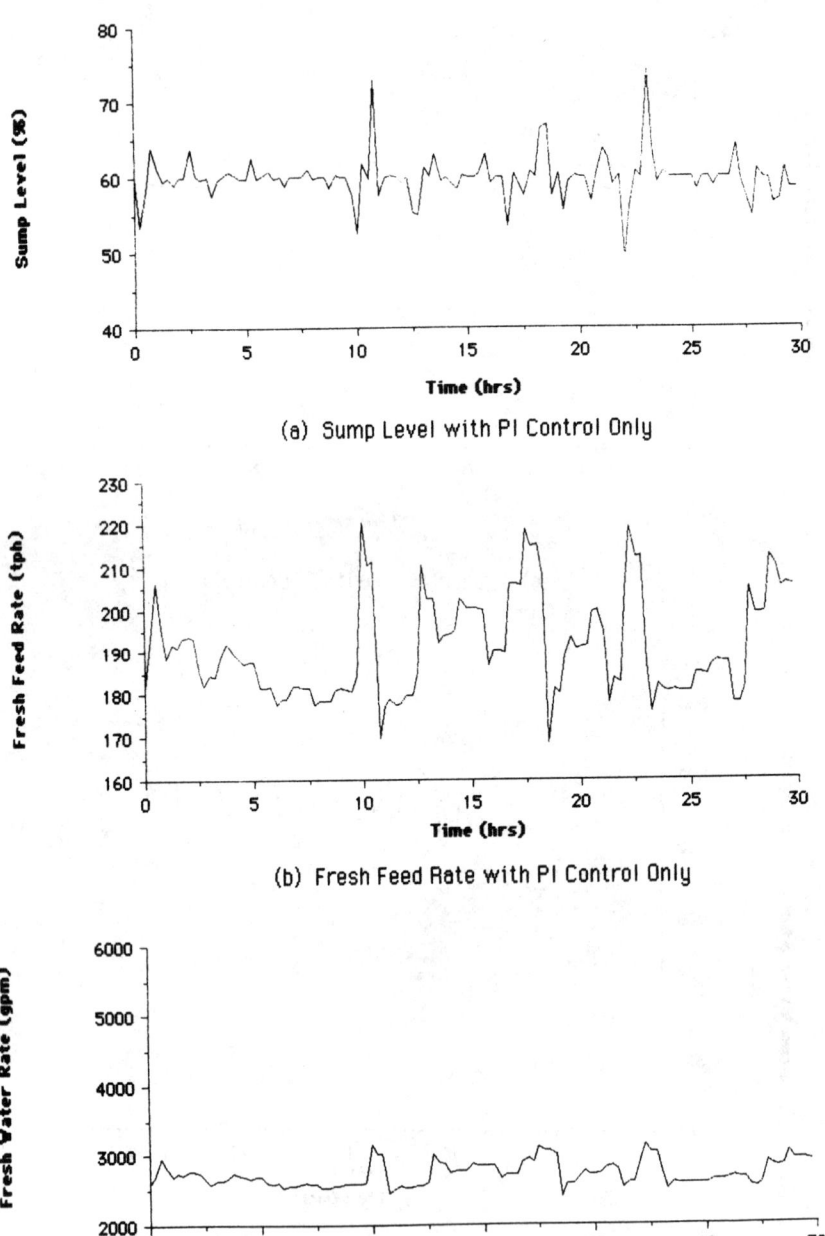

(a) Sump Level with PI Control Only

(b) Fresh Feed Rate with PI Control Only

(c) Ratioed Fresh Water with PI Control Only

Figure 5: Sump level, feed rate, and fresh water addition with PI control

110

Expert System Results

The expert system was made up of 24 rules. The general rule format is of the IF-THEN-ELSE form. Table 1 shows some example rules used for the simulation. The two main rules formulate the goals of the expert system. Each of these rules can be broken down further into sub-rules as shown in the table. Each sub-rule can also branch out. In the table, rules are underlined to show how the sub-rules link to their parent rules. Note that most of the rules compare current values of the operating variables to high or low limits. These limits can be set by management as fixed values, or they can be adjusted by the expert system to optimize the mill performance.

Table I. Example Rules for the Expert System

Main Rules

IF a control action may be taken **AND** the fresh feed rate is above the low limit **THEN** decrease the size setpoint; **ELSE** take no action **OR** (if control action may be taken and feed rate is below the low limit) increase the size setpoint.

IF the estimated mill percent solids are high **THEN** increase the fresh water ratio; **ELSE** decrease the fresh water ratio.

Sub Rules

    IF the circuit is on **AND** the controller is on **AND** there is no emergency **THEN** take a control action; **ELSE** wait for circuit and/or controller to turn on **OR** correct emergency.

        IF the sump level is OK **AND** the sump water controller output is OK **AND** the cyclone pressure is OK **AND** the feed water is OK **THEN** there is no emergency; **ELSE** there is an emergency which must be corrected.

            IF the sump is not cavitating **AND** the sump is not nearing overflow **THEN** the sump level is OK; **ELSE** a corrective action must be taken.

The objective of the expert system was to monitor the process and, based on the performance of the control loops, make setpoint changes that take advantage of the ore changes and minimize the final grind size while not letting the throughput rate decrease below 200 tph. An additional feature that the expert system used was the calculated mill percent solids that were estimated continuously by an on-line Kalman Filter model system (5). Based on the calculated mill solids the water ratio setpoint was also changed as necessary to maintain a percent solids of 68 percent. The mean cyclone overflow particle size was 24.1 percent plus 100 mesh with a standard deviation of 1.71. The throughput averaged 200.0 tph with a standard deviation of 6.9 tph. Figures 6 and 7 show the results achieved by the expert control strategy.

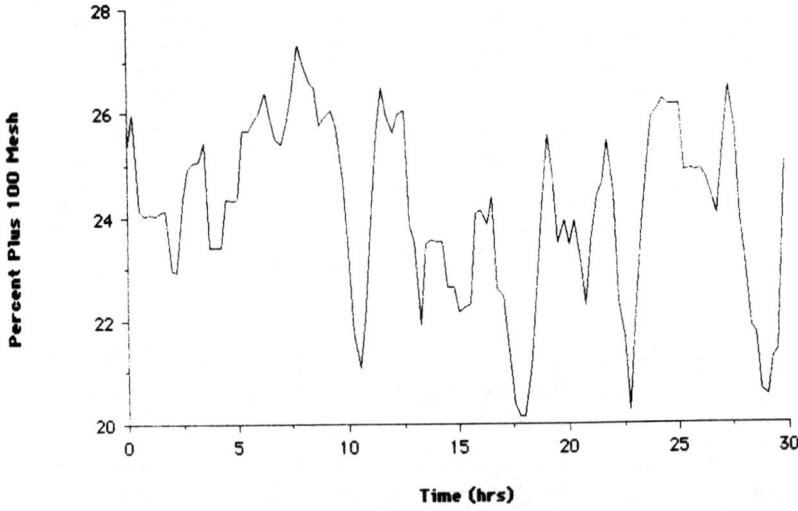

(a) Product Size with Supervisory Control

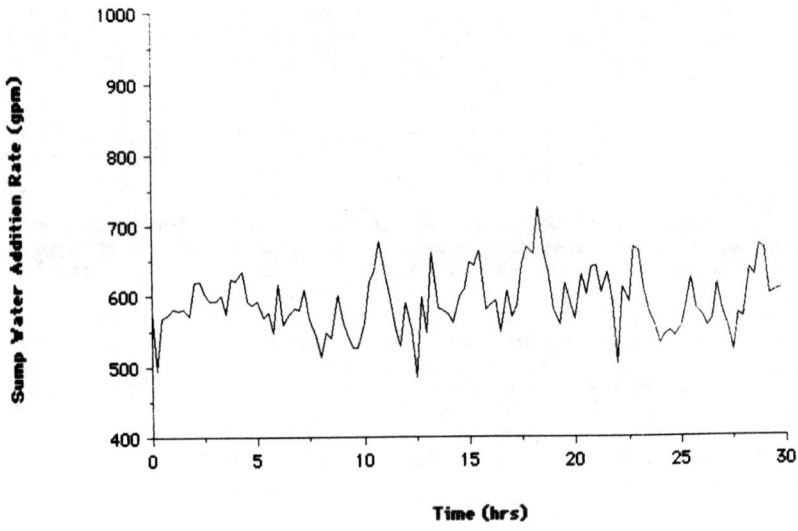

(b) Sump Water Addition with Supervisory Control

Figure 6: Size and sump water addition with Expert Control

112

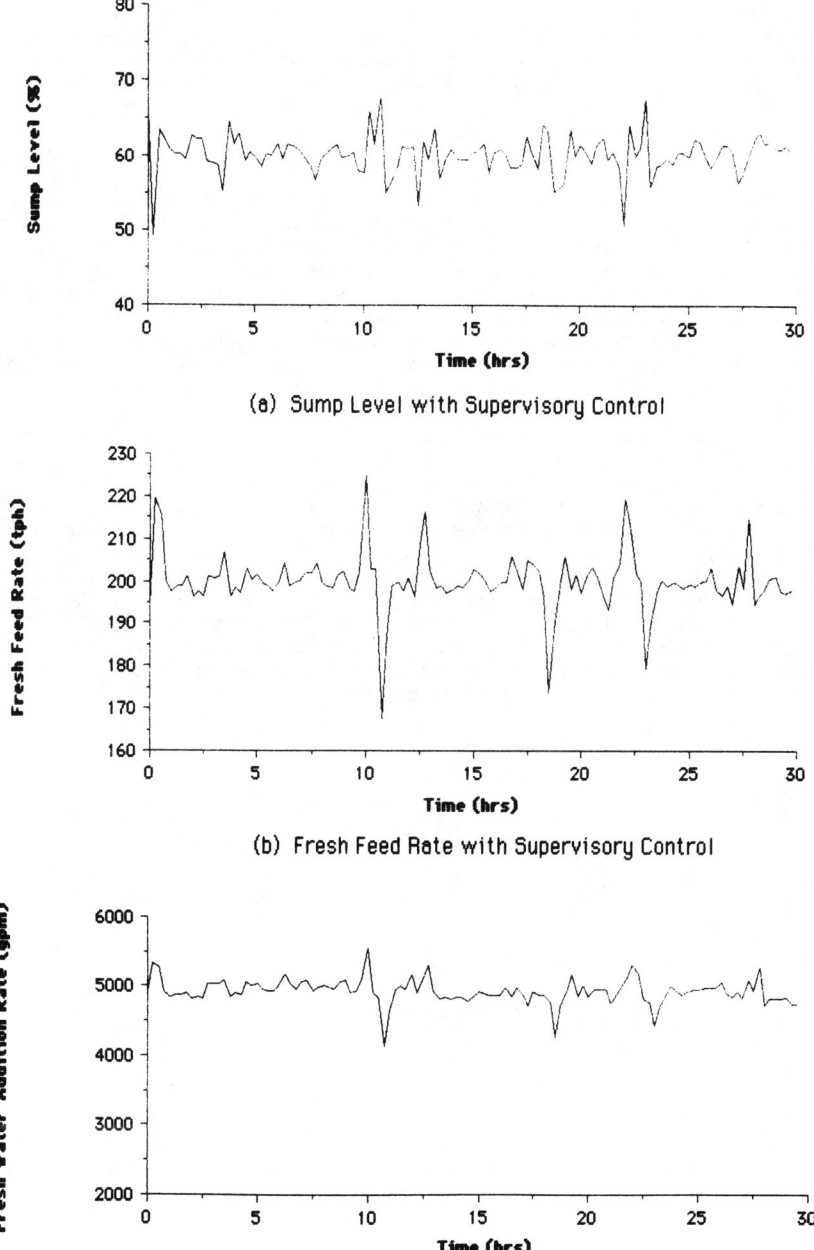

(a) Sump Level with Supervisory Control

(b) Fresh Feed Rate with Supervisory Control

(c) Fresh Water Addition with Supervisory Control

Figure 7: Sump level, feed rate and fresh
water addition with Expert Control

113

Discussion of Results

As the simulation study shows, the expert system performed significantly better than the standard stabilizing control loops. Even under sub-optimal conditions, the average throughput increased by 10 tph while at the same time the product became finer by one percent. The reasons for this are both obvious and subtle. Figures 8 and 9 show two histograms of the feedrate and cyclone particle size. Here again the improvement achieved by the control strategy using the expert system is readily seen.

The obvious results are that as ore conditions vary setpoint changes could be made to take advantage of softer ore or to compensate for harder ores. In fact, comparing Figures 3 and 6a shows that the product size "traces" the hardness disturbance. The reasons that operators do not perform as well as the computer control strategy are due to their incomplete set of "rules of thumb" which often times include rules that are incorrect or that are applied incorrectly or incompletely. Another problem is the fact that one operator is usually responsible for more than one grinding line and problems in one area can result in inattention to other areas.

The expert control strategy used for this study was relatively easily setup and evaluated. Adding or deleting rules to change the performance of the strategy is simpler than the corresponding changes to a FORTRAN program would have been. The improvement in performance of the expert system over the stabilizing control loops was enhanced considerably by incorporating the Kalman Filter model system that estimated the mill percent solids continuously. It is intuitive that grinding efficiency is effected by percent solids and, in fact, laboratory test work has shown this to be true. Because variations in ore hardness and feed size distribution naturally produce variations in mill percent solids the ability to accurately calculate mill percent solids and then control it at its optimum level greatly improves the grinding performance.

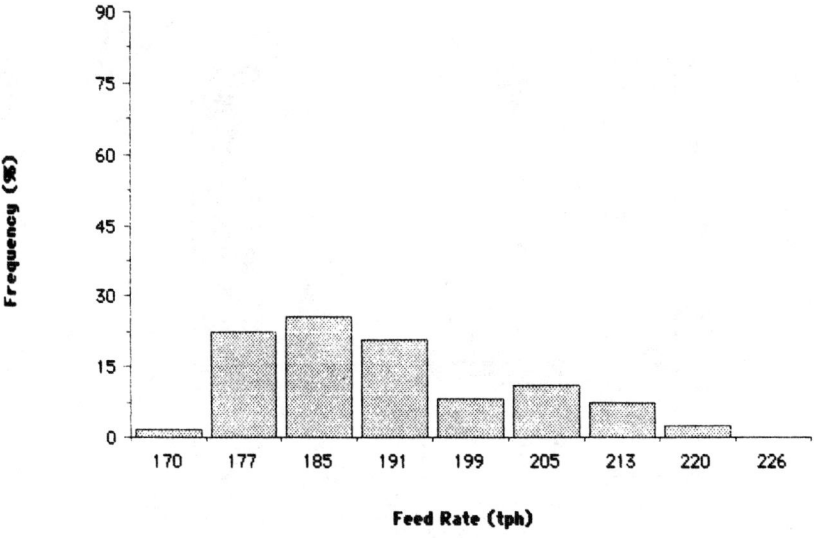

(a) Feed Rate Histogram with PI Control Only

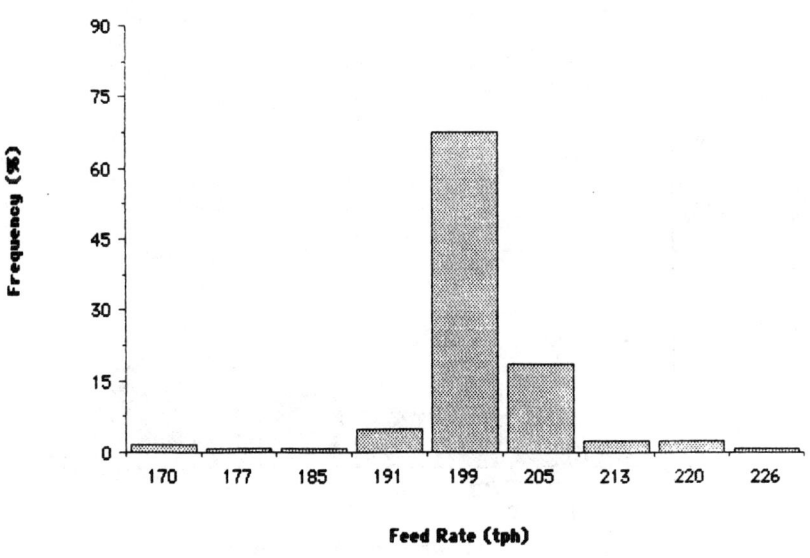

(b) Feed Rate Histogram with Supervisory Control

Figure 8: Histograms for feed rate using the two strategies

115

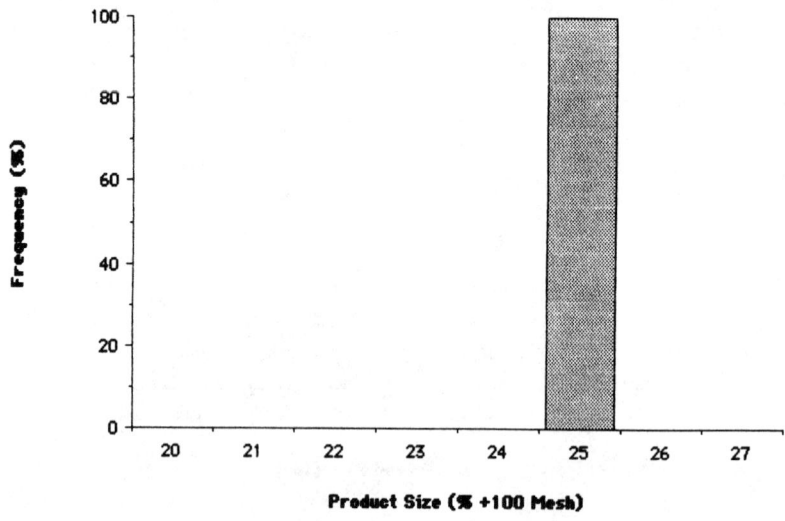

(a) Size Histogram with PI Control Only

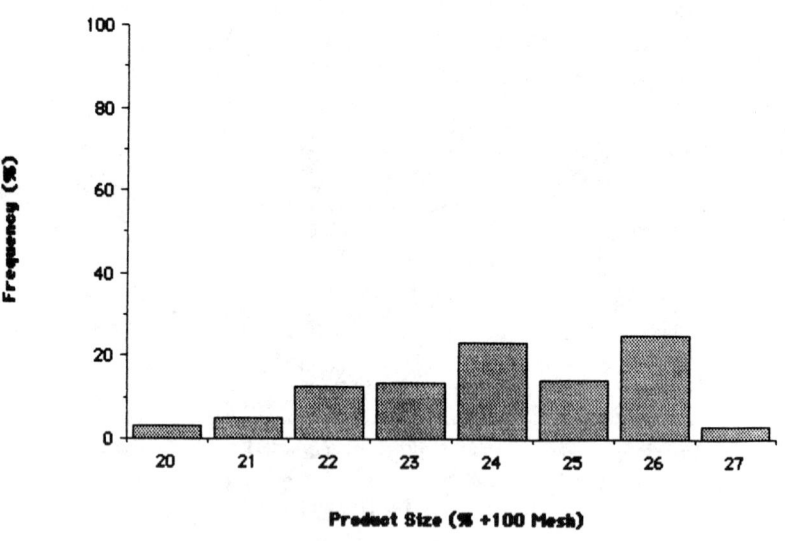

(b) Size Histogram with Supervisory Control

Figure 9: Histograms for size using the two strategies

116

Conclusions

The need for supervisory computer control is obvious and can result in substantial benefits. It has been difficult in the past to continuously achieve these benefits because of the problems associated with traditional programming languages. Many of these problems can be overcome by an appropriately designed expert control system. Many problems, however, still have to be overcome before expert systems of the type described here will be used routinely in our industry. They do offer great promise and further development will result in expert control systems that will be suitable for real-time control in the near future.

References

1. E. Wang, "Brain Waves," AI Expert Magazine, 2 (1) (1987), 5.

2. R. Forsyth, ed., Expert Systems - Principles and Case Studies. (London: Chapman and Hall, 1984), 9.

3. R.I. Levine, D.E. Prang, and B.A. Edelson, 1986, Comprehensive Guide to AI and Expert Systems (New York, NY: McGraw Hill, 1986), 4.

4. K. Rajamani and J.A. Herbst, "A Dynamic Simulator for the Evaluation of Grinding Circuit Control Strategies," Proceedings of the Particle Technology 1980, ed. K. Schonert (Amersterdam: Dechema, 1980), 64-81.

5. P.S. Maybeck, Stochastic Models, Estimation and Control, vol. 1. (New York, NY: Academic Press, 1981), 206-226.

FLOTATION CELL SIMULATION

Session Chairmen
Lynn B. Hayles, Control International, Inc.
Greg Adel, Virginia Polytechnic Institute
State University

This session describes model development from phenomenalogical data for a semi-selective flotation circuit, as well as model implementation to analyze reagent behavior. Application of a flowsheet simulation package involving predictive and optimazation capabilities is also discussed.

A NEW TWO-PHASE DYNAMIC MODEL OF FLOTATION

Ligang Chang and Zhen Su
Mineral Engineering Department
Central South University of Technology
Changsha, Hunan
P.R. China

and Weibai Hu
Department of Metallurgy and Metallurgical Engineering
University of Utah
Salt Lake City, Utah 84112

Abstract

A new two-phase dynamic model has been developed from a detailed phenom-enological circuit description based on the concept of flotation rates and drainage rates. Systematic analysis was made about the model, and it was tested with the data from a semi-selective flotation circuit of a Chinese concentrator.

121

Introduction

In the beginning of the 1960s, Arbiter and Harris postulated a two-phase model to analyze the dynamics of flotation (1). They divided the floating materials into four groups, namely, pulp, froth, concentrate, and tailings. They assumed the existence of material exchange among the four groups. Then a dynamic equation could be derived. However, such a theoretical model was not used in practice. Niemi et al. tried to develop dynamic models for commercial flotation cells. Their models, though simple in form, provided the basis for the development of the optimal control schemes and gained a lot of positive results (2-9). All of these models are the two-phase models with some modifications. Sadler tested the two-phase model and three other models under continuous transient conditions (10). He reported that the data obtained were inadequately described by the two-phase model and that a variation of the model referred to as Bubble Surface Area as well as a distributed rate constant model offer excellent alternatives in the dynamics of his particular system. In this flotation control simulation, Fewings and Lynch also used a model similar to those mentioned above (11). Although much has been done in this respect, but due to the complexity of the process and the variety of ores, most previous results are not compatible.

In this paper, some modifications of the two-phase model have been suggested. The results show that the new model has satisfactory dynamic characteristics under some conditions. The model was tested against the data from a semi-selective flotation circuit of a Chinese concentrator.

Modeling

We consider a simple flotation process consisting of only one cell as shown in Figure 1. Assuming

(1) There are two phases in the cell, the froth layer and the pulp phase, and the phases are each ideally mixed,

(2) The floating from the pulp and drainage from the froth are both first-order kinetics.

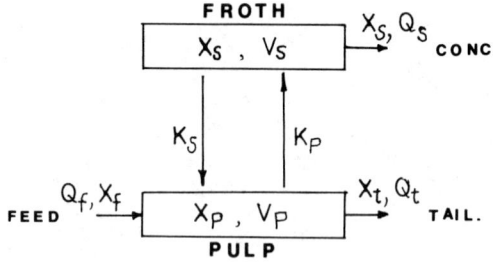

Figure 1. Illustration of two-phase model.

(3) There are many kinds of materials which have different flotation constants α_i and drainage rates β_i, and let

$$K_p = \text{diag} (\alpha_1, \alpha_2, \ldots, \alpha_n)$$
$$K_s = \text{diag} (\beta_1, \beta_2, \ldots, \beta_n)$$

where n is the number of kinds of materials.

(4) The grade of some kind of material in the concentrate is equal to that in the froth.

Based on the principle of population balance, dynamic balance equations can be written for each of the phases:

for the pulp:

$$\frac{d(V_p X_p)}{dt} = -K_p V_p X_p - Q_t X_t + k_s V_s X_s + X_f Q_f \qquad (1)$$

for the froth layer:

$$\frac{d(V_s X_s)}{dt} = K_p V_p X_p - K_s V_s X_s - Q_s X_s \qquad (2)$$

where Q_f, Q_t, and Q_s, respectively, are the feed, tailing, and concentrate volume flow rates; X_p, X_t, X_s, and X_f are the volume concentrations of the different flotation material in, respectively, the pulp, tailing, froth layer, and feed; V_p is the pulp volume, and V_s the froth volume.

We can suppose that V_p and V_s are constants; then it follows from Equations (1) and (2) that

$$\frac{dX_p}{dt} = - K_p X_p - \frac{C_Q}{C_t} X_t + C_v K_s + \frac{1}{C_t} X_f \qquad (3)$$

$$\frac{dX_x}{dt} = \frac{1}{C_v} K_p X_p - [K_s + \frac{1 - C_Q}{C_v C_t} I]X_s \qquad (4)$$

where $C_v = V_s/V_p$; $C_Q = Q_t/Q_f$; and $C_t = V_p/Q_f$.

Since there are different functional relations between the grades of the pulp and tailing, different models can be obtained after taking account of these differences, so that Equations (3) and (4) represent a great range of functions.

We consider only the simplest condition, that is, let X_p and X_t be as follows:

$$X_t = CX_p$$

where C is a constant, $0 \leq C \leq 1$.

With the above-mentioned assumption, we obtain

123

$$\frac{dX_t}{dt} = -(K_p + \frac{CC_Q}{C_t} I)X_t + CC_vK_sX_s + \frac{C}{C_t} X_f \tag{5}$$

$$\frac{dX_s}{dt} = \frac{1}{C_v} K_pX_t - (K_s + \frac{1 - C_Q}{C_vC_t} I) X_s \tag{6}$$

Equations (5) and (6) are the two-phase dynamic model used for theoretical analysis and simulations.

Model Analysis

Rewriting Equations (5) and (6) using matrix notation, we have

$$\frac{dX}{dt} = AX + X_f' \tag{7}$$

where
$$X = \begin{vmatrix} X_t \\ X_s \end{vmatrix} \quad ; \quad A = \begin{vmatrix} -K_p + aI & bK_s \\ K_p/b & -K_s + dI \end{vmatrix}$$

$$X_f' = \frac{C}{C_t} \begin{vmatrix} X_f \\ 0 \end{vmatrix} \quad ; \quad a = -\frac{CC_Q}{C_t} \; ; \quad b = CC_v; \quad d = -\frac{1 - C_Q}{C_vC_t}$$

For the general solution of Equation (7), using the technique in matrix and determinant theory, we obtain the eigen function of matrix A:

$$| A - \lambda I | = | \lambda^2 I + \lambda[K_s - (a + d) I + K_p] + adI - aK_s - dK_p |$$

where I is the unit matrix. Therefore, the eigen values of matrix A are

$$\lambda k_{1,2} = \frac{a + d - (\alpha_K + \beta_K) \pm \sqrt{(a + d - \alpha_K - \beta_K)^2 - 4(ad - d\alpha_K - a\beta_K)}}{2} \tag{8}$$

Because $a < 0$, $d < 0$, $\alpha_K > 0$, and $\beta_K > 0$, then $a + d < \alpha_K + \beta_K$, and $ad > d\alpha_K + a\beta_K$.

It is evident that A has 2n's distinct negative eigen value λk_1, λk_2 (k = 1, 2, ..., n). Thus the differential Equation (7) has stable solutions. With computation, we obtain the solution of Equation (7):

$$X = e^{At} [X_o + \int_o^t e^{-Au} \cdot X_f'(u)du] \tag{9}$$

where $X_o = X(0)$ is the initial state when t = 0 and

$$e^{At} = M \text{ diag } \{e^{\lambda 11^t}e^{\lambda 12^t}, \ldots, e^{\lambda n1^t}, e^{\lambda n2^t}\}M^{-1} \tag{10}$$

124

$$M = \begin{vmatrix} 1 & 1 & 0 & 0 & \cdots & 0 & 0 \\ -\dfrac{a-\lambda_{11}}{d-\lambda_{11}} & -\dfrac{a-\lambda_{12}}{d-\lambda_{12}} & 1 & 1 & \cdots & 0 & 0 \\ 0 & 0 & -\dfrac{a-\lambda_{21}}{d-\lambda_{21}} & -\dfrac{a-\lambda_{22}}{d-\lambda_{22}} & \cdots & \vdots & \vdots \\ \vdots & \vdots & 0 & \vdots & \cdots & 0 & 0 \\ \vdots & \vdots & 0 & \vdots & \cdots & 1 & 1 \\ 0 & 0 & 0 & 0 & \cdots & -\dfrac{a-\lambda_{n1}}{d-\lambda_{n1}} & -\dfrac{a-\lambda_{n2}}{d-\lambda_{n2}} \end{vmatrix} \qquad (11)$$

Also, consider several special conditions as follow:

1. Input is stable and C_t is a constant; obtain from Equation (9)

$$X = e^{At}(X_o + A^{-1} X_f') - A^{-1}X_f' \qquad (12)$$

where $A^{-1} = M \, \text{diag} \, (1/\lambda_{11}, \, 1/\lambda_{12}, \, \ldots, \, 1/\lambda_{n1}, \, 1/\lambda_{n2}) \, M^{-1}$ (13)

2. C_t keeps constant, and a long time after X_f becomes stable, obtain from Equation (12) directly

$$X = -A^{-1}X_f' \qquad (14)$$

Solving Equation (14), we obtain

$$X_t = - \left[\frac{1}{a} I + \frac{d}{a^2} (-K_s + dI - \frac{d}{a} K_p)^{-1} K_p\right] \frac{C}{C_t} X_f \qquad (15)$$

$$X_s = \frac{1}{ab} (-K_s + dI - \frac{d}{a} K_p)^{-1} K_p \frac{C}{C_t} X_f$$

Equation (15) is useful for analysis of flotation processes.

Simulation Studies

In order to simulate the plant conditions using Equations (5) and (6), an appropriate model for predicting must first be found. Since the functional relation between feed grades X_f and time t cannot be defined for most industrial flotation processes, it is difficult to solve Equation (9) directly.

An alternative equation was used for an approximate solution.

Rewriting Equations (5) and (6),

125

$$\frac{d}{dt} X(t) = f[X(t)]$$ (16)

Now for the discretization we obtain from Equation (16) that

$$X_{K+1} - X_K = f[X(t)]dt$$

If the sample interval $t_{K+1} - t_K = T$ is very small, then $f[X(t)]$ can be developed approximately at intervals $[t_K, t_{K+1}]$:

$$f[X(t)] = f(X_K) + A(X_K) f(X_K) (t - t_K)$$ (17)

where $X_K = X(t_K)$ and $A(X) = \partial f(X)/\partial X$. Rewriting Equation (16) and integrating, we obtain

$$X_{K+1} = X_K + f(X_K)T + A(X_K) f(X_K)T^2/2$$ (18)

Equation (18) is the model used for simulation.

The data for simulation were taken from the semi-selective flotation circuit of the Feng Huang Shan copper ore concentrator. The grades of semi-selective feed, concentrate, and tailing were measured with an on-line x-ray analyzer. Equation (18) was used to predict these grades. The data was taken for seventeen hours. The grades have all been transformed into weight percentage. Figures 2 and 3 illustrate the variance of the grades with time.

Results from analysis of the measurements and the predictions are shown in Table 1.

It can be seen from Table 1 that the predictions and the measurements are very close, and the relative deviations for grade predictions and measurements fall into the range acceptable by the mineral-processing experiment.

Discussion

These equations were obtained in the case of a single cell, and good results have been obtained, which are shown in Figures 2 and 3. However, flotation processes usually consist of several cells. There are two ways to consider the case of multiple cells. One is to consider these cells as one big cell, and the model becomes an aggregate one with the same form as Equation (7). Another is to derive the multicell model which can be obtained by linking the single-cell models together.

Table 1. Statistics of Deviations

	Absolute Deviation	Relative Deviation
Concentrate	0.33	1.2%
Tailing	0.05	3.7%

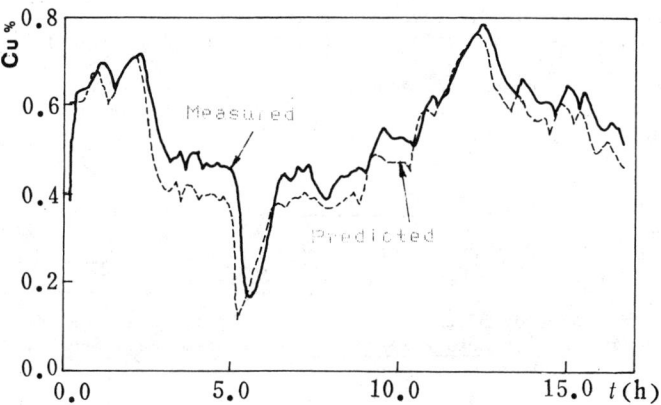

Figure 2. Cu % in tailing vs. time.

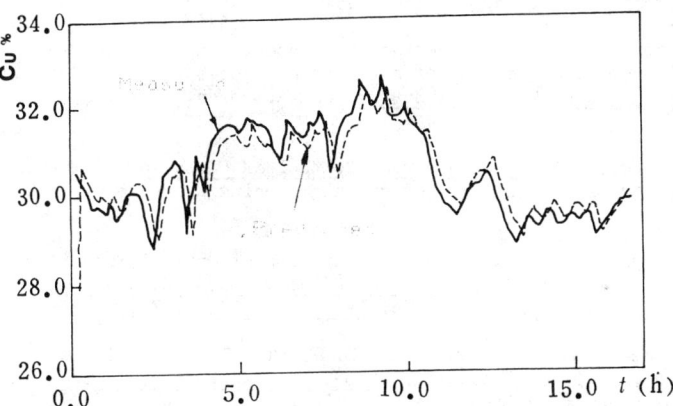

Figure 3. Cu % in concentrate vs. time.

The manipulating variables are not contained in the model, so it cannot be used for control purposes yet. These variables should be combined with the model's parameters.

In addition, some parameters are supposed to be constants in the simulation. In fact, they are variable in the flotation processes because of disturbances. These should be considered in further research.

Conclusions

A great range of two-phase dynamic flotation models was obtained based on the concept of flotation and drainage rates and the principle of population balance. After assuming a linear relationship between pulp grades and tailing grades, the model was further analyzed theoretically, and two special solutions were obtained. Although Equations (12) and (15) were obtained under ideal conditions and it is difficult for quantitative simulations, they are helpful for qualitatively analyzing the phenomenon of flotation.

Equation (18) is proper for recurrence estimating qualitatively. The estimated results compared with plant data showed that the model can represent the variation of the process.

References

1. N. Arbiter and C. C. Harris, "Flotation Kinetics," Froth Flotation, ed. D. W. Fuerstenau (AIME, New York, 1962), 215-262.

2. A. J. Niemi, "A Study of the Dynamic and Control Properties of Industrial Flotation Processes," Acta Polytech. Scand., Chem. Met. Series, 48 (1966).

3. A. J. Niemi, "On the Dynamics of a Pneumatic Flotation Cell," Acta Polytech. Scand., Chem. Met. Series, 49 (1966).

4. A. J. Niemi and V. E. Paakkinen, "Simulation and Control of Flotation Circuits," Automatica, 5 (1969), 551-561.

5. P. I. Eerola and V. E. Paakkinen, "Introduction of Theoretical Aspects into the Computer Control of a Flotation Process," A Decade of Digital Computing in the Mineral Industry, ed. A. Weiss (SME/AIME, 1969), 827-851.

6. A. J. Niemi, Optimal Feedforward Control of Flotation Cell (Helsinki University of Technology, Control Engineering Laboratory, ISBN 951-750-067-X, 1971).

7. A. J. Niemi and J. Maijanen, "Computation of Optimal Feed Forward Control for Flotation Cell," Symposium on Dynamic Modelling and Computer Control of Technological Processes, Tbilisi, Nov. 1973 (1973), 352-368.

8. A. J. Niemi, J. S. Maijanen, and M. T. Nihtila, "Singular Optimal Feedforward Control of Flotation," Proc. of IFAC/FORS Symposium on Optimization Methods -- Applied Aspects, Varna, Bulgaria (74), 277-283.

9. A. J. Koivo and H. N. Koivo, "Optimal (Singular), Control of a Flotation Cell in Mineral Processing," Int. J. Control, 23 (2) (1976), 217-228.

10. L. Y. Sadler, "Dynamic Response of the Continuous Mechanical Froth Flotation Cell," Trans. SME-AIME, 254 (1973), 336.

11. J. H. Fewings et al., "The Dynamic Behaviour and Automatic Control of the Chalcopyrite Flotation Circuit at Mount Isa Mines Limited," XIII IMPC, Warsaw, vol. 2 (1979), 405-432.

USE OF KINETIC MODELS TO EVALUATE NEW FLOTATION

COLLECTING REAGENTS

Edward C. Dowling, Jr.[*], Richard R. Klimpel[+] and F.F. Aplan[*]

*Mineral Processing Section
The Pennsylvania State University
University Park, PA   16802

+1776 Building
The Dow Chemical Company
Midland, MI   48640

ABSTRACT

Kinetic flotation models can be used in many process analysis, control and development applications. In this paper a kinetic model, previously identified as being most reliable in describing the flotation of sulfide ores (Dowling, Klimpel and Aplan, Schuhmann Symposium, 1986), is used to evaluate flotation reagent performance. Using flotation rate and recovery as the criteria of merit, flotation tests were made using copper sulfide ores with the goal of evaluating various operational conditions and a broad variety of collectors and frothers.

INTRODUCTION

Kinetic flotation models were initially developed many years ago by those investigators that recognized the great utility of these analytical tools, (1-3). More recently, the availability of relatively inexpensive information processing systems and economic pressures for efficient resource management have stimulated an increased interest in flotation modeling (4-7). Although there are several model types, kinetic models are of particular interest due to their wide range of application. Models of this type can be used to develop control systems, characterize or predict plant performance (on or off line), assist engineering design simulations and scale-up and evaluate small scale tests (8-13). The present study will focus on the use of kinetic models to evaluate flotation reagents.

The work reported here is the result of an ongoing research program having the ultimate goal of improving industrial sulfide flotation practice. A specific goal is to identify chemical reagent systems superior to those currently used. To this end, over one hundred collectors were first individually tested on sulfide minerals using microflotation techniques (14-18). These initial tests were evaluated using a simple first-order flotation rate and near ultimate mineral recovery approach as the criteria of merit. This approach identified a number of promising reagents. The present paper uses a more sophisticated approach employing a statistically reliable flotation model to evaluate and report the results of follow-up laboratory batch-scale tests for a variety of reagent systems and conditions.

EXPERIMENTAL METHODS

Froth flotation is recognized as an interactive engineering system composed of three factors (chemical, operational and equipment), each of which contributes to the overall flotation performance (19). Industrial operators know from experience that a certain change in any factor (pH, reagent levels, aeration rate, percent solids, etc.) can improve or worsen flotation performance. Consequently, over time, operators will typically balance the overall contribution of each factor to improve circuit performance. With this in mind it is important to realize that flotation tests comparing differences in reagent performance must be designed to test for these differences only. Consequently, special care has been taken in sample preparation, testing procedures and data analysis methods.

Sample Preparation

Flotation tests were made with porphyry copper ores from Magma Copper Company's Pinto Valley, AZ, and Kennecott Copper Company's Arthur, UT, mills. The Pinto Valley ore was evaluated in our laboratories and by W. Mueller of Newmont Exploration Ltd., Danbury CT. This ore is a quartz monzonite porphyry containing approximately equal amounts of chalcopyrite and pyrite as the principle sulfide minerals along with a minor amount of molybdenite. Considerable liberation of chalcopyrite from gangue was noted at a nominal -48 mesh grind. This ore averaged 0.54% copper with only 0.006% oxide copper. Simple mineralogy makes this ore ideal for study.

The Utah ore, also of the porphyry type, has a somewhat more complex sulfide mineralogy, containing significant amounts of three copper sulfide minerals ($\sim$1.5% total, of which chalcopyrite $\sim$75%, bornite $\sim$20% and chalcocite $\sim$5%), pyrite ($\sim$3%) and a trace of molybdenite. Reasonable liberation occured at a grind of nominal -65 mesh and heads for the sample used here averaged 0.50% copper. The mineralogical results noted here are in good agreement with the published literature (20,21). The proportion of each ore selected for study was carefully sampled using standard riffling procedures, stage crushed (to avoid oxidation of fines) to nominal -10 mesh (∿2 mm) and split into 500 g test samples.

Test Procedures

Ore grinding was done in an iron rod mill, and soda ash (2.0 kg/t) was added to the grind as a pyrite depressant for those tests made at pH 10.5. Two grinding times of six minutes (2.1% +48, 42.7% -200 mesh) and four minutes (11.5% +48, 31.6% -200 mesh) were used for the Pinto Valley ore. The short grinding time was intentionally used in order to study the influence of various reagent systems on the recovery of coarse particles. A five minute grind (1.6% +65, 51.9% -200 mesh) was selected for the Utah ore.

Flotation tests were made using a Galligher LA500 flotation machine (500 g cell) using an automatic froth removal system and close air control to minimize human bias. Immediately after the ground material was transferred to the cell, pulp level was adjusted to the appropriate level and the pH adjusted to either 5.0 or 10.5, using hydrochloric acid or milk of lime additions. After the final reagent addition, the pulp was conditioned for one minute. Collector additions were made both to the grinding mill and directly to the flotation cell. It will be subsequently shown that the point of collector addition is often an important operating factor. Collector additions were made to the mill by carefully weighing the appropriate amount of collector on to a microscope glass slide cover and adding this to the grinding charge. Additions to the flotation cell were made by adding a dilute solution of the collector by pipet. All collectors used in this study were purified before use (14,17). Table I gives the name, abbreviation and formula of the collectors and frothers (22,23) used here. While the bulk of the testing was done using DF-200, other frothers were also evaluated, and these were added directly to the flotation cell by microsyringe.

Flotation rate studies were made in all cases and zero time was defined as the instant the froth began to flow freely over the cell lip. Normally, five concentrates were collected at flotation times of 30, 60, 120, 240 and 480 seconds. However, only four concentrates were collected at flotation times of 30, 60, 120 and 480 seconds for the Pinto Valley tests done at the coarse grind (4 minutes) in order to insure that there was sufficient material for analysis. Because mineralogical results showed that all of the sulfur contained in the Pinto Valley ore was present as chalcopyrite or pyrite (not counting a trace of molybdenum), pyrite assays were determined by analyzing for total copper and total sulfur and attributing all residual sulfur not accounted for by the chalcopyrite stoichiometry to pyrite. In this way the influence of the reagents on pyrite flotation could be evaluated. Metallurgical balances and time-recovery profiles were calculated from the primary data. Further details on the procedure are to be found elsewhere (11).

Analysis Procedures

As a prelude to these tests, the authors made a systematic comparison of a number of kinetic models, described in the literature, for their reliability in precisely describing sulfide flotation in both laboratory and plant scale settings (11-13). The models were compared using the requirements of goodness of fit and discreteness (narrow range of statistical significance) of model parameters. Results of this model evaluation identified a two parameter, first-order model with rectangular distribution of floatabilities as being especially reliable. The mathematical form of this model [6,10] is given by Equation 1 which describes flotation recovery,

Table I. Reagents Used

Name	Abbreviation	Structural Formula
Potassium Ethyl Xanthate	KEtX	$CH_3CH_2OC(S)SK$
Potassium Propyl Xanthate	KPX	$CH_3(CH_2)_2OC(S)SK$
Potassium Isopropyl Xanthate	KIpX	$(CH_3)_2CHOC(S)SK$
Potassium Amyl Xanthate	KAmX	$CH_3(CH_2)_4OC(S)SK$
Potassium Octyl Xanthate	KOcX	$CH_3(CH_2)_7OC(S)SK$
Potassium Dodecyl Xanthate	KDdX	$CH_3(CH_2)_{11}OC(S)SK$
Diisopropyl Dixanthogen	$(IPX)_2$	$[(CH_3)_2CHOC(S)S]_2$
Sodium Diisopropyl Dithiophosphate	IpDTP	$[(CH_3)_2CHO]_2P(S)SNa$
N-Methyl-O-Ethyl Thionocarbamate	NMe-OEtTC	$CH_3NHC(S)OCH_2CH_3$
N-Ethyl-O-Ethyl Thionocarbamate	NEt-OEtTC	$CH_3CH_2NHC(S)OCH_2CH_3$
N-Isopropyl-O-Ethyl Thionocarbamate	NIp-OEtTC	$(CH_3)_2CHNHC(S)OCH_2CH_3$
N-Butyl-O-Ethyl Thionocarbamate	NBu-OEtTC	$CH_3(CH_2)_3NHC(S)OCH_2CH_3$
N-Ethyl-O-Isopropyl Thionocarbamate	NEt-OIpTC	$CH_3CH_2NHC(S)OCH(CH_3)_2$
Z-200		Commercial form of NEt-OIpTC
bis-Isopropyl Thionocarbamate	bIpTC	$(CH_3)_2CHOC(S)NHNHC(S)OCH(CH_3)_2$
sec-Butyl bis-Isopropyl Thioncarbamate	sBbIpTC	$(CH_3)_2CHO(S)CNH(CH_2)_2CH(CH_3)-$ $NHC(S)OCH(CH_3)_2$
Hexyl bis-Isopropyl Thionocarbonate	HbIpTC	$(CH_3)_2CHOC(S)NH(CH_2)_6-$ $NHC(S)OCH(CH_3)_2$
Isopropyl Xanthogen Ethyl Formate	IpXEF	$(CH_3)_2CHOC(S)SC(O)OCH_2CH_3$
Benzyl Xanthogen Ethyl Formate	BeXEF	$C_6H_5CH_2OC(S)SC(O)OCH_2CH_3$
Isopropyl Xanthagen Phenyl Formate	IpXPhF	$(CH_3)_2CHOC(S)SC(O)OC_6H_5$

1-Hydroxyethyl-2-Heptadecenyl Glyoxalidine   Amine 220

$$CH_3(CH_2)_7CH=CH(CH_2)_7 - C\begin{array}{c} N \underline{\quad\quad} CH_2 \\ \| \quad\quad\quad | \\ \quad\quad CH_2 \\ N \\ | \\ CH_2CH_2OH \end{array}$$

Experimental Collector A	Proprietory thionocarbamate

Frothers

Methyl Isobutyl Carbinol (4-methyl-2-pentanol)	MIBC	$(CH_3)_2CHCH_2CH(OH)CH_3$
Dowfroth 200[*]	DF-200	$CH_3(C_3H_6O)_nOH$ $n\sim3$
Dowfroth 250[*]	DF-250	as above $n\sim4.3$
Dowfroth 1012[*]	DF-1012	as above $n\sim6.3$
Dowfroth 1263[*]	DF-1263	Butylene oxide capped DF-250
Experimental 35003.02[*]		$CH_3(CH_2)_5(C_3H_6O)_2OH$
Experimental 35004.00[*]		Polypropylene glycol ether

[*] references (22,23)

r, at any time, t, as a function of 'ultimate' recovery, R_∞, and flotation rate constant, K, (min.$^{-1}$) parameters.

$$r = R_\infty[1-1/Kt[1-\exp(-Kt)]] \qquad (1)$$

A generalized parameter estimation computer program was used to determine optimal parameter values (24). The criterion used for parameter estimation was the minimization of the absolute squares of deviations between the observed and predicted recovery values. The computationally efficient conjugate gradient method was employed to approach this objective function (11,25). For convenience, the authors have chosen to present the predicted ultimate recovery, R_∞, in the form of fractional recovery rather than the more common percent recovery.

CHARACTERIZATION OF REAGENT PERFORMANCE

The following sections illustrate the utility of using a kinetic model to characterize flotation results by describing the performance of reagent types as a function of concentration, hydrocarbon chain length, position of the hydrocarbon grouping in the collector molecule, pH level and point of reagent addition.

Effect of Collector Concentration

In order to examine the influence of collector concentration on flotation, a number of tests were made using the Utah ore. Figure 1 shows how ultimate recovery (R_∞) and flotation rate constant (K) change as the amount of KAmX collector is varied from 0.01 to 0.25 kg/t.

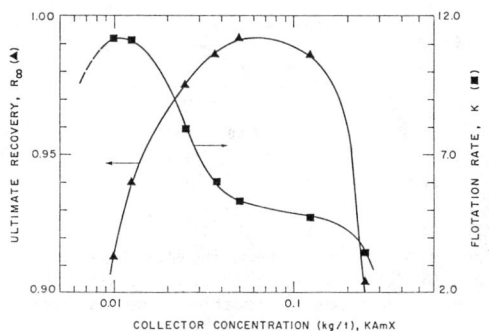

Figure 1. Effect of Collector (KAmX) Concentration on the Flotation of Utah Copper Ore at pH 10.5. MIBC:0.081 kg/t.

Note, particularily, that K is the more sensitive variable as it changes nearly 6-fold while R_∞ is changed by less than 10%. As collector concentrations increased from zero to levels necessary to achieve 'good' fractional recovery (+0.90), both K and R_∞ increase. From Figure 1, it is seen that increases in anticipated R_∞ to values greater than about 0.90 are achieved at the expense of rate of recovery in the collector range ∿0.02 to ∿0.07 kg/t KAmX. This result has been noted by several investigators (19,25,26) and has been termed the 'R/K-tradeoff' (19). In terms of practical flotation plant performance, this means that if a plant is operating in a rate-limited condition, the addition of more collector would lower the recovery in a bank of cells of fixed number even though conventional wisdom would expect an increase in recovery.

133

However, the recent work of Ackerman et al. (16) has indicated that the situation is rather more complex, and they find, under various conditions of collector concentration, hydrocarbon chain length and the mineral to be floated, that recovery and rate may vary in consort or separately. Figure 1 shows, for example, that above ∿0.1 kg/t both recovery and rate decrease, the former precipitously. The maximum collector addition of 0.25 kg/t corresponds to a concentration of ∿1.8x10^{-4} mol l^{-1}, and at this high concentration the reduction in R_∞ and K may be due to the onset of reverse collector adsorption.

Effect of Collector R-Chain Length

This aspect was explored using both xanthates and thioncarbamates to float the chalcopyrite containing Pinto Valley ore. Figure 2 shows that both ultimate recovery (R_∞) and, most especially, the flotation rate constant (K) increase when the xanthate hydrocarbon or R-chain length is increased from ethyl (C_2) to amyl (C_5) with a maximum in both parameters occuring at about amyl (C-5).

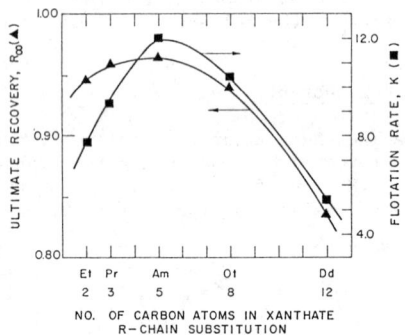

Figure 2. Effect of Xanthate Hydrocarbon (R) Chain Length on the Flotation of Pinto Valley Copper Ore at pH 10.5. Collector:0.02 kg/t, DF-200:0.081 kg/t.

A further increase in the R-chain length is accompanied by a decrease in both R_∞ and K. Considering the maximum in flotation performance and that collector costs usually increase with increasing R-chain length, there appears to be no advantage in using xanthates having more than about five carbon atoms in its R-chain group. These data confirm the microflotation results of Ackerman et al. (16), excepting that they found recovery remained constant while rate decreased for chalcopyrite flotation with dodecyl xanthate. They have provided a detailed explanation for the observed phenomena (16).

The results of using N-substituted O-Ethyl-Thionocarbamates (see Figure 3) indicate that as the N-alkyl substituent length is increased from methyl to butyl, both R_∞ and K continually increase. The total number of carbons reported in the hydrocarbon chain are the sum of both the N- and O- substituents. At least for the thionocarbamates used here, collector effectiveness increases up to C-6.

In general, it is observed that while the flotation rate typically follows the ultimate recovery parameter (i.e. high recoveries are accompanied by high rates, low recoveries by low rates), as a maximum in recovery is approached, differences between collector performance are best distinguished using the flotation rate criterion. Reasons for the effect of

134

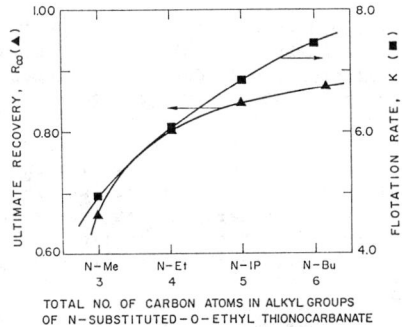

Figure 3. Effect of Alkyl Substitutions in N-substituted-OEthyl Thionocarbamates on the Flotation of Pinto Valley Copper Ore at pH 10.5. Collector:0.02 kg/t, DF-200:0.081 kg/t.

xanthate hydrocarbon chain length on sulfide mineral flotation have been detailed elsewhere (16). The response of thionocarbamates of varying chain length as determined by microflotation are similar to those given here, excepting that Ackerman et al. (17) found a slight decrease in flotation with the N-Isopropyl form, and they attributed this to steric interference. Perhaps the difference is due to the concentration variable since ∿50% more collector is used here than was used in the microflotation tests. Alternatively, the noted difference may be due to a reagent solubility problem and its point of addition (see below).

Effect of Interchanging Hydrocarbon Groups

A few additional tests were made in order to study the effect of interchanging the hydrocarbon groups of thionocarbamate collectors from N-Isopropyl-O-Ethyl to N-Ethyl-O-Isopropyl Thionocarbamate. The data of Table II indicates that the difference in rate and recovery parameters, though slight, is statistically significant. Microflotation experiments (17,27) also showed little difference between these two reagents in chalcopyrite floatability. This experiment supports the observation that, at least up to C-5, the hydrocarbon chain of a thionocarbamates length may be approximated by the sum of the two alkyl substituent groups.

Table II. Effect of Interchanging Hydrocarbon Groups in Thionocarbamates Pinto Valley ore. Collector added to grinding mill, frother: DF-200 (0.081 kg/t)

Collector	Conc.(kg/t)	R_∞	K^*
NIp-OEtTC	0.02	0.847	6.84
NEt-OIpTC	0.02	0.876	7.73

* units - min.$^{-1}$

135

Effect of pH

It is widely understood that sulfide flotation is very dependent on pH, and the extensive work of Ackerman et al. (14-17), using microflotation, has shown that for most collectors, sulfide mineral flotation is better at pH 5 than at pH 8.5 or 10.5, but that selectivity against pyrite is generally better in an alkaline circuit. A series of tests were undertaken to quantify this effect using the Pinto Valley chalcopyrite ore, standard laboratory flotation procedures and several well-known commercial collector types. The individual collectors were studied at two pH levels (5.0 and 10.5) and the results were characterized with respect to both copper and pyrite flotation recovery and flotation rate (see Table III). Selectivity against pyrite during chalcopyrite flotation is also characterized using recovery and rate ratios.

Examination of Table III clearly shows that all xanthates are totally ineffective when added to the grinding mill at pH 5. This is in sharp contrast to the microflotation results of Ackerman et al. (14). Differences between the 'pure' microflotation system and the 'dirty' standard flotation system used here include a longer contact time in acid solution and a plethora of ions in the water (especially iron) in the present system. The discrepancy between the two systems is undoubtedly related to the xanthate decomposition in an acid circuit (28). As will be subsequently shown, the point of addition is a very important operating parameter. At pH 5, the recovery of copper by xanthate is always low, but it is seen to increase slightly with increasing R-chain length. This indicates that the higher molecular weight xanthates are probably more acid stable (28).

At pH 10.5, all of the xanthates are good copper sulfide collectors, giving excellent R_∞ and moderate to high K values. It is common flotation practice to use alkaline pH levels to improve the selectivity between copper sulfide minerals and pyrite. Surprisingly, examination of Table III shows that these xanthate collectors recovered significant amounts of pyrite. Additionally, the xanthates having longer R-chains are seen to be stronger collectors and somewhat less selective toward pyrite. The poor ability of xanthates to discriminate against pyrite would be a mitigating factor in their use.

Although dixanthogens are not common commercial collectors, this species is reported to be responsible for the hydrophobic coating on chalcopyrite (29). Reference to Table III shows that while this collector gives only moderate copper recoveries, $(IpX)_2$ is much more selective against pyrite than are the normal xanthates tested here in spite of its oily nature which presumably should make it unselective toward any sulfide. Its use as a copper sulfide collector has been fully discussed elsewhere (14).

Sodium Diisopropyl Dithiophosphate was also added to the grinding mill and gave poor recoveries at pH 5, again indicating a stability problem in acid solution. At pH 10.5, it gives good copper recovery at a moderate rate, and, as shown by the recovery and rate ratios, this collector provides excellent selectivity between copper sulfides and pyrite. Dithiophosphates are known to be selective against pyrite (14,30).

Table III also shows that both the thionocarbamate (Z-200, commerical NEt-OIpTC) and xanthogen ethyl formate (IpXEF) exhibit excellent acid circuit stability, and both give high R_∞ values at pH 5.0. While Z-200 produced only moderate flotation rates at both pH levels, this collector gives a relatively low R_∞ value at pH 10.5 as has been noted previously (14,18). However, it is very selective toward pyrite at pH 10.5, (18), and this is confirmed by the relatively high, recovery and rate ratios. On the

Table III. Effect of pH Using Commercial Collectors (0.02 kg/t to grinding mill). Pinto Valley ore, frother: DF-200 (0.081 kg/t)

Collector	pH	$R_{\infty,Cu}$	R_{∞,FeS_2}	K_{Cu}^*	$K_{FeS_2}^*$	$R_{\infty,Cu}/R_{\infty,FeS_2}$	K_{Cu}/K_{FeS_2}
KEtX	5.0	0.402	0.062	3.61	1.43	**	**
KEtX	10.5	0.945	0.743	7.67	6.07	1.27	1.28
KIpX	5.0	0.429	0.081	3.96	3.43	**	**
KIpX	10.5	0.961	0.819	11.1	6.95	1.17	1.59
KAmX	5.0	0.511	0.086	7.14	2.61	**	**
KAmX	10.5	0.963	0.824	12.0	7.49	1.17	1.60
$(IpX)_2$	5.0	0.800	0.167	7.11	3.08	4.79	2.31
$(IpX)_2$	10.5	0.774	0.092	3.41	3.42	8.41	0.99
NaIpDTP	5.0	0.624	0.295	7.83	7.48	**	**
NaIpDTP	10.5	0.923	0.243	5.60	0.65	3.80	8.62
Z-200	5.0	0.933	0.674	7.08	4.63	1.38	1.53
Z-200	10.5	0.876	0.182	7.73	3.53	4.81	2.19
IpXEF	5.0	0.961	0.896	14.8	13.4	1.07	1.10
IpXEF	10.5	0.969	0.904	13.9	7.44	1.07	1.87

* units - min.$^{-1}$
** low copper recovery, <0.62 makes computation meaningless

other hand, IpXEF is seen to give very high R_∞ and K values for both chalcopyrite and pyrite at both pH levels. Overall, the xanthogen formate tested here appears to be a very strong sulfide mineral collector.

Except for the decomposition problem with xanthate at pH 5.0, the results agree well with microflotation tests on chalcopyrite and pyrite (14).

Effect of Reagent Point of Addition

Reagent point of addition (to the grinding mill or to the flotation cell) was found to have a significant impact on flotation performance. As an elementary premise, this could be attributed to solubility problems or to time effects controlling the extent of collector decomposition by acid or by deleterious ions in solution.

The xanthate results given in Table IV indicate that the xanthate is an effective collector at pH 10.5 but a poor collector at pH 5.0. While both forms of collector addition made at pH 10.5 level result in similar flotation rates, a slight, but statistically significant, higher ultimate recovery (R_∞) is achieved when the xanthate is added directly to the flotation cell. One may conclude from this that adding xanthate to the grinding mill affords more time for it to react, unfruitfully, with metal ions, or their precipitates. The problem is especially serious in even weakly acidic solutions (pH 5.0).

Table IV. Effect of Point of Addition. Pinto Valley ore, frother: DF-200 (0.081 kg/t)

Collector	Conc.(kg/t)	Addition Point	pH	$R_{\infty,Cu}$	K_{Cu}^*
KAMX	0.02	Grind	5.0	0.511	4.14
KAMX	0.02	Grind	10.5	0.963	12.0
KAMX	0.02	Cell	5.0	0.799	8.43
KAMX	0.02	Cell	10.5	0.976	11.8
bIpTC	0.02	Grind	5.0	0.392	4.35
bIpTC	0.02	Grind	10.5	0.914	7.35
sBbIpTC	0.02	Cell	10.5	0.613	4.13
sBbIpTC	0.02	Grind	10.5	0.974	9.78

* units: min.$^{-1}$

At the lower pH, the slightly acidic environment and the presence of interfering ions may account for the lower recovery. A similar reagent decomposition problem is noted for the bis-thionocarbamate, bIpTC. Harris (31) has indicated that Fe^{+3} can react with this reagent to cause the formation of a double bond between the nitrogens, as shown:

$$
\begin{array}{ccc}
\text{S H H S} & Fe^{+3} & \text{S} \quad \text{S} \\
\text{\char34\ \ \textbar\ \ \textbar\ \ \char34} & & \text{\char34} \quad \text{\char34} \\
\text{IpOC-N-N-COIp} & \rightarrow & \text{IpOC-N=N-COIp}
\end{array} \qquad (2)
$$

Table IV shows that both R_∞ and K are significantly lower at pH 5.0 than at 10.5 indicating decomposition is much greater in the slightly acidic solution. A low pH should probably be avoided with this reagent or, alternatively, a sequesterant used.

While reagent additions to the grinding mill can lead to reagent decomposition problems, collector additions made to the grinding mill should be advantageous for relatively insoluble collectors (32). Table IV shows that adding the relatively insoluble collector sBbIpTC to the grinding mill greatly improves the recovery over that noted when this reagent is added directly to the flotation cell at the same pH (10.5). The improved performance is probably due to the vigorous mixing and emulsification of the collector inside the mill as well as freshly created mineral surfaces for enhanced collector attachment. Alternatively, the insoluble collectors could be added as an emulsion or in a water soluble carrier such as a frother (32).

COMPARATIVE REAGENT PERFORMANCE

Some potentially good collectors identified in the previous microflotation studies (15,27,34) were compared, under standard flotation conditions (Table V), to two popular commercial collectors. Several of the bis-thionocarbamates were previously found to be attractive collectors (33), and results of tests with three of these are given in Table V. The bIpTC shows good pyrite rejection, and from Figure 3 one would expect this collector, with two hydrocarbon chains totaling C-6, to be a good collector. However, it is inferior to another C-6 compound NBu-OEtTC in terms of both R_∞ and K for Cu (see Figure 3). A much superior copper recovery ($\sim$0.91) with bIpTC at pH 10.5 by microflotation (33), however, indicates this collector is worthy of further study, e.g. at a slightly higher concentration. The two bis-IpTC compounds with alkyl groups between the two mirror image halves (HbIpTC and sBbIpTC) were also evaluated here since they also had been found to be good chalcopyrite collectors by microflotation (33). At least at the concentration used here, the sBbIpTC is the superior collector of the two, though the ability of the HbIpTC to reject pyrite is particularly attractive and warrants further study. A comprehensive flotation study of the bis-thionocarbamates as sulfide mineral collectors has been given elsewhere (33).

Also shown in Table V are the results obtained using xanthogen formates. BeXEF is a good copper collector and shows good selectivity against pyrite, whereas IpXPhF floats both sulfides well. The IpXPhF could thus find use in flotation systems where all sulfides must be floated in a bulk concentrate (e.g. precious metal recovery). The use of xanthogen formates has been reported in great detail elsewhere (14,34).

The glyoxalidine chelate forming reagent (Amine 220) was evaluated since on an equal molar basis it proved to be as effective as most sulfhydral collectors by microflotation (15). When tested at pH 10.5 at the 0.02 kg/t level, its recovery and rate were not quite as good as expected. Increasing the collector concentration to 0.035 kg/t, to put it on an

Table V. Comparative Experimental Collector Performance. Pinto Valley ore, 0.02 kg/t collector to grinding mill, frother: DF-200 (0.081 kg/t)

Collector	pH	$R_{\infty,Cu}$	R_{∞,FeS_2}	K^*_{Cu}	K^*_{FeS2}	$R_{\infty,Cu}/R_{\infty,FeS2}$	K_{Cu}/K_{FeS2}
			Experimental Collectors				
bIpTC	5.0	0.392	0.065	4.35	3.01	6.03	1.44
bIpTC	10.5	0.740	0.164	0.18	0.11	4.57	1.64
HbIpTC	5.0	0.403	0.061	4.39	3.10	6.61	1.42
HbIpTC	10.5	0.643	0.059	2.89	3.01	10.9	0.96
sBbIpTC	5.0	0.835	0.218	7.43	4.95	3.83	1.50
sBbIpTC	10.5	0.974	0.674	9.78	2.66	1.45	3.68
BeXEF	5.0	0.925	0.302	8.58	6.28	3.06	1.37
BeXEF	10.5	0.902	0.372	5.98	3.29	2.42	1.82
IpXPhF	5.0	0.961	0.834	12.7	13.3	1.15	0.95
IpXPhF	10.5	0.973	0.874	13.6	9.84	1.11	1.30
Amine 220**	10.5	0.795		3.17			
Amine 220**	10.5	0.938		10.4			
			Conventional Collectors				
KIpX	10.5	0.961	0.819	11.1	6.95	1.17	1.59
KAmX	10.5	0.963	0.824	12.0	7.49	1.17	1.60

* units: min.$^{-1}$
** 0.035 kg/t ($\sim$ equal molar conc.)

approximately equimolar equivalent to the conventional sulfhydral collectors, produced excellent results. These data, together with those of Figure 1, clearly indicate the strong influence collector concentration may sometimes have on flotation recovery and rate. The most rational means of comparing collectors is probably to use an equal molar basis, but any comprehensive reagent test program should include concentration as a major variable.

Collectors of the types given here merit considerable further study.

Acid Circuit Collectors

Approximately 400,000 tons of lime valued at about $10,000,000 were used in 1980 for pH control in flotation processing of copper ore (35). A significant economic improvement could be made in flotation reagent costs if the copper sulfide could be recovered by a collector that worked well in a neutral to slightly acid circuit and also showed a good ability to reject pyrite. In addition to reducing the use of lime for pH control, this flotation scheme would decrease plant maintenance costs by lessening the build-up of lime and gypsum deposits in pipes, pumps, sumps, launders and other equipment. Examination of Table V indicates that BeXEF is a strong potential candidate here. Other reagents which perform a similar function are being reported elsewhere (27,33,34).

Specific Mineral Affinity

Bornite (Cu_5FeS_4) is generally more difficult to float than the other common copper sulfide minerals (14). Considering the large amount of copper contained in bornite ($\sim$63% Cu) as compared to the predominant copper sulfide mineral, chalcopyrite ($\sim$35% Cu), a small bornite loss could account for a significant loss in the overall recovery of copper. An experimental collector (Collector A), developed by the Dow Chemical Co., showed promise by microflotation testing for bornite flotation. Conventional flotation tests with the Utah Copper ore (which contains about 20% of the copper sulfides as bornite) gave R_{∞} of 0.93 with KAmX and 0.88 with Collector A using 0.01 kg/t each. Further confirmation of bornite flotation was made by

139

concentrating the sulfide minerals in the froth obtained from the first 30 seconds of flotation by fractionation in bromoform at 2.90 sp. gr. The heavy fraction was then briquetted and examined in polished section. About 25% of the material that floats in the first thirty seconds with Collector A is bornite (see Figure 4a), whereas only ∿5% of the comparable material floated with the KAmX is bornite (see Figure 4b). Further, much of the bornite floated by Collector A is coarse grained (Figure 4a).

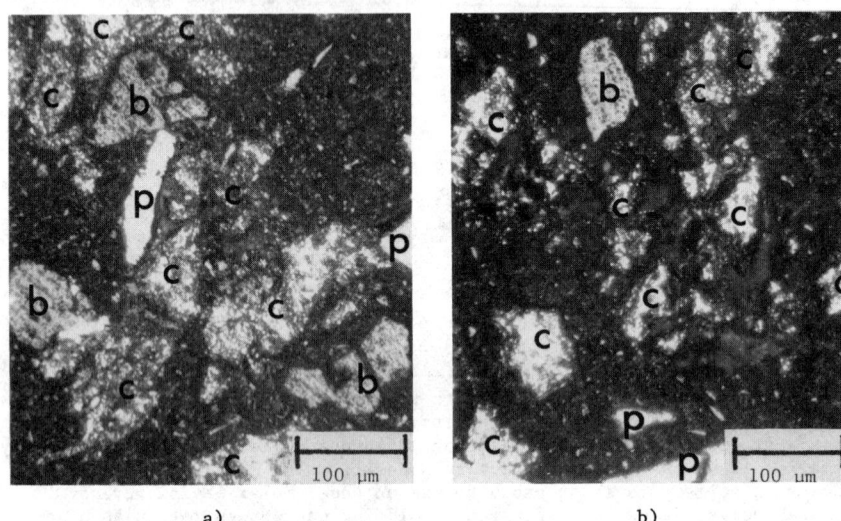

a)   b)

Figure 4. Heavy Mineral Concentrate from first 30 sec. of Flotation. Utah Copper Ore. MIBC frother 0.081 kg/t
   a) Experimental Collector A   b) KAmX

Influence of Frothers

Another specific goal of this project was to identify reagent systems that can effectively recover valuable copper sulfide minerals at a relatively coarse (+65 mesh) size. One of the best ways to reduce grinding costs is to minimize the degree of grinding in the first place (36,37). To accomplish this goal will require reagents with a superior ability to float coarse particles. Recently, a series of new frothers each designed to float coarse particles, or to float fine particles, or to float particles rapidly or slowly has been reported (22,23). Accordingly, tests were designed to evaluate these four frother mutations: coarse, fine, fast, slow. For this purpose, the Pinto Valley ore was used.

Two xanthate collectors (KIpX-strong, KEtX-weak) were used in the amount of 0.02 kg/t along with the frother of interest (0.081 kg/t) at pH 10.5. Results of these tests for the coarse (+65 mesh), fine (-65 mesh) and combined material are given in Table VI. As a point of reference, metallurgical balances indicate the coarse fractions contained on average about 22% of the total copper in the combined flotation feed. Consequently, comparison of the various R_∞ and K parameter values should be made with this fact in mind.

Table VI indicates that KEtX gives lower R_∞ values than does KIpX for all size fractions and all frothers, and this is achieved at a generally slower rate. In one or two instances, the flotation rates appear to be higher using KEtX than with KIpX. This is an artifact related to poor recovery. Examination of Equation 1 indicates that the ratio of recoveries

140

TABLE VI. The Effect of Frother Variations. Collector (0.02 kg/t) to cell, frother: DF-200 (0.081 kg/t)

Frother Type

Collector	Size	Coarse DF 1263		Coarse 35003.02		Fine MIBC		Fine 35004.00		Slow DF-200		Fast DF-1012	
		K^*	R_∞	K^*	R_∞	K^*	R_∞	K^*	R_∞	K^*	R_∞	K^*	R_∞
A. On Overall Basis													
KIpX (strong)	+65 Mesh	10.0	0.199	9.49	0.172	6.63	0.095	6.31	0.044	6.30	0.153	16.4	0.108
	−65 Mesh	14.4	0.695	13.5	0.690	16.4	0.688	15.8	0.754	16.4	0.702	26.2	0.692
	combined	18.3	0.894	17.2	0.862	12.8	0.783	12.0	0.798	12.7	0.855	24.8	0.800
KEtX (weak)	+65 Mesh	10.1	0.146	9.21	0.130	6.27	0.048	–	–	6.30	0.119	11.3	0.071
	−65 Mesh	12.5	0.659	12.0	0.657	16.6	0.650	13.2	0.720	17.0	0.639	23.9	0.642
	combined	16.7	0.805	15.7	0.787	12.8	0.698	13.2	0.720	13.0	0.758	20.5	0.713
B. On Basis that Each Size Fraction is 100%													
KIpX	+65 Mesh		0.892		0.771		0.426		0.197		0.686		0.484
	−65 Mesh		0.895		0.888		0.885		0.970		0.903		0.891
KEtX	+65 Mesh		0.655		0.583		0.215		–		0.534		0.318
	−65 Mesh		0.848		0.846		0.837		0.927		0.822		0.826

units: min.$^{-1}$

141

(r/R_∞) will influence the value determined for the flotation rate constant. Generally, the results of using KIpX with these various frothers follow a similar, but superior, pattern to those where KEtX was used as collector. Note that there are various frother-collector combinations where the weaker collector can produce results similar to the strong collector. This reinforces the concept that the flotation system is an interactive one and various machine, operational and reagent combinations can produce similar results (19). In the present case the use of the appropriate frother can mitigate some of the deleterious effects of a weak collector (see e.g. the improved flotation of coarse chalcopyrite using DF-1263 as compared to the conventional MIBC using KEtX collector).

Table VIB shows the effectiveness of the coarse frothers, DF-1263 and 35003.02, for the flotation of +65 mesh chalcopyrite. When used in conjunction with KIpX, DF-1263 gave a coarse chalcopyrite recovery ($R_\infty \sim$ 0.89) essentially as good as that achieved with the fine -65 mesh material.

Reference to Table VI shows that Frother 35004.00, specifically designed for fine particle flotation, gave the best (by far) fine chalcopyrite recovery using either collector, though it also gave the worst coarse copper recovery. In fact, no coarse copper was recovered when the weak collector was used with this frother. While MIBC can be rated as a fine particle frother, it is essentially no better than the two coarse frothers for the recovery of -65 mesh copper sulfide.

It is recognized that the fine frother, 35004.00, would be particularly useful in floating the finer size particles in split (sized) feed flotation systems, or when used in the presence of many fine particles (e.g. rougher flotation), or when used in conjunction with a coarse particle frother. The coarse particle frothers could be used on coarse flotation from split feed flotation, or in situations where most of the fine, valuable particles have been removed (e.g. scavenger flotation), or in combination with a fine frother.

The data of Table VI also indicates that some frothers float particles slowly while other frothers float them rapidly. The fast frother (DF-1012) is especially interesting. It floats the chalcopyrite faster than any other frother, in each case, and it often improves the flotation rate by 50 to 100%. Table VI, also shows that there appears to be a crude relationship between flotation rate and particle recovery by size (i.e. high rate - coarse particle recovery, low rate - fine particle recovery) for those frothers with the polypropylene oxide backbone (22, 23). However, examination of the results for DF-200 and DF-1012 indicate that the successful flotation of coarse particles is not solely rate dependent and that something more complex is probably involved. For example, the slow frother, DF-200, gives higher coarse particle recoveries than does the fast frother, DF-1012. This occurs even though DF-1012 gives the highest flotation rate for all sizes of any frother tested here. The superior coarse flotation results noted for DF-1263 and 35002.02 may be due in part to the carbon chains attached to the polypropylene oxide backbone structure (see Table I). It is speculated that there may be some R-chain interaction between these carbon chains and those of the collector. Tests are currently planned to study this aspect.

SUMMARY AND CONCLUSIONS

This study demonstrates the utility of using the kinetic approach to describe and compare laboratory flotation results. A statistically reliable kinetic flotation model is used here to characterize reagent performance as a function of concentration, hydrocarbon chain variations, pH and point of addition, as well as to compare the effectiveness of a broad variety of collectors and frothers.

1. The laboratory flotation studies reported here generally confirm previous microflotation results (14-18), excepting where complications (such as reagent decomposition), not found in the pristine microflotation environment, occur.

2. Flotation recovery and rate are often critically dependent on collector concentration. This variable must be given careful consideration in any comprehensive reagent test program.

3. For the flotation of chalcopyrite bearing ore with xanthate, both recovery and rate were found to go through a maximum at about a hydrocarbon chain length of five, under the test conditions used here.

4. For thionocarbamates, however, both recovery and rate continually increase up to a C-6 hydrocarbon chain length.

5. The point of collector addition to the flotation system, whether to the grinding circuit or to the flotation cell, can often have a pronounced effect on the results. The effect of adding them to the grinding mill may be favorable if the reagent is relatively insoluble and unfavorable if the reagent can be easily decomposed (as by acidic or catalytic ion containing solutions).

6. Several new collectors, that appeared to be favorable for sulfide mineral flotation by microflotation testing (14-17,27,33,34), were evaluated by conventional flotation testing procedures. These included reagents that gave favorable results for the flotation of copper sulfides, for the flotation of all sulfides and for the discrimination against pyrite concurrent with copper sulfide flotation. These new collectors included various thionocarbamates, xanthogen formates and a glyoxalidine. All of these classes merit further study.

7. Several new frothers, specifically developed for either coarse or fine particle flotation, have been evaluated for the flotation of copper ore. One of the new coarse frothing agents proved to be especially good for the flotation of 28x65 mesh chalcopyrite from its ore. Use of such a frother could lead to a substantial reduction in grinding costs for flotation. An experimental fine particle frother gave somewhat improved results for flotation of -65 mesh particles over that of MIBC.

8. Specially developed frothers for the fast or slow flotation of particles were found to be equivalent to slightly better than the traditionally used MIBC. One of them is significantly better than MIBC for the flotation of +65 mesh particles.

ACKNOWLEDGEMENTS

The authors wish to acknowledge the financial support of the National Science Foundation (NSF grants CPE 8111792 and CPE 8442865), for the work reported above, which is part of a cooperative research program between Dow Chemical Company, Newmont Mining Corporation and The Pennsylvania State University. They also wish to thank Drs. P.K. Ackerman and G. H. Harris for preparation of reagents used here. Pinto Valley Copper Co. and Kennecott Copper Corporation are acknowledged for supplying the ores used. Mr. Dowling also wishes to thank the Department of the Interior, Minerals Research Institute Program (USBM grant #G1164142), and the Pennsylvania Mining and Mineral Research Institute for the fellowship support received during the latter part of this research program.

REFERENCES

1. Garcia-Zuniga, H. "Flotation Recovery is an Exponential Function of Time," Bol. Soc. Nac.Min., 47(1935), 83.

2. Schuhmann, R., "Flotation Kinetics I., Methods for Steady State Study of Flotation Problems," J. Phys. Chem. 64(1942), 891-902.

3. Kelsall, D.F., "Application of Probability in the Assessment of Flotation Systems," Trans. I.M.M., 70(1961), 191-204.

4. Arbiter, N. and Harris C.C., "Flotation Kinetics," Froth Flotation-50th Anniversary Volume, ed., D. W. Fuerstenau, (New York:AIME, 1962), 215-246.

5. Lynch, A. J., Johnson, N. W., Malapig, E. V. and Thorne, C. G., Mineral and Coal Flotation Circuits, Developments in Mineral Processing 3, (Amsterdam: Elsevier, 1981).

6. Huber-Panu, I, Ene-Danalache, E. and Cojocariu, D. G., "Mathematical Models of Batch and Continuous Flotation," Flotation - A. M. Gaudin Memorial Volume, ed., M. C. Fuerstenau, (New York: AIME, 1976), 675-724.

7. Thorne, G. C., Malapig, E. V., Hall, J. S. and Lynch, A. J., "Modelling of Industrial Sulfide Flotation Circuits," Flotation- A. M. Gaudin Memorial Volume, ed. M. C. Fuerstenau, (New York: AIME, 1976), 725-750.

8. Agar, G. E., Stratton-Crawley, R. and Bruce, T. J., "Optimizing the Design of Flotation Circuits," CIM Bul., 73(824)(1980), 173-181.

9. Mular, A. L., Chen, Z. N. and Cheng, K. K., "Digital Simulation of a Flotation Circuit for Design--Case Study," Design and Installation of Concentration and Dewatering Circuits, eds., A. L. Mular and M. A. Anderson, (Littleton, CO:SME, 1986), 588-603.

10. Klimpel, R. R., "Selection of Chemical Reagents for Flotation," Mineral Processing Plant Design, eds., A. L. Mular and R. B. Bhappu, (New York: AIME, 1980), 907-934.

11. Dowling, E. C., Klimpel, R. R., and Aplan, F. F., "Model Discrimination in the Flotation of a Porphyry Copper Ore," Min. and Met. Proc., 2(2) (1985), 87-101.

12. Dowling, E. C., Klimpel, R. R., and Aplan, F. F., "Model Discrimination for the Flotation of Base Metal Sulfide Ores-Circuitry and Reagent Variations," Design and Installation of Concetration and Dewatering Circuits, eds., A. L. Mular and M. A. Anderson, (Littleton, CO:SME, 1980), 570-587.

13. Dowling, E. C., Klimpel, R. R., and Aplan, F. F., "Use of Kinetic Models to Analyze Industrial Flotation Circuits," The Reinhardt Schuhmann International Symposium on Innovative Technology and Reactor Design, eds., D. R. Gaskell, J. P. Hager, J. E. Hoffman and P. J. Mackey, (Warrendale, PA:TMS, 1986), 533-552.

14. Ackermann, P. K., Harris, G. H., Klimpel, R. R. and Aplan, F. F., "Evaluation of Flotation Collectors for Copper Sulfides and Pyrite, Part I-Common Sulfhydral Collectors," Int. J. Min. Proc., (in press).

15. Ackermann, P. K., Harris, G. H., Klimpel, R. R. and Aplan, F. F., "Evaluation of Flotation Collectors for Copper Sulfides and Pyrite, Part II-Non-Sulfhydral Collectors," Int. J. Min. Proc., (in press).

16. Ackermann, P. K., Harris, G. H., Klimpel, R. R. and Aplan, F. F., "Evaluation of Flotation Collectors for Copper Sulfides and Pyrite, Part III-Effect of Xanthate Chain Length and Branching," Int. J. Min. Proc., (in press).

17. Ackermann, P. K., Harris, G. H., Klimpel, R. R. and Aplan, F. F., "Effects of Alkyl Substituents on Performance of Thionocarbamates as Copper Sulfide and Pyrite Collectors," Reagents in the Minerals Industry, (I.M.M.,1985), 69-78.

18. Ackermann, P. K., Harris, G. H., Klimpel, R. R. and Aplan, F. F., "Importance of Reagent Purity in Evaluation of FLotation Collectors," Trans. I.M.M., 95(1986), C165-C168.

19. Klimpel, R. R., "The Effect of Chemical Reagents on the Flotation Recovery of Minerals," Chem. Eng., 91(18)(1984), 75-69.

20. Dayton, S., "The Quiet.Revolution in Mineral Processing Cu/Mo Recovery: A Flowsheet Study at Pinto Valley," Eng. Min. J., 131 (6)(1975), 90-97.

21. Tveter, E. C. and McQuiston, Jr., F. W., "Plant Practice in Sulfide Mineral Flotation," Froth Flotation-50th Anniversary Volume, ed., D. W. Fuerstenau, (New York: AIME, 1962), 383-426.

22. R. R. Klimpel and R. D. Hansen, "Chemistry of Fine Coal Flotation" Fine Coal Processing, ed., S. K. Mishra and R. R. Klimpel (Park Ridge, NJ: Noyes Pubs., 1987) 78-109.

23. R. R. Klimpel and R. D. Hansen, "Frothers," Reagents in the Mineral Industry (New York: Marcel Dekker, in press), 385-409.

24. Blau, G. E.,, Klimpel, R. R., and Steiner, E. C., "Equilibrium Constant Estimation and Model Distinguishability," Ind. Eng. Chem. Fund., 11(3)(1972), 324-332.

25. Fletcher, R. and Powell, M., "A Rapidly Convergent Descent Method for Minimization," Brit. Comp. J., 6(1963), 163.

26. Gaudin, A. M., Schuhmann, R., and Schlechten, A. W., "Flotation Kinetics. II. The Effect of Size on the Behavior of Galena Particles," J. Phys. Chem, 64(1942), 902-910.

27. P. K. Ackerman, G. H. Harris, R. R. Klimpel and F. F. Aplan, "Evaluation of Flotation Collectors for Copper Sulfides and Pyrite: Part IV Effect of Branching on Thionocarbamates," manuscript in preparation.

28. Harris, G. H., "Xanthates," Kirk-Othmer Encyclopedia of Chemical Technolgy, 3rd ed., 24(1984), 645-661.

29. Allison, S. A., Goold, L. A., Nicol, M. S., and Granville, A., "A Determinazation of the Products of Reaction Between Various Sulfide Minerals and Aqueous Xanthate Solutions and Correlation of the Products with Electrode Rest Potentials," Met. Trans., 3(1972), 2613-2618.

30. Staff, American Cyanamid Co., Mining Chemicals Handbook, Mineral Processing Notes No. 26, (American Cyanamid Co., 1976), 4.

31. Harris, G. H., private communication with authors, January 1985.

32. Taggart, A. F., Handbook of Mineral Dressing, (New York: Wiley, 1945). Sect. 12.

33. P. K. Ackerman, G. H. Harris, R. R. Klimpel and F. F. Aplan, "Evaluation of Flotation Collectors for Copper Sulfides and Pyrite: The bis-Thionocarbamates, to be presented at AIME Annual Meeting, Phoenix, Jan. 26-29, 1988.

34. P. K. Ackerman, G. H. Harris, R. R. Klimpel and F. F. Aplan, "Evaluation of Flotation Collectors for Copper Sulfides and Pyrite: The Xanthogen Formates" AIME Annual Meeting, Denver, Feb, 1987. Manuscript in preparation.

35. Staff, U. S. Bureau of Mines, Minerals Yearbook, 1980, (Washington, D. C.: U.S. Govt. Ptg. Off., 1981).

36. F. F. Aplan, "The Future of Mineral Beneficiation - The Impact of Scientific Studies," Mineral Resources of Australia, ed., D. F. Kelsall and J. T. Woodcock. (Parkville, Victoria: Australian Academy of Technological Sciences, 1979), 171-189.

37. R. R. Klimpel and R. D. Hansen, "The Interaction of Flotation Chemistry and Size Reduction in the Recovery of a Porphry Copper Ore", to appear in Int. J. Min. Proc.

SIMULATION AND OPTIMISATION OF GRAVITY

CONCENTRATION CIRCUITS

R.I. Mackie and P. Tucker

Minerals and Metals Division
Warren Spring Laboratory, Department of Trade and Industry
Gunnels Wood Road, Stevenage, Hertfordshire SG1 2BX, UK

Abstract

A flowsheet simulation package has been written for use in the mineral processing industries. The simulator includes a number of novel features: (i) fully predictive models of gravity concentration and hydraulic classification devices are included; (ii) the simulator can handle constraints imposed upon the flow to the devices in the circuit; (iii) an optimisation routine is included, enabling the simulator to select the optimum operating conditions for a particular application. A serial solution method is used in simulator, and a pattern search method is used in the optimisation routine. The simulator can run on either minicomputers or PC's, and is designed to be of use both at the design stage and in the day to day running of a mineral processing plant. The paper describes the mathematical and computational aspects, and the use of the simulator as applied to European mineral processing operations.

Introduction

A flowsheet simulator is a computer/mathematical model which parallels on a computer what will happen on a real processing plant. Simulation offers many potential advantages to both the plant operator and the plant designer, and indeed simulation has been well established in the chemical industries for several years now. For instance, simulation can be used to investigate alternative plant configurations, optimise the performance of existing plant, and as a useful tool in answering a wide range of questions that need to be asked at the design stage (e.g. choosing the type of devices, estimating costs etc).

Despite these potential benefits and simulations proven worth in the chemical industries, simulation has been slow to make an impact on mineral processing. One reason for this is undoubtedly the inherent complexity of the unit processes involved in mineral processing, for simulation requires good mathematical models of the unit processes if it is to be applied successfully. In recent years good models have begun to appear in many areas of mineral processing, namely comminution (Austin et al (1), Lynch (2)), classification (Plitt (3), Karra (4)) and gravity separation (Tucker et al (5), Mackie et al (6)). However, until recently simulation has been confined to specific parts of a flowsheet, usually either comminution or flotation. The first general simulator for mineral processing was designed by Ford and King (7). This paper describes a new general purpose simulator (GSIM) developed at Warren Spring Laboratory (WSL). GSIM includes a number of novel features: (i) the inclusion of proven models for gravity concentration and hydraulic classification devices; (ii) the ability to handle constraints imposed upon the flow to the devices; (iii) an optimisation routine enabling the simulator to choose the best operating conditions for a particular application; and (iv) the simulator can be used on micro computers such as the IBM PC (512 K memory).

General Principles of Simulation

A flowsheet simulator requires three sets of information:

1. Definition of the flowsheet

    This information tells the simulator what the circuit looks like, and where the various streams are coming from and going to. In GSIM each of the devices is given a number, as is each feed stream entering the circuit, and each circuit output (e.g. concentrate, tailings etc). Then the output streams from each device are given a local numbering, e.g. for a spiral the concentrate stream would be numbered 1, the middlings 2, and the tailings 3. Finally the destination of each feed stream and of each device output stream is given. This will either be the number of another device or of a circuit output. This information is supplied by the user, and is entered via user-friendly input routines. GSIM carries out checks on the data for obvious inconsistencies, e.g. devices having no streams feeding them.

2. Feed definition

    The simulator needs to know what the feed to the circuit is. For the mathematical models currently included in GSIM the key characteristics of particles are the size and SG (specific gravity). Therefore the size/SG distribution needs to be given, this is in addition to the flowrate and pulp density. In most applications it is the grade and recovery of one or more minerals

148

that is of interest to the engineer, and therefore in GSIM there is
also the option of supplying assay values to the individual size/SG
fractions. For any circuit there may be more than one feed stream,
and the assay values may differ from feed stream to feed stream.
Although GSIM currently is designed primarily for gravity separation circuits, it is envisaged that other processes such as
magnetic separation will be included. When this is done other
quantities such as magnetic susceptibility will be needed in the
feed definition.

3. Device performance

    The most flexible method is to use a modular approach, where
device modules are available for each type of unit process. Each
device module contains a mathematical description of the device
performance. The modular approach allows the simulator to deal
with any configuration of circuit, the simulator calling the
relevant device module when necessary. This approach also allows
new device modules to be added very easily.

    The device modules are by far the most important aspect of a
simulator, for the results of simulation are only as good as the
device modules used. Therefore this aspect will be expanded upon a
little.

Device Modules

    As mentioned above, each device module contains a mathematical
description of a particular unit process. When the simulator calls a device
module it supplies the module with certain information, namely the feed
rate, pulp density and size/SG distribution. In addition most device
modules will also need to know certain operating conditions. The exact
nature of this information will vary from device to device, for example for
a spiral concentrator the module will need to know the cutter settings and
wash water levels, for the Plitt (3) cyclone model the cyclone geometry is
required. The device module then calculates the feed rate, pulp density and
size/SG distribution for each of its output streams.

    WSL has developed several models for gravity concentration devices and
hydraulic classifiers. While size and SG are the most important particle
properties, there are other characteristics, such as shape, whose effect
cannot be easily described mathematically. These factors are taken account
of by means of model parameters which are used to tune the models to
particular applications. Further details of the models and their application can be found in Tucker et al (8), Tucker (9), and Mackie et al (6).

    In addition several literature models have been included in GSIM. In
some cases the models have required modification in order to enable them to
predict what happens to each size/SG fraction. For instance, Austin's (10)
grinding model predicts the size distribution of the mill product, but not
the size/SG distribution. There is no generally accepted model of
liberation for grinding mills, so the following expedient was adopted:

    Given the feed to the mill, Austin's model is used to predict the size
distribution of the product. Assume initially that the SG distribution for
each size fraction is the same as that of the same size fraction in the
feed. Then an estimate of the mass flow in each size/SG fraction can be
calculated.

    Suppose the estimated mass flow in the jth density fraction of the ith

149

size fraction is z_{ij}.

A better estimate of the mass flows, y_{ij}, is then obtained by minimising the function

$$E = \sum_{ij} \left(y_{ij} - z_{ij}\right)^2 / z_{ij}^2 \qquad (1)$$

subject to the constraints

$$\sum_j y_{ij} = Y_i \qquad (2)$$

where Y_i is the total mass flow for size fraction i predicted by Austin's model, and

$$\sum_{ij} a_{ij} y_{ij} = A \qquad (3)$$

where A is the total mineral content of the feed, and a_{ij} are the mineral assays for each size/SG fraction of the feed. Equation (2) ensures that the size distribution remains unchanged, and eqn (3) ensures that the total amount of the mineral is conserved. Solving the minimisation problem then gives improved estimates of the SG analysis of the mill product.

It is realised that this method is not ideal, for instance it is assumed that the mineral assays of the size/SG fractions in the product are the same as the corresponding feed fractions. However, the method does give reasonable results and the calculations are quick. Table 1 shows a comparison of the measured and calculated SG analysis of the product from the South Crofty rod mill (using the above method). The measured and calculated SG distributions are in good agreement. The calculated amount of tin in the lighter SG fractions is lower than the measured values because of changes in the assay values between the feed and the product.

Table I. Measured and Calculated SG Analysis of Mill Product

Weight distribution (%) by SG

SG	Feed	Product	
		Meas	Calc
−3.3	96.09	94.83	95.08
+3.3 −4.0	2.44	1.78	1.76
+4.0	1.47	3.39	3.16

Tin distribution by SG

SG	Feed	Product	
		Meas	Calc
−3.3	38.25	7.33	4.87
+3.3 −4.0	24.38	9.34	5.05
+4.0	37.37	83.33	90.08

Mathematics of Simulation

The mathematics of flowsheet simulation have received a great deal of attention from researchers in chemical engineering, and a good review can be found in Westerberg et al (11). However, since flowsheet simulation is relatively new to mineral processing a brief overview will be given here.

If there are no recycle streams in the circuit its solution is straightforward. All that needs to be done is to determine the correct precedence ordering, i.e. the order in which the devices should be calculated. The principle of precedence ordering can be illustrated by reference to Fig. 1. Clearly the correct calculation order is 1-2-3, and to calculate the devices in the order 2-1-3 would be nonsense.

In general recycle streams are present, and this means that, a priori, the flow in certain streams is not known. For instance, in a closed grinding circuit although the feed to the circuit is known, the total feed to the mill (circuit feed plus classifier oversize) is not known initially. Therefore iterative techniques have to be used. Solution is simplified by identifying the loops within the circuit, and condensing the loops into groups such that there is no recycle between groups. This concept can be understood by reference to Fig. 2. In Fig. 2 there is no recycle between the grinding and spiral circuits, therefore in solving the system the grinding circuit can be solved first, and then the spiral circuit, rather than trying to solve the whole circuit at once. Identifying the constituent loops of a circuit is called partitioning the circuit. Many algorithms exist for doing this, and most establish the correct precedence ordering as well. The simplest ones use path searching methods (e.g. Sergeant and Westerberg (12)), and such a method is used in GSIM.

In solving each loop the flow in certain streams an initial guess of the flow in certain streams has to be made, and then interative methods used to calculate the actual flow. So the solution of a loop involves two steps:

1. Choosing which streams to "guess".
2. Iterative solution of the loop.

The first step is called "tearing", and the chosen streams are called "tear" streams, and these streams are said to be "torn". For any given loop there will be several different tear sets which enable the loop to be solved. Kehat and Shacham (13) give a summary of many of the different methods of tearing a circuit, and present their own method. A similar method is used in GSIM.

The iterative methods used to solve the loop fall into two main classes: (i) simultaneous solution; and (ii) serial solution. The simultaneous solution methods usually converge more quickly, but the serial solution methods use less computer storage space and tend to be more robust. For this reason the serial solution method is used in GSIM, and is accelerated using the bounded Wegstein (14) method. This method works as follows:

Suppose that the latest estimates of the flows in the tear streams are z_i. Using these values new estimates, r_i, of the flows in the tear streams are obtained. For the basic serial solution method the new estimate, z_{i+1}, of the tear variables are taken to be

$$z_{i+1} = r_i \qquad (4)$$

In order to accelerate the convergence z_{i+1} can be defined as

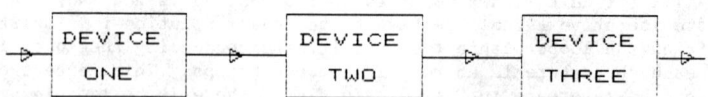

Figure 1. Precedence Ordering

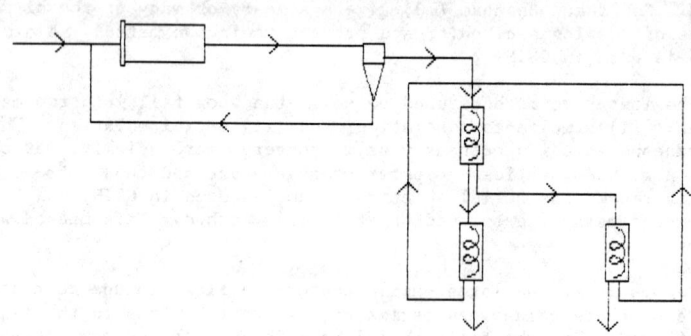

Figure 2. Partitioning a Flowsheet

$$\underline{z}_{i+1} = Q_i \; \underline{z}_i + \left(1 - Q_i\right) \underline{r}_i \qquad (5)$$

where Q_i is a diagonal matrix whose elements q_{ij} are defined by

$$q_{ij} = a_{ij}/\left(a_{ij} - 1\right) \qquad (6)$$

where

$$a_{ij} = \left(r_{ij} - r_{i-1j}\right)/\left(z_{ij} - z_{i-1j}\right) \qquad (7)$$

r_{ij} and z_{ij} being the jth elements of $\underline{r}_i$ and $\underline{z}_i$ respectively. Instabilities are avoided by putting limits on the values of the q_{ij}, and by not using the acceleration at every iteration.

Enhancements to Simulation

This completes the description of the basic simulator. The basic simulator can predict the performance of a given circuit under a given set of operating conditions. While such a simulator is a very useful design tool, its usefulness can be further enhanced by the addition of two features:

1. The ability to handle constraints upon the flow to the devices.
2. The ability to choose the operating conditions in order to optimise a given circuit.

Flow Constraints

A device can either be a single unit device (e.g. a mill or a hydro-sizer), or a multiple unit consisting of several unit devices operating under exactly the same operating conditions (e.g. a bank of spirals). Clearly the number of units in a device affects the flow per unit, and in general the performance of a device is dependent upon the flowrate per unit. Therefore the designer will usually wish to choose the number of units in a device so as to attain a desired flow per unit. Similarly device performance is also dependent upon the pulp density of the feed, so either water will need to be added to or extracted from (dewatering) the feed in order to meet a specified target. The number of units per device and the degree of water addition or dewatering needed could be determined by trial and error doing repeat simulations of a circuit. However, GSIM has been designed so that the constraints on the flow per unit and the pulp density are satisfied automatically. This is done as follows:

A target solids mass flow per unit, F_i, and a target pulp density, ϕ_i, are specified for each device, and the number of units in each device, n_i, and the water input, w_i, are considered to be variables.

The serial solution is then carried out in the normal way, except that before calculating the flows from each device the values of n_i and w_i are chosen by the simulator to ensure that the flow constraints are satisfied, i.e. suppose that when the flows from device i are to be calculated the total solids mass flow to the device is M_i and the pulp density α_i. The n_i and w_i are set to the values

$$n_i = M_i/F_i \qquad (8)$$

$$W_i = M_i\left(1/\phi_i - 1/\alpha_i\right) \qquad (9)$$

The flow constraints are then satisfied.

153

When the serial solution has converged n_i and w_i will have the values necessary in order to give the required flow conditions. In general n_i will not have an integer value. Therefore the values of n_i are rounded to the nearest integer and the calculation is repeated, but this time only the w_i are varied. Of course the flow per unit device constraint will not be satisfied exactly, but in most cases the flow per unit is close to the ideal. Positive values of w_i indicate that water needs to be added to the feed to device i, and negative values indicate that dewatering needs to take place.

Far from making the solution more difficult, imposing flow constraints actually reduces the calculation time in many cases. The reason for this is that several of the mathematical models take account of the flowrate and pulp density, but not the size/SG distribution. When flow constraints are imposed the flow per unit (though not the total flow to the device) and the pulp density remain the same from one iteration to the next. Therefore the transfer coefficients for the device need to be calculated only once.

The flow constraint facility could be helpful at the design stage, indicating the number of unit devices needed at each stage, and predicting the degree of water addition or dewatering needed at various parts of a flowsheet.

Flowsheet Optimisation

As well as predicting the performance of a circuit under a given set of conditions, one of the prime uses of flowsheet simulation is to assist in determining the optimum operating conditions. Basic flowsheet simulation can be used to investigate the performance of a circuit under different sets of conditions. However, even for quite simple circuits the number of operating conditions that can be altered can be very large. For instance, consider a rougher/cleaner/scavenger circuit made up of GEC spiral concentrators. For each spiral the operating conditions that can be varied are: the flow per unit spiral, the feed pulp density, the wash water level, and fifteen port settings. So for the whole circuit there are 54 variables. To examine all the combinations, even using simulation, would be impossible. The problem is overcome in GSIM by incorporating an optimisation routine into the simulator.

In GSIM two options are available for defining the objective of the optimisation:

1. Achieve the maximum recovery, R_{max}, of a mineral while maintaining a grade of at least G_{min}.

2. Achieve the maximum grade, G_{max}, of a mineral while maintaining a recovery of at least R_{min}.

Each variable (e.g. flow per unit, wash water level etc) can either be set to a definite value, or a range over which it can vary may be specified. The optimisation routine then chooses the values of the variables which can be varied so as to give the best performance.

Some of the mathematical models are very complex, so that optimisation methods which rely on derivatives are not a practical option, and as described above the number of conditions that can be varied may be very large. Therefore a search method that does not use derivatives, and that can cope with large numbers of variables is required. The one chosen was the pattern search method, which works as follows:

154

Let $\underline{x}$ be the vector of the operating conditions that can be varied. Initially set the variables to the midpoints of their specified ranges.

$$x_i = \left(x_{imin} + x_{imax}\right)/2 \qquad (10)$$

and the performance of the circuit is calculated under these conditions. Then the first variable is increased by a step size h_1, i.e.

$$x_1{}^* = x_1 + h_1 \qquad (11)$$

and the performance is recalculated. If this improves the performance x_1 is set to this new value. Otherwise x_1 is decreased, i.e.

$$x_1{}^* = x_1 - h_1 \qquad (12)$$

and the performance of the circuit is calculated again. If the performance has improved, x_1 is set to this new value, otherwise x_1 retains its original value and the step size, h_1, is decreased.

When this has been done once for all the variables, all the variables are changed at once in the direction indicated by the individual changes, i.e. if x_i was increased, then x_i is increased again; if x_i was decreased then x_i is decreased again. So if $\underline{x}^0$ is the vector of variables after all the individual changes, a new vector, $\underline{x}^1$ is given by

$$\underline{x}^1 = \underline{x}^0 + \underline{H} \qquad (13)$$

where

$$H_i = \begin{bmatrix} +h_i & \text{if } x_i \text{ was increased} \\ 0 & \text{if no change was made to } x_i \\ -h_i & \text{if } x_i \text{ was decreased} \end{bmatrix}$$

When this has been done the performance of the circuit is calculated. If performance has improved the process is repeated, otherwise the process of changing one variable at a time is repeated.

The whole process is repeated until performance can no longer be improved. Note that the flow constraint facility is used in the calculation of the flowsheet performance as two of the conditions that can be altered are the flow per unit and the feed pulp density. This also means that the number of units per device has to be rounded up to the nearest integer and the performance recalculated.

If the initial grade (recovery) is less than G_{min} (R_{min}) the optimisation routine attempts to attain the specified minimum grade (recovery), and performance is said to improve if the grade (recovery) increases. Once the minimum has been attained the performance is said to improve if the recovery (grade) increases without the grade (recovery) falling below G_{min} (R_{min}).

Example

The general advantages of using flowsheet simulation have been described in the introduction. The following example of the use of GSIM illustrates some of the ways in which flowsheet simulation can be of benefit.

Suppose that it is desired to design a spiral circuit for a tungsten

155

ore, and that the size/SG/assay distribution of the feed to the proposed circuit is known.

Initially a simple rougher/scavenger/cleaner circuit was investigated. Only a concentrate and tailings cut was taken from each spiral, the scavenger concentrate and cleaner tailings being recycled back to the rougher spirals. Flowsheet optimisation was used to predict the best performance obtainable from the circuit, and it was found that 88% of the tungsten could be recovered at a grade of 55%. The circuit together with the grades of the flows in each stream are shown in Fig. 3 and Table II.

It was then decided to investigate the advantages of taking middlings cuts from each of the spirals. Middlings rougher and cleaner bansk of spirals were introduced, the complete circuit being shown in Fig. 4 and Table III. Again flowsheet optimisation was used to predict the best performance, and it was predicted that 86% of the tungsten could be recovered at a grade of 65% a significant improvement on the first circuit. However, the results also showed that the grades (see Table III) for the middlings and concentrate streams of the cleaners and middlings cleaners were all about the same. This suggested that the middlings cuts from these two spirals were superfluous. Therefore it was decided to take only a concentrate cut from these two sets of spirals, and the circuit is shown in Fig. 5 and Table IV. This time it was predicted that 82% of the tungsten could be recovered at a grade of about 63%. This is slightly lower than for the second circuit, but the flowsheet is significantly simpler and this would reduce operating costs.

All these calculations were carried out in about half an hour on the computer. This demonstrates how a wide variety of options can be investigated before embarking on the physical testwork.

Conclusions and Summary

A general purpose simulator has been developed. The basic simulator has been enhanced so that (i) it can deal with constraints imposed upon the flow to the various devices in the circuit, and (ii) it can optimise the operating conditions in order to obtain a desired result.

The simulator has been used in several consultancy contracts for European companies, and the simulator will operate both on mini computers, and on micro computers such as the IBM PC. GSIM has been used on several occasions to assess the feasibility of concentrating new ore prospects. The optimisation facility in GSIM enabled predictions of the size of circuit, and the best operating conditions to be made. GSIM is also currently being used by the Wheal Jane tin mine in Cornwall where a new hydrosizer is being installed (Mackie et al (6)). GSIM has the capability to enable the hydrosizer and shaking table performance to be studied simultaneously. In addition GSIM is being used in the study of coal spiralling. Detailed testwork on a single coal spiral was carried out, and the results used to develop a mathematical model. This model was then installed in GSIM to predict the optimum performance of various spiral circuits, e.g. rougher/cleaner/scavenger, rougher/scavenger etc.

The simulator represents an important step towards being able to simulate the performance of a complete mineral processing flowsheet, and is already of considerable practical use in the detailed analysis of gravity circuits. In its present stage of development the simulator contains models for gravity concentration, hydraulic classification and grinding. To extend its applicability, models need to be included for crushing, flotation, and magnetic separation. In some cases (notably magnetic separation) this will

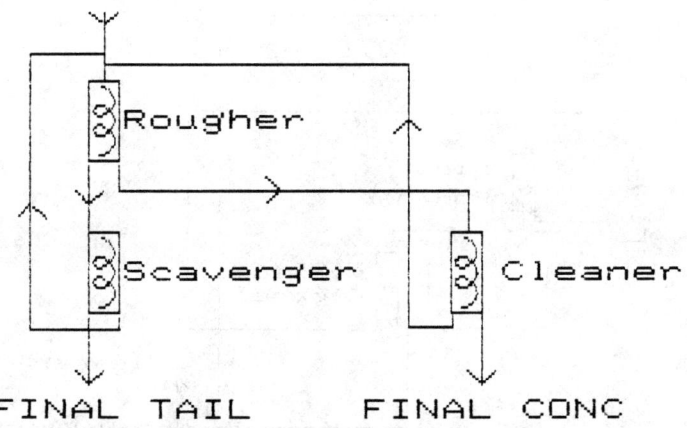

Figure 3. Rougher/Cleaner/Scavenger Circuit

Table II. Tungsten Grades for Rougher/Cleaner/Scavenger Circuit

Spirals	Tails	Mids	Conc
Roughers	0.51		19.06
Scavengers	0.14		4.74
Cleaners	6.31		55.05

	Grade	Recovery
Final Conc	55.05	87.9

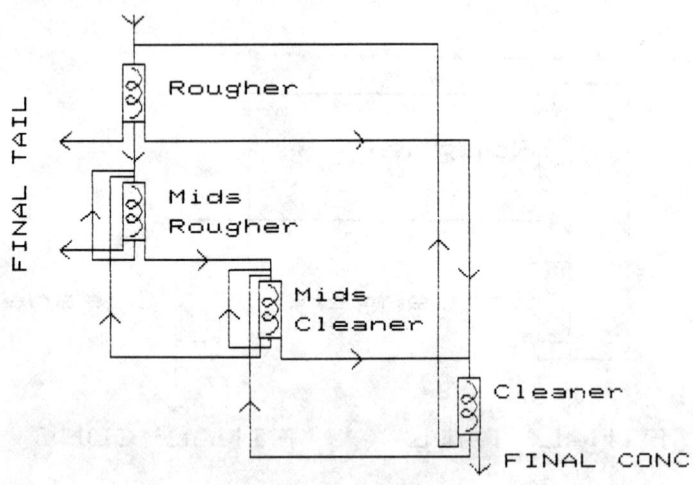

Figure 4. Mids Retreatment Circuit

Table III. Tungsten Grades for Mids Treatment Circuit

Spirals	Tails	Mids	Conc
Roughers	0.11	7.78	18.13
Mids Roughers	0.68	22.46	42.33
Mids Cleaners	10.35	63.39	65.62
Cleaners	5.99	63.67	65.49

	Grade	Recovery
Final Conc	65.50	86.2

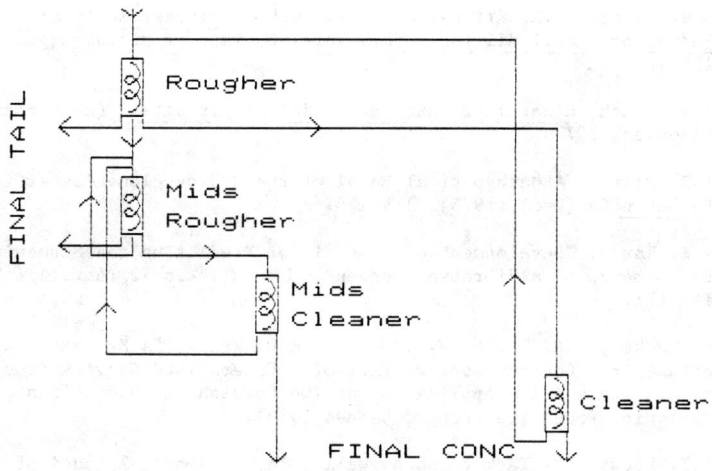

Figure 5. Simplified Mids Retreatment Circuit

Table IV. Tungsten Grades for Simplified Mids Treatment Circuit

Spirals	Tails	Mids	Conc
Roughers	0.16	10.67	21.84
Mids Roughers	0.88	44.31	57.99
Mids Cleaners	50.06		64.84
Cleaners	19.15		54.29

	Grade	Recovery
Final Conc	63.40	82.1

require the development of new mathematical models. In other areas it is important that further work is done in improving the predictive capabilities of the models.

References

1. L.G. Austin, R.R. Klimpel and P.T. Luckie, Process Engineering of Size Reduction: Ball Milling. (New York, Society of Mining Engineers, 1984).

2. A.J. Lynch, Mineral Crushing and Grinding Circuits. (Amsterdam, Elsevier, 1977).

3. L.R. Plitt, "A Mathematical Model of the Hydrocyclone Classifier", CIM Bulletin 69 (Dec) (1976), 114-123.

4. V.K. Karra, "Development of a Model for Predicting the Screening Performance of a Vibrating Screen", CIM Bulletin 72 (April) (1979), 167-171.

5. P. Tucker, K.A. Lewis, W.J. Hobba and D. Wells, "A Mathematical Model of a Spiral Concentrator as Part of a Generalised Gravity Process Simulation and Its Application at Two Cornish Tin Operations", (XVth Int. Min. Processing Cong., Cannes 1985).

6. R.I. Mackie, P. Tucker and A. Wells, "A Mathematical Model of the Stokes Hydrosizer", Accepted for publication in IMM Trans (Sect C), 1987.

7. M.A. Ford and R.P. King, "Simulation of Ore Dressing Plants", Int. J. of Mineral Processing, 12 (4), (1984), 285-304.

8. P. Tucker, K.A. Lewis and M.P. Hallewell, "Computer Modelling as an Aid to Optimisation of the Primary Sprial Circuit at South Crofty Ltd", Accepted for publication in IMM Trans (Sect C), 1987.

9. P. Tucker, "An Approach to Modelling Industrial Unit Processes: Application to a Spiral Concentrator for Minerals", Applied Mathematical Modelling, 9 (Oct) (1985), 375-379.

10. L.G. Austin, P. Bagga and M. Celik, "Breakage Properties of Some Materials in a Laboratory Ball Mill", Powd. Tech., 28 (1981), 235-244.

11. A.W. Westerberg et al, Process Flowsheeting, (Cambridge University Press, 1979).

12. R.W.H. Sargent and A.W. Westerberg, "SPEED-UP in Chemical Engineering Design", Trans. Inst. Chem. Eng., 42 (1964), 190-197.

13. E. Kehat and M. Shacham, "Chemical Process Simulation Programs. II Partitioning and Tearing System Flowsheets", Process Technol., 18 (3) (1973), 115-118.

14. J.H. Wegstein, "Accelerating Convergence of Iterative Processes", Commun. Assoc. Comput. Mach., 1 (1958), 9-13.

THE USE OF MATHEMATICAL MODELING TO ANALYZE COLLECTOR

DOSAGE EFFECTS IN FROTH FLOTATION

Richard R. Klimpel

Mining Chemicals Research and Development
The Dow Chemical Company
Central Research Laboratories
Midland, Michigan 48674

Abstract

Recent work on developing improved procedures for the use of batch laboratory flotation data to indicate full-scale continuous plant operational trends has clearly shown that controlling collector dosage is a key variable. In particular, a phenomenon denoted as the R/K trade-off plays an important role, especially with regard to influencing the rate at which flotation occurs. In simple terms, the R/K trade-off describes the observation that in the laboratory, as collector dosage is initially increased from starvation dosages, the ultimate (equilibrium) recovery at long flotation time and the rate at which this recovery is achieved at shorter times both increase. However, as further collector is added, eventually the rate of recovery goes through a maximum value and starts to decrease at high dosage levels while the equilibrium recovery generally continues to increase. This time effect also makes its influence felt at the plant level as the changing rate of flotation may or may not be important depending on the cell capacity that is available along with operating parameters such as feed rate, particle size, percent solids, and slurry temperature. This paper shows that, if residence time data on a bank of cells is known and when appropriate laboratory rate information on the same ore slurry is experimentally measured and quantified, it is possible to mathematically predict with reasonable accuracy where the R/K trade-off will occur as a function of collector dosage and what the magnitude of the recovery loss will be.

Introduction

Froth flotation is industrially one of the most widely used and economic means of concentrating metal containing ores (1,2). There are numerous reference sources that have been published on the chemical reagent aspects of froth flotation (3-10). In reality, the actual chemical reagents being used commercially have changed little over the last 20 years or so (11). However, it is clear there still remains a host of problems to be solved, especially in the plant scale optimization of the "flotation system" which involves highly interactive components of reagent chemistry, equipment, and operating parameters as shown in Figure 1. As with any system, the key to operating the flotation system successfully (economically) is to understand those factors or variables (including variable interactions) which have a tendency to dominate the overall response of the system. Detailed knowledge of any given variable's influence is often of limited value in such a system, especially one that is as self-compensating as the flotation system has shown itself to be in various industrially oriented test programs (1,10,12-21).

Despite the complexity involved in the flotation system with taking variable feed ores and complex water chemistry, measuring batch laboratory data, and then predicting with reasonable accuracy the related industrial scale response, numerous test experiences by this author and his coworkers (1,10,12-21) did repeatedly demonstrate several important "industrial" interactions. These are: 1) it is necessary to accurately measure time-recovery profiles (see Figure 2) at the laboratory level in order to simulate the effect of reagents at the plant level; 2) the time equivalency of matching laboratory results to inferring plant scale operation is not generally at long laboratory flotation times (thus both rate and equilibrium recovery are important); 3) different chemical structures and changes in dosages of these reagents cause very consistent and predictable changes in rate and equilibrium recovery trends; 4) many reagent types and other system factor combinations cause the associated experimental time-recovery profiles to cross as in Figure 2 implying that these conditions can either favorably or negatively affect recovery in a fixed plant configuration; 5) the overdosing of frothers and especially collectors leads to an important plant phenomenon denoted as the R/K trade-off; 6) the influence of frother chemical structure is very important on the recovery/selectivity associated with both coarse and fine particles; 7) the selectivity of existing commercial collectors over iron sulfides in sulfide/mineral flotation is often more dependent on dosage/pH control with a given collector than the inherent selectivity resulting from chemical structure changes within existing classes of commercial collectors; and 8) several families of new frothers and collectors have been invented with initial tests showing enhanced performance over existing commercial reagents (10,20,21).

It is item 5) of the above list that will be the subject of this paper. The interaction of flotation rate with collector dosage is a general observation that seems to apply to almost all ore types and plant configurations studied, only the location and magnitude changes from plant to plant. The R/K trade-off describes the observation that in the laboratory, as collector dosage is initially increased from starvation dosages, the ultimate (equilibrium) recovery at long flotation time and the rate at which this recovery is achieved at shorter times both increase. However, as further collector is added, eventually the rate of recovery goes through a maximum value and starts to decrease at high dosage levels while the equilibrium recovery generally continues to increase. Exactly why from a detailed surface chemistry viewpoint such a phenomenon occurs is not known exactly. Probably it is related to the manner in which molecules (collectors) having both a polar and nonpolar character arrange themselves relative to a mineral surface with increasing dosage. The orientation of

162

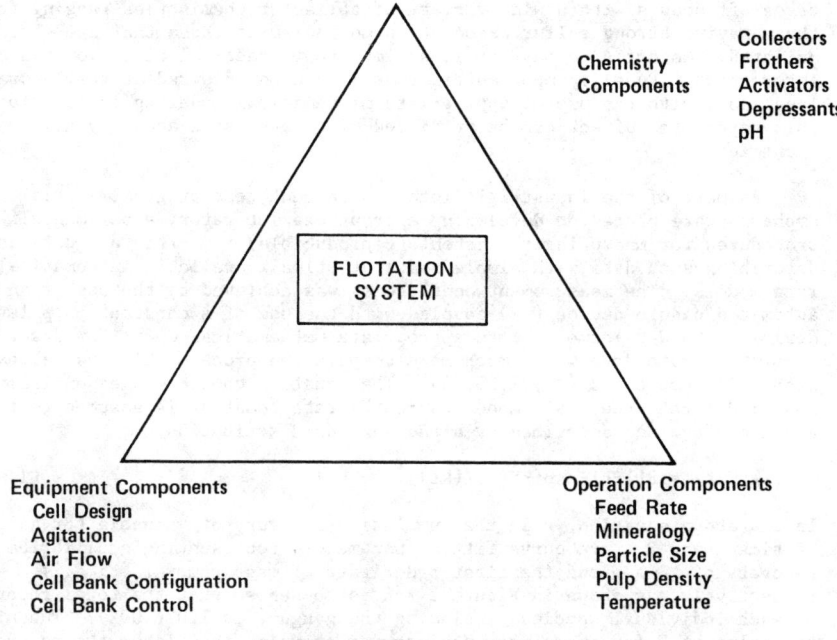

| Chemistry Components | Collectors
Frothers
Activators
Depressants
pH |

FLOTATION SYSTEM

Equipment Components
  Cell Design
  Agitation
  Air Flow
  Cell Bank Configuration
  Cell Bank Control

Operation Components
  Feed Rate
  Mineralogy
  Particle Size
  Pulp Density
  Temperature

Fig. 1   The Process of Flotation Illustrated as a
          Three—Cornered Interactive System

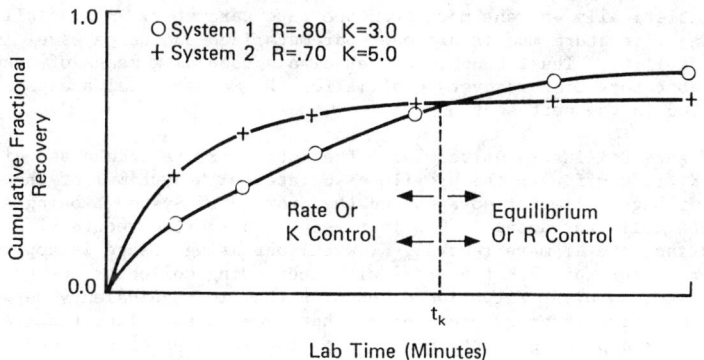

○ System 1 R=.80 K=3.0
+ System 2 R=.70 K=5.0

Rate Or K Control   Equilibrium Or R Control

Cumulative Fractional Recovery

t_k

Lab Time (Minutes)

Fig. 2   An Illustration of Two Laboratory Time-Recovery
          Profiles Run Under Different Conditions

selected portions of the collector into the liquid phase, the amount of
surface coverage by the collector, and the formation of micellar types of
structures all play a role. It is also interesting to note that this R/K
trade-off occurs with a wide variety of collector chemistries ranging from
those having strong sulfur association mechanism to those that are chelation
oriented. Needless to say, it is an important practical flotation system
interaction. Equally important is that it can be measured at the laboratory
level and, with the use of appropriate mathematical modeling as described in
this paper, the effect can be predicted with reasonable accuracy at the
plant level.

As part of the industrial flotation reagent test program, initial
emphases were placed on developing appropriate laboratory experimental
procedures for measuring consistent (reproducible) time-recovery data and
describing such data with simple and statistically reliable mathematical
rate models. The measurement consistency was achieved by the use of an
automated paddle device (22) coupled with the use of a constant pulp level
device. The development of an appropriate mathematical model for describing
laboratory rate data was a much more complicated process which has already
been well documented (13,22,23,24). The equation chosen was essentially a
modified first order rate model where the rate constant is assumed to follow
a range of values described by a rectangular distribution:

$$r(K,t) = R[1-(1-\exp(-Kt))/(Kt)] \tag{1}$$

In the above equation, r is the cumulative recovery of valuable (or gangue)
at time t and R,K are curve-fitting parameters representing equilibrium
recovery at $t \rightarrow \infty$ and the first order rate of mass removal (time^{-1}),
respectively, as shown in Figure 2. R is chosen so that the total recovery
of each individual species, including the gangue, is 1.0 (100%). The higher
the value of K for any given species, for example, the higher the recovery
of this species at shorter times of flotation. The major reason for
selecting Equation 1 over other mathematical forms was its tendency, when
evaluated using statistical model discrimination concepts, to maximize the
ability to distinguish differences between two time-recovery profiles each
run under a different set of conditions (for example, such as differing
collector dosage). It was clearly found that fitting an ordinary laboratory
flotation time-recovery profile run under one set of conditions simply did
not statistically warrant more than one rate parameter. Almost all of the
existing literature models are over-paramaterized and hence have diminished
predictability. Thus, Equation 1 was always used as a means of quantifying
each laboratory time-recovery profile for the mathematical analysis to be
described in the next section.

Figure 2 illustrates very well the laboratory rate data associated with
the R/K trade-off with the profile associated with System 1 of the figure
being at high collector dosages and the profile of System 2 being at
moderate collector dosage. The increasing equilibrium recovery, R, and
decreasing rate of mass removal, K, with increasing dosage is apparent.
Whether or not this R/K trade-off with increasing collector dosage is
important in a plant situation depends on the time-equivalency between the
laboratory time-recovery profile and that time in the plant time-recovery
profile corresponding to the end of a particular bank of continuous
flotation cells. Thus, if the plant cell capacity for current conditions is
ample in size, then the laboratory time equivalency corresponding to the end
of the flotation bank will be at longer laboratory times. In this
situation, conditions that cause a change in the laboratory R value will
quite likely give the same qualitative change in the plant bank of cells
regardless of the possible changes in the rate, K. This is the normal
assumption made in classical reagent testing and is the case of a plant that
is operating within its intended design criteria.

164

On the other hand, if the plant bank of cells is being pushed in any number of ways, such as increased throughput, changes in percent solids, coarser particle sized feed, colder water (slurry) temperatures, etc. (1,12-19) then the time equivalency between plant and laboratory will move to smaller values on the laboratory time scale. Once this is done, it is obvious from Figure 2 that the value of K will become more and more important regardless of the final equilibrium R values. In this situation, anything that decreases K significantly will directly influence the plant leading to lower observed recovery at the end of the cell bank. All of this is very logical and typical of almost all chemical engineering process behavior (1,10,12,13). In over half of all of the flotation plant studied by this author in detail, some type of R/K trade-off balance was occurring with the time equivalency being close to the value of Figure 2. Thus, a loss or gain of several percent of valuable recovery could easily be experienced if operators were not careful with the R/K trade-off.

The R/K trade-off that occurs simply with collector dosage was one of the more unexpected flotation system interactions identified. Prior to this work, there was essentially nothing in the published literature on this effect and typically one thinks of rate limitations as being more physical oriented (for ample, excessive mass flow or too little equipment capacity). Yet, having too much collector being added to an otherwise constant set of conditions in an industrial bank of cells has an effect equal to removing cell capacity, etc.

Residence Time Measurement and Description

The concept of residence time distribution (RTD) is well known (25,26) and describes the distribution of times for all the fractions of material as they pass through a reactor. For the situation of a flotation bank of cells, the situation is somewhat more complex because the "reactor" has two product streams, tailings and froth concentrates. Previous studies (27,28) on mineral flotation have shown that the froth concentrate RTD effect is minimal so that measuring the RTD on a bank of cells can be reasonably approximated by simply following the tailings RTD. This assumption was also made in this study.

Normally, with measuring RTD in particulate systems such as grinding and flotation, there is an additional problem of measuring the transfer of two phases, water and solid. There are two general approaches utilized (26) consisting of either irradiating a specific fraction of material in the feed and following the radioactivity in the product over time or adding a salt tracer such as lithium chloride and following the appearance of the salt in the product by chemical means. The first method, while clearly more difficult to use in practice as it requires special equipment, extra safety precautions, etc., is more preferred as it traces the solid directly. The salt method, while easier to implement, gives much more questionable data as it is tracing the water flow with the assumption that the solid behavior is similar to the water. Experiences in coal flotation (10) and in size reduction (26) indicate that one has to be very careful of assuming water/ solid RTD similarity.

In this study, two different techniques were used to measure RTD on a 10 cell (500 cubic foot cells) industrial rougher bank of mechanical flotation machines processing a chalcopyrite ore. The first was to inject lithium chloride into the conditioning cell just prior to flotation and then collecting appropriate samples at regular intervals using atomic adsorption. The second was to add to the conditioning cell a small amount of silica-like material having a size distribution and gravity similar to that of the feed ore. This material had been specially treated with a strongly adsorbed proprietary surfactant that contained a chemical moiety that allowed for its measurement with appropriate chemical analysis equipment.

The data were analyzed as has been previously outlined (26) to determine an experimental residence time distribution ($\emptyset$) and a total mean residence time (τ) for the bank of cells. Each tracing technique was repeated twice under constant conditions, the results composited for each technique, and the subsequent distributions fitted to a number of different mathematical equations (26). Based primarily on its simplicity, a functional form similar to one recently used to describe coal flotation equipment was used (29). Thus, each distribution was fitted to the functional form:

$$\emptyset'(t') = a(t')^n \exp(-bt') \qquad (2)$$

which is a dimensionless representation of the actual residence time distribution data, $t' = t/\tau$ and $\emptyset' = \tau\emptyset$. The constants a, n, and b are fitting parameters determined so as to give the best least squares fit of $\emptyset'(t')$ calculated versus $\emptyset'(t')$ measured. When n and b are equal to the number of cells in the bank, this equation reduces to the ideal-mixers-in-series expression for the RTD.

Figure 3 gives the fitted residence time distributions for both the solid and liquid (salt) tracer techniques with the ideal mixers in series curve given for comparison. The experimental points are not given for the sake of clarity but the fit is good enough that there are no significant departures of the curve from the data. It can be seen that the two experimental RTD's are both less than ideal indicating some short circuiting through the bank of cells. It is also interesting to note that the liquid tracer RTD is indicating a less ideally mixed RTD than is the solid tracer. The accuracy of the experiments and technique is such that one can be quite confident of the trend just mentioned. In addition, the liquid tracer mean residence time was less than that of the solid tracer indicating that the liquid phase is moving somewhat faster than the solids through the bank. Apparently, this rather surprising result implies that the solid phase is actually receiving better mixing than would be indicated from simply viewing the water phase alone. In addition, the solids recovery predictive capability using the salt tracer is much poorer which will be shown in the next section.

Development of Mathematical Equations

A basic assumption behind Equation 1 for fitting laboratory time-recovery profiles is that mass is being removed from the cell in a modified first-order fashion (13). Using concepts from chemical engineering process modelling (26), if the batch fully mixed laboratory device operates in a reasonably similar fashion to a larger scale continuous industrial device, then the relationship between the laboratory and plant at the end of the plant device can be given by:

$$C/C_o = \int_0^\infty f(K,t)\emptyset(t)dt \qquad (3)$$

where C/C_o is the fraction of a specified constituent of the initial feed showing up in the tails product, $\emptyset(t)$ is the RTD of the larger device (for the tails product) and $f(K,t)$ is the small scale batch flotation rate model for the tailings portion. Selecting $f(K,t)$ is critical in this approach and the logical choice was the rate expression of Equation 1 put into a tailings product form:

$$f(K,t) = [1/(Kt)][1-\exp(-Kt)] \qquad (4)$$

166

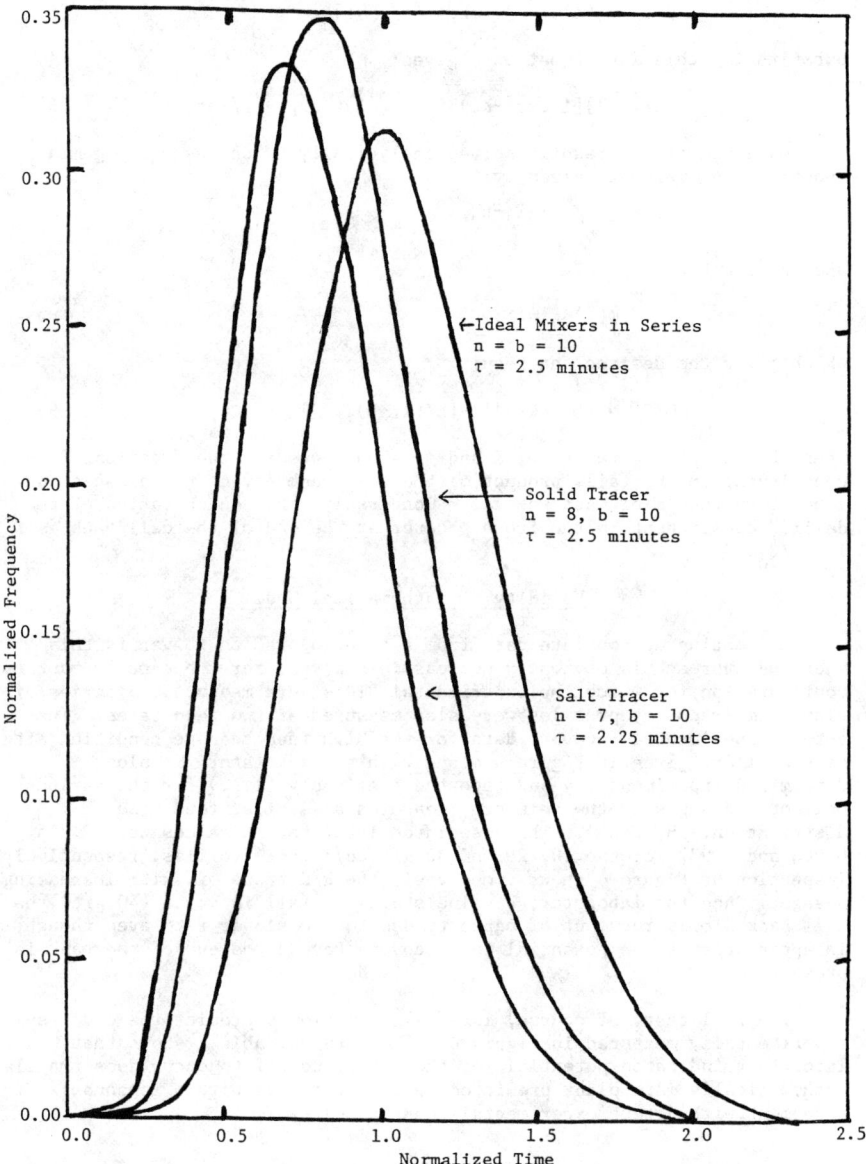

Figure 3   Fitted Residence Time Distributions

167

Then putting Equation 2 into its absolute form (29) gives:

$$\emptyset(t) = (a/\tau^{n+1})t^n \exp[-(b/\tau)t] \tag{5}$$

Substituting this into Equation 3 gives:

$$C/C_o = \int_0^\infty [1/(Kt)][1-\exp(-Kt)](a/\tau^{n+1})t^n \exp[-(b/\tau)t]dt \tag{6}$$

This equation can be readily solved analytically if one uses the gamma function $\Gamma(n)$ which is given by:

$$\Gamma(m) = \int_0^\infty t^{m-1}e^{-t}dt \qquad m > 0 \tag{7}$$

and the identity:

$$\int_0^\infty \emptyset(t)dt = 1 \tag{8}$$

which gives the desired end result:

$$C/C_o = [b/(nK\tau)]\{1-[b/(k\tau+b)]^{n-1}\} \tag{9}$$

Thus, Equation 9, given b, n, K and τ, should predict the fraction of some constituent in the tails product of the cell bank having the associated laboratory rate K of mass removal. Conversely, the concentration of the desired constituent in the froth product at the end of the cell bank is $1 - C/C_o$.

Validity of Mathematical Model

Collecting appropriate data from a plant operation to verify this modeling approach is obviously not easy. However, for the same 10 bank cell configuration for which the experimental RTD's were measured, a series of plant time-recovery profiles were also measured at two feed rates. The actual experimental recovery data for the high feed tonnage condition after each cell are given in Figure 4 along with the laboratory K values determined experimentally and reported previously (1,12) for the various collector dosages on the same ore (measured at similar feed size distribution, pH, etc.). The associated laboratory R values were 0.856, 0.908 and 0.937 for the 10, 20 and 40 g/t collector profiles, respectively. Inspection of Figure 4 shows immediately the R/K trade-off with increasing dosage. Once the laboratory K value starts to fall in value (40 g/t) the 10 cell bank simply runs out of capacity due to the slower rate even though it is apparent that the potential for recovery beyond the end of the bank is present.

The real test, of course, is: will Equation 9 predict the end result experimentally measured in Figure 4? The data of Table I shows that laboratory indicated potential for the R/K trade-off to occur does translate mathematically into plant predicted results that are within reasonable agreement with direct experimental plant results.

168

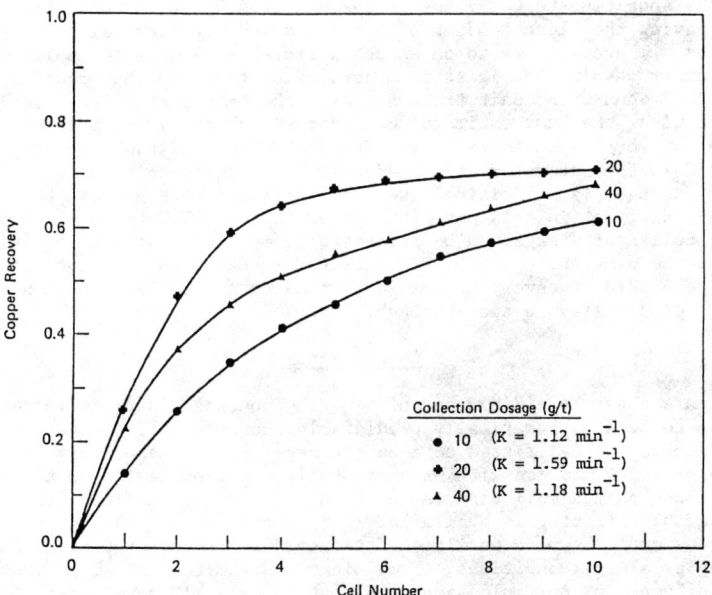

Fig 4   The Influence of Potassium Amyl Xanthate Collector
Addition on Rougher Bank Performance Holding all
Other Factors Constant Including MIBC Frother Dosage
of 30g/ton

Table I. Prediction of Copper Recovery in a 10 Cell Flotation Bank Using
RTD Distributions and Laboratory Measured Rate Values

Feed Rate	K	τ	n	b	Predicted C/C_o	Predicted $1 - C/C_o$	Experimental Recovery
					Using Solid Tracer RTD		
High	1.59	2.5	8	10	.284	.716	.705
"	1.18	"	"	"	.354	.646	.658
"	1.12	"	"	"	.367	.633	.611
Low	1.59	4.0	"	"	.190	.801	.822
"	1.18	"	"	"	.247	.753	--
"	1.12	"	"	"	.258	.742	--
					Using Salt Tracer RTD		
High	1.59	2.25	7	10	.336	.664	.705
"	1.18	"	"	"	.407	.593	.658
"	1.12	"	"	"	.420	.580	.611
					Using Ideally Mixed RTD		
High	1.59	2.5	10	10	.239	.761	.705
"	1.18	"	"	"	.306	.694	.658
"	1.12	"	"	"	.318	.682	.611

Inspection of Table I shows that the predicted bank recovery is reasonably sensitive to the value of the measured RTD with the solid tracer clearly giving the closest match of predicted and experimental bank recoveries. There appears to be enough difference between the solid and liquid tracer results (at least in mineral flotation) so that caution must be exercised when using salt tracers. As might be expected, the ideally mixed RTD gives the best estimated Cu recovery but is giving predictions that are not very close to reality. The change in cell bank recovery with changing K in the laboratory model (due in these tests only to variations in collector dosage) is surprisingly well quantified. Thus, it can be concluded, at least in this one plant test, that the R/K trade-off due to changing collector dosage can be successfully modelled mathematically. The fact that the same solid tracer RTD measured at the high feed rate can also be used to predict recovery at the lower feed rate with reasonable agreement also adds credibility to the approach.

<u>Conclusions</u>

In this paper, the R/K trade-off interaction was described in detail and shown to be a mathematically predictable phenomenon if one uses an appropriate rate model fitted to accurate experimental laboratory time-recovery profiles coupled with experimentally measured residence time distributions on the cell bank in question. As so little flotation RTD data involving solid tracing is available, it is not known yet whether such predictions could be made easily on other cell sizes and/or configurations and changing slurry conditions. Preliminary unreported investigations (30) on the influence of particle size on flotation cell RTD have shown, for example, that there can be a strong RTD influence of coarse particles under some operating conditions. In addition, as cell size increases to 1000 cubic feet or above, preliminary data (30) suggest some significant departures from ideally mixed RTD's which could cause problems in using laboratory measured K values for prediction. At any rate, this paper has shown some encouraging first modelling results on the important rate/recovery trade-off in flotation.

References

1. R. R. Klimpel, "The Industrial Practice of Sulfide Mineral Collectors", Chapter 21 in B. Moudgil and P. Somasundaran, eds., Reagents in Mineral Technology (New York: Marcel Dekker, 1987), 663-682.

2. R. R. Klimpel, "Froth Flotation", Chapter in Encyclopedia of Physical Science and Technology, 5, (1987), 614-625.

3. K. L. Sutherland and I. W. Wark, Principles of Flotation (Melbourne: Aust. I.M.M., 1955).

4. V. I. Klassen and V. A. Mokrousov, An Introduction to the Theory of Flotation (London: Butterworths, 1959).

5. D. W. Fuerstenau, ed., Flotation - 50th Anniversary (New York: AIME, 1962).

6. M. C. Fuerstenau, ed., Flotation - A. M. Gaudin Memorial (New York: AIME, 1976).

7. J. Leja, Surface Chemistry of Froth Flotation (New York: Plenum Press, 1982).

8. P. Somasundaran, ed., Advances in Mineral Processing (Littleton, Colorado: SME, 1986).

9. B. Moudgil and P. Somasundaran, eds., Reagents in Mineral Processing (New York: Marcel Dekker, 1987).

10. S. K. Mishra and R. R. Klimpel, eds., Fine Coal Processing (Park Ridge, New Jersey: Noyes, 1987).

11. R. D. Crozier, "Flotation Plant Reagents", Mining Magazine, Sept. 1984, 202-219.

12. R. R. Klimpel, "Use of Chemical Reagents in Flotation", Chemical Engineering, 21 (18) (1984) 75-79.

13. R. R. Klimpel, "Selection of Chemical Reagents for Flotation", Chapter 45 in A. Mular and R. Bhappu, eds., Mineral Processing Plant Design, 2nd. edn. (New York: AIME, 1980), 907-934.

14. R. R. Klimpel, "The Engineering Characterization of Flotation Reagent Behavior", in Mt. Isa Mill Operators Conference (Melbourne: Aust. I.M.M., 1982) 297-311.

15. R. R. Klimpel, R. D. Hansen, and W. C. Meyer, "The Engineering Characterization of Flotation Reagent Behavior in Sulfide Ore Flotation", paper IV-17 in SIV Int. Min. Proc. Cong. (Toronto: CIM, 1982).

16. R. R. Klimpel, "Froth Flotation: The Kinetic Approach", in L. F. Haughton, ed., Proceedings of Mintek 50 (Randburg, South Africa: Council for Mineral Tech, 1985), 385-392.

17. R. R. Klimpel, R. D. Hansen and B. S. Fee, "Some Recent Advances in New Reagents for Sulfide Mineral Flotation in S. Castro, ed., II Latin America Congress of Flotation (Concepcion, Chile: U. of Concepcion, 1985), Vi 3.1-Vi 3.27.

18. R. R. Klimpel and R. D. Hansen, "Frothers", Chapter 12 in B. Moudgil and P. Somasundaren, eds., Reagents in Mineral Technology (New York: Marcel Dekker, 1987), 385-409.

19. R. R. Klimpel and R. D. Hansen, "The Interaction of Flotation Chemistry and Size Reduction in the Recovery of a Porphyry Copper Ore", Int. Journal of Mineral Proc., in press.

20. R. R. Klimpel, R. D. Hansen, and B. S. Fee, "Recent Advances in New Frother and Collector Chemistry for Sulfide Mineral Flotation", XVI Int. Min. Proc. Congress, Stockholm, 1988, in press.

21. R. R. Klimpel, R. D. Hansen, G. Garcia-Huidobro, J. Broitman, "New Frother and Collector Chemistry for Sulfide Mineral Flotation", Copper 87, Vina del Mar, Chile, 1987, in press.

22. W. C. Meyer and R. R. Klimpel, "Rate Limitations in Froth Flotation", Annual Bound Vol. Trans. AIME, 274 (1983) 1852-1858.

23. E. C. Dowling, R. R. Klimpel, and F. F. Aplan, "Model Discrimination in the Flotation of a Porphyry Copper Ore", Minerals and Metallurgical Proc., 2(2) (1985) 87-101.

24. E. C. Dowling, R. R. Klimpel, and F. F. Aplan, "Model Discrimination in the Flotation of Base Metal Sulfide Ores", A. Mular and M. Anderson, eds., Design and Installation of Concentration and Dewatering Circuits (Littleton, Colorado: SME, 1986), 570-587.

25. P. V. Danckwerts, "Continuous Flow Systems", Chemical Eng. Science, 2(1) (1953), 1-13.

26. L. G. Austin, R. R. Klimpel, and P. T. Luckie, Process Engineering of Size Reduction (Littleton, Colorado: SME, 1984).

27. E. T. Woodburn, R. P. King, and R. P. Colborn, "The Effect of Particle Size Distribution on the Performance of a Phosphate Flotation Process", Metallurgical Transactions, 2 (1971), 3163-3174.

28. R. P. Gardner, "The Use of Radiotracer Techniques in the Development of a Phenomenological Model for Continuous Flotation Processes", (Paper 85-148 presented at SME Annual Meeting in New York, February, 1985).

29. D. R. Van Orden, "Simulation of Continuous Flotation Cells Using Observed Residence Time Distributions", in P. Somasundaran, ed., Advances in Mineral Processing (New York, Marcel Dekker, 1987), 714-725.

30. R. R. Klimpel, paper in preparation, 1987.

SIMULATION OF COPPER PORPHYRY ORE FLOTATION

BY A TRANSFORMED LIBERATION MODEL

C.L. Lin, A. Cortes and J.D. Miller

Department of Metallurgy and Metallurgical Engineering
University of Utah
Salt Lake City, Utah

Abstract

The mineralogical composition of particles sampled during various stages of flotation was measured using polished sections together with an image analysis technique. The measured linear (or areal) grade distributions were transformed into the corresponding volumetric grade distributions. The particle size and mineralogical information thus obtained were combined with kinetic parameters estimated from King's flotation model using batch flotation experiments. This model was used to simulate the performance of continuous pilot-plant scale flotation.

Introduction

Separation of mineral particles in the flotation process has been studied extensively over the past decade. Although many fundamental details, such as the interaction between the hydrophobic particles and air bubbles, remain to be investigated, there is enough information now available to develop a simple, realistic model to simulate the flotation behavior of various ores. After considering the different flotation models developed in the last decade, the distributed-rate-constant model (1-3) from King's group has been tested and found to be useful in the simulation of the steady-state behavior of plant operations. Distributions of the rate constants over the entire particle population are considered as the primary data in the simulation. The distribution of these flotation rate constants needs to be specified for each particle type. For different mineral systems, the flotation behavior of a particular particle type can be characterized and classified with respect to its floatability. Determination of the mineralogical and liberation characteristics plays an important role in estimating the parameters for the distributed-rate-constant flotation model.

In this study, a transformed liberation model (4,5) for the determination of the mineralogical composition of the particles was used to estimate the parameters for the distributed-rate-constant model. The kinetic parameters from a copper porphyry ore flotation separation were estimated using both graphical and computer techniques. MODSIM (6), a simulator for ore dressing plants developed by King, together with the parameters derived from the transformed liberation model, were used to simulate a pilot-plant scale copper ore flotation operation and the utility of the flotation model was demonstrated.

Model Formulation

Transformed Liberation Model

The liberation parameter (degree of liberation in each size range) plays an important role in simulations using the distributed-rate-constant flotation model. This parameter can be established on the basis of the mineralogical composition distribution. Microscopic examination of polished sections of the population of particles is one of the most preferred methods in assessing the extent of liberation. For two-dimensional information gathered from the examination of polished sections, biassed and incomplete mineralogical composition data are expected (7). The true degree of mineral liberation should be based on the volumetric grade distribution of the particle population. The measured grade distributions must be converted into three-dimensional information in order to predict the true mineral liberation. A transformed liberation model (4,5) that converts the one- and two-dimensional data into three-dimensional information has been developed and tested. For a monosize sample, the relationship between linear grade distribution $f(g_1)$ and volumetric grade distribution $p(g)$ can be established, provided the information about the transformation function $H(g_1|g,Nn)$, a conditional probability function, is available. The desired transformation can be expressed as:

$$f(g_1) = \int_0^1 H(g_1|g,Nn)p(g)dg \tag{1}$$

The transformation function has been established from extensive computer simulation (8). Based on this approach, the transformation equation (equation 1) has been solved and tested against experimental data obtained through depth profile measurements for different ore samples (4,5).

In practice, particles of specified size d are mounted in a resin matrix and the linear grade distribution from the polished section is determined as $f(g_1)$ of equation 1. Then, the volumetric grade distribution $p(g)$ is estimated by solving the transformation equation. This transformed liberation model provides a method for the classification of the ore particles into different g-classes (grade classes based on mineral content) which is one of the important parameters of the distributed-rate-constant flotation model.

Flotation Model

The distributed parameter kinetic flotation model developed by R.P. King is used in this study. The model is described briefly below. Detailed development and description of the model can be found in the previously cited literature (1-3).

The distributed-rate-constant model begins with the assumption that the flotation cell consists of a perfectly mixed pulp through which a cloud of bubbles rise. Particles that become attached to the bubbles are transported into the froth phase. A portion of the particles that enter the froth is assumed to return to the pulp while the rest is removed from the cell.

Three distinct properties of the particles are considered for model development; size (d), mineralogical composition or grade (g), and surface activity or rate constant (k). Individual particles from the entire population are classified according to these three parameters. For model development, it is assumed that the flotation of a particular particle type (size, grade, surface activity) obeys a first order rate law of the following form:

$$r(k,g,d) = \gamma k\phi(d)ASWf_T(k,g,d) \tag{2}$$

Where:

$r(k,g,d)$ = rate of flotation of particles characterized by property values k, g, and d.

$f_T(k,g,d)$ = distribution of particles from an entire population characterized by rate constant k, grade g and size d. $f_T(k,g,d)$ can be specified as separate conditional distribution functions; i.e.,

$$f_T(k,g,d) = f_T(k|g,d) \; f_T(g|d) \; f_T(d).$$

$f_T(d)$ = the particle size distribution which can be determined easily.

$f_T(g|d)$ = the volumetric grade distribution of size d, estimated from the transformed liberation model.

γ = the fraction of particles entering the froth phase that is removed over the froth lip.

A = total bubble surface area per unit volume of pulp.

S = fraction of the surface area of a single bubble not covered by adhering particles.

W = hold-up mass of particles in the cell.

$\phi(d)$ = the fractional efficiency of impaction, adhesion and levitation for particles of size d given by:(9)

$$\phi(d) = 2.33(\sqrt{\varepsilon}/d)(1-(d/\Delta)^{1.5})\exp(-\varepsilon/d^2) \tag{3}$$

Where:

Δ = largest size of particle that will float.

ε = tubulence intensity parameter that depends on the hydrodynamic conditions in the cell.

The mass balance for any particular class of particles is then,

$$M_I f_I(k,g,d) - M_T f_T(k,g,d) = M_F f_F(k,g,d)$$
$$= \gamma k\phi(d)AS_{AV}Wf_T(k,d,g) \tag{4}$$

Where M is the total mass flowrate of particles and I, T and F denote feed, tailing and froth streams respectively.

$$M_T f_T(k,g,d) = \frac{M_I f_I(k,g,d)}{1 + \gamma k\phi(d)AS_{AV}W/M_T} \tag{5}$$

176

The performance of the flotation process can be simulated by computing and solving the mass balance equation for each particle type together with the mass balance equation for water. Since the equations are nonlinear, as indicated by King, a Newton-Raphson scheme was developed for a numerical solution based on iterations involving the stage holding time.

Experimental and Simulation Techniques

Batch and Continuous Pilot-Plant Flotation

To test the predictive capability of the models, batch and pilot-plant scale flotation tests were conducted using copper porphyry ore as feed material. Batch flotation was performed in a one cubic foot cell and the results were used to estimate the flotation model parameters previously described. The experimental conditions and details for the batch test are shown in Table I. To verify the model, an integrated pilot-plant scale ball mill/flotation cell circuit was operated. Although the flotation conditions differ somewhat, as is evident from Table II, experimental results suggests that this difference is not of great significance. Only the flotation results are discussed in this study. The pilot-plant circuit configuration and its operational conditions are shown in Figure 1 and Table II respectively. The products from the grinding mill are discharged to a flotation feed sump which feeds a one cubic foot rougher cell. The concentrate from the rougher bank becomes the first concentrate while the tails become feed to the one cubic foot scavenger cell. The tails from the scavenger cell are the final tails of the circuit. The concentrate from the scavenger bank was not recycled.

Table I. Experimental Conditions for Batch Flotation

Collector	Sodium Isopropyl Xanthate 0.4 g/100 ml
	Add 1 ml to grinding
	Add 1 ml to flotation
NaCN	0.1 g/100 ml
	Add 1 ml to grinding
Soda Ash	0.5 g
500 g/m ore, 1,000 ml water, grinding for 15 minutes	
Frother	MIBC
pH	8-9

Table II. Pilot Plant Circuit Conditions

Reagent

Collector	Minerec 2030	0.04 lb/ton
Frother	MIBC	0.03 lb/ton
Modifier	Lime	pH 10

Flotation Cell (manufactured by Hazen Quinn)

Each cell 1 cubic foot size

Aeration rate 30.0 liter/min.

pH adjusted by lime addition

CIRCUIT OF PILOT—SCALE FLOTATION TEST

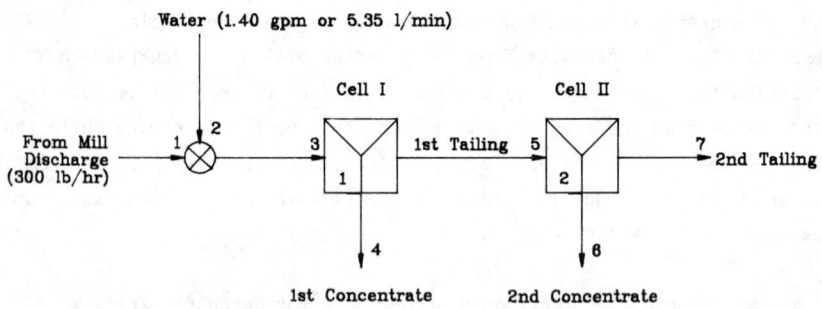

Figure 1. Pilot-scale flotation circuit.

Copper assay, size and grade distribution analyses were conducted for all the samples taken from batch and pilot plant scale tests. The size analysis was done with a standard set of Tyler sieves. Out of each sample a portion was cut using an automated sample splitter. The samples were mounted in a resin matrix for grade distribution analysis. Both X-ray probing and optical image analysis (IBAS II system) were used to measure the linear grade distributions of these samples(4). The transformed liberation model described previously was then used to estimate the corresponding volumetric grade distribution. For pilot-plant testing, copper analysis of each stream was done and the analyses were used to check the mass balance of the flotation circuit.

Parameter Estimation

In order to use any model effectively, the model parameters must be known and easily measured. For each size class, linear grade distributions were measured and used to estimate the corresponding volumetric grade distributions. Detailed procedures are described in previous papers(4,5). The parameters related to grade (g-class) for the distributed-rate-constant model were established based on these volumetric grade distributions.

Graphical techniques (9) can be used to estimate the particle size parameters such as ε and Δ for the distributed-rate-constant flotation model. Parameters, such as k, f(k|g) and γ can be determined by examining the behavior of the individual particle size classes. These parameters are related to the recovery of the individual mineral species; i.e., the recovery based on the mineralogical composition distribution. Besides graphical techniques, the flotation parameters can be estimated by using a computer program that extracts the desired information from experimental data. One such program, ESTIFLOTE(10), was used to estimate the flotation parameters in this study. The test results obtained from batch flotation experiments were used for the determination of the desired parameters. For a particle belonging to a particular size class and having a specified grade, its floatability is determined by a distinct combination of three flotation rate constants. These rate constants were obtained by using the program estimator.

Simulation of the Flotation Circuit by MODSIM

Simulation can be used to improve the performance of a mineral processing circuit with a minimum amount of plant scale testing. The parameter values determined from batch tests were used in the pilot plant flotation circuit simulation. Performance of the entire pilot plant flotation circuit were simulated using MODSIM, a simulator for ore dressing plants developed by King. Experimental test runs were conducted to verify the simulation and for final tuning of the flotation model parameters.

Results and Discussion

Parameter Estimation

The size distribution data from a batch flotation test are shown in Table III. Examination of the data presented shows that at the initial stage of flotation, a higher proportion of fine particles report to the concentrate. In contrast, coarse particles are transported to the froth phase only after an extended time of flotation. However, the difference between overall particle size distributions of concentrates at different flotation times is not particularly significant.

Table III. Batch Flotation Test Size Distribution Data

Size (Mesh)	A (15 Sec) Wt%	B (30 Sec) Wt%	C (1 Min) Wt%	D (2 Min) Wt%	E (4 Min) Wt%	F (8 Min) Wt%	G (16 Min) Wt%	Tail Wt%
+70	0.03	0.04	0.05	0.08	0.09	0.26	0.28	0.87
70x100	0.05	0.02	0.13	0.23	0.35	0.52	0.87	4.09
100x150	0.24	0.22	0.61	0.83	0.91	1.15	2.01	11.19
150x200	1.20	2.54	4.01	4.87	4.40	3.59	3.97	18.43
200x270	5.12	7.83	8.89	8.77	8.12	6.09	5.22	14.94
270x400	7.16	8.73	8.17	6.99	6.46	4.92	4.35	11.00
-400	86.20	80.62	78.15	78.04	79.52	83.42	83.09	39.47

The volumetric grade distributions of three different size classes (100x150, 200x270 and 270x400 mesh) from the concentrate after different flotation times have been estimated by using the transformed liberation model. These are illustrated in Figures 2 to 4, respectively. One should note that the flotation behavior is influenced not only by the particle size, but also by the effect of the particle liberation characteristics. Figure 5 shows the flotation behavior characteristics in terms of various particle types. In the initial stages of batch flotation, a higher percentage of liberated chalcopyrite is floated, however, the percentage decreases as the flotation time is increased. For example, let us examine the 270x400 mesh flotation curve. After one minute of flotation time, approximately 30% of free chalcopyrite was observed. This quantity decreased to about 10% after eight minutes of flotation. Furthermore, the liberation characteristics of different size classes is clearly shown in the figure. Finally, a higher percentage of liberated chalcopyrite in the smaller size class is observed. Such data for both size and grade distribution are used to estimate the flotation parameters described later.

The removal rate of water in the batch test of each concentrate collection time is shown in Figure 6. Copper assay has been done for each concentrate and the corresponding copper recovery based on the flotation time is presented in Figure 7. As indicated, after 16 minutes of flotation, 99% of the total copper content has been recovered.

The flotation parameters were estimated using both graphical and computer methods. The flotation rate constants for different sizes and mineralogical compositions, are shown in Figures 8 to 10. These data clearly indicate that these flotation rate constants are distributed with the finer particles floating at the highest rate. In actual flotation

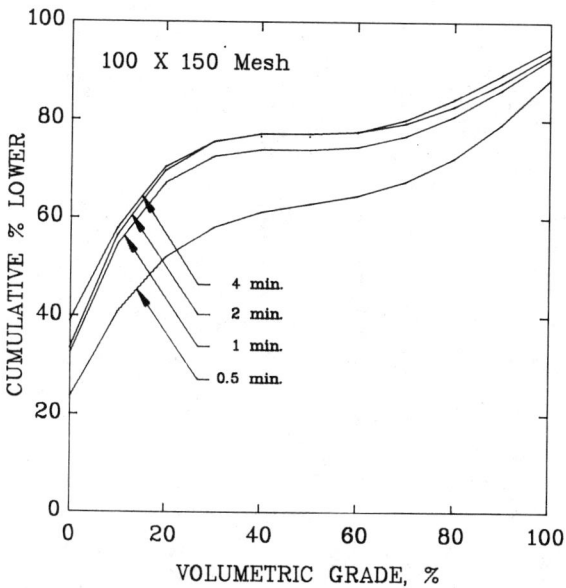

Figure 2. Volumetric grade distributions of concentrate at different flotation times for the 100x150 mesh size interval.

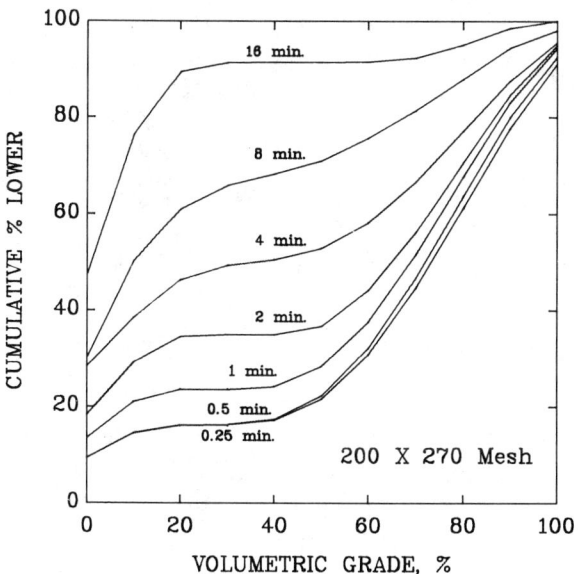

Figure 3. Volumetric grade distributions of concentrate at different flotation times for the 200x270 mesh size interval.

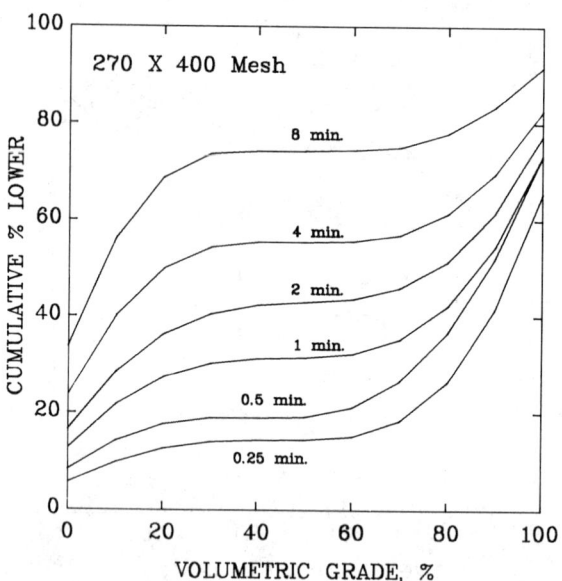

Figure 4. Volumetric grade distributions of concentrate at different flotation times for the 270x400 mesh size interval.

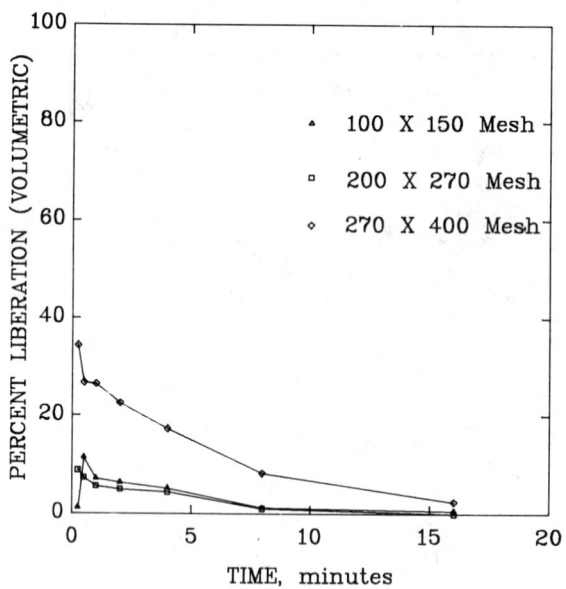

Figure 5. Flotation rate curves of free chalcopyrite for different particle sizes in batch flotation.

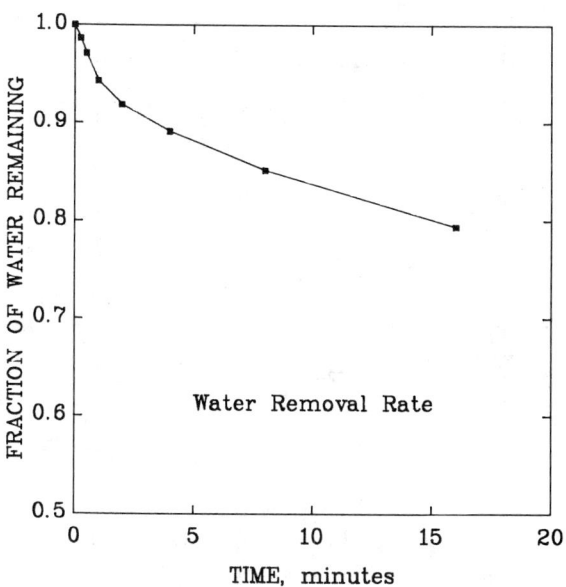

Figure 6. Water removal rate curve for batch flotation.

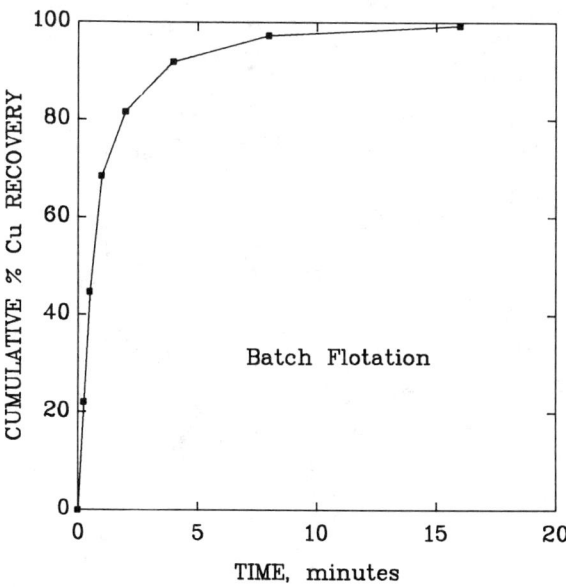

Figure 7. Copper recovery versus time for batch flotation.

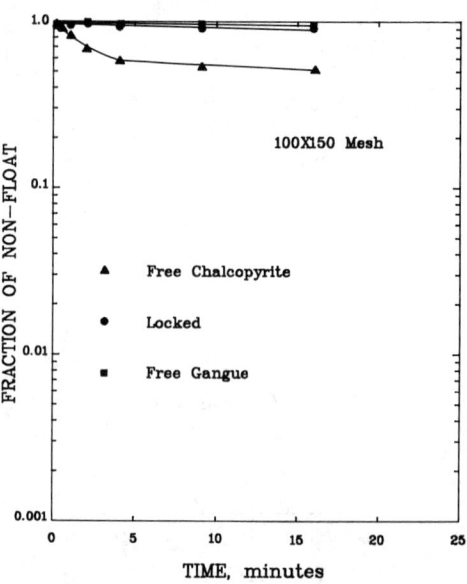

Figure 8. Flotation rate curves of free chalcopyrite, middling and gangue particles of 100x150 mesh size for batch flotation.

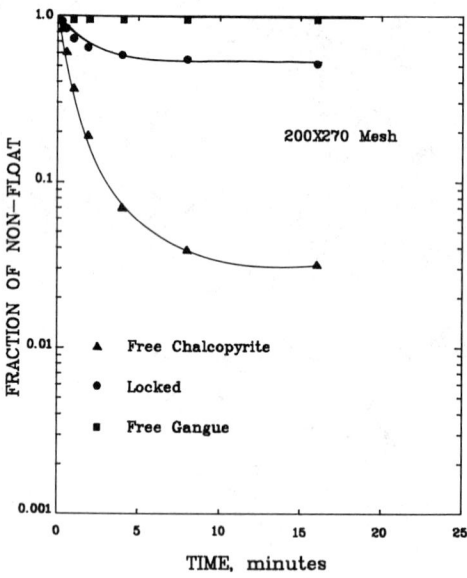

Figure 9. Flotation rate curves of free chalcopyrite, middling and gangue particles of 200x270 mesh size for batch flotation.

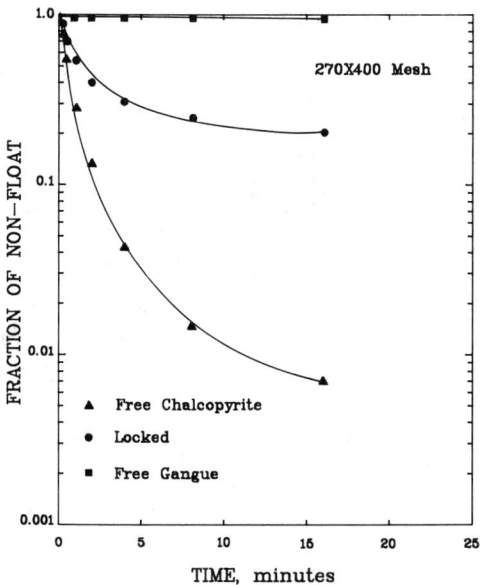

Figure 10. Flotation rate curves of free chalcopyrite, middling and gangue particles of 270x400 mesh size for batch flotation.

systems, the flotation rate constant, k, serves as the basis for quantifying the floatability of mineral particles. These rate constants are used as the primary data in the final tuning of the MODSIM simulation of the pilot plant circuit.

The flotation model parameters were further estimated using the computer program ESTIFLOTE(10). The manual for this parameter estimation is easy to follow and the computer program was used to improve the fit of several parameters to the data at one time. The procedure was repeated for all parameters until the parameters had converged and residual error could not be further reduced. The final values of each step and the reduction in residual error are listed in Table IV. These values serve as the basis for the pilot-plant circuit simulation.

Table IV. Flotation Parameters Estimation

Step	Parameter	Initial	Final	SSQ Error
1	k_2 k_3	1.0E-4 1.0E-6	1.0E-4 1.05E-6	7.295E-1
2	ε_1 ε_2 ε_3	1.12E-9 1.00E-5 3.59E-6	2.031E-9 1.0E-5 3.59E-6	1.630E-1
3	Δ_1 Δ_2 Δ_3	1.28E-3 4.50E-3 4.50E-4	No Change	
4	$f(k_1\|g_1)$ $f(k_2\|g_1)$ $f(k_3\|g_1)$	1.0 0 0	0 0.2412 0.7588	1.702E-2
5	$f(k_1\|g_2)$ $f(k_2\|g_2)$ $f(k_3\|g_2)$	0 1.0 0	No Change	
6	$f(k_1\|g_3)$ $f(k_2\|g_3)$ $f(k_3\|g_3)$	0 0 1.0	No Change	
7	k_2 k_3	1.0E-4 1.05E-6	8.399E-4 1.0E-7	1.098E-2

Pilot-Plant Circuit Simulation and Verification

The distributed-rate-constant flotation model was used to predict the behavior of the pilot plant scale test. The details of the pilot plant operation have been given previously in the experimental section. Initially, the parameters used in the prediction were obtained from the batch flotation test and are listed in Table V. The predicted results from these parameters are given in Table VI as simulation No. 1. A significant difference between the predicted and experimental responses is observed. This discrepancy may be caused by the difference in reagents used in the pilot plant and batch flotation tests. Based on these limited data, it cannot be concluded that the model is able to predict the performance of the pilot plant from batch data.

Further simulations for the pilot plant circuit were done using the MODSIM package. The performances of the two flotation cells were simulated sequentially. The simulation results are listed in Table VI. Simulations 2

Table V. Initial Parameters Values for Cells I and II

Parameters	Cells I and II
1. Cell Volume (m^3)	0.0245
2. Aeration Rate (m^3/s)/m^3	0.0272
3. Froth Transmission Coefficient (γ)	1.0
4. Bubble Size (m)	0.0012
5. Estimated Solids Residence Time (s)	300
6. % Solids in Concentrate	25
7. Epsilon ε	2.031E-9
8. Colby's Delta Δ	1.28E-3
9. Rate Constant k_1 (1/s)	1.0E-5
10. Rate Constant k_2 (1/s)	8.399E-4
11. Rate Constant k_3 (1/s)	1.0E-7

to 9 in Table VI are the simulation results for the rougher flotation cell (cell I). A different set of distributed rate constants was used for the scavenger stage (cell II) and the simulation results are given in simulations 10 to 13 of Table VI The overall quality of predicted grades and recoveries from simulation 11 is presented in Table VII. Also, the experimental results from the flotation products are listed in Table VII for comparison. The size distributions of different streams from both experimental and simulated results are given in Table VIII. These results illustrate that the performance of pilot-plant operation can be simulated successfully.

Conclusions

A distributed-rate-constant flotation model is used to simulate both batch and pilot plant operations. A transformed liberation model is used to measure the volumetric grade distribution of the sample from different streams of flotation products. These volumetric grade distributions together with the size distribution and the recovery of solids were used to estimate the parameters for the distributed-rate-constant flotation model. The prediction and estimation was limited to only a few experimental data sets. On the basis of these limited data and differences in experimental conditions, estimated parameters from batch experiments cannot be used to

Table VI. Simulation Results of Pilot Plant Circuit

Sim. No.	f(k_i|g1) i=1	2	3	f(k_i|g2) 1	2	3	f(k_i|g3) 1	2	3	Y	*A	Conc. I Solid Flow	Grade	Recov.	TAIL I (Feed II) Solid Flow	Grade	Recov.	Conc. I Solid Flow	Grade	Recov.	TAIL II Solid Flow	Grade	Recov.
Exp.	0	24.1	75.8	0	100	0	0	0	100			.0130	44.80	93.00	.1234	0.32	7.00	.0148	8.28	5.11	.1086	0.09	1.87
1	0	97	0	0	100	0	0	5	95	1	25	.0058	26.09	26.19	.1306	3.27	73.81	.0008	8.90	1.20	.1298	3.24	72.61
2	0	97	0	0	100	0	0	2	98	1	25	.0160	34.41	94.97	.1204	0.24	5.03						
3	0	97	0	0	100	0	0	2	98	1	35	.0123	44.74	94.84	.1241	0.24	5.16						
4	0	97	0	0	100	0	0	2	98	1	50	.0122	44.88	94.57	.1242	0.25	5.43						
5	0	97	0	0	100	0	0	2	98	1	75	.0122	44.92	94.49	.1242	0.26	5.51						
6	0	97	0	0	100	0	0	2	98	0.6	25	.0118	45.55	92.57	.1246	0.34	7.43						
7	0	97	0	0	100	0	0	2	98	0.4	25	.0113	45.97	89.99	.1251	0.46	10.01						
8	0	97	0	0	100	0	0	2	98	0.8	25	.0120	45.15	93.96	.1244	0.28	6.04						
9	0	97	0	0	100	0	0	2	98	0.9	25	.0122	44.94	94.44	.1242	0.27	5.56						
	k1			k2			k3			Y	*A												
10	1.0 E-5			8.399 E-4			1.0 E-6			1	25	(Same as simulation 6)						.0067	4.89	5.62	.1180	0.09	1.81
11	1.0 E-5			8.399 E-4			1.0 E-6			0.6	25							.0042	7.22	5.29	.1204	0.10	2.14
12	1.0 E-5			8.399 E-4			1.0 E-6			0.4	25							.0123	2.40	5.08	.1124	0.12	2.34
13	1.0 E-5			8.399 E-4			1.0 E-6			1	25							.0296	1.13	5.78	.0951	0.10	1.64

*A = % Solid in Concentrate

Table VII. Chalcopyrite Grade and Recovery Prediction

	Recovery (%)		Grade (%)	
	Observed	Predicted	Observed	Predicted
First Cell Feed	100.00	100.00	4.48	4.24
First Cell Concentrate	93.00	92.57	44.80	45.55
First Cell Tail	7.00	7.43	0.32	0.34
(second cell feed)				
Second Cell Concentrate	5.11	5.29	8.38	7.22
Second Cell Tail	1.89	2.14	0.09	0.10

Table VIII. Comparison of Experimental and Simulation (No.11) Results with Respect to Size for Pilot Plant Flotation of Low Grade Copper Porphyry Ore

Size (micron)	First Conc. (%)		First Tail (%)		Second Conc. (%)		Second Tail (%)	
	Observed	Predicted	Observed	Predicted	Observed	Predicted	Observed	Predicted
+149	5.04	0.93	0.40	0.83	1.28	1.03	11.03	0.82
149x105	7.81	1.25	0.48	1.17	1.45	2.18	8.64	1.14
105x53	21.34	14.08	13.69	14.46	10.22	34.96	22.97	13.73
-53	65.81	83.74	85.43	83.54	87.05	61.83	57.36	84.31

189

predict the behavior of the pilot plant circuit. However, the parameters for the pilot plant operation can be estimated using MODSIM and the simulation results can describe the performance of the pilot plant operation successfully.

Acknowledgement

Authors would like to express their appreciation to Drs. Prisbrey, Rajamani, and Herbst for their assistance in this endeavor. This research done at the University of Utah Generic Center on Comminution.

References

1. R.P. King, "A Model for the Design and Control of Flotation Plant," 10th International Symposium on Application of Computers in the Minerals Industry, IMM, ed. by Salamon and Lancaster, (1972), p. 341.

2. R.P. King, "A Computer Programme for the Simulation of the Performance of a Flotation Plant," National Institute for Metallurgy, Johannesburg, S. Africa, Report No. 1436, (1973).

3. R.P. King, "Simulation of Flotation Plant," Mining Engineering/AIME, 258, (1975), p. 286-293.

4. C.L. Lin, J.D. Miller, and J.A. Herbst, "Solution to the Transformation Equation for Volumetric Grade Distribution from Linear and/or Areal Grade Distributions," Powder Technology, 50, (1987), p. 55-63.

5. J.D. Miller and C.L. Lin, "Treatment of Polished Section Data for Detailed Liberation Analysis," presented at Conference on Recent Developments in Comminution, Dec. (1985), Hawaii, to be published in Int'l J. of Mineral Processing, (1987).

6. R.P. King, "A User's Guide to MODSIM," Dept. of Metallurgy, Report GEN/1/86, University of the Witwatersrand, Johannesburg, S. Africa.

7. J.D. Miller and C.L. Lin, "Comparison of Linear and Areal Grade Distributions as Estimates of the Volumetric Grade Distribution in Liberation Analysis," Particulate and Multiphase Processes. T. Ariman, et al. ed., Hemisphere Publishing Corp., (1987), p. 421-432.

8. J.E. Sepulveda, J.D. Miller and C.L. Lin, "Generation of Irregularly Shaped Multiphase Particles for Liberation Analysis," XV IMPC, Cannes, France, 1, (1985), p. 120.

9. R.P. Colburn, E.T. Woodburn and R.P. King, "The Effect of Particle Size Distribution on the Performance of a Phosphate Flotation Process," Metall. Trans., 2, (1971), p. 3163-3174.

10. R.P. King and M.H. Moys, "Estimation of the Parameters in the Distributed Constant Flotation Model, "National Institute for Metallurgy, Johannesburg, S. Africa, Report No. 1567, (1974).

COLUMN FLOTATION

Session Chairman
Donald G. Foot, US Bureau of Mines
Salt Lake City Research Center

This session examines advances in continuous column flotation technology for sequential recovery of mineral values from flourite concentrates and equipment innovation for uniform reagent disperison and froth control in packed column flotation of copper sulfide ores. Models describing the dynamic and steady state performance of flotation columns, and the interactive computer control for automating laboratory columns are presented.

CONTINUOUS COLUMN FLOTATION RECOVERY

OF MULTIPLE PRODUCTS FROM A FLUOUITE ORE

M.R. Peterson, L.J. Duchene, and D.G. Foot, Jr.

Department of the Interior, U.S. Bureau of Mines
Salt Lake City Research Center
729 Arapeen Drive
Salt Lake City, UT 84108

Abstract

The Bureau of Mines investigated column flotation for recovery of high grade fluorite concentrate and byproduct concentrates from the Fish Creek fluorite deposit in Nevada. The recovery scheme consisted of (1) grinding to minus 48 mesh, (2) fluorite rougher and cleaner flotation, (3) desliming at 20 µm, (4) muscovite flotation, and (5) silicate rougher flotation. Acidgrade fluorspar, mica, silica glass sand, and low-grade beryl concentrate were produced in a 100-1b/h continuous column flotation unit (CCFU). Best results achieved were 96.6-pct fluorite recovery in a first cleaner concentrate containing 99.7 pct CaF_2. Beryl recovery was as high as 92 pct at a grade of 5.3 pct BeO. The silica glass sand assayed 96-99 pct SiO_2, and was recovered as the rougher beryl flotation tailings. Collector reagent consumption was 24 pct lower for fluorite and mica flotation steps than continuous pilot-scale conventional flotation.

COMPUTER CONTROL OF A LABORATORY FLOTATION COLUMN

P.J. McDonough, J.D. McKay, and D.G. Foot, Jr.

Department of the Interior
U.S. Bureau of Mines
729 Arapeen Drive
Salt Lake City, UT 84108

Abstract

As part of its research program on column flotation, the Bureau of Mines designed and installed a distributed computer control system for automation of laboratory flotation columns. The system uses a five-loop simple interactive control strategy for flow control of (1) feed, (2) tailings, (3) wash water, (4) bubble generation air, and (5) bubble generation water, using an external bubble generation system. The system consists of (1) five stand-alone controllers with menu selectable configuration, (2) a PC-AT style computer with 40 mb hard drive, (3) enhanced color graphics monitor, and (4) an extensive software package which permits dynamic flowcharting, concurrent multiple data point trending, keyboard access to controllers, and storage and manipulation data. Selection of peripherals such as pumps, actuated valves, level sensors, pressure transmitters, and flowmeters is discussed.

PACKED COLUMNS FOR COPPER SULFIDE FLOTATION

David C. Yang and Chin Li
Institute of Materials Processing
Michigan Technological University
Houghton, Michigan 49931

Abstract

A packed column flotation system, originally developed for
processing finely disseminated taconites, is applicable to a wide variety
of ores and coals. The system combines a controlled dispersion reagent
scheme with an innovation in flotation machines. The salient design of
this new machine is its in-line packing which supports a controllable
froth bed height and permits intimate particle-bubble contact. The
machine is equipped with no moving parts and no air sparger. Fine
bubbles are generated and evenly distributed while air passes through the
small circuitous passages conforming to the structural pattern of packing
elements.

This paper presents the progress made in applications of the packed
column to copper sulfides. A recent feasibility study at the White Pine
concentrator of the Copper Range Company indicates that one stage of the
packed column has the potential of replacing both the conventional
cleaning and recleaning stages with much higher recovery and better grade
of the concentrate. The tests of the system on fine coal were also
encouraging. An attempt is made to modify a first order rate equation to
describe the performance of the packed column machine. A preliminary
model established is found to fit the data reasonably well. However, a
more detailed study is recommended before the model can become generally
acceptable.

Introduction

White Pine Copper

The Copper Range Company is located in the western part of Michigan's Upper Peninsula and south of Lake Superior. The copper mineralization is confined to a narrow stratigraphic interval near the contact of the Nonesuch shale and the underlying Copper Harbor formation, both of late Cambrian age. The ore contains finely disseminated chalcocite, native copper, and native silver as a trace element (1).

Due to economic hardship, White Pine Copper mine was idled for more than two years until it was reopened in November of 1985 based on an employee ownership structure. Despite the continued depressed copper market and price, the company is facing up to the challenge with renewed efforts to restore its competitive position.

Technological advances could improve productivity and bring the company back to profit. A new flotation system recently developed at Michigan Technological University (MTU) is considered a promising approach to meet such a challenge (2). Sufficient data from the tests on various ores and coals have been obtained to date. It has been demonstrated that the new flotation machine is especially suited for separation of fine particles (3, 4, 5). Normally, one stage of the new machine can replace up to eight stages of conventional mechanical cells with equal or better metallurgical performance. This paper focuses on the first step of cooperative efforts between MTU and the Copper Range Company to improve its White Pine operation.

Experimental

Tube Flotation Machine

A schematic representation of a static tube flotation machine, or commonly referred to as a packed column, is shown in Figure 1. A pulp inlet is provided for introducing a conditioned feed slurry into the middle section of the tube. Air is injected into the bottom of the tube while wash water is introduced into the top. The froth product and tailings are discharged through two separate outlets located at the top and bottom of the tube, respectively. The tube is entirely or partially filled with packing which provides a large number of small flow passages in a winding pattern. Since the tube is packed, no provision is needed to generate or disperse fine air bubbles. Instead, injected air is evenly distributed and broken up into fine bubbles upon rising through the packing elements. A steady stream of fine bubbles of relatively uniform size can then be supplied inside the tube. The tube flotation machine has many advantages which include plug flow, water washing, and minimum water and energy usage. More important is the fact that the packing elements in the new machine can support a much deeper froth bed. This froth bed can be used to control flotation cleaning by providing suitable time for froth cleaning.

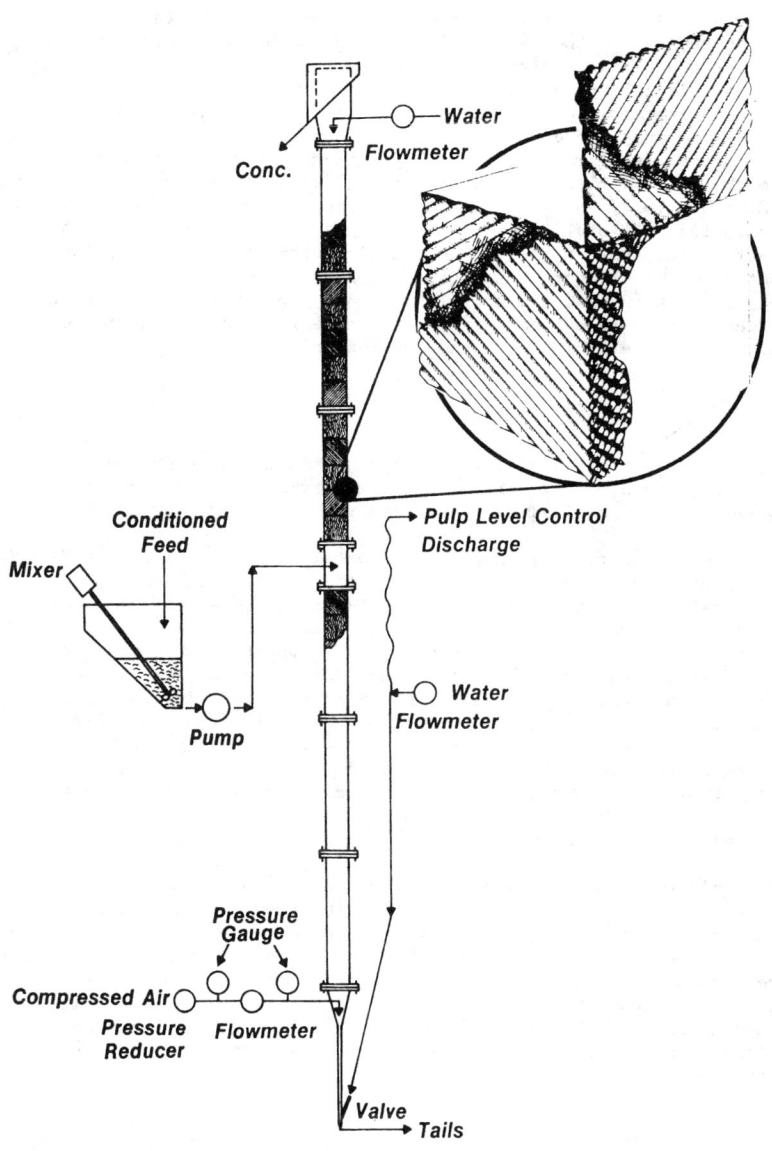

Figure 1 - Static tube flotation machine and its packing.

White Pine Flotation Circuit

Figure 2 is a simplified flowsheet used at the White Pine Copper concentrator. Cyclones are used to close the primary grinding circuit. The overflow is fed to the primary flotation circuit while the underflow goes back to the primary grinding circuit. The tailings from the primary flotation are deslimed in deslime cyclones before being subject to a secondary (sand) flotation. The slime fraction (slime tails) combines with the tailings from the secondary flotation circuit (sand tails) to form the final tailings which are sent to a tailings pond. Concentrates from both primary and sand flotation circuits along with the tailings from recleaner flotation are reground in a tube mill which is closed by two-stage regrind cyclones. The overflow from the regrind cyclones is floated in the cleaner flotation, tailings from which are recirculated back to the primary cyclones. The cleaner concentrate goes to the recleaner, from which the final concentrate is produced while the tailings recirculate back to the regrind mill.

Testing Procedure

Parallel comparison tests. The experimental work carried out for this project can be categorized into two parts: (1) Feasibility study and (2) On-site parallel comparison tests.

A series of tests were first conducted at the Institute of Materials Processing (IMP) of MTU on samples cut from the plant cleaner concentrate, i.e., the feed to the recleaner. These were then followed by a series of confirming tests including the finding of more data points to extend the range of grade-recovery curve toward higher quality products. In the course of this investigation, great care was exercised in obtaining representative samples for testing. Following a plant inspection, a decision was made to tap samples from the feed line to the two recleaners. A hose was connected to a valved tap diverting a small portion of the slurry from the feed line into one of the recleaner feed boxes, from which representative feed samples were cut into individual containers premarked for volume control. These containers were sealed to keep the samples from oxidizing.

Tests were conducted the very next day at IMP using a laboratory tube unit (Figure 3). The main body of the tube unit comprises two 4-foot packed glass tubings of 2-inch outside diameter with several inlet and outlet ports. The individually measured charges (4-liter in volume) were dumped batchwise into the recirculation feed cone and continuously fed to the tube machine at a controlled rate (i.e., 100 ml/min). The on-site comparison tests were carried out in a similar fashion. No additional reagents were added in both the laboratory and on-site comparison tests.

Two-stage substitution. Our next attempt was to replace both cleaner and recleaner flotation circuits with a one-stage circuit using the tube flotation machine. Since overflow from the second regrind cyclones flows through a closed launder into a cleaner feed box, which is not easily accessible, a special device was used for sampling. The idea was to create a stream of the slurry representative of the cleaner feed circulated in a closed loop using a pump connected to two PVC pipes which were inserted into the cleaner feed box through a gap located in the center of the feed box. Feed samples were tapped into the premarked container from a valved side line.

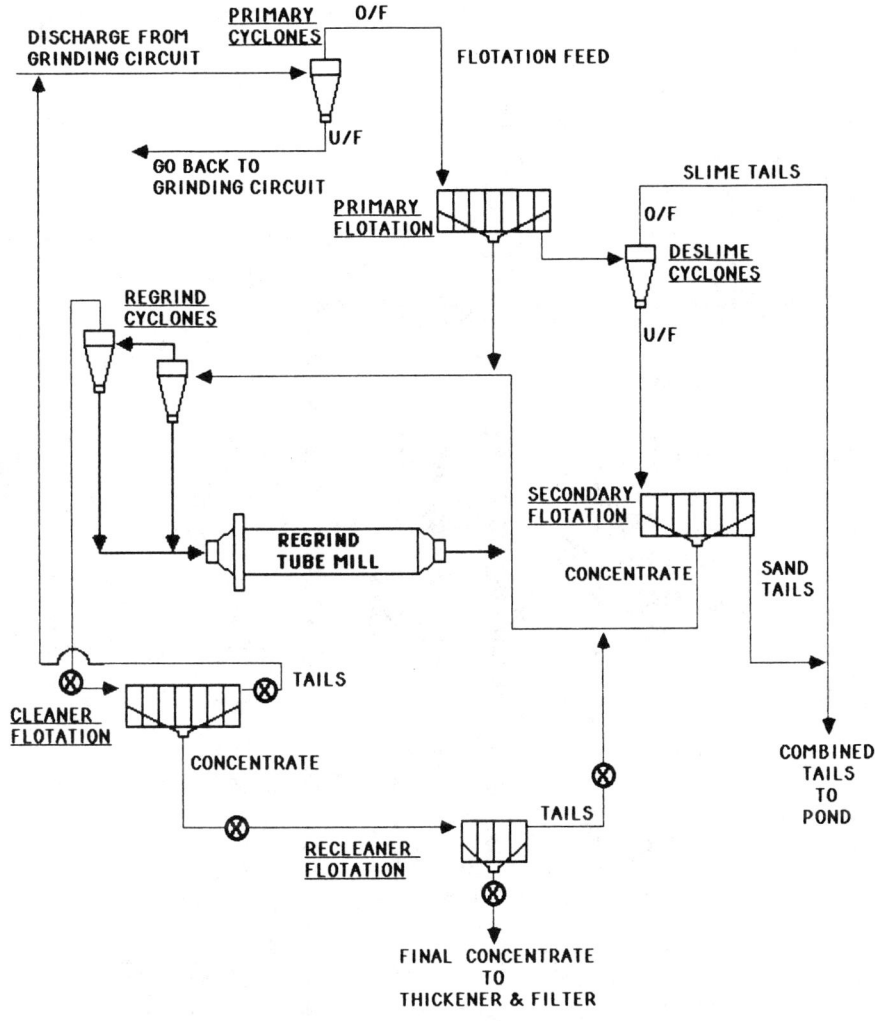

DISCHARGE FROM
GRINDING CIRCUIT

PRIMARY
CYCLONES

O/F

FLOTATION FEED

GO BACK TO
GRINDING CIRCUIT

U/F

PRIMARY
FLOTATION

SLIME TAILS

O/F

DESLIME
CYCLONES

U/F

REGRIND
CYCLONES

SECONDARY
FLOTATION

REGRIND
TUBE MILL

CONCENTRATE

SAND
TAILS

TAILS

CLEANER
FLOTATION

CONCENTRATE

COMBINED
TAILS
TO
POND

TAILS

RECLEANER
FLOTATION

FINAL CONCENTRATE
TO
THICKENER & FILTER

⊗ five-point sample set

Figure 2 - Flotation flowsheet of White Pine concentrator.

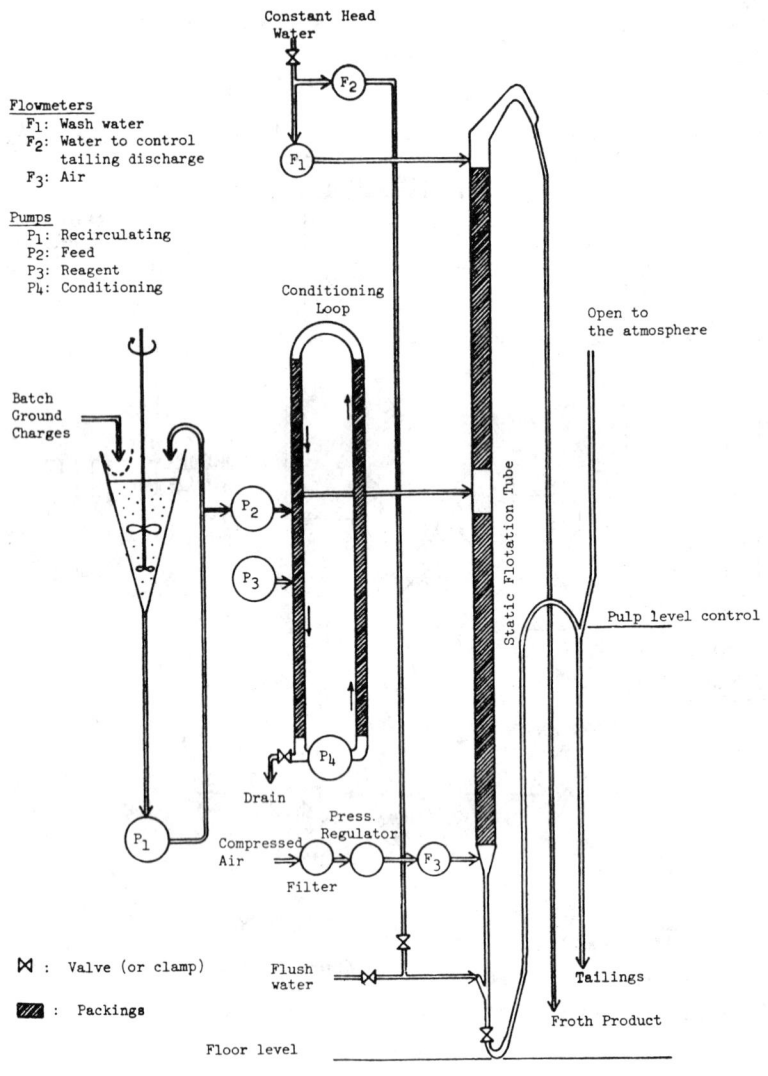

Figure 3 - A sketch of the laboratory tube flotation setup.

202

According to the mill superintendent, NaSH is added at 0.08 lbs/ton to the first cell of the plant cleaner flotation bank to increase the copper recovery. Tests were conducted to compare the effect of the addition of NaSH to that of no addition. The required NaSH was injected into the feed line using a syringe pump.

Incidentally, the effect of solids content was also studied in two different time periods when the mill was running full load and when it was running half load during the maintenance routine of its grinding circuit. On the half section load, pulp density of the flotation feed was reduced to about half of the normal running condition.

Sample and Data Acquisition

All the feed samples were measured batchwise using the premarked 4-liter container and poured sequentially into the recirculation feed cone. The slurry was then pumped continuously from a side line of the recirculation loop into the tube flotation machine. A wide range grade-recovery curve can be obtained by sampling from different operating conditions by varying air, water, pulp level, feed rate, etc. Both concentrate and tailings were collected for the same time period after a steady-state condition had been reached. Samples were then filtered, dried, and analyzed for copper content. Data obtained from tests of the recleaner feed were compared with those provided by the company. As for the cleaner test, a five-point sample set was taken from the plant circuit (Figure 2) during each testing period in order to establish a reference standard.

Results and Discussions

Recleaner Feed - Direct Comparison

The result obtained from the laboratory tests on recleaner feed are listed in Table I. The data is quite reliable judging by its good consistency, although grades of all calculated heads are somewhat higher than the head analysis of the plant sample. The difference was possibly due to the fluctuation of the mill and/or the segregation of solid particles when testing samples were taken. Table II presents the results of the plant on-site comparison tests of which the grade of calculated heads are very close to the head grade. All available data was plotted on a graphic paper of % Cu Recovery versus % Cu in Concentrate for comparison with the data point provided by the company (Figure 4). The extension of the laboratory results to a higher grade end by the plant on-site tests was considered quite successful. The smoothness of the curve also indicates the reliability of the data. It can be projected that through one-to-one comparison the tube flotation system is far superior to the existing recleaner flotation circuit. In terms of metallurgical performance, more than 90% copper can be recovered based on present product grade of about 26% Cu. On the other hand, a concentrate with over 40% copper content can be obtained by keeping the same copper recovery at about 74%.

Cleaner Feed - Two-Stage Substitution

Pulp densities of full and half section tests are about 9 and 5

Table I. Results of Laboratory Tube Flotation Test
on White Pine Recleaner Feed

	Cal. Head	Concentrate			Tails
Test No.	%Cu	%Cu	%Wt	%Cu Rec.	%Cu
86117	15.58	19.97	76.54	98.12	1.25
	15.21	21.59	68.43	97.12	1.39
	15.88	22.84	67.44	97.01	1.46
	15.25	27.98	51.39	94.32	1.78
	15.35	28.29	51.12	94.24	1.81
	15.05	28.66	49.34	93.97	1.79
Plant					
Samples	12.77*	25.98	33.24	73.68	4.62

* Head assay

Table II. Results of On-Site Tube Flotation Test
on White Pine Recleaner Feed

	Cal. Head	Concentrate			Tails
Test No.	%Cu	%Cu	%Wt	%Cu Rec.	%Cu
86272	14.47	33.41	39.08	90.23	2.32
	14.56	33.33	39.36	90.13	2.37
	14.07	35.96	34.29	87.63	2.65
	13.77	36.10	33.24	87.15	2.65
	14.11	36.50	33.67	87.11	2.74
	13.99	36.62	33.36	87.33	2.66
Plant					
Samples	14.01*	25.98	33.24	73.68	4.62

* Head assay

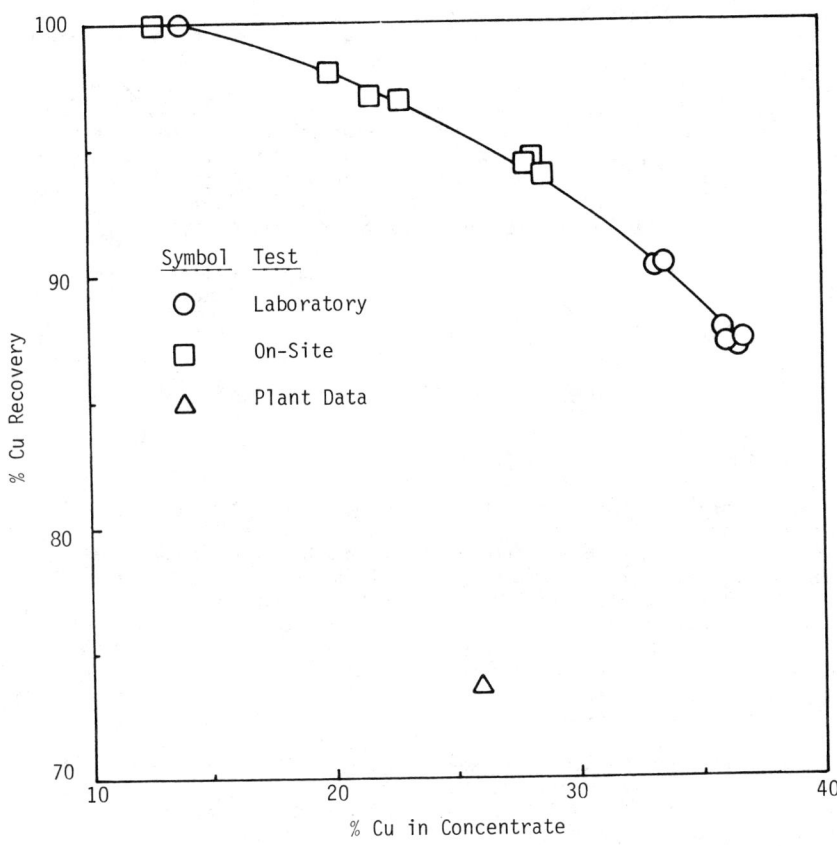

Figure 4 - Comparison of the tube flotation with the
plant data on White Pine recleaner feed.

percent solids, respectively. The results of the full section tests are listed in Table III, at which time no NaSH was added. Table IV shows the results of the half section tests including those with and without additon of NaSH. Figures 5 and 6 illustrate the plant material balance calculated from the data of the five-point sample sets obtained during full and half section testing periods, respectively. The calculated heads of the full section test show fairly good consistency but lower than the plant data. The plant data in the half section test is lower than the calculated heads which show a downward trend. This means the fluctuation of the mill is significant especially for the half-section condition. Figure 7 is a comparison between all available test results and the plant data. The curves for both full and half section tests are quite smooth and thus it is believed that the results are reliable. Obviously, the static tube flotation machine can replace both cleaner and recleaner circuits with much improved metallurgical performance.

Lowering the pulp density can improve the grade of the final concentrae while decreasing the copper content in the final tailings. In other words, more copper could be recovered by decreasing the feed density. However, there is no evidence showing that better results can be obtained by adding NaSH.

Design and Scale-up Considerations

Control Strategies

A laboratory tube flotation machine can be directly controlled by varying the "machine operating variables" which at least include elevation (or height) of tailing discharge head, air flow, wash water and feed rate. Each variable actually affects one or more of the "effective parameters" (e.g., slurry residence time, bubble surface area, bubble diameter, etc.) which in turn control the material distribution inside the tube flotation machine. The effect of these "machine operating variables" has been studied recently on coal flotation (5), but since the principles addressed are common in almost all cases, some highlights will be presented.

Elevation of the discharge head. A discharge line is used in the laboratory tube unit to control the slurry hold-up inside the flotation machine (refer to Figure 3). The elevation of the discharge head is defined as the distance from the air inlet to the highest point of the tailings discharge exposed to the atmosphere. When there is no air passing through the tube, the liquid level inside the tube flotation machine is the same as the elevation of discharge head due to equivalent hydrostatic pressure. Under aerating conditions, the discharge line is considered to have a similar effect on the slurry hold-up inside the tube even if it is not clearly observable. Higher elevation of the discharge head results in higher slurry hold-up inside the tube and hence longer residence time of the slurry. Since lacking reliable measurement of the slurry residence time, at least for existing facility, it is reasonable to accept the height of the discharge head as an indirect parameter of the slurry residence time. Figure 8 shows some typical results from tests in which the height of the discharge head was varied. The trend clearly indicates that increasing elevation of the discharge head or prolonging residence time increases not only the weight recovery but also the ash content of the concentrate. There appears to be an optimum range of this important control parameter to obtain desired results.

206

Table III. Results of On-Site Tube Flotation Test
on White Pine Cleaner Feed
(Full Section)

Test No.	Cal. Head %Cu	Concentrate			Tails
		%Cu	%Wt	%Cu Rec.	%Cu
86274	3.84	10.50	31.36	85.71	0.80
	3.74	12.07	25.72	83.10	0.85
	3.72	12.35	25.02	83.07	0.84
	3.91	14.72	21.43	80.71	0.96
	3.78	14.76	20.44	79.80	0.96
	3.89	24.39	11.54	72.29	1.22

Table IV. Results of On-Site Tube Flotation Test
on White Pine Cleaner Feed
(Half Section)

Test No.	Cal. Head %Cu	Concentrate			Tails
		%Cu	%Wt	%Cu Rec.	%Cu
86278	6.00	29.59	16.93	83.52	1.19
	5.79	32.45	14.53	81.41	1.26
	6.46	34.03	15.85	83.46	1.27
	6.00	34.87	13.60	78.99	1.46
	4.63*	12.68	33.12	90.75	0.64
	5.00*	15.80	28.58	90.29	0.68
	4.53*	19.46	20.08	86.24	0.78
	5.24*	28.06	15.66	83.89	1.00

* NaSH is added as a modifier.

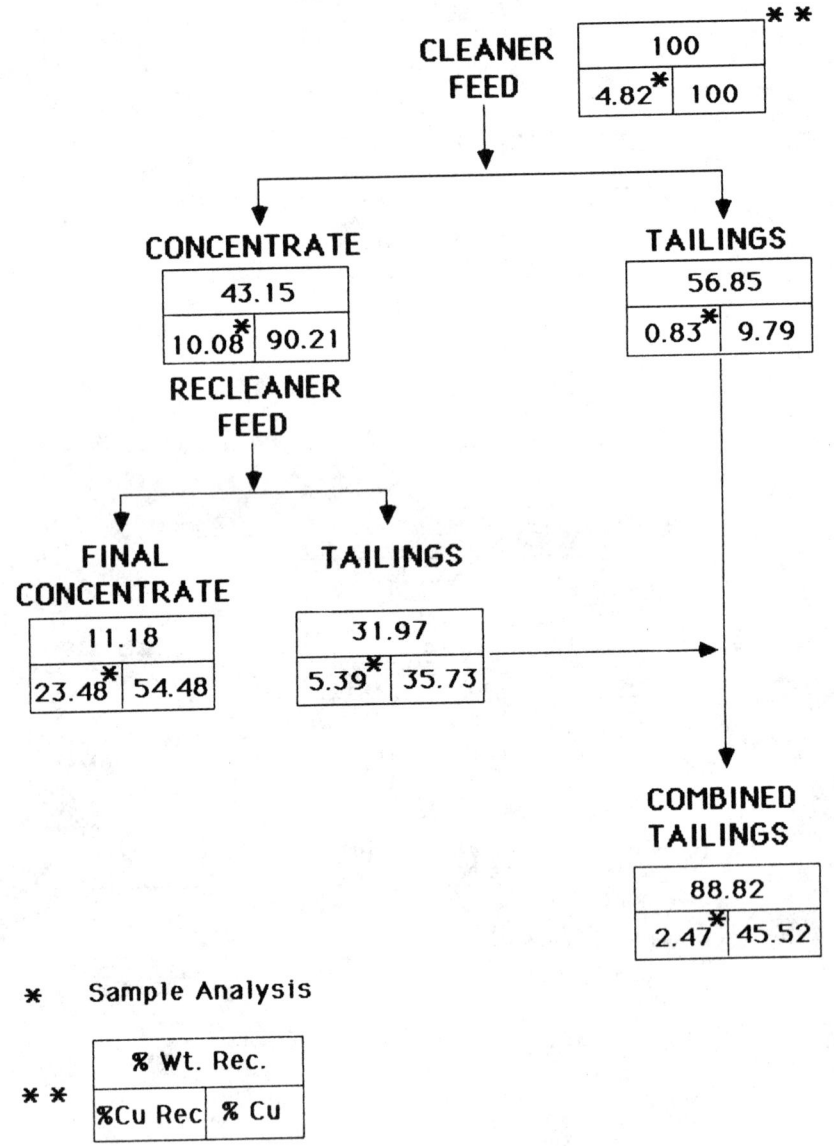

CLEANER FEED

100	
4.82*	100

**

CONCENTRATE

43.15	
10.08*	90.21

TAILINGS

56.85	
0.83*	9.79

RECLEANER FEED

FINAL CONCENTRATE

11.18	
23.48*	54.48

TAILINGS

31.97	
5.39*	35.73

COMBINED TAILINGS

88.82	
2.47*	45.52

*   Sample Analysis

	% Wt. Rec.
%Cu Rec	% Cu

**

Figure 5 - Material balance from the plant data
during full section operation.

208

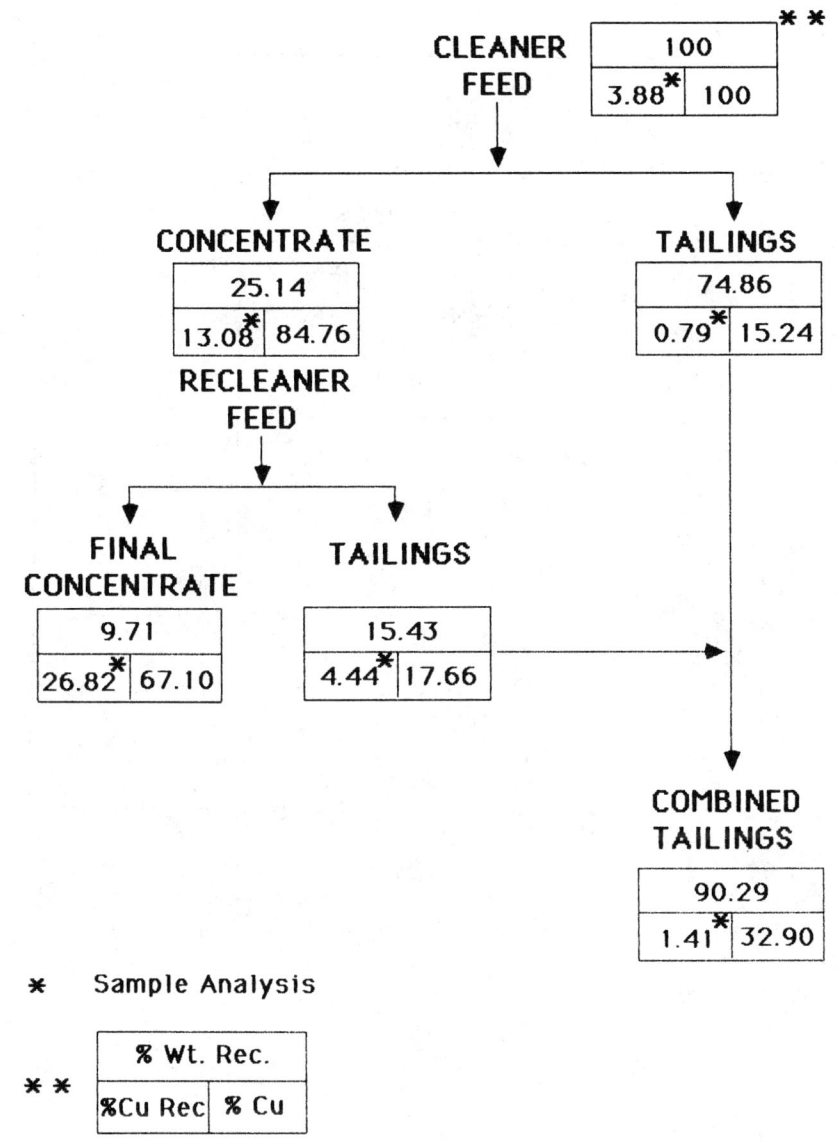

Figure 6 - Material balance from the plant data
during half section operation.

209

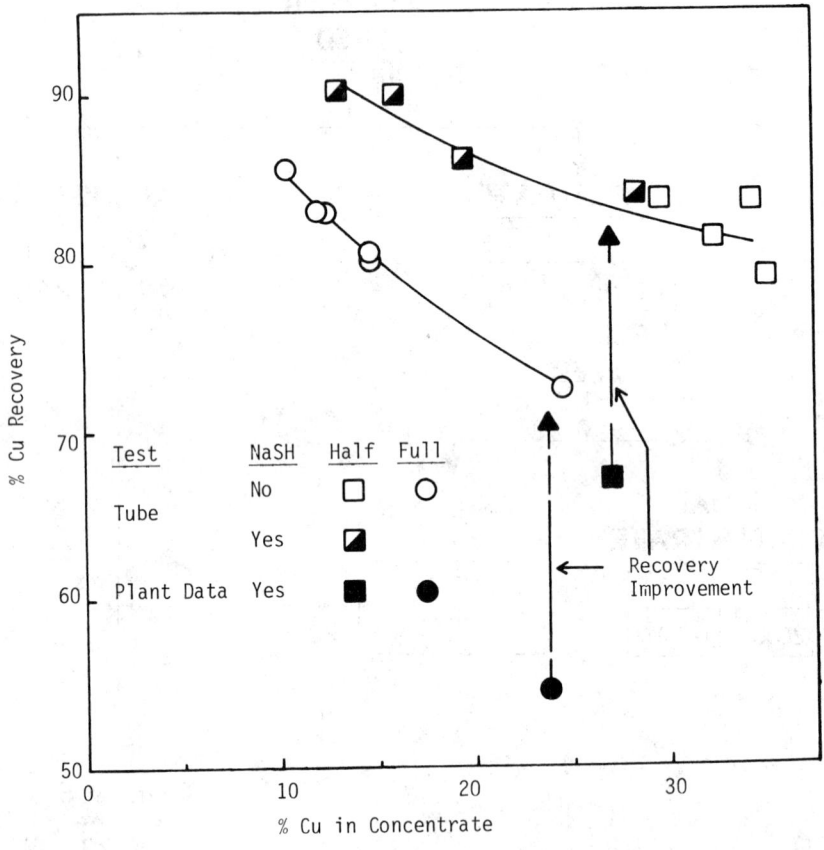

Figure 7 – Comparison of the tube flotation with the
plant data on White Pine cleaner feed.

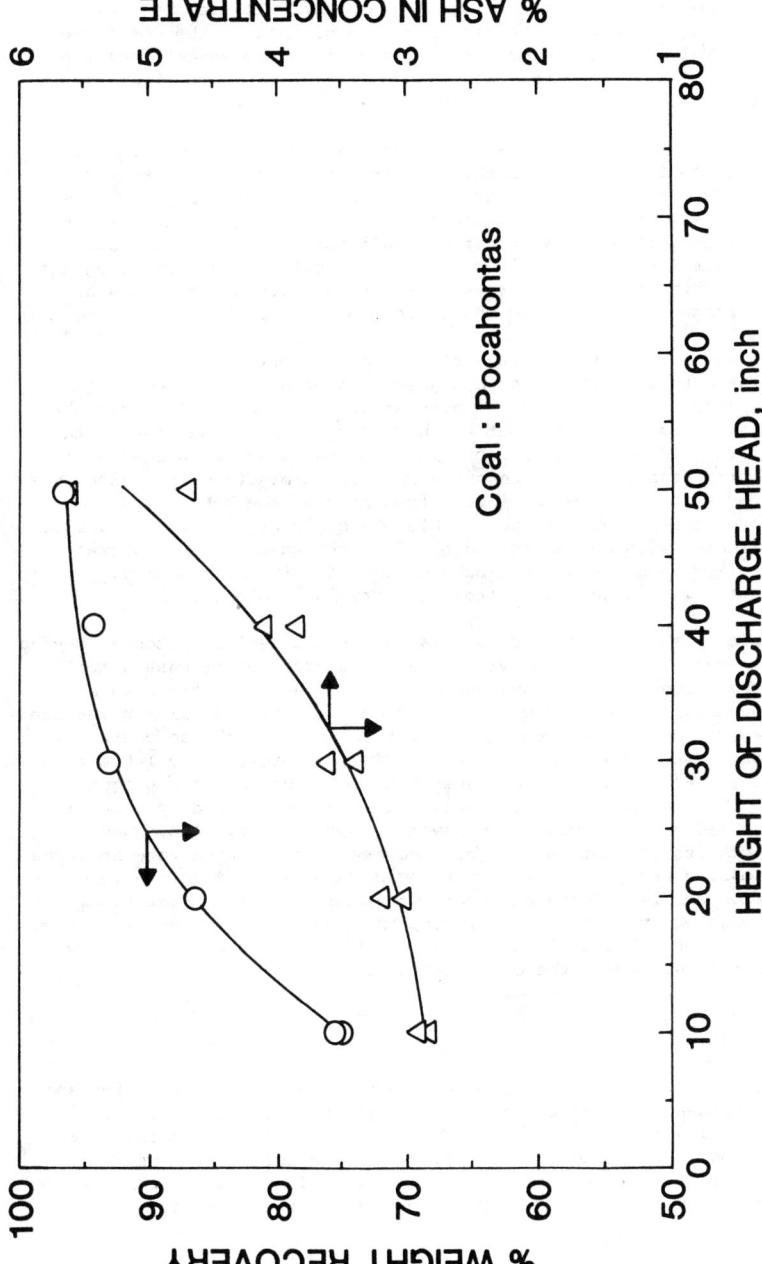

Figure 8 – Effects of discharge head on tube flotation.

Air flow rate. Since the cross-sectional area of the tube machine might be different, air flow rate cannot be used alone as a standard. The term, superficial air velocity, was introduced to overcome this problem. It is defined as the air flow rate divided by the effective cross-sectional area of the machine. For a fixed cross-sectional area and constant bubble diameter, increasing air flow rate also increases bubble population and thus available bubble surface area.

Figure 9 shows some typical data obtained from a series of flotation tests with changing air flow rate. Within the experimental range of the present study, the weight recovery increased with increasing air flow rate to some extent, which is possibly the effective capacity for the tube flotation machine at a specified condition. However, the ash content in the froth product increased continuously with increasing air flow rate. This is due to increasing water recovery and the flow of entrained gangue minerals into the froth product.

Wash water. It is well understood that the entrained gangue minerals have to traverse through a great number of interconnecting tortuous paths in the packing and also around the bubbles to reach the tailings discharge end. Without froth washing or spraying, the froth becomes more arid as it rises along the flow paths while bubbles shrink as the slurry drains away. The entrained gangue minerals then have less chance to drain away from the interstices between the bubbles. Furthermore, the probability for bubbles to coalesce with each other or burst increases with decreasing bubble film thickness. The wash water keeps the flow channels between bubbles open and in the meantime, displaces the entrained slurry from the uprising froth bed.

Figure 10 shows some results obtained from a series of tests varying wash water rate from 0 to 220 ml/minute. The addition of wash water affects the weight recovery and ash content of the concentrate in a complex manner. Increasing wash water from 0 to about 40 ml/min resulted in decreasing both ash content and weight recovery in the concentrate. Further increase in wash water rate from 40 to approximately 140 ml/min lowered the ash content in the concentrate but increased the weight recovery. This complex effect of wash water may be caused by a partial depression of the clean coal with lower hydrophobicity. Above 140 ml/min, both ash content and weight recovery of the concentrate increase with increasing wash water rate until reaching a limited void volume for slurry drainage, i.e., machine saturation point at a given air rate. When the wash water exceeds this limit, the excess wash water will flood out from the concentrate discharge outlet which would cause an adverse effect on the quality of the clean product.

Modeling

A model of the flotation process can be very useful for design and control purposes. For design purposes it can be used to predict the performance of a large plant from laboratory and pilot plant data. For control purposes the model can be used to assess the utility of any proposed operating modification to an existing plant, and it can also be used to determine optimal control settings for best operating performance.

In 1962 Arbitar and Harris proposed a simple model for the cumulative recovery of an n-cell conventional flotation bank (6),

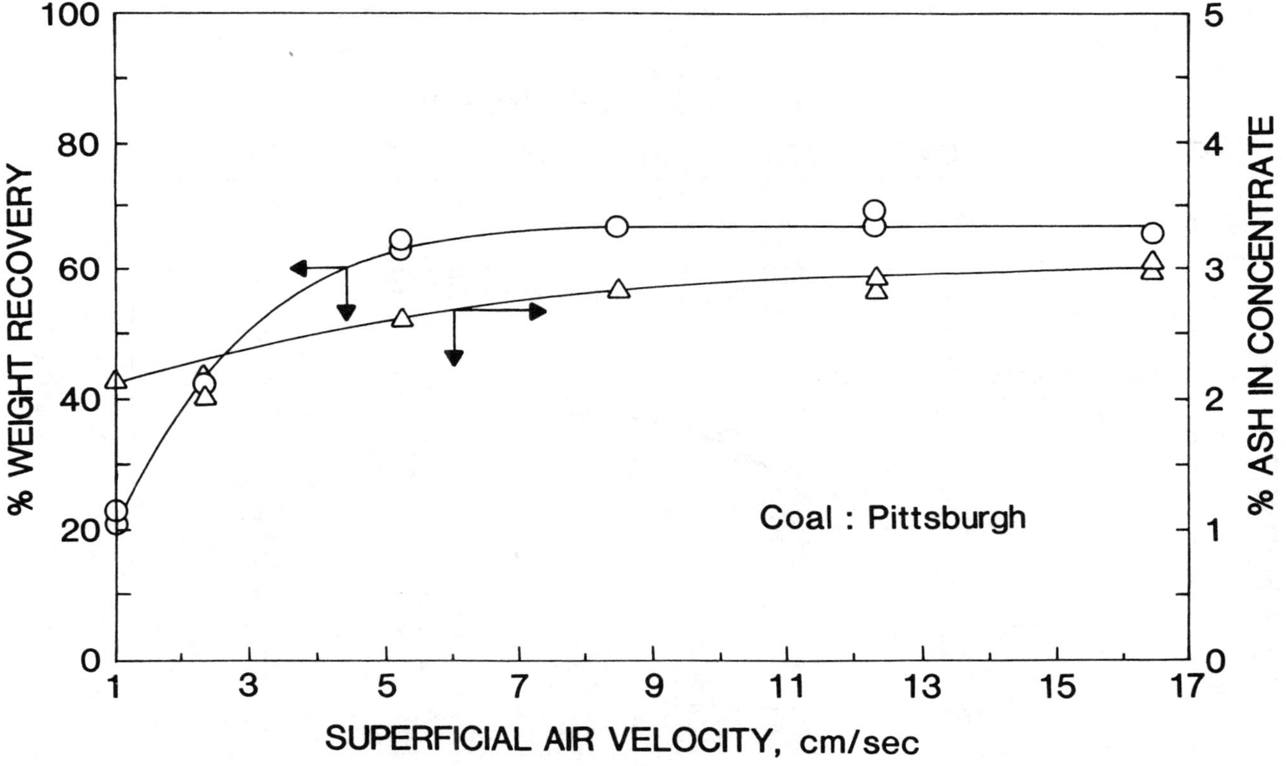

Figure 9 - Effects of air flow rate on tube flotation.

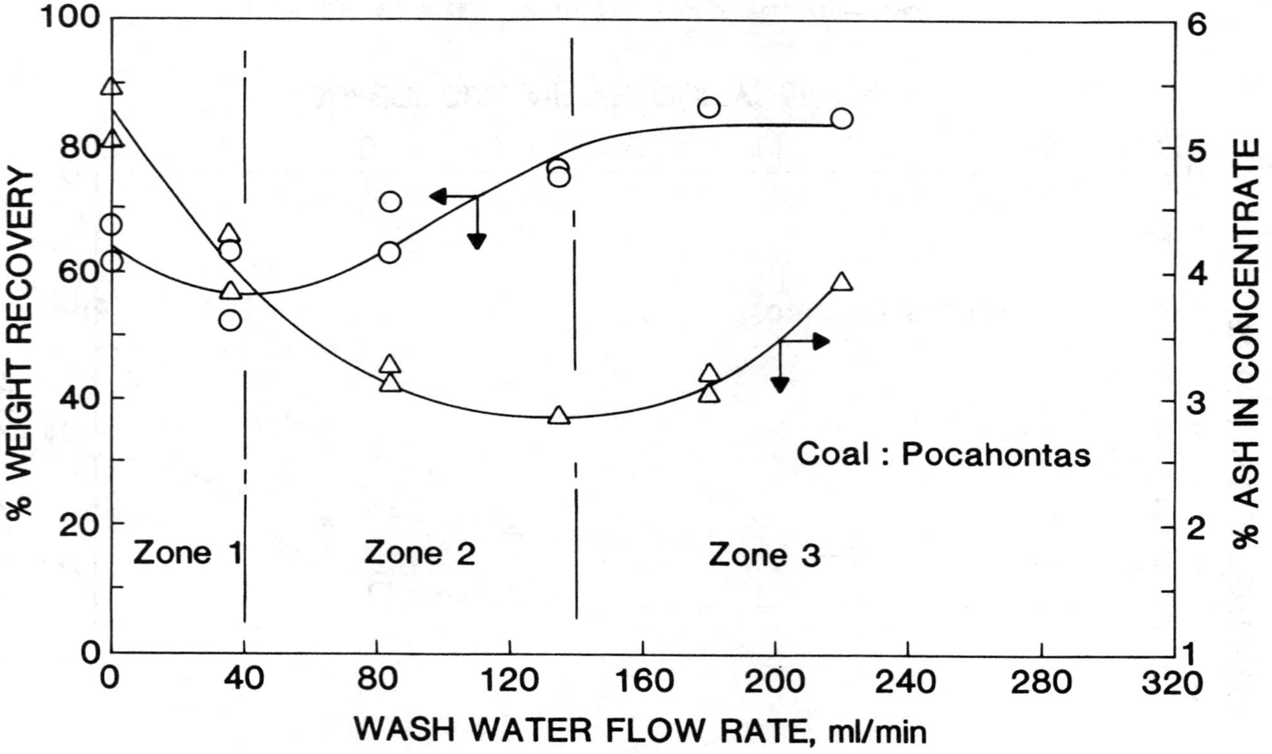

Figure 10 - Effects of wash water on tube flotation.

$$Rn = 1 - EXP (- K t n)$$ (1)

where K = the first order rate constant, and t = the nominal residence time of the slurry. Since the number of conventional cleaning or scavenging stages can be regarded as a length equivalent in a tube flotation machine, it is reasonable to assume the tube flotation machine as a stacking of a number of short flotation columns together. Each unit can be considered to be a perfectly mixed phase having a relatively short resident time. Therefore, an attempt was made to develop a preliminary model to fit the tube flotation machine based on the similar approach taken to derive Equation 1.

As already discussed, superficial air velocity and slurry residence time are two of the major controlling variables to affect the performance of coal flotation. As a first approximation, Equation 1 can then be written into the following form:

$$R = 1 - EXP (- K' Vs t)$$ (2)

where K' = the rate constant, and Vs = the superficial air velocity. As the measurement of the slurry residence time is not reliable for the existing setup, the height of the discharge head is considered instead. The validity of Equation 2 can be checked by a graphic method to see whether a linear relationship exists between the superficial air velocity or the height of discharge head with the logarithm of the fraction of floatable particles remaining unfloated. Figures 11 and 12 are such plots for the superficial air velocity and the height of the discharge head, respectively. In both cases, good straight lines can be plotted except that there is a sharp change in the slope when air velocity exceeding 5 cm/sec (Figure 11). This indicates that first order rate kinetics can be applied to the tube flotation machine, at least within the operating range below the machine flooding point.

Besides the residence time and the superficial air velocity, bubble size may be a critical factor in conventional columns for fine coal recovery. This parameter was regarded as a constant in the present work due to the unique packing design which requires neither moving parts nor air sparger. Other control variables such as frother concentration and corrugation of the packing elements would certainly play significant roles in the packed column system. Despite being held constant or neglected, these variables should be considered in the further study.

Summary

The key points of the present work can be summarized as follows:

1. The new tube flotation system is suitable for processing copper sulfide ore.

2. Single stage tube flotation system can replace both cleaner and recleaner circuits with better metallurgical performance.

3. No special arrangement or modification is needed to use this new flotation system.

4. Better results can be obtained by decreasing pulp density.

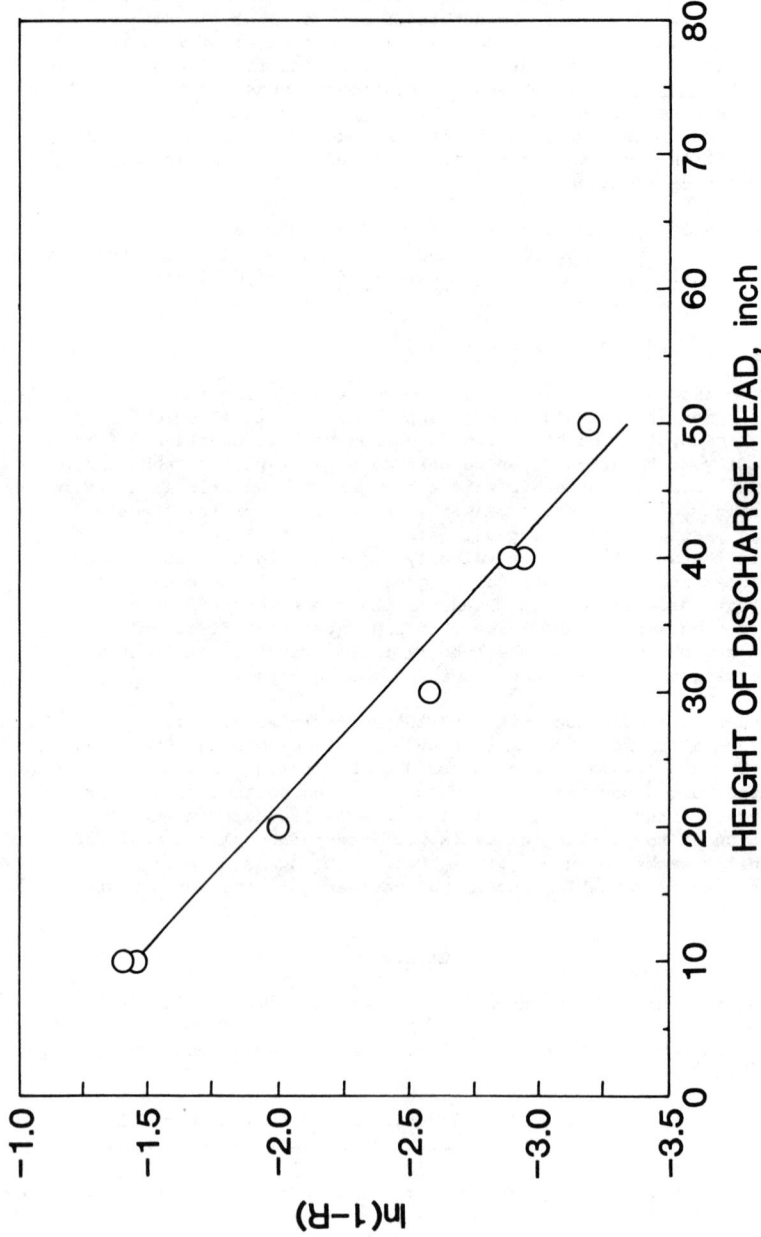

Figure 11 - Linear relationship of ln(1-R) with superficial air velocity.

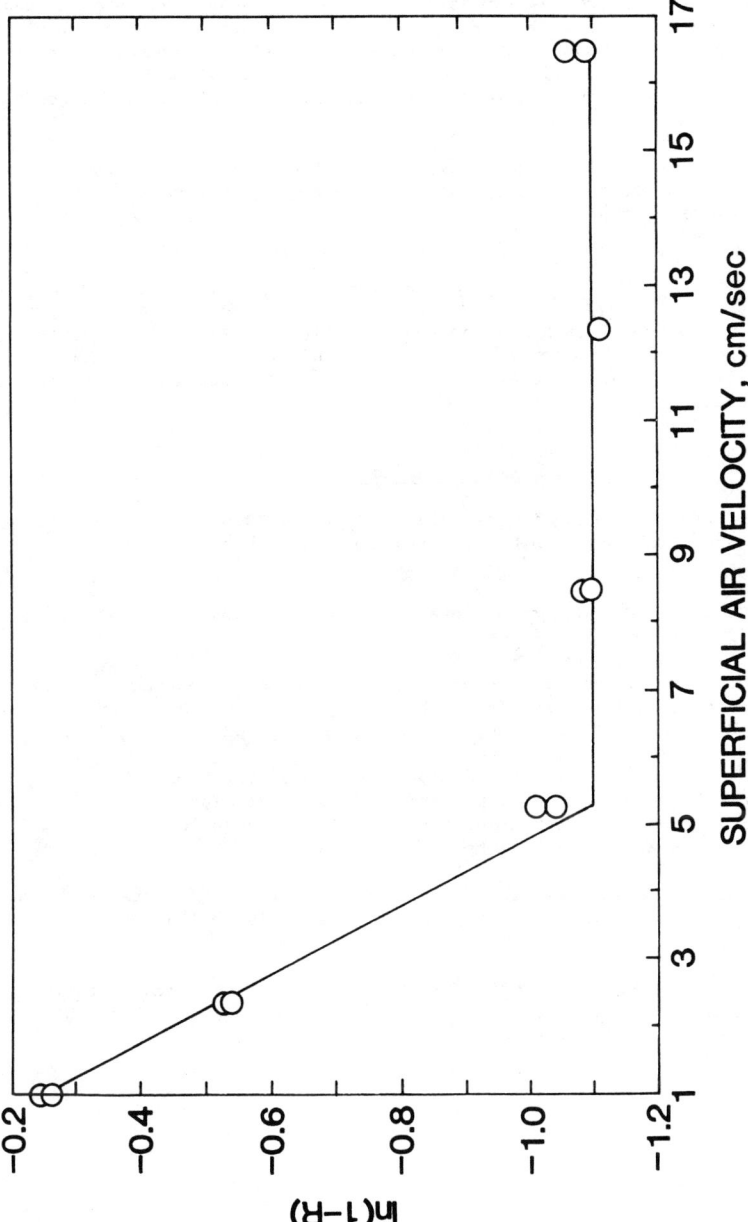

Figure 12 – Linear relationship of ln(1-R) with height of discharge head.

217

5. The effectiveness of NaSH in the recovery of copper has to be reevaluated. Some other reagents might be considered superior.

6. Three "control variables," elevation of the discharge head, wash water, and air flow rate were studied. The elevation of the discharge head is closely related to the slurry residence time while the air flow rate determines the population of air bubbles. The wash water has a significant effect on the grade of the final product.

7. Two "effective parameters," superficial air velocity and slurry residence time, were found to play significant roles in the packed column system.

Acknowledgements

The help from Messrs. Herman Ponder, Allen Trujillo, William Vlasak, and Joseph Rickard of Copper Range Company and Mr. Robert Trevethan of the Institute of Materials Processing, Michigan Technological University is gratefully acknowledged.

References

1. A.D. Brown, "Zoning in the White Pine Copper Deposit, Ontonagon County, Michigan" (Ph.D. Dissertation, University of Michigan, 1968).

2. D.C. Yang, "Column Froth Flotation," June 3, 1986, U.S. Patent 4,592,834.

3. D.C. Yang, "Static Tube Flotation for Fine Coal Cleaning," (Paper presented at the Sixth International Symposium on Coal Slurry Combustion and Technology, Orlando, Florida, June 26, 1984), 16.

4. D.C. Yang and C. Li, "A New Flotation System for Producing Superclean Coal," (Paper presented at the Sixth International Workshop on Coal-Liquid and Alternate Fuels Technology, Halifax, Nova Scotia, Canada, October 1, 1986), 15.

5. C. Li, "Process Analysis of A Static Tube Flotation System for Fine Coal Cleaning" (M.S. Thesis, Michigan Technological University, 1987).

6. N. Arbiter and C.C. Harris, "Flotation Kinetics," Froth Flotation 50th Anniversary Volume, D.W. Fuerstenau, editor, (New York, NY: SME/AIME, 1962), 215-246.

MODEL-BASED DESIGN OF COLUMN FLOTATION

M. J. Mankosa, G. T. Adel, G. H. Luttrell and R. H. Yoon

Department of Mining and Minerals Engineering
Virginia Polytechnic Institute and State University
Blacksburg, Virginia 24061

Abstract

Using the concept of the population balance, two models have been
developed to describe the dynamic and steady-state performance of a flotation
column. Flow conditions in the first model are represented by a series of
first order differential equations describing completely mixed zones, while
flow conditions in the second model are represented by a single partial
differential equation describing axially-dispersed plug flow. Separate
equations are used in both models to represent the behavior of three distinct
phases: air bubbles, unattached particles and bubble-particle aggregates.
Each model considers transport terms describing the movement of air bubbles
and solid particles due to fluid flow and bouyancy or gravity, and rate terms
describing the abrupt disappearance or appearance of material from one phase
into another due to bubble-particle attachment. The rate expressions used in
both models have been derived from the first principles of bubble-particle
adhesion. Dynamic and steady-state simulations are presented and are
compared to experimental results obtained from bench-scale column flotation
tests.

Introduction

Since their development in the early 1960's, flotation columns have often been reported to give improved recovery and product grade with a savings in floor space and capital cost over conventional flotation banks. This has been especially true when the processing of fine particles is involved. However, until recently, the acceptance of the flotation column by industry has been slow. Part of the reason for this slow acceptance can be attributed to the natural resistance to change that comes with the introduction of any new process. But perhaps a more important reason is the lack of operational and scale-up data which can be directly applied to the design of a flotation column.

Much of the early expertise in column design and operation was gained from in-plant testing. For example, tests conducted from 1963-65 at the Opemiska copper mine in Canada (1,2,3) revealed problems with plugging of the porous diffuser used for bubble generation. This led to a redesign of the bubble generator to one which could be easily replaced without shutting down the column. In addition, it was found that the metallurgical performance of the column could be greatly improved by increasing the column height to reduce internal mixing.

The first real attempt to develop fundamental design criteria for column flotation was presented by Sastry and Fuerstenau (4) in the form of a mathematical process model. Through the use of dimensionless numbers, they were able to relate several of the column operating parameters to the concentration gradient in the column and the internal mixing. Unfortunately, this model contained several theoretical parameters which were difficult to obtain experimentally, and therefore, no validation of the model was presented.

With the renewed interest in column flotation over the past few years, an increasing effort has been underway to develop a fundamental basis for the design and scale-up of flotation columns. Included in this effort is the recent work by Dobby and Finch (5,6) on column mixing and scale-up, and the dynamic column model developed by Luttrell et al. (7). The purpose of this paper is to present two population balance models describing the column flotation process. The first is based on a series of completely mixed zones and the second is based on the concept of axially-dispersed plug flow. Steady-state and dynamic simulation results are presented to illustrate the use of these models in column design, and the importance of internal mixing is discussed.

Model Development

Mixed Zone Model

In a first attempt to describe the flow patterns in a flotation column, the column has been represented as a series of well-mixed zones, the number of which depends on the desired height of the column. For each zone, a mass (or volume) balance has been carried out for each particulate phase present in the column. These phases include air bubbles, unattached particles and bubble-particle aggregates. Particulate solids have been further classified as either valuable or gangue material, with each having a different flotation rate constant and a different settling velocity. Perfect liberation of the valuable mineral from the gangue has been assumed, although it is possible to incorporate composite particles into the simulation. It has also been assumed that the size of particles and bubbles can be reasonably represented by a single mean value for each. Other factors such as particle agglomeration, bubble coalescence, bubble loading and bubble-particle detachment have not been included at this time. A summary of the model derivation is discussed below.

Volumetric Flow Balance. The schematic shown in Figure 1 illustrates the volumetric flow balance around each zone of the column. The five flows considered here include feed (Q_F), tailings (Q_T), gas (Q_G), wash water (Q_W) and product (Q_P). Of these flows, Q_F, Q_G and Q_W are input parameters, while Q_T and Q_P must be calculated. The feed flow (Q_F) is assumed to split into Q_P and Q_T as it enters the feed zone, after which these flows are translated throughout the upper and lower parts of the column, respectively. Likewise, Q_W is translated throughout the upper part of the column and Q_G throughout the lower part of the column. By using this approach it is possible to approximate mixing within the column.

The unknown flows of Q_P and Q_T are approximated by assuming a steady-state model of the froth and by conducting an overall volume flow balance. From a steady-state air balance around the froth zone (Figure 2), an expression for Q_P is given as:

$$Q_P = (U_b A \varepsilon_{i-1} + Q_P \varepsilon_{i-1})/\varepsilon_F \qquad (1)$$

where U_b represents the bubble rise velocity, A the cross-sectional area of the column, ε_{i-1} the volume fraction of air in the transition zone and ε_F the volume fraction of air in the froth. At present, the value of ε_F is obtained by considering that each bubble carries a "sheath" of slurry into the froth. The average thickness of the sheath (T) depends on parameters such as surface tension, superficial gas velocity, average bubble diameter, etc. From a simple geometric analysis, it can be shown that:

$$\varepsilon_F = \frac{D_b^3}{D_b^3 + 6D_b^2 T + 12D_b T^2 + 8T^3}. \qquad (2)$$

In the present work, T has been determined experimentally (7).

Once Q_P is known, Q_T is obtained from the overall volumetric flow balance given as:

$$Q_T = Q_G + Q_W + Q_F - Q_P. \qquad (3)$$

Thus, Equations [1] through [3] are used as supplementary equations to solve the air and solids balances discussed in the following sections.

Volumetric Air Balance. The terms necessary to describe the air volume balance between zones are shown schematically in Figure 2. The net change in the air content with time is determined by the difference of the volume of air per unit time entering a zone and the volume of air per unit time leaving a zone. For example, the volumetric flow of air which enters the reject zone consists of the air feed rate (Q_G) and the air carried into the zone from above by the downward flow of pulp in the column ($Q_T \varepsilon_{i+1}$). Flows leaving the zone include the volume flow of air to reject ($Q_T \varepsilon_i$), the volume flow displaced by the air which enters the zone ($Q_G \varepsilon_i$) and the flow due to the bouyancy of the bubbles ($U_b A \varepsilon_i$). Each of these volume flows is converted to fractional air content by dividing by the total volume of the zone (V_z). A balance of terms around the reject zone dictates that:

$$d\varepsilon_i/dt = [(Q_G + Q_T\varepsilon_{i+1})/V_z] - [((Q_T + Q_G + U_bA)\varepsilon_i)/V_z] \qquad (4)$$

which provides an expression for the change in the fractional air content in the reject zone with time. Volumetric balance equations for the other zones can be derived in the same manner.

Unattached Solids Balance. The terms considered in the unattached solids balance are represented schematically in Figure 3. In developing this balance, it has been assumed that there is a net amount of slurry which moves with the air in the column. The ratio of slurry to air (F) is determined from the volume fraction of air in the froth as follows:

$$F = \frac{1 - \varepsilon_F}{\varepsilon_F}. \qquad (5)$$

In order to determine the volume flow of slurry at any point within the column, F is multiplied by each term which describes the flow of air. Once the volume flow of slurry is known, it is multiplied by the concentration of unattached solids in each zone to determine the mass flow rate of unattached solids.

A second type of unattached solids flow which must be considered is that contribution which is carried down the column directly by Q_W and Q_T. These flows must be considered separately since they can be present even if Q_G and Q_p are zero. The mass per unit time carried by either of these flows is determined simply by multiplying the flow by the concentration of unattached solids in the appropriate zone.

In addition to the flow terms, there is a settling term and a rate term which must be considered. The settling term is very similar to the rise term used in the air balance. In this case, the Stokes equation for particle settling has been used. The rate term accounts for particles which suddenly disappear from the unattached phase and appear in the attached phase. It is the product of the attachment rate constant (k_a) and the mass of unattached particles in a given zone ($M_{f,i}$). The attachment rate constant is determined from:

$$k_a = \frac{6PQ_G}{\pi D_b D_c^2}, \qquad (6)$$

where P is the probability of collection, Q_G is the air flow rate, D_b is the diameter of the bubble and D_c is the diameter of the column. In the present work, P has been evaluated as a function of particle size, bubble size and hydrophobicity using a fundamental hydrodynamic analysis detailed elsewhere (8).

Upon converting particle concentration to mass (i.e., $C_{f,i} = M_{f,i}/V_z$), a mass balance around the reject zone yields:

222

$$dM_{f,i}/dt = [(Q_T\varepsilon_{i+1}F + Q_T + U_pA)/V_z]M_{f,i+1}$$

$$- [((Q_T + Q_G + U_bA)\varepsilon_iF + Q_T + U_pA)/V_z - k_a]M_{f,i} \quad (7)$$

A similar procedure can be used in handling other zones.

Attached Solids Balance. Figure 4 is a schematic representation of terms necessary to describe the movement of particles attached to air bubbles. Once again, consider the balance of terms around the reject zone. The mass flow of attached particles leaving the reject zone consists of those carried out by the tailings flow ($Q_TC_{a,i}$) and those carried upward by the bubbles and gas flow ($\{Q_G + U_bA\}C_{a,i}$). Attached particles enter the zone either by the downward flow of pulp in the column ($Q_TC_{a,i+1}$) or by bubble-particle attachment ($k_aM_{f,i}$). A balance of terms around the reject zone yields:

$$dM_{a,i}/dt = k_aM_{f,i} + [Q_TM_{a,i+1} - (Q_T + Q_G + U_bA)M_{a,i}]/V_z. \quad (8)$$

In the present work, particle detachment has not been considered because fine particles have a very low probability of detachment due to their low intertial force. Expressions for the other zones are obtained in a similar manner.

Solution Methodology. Before solving the mixed zone model, it is necessary to obtain values for two unknown model parameters, probability of collection (P) and froth film thickness (T). The experimental methods for determining each of these values have been discussed elsewhere (7).

The dynamic solution to the mixed zone model was obtained by applying the numerical Euler method. This technique has the advantage that it is simple to program; however, it can lead to large discretization errors if the step size is not sufficiently small. On the other hand, as the step size is decreased, the cumulative round-off error tends to increase. The cumulative error, however, did not appear to be substantial when the steady-state results of a dynamic simulation were compared to the true steady-state solution.

The steady-state solution to the mixed zone model was obtained by setting the left-hand side of each differential equation to zero, and solving the resulting set of simultaneous algebraic equations. A resubsitution technique was used for this purpose. Using this approach, a solution could be obtained in less than one minute with compiled BASIC on an IBM-PC.

In presenting the simulation results, the total mass flow of valuable mineral was determined by the summation of the mass of unattached and attached mineral reporting to either the product or reject streams. Combining this solution with a similar analysis for gangue allowed both the grade and recovery of the product and reject streams to be calculated. Other values, such as percent solids in the product and reject streams, air and liquid flow rates, etc. can also be determined.

Dispersion Model

In the mixed zone model, mixing in the column is approximated by the opposing flows of Q_G and Q_T in the bottom of the column, and Q_P and Q_W in the top of the column. Unfortunately, these flows must be varied in order to make a change in the mixing conditions. It is quite commonly known, however,

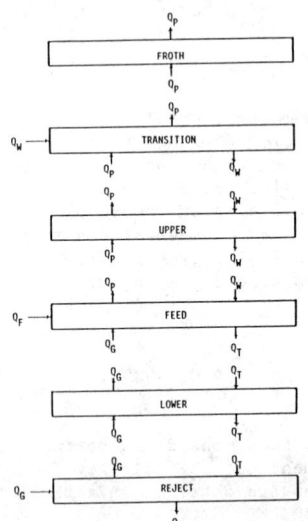

Figure 1 - Schematic representation of the volumetric flows within the flotation column.

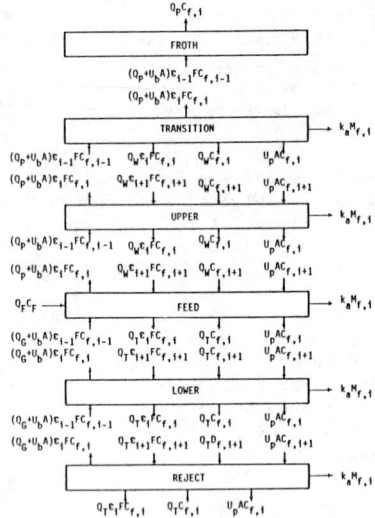

Figure 3 - Schematic representation of the transport of unattached solids within the flotation column.

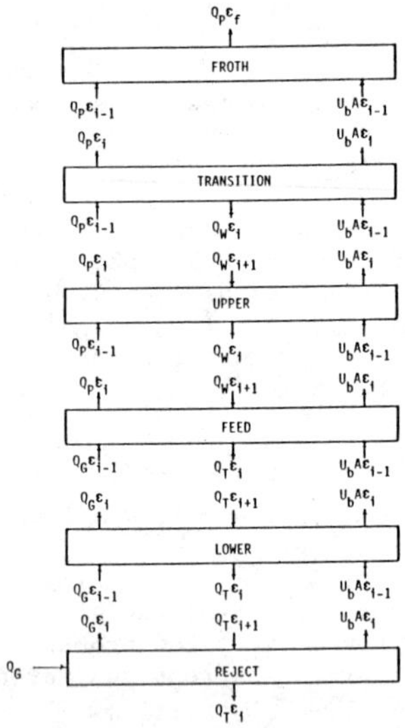

Figure 2 - Schematic representation of the transport of air within the flotation column.

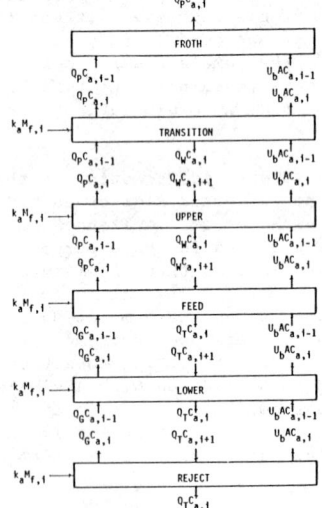

Figure 4 - Schematic representation of the transport of attached solids within the flotation column.

that column geometry can affect mixing without the need to change any of the external flows. In order to incorporate mixing on a more fundamental basis, the axially-dispersed plug flow model was used. The general form for this model is given as:

$$\frac{\partial C}{\partial t} = D_\ell \frac{\partial^2 C}{\partial z^2} - U_z \frac{\partial C}{\partial z}, \tag{9}$$

where C represents the concentration of particulates in the phase of interest, U_z represents the superficial velocity of the particulates and D_ℓ represents the axial dispersion coefficient. It is important to note that this is a partial differential equation which is continuous both in time and in the z-direction (length) of the column. Thus, the discretization error associated with the mixed zone model is eliminated.

Since the superficial velocity (U_z) changes above and below the feed point, it is necessary to write two equations for each phase. For the air phase the model equations are given by:

$$\frac{\partial \varepsilon}{\partial t} = D_\ell \frac{\partial^2 \varepsilon}{\partial z^2} - [U_b - (Q_T - Q_G)/A] \frac{\partial \varepsilon}{\partial z} \tag{10}$$

and

$$\frac{\partial \varepsilon}{\partial t} = D_\ell \frac{\partial^2 \varepsilon}{\partial z^2} - [U_b + (Q_P - Q_W)/A] \frac{\partial \varepsilon}{\partial z}. \tag{11}$$

The same form can be used to describe the unattached solids phase by replacing U_b with U_p and adding a rate term for disappearance into the attached solids phase. These two equations are as follows:

$$\frac{\partial C_f}{\partial t} = D_\ell \frac{\partial^2 C_f}{\partial z^2} + [U_p + (Q_T - Q_G)/A] \frac{\partial C_f}{\partial z} - k_a C_f \tag{12}$$

and

$$\frac{\partial C_f}{\partial t} = D_\ell \frac{\partial^2 C_f}{\partial z^2} + [U_p - (Q_P - Q_W)/A] \frac{\partial C_f}{\partial z} - k_a C_f. \tag{13}$$

Finally, the description for the attached solids is given by:

$$\frac{\partial C_a}{\partial t} = D_\ell \frac{\partial^2 C_a}{\partial z^2} - [U_b - (Q_T - Q_G)/A] \frac{\partial C_a}{\partial z} + k_a C_f \tag{14}$$

and

225

$$\frac{\partial C_a}{\partial t} = D_\ell \frac{\partial^2 C_a}{\partial z^2} - [U_b + (Q_P - Q_W)/A]\frac{\partial C_a}{\partial z} + k_a C_f. \tag{15}$$

In order to solve this model, it is once again necessary to make use of auxilliary equations [1 - 3] for determining Q_P and Q_T. It is also necessary to determine the collection probability (P) and the froth film thickness (T) as discussed earlier. The other parameter which can be experimentally determined is D_ℓ. In this work, a tracer of salt solution was used to determine the residence time distribution in a flotation column. From the resulting RTD curve, the mean residence time and the axial dispersion coefficient were obtained. The equations used for fitting the RTD data are discussed in more detail in the following section. Finally, the model equations were solved numerically using the technique of finite differencing.

Results and Discussion

Steady-State Simulations

Using the steady-state solution to the mixed zone model, a series of simulations were conducted to determine the effects of various operating and design parameters on the product recovery and ash obtained during the flotation of fine coal. The conditions under which these simulations were conducted are given as follows:

Superficial Feed Velocity	=	15.4 cm/min
Superficial Gas Velocity	=	19.7 cm/min
Feed Percent Solids	=	5.3%
Feed Percent Ash	=	36.4%
Mean Particle Size	=	5.5 μm.

The unknown parameters of froth film thickness (F) and particle collection probability for coal (P), as determined from experimental data, were 4.9 microns and 0.00028, respectively (7). The particle collection probabilty for mineral matter was assumed to be zero.

As a result of several simulations, it was determined that column length to diameter ratio, bubble diameter and wash water addition rate had the largest influence on product grade and recovery. The effects of these three parameters are illustrated in Figures 5 - 8. Figure 5 shows the relationship between recovery and bubble diameter (D_b) for various column length-to-diameter ratios (L/D) in the absence of wash water. As can be seen, recovery increases significantly as bubble size is reduced. This is a direct result of the increase in the number of bubbles and the probability of collection (P) with decreasing bubble size. The latter effect has been discussed in detail elsewhere (9). An increase in recovery is also observed as L/D increases, although this increase is not as significant as that produced by decreasing the bubble size. This effect is primarily a result of the increased residence time which provides more opportunity for bubble/particle collision.

A corresponding relationship to that discussed above is illustrated in Figure 6 for product ash. As shown, the ash content tends to decrease with decreasing bubble size down to a diameter of approximately 250 microns, after which, the ash content tends to increase. The intial decrease in ash content with decreasing bubble diameter is largely due to the increase in P discussed previously. Since the mineral matter in this simulation has been considered to have no floatabilty, an increase in P should increase the selectivity. Below a bubble size of approximately 250 microns, however, the increased water recovery due to the large number of bubbles increases the nonselective entrainment of mineral matter and the ash content of the product increases. Thus, the simulations predict that operating at a bubble size between 200 and 300 microns should give the best conditions for obtaining a low ash product at a relatively high recovery. The results shown in Figure 6 also suggest that increasing the column L/D ratio has a significant effect on the product ash. In Fact, under the conditions employed for this simulation, increasing the L/D ratio has a more significant effect on product quality than on recovery.

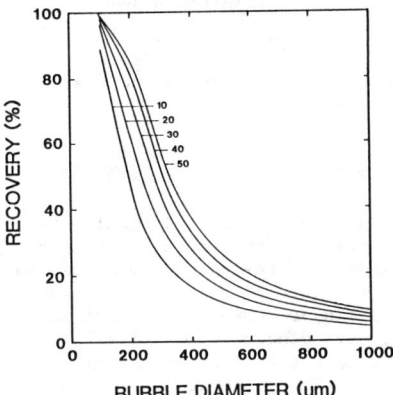 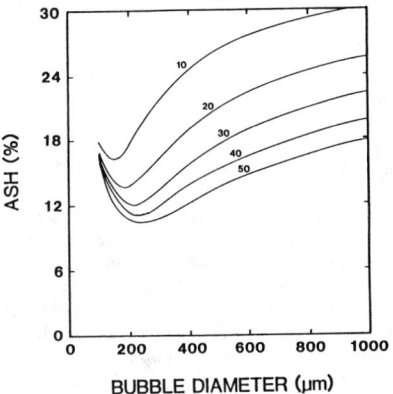

Figure 5 - Recovery as a function of bubble diameter for different L/D ratios in the absence of counter-current wash water.

Figure 6 - Product ash as a function of bubble diameter for different L/D ratios in the absence of counter-current wash water.

When countercurrent wash water is added just below the froth/pulp interface, little change is seen in the shape of the recovery curve as shown in Figure 7, although the recovery values are somewhat lower. In this case, wash water was added at a superficial velocity of 2 cm/min which lowered the recovery values approximately 5%. The effect of wash water on product ash, however, is considerably more significant. As shown in Figure 8, the use of countercurrent wash water reduces the product ash content for larger values of D_b and L/D to less than 1%. If a typical superficial wash water velocity of approximately 20 cm/min is used, nearly all ash entrainment can be suppressed.

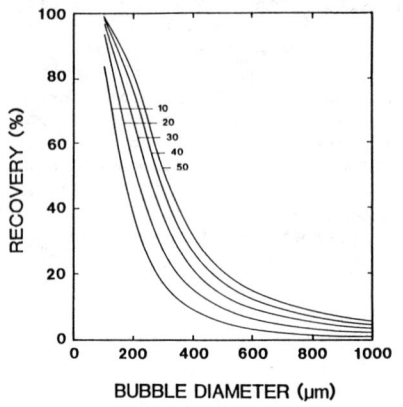

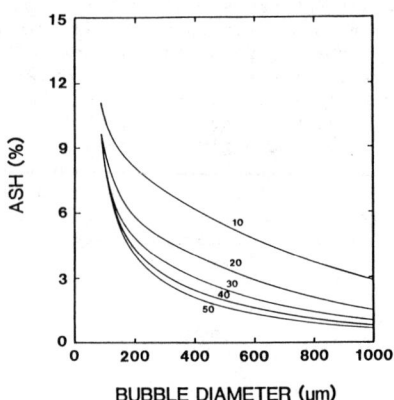

BUBBLE DIAMETER (μm)

BUBBLE DIAMETER (μm)

Figure 7 - Recovery as a function of bubble diameter for different L/D ratios in the presence of counter-current wash water.

Figure 8 - Product ash as a function of bubble diameter for different L/D ratios in the presence of counter-current wash water.

Dynamic Simulations

The dynamic solution to the mixed zone model was used to determine the time required to achieve steady-state from a start-up condition. This can also be used as an indication of the time needed for the column to stabilize when it encounters a disturbance. As shown in Figure 9, the recovery appears to reach steady-state quite rapidly (i.e. < 5 min.). However, the product ash content requires nearly 20 minutes to achieve steady-state. The simulated results were found to be in good agreement with experimental results produced in a 1-inch laboratory column. The implication of this finding is that if the column is operated for a short time, biased results can be produced which will show a much cleaner product than can be achieved under steady-state conditions. Thus, it is important to monitor the product grade rather than recovery when determining steady-state for flotation column testing.

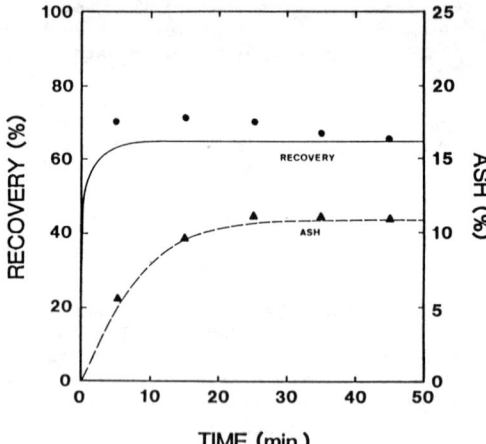

TIME (min.)

Figure 9 - Simulated (lines) and experimental (points) recovery and product ash as functions of time.

228

Column Mixing

As mentioned previously, one of the main disadvantages of the mixed zone model is the absence of a term which incorporates mixing from a fundamental perspective. However, the state of mixing in a column can play a major role in determining the recovery. Figure 10 shows recovery as a function of the product of flotation rate constant and mean residence time ($k\tau_m$). This latter term is a dimensionless quantity which combines the fundamentals of particle collection with the capacity of the column. The recovery values for the extreme conditions of plug flow and perfect mixing were determined from standard equations available in the literature (10). As shown, plug flow conditions provide nearly 20% improvement in recovery over perfectly mixed conditions when the value of $k\tau_m$ is low. Only at high values of $k\tau_m$ do the two curves begin to come together. Thus, it appears advantageous to make the column as plug flow as possible.

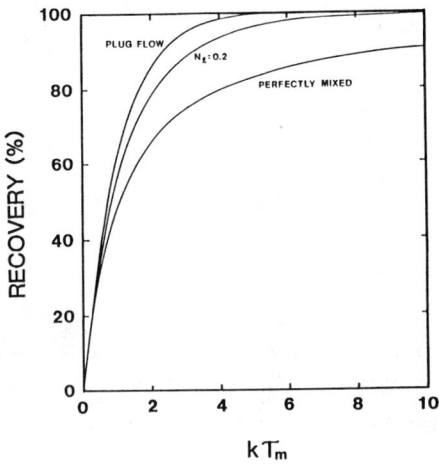

Figure 10 - Recovery versus $k\tau_m$ for perfectly mixed, plug flow and axially-dispersed flotation columns.

In order to incorporate mixing into the column model, the axially-dispersed plug flow model was used. This form allows mixing to be represented as a dispersion number or dispersion coefficient which can be determined from a residence time distribution. Figure 11 shows experimentally determined residence time distributions for a 2-inch diameter microbubble flotation column being tested at Virginia Tech. The ordinate is represented by dimensionless concentration (C/C_o) and the abscissa is represented by dimensionless residence time (τ/τ_m). As shown, the column becomes more plug flow as L/D ratio increases; however, there appears to be a substantial amount of mixing present for all cases shown in Figure 11.

The dispersion numbers (N_ℓ) for each of the residence time distributions were determined by fitting a standard mixing equation to the curves in Figure 11. In this case, the equation which applies is given by (11):

$$\frac{C}{C_o} = \frac{1}{2}[1 - erf\{\frac{1}{2}(\frac{1}{N_\ell})\frac{1 - \tau/\tau_m}{\tau/\tau_m}\}]. \tag{16}$$

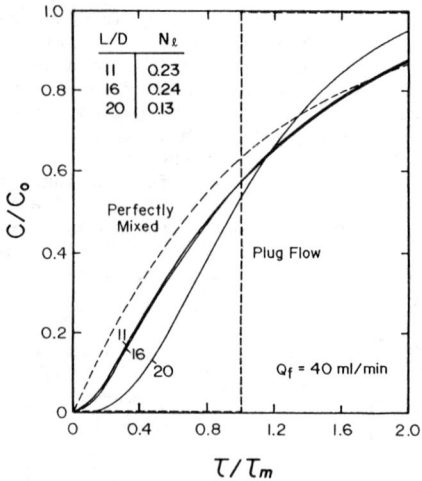

Figure 11 - Effect of L/D ratio on residence
time distribution in a flotation column.

As shown in Figure 11, the dispersion number becomes more plug flow as L/D
increases, although a decrease in the dispersion number of nearly two orders
of magnitude would be required in order to make the RTD appear close to the
plug flow condition. On the other hand, when a typical dispersion number of
0.2, as determined experimentally for the microbubble flotation column, is
plotted on the recovery vs. $k\tau_m$ diagram (Figure 10), the curve is much closer
to the plug flow condition. In this case, the recovery for the intermediate
mixing condition was determined from an equation available in the literature
(12). Thus, it appears that a small change in mixing can have a large effect
on the recovery in a column.

Model Comparison

    Using a dispersion number of 0.2, the axial dispersion model was used to
simulate the air phase in the column. The dispersion coefficient (D_ℓ) for
the model was determined from the relationship (12),

$$D_\ell = N_\ell u_\ell L \tag{17}$$

where u_ℓ is the interstitial liquid velocity and L is the length of the
collection zone in the column. Figure 12 shows the steady-state result of
this simulation as compared to the result obtained with the mixed zone model.
The two models appear to agree quite closely although the dispersion model
shows a more gradual change along the length of the column since it does not
have the discretization error present in the mixed zone model. Furthermore,
mixing in the dispersion model can be easily varied without the need to
change other flows in the column. Similar comparison simulations are in
progress for the unattached and attached solids phases.

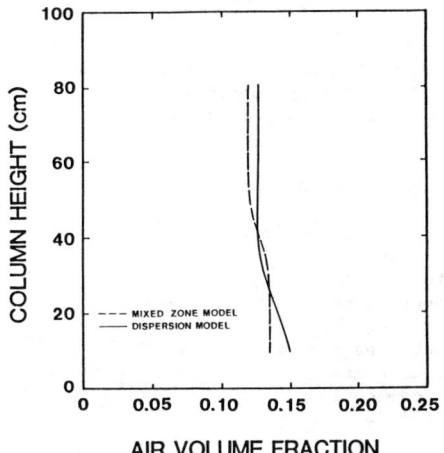

AIR VOLUME FRACTION

Figure 12 - A comparison of the air profiles produced
using the mixed zone model and the dispersion model.

Conclusions

1. Two population balance models for column flotation have been
   presented. The first model (mixed zone model) can be solved with a
   relatively simple numerical procedure but contains discretization
   error. The second model (dispersion model) incorporates fundamental
   mixing theory and eliminates much of the discretization error of the
   mixed zone model.

2. Steady-state simulations of fine coal cleaning in a flotation column
   indicate that the use of small bubbles and tall columns produces the
   highest recoveries, with bubble size having the most significant
   effect. Countercurrent wash water provides the best means for
   eliminating the entrainment of mineral matter, although under the
   conditions employed here, column height has a significant effect on
   product ash.

3. Dynamic simulations of fine coal cleaning in a flotation column
   indicate that the recovery reaches an apparent steady-state much
   sooner than the product ash. Care must be taken when collecting
   samples from a flotation column to ensure that true steady-state has
   been attained.

4. Mixing in the column can play an important role in determining the
   column performance. Theoretical predictions indicate that recovery
   is improved as the flow pattern in the column becomes closer to plug
   flow.

Acknowledgements

The authors would like to acknowledge the United State Department of
Energy for the support of this work through the University Coal Research
Program (Grant No. DE-FG22-83PC60806) and Contract No. DE-AC22-86PC91221.
They would also like to thank Beth Dillinger for assisting in the preparation
of the manuscript.

Nomenclature

A - column cross-sectional area
C - concentration of particulate of interest
C_a - concentration of attached particles
C_f - concentration of unattached particles
D_b - bubble diameter
D_c - column diameter
D_ℓ - liquid axial dispersion coefficient
ε - volume fraction air
ε_f - volume fraction air in the froth
F - slurry to air ratio
k_a - attachment rate constant
L - length of recovery zone
M_a - mass of attached particles
M_f - mass of unattached particles
N_ℓ - liquid dispersion number
P - probability of collection
Q_F - volumetric feed flow rate
Q_G - volumetric gas flow rate
Q_P - volumetric product flow rate
Q_T - volumetric tailings flow rate
Q_W - volumetric wash water flow rate
t - time
T - sheath thickness
τ - residence time
τ_m - mean residence time
U_b - bubble rise velocity
U_ℓ - interstitial liquid velocity
U_p - particle settling velocity
U_z - superficial velocity of particulates
V_z - zone volume

References

1. P. Boutin and D. A. Wheeler, "Column Flotation Development Using an 18 Inch Pilot Unit," Can. Mining J., 94 (1967) 101.

2. P. Boutin and D. A. Wheeler, "Column Flotation," World Mining, 67 (1967) 50.

3. D. A. Wheeler, "Big Flotation Column Mill Tested," Engrg. Mining J., (1966) 98.

4. K. V. S. Sastry and D. W. Fuerstenau, "Theoretical Analysis of a Countercurrent Flotation Column," Trans. SME-AIME, 247(1)(1970), 46-52.

5. G. S. Dobby and J. A. Finch, "Mixing Characteristics of Industrial Flotation Columns," Chemical Engineering Science, 40(7)(1985), 1061-1068.

6. G. S. Dobby and J. A. Finch, "Flotation Column Scale-Up and Modeling," Can. Inst. Mining Bulletin, 79(889)(1986), 89-96.

7. G. H. Luttrell, G. T. Adel and R. H. Yoon, "Modeling of Column Flotation," (116th Annual Meeting of SME-AIME, Denver, Colorado, 1987), Preprint No. 87-130, 9.

8. G. H. Luttrell, "Hydrodynamic Studies and Mathematical Modeling of Fine Coal Flotation" (Ph.D thesis, Virginia Polytechnic Institute and State University, 1986).

9. R. H. Yoon and G. H. Luttrell, "The Effect of Bubble Size on Fine Coal Flotation," Coal Preparation, An International Journal, 2 (1986), 179.

10. A. J. Lynch et al., Mineral and Coal Flotation Circuits - Their Simulation and Control, (Amsterdam: Elsevier, 1981), 42.

11. J. M. Smith, Chemical Engineering Kinetics, (New York, NY: McGraw-Hill, Inc., 1970), 257.

12. O. Levenspiel, Chemical Reaction Engineering, (New York, NY: John Wiley & Sons, Inc., 1972), 286.

SENSORS AND INSTRUMENTATION

Session Chairmen
John L. Watson, University of Missouri-Rolla
Keith Prisbrey, University of Idaho

This session reviews the demanding service requirements for sensing devices and instrumentation in mineral processing plant environments. On-line reagent analysis in flotation systems, in-situ monitoring of chemical parameters in sulfide circuits, in process particle size characterization, and application of acoustic sensors for both pulp leel maintainence and grinding mill optimization are thororoughly covered in field testing reports.

MEASUREMENTS OF AQUEOUS SOLUTIONS CHARACTERISTICS

IN MINERAL PROCESSING PLANTS

G. Barbery* and J.L. Cécile**

*Department of Mining and Metallurgy, Laval University
Québec, Canada G1K 7P4

**Mineral Technology Department, BRGM
B.P.6009 45069 Orléans Cédex, France

Abstract

Aqueous solution characteristics (pH, Eh, concentration in various reagents..) are of primary importance in processing plants where reagents are added to modify mineral properties. Flotation of sulphide ores and gold ore cyanidation are taken as examples of processes for which solution composition is desired for process stabilization and optimal reagent addition. A review is given of the state of development of in-line intruments to be used for these processes. Oxidation potential and xanthate residual concentration can be measured with instruments available from some manufacturers, either in slurries or in clear solutions extracted from the slurries. Measurement of free cyanide concentration in gold processing plants is now possible with some techniques which all involve extraction of a clear solution and either spectometry or potential measurements. The needs of extensive instrument testing and development of process control strategies that incorporate these measurements are pointed out.

Introduction

A large number of mineral separation processes involve modifications of mineral properties due to the reaction of the minerals with reagents contained in aqueous solutions. Flotation and hydrometallurgical processing are typical examples that will be considered in the present paper. The determination of the characteristics of the aqueous solution composition at various points in such processes is an obvious application of instrumentation in mineral processing plants, since the extent and the rate of the modification depends upon the concentration of active species in solution. For some species, proton activity for example, practicing engineers carry out measurements without major problems. In the general case however, and for applications as major as cyanide concentration control in gold processing plants, progress in industrial applications over the past twenty years has been slow. The present paper reviews the state of the art in two fields of activity: sulfide flotation reagent control and free cyanide determination in gold processing plants.

Reagents control in the flotation of sulphide ores

The authors have presented a paper at the Arbiter Symposium in 1986 (1), in which they reviewed the various methods in use for reagent control in flotation. In particular they made an analysis of the development of automatic control of flotation; they pointed out to the fact that the developments which have taken place in the Western world in that field were related to the availability of reliable solid composition analyzers. Mineral processing engineers working in plants instrumented with such analyzers have a means to monitor the evolution in the solids assays at various points in their circuits; fairly complex automatic control schemes have developed for flotation, some of which include actions to be taken for reagents additions based on the actual circuit response. Both feed forwards and feed back techniques have been developed, see references in (1). The applications of some of the methods give very good results in practice, but some authors, for example Lynch et al. (2), indicate that the determination of the proper action to take for reagent addition is not simple, especially for ores in which various types are present, each requiring its own combination to obtain selectivity.

It is thus not too surprising that various researchers have considered aqueous solution monitoring in order to maintain and optimize the conditions of selectivity. Woodcock (3) has presented a review of the methods available for reagent measurements in flotation slurries. Russian workers, and in particular Abramov (4), have developed complex strategies for reagent addition control based on aqueous solution characterization. Companies such as Outokumpu (5), BRGM (6) and Licensintorg (7) are developing, or are making available to operating mining companies, systems for aqueous solutions monitoring.

The advantages to be gained by solution analysis, compared to other control strategies, have been made clear by Jones and Woodcock (8) and Barbery and Cécile (1). In particular the paper by Jones and Woodcock (8) presents a detailed analysis of the information contained in the evolution of solution characteristics as a function of ore changes. In front of all the positive elements which favour the analysis of solutions, few methods are reliable enough to have been developed at industrial scale. Obviously there are major problems in the developments, and some of the features of equipment limitations and problems will be assessed here.

As a general principle, the equipments should be made available for direct application in slurries. Few instrumental methods are amenable to this type of operation in the abrasive slurries that are characteristic of mineral processing plants. In fact, pH is the only solution feature which is measured routinely with few problems. Electrode fouling and cleaning is usually not too difficult with glass electrodes, and the measurements are considered to be reliable.

Another measurement that can be carried out simply with electrodes in slurries is Eh, and a great deal of activity is being developed, both a theoretical level (9) and in plants (5, 6) to quantify the Redox characteristics of slurries, or solutions in contact with solids in flotation streams, using either a mineral electrode or a platinum electrode. Slurry measurements are presented in a paper by Outokumpu (5). A technical pamphlet has been prepared by the same company to present the resulting instrument, called ELEXAN 80P for Eh measurement (10). The method is described in the paper (5) as a modification of a voltametric equipment (Voltameter 470 from the same company, which works on clear solutions, 10), but seems to be closer to potentiometry using special electrodes. No information is given in the paper, but the company states that the electrodes are made of a mineral clearly involved in the process, e.g. a chalcopyrite electrode would be used in chalcopyrite flotation (11). The company claims that it has some proprietary cleaning system for the electrode (10) based on the application of ultrasounds. The cleaning is carried out every 30 seconds.

Similar methods are being tested by the US Bureau of Mines: electrodes of platinum or sulphide minerals are immersed in slurries (12). The meaning of Eh measurements in slurries is a delicate subject, especially when the solution in contact with the slurry contains several redox systems. Rand and Woods (13) have pointed out to the great difficulty in relating electrode potential to species activities of Redox couples in complex systems. In the presence of conducting mineral particles, such as exist in sulphide flotation streams, the electrode might take the potential of the particles; this is a mixed potential which depends on the mineralogical composition of the particles in the slurry due to galvanic interactions.

The measurement of xanthate residual concentration is of major interest, since, in addition to pH and Eh, it represents information on adsorption mechanisms. Various attempts have been made to quantify this parameter. Barbery and Cécile (1) and Jones and Woodcock (8) have reviewed the various methods that have been put forwards to carry out the measurement in flotation slurries. At the present state of development, the following companies are developing xanthate sensors for in-line industrial use:

- Licensintorg, from the Soviet Union is offering a "Method and device for xanthate batching control" (7). The measurement principle for xanthate concentration is a potentiometric measurement in slurry with a sulphide electrode (probably an argentite electrode). The system incorporates Eh measurement by potentiometry. Xanthate concentrations in the range 0.25-30 mg/l (for butyl xanthate) are claimed. Optimization of xanthate addition can be made under two modes: stabilization (constant residual concentration), or optimization (xanthate residual is kept at a value which depends on oxidation potential). Abramov (4,14) has described various schemes for the optimization and control of solution parameters. The implications of such developments in Soviet Union are probably that a reliable instrument is available and that the technology may be acquired by companies on a world wide basis.

- Outokumpu, as described above, has put on the market the Elexan 80X (10) for the measurement of xanthate concentration. The electrode composition is not given, but the company states that it has tried various compositions (11). The measurement is made by collection voltametry. Although the original paper (5) indicates the value of the method, the company is evaluating the

239

performance of the instrument for potential and xanthate measurement in three Finnish mines (11). The claimed performance are the following: in slurry measurement, concentration range 0.5-100 mg/l, 60s measurement time, 5% accuracy. Electrode cleaning is the same as that described for potential measurement. The performance obtained is stated to be satisfactory, but there are still design problems, and the product is not marketed for control purposes. The company is ready to supply the equipment and guarantee the measured values, but does not assist customers in the use of the results(11).

- AMDEL (Australia) has been working for quite a few years on the development of a xanthate monitor, based on the initial work of Sullivan and Woodcock (15). Although various papers published in 1984 (8, 16) mention the release of the instrument on the commercial market, there has been apparently no further development in this field. Since the instrument was based on UV-VIS spectrometry, problems were probably encountered in the filtration of a clear solution from flotation slurries.

- BRGM-Instruments in France is developing a modular instrument for aqueous solution characterization in mineral processing streams, called PIRANA[R] (17). A version of this instrument is dedicated to xanthate concentration measurements. The basic feature of the equiment is a patented tangential filter, which extracts a clear solution from the slurry stream. A flowrate of 0.2 to 1 m³/h of slurry at a solid concentration of up to 65% by mass is used; the pressure differential in the filter unit (0.05MPa) is such that a 1% representative sample of the clear solution in the slurry is extracted (6). The clear solution characteristics are 99% transmission coefficient in white light. Conventional analytical techniques can be used on the clear solution stream. In the basic version of the instrument, pH, temperature and Eh are measured. In the xanthate measuring version, the clear solution from the electrochemical cell is sent to a UV-Vis spectrometer in which xanthate measurement is carried out. Figure 1 is a photograph of the complete instrument; the analytical instrumentation box is fitted on top of the filter unit. The filter sheet has to be changed on average once a week.

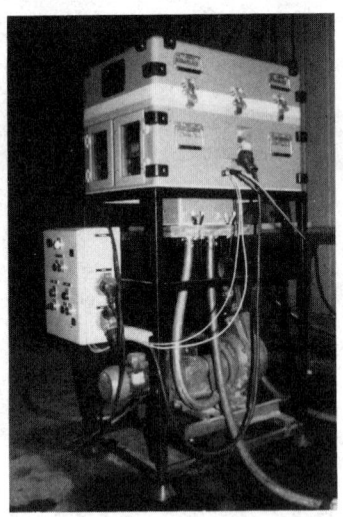

Figure 1 - PIRANA[R] sensor for pH, Eh,
temperature and xanthate measurements
in sulphide flotation slurries.

Replacement of the sheet takes 5 minutes. The equipment has been tested at pilot and industrial scale plants (6). Detection limit for xanthate is given at 0.5 umol/l. The commercial version of the instrument is available from BRGM-Instruments.

From the list of recent developments given above, it is clear that there are difficulties in solving all the problems involved, from measurement principle to the complete reliable industrial sensor. Moreover, the analytical instrument is not the final product, which should be the stabilization and optimization module for reagent addition and modulation of Eh potential. Apart from the equipment marketed by Licensintorg in USSR, it seems that there is not sufficient data available to provide the adequate information to potential users. Even at fundamental level, as Trahar (e.g. 18) has pointed out at various times, the information developed by electrochemical fundamental approaches to sulphide flotation is still limited to single sulphides, and does not lend itself to complete and general applications. Abramov (4, 14) is the only researcher who dares to give general relationships for various mineral separations. For further developments in the application of aqueous solution analysis in flotation control, and especially in the field of optimization, extensive work should be carried out in order to document the gains to be made using such instruments.

Measurement of cyanide concentration in gold processing plants

Hydrometallurgical processing of gold ores is almost universally carried out by cyanidation. Chemical reactions and reaction kinetics have been well studied, see for example Habashi (19). Optimization of cyanide addition is a problem in gold processing plants, since the consequences of having too little or too much cyanide are costly, either due to gold losses, or as indirect costs due to excessive cyanide consumption or destruction requirements. The control of cyanide addition is carried out in plants according to the well known titration technique with silver nitrate in the presence of potassium iodide. The method is manual, fairly tedious, and the determination of the end point is not always clear cut. Operators of gold processing plants are familiar with the method, and the instrumental developments which have taken place recently aim at improving the results that can be obtained with this procedure, by providing either an instrumental off-line method of measurement, or, more interestingly, an in-line or continuous industrial method or sensor.

Solution compositions in gold ores processing are complex. Oseo-Asare et al. (20) have presented a synthetic view of the reactions that can take place in the presence of various contaminants (such as iron and base metal salts resuting from sulphide minerals, almost always present in gold ores). The problems involved in analyzing complex cyanide solutions have been studied by various research groups, especially from the point of view of environmental studies. At the Workshop "Cyanide from mineral processing" (21) a paper was presented on "Analytical procedures for determining cyanide concentrations". Cyanide concentrations can be expressed as total cyanide, weak acid dissociable cyanide, cyanide amenable to chlorination and free cyanide. Usually, for the determination of cyanide in gold processing plants, free cyanide is the desired feature, since it is the concentration or activity which is involved in kinetic equations or in equilibria. It is this value which is measured in silver nitrate titration. The concentration measurement of all cyanide species in gold plant effluents is outside the scope of the present paper.

Instrumental method to measure free cyanide in gold plant solutions off-line have been tested over the years. The most important work carried out

241

recently appears to be that done at MINTEK in South Africa. A number of research reports have been published on the subject by Pohlandt-Watson (22, 23, 24). Flow Injection Analysis (FIA) (25) is demonstrated to be a very versatile method for the determination of free cyanide in the presence of ionizable metal cyanide complexes, the determination of cyanides derived from free cyanides and weak metal complexes, and the determination of "total" cyanides. Lynch (26, 27) has developed a method which combines flow-injection analysis and specific cyanide electrode. For such a procedure, as checked by Pohlandt-Watson (23), concentrations in free cyanide can be determined very reliably beween 20 and 1000 mg/l, a typical range in gold processing plants. Lynch obtained a rate of measurement of up to 120 per hour when the carrier solution contains traces of cyanide, a result that could not be reproduced by Pohlandt-Watson, who obtained a maximun of 60. Interferences have been shown to exist using cyanide electrodes: sulfide ions is given as a very deleterious constituent of solutions, since it prevents measurements to be carried out. The recommended practice would be to remove it from the solution stream by addition of a lead salt in the carrier solution. Thiocyanate and thiosulfate also interfere markedly with the cyanide electrode (23). These ions are present in large quantities in some gold processing operations. No study appears to have been carried out to remove their influence. The references mentioned above present work carried out on South African ores. Gold ores in South Africa are simple compared to the ores of, say, Canada, and direct transfer of analytical procedures should not be done without complete checks on the interference of various solutions constituents.

The conventional titration technique can also be automatized or replaced by a procedure which is amenable to in-line measurement. Various systems have been put forwards in which the method is spectrometry or potentiometry. It is in that field that the most interesting developments have taken place recently for plant applications.

An industrial instrument has been developed in South Africa for measuring free cyanide in plant solutions. It is described in a recent paper (28), and the instrument, CYANOSTAT, is distributed world wide by the Swiss company Polymetron (29). The principle of the instrument, developed by Gold Fields of South Africa since 1982, is to add to a clear solution containing cyanide, cupric ions in excess, and to measure the excess copper by complexing with EDTA. Copper-EDTA complex develops a blue color, and its concentration can be measured by spectrometry. The complete system thus consists of three parts: extraction of a clear solution from the slurry, reagents addition-mixing, microprocessor to process data and present results. Multi streams (up to six) can be processed in the same instrumental unit. Sample extraction and filtering is carried out in 2m long probes fitted with hollow polyehylene filter cartridges. Sampling is done in a batch mode consisting of four operations: a) the probe is put under vacuum and clear solution is withdrawn in to a filtrate receiving tank; b) compressed air is then admitted to the top of the filter probe finalizing filtrate transfer to the tank, and blowing off filter cake; c) primary filtrate is refiltered from the tank on its way the the analyzer sample holder d) a peristaltic pump withdraw a sample for the analysis. The critical part of the sample extraction unit is clearly the filter; cartridge replacement has to be carried out on average every week.

The CYANOSTAT system has been installed in a number of plants in South Africa, Australia and Papua New-Guinea. Links with Proscon controller from Outokumpu are described for one operation (28). The number of installations was the following, at the end of 1986: seven in South Africa, two in Autralia and one in Papua-New Guinea. Typical costs are given in (28); a capital cost of about $30,000 and a pay-back period of 3-4 months based on cyanide consumption reduction in a 6,000 tpd plant processing a typical South African gold ore.

Witteck Development Inc. (30) is developing an in-line sensor based on potentiomentry. The method is based on initial work by CANMET in the 1960's (31,32,33). The principle of the method is the following: a clear solution is extracted from a slurry, using a very simple porous polyethylene tube filter in which the slurry is made to circulate at slight over pressure (0.05 MPa) (31); metering pumps send the solution to be quantified and a titrating solution (containing silver nitrate, sodium hydroxide and ammonia) to a mixing container; the mixed solutions flow to the measuring section in which the potential between a silver electrode and a reference electrode is measured. A complete instrument based on this method was developed by CANMET in the early 60's (31); it was installed in Canadian plants and the results showed the value of the system. The filtering part is a good contribution to the problem of clear solution extraction from slurries. It has been tested at pilot and industrial scale; for a 0.5 inch diameter, 6 inches long filter in a 30 percent slurry, under 0.04 MPa pressure, a clear solution flows at a rate of 10ml per minute, sufficient for the subsequent instrument. The filter is said to be self cleaning, and retains its characteristics for up to one month (34).

Witteck Development Inc. has carried out recently the development and testing of an instrument based on similar principles (35). The on-line continuous cyanide monitor has been tested for weeks in Canadian gold processing plants, in early 1987. It has proven insensitive to the level of impurities found in these plants (especially with respect to base metal salts, oxidized sulphur species..). It is being currently commercialized.

BRGM-Instruments (17) is developing a version of the PIRANA[R] instrument for cyanide measurements.

Conclusions

The value of solution characterization in mineral processing plants has been demonstrated by various authors, in order to make use of the chemical information contained in these measurements, and to relate them to process control and optimization. Developments in the field are taking place rather rapidly, due to the incentives in process stabilization, reagent consumption reduction and innovations in analytical instrumentation. Although some organizations are developing sensors for in slurry measurements, which are the simplest to carry out, only one system, made in Soviet Union seems to have solved the difficulties of potential and residual xanthate measurements in flotation slurries. Progress has beeen made in the development of slurries filters, which are requisite to obtain a clear solution when the measurement method is based on spectrometry, or when it requires metering pumps. Various designs are available which can provide high quality solutions and a reasonable life in slurry. Clear solution characterization can use almost any instrumental method of analysis, such as spectrometry for xanthate concentration measurement, spectrometry for free cyanide measurement, and potentiometry for free cyanide measurement. Further progress is required, both in testing and demonstrating the reliability of the instruments, and to provide mineral processing engineers with a methodology to use the information for process optimization.

References

1. G. Barbery and J. L. Cécile "Instrumentation for reagent control in flotation: present status and recent developments" Advances in Mineral Processing - the Arbiter Symposium (New-York, NY,AIME, 1986), 726-739.

2. A. J. Lynch et al., Mineral and coal flotation circuits, their simulation and control (Amsterdam, Elsevier, 1981).

3. J. T. Woodcock, "Automatic control of chemical environment in flotation plants" Proceedings XIIIth International Mineral Processing Congress (Amsterdam, Elsevier, 1980), 1485-1512.

4. A. A. Abramov Technologia obogashenia rud tsvetye mettalov (Moscow, USSR, Nedra, 1983).

5. S. Haimala et al., "New controlled flotation methods developed by Outokumpu Proceedings XVth International Mineral Processing Congress (France, GEDIM, 1985), 3 88-98.

6. J. L. Cécile et al. "A new system for the measurement and control of the physico-chemical parameters in mineral processing plants" submitted for presentation at the XVIth International Mineral Processing Congress, Stokholm, Sweden, June 1988.

7. Licensintorg, 11 Minskaya ul., 121108 Moscow, USSR, pamphlet describing a "Method and device for xanthate batching control", 1985.

8. M. H. Jones and J. T. Woodcock, "Application of pulp chemistry to regulation of chemical environment in sulphide mineral flotation" Principles of Mineral Flotation - the Wark Symposium, (Victoria, Australia, Australasian Institute of Mining and Metallurgy, 1984), 147-183.

9. K. S. Forssberg (Editor), Flotation of Sulphide Minerals (Amsterdam, Elsevier, 1985).

10. Outokumpu Electronics, P.O. Box 85, SF-02201, Espoo, Finland, pamphlets describing ELEXAN 80 and Voltameter 470, 1987.

11. Mike Schmidt, private communication with G. Barbery, 22 July 1987, Outokumpu Equipment Canada Ltd.

12. J. R. Pederson Bureau of Mines Research 1986 (Washington, DC, United States Department of the Interior, 1987), 52-53.

13. D. A. Rand and R. Woods "Eh measurements in sulphide electrode slurries" International Journal of Mineral Processing 13 (1984) 29-42.

14. A. A. Abramov et al., "Improvement of technology for beneficiation of polymetallic ores having variable composition" Proceedings XVth International Mineral Mineral Processing Congress (France, GEDIM, 1985), 2 314-327.

15. J. V. Sullivan and J. T. Woodcock "A simple on-stream xanthate monitor" Proceedings Australasian Institute of Mining and Metallurgy, 248 (1973) 1-7.

16. B. E. Ashston, "On-Stream Analysis", Principles of flotation- the Wark Symposium (Australia, Victoria, Australasian Institute of Mining and Metallurgy, 1984), 285-299.

17. BRGM-Instruments, BP 6009, 45060 Orléans Cedex 02 France, pamphlet

describing PIRANA™

18. W. J. Trahar, "The influence of pulp potential in sulphide flotation",
Principles   of   flotation-   the   Wark   Symposium   (Australia,   Victoria,
Australasian Institute of Mining and Metallurgy, 1984), 117-135.

19. F. A. Habashi, "Principles of Extractive Metallurgy" (New-York, NY,
Gordon and Breach, 1970), vol.2 23-39.

20. K. Oseo-Asare et al., "Solution chemistry of cyanide leaching systems" in
"Precious   Metals   :   Mining,   Extraction   and   Processing"   (New-York, NY,
AIME-TMS, 1984) 173-197.

21. "Cyanide from Mineral Processing", Workshop Proceedings (Salt Lake City,
Utah, College of Mines and Minerals Industry, 1983).

22. C. Pohlandt, "The determination of cyanide in hydrometallurgical process
solutions and effluents by ion chromatrography" MINTEK Report M128, 1984.

23.   C.   Pohlandt-Watson, "A simplified flow-injection method for the
determination of free cyanide inp/rocess solutions" MINTEK Report M275, 1986.

24.   C. Pohlandt-Watson, "A revised ion-chromatography method for the
determination of free cyanide" MINTEK Report M283, 1986.

25. J. Ruzicka and E. H. Hansen, "Flow injection analyses Part 1. A new
concept of fast continuous analysis" Analytica Chemica Acta" 78 (1975)
145-157.

26. T. P. Lynch, "Determination of free cyanide in mineral leachates" The
Analyst (London) 109 (1984) 421.

27. T. P. Lynch et al., "Application of flow injection procedures to the
analysis of mineral process solutions" Extraction Metallurgy '85 (London,
IMM, 1985) 93-116.

28. R. J. Wyllie, "Cyanostat : at last a new on-line control for precious
metals leach plants", Engineering and Mining Journal, January 1987, 32-33.

29. Polymetron, CH-8634 Hombrechtikon, Switzerland.

30. Witteck Development Inc., 2640 South Sheridan Way, Missisauga, Ontario,
Canada L5J 2M4.

31. W. A. Gow et al. "Instrumentation in the cyanide process", Canadian
Mining and Metallurgical Bulletin, 59 (1966) 872-880.

32. K. D. Downes, "Control in the cyanide process" Canadian Mining Journal 85
(1964) (10) 92-95.

33. J.C. Ingles, "Measurement of free cyanide concentration by continuous
potentiometric titration" Mines Branch Research Report R 127, 1964.

34. Fred Kelly, personal communication with G.Barbery, 24 July 1987, CANMET,
Ottawa, Canada.

35. Doug Bartlett, personal communication with G.Barbery, 21 July 1987,
Witteck Development Inc.

DEVELOPMENT OF AN IN SITU PROBE FOR MONITORING REDOX CONDITIONS

IN COMMERCIAL PROCESSING PLANTS

C. R. Neuharth, C. S. O'Dell, and G. W. Walker

U.S. Department of the Interior
Bureau of Mines
Washington, D.C.   20241

Abstract

A number of sulfide mineral-thiol collector interactions have been identified as electrochemical or electron transfer processes. Laboratory studies have confirmed that these interactions can be controlled by variation of the mineral potential, using either a directly applied potential or redox-controlling reagents, to induce or depress flotation. Control of the mineral potential in the laboratory has proved to be a very precise means of achieving flotation.   If similar results can be achieved on a commercial scale, more efficient recovery of domestic critical and strategic minerals might be possible.

In order to develop an understanding of the relationship between redox potential and recovery in commercial flotation plants, a portable system for in situ monitoring of the pulp conditions in sulfide flotation circuits has been developed.   The multichannel capability of this data acquisition system provides for the simultaneous monitoring of redox potentials on a variety of metal and mineral electrodes.   The system also provides for the monitoring of such parameters as pH, dissolved oxygen content, and soluble ion concentration.   The results of probe tests at two commercial processing plants will be discussed in terms of the differences in mineral and metal electrode potentials under similar conditions; the effects of reagent, pH, and dissolved oxygen on these potentials; and the viability of using potential as a commercial flotation control parameter.

Introduction

For over 75 years, froth flotation has been used to recover metal values from complex ores. The steadily decreasing grade of domestic ore deposits necessitates the optimization of mineral processing techniques if the United States is to maintain its current levels of production and compete economically with foreign producers. Improving efficiency will require a more complete understanding of the mechanisms involved in froth flotation.

Fundamental research into sulfide mineral flotation has shown that the identified interactions between sulfide minerals and thiol collectors are generally electrochemical, charge transfer processes (1-4). Control of flotation response with thiol collectors has been achieved with applied potentials (5-7) and with addition of redox-controlling reagents (8). In cases involving copper, copper-iron, or iron sulfides, such as chalcocite, bornite, chalcopyrite, and pyrite, Richardson and Walker (5) have reported that: (1) the potential can be used quasi-reversibly to initiate or depress flotation; (2) the flotation potential increases with increasing iron to copper ratio in the mineral lattice; and (3) the potential regions where the individual minerals float are reasonably well defined and distinct. With mixed bed systems (9-11), mineral separations can be achieved, but mineral dissolution and galvanic interactions can lead to loss of selectivity in flotation.

Control of the electrochemical potential in flotation circuits might be a way of achieving better recoveries and grades from low-grade, complex ores. While the redox conditions of flotation pulps are known to cause potential differences between the various sulfide minerals in the laboratory, the measurement of similar potential differences in industrial circuits has not been investigated to any extent. In the limited cases where redox potentials have been measured (8) or where potential is being used as a control parameter (i.e., Duval Sierrita's closed molybdenum flotation cells), platinum and gold electrodes were used primarily. These noble metal electrodes measure "pulp potential" (12) rather than mineral potential; however, it is arguable whether pulp potential rather than mineral potential is suitable for controlling the separation of sulfide minerals (13). In order to determine if and how metal and mineral electrodes differ in their response to changes in pulp conditions, a portable, multichannel system for monitoring in industrial flotation circuits has been developed. The purpose of this report is to detail the design, development, and current applications of such a system.

Monitoring System

The Bureau-designed monitoring system (fig. 1) consists of (1) a probe containing the various sensing electrodes, (2) signal conditioners, (3) a programmable data acquisition and temporary data storage system, and (4) a computer for permanent data storage and data handling.

The probe heads (fig. 2) were fabricated from plexiglass and threaded to accommodate a 1 1/4-in (dia.) PVC pipe which housed the electrical cables and provided mechanical support when the probe was submerged in the flotation cell. At the bottom, the probe had removable double plates with O-ring fitted openings for the pH and dissolved oxygen (DO) electrodes. The removable side plates were drilled to accept metal/mineral electrodes. All the plates were attached to the probe head with stainless steel or nylon screws and a viton rubber gasket to achieve water-tight seals. The pH electrode was a combination glass electrode with a gel-filled double junction silver/silver chloride reference

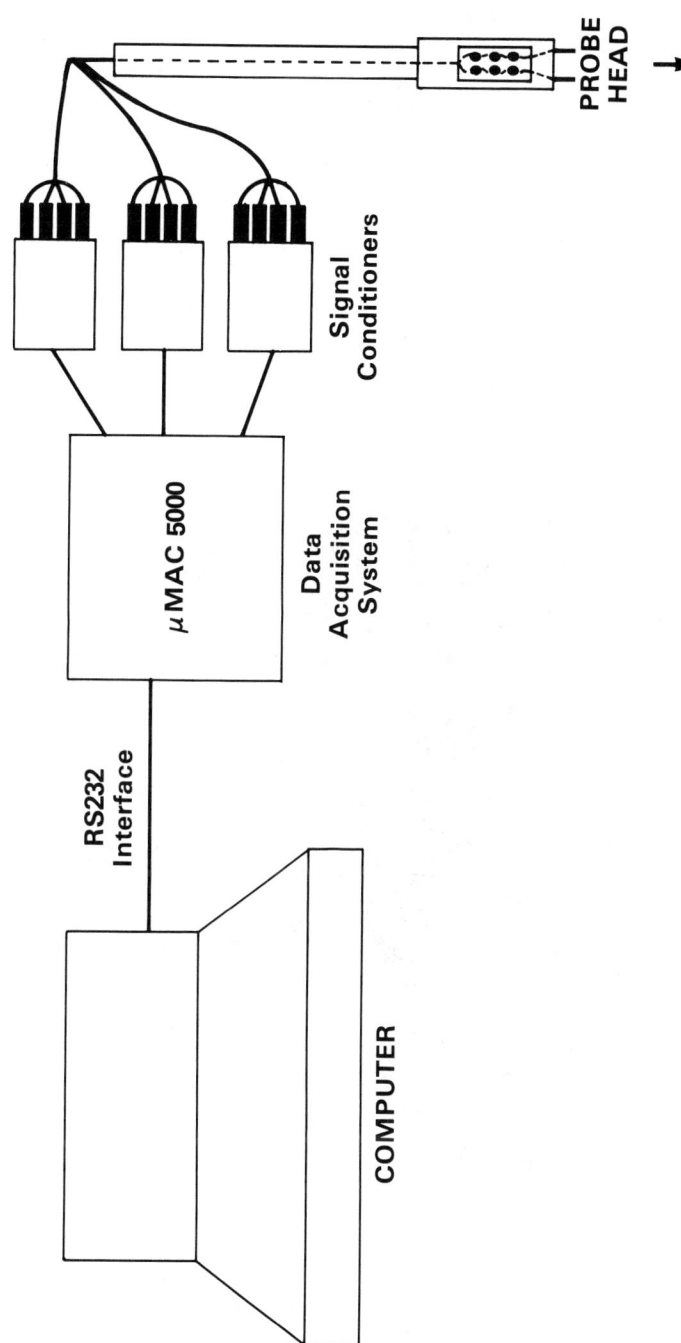

Figure 1 — Bureau-designed monitoring system.

TO MICROPROCESSOR

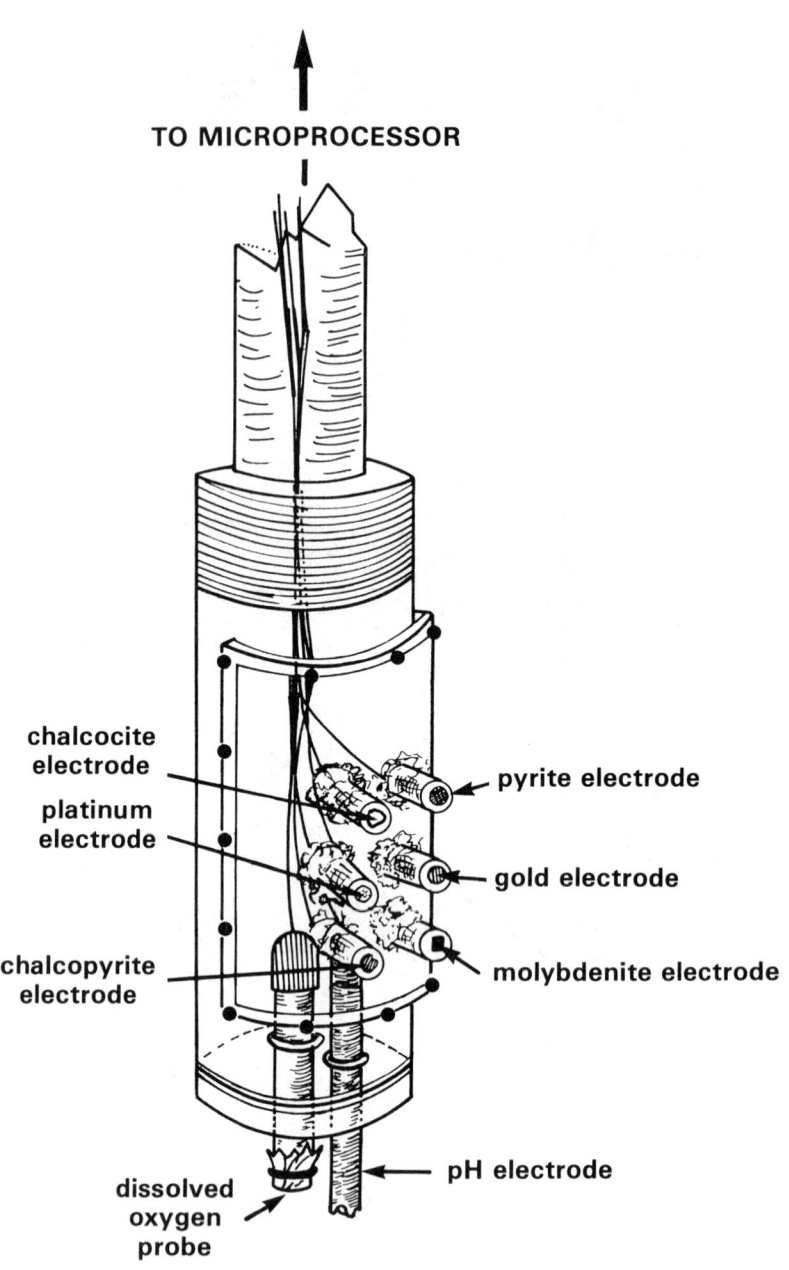

chalcocite electrode

platinum electrode

chalcopyrite electrode

pyrite electrode

gold electrode

molybdenite electrode

dissolved oxygen probe

pH electrode

Figure 2 - Single probe head containing
pH, DO, and metal/mineral electrodes.

(Markson Scientific R-739). A standard Clark cell (Lazar Model DO-166) was used to measure the DO. Since the silver anode in the DO cell is susceptible to contamination from hydrogen sulfide diffusing through the Teflon membrane, care must be taken when using these electrodes in flotation circuits where sulfide ions are present. Other DO electrodes were tried, but the Lazar electrode appeared to be the least susceptible to malfunctioning when exposed to the suspended particles in flotation pulps. The metal/mineral electrodes were fabricated by first attaching copper wire to the backs of the electrodes with Electrodag silver paint (Acheson Colloids Co.) and then encasing all but the electrode face in Arlidite epoxy. The Electrodag paint remained conducting for long periods but had very little structural stability. Bonding of the electrodes to the side plates with silicone adhesive provided a flexible water-tight seal and permitted easy exchange of the electrodes.

The signal conditioners consisted of high-impedance, operational amplifiers ($>10^{14}$ ohms, such as the Burr-Brown OPA 104CM or Analog Devices 545), as voltage followers with separate 20k ohm potentiometers for zero and gain adjustments and a passive RC noise filter on the output side. Ground looping difficulties were avoided by using individually isolated power supplies (i.e., Burr-Brown 724 DC-DC convertor) to each operational amplifier. Stable pH signals were achieved with a battery powered unity gain operational amplifier connected to the pH electrode inside the probe head; the conditioned signal could then be transmitted over the cable without appreciable noise pickup or ground looping. Fifty foot lengths of cable have been used without difficulty permitting simultaneous monitoring with the probes in flotation cells up to 80 feet apart. With this degree of signal isolation only one reference electrode was needed in each probe head simultaneously measure pH and up to eleven metal/mineral electrode potentials. The potentials of the sulfide mineral electrodes, including ones with extremely low conductivities, such as molybdenite, were measured without difficulty.

Individually shielded twisted-pair cables provided electrical continuity between the probe electrodes and the signal conditioners housed next to the data acquisition system in the main box. Type X pin and socket connectors were used to connect the electrodes to the cable inside the probe head while standard stereo phone plug connectors were used for cable connection to the signal conditioners.

Data acquisition and temporary data storage was handled using a programmable uMac-5000 unit (Analog Devices) with 128k of RAM and 3 voltage input modules (QMX04's). The uMac-5000 provided ~1000V channel isolation, 12 bit A/C conversion, and $>10^9$ input impedance. This impedance was not sufficient to prevent ground looping when only one reference electrode was used for each probe head, thus necessitating the need for the signal conditioners described earlier. With the gain and zero adjustments on the signal conditioners and programmed calculations, each channel could be calibrated in scientific units for either potential, pH, or DO. The system was configured to monitor 12 individual channels with 1000 points of data per channel; the number of data points per channel can be increased by reducing the number of channels. If required, expansion of the number of channels is possible with this unit. The temporary storage of data in the uMac-5000 was protected against loss due to power outages by battery backup. Communication between the uMac and a portable computer was via an RS-232 cable with data communication software running on the computer. Both Hewlett-Packard 85 and Hewlett Packard 110 (Portable Plus) computers have been used, although any computer with a serial port and data communication software will control the uMac-5000.

The software developed to control programming for the uMac allowed for an initial calibration stage of all the channels for potential and then for calibration of pH and DO electrodes. The sampling rate was user definable with signal averaging over the last 25 seconds of each sampling period. A comment statement along with the date, starting time, channel names, and number of readings were stored as header information in each data file. The software also allowed data collection to be paused, with channel recalibration as needed, and then resumed or restarted. At the end of a test, the data was transferred from random access memory in the uMac-5000 through the portable computer and to disk in a file format readable by Lotus 1-2-3. In addition the Lotus software was used to massage the data and to generate time plots for each electrode channel. The data files and graphics files could be transferred via modem for evaluation by personnel at the research center to aid in discussion and planning of the next day's monitoring at the processing plant.

Applications

Initial Chino Visit

In the initial study, the pulp conditions at the Chino Concentrator in Hurley, NM, were monitored using platinum, conductivity, DO, and pH electrodes. The objective of this visit was to test the operation of the newly designed monitoring system and to determine if these electrodes would respond to changes in plant operating conditions, such as variations in lime, collector, and frother dosages. The tests showed that under certain conditions the platinum electrode was sensitive to changes in the rate of the amylxanthate collector addition. Comparison of pH and platinum responses (fig. 3) showed that both electrodes tracked an

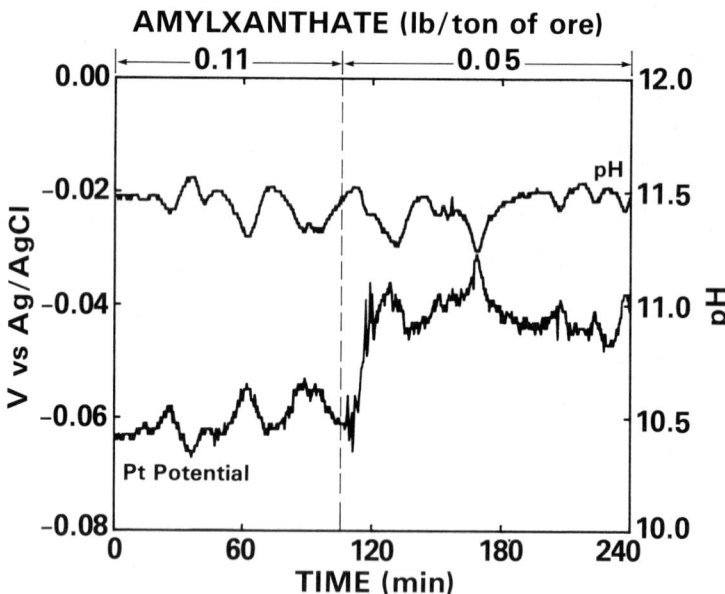

Figure 3 - Electrode response to variation in collector concentration at Chino.

oscillating condition throughout the test, which was believed to be the normal fluctuations caused by the plant's feedback circuit controlling the pH of the flotation pulp. At -120 min on the plot, the platinum electrode responded to a change from 0.11 to 0.05 lbs/ton-of-ore in collector dosage. As expected, the pH electrode did not reflect this change. In another test, the collector dosage was varied sequentially from 0.05 to 0.08 to 0.11 lbs/ton, but a significant change in platinum potential could not be observed. Based on a calculation of the expected Nernst potential change, these changes in xanthate dosage should have been observed. The lack of response was believed to reflect platinum's sensitivity of oxygen, which in the oxygen saturated conditions of a flotation cell masked any small potential changes. As a result, only the effects of gross xanthate changes were observed.

Duval Visit

Monitoring was confined to the molybdenum circuit (fig. 4), which consisted of two banks of rougher cells, two banks of cleaner cells, three column recleaner cells, one column re-recleaner cell, and one bank of scavenger cells. Separation was accomplished using sodium sulfide as the depressant for the copper minerals. Feed was comprised of underflow from the copper-molybdenum concentrate thickener. Tailings from the molybdenum roughers made up the final copper concentrate.

Duval had originally used an iron-cyanide complex as their copper depressant, but at the time of the plant visit, both banks of rougher cells and one bank of cleaner cells had been modified to operate as closed flotation cells and sodium sulfide was being used as the copper depressant. The closed systems were initially charged with air (approximately 20% oxygen and 80% nitrogen); however, the oxygen was quickly consumed in a homogeneous reaction with the sodium sulfide. This reaction produced a recirculating flotation gas that was virtually pure nitrogen and permitted the economical use of the sodium sulfide depressant. Sodium sulfide addition in the rougher and closed cleaner circuits was controlled via redox potential monitored on a platinum electrode, while reagent additions in the open cells and columns were controlled by the ore feed rate. The primary objectives in monitoring at Duval were to determine electrode response under low oxygen conditions and to compare the redox conditions in open and closed flotation cells.

The probes, along with the data acquisition system, were tested at various points throughout the molybdenum circuit. Generally the data acquisition system worked quite well; however, some problems were encountered because of poisoning of the pH reference electrodes due to the high concentrations of sodium sulfide being used and contamination of the DO probe by fine mineral particles working in and around the electrode membrane with time. Changing to the gel-filled double junction reference electrode described in the System section eliminated the pH and reference difficulties. Careful installation of the DO membrane using two small diameter o-rings to secure the membrane prevented mineral particle contamination. However, sulfide contamination of the silver electrode can, as mentioned, be a problem.

Results from one of the tests conducted at Sierrita can be seen in fig. 5. In this test, the probes were placed in the #1 and #3 recleaner column cells, while changes were made in the sodium sulfide addition rate and the volume of air fed to the columns. The flotation gas could be varied for individual columns. Sulfide addition was to the recleaner circuit as a whole and not to the individual column cells. As expected, both of these variables had significant effects on the potentials of the metal and mineral electrodes as well as the DO level in the pulp. The

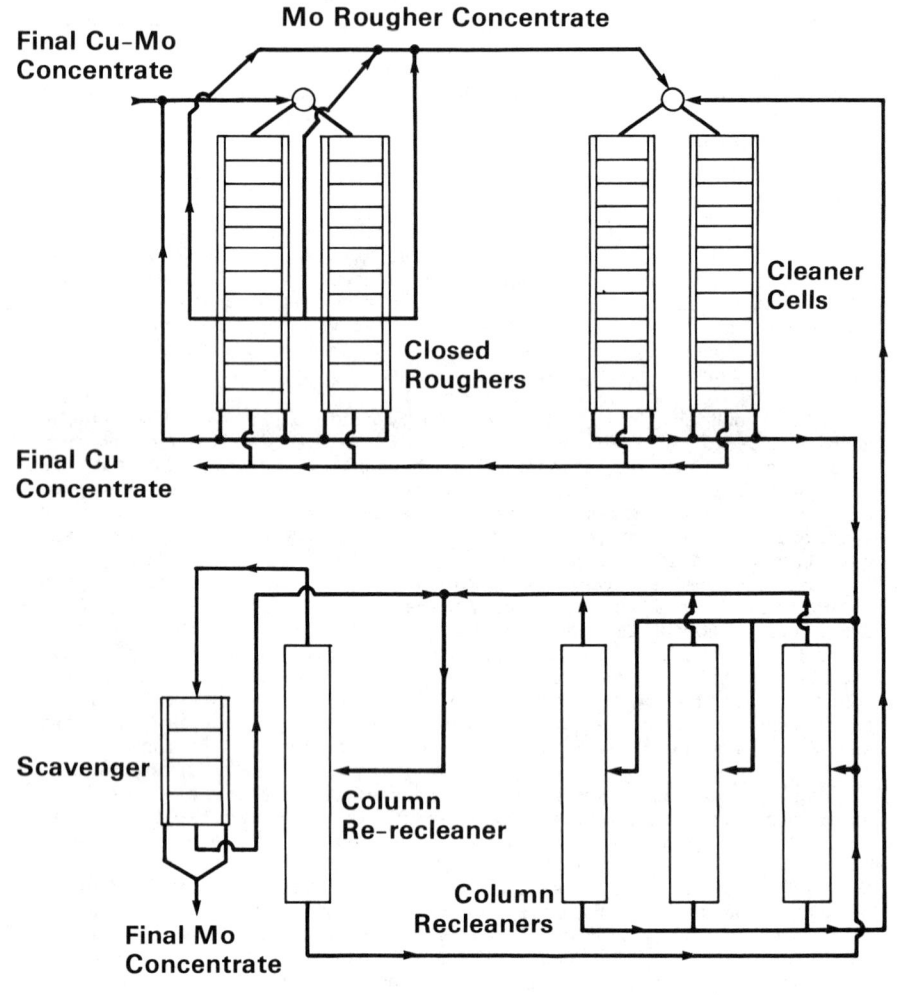

Final Cu-Mo Concentrate

Mo Rougher Concentrate

Cleaner Cells

Closed Roughers

Final Cu Concentrate

Scavenger

Column Re-recleaner

Column Recleaners

Final Mo Concentrate

Figure 4 - Duval's molybdenum flotation circuit.

Na₂ S ADDITION RATE

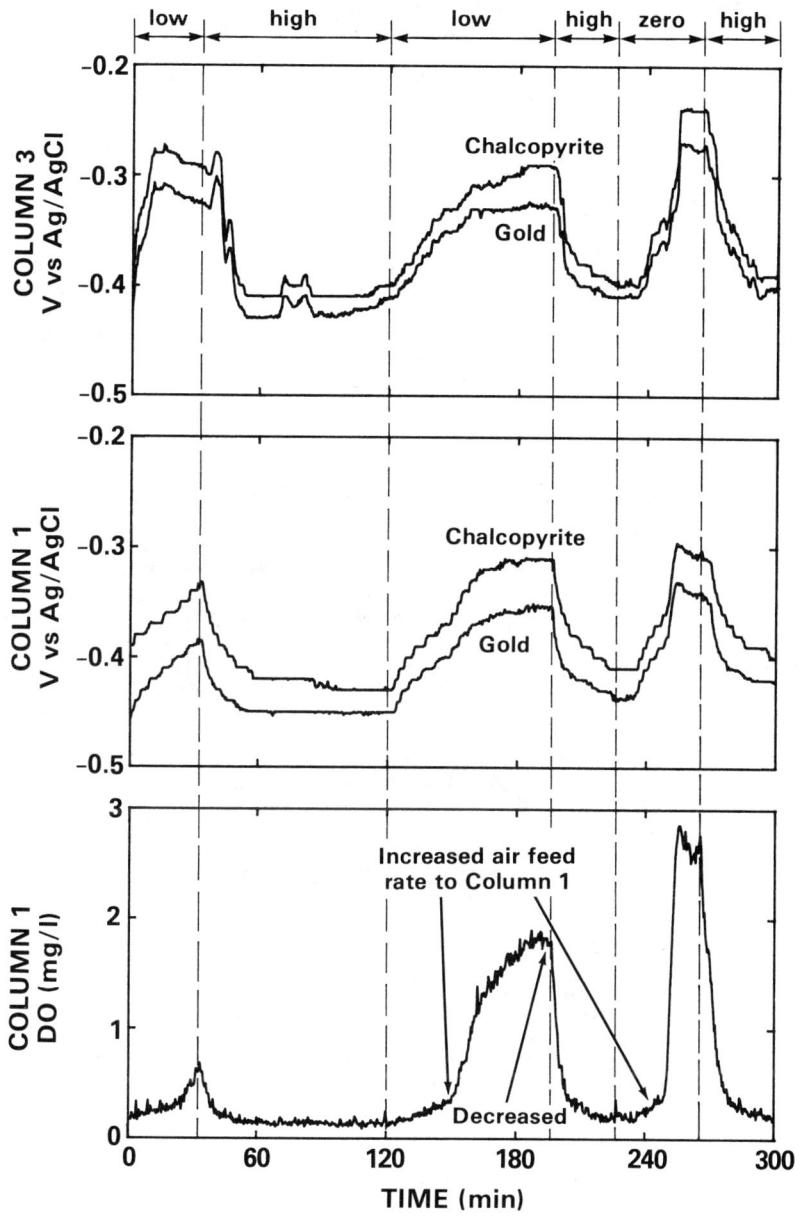

Figure 5 – Electrode responses to variations
in Na₂S and air feed rates at Duval.

high sodium sulfide concentrations necessary for separation limited DO to negligible amounts in the open column cells; however, in the closed conventional cells, even lower DO levels were found with half the sodium sulfide dosage used in the column cells. These results indicated that the metal and mineral electrodes as well as the DO probe were capable of tracking the variations in pulp conditions resulting from changes in the addition rate of a reducing agent such as sodium sulfide. Platinum electrodes were used at Duval in the sulfide addition control loop, because gold electrodes had been found to foul with time. This raised an interesting paradox, because operators at the Twin Butte's concentrator, adjacent to Duval, had found that platinum electrodes fouled more readily than gold electrodes. This situation points out the importance of considering a number of metal and mineral electrodes when considering potential as a control parameter.

The shortness of this visit did not provide ample time to make a firm judgment concerning the operation of mineral electrodes versus metal electrodes. However, in the Duval operation, where the conditions are kept very reducing throughout the molybdenum circuit and only one mineral is being recovered, a single electrode, either metal or mineral, should be satisfactory. The primary concern would appear to be the longevity of operation.

Second Chino Visit

The Santa Rita deposit is composed of a stock ore containing primarily pyrite and chalcocite and a sedimentary ore containing chalcopyrite, pyrite, magnetite, and sulfate mineral. Even with blending techniques, the composition of the ore fed to the plant is highly variable, and control of pH alone is not always effective in producing a stable grade of concentrate. Monitoring of the primary flotation circuit (fig. 6) at the Chino concentrator was conducted to determine if pulp potential might be a more efficient method of controlling flotation parameters. The four main objectives addressed during the monitoring visit were:

1) Long-term monitoring to determine the stability, reproducibility, and lifetime of a variety of metal and mineral electrodes;

2) Monitoring to obtain a profile of the average pulp conditions along a row of rougher and middling flotation cells under normal operating conditions;

3) Monitoring along the rougher flotation cells while changing pH and collector concentration to determine the effects of these changes on the potential of the various metal and mineral electrodes; and

4) Monitoring around the sodium hydrosulfide addition point to determine the effects of this reducing agent on the potentials of the various metal and mineral electrodes and the dissolved oxygen level in the pulp.

The long-term probe, containing only metal and mineral electrodes, was left unattended, except for periodic monitoring of the potentials, in the #4 rougher cell for nine days. The electrodes behaved normally (i.e., the potential readings mirrored pH changes similar to fig. 3 data) for four days, then exhibited some potential shifting relative to each other. When the probe was removed from the pulp on the ninth day, a thick black scale covered the entire probe including the electrodes.

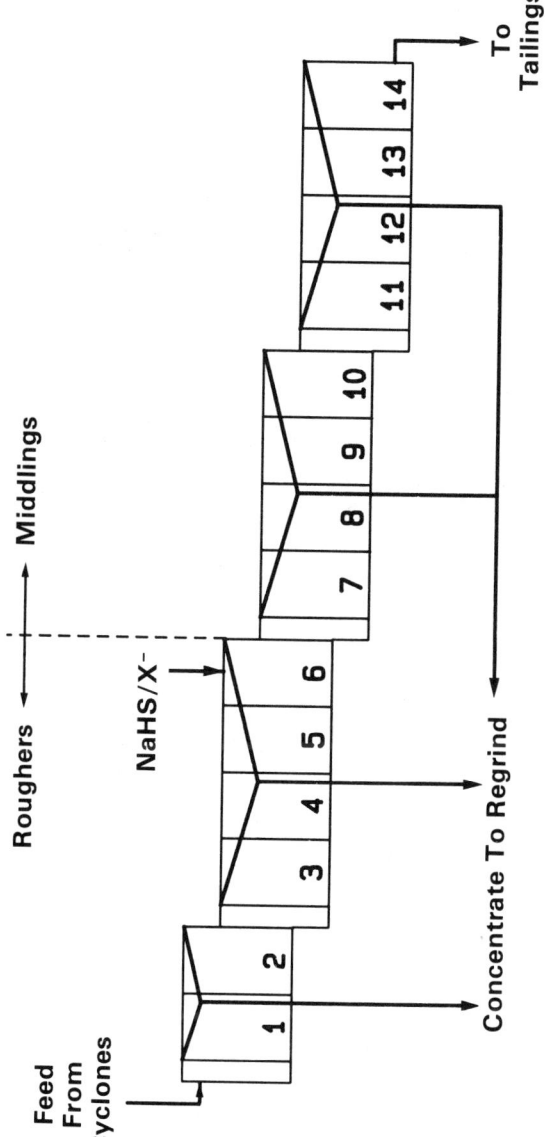

Figure 6 – Chino's primary flotation circuit.

This scale, which Chino finds regularly on their pH electrodes after three day's exposure to the pulp, has been identified by X-ray diffraction and infrared spectroscopy to consist primarily of calcite with magnetite as the minor fraction. The high resistivity of the scale may have effected the potential readings of the electrodes after four days. When these electrodes were compared with uncoated electrodes in the laboratory, the magnitudes of their potential shifts to a perturbation were generally equivalent to shifts on clean electrodes, but their responses were much slower. These calcite deposits were easily removed with the same weakly acidic solutions routinely employed by Chino to clean their pH electrodes.

A profile of a complete row of rougher and middling cells was obtained over a 6-hour period under normal operating conditions (i.e., all of the reagent additions, with the exception of lime to the grinding circuit, were held at a constant level). Due to the variability in the natural pH of the slurry and the separation between the lime addition point (SAG mill) and the pH monitor (head of primary flotation circuit), the control loop would sometimes oscillate by as much as 0.5 pH units. As would be expected, the potentials of the metal/mineral electrodes mirrored these pH fluctuations while DO levels remained constant. The electrodes exhibited very real differences in potential under the same solution conditions. While the establishment of electrode potential appears to be a complex process, in relative terms, the potential sequence was generally platinum > gold ~ pyrite > chalcopyrite > chalcocite > molybdenite. The gold and platinum electrodes exhibited the greatest variability; platinum, with its high oxygen sensitivity, was more anodic. It is interesting to note that the pyrite > chalcopyrite > chalcocite sequence is the same as their flotation potential sequence observed in the laboratory.

After obtaining the background data from the profile experiments, the more complex objectives (3 and 4) were addressed. When pH and collector concentration were varied, the mineral electrodes continued to show favorable response to the changes in pH; however, no potential changes that could be related to variations in collector concentration were detected (fig. 7). This does not contradict the platinum-xanthate results from the first trip because Chino had subsequently changed from amylxanthate to a blend (Minerec 2694) of isopropyl thionocarbamate and isobutyl xanthogen ethyl formate in an effort to improve pyrite rejection at lower pH. In the test shown in fig. 7, only the thionocarbamate was used, and it is possible that the absence of a potential response is due to the fact that thionocarbamate-mineral interactions are not charge-transfer reactions or that the magnitude of the charge-transfer reaction was too small to measure.

Tests were conducted to determine the effects of sodium hydrosulfide additions on the potentials of the metal/mineral electrodes and DO. At the time of the monitoring visit, Chino was adding sodium hydrosulfide to the #6 rougher cell to lower the pulp potential and improve chalcopyrite recovery in the middlings stage (i.e., cells 7-14). The effects of this reagent on the potentials of the mineral electrodes as well as on the DO level in cell #6 were very evident (fig. 8), but were found to be short lived. The mineral potentials and DO level were much lower than in the proceeding cells; however, the values recorded in cells #7-11 both before and during sulfide addition to cell #6 were virtually identical. The shifts in DO and potential were easy to correlate with the changes in sulfide addition rates; however, the responses of the same electrodes during periods of constant sulfide addition were not as easily under-stood. Between 165 and 230 min in fig. 8, DO was constant while all the potentials were decreasing and even more surprising was the fact that

LIME ADDITION RATE

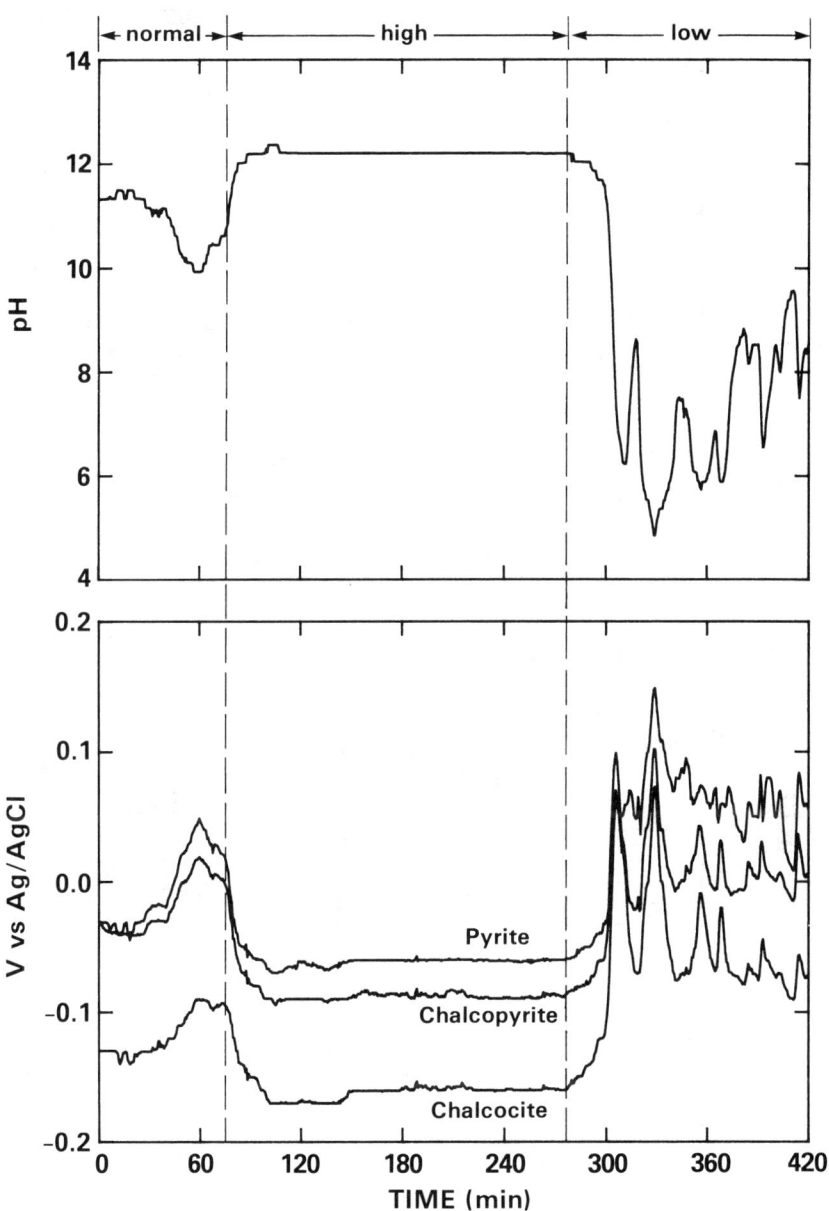

Figure 7 - Electrode responses to variations in pH and collector concentration at Chino.

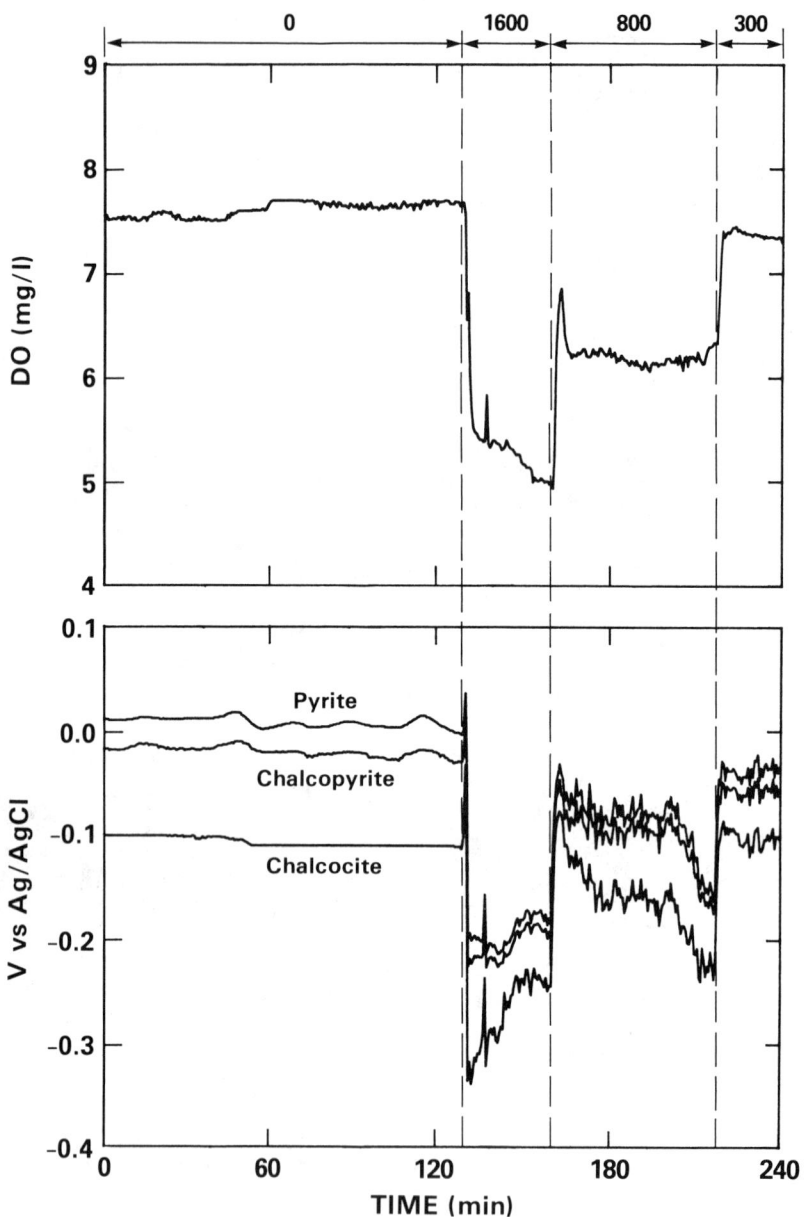

Figure 8 – Electrode responses to variations in NaHS/X⁻ feed rate at Chino.

when DO was decreasing (i.e., between 140 and 165 min), the potentials were increasing. While these observations cannot be explained by sulfide concentration alone, when the normal fluctuations in pH in the plant (not shown) are considered, then the potential shifts are understandable.

Not only do these results illustrate the short-ranged effects of sulfide additions in an oxygen saturated system , they also point out the complex nature of pulp potential. Just as pH alone cannot be used to produce consistently high recoveries and grades, potential by itself will probably not be an effective means of controlling flotation. Instead a combination of factors, such as potential, pH, and DO will need to be monitored and controlled.

Summary

The monitoring system described in this paper has the ability to measure various in situ parameters in industrial scale flotation cells. With this portable system, pH, DO, and metal/mineral electrode potentials have been measured. It should be noted that any quantity which can be reduced to an electrical signal could be monitored and possibly used to control a flotation circuit. The high degree of signal conditioning in the present system permits monitoring with a single reference electrode in each flotation cell and allows values for low conducting sulfide mineral electrodes such as molybdenite to be recorded.

Results of the monitoring visits have revealed that:

1) Metal and mineral electrodes can be operated for extended periods in flotation cells, suggesting that these electrodes could be used in a control loop system to adjust pulp or mineral potential. In carbonate-rich environments where calcite buildup is a problem, electrodes will require periodic cleaning to prevent fouling; however, the cleaning procedure will generally be the same as that used to remove the calcite from the pH electrodes currently used in commercial processing plants.

2) The various metal and mineral electrodes exhibit different electrochemical potentials. Response of an electrode varies with system changes such as lime addition, DO, and perhaps collector additions. While gold and platinum are generally the most sensitive to condition changes, the mineral electrodes appear to display sufficient responses to be considered for use as control electrodes.

3) Potentials for the mineral electrodes in flotation circuits fall within the potential range where mineral-collector charge transfer processes have been observed in the laboratory. Control of the potential by reducing and oxidizing reagents could provide control of the flotation response in plants similar to that achieved in laboratory systems.

Continued on-site monitoring should provide the link between flotation theory as studied in laboratory situations and current techniques used in commercial concentrators and may lead to development of more efficient methods of process control for the industry.

REFERENCES

1. R. Woods and P. E. Richardson, The Flotation of Sulfide Minerals--
   Electrochemical Aspects. Advances in Mineral Processing, ed. by P.
   Somasundaran, Society of Mining Engineers, Inc., Littleton, CO,
   1986, pp. 154-170.

2. S. Chander, Oxidation/Reduction Effects in Depression of Sulfide
   Minerals - A Review. Minerals and Metallurgical Processing, v. 4,
   1985, pp. 26-35.

3. A. Kowal and A. Pomianowski, Cyclic Voltammetry of Ethyl Xanthate on
   a Natural Copper Sulphide Electrode. J. Electroanal. Chem. and
   Interfacial Electrochem., v. 46, 1973, pp. 411-420.

4. O. Lam, M. Lamache, D. Bauer, and J. L. Cecile, Etude analytique des
   reactions lors de l'oxydation de la galena au contact de solution de
   xanthate a pH 9. Paper in XVth International Mineral Processing
   Congress, (Cannes, France, June 2-9, 1985). Edition Gedine, St.
   Etienne, France, v. 2, 1985, pp. 154-166.

5. P. E. Richardson and G. W. Walker, The Flotation of Chalcocite,
   Bornite, Chalcopyrite, and Pyrite in an Electrochemical-Flotation
   Cell. Paper in XVth International Mineral Processing Congress,
   (Cannes, France, June 2-9, 1985). Edition Gedim, St. Etienne,
   France, v. 2, 1985, pp. 198-210.

6. C. S. O'Dell, G. W. Walker, and P. E. Richardson, Electrochemical
   Reactions in the Chalcocite-Xanthate system. J. Applied
   Electrochem., v. 16, 1986, pp. 544-554.

7. R. Woods, Electrochemistry of sulfide flotation. Principles of
   Mineral Flotation. The Australian Institute of Mining and
   Metallurgy, Parksville, Victoria, Australia, 1984, pp. 91-115.

8. W. J. Trahar, The influence of pulp potential in sulphide flotation.
   Principles of Mineral Flotation. The Australian Institute of Mining
   and Metallurgy, Parksville, Victoria, Australia, 1984, pp. 117-135.

9. J. E. Gebhardt, N. F. Dewsnap, and P. E. Richardson, Electrochemical
   Conditioning of a Mineral Particle Bed Electrode for Flotation.
   BuMines RI 8951, 1985, 10 pp.

10. J. P. Guy and W. J. Trahar, The Effect of Oxidation and Mineral
    Interaction on Sulphide Flotation. Flotation of Sulphide Minerals,
    ed. by K. S. E. Forssberg, Elsevier, Netherlands, 1985, pp. 61-79.

11. C. R. Neuharth, G. W. Walker, and P. E. Richardson, Electrochemical
    Aspects of Galena Flotation. Paper presented at 1986 Fall AIME
    Meeting, St. Louis, MO, September 7-10, 1986.

Evaluation of AE Sensor for Level Control in Flotation Cells

Matthew D. Light and Keith Prisbrey

Department of Metallurgical and Mining Engineering
University of Idaho, Moscow, Idaho 83843

Abstract

Acoustic Emission (AE) sensing for flotation level control
is attractive because of the non-intrusive nature of the
sensors. Where most other sensors are exposed to abrasive
slurry streams and fouling environments an AE sensor can be
attached to the side or frame of the flotation cell and be
relatively immune. Slight changes in pulp level result from
large changes in water mass, which in turn affects the
resonating frequencies of the system. Data collected on
industrial and laboratory cells shows that these shifts in
resonant frequency from background AE give pulp level changes
which are accurate to within a centimeter. The repetitive AE
measurements are noisy, and it is necessary to use a Kalman
filter on the measurements. We discuss the problems in
development of a complete hardware and software package which
is capable of periodic recalibration.

Introduction

Despite the plethora of approaches to level measurement, including float, displacer, differential pressure, air bubbler, capacitance, ultrasonic, and nuclear (1-6), there is still need for inexpensive applications in flotation cell devices. In the very harsh environmental conditions, seals can be troublesome, buildups and deposits occur on floats and displacers, and maintenance costs are too high. Acoustic emission (AE) sensors are potentially more robust, and avoid the fouling environment.

The AE sound responds to pulp level. Changes in both resonating frequency and amplitude correlate to pulp level state. However, the AE signal has noise associated with it which requires filtering.

We do not introduce state equations for pulp level to develop the control algorithm, but rather to relate them to the AE signal. The AE signal is random, and always has some element of chance associated with it. The state equations are used in a random signal processing problem for sensing the level with the use of a Kalman filter (7). The state equations are developed from the material balance on a flotation cell as follows:

$$F - P = d/dt \ (rAH) \tag{1}$$

where

    r = liquid density
    A = cross-sectional area of tank
    F = Feed flow
    P = discharge flow
    H = pulp level

The important mathematical assumption is to consider Equation 1 as a stochastic state equation, and to consider the feed disturbance variable, F, as a white noise source, i.e. as having a uniform power spectral density function. The reason for this assumption is that feed fluctuates wildly due to ball mill and hydrocyclone amplified surges. The control variable is discharge flow, P. It is next deleted from Equation 1, for the purposes of this paper, in order to focus on the optimal estimation problem. After the foundation of pulp level estimation has been developed, optimal control of pulp level can be developed as a separate problem using, for example, variable position dart plugs to adjust P.

Letting the pulp level become the state variables,

$$x_1 = H$$
$$x_2 = dH/dt$$

the state equations become:

$$\frac{d}{dt}\begin{bmatrix} x1 \\ x2 \end{bmatrix} = \begin{bmatrix} 0 & 1 \\ 0 & -rA \end{bmatrix} \begin{bmatrix} x1 \\ X2 \end{bmatrix} + \begin{bmatrix} 0 \\ 1 \end{bmatrix} F \qquad (2)$$

or, in matrix notation,

$$dX/dt = A X + B F \qquad (3)$$

Using the transfer-function formalism necessary for a discrete Kalman filter, the transfer function, C, is derived from

$$C = L^{-1} (sI - A)_t = \Delta t \qquad (4)$$

L^{-1} = inverse Laplace transform

which gives

$$C = \begin{bmatrix} 1 & (1 - e^{(-rA \, \Delta t)}/rA \\ 0 & e^{(-rA \, \Delta t)} \end{bmatrix} \qquad (5)$$

and state equation in matrix form is

$$X_{k+1} = C_k X_k + Q_k \qquad (6)$$

where :

$$Q_k = E [F_k F_k']$$
(expected value of the white noise disturbance)

We use a vector of measurement variables composed of AE power spectra features, Z, and relate them via a linear transformation, J, to the state variables:

$$Z_k = J X_k + R \qquad (7)$$

R = matrix of measurement variances, assumed to be uncorrelated with Q.

Our experimental procedure was to measure AE on flotation cells at different pulp levels in both the laboratory and industrial setting and examine the spectra for measurement features. We developed the linear regression equations and error terms necessary and programmed the state equations into a Kalman filter. The Kalman filtering was successful enough to estimate pulp level, even with very noisy AE signals.

Experimental

AE data was digitized with a Nicolet digital oscilloscope on loan for this project from Kennecott Research. Continuous industrial data was collected by changing the position of the sensor with respect to the pulp level during steady state operation. Batch data was collected on a laboratory flotation cell by varying the pulp level while maintaining the sensor position.

We used a piezoelectric AE sensor to digitize waveforms at 0.05 milliseconds/sample point. The AE sensor was coupled to the outside of flotation cells using petroleum jelly. The background AE was collected for evaluation, which was generated by the impeller, motor noises, and mechanical motion within the flotation cell.

A 100 KHz filter kept us on the correct side of the Nyquist frequency in order to avoid aliasing errors. When AE events exceeded a trigger threshold, the oscilloscope captured 16K sample points, taking about 4 seconds per waveform "sample" in order to store it on a floppy disk.

Twenty such samples were collected at each flotation level. We used an FFT to obtain Fourier coefficients of each sample which were then transferred to a personal computer. The Fourier coefficients were used to estimate the power spectral density function according to the following relation[7]:

$$E \ [1/T \ |F \ \{ \ X_t \ \}|^2 \ ] \qquad (8)$$

where E = Expected Value

$F\{ \ \}$ = Fourier Transform

X_t = sound wave of record length T

The quantity inside the brackets of Equation 8 is the periodogram. The ensemble of twenty periodograms was averaged to yield the power spectral density function for each pulp level. Averaging may not be essential in the analysis of deterministic signals, but it is for random signals.

The periodograms were examined for features that would allow unique association with pulp level. Our earlier work (8) has shown that both time domain and frequency domain features can be successful. For example, a group of frequency domain features, S1..S4, were found to correctly classify the periodograms in a linear discriminant function of form:

$$\text{State Variable} = a_1\ S_1 + a_2\ S_2 + a_3\ S_3 + a_4\ S_4 \qquad (9)$$

In addition to visual examination of the power spectra, we used stepwise discriminant analysis (PROC STEPDISC in the mainframe computer SAS package) on the ensemble of periodograms to find those features in the frequency domain most useful for discriminating flotation pulp level. After determining the best AE features, we developed linear regression equations relating AE features as dependent variables to the state variables (Equation 7). We then coupled these with Equation 6 in a Kalman filter to estimate pulp level. This required values of Q in Equation 6 and R in Equation 7. The assumption of Q was based on a "white noise" feed fluctuation causing plus or minus 1 inch in pulp level. R was experimentally determined from the AE measurements.

Results

Experimental data from both laboratory and industrial flotation shows an enormous amount of information in the AE power spectral density function about the state of the system. That this information exists is shown using linear discriminant functions of the form of Equation 9. However, since the information is extremely noisy, filtering is necessary. A Kalman filter gives satisfactory results for pulp level when the AE sensor is positioned to minimize the noise (i.e. minimize R in Equation 7).

Figure 1 shows distinctly different power spectral densities for each of four levels in the batch flotation cell. Pattern recognition processes can begin with visual examination, which shows that the frequency of maximum power is different for different levels (26600 Hz for pulp levels 1 and 26600 vs. 26900 Hz for pulp levels 3 and 4). This indicates part of the overall shift in resonating frequency due to changes in pulp mass holdup--similar to hearing a change in the tone of a glass of water as it is tapped with a spoon after filling to different levels. The frequency of maximum power can be a pattern recognition feature.

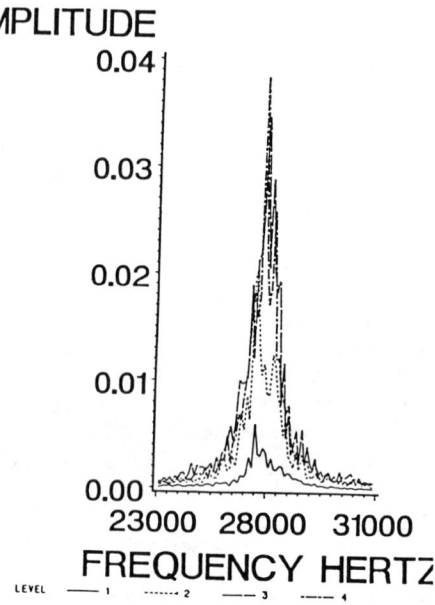

POWER SPECTRAL DENSITY

Figure 1. AE Power Spectra Densities for Four Flotation Pulp Levels.

Other pattern recognition features can be chosen. The power (amplitudes) of the 26600 Hz and the 26900 Hz frequencies are related to pulp level (Figure 2).

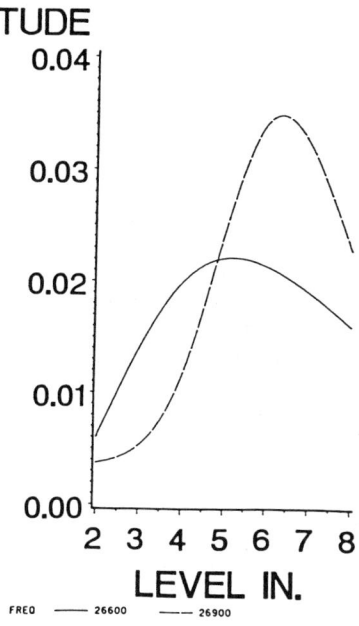

POWER SPECTRAL AMPLITUDES VS. LEVEL

AMPLITUDE

FREQ ——— 26600 ——— 26900

Figure 2.   Relation of Two Power Spectral Density Features to
Pulp Level.

The PROC STEPDISC search for more features was conducted
on 78 periodograms taken at four pulp levels. The best in
order of importance were the power at 25700, 25800, 26500,
25900, and 25600 Hz. Table 1 shows the accuracy of classifying
the 78 periodograms to their correct pulp level category using
these five features.

===
Table 1.   Classification Summary Using A Linear
Discriminant Function of Power at Five Key Frequencies.

| Classified from | Percents Classified To Pulp Level | | | |
Pulp Level	1	2	3	4
1	90	5	0	5
2	31	47	16	4
3	10	15	60	15
4	10	10	32	48

===

A different set of AE periodograms was collected from industrial flotation cells, only the position of the sensor was changed with respect to the steady state pulp level. The results were similar, with high classification accuracy for certain positions of the sensor, and lower accuracy for the others.

The accuracy achieved by the linear discriminant analysis suggests that the periodograms contain enough information to estimate pulp level using a Kalman filter, despite some misclassification and noise. To determine how fast a Kalman filter could sense a step change in pulp level we started at an initial pulp level and "fed" the filter features from a sequence of ten experimental periodograms taken at a pulp level four inches away (called the "zero" level). We calculated the discrete Kalman filter gain, L, and updated the estimate of the pulp level recursively with each new periodogram (Z).

$$X_{new} = X_{old} + L(Z - J X_{old}) \tag{10}$$

where

$$L = PJ'(JPJ' + R)$$

and

$$P_{new} = (I - KJ)P_{old}$$

Figure 3 shows how fast the Kalman filter could estimate the actual pulp level from an initial condition four inches away, given two sequences of periodograms, one more noisy than the other. The degree of noise depends on the relative position of the sensor and pulp level. Figure 4 shows one of the pattern recognition features (power at 26600 Hz) from each of the two sequences of periodograms in order to indicate the difference in noise. In both cases the Kalman filter responds rapidly. However, in the noisy case the estimated pulp level occasionally deviates from the actual pulp level by plus or minus two inches, which is unacceptable for control purposes. In the less noisy case the estimated pulp level deviates at most about a quarter of an inch, which is acceptable for control. Since both degrees of noise were experimentally determined, depending on the position of the AE sensor with respect to pulp level, we conclude that an AE sensor may be industrially feasible, and the sensor should be positioned several inches above pulp level on the outside of the flotation cell.

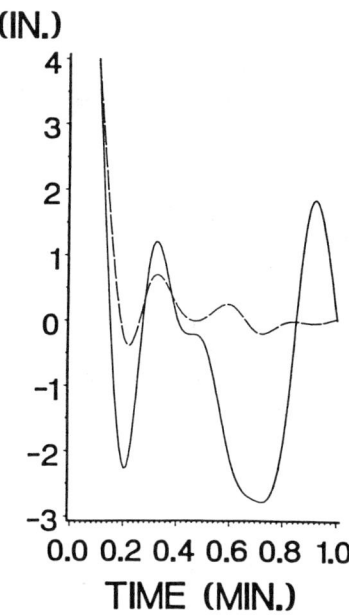

Figure 3. Kalman Filter Estimate of Pulp Level Using a Time Ensemble of More and Less Noisy AE Features.

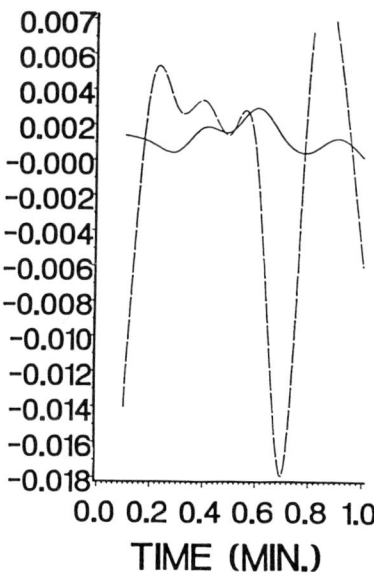

Figure 4. Typical AE Features (Fourier Coeffients) in millivolts from more and less Noisy Ensembles.

271

The nine measurement features consisted of seven powers at key frequencies and two frequencies at maximum powers. A time interval of 0.1 minute was experimentally about the "best case" 10 MHz 80286 processing speed of digitizing an AE sound wave, performing an FFT, extracting the measurement features, and updating the pulp level estimate with the Kalman filter using 10 MHz 80286 processing speeds. Hence Figure 3 shows that about a quarter of a minute is needed for the Kalman filter to correctly estimate pulp level after a step change. If only five measurement features were used instead of the nine, not much time is saved from the 0.1 minute per iteration. However, the Kalman filter takes much longer (a minute and a half) to respond to a step change in pulp level of eight inches (Figure 5). Using computer processing power and more measurement features is profitable.

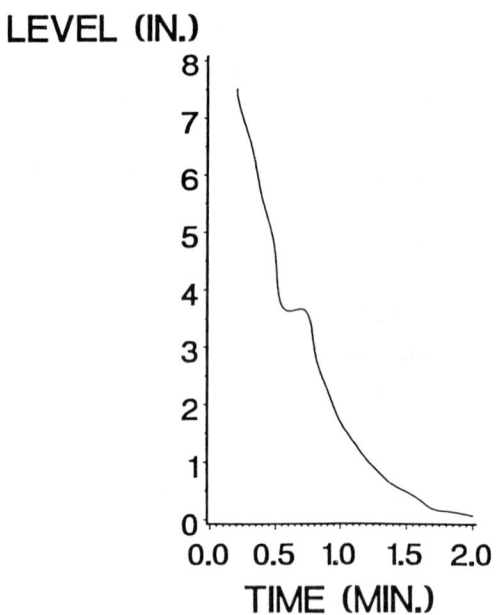

Figure 5. Kalman Filter Estimate of Pulp Level using only Five AE Features in a Time Ensemble.

Conclusion

Using random signal processing techniques, AE sensors give reasonable estimates of pulp level. The optimal estimation of pulp level with a Kalman filter can be integrated into a control unit which will contain computer coupled AE sensors. We have presented results on data collected at one time and analyzed separately at a later time. The next step is to have real time estimation. This kind of test needs significant hardware and software development. Another problem is periodic recalibration. The program software must allow updating of the regression equations (Equation 7).

The hardware and software requirements for this estimation and control sequence are relatively inexpensive. An industrially hardened personnel computer may be satisfactory with its high level languages, microprocessor speeds, and ability to handle matrix manipulations and fast Fourier transforms. A contained spectrum analyzer that can directly receive the analog signal from the AE sensor would help. If these methods prove too slow, large scale integrated circuits and/or the utilization of IC's in combination, such as using array processors, FFT chips, etc., would be necessary. Computer control has proven itself as reliable, maintainable, and cost effective. With more accurate process models, this method of control can bootstrap its way to better product quality, higher production and lower costs.

References

1. C. L. Smith, "Liquid-Measurement Technology", Practical Process Instrumentation and Control, McGraw-Hill (New York, Chemical Engineering, 1980), 116-125.

2. D. E. Zientara, "Measuring Process Variables", Practial Process Instrumentation and Control, McGraw-Hill (New York, Chemical Engineering, 1980), 105-115.

3. H. P. Walsh, "An Instrument to Detect the Position of the Froth/Pulp Interface During Flotation", NIMTEK Rep. No. M4, Council for Mineral Technology, Randburg, South Africa, 1982.

4. V. G. Artem'ev, et al. "Foam Thickness and Pulp Level Measuring Device for Flotation Machines, Tsvetn. Met. 7, (1979) 104-105.

5. P. Thwaites, "Continued Development of Copper Flotation Control at the Kidd Creek Concentrator", Can. Min. Metall. Bull. 767, (1983) 41-46.

6.   P. Gajowski, and W. Penar, "Automation in the Processing of Non-Ferrous Metals Ores", _Pol. Tech. Rev. 1_, (1984) 2-4.

7.   R. G. Brown, _Introduction to Random Signal Analysis and Kalman Filtering_, John Wiley, New York, 1983.

8.   T.P. Harring, P.G. Doctor, and K. Prisbrey, "Analysis of Acoustic Emission Spectra of Particle Breakage in a Laboratory Cone Crusher", _Trans. AIME/SME 170_, (1982) 1878-1882.

SOUND LEVEL PARAMETER DETERMINATION AND APPLICATION IN CONTINUOUS LABORATORY BALL MILLING

John L. Watson and Michael P. Tonelis
Department of Metallurgical Engineering
University of Missouri-Rolla
Rolla, MO 65401.

Abstract

The size reduction of many materials by rod and ball milling is an expensive and inefficient process. Any improvement in mill operation would result in considerable economic savings, and the use of mill sound, as a control parameter, represents one possible method by which improvements may be achieved. This research investigated the relationships between mill sound parameters and various mill operating parameters. The sound pressure levels at 1 kHz were monitored for laboratory continuous ball mill using a microcomputer/data acquisition system and the data were analysed by means of computer software. The feed materials used in the laboratory were dolomite, magnetite and traprock, and the major mill parameters investigated were pulp density, feed rate and product sizing.

The results illustrated that mill sound parameters can reflect mill state variables without disrupting the mill circuit, and without exposing the measurement sensors to the harsh environment normally associated with milling operation instrumentation. It was shown that the internal mill pulp viscosity can be inferred from mill sound level measurements and that an optimum pulp viscosity exists for the fine grinding of magnetite under laboratory conditions. The next stage in the research will be full scale industrial testing of the mill sound concept and it is anticipated that the inclusion of a mill sound parameter, in modern mill control alogorithms, could effect improvements in mill operation.

Introduction

Considerable research has been undertaken to lessen mill noise to prevent possible damage to the hearing of mill operators, but mill noise has also been used as a mill control parameter. For many years mill sound pressure level has been monitored and feed rate control (1,2) has been based upon this noise measurement. The use of mill sound pressure level, as a parameter to reflect mill performance, has many advantages in that the sensor is non-invasive, and as such it is not subject to the harsh internal environment of a mill, and it does not interfere with the grinding process in any way. Recent research (3,4,5) has demonstrated that the use of mill noise need not be limited to feed rate control, and that the measurement of sound levels emanating from a batch ball mill can be correlated with a variety of mill and mill charge parameters. The basic premise of this research (3) is that mill noise is essentially a function of the collisions of the steel balls with the mill wall. The mill speed, ball load, and ball sizing will obviously influence these collisions and their influence has been detected in the mill sound levels. In addition it has been shown that the particle size distribution in the mill also influences the noise levels (4). Coarser particles are capable of blocking the ball / wall collisions and hence reduce the sound levels, while finer particles cannot prevent the collisions and thus noise levels increase. The residence time of a particle in a mill will determine its final size and obviously finer particles will result from longer residence times and thus residence times will influence mill noise. In general a higher feed rate will produce shorter mill residence times and hence a coarser product and a quieter mill. The particle sizes capable of blocking ball/wall collisions are a function of ball size and mill geometry. Further research (5) using pulps has revealed that pulp density and viscosity also strongly affect the mill noise levels and various rheological regimes within the mill have been identified and determined by mill noise levels. It has been demonstrated that the mill sound levels reflect changes in pulp viscosity, which result from particle number and size variations within the pulp. The pulp viscosity, for certain ores, can change from dilatant to pseudoplastic and finally to pseudoplastic with yield as the pulp density increases and particle size distribution varies. It has been shown, for such ores, that grinding may be maximized, for a constant volume pulp, in the pseudoplastic regime. Overall it is considered that mill noise for wet grinding is a function of -

    a) the number and size of particles in the mill.
    b) the viscosity of the pulp in the mill.

It is apparent that ore character and pulp feed rate and density will greatly affect a) and b) and hence be the major mill parameters which control mill sound levels.

It is considered that mill sound level parameters have the potential to provide valuable on-line operating data, which can supplement existing mill data to improve control. One possible example of sound usage relates to a change in ore type, which in a typical modern mill is not detected until cyclone overflow parameters change. However it appears that such ore changes may well be detectable by sound measurements in the rod mill, if the mill operating conditions remain the same. Some work in this field has been undertaken in the USSR (6) and in Finland (7), but little research has been published in the last ten years.

The recent research (3,4,5), reviewed above, was carried out in the laboratory using dry and wet batch ball milling, and a variety of minerals and ores were treated. However very little industrial grinding is done in the batch mode and therefore if the work is to have any industrial application, it is necessary that data be produced from continuous milling tests. Therefore the research reported in this paper represents the sound pressure level responses of a continuous laboratory mill to a variety of feed conditions, including feed rate and pulp density. In addition the mill product size distributions are presented and the collected mill data analysed to determine any meaningful relationships between the various measured parameters.

Experimental

A 30 X 30 cm continuous ball mill was fed with ore from a vibrating feeder and storage hopper. The mill was operated both dry and wet, and water was added through a rotameter for wet milling. The mill volume was 22 liters and the standard charge was 235 steel balls of 2.5 cm diameter and weighing 16 kg. A sound level detection system (GenRad 1988 Precision Sound level Indicator and Analyser) was linked to an Apple II computer through an Isaac data acquisition system to permit the mill noise level to be monitored and recorded. The milling circuit is illustrated in Fig. 1. together with the noise monitoring circuit. Typically noise data at a frequency of 1 kHz was collected over a period of 25 seconds and then averaged and stored to represent a 30 second grinding period. Mill product samples were taken manually over a 60 second period at various time intervals throughout a grind. Such samples were then sized by wet screening to provide a measure of the effectiveness of the grind. Full details of the experimental circuit are available elsewhere (8)

The feed materials used for the research included dolomite from the University of Missouri-Rolla mine, and magnetite ore and trap rock waste from the Pea Ridge Mine, Sullivan, MO. Typical feed size was -4.7 + 2.8 mm and feed rates of 15 to 760 g/min were used. The following sets of tests using various feed materials were undertaken -

a) Reproducibility tests for dry grinding.
   Dolomite ore at feed rates of 60, 225 and 760 g/min.
b) Dry grinding at various steady state feed rates.
   Dolomite, magnetite and traprock at feed rates of 100 g/min
c) Dry grinding at combinations of feed rates.
   Dolomite, magnetite and traprock at sequential feed rates of 15, 85, 225, 85 and 15 g/min for hour intervals.
d) Steady state wet grinding tests at various pulp densities.
   Magnetite ore at 100 g/min at pulp densities of 60 - 85% solids by weight.
e) Wet grinding tests at combinations of feed rate.
   Dolomite at 100, 200, 300, 200 and 100 g/min in a 60% solids by weight pulp for hour intervals.
f) Computer controlled pulp density tests.
   Dolomite at 100g/min with pulp densities of 40 - 70%.

Test f) refers to tests performed at the end of the project, when the feed system to the mill was revised, and computer control introduced for ore and water as illustrated in Fig. 2. The major objective in these tests was to investigate the effect of computer manipulation of pulp density on the mill operational parameters and the mill noise.

Results and Discussion

a) Reproducibility Tests.

Figures 3,4, and 5 illustrate the mill sound pressure levels for three

277

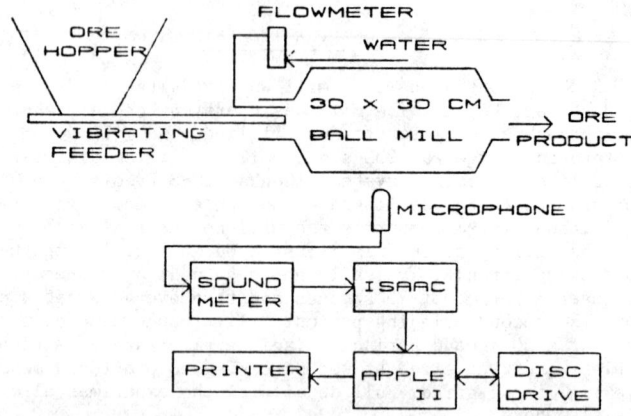

Figure 1. Laboratory Circuit for Mill Noise Data Collection.

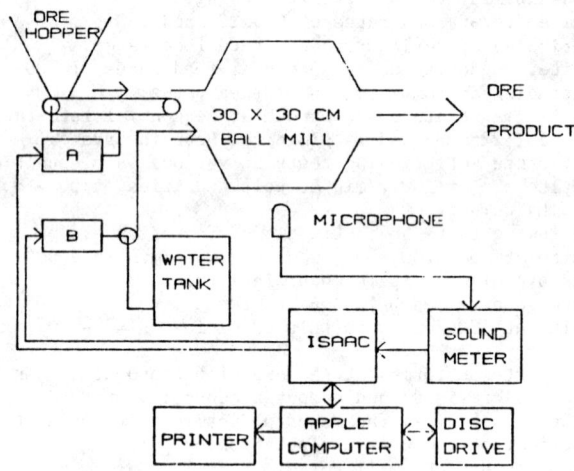

Figure 2. Laboratory Circuit for Mill Noise Data Collection with Computer Control of Mill Feed.

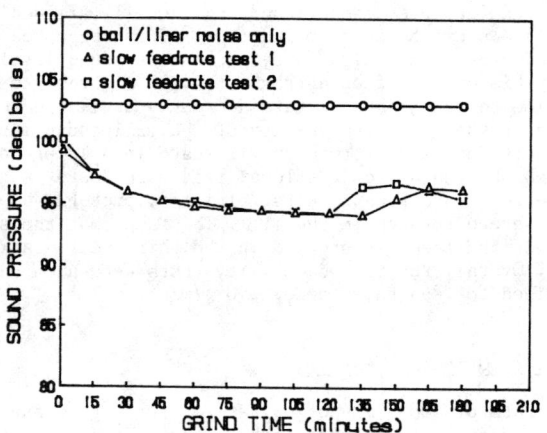

Figure 3. Mill Sound Pressure Level Variations with Grind
Time for a 60 g/min Dolomite Feed.

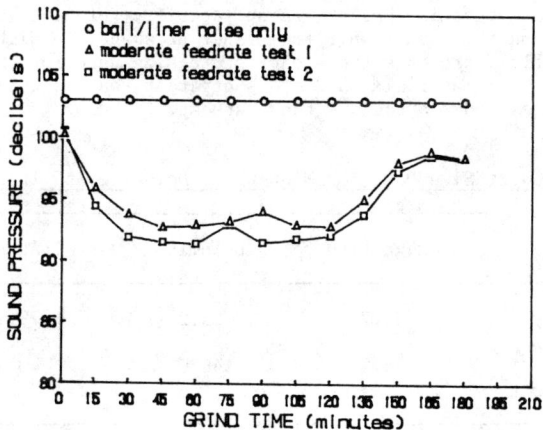

Figure 4. Mill Sound Pressure Level Variations with Grind
Time for a 225 g/min Dolomite Feed.

279

feed rates for a three hour grind. The horizontal plot in each figure
represents the mill sound level with no ore charge. The empty mill was fed
with the dolomite at 60, 225 and 765 g/min and in each case the mill noise can
be seen to decrease as ore accumulates in the mill and to then level off.
After two hours the feed was discontinued and the mill allowed to empty and
the mill sound levels can be seen to increase. It is suggested that mill
steady state conditions are indicated by a constant sound level and for the
lower feed rate this occurs after approximately 70 minutes. As the feed rate
increases the time to reach steady state decreases until the feed rate exceeds
the capabilities of the mill. The ability of the mill sound level to identify
steady state conditions is confirmed by mill data in a later section. The
reproducibility of the data from the lower feed rate tests is good but the
higher feed rate test gave poor results initially, probably due to over
feeding causing severe surging in the mill. The fact that the sound levels did
not increase, when the feed was stopped in the third test, indicates an
overloaded mill. Overall the reproducibility tests were acceptable and future
tests were confined to feed rates below 250 g/min.

b) Steady State Dry Grinding Tests.

For a feed rate of 100 g/min of magnetite ore, Fig. 6 shows the mill
sound levels and product sizing for a 5 hour test. It appears that the mill
operation has settled down after 90 minutes according to the noise levels, but
the size data suggests that steady state conditions were not achieved for a
further 30 minutes. Figure 7 illustrates the discharge rate of the mill which
confirms that steady state conditions are achieved, at least in terms of mass
transport, after 90 minutes. This shows that mill sound levels are capable of
signifying the attainment of steady state conditions.

Grinding tests were then carried out for three materials at feed rates of
100 g/min and Fig. 8 shows typical results. It can be seen that steady state
conditions are obtained at 90 minutes for all three materials, and that
magnetite produces the highest sound levels with traprock giving the lowest.
This suggests that magnetite has the least capability of blocking the ball
wall collisions and this can result from finer breakage. ie higher
grindability. Table I presents the steady state size date for the ground
products of the three materials, and it is apparent that magnetite is the
finest of the three and traprock is the coarsest.

Table I. Mill Product Size Data for Magnetite, Dolomite and Traprock.

Size	Magnetite	Dolomite	Traprock
% - 149 microns	89.7	85.2	79.2
% -37 microns	46.9	46.0	40.2

Thus traprock gives a coarse product, which is capable of blocking the noise
producing ball/wall collisions and hence giving the lowest mill sound levels.
Previous results (8) have also shown that magnetite does have the highest
grindability and traprock the lowest. Thus the results for continuous grinding
follow closely those from batch grinding in indicating ore character.

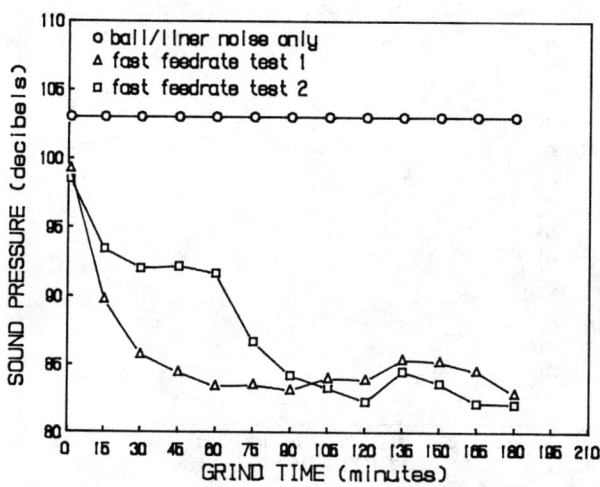

Figure 5. Mill Sound Pressure Level Variations with Grind
Time for a 765 g/min Dolomite Feed.

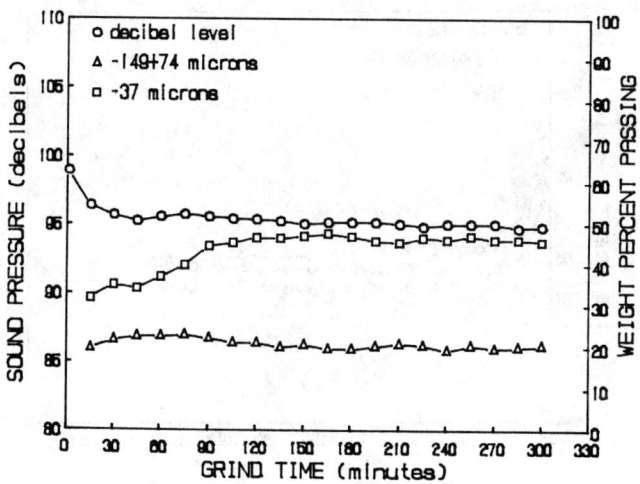

Figure 6. Mill Sound Pressure Level and Particle Size Variations
with Grind Time for a 100 g/min Magnetite Feed.

281

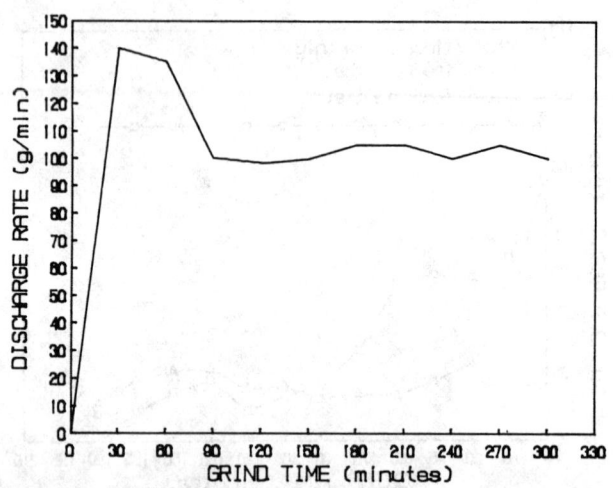

Figure 7. Mill Discharge Rate Variations with Grind Time for
a 100 g/min Magnetite Feed.

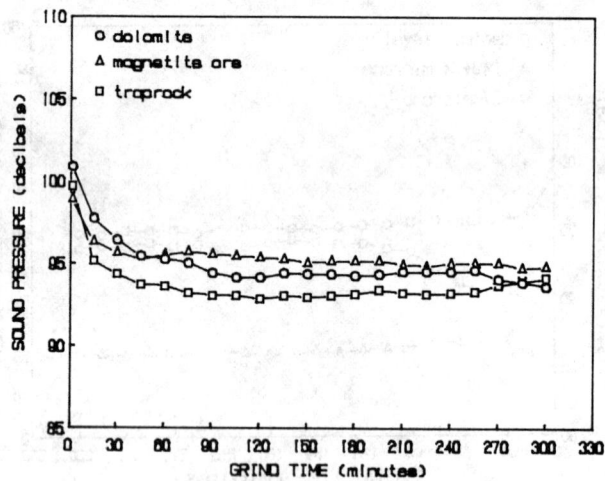

Figure 8. Mill Sound Pressure Level Variations with Grind Time for
Dolomite, Magnetite and Traprock Feeds of 100 g/min.

c) Variable Feed Dry Grinding Tests.

To investigate the effect of a variable feed rate, dolomite was fed at a rate which was increased from 15 to 85 to 225 g/min at hourly intervals, and then decreased at hourly intervals in a six hour test. Figure 9 shows the mill sound level variations, together with product size data. The mill sound initially falls with the new feed and appears to indicate that steady state conditions are not attained at the time of the first feed increase. The second hour does show steady state as does the third hour, with the sound levels decreasing each time. The second half of the test tends to be a mirror image of the first half. From Fig. 9 it is apparent that the product fineness (-37 microns) closely follows the mill sound level which in turn reflects the changing feed rate. The coarser material in the product shows a less rigid inverse relationship with sound and feed rate. Again the results may be explained in terms of ball-wall collisions with the lower feed rate permitting longer particle residence times in the mill, which result in more breakage and hence finer particle sizing. This reduces the ability of the particles to block the noise producing collisions and hence low feed rates are associated with higher mill sound levels. As the feed rate increases so does the coarseness of the particle size distribution and thus the noise levels fall.

The results of varied feed rate tests for dolomite, magnetite and traprock are contrasted in Fig. 10. While each plot does reflect the changing feed rate, it is apparent that steady state conditions are not attained and that the feed rate variations affect the mill sound levels for each material to a differing extent. This is probably related to the density and number of particles present in the mill and could reflect some feed rate inaccuracies. The results of these grinding tests do however confirm the ability of mill sound to reflect the fineness of grind in a continuous mill, as shown in Table II.

Table II. Mill Product Size Data for Magnetite, Dolomite and Traprock at Various Feed Rates.

Size	Magnetite		Dolomite		Traprock	
	60 min (15g/min)	120 min 85g/min	60 min 15g/min	120 min 85g/min	60min 15g/mi	120 min 85g/min)
% -149 micron	94.6	92.8	96.8	90.2	95.3	87.1
%-37 micron	79.3	55.0	69.4	47.1	76.0	50.0

The effect of increased feed rate is to decrease the breakage for all materials as expected. At grind times of 60 and 120 minutes the dolomite displays the lowest mill sound levels and accordingly it should have the coarsest size. This is true at -37 microns but not at -149 microns. This suggests that the -35 micron material possibly has a greater effect on mill sound levels than the material in the size range 75 to 149 microns. Magnetite and traprock show similar mill sound levels and in general their sizings are also similar. However it is apparent that departure from steady state conditions and variable feed rates complicate the noise relationships considerably.

Overall the dry grinding tests indicate that mill noise is controlled by the

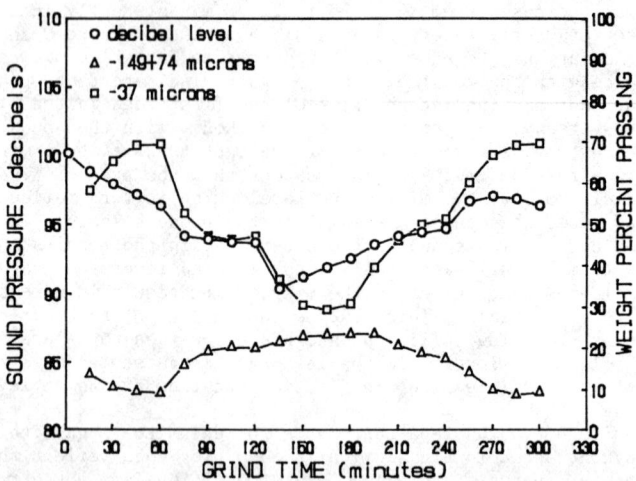

Figure 9. Mill Sound Pressure Level Variations with Grind Time for a Stepped Dolomite Feed Rate.

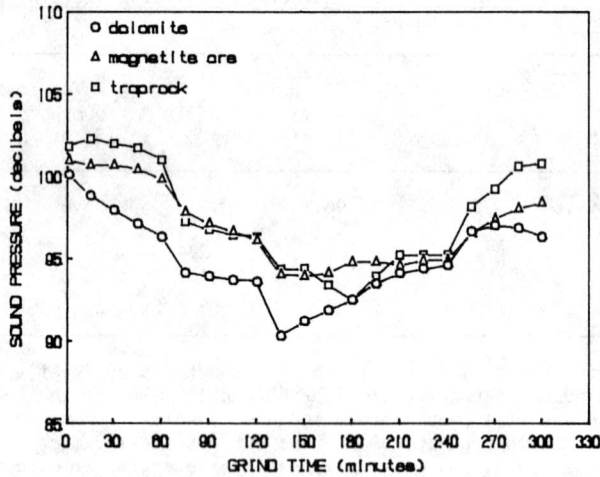

Figure 10. Mill Sound Pressure Level Variations with Grind Time for Stepped Dolomite, Magnetite and Traprock Feed Rates.

size and number of particles in the mill and these parameters in turn are controlled by ore character and feed rate.

d) Wet Grinding Tests.

The introduction of water into the grinding process alters the mass transport and grinding mechanisms a great deal. The mill sound levels will now be a function of pulp viscosity in addition to ore character and feed rate. Figure 11 shows the variation of sound level and product size data with time for feed of 100 g/min of magnetite at a pulp density of 60% solids by weight. It appears that the mill takes approximately 75 minutes to attain steady state and this is confirmed by the product pulp density and discharge rates shown in Fig. 12. The production of fine particles in this case is approximately 5% greater than that achieved for the same feed rate in dry grinding.

The effect of decreasing the mill water, and hence increasing the pulp density, on mill noise is shown in Fig. 13. It can be seen that steady state conditions are attained in the same time for each plot, and that mill sound levels are not simply related to pulp density. As the density increases from 60 to 65% the mill sound levels fall, even though the magnetite feed rate remains constant, with a thicker pulp preventing ball-wall collisions. At 68% (not shown for clarity) and 70% solids the mill is noisier than at 65% solids and this is the opposite of what would be expected from the foregoing arguments. However this phenomenon was also observed in batch grinding (5), and there it was explained in terms of a viscosity decrease due to the pulp rheology changing from dilatant to pseudoplastic. Above 70% solids the mill sound levels fall rapidly due to the pulp thickening up to such an extent that pseudoplasticity with yield occurs and the balls become trapped in the pulp.

The relationship between sound and pulp density is represented in Fig. 14 for seven tests at three grind times, and it can be seen that there are three distinct regions present. Observation of the pulp conditions suggested that below 65% the pulp could be classified as dilatant, while above 65% the pulp displays pseudoplasticity and above 70% the onset of yield was apparent. The sizing data for these tests is presented in Table III for 60 and 120 minute grind times.

Table III. Mill Product Size Data for Magnetite at 60 and 120 Minute Grind Times for Pulp Densities of 60 -75 % Solids by Wt

Pulp Density %solids by wt.	60 min % -37 micron	120 min % -37 micron
60	50.1	51.4
63	45.4	51.3
65	40.9	37.7
68	47.3	60.7
70	50.6	50.6

From Table III and Fig. 14 it can be seen that the mill sound levels and the production of fine particles are closely related over the range of pulp densities tested. As pulp density increases the thicker, more viscous pulp will tend to protect the particles from the grinding action of the balls and hence less grinding and mill noise would be expected. However the above data shows that, between 65 and 70% solids, the pulp permits more noise and more

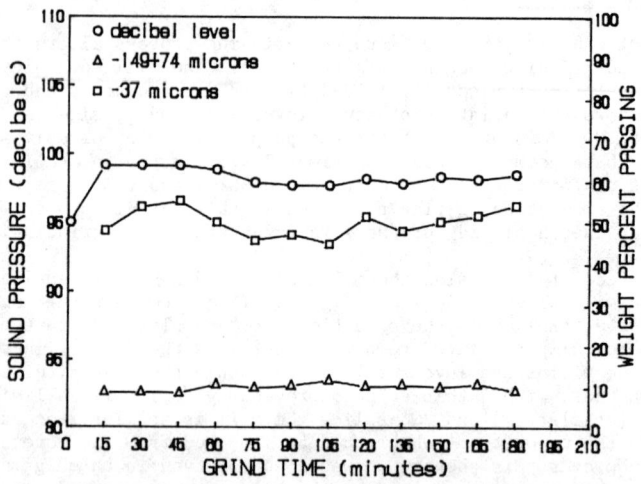

Figure 11. Mill Sound Pressure Level and Particle Size Variations
with Grind Time for a 60% Solids Magnetite Pulp.

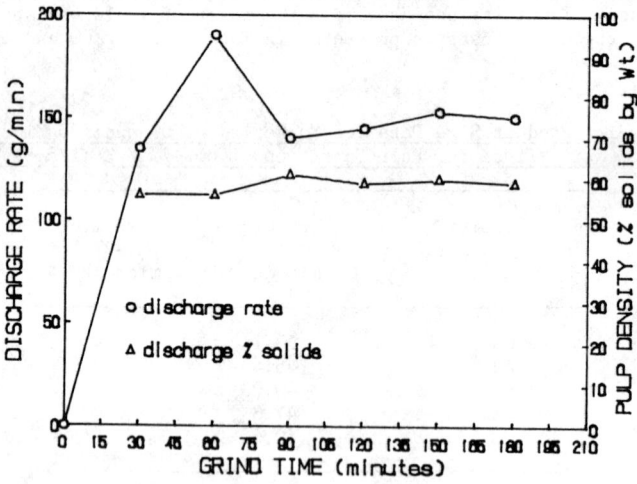

Figure 12. Mill Discharge Rate and Pulp Density Variations with
Grind Time for a 60% Solids Magnetite Pulp.

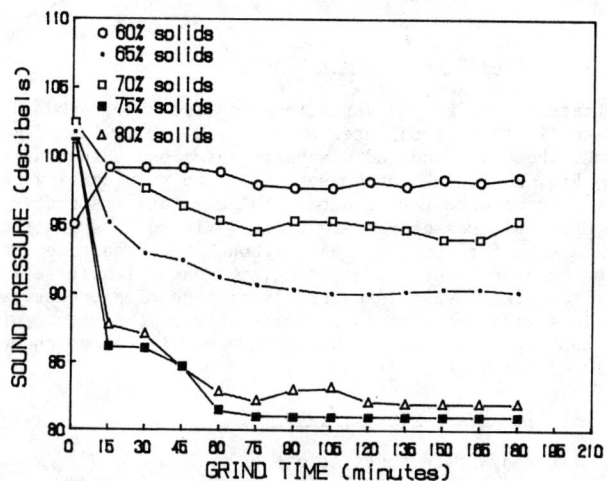

Figure 13. Mill Sound Pressure Level Variations with Grind Time for Magnetite Pulps.

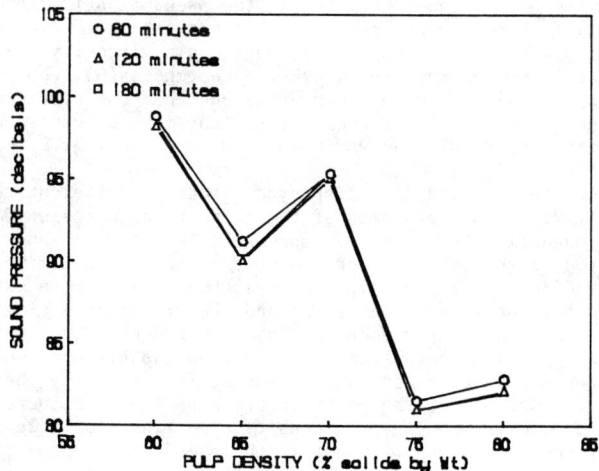

Figure 14. Mill Sound Pressure Level Variations with Pulp Density for a Magnetite at Increasing Grind Times.

grinding and this region corresponds to the pulp observations of pseudoplasticity. It therefore appears for magnetite that a rheological change permits increased grinding due to decreased pulp viscosity and that this change may be identified by an increase in mill sound levels.

e) Wet Grinding Tests with Feed Rate Variations.

To illustrate the effect of variations in feed rate, a 60% dolomite pulp was fed to the mill at ore feed rates of 100, 150, 300, 150 and 100 g/min over a 6 hour period. The mill feed and discharge rates and pulp density are illustrated in Fig. 15 and it can be seen that the pulp density achieves steady state at approximately 90 minutes, but discharge rate does not appear to attain the appropriate steady state values. Figure 16 shows the mill noise and product size data for the test, and although the feed rate transitions are not as sharp as they were for dry grinding, the data clearly reflects them. Again higher feed rates result in shorter residence times, decreased breakage, coarser particle sizing and thus lower noise levels. The effect of changes in particle size on pulp viscosity in this test may well oppose the particle size effect on mill sound levels.

f) Computer Controlled Pulp Density Tests.

A series of steady state tests at a feed rate of 90-95 g/min of dolomite and varying water addition produced the mill noise / pulp density relationship illustrated in Fig. 17. It is apparent that mill noise is strongly related to pulp density in the region 45 - 70% solids. A plot of the product size factor (represented by wt% greater than 104 microns / wt% less than 75 microns ie. coarseness) against pulp density and mill noise is shown in Fig. 18. It is apparent that particle coarseness decreases with pulp density and increases with mill sound level. Together Figs. 17 and 18 suggest that a quieter mill is produced by a thicker pulp and a finer product results. This relationship is, of course, limited by the onset of yield in the pseudoplastic viscosity regime (ie 67% solids). In these tests for dolomite feeds, the data did not indicate the relationships apparent for magnetite, where pulp viscosity regimes were evident. This confirms the results of other researchers (9), who found that only certain ores display rheology boundaries and that the particle size distribution and the double layer chemistry of the mineral surfaces determine the ability of the ore to display variations in pulp rheology.

In addition the dolomite data displayed a totally different relationship for mill sound level with coarseness of the grind than that seen for magnetite. For magnetite the mill sound pressure levels were shown to be inversely related to the mill product coarseness, but for dolomite the relationship is direct over the pulp range plotted. This may be explained in terms of the number and size of particles and their residence time within the mill and their effect on pulp viscosity. The higher density and grindability of magnetite will permit finer particle production capable of precipitating viscosity changes as the pulp density increases. The viscosity changes are more effective in controlling the particle breakage then the increased residence time due to increasing pulp density. The finer particle sizes would be cushioned by the viscous pulp except in the pseudoplastic region, where pulp viscosity falls and particle breakage improves. Thus these changes result in the viscosity controlled relationships exhibited in Fig. 14. For dolomite, it is difficult to understand why the particle breakage increases over the pulp density range 40-70% solids. Again an increasing pulp density at constant solids feed rate will produce an increased residence time, which will permit increased breakage. The corresponding mill sound level decrease may simply be explained in terms of increased pulp viscosity. It is apparent that the increased breakage is not capable of influencing the mill noise due to the

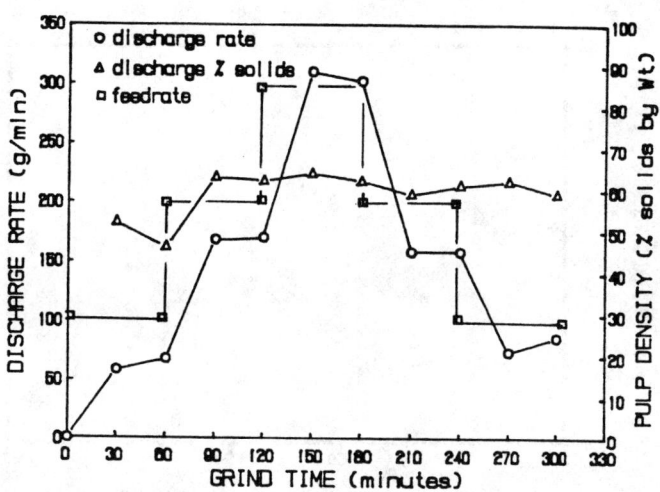

Figure 15. Mill Discharge Rate and Particle Size Variations with Grind
Time for a Stepped Dolomite Feed Rate (60% Solids).

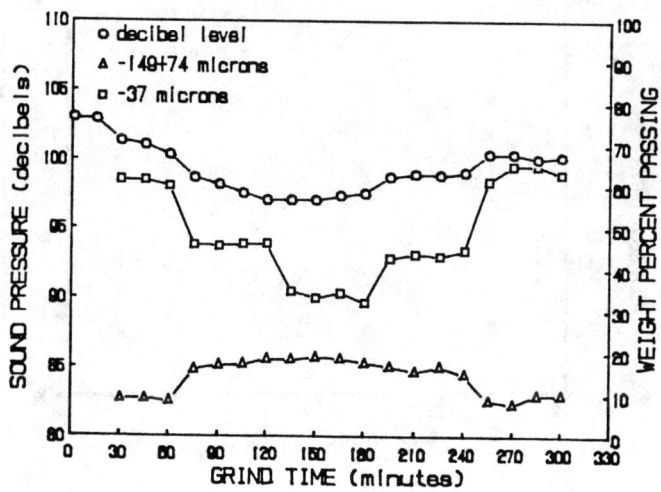

Figure 16. Mill Sound Pressure Level and Particle Size Variations with
Grind Time for a Stepped Dolomite Feed Rate (60% Solids).

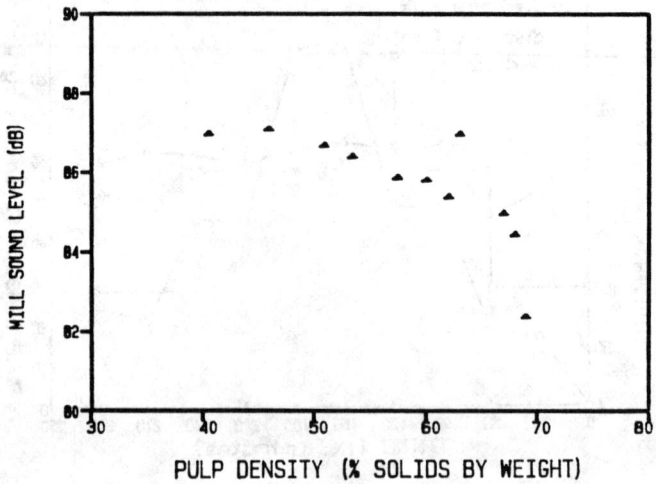

Figure 17. Mill Sound Pressure Level Variation with Pulp Density for a 100 g/min Dolomite Feed.

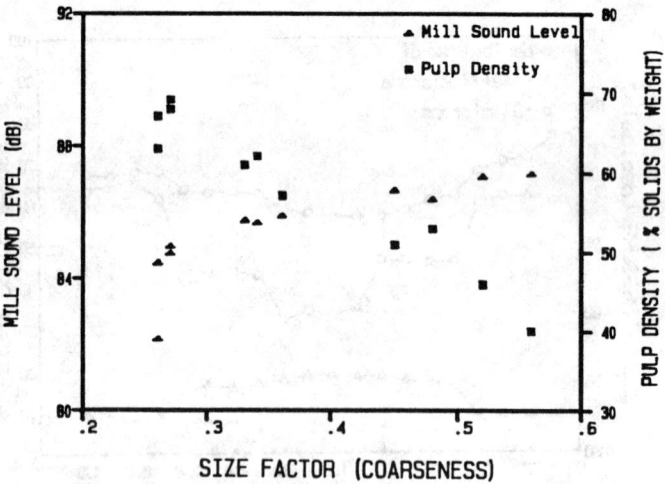

Figure 18. Mill Sound Pressure Level and Pulp Density Variations with Product Coarseness for 100 g/min Dolomite Feeds.

overwhelming effect of increased pulp viscosity. One possible explanation for the increased breakage may be the interaction of particles in the pulp causing them to agglomerate and to be readily exposed for collision with the balls, and this would further provide noise reduction. Thus for wet milling it appears that mill noise levels are more influenced by pulp viscosity then directly by particle sizing and that pulp viscosity can be both beneficial and detrimental in terms of grind fineness, depending upon the ore character and particle size distribution.

Utilizing the ability of the computer to continually vary the water addition rate, a 4.5 hr test was undertaken with the ore rate at 108 to 112 g/min and the water rate being reduced from 100g/min at a rate of 0.25g/min/min. The mill sound level variations are shown in Fig. 19, together with the non steady state pulp density and water rate. Again the pulp density and mill sound levels can be seen to be inversely related and this is confirmed in Fig. 20. At a pulp density of 67% solids by weight the relationship breaks down, which suggests the state of the mill abruptly changes and this is linked to the onset of pseudoplasticity with yield. Figure 21 displays the mill noise variation with product coarseness and again a strong linear relationship is seen. This may again be explained in terms of pulp viscosity increases due to increased pulp density. The finer grind results from longer residence times as the pulp rate decreases with increasing pulp density, and this is verified by the pulp density-coarseness plot in Fig. 21.

It is apparent from these tests, where solids feed rate was kept constant and hence the pulp density was varied by means of the water rate, that mill sound levels can be related to pulp viscosity and to a simple grinding parameter. The marked difference in mill sound level and product coarseness response with pulp density, for magnetite and dolomite, indicates the ability of mill noise to reflect pulp character. Finally the fact that these relationships exist for continuous grinding is very encouraging with respect to the application of this research topic to industrial mills.

Conclusions

1 It has been demonstrated that mill noise levels from a continuous ball mill are capable of reflecting mill state variables in a similar manner to previous results obtained for a batch ball mill. This means that the noise parameter premise is applicable to full scale grinding operations and that noise sensors could well have the capability of providing additional on-stream data upon which to base the operation and control of grinding mills. The monitoring of mill noise is extremely simple to accomplish as the sensors are non-invasive and hence not subjected to the rigors of the mill, and do not interfere with the operation of the mill. In addition the instrumentation is not expensive and may be easily integrated into existing control systems.

2 Specific areas in which a mill noise parameter could have industrial application are -

    a) Mill feed rate - mill sound levels can indicate approaching grind out or mill overload conditions.
    b) Mill pulp viscosity - mill sound levels will reflect the internal pulp viscosity and permit the mill to be operated in the desired viscous regime to optimize the grind.
    c) Mill feed character - mill sound levels are capable of indicating changes in feed character on-stream when used in combination with other mill parameters.
    d) Mill product size distribution - mill sound levels are

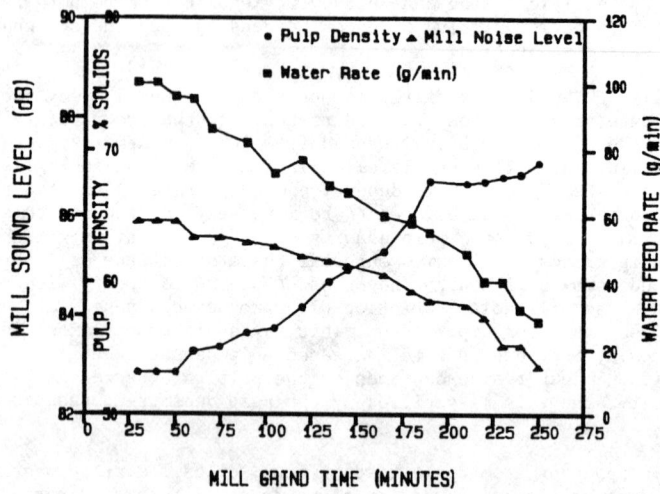

Figure 19. Mill Sound Pressure Level, Water Feed Rate and Pulp Density
Variations with Grind Time for 100 g/min Dolomite Feeds.

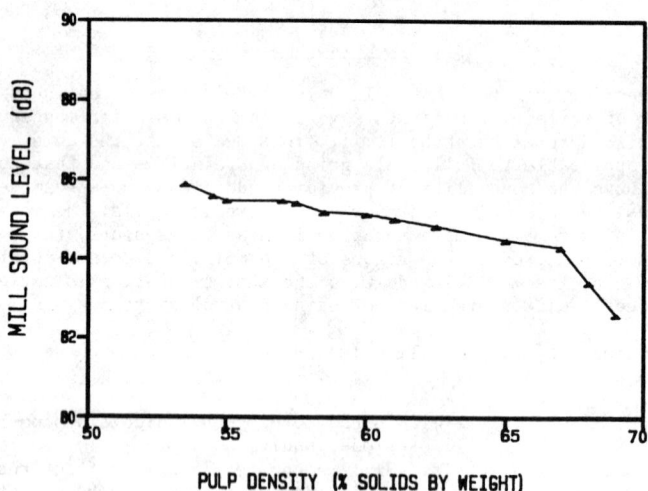

Figure 20. Mill Sound Pressure Level Variations with Continuously
Varying Pulp Density for a 100 g/min Dolomite Feed.

292

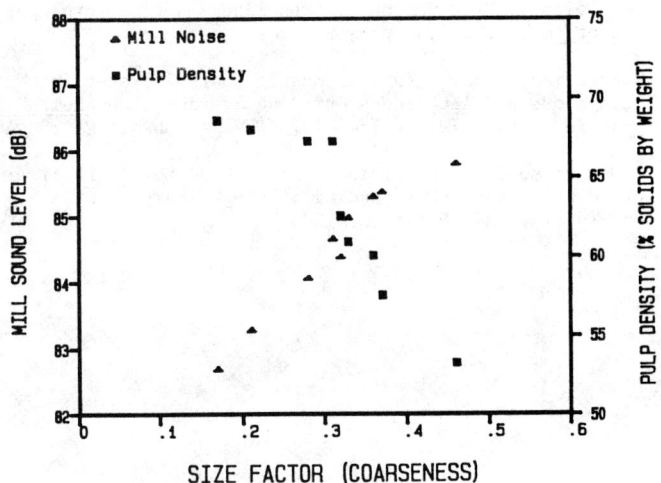

SIZE FACTOR (COARSENESS)

Figure 21. Mill Sound Pressure Level and Pulp Density Variations with Product Coarseness for a 100 g/min Dolomite Feed.

sensitive to the size distribution of the ore charge for dry grinding and hence the grind size could be monitored using mill noise in conjunction with existing mill parameters. For wet milling the size effect on mill sound levels is overwhelmed by the pulp viscosity effect.

3 This research, together with previous research, has provided justification for the mill noise concept to be applied in an industrial situation. Therefore the next stage will be to collect data from an industrial operation to determine the applicability of mill noise as a control parameter. In addition further laboratory work will be undertaken to close the control loop and to use mill sound levels to permit pulp density manipulation for optimum grinding.

References

1   H.Hardinge, "The Electric Ear, a Device for Automatically Controlling the Operation of Grinding Mills by Their Sound", AIME Trans. , (134), 1939.

2   J.O.Bernt, "Automatic Control of Feed Rates to Grinding Mills in the Cement Industry", ISA Annual Conf., New York, NY, Oct, 1964.

3   J.L.Watson, "An Analysis of Mill Noise", Powder Tech. , Vol 41, No 1, p83-89, 1985.

4   J.L.Watson and S.D.Morrison, "Indications of Grinding Mill Operation by Mill Noise Parameters.", Part. Sci and Tech. , Vol 3, No 1, p49-63, 1985.

5   J.L.Watson and S.D.Morrison, "Estimation of Pulp Viscosity and Grinding Mill Performance by Means of Mill Noise Measurements." Journal of Mineral and Met Proc. , AIME-SME Vol 4, p216-221, Nov, 1986.

6   I.I.Belyaev, "Automation of Nephelite and Limestone Crushing",
    Tsvet. Metal. , (2), p53-59, 1962.

7   H.Aurasmaa et al., "An Audiometric Control System for Wet Semi-
    Autogenous and Autogenous Grinding Systems", Dechema-Mono-
    graphien , Band 79, Nr 1549-1575, Verlag Chemie, p435-446, 1976.

8   S.D.Morrison, "An Analysis of Mill Sound Levels to Aid in
    the Control and Understanding of the Grinding Process",
    PhD Thesis, UMR, December, 1986.

A BASIS FOR ON-LINE ESTIMATION OF COARSE PARTICLE SIZE DISTRIBUTIONS

J. A. Herbst, K. Y. Lee, and F. Alba J.

Generic Mineral Technology Center in Comminution
University of Utah
Salt Lake City, UT 84112

Abstract

A new method is presented for converting linear intercept measurements
made on projected images of particles to the corresponding mass distribution
based on mesh size. The method is general and can be applied to particles
of irregular shape. The measurement technique which uses an image analyzer
is described and the numerical method for the transformation is provided.
The method was confirmed experimentally by successfully reconstructing the
mesh size distribution for several different particle populations. This
non-contact technique is suitable for the on-line measurement of coarse
particle size distributions.

Introduction

Particle size reduction (comminution) is a basic unit operation of mineral processing as well as in the chemical, cement, stone and coal industries, etc. The size distribution of feed and products are key variables for size reduction devices. It is well-known that plant performance for these types of operations is very sensitive to the size distribution of the material processed. The objective of automatic control systems for such plants is often to maintain operating conditions that result in either maximum throughput at a constant product size or maximum throughput at the finest product size possible.

Even though the importance of a knowledge of the size distribution of the product of a comminution device is widely recognized, control strategies based on its direct measurement are not very common in industry. Accurate and robust on-line size analyzers are commercially available for fine materials (< 1mm) but coarse material size measurement has received relatively less attention. In crushing, for example, Herbst and Rajamani pointed out that the crusher product size should be treated as a principal measured and controlled variable (1). However, the actual implementation of such a strategy requires a reliable on-line instrument for coarse particle size measurement. For this reason, on-line measurement of the particle size distribution of coarse material as it moves along a plant conveyor belt has received increasing attention during this decade.

A direct extension of conventional off-line techniques to on-line analysis, such as automatic sampling, sieving and weighing, would be very time consuming and inaccurate due to sampling error, unless an impractically large sample size was used (2). A more suitable approach consists of a non-contact surface measurement employing optical imaging and electronic image processing techniques. This technique overcomes the material handling and sampling statistics problems and is particularly appropriate for continuous on-line measurement.

The first commercial system of this type, based on light-shadow profiles measurement, was developed at the Julius Kruttschnitt Mineral Research Centre in Queensland, Australia (2), with a first prototype manufactured by FOXBORO and currently commercialized by ARMCO AUTOMETRICS (3). A second system, recently described in the literature (4), has been developed by CONTAC INGENIEROS LTDA., Chile. This device is based on a similar concept to the ARMCO AUTOMETRICS unit, but it uses image analysis technology instead of the "light-shadow" method.

There is an inherent drawback associated with both mentioned systems due to material segregation occurring on the belt. Since these instruments scan only the top surface (e.g. the Autometric's sizer scans only the middle line on the surface), vertical and horizontal material segregation constitute a significant problem causing the instrument to deliver a biased estimate of the actual distribution. In many cases, however, this segregation can be eliminated or at least maintained approximately constant so that the performance of a control strategy based on such a measurement is acceptable even though the measurement itself is not representative of the actual size distribution.

A more fundamental problem associated with the current systems, for which a solution is proposed in this paper, is that these instruments really measure the distribution of chord-lengths cut randomly across particles during scanning, not the sieve size distribution of particles on the belt. This fact is explicitly recognized by both Vignos, et al. and Yacher, et al. in their publications (2,4). Although they suggest the possibility of correcting the linear intercept information to estimate the sieving data, they imply that, for control purposes, the chord-length distribution may be adequate.

This last statement may be debatable, but perhaps the greatest weakness of the simple linear intercept distribution approach is that sieve analysis has become a size standard for mineral processing plant operators and engineers. This provides considerable incentive to provide a sieve size output from a coarse particle size analyzer.

Taking these considerations into account and based on significant theoretical work on statistical stereology carried out in the last twenty years (5,6,7,8), a project was initiated at the Comminution Center of the University of Utah to develop an on-line instrument which directly delivers the sieve sizing information. The emphasis of this project is on the development of a computational technique, easy to implement in a microprocessor or a personal computer, to convert the linear intercept information delivered by either the Armco Autometrics MSD-95 system or an image analyzer to the sieve information.

Theoretical Background

The problem being addressed here can be stated concisely as one of estimating the mass distribution based on mesh size (or the volume distribution if constant density is assumed) from the on-line linear intercept (chord-length) information.

It is apparent that since chords are cut at random through particles to obtain linear intercept information, that there is no deterministic relationship between a single observation of chord length and the volume of a particle. It can, however, be shown from statistical stereology and geometrical probability considerations, that both chord-length and volume distributions for a population of particles are related through an integral equation (6,7,8,9).

The basic stochastic model proposed by King (10) has been employed in this work and experimentally tested. King states that, under the very general assumption that the distribution of shape parameters does not vary with size in a particle population, the following integral relationship between linear intercept and volumetric (mass) density functions is valid:

$$p(L) = \mu_n \int_0^\infty p(L|D) \frac{f(D)}{L_{D,n}} dD \qquad (1)$$

where:

$p(L)$:   linear intercept density function by length

$f(D)$:   volumetric (mass) density function based on mesh size D

$p(L|D)$:   conditional probability density function for linear intercepts by length conditioned on their being monosize particles of mesh size D

μ_n: n^{th} moment of $p(L)$ defined by:

$$\mu_n = \int_0^\infty L^n p(L) dL \qquad (2)$$

297

$L_{D,n}$: n^{th} conditional moment of p(L) defined by:

$$L_{D,n} = \int_0^\infty L^n p(L|D)dL \qquad (3)$$

Equation 1 is the basic working equation and the terms $p(L|D)$, μ_n and $L_{D,n}$ can all be measured experimentally or calculated from measured values. However, the calculation requires that the geometry of the particles is well understood and this is obviously not the case in most practical situations.

A numerical and an analytical solution were derived by King (10,11) for the transformation from chord-length measurements made on polished sections of irregularly shaped particles to the mass distribution based on mesh size. The results are satisfactory, but the calculation procedure is very complicated and the method is suitable only for particle populations with a limited range for the shape factor.

The problem of finding f(D) requires the inversion of equation 1. Equation 1 itself can be recognized as a Fredholm integral of the first kind (12) with a kernel function given by:

$$H(L,D) = \frac{\mu_n}{L_{D,n}} p(L|D) \qquad (4)$$

The kernel is a function of both the mesh size and the linear intercept. Conceptually, a single particle can produce a typical linear intercept distribution on its projected image. A group of particles with the same mesh size (i.e., monosize particles) will have the same linear intercept distribution as is produced by a single particle. So the kernel function H(L,D) can be regarded as the relationship between a monosize particle distribution with mesh size D and the linear intercept distribution it produces. The linear intercept distribution of a polydispersed particle population can then be constructed by summing the contributions to the chord-length distribution due to each volumetric size fraction as dictated by the kernel function.

Since, as stated, the kernel can only be theoretically calculated for particle populations of very restricted geometrical properties, a more practical approach for obtaining the kernel H(L,D) involves experimentally measuring the linear intercept distribution for a sample of monosize material of mesh size D.

Assuming the kernel is known, equation 1 can then be discretized and written in matrix form as follows:

$$\underline{p} + \underline{e} = \underline{\underline{H}}\, \underline{f} \qquad (5)$$

where $\underline{\underline{H}}$ is the kernel matrix (transformation matrix) for the discrete density vectors $\underline{f} = (f_1, f_2, ... f_m)^T$ and $\underline{p} = (p_1, p_2, ... p_m)^T$ and $\underline{e} = (e_1, e_2, ... e_m)^T$ is the corresponding experimental error vector. Phillips (13) discussed the problem of stability associated with the calculation of the exact numerical solution for this equation. He provided a simple smoothing technique with constrained error to stabilize the solution.

Twomey (12) extended Phillips' method to use a controlled amount of smoothing in solving the matrix system. The solution is given as:

$$\underline{f} = (\underline{\underline{H}}^T \underline{\underline{H}} + \gamma \underline{\underline{Q}})^{-1} \underline{\underline{H}}^T \underline{p} \tag{6}$$

where $\underline{\underline{Q}}$ is the smoothing matrix for function $\underline{f}$ and γ is an undetermined Lagrangian multiplier. Several different smoothing techniques were mentioned by Twomey and the effect of errors was discussed to determine the best value of γ (14). Successful applications of this inversion technique have been reported by authors in a variety of fields (15,16,17).

Experimental Work

Measurement of the Linear Intercept

The method proposed in this paper was tested experimentally to confirm that it is capable of accurately estimating the sieve size distribution from the linear intercept distribution. The sample material used here is crushed onyx from a Southern Utah quarry. These irregularly shaped particles are bright so that they form a significant contrast to the black surface of the belt and the contours are easily distinguishable. Photographs of a dispersed monolayer of particles randomly distributed on a static belt were taken with a 35mm camera. Then an image analyzer with a video camera was used to obtain the digitized grey-level images from these photos for later processing. The fully automatic image analyzer--Kontron IBAS II system--is shown in Figure 1. A sequence of user-interactive image processing modules

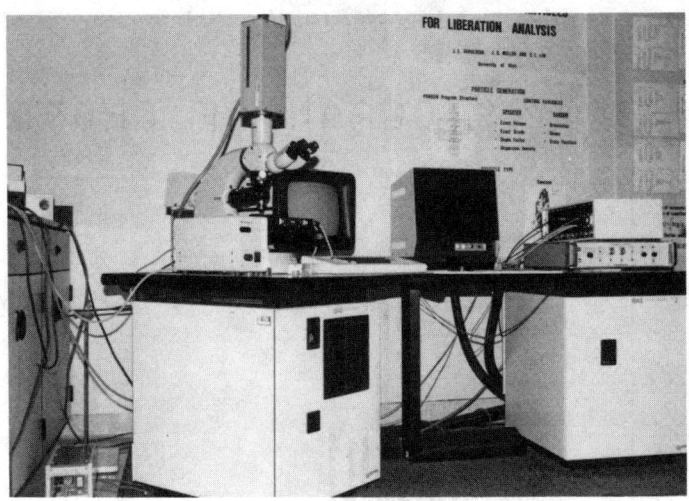

Figure 1 - Kontron IBAS II Image Analyzer

built in the system's software were used to extract the features of interest
from the grey-level image and the images are displayed on a 512 x 512 pixels
color monitor during processing. A wide variety of software is available
for the enhancement, editing, segmentation, and noise filtering of the
stored images. Generally speaking, different material requires a different
sequence of functions. The extraction process usually results in a binary
image in which the features are represented as white areas on a black
background. The measurement of the distribution of random chord lengths was
then performed on the binary image (18). Note here that the length
distribution rather than the number distribution is employed for the linear
intercept measurements.

<u>Determination of the Kernel Matrix</u>

The conditional function $p(L|D)$ is best obtained by experimental
measurement on a sample of material all of which has the same mesh size D.
In practice, it is impossible to prepare such a sample and it is necessary
to examine a set of samples prepared by sieving fractions with contiguous
narrow screen size intervals and then obtain the linear intercept
distribution for each "monosize" sample with mean mesh size D_i. Each one of
these distributions constitutes the conditional function $p(L|D_i)$
corresponding to size D_i and becomes the ith column of the kernel
matrix $\underline{\underline{H}}(L,D)$ as per equation 4. A computer printout histogram of the
function $p(L|D_i)$ based on number distribution for a typical "monosize"
sample is shown in Figure 2 and the function $p(L|D_i)$ based on length
distribution can then be calculated from these information.

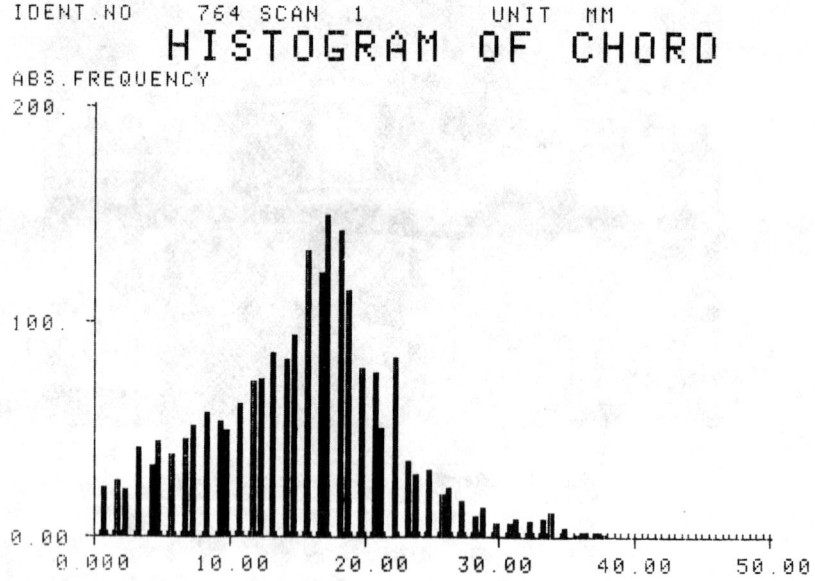

Figure 2 - Histogram of Chord-Length Distribution for a
Typical Monosize Sample

300

Validation of the Mathematical Model

     Several polydispersed particle populations having significantly different mass distributions were prepared and are shown in Figure 3, 4, and 5. The size ranges are from .425 mm to 45 mm. These samples were treated the same way as just mentioned and the linear intercept distribution as well as its nth moment for each sample were obtained. The average experimental error on the measurement of the linear intercept distribution is about 2% absolute for each size fraction of a typical sample.

     Forward calculation of the linear intercept distribution using the model with the predetermined kernel matrix and sieving results was performed in a HP-A700 microcomputer. This predicted linear intercept distribution was compared with the one obtained by image analysis on the actual sample. Precision of the kernel matrix determination was improved as necessary to obtain an accuracy for the mathematical model of about 2% absolute error for each size fraction. A final number of size classes for the sieving and linear intercept distribution of 10 was considered acceptable.

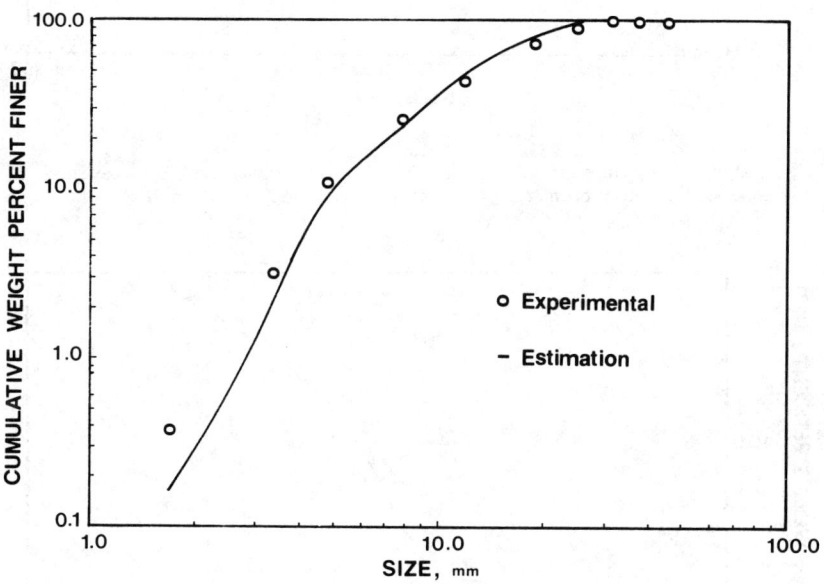

Figure 3 - Comparison of Mass Distribution Obtained by Sieving and Estimation from the Inversion Technique for Sample 1

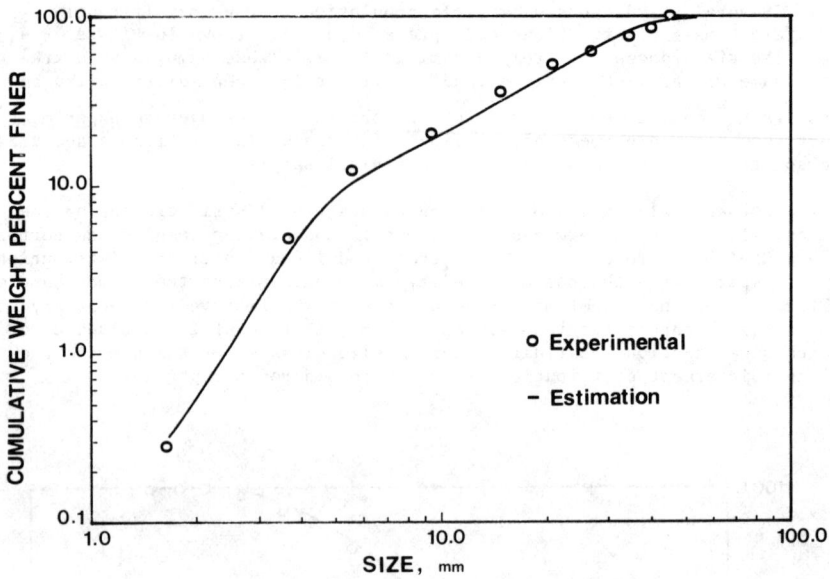

Figure 4 - Comparison of Mass Distribution Obtained by Sieving and Estimation from the Inversion Technique for Sample 2

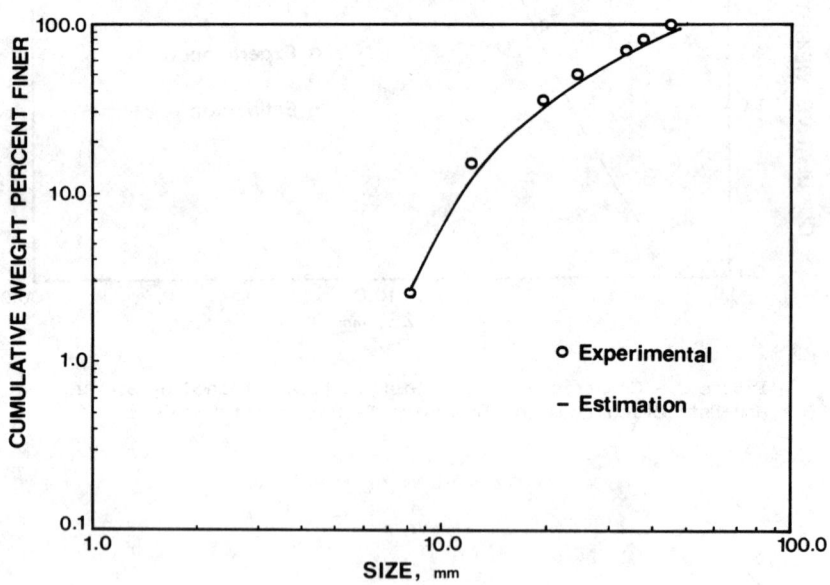

Figure 5 - Comparison of Mass Distribution Obtained by Sieving and Estimation from the Inversion Technique for Sample 3

Estimation of the Volumetric Distribution

A FORTRAN program which implements the constrained linear inversion technique was coded on an HP-A700 computer. The mesh size distributions for a typical sample were estimated for several values of γ. The residual $|\underline{Hf}'-\underline{p}|$, where $\underline{f}'$ is the estimated solution, and the error

$|\underline{f}'-\underline{f}|$, where $\underline{f}$ is the known sieving size distribution, were also calculated for each γ. Then the most appropriate value for γ was decided using the error analysis described by Twomey (14). A globally suitable value of γ was also chosen by comparison of the error and residual for all cases.

Results and Discussion

The function $p(L|D)$ was obtained by the procedure mentioned above and plotted in Figure 6 for four of the ten different classes of linear intercept L and mesh size D. Plotted here is the discrete form of the conditional probability function $\underline{p}(L|D_i)$ for size D_i where

$$p_j'(L|D_i) = \int_{L_j}^{L_{j+1}} p(L|D_i)dL \quad \text{for each linear intercept class j.}$$

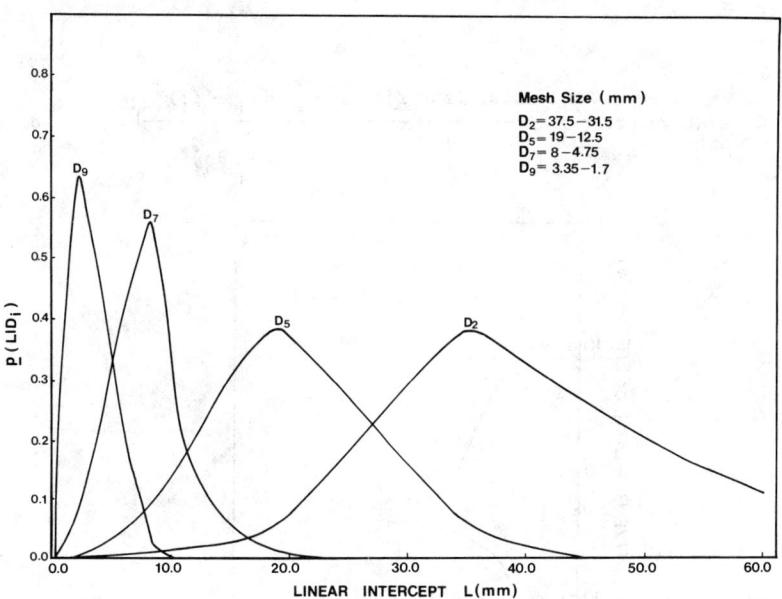

Figure 6 - Discrete form of Conditional Probability Function $p(L|D_i)$ determined for four monosize sieve samples

It is obvious that the kernel function (given by Equation 4) is flat for large particle size and is sharply peaked for small particle size. This means the kernel value changes strongly at small particle size so extra care is necessary to reduce the variation of $p(L|D)$ and increase the resolution in this region. The remedy is to add more mesh size and linear intercept classes in the small size region but this is also subject to the physical limitation of the sieve sizing system. An acceptable result was achieved by comparing the consistency of the forward calculation with the measured linear intercept distribution when the number of classes for both mass and chord-length distributions was increased from 7 to 10. Table I shows the forward calculation and the measurement of linear intercept distribution for the three populations.

Although there is no theoretical value for the order of the nth moment, the first, the second, and the third moment were all tested. It was found experimentally that the first moment fits the model the best; i.e., n equal to 1 was adopted in the following calculation.

The result of an error analysis for Sample 3 is shown in Figure 7 by plotting the r.m.s. relative residuals and errors vs. γ varying from 10^{-5} to 10. The definitions of these two terms are given by the following equations:

$$\text{r.m.s. relative residual} = \frac{\text{r.m.s. } (\underline{p} - \underline{p}')}{|\underline{p}|} = \left[\frac{\sum_{i=1}^{m} (p_i - p_i')^2}{m \sum_{i=1}^{m} p_i^2}\right]^{1/2} \quad (7)$$

$$\text{r.m.s. relative error} = \frac{\text{r.m.s. } (\underline{f} - \underline{f}')}{|\underline{f}|} = \left[\frac{\sum_{i=1}^{m} (f_i - f_i')^2}{m \sum_{i=1}^{m} f_i^2}\right]^{1/2} \quad (8)$$

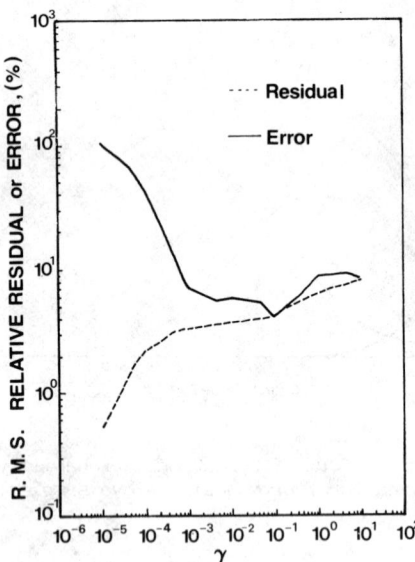

Figure 7 – R.M.S. Relative Residuals and errors vs. for Sample 3

where r.m.s. is the abbreviation of root mean squares and m is the number of size classes. The best value of γ in this case is chosen at γ = 0.1 where the error has the minimum value. Though the residual still decreases with decreasing γ, a larger γ is suitable because the error soon begins to increase with decreasing γ values and the residual at that point is still not less than the estimated overall error (experimental, quadrature, etc.) which is about 4% absolute for each size fraction in p in this case. If γ is overestimated, a useful but oversmoothed solution is obtained; too small a value of γ which is too small can produce instability (14). A survey of best γ values for several different samples of this kind of material results in an optimal global value of γ equal to 0.1.

Table 1. Comparison of the Measured (p_m) with the Forward Calculated (p_{fc}) Linear Intercept Distribution for Three Different Samples

Chord Length (mm)	Sample 1		Sample 2		Sample 3	
	p_m (%)	p_{fc} (%)	p_m (%)	p_{fc} (%)	p_m (%)	p_{fc} (%)
70–50	0.00	0.03	3.05	2.21	6.89	3.95
50–40	0.00	0.78	3.96	5.24	6.69	9.35
40–30	1.33	3.12	7.93	8.40	16.39	14.98
30–20	11.96	10.52	15.54	12.39	20.20	22.09
20–15	14.94	13.82	9.36	11.30	17.92	20.06
15–10	20.07	15.68	11.95	11.70	18.34	19.08
10–6	24.58	22.64	12.64	15.19	10.78	10.98
6–3	17.78	23.14	21.29	23.27	2.22	4.74
3–1	8.47	11.69	13.13	14.63	0.57	0.83
–1	0.87	1.69	1.16	1.88	0.001	0.05

With γ = 0.1, the estimated mass distributions resulted from the inversion technique for Samples 1, 2, and 3 were obtained and plotted in Figure 3, 4, and 5 for comparison with the sieving distributions. It is apparent from these figures that the estimates are quite satisfactory. The r.m.s. absolute error for each case are 4.24%, 3.33%, and 1.63%, respectively.

Conclusions

A convenient stereological method has been developed for the construction of the mass distribution based on mesh size of a particle population from the linear intercept measurement on its projected image. The method has been shown to be effective and accurate for several populations.

Considerable effort is required to determine precisely the conditional linear intercept distribution $p(L|D)$ from several closely sized fractions of material. The function $p(L|D)$ would have to be initially determined for the kind of material processed and then employed in all subsequent determinations.

A constrained linear inversion technique has proven to be capable of transforming the measured linear intercept information to mass distribution based on mesh size correctly. The computation time required for the inversion is small in comparison with the time taken to generate the measurement of the linear intercept distribution from the image analyzer.

Further effort is required to apply this measurement technique in an on-line fashion. Off-line measurements from the image analyzer (Kontron IBAS II) will be replaced by the on-line signals coming from a system like the MSD-95 or a simple on-line image analyzer.

Acknowledgment

The authors wish to acknowledge the United States Bureau of Mines for support of this research, under the Generic Mineral Technology Center Program in Comminution, Contract Number G1125149.

References

1. J. A. Herbst and K. Rajamani, "The Application of Modern Control Theory to Mineral Processing Operations," Proceedings of the 12th CMMI Congress, (S. Afr. Inst. Min. Metall., Johannesburg, 1982).

2. J. H. Vignos, L. Elber, and E. Gallagher, "Coarse Particle Size Distribution Transmitter," On-Stream Characterization and Control of Particulate Processes, J. A. Herbst, and K. V. S. Sastry, ed., (Engineering Foundation, New York, 1978).

3. Interim Operating Manual of MSD-95, (Armco Autometrics, Boulder, Colorado, November 1982).

4. L. Yacher et al., "Coarse Particle Size Distribution Analyzer," Instrumentation in the Mining and Metallurgy Industries, Vol. 12, H. R. Cooper, ed., (Instrument Society of America, North Carolina, 1985).

5. E. E. Underwood, Quantitative Stereology, (Addison-Wesley, Reading, MA, 1970).

6. E. R. Weibel, Stereological Methods, (Academic Press, New York, 1979).

7. M. G. Kendall and P. A. P. Moran, Geometrical Probability, (Charles Griffin, London, 1963).

8. L. A. Santalo, Integral Geometry and Geometrical Probability, Encyclopedia of Mathematics and its Applications, Vol. 1, G. C. Rota, ed. (Addison-Wesley, Reading, MA, 1976).

9. P. L. Goldsmith, "The Calculation of True Particle Size Distributions from the Sizes Observed in a Thin Slice," Brit. J. Appl. Phys., 18(1967).

10. R. P. King, "Determination of the Distribution of Size of Irregularly Shaped Particles from Measurements on Sections or Projected Areas," Powder Technology, 32(1982).

11. R. P. King, "Measurement of Particle Size Distribution by Image Analyzer," Powder Technology, 39(1984).

12. S. Twomey, "On the Numerical Solution of Fredholm Integral Equations of the First Kind by the Inversion of the Linear System Produced by Quadrature," J. Assoc. Comput. Mach., 10(1963).

13. D. L. Phillips, "A Technique for the Numerical Solution of Certain Integral Equations of the First Kind," J. Assoc. Comput. Mach., 9(1962).

14. S. Twomey, Introduction to the Mathematics of Inversion in Remote Sensing and Indirect Measurements, (Elsevier, New York, 1977).

15. L. C. Chow and C. L. Tien, "Inversion Techniques for Determining the Droplet Size Distribution in Clouds: Numerical Examination," Applied Optics, 15(2)(1976).

16. C. D. Capps, R. L. Henning and G. M. Hess, "Analytic Inversion of Remote-Sensing Data," Applied Optics, 21(19)(1982).

17. C. L. Lin, "Measurement and Prediction of Mineral Liberation During Grinding," (Ph.D. thesis, University of Utah, 1986).

18. Kontron IBAS II Reference Manual, Rel. 4.2., (Kontron Bildanalyse, 1984).

HYDROMETALLURGY

Session Chairman
Milton E. Wadsworth, University of Utah

This session addresses microprocessor technology as a tool to enhance metallugical research laboratory capabilities and commonly available spreadsheet analysis to evaluate processing options. Other presentations investigate mathematical models applied to ion-exchange, pressure leaching and complete plant design.

SOLUTION MINING SYSTEMS

Milton E. Wadsworth

Department of Metallurgy and Metallurgical Engineering
University of Utah
Salt Lake City, Utah 84112

Abstract

Preliminary to the modeling of solution mining systems, it is necessary to define the physical and chemical conditions for various configurations. It may be shown that, based upon either the physical chemistry of the mineral phases present or upon the geological positioning, definitive solution mining systems may be described which relate to chemistry, the oxidation potential or the insitu mineral site, the position of the water table, and the disposition of waste rock. Factors relating to the management of these systems, which might be integrated into models, are reviewed mainly for base metal sulfides. Similar concepts may be extended to gold and uranium. Noteworthy in the modeling of solution mining systems is the complex overlapping of multiple technologies including surface chemistry, rate processes, hydrology, mining and chemistry of both metal-containing and nonmetallic mineral phases.

311

In defining solution mining systems, it is important to consider both geologic and chemical factors which may influence the extraction of metal values during the leaching cycle. The modeling of a solution mining system is related to the character of the target minerals deposited in cracks, fissures, veinlets, and pores of the rock mass. Figure 1 illustrates an idealized primary sulfide deposit containing copper and iron which has been subjected to variable weathering. Above the water table the oxidation potential is sufficiently high for oxidation of sulfides to occur. Oxidized mineral zones are often depleted by the leaching action of supergene water. The oxidized or vadose zone shown in Figure 1 illustrates the depletion of copper and iron by the downward percolation of supergene water. If pyrite is present in abundance, acidic solutions solubilize base metal sulfides providing downward mobility of metal sulfate solutions. If acidic conditions cannot be maintained because of low abundance of pyrite, carbonate minerals may form in the vadose zone. Secondary enrichment, illustrated as the supergene zone, results from the secondary conversion of primary sulfides by the action of supergene water containing base metals in regions of low oxidation potential. The conversion of chalcopyrite ($CuFeS_2$) to chalcocite (Cu_2S) and covellite (CuS) is an example. Also pyrite may be converted by these reactions to chalcocite. As shown in Figure 1, the copper content in the enriched zone is appreciably higher than that in either the weathered zone or in the primary hypogene zone. Below the zone of secondary enrichment is the primary or original sulfide ore, which, as shown in Figure 1, contains copper at some intermediate value. There are three zones of variable mineralization. These are the oxidized zone well above the water table, the secondary supergene zone at and immediately below the water table, and the deeper hypogene zone, all of which present differing chemistry and mineralization and thus differing approaches for design of solution mining systems. Other examples of enrichment by weathering are nickel laterite and bauxite deposits. In some cases the dissolved metal may be highly mobile and move over great horizontal distances. Uranium is a typical example where, because of the presence of carbonate, stable uranium carbonate complexes form. These soluble

PERCENT METAL CONTENT

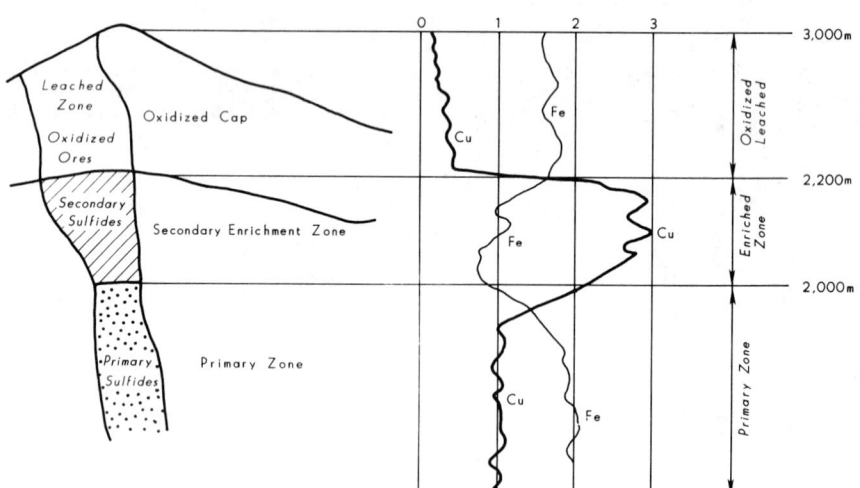

Figure 1. Primary copper sulfide deposit showing oxidized, enriched and primary ores.

complexes may travel great distances until precipitation or reduction immobil-
izes the uranium in so-called roll-front-type deposits. Such deposits have
been important in uranium solution mining in recent years.

Hydrometallurgy is the only economically feasible method for extracting
metals from dilute phases. Table 1 illustrates the range of dilution in the
solids treated and the aqueous effluents produced. The economic incentive
results from the effectiveness of separation and the ability to treat coarse
rubblized rock and/or fractured material without comminution.

Eh-pH Relationships

Figure 2 illustrates E_h-pH relationships for the copper, iron water system
as it relates to the various copper sulfide minerals. As was observed in Figure 1 from the geology of a typical weathered copper deposit, the three zones
are clearly identifiable from E_h-pH data. The secondary enriched zone corresponds to the region of approximately 0.2 to 0.6 volts at pH = 0, to 0.0 to -0.4
volts at pH 10, with the vadose zone above this range and the hypogene zone below. The hatched line indicates the region in which solution mining solutions
must be maintained to provide metal mobility during solution mining. The
leaching of oxidized ores presents special chemical problems since the leaching
process is acid-consuming. The leaching of sulfides, however, differs in that
the leaching process itself may be acid-producing and may maintain very broad
and uniform reaction zones within a deposit depending upon conditions of aeration. Also shown in Figure 2 is the very limited region in which autotrophic
bacteria can remain active. They require a high oxidation potential and are
active predominantly in the pH range of 1.5 to 3.0. It should be noted that
these bacteria favor high Fe^{III}/Fe^{II} ratios, catalyzing the oxidation of pyrite, sulfur, and ferrous ions.

The active Eh-pH range for these autotrophic bacteria also favors hydrolysis and precipitation of iron as ferric oxide or in the form of basic iron sulfates (jarosites). These hydrolysis reactions are acid-producing and coupled
with oxidation of pyrite, under conditions of good aeration, provide open porosity, uniform oxidation potential, and stable pH values in the range of 2.0 to
2.4.

Table 1. Typical Ore and Solution Grades

Operation	Grade of Solids	Grade of Solution	Method of Recovery
Copper Dump Leaching	<0.2% Cu <2000 g/mt <4 lb/st	~0.1% 1000 g/mt 2 lb/st	Cementation or Solvent Extraction
Heap Leaching of Gold	$9x10^{-5}$-$4.2x10^{-4}$% Au 0.9-4.2 g/mt 0.03-0.13 oz/st	$6x10^{-5}$-$3.8x10^{-4}$% 0.6-3.8 g/mt 0.02-0.12 oz/st	Adsorption on C, Elution Electrolysis Precipitation
Insitu Extraction of Uranium	0.05-0.5% U_3O_8 500-5000 g/mt 1.0-10 lb/st	$5x10^{-4}$-0.02% 5-200 g/mt 0.01-0.04 lb/st	Anion Exchange Resins, Elution Precipitation

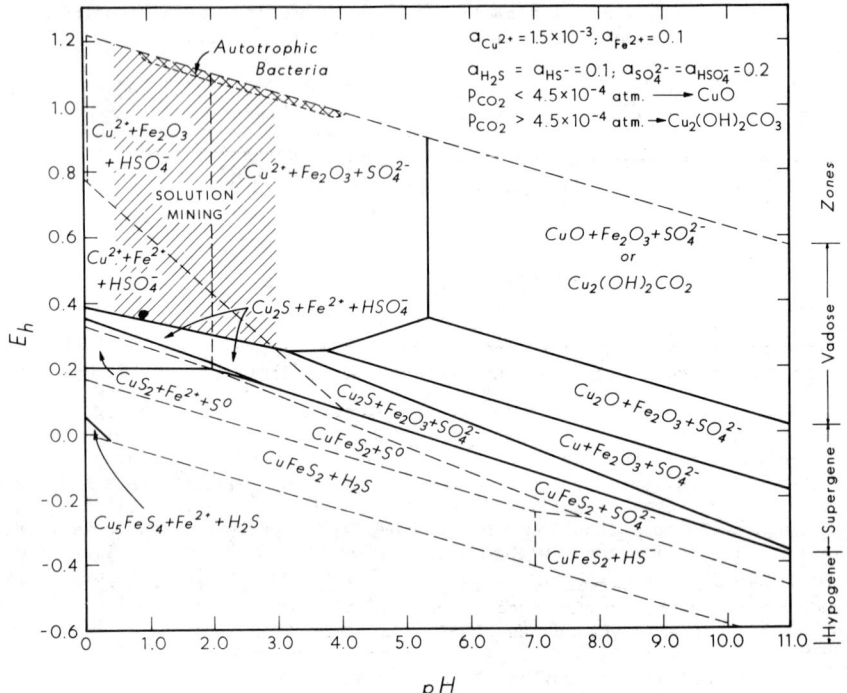

Figure 2. E_h-pH diagram for the copper, iron water system showed Vadose, Supergene and Hypogene zones.

Figure 3 illustrates E_h-pH relationships for the uranium, carbonate, water system (1). Uraninite is the stable oxide formed under reducing potentials, 0.0 to -0.4 volts at pH 9, and would correspond to the secondary mineralization zone of Figure 2. The stable U^{VI} ion is the complex carbonate ion $UO_2(CO_3)_3^{4-}$. The hatched region of Figure 3 illustrates the range employed for insitu extraction of uranium. Acid leaching of uranium is possible and is based upon the formation of stable uranium sulfate complexes. Acid leaching is more complex because of the many side reactions which occur.

Generic Types

As previously noted from both geological evidence and Eh-pH relationships, three general zones for solution mining are apparent. These are regions of high oxidation, regions of secondary enrichment, and primary base mineralization. The presence of aqueous solutions was instrumental in the original deposition, and its historical positioning controls the oxidation potential within a given deposit. Thus the depth of the deposit and its positive relative to the water table are important factors. Deep deposits may not be accessed economically by underground mining methods and may have to be approached by insitu extraction techniques. On the other hand, shallow deposits may be accessible by combined mining and leaching. The leaching of roll-front uranium deposits is typical of primary sandstone-type materials of sufficient porosity that direct leaching without rubblization or hydrofracting is possible. Deep-seated hard-rock uranium, common throughout the world, may require rubblization for effective leaching in the future.

314

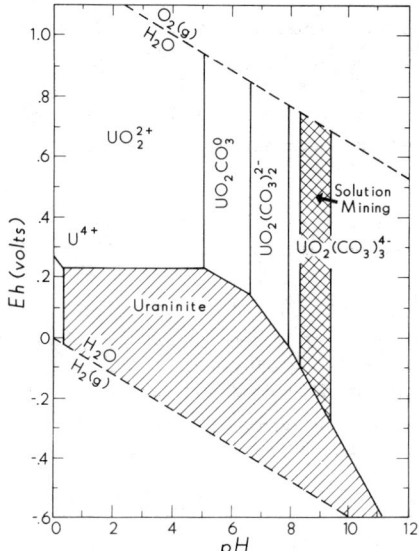

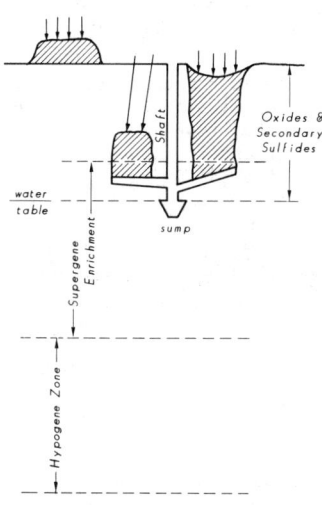

DUMPS AND DEPOSITS ABOVE
WATER TABLE

Figure 3. E_h-pH diagram for
          uranium-water,
          showing solution
          mining region.

Figure 4. Deposits showing dumps on
          surface and newly or
          previously mined areas
          below surface.

The percentage of copper produced by leaching will probably increase in
the future as the average grade of major deposits decreases. The distinction
between ore and waste will require an even more definitive assessment as energy
costs for milling increase (2), probably favoring a greater percentage production of copper by leaching. As an open-pit operation continues, the stripping
ratio will increase to the point that underground mining or some other alternative must be considered if increased recovery is to be achieved. As a rough
estimate, approximately an equal amount of copper will remain, after conventional mining, in low- to medium-grade zones including regions of "halo" mineralization, deep-seated medium-grade ores, and unleached residues in waste
dumps. This is a worthy target for extraction by dump leaching and insitu
solution mining.

Deposits above the Water Table

Figure 4 indicates the position of oxides and/or secondary sulfides located near and above the water table prepared for leaching. Dumps located on
the surface are made up of materials removed from open-pit mining or from the
removal of materials during block caving. The earliest successful examples of
solution mining were carried out in rubblized low-grade ore remaining in the
deposit after mining (3). The amount of such materials is expected to be very
great in the future. Waste rock removed from open-pit mining contains less
than approximately 0.2 percent copper and, for typical stripping ratios of 3
to 1, may actually contain as much copper as ore being shipped to the mill.
Therefore it is quite apparent that the inventory of copper contained in low-
grade rock in typical southwestern U.S.-type operations is considerable and re-

presents a major source for recovery of copper by solution mining. Economically unminable ore by conventional methods represents a major source of copper
for the future. Open-pit mining must cease when an excessive stripping ratio
is required. At this point, it is not unusual to expect less than 50 percent
recovery of the total copper originally in such a deposit. Therefore, the application of solution mining in previously mined operations is expected to be a
very important technology for the future. Such dumps or deposits are depicted
in Figure 4 and refer either to rubblized materials, porous materials, or fractured zones through which solutions can penetrate. The important feature of
these deposits is that the chemistry is quite similar. For sulfides it would
be expected that autotrophic bacteria would contribute significantly to the
leaching and would have to be considered in any of the modeling applications.
Most effective management of solutions underground is achieved by collection in
sumps at the water table.

The management of solutions, solution chemistry, and the importance of
autotrophic bacteria are similar for dumps and insitu deposits above the water
table. Effective leaching depends upon a sequence of processes. For these
types of deposits, these are

        * Effective air circulation
        * Good bacterial activity
        * Uniform solution contact.

Figure 5 illustrates the general flow of solution to a large copper dump
with collection of solution in a holding basin. Solutions are introduced on
the surface by one of several methods, mostly commonly ponding, trickle leaching, or spraying (4). Ponding is used but the trend is to the use of trickle
leaching or sprays. In ponding, channeling can cause excessive dilution of
effluent. Trickle leaching is carried out by using a network of perforated PVC
pipe, as shown in Figure 5, or spraying using low-pressure multiple sprays or
single high-pressure sprays. Trickle and spray application provide uniform
controlled application of leachant. Sprays suffer the highest evaporation
loss, followed by trickle application. During winter months, ice formation may

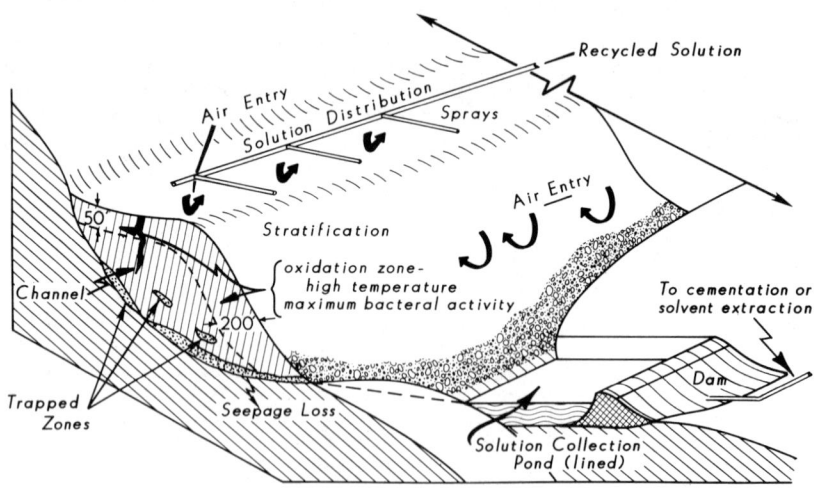

Figure 5. Cross-section of copper leach dump illustrating solution management.

require alternate methods of solution application. As shown in Figure 5, solutions from the toe of the dump flow down natural channels to a collection pond. Solutions are then pumped to separate circuits for copper extraction.

Following extraction, solutions are recycled or enter a containment pond where some aeration occurs. It should be noted, however, that the iron balance for the greater part is achieved by precipitation of iron salts throughout the dump itself. In general, intermittent leaching with alternate leach and rest cycles is preferred to continuous leaching. This practice conserves energy consumed in pumping and is effective, since pore leaching continues during the rest period, under conditions of good aeration, building up dissolved metal values in the contained liquid phase. Continuous leaching without the rest cycle, because of the large volume of water, extracts large quantities of heat from the dump (up to one-half of the exothermic heat of reaction), adversely affecting leaching rates (4).

Figure 6 illustrates oxide or secondary sulfide ores, produced by block caving, prepared for leaching underground as proposed by Roman (5). This configuration is similar to dump leaching and, in the case of sulfides, would be enhanced by autotrophic bacteria. An alternative would be to rubblize using explosives placed to provide a predesigned ore fragment size distribution. Figure 7 illustrates a multiple cell insitu leaching system proposed by Wells (6). During adjacent underground mining, and area is proposed to be treated by circulation of leaching solutions through a series of cells containing ore rubblized by block caving or controlled explosive fragmentation. Depleted cells would be bypassed as new cells are developed. In this proposal, aeration cycles would be alternated with flooding and drainage cycles.

Conditions promoting active bacterial action would be possible for these underground leaching systems similar to those obtained in dump leaching. The insitu leaching systems have the advantage of providing more positive control over aeration. This would enhance bacterial activity as well.

When bacteria are present and active under conditions of uniform aeration and insitu acid generation, the oxidation potential will be uniform and leaching will progress uniformly through the deposit because of the uniform generation of ferric ion Fe^{III} which is the active lixiviant. Large fragments may be too large for effective leaching to occur in the active leaching period of months to years for sulfide ores.

Leaching of oxide copper ores depends virtually exclusively upon the water drive reaction of an oxide dissolving in acidic solutions without oxidation or reduction. For this reason the modeling is totally different from sulfide leaching. A reaction zone progresses through the deposit and the acidic lixiviant is consumed. Excessive acids may cause severe acid consumption through alternate gangue reaction. Chae (7) identified three separate acid-consuming reactions during the leach of oxide copper ores. Insufficient acid will not support adequate leaching over long leaching paths. Excessive acid may produce strong acid effluents which cannot be treated by solvent extraction without neutralization. Thus the leaching of oxides is critically depth-dependent and calls upon a configuration of solution management quite different from those required for the leaching of sulfides. Oxides have the advantage that leaching is often faster in given rock fragments. Initial rapid release of metal values is often observed followed by slower release and increased reaction with gangue minerals. Accordingly strong leaching solutions may be effective initially but should be followed with less-active solutions to minimize gangue mineral reactions.

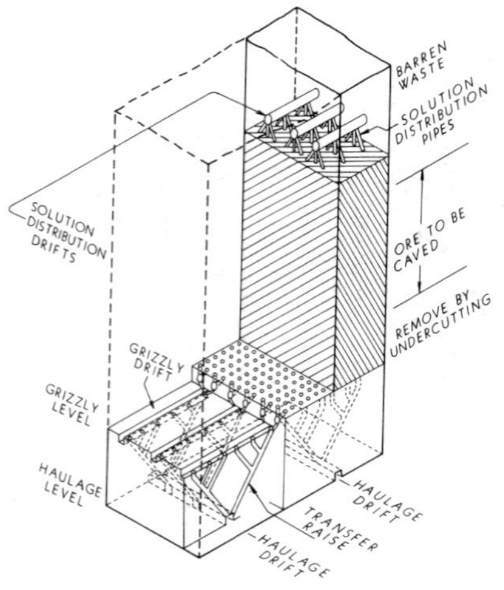

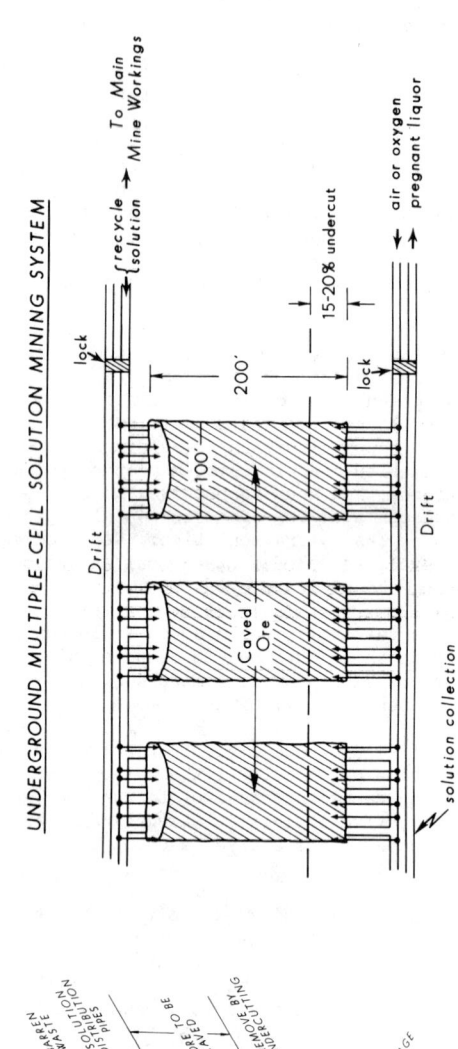

Figure 7. Multiple underground leach cells concurrent with underground mining [Wells (6)].

Figure 6. Underground leaching of caved ore [Roman (5)].

Deposits Immediately Below the Water Table

This regions corresponds to readily accessible regions by mining and shallow drilling. Such deposits would include the secondary enrichment zone of copper sulfide deposits, uranium and vanadium in roll-front deposits and many other metal deposits, including gold and silver. Figure 8 illustrates three conditions which may be considered. The first is illustrated with an artificial water table produced by mining and dewatering. The deposit may be leached following procedures applicable to deposits above the water table. Dewatering would result in continuous dilution of the leach effluent in this case. The second case is one in which a zone is rubblized by mining or fractured in some other way, followed by flooding. Oxidant such as air or oxygen may then be injected in the bottom to provide solution mixing action. Solution would be continuously or intermittently removed for metal recovery. The third case is one using the more conventional secondary oil-recovery technology successfully applied to uranium insitu extraction. Here injection wells and solution recovery or production wells are placed in a grid-like configuration for controlled solution management.

Figure 9 illustrates the application of well-injection technology for uranium recovery from porous, confined roll-front deposits (8). Injection and production wells are shown in a grid pattern. This method is well suited to uranium recovery because of the relatively ideal confinement of the uranium-bearing strata.

Deep Deposits

Deep deposits, for example more than 200 meters below the water table, may or may not be accessible by mining. Examples are illustrated in Figure 10. Adequate permeability may require rubblization or hydrofracting or may depend upon the natural porosity which may be chemically enhanced during leaching. Researchers at Lawrence Livermore (9) proposed the use of nuclear devices for springing and block caving rubblization of deep primary copper deposits. The advantage of an openly porous deposit comes from the improved oxygen solubility due to the hydrostatic head. At atmospheric pressure, water in equilibrium with air contains approximately 7 ppm O_2 and 35 ppm for equilibrium with oxygen. For each 9.75 meters these amounts would increase incrementally. The graph on the left shows the concentration of oxygen at various depths for water in equilibrium with air and pure oxygen. The rate of leaching is proportional to the activity of oxidant. If the oxidant is oxygen, then the hydrostatic head will greatly increase the rate of leaching. This was demonstrated for hypogene copper porphyry ore by the Lawrence Livermore simulations (10).

The major problem with deep deposits will be solution management and the development of suitable permeability. The systems can readily be operated at elevated temperature and maintain markedly improved oxidant activity. The rates of reaction would depend upon the direct chemical attack by oxygen rather than by ferric ion. The latter would be readily precipitated out at elevated temperatures and high oxygen activity. The dissolution would proceed by direct chemical reaction. The autotrophic bacteria important to dump and shallow insitu systems would not survive under these conditions.

Modeling

Virtually all researchers agree that the leaching of matrix mineral ore fragments involves penetration of solution into the rock pore structure. The kinetics thus involve diffusion of lixiviant into the rock where reaction with individual mineral particles occurs. The kinetics are complicated by changing

319

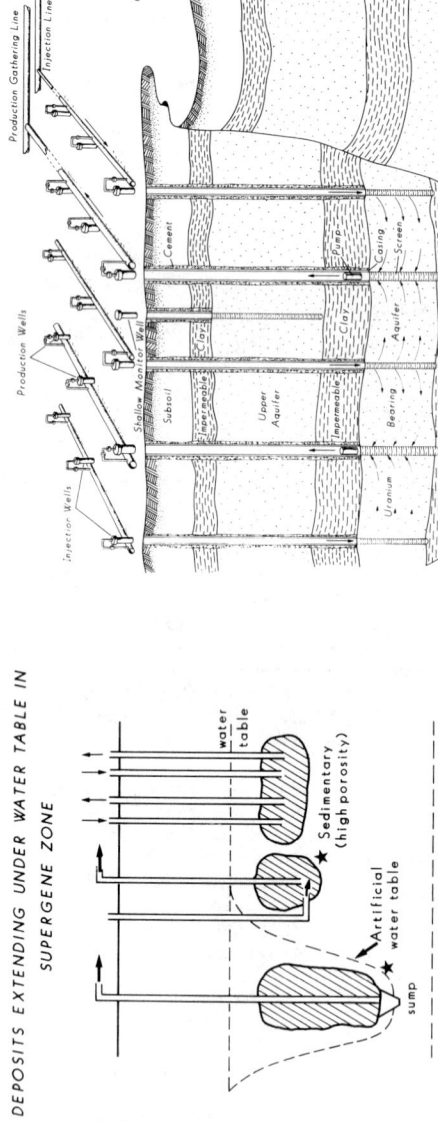

Figure 9. Example of injection and production wells for uranium recovery by insitu solution mining.

Figure 8. Example of deposits at or immediately below water table.

320

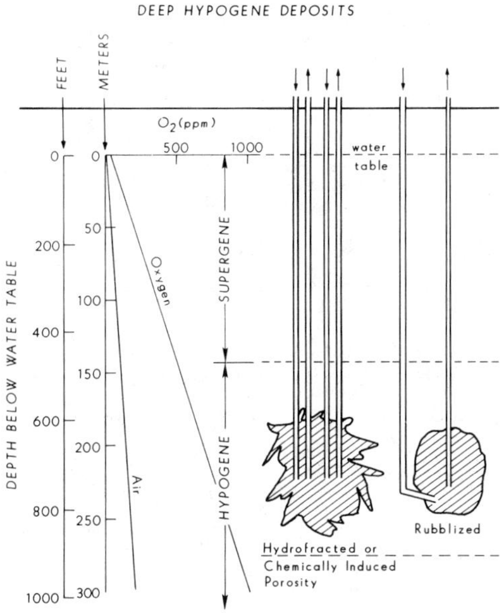

DEEP HYPOGENE DEPOSITS

Figure 10.   Solution mining of deep deposits showing
enhanced oxygen solubility with depth.

porosity, pH, and solution concentration. Bartlett (11) applied the continuity
equation to a system as described for a deep rubblized deposit. Madsen and
Wadsworth (12) applied the continuity equation to the leaching of partially
enriched copper ores and incorporated intrinsic mineral kinetics for five
different copper sulfides. Recently Lin et al. (13) have shown that, for
oxygen-mineral reactions with chalcopyrite and pyrite, electrochemical reaction
kinetics are required to explain observed intrinsic kinetics.

The shrinking-core concept is a special case used successfully by several
investigators to model the leaching of matrix mineral ore fragments (10,14-18).
Figure 11 illustrates a matrix mineral ore fragment showing a reacted zone, a
zone of partially reacted sulfides, and an unreacted core for copper sulfide
mineralization. Also shown is the concentration profile for copper and oxidant. The oxidant would normally be ferric ion complexes Fe^{III} or, for deep
solution mining, oxygen. If the reaction zone is relatively narrow, the
shrinking-core simplification may be used. As illustrated on the right of
Figure 11, trickle leaching would provide a flowing film which would transfer
oxygen with bacterial-supported oxidation of Fe^{II} to Fe^{III}. Thus, with good

321

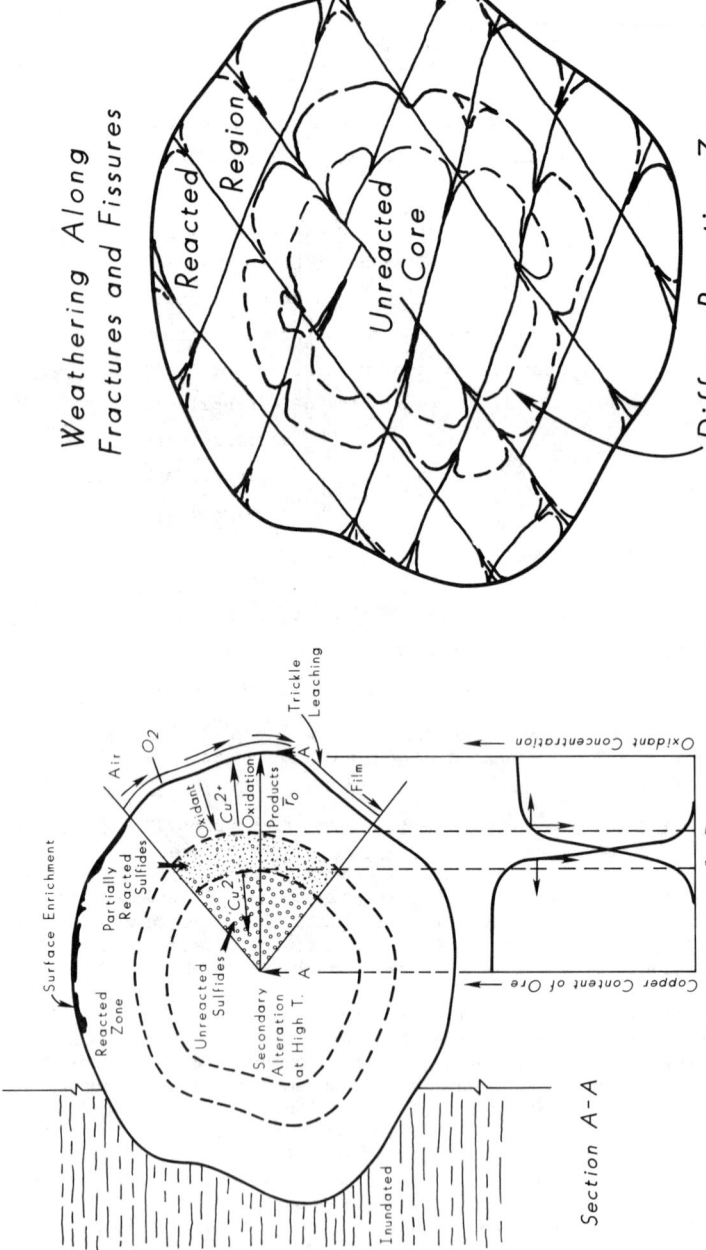

Weathering Along Fractures and Fissures

Reacted Region

Unreacted Core

Diffuse Reaction Zone

Figure 12. Ore fragment after long exposure to solutions, showing induced weathering along fractures due to acidic solutions.

Figure 11. Ore fragment showing reaction zone and concentration profiles.

aeration, steady-state generation of Fe^{III} would be achieved over broad regions of the deposit. Leaching occurs by diffusion of Fe^{III} into the fragment. In deep solution mining with the fragments inundated (left side of Figure 11), dissolved oxygen must transfer to the ore surface and diffuse inwardly for the reaction to proceed.

A complication which influences the long-term recovery is the induced weathering of the ore matrix fragments which results from the acidic solutions. Figure 12 illustrates the effect of weather by dissolution of gangue constituents and penetration down existing cracks in the ore. Braun et al. (10) considered the weathering as a systematic increase in interfacial area.

Cathles and Apps (16) modeled dump leaching and included in the model solution flow, air convection, and heat flow. Madsen et al. (14) and Braithwaite (17) used a simplified continuity equation proposed by Sohn and Szekely (18) for copper sulfide ore using finite difference approximations. Schechter et al. (19,20,21) have modeled the leaching of uranium ores contained in porous deposits using equations of continuity for both carbonate and acid ferric sulfate leaching.

A leaching model proposed by Braun et al. (10) assumed uniform distribution of oxygen and temperature by bubbling through rubblized ore. Gao et al. (22,23) have proposed unsteady-state, one-dimensional models for single-phase and two-phase flow applied to a deep chimney of copper porphyry ore having variable chalcopyrite to pyrite ratios. Their model used the observed leaching rate data observed by Braun et al. (10) in their calculations.

The single-phase one-dimensional flow model of Gao et al. (22) will be used as an example. If i refers to the lixiviant species in solution, the continuity equation is

$$\frac{\partial}{\partial t} (\epsilon C_i) + V_z \frac{\partial C_i}{\partial z} - \frac{\partial}{\partial z} (D_{ai} \frac{\partial C_i}{\partial z}) = \dot{R}_i \tag{1}$$

where
- ϵ = porosity of rubblized ore
- C_i = liquid concentration of species i
- V_z = superficial velocity
- z = distance in direction of flow in deposit from position of entry
- t = time
- D_{ai} = axial dispersion coefficient
- $\dot{R}_i$ = net rate of production per unit volume of chimney

Using the rate equation developed by Braun et al. (10), the rate of consumption of oxygen per unit volume of chimney may be calculated and coupled to the mass-transfer process of oxygen from the bulk liquids to the external surfaces of the ore fragments. From the data of Braun et al.,

$$\dot{R}_{O_2} = - \sum_{j=1}^{N} q_j \frac{4\pi r_j^2}{\phi_j} \times [\frac{C_{O_2}}{(\frac{r_j}{R_j})^2 \frac{1}{k_r} - \frac{1}{\sigma G\beta} + \frac{1}{D_e} \frac{r_j}{R_j} (R_j - r_j)}] \tag{2}$$

where
- β = $3\rho_r \phi k / r_p \rho_p$
- C_{O_2} = bulk liquid oxygen concentration
- R_j^2 = initial spherical radius of the jth fragment size fraction
- r_j = radius of unreacted core for the jth size
- r_p = average spherical copper sulfide radius

323

q_j = average number of particles of the jth size
per unit volume of the chimney
ϕ_j = the sphericity shape factor for the jth fraction
σ = the stoichiometry number
D_e = effective diffusivity
G = ore grade in wt. fraction of $CuFeS_2$
ρ_p = density of chalcopyrite
ρ_r = bulk rock density
δ = thickness of the reaction zone
k_t = global rate constant for chalcoyprite and pyrite combined

The rate of change of the unreacted core radius for the jth size fraction
is

$$\frac{dr_j}{dt} = - \frac{M_w}{32\phi_j\rho_r G} \left[\frac{C_{O_2}}{(\frac{r_j}{R_j})^2 \frac{\sigma}{k_t} - \frac{1}{G\beta} + \frac{\sigma}{D_e} \frac{r_j}{R_j} (R_j - r_j)} \right] \tag{2}$$

where M_w is the molecular weight of chalcopyrite. Follow the findings of Braun
et al., the higher surface grade was accounted for by partitioning G for a surface layer. Also, the weathering correction giving the change of ϕ_j proposed
earlier (10) was used where

$$\phi_j = [\phi_{jo}^2 - \frac{2}{3}\lambda(R_j^3 - r_j^3)]^{1/2} \tag{3}$$

Using calculated external mass-transfer coefficients, the rate of reaction was
calculated using numerical methods. The rate was coupled with the energy conservation equation for axial conduction to determine the temperature-time relationship,

$$[(1 - \epsilon)\rho_s c^5 + \epsilon\rho_\ell c^\ell) \frac{\partial T}{\partial t} + \rho_\ell c^\ell V_z \frac{\partial T}{\partial z} = \dot{Q} \tag{4}$$

where ρ_ℓ and ρ_s are the densities of liquid and ore, and $\dot{Q}$ is the rate of heat
generated per unit volume of chemistry. The total reaction for chalcoyprite
($CuFeS_2O$ and pyrite (FeS_2) were considered to be

$$CuFeS_2 + \frac{15}{4} O_2 + \frac{3}{2} H_2O = Cu^{+2} + \frac{1}{3} Fe_3(SO_4)_2(OH)_5 \cdot 2H_2O + \frac{1}{3} S + SO_4^{-2} \tag{5}$$

and $$FeS_2 + \frac{15}{4} O_2 + \frac{17}{6} H_2O = \frac{1}{3} Fe_3(SO_4)_2(OH)_5 \cdot 2H_2O + \frac{4}{3} SO_4^{-2} + \frac{8}{3} H^+ \tag{6}$$

Table II presents a summary of copper extraction-time calculations for various
beginning chalcopyrite to pyrite ratios. An actual run of mine size distributions of ore fragments was used in the calculations.

Conclusions

Hydrometallurgy provides a unique method for the recovery of metal values
from dilute and remote ore bodies. The application to low-grade waste rock and
insitu ore bodies will, in the future, assume a role of increasing importance.
Modeling will be important for estimation of recovery-time relations for a
variety of needs. These models must include factors related to the geology,
chemistry, natural or generated porosity and hydrology (24) for a large variety

Table II. Summary of Insitu Leaching Model Calculation

Case No.	Ore Grade Wt Pct CuFeS$_2$	Py/Cp Molar Ratio	Shape Factor	Chimney Porosity	Initial Chimney Temp. °C	Flow Rate (m³/m²-day)	Days to Reach 90°C	250	500	750	1000	Total Copper Extracted at 1000 Days (kg)	Days for Copper Loading in Exit Stream to Leach 10 kg/m³	Max. Pres. Drop (psi)	Total Oxygen Supplied for 1000 Days (kg-mole/Square Meter of Chimney)
1a	2	2	0.65	0.15	60	10	159	12.1	24.9	35.8	43.8	2415	214	0.059	429
1b						15	126	16.5	31.8	41.5	47.5	2621	168	0.093	464
1c						20	112	18.8	34.8	43.2	48.7	2685	144	0.131	476
1d						30	103	22.6	36.7	44.3	49.4	2726	128	0.223	483
1e						variable	103	23.9	37.4	44.8	49.9	2752	125	0.257	487
2a	2	2	0.65	0.25	60	10	182	12.7	26.7	37.7	45.1	2195	356	0.010	391
2b						20	138	19.3	35.0	43.2	48.7	2371	252	0.022	422
2c						30	131	21.7	36.3	44.0	49.3	2396	230	0.038	427
2d						variable	133	22.1	36.5	44.3	49.5	2409	217	0.051	429
3a	1.5	2	0.65	0.25	60	20	186	18.6	36.1	45.0	50.7	1852	329	0.022	331
3b						variable	187	19.1	36.5	45.3	51.1	1865	315	0.023	333
4a	3.3	2	0.65	0.25	60	20	95	16.0	29.7	38.3	43.7	3509	186	0.022	623
4b						variable	73	23.5	34.8	41.4	45.9	3685	117	0.130	655
5a	2	1	0.65	0.25	60	20	212	17.3	36.1	45.4	51.4	2499	269	0.022	298
5b						variable	217	17.0	36.2	45.5	51.6	2511	267	0.017	299
6a	2	0.5	0.65	0.25	60	20	296	12.3	34.7	45.4	52.2	2540	316	0.022	228
6b						variable	308	11.3	33.9	45.2	52.2	2540	326	0.007	228
7a*	2	2	0.65	0.25	60	20	260	7.8	21.0	30.0	36.9	1802	462	0.006	318
7b*						variable	272	7.1	20.4	29.5	36.6	1783	470	0.007	315
8a	2	2	0.35	0.25	60	20	95	26.1	44.1	52.9	58.5	2848	186	0.022	506
8b						variable	72	35.1	48.2	55.4	60.3	2934	118	0.110	522
9a**	2	2	0.65	0.25	60	20	75	11.3	22.4	32.5	40.4	5284	161	0.022	936
9b**						variable	33	26.4	37.4	43.7	48.2	6299	53	0.517	1110
10a	5.37	2	0.65	0.25	60	20	79	11.1	21.8	30.5	36.9	4875	164	0.022	855

*Uniform particle size distribution (d_p = 0.092 m)

**Ore grade follows a log distribution: $\log G = -0.21 \log d_p - 1.096$ (d_p in centimeters. G = pct CuFeS$_2$/100)

of systems. Insitu solution mining has potential for application for many metals. It has already been demonstrated favorably for copper and uranium. Based upon economics, physical and chemical characteristics, and environmental control, Potter et al. (24) have listed eight prime candidates in addition to copper and uranium. These are:

Aluminum	Molybdenum
Gold-Silver	Cobalt-Nickel
Lead-Zinc	Vanadium
Manganese	Tungsten

Conceptual designs were presented for aluminum, gold-silver, manganese, and cobalt-nickel.

Many problems exist in applying insitu technology to the broad variety of sources available. These problems relate mainly to unknown factors including geologic characteristics, fundamental mineral and solution chemistry, methods for generating porosity, and fluid management. Interfacing hydrology with extraction rate processes becomes increasingly difficult as we move from uniformly porous to rubblized material. Engineering parameters introduced through rubblization, hydrofracting or chemically induced porosity become integral parts of any modeling effort. A study by a task force under the University of Utah Research Institute (UURI) (26) recently presented a four-volume analysis of the state-of-the-art of insitu leaching and solution mining. The task force consisted of 21 consultants and covered geological characterization, solution chemistry, fracturing and rubblization, and fluid flow management. The task force listed twenty topics of highest priority for research needed to develop new knowledge and technology in understanding, determining, measuring, and/or controlling the following:

Geological Characterization

* Primary porosity and permeability
* Prediction of failure geometry within ore zone
* Location, size and distribution of ore minerals
* Quantitative mineral and chemical composition through ore body
* Three-dimensional patterns of structural elements in deposit

Chemistry

* Rate and release sequence of metals from host rocks
* Permeability degradation or enhancement
* Lixiviant consumption and composition
* Environmental problems
* Removal of metals from solution by gangue minerals

Fracturing and Rubblization

* Hydrofracturing
* Fragmentation prediction and control
* Mine layout design for solution mining
* Modified insitu methods
* Blasting to enhance permeability

Fluid-Flow Management

* Transport models in fractured and rubblized media
* Well testing
* Well pattern design and production/injection schemes
* Coupling of chemical and fluid-flow models
* Regional-subregional hydrological parameters

It is quite obvious that successful design and operation of insitu solution mining will require the combined cooperative efforts of many technologies and extensive background information. These efforts must be economically and environmentally sound for future development. A coordinated research effort is justified on the basis of the critical and strategic importance of the vast resource of metal commodities in dilute and remote deposits in the earth's crust. Modeling, based upon reliable background information, is central to our ability to predict success or failure of these efforts in the future.

References

1. D. Langmuir, "Uranium-Solution Equlibria at Low Temperatures with Applications to Sedimentary Ore Deposits," Geochimica et Cosmochimica Acta, 42 (1978), 547-569.

2. M. E. Wadsworth and C. H. Pitt, "An Assessment of Energy Requirements in Proven and New Copper Processes," DOE/CS/40132 (1980); also NTIS PC A17/MF A01.

3. J. B. Fletcher, "In-Place Leaching at Miami Mine, Miami, Arizona," Soc. of Min. Engrs., 250 (1971), 310-314.

4. J. S. Jackson and B. P. Ream, "Solution Management in Dump Leaching," Leaching and Recovering Copper from As-Mined Materials, ed. W. J. Schlitt [Soc. of Min. Engrs., 1980), 79-94.

5. R. J. Roman, "Simulation of an In Situ Leaching Operation as an Aid to the Feasibility Study" (New Mexico Bureau of Mines and Mineral Resources, Socorro, New Mexico, 1973).

6. H. M. Wells, "Mining for Multiple Cell Insitu Leaching and Solution Management," private communication of work in progress, Department of Mining Engineering, University of Utah, Salt Lake City, Utah (1985).

7. D. G. Chae, "The Modeling of the Leaching of Copper Oxide Ores" (PhD Thesis, University of Utah, 1980), 1-60.

8. A. L. Bishop, "Non-Production Zone Excursions," Forth Annual Uranium Seminar (Soc. of Min. Engrs., 1980), 75-80.

9. A. E. Lewis, R. L. Braun, J. C. Sisemore and R. G. Mallon, "Nuclear Solution Mining -- Breaking and Leaching Considerations," Solution Mining Symposium, eds. F. F. Aplan et al. (SME and TMS of AIME, 1974), 56-75.

10. R. L. Braun, A. E. Lewis, and M. E. Wadsworth, "In-Place Leaching of Primary Sulfide ores: Laboratory Leaching Data and Kinetics Model," Met. Trans., 5 (1974), 1717-1726.

11. R. W. Bartlett, "Pore Diffusion-Limited Metallurgical Extraction from Ground Ore Particles," Met. Trans., 3 (1972), 913-977.

12. B. W. Madsen and M. E. Wadsworth, "A Mixed Kinetics Dump Leaching Model for Ores Containing a Variety of Copper Sulfide Minerals," U.S.B.M., RI 8547 (1981).

13. H. K. Lin, H. Y. Sohn, and M. E. Wadsworth, "The Kinetics of Leaching of Chalcopyrite and Pyrite Grains in Primary Copper Ore by Dissolved Oxygen," unpublished (Department of Metallurgy and Metallurgical Engineering, University of Utah, Salt Lake City, Utah, 1987).

14. B. W. Madsen, M. E. Wadsworth, and R. D. Groves, "Application of a Mixed Kinetics Model to the Leaching of Low Grade Copper ore," Trans. Soc. Min. Engrs., 258 (1974), 69-74.

15. R. J. Roman, B. R. Benner, and G. W. Becker, "Diffusion Model for Heap Leaching and its Application to Scale-Up," Trans. Soc. Min. Engrs., 256 (1974), 247-256.

16. L. M. Cathles and J. A. Apps, "A Model of the Dump Leaching Process that Incorporates Oxygen Balance, Heat Balance, and Air Conductivity," Met. Trans. B, 6B (1975), 617-624.

17. J. W. Braithwaite, "Simulated Deep Solution Mining of Chalcopyrite and Chalcocite" (PhD Thesis, University of Utah, Salt Lake City, Utah, 1976).

18. H. Y. Sohn and J. Szekely, "A Structural Model for Gas-Solid Reactions with a Moving Boundary -- Part III," Chem. Eng. Sci., 27 (1972), 763-778.

19. P. M. Bommer and R. S. Schechter, "Mathematical Modeling of In Situ Uranium Leaching," Soc. Pet. Eng. J., 19 (1979), 393-400.

20. M. I. Kabir, L. W. Lake, and R. S. Schechter, "A Minifield Test of In Situ Uranium Leaching -- Interpretation of Results," Interfacing Technologies in Solution Mining, eds. W. J. Schlitt and J. B. Hiskey (SME and SPE of AIME, 1982), 75-87.

21. A. Tatom, R. S. Schechter, and L. W. Lake, "Factors Influencing the In Situ Acid Leaching of Uranium Ores," Interfacing Technologies in Solution Mining, eds. W. J. Schlitt and J. B. Hiskey (SME and SPE of AIME, 1982), 131-147.

22. H. W. Gao, H. Y. Sohn, and M. E. Wadsworth, "A Mathematical Model for the Solution Mining of Primary Copper Ore: Part I. Leaching by Oxygen-Saturated Solution Containing No Gas Bubbles," Met. Trans. B, 14B (1983), 541-551.

23. H. W. Gao, H. Y. Sohn, and M. E. Wadsworth, "A Mathematical Model for Solution Mining of Primary Copper Ore: Part II. Leaching by Solution Containing Oxygen Bubbles," Met. Trans. B, 14B (1983), 553-558.

24. C. R. McKee, R. H. Jacobson, S. C. Way, M. E. Hanson, and K. Chong, "Design Criteria for In Situ Mining of Hard Rock Deposits," Interfacing Technologies in Solution Mining, eds. W. J. Schlitt and J. B. Hiskey (SME and SPE of AIME, 1982), 103-121.

25. G. M. Potter, C. Chase, and P. G. Chamberlain, "Feasibility of In Situ Leaching of Ores Other than Copper and Uranium," Interfacing Technologies in Solution Mining, eds. W. J. Schlitt and J. B. Hiskey (SME and SPE of AIME, 1982), 123-130.

26. In Situ Leaching and Solution Mining Evaluation of State of the Art: Summary Report and I. Geological Characterization, II. Chemistry, III. Fracturing and Rubblization and IV. Fluid Flow Management ed. P. M. Wright (Earth Science Laboratory, University of Utah Research Institute, 391 Chipeta Way, Salt Lake City, Utah, 1983).

NUMERICAL PREDICTION OF COBALT SORPTION IN A

CONTINUOUS ION-EXCHANGE COLUMN

K. S. Gritton

U. S. Bureau of Mines, Salt Lake City Research Center,
729 Arapeen Drive, Salt Lake City, Utah, 84108

Abstract

The Bureau of Mines investigated the feasibility of numerically
modeling the sorption of cobalt, a vulnerable, strategic, and critical
metal, from spent copper leach solutions in a multiple-compartment
ion-exchange (MCIX) column. Equilibrium resin loadings, cobalt-sorption
kinetics, and resin-fluidization characteristics were determined in simple
laboratory tests, and a numerical model was developed to predict cobalt
extraction and the MCIX-column resin inventory based upon the test results.
The model was implemented on a microcomputer to approximate the solution to
the set of differential equations which govern model performance. Results
compared very well to data from 12 pilot-scale MCIX-column experiments.
The average difference between predicted and actual cobalt extractions was
3.2 pct, while the maximum difference was 9.3 pct. The actual and
predicted resin inventories differed by an average of 5.8 pct, and the
maximum difference was 12.0 pct.

[1]Chemical engineer, Salt Lake City Research Center, Bureau of Mines, U.S.
Department of the Interior, Salt Lake City, UT.

Introduction

The Bureau has developed a process using ion exchange to extract cobalt, a strategic and critical metal, from domestic copper leach solutions (1-3). These solutions are produced by dump or heap leaching of low-grade ores with dilute sulfuric acid and contain significant amounts of readily accessible cobalt. Although the cobalt concentrations in these streams are only 15 to 30 ppm, potential annual recoveries exceed 2 million lb.

The Bureau's process used ion exchange to extract the cobalt from the leach solution, followed by solvent extraction to purify and concentrate the cobalt. Metallic cobalt was produced by electrowinning. Economic considerations dictated that ion-exchange resin inventories in the Bureau's process be minimized. Thus, the ion-exchange investigation was directed toward the use of a continuous, fluidized bed, ion-exchange column. The Bureau-developed MCIX column was chosen for the cobalt recovery studies since previous work using this column to extract uranium from low-grade solutions had demonstrated its utility in reducing resin inventory when compared to fixed-bed systems (4). Ion-exchange resin Dow 4195.02[3] was used during the cobalt recovery investigation.

[3]Reference to specific products does not imply endorsement by the Bureau of Mines.

Although modeling of fixed-bed ion-exchange systems has been the subject of intensive investigations (6-7), results from these studies are not generally applicable to fluidized-bed systems. Previous approaches to modeling fluidized systems included approximation of a continuous ion-exchange column as a series of agitated resin slurries (8). This approach, however, required pilot-plant data prior to complete model definition. Other modeling approaches required knowledge of terminal concentrations of the solution and resin phases (9) or did not allow for possible changes in resin volume as metals were sorbed (10). More general modeling approaches have been published (11-12), but restrictive assumptions concerning reaction kinetics and mass-transfer rate expressions inhibited general application of these methods. Thus, the Bureau investigated a general method to numerically model the MCIX column.

The objective of the Bureau's numerical studies was to provide a dependable method of predicting MCIX column performance using minimal laboratory data. The Bureau-developed model used data from simple laboratory tests to establish equilibrium and kinetic parameters and resin-fluidization characteristics. Test results were then used in the numerical solution of a series of ordinary differential equations to predict MCIX column performance, and model predictions were compared to results from pilot-scale MCIX column experiments.

Cobalt Sorption Studies

Description of the MCIX Column

The Bureau-developed MCIX column consisted of a series of vertically arranged fluidized beds of ion-exchange resin. Stages, or sections, of the column were separated by orifice plates having an orifice area comprising 6 pct of the cross-sectional area of the column. A schematic of the MCIX column is depicted in figure 1. Leach solution was introduced into the bottom of the column, and the upflow of solution fluidized the resin in each compartment. This fluidization was maintained except during brief intervals when resin was withdrawn from the bottom of the column. During

330

these withdrawal periods, solution flow was interrupted, resin and solution
in each stage were transferred to the next lower stage, and fresh resin was
added to the top stage of the column. At the completion of the resin
discharge period, solution flow was restored, and resin within the column
was again fluidized.

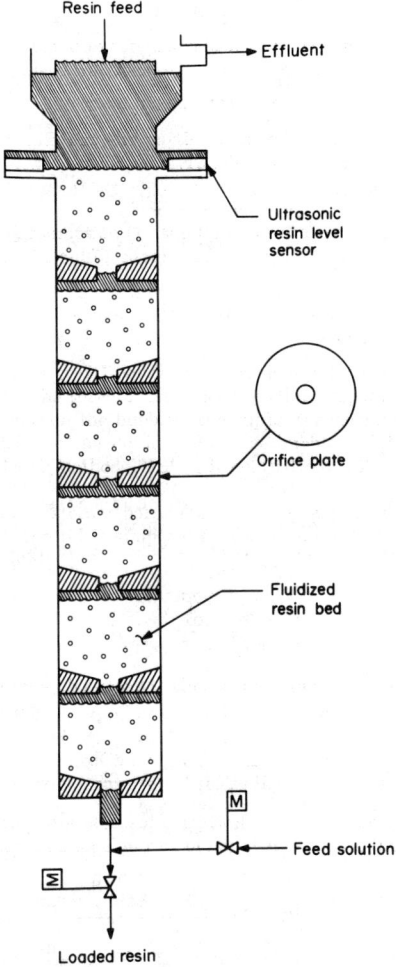

Figure 1 - Multiple compartment
ion-exchange (MCIX) column.

Resin, discharged from the column, was eluted with H_2SO_4 and NH_4OH to
produce metal-rich eluate solutions and a barren, base-form resin which was
recycled to the MCIX column (1-3, 13).

Description of the Ion-Exchange Resin

The resin used in the cobalt sorption studies, Dow resin 4195.02, is
classified as a chelating resin. The resin consists of a three-
dimensional network of copolymerized styrene and divinylbenzene with

attached bispicolylamine functional groups (14-15). The resin, as received in the acid form, was minus 20 plus 50 mesh, and was screened to minus 20 plus 28 mesh for use in the sorption studies. The minus 20-plus 28-mesh size fraction amounted to 50 pct of the as-received resin. The volume change associated with changing from the acid or metallic forms to the base form of the resin was minus 30 pct for the minus 20- plus 28-mesh resin.

Description of the Solution

The solution used in the cobalt sorption investigation was an effluent from a domestic copper cementation plant. The solution pH was 3.1, and the approximate metals content of the solution was, in grams per liter, 0.03 Co, 0.03 Ni, 0.06 Cu, 2.0 Fe, 0.2 Zn, 4.5 Al, 7.2 Mg, and 0.4 Mn.

Determination of the Equilibrium-Loading Isotherm

Although sorption studies emphasized removal of cobalt from the cementation effluent, Dow 4195.02 also sorbed nickel, copper, iron, and zinc from this same solution. Single-component equilibrium isotherms for the above metals in acidic sulfate solutions have been reported (14), but of more interest in the modeling investigation was the loading isotherm for Dow 4195.02 in equilibrium with cobalt in the multicomponent domestic copper leach solution.

The equilibrium-loading isotherm for cobalt was determined by contacting cementation plant effluent with barren, base-form resin at aqueous-to-resin (A:R) volume ratios ranging from 1:1 to 300:1, at a constant pH of 3.1, in a stirred vessel at room temperature for 24 h. Preliminary research had indicated that 24 h was sufficient time for the resin and solution to reach equilibrium. Resin loadings were determined from a material balance using differences between fresh and spent solution assays. Equilibrium isotherms varied somewhat depending upon the exact composition of a particular sample of solution, but a typical equilibrium isotherm for cobalt is depicted in figure 2.

The relative affinity of Dow 4195.02 for a number of the metallic cations in the copper leach solution has been reported as $Cu^{2+} > Ni^{2+} > Fe^{3+} > Zn^{2+} > Co^{2+} > Fe^{2+}$ (14). The resin had no significant affinity for other cations, such as aluminum, magnesium, and manganese in the leach solution. The difference in affinity of the resin for various cations in the leach solution suggested that sorbed cobalt ions could be crowded from the resin by ions for which the resin had a higher affinity. Crowding effects were not quantified in the modeling procedures since equilibrium isotherms were generated using actual cementation plant effluent, thus, the equilibrium isotherm depicted in figure 2 accounts for crowding implicitly.

Although fresh cementation effluent solution contained iron, primarily in the ferrous state, oxidation to the ferric state was rapid, and many of the solutions used in the investigation contained primarily ferric iron. The effects of the iron oxidation were also accounted for implicitly, since both fresh and aged solutions were used in generating the equilibrium isotherms.

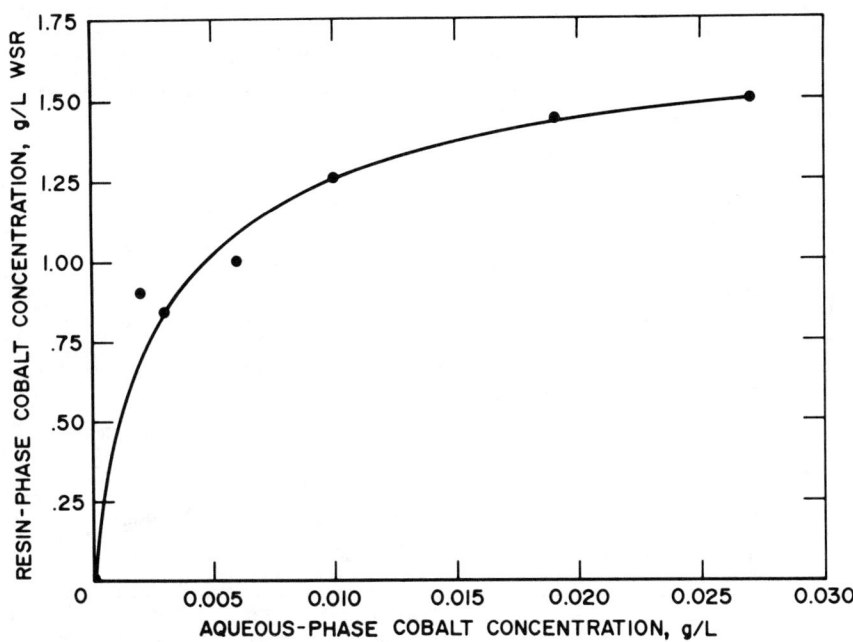

Figure 2 - Equilibrium-loading isotherm for cobalt in pH 3.1 copper
leach solution using Dow 4195.02.

Cobalt Sorption Kinetics

Since solution residence time in the MCIX column was only 7.5 to
28 min, equilibrium between resin and solution was not achieved during the
time between resin discharges in the MCIX column. Consequently, loading
kinetics had a significant effect on extraction efficiency. Studies were
conducted to determine the rate at which cobalt was extracted from the
spent copper leach solution by resin 4195.02.

Cobalt sorption rates for the resin-cementation effluent system were
determined in simple batch tests. Although sorption rates could be
determined by a number of analytical procedures, a technique using a
radiotracer was implemented since the very small sample sizes, typically 1
to 5 mL, required for analysis, did not alter A:R ratios significantly
during the tests. A trace amount of ^{60}Co, .002 x 10^{-6} Ci, dissolved in
0.1N HCl, was added to 3 L of pH 3.1 cementation plant effluent. This
solution was contacted with barren, base-form resin at A:R ratios between
30 and 60. These ratios spanned the practical operating range of A:R
ratios for the MCIX column using Dow resin 4195.02. Sufficient agitation
was provided to ensure complete resin fluidization. Aqueous-phase cobalt
concentration was determined by the rate at which gamma radiation was
emitted by samples of solutions taken at various times during the batch
experiments. Gamma radiation emission was monitored using a NaI
scintillation detector with background correction. Emission rates were
low, and anticoincidence correction was not necessary. Cobalt
concentration versus time was plotted for each A:R ratio, and sorption
rates were determined by graphical determination of the slopes of the
curves. The rates of sorption, in grams per liter wet-settled resin (WSR)

per minute, were estimated as the negative of the slopes of the curves at given times, multiplied by the A:R ratios. Data obtained at an A:R ratio of 40:1 are depicted in figures 3 and 4.

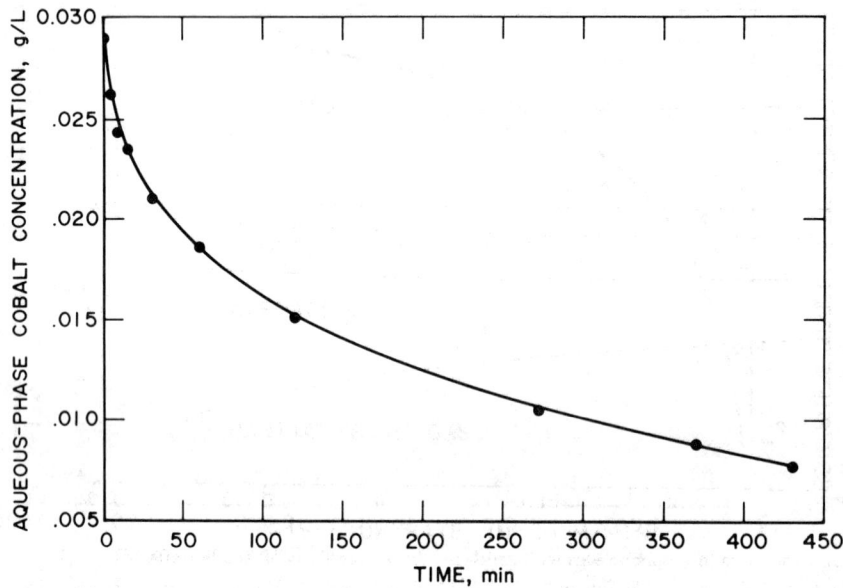

Figure 3 - Batch kinetic data, aqueous-phase cobalt concentration versus time curve, A:R = 40:1.

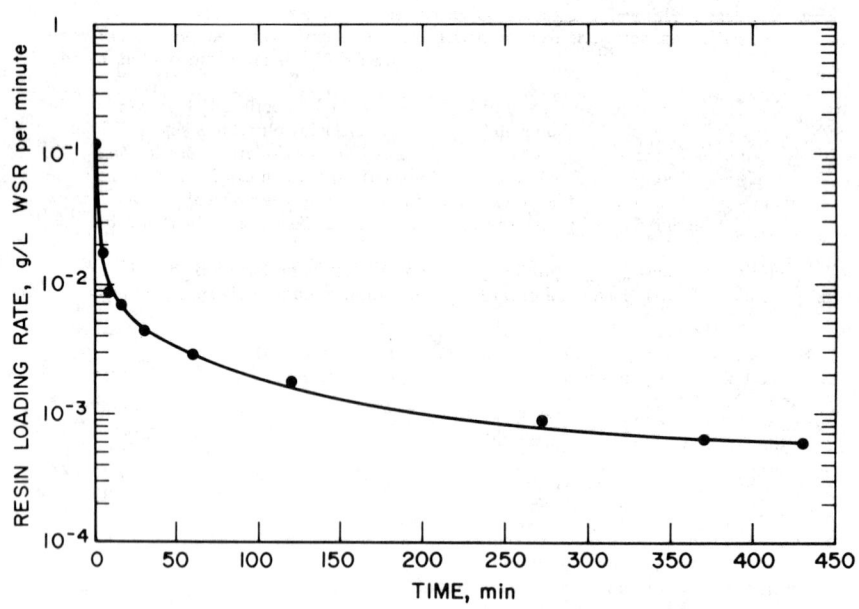

Figure 4 - Loading-rate data, A:R = 40:1.

Initial cobalt sorption was rapid, and the concentration of cobalt in solution declined throughout the duration of the tests, although sorption rates changed only slightly after the first 2 h of the tests. The results depicted in figures 3 and 4 were typical of those obtained at other A:R ratios.

Resin-Fluidization Studies

Fluidization characteristics of Dow 4195.02 were determined by adding a measured amount of resin to a 3-ft-high by 2-in-ID glass tube and fluidizing the resin with an upflow of solution at various liquid flow rates. Solution was introduced into the bottom of the tube through an orifice plate similar to those used between MCIX stages. The expansion of the resin bed was determined as a function of the superficial liquid flow rate by comparison of the height of the fluidized bed to the height of the WSR bed prior to fluidization.

This experimental procedure allowed investigation of the effects of changes in resin density due to loading on the fluidization characteristics of the resin. Water was used to fluidize the barren resin in laboratory tests since its viscosity was not significantly different from that of the leach solution, and the base-form resin would have sorbed metals from the leach solution. Both water and leach solutions were used to fluidize the loaded resin with no significant differences in fluidization characteristics.

Data from the fluidization experiment, using loaded resin and leach solution, are depicted in figure 5. The curve in figure 5,

$$\text{percent expansion} = 9.331(Q)^{1.434}, \tag{1}$$

where Q is the liquid flow rate in gallons per minute per square foot, was established by the application of a least-squares procedure. The equation for the curve was used to determine the MCIX column resin inventory in the modeling studies.

MCIX-Column Modeling Studies

The objectives of MCIX modeling research were (1) to develop a numerical model that could correctly predict cobalt extraction and resin inventory at a specified set of MCIX column operating conditions, and (2) to determine the requisite laboratory data for making such a prediction.

Mathematical Basis for the Model

General Kinetic Model. The rate at which Co^{2+} ions were extracted from solution was dependent on the concentration of cobalt in solution, the concentration of cobalt on the resin, and the equilibrium capacity of the resin in contact with the cobalt-bearing solution. Thus, the extraction rate was expressed by an ordinary differential equation of the form

$$(dY/dt) = k_1 \cdot C_a{}^\alpha \cdot (1 - Y/Y^*)^\beta, \tag{2}$$

where Y was the concentration of cobalt on the resin in grams per liter WSR, k_1, α, and β were experimentally determined constants, and Y^* was the concentration of cobalt on the resin which would be in equilibrium with pH 3.1 cementation-plant effluent containing C_a grams per liter of cobalt.

335

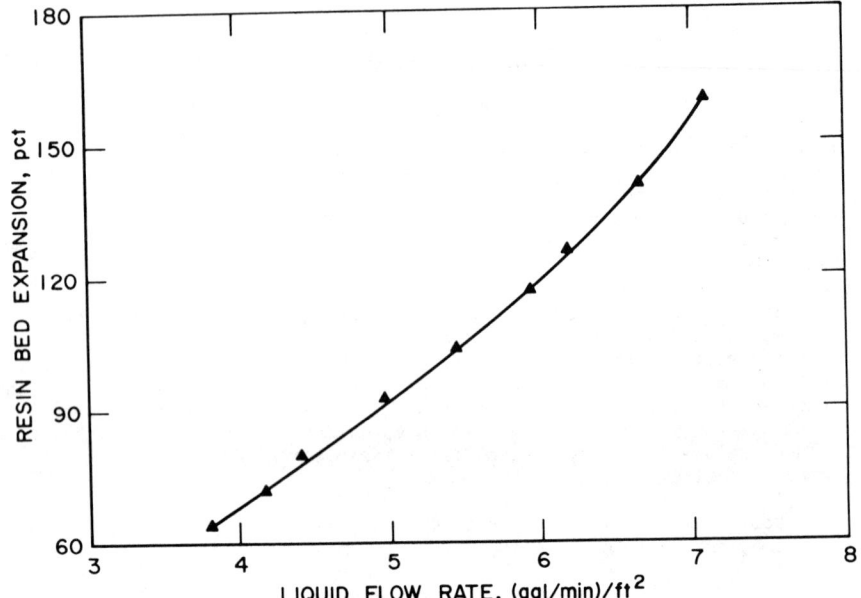

Figure 5 - Fluidization of Dow 4195.02, bed expansion versus supervicial liquid flow rate.

The equilibrium-resin concentration, Y^*, was expressed as a function of solution composition according to the relationship

$$Y^* = A(C_a) / (B(C_a) + 1),\qquad(3)$$

where A and B were determined empirically using equilibrium data and curve-fitting procedures. The values for A and B, which described the isotherm depicted in figure 2, were 550 and 325, respectively, but values of A and B, which best described the equilibrium resin loading of cobalt, differed for various samples of cementation plant effluent due to variations in solution composition.

The values of k_1, α, and β, in equation 1, were obtained from previously described batch loading-rate tests. Loading rates were determined from plots of the experimental data from the batch experiments, and equation (1) was approximated by the linearized form

$$\ln(-dC_a/dt) = \ln(k_1) + \alpha \cdot \ln(C_a) + \beta \cdot \ln(1 - Y/Y^*).\qquad(4)$$

A multiple-linear regression procedure yielded approximate values for k_1, α, and β. Correlation coefficients from the regressions were typically 0.8. Attempts to improve the accuracy of the values of k_1, α, and β using nonlinear least-squares procedures were unsuccessful.

Numerical Loading-Rate Equations. A cobalt material balance on the resin phase between discharges was approximated using an Euler method to solve equation (2) for each stage of the MCIX column. The numerical expression used in the model was of the form

$$Y_i^{t+\Delta t} = Y_i^t + [k_1 \cdot (C_a^t)_i^\alpha (1 - Y_i^t/Y_i^*)^\beta] \Delta t, \tag{5}$$

where the subscript i denotes the MCIX stage number, the superscript t denotes time, and Δt is the time increment used in approximating the solution to equation (2). The solution to the set of equations represented by equation (5) required initial conditions for both the resin and solution-phase cobalt concentrations in each section of the column. For simplicity, the following initial conditions were chosen:

$$Y_i^0 = ((N-i+1)/N) \cdot 0.75 \cdot Y_M, \quad i = 1,2, \ldots ,N, \tag{6}$$

and,

$$(C_a)_i^0 = (\text{feed concentration}) \cdot ((N-i+1)/N), \quad i = 1,2, \ldots ,N, \tag{7}$$

where N was the total number of MCIX stages, and Y_M was the maximum loading of cobalt on the resin as determined from the equilibrium isotherm using the initial leach solution concentration as the abscissa. The modeling of the column was performed such that any reasonable set of initial conditions for the resin and solution concentrations would be adequate, but the above set of initial conditions accelerated convergence of the calculations.

Since local concentration gradients and mixing transients in a single MCIX stage are highly complex, average resin and aqueous-phase concentrations were assumed to prevail, and equations (2), (4), and (5) were applied to each MCIX stage with Y_i, $(C_a)_i$, and Y_i^* taken as average quantities for each stage during the current time interval.

Determination of Resin Inventory. The resin inventory in the column was determined by correlating laboratory measurements of the resin bed expansion at various superficial liquid velocities. The resin inventory was then determined from the following expression:

$$\text{Resin inventory} = \sum_{i=1}^{N} V_i / (1 + (\text{pct bed expansion})_i/100), \tag{8}$$

where

$$\text{percent bed expansion} = a \cdot Q^b \cdot Y_i^c, \tag{9}$$

N was the number of MCIX stages, V_i was the volume of a single MCIX stage, Q was the liquid flow rate per unit of cross-sectional column area, and a, b, and c were empirical constants estimated by application of a least-squares procedure to the data from the resin-fluidization tests, depicted in figure 5.

Determination of the Loading Cycle Length. The amount of time between resin discharges in the MCIX column was determined based on the amount of loaded resin discharged, the liquid flow rate, and the A:R ratio. One hundred percent of the resin and solution in each stage were assumed to transfer to the next lower stage during a discharge, although the model allowed for less than a full compartment to be discharged during a simulation. The cycle length was determined to be the volume of WSR in the bottom compartment, multiplied by the fraction of resin and solution in

337

each stage to be transferred during a discharge, divided by the quotient of the liquid feed rate and the A:R ratio.

Since the volume of displaced solution, due to expansion of the loaded resin, was insignificant compared to the amount of solution passing through the column during the loading cycle, the displaced solution was not considered in determination of the loading cycle length.

Modeling Algorithm

The step-by-step procedure used in the modeling calculations may be summarized as follows:

1. Specify the number of MCIX stages, the column dimensions, A:R ratio, liquid feed rate and feed concentration, the fraction of resin and solution in each stage to be transferred during a resin discharge, and the number of discharge cycles to be simulated. The number of cycles required for the model to reach steady state was approximately four times the number of MCIX stages.

2. Set initial resin and solution compositions.

3. Estimate the stage-wise and total resin inventories.

4. Estimate the length of the loading cycle.

5. Calculate the equilibrium resin-phase concentration and sorption rates for each stage during the current time increment.

6. Calculate the amount of metal sorbed during the current time increment and the new resin compositions for each stage.

7. Calculate new solution compositions based upon the amount of metal sorbed onto the resin and the flow of solution through each stage during the current time increment.

8. Check to see if a loading cycle was completed. If so, check to see if the specified number of discharge cycles have been completed. If so, then the simulation is complete. Calculate the desired instantaneous and total metal extractions. Print extractions, resin and solution concentration profiles, then terminate program execution. If the specified number of cycles were not complete, transfer the solution and resin in each stage to the next lower stage, add barren resin and spent solution to the top compartment, increment the current time, and proceed to step 5.

The modeling algorithm is depicted as a simplified flowchart in figure 6. Computer code was written in FORTRAN to implement the algorithm on a microcomputer, and model results were compared to data obtained from a pilot-scale MCIX column. A copy of the FORTRAN source code may be obtained from the author.

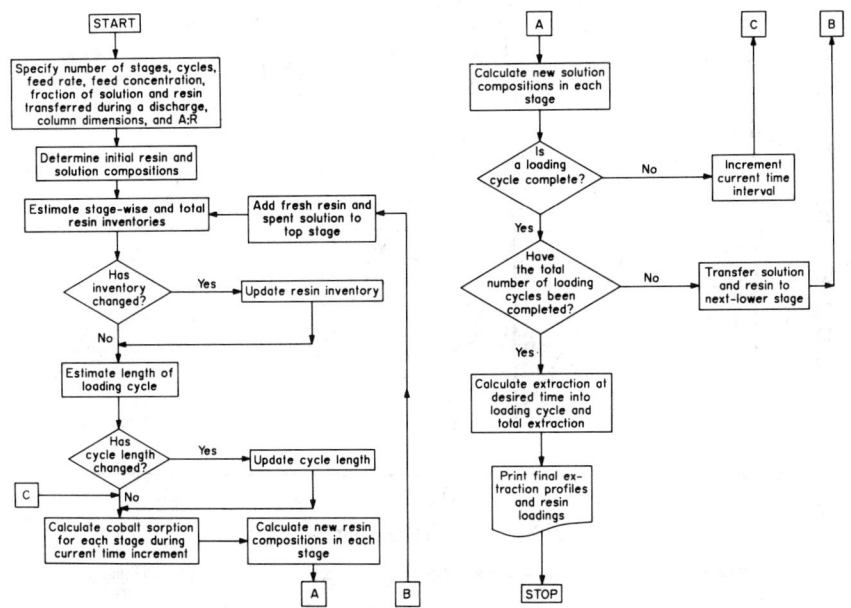

Figure 6 - Flowchart of algorithm for modelling the MCIX column.

Dynamic Behavior of the MCIX Column

The percentage of the cobalt in the feed solution extracted by the resin in the column varied throughout the length of a loading cycle. This was due to the cyclic mode of operation of the MCIX column. Resin was transferred periodically, while solution moved continuously. Thus, at the beginning of a cycle, the resin in a particular MCIX stage had less metal on it than at the middle or end of the cycle. Sorption rates and metal extractions were higher early in the cycle, and as the cycle progressed, sorbed metal reduced the difference between the equilibrium resin-phase metal concentration, Y^*, and the average concentration of metal on the resin, Y, in a particular stage. Since the difference between the resin loading and equilibrium loading constituted the driving force for mass transfer, the extraction rates decreased as the cycle progressed. This decrease in the mass-transfer driving force was only slightly compensated by the increased average solution-phase concentration which resulted from the lower extraction rates, and the trend during the loading cycle was one of decreased extraction. A plot of the predicted, transient performance of a 15-compartment MCIX column, operated at an A:R ratio of 40:1 and a solution flow rate of 5.5 gpm/ft², is depicted in figure 7. Instantaneous extraction is depicted by the solid line, while overall extraction is represented by the dashed line.

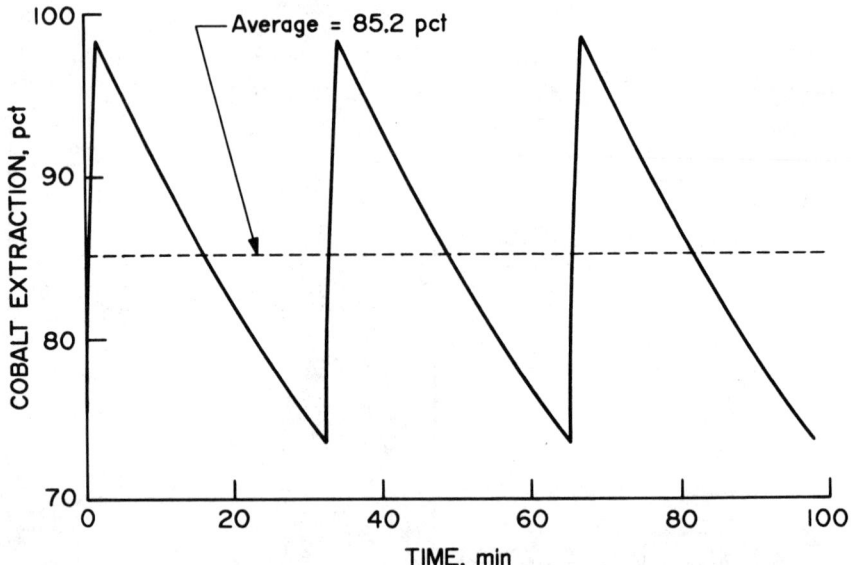

Figure 7 - Dynamic extraction behavior of the MCIX column.

Figure 7 suggests that shorter cycle times would result in greater total recoveries since the valleys in the instantaneous recovery curve would be more shallow. While this is true, the net result of faster cycle times would be incomplete utilization of the ion-exchange resin, i.e., the A:R ratio would be decreased due to the increased resin flow rate, and resin discharged from the column would be only partially loaded. While cobalt recoveries might be greater at lower A:R ratios, resin inventories would be unreasonably large, and resin costs would be correspondingly high.

Comparison of Model Predictions to MCIX Column Data

Comparison of Actual-to-Predicted Optimum Operating Conditions

Optimum operating conditions and cycle times for a MCIX column must be determined by a balance between product values and capital and operating costs. Due to the high cost of the ion-exchange resin used in the Bureau's cobalt recovery studies, the Bureau's 2-in-ID MCIX column was operated to obtain maximum cobalt recovery per unit of ion-exchange resin. Maximum cobalt extraction per unit of resin was achieved in a column consisting of ten 1-ft compartments, at an A:R ratio of 40:1, using a solution feed rate of 5 (gal/min)/ft² (3). Modeling studies predicted maximum cobalt recovery per unit of resin would occur at the same set of operating conditions.

Comparison of Actual-to-Predicted Cobalt Extractions and Resin Inventories

Due to the dynamic behavior of the MCIX column, actual and predicted metal extractions must be compared at specified times during a loading cycle. During most of the cobalt sorption studies, MCIX-column extraction profiles were taken 20 min into the loading cycle. Thus, model predictions of extractions were calculated 20 min into the loading cycle for comparison.

For those conditions resulting in a cycle time of less than 20 min, comparisons were made at the end of the loading cycle. Actual and predicted extractions for various MCIX column experiments are summarized in Table I.

The MCIX column stages used in the 10- and 15-compartment experiments were 1-ft high by 2-in ID, while the stages used in the 5-compartment experiment were 2-ft high by 2-in ID. Data and predictions agreed well for both the resin inventory and cobalt extractions. The average difference between predicted and actual cobalt extractions was 3.2 pct, while the maximum difference was 9.3 pct. The actual and predicted resin inventories differed by an average of 5.8 pct, and the maximum difference was 12.0 pct.

Table I. - Comparison of MCIX Column Data With Model Predictions 20 min Into the Loading Cycle

MCIX stages	A:R[1]	Flow rate, (gal/min) /ft²	Resin inventory			Cobalt extraction, pct	
			Actual, mL	Predicted, mL	Difference,[2] pct	Actual	Predicted
15	40	4.0	6,103	5,508	9.7	100.0	100.0
15	40	5	5,047	4,777	5.3	100	99.8
15	40	5.5	4,340	4,460	2.8	92.3	98.3
15	40	6	4,080	4,172	2.3	82.8	81.7
15[3]	40	7	3,756	3,672	2.2	72.4	73.6
15	50	5.5	4,580	4,460	2.6	84.6	82.2
10[4]	30	6	3,090	2,781	10	69.7	66.1
10	40	4	3,988	3,672	7.9	100	97.6
10	40	5	3,322	3,184	4.1	96.2	89.4
10	40	6	2,970	2,781	6.4	72.7	63.4
10	80	5	3,620	3,185	12	45.5	49
5	40	5	3,306	3,184	3.7	96.2	94.5

[1]Aqueous-to-resin volume ratio. [3]Cycle time: 16.9 min.
[2]Absolute value. [4]Cycle time: 16.8 min.

Comparison of Actual-to-Predicted Extraction Profiles

Stage-wise extraction profiles, indicating model predictions and actual MCIX column data, are depicted for various MCIX column experiments in figures 8 to 12. Although extraction profiles might be of less significance than predictions of quantities which influenced process economics, i.e., total extraction and resin inventory, the profiles aid in evaluating the ability of the model to reflect actual conditions within the MCIX column. Compartment numbers in figures 8 to 12 correspond to the sequence of compartments in the MCIX columns with 1 being the bottom compartment. Model predictions were usually slightly lower than the actual extraction profiles in the lower and middle sections of the column. This trend may be due to the fact that only a first-order numerical method, i.e., Euler's method, was used in approximating the solution to the loading-rate equations. A higher order solution method, e.g., a Runge-Kutta or a predictor-corrector method, may result in more accurate extraction profiles. In all cases, however, model predictions were within a few percent of actual cobalt extractions.

341

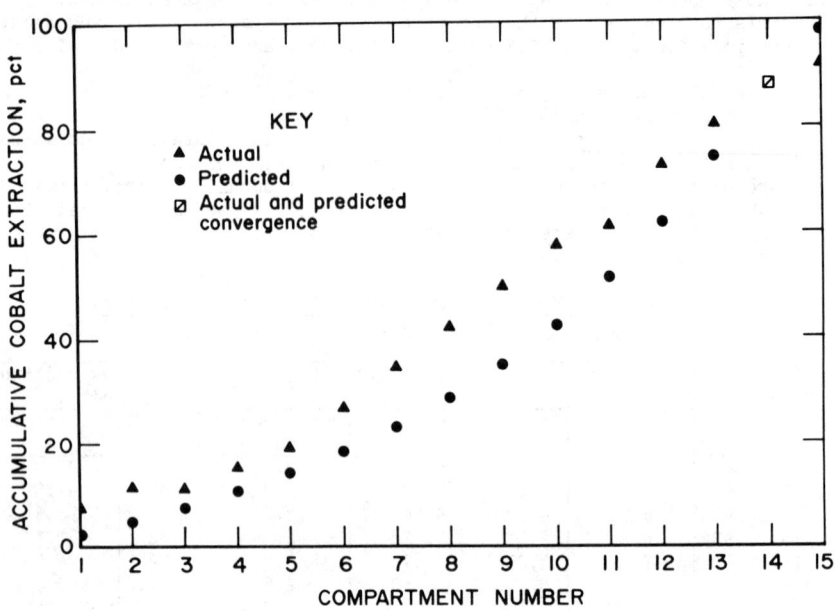

Figure 8 - Predicted and actual extraction profiles for a 15-stage
MCIX column 1-ft compartments, A:R = 40:1, 5.5 gpm/ft².

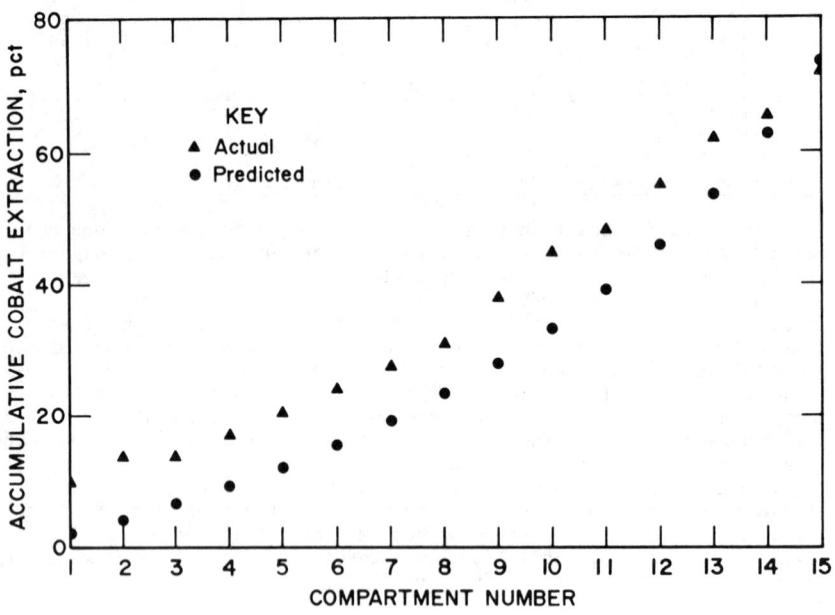

Figure 9 - Predicted and actual extraction profiles for a 15-stage
MCIX column, 1-ft compartments, A:R = 40:1, 7.0 gpm/ft².

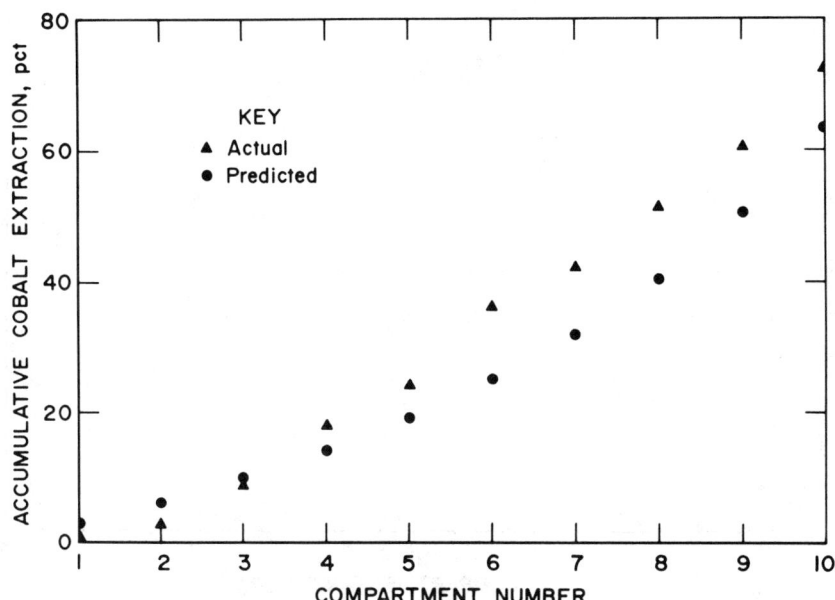

Figure 10 - Predicted and actual extraction profiles for a 10-stage
MCIX column, 1-ft compartments, A:R = 40:1, 6.0 gpm/ft².

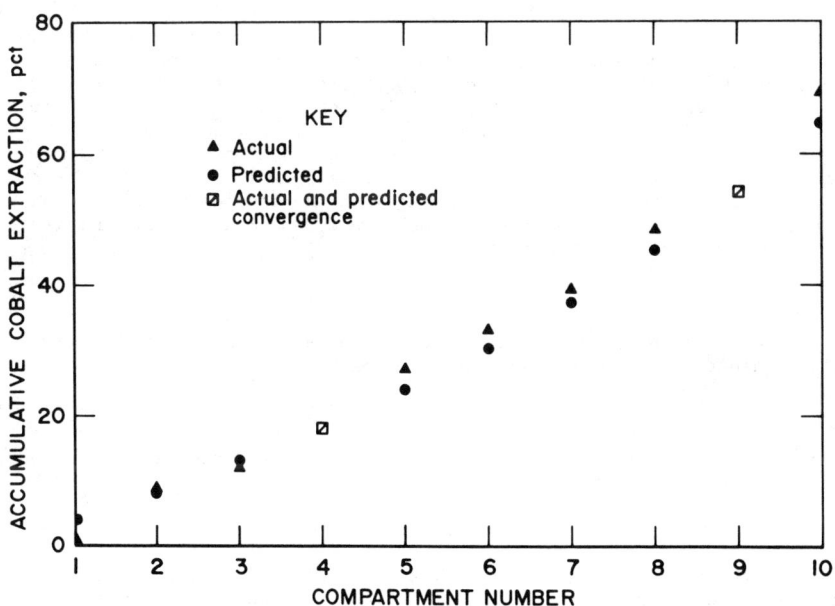

Figure 11 - Predicted and actual extraction profiles for a 10-stage
MCIX column, 1-ft compartments, A:R = 30:1, 6.0 gpm/ft².

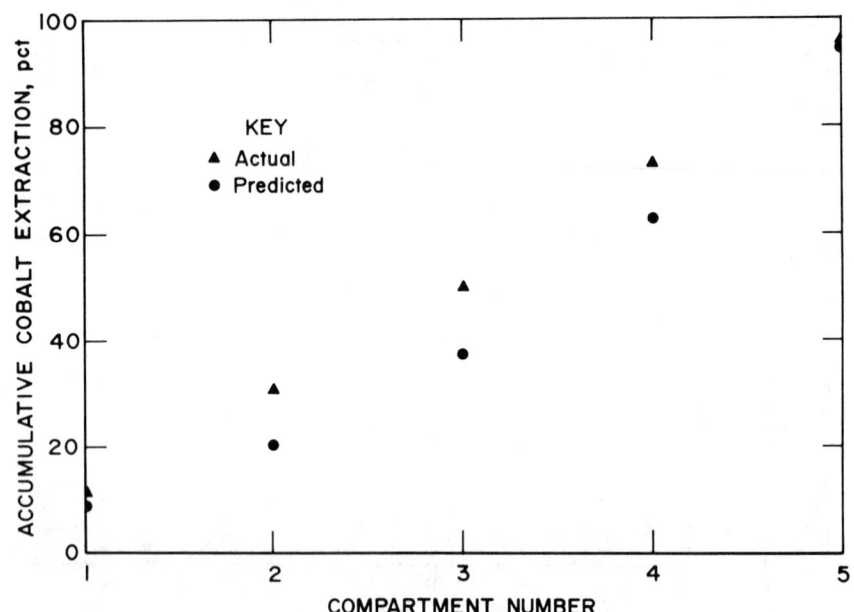

Figure 12 - Predicted and actual extraction profiles for a 5-stage
MCIX column, 2-ft compartments, A:R = 40:1, 5.0 gpm/ft².

Summary

The Bureau-developed numerical model of the MCIX column accurately
predicted MCIX column performance on the basis of simple laboratory test
results. Eight parameters, determined in simple laboratory tests, were
necessary for complete model definition. Equilibrium resin and
aqueous-phase cobalt concentrations were determined by contacting barren
resin with cementation effluent at A:R ratios from 1:1 to 300:1. The resin-
phase concentration, Y^*, was plotted against the aqueous-phase
concentration, C_a, and constants A and B were determined to express the
equilibrium-loading isotherm as

$$Y^* = A(C_a)/(B(C_a) + 1). \tag{10}$$

Cobalt extraction rates were assessed, and constants k_1, α, and β were
determined to approximate the extraction rate as

$$(dY/dt) = k_1 \cdot C_a{}^\alpha \cdot (1 - {}^Y/Y^*)^\beta. \tag{11}$$

Finally, resin fluidization tests were conducted, and the resin-bed
expansion was correlated by

$$\text{percent bed expansion} = a \cdot Q^b \cdot Y^c. \tag{12}$$

These eight parameters, A, B, k_1, α, β, a, b, and c, were used in a
numerical model to predict metal extractions and resin inventories in the
MCIX column under a number of different operating conditions.

Results compared very well to data from 12 pilot-scale MCIX column experiments. The average difference between predicted and actual cobalt extractions was 3.2 pct while the maximum difference was 9.3 pct. The actual and predicted resin inventories differed by an average of 5.8 pct, and the maximum difference was 12.0 pct.

References

1. Jeffers, T. H., and M. R. Harvey. Cobalt Recovery From Copper Leach Solutions. BuMines RI 8927, 1985, 12 pp.

2. Jeffers, T. H., K. S. Gritton, and P. G. Bennett. Recovery of Cobalt From Spent Copper Leach Solution Using Continuous Ion Exchange. Paper in Recycle and Secondary Recovery of Metals, (Int. Symp. on Recycle and Secondary Recovery of Metals). The Metallurgical Society Inc., Warrendale, PA, 1985, pp. 609-621.

3. Jeffers, T. H., K. S. Gritton, P. G. Bennett, and D. C. Seidel. Recovery of Cobalt From Spent Copper Leach Solution Using Continuous Ion-Exchange. BuMines RI 9084, 1987.

4. Ross, J. R., and D. R. George. Recovery of Uranium From Natural Mine Waters by Countercurrent Ion Exchange. BuMines RI 7471, 1971, 17 pp.

5. Kirk, W. Cobalt. Sec. in BuMines Mineral Commodity Summaries 1987, pp. 38-39.

6. Morbidelli, M., A. Servida, G. Storti, and S. Carra. Simulation of Multicomponent Adsorption Beds. Model Analysis and Numerical Solution. Ind. Eng. Chem. Fundam., v. 21, 1982, pp. 123-131.

7. Barba, D., G. Del Re, and P. U. Foscolo. Numeric Simulation of an Ionic Exchange Bed. Paper in Proceedings of the 7th International Symposium on Fresh Water From the Sea, v. 2, European Federation of Chemical Engineers, Athens, Greece, 1980, pp. 13-23.

8. Chen, J. W., F. L. Cunningham, and J. A. Buege. Computer Simulation of Pilot-Scale Multicolumn Adsorption Processes Under Periodic Countercurrent Operation. Ind. Eng. Chem. Process Des. Develop., v. 11, No. 3, 1972, pp. 430-434.

9. Slater, M. J. Continuous Ion Exchange in Fluidized Beds. The Can. Jour. Chem. Eng., v. 52, Feb. 1974, pp. 43-51.

10. Dodds, R., P. I. Hudson, L. Kershenbaum, and M. Streat. The Operation and Modelling of a Periodic, Countercurrent, Solid-Liquid Reactor. Chem. Eng. Sci., v. 28, 1973, pp.1233-1248.

11. Wright, R. S. The Prediction of Concentration Profiles for a NIMCIX Column Absorbing Uranium From Aqueous Solution. Atomic Energy Board, South Africa, Report PER 26, May 14, 1979, 16 pp.

12. Ford, M. A. The Simulation and Process Design of NIMCIX Contactors for the Recovery of Uranium. Ion Exchange Technology, ed. by D. Naden and M. Streat. Messrs. Ellis Horwood Ltd, 1984, pp. 668-678.

13.   Bennett, P. G., G. R. Palmer, and D. C. Seidel.   Application of the
Pachuca Reactor for Elution of Metal Values From Loaded Ion-Exchange Resin.
Paper in The Reinhardt Schuhmann International Symposium on Innovative
Technology and Reactor Design in Extraction Metallurgy, (TMS-AIME Annu.
Meeting, Colorado Springs, CO, Nov. 9-12, 1986).   The Metallurgical Society
Inc., Warrendale, PA, 1986, pp. 335-346.

14.   Grinstead, R. R.   Selective Absorption of Copper, Nickel, Cobalt and
Other Transition Metal Ions From Sulfuric Acid Solutions With the Chelating
Ion Exchange Resin XFS 4195.   Hydrometallurgy (Amsterdam), v. 12, 1984, pp.
387-400.

15.   Grinstead, R. R., and W. A. Nasutavicus.   Water Insoluble Chelate
Exchange Resins Having a Crosslinked Polymer Matrix and Pendant Thereto a
Plurality of Methyleneaminopyridine Groups.   U. S. Pat. 4,031,038, July 4,
1978.

The Mathematical Modelling of the Zinc Pressure Leach

D. B. Dreisinger and E. Peters

Department of Metals and Materials Engineering
University of British Columbia
Vancouver, B.C. V6T 1W5

Abstract

The zinc pressure leach process developed by Sherritt Gordon Mines and used by Cominco and others has been mathematically modelled. The model considers the following reaction sequence; 1) oxygen in the gas phase is absorbed into the leach solution, 2) ferrous iron in solution is homogeneously oxidized to ferric ion and 3) ferric ion leaches the sulphide mineral. The data necessary to develop the model included gathering particle size data for the feed to the autoclave and obtaining rate constants for the chemical steps which were either determined in laboratory studies or estimated based on reported data from the Cominco autoclave. The objective of the modelling was to develop a preliminary mathematical framework for analyzing the process and to identify potentially important laboratory studies which might give new information about the process and how it might be optimized.

In the first modelling exercise, the Cominco operating data were forced to fit the model. This exercise showed that 1) each of the three steps mentioned above were important in rate control in the first compartment and 2) that the rate at which ferric ion leaches the zinc sulphide falls dramatically from stage 1 to stage 4. In the second exercise the model was used to predict process performance at enhanced feedrates and for a three stage autoclave configuration.

Introduction

In 1977, Sherritt Gordon Mines Ltd. and Cominco Ltd. demonstrated a direct leach of zinc sulphide concentrates in dilute sulphuric acid in a pilot plant, using oxygen pressure leaching technology. The process was installed at Trail, B.C. alongside a conventional roast-leach plant in 1981. One autoclave, having a rated capacity of 190 TPD of zinc concentrate has been operating since 1981.

347

The most recent published description of this plant is provided by Martin and Jankola(1) who supplied data on operations at 97% and 127% of rated capacity. It should be noted that there is evidence that the autoclave can be run at up to 200% of rated capacity(2). For the purposes of this paper, however, only the published data at 127% of capacity will be used for modelling. Figure 1 shows a schematic flowsheet for the zinc pressure leach while Table 1 presents the operating data abstracted from Martin and Jankola.

Table 1. Selected Operating Data for the Cominco Autoclave (1).

		Assay g/l or %					Distribution %		
Leach Input	Zn	Pb	Fe	S	H_2SO_4	Zn	Pb	Fe
Zinc Conc.	48.6	5.7	11.6	32	-	100	100	100
Feed Acid	50	-	-	-	165	-	-	-
Leach Outputs								
$ZnSO_4$ Soln.	115	-	5.0	-	30	98	-	33
Flot. Conc.	1.8	0.6	0.8	95.2	-	1	1	2
Flot. Tail.	1.4	16.0	21.3	16.5	-	1	99	65

The chemistry of the leach can be approximated by equations 1-3:

$$ZnS + H_2SO_4 + 0.5O_2 \ ----> \ ZnSO_4 + H_2O + S \tag{1}$$

$$PbS + H_2SO_4 + 0.5O_2 \ ----> \ PbSO_4(s) + H_2O + S \tag{2}$$

$$FeS + 0.5H_2SO_4 + 0.75O_2 + 0.167PbSO_4 + 0.5H_2O$$

$$----> \ 0.333Pb_{0.5}Fe_3(OH)_6(SO_4)_2 + S \tag{3}$$

The Cominco leach autoclave contains four compartments, representing the equivalent of four stirred tanks in series. Oxygen is supplied by means of spargers in the first two compartments, with agitators operating so as to disperse gas bubbles, mix the slurry, and pump additional gas into the slurry from the plenum space. More than 80% of the overall leaching reaction takes place in the first compartment, which therefore takes the main burden for oxygen absorption. The first compartment extracts 83.5% of the zinc from the solids in 27.8 minutes of mean residence time at 127% of rated capacity. The corresponding oxygen absorption rate is calculated to be about 22.35 moles/m³.min (0.50 std m³ of gas per m³ of slurry per minute at 150 C).

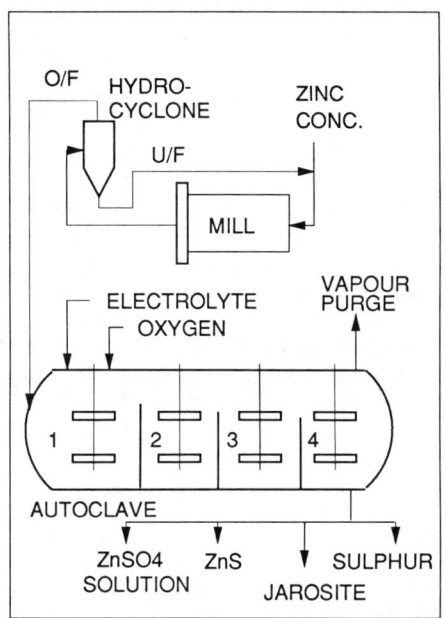

Figure 1. The Cominco Zinc Pressure Leach Flowsheet.

It has been conventional wisdom to consider heavily agitated leaching systems to be kinetically controlled at the mineral-solution interface, even when a gaseous reagent is involved. In the case of gas absorption rates as high as those quoted above for the first compartment, the resistance at the gas-liquid interface may be significant. In fact it has already been observed that leaching rates in the first compartment are altered by sparging rate changes. Further, there is compelling evidence that the presence of ferric ions in solution is necessary for high leaching rates: i.e. iron acts as a catalyst by being oxidized "homogeneously" to ferric ion by dissolved oxygen, and reduced by zinc sulphide mineral as leaching proceeds. There are thus at least three steps in the leaching process, involving reactions taking place at different sites. Equations 4-6 provide mathematical statements for the rate at which each step proceeds.

Step 1 $O_2(g) \longrightarrow O_2(aq)$

$$R_1 = R(O_2) = k_1([O_2]^*-[O_2]) \tag{4}$$

Step 2 $4Fe^{2+} + 4H^+ + O_2 \longrightarrow 4Fe^{3+} + 2H_2O$

$$R_2 = 4R(O_2) = k_2[Fe^{2+}]^2[O_2] \tag{5}$$

Step 3 $2Fe^{3+} + ZnS \longrightarrow 2Fe^{2+} + Zn^{2+} + S$

$$R_3 = 2R(O_2) = k_z A_{ZnS}[Fe^{3+}] \tag{6}$$

The rate expressions for each of the three steps are based on assumed mechanisms. These mechanisms will be discussed in a later section.

Since the first compartment is normally operated at steady state, it is clear that the three steps have the same rate when expressed as $R(O_2)$, the oxygen consumption rate, and that the values of all variables contained in equations 4-6 are in fact, constant and established by the operating conditions of temperature, pressure, agitation, pulp density, feed solution composition, mean residence time and feed particle size distribution. A mathematical model of the system will identify the effect of such dependent parameters as dissolved oxygen content $[O_2]$, iron concentration and distribution as ferrous and ferric ion on the net percent zinc extraction of the stage as well as the solution composition and particle size distribution of the overflow entering the second compartment. A similar model for the second stage would use first stage exit properties as second stage input parameters; this may then be repeated for the third and fourth stages.

Mathematical Description

Reaction Stoichiometry

Equations 1-3 are a good representation of the overall pressure leach chemistry but are inadequate in describing the first stage leach reactions. The first compartment remains considerably more acid than the following compartments, and also considerably more reducing (because of the presence of a larger load of unreacted zinc sulphide concentrates). The result is that little or no jarosite is formed in the first compartment and that the iron that is leached from concentrates is essentially all in solution as a mixture of Fe^{2+} and Fe^{3+}. If it is assumed that no jarosite is formed in the first compartment, the steady state solution composition can be estimated as a function of % zinc extraction as shown in Table 2.

Table 2. First Compartment Solution Composition.

Assuming: 136.5 kg of solids/m³ of solution. The acid concentration is based on a ferrous/ferric ratio of 1.0 and 2% of the sulphur is oxidized to sulphate.

Species	Feed Acid	70%	Extraction 80%	90%
H_2SO_4	165	70.3	56.8	43.3
Zn	50	96.4	103.0	109.6
Fe	0	11.1	12.7	14.3

In view of the leaching of iron (assumed to be present in the concentrate as marmatite) and Pb present as galena, and the oxidation of some sulphur to sulphate and iron to ferric ion, it is necessary to adjust the total oxygen demand of the system by a factor of 1.454 above that required solely for zinc leaching. This factor has been used in all subsequent calculations.

Equations 4, 5 and 6 cannot be solved without a knowledge of the rate constants k_g (gas absorption), k_2 (a third-order homogeneous rate constant for ferrous oxidation), k_z (the heterogeneous rate constant for Fe^{3+} attack of the sulphide mineral) and A_{ZnS} (the steady state sulphide mineral surface area). In addition, these equations contain three internal variables, $[O_2]$, $[Fe^{2+}]$ and $[Fe^{3+}]$, which are at steady state in each compartment. It has already been suggested that the first compartment may form little or no jarosite, in which case there is an additional equation which may be employed.

$$[Fe^{2+}] + [Fe^{3+}] = C_F \tag{7}$$

C_F represents the total concentration of iron leached.

Supplemental Assumptions

The zinc pressure leach is successful because a detergent was found (calcium lignin sulphonate) that prevents elemental sulphur formed by the reaction from wetting and protecting unreacted zinc sulphide mineral. In the absence of any protective reaction product, it is reasonable to suppose that the mineral particles shrink at a constant rate.

$$\frac{-dr}{dx} = k_l = k_z[Fe^{3+}]V_m \tag{8}$$

where r is the particle radius and V_m the molar volume of the mineral per mole of Fe^{3+} reacting.

It is also possible for dissolved oxygen to react directly with the mineral:

$$2H^+ + ZnS + 0.5O_2(aq) \longrightarrow Zn^{2+} + 2H_2O + S \tag{9}$$

At autoclave conditions, the maximum concentration of oxygen in solution will be directly proportional to the partial pressure of oxygen in the autoclave (7.5 atm). (For the purposes of this paper, the dissolved oxygen concentration will be reported in atm., not moles/m^3. The calculations are unaffected by the proportionality constant required to convert from atm. to moles/m^3.) Assuming a proportionality constant of 0.5 to 1.0 moles/(m^3.atm.) the maximum concentration of oxygen is 3.75 to 7.5 moles/m^3. The value of the steady state oxygen concentration is lower. Since total iron concentrations (12-15 gpl) are much higher at 215 - 270 moles/m^3, the leaching rate due to dissolved oxygen would be expected to be less than 10% of the Fe^{3+} rate on mass transfer considerations alone. On the basis of reactivity, oxygen is known to be particularly unreactive, except on selected catalytic surfaces. It appears that it is a reasonable assumption that the contribution of direct oxygen leaching (equation 9) can be safely neglected.

One other possibility which warrants consideration is that the reaction may have an additional catalytic step involving $H_2S(aq)$.

$$2H^+ + ZnS \longrightarrow H_2S + Zn^{2+} \tag{10}$$

Ferric ion would then oxidize the H_2S to elemental sulphur.

$$2Fe^{3+} + H_2S \longrightarrow 2Fe^{2+} + 2H^+ + S \qquad (11)$$

This mechanism has been reported by Locker and Debruyn (3) and Crundwell and Verbaan (4) for acid sulphate solutions and by Majima (5) for chloride solutions. In general, the non-oxidative leach (equation 10), when studied in the absence of Fe^{3+}, is much slower than the ferric ion leach. The initial non-oxidative leaching rate from the work of Crundwell and Verbaan for 0.5 M H_2SO_4 solution was extrapolated to zinc pressure leaching temperatures using their quoted activation energy. A leaching rate of approximately 0.06 μm/min was calculated. This is obviously much too slow for dissolving a 44 μm particle in less than 2 hours (zinc pressure leach performance). On this basis the non-oxidative leach (equation 10) has been neglected.

Selection of Input Constants

Equations 4-7 completely describe the first stage of the pressure leach. The input parameters are k_g, k_2, k_z and A_{zns}, while the values of $[O_2]$, $[Fe^{2+}]$ and $[Fe^{3+}]$ can be calculated along with $R(O_2)$. These can then be compared with measured values from the industrial autoclave for $[Fe^{2+}]$, $[Fe^{3+}]$ and $R(O_2)$. It is not possible to measure the dissolved oxygen concentration in the autoclave, so this value can only be calculated.

The Value of k_g for Oxygen Mass Transfer. k_g is a constant that combines the gas-liquid interfacial area with the mass transfer coefficient, i.e.

$$k_g = k_{g/l} \cdot A_{g/l}, \ (min^{-1}) \qquad (12)$$

The first compartment of the Cominco autoclave has a dual impellor configuration with an oxygen sparger positioned under the lower impellor. There are no reliable correlations available for predicting a value of k_g for the particular agitation configuration, temperature and solution conditions employed in zinc pressure leaching. A starting value of k_g must therefore be extracted from the operating data.

The maximum value of $R(O_2)$ if equation (4) is rate determining is given by:

$$R(O_2)_{max} = VMR(1) = k_g[O_2]^* \qquad (13)$$

where the term "VMR(1)" refers to the virtual maximum rate for oxygen absorption, based on a zero value for $[O_2]$, the steady state oxygen concentration of the slurry. Since observed oxygen absorption rates are of the order of 20-25 moles/(m^3.min), it is reasonable that VMR(1) is in the range of 25 to 100 moles/(m^3.min), and at an oxygen solubility of 7.5 atm. this leads to a value for k_g of 3.33 to 40 min^{-1} for the operating first compartment of the autoclave (depending on the value of $[O_2]$, assumed to be between 0 and 5.0). The second stage of the autoclave has a similar configuration to the first stage (sparger and dual agitator). The value of k_g is therefore probably similar for the second stage. The third and fourth stages do not have spargers and therefore rely on oxygen pumping by the agitators from the plenum. The requirement for oxygen in these stages is quite low as most reaction occurs in the first two stages. For the purposes of modelling it will therefore be assumed that solution in the final two stages of the autoclave are saturated in oxygen.

The Value of k_2 for Ferrous Oxidation by Dissolved Oxygen. The rate of ferrous oxidation with dissolved oxygen has been studied by numerous workers (for example (6)-(11)) who have generally presented their results in the form of a rate law for the termolecular reaction between two ferrous ions and dissolved oxygen. This rate law was presented in equation (5). The studies were generally carried out at temperatures lower than the zinc pressure leach (25 - 135 C) under a variety of solution conditions. When rate constants were extrapolated to pressure leach temperature and solution conditions, the constants showed significant variation. This variation prompted a study of the rate of ferrous oxidation under zinc pressure leach conditions by the present authors. This study will be reported elsewhere (12). A rate constant of $k_2 = 2.54 \times 10^{-3}$ moles^{-1}.m^3.min^{-1}.atm.$^{-1}$ was determined for an average pressure leach solution. Using this rate constant, the virtual maximum rate for Fe^{2+} oxidation was calculated by assigning $[Fe^{2+}] = C_F$, and $[O_2] = [O_2]^*$, i.e. all iron in solution is reduced to the ferrous state and the solution is saturated with oxygen.

$$VMR(2) = 0.25k_2(C_F)^2[O_2]^*$$ (14)

The value of VMR(2) is dependent on iron extraction, which determines the value of C_F (total iron extraction). Values for three extractions are shown in Table 3.

Table 3. Virtual Maximum Rates Calculated for Ferrous Oxidation.

Temperature = 150 C and $[O_2]^* = 7.5$ atm.

Iron Extraction %	70	80	90
Iron Concentration ,gpl	11.1	12.7	14.2
Value of C_F, moles.m^{-3}	198	227	255
VMR(2), moles O_2.m^{-3}.min^{-1}	187	245	310

The Value of k_z, the Rate Constant for ZnS Oxidation by Fe^{3+}. The value of k_z is either a mass transfer coefficient or a heterogeneous rate constant for Fe^{3+} ions on the ZnS particle surface. If it is a diffusional mass transfer coefficient, it's value can be estimated from Harriott's equation (13) for mass transfer to or from a particle.

$$k_z = \frac{D}{2r}(2 + 0.6Re^{0.5}Sc^{0.33})$$ (15)

For the maximum particle size in the range being leached (0-60 μm) the value of k_z would increase from about 0.0031 m/min at room temperature to about 0.0168 m/min at 150 C.

In order to test this value of k_z we must calculate the rate of mineral dissolution under the mass transfer controlled conditions. We will consider the first compartment case. The concentration of iron in solution and its distribution between the ferrous and ferric states will determine the rate of leaching. Leaching rates have been calculated using a mass transfer coefficient of 0.0168 m/min and assuming that half the iron is ferric for zinc extractions of between 70 and 90%. These rates are listed in Table 4.

Table 4. Mass Transfer Limited Leaching Rates for ZnS Leached with Fe^{3+}.

Oxidation at 150 C and 7.5 atm. of oxygen.

Zinc Extraction, %	70	80	90
Iron Concentration, moles/.m³	198	227	255
Leaching Rate, μm/min	19.7	22.6	25.4

It is apparent from Table 4 that the leaching rates predicted by the mass transfer of ferric ion to the particle surface are much too high to account for the kinetics of the Cominco autoclave.

The best estimate of the leaching rate in the first compartment of the commercial autoclave is about 0.4 μm/min (i.e. particle disappears at rate of 0.8 μm/min) from Martin and Jankola's work. There are two possible mechanisms to account for such a slow rate. The first is an electrochemical mechanism with the ferric-ferrous couple carrying out the oxidizing work. Peters (14) has shown in previous work that although ZnS is one of the least conductive metal sulphides it will still leach electrochemically. The second possibility is that a slow surface reaction involving Fe^{3+} may limit the rate. It is not possible at this point to discriminate between an electrochemical or a surface reaction. However, the rates of both reactions will depend on the ferric concentration. We will therefore assume that the rate is first order in ferric ion and assign an estimated value to k_z. Table 5 summarizes values of k_z for a leach rate of 0.4 μm/min under different extraction conditions.

Table 5. Calculated Values of k_z for Various Amounts of Zinc Extraction.

Assume that half the iron is in the ferric state and the mineral dissolves at the rate of 0.4 μm/min. Reaction at 150 C.

Zinc Extraction, %	70	80	90
C_F, moles/m³	198	227	255
$[Fe^{3+}]$, mole/m³	99	114	128
k_z, m/min	3.79 x 10⁻⁴	3.31 x 10⁻⁴	2.94 x 10⁻⁴
$[Fe^{3+}].k_z$, moles/m³.min⁻¹	0.03752	0.03752	0.03752

Estimation of A_{ZnS}. The surface area of ZnS particles is spread over particles of initial sizes that have been exposed to widely varying residence times in the first compartment. To make a useful area estimation, it is necessary to know a) the particle size distribution of the feed particles, b) the residence time distribution of the material in the compartment and c) the rate at which particles shrink. It is also necessary to make some assumptions about particle shapes and leaching morphology.

The following considerations permit a fairly useful theoretical calculation to be made:

1. The feed to the zinc autoclave is concentrate that is ground in a ball mill, with a hydrocyclone classifier that recycles material larger than 325 mesh (+44 μm) to the mill.

There are basically four common functions which are used to fit particle size data: the Rosin-Rammler-Bennett (15) equation; the Gates-Gaudin-Schuhmann (16) equation; the log probability distribution (17) and; the Gaudin-Meloy (18) equation. These functions and their applications are summarized in Perry (19).

Particle size data has been made available by Cominco (2) for a typical feed material to the autoclave. The Rosin-Rammler-Bennett equation was fitted to this data using weighted non-linear least squares estimation. The Rosin-Rammler-Bennett formula was found to give the best fit of the four options mentioned above. Equation (16) shows the general form of the function while equation (17) provides the fitted form. Figure 2 illustrates the fit of the function to the data.

$$Y = 1 - \exp(-(x/x')^n) \tag{16}$$

$$Y = 1.026 - \exp(-(x/22.76)^{1.563} \tag{17}$$

Y is the cumulative fraction of material found below a size x. It is also possible to differentiate equation (17) to give dY/dx, the probability of finding a given differential unit of material in a size range between x and x+dx.

$$dY/dx = 1.181 \times 10^{-2}.x^{0.563}.\exp(-(x/22.76)^{1.563}) \tag{18}$$

For this paper the differential will be referred to as the Psi function ($\Psi(u_o)$), where u_o replaces x as the initial particle size.

2. The residence time distribution function in a perfectly mixed reactor is given by the equation:

$$\Phi(t) = \exp(-t/t_{ave}) \tag{19}$$

where t_{ave} is the mean residence time for the slurry in the stage. This function is normalized with respect to the variable (t/t_{ave}). In order to normalize it in terms of the variable "t" (time) the function must be divided by t_{ave}. Therefore the probability of a residence time t in a reactor is given by:

$$P(t) = \frac{1}{t_{ave}}.\exp(t/t_{ave}) \tag{20}$$

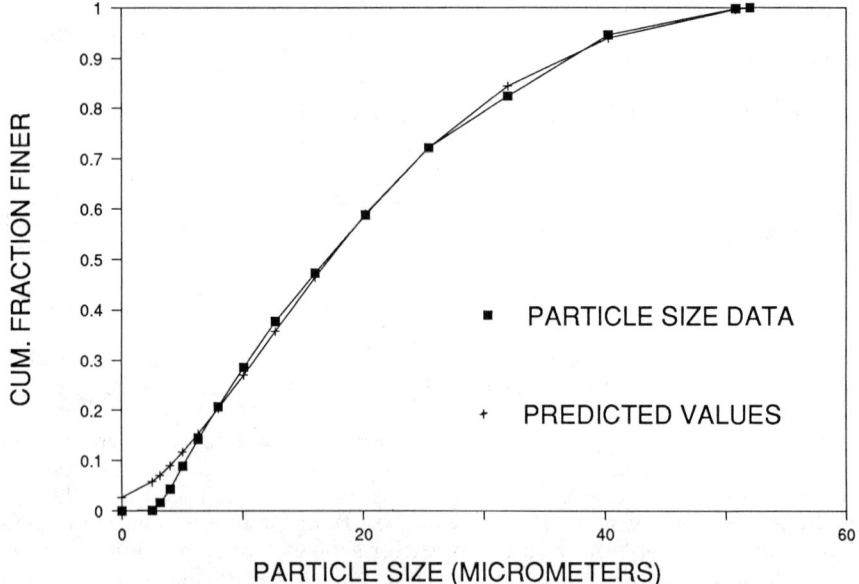

Figure 2.   Particle Size Distribution Data for the Feed to the Pressure Leach. Predicted values from Rosin-Rammler-Bennett Distribution Function.

In view of equation (8), it is possible to establish the size of a particle after a residence time t:

$$u = u_o - 2k_l t \tag{21}$$

The volume fraction (or weight fraction) of a particle remaining unleached in time t is therefore:

$$\frac{V}{V_o} = \frac{W}{W_o} = \left(\frac{u}{u_o}\right)^3 = (1 - \frac{2k_l t}{u_o})^3 \tag{22}$$

The fraction of all particles remaining unleached at a mean residence time of t is therefore:

$$\frac{V}{V_o} = \int_{2k_l t}^{u_o(max)} \int_0^{\frac{u_o(max)}{2k_l}} \left(1 - \frac{2k_l t}{u_o}\right)^3 \Psi(u_o) \frac{\exp\left(-\frac{t}{t_{ave}}\right)}{t_{ave}} dt du_o \tag{23}$$

For particles whose effective diameter is u_o in the feed, and whose surface area is πu_o^2 (assume spheres), a numerical evaluation of equation (23) leads to the results shown in Figure 3. Note that the area under the Psi function represents the fraction unleached. If we know the probability of material falling in a certain size range (Psi vs u) we are also able to calculate the area per unit weight. The feed material has an effective surface area of about 144 m²/kg of feed or 1.96 x 10⁴ m²/m³ of slurry. In 27.8 min of residence time at a linear leach rate of 0.418 um/min the area at steady state is 16.9 m²/kg of feed or 2306 m²/m³ of slurry. The zinc extraction is 83.5 % but the surface area depletion is closer to 90 %. This is of course due to the disappearance of much of the fine material. Figure (3) shows how the normalized values of the Ψ function have shifted to a larger particle size. The steady state values of Ψ and area provide feed values for the second compartment of the leach.

The value of A_{ZnS} assigned to equation (6) is therefore 2306 m²/m³ of slurry at 83.5 % zinc extraction. Further calculations at other extractions indicate that between 70 and 100 % zinc extraction, the steady state value of A_{ZnS} is approximately linear. This is shown in Figure 4 and summarized in equation 24.

$$A_{ZnS} = 13979(\text{Fraction Unleached}) \tag{24}$$

Combining the estimated value of k_z and A_{ZnS}, we can compute the virtual maximum rate for the zinc leach by ferric ions by assigning $[Fe^{3+}]=C_F$.

$$VMR(3) = 0.5k_zA_{ZnS}C_F \tag{25}$$

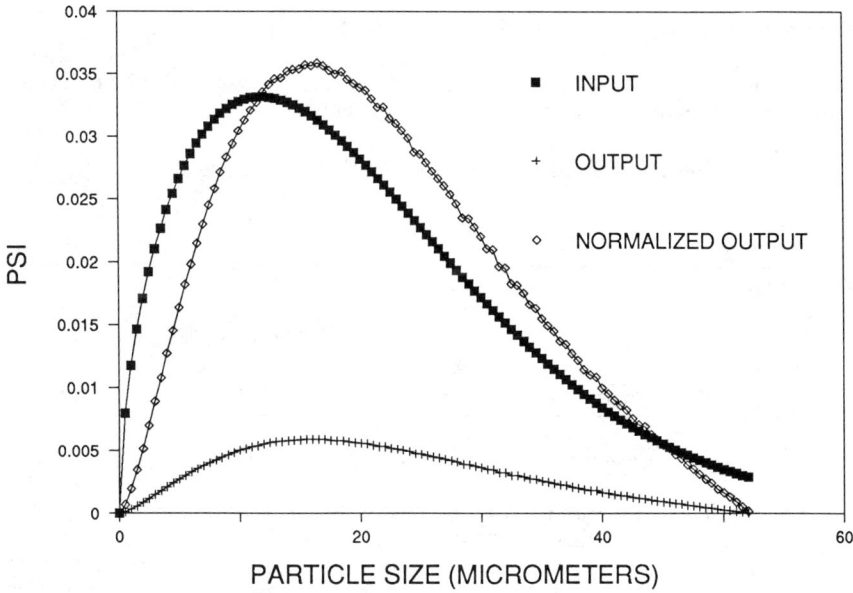

Figure 3. Psi Distribution for First Stage Leach. $k_1 = 0.418$ and $t_{ave}=27.8$ min.

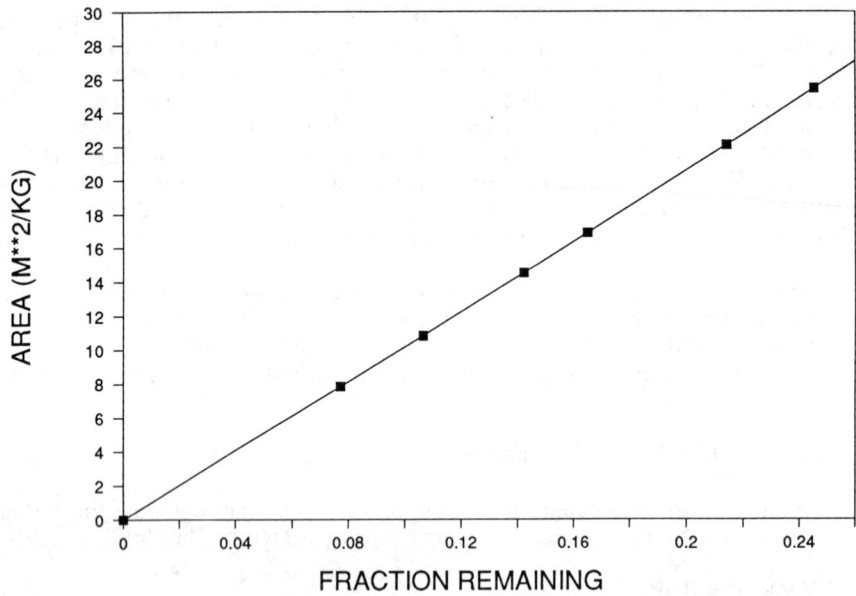

Figure 4. Area versus Fraction Unleached for First Stage Leach.

This yields values of VMR(3) at different zinc extractions as shown in Table 6. An average k_z value of 3.306×10^{-4} m/min was chosen.

Table 6. Virtual Maximum Rates for Fe^{3+} Leaching of ZnS.

	% Extraction		
Variable	70	80	90
k_z, m/min	3.306×10^{-4}	3.306×10^{-4}	3.306×10^{-4}
A_{ZnS}, m²/m³	4194	2796	1398
C_F, moles/m³	198	227	255
VMR(3), moles O_2/(m³.min)	137.3	104.9	58.9

Modelling Stage 1 of the Zinc Pressure Leach

Stage 1 deserves special attention because more than 80% of the leaching is typically carried out in this compartment. In the modelling of stage 1 we would like to know, what is the relative importance of each of the reaction steps? Also, by varying the rate constants (k_g, k_2, k_z) can we achieve significant increases in extraction and/or throughput?

In order to model the leach, equations 4, 26-28 must be solved for $R(O_2)$, $[Fe^{2+}]$, $[Fe^{3+}]$ and $[O_2]$.

$$R(O_2) = k_g([O_2]^* - [O_2]) \tag{4}$$

$$R(O_2) = 0.25k_2[Fe^{2+}]^2[O_2] \tag{26}$$

$$R(O_2) = 0.5k_z A_{ZnS}[Fe^{3+}] \tag{27}$$

$$C_F = [Fe^{2+}] + [Fe^{3+}] \tag{28}$$

C_F and A_{ZnS} are calculated as a function of recovery in equations 29 and 30.

$$C_F = 283.5*(\text{Fraction Leached}) \tag{29}$$

$$A_{ZnS} = 13989x(1 - \text{Fraction Leached}) \tag{30}$$

Equations 4, 26-28 have been rearranged to give one equation in $[Fe^{2+}]$.

$$0.25k_2[Fe^{2+}]^2\left([O_2]^* - \frac{0.5(C_F - [Fe^{2+}]k_z]}{k_g}\right) - 0.5(C_F - [Fe^{2+}])k_z = 0 \tag{31}$$

Equation 31 has been solved for $[Fe^{2+}]$ given values of fraction leached between 0.70 and 0.95 (70 -95 % recovery) using the bisection root finding method. The value of $[Fe^{2+}]$ was then used to calculate values of $R(O_2)$, $[Fe^{3+}]$ and $[O_2]$. Table 7 summarizes a set of these calculations for a given set of rate constants. Also included in the table is a line (second line) representing actual $R(O_2)$ values determined from zinc extraction and residence time parameters. The calculated values (last line) are to be compared with these actual values. The actual value of $R(O_2)$ remains lower than the calculated value for 70 - 90 % extraction but is higher for 95 % extraction. Therefore this particular set of rate constants would predict an extraction of between 90 and 95 % in the first stage of the leach.

Table 7.   Actual vs. Calculated Rates at Various Extractions for the First Stage

Residence time of 27.8 min and $[O_2]^* = 7.5$ atm.

Zinc Extraction, %	70	80	90	95
$R(O_2)$, moles O_2.m^{-3}.min^{-1}	18.97	21.68	24.39	25.75
Input Parameters				
k_g, min^{-1}	5	5	5	5
k_2, mole^{-1}.m^3.min^{-1}.atm.$^{-1}$	2.54×10^{-3}	2.54×10^{-3}	2.54×10^{-3}	2.54×10^{-3}
k_z, m.min^{-1}	3.306×10^{-4}	3.306×10^{-4}	3.306×10^{-4}	3.306×10^{-4}
A_{ZnS}, m^{-1}	4194	2796	1398	690
C_F, moles.m^{-3}	198	227	255	269
Residence Time	27.8	27.8	27.8	27.8
Calculated Values				
$[O_2]$, atm.	1.82	1.70	2.16	3.49
$[Fe^{2+}]$, moles.m^{-3}	157	164	139	95
$[Fe^{3+}]$, moles.m^{-3}	41	63	116	174
$R(O_2)$, moles O_2.m^{-3}.min^{-1}	28.40	29.02	26.69	20.07

Equation 31 has been used to study the effect of varying each of the rate constants. To study this variation the fraction leached was varied for each set of rate constants until the actual and calculated values of $R(O_2)$ converged. Figure 5 to 7 illustrate the effect of varying k_g, k_2 and k_z on fraction leached, $[O_2]$, $[Fe^{2+}]$ and $[Fe^{3+}]$. The constants were varied one at a time with the starting values $k_g = 3.70$ min^{-1}, $k_2 = 2.54 \times 10^3$ mole^{-1}.m^3.min^{-1}.atm.$^{-1}$ and $k_z = 3.42 \times 10^{-4}$ m.min^{-1}. (These are the fitted constants for 83.5% extraction in the first stage).

Figure 5 shows that the fraction leached increases steeply as a function of k_g and then levels off. The value of $[Fe^{2+}]$ moves roughly opposite to $[O_2]$ as k_g increases. $[Fe^{3+}]$ increases over the whole range.

Figure 6 shows that fraction leached is strongly affected by k_2. $[Fe^{2+}]$ and $[O_2]$ decrease with increasing k_2 while $[Fe^{3+}]$ increases. Figure 7 shows that fraction leached increases rather slowly with k_z. $[Fe^{3+}]$ and $[O_2]$ decrease with increasing k_z while $[Fe^{2+}]$ increases.

The conclusion of this exercise in modelling stage 1 of the zinc pressure leach is that the reaction rate is probably controlled by each of the three steps of oxygen absorption, ferrous oxidation and ferric leaching of ZnS. An improvement in any of the three rate constants k_g, k_2 and k_z would result in an overall improvement in zinc pressure leach performance. In comparing the steepness of the fraction leached graphs (c), the greatest benefit seems to be in increasing the value of k_g, the second greatest, the value of k_2 and the least benefit in extraction was projected by increasing k_z.

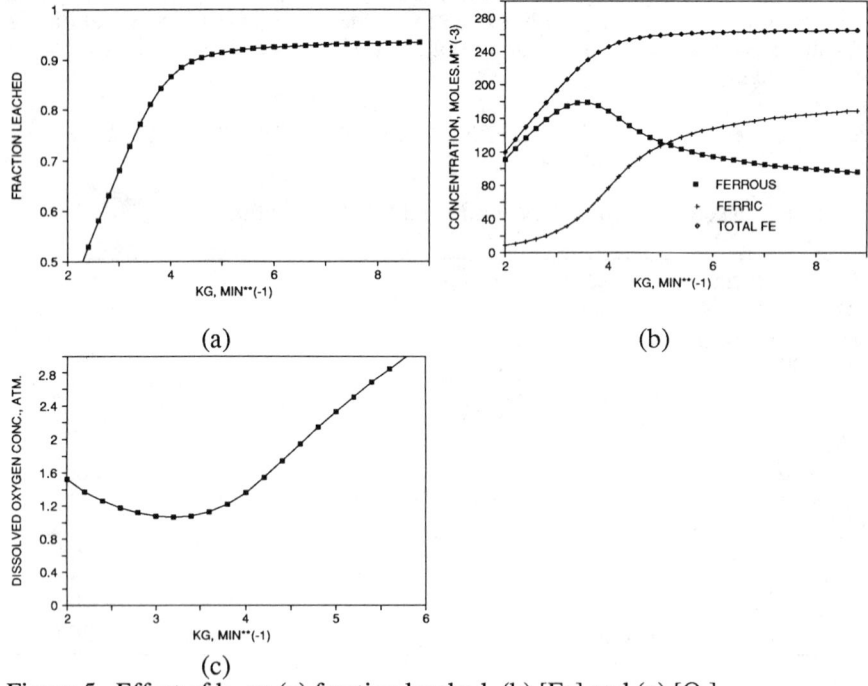

(a)

(b)

(c)

Figure 5. Effect of k_g on (a) fraction leached, (b) [Fe] and (c) $[O_2]$

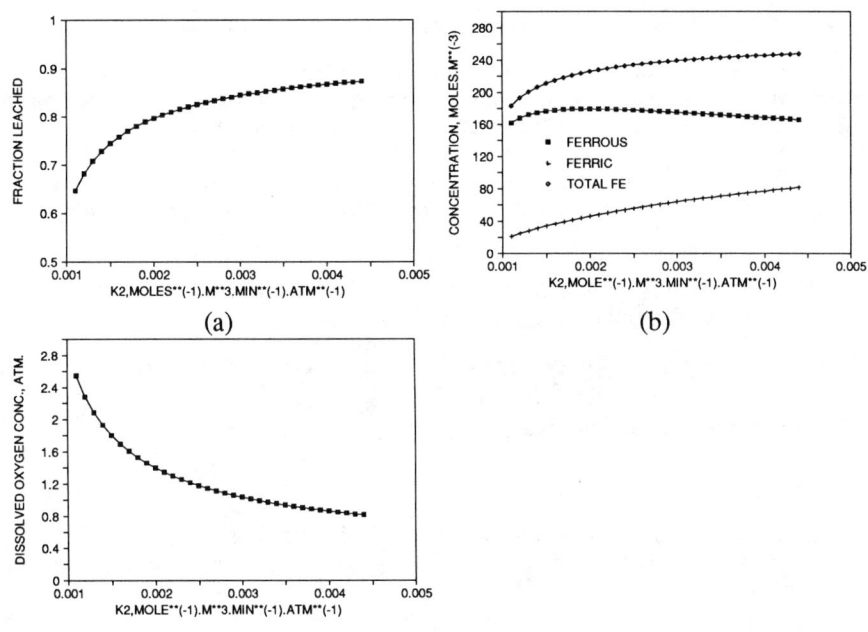

Figure 6. Effect of k_2 on (a) fraction leached, (b) [Fe] and (c) [O$_2$]

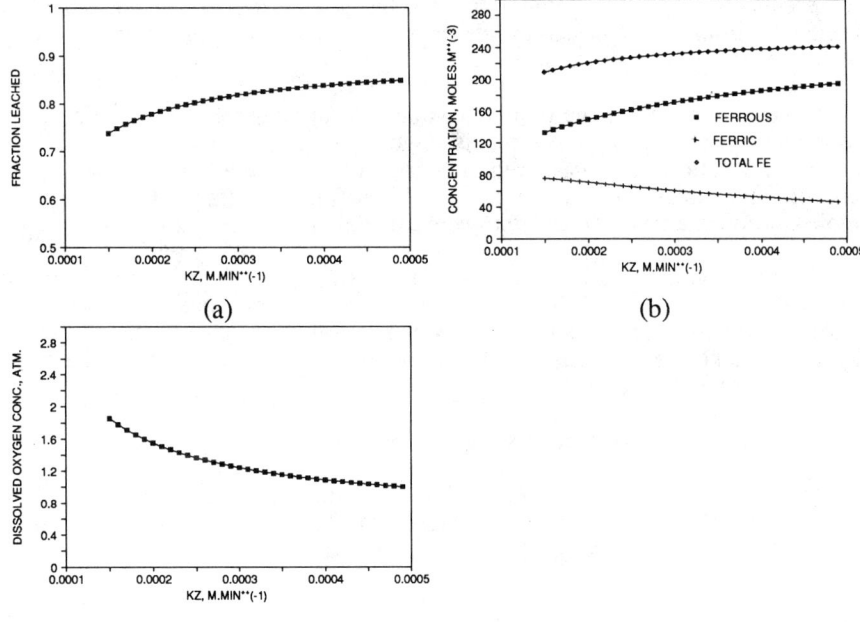

Figure 7. Effect of k_z on (a) fraction leached, (b) [Fe] and (c) [O$_2$]

Modelling of Stages 2-4 of the Zinc Pressure Leach

The modelling of stages 2-4 has been carried out using the output Psi distribution from the previous stage (starting with stage 1) to model each subsequent stage. The extractions reported by Martin and Jankola for leaching at 127 % of capacity have been used as targets in the modelling procedure. Values of k_l were determined based on the integration of equation 23. Equations 4, 26-28 were then used to determine steady state values of $[Fe^{2+}]$, $[Fe^{3+}]$ and $[O_2]$.

The precipitation of jarosite in stages 2-4 has been taken into account by considering the precipitation reaction following.

$$3Fe^{3+} + 0.5PbSO_4 + 1.5SO_4^= + 6H_2O \longrightarrow Pb_{0.5}Fe_3(OH)_6SO_4 + 6H^+ \qquad (32)$$

At equilibrium for reaction (32),

$$\log[Fe^{3+}] = 0.333\log K - 2pH - 0.5\log[SO_4] \qquad (33)$$

The usual exit solution from the autoclave is approximately 30 gpl H_2SO_4 and 5 gpl Fe^{3+}. Assuming that these values are close to equilibrium we can write equation (33) as,

$$\log[Fe3+] = 0.621 - 2pH \qquad (34)$$

For modelling purposes we have assumed that equation (34) adequately represents the jarosite precipitation in stages 2-4. That is, we are using an pseudo-equilibrium expression to describe a process that is probably occurring relatively close to equilibrium.

Table 8 summarizes the model calculations for stages 1-4. The value of k_g for stage 2 was assumed the same as stage 1, i.e. $k_g = 3.70$ min^{-1}. For stages 3 and 4 the solution was considered to be saturated in oxygen due to low oxygen consumption rates in these stages. k_2 was kept constant at 2.54×10^{-3} mole^{-1}.m^3.min^{-1}.atm.$^{-1}$. Values of k_z were calculated using equation 27.

Figure 8 summarizes the normalized Psi distributions for the input feed and the output from stages 1-4. It is apparent that the fine material in the feed disappears rather quickly. By stage 4 the distribution appears to be almost symmetric about the middle of the particle size range as it would be for a Gaussian probability curve.

Table 8 has the following interesting features.

1. The value of k_l, the linear leaching rate constant, decreases from stage 1 to stage 4 by almost one order of magnitude. This is reflected in a similar decrease in k_z. The leaching rate should be increasing as the

Table 8. Computed Steady-State Dynamic Properties in a Four Compartment Autoclave during Zinc Pressure Leaching

$[O_2]^* = 7.5$ atm.

Compartment	1	2	3	4
Zinc Extraction, %	83.5	95.3	97.0	97.8
Residence Time, min	27.8	19.6	19.1	20.4
$R(O_2)$, moles $O_2.m^{-3}.min^{-1}$	23.4	4.69	0.69	0.30
k_g, min^{-1}	3.70	3.70	-	-
k_2, $mole^{-1}.m^3.min^{-1}.atm.^{-1}$	2.54×10^{-3}	2.54×10^{-3}	2.54×10^{-3}	2.54×10^{-3}
k_z, $m.min^{-1}$	3.42×10^{-4}	1.03×10^{-4}	2.40×10^{-5}	1.54×10^{-5}
k_1, $\mu m.min^{-1}$	0.418	0.336	0.0782	0.047
A_{ZnS}, m^{-1}	2306	594	371	270
$[O_2]$, atm.	1.17	6.23	7.5	7.5
$[Fe^{2+}]$, $moles.m^{-3}$	177.3	34	12	8
$[Fe^{3+}]$, $moles.m^{-3}$	59.4	153	150	147
Jarosite Ppted., $moles.m^{-3}$	0	83	113	123
Total Fe Leached, $moles.m^{-3}$	237	270	275	277
$[H_2SO_4]$, $moles.m^{-3}$	550	401	395	392

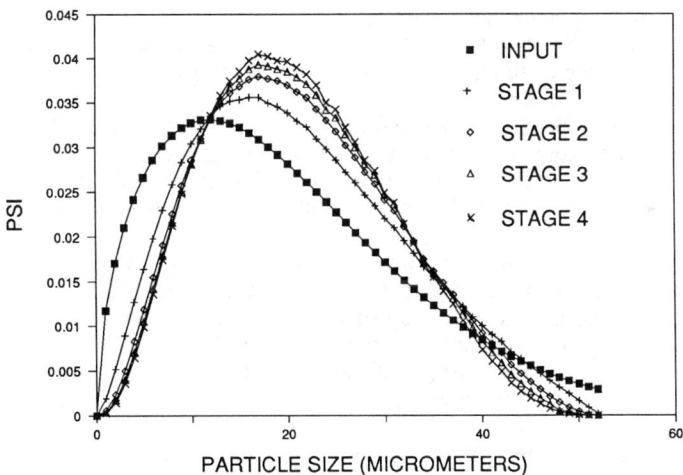

Figure 8. Normalized Psi Distributions for all Four Stages of the Leach.

feed moves through the autoclave because the particles are encountering progressively more oxidizing conditions. The possible explanations for this decreasing rate are that a) there may be some refractory zinc minerals present which leach rather slowly or b) the detergent that keeps the liquid sulphur from coating ZnS is losing it's effectiveness in later stages, thereby lowering the leaching rate or, c) the particle size distribution data used for calculations may be in error, i.e. there may have been more coarse material present than reported. If, in fact, the detergent has lost it's effectiveness it may be possible to add detergent to stages 3 and 4 to give increased leaching rates.

2. The bulk of the leaching work is carried out in stage 1 with almost no work being done in stage 4. Stage 4 contributes an incremental amount of zinc recovery but may be useful for optimizing jarosite precipitation and sulphur coalescence.

3. Most of the jarosite precipitates in compartment 2 of the autoclave. This is in agreement with industrial data.

4. The ferric to ferrous ratio increases dramatically from less than 1.0 in stage 1 to over 10 in stage 4. This highly oxidizing condition may be responsible for the destruction of detergent in stages 3 and 4.

Model Predictions for Modified Operating Conditions

The goal of any change in operating conditions would be to increase the feedrate to the autoclave and/or increase the fraction leached. Three sets of hypothetical calculations have been carried out to predict autoclave performance under the following conditions. At this point in the modelling exercise heat balances have not been calculated for modified conditions.

1. A feedrate of 200 % of design for a four stage autoclave was used. The linear leaching rate constants, k_l, were kept at the same values as determined for the data at 127 % of design. The value of k_g was raised to 10 min^{-1} for stage 1 in order to support the rate of oxygen absorption. Values of k_z were calculated and summarized.

2. A feedrate of 200 % of design for a three stage autoclave was used. Stages 1 and 2 were combined. (As an aside, in the Cominco autoclave the agitator positions are fixed so the only simple way of reducing the number of stages is to combine 1 and 2 and have 2 agitators in the single compartment). The value of k_l for the new stage 1 was the average of the original values for stages 1 and 2. k_l values for stages 3 and 4 were as determined in the modelling of the data at 127 % of design.

3. A feedrate of 200 % of design for a three stage autoclave was used. The value of k_l for stages 1-3 was set at 0.418, the predicted values for stage 1 from the data at 127 % of design.

Tables 9 to 11 summarize the results for the three different hypothetical sets of operating conditions.

Operating condition (1), as would be expected, lowers the zinc extraction from 97.8 % (at 127 % of design) to 94.7 % according to the model. The value of k_g had to be increased to 10 for stage 1. This indicates that if the autoclave feedrate is to be increased to 200 % of design a significant increase in the gas-liquid mass transfer characteristics of stage 1 must be affected. A change in agitator design or power input or an increase in sparging rates may be able to accomplish this.

Operating condition (2) lowers the zinc extraction to 90.3 %. Once again, this decrease would be expected due to the residence time distributions for a three compartment autoclave being less favourable. The attractive feature of this configuration is the decrease in the required value of k_g for stage 1. The increased volume of the combined stage 1 lowers the volumetric leaching rate in this compartment and thus allows the standard value of $k_g = 3.70$ min^{-1} to suffice. Very little jarosite precipitates in the three stages due to a higher exit acidity.

The operating condition (3) was intended to simulate what would happen if a uniform leaching rate was employed for all stages. As was previously discussed, the addition of detergent in later stages may affect this result of a uniform leaching rate. The calculated extraction for this configuration was 98.3 % at 200 % of design. The value of k_g was 3.70 for the model calculations. As in operating condition (2), the result of combining two stages into one was to lower the volumetric leaching rate in compartment 1. Therefore no extra measures would be required to enhance gas-liquid mass transfer in operating condition (3).

Table 9. Computed Steady-State Dynamic Properties in Four Compartments during Zinc Pressure Leaching

$[O_2]^* = 7.5$ atm., 200 % of design rate.

Compartment	1	2	3	4
Residence Time, min	17.45	12.45	12.13	12.95
k_1, μm.min^{-1}	.418	0.336	0.0782	0.047
k_g, min^{-1}	10	3.7	-	-
Zinc Extraction, %	76.5	90.1	93.43	94.71
$R(O_2)$, moles O_2.m^{-3}.min^{-1}	33.8	9.07	1.55	0.77
$[O_2]$, atm.	4.12	5.05	7.5	7.5
$[Fe^{2+}]$, moles.m^{-3}	113.2	53.2	18.1	12.7
$[Fe^{3+}]$, moles.m^{-3}	103.2	171.3	166.6	161.2
A_{ZnS}, m^{-1}	3316	1145	818	652
Jaro. Ppted., moles Fe.m^{-3}	0	33.5	80.2	95
$[H_2SO_4]$, moles.m^{-3}	630	42.3	417	410
k_2, m.min^{-1}	1.97×10^{-4}	9.22×10^{-5}	2.28×10^{-5}	1.46×10^{-5}

Table 10. Computed Steady-State Dynamic Properties in a Three Compartment Autoclave during Zinc Pressure Leaching

$[O_2]^* = 7.5$ atm., 200 % of design rate.

Compartment	1	2	3
Residence Time, min	30.1	12.13	12.95
k_1, μm.min^{-1}	0.377	0.0782	0.047
k_a, min^{-1}	3.7	3.7	-
Zinc Extraction, %	83.2	87.9	90.3
$R(O_2)$, moles O_2.m^{-3}.min^{-1}	21.54	3.02	1.46
$[O_2]$, atm.	1.678	6.68	7.5
$[Fe^{2+}]$, moles.m^{-3}	142.2	26.7	17.5
$[Fe^{3+}]$, moles.m^{-3}	93.6	194.0	183.1
A_{ZnS}, m^{-1}	2352	1590	1237
Jaro. Ppted., moles Fe.m^{-3}	0	28.3	55.4
$[H_2SO_4]$, moles.m^{-3}	548	450	437.4
k_z, m.min^{-1}	1.96×10^{-4}	1.96×10^{-5}	1.29×10^{-4}

Table 11. Computed Steady-State Dynamic Properties in a Three Compartment Autoclave during Zinc Pressure Leaching

$[O_2]^* = 7.5$ atm., 200 % of design rate.

Compartment	1	2	3
Residence Time, min	30.1	12.13	12.95
k_1, μm.min^{-1}	0.418	0.418	0.418
k_a, min^{-1}	3.7	3.7	-
Zinc Extraction, %	84.50	94.74	98.3
$R(O_2)$, moles O_2.m^{-3}.min^{-1}	21.89	6.55	2.12
$[O_2]$, atm.	1.58	5.73	7.5
$[Fe^{2+}]$, moles.m^{-3}	147.5	42.4	21.1
$[Fe^{3+}]$, moles.m^{-3}	92.1	154.5	141.5
A_{ZnS}, m^{-1}	2157	666	215
Jaro. Ppted., moles Fe.m^{-3}	0	71.7	116
$[H_2SO_4]$, moles.m^{-3}	531	402	384
k_z, m.min^{-1}	2.20×10^{-4}	1.27×10^{-4}	1.39×10^{-4}

Summary

A preliminary mathematical model of the zinc pressure leach has been developed. The objective in developing the model was to gain an understanding of the important features of the process and to identify potentially important laboratory studies which might provide new information about the process. Features of the process include;

1. Stage 1 of the autoclave performs most of the leaching work (>80%). In the first stage, the rates of oxygen absorption, ferrous oxidation by dissolved oxygen and ferric ion leaching control the overall rate of leaching.
2. Rate control in stages 2-4 shifts to the ferric ion leaching step, away from oxygen absorption and the ferrous oxidation. In addition, the rate constant for ferric leaching (k_z) falls dramatically from stages 1-4. It was speculated that this was caused by 1) the presence of refractory zinc minerals, 2) a loss in detergent activity causing sulphur to encapsulate the remaining zinc, or 3) the particle size distribution data may have been in error causing a misestimate of the value of k_z.
3. Calculations carried out with the model for higher feedrates (200 % of design) and altered autoclave configurations (stages 1 and 2 combined to give a 3 stage autoclave) showed that the leach could be run at high recoveries and throughput if a more uniformly high ferric ion leaching rate could be achieved. No increase in k_g was required to affect this result.

Experimental studies which arise from the model development include;

1. The precise determination of the particle size distribution of the feed material to the autoclave. Small errors in the distribution, especially at the coarse particle sizes, could result in large errors in the fraction unleached.
2. The study of gas-liquid mass transfer remains a compelling area of investigation. Degraaf (20) has carried out studies at UBC related to finding the optimum agitator design and sparging rates for the first stage of the Cominco autoclave and these studies are continuing.
3. The homogeneous oxidation of ferrous ion by dissolved oxygen has been studied in our laboratories and the results will be published shortly (12).
4. The rate of ferric ion leaching of ZnS and the role of the detergent need to be examined. This must include studies on the action of the detergent on the sulphur-mineral interface and the rate of degradation of detergent under various oxidizing conditions. The precise rate law governing ferric ion leaching of ZnS must also be investigated.

In conclusion then, the mathematical modelling of the zinc pressure leach has provided a framework for ongoing study and possible process enhancement. The authors look forward to providing important updates on the progress of the studies outlined above.

Acknowledgements

The authors would like to thank Cominco Ltd. and the Natural Science and Engineering Research Council (NSERC) for providing the authors with a NSERC University-Industry Research Grant to conduct this research.

References

1. M.T. Martin and W.A. Jankola, "Cominco's Trail zinc pressure leach operation", CIM Bull., 78(876)(1985), 77-81.

2. E.G. Parker and D.J. McKay, private communication with authors, Cominco Ltd., 1987.

3.   L.D. Locker and P.L. Debruyn, J. Electrochem. Soc., 116(12)(1969), 1659.

4.   F.K. Crundwell and B. Verbaan, "Kinetics and Mechanism of the Non-oxidative Dissolution of Sphalerite (Zinc Sulphide)", Hydrometallurgy, 17(1987), 369-384.

5.   H. Majima, Y. Awakura and N. Miskaki, Met. Trans. B., 12B(1981), 645.

6.   T. Chmielewski and W.A. Charewicz, "The Oxidation of Fe(II) in Aqueous Sulphuric Acid Under Oxygen Pressure", Hydrometallurgy, 12(1984), 21-30.

7.   D.R. McKay and J. Halpern, "A Kinetic Study of the Oxidation of Pyrite in Aqueous Suspension", Trans. Met. Soc. AIME, 212(1958), 301.

8.   R.J. Cornelius and J.T. Woodcock, Proc. Aust. Inst. Min. Metal., 185(1958), 65.

9.   C.T. Mathews and R.G. Robins, Proc. Aust. Inst. Min. Metal., 242(1972), 47.

10. I. Bielpolskij and N. Urusov, Zh. Prikl. Khim., 21(1948), 903.

11. J.R. Pound, "The Oxidation of Solutions of Ferrous Salts", J. Phys. Chem., 43(1939), 955-967.

12. D. Dreisinger and E. Peters, "The Oxidation of Ferrous Sulphate Under Zinc Pressure Leach Conditions", to be submitted to Hydrometallurgy.

13. P. Harriott, "Mass Transfer to Particles: Part 1. Suspended in Agitated Tanks", A.I.Ch.E.J., 8(1962), 93-102.

14. E. Peters, "The Physical Chemistry of Hydrometallurgy", International Symposium on Hydrometallurgy, D.J.I. Evans and R.S. Shoemaker, eds. (Chicago, IL., AIME, 1972), Chapter 10.

15. P. Rosin and E. Rammler, J. Inst. Fuels, 7(1933), 29-36.

16. R. J. Schuhmann Jr., A. I. Min. Metal. Pet. Eng., Tech. Pap. 1189 (1940).

17. T. Hatch and S.P. Choate, J. Franklin Inst., 207(1929), 369.

18. A.M. Gaudin and T.R. Meloy, Trans. A.I.M.M. Pet. Eng., 223(1962), 40-50.

19. R.H. Perry et al, eds., Perry's Chemical Engineers Handbook, Sixth Edition, (New York, NY: McGraw Hill, 1984), 8-5.

20. K.B. Degraaf, "An Investigation of Gas/Liquid Mass Transfer in Mechanically Agitated Systems", <u>M.A.Sc. Thesis</u>, University of British Columbia, (1984).

MODERNIZED HYDROMETALLURGICAL PROCESS DEVELOPMENT TECHNIQUES

K. G. Tan

CANMET, Energy, Mines and Resources Canada
555 Booth Street, Ottawa, Ontario, K1A 0G1, Canada

Abstract

CANMET is involved in the development and testing of new and improved
processes for the metals industry. Recent implementation of microprocessor
controlled instrumentation and data analysis in a hydrometallurgical
laboratory is discussed and evaluated. The immediate advantage of laboratory
modernization is masked by the necessity for diversification (steep learning
curves) and for processing larger amounts of data. The access to more
accurate and comprehensive process information, however, allows the completion
of projects in relatively short times, once the appropriate tools have been
developed. These findings are illustrated by a selection of examples,
indicating that modern techniques serve to enhance all facets of fundamental
and applied research. Prerequisites for success are still careful experimentation, innovation and basic research, but the tools have grown and continue
to expand. The stepwise improvement and accumulation of experimental data
also should provide a basis for the development of realistic mathematical
models capable of representing the more complicated situations often
encountered in hydrometallurgical process industries.

Introduction

Microprocessors have added a new dimension to the scope and efficiency of industrial process development work carried out in metallurgical laboratories (1). The successful implementation of microcomputers and microprocessor controlled equipment should involve all aspects of laboratory operations, including inventory control, planning, interfacing, process control, automation, data evaluation, literature research, reporting, and communication. It is important to become familiar with all available options on a particular piece of equipment or a combination of instruments to achieve adequate efficiency, flexibility, and reliability of potential customized applications.

The problems faced by laboratory personnel are manyfold, regardless of previous experience with mainframe computers and sophisticated analytical equipment. The diversity of available products is further complicated by different features, options, accessories, operating languages, and electronic details. Selection must be based on the degree of flexibility and compatibility with the available resources and foreseeable requirements. Adequate support must be solicited from specialists within or outside the organization, but the best approach is to grow with the tools in a prudent and results oriented manner. Thus, new equipment and techniques should be applied gradually to basic areas of operation so that the expenditure of capital and effort is directly reflected by tangible improvements in the quality and/or quantity of output. Confidence must be gained in a thorough and well balanced manner, and planning further ahead is essential. By keeping up with new developments in basic components and sophisticated software, the expansion of laboratory tools and increasing levels of expertise may be attained in a logical and budget-wise fashion.

At CANMET, diversified research is carried out in several sections which are differently exposed to various levels of microprocessor technology. This presentation reflects the experience of one solution chemistry group, where a microcomputer was introduced at the start of 1983. A summary of the progress made is presented, mainly to assess the needs for future developments, but also to aid other groups undergoing similar evolutions. The exposure at this forum also may be rewarding in introducing some of CANMET's current activities, and in establishing contacts in the broad and active field of hydrometallurgical process research.

Thiosalt Oxidation Studies

An example typical of the pre-microprocessor era may illustrate the classical techniques still widely used to generate data in many hydrometallurgical laboratories. A long-term study addressed the seasonal pH changes of natural drainage areas in the vicinity of many complex-sulphide processing mills that were attributed to the delayed acid-generating capacity of incompletely oxidized sulphur species called thiosalts (2). Once discharged into the environment, these thiosalts are oxidized by bacteria; e.g., according to:

$$Na_2S_xO_6 + (3x-5)/2 \ O_2 + (x-1) \ H_2O \ ----->$$

$$Na_2SO_4 + (x-1) \ H_2SO_4 \qquad (1)$$

Although the thiosalt levels in treated mill effluents seldom exceed 2 g/L, typical discharge volumes of single plants are in the order of 10 m^3/min (3). Because of the harsh Canadian climate, significant bacterial action starts only in late spring, and the acid generating capacity of accumulated thiosalts is released within a relatively short time. Various groups at CANMET studied this problem between 1976 and 1985, and reference is made to a few recent articles (2-5) for details on the background, scope and experimental procedures. Much of the work was based on analytical determinations of the indivi-

372

dual thiosalt species in process streams, natural waters, synthetic solutions, and in oxidizing media. Typical pieces of equipment used were standard laboratory apparatus including metering pumps, pH meters, titrators, a Beckman DK-2 spectrophotometer, and an ozone generator. Mass balances around grinding and flotation circuits were calculated manually to determine the generation of thiosalts under prevailing conditions (4). Oxidation tests and kinetic measurements yielded simplified chemical reaction models with limited predictive powers, and provided the basis for the technical-economic evaluation of various industrial treatment options (3,5). The Pourbaix diagram for metastable thiosalts (Fig. 1) summarizes some of the well known reaction paths such as:

- 2 H_2S + SO_2 ----> 3/8 S_8 + 2 H_2O (Claus Reaction);

- H_2S + excess SO_2 (aq.) ----> $H_2S_xO_6$ (Wackenroder Liquids);

- 1/8 S_8 + Na_2SO_3 (aq.) ----> $Na_2S_2O_3$ (Thiosulphate Formation);

- 2 $H_2S_2O_3$ + oxidant ----> $H_2S_4O_6$ + H_2O (Iodometry);

- stepwise oxidation and disproportionation reactions of thiosalts in alkaline and acidic media (6); and

- nucleophilic sulphur chain building and degradation reactions (7).

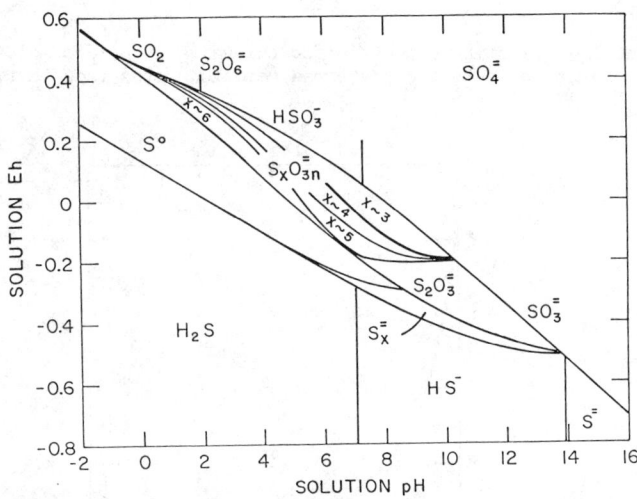

Figure 1 - Idealized Pourbaix diagram of thiosalts at 25°C.

The oxidation schematic in Fig. 2 shows some of the complexity of thiosalt chemistry in acidic to near neutral solutions. The most direct reaction paths possible are considered (i.e., involving the transfer of one or two electrons only), and the ease of oxidation is indicated by standard potentials estimated from the available literature. Starting with the oxidation of thiosulphate, tetrathionate is usually the main product but sulphite also may be formed, especially in the presence of strong oxidants. Tetrathionate is further oxidized to trithionate plus sulphite since the direct oxidation to sulphite alone is not favoured kinetically (high reaction order). The non-oxidative nucleophillic degradation of tetrathionate by thiosulphate or sulphite (not shown) is an alternative route to trithionate. The main point is,

373

however, that the further oxidation of trithionate is the rate determining step. This explains the large accumulation of dissolved trithionate that is often observed with all but the strongest oxidants (ozone).

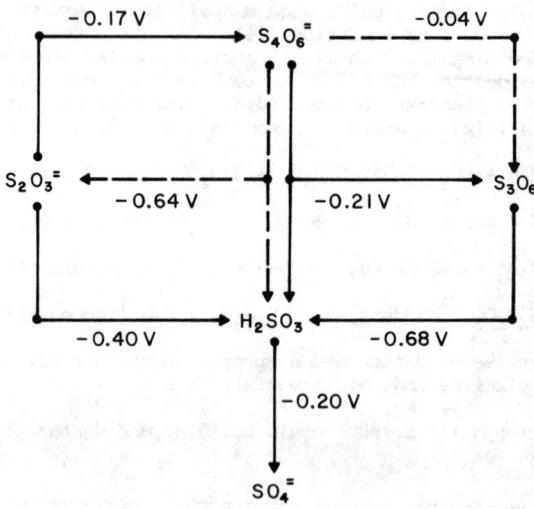

Figure 2 - Oxidation schematic for thiosalts in acidic solutions; solid lines indicate the preferred reaction paths (see text).

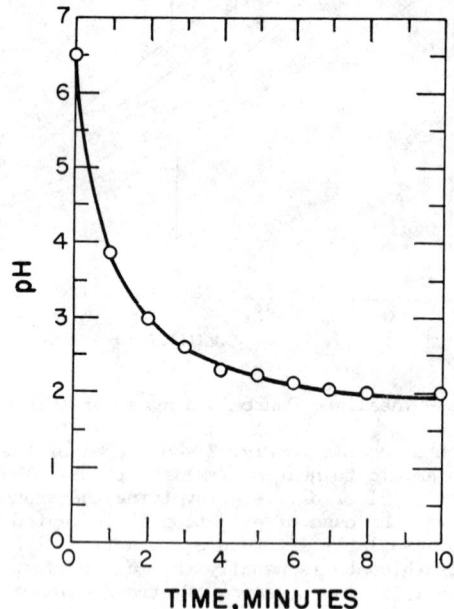

Figure 3 - Low temperature oxidation of an industrial effluent with Fenton's reagent, as monitored by the solution pH.

374

Selection of industrial process conditions should target the destabiliza-
tion of trithionate by the use of catalysts and higher acidities. The use of
hydrogen peroxide catalysed by ferrous salts, for example, produces radical
hydroxyl and perhydroxyl intermediates (Fenton's Reagent). In acidic solutions, these highly oxidative radicals rapidly and completely oxidize the
thiosalts to sulphate, even at temperatures as low as $6^{\circ}C$. The course of
reaction is characterized by decreasing pH related to the generation of H_2SO_4
in analogy to Eqn. 1. As shown in Fig. 3, the entire thiosalt content of a
typical industrial effluent was eliminated in less than 10 minutes, but about
3.7 mol H_2O_2 were consumed per mol of thiosalt.

Lead Chloride Solubility Studies

Since 1979, another CANMET project has focussed on the development of a
ferric chloride leaching process for complex sulphide concentrates (8). One
element involved the measurement of lead chloride solubility in a variety of
aqueous chloride systems (9). Extensive solubility and solution density data
generated in this study were evaluated on a mainframe computer, with increasing interaction of a microcomputer. The results for binary systems were
interpreted in terms of solubility models, and complete $PbCl_2$-solubility surfaces were generated as a function of the temperature and co-electrolyte concentrations (9,10). The solubilization of $PbCl_2$ in aqueous chloride media is
associated with the stepwise formation or dissociation of various chloro-com-
plexes (e.g., $Me^{m+} = Fe^{3+}$, Pb^{2+}) according to:

$$(MeCl_{n-1})^{m-n+1} + Cl^- ===== (MeCl_n)^{m-n} \qquad (2)$$

Simple thermodynamic models (10) predict that the $PbCl_2$ solubility is determined by the free chloride concentration at a given temperature and solution
composition:

$$S = A/c^2 + B/c + C + Dc + Ec^2 \qquad (3)$$

The concentration of $PbCl_2$ (non-complexed or undissociated species) is given
by C, and the remaining parameters are related directly to the stepwise formation constants K_n for $Me^{m+} = Pb^{2+}$, in Eqn. 2. The study showed that Eqn. 3
fits the measured solubility data, but non-meaningful values for the parameters A to E were obtained, even in the absence of other complexing cations
(10). It was shown that c should be replaced by the chloride activity in
concentrated solutions, and that the precision and reliabilty of measurements
must be improved. The application of more sophisticated solution models (11)
requires additional data on coordination numbers, stability constants,
solution speciation, partial molar volumes, hydration numbers, and activity
coefficients.

Experimentally, bulk electrolyte solutions were mixed with excess $PbCl_2$
crystals and the mixture was equillibrated in a 2 L vessel, at different temperatures. A pycnometric sampling method was employed. Saturated solution
density determinations were used to assess the reliability of the sampling
procedures, and the measurements were based on molar concentration units. The
density of the $PbCl_2$-saturated chloride solutions decreased linearly with
increasing temperatures, and in many cases a linear additivity rule for co-
electrolyte contributions was observed (9,10). Solubility curves as a
function of temperature were approximated by an exponential equation:

$$S = At.e^{Bt} + C \qquad (4)$$

The solubility isotherms for individual co-electrolyte compositions then were
fitted to Eqn. 3, by inserting the initial (lead-free) chloride concentration
c_o instead of c. Assuming that dS/dc_o approximates dS/dc, c is estimated from:

$$c = c_o/(1 + dS/dc) \qquad (5)$$

375

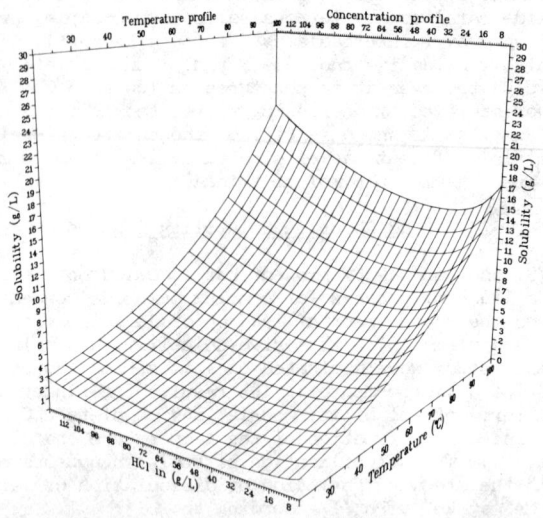

Figure 4 – Lead chloride solubility surface
in 0.1 M $FeCl_3$ and 0 to 3 M HCl.

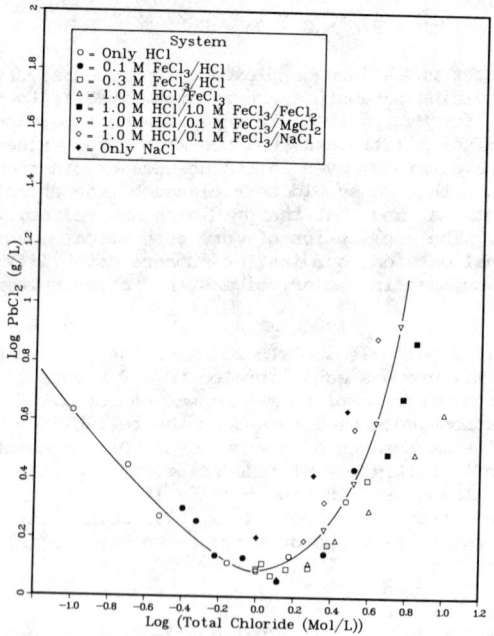

Figure 5 – Lead chloride solubility at 25°C as a
function of total chloride concentration.

The PbCl$_2$-solubility surface in 0.1 M FeCl$_3$ and HCl is shown in Fig. 4. The trough below 1 M HCl reflects the transition between the solubility depression by the common ion effect and the solubility enhancement by complexation with increasing chloride concentrations. The endothermic heats of solution are evident from the inclination of the surface along the temperature axis. The PbCl$_2$ solubility (S) provides direct information about the activity of free chloride ions (a$_{Cl}$) that is not readily available from other sources.

In Fig. 5, the relationship between S and the total chloride concentration (c$_t$=c$_o$+2S) is displayed for a number of aqueous systems at 25°C. Increasing a$_{Cl}$ at low c$_t$ would depress S (common ion effect), and the opposite effect is expected at high c$_t$. The presence of different co-electrolytes affect a$_{Cl}$ through dissociation and complexation, and through diminishing the activity of water by hydration. In the complexation region (log c$_t$ > 0.1), a$_{Cl}$ decreases in the following sequence: NaCl > NaCl + FeCl$_3$ > MgCl$_2$ + HCl + FeCl$_3$ > HCl + FeCl$_3$, where FeCl$_3$ < 0.3 M. At higher ferric concentrations, withdrawal of free chloride is no longer described by Eqn. 2. Similar plots were obtained at higher temperatures, and refinement of the measurements could yield more quantitative models of complex hydrometallurgical solutions.

Copper Sulphate Solubility Studies

Anode passivation in copper electrorefineries often is associated with the formation of impervious layers of CuSO$_4$.5H$_2$O that protect the anode from further corrosion. Because of the high solubility of copper, reliable measurements in industrial solutions are problematic. In concentrated aqueous solutions, water is no longer in excess, and classical thermodynamic solution models are valid no longer. Changes in water concentration due to evaporation and hydration of soluble and crystallized species become important, and internal consistency checks should be employed to determine the extent of such occurences.

In a study currently in progress, the experimental apparatus was miniaturized and modernized to avoid massive additions of CuSO$_4$ and to eliminate major sources of systematic error. Thus, a battery of 10 mL flasks with a narrow side arm to accommodate volume expansions was placed in a shaker bath, and samples were taken by an automatic dilutor and rinsed into sealed weighing vials. Elimination of the headspace prevented evaporative losses, and the reliability of the sampling was checked by weight and density measurements, as well as by periodic calibration tests. The initial solution composition was determined by weight to avoid the uncertainties of volume measurements, and to obtain concentration units which are invarient with temperature and solubility changes. This involved maintaining a rigorous water balance and taking contributions from hydrated solids into account. The co-electrolyte levels (H$_2$SO$_4$ and NiSO$_4$) and saturated solution densities were used as internal reliability checks, and all major operations (equilibration, sampling and analysis) were carried out in triplicate. As a result of these improvements, the analytical determination of the electrolytes by titration and by atomic absorption spectrophotometry was identified as the largest remaining source of error. The total sulphate concentration now is measured by liquid ion-exchange chromatography, and individual cations will be determined in a similar manner. Direct density measurements in the 'bulk' solutions will be made using calibrated density beads, and accurate data may then be used to gain further insight into the molecular structure of concentrated solutions.

Mini-Plant Chlorination

Another aspect of CANMET's ferric chloride leach process involves the chlorination of ferrous chloride solutions to regenerate the lixiviant (12). A mini-plant reactor column was designed and operated in a co-current contacting mode under semi-adiabatic conditions. Mass transfer rates were measured by controlling precisely the inlet liquid and gas flowrates, by monitoring temperatures and pressures, and by sampling at several points along the column. Raw data from the data logger and analytical results were plotted, converted and evaluated immediately after each run using home-made Basic programs. A compact UV/VIS spectrophotometer was interfaced with a microcomputer to provide rapid analytical support in addition to standard ferrous titrations (13). The whole project divided into several shorter terms was completed within one year to the extent that process optimization and integration with upstream and downstream operating steps could be considered.

A prototype reactor was designed and tested, and further improvements to the design yielded reliable kinetic data. The basic equations for mass transfer calculations using standard notation (14) are:

$$Cl_2 \text{ (g)} + 2 \text{ FeCl}_2 \text{ (aq.)} \longrightarrow 2 \text{ FeCl}_3 \text{ (aq.)} \tag{6}$$

$$- p_A/r_A = 1/k_G a + H_A/k_L a.E + H_A/kC_B f_L \tag{7}$$

The subscripts refer to the reagents (A=Cl_2, B=$FeCl_2$) and to the fluid phases (G=gas, L=liquid); p_A = pressure, r_A = rate, k_i = rate constants, a = packing area, H_A = Henry's law constant, E = solubility enhancement factor, C_B = concentration, and f_L = fractional volume flow. Other notations used below are the volumetric gas flow rate G_A and the reactor volume V_R.

The reaction according to Eqn. 6 is very rapid, irreversible, and exothermic, and the Cl_2 is well soluble in aqueous solutions. At ferrous concentrations above 0.01 mol/L, the overall mass transfer is expected to be gas film controlled; i.e., a pseudo first-order reaction occurs at the gas-liquid interface. Mass balance considerations according to Eqn. 6 establish the following relationships, in terms of molar gas and liquid reagent flows, at any point along the length of the column:

$$- r_A = p_A k_G a = dG_m/dV_R = 1/2 \, dL_m/dV_R \tag{8}$$

The literature (15) predicts a direct dependence of $k_G a$ on the volumetric gas flow rate G_A, such that

$$\ln(-r_A) = A + B.\ln(G_A) \tag{9}$$

with A = - 0.233 and B = 0.70. The residual gas flow rates were calculated from titrated ferrous concentrations, and were plotted as a function of the cumulative reactor volume V_R. Regressional analysis yielded excellent fits in the form of

$$G_A = (a + bV_R)/(1 - cV_R) \tag{10}$$

Mass transfer rates were calculated by differentiation (Eqn. 8 and 10), and were plotted against G_A as shown in Fig. 6. Most of the data points are clustered around the solid line (approximate best overall fit):

$$\ln(-r_A) = - 3(G_A - G_O)^2 \tag{11}$$

where the optimum inlet gas flow, G_O, approximates the target flow of 1 L/min.

Much of the scatter is caused by low reagent flow rates (dashed line, filled squares); i.e., underloading of the column. It was shown that the effects of column material and diameter, inlet temperature, reagent concentrations, and stoichiometric ratios are quite small, and that pressurization and loading beyond a critical gas flow rate are advantageous. This was demonstrated by plotting selected subgroups of the test results displayed in Fig. 6, in a similar but separate manner. To examine the effect of column loading at

80°C, for example, the results of three tests are summarized in Fig. 7. The reactor orientation is reversed in the graph as the inlet (bottom) is characterized by high Cl_2 flow rates (G_A) which approach zero near the outlet (top) of the column. The target Cl_2 flow (dash-dotted line) yielded the best results with high residual driving forces ($-r_A$) at the indicated sampling points. The distance between like symbols for a given chlorine flow (G_A) represent the 'bite' in reagents achieved by equivalent lengths of the column. The performance gradually deteriorated at lower column loadings (dotted vs dashed lines) when pure Cl_2 was used.

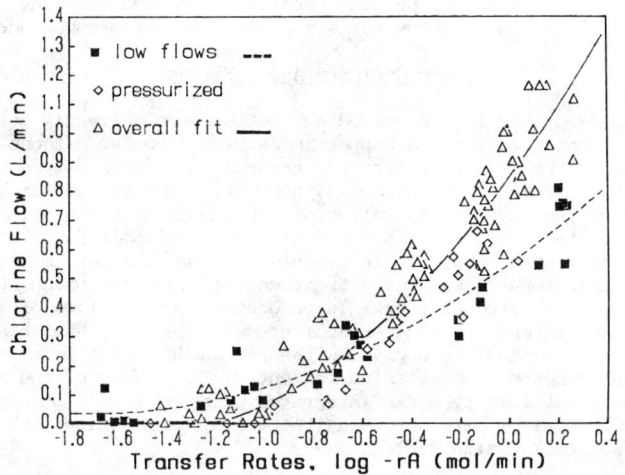

Figure 6 - Operating curves for the chlorination of $FeCl_2$ solutions under a wide variety of conditions.

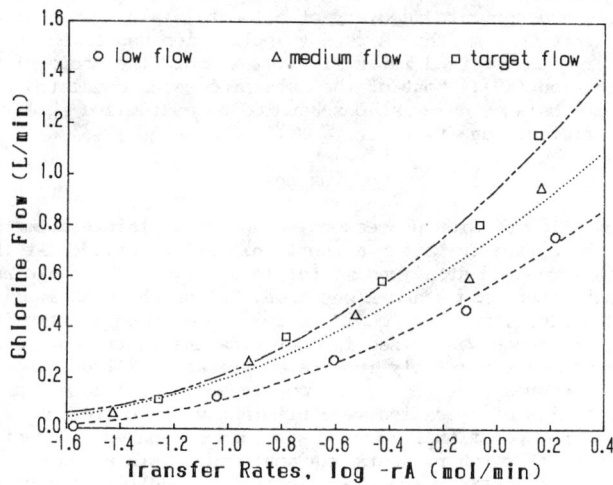

Figure 7 - The effects of column loading at high inlet temperatures.

The developed method may be used for many fundamental and practical purposes. The operating conditions must be controlled precisely and the effects of individual parameters can be characterized by a unique family of operating curves. Mechanistic details about individual mass transfer steps may then be elucidated. In using pure Cl_2 gas, for example, the absorption of Cl_2 at the gas/liquid interface is not limited by diffusion since a diluent gas is absent. It would be a function of bubble size, internal pressure, surface tension, and the chemical enhancement factor only. The effects of these and other parameters such as pH, viscosity, temperature profile, size factors, geometries, catalysts, and operating modes will provide useful information. The introduction of predictive process control, automated sampling, and online analytical techniques, however, is a prerequisite for such endeavours.

Copper Refinery Anode Slimes

The tools described in the preceding sections are currently being used in process development work for the copper industry. A copper cathode output of several hundred thousand tonnes per year generally produces only a few tonnes per day of anode slimes. Both products, however, often are comparable in value because of the high concentration of noble metals in the slimes. As a result, only limited amounts of this material are available for testwork outside the plant, even though large amounts of copper, selenium, tellurium, and other nuisance impurities create treatment difficulties which should be addressed via process studies (16). To reduce the lock-up time of valuable metals in process inventories, ammoniacal pressure leaching was chosen as a fast and efficient method for upgrading the raw anode slimes by selective leaching of their copper and selenium contents. The safety of the method (17) and the economic feasibility of an integrated process were to be demonstrated. These project objectives were met using the rapid optimization technique provided by factorial design (18).

The method was explored using a particular type of complex feed material containing 20 to 30% each of Ag, Cu, Pb and Se. The range for parametric variations was selected on the basis of a few preliminary tests, and a 2^4 factorial representation exploring 8 major effects was initiated. Similar testwork on the primary leach products established suitable downstream steps for the complete and separate recovery of all metal values in a direct and simple manner. Within 4 months, a patent application was filed and subsequent studies on other types of feed material have extended the scope of the original application (19). Most of the data processing consisted of chemical analyses and mass balance calculations handled by customized spreadsheet programs, in an efficient manner.

Conclusions

The introduction of microprocessor technology to this hydrometallurgical laboratory has been fruitful after a short initiation period. At the start, new instruments were used at a fundamental level; i.e., for basic and routine operations, most often in a stand-alone mode. Although these applications often duplicated functions which could be performed by well established traditional means, experience was gained in operating the instruments and designing customized procedures for a fully integrated operation. Thus, the reliability of tools and procedures were tested thoroughly and manual data transfer steps during various stages of operation were eliminated. A better overview of the merits of different operational modes was obtained, and more direct control over the course of research projects was achieved. This aided the gradual upgrading of the applications and the selection of additional equipment and accessories for further expansion. Major efforts towards improving experimental techniques and laboratory procedures are still in progress, and fundamental insight on all aspects of the work was obtained in the process.

The precise control over operating conditions and the continuous monitoring of many output variables in a digital format are major objectives in scientific research. These tedious functions can be performed conscientiously with the help of microprocessor controlled instrumentation. Sources of experimental error then become more obvious, and they can be systematically traced and eliminated to a large extent. Results and experience gained can be used for fundamental and practical purposes, and an accumulating mass of reliable data can be evaluated in an efficient manner. All this, however, requires additional effort to learn about the available equipment and technology, to plan and design customized installations, and to maintain the consistency of increasingly automated procedures. Laborious manual operations are replaced by the constant updating and interphasing of laboratory tools and procedures that included frequent reliability checks on new applications. In this sense, microprocessors are not so much time-saving devices, but tools that increase the scope and versatility of scientific research. A high degree of multi-disciplinary skills is acquired, but fool-proof error indicators must be designed to include support from less specialized personnel. The rewards are manyfold, especially in minimizing the need for personnel, manual intervention, and unnecessary hold-ups in operations. The ultimate aim, of course, is the increase in quality and quantity of output plus a detailed understanding of processes at more sophisticated levels.

Microprocessors also are powerful learning tools in all aspects of research. Access to extensive raw industrial data is a bonus which often creates confusion. The best approach is to install modern techniques capable of sorting the data and eliminating many uncertainties. This may allow a better understanding of older results and their use to complete the overview. The postulation of simplified models is extremely helpful at this stage, and more rigorous mathematical modelling may be attempted on refined experimental data to yield a better understanding of many details. The implementation of microprocessors as data loggers and process controllers in industrial operations may contribute to this objective. The long term effects of varying operating conditions can be recorded and scrutinized to serve as a model for further experimentation and optimization. Access to such data is invaluable to management, process engineers and research scientists. The weak links remain, however, the confusion created by an overwhelming and often contradictory mass of data, and the uncertainty about relevant details including the reliability of many chemical analyses. It is in these areas that chemical laboratories such as CANMET's should strive for improvement.

Acknowledgements

The author would like to thank many co-workers for their support and assistance during the course of this work, in particular: Dr. J.E. Dutrizac, K. Bartels, P. Bedard, I. Cottrell, E. Rolia, and C. J. Weatherell.

References

1. "Programme, 36th Canadian Chemical Engineering Conference, 5-8 October 1986, Sarnia, Ontario," Can. Chemical News, 38(6)(1986), centrefold.

2. M. Wasserlauf and J. E. Dutrizac, "CANMET's Project on the Chemistry, Generation and Treatment of Thiosalts in Milling Effluents," Can. Metall. Q., 23(3)(1984), 259-269.

3. "Thiosalt Control," Impurity Control & Disposal, ed. A. J. Oliver (Montreal: The Can. Inst. of Mining and Metall., 1985), papers 27-31.

4. E. Rolia and K. G. Tan, "The Generation of Thiosalts in Mills Processing Complex Sulphide Ores," Can. Metall. Q., 24(4)(1985), 293-302.

5. K. G. Tan and E. Rolia, "Chemical Oxidation of Thiosalt-Containing Milling Effluents," Can. Metall. Q., 24(4)(1985), 303-310.

6. M. Goehring, "Chemie der Polythion Sauren," Fortschr. chem. Forsch., 2 (1952), 444-483.

7. M. Schmidt, "Reactions of the Sulphur-Sulphur Bond," Elemental Sulphur, ed. B. Meyer (London: Interscience, 1965), 306-326.

8. J. E. Dutrizac, "The Leaching of a Pyritic Zn-Pb-Cu-Ag Bulk Concentrate in Ferric Chloride Media," (Report MRP/MSL 81-58(TR), CANMET, 1981).

9. K. Bartels, P. L. Bedard, and K. G. Tan, "Solubility of Lead Chloride in Ferric Chloride Leaching Media," (Report MRP/MSL 85-7(TR), CANMET, 1985).

10. K. G. Tan, K. Bartels, and P. L. Bedard, "$PbCl_2$ Solubility and Density Data in Binary Aqueous Solutions," Hydrometallurgy, 17 (1987), 335-356.

11. A. E. Martell, ed., Coordination Chemistry, Vol. I, (New York: Van Nostrand Reinhold Company, 1971), 394-541.

12. K. G. Tan and P. L. Bedard, "Development of a Mini-Plant Chlorinator for Ferric Chloride Leach Processes," (Report MSL 86-98(OP), CANMET, 1986).

13. P. L. Bedard, F. Richer, and K. G. Tan, "Interfacing of the Pye Unicam PU8610 UV/VIS Spectrophotometer with the Osborne 1 Microcomputer," (Report MSL 86-170(TR), CANMET, 1986).

14. K. G. Tan and P. L. Bedard, "Development of a Mini-Plant Chlorinator for Ferric Chloride Leach Processes," (Paper presented at the 36th Canadian Chemical Engineering Conference, Sarnia, Ontario, 7 October 1986), 19a.

15. R. H. Perry and C. H. Chilton, ed., Chemical Engineers' Handbook, (New York: McGraw-Hill, 1973), Eq. 18-55 and Eq. 18-68.

16. R. A. Zingaro and W. C. Cooper, ed., Selenium, (New York: Van Nostrand Reinhold, 1974), Chapter 2.

17. K. G. Tan, "Fulminating Gold and Silver," (Report MSL 87-59(J), CANMET, 1987).

18. W. E. Duckworth, Statistical Techniques in Technological Research, (London: Methuen & Co., 1968), Chapters 4-8.

19. K. G. Tan, P. L. Bedard, and C. J. Weatherell, "Selective Ammoniacal Pressure Leaching of Noble Metal Concentrates," (Patent application in progress, CANMET, 1987).

HYDROMETALLURGICAL PROCESS SIMULATION

USING SPREADSHEET PROGRAMS

Rein Raudsepp

Bacon Donaldson & Associates Ltd.
2036 Columbia Street
Vancouver, B.C., Canada   V5Y 3E1

Abstract

Spreadsheet programs available for personal computers are an ideal medium for the production of comprehensive process mass balances. Spreadsheets make it possible to readily simulate hydrometallurgical processes from laboratory data and so evaluate the effects of water balance, recycles and process options. The paper discusses the efficient and rapid set-up of process spreadsheets, troubleshooting and the expedient extraction of results. A comprehensive example using the program Lotus 1-2-3 is given.

Introduction

In the development of a hydrometallurgical process the proposed unit operations must eventually be fit together and the entire flowsheet evaluated as an entity. The earlier this evaluation is done, the more efficient the subsequent development can be. Manual calculation of a process mass balance can be a major undertaking especially if there are a large number of components and recycle streams. The evaluation of a range of process conditions and flowsheets options adds to the effort. The use of a personal computer and a spreadsheet program, such as Lotus 1-2-3, can speed the calculation time and so, allow a much more detailed examination.

This paper sets out a systematic approach to the formulation of a process spreadsheet including:

a)   organization of the information available
b)   the structure of the process spreadsheet
c)   process flow calculations
d)   entering information
e)   error tracing
f)   spreadsheet convergence
g)   extracing results

These points will be illustrated using a simple example: a hypothetical Chalcopyrite Upgradge Process. In this process, a pyrrhotite/chalcopyrite concentrate is decomposed to upgrade its precious metal content.

Information Organization

The first step is to organize the available process information into a complete process flowsheet which can be translated into a process spreadsheet:

1.   The basic unit operations must be identified and arranged into a rough process flowsheet.

2.   For each unit operation process reactions must be written. The reactions must be written expressed in a consistent set of chemical components that provide meaningful information. For example, acid can be described by the component H^+, HCl or H_2SO_4. The first example would require a counter anion to maintain the solution charge balance. The other two would require Cl^- or SO_4^{2-} to be listed as components. The set of reactions and component must be comprehensive. If water evaporates, both aqueous and gaseous water components are required and a process equation to link them. If a soluble salt is added, a reaction is required to dissolve it. The list of components should also include spectators which do not react. For example, if the oxygen in air is a reactant, nitrogen must also be a component.

3.   The process flowsheet must then be upgraded to show all the process unit operations including solid/liquid

separation steps and all the process flows. Flows made up of spectator components must also be included. A unit operation which uses the oxygen from air will have to vent the nitrogen. The process flows should be numbered.

4. Using the process flowsheet and reactions, component flow diagrams should be made up to show how each component deports. In the process spreadsheet each component must be controlled and the component flow diagrams are invaluable aid in determining the control strategy.

Process Spreadsheet Structure

The information in the process flowsheet and reactions can now be translated into a spreadsheet. The spreadsheet is a grid of cells into which information is entered. The information in each cell can be text which is displayed or amathematical formula for which the current value is displayed. The formula can use the value of other cells and so input data can be entered as specific process streams and the entire process mass balance can be calculated from this input. The framework of a spreadsheet program, like Lotus 1-2-3, has powerful features to assist in the handling of the information in the cells.

A functional spreadsheet structure is as follows:

1. The process spreadsheet is set out with the process flows as columns ordered in an appropriate manner. Flows can appear several times to help visual the progression of the calculations or to show balances around unit operations.

2. The process reactions in each unit operation are also set out as columns: with one or more reactions per column. The components which are produced in the reaction(s) are positive; those which react: negative. The net sum of the components in a reaction must be zero. The streams which enter a unit operation can be combined into one column, though they may be physically separate. The streams that exit can also be combined. This format: input, reaction and output, clearly shows the chemical changes within the unit operation.

3. The process flow parameters are set out as the rows of the flowsheet. An effective format to follow down each column is:

    a) process flow name usually as a abbreviation
    b) flow number from the flowsheet
    c) the phase(s) in the flow: liquid, solid and/or gas
    d) the mass flows: total, aqueous, solid and gas
    e) the volume flows: total, aqueous, solid and gas
    f) the specify gravities: overall, aqueous, solid and gas
    g) flow temperature and pressure

h) the mass flows of each component broken down into aqueous, solid and gas components
i) the concentration of each component: aqueous components in g/l; solid and gas components in %.

4. In working with the spreadsheet it is useful not to enter the input data directly into the process streams and reactions but rather to transfer the information from an input section. In this area, all the input variables appear and are labelled. The input section can be split into two: one section which contains reference information which is fairly constant such as the analysis of the concentrate used; and, the other which contain process parameters which are to be varied more frequently.

It is very useful to have the molecular weights of all the components and their constituent elements as input variables. These can be labelled with appropriate names and used in the cell formula.

The spreadsheet rows and columns should be appropriately titled and the input sections set up with all the input data labelled.

Process Flow Calculations

The manner in which information entered into the spreadsheet is manipulated is crucial to the success of the calculations. There are three types of operations:

1) In stream addition two process flows with all their components are added together.

2) In stream splitting a flow with all its components is split as determined by a criterion on a key component(s).

3) A chemical reaction converts one set of components into another. The extent of reaction is determined by a criterion on a key component. The reactant and products of reactions may provide the criteria for other reactions and stream splits.

Throughout these operations each component must be directly controlled. Any component not controlled will not have stable values and will lead to a mass balance that cannot close. Control directly from input data is most effective. Dependence on intermediate calculated values is much less effective and may lead to an iterative calculation which cannot converge. Component control strategies should be developed using the component flowsheets and written out for future reference. [It should be remembered that for any part of the flowsheet the flow out has to equal flow in plus changes due to reactions.]

Entering Information

The spreadsheet is built up column by column as process flows and reactions are added. Input data is converted to component masses upon which the process operations are performed. Recycle streams which are yet to be fully calculated can be initially estimated and a fixed value inserted. If it is possible to calculate a component by several means the one most directly linked to the input data should be used. Automatic spreadfast recalculation with each cell entry should be disabled. The spreadsheet should be saved frequently during development and prior to calculation to limit losses due to errors.

For solids and gases component concentrations can be readily calculated as weight percentages. For aqueous streams, the water concentration can be arbitrarily fixed; a suitable value between 900 and 1000 g/l can be chosen. This makes the volume of an aqueous flow only a function of the mass of water.

Error Tracing

Errors in component "bookkeeping" are easy to make and can be tracked down by three means.

1) The total mass flow for chemical reactions must be zero.

2) An element tracking table can be set up. The components in each column can be broken down to their constituent elements and set up much like the component mass balance. The total of the constituent elements must equal the component total.

3) The sum of all the input streams must equal the sum of the output streams for the total flows and each constituent element. The element table can be used for this purpose. The input streams can be multiplied by 1; the output: by -1 and, all internal streams by 0. The sum of all the columns should be zero. [A portion of a spreadsheet e.g., the element table can easily be recalculated by copying it onto itself repeatedly.]

Spreadsheet Convergence

Spreadsheets that contain recycle streams must be iterated, usually 20 to 25 iterations is adequate. Convergence of the iteration can best be gauged by looking at the sum of the flows into and out of the process. The sum should get progressively smaller with each recalculation. If the spreadsheet does not converge, irrepairable damage can be done. It is essential to have an uncalculated version saved which can be retrieved.

The main reason for a lack of convergence, other than errors, is the lack of direct component control. "Calculation loops" in the spreadsheet can make the value of a calculated component in a specific cell swing in response to other calculated values including that of the specific cell. The problem can be insidious. One means to track down the source of the problem is to replace calculated values with fixed values and see if this forces convergence.

Spreadsheets can also diverge though no errors exist due to large changes in input parameters which create large changes in specific streams or by input parameters which create an impossible situation, ie, negative flows. This problem can be checked by writing conditional component values. Lotus 1-2-3 permits a cell to have two values depending on a conditional statement. By this means recycle streams can be fixed if their true calculated value goes negative or becomes excessive.

Extracting Results

The final calculated spreadsheet can be readily printed. Each area of the spreadsheet should be labelled to simplify the printing commands. If a large number of cases have to be produced, several options are available. A macro program can be written to change the required variables, calculate the spreadsheet and print the output. Alternatively Lotus 1-2-3 has a Data Table facility. This enables values for 1 or 2 variables to be input and the program automatically calculates the spreadsheet and outputs specified results on the form of a table. The results of the one variable Data Table can readily be graphed for an effective presentation.

Process Simulation Example

To illustrate the process simulation procedure, the development of a process spreadsheet for a hypothetical Chalcopyrite Upgrade Process (CUP) is given below.

In the CUP process a chalcopyrite/pyrrhotite/gangue PGM concentrate is first treated with an acid preleach to convert the pyrrhotite to H_2S and soluble iron. The preleach slurry is then leached with a mixture of ferric iron and cupric copper to decompose most of the chalcopyrite. The product cuprous copper is precipitated with H_2S to form chalocite. The product solution is then oxidized to regenerate the oxidized iron and copper for leaching. Iron is precipitated as sodium jarosite. A solution stream is taken off and neutralized and this forms the solution bleed for water control and the recycle solution for the preleach. The residue from this process will be treated for sulphur removal and PGM recovery. Figure 1 shows a rough process flowsheet. A complete process flowsheet has 31 streams.

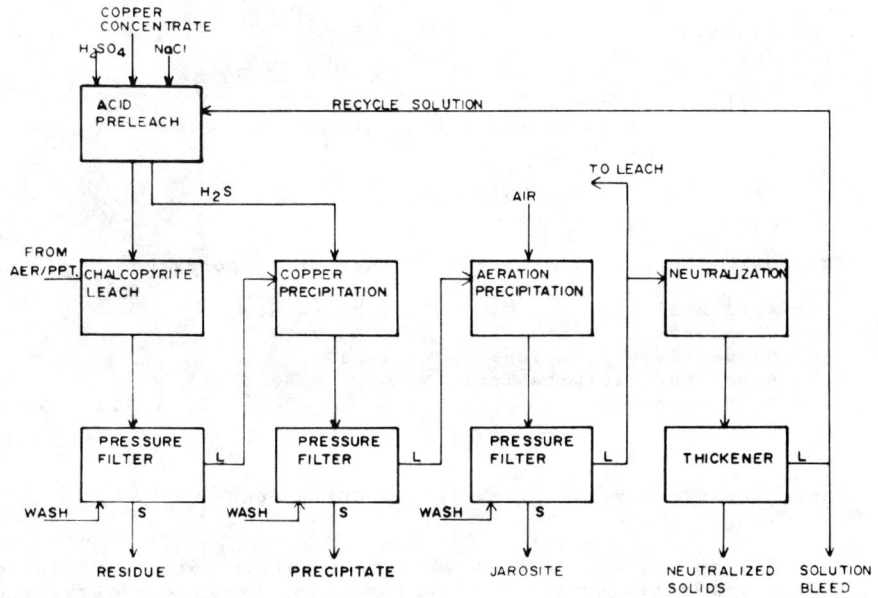

Figure 1 - A rough process flowsheet for the
Chalcopyrite Upgrade Process.

A complete set of process reactions follows:

R1: $FeS + 2H^+ = H_2S + Fe^{2+}$

R2: $CuFeS_2 + 3Fe^{3+} = Cu^+ + 4Fe^{2+} + 2S$

R3: $FeS + 2Fe^{3+} = 3Fe^{2+} + S$

R4: $Cu^+ + Fe^{3+} = Cu^{2+} + Fe^{2+}$

R5: $2Cu^+ + H_2S = Cu_2S + Fe^{2+}$

R6: $4Fe^{2+} + O_2 + 4H^+ = 4Fe^{3+} + 2H_2O$

R7: $4Cu^+ + O_2 + 4H^+ = 4Cu^+ + 2H_2O$

R8: $Na^+ + 3Fe^{3+} + 2SO_4^{2-} + 6H_2O = 4NaFe_3(SO_4)_2(OH)_6 + 6H^+$

R9: $Cu^{2+} + 2NaOH = 2Na^+ + Cu(OH)_2$

R10: $H^+ + NaOH = Na^+ + H_2O$

R11: $2H_2S + O_2 = 2H_2O + 2S$

R12: $H_2O(aq) = H_2O\ (gas)$

R13: $NaCl = Na^+ + Cl^-$

A complete set of components including spectators is:

Aqueous: Cu^+, Cu^{2+}, Fe^{2+}, Fe^{3+}, H^+, Na^+, Cl^-, SO_4^{2-}, H_2O

Solid: $CuFeS_2$, FeS, Cu_2S, SiO_2, S, $NaCl$, $NaFe_3(SO_4)_2(OH)_6$ [sodium jarosite], $NaOH$, $Cu(OH)_2$

Gas: O_2, N_2, H_2S, H_2O

The following reactions occur in each unit operation.

Acid Preleach:	R1 + R12 + R13
Chalcopyrite Leach:	R2 + R3 + R4
Copper Precipitation:	R5 + R12
Aeration/Precipitation:	R6 + R7 + R8 + R12
Neutralization:	R6 + R7 + R8 + R9 + R10 + R11 + R12

There are two stream splits in the CUP process.

1. The solution from Neutralization splits into the Solution Bleed which consists of the net input minus the reacted: Na^+, Cl^-, SO_4^{2-} and H_2O. The Recycle Solution has to have a specified volume and so the water flow to that unit is set. The other components follow the water. Through selection of specific parameters, it is possible to have a negative H_2O flow in the water bleed. In this case the recycle stream is forced to have a zero value by conditional cell values.

2. The solution from Aeration/Precipitation splits as defined by an input variable.

The solid liquid separation steps are controlled by % solids and the wash ratio, defined as the mass of wash water per mass of contained solution in the filter cake at the final % solids. The washes were complete - the final cakes contained no dissolved solids.

A complete list of the component control strategies would be extensive. Three examples are presented:

Cu^+: Produced in the Chalcopyrite Leach (R2); partially precipitated in Copper Precipitation (R5); converted to Cu^{2+} in Aeration Precipitation (R7).

Cu^{2+}: Produced in Aeration Precipitation (R7); portion to Neutralizatio where totally precipitated (R9); remainder reacts in Chalcopyrite Leach (R4).

SO_4^{2-}: Added in Acid Preleach with H_2SO_4 (in sufficient quantity to keep Aeration/Precipitation neutral); reacts in Aeration/Precipitation and Neutralization

(R8); recycles to Acid Leach, remainder out of process in solution bleed.

A portion of the spreadsheet encompassing the Acid Preleach is shown in Figure 2 on the following page.

In a preliminary study the characteristics of the flowsheet were determined and the following conclusions could be made:

1) The sulphuric acid use was excessive due to the need to keep the leach neutral. As a result the sulphate concentrations in the process solutions were high.

2) As the ratio of Aeration/Precipitation solution sent to Neutralization versus the Chalcopyrite Leach was increased, the mass flow to the leach decreased (as the flow to Neutralization is constrained to be relatively constant). This resulted in an increase in the copper and iron concentrations in the leach. The increase in iron concentration also meant more iron was sent to Neutralization and relatively less iron could be removed in Aeration/Precipitation.

3. The process water balance was reasonable primarily due to the evaporation of water from Aeration/Precipitation.

The net conclusion is that the cup process has serious problems which can only be overcome by flowsheet changes. It is unlikely that a satisfactory set of operating parameters can be selected.

Conclusion

The method presented for flowsheet simulation offers a ready means of producing detailed mass balances for complex hydrometallurgical processes. Through organization of the information available and the extensive calculation capabilities of spreadsheet programs the positive and negative attributes of a flowsheet become apparent and subsequent development can be guided.

Stream	Conc	H2SO4	NaCl	Rcy Soln	APL In	APL Rctn	APL Out	APL H2S	APL Sly
No	1	2	3	4	1+2+3+4	R1+R12 +R13	5+6	5	6
Phase(s)	SL	L	S	L				G	SL
Mass Flow-kg/min									
Total	117.65	72.57	10.00	302.67	502.89	.00	502.89	14.19	488.70
Aqueous	17.65	72.57	0.00	302.67	392.89	20.81	413.70	0.00	413.70
Solid	100.00	0.00	10.00	0.00	110.00	-35.00	75.00	0.00	75.00
Gas	0.00	0.00	0.00	0.00	0.00	14.19	14.19	14.19	0.00
Vol Flow-l/min									
Total	41.64	39.44	4.62	177.60				15469	213.82
Aqueous	17.65	39.44		177.60					195.27
Solid	23.99		4.62						18.55
Gas								15469	
Spec Grav-kg/l									
Overall	2.826	1.840	2.165	1.704				0.001	2.286
Aqueous	1.000	1.840		1.704					2.119
Solid	4.169		2.165						4.042
Gas								0.001	
Temp deg C	amb	amb	amb	85		80		80	80
Press psig	0	0	0	0		0		0	0
Aqueous-kg/min									
Cu+									
Cu++									
Fe++						15.88	15.88		15.88
Fe+++									
H+		1.42			1.42	-0.57	0.84		0.84
Na+				45.28	45.28	3.93	49.21		49.21
Cl-				16.72	16.72	6.07	22.78		22.78
SO4--		67.53		71.95	139.48		139.48		139.48
H2O	17.65	3.63		168.72	190.00	-4.50	185.50		185.50
Solid-kg/min									
CuFeS2	70.00				70.00		70.00		70.00
FeS	25.00				25.00	-25.00	0.00		0.00
Cu2S									
SiO2	5.00				5.00		5.00		5.00
S									
NaCl			10.00		10.00	-10.00			
Na Jarosite									
NaOH									
Cu(OH)2									
Gas-kg/min									
O2					0.00		0.00		
N2					0.00		0.00		
H2S					0.00	9.69	9.69	9.69	
H2O						4.50	4.50	4.50	

Stream	Conc	H2SO4	NaCl	Rcy Soln	APL In	APL Rctn	APL Out	APL H2S	APL Sly
No	1	2	3	4	1+2+3+4	R1+R12 +R13	5+6	5	6
Phase(s)	SL	L	S	L				G	SL
Aqueous-g/l									
Cu+									0.0
Cu++									0.0
Fe++									81.3
Fe+++									0.0
H+		2.0%							4.3
Na+				254.9					252.0
Cl-				94.1					116.7
SO4--		93.0%		405.1					714.3
H2O	1000.0	5.0%		950.0					950.0
Solid-%									
CuFeS2	70.0%								93.3%
FeS	25.0%								0.0%
Cu2S									
SiO2	5.0%								6.7%
S									
NaCl			100.0%						
Na Jarosite									
NaOH									
Cu(OH)2									
Gas-%									
O2									
N2									
H2S								68.3%	
H2O								31.7%	
Elements-kg/min									
H	1.97	1.82	0.00	18.88	22.68	.00	22.68	1.08	21.60
N	0.00	0.00	0.00	0.00	0.00	0.00	0.00	0.00	0.00
O	18.63	48.21	0.00	197.78	264.62	0.00	264.62	3.99	260.62
Na	0.00	0.00	3.93	45.28	49.21	.00	49.21	0.00	49.21
Si	2.04	0.00	0.00	0.00	2.04	0.00	2.04	0.00	2.04
S	33.58	22.54	0.00	24.02	80.13	.00	80.13	9.12	71.02
Cl	0.00	0.00	6.07	16.72	22.78	.00	22.78	0.00	22.78
Fe	37.18	0.00	0.00	0.00	37.18	.00	37.18	0.00	37.18
Cu	24.24	0.00	0.00	0.00	24.24	0.00	24.24	0.00	24.24
Total	117.65	72.57	10.00	302.67	502.89	.00	502.89	14.19	488.70
Component Total	117.65	72.57	10.00	302.67	502.89	.00	502.89	14.19	488.70
Difference	0.00	0.00	0.00	.00	.00	.00	.00	0.00	.00

Figure 2 - Portion of CUP spreadsheet

THE USE OF MASS BALANCE MODELS IN THE PROCESS DESIGN

OF HYDROMETALLURGICAL PLANTS FOR ZAMBIA CONSOLIDATED COPPER MINES LTD

G.K. Chibuye, I.M. Johnson and E.P. Smithson

Nkana Division, ZCCM Ltd., Kitwe, Zambia; ZES Ltd.,
Ashford, Kent, UK; Technical Services, ZCCM Ltd.,
Kalulushi, Zambia.

Abstract

In recent years Zambia Consolidated Copper Mines Ltd., (ZCCM) has designed, constructed and brought into operation two major new hydrometallurgical plants. The plants incorporate complex circuits for the production and purification of leach liquors from which metals are extracted by electrowinning. Mass balance models were used during the design phases of the two projects. In both cases the models assisted in the development of process flowsheets for the new circuits. Simple techniques were used to build the steady state mass balance models. However the simplicity of the formulation did not prevent the models from playing a key role during the project design stages: the results generated from them became the reference base for process design decisions. The models have been widely applied. Process investigations using them have included impurity deportment, production optimisation, alternative sub-circuit operations and water retention and conservation.

Introduction

Zambia Consolidated Copper Mines Limited, (ZCCM), has built and brought into operation two major new hydrometallurgical extraction plants during the past five years. A plant for the production of copper and cobalt metals from mixed sulphide concentrate was completed at the Nkana division of ZCCM in 1982. This plant is now producing copper and cobalt at annual rates of 11000 and 2250 tonnes respectively. More recently a major expansion and restructuring of the Tailings Leach Plant facility at Nchanga division has been completed. The additional material treated by this expansion is from old tailings dams, which contain residual 'oxide' copper from the inefficient oxide flotation process. In its first full year of operation the plant produced 30000 tonnes of electrowon copper from old tailings.

During the early design phases of both plants it was recognised that manual methods would be uneconomic for flowsheet development because of the volume of calculations required. Accordingly computer based models were constructed for determining mass balances for the respective circuits. The models were developed by Zambia Engineering Services Limited, (ZES), a British based subsidiary company of ZCCM. ZES was assigned responsibility for the management and design of both projects. Mass balance modelling is a standard process design technique employed within ZES.

The mass balance models were used extensively in investigations of process flow volumes and compositions. The responses of the circuits to changes in control parameters and variations in feed composition were studied. Conclusions on the optimisation of plant operations were derived from this work.

Nchanga Tailings Leach Plant

Plant Description

Current arisings of tailings from the Nchanga copper concentrator and old tailings slurried with high pressure water are pumped to the plant. In each case the dry solid content of the material flow is approximately 750000 tonnes per month.

The feed slurries are dewatered by thickening and filtration on horizontal belt filters, (the preleach filters). This enables a water balance to be achieved in the subsequent leach circuit. The water that is extracted from the feed slurries is recirculated to the tailings reclamation operation, (figure 1).

The preleach filtercake is fed to the primary leaching stage where sulphuric acid is added to control the pH of the leach discharge. A clear leach solution, containing typically 5 gpl copper at pH2, is obtained by thickening and clarification of the leach discharge. The leach solution is treated in a solvent extraction process to produce an advance electrolyte containing 50 gpl copper, approximately. Copper metal is extracted from the electrolyte in conventional electrowinning tankhouses.

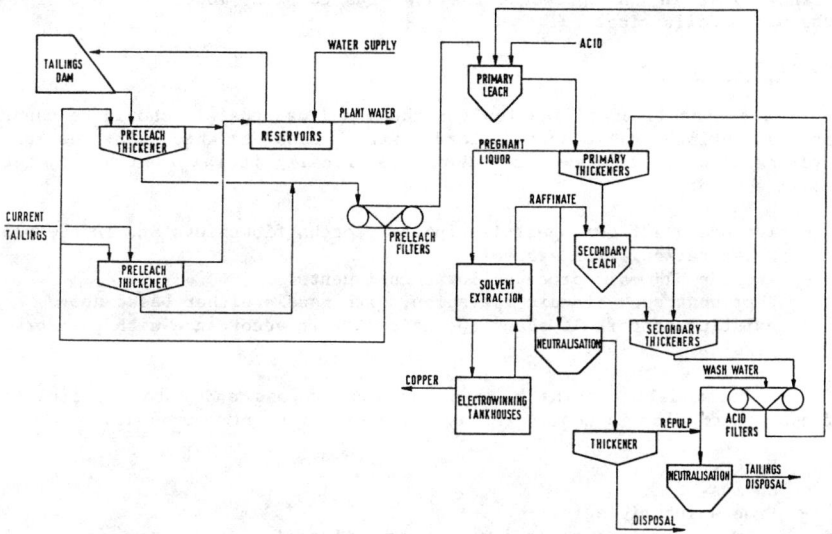

The acid bearing raffinate from the solvent extraction plant is partially routed to the secondary leach pachucas for further leaching of the solids and partially discarded via a limestone/lime neutralization circuit to maintain volume balance. After neutralization the water is largely recovered for reuse. The thickened solids from the primary leach stage are also fed to secondary leach where additional copper is extracted into solution from the solid material. The residual solids from secondary leach are thickened, and then filtered and washed with water on a second set of horizontal belt filters, (the acid filters). The washed filtercake is neutralized and discarded and the copper bearing solutions are recirculated within the plant.

The overall process contains a number of major sub-circuits. Within each of these the solution medium is recirculated and retained. Water balances and the retention of water within the plant were a major consideration in the development of the process design.

The Nchanga Tailings Leach Plant has been built in stages. The recently completed Stage 3 Project was a major extension to an existing plant that had been in operation for a decade (1). The Stage 3 project doubled the capacity of the plant and brought about a significant change in plant operations. The introduction of filtration within plant operations has greatly improved the control and retention of process liquors within the circuit. Certain unit operations have been altered as a result of the Stage 3 project. For example thickeners that were previously used in a 5-stage counter current decantation mode are now employed in two parallel lines of 2-stage counter current operations, with one standby thickener between them. The residue from both thickener trains is fed to the new filter plant. Thus the process circuit for the plant has been altered significantly with the advent of Stage 3. Accordingly the project included the drawing up of new process flowsheets for the whole plant. This in turn promoted the use of a computer model in the design of the new metallurgical balance.

Model Formulation

The overall specification for the model was that it should be capable of generating an accurate steady state mass balance for the plant hydrometallurgical processes. More specifically it was required that the model should:

- Include all flows specified in the process flowsheets and in the plant water reticulation system.
- Include all main process flow constituents.
- Represent each circuit operation, with models either based upon existing copperbelt plant operations or in accordance with testwork results.

The model contains a total of 134 process and water reticulation flows. Each flow is considered as a mixture of eight components:

- Water.
- Solids.
- Free sulphuric acid.
- Sulphate ions associated with gangue minerals.
- Unleached acid soluble copper.
- Acid insoluble copper.
- Copper in solution.
- Copper hydroxide.

The specification of a process flow enables a complete mass balance for copper to be calculated across the circuit. It also allows reagent requirements, (acid, lime), to be calculated through the determination of a sulphate ion balance.

The Nchanga Tailings Leach Plant contains the following types and numbers of unit processes.

```
Thickeners                   7
Filters                      2 operations, (26 filters)
Reaction cascades            4 operations, (6 sets of pachucas)
Solvent extraction plant     1
Electrowinning tankhouses    1 operation   (2 tankhouses)
```

In addition to this there are numerous tanks and sumps, which act as flow mixers. There are also many flow splits where offtakes from a flow are routed to other destinations. All of these processes were included in the model with the exception of the tankhouses. The solvent extraction process acts as a boundary to the leaching and hydrometallurgical circuit and hence was a natural limit to the scope and extent of the model.

The major difficulty in the formulation of the model was the specification of the leaching processes. Testwork results and experience of operating the Stage 2 plant circuit had established that all acid would

be consumed in the leach pachucas, with the size of tank, (and
corresponding residence time), envisaged. The modelling problem was one
of specifying the respective quantities of acid consumed in leaching copper
and in reacting with gangue minerals. These reactions are continuous
within the cascade; the rates at which they occur depend upon many factors.
The concentration of acid and the grade, mineralogy and physical
availability of the reactants within the solid material are prime
influences.

Considerable testwork has been done to establish rate reaction curves
for copper leaching and to determine gangue acid consumption, (GAC), values
for typical plant feeds. The work has shown that GAC values vary
considerably as the mineralogy of the mined ore at the division changes.
Testwork results also confirm that gangue mineral leaching will continue
for a long time if free acid is present. Hence, while copper oxides will
react relatively quickly and largely to completion, gangue minerals will
continue consuming acid. Typical curves for the two types of reaction are
shown in figure 2. Clearly the optimum plant operation would be achieved
when leaching terminates at the point Y in figure 2. This point indicates
the position where the marginal copper revenue from additional production
equals the marginal acid cost from increased GAC. As other leach
conditions were fixed by other process considerations this largely
determined the leach residence times.

FIGURE. 2   LEACHING REACTION CURVES

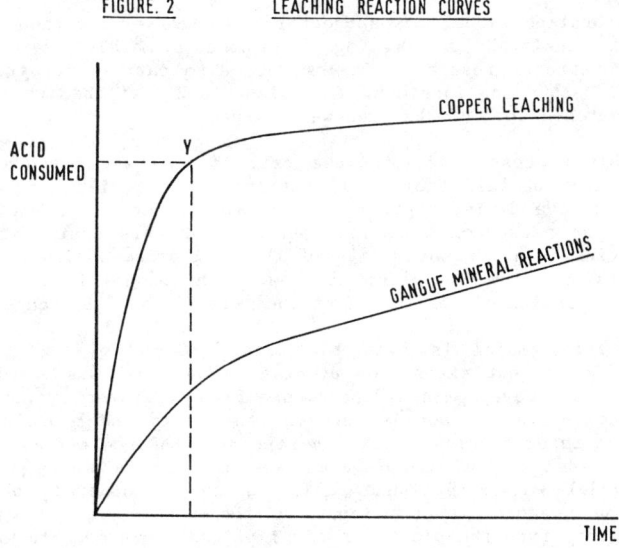

It was fortunate for the model building work that Nchanga division
declares GAC values for plant feed materials on a routine basis from
testwork at the standard leach conditions. The declared values are
estimates from testwork corresponding to the point Y in figure 2: the
optimum rate of acid consumption in relation to copper extraction. The
existence of 'accepted' GAC values for terminal conditions enabled a
complicated model formulation to be avoided. Instead of modelling

reactions directly and using as data the acid added to primary leach, the model works in the opposite manner. GAC values for each feed material are input to the model as data, together with a value of the extraction efficiency for acid soluble copper. The model calculates the acid that must be added in order to achieve the specified terminal conditions after the two stages of reaction.

The formulation used for the leaching calculations is extremely simple In practice it has provided acceptable results. The use of established operating parameters in the calculations gave ZCCM personnel confidence in the model results.

An iterative calculation method is applied in the program to achieve a mass balance. The composition of process flows are calculated from the feed input stage through the circuit. The volume and composition of recirculating flows are estimated to enable a once-through calculation to be completed. Discrepancies occur between the estimated values and subsequently calculated values for the composition and size of the recirculating flows. These discrepancies in the balance are reduced in successive program iterations. Finally a mass balance is achieved for the whole circuit. This method of calculating mass balances for circuits with recirculating flows has been described by Westerburg (2).

Model Verification

The mass balance model was subject to careful verification. In the main the work consisted of checking unit process models against plant operations or testwork results. Checks were also carried out to confirm that the model output was indeed a mass balance and that the circuit logic of the model matched that of the process flowsheet.

The unit process model for the belt filters was based upon the operating results obtained from a pilot filter. This piece of equipment was installed at the division prior to the main project. It was used to familiarise plant personnel with the operation of this equipment and to provide operating data for the design of the full process plant. The key parameters obtained from the pilot plant were the wash efficiency of the filter and the fraction of moisture contained within the filtercake.

A detailed model is available within ZCCM for predicting the performance of a solvent extraction process. The model was developed by the Research and Development department for interpreting laboratory testwork results. It was decided however that this model would not be appropriate for incorporating in the overall mass balance model. Instead a statistical analysis of plant data was made from which a simple performance model was established. The analysis compared the copper content of the pregnant liquor input to the plant with that of the raffinate returned from the plant. A simple linear relationship was shown to exist. This is another example of a simple formulation being employed in the model, rather than a more complicated alternative.

As indicated in an earlier section, the modelling of the leaching process was a major consideration in the construction of the model. The modelling approach was subject to careful scrutiny. ZCCM personnel reviewed the model and the results obtained from it. They confirmed that the predicted extent of leaching in the respective cascades was in accordance with their Stage 2 plant experience and their expectations of the new circuit. Solution tenors were also confirmed as being within the expected range. The leaching model was accepted on this basis.

398

Model Application

The primary use of the model was in establishing mass balance flows for the process flowsheets. The engineering design of the plant was based upon the model generated flowsheet specifications of process flows. Hence the model made a major contribution to the project design, in particular with respect to piping and equipment sizing.

The model was used to examine process flowsheet modifications. One change altered a water supply line from the current preleach thickener to a water reservoir. The modification also split preleach filtrate between the current and reclamation preleach thickeners. The model runs that were carried out to examine this modification showed that water reticulation would be improved by the changes. The overflow volumes of the two preleach thickeners were in better balance in the modified circuit.

Model runs have been carried out to examine the effect of varying the volume of wash water used on the acid filters. Changes in this flow volume have two direct effects on the recovery of copper from solution in the circuit. Increasing the wash volume displaces additional copper from acid filtercake. The loss of solution copper via this material leaving the circuit is reduced. Increases in the wash volume also give rise to corresponding increases in the volume of raffinate that must be bled out of the process circuit in order to maintain a volume balance. The raffinate bleed contains copper in solution and this is lost from the circuit. Increasing the raffinate bleed volume also increases acid losses from the circuit; both acid and lime consumption rise as a result. The model was used to determine an optimum wash operation, taking into account these various loss factors.

The model has been used to examine the water input requirements of the circuit. A number of different operating conditions were modelled. For each case changes in the control settings for the plant were tested. The work established methods of reducing the net requirements for supplementary water. These and similar exercises also established the maximum flow volumes that could be expected in certain key pipelines.

The model is currently being used in production planning exercises. The studies are examining means of maximising copper output under conditions of restricted acid and lime supply.

Nkana RLE Cobalt Plant

Plant Description

The plant treats a mixed sulphide concentrate to produce both cobalt and copper metals. It has a design capacity to treat up to 130000 tpa of concentrates and produce a maximum of 18400 tpa copper and 4300 tpa cobalt (3).

399

The concentrate feed is input to a fluosolids roaster which produces a calcine containing soluble metallic sulphates. The roaster off-gas is scrubbed to remove entrained solids and routed to a sulphuric acid production complex. The calcine is quenched and the metals contained in it are leached into solution. Thereafter the treatment route is via a complicated series of hydrometallurgical circuits. A simplified diagram of the extraction process is given in figure 3.

FIGURE.3   <u>NKANA COBALT PLANT SUB-CIRCUITS AND MAJOR OPERATIONS</u>

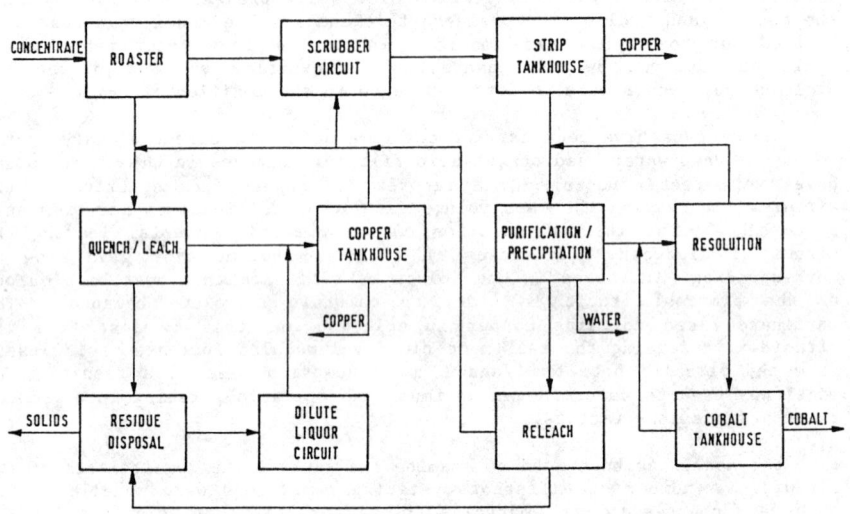

The leach solution first enters the copper circuit. The solution is clarified and copper metal is partially recovered in an electrowinning tankhouse. A bleed of tankhouse spent solution is taken via the roaster gas scrubbing circuit to an electro strip tankhouse. In this tankhouse almost all the copper is removed whilst the other leached metals remain in solution. This stripped electrolyte stream is a cobalt rich solution which is then fed to the purification and precipitation circuit.

The purification of cobalt sulphate solution is achieved in a series of selective precipitation stages. Slaked lime is used as the precipitation agent. The pH is carefully controlled in each of the reaction cascades to achieve the desired separation of the metals in the incoming solution. Thickeners and rotary drum filters are used for solids/liquid separation after each cascade. In the final precipitation stage cobalt hydroxide is formed and excess water is removed from the process circuit.

The cobalt hydroxide formed in the final purification stage is repulped in spent electrolyte from the cobalt tankhouse. The resulting advanced electrolyte is clarified and conditioned before passing to the cobalt tankhouse where cobalt metal is recovered by conventional electrowinning.

The impurity content of the finished cobalt metal is constantly monitored and controlled within set specifications. The objective of maximising recovery conflicts with minimising the impurity content of the finished product: some important impurities are metals with similar reaction characteristics to those of cobalt.

Process control is achieved primarily via the pH levels in the cobalt circuit precipitation stages and by controlling tenors of certain streams in other circuits. Some examples of the latter are:

- Copper, cobalt and acid in advanced copper electrolyte
- Copper and cobalt in the scrubber circuit
- Cobalt electrolyte tenor

Variations in the settings of key control parameters can have significant effects on the steady state mass balance throughout the plant.

Model Formulation

The model contains 170 process flows. Each flow is considered as a mixture of ten constituents:

- Water
- Solids
- Sulphuric acid
- Gypsum
- Copper sulphate, copper hydroxide
- Cobalt sulphate, cobalt hydroxide
- Zinc sulphate, zinc hydroxide

This specification is the minimum capable of providing a mass balance for the process and of determining how recoveries could be maximised within metal purity specifications. (Zinc is the most significant impurity within the circuit that is controlled by pH and flowrate/tenor settings).

The types and numbers of unit operations incorporated in the process flowsheet are as follows:

Flow mixers	49
Flow splits	25
Thickeners	12
Filters	14
Reaction Cascades	17
Tankhouses	4

A model was developed for each of the above unit process types. The unit process models operate as simple flow partitions. Each component is treated separately; interactions such as co-precipitation were ignored in the formulation. Careful attention was applied to the modelling of the filter wash operations and the reaction cascade processes. The effect of applying a wash on a filter is shown in figure 4.1. The dilution of the entrained liquor in the filtercake depends upon the relative wash volume. 'Curves' were also applied in the model to define the precipitation of metals in the reaction cascades. An illustration of a set of pH - precipitation curves is given in figure 4.2.

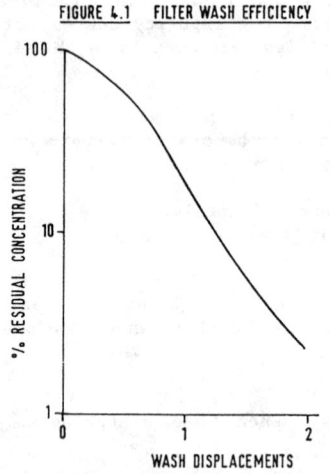

FIGURE 4.1   FILTER WASH EFFICIENCY

% RESIDUAL CONCENTRATION

WASH DISPLACEMENTS

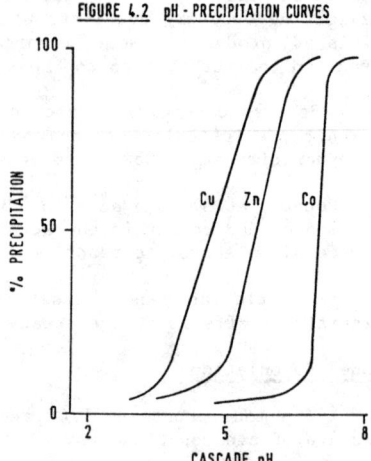

FIGURE 4.2   pH - PRECIPITATION CURVES

% PRECIPITATION

Cu   Zn   Co

CASCADE pH

   Both the filter wash efficiency curve and the pH – precipitation curves were specified in the model by sets of straight line approximations.

   The overall mass balance model was constructed from the unit process 'building blocks'. This method of developing the model proved to be extremely effective. It also has the advantage that changes in process flow routings can be incorporated in the model very easily.

   The program calculates a mass balance by the approximation and iteration method that was described earlier for the Nchanga Tailings Leach model. The balance obtained is a simulation of process operations for a specified set of plant conditions. Variations in control parameter values will alter the mass balance. However, the calculation procedure is not capable of optimising the balance by varying control parameters. In the absence of this facility optimisation studies have been carried out by controlled sequences of model runs.

Model Verification

   The same verification methods that were used for the Nchanga Tailings Leach model were applied. In general all unit process models were verified against either existing installations or testwork results. Particular care was taken with the pH/precipitation relationships for the reaction cascades. In one case operating data from an existing cobalt plant and laboratory testwork results were significantly different. The model verification work prompted an investigation of the differences and a resolution of the problem.

   A cobalt extraction plant has been in operation at the Nkana Division of ZCCM since the early 1960's. The existence of the 'old Nkana' cobalt circuit enabled sub-circuits within the model to be verified against plant data. These exercises confirmed impurity deportment parameters for cobalt circuit processes and gave a general confirmation of model validity. In one particular exercise the model of cobalt hydroxide precipitation and the cobalt electrolyte circuit were compared with plant operations. The known problems of water balance and tenor control within this sub-circuit were clearly demonstrated in the model. The exercise gave one process engineer, with experience of running the existing plant, a clear insight into why the problems arose and how they could be alleviated. The exercise also confirmed the values of sensitive process parameters, (hydroxide filter performance, effective availability of lime), that should be applied in the model.

Model Application

The model was used extensively during the project design stages to determine process flow volumes and compositions. As with the Tailings Leach Plant model, the prime requirement was that of generating a standard mass balance for the plant. This was achieved; the material movements defined in the project flowsheets were established in model runs. Currently the model is being widely used in process investigations and in sub-circuit and plant optimisation studies.

As the project developed it became apparent from later geological data that the levels of impurities in future concentrates would be higher than originally predicted. Alterations in the process balance were made in order to maintain the purity of finished metal whilst also maximising its recovery. The operating load upon the purification circuit became greater since, in order to control the impurities to desired levels, cascade pH settings were raised resulting in increased circulating loads. Manual calculation methods could not have provided a practical alternative to the facilities provided by the model in generating new process balances for the changing circumstances.

The copper input to the strip tankhouse increased as the process balance changed and recirculating loads within sub-circuits rose. These developments occurred after the general design of the tankhouse had been fixed and detailed design was well advanced. Redesign would have been costly and would have delayed the completion of the project. As an alternative, the model was used to investigate how copper deportment could be altered to accommodate the capacity limitation of the strip tankhouse. The following changes were considered in the investigations:-

-   Re-routing of flows in the copper circuit.
-   Cobalt circuit adjustments to reduce recirculating loads.
-   Alterations in controlled volume flows/controlled tenor levels.

The model runs showed that copper deportment could be altered significantly. It was concluded that the capacity constraint could be met and that a redesign of the strip tankhouse could be avoided.

The model was used for a number of investigations of the scrubber circuit operation. The manner in which this circuit is connected to other sub-circuits can be altered. The secondary roles performed by the circuit vary according to the interconnections. For example the circuit can be used to remove selenium from the plant in conjunction with the strip tankhouse. However, this is not acceptable as a continuous mode of operation. The operating roles of the scrubber circuit were assessed in steady state mass balances for the different circuit configurations. It was concluded that the operation of the circuit should be alternated between two modes of use.

403

Since the commissioning of the plant in 1984 other model investigations of the scrubber circuit have been requested by plant management. The feed liquors to the circuit were varied considerably in the exercises. Both the direct effects of the changes within the scrubber circuit and the allied effects in the remainder of the plant were assessed. The results clearly defined the implications of the alternative operating strategies.

The level of zinc in finished cobalt metal was a major process design consideration. The deportment of zinc in the cobalt circuit is controlled by the pH levels of two of the precipitation cascades and the operation of the releach circuit. The interactions of control parameter settings within these operations were assessed using the model. In an extension of this work the possibility of taking a partially purified stream from the new cobalt plant to the old Nkana cobalt circuit was examined. This arrangement enabled both a high quality and a standard quality cobalt metal to be produced. The scheme matched production output with the known markets for cobalt. As a result of this work the decision was made to build the facilities that would enable the circuits of the new and old cobalt plants to be operated in this manner. The necessary engineering design work was completed and the required equipment installed.

The maximisation of metal recovery was an underlying objective throughout the development of the process balance and during the investigations described above. Many of the circuit control parameters have interacting effects. Often specific objectives could be achieved by different arrangements and combinations of the control settings. In these circumstances the maximisation of metal recovery defined the optimum combination within a set of feasible solutions.

At present there are market requirements for special grades of cobalt metal. In response to this the model is being used to examine in detail the deportment of impurities other than zinc. Hence the model is currently playing an important role in matching the finished product to customer needs.

Summary and Conclusions

Mass balance modelling has been a major tool used in the development of process flowsheets in two large hydrometallurgical projects carried out by ZES Limited.

In both projects the applications in which the models were used went far beyond the original intention and justification for their development. 'What if' questions inevitably arise when future plant operations are under consideration. The wide diversity of the questions that were posed during the two projects has been outlined above. The models were able to supply the quantitative analysis necessary to answer those enquiries. In many cases the results led directly to engineering design decisions.

The cost of building each of the models was of the same order of magnitude as calculating a mass balance by manual methods. Thereafter subsequent mass balances were produced by the models at a fraction of that cost. Hence the technique has proven to be extremely cost effective.

A major feature of both models is the simplicity of their formulation. This has proven to be an advantageous characteristic. It has enabled the models to be easily understood and has ensured results that can be easily reconciled to input data. This has prompted user confidence in model output and predictions. In some cases an extrapolation or interpretation of model results has been necessary to allow for a more complicated situation than that specified in the model formulation. Experience has shown that in these circumstances users are willing to apply their specialised knowledge to interpret results from a model in which they have confidence.

The authors would like to emphasise this latter point. A simple verifiable model used in conjunction with expert knowledge and experience is a powerful combination. Recent experience in the design of two new hydrometallurgical plants for the Zambian copper industry gives testimony to this conclusion.

Acknowledgement

The authors wish to gratefully acknowledge the permission of Zambia Consolidated Copper Mines Limited to publish this paper.

References

1. P.R. Hampsheir, "Design and Construction of the Nchanga Copper Tailings Leach Plant Stage 3", Proceedings of the Institution of Mechanical Engineers, 200 (131) (1986).

2. A.W. Westerburg et al., Process Flowsheeting (Cambridge University Press, 1979), Chapter 6.

3. G.L. Willis et al., "The New Nkana Cobalt Plant for Zambia Consolidated Copper Mines Limited" (Paper presented at the 115th TMS Annual Meeting, New Orleans, March 1986).

FERROUS

Session Chairmen
D. Ablitzer, Ecole des Mines, Nancy
Walter E. Wahnsiedler, Alcoa Technical Center

One of these two sessions focusses on iron making. Numerical modeling techniques study the influence of process variables on ore pellet induration. Reduction of pellets using an un-reacted-core model, simulation of combustion chamber phenomena, comparison of actual desulfurizing behavior with predictions of kinetic algorithm, and illustration of mass transfer effects in a bottom blown iron melt using a two-fluid mathematical formulation are also described.

On the steel making side, another session covers the segregation effects and tracer measurements of turbulent flow in tundishes. Steel refining papers include quantitative analysis of gas-liquid interactions, mixing in ladle metallurgy, and predictions about heat and mass transfer associated with high melting point ladle metallurgy, and predictions about heat and mass transfer associated with high melting point ladle additions.

THE MODELLING OF FLUID FLOW, TRACER DISPERSION

AND INCLUSION BEHAVIOR IN TUNDISHES

O.J Ilegbusi and J. Szekely

Department of Materials Science and Engineering
Massachusetts Institute of Technology
Cambridge, MA 02139, U.S.A.

Abstract

Fluid flow, tracer dispersion and flotation of inclusion particles are modelled for various tundish systems. The system behavior is considered from both order-of-magnitude analysis and numerical computation. The results show that there are significant differences between the behavior of tracer and that of inclusion particles, due principally to the finite rising velocity of the latter.

Introduction

Continuous casting is one of the major recent advances in steel processing technology, which is now being generally adopted. In this process molten steel is poured from a ladle through a tundish (an intermediate vessel) into molds, where solidification takes place, yielding continuous slabs or billets as sketched in Fig. 1.

In most conventional applications, this tundish is a relatively shallow trough (about 4-8 meters long, a meter wide and a meter deep), into which molten steel is poured at one location, and withdrawn at one or more locations, to provide the feed to the mold of the continuous caster. A typical tundish arrangement is sketched in Fig. 2.

Initially, the main role of the tundish was to distribute the liquid metal and to act as a buffer, ensuring uniform metal flow. More recently, it has been appreciated [1-5] that tundishes may play a key role in affecting the quality of continuously cast steel products. This is due to several factors, which include the following:

-- Non-metallic impurities, termed inclusions, may have a chance to float out, and thus may be partially removed in the case of properly designed tundishes.

-- By the same token, the temperature fields may be homogenized, and temperature fluctuations may be minimized.

-- Flow fluctuations, vortexing, and turbulence may be minimized in the exit regions with proper tundish design; and finally,

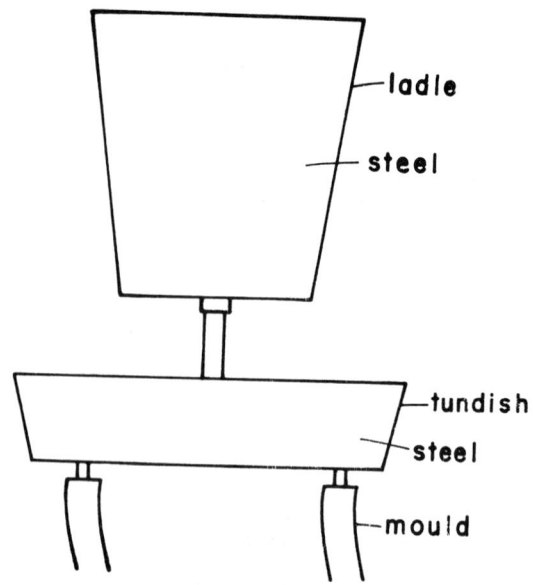

Fig.1: Continuous casting arrangement

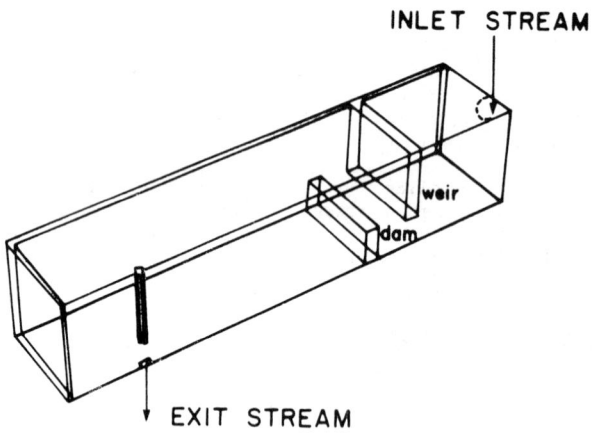

Fig.2: Typical tundish arrangement

-- The temperature of the exit stream may be adjusted through auxiliary heating, and alloying additions may also be made in newer tundish designs.

Recognition of these factors has led to intensive physical modelling efforts [1-3]. Indeed, most major steel corporations have massive water models of tundish systems. Most of this work, however, has been confined to performing tracer tests to assess the performance of different tundish designs, and relatively little insight has been sought into the more fundamental aspects of fluid flow behavior in tundish systems.

In recent years, the ready availability of large-scale computational facilities has stimulated realistic mathematical modelling efforts into tundish behavior. It has to be recognized that the flow in tundishes is inherently three-dimensional, so that a major computational effort is needed; furthermore, the intelligent display and assimilation of the large number of data obtained (say for over 2-4,000 grid points) will also require some creativity and imagination.

As a result of work such as [4]-[7], a much better general understanding of these systems is emerging.

The work presented up to the present has addressed the question of velocity distribution and tracer dispersion and with one very recent exception [7], could not consider the most critical problem of inclusion removal, although tracer behavior will be an indication of the expected inclusion removal efficiency.

The purpose of this paper is to provide a general insight into tundish behavior through the complementary use of asymptotic considerations and three-dimensional flow modelling. In this work we shall specifically address the question of inclusion removal, and will discuss the use and limitations of tracer techniques for addressing problems of this type.

Some General Asymptotic Considerations

If we consider a typical tundish, such as that sketched in Fig. 2, the linear melt velocity at the inlet nozzle (which may range in diameter from 50-100 mm) is of the order of 2-10 m/s. The exit nozzles have a diameter of 50-75 mm; thus, the corresponding melt velocities there will be of the same order of magnitude. The Reynolds numbers are of the order of 10^5-10^6, which correspond to highly turbulent conditions.

In contrast, in the body of the tundish, linear melt velocities are of the order of a few centimeters/second, which under equilibrium conditions would result in mildly turbulent or transitional flow, with Reynolds numbers generally in the thousands. Indeed, there are many tundish designs, particularly for billet casters, in which the equilibrium flow would be laminar.

If we are concerned with the behavior of inclusions, two factors must be considered. In the vicinity of the entry of the pouring stream, there is a very high rate of energy dissipation (typically on the order of 200 watts/ton). Indeed, this rate of energy dissipation is comparable to that found in ladle treatment facilities employing induction stirring or vacuum degassing [8,9]. It follows that significant agglomeration of the inclusion particles may occur in this region.

411

In the bulk of the tundish, we wish to encourage flotation of inclusion particles. In a quiescent fluid, the terminal rising velocity of inclusion particles is given by Stokes's Law, which may be written as:

$$U_T = \frac{2R_p^2(\rho_p - \rho_f)\,g}{9\mu}$$ (1)

Thus, the time required to float out a 20- and a 100-micron particle having a density some 0.4 times that of steel would be of the order of 7500 and 300 seconds, respectively. The rising velocity of the larger particles may be comparable to the mean velocity in the tundish (outside the pouring region), and would be significantly larger than the fluctuating turbulent velocity. However, the rising velocity of the smaller (say below 30 micron) particles may be smaller than the fluctuating velocities taken as 1/10 of the mean; so for these, the flotation process could be adversely affected by turbulence.

Let us now consider the residence or retention time of the melt in the tundish. The nominal residence time is given as:

$$t_R = \text{tundish volume/volumetric flow rate}$$ (2)

For most tundish applications, this value tends to range from about 2 to 10 minutes. However, for all tundish systems examined, the "breakthrough time," i.e. time required for the first appearance of a tracer at the exit, counted from the moment the addition was made at the inlet, is much shorter, say of the order of one-third of this value.

This departure from "ideal behavior" may be due to by-passing, i.e. the existence of "dead volumes," and to highly mixed regions of the system.

As an illustration, Fig. 3 shows a sketch of three idealized behavior types: (a) plug flow, when $t = t_R$; (b) complete mixing, when some of the tracer will appear at the exit immediately after it has been added; and (c) the most realistic case, a combination of plug flow, completely mixed flow, and "dead zone."

Most tundishes correspond to this last case; it is reasonable to expect that the region where the pouring stream enters will be well mixed, and that the remainder will be divided in some fashion between plug flow and dead regions. In many poorly-designed tundishes, these dead regions may take up the majority of the whole tundish volume.

If the whole system were in plug flow, there would be an abrupt transition between the size of the particles retained, on the one hand, and removed by flotation on the other. The critical cut-off diameter could be calculated from equation (1) and the relation

$$U_T = Q/bL$$ (3)

where Q is the volumetric flow rate, b is the tundish width, and L is the half-length. This equation expresses a balance between the rising time and the nominal residence time.

For perfectly mixed systems, there is no such cut-off diameter, but the inclusion removal rate, f, could be calculated from the relation:

412

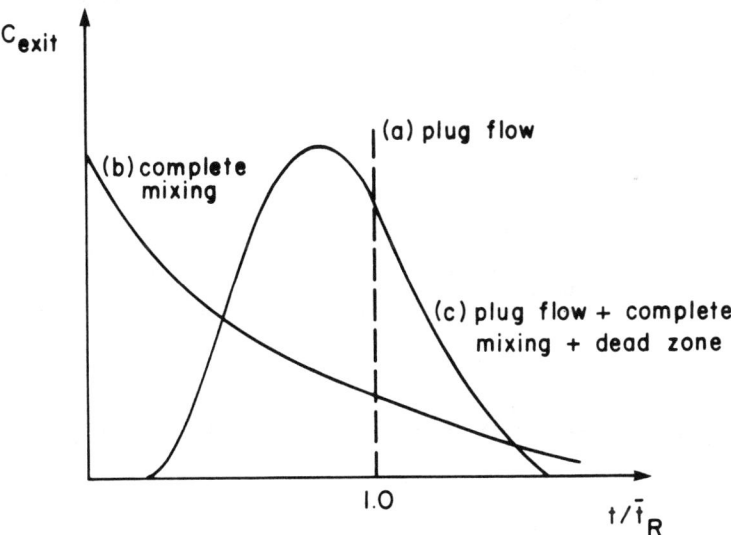

Fig. 3 Idealised types of tundish behavior

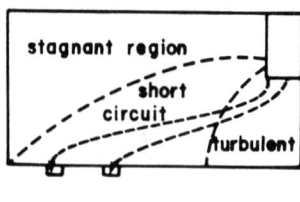

(a) No flow control.

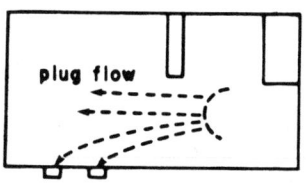

(b) Flow control with shallow weir.

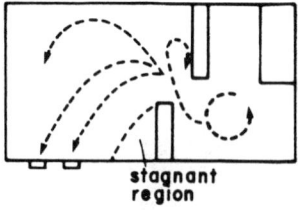

(c) Flow control with one weir and one dam.

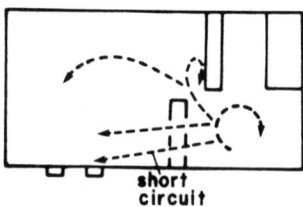

(d) Flow control with one weir and one slotted dam.

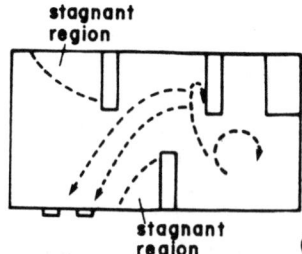

(e) Flow control with two weirs and one dam.

Fig.4: Tundish installations showing different flow regimes

$$f = 1 - \frac{1}{1 + U_T bL/Q} \tag{4}$$

Finally, the presence of dead volumes would reduce the actual residence time in the system.

Fig. 4 shows sketches of several tundish installations, indicating the anticipated "completely mixed" and "dead zones"; the possible short-circuiting flow paths are also indicated on these plots.

We should note here that in the absence of inserts (i.e., dams or weirs), significant short-circuiting will be inevitable. Furthermore, tundishes with a small separation between the inlet and outlet ports would be undesirable. The use of baffles or weirs could greatly improve matters, but their locations can be critical.

Finally, a comment should be made on tracer dispersion. In general, a tracer will be dispersed due to two factors, namely eddy diffusion and bulk flow. The relative importance of these may be assessed with the aid of the Peclet number, defined as:

$$N_{Pe} = \frac{LU}{D_{eff}} \tag{5}$$

where D_{eff} is the effective diffusivity.

Order-of-magnitude estimates for eddy diffusivity, D_{eff}, show that it may be of the order of 1-50 cm^2/s in the vicinity of the pouring stream, giving a Peclet number of 0.1 - 0.01. This indicates significant diffusive mixing. In contrast, the Peclet number will be very much higher in the remainder of the tundish, signifying that tracer transport there will be due to convection. In this regard, the existence of short-circuiting or by-pass streams such as those sketched in Fig. 4, could have a particularly adverse effect.

These asymptotic considerations are helpful because they allow us to make some useful intermediate conclusions, which may be summarized as follows:

-- The flow is turbulent at the inlet and at the exit regions, but will be only mildly turbulent or transitional in the bulk.

-- The breakthrough time for a tracer will be much smaller than the nominal residence time; this is inevitable, but one should be able to do much better than the 1/3 - 1/5-or-less ratio currently experienced. A target value of between 1/2 and 1/3 might be a realistic expectation.

-- The time required to float out the finer (say below 20-30 micron) particles from a quiescent melt would be much longer than the realistically achievable residence times. However, the time required to float out particles larger than ~60-100 microns should be feasible. Floating out the finer particles may be possible only through the appropriate adjustment of the melt velocities. Here we should stress that there is only a partial correspondence between the behavior of the tracers and of the inclusion particles. The very fine inclusion particles, say below about 10 microns in diameter, will act like the tracers in that their rising velocity is quite small. However, the larger inclusion particles will have a finite rising velocity, so that the behavior of a

415

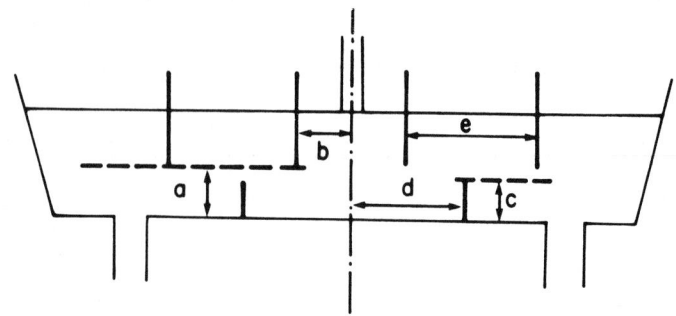

Fig.5: Schematic sketch of tundish arrangements considered

Table 1:

PRINCIPAL INPUT AND COMPUTATION PARAMETERS

Tundish length (X') = 6.79 m

Tundish width (Y') = 0.65 m

Melt depth (Z') = 0.75 m

Inlet stream velocity = 7.67 m/s

Inlet nozzle diameter = 54 mm

Outlet nozzle diameter = 54 mm

Inclination of walls (case 4) = 8°

Ratio of particle to fluid density = 0.5

Grid points along x,y,z = 25,11,15 (non-uniform)

CPU time for velocity field = 4 hrs on MicroVAX VT220

CPU time for concentration field = 8 hrs

CPU time for particle concentration = 30 mins

tracer will give only a rough indication as to the extent to which these particles may be removed or floated out in a given tundish arrangement. This is a very important point, which will be discussed subsequently.

-- Finally, the use of proper design and the deployment of flow control devices is essential for most applications. How these inserts should be deployed is often far from obvious, and will require quantitative assessment through both water modelling and machine computation.

The Mathematical Formulation[*]

Differential Equations

Fig. 5 shows the actual geometries considered in modelling the fluid flow phenomena in a tundish. In essence, we assumed a "straight flow" tundish, with one inlet and two outlets. As is seen, provision was made to have one dam and up to two weirs; allowance was also made to have slots in the dam. The input parameters employed are summarized in Table 1, and the exact location of the flow control devices is summarized in Table 2. It should be noted that the case examined here are to be taken as illustrative examples; other geometries could be readily examined.

From a fluid flow standpoint, the problem is to calculate the three-dimensional turbulent (transitional) velocity fields in a vessel with a specified inlet and outlet streams modified by flow control devices, together with the dispersion of a tracer and the behavior of the inclusion particles.

The fluid flow problem may be stated by writing down the turbulent Navier-Stokes equations; thus, we have:

Equation of continuity:

$$\nabla \cdot \rho \underset{\sim}{u} = 0 \tag{6}$$

Equation of motion:

$$\underset{\sim}{u} \cdot \nabla \rho u = -\nabla P - \nabla \cdot (\mu_{eff} \nabla \underset{\sim}{u}) \tag{7}$$

Turbulence model:

$$\mu_{eff} = \mu + \mu_t \tag{8}$$

where μ is the molecular viscosity, and μ_t is the turbulent viscosity, calculated from the turbulent energy K and its rate of dissipation ϵ, as given in Appendix 1.

Tracer Conservation

$$\frac{\partial c}{\partial t} + \underset{\sim}{u} \cdot \nabla C = -\nabla D_{eff} \nabla C \tag{9}$$

where

$$D_{eff} = D + D_t \tag{10}$$

---

[*] All symbols are defined in the Nomenclature.

$$\text{and} \quad \frac{\mu_t}{\rho D_t} = 1 \tag{11}$$

Particle Transport

The transport of particles, unlike the tracer, will be inherently steady in a time-averaged statistical sense. An equation similar to [8] is thus employed, but without the transient term. In addition, diffusion is cut off by using a high Schmidt number.

In effect, C_p, the concentration of the particles, is interpreted as the number of particles per unit volume, and is calculated after the flow and turbulence fields have been computed.

Boundary Conditions

A no-slip boundary condition is imposed at the solid walls by using a wall-function approach [10]. Thus, the form of the drag law employed at the first computational node nearest to the surface is:

$$\frac{U}{\sqrt{\tau_s/\rho}} = \frac{1}{\kappa} \ell n \left[\frac{E \ y \ \sqrt{\tau_s/\rho}}{\mu} \right] \tag{12}$$

where τ_s is the surface shear stress.

The free surface is assumed to be stress-free, and the appropriate melt velocities are specified at the inlet. The fluid is assumed to leave the tundish with zero velocity components in the cross-stream directions.

It is assumed that the particles are absorbed at the free surface in accordance with the ideal relationship:

$$q_s = U_T C_{p_s} \tag{13}$$

where q_s and the C_{ps} are the upward particle flux density and the particle concentration at the surface, respectively.

Solution Procedure

The method of solution is that embodied in the PHOENICS computational code [11].

In generating the results, we considered only one-quarter of the tundish, utilizing the planes of symmetry available. A 25 x 11 x 25 non-uniform grid structure was used. This arrangement was selected after testing for the grid sensitivity of the computed results.

The flow control devices and the angled walls were represented by the partial "blocking" of nodes.

Typically, about 200 sweeps were needed for the velocity fields, and a smaller number of sweeps was employed in computation of the concentration fields. Computation required about 4 hours, and about 12 hours of CPU time on a MicroVAX II, respectively, for the velocity and the tracer concentration fields, while the particle concentration took about 30 minutes. The tracer calculations were transient and were carried out up to about 8 minutes of real-time operation.

One should stress the need to check for the internal consistency of the results by ensuring that both the fluid and tracer conservation relationships have been fully satisfied. These issues can become particularly critical when one is blocking off certain nodes. In the present case, the discrepancy was always less than 1%. Finally, the dependence of the particle flux density on the concentration (Equ 12) demands linearization of the source terms in the numerical scheme. The principal input and computational parameters are summarized in Table 1, while Table 2 gives the location of flow control devices for the various systems considered.

Computed Results

In presenting the computed results, we shall concentrate our efforts on tracer dispersion and on the description of particle behavior, since extensive velocity data have been published elsewhere [12]. Reference will, however, be made to the general nature of the flow field for given applications, because it is this very flow field which will govern the overall behavior of the system.

Figs. 6(a) and (b) show selected tracer concentration isopleth maps in the absence of flow control. Under these conditions, there are no baffles or dams, and the incoming stream is restrained only by the bounding walls. The very rapid dispersion of the tracer is readily apparent, and after the passage of 40 seconds, corresponding to $t/t_R \sim 1/3$, the tracer will start appearing at the outlet.

Figs. 7(a) and (b) show the corresponding plots with a dam and weir flow control. It is seen that the tracer is confined during an initial time period, and is retained in the system for a significantly longer time.

Not all inserts or flow control devices are equally helpful, however. Figs. 8(a) and (b) show the behavior of a slotted dam, which is seen to lead to very rapid tracer dispersion.

The results of the tracer data are summarized in Fig. 9, and also in Table 3. This collection of data includes three additional cases. It is of interest to note that significant by-passing appears to occur in all of these cases, and there appear to be major regions which are completely mixed. The plug flow region is seen to be quite small in all the cases considered.

As seen in Table 3, the theoretical predictions are in quite good agreement with measurements.

Let us now turn our attention to the behavior of the inclusion particles. In considering the history of the inclusion particles, we must note that, in contrast to the tracers, we are dealing with steady-state behavior. For a perfect plug flow, one would expect the concentration isopleths to be parallel planes giving progressively smaller inclusion levels as we progress toward the free surface. All these plots show inclusion concentration levels that are normalized with respect to the inlet value.

Figs. 10(a), (b) and (c) show the 0.4, 0.5 and 0.6 isopleths, respectively, for the 100-micron particles in the absence of flow control, and the depletion of these is readily apparent.

Table 2:

LOCATION OF FLOW CONTROL DEVICES [see Fig. 5]

CONFIGURATION	a (mm)	b (mm)	c (mm)	d (mm)	e (mm)
1 No flow control	---	---	---	---	---
2 Flow control with 1 weir, 1 dam	229	762	229	1067	---
3 Flow control with 1 weir, 1 slotted dam	229	762	229	1067	---
4 Flow control with 2 weirs, 1 dam	229	762	229	1067	888
5 Flow control with 1 weir, 1 dam and allowance for wall inclination	229	762	229	1067	---

Table 3:

COMPARISON OF PREDICTED RETENTION TIME WITH DATA

Configuration	Data [3] (sec)	Prediction (sec)
No flow control	20	27
Flow control 1 weir, 1 dam	80	62
Flow control Dam with slots	--	33
Flow control 2 weirs, 1 dam	--	56
Flow control Allowance for slope	--	50

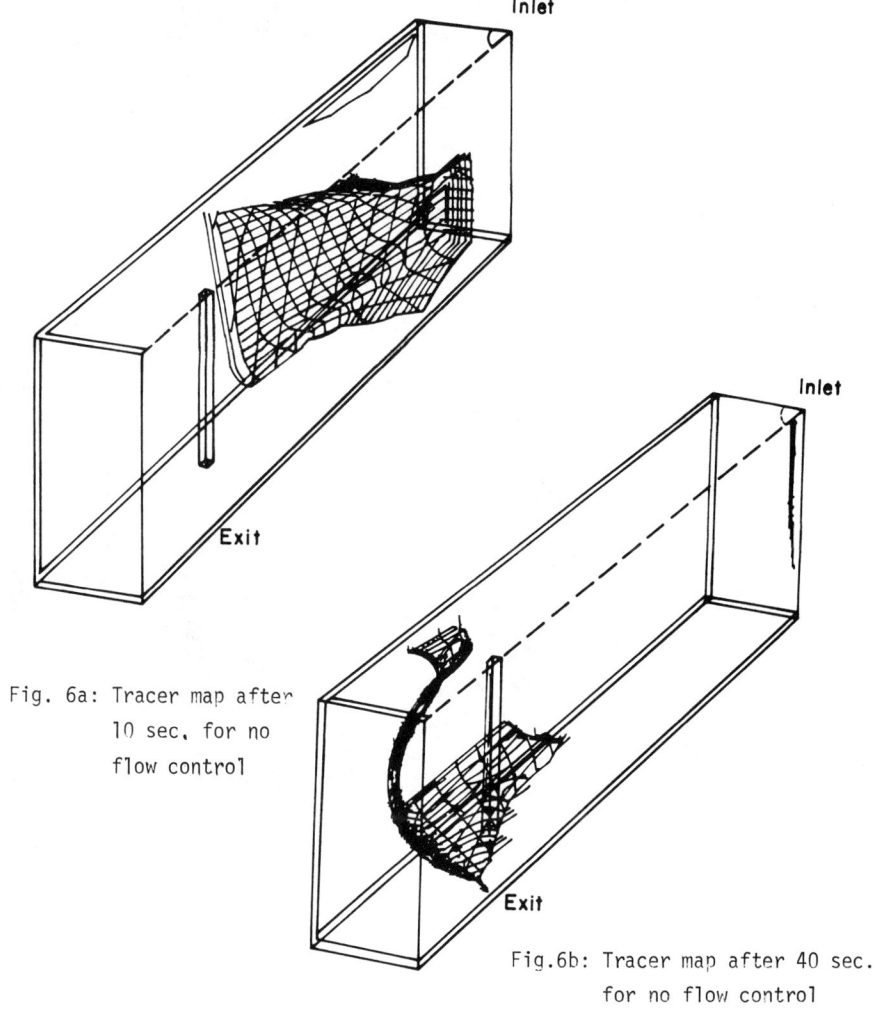

Fig. 6a: Tracer map after
10 sec. for no
flow control

Fig.6b: Tracer map after 40 sec.
for no flow control

421

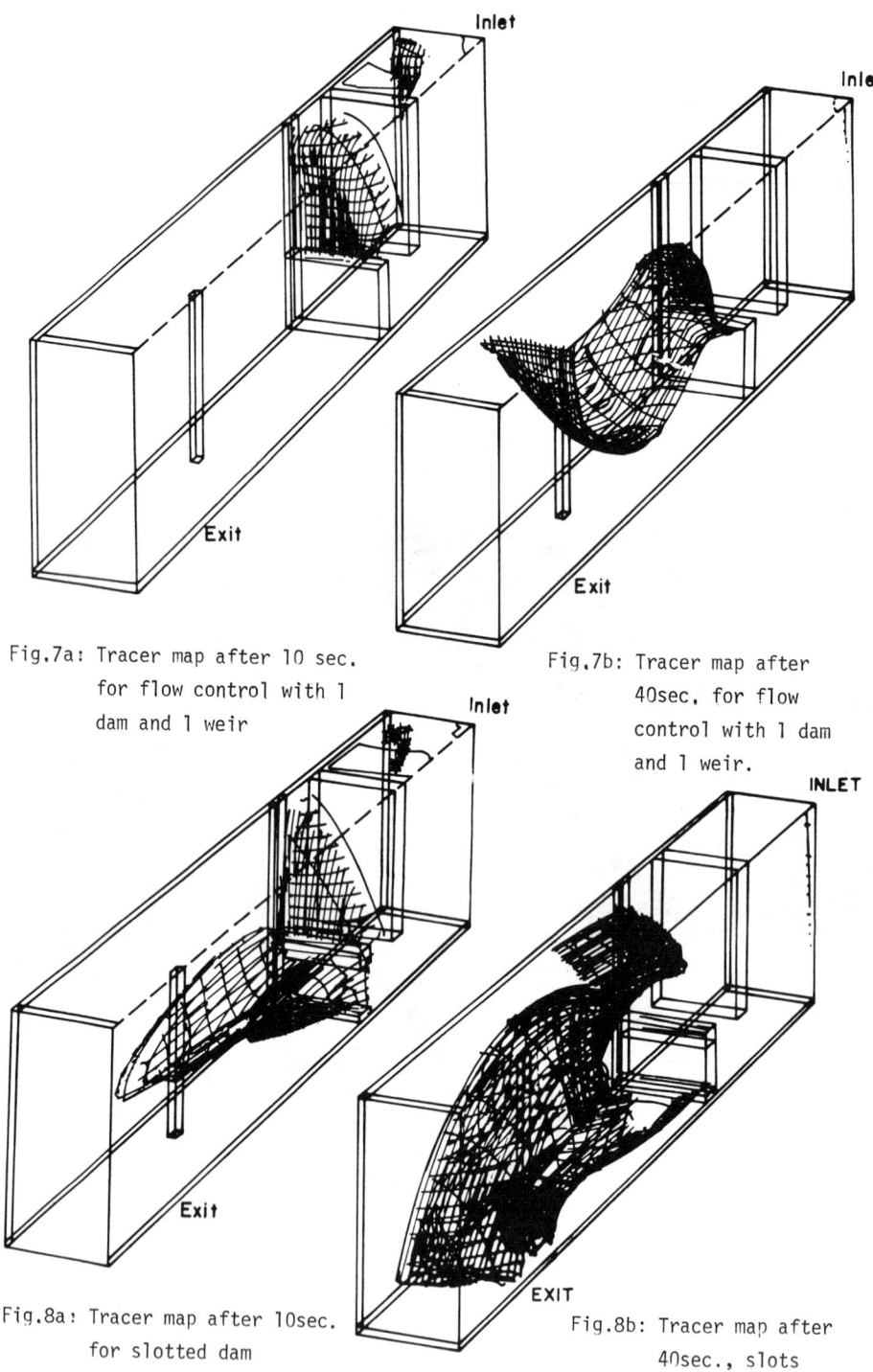

Fig.7a: Tracer map after 10 sec. for flow control with 1 dam and 1 weir

Fig.7b: Tracer map after 40sec. for flow control with 1 dam and 1 weir.

Fig.8a: Tracer map after 10sec. for slotted dam

Fig.8b: Tracer map after 40sec., slots

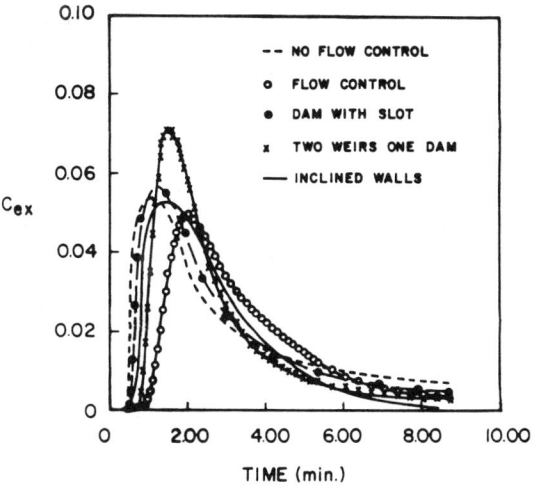

Fig.9: Exit tracer concentration as a function of time

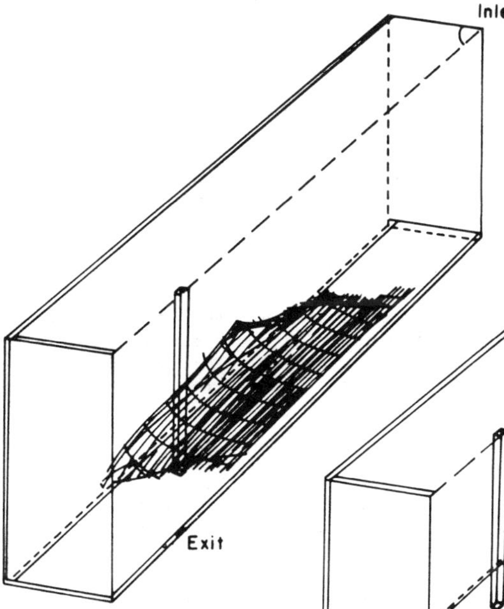

Fig.10a: Conc. map of 0.4 for a 100μm particle for no flow control

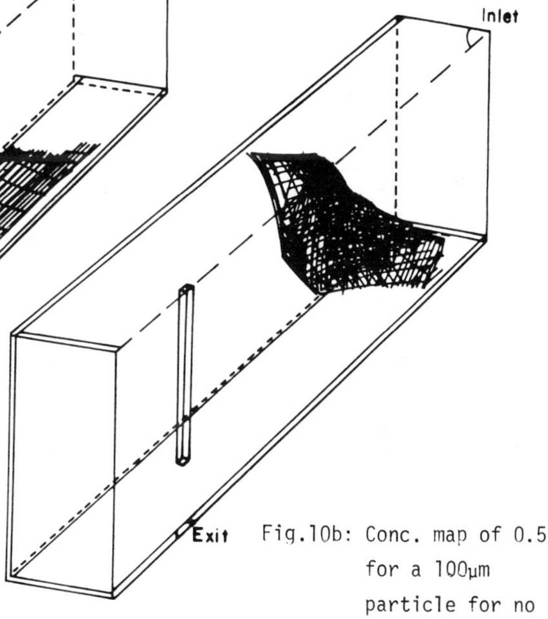

Fig.10b: Conc. map of 0.5 for a 100μm particle for no flow control

The situation is somewhat different in the presence of flow control, as seen in Fig. 11(a), (b) and (c). As shown in 11(a), the concentration isopleth corresponding to C = 0.4 is now at a higher vertical level, indicating a more effective rate of inclusion removal.

Similar results have been obtained for the smaller particles, but these are not reproduced here.

Fig. 12 shows a summary of these computed results on a plot of the percentage inclusion removal, as a function of the particle rising velocity and, hence, particle size for various tundish arrangements. Also given, for the sake of comparison, is the behavior of a completely mixed system and the behavior of a system in plug flow.

This plot is instructive because it shows that had plug flow been attainable, the inclusion removal efficiency would have been increased very markedly. Even complete mixing would provide a better performance than any of the "real" systems considered. The reason for this is simple; in the real systems, a significant fraction of the tundish is a dead, inactive region. Fig. 12 is also helpful because it clearly indicates the effect of changing the internal, i.e. the dam and weir, configurations.

The subtleties involved are further underlined upon further inspection of Fig. 12, which indicates that the absence of flow control arrangements may actually promote the removal of finer particles, when compared to the performance of dams and weirs. The reason for this is simple; in the absence of flow control, there is a larger recirculating flow field, which could bring the finer particles to the top surface.

As a practical matter we should state, however, that the two overriding factors that have to determine the tundish performance as a means for separating the inclusion particles are the following:

(a) The effective tundish volume, as affected by the length and the width, but not the depth of the tundish, in relation to the volumetric flow rate; and

(b) The attainment of a situation such that dead volumes are minimized.

It is of interest to compare the theoretical predictions with measurements; Fig. 13 shows a comparison of the computed results with measurements reported by Tanaka [13], using a water model system. The agreement between predictions and measurements is excellent, providing support for the validity of the modelling approach described here.

Discussion

In the paper we discussed the modelling of fluid flow phenomena, tracer dispersion and the flotation of inclusion particles in tundish systems. We have shown that through the combined application of order-of-magnitude analyses and machine computation, a useful insight may be obtained into the behavior of tundishes, and that a methodology exists for the optimal design of these systems.

The principal conclusions emerging from this work may be summarized as follows:

(1) Due to the inherent nature of these systems, the flow is highly turbulent in the vicinity of the inlet regions, and will become quasi-laminar in the bulk of the tundish. As a result, there will be exten-

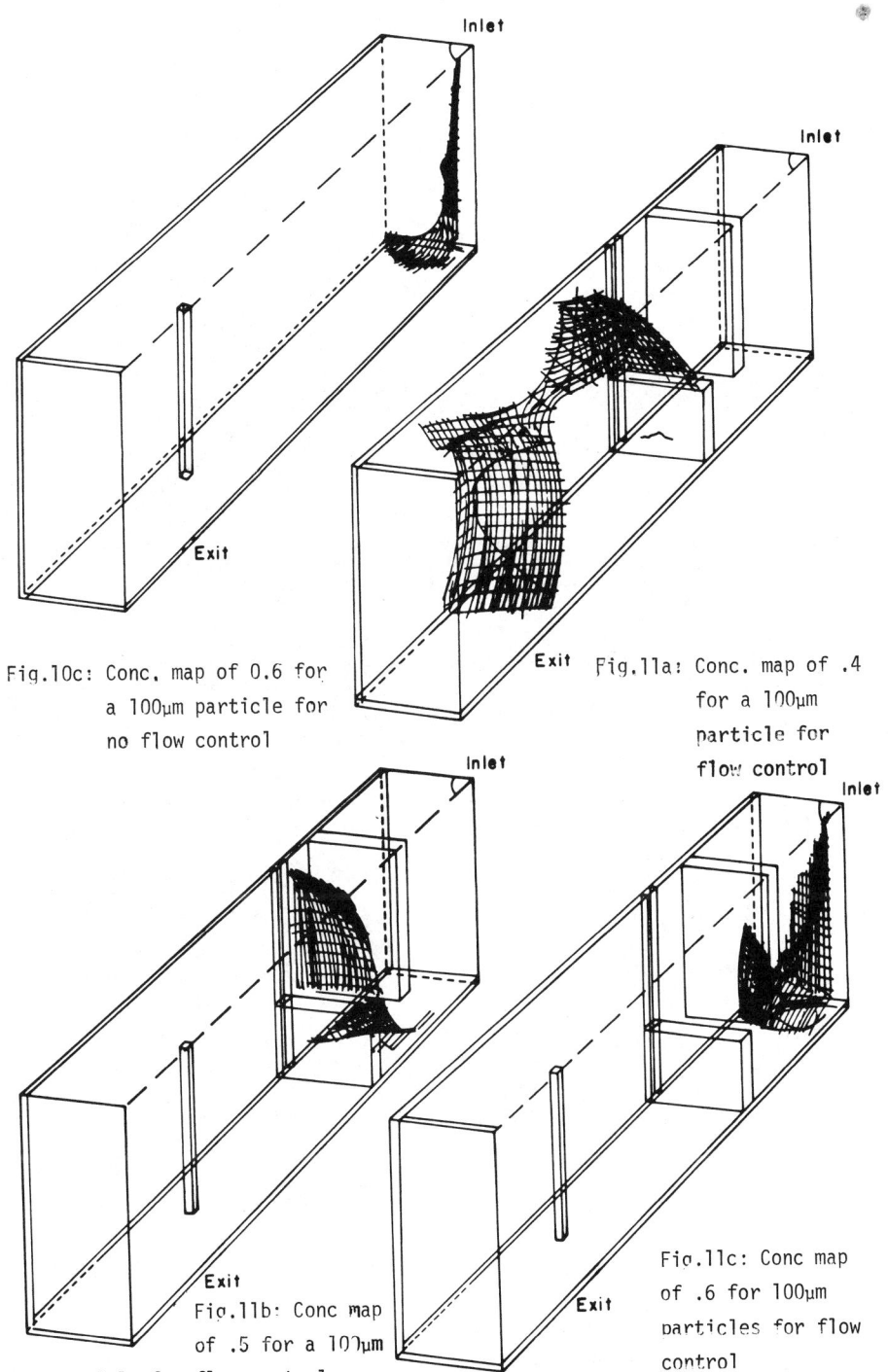

Fig.10c: Conc. map of 0.6 for a 100μm particle for no flow control

Fig.11a: Conc. map of .4 for a 100μm particle for flow control

Fig.11b: Conc map of .5 for a 100μm particle for flow control

Fig.11c: Conc map of .6 for 100μm particles for flow control

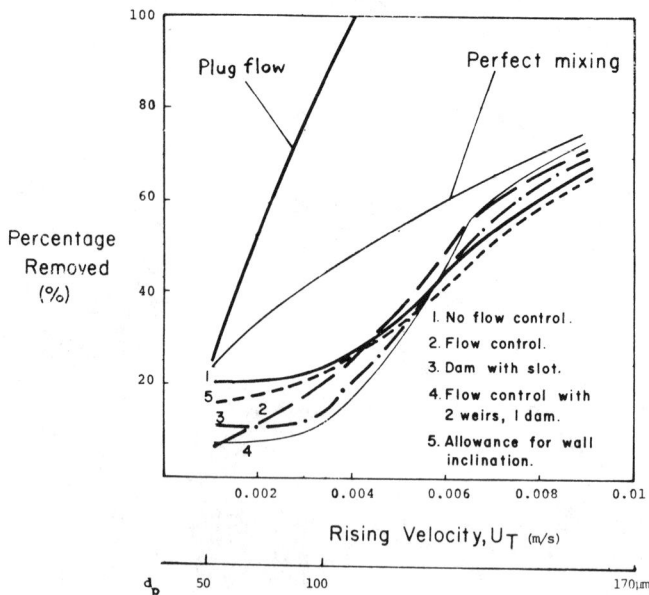

Fig.12: Percentage of inclusion removed as a function of rising velocity

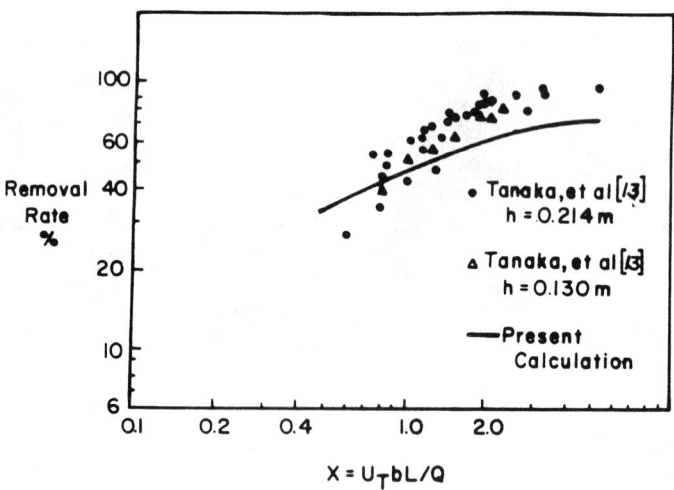

Fig.13: Comparison of predicted removal rate with measurements for
for tundish without flow control

sive mixing in the inlet region, but transport of the tracer and the inclusion particles will occur primarily by convection elsewhere.

(2) Both experimental and theoretical considerations suggest that significant by-passing will occur, and that the establishment of relatively stagnant regions will be difficult to avoid. A good indication of this behavior is the fact that the breakthrough times were 1/5-1/3 of the nominal residence time within the system.

(3) A limited parametric study of the role played by the location of the dams and weirs suggests that while these will play a role in modifying the behavior of the system, the overriding factor appears to be the nominal residence time in the system and the elimination, or at least minimization, of dead regions.

(4) An important new development has been the fact that the behavior of the inclusion particles could be considered explicitly. It has been found that while there is a parallel between tracer behavior and that of the inclusion particles, there are also significant differences, brought about by the finite rising velocity of the particles, particularly of the larger size.

The computational techniques described here allow us to calculate the inclusion removal rate for specific tundish arrangements. Thus, these methods should provide the means for the computer-aided design of tundish installations.

Acknowledgement

The authors wish to thank Dr. Karl-Hermann Tacke, previously of CONCAST A.G., for providing us with an advance copy of reference [7].

References

1. A.T. Hart and C.J. Cripps-Clark, Proc. ICSTIS, sec. 2, p. 69 (1971).

2. W.J. Maddever, A. McLean and G.E. Forward, Canadian Met. Quart. 12, 79 (1973).

3. F. Kemeny, D.J. Harris, A. McLean, T.R. Meadowcroft and J.D. Young, Proc. Second Process Technology Conference, Continuous Casting of Steel (Chicago, Feb. 23-25, 1981), pp. 232-245. Warrendale, PA: Iron and Steel Society-AIME (1981).

4. J. Szekely and N. El-Kaddah, Proc. SCANINJECT IV, Lulea, Sweden (1986).

5. Y. He and Y. Sahai, Proc. 5th International Iron and Steel Congress (Washington, DC, April 2-6, 1986). Warrendale, PA: Iron and Steel Society-AIME (1986).

6. Y. Sahai and R. Ahuja, Ironmaking and Steelmaking 13, 241-247 (1986).

7. K-H. Tacke and J.C. Ludwig, Archiv fur das Eisenhuttenwesen, in press (1987).

8. K. Nakanishi and J. Szekely, Trans. ISIJ 15, 522 (1975).

9. K. Shirabe and J. Szekely, Trans. ISIJ 23, 465-474 (1983).

10. S.V. Patankar and D.B. Spalding, International Journal of Heat and Mass Transfer 15, 1786 (1972).

11. D.B. Spalding, Mathematics and Computers in Simulation, XIII, 267-276 (1981).

12. J. Szekely, O.J. Ilegbusi and N. El-Kaddah, Journal of Physico-Chemical Hydrodynamics, in press (1987).

13. S. Tanaka, M. Lye, M. Salcudean and R.I.L. Guthrie, Proc. 24th Annual Conference of Metallurgists, 15th Annual Hydrometallurgical Meeting, Canadian Institute of Metallurgy (Vancouver, BC, August 18-21, 1985) (1985).

14. B.E. Launder and D.B. Spalding, Mathematical Models of Turbulence. London: Academic Press (1972).

Nomenclature

b	width of tundish
c	tracer concentration
c_p	number of particles per unit volume
D	diffusivity
E	sublayer resistance factor in the law of the wall
f	inclusion removal rate
g	gravitational acceleration
K	specific turbulence energy
ℓ	characteristic length
L	length of half tundish
Q	volumetric flow rate per strand
R_p	radius of particle
t_R	nominal residence time
$\underset{\sim}{u}$	velocity vector
U_T	terminal velocity of particle
V	characteristic velocity
x	normalized longitudinal distance
X	dimensionless parameter $U_T bL/Q$
ρ	density
ρ_f	density of fluid

428

ρ_p	density of particle
ϵ	specific rate of turbulence energy dissipation
κ	von Karman constant
μ	molecular viscosity
μ_{eff}	effective viscosity
μ_t	turbulent viscosity
σ	Prandtl number
τ_s	surface shear stress

Appendix 1: Calculation of the Turbulent Viscosity

The turbulent viscosity, μ_t, is calculated from two characteristic parameters of turbulence, namely the energy K and its rate of dissipation, ϵ, thus:

$$\mu_t = C_\mu \rho K^2 / \epsilon \tag{A1}$$

$$\rho \underset{\sim}{u} \cdot \nabla K = -\nabla \frac{\mu_t}{\sigma_K} \nabla K + S \tag{A2}$$

$$\rho \underset{\sim}{u} \cdot \nabla \epsilon = -\nabla \frac{\mu_t}{\sigma_\epsilon} \nabla \epsilon + S_\epsilon \tag{A3}$$

The source terms S_K and S_ϵ are expressed as:

$$S_K = G_K - \rho \epsilon \tag{A4}$$

and $$S_\epsilon = \frac{\epsilon}{K} (C_1 G_K - C_2 \rho \epsilon) \tag{A5}$$

where G_K is the generation of K, and is given in Cartesian tensor form through the velocity gradients as:

$$G_K = \mu_t \left(\frac{\partial U_i}{\partial x_j} + \frac{\partial U_j}{\partial x_i} \right) \frac{\partial U_i}{\partial x_j} \tag{A6}$$

The values of the constants C_1, C_2 and C_μ are those employed in [14].

COMPUTER SIMULATION OF MELT FLOW CONTROL DUE TO

BAFFLES WITH HOLES IN CONTINUOUS CASTING TUNDISHES

Y. Sahai

Department of Metallurgical Engineering
The Ohio State University
Columbus, Ohio 43210

Abstract

The understanding of the flow of molten steel through continuous casting tundishes is a prerequisite for its optimum designing. In this work, the changes in fluid flow due to insertion of a baffle with holes has been simulated. The effects of horizontal and a combination of horizontal and inclined holes have been studied. The mathematical model involves solution of the three-dimensional, Navier-Stokes' equation. The role of these changes on the residence time distribution of fluid and consequently on the separation of nonmetallic inclusions and cleanliness of steel has been discussed.

I. Introduction

Continuous casting of steel has become an important step in the manufacture of steel in the past couple of decades. The share of continuously cast steel has increased significantly over this period. A tundish is an intermediate vessel between the ladle and the mold which acts as a distributor of steel from the ladle to the continuous casting strands. It is now well known that during melt flow through the tundish the nonmetallic inclusions could be floated out, influencing the quality of the cast steel. Generally, the time available for the separation of nonmetallics in the tundish is very short. The inclusion flotation and hence their removal can be maximized if the melt flow can be so modified as to result in having minimum stagnant volume and optimum circulation conditions in the tundish. For this purpose, flow control devices such as dams, weirs, baffles, etc., are used. The effectiveness of such devices has been studied by physical (1-10) and/or mathematical (11-14) modeling. Most major steel companies study the changes in fluid flow due to flow control devices by physical modeling in large plexiglas models using water as working fluid. Flow visualization with dye or other tracers and residence time distribution studies in such physical models have provided very useful information. However, the detailed information of velocity and turbulence field during fluid flow is difficult to obtain in water modeling experiments.

In recent years mathematical modeling has emerged as a tool to simulate melt flow in tundishes. It is now possible to simulate three-dimensional fluid flow in any shape of a tundish with flow control devices and obtain detailed information of the turbulent fluid flow. With the increasing power of computing, it is becoming easier to assess the effectiveness of any flow control device or of their combinations and develop an optimum design of a tundish.

Baffle with holes is one of the physical flow control devices which is placed to cover the entire cross section of a tundish to affect the flow of molten steel. These baffles are finding increasing application in tundishes for producing quality steel. In this work, the changes in fluid flow due to insertion of a baffle with holes has been simulated. The effects of horizontal and a combination of horizontal and inclined holes have been studied. The role of these changes on the residence time distribution and consequently on the separation of nonmetallic inclusions and cleanliness of steel has been assessed.

II. Tundish and Baffle Design

A straight trough type, two strand tundish as shown in Fig. 1, is used in these calculations. The important parameters of the tundish are given in Table I. Two baffles used in the simulation are shown in Figs. 2(a) and (b).

Table I. Important Parameters of Tundish

Tundish Length (half length) at the Bottom	4.10 m
Tundish Width at the Bottom	0.60 m
Depth of Melt	1.2 m
Angle on Tundish Walls (to the vertical)	12.5°
Number of Strands	2
Distance of the Baffle from the Axis of Jet	1.34 m
Velocity of Incoming Jet from Ladle to Tundish	1.92 ms^{-1}

Baffle (a) has four rows of horizontal holes, whereas baffle (b) has upper two rows of horizontal holes and lower two rows of holes inclined at $45°$. Fluid flow has been simulated with no baffle, with baffle (a) or with baffle (b).

III. Mathematical Model

Referring to Fig. 3, the governing equations representing the mathematical model are:

Equation of Continuity

$$\frac{\partial}{\partial x_i} (\rho u_i) = 0 \qquad [1]$$

Momentum Balance Equations

$$\frac{\partial(\rho u_i u_j)}{\partial x_j} = -\frac{\partial p}{\partial x_i} + \frac{\partial}{\partial x_j}\left[\mu_{eff}\left(\frac{\partial u_i}{\partial x_j} + \frac{\partial u_j}{\partial x_i}\right)\right] \qquad [2]$$

A. Turbulence Model

For modeling turbulence, the K-ε two-equation model of Launder and Spalding (15) was used. The governing transport equations for turbulent kinetic energy K and its dissipation rate ε can be represented as:

Turbulent Kinetic Energy

$$\frac{\partial}{\partial x_i}\left(\rho u_i K - \frac{\mu_{eff}}{\sigma_k}\frac{\partial K}{\partial x_i}\right) = G - \rho\varepsilon \qquad [3]$$

Dissipation Rate of Turbulence Energy

$$\frac{\partial}{\partial x_i}\left(\rho u_i \varepsilon - \frac{\mu_{eff}}{\sigma_\varepsilon}\frac{\partial \varepsilon}{\partial x_i}\right) = (C_1\varepsilon G - C_2\rho\varepsilon^2)/K \qquad [4]$$

where

$$G = \mu_t \frac{\partial u_j}{\partial x_i}\left(\frac{\partial u_i}{\partial x_j} + \frac{\partial u_j}{\partial x_i}\right) \qquad [5]$$

$$\text{Effective viscosity, } \mu_{eff} = \mu_1 + \mu_t \qquad [6]$$

$$\mu_t = C_D\rho K^2/\varepsilon \qquad [7]$$

Following the recommendations of Launder and Spalding (16), the five constants appearing in Eqs. [3] to [7] take the value given in Table II.

Table II. Values of Constants in K-ε Turbulence Model

C_1	C_2	C_D	σ_K	σ_ε
1.43	1.92	0.09	1.00	1.30

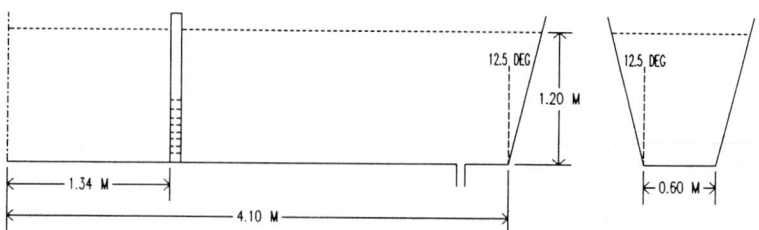

Figure 1 - Tundish with a baffle and its dimensions.

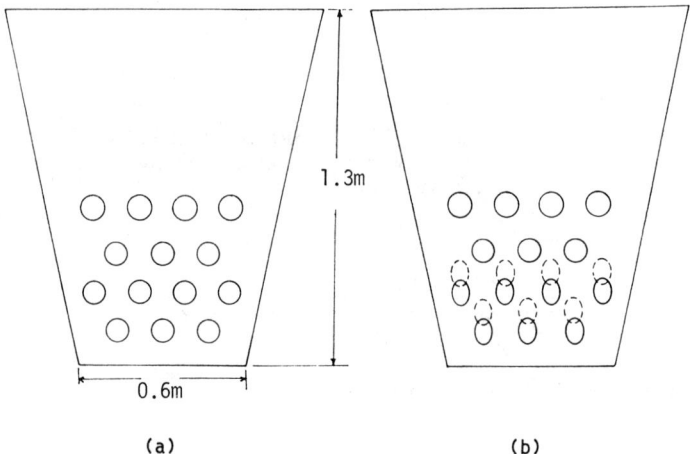

(a)   (b)

Figure 2 - Baffle with (a) horizontal holes, (b) mixed holes.

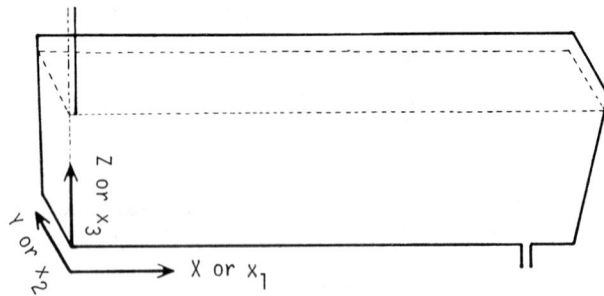

Figure 3 - The coordinate system with respect to the tundish.

B. Boundary Conditions

Close to solid walls, including the baffle, the variation in flow properties are much steeper than within the bulk fluid, and levels of turbulence Reynolds number are sufficiently low for molecular viscosity to influence the production, dissipation, and diffusion of K and ε , and local anisotropy no longer prevails. Consequently, the momentum (u_1, u_2, u_3) and scalar (K, ε) transport properties are modeled using wall functions (15). Also, no slip boundary conditions were imposed at the solid walls.

At the free surface, which was considered flat, and at the symmetry planes the normal velocity components and the normal gradients of all other variables (momentum and scalar transport properties) were used as zero. At the jet entry, the velocity perpendicular to the free surface was calculated from the volumetric flow rate and the area of nozzle, and a flat velocity profile was assumed. This simulates submerged jet with minimal submersion depth to avoid any air entrainment. Similar boundary conditions were imposed at the outlet nozzle.

C. Numerical Solution Procedure

Finite difference equations were derived from Eqs. [1] to [4], using an implicit finite difference procedure referred to as SIMPLE (Semi-Implicit Method for Pressure Linked Equations). The computer code for three-dimensional geometry was developed based on the TEACH-2E code of Gosman et al. (17).

The flow domain corresponding to a symmetrical quarter of the tundish was considered for the computations. The domain was divided into a nonuniform grid of 30 (longitudinal) x 10 (vertical) x 12 (transverse) for no baffle case and 30 (longitudinal) x 12 (vertical) x 15 (transverse) for flow with a baffle.

D. Tracer Dispersion

The dispersion of tracer introduced into the melt flowing through tundish may be described by the mass transport equation:

$$\frac{\partial}{\partial t}(\rho C) + \frac{\partial}{\partial x_i}(\rho u_i C) = \frac{\partial}{\partial x_i}\left(\rho D_e \frac{\partial C}{\partial x_i}\right) \qquad [8]$$

This is a time dependent equation in which C represents the concentration of tracer and D_e is the effective mass-diffusion coefficient, which would depend upon the fluid flow field. D_e is related to effective viscosity, μ_{eff}, in the following manner:

$$\frac{\mu_{eff}}{\rho D_e} \sim 1 \qquad [9]$$

The boundary conditions required for the solution of Eq. [8] have to express the physical constraints that all the bounding surfaces are impervious to the tracer. Mathematically, this corresponds to zero flux at all the bounding surfaces.

IV. Results and Discussion

A. No Flow Control

The predicted velocity field is shown in Figs. 4, 5 and 6. Figure 4 shows flow in selected longitudinal, vertical planes (X-Z planes). Figure 4(a) is near the middle of the tundish showing the entering metal stream and the exit to mold. Figs. 4(b), (c) and (d) show flow fields in other parallel planes moving towards the inclined wall. Figure 4(d) is near the inclined wall. Since the wall is inclined, the lower part of the vertical plane lies outside the tundish and consequently flow is not shown in this region. Figure 5 shows flow fields in the longitudinal, horizontal planes (X-Y planes) at different elevation. Figure 5(a) shows flow near the free surface whereas Figs. 5(b), (c), (d) and (e) show flow fields at increasing depths in the tundish. Figure 5(e) is near the bottom, where exit to the mold is shown. Finally, Fig. 6 shows flow in transverse, vertical planes (Y-Z planes). Figure 6(a) is the predicted flow near the axis of symmetry, where jet enters the tundish and Figs. 6(b), (c), (d), (e) and (f) are flows in transverse, vertical planes at increasing distances towards the exit to mold. Figure 6(f) is vertical plane through the exit to mold. Although the computations were performed only for a symmetrical quarter of the tundish, Figs. 5 and 6 show flow on both sides of the jet for better visual representation.

From Figs. 4 and 5, it can be seen that the entering metal jet penetrates to the bottom of the tundish where it flows downstream and towards the long walls of the tundish (see Fig. 5(e)). Most of the metal then flows up towards the free surface. Part of this ascending metal moves downstream in the direction of the exit and some liquid recirculates back towards the incoming jet (see Figs. 5(a), 4(c) and 4(d)). The incoming jet entrains fluid around it; in fact, Fig. 6(a) shows the entire plane has downward flow. Figure 6(b) shows a recirculation near the bottom, whereas further downstream metal flows up towards free surface (Fig. 6(c)). In Fig. 6(e) flow is predominantly towards the bottom and, finally, Fig. 6(f) shows flow near the exit to mold and a recirculation in the upper half of the section. There exists a recirculation near the end wall, as evident in Figs. 4(b), (c) and (d). It is clear from Fig. 6 that the overall velocities drop significantly in the transverse vertical planes as the distance from the incoming jet increases and one moves towards the exit nozzle. Incoming jet velocity is 1.92 ms^{-1} and, in general, the overall velocities in the tundish varies from 2 to about 10 cm per second.

B. Baffle With Horizontal Holes

Baffles with holes are industrially employed in tundishes to obtain cleaner steel. These baffles are inserted in the cross section of the tundish and the metal flows through the holes of the baffle. In this simulation, one baffle was considered at 1.34 m from the axis of incoming jet (see Fig. 1). The baffle, as shown in Fig. 2(a), has four rows of horizontal holes with a total of 14 holes. The presence of this baffle limits the turbulence created due to the plunging jet, to the upstream side of the baffle. This baffle also prevents the flow of any ladle slag to downstream side, thus avoiding entrainment of slag inclusions to cast steel. It may be noted that the holes are only in the lower half of the cross section. This prevents any strong flow currents near the free surface of the tundish and avoids breaking of the top layer of the tundish powder/flux, thus minimizing any chances of the reoxidation of metal. The second reason for the holes in the lower half is to avoid any transfer of ladle slag to the downstream side during ladle changes when the level of metal is lowered.

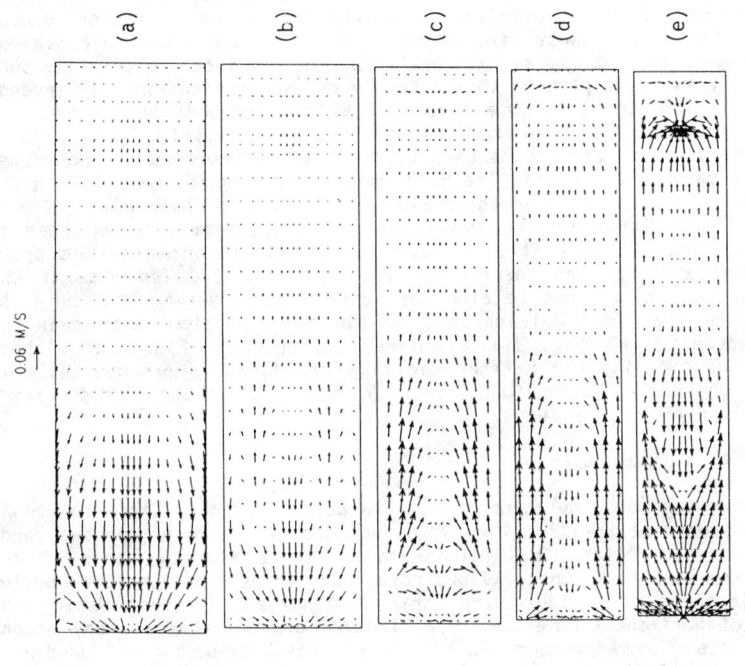

Figure 5 - Predicted velocity field in longitudinal, horizontal planes at different elevations with no baffle (a) near the free surface and (e) near the bottom.

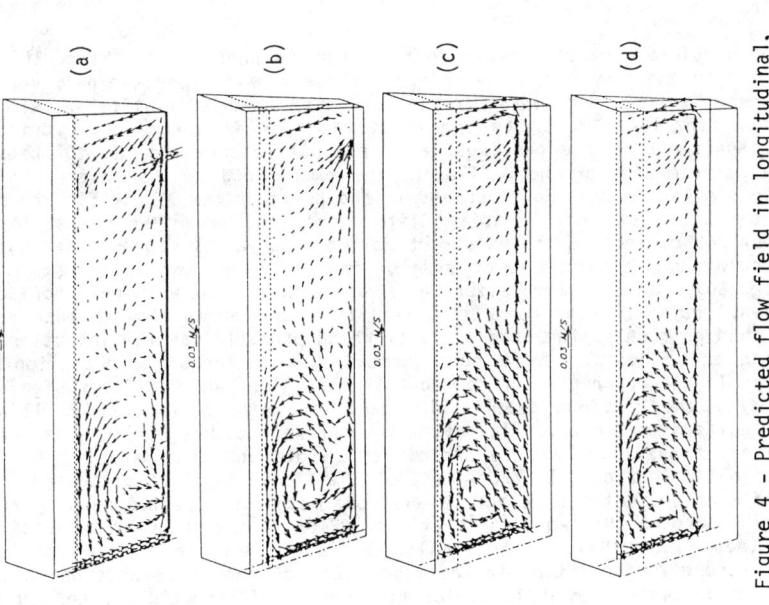

Figure 4 - Predicted flow field in longitudinal, vertical planes with no baffle in the tundish (a) near the plane of symmetry and (d) closer to the inclined wall.

The predicted velocity field is shown in Figs. 7, 8 and 9. As in the previous cases, Fig. 7 shows flow in selected longitudinal, vertical planes. Figure 8 shows flow in longitudinal, horizontal planes at different elevations and Fig. 9 represents flow of metal in transverse, vertical planes. The presence of this baffle significantly changes the flow field in the tundish. The incoming jet plunges to the bottom of the tundish where it spreads radially. The jet entrains some liquid from its surroundings which causes formation of recirculation on the upstream side of the baffle. Thus, part of the incoming liquid recirculates back to the jet and the rest of the molten metal flows through the holes. The holes seem to streamline the flow in the downstream region, as can be seen in Fig. 8. The flow in this region is more like plug flow. After exiting the holes, the downstream flow enlarges in the entire cross section, thus creating a recirculation behind the upper half of the baffle on the downstream side (see Fig. 7). The flow in the transverse, vertical planes is also very different. A strong recirculating vortex can be seen in the lower part of the cross section, even near the plane of incoming jet (Fig. 9(a)), unlike flow shown in Fig. 6(a) without baffle. After crossing the baffle, the flow of liquid is predominantly in the X or longitudinal direction. Velocity components in the Y or Z directions are significantly reduced.

C. Baffle With Mixed Holes

The next case of simulation included a baffle shown in Fig. 2(b). This baffle has lower two rows inclined at an angle of 45 degrees, whereas upper two rows are horizontal. The approximate positions and the sizes of the holes are the same as in the previous case. Many industrial tundishes employ such combinations of inclined and horizontal holes. In this design the upper set of horizontal holes are not directed upwards to avoid any strong flow currents near the free surface. The potential benefit considered with the lower set of 45 degrees inclined holes is in directing more fluid towards the free surface, thus providing a better chance of nonmetallics flotation.

The predicted velocity field in this case is presented in Figs. 10, 11 and 12. As in the previous cases, Fig. 10 shows flow in longitudinal, vertical planes, Fig. 11 represents flow in horizontal planes at different elevation and, finally, Fig. 12 is the predicted flow in transverse, vertical planes. Again, as in the previous case, the baffle contains the turbulence to the upstream side of the baffle and the flow field in the upstream side is quite similar to the one discussed in the previous case. The flow on the downstream side, however, is quite different in the later two cases. Since the lower set of inclined holes direct molten metal at 45 degrees, it causes a recirculation behind the lower side of the baffle on the downstream side (see Fig. 10). This recirculation was almost nonexistent with only horizontal holes. Further downstream, this recirculation causes flow to bend down rather than go towards the free surface, which probably was the intention of inclining these holes. As a consequence, a very large recirculation is developed in the upper part of the downstream region, which differs significantly from the previous case. Flow near the free surface, Fig. 11(a), shows a uniform movement of fluid from the end wall to the baffle. At lower elevations, Figs. 11(c) and (d), flow is uniform and predominantly in the longitudinal direction. In fact, most of the liquid is short-circuited from the baffle holes to the exit nozzle and some metal is directed back towards the baffle through the two recirculations. Flow in the transverse, vertical planes (see Fig. 12) is quite similar to that described in the previous case; a strong recirculation in the lower part of the cross section on the upstream side and extremely low velocities in Y or Z directions after liquid passes through the holes.

438

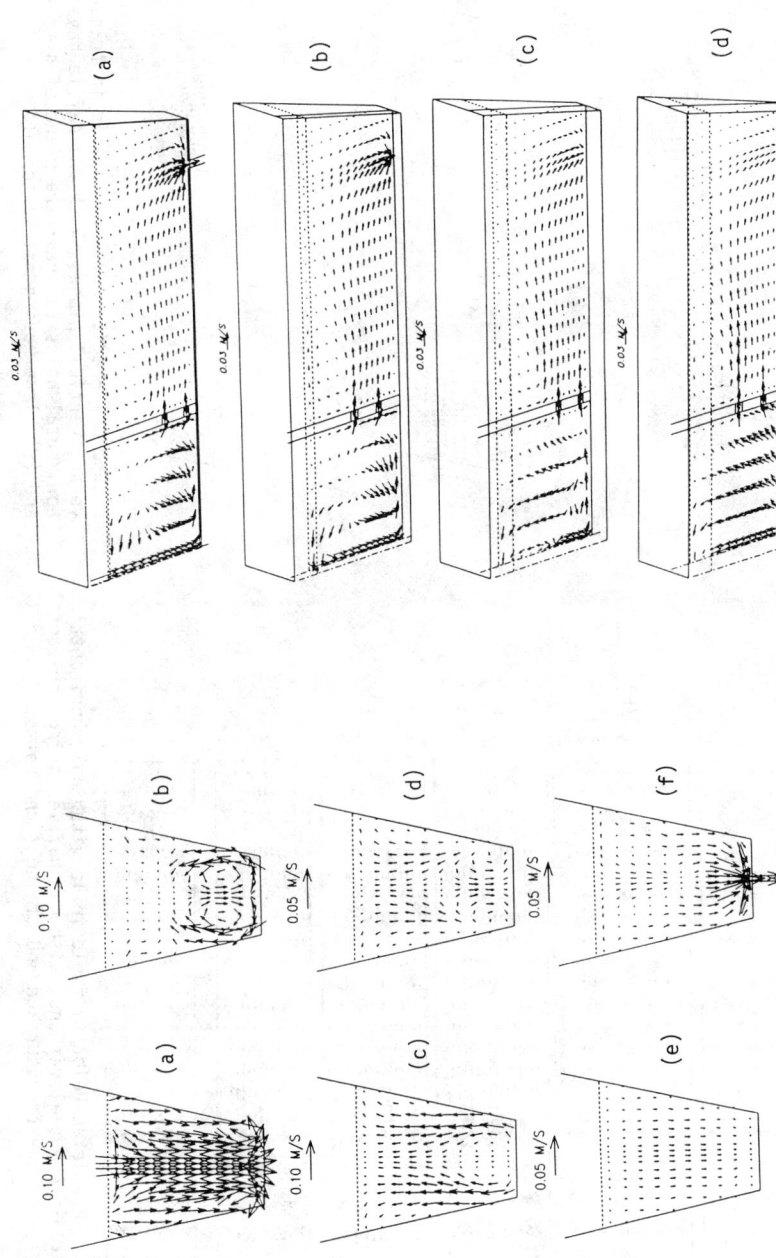

Figure 7 - Predicted flow field in longitudinal, vertical planes, with one baffle having horizontal holes, in the tundish (a) near the plane of symmetry and (d) closer to the inclined wall.

Figure 6 - Predicted velocity field in transverse, vertical planes with no baffle (a) near the plane of symmetry and (f) plane having nozzle to mold.

439

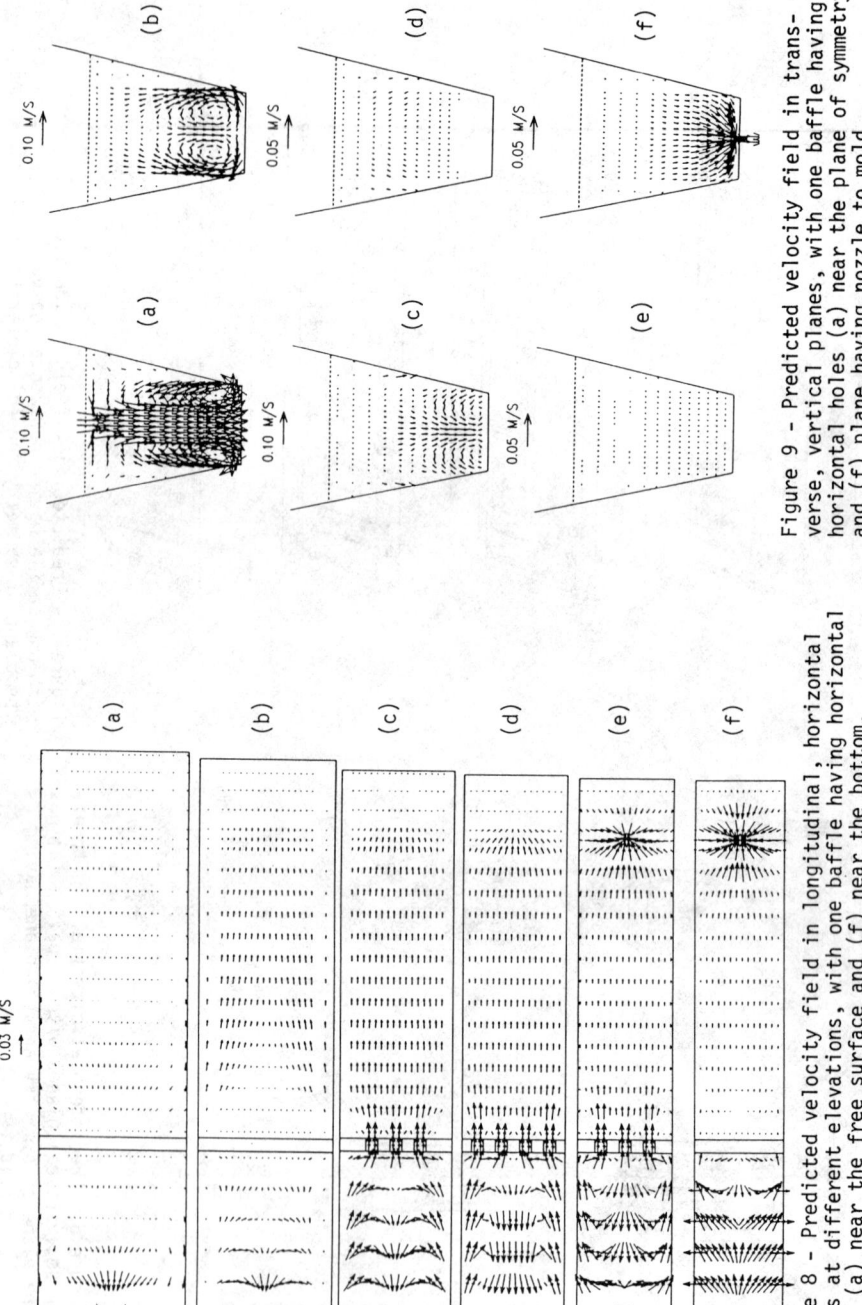

Figure 9 - Predicted velocity field in transverse, vertical planes, with one baffle having horizontal holes (a) near the plane of symmetry and (f) plane having nozzle to mold.

Figure 8 - Predicted velocity field in longitudinal, horizontal planes at different elevations, with one baffle having horizontal holes (a) near the free surface and (f) near the bottom.

0.03 M/S

0.10 M/S

0.05 M/S

440

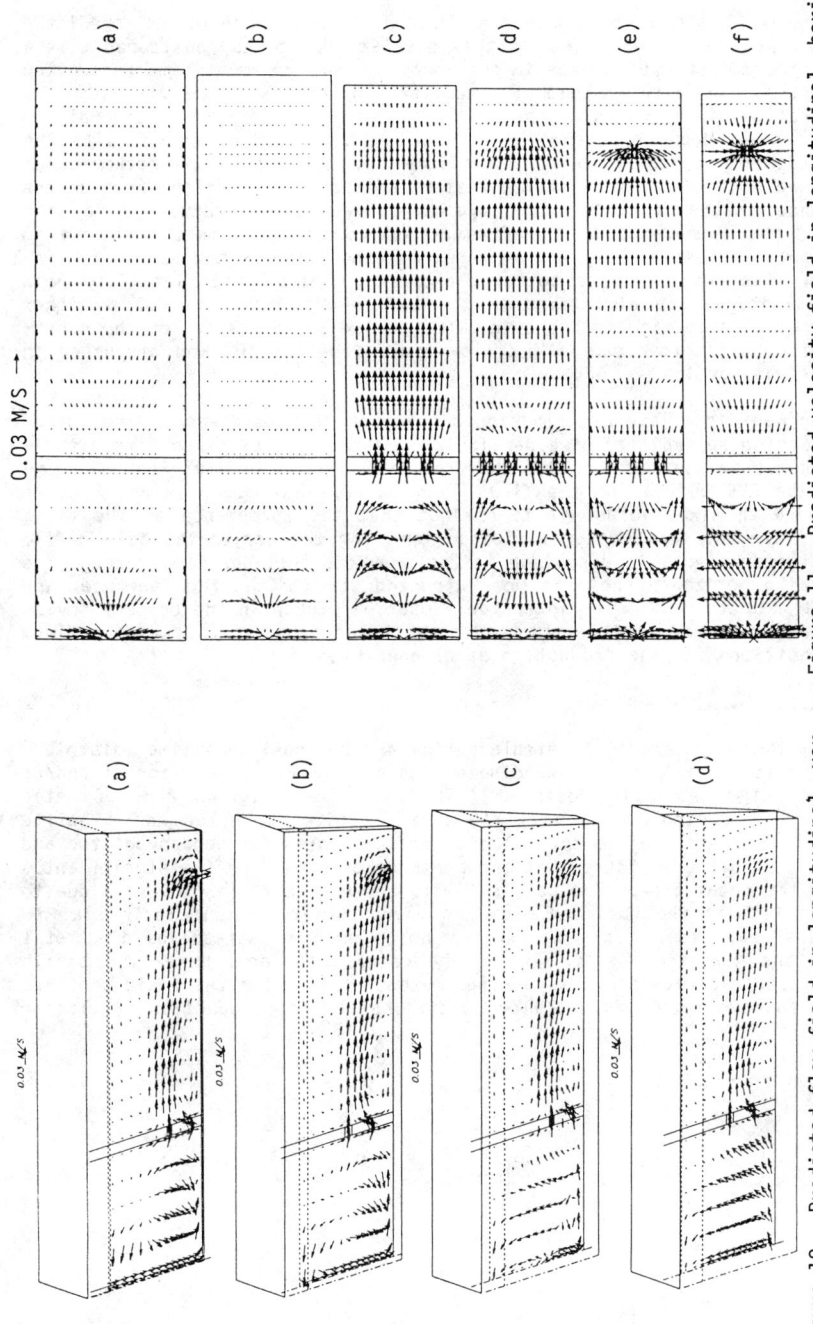

0.03 M/S →

(a)
(b)
(c)
(d)
(e)
(f)

Figure 11 - Predicted velocity field in longitudinal, horizontal planes at different elevations, with one baffle having mixed holes (a) near the free surface and (f) near the bottom.

(a)
(b)
(c)
(d)

Figure 10 - Predicted flow field in longitudinal, vertical planes, with one baffle having mixed (horizontal and inclined) holes, in the tundish (a) near the plane of symmetry and (d) closer to the inclined wall.

441

Tracer Dispersion and Metal Cleanliness

Figure 13 shows the Residence Time Distribution (RTD) as predicted using Eq. [8] for three cases. In this simulation, it was considered that a pulse injection of dye is made in the incoming jet stream at time zero. The dye disperses and flows through the tundish. In this plot, the dimensionless concentration of dye reaching the exit nozzle is plotted against the dimensionless time. Upon comparing the curves for the no flow control with the baffle with horizontal holes, a significant improvement in the retention of dye can be noticed. With baffle, the start of the concentration curve or the appearance of the first trace of dye is delayed significantly. Also, the time for the appearance of the maximum concentration is more than the no flow control case. Thus, the average residence time of the metal is increased, which will provide better chances of nonmetallics flotation. Both these advantages are diminished in the baffle with mixed holes. A higher concentration appearing in a shorter time period is indicative of the short-circuiting, which is clear in its flow field (see Fig 10) and discussed in the previous Section.

The Residence Time Distribution should not be considered in isolation. For assessing suitability of a particular design, the residence time distribution curve in conjunction with the flow field should be analyzed. On comparing the RTD curves for the flows with baffles having horizontal holes only and with mixed holes, it is obvious that the appearance of the first trace of dye and the average residence time for the horizontal hole baffle are at larger time. As discussed in the previous section, flow through the baffle with horizontal hole is more directed towards the free surface, and may have a better chance of inclusion flotation. Thus, in the present investigation the use of baffle with horizontal holes appears to provide optimum flow conditions for the production of cleaner steel.

Summary and Conclusions

A mathematical model to simulate flow and the residence time distribution of molten metal through tundishes with baffles having horizontal and/or inclined holes has been presented. In this investigation, flow of metal with no flow control, with a baffle having horizontal holes and a baffle having a combination of horizontal and vertical holes has been predicted and analyzed. It has been found that the presence of such a baffle significantly changes the metal flow. The baffle limits the high turbulence levels due to plunging jets to the upstream side and prevents flow of any ladle slag to the downstream side. In general, the holes seem to streamline the metal flow in the downstream side. Among the cases considered here, the baffle with horizontal holes provides maximum retention time for the fluid and thus appears to have optimum fluid flow conditions for the production of cleaner steel.

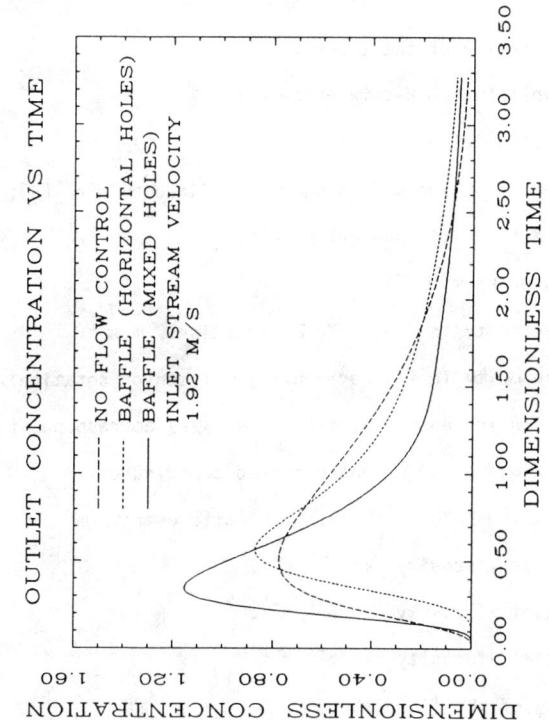

OUTLET CONCENTRATION VS TIME

- --- NO FLOW CONTROL
- ······ BAFFLE (HORIZONTAL HOLES)
- ── BAFFLE (MIXED HOLES)
- INLET STREAM VELOCITY
  1.92 M/S

Figure 13 - Theoretically predicted residence-time distribution of molten steel flowing through the tundish with no flow control and with a baffle having horizontal or a combination of horizontal and inclined (mixed) holes.

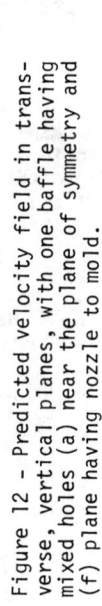

Figure 12 - Predicted velocity field in transverse, vertical planes, with one baffle having mixed holes (a) near the plane of symmetry and (f) plane having nozzle to mold.

443

List of Symbols

C   Concentration of the tracer.

C_1, C_2, C_D Constants in the K-ε turbulence model.

D_e Eddy diffusivity, $m^2 \ s^{-1}$.

G   Generation of turbulence energy, defined by eqn. [5].

K   Turbulent kinetic energy, $m^2 \ s^{-2}$.

p   Pressure, $kg \ m^{-1} \ s^{-2}$.

u_i Mean velocity in i = X, Y, Z directions, $m \ s^{-1}$.

u_j Mean velocity in j direction, (j = i tensor notation), $m \ s^{-1}$.

X_i Direction in tensor notation, i = 1,2,3 corresponds to X,Y,Z.

X,Y,Z   Length, width and height directed coordinates.

ε Dissipation rate of turbulent kinetic energy, $m^2 \ s^{-3}$.

μ_l Molecular viscosity, $kg \ m^{-1} \ s^{-1}$.

μ_t Turbulent viscosity, $kg \ m^{-1} \ s^{-1}$.

μ_{eff} Effective viscosity, $kg \ m^{-1} \ s^{-1}$.

ρ Density of the fluid, $kg \ m^3$.

$\sigma_K, \sigma_\varepsilon$ Schmidt number for K and ε .

Acknowledgements

The work has been supported by the National Science Foundation, Grant No. MSM–8602523 and by the Magneco/Metrel, Inc. Magneco/Metrel, Inc. also provided the baffle designs. The author gratefully acknowledges help from Professor Youduo He in computational work and from David Blevins in plotting all the figures.

References

1. M. Hashio, M. Tokuda, M. Kawasaki and T. Watanabe: 2nd Process Technology Conference, Chicago, ISS-AIME, 1981, pp. 65-73.

2. L. Heaslip, A. McLean, J. Mitchell, A. Hohulin, N. Woost and J. Buchannan: 2nd Process Technology Conference, Chicago, ISS-AIME, 1981, pp. 54-64.

3. G. L. Dressel, D. R. Shrader and D. A. Dukelow: Steelmaking Proceedings, ISS-AIME, Vol. 66, Atlanta, 1983, pp. 205-210.

4. F. Kemeny, D. J. Harris, A. McLean, T. R. Meadowcroft and J. D. Young: 2nd Process Technology Conference, Chicago, ISS-AIME, 1981, pp.232-245.

5. N. A. McPherson and S. Henderson: Ironmaking and Steelmaking, Vol. 10, No. 6, 1983, pp.259-268.

6. E. M. Rehlaender: M.A.Sc. Thesis, Univ. of Toronto, 1983.

7. K. Hamagami, K. Sovimachi, M. Kuga, T. Koshikawa and M. Saigusa: Steel-making Proceedings, ISS-AIME, Vol. 65, Pittsburgh, 1982, pp.358-364.

8. A. Yamagami, M. Tate, A. Masui and H. Uchibori: Steelmaking Proceedings, ISS-AIME, Vol. 64, Toronto, 1981, pp. 234-244.

9. Y. Nuri, K. Umezawa, T. Ohashi, R. Itoh, R. Mizoguchi and Shin-Ichi Yokoi: 106th ISIJ Meeting, Oct. 1983, L. No. S989, Abstracted in Trans. ISIJ, Vol.24, No. 1, 1984, p.B16.

10. M. Kitamura: Ironmaking and Steelmaking, Vol. 10, No.2, 1983, pp.82-90.

11. T. DebRoy and J. A. Sychterz: Metal. Trans. B, Vol. 16B, 1985.

12. N. El-Kaddah and J. Szekely: paper presented at the 114th AIME Annual Technical Meeting at New York, Feb. 1985. Abstracted in J. of Metals, Dec. 1984, p. 15.

13. S. Tanaka, M. Lye, M. Salcudean and R.I.L. Guthrie: Proceedings of the Int. Sym. on the Continuous Casting of Steel Billets, Vancouver, 1985, p. 142.

14. Y. He and Y. Sahai: Metal. Trans., Vol. 18B, 1987.

15. B. E. Launder and D. B. Spalding: Computer Methods in Applied Mechanics and Engineering, 1974, Vol. 3, p. 269.

16. B. E. Launder and D. B. Spalding: Heat Transfer Section Report No. HTS/73/2, Imperial College, London, 1973.

17. A. D. Gosman, B. E. Launder: TEACH-2E, Internal Report, Mech. Eng. Dept., Imperial College, University of London, 1976.

445

PHYSICAL AND MATHEMATICAL MODELS FOR LADLE

METALLURGY OPERATIONS

R.I.L. Guthrie

Macdonald Professor of Metallurgy
Department of Mining and Metallurgical Engineering,
F.D.A. Bldg., McGill University,
3450 University St., Montreal H3A 2A7
Quebec, Canada

Abstract

A series of processing steps carried out in ladle metallurgy operations is described, in order to illustrate the way in which mathematical models can be used to describe liquid flows and attendant heat and mass transfer phenomena through solution of turbulent two, and three, dimensional forms of the turbulent Navier-Stokes equation. Such solutions then allow one to show how the melting, and mixing in, of ladle alloy additions can be treated when they are made during, or following, furnace tapping operations.

It is shown how mixing rates of these additions can be a function of bubbler position, the number of porous plug elements, and/or their disposition. The value of physical models to verify the numerical solutions is emphasized, particularly in regard to the proper prescription for turbulent viscosities used in these mathematical models.

Introduction

The purpose of this paper is to review some recent work by the author, and co-workers, in the modelling of ladle metallurgy processing operations (2, 4-23). While the physical modelling of metallurgical processing operations dates back to at least the early 1960's (e.g. Ref. 1), mathematical models have only recently come of age (i.e. mid 1970's). This progress has been largely achieved through advances in the computing power and speed of digital computers which are now available at moderate cost. Concurrent with these advances have been the numerical algorithms, and special computational procedures, needed to solve transient, three dimensional forms of the turbulent Navier Stokes, (or Reynolds Stress) equation and their equivalent heat and mass counterparts. A summary of "Recent Advances in the Hydrodynamics of Metallurgical Processing" up to 1982 is presented by Y.Sahai and this author in Volume IV of Advances in Transport Processes (2).

The present paper is limited to a discussion of metallurgical processing operations carried out in ladles. These begin with tapping steel from an electric, or basic oxygen, steelmaking furnace (B.O.F.), into the type of steelworks ladle shown in Figure 1. During the course of this tapping, alloy additions in lump form can be fed through chutes set above the filling ladle, so that they fall into the churning bath of molten steel. The additions melt and/or dissolve into the steel bath, so that the bath chemistry can hopefully be adjusted to that specified for the grade of steel being produced. Towards the end of tapping operations, some steelmaking slag lying on top of the raw steel within the furnace, is almost inevitably carried over into the ladle. As this slag is oxidising, any steel deoxidation procedures will be affected by its presence.

Practically all ladle metallurgy operations follow this basic route. Sometimes less crude alloy addition procedures are subsequently made (e.g. wire feeding, submerged injection of powders, etc.) at alloy trimming stations, in order to achieve closer control of the final product chemistry.

The second step, now standard in most steel shops, is to bubble gas through a lance set deep within the liquid steel, or through a porous plug element set in the bottom of the ladle. Over the correct range of flows, this practice can eliminate chemical and thermal gradients within the steel by thoroughly mixing the vessels contents, while minimizing slag entrainment and temperature loss. The procedure can also rinse the liquid steel of condensed oxide inclusions, which nucleate following aluminum, or ferro-silicon additions to the steel. These additions are used to getter dissolved oxygen and to condense it in the form of fayalite or alumina type inclusions.

For higher quality steels, secondary processing is needed in order to remove harmful residual elements to parts per million levels. For instance, arctic oil rig platforms require that the ductile/brittle transformation temperature be lowered as far as possible, and certainly to -20°C. The special properties sought can be approached by reducing the concentration of residual elements ($\underline{S}$, $\underline{P}$, $\underline{O}$, $\underline{N}$, $\underline{H}$, ...) within the steel bath, prior to teeming into the continuous casting machine. This reduces the number of inclusions, while the remainder must be converted into harmless refractory type spheroids prior to hot rolling, so that they do not subsequently form elongated stringers. For plate, pipe, rail steels, and other thick sections, dissolved hydrogen retained within the matrix can precipitate at

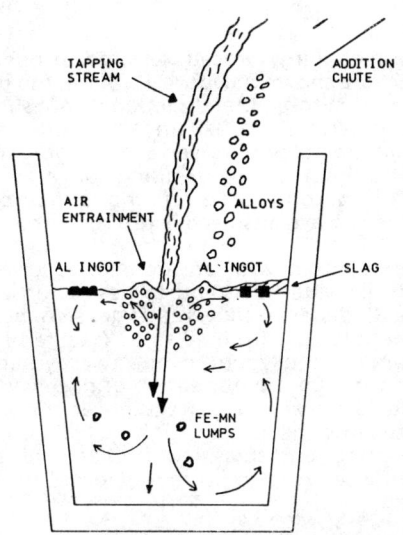

Figure 1   Schematic of tapping a furnace into a
             steelworks teeming ladle, showing alloy
             additions and slag carryover.

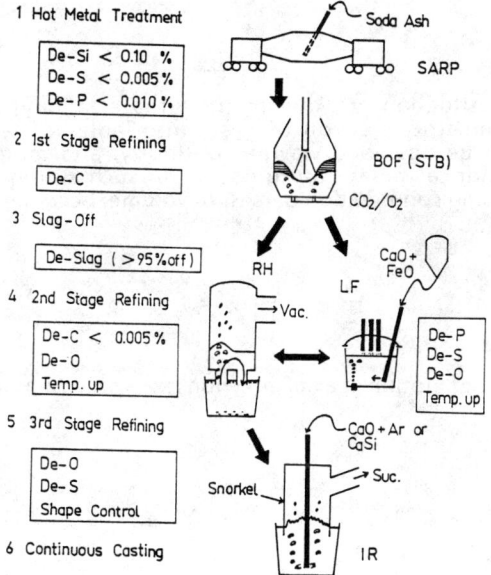

Figure 2   An example of current processes for
             making clean steels (Sumitomo Metals
             Industries, Kashima Works).

449

discontinuities within the steel (e.g. inclusions), and lead to hydrogen induced cracking (HIC).

Typical ladle metallurgy procedures used to donate superior properties to the final steel product are shown in Figure 2. There, the particular process routes currently in operation at Sumitomo Metals, in Japan, are shown .[3] Thus hot metal is first dephosphorised with soda ash in a torpedo car/ladle, and is then decarburised in a combination blown converter employing top and bottom blowing of oxygen. After tapping the steel, any oxidising converter slag, which carries over to the ladle, is removed. Second stage refining is then conducted using an RH vacuum degassing unit to remove dissolved hydrogen nitrogen, oxygen and carbon.

A ladle furnace fitted with an arc heating unit where fluxes can be injected can be used to provide heat, and to desulphurise the steel. The use of vacuum (~1 Torr) in the RH degasser allows initial levels of hydrogen of 4-5 ppm (say) to be lowered to 2ppm, nitrogen from 40 to 32 (say), and oxygen from 300 ppm (say), to perhaps 10 ppm (total oxygen), over a twenty minute processing period. The oxygen is removed through the nucleation of dissolved carbon and oxygen to form carbon monoxide, the reaction proceeding at the interfaces of ascending bubbles of argon injected into the 'up-leg' of the RH degasser, and the steel/vacuum interfaces of the bath surface and splashing droplets. Initial carbon levels of 200-300 ppm, can typically be reduced to 'ultra low levels' of 20 ppm, for 'interstitial free' steels, by degassing.

This cursory glance at primary and secondary ladle metallurgy operations illustrates a wealth of procedures and phenomena to which mathematical and/or physical modelling can be applied. Following a statement of the basic differential equations governing the flow of heat, mass, and momentum in such systems, therefore, a series of examples is provided, which hopefully demonstrate the value such analyses can have in terms of process design and operation.

Theory

For the situation of axisymmetric furnace tapping into a ladle, or of submerged axisymmetric injection of gases into ladles of typical circular cross section, flows can be described via the continuity, and momentum, equations expressed in cylindrical polar co-ordinates. In such systems, the equation of continuity (expressing conservation of mass or volume) becomes

$$\frac{\partial u}{\partial z} + \frac{1}{r}\frac{\partial}{\partial r}\left(rv\right) = 0 \tag{1}$$

while the equation of motion in axial direction is:

$$\frac{\partial}{\partial z}\left(\rho uu\right) + \frac{1}{r}\frac{\partial}{\partial r}\left(\rho ruv\right) = -\frac{\partial P}{\partial z} + \frac{\partial}{\partial z}\left(\mu_{eff}\frac{\partial u}{\partial z}\right) + \frac{1}{r}\frac{\partial}{\partial r}\left(r\mu_{eff}\frac{\partial u}{\partial r}\right) + S_u \tag{2}$$

where

$$S_u = \frac{\partial}{\partial z}\left(\mu_{eff}\frac{\partial u}{\partial z}\right) + \frac{1}{r}\frac{\partial}{\partial r}\left(r\mu_{eff}\frac{\partial v}{\partial z}\right) - \rho g \tag{3}$$

S_u is known as the source term for the u momentum equation. In a similar fashion, the equation of motion in the radial direction becomes:

$$\frac{\partial}{\partial z}\left(\rho uv\right) + \frac{1}{r}\frac{\partial}{\partial r}\left(\rho rvv\right) = -\frac{\partial P}{\partial r} + \frac{\partial}{\partial z}\left(\mu_{eff}\frac{\partial v}{\partial z}\right) + \frac{1}{r}\frac{\partial}{\partial r}\left(r\mu_{eff}\frac{\partial v}{\partial r}\right) + S_v \tag{4}$$

where

$$S_v = \frac{\partial}{\partial z}\left(\mu_{eff}\frac{\partial u}{\partial r}\right) + \frac{1}{r}\frac{\partial}{\partial r}\left(r\mu_{eff}\frac{\partial v}{\partial r}\right) - \mu_{eff}\frac{v}{r^2} \tag{5}$$

It is to be noted that all the transport equations written in this way take the general tensorial form:

$$\frac{\partial}{\partial x_i}\left(\rho u_i \phi\right) = \frac{\partial}{\partial x_i}\left(\Gamma \frac{\partial \phi}{\partial x_i}\right) + S_\phi \tag{6}$$

in which convective terms balance diffusive and 'source' terms on the right side of the equation. By casting all the equations in this form, where ϕ represents (u, v, w, k, $\dot{\epsilon}$, T, C, ...) an efficient iterative solution routine then becomes possible.[24]

In solution procedures carried out by the author, and co-workers, gas injection problems have been treated as pseudo one phase flow phenomena, in which the gas-liquid metal plume is treated as a region of lower density steel. Typical gas voidages of 5% within a rising gas-liquid plume can be accounted for by introducing a term, $\rho_L g\alpha$, involving the gas voidage, α, on the right side of equation 3 for S_u, the 'source term' for the u momentum equation. The approach is simple and correctly emphasizes the importance of buoyancy, versus shear, forces in gas driven recirculating flows. It has been confirmed through many experiments to be an effective way of treating such problems. Nonetheless, the approach requires that the plume dimensions and gas voidage be specified "a priori".

In an alternative, but more computationally demanding procedure, by Boysan et al [25], a flow field is first deduced using the Eulerian scheme implied by equations 1-5. Successive bubbles are then introduced into the system, using a Lagrangian framework. This allows spatial variations in plume voidage to be computed as a function of bulk flow patterns. Through successive iterations between Eulerian and Lagrangian frames of reference therefore, plume geometries can be deduced as part of the numerical solution procedures. Such an approach can predict, conditions for which the plume is 'bent' inwards, or outwards, as a result of interactions with the bulk flow fields. As demonstrated later, such situations can arise in three dimensional flows in ladles.

In modelling turbulent effective viscosities, it is well known that the scale of turbulence is far smaller than the finest grid systems that can be handled by present day computers in arriving at numerical solutions to the discretised forms of equations 1-5.

The author, and co-workers, have tested a variety of prescriptions for mimicing the effect of turbulent fluctuating velocities on bulk flows and mixing rates. These range from ad hoc specifications of local turbulent viscosities, to their own algebraic models (11,12) and the popular k-$\dot{\varepsilon}$ model of turbulence. In this latter model, the governing transport equations for turbulence kinetic energy, k, and its dissipation rate, $\dot{\varepsilon}$, can then be represented in terms of cylindrical polar coordinates according to (26):

Equation of turbulence kinetic energy:

$$\frac{\partial}{\partial z}\left(\rho u k\right) + \frac{1}{r}\left(\rho r v k\right) = \frac{\partial}{\partial z}\left(\frac{\mu_{eff}}{\sigma_k}\cdot\frac{\partial k}{\partial z}\right) + \frac{1}{r}\frac{\partial}{\partial r}\left(r\frac{\mu_{eff}}{\sigma_k}\cdot\frac{\partial k}{\partial r}\right) + S_k \qquad (7)$$

where S_k, the net source term, can be represented as:

$$S_k = G - \rho\dot{\varepsilon} \qquad (8)$$

where G, the generation of kinetic energy of turbulence term, is given by

$$G = \mu_t\left\{2\left[\left(\frac{\partial u}{\partial z}\right)^2 + \left(\frac{\partial v}{\partial r}\right)^2 + \left(\frac{v}{r}\right)^2\right] + \left(\frac{\partial u}{\partial r} + \frac{\partial v}{\partial z}\right)^2\right\} \qquad (9)$$

Equation of dissipation rate of turbulence kinetic energy:

$$\frac{\partial}{\partial z}\left(\rho u \varepsilon\right) + \frac{1}{r}\frac{\partial}{\partial r}\left(\rho r v \varepsilon\right) = \frac{\partial}{\partial z}\left(\frac{\mu_{eff}}{\sigma_\varepsilon}\cdot\frac{\partial \varepsilon}{\partial z}\right) + \frac{1}{r}\frac{\partial}{\partial r}\left(r\frac{\mu_{eff}}{\sigma_\varepsilon}\cdot\frac{\partial \varepsilon}{\partial r}\right) + S_\varepsilon \qquad (10)$$

where

$$S_\varepsilon = \frac{C_1 \varepsilon G}{k} - \frac{C_2 \rho \varepsilon^2}{k} \tag{11}$$

The auxiliary relationships are:

$$\mu_{eff} = \mu_\ell + \mu_t \tag{12}$$

where

$$\mu_t = \frac{C_\mu \rho k^2}{\varepsilon} \tag{13}$$

and

$$\rho = \alpha \rho_G + (1-\alpha) \rho_L \tag{14}$$

for gas injection procedures.

The five constants, in this turbulence model, together with recommended numerical values, are given in Table 1. The interested reader is referred to Reference (27) for further details and explanations.

TABLE 1   RECOMMENDED VALUES OF CONSTANTS
FOR k-ε MODEL OF TURBULENCE

C_1	C_2	C_μ	σ_k	σ_ε
1.44	1.92	0.09	1.0	1.3

Dispersion of alloy additions/tracers

In the presence of a steady state two-dimensional flow field, to which a tracer 'i' is added at time zero, the appropriate statement for conservation of mass i, in terms of cylindrical polar coordinates, as it spreads through the system is:

Mass Conservation

$$\frac{\partial}{\partial t}\left(\rho m_i\right) + \frac{\partial}{\partial z}\left(\rho u\, m_i\right) + \frac{1}{r}\frac{\partial}{\partial r}\left(\rho rv m_1\right) = \frac{\partial}{\partial z}\left(\Gamma_e\, \frac{\partial m_i}{\partial z}\right) + \frac{1}{r}\frac{\partial}{\partial r}\left(r\Gamma_e\, \frac{\partial m_i}{\partial r}\right) \tag{15}$$

where the effective exchange coefficient, Γ_e, is defined by

$$\Gamma_e = \frac{\mu_\ell}{\sigma_\ell} + \frac{\mu_t}{\sigma_t}$$

σ_ℓ and σ_t were assumed to have a value of unity.

In the case of alloy additions, the dispersing alloy generally has a density which is either higher, or lower, than that of the steel bath. It is shown elsewhere (34), that this can significantly affect metal flows in procedures such as aluminum wire feeding into a ladle. Mathematical models of such phenomena require an additional buoyancy term equivalent to that used for S_u in equation 3.

Three dimensional flows

While many flows within ladles can be idealised by assuming axisymmetric conditions, and while some procedures are indeed carried out in this manner, a major fraction of the flows in ladle metallurgy processing operations have a three dimensional character. With the larger storage capacities of today's computers and their faster processing times, three dimensional solutions are now becoming relatively common-place. These require the introduction of convective and diffusive terms for the angular direction. Thus, the differential equations for steady-state, turbulent flow, and associated heat and mass transport phenomena can be written in three-dimensional, incompressible, cylindrical and ensemble-averaged form.

Continuity Equation

For incompressible flow,

$$\frac{\partial u}{\partial z} + \frac{1}{r}\frac{\partial}{\partial r}\left(rv\right) + \frac{1}{r}\frac{\partial w}{\partial \theta} = 0 \tag{16}$$

Momentum Conservation Equation

Axial direction

$$\frac{\partial}{\partial z}\left(\rho uu\right) + \frac{1}{r}\frac{\partial}{\partial r}\left(r\rho vu\right) + \frac{1}{r}\frac{\partial}{\partial \theta}\left(\rho wu\right) - \frac{\partial}{\partial z}\left(\mu_{eff}\frac{\partial u}{\partial z}\right)$$

$$- \frac{1}{r}\frac{\partial}{\partial r}\left(r\mu_{eff}\frac{\partial u}{\partial r}\right) - \frac{1}{r}\frac{\partial}{\partial \theta}\left(\mu_{eff}\frac{1}{r}\frac{\partial u}{\partial \theta}\right) = -\frac{\partial P}{\partial z} + S_u \tag{17}$$

where

$$S_u = \frac{\partial}{\partial z}\left(\mu_{eff}\frac{\partial u}{\partial z}\right) + \frac{1}{r}\frac{\partial}{\partial r}\left(r\mu_{eff}\frac{\partial v}{\partial z}\right) + \frac{1}{r}\frac{\partial}{\partial \theta}\left(\mu_{eff}\frac{\partial w}{\partial z}\right) - \rho g \tag{18}$$

Radial direction

$$\frac{\partial}{\partial z}\left(\rho uv\right) + \frac{1}{r}\frac{\partial}{\partial r}\left(r\rho vv\right) + \frac{1}{r}\frac{\partial}{\partial \theta}\left(\rho wv\right) - \frac{\partial v}{\partial z}\left(\mu_{eff}\frac{\partial v}{\partial z}\right)$$

$$- \frac{1}{r}\frac{\partial}{\partial r}\left(r\mu_{eff}\frac{\partial v}{\partial r}\right) - \frac{1}{r}\frac{\partial}{\partial \theta}\left(\frac{\mu_{eff}}{r}\frac{\partial v}{\partial \theta}\right) = -\frac{\partial P}{\partial r} + S_v \tag{19}$$

where

$$S_v = \frac{\partial}{\partial z}\left(\mu_{eff}\frac{\partial u}{\partial r}\right) + \frac{1}{r}\frac{\partial}{\partial r}\left(r\mu_{eff}\frac{\partial v}{\partial r}\right) + \frac{1}{r}\frac{\partial}{\partial \theta}\left[r\mu_{eff}\frac{\partial}{\partial r}\left(\frac{w}{r}\right)\right]$$

$$+ \left[\frac{\rho w^2}{r} - \frac{2\mu_{eff}}{r^2}\left(v + \frac{\partial w}{\partial \theta}\right)\right] \tag{20}$$

Angular direction

$$\frac{\partial}{\partial z}\left(\rho uw\right) + \frac{1}{r}\frac{\partial}{\partial r}\left(r\rho uw\right) + \frac{1}{r}\frac{\partial}{\partial \theta}\left(\rho ww\right) - \frac{\partial}{\partial z}\left(\mu_{eff}\frac{\partial w}{\partial z}\right)$$

$$-\frac{1}{r}\frac{\partial}{\partial r}\left(r\mu_{eff}\frac{\partial w}{\partial z}\right) - \frac{1}{r}\frac{\partial}{\partial \theta}\left(\frac{\mu_{eff}}{r}\frac{\partial w}{\partial \theta}\right) = -\frac{1}{r}\frac{\partial P}{\partial \theta} + S_w \tag{21}$$

where

$$S_w = \frac{\partial}{\partial z}\left(\frac{\mu_{eff}}{r}\frac{\partial u}{\partial \theta}\right) + \frac{1}{r}\frac{\partial}{\partial r}\left[\mu_{eff}\left(\frac{\partial v}{\partial \theta} - w\right)\right] + \frac{1}{r}\frac{\partial}{\partial \theta}\left(\frac{\mu_{eff}}{r}\left(\frac{\partial w}{\partial \theta} + 2v\right)\right)$$

$$+ \frac{\mu_{eff}}{r}\left(\frac{\partial w}{\partial r} + \frac{1}{r}\frac{\partial v}{\partial \theta} - \frac{w}{r}\right) - \frac{\rho vw}{r} \tag{22}$$

Energy Conservation

$$\frac{\partial}{\partial t}\left(\rho T\right) + \frac{\partial}{\partial z}\left(\rho uT - \Gamma_{T.eff}\frac{\partial T}{\partial z}\right)$$

$$+ \frac{1}{r}\frac{\partial}{\partial r}\left[r\left(\rho vT - \Gamma_{T.eff}\frac{\partial T}{\partial r}\right)\right] + \frac{1}{r}\frac{\partial}{\partial \theta}\left(\rho wT - \Gamma_{T.eff}\frac{1}{r}\frac{\partial T}{\partial \theta}\right) = 0 \tag{23}$$

In this equation it was assumed that thermal (temperature) homogenization resulted from both convective and diffusive heat transfer phenomena. The effective diffusivity concept was used to represent the combined effects of molecular and turbulent thermal diffusion. The exchange coefficient, Γ, is given by

$$\Gamma_{T.eff} = \frac{\mu_t}{\sigma_{T.t}} + \frac{\mu_\ell}{\sigma_{T.\ell}} \tag{24}$$

Conservation of chemical species

$$\frac{\partial}{\partial t}\left(\rho c\right) + \frac{\partial}{\partial z}\left(\rho u c - \Gamma_{C.eff}\,\frac{\partial c}{\partial z}\right) + \frac{1}{r}\frac{\partial}{\partial r}$$

$$\left[r\left(\rho v c - \Gamma_{C.eff}\,\frac{\partial c}{\partial r}\right)\right] + \frac{1}{r}\frac{\partial}{\partial \theta}\left(\rho w c - \Gamma_{C.eff}\,\frac{1}{r}\frac{\partial c}{\partial \theta}\right) = 0 \tag{25}$$

Motion of alloy additions

Before alloy additions disperse into the steel within the ladle they remain as discrete objects, and are therefore subject to Newton's laws of motion. For additions such as aluminum and ferro-manganese, a chilled layer of steel forms over a molten core of alloy.[28] Ignoring changes in mass for such additions, an appropriate force balance on a particle within the fluid is (4):

$$\frac{4}{3}\,R_p^3\,\rho\,\frac{d\mathbf{U}_p}{dt} = \frac{4}{3}\,R_p^3\,g\left(\rho_p - \rho\right)$$

$$-\frac{C_D}{2}\,R_p^2\,\rho\,\mathbf{U}_r\left|\mathbf{U}_r\right| - C_A\,\frac{4}{3}\,R_p^3\,\rho\,\frac{d\mathbf{U}_p}{dt}$$

where,

$$\mathbf{U}_p = \frac{d\mathbf{x}}{dt}$$

In equation 26, $\mathbf{U}_p$ is the instantaneous velocity vector of the particle while $\mathbf{U}_r$, is the relative velocity between the particle and the bulk fluid. It is through the drag term (i.e., $F_D \propto U_r^2$) that the liquid's motion within the vessel exerts an influence on the trajectory of submerged particles. Consequently, the distribution of flow parameters within the vessel must first be specified, before subsurface trajectories can be predicted via Equations 26 and 27.

Numerical procedures

Discretisation equations derived from Equations 1 through 10 were solved using an implicit finite difference procedure, referred to as SIMPLE (29)

(Semi Implicit Method for Pressure Linked Equations). For analysis of the gas liquid region, the GALA (30) procedure was incorporated into the SIMPLE algorithm. In this, the physical properties of fluid mixture in a cell in the two phase region were averaged on a volumetric basis. This required the conventional mass continuity equation to be replaced by a volume continuity equation, such that the volume of fluids entering a volume element equalled the total volume of fluids leaving.

Evidently, there can be significant discrepancies between the total mass inflow and outflow to cells on the edge of the plume, when modelling a two phase flow system using a single phase approximation. The use of GALA allowed computation of hydrodynamic variables over the entire flow domain.[18]

For the majority of examples illustrated in the next section, variable grid networks, typically 24 x 16, were chosen for obtaining numerical solutions. Computations were performed on McGill's Amdahl V7 machine. A typical 2D execution required 700 iterations, corresponding to an execution time of about 162s. A convergence criterion was set (⟨0.005) on all variables, and computations were carried out until the absolute sum of residuals on u, v, and volume continuity fell below this stipulated value. By way of example, a schematic of the C.A.S. alloy addition procedure, illustrating central injection of gas, together with the grid network used for its mathematical representation, is shown in Figure 3.[18]

One way coupling between (i) the fluid flow and dispersion equations and (ii) the fluid flow and particle motion equations was assumed. This meant that the fluid flow equations were solved first. The converged flow fields were then used to obtain transient concentration profiles and mean trajectories of spherical particles within the flow domains.

The numerical time step integration in the mass conservation equation (viz., Eq. 15) was approximated by a fully implicit marching integration procedure (27), while for the representation of total flux, i.e., convection + diffusion, an hybrid differencing scheme (28) was adopted. Equation 15, though a linear differential equation, was solved iteratively, using a line by line solution scheme, a convergence criterion having been set so that the absolute sum of residuals for m_i fell below 10^{-6}. About 12 to 16 iterations were required to satisfy this criterion.

Numerical solutions to the particle motion equations (viz., Equations 26 and 27) were obtained via a fourth order Runge-Kutta-Gill method. Embodying the converged velocity field with Equations 26 and 27, spherical particles' subsurface trajectories were predicted at every 0.01 seconds, following their entry into the system. Computations were continued until the particle had either resurfaced, or settled to the bottom of the ladle.

458

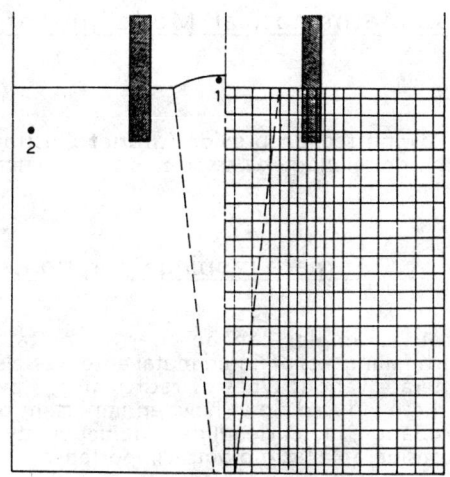

1 Tracer addition location; 2 Probe location

Figure 3   Schematic of C.A.S. alloy addition system, illustrating central gas injection and grid networks used for mathematical representation (vessel diameter = 1.12m, liquid depth = 0.93m).

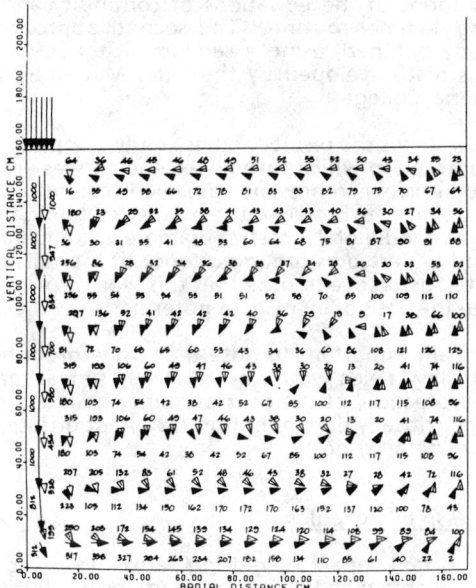

Figure 4   Comparison of velocities predicted on the basis of turbulence theory and equivalent velocities assuming laminar flow conditions at filled height of 1.6m.

Illustrations of Mathematical Modelling of Ladle Metallurgy Operations

The following section provides a brief illustration, and summary, of modelling research into phenomena associated with primary processing operations in ladles.

Furnace Tapping Operations

Numerous examples can be cited of metallurgical processes and procedures in which falling jets of liquid metal enter vessels such as furnaces, ladles or molds and generate various types of recirculatory flows. As the nature and characteristics of these induced flows have an important bearing on alloy mixing, heat transfer, reoxidation, metal cleanliness, homogeneity, and so on, the general subject area is of considerable technological importance.

In an idealised simulation of an actual tapping operation from a Basic Oxygen Furnace, (5,6), it was assumed that a vertical jet of steel plunges axially into a cylindrical ladle with an entry velocity of 10 m/s. Two completely independent programs were written for numerical solutions. The first was based on an early version of the MAC (marker and cell) method, first introduced by the Los Alomos Laboratories modelling group. This method relies on a finite difference scheme using discretised forms of the equations of continuity and motion and used an explicit, marching in time, routine. The second approach followed numerical procedures already outlined, using a semi-implicit T.D.M.A. coupled to a Gauss-Siedel routine, as first developed by the Fluid Modelling Group of Mechanical Engineers of Imperial College.

Figure 4 shows the results of these early computations. The black arrows represent computations in which turbulence energy was accounted for via the k-ε equation, the open arrows represent equivalent computations based on laminar flow. The striking similarity between the two recirculating flows predicted for a 25 cm dia. jet entering a ladle of 3.3 metres diameter, at a filling height of 1.6 metres, is remarkable, and underscores the importance of inertial versus turbulent viscous forces. Indeed, it is instructive to note that this type of recirculating flow would be predicted even on the basis of inviscid, or potential, flow theory.

It is equally important to note that while predicted trajectories of alloy additions in such flow fields would be relatively similar,that once any entrained objects lost the 'protection' of their surrounding shells of steel, dispersion of their contents would be critically dependent on turbulent diffusion phenomena. This is considered in more detail later.

Particle trajectories in filling ladles

An important aspect of getting ferro-alloys and aluminum to melt in molten steel in a reliable manner, is to try and ensure that they melt and disperse subsurface, out of harms way of any oxidising slags, or interactions with air.[7]

Figure 5A shows computations and experimental data using a smaller scale water model simulation, wherein the correct density ratio between addition and liquid was maintained. Thus wooden spheres with a specific gravity of 0.4 (roughly, corresponding to spheres of aluminum, γ = 2.3/7.0 = 0.33, and ferrosilicon, γ = 4.0/7.0 = 0.57), are predicted to resurface almost immediately after projection into any part of such recirculating flows. On the other hand, heavier particles, (γ = 0.8), (roughly corresponding to the apparent densites of ferro-manganese, silico-maganese, etc., additions), have a chance of being drawn downwards into the recirculating flow, provided they are projected into the bath, reasonably close to the plunging jet. Figure 5B shows predictions for a ladle of 4 metres diameter, at a filling height of 1.5 metres. Figure 5C illustrates that these predictions are based on steady flows. In reality, turbulence can introduce stochastic effects, as demonstrated by the experimental data appearing there.

It is worth noting that these predictions do not allow for air becoming entrained as the plunging jet enters the liquid bulk. This can be significant, reversing steel flows in that vicinity, and leading to undesirable, increases in dissolved nitrogen levels of up to 20 ppm typically.

While these simulations point to the deficiences of current alloy addition making procedures, the most critical problem in practice, is the variable amount of slag carryover from the furnace into the ladle, and resultant slag-solute interactions.[19]

Gas Stirred Ladles

Flow Patterns

Once, teeming ladles have been filled with molten steel, the next step is to stir the vessel's contents to ensure chemical and thermal homogeneity. Figure 6 illustrates the computed flow field generated in a 250 tonne ladle, when a gas flow of 0.25 Nm^3/min is injected axially through a porous plug, (or deeply submerged lance). The main point to note is the high velocities within a rising plume of argon bubbles and entrained steel. The bubbles comprise about 5-10% of the nett volume of the plume. The plume expands due to entrainment of adjacent liquid, such that its the ascent velocity remains almost constant with height, and typically ranges from, 1-1.5 m/s. High radial outflows towards the ladle's rim, together with strong downflows near the upper sidewalls, can lead to slag droplet entrainment and refractory erosion. The last parts of the vessel's contents to become mixed, in are to be found deep in the ladle, near the sidewall and plug regions.

Algebraic Approximation for Plume Velocities, Turnover Rates and Mixing Times

While computational procedures are valuable in that they can provide precise details of the flows generated, and thereby lead on to computations of alloy addition trajectories, mixing-in times, inclusion floatout behaviour, thermal de-stratification, etc., it is useful to present a few simple algebraic equations deriving from these studies, so as to provide guidance for steelmakers. Thus, Sahai and Guthrie (9,10) have shown that the mean rising velocity within a gas-liquid plume approximates the expression (S.I. Units):

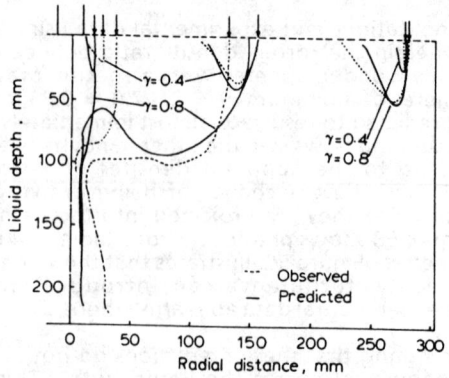

Figure 5A   Trajectories of 10mm diameter wooden spheres in a 0.15 scale
model of a 250 ton teeming ladle, during idealised furnace
tapping operations - vertical submerged entry jet.

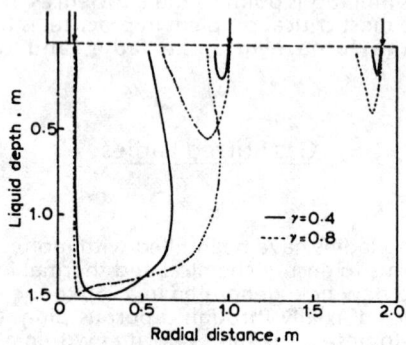

Figure 5B   Predicted mean trajectories of 67mm (2.6") lumps (spheres) of
alloy additions (γ = 0.4 and 0.8).

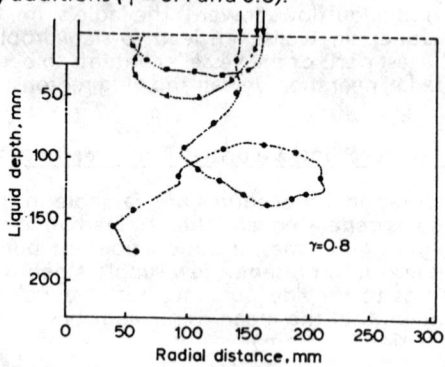

Figure 5C   individual trajectories of particles in water model, illustrating
sensitivity to injection location, and bath turbulence.

462

$$U_p = 4.19 \left(\frac{L^{\frac{1}{4}} Q^{\frac{1}{4}}}{R^{\frac{1}{4}}} \right) \qquad (35)$$

where Q is the flow of gas at mean ladle height and temperature (m³/s), L is the depth of liquid steel, (m) and R is the mean radius of the ladle (m).

In a more recent paper (21), Mazumdar and Guthrie go on to show that the 95% mixing-in time for a neutrally buoyant addition (or tracer) added to the surfacing plume of liquid in an axisymmetrically stirred ladle, can be approximated by the empirical expression (S.I. Units).

$$\tau_m = 25.4 \frac{R^{7/3}}{\left(\beta Q \right)^{\frac{1}{4}} L} \qquad (36)$$

Since the energy input to the ladle derives almost exclusively from the potential energy supplied by rising gas bubbles, the energy input per unit mass of liquid becomes

$$\varepsilon_m \simeq \frac{\rho_L g \beta Q L}{\rho_L \pi R^3 L} \qquad (37)$$

where β is the fractional depth of lance submergence (β = 1 for porous plugs), allowing equation (25) to be rewritten in the form

$$\tau_m = 37 \varepsilon_m^{-1/3} R^{5/3} L^{-1} \qquad (38)$$

In computing mixing times, it is important that there be a clear definition of what is meant by τ_m. The authors have suggested that the 95% mixing time be used as a suitable standard, and that it be defined as that time at which *all* the local concentrations of tracer addition have reached 95% of the bulk well mixed value (which is theoretically reached at time infinity).

Numerical Solutions

Thus Figure 7 illustrates the variation in local 95% mixing times. These numerical solutions were obtained (21) through the procedures presented earler

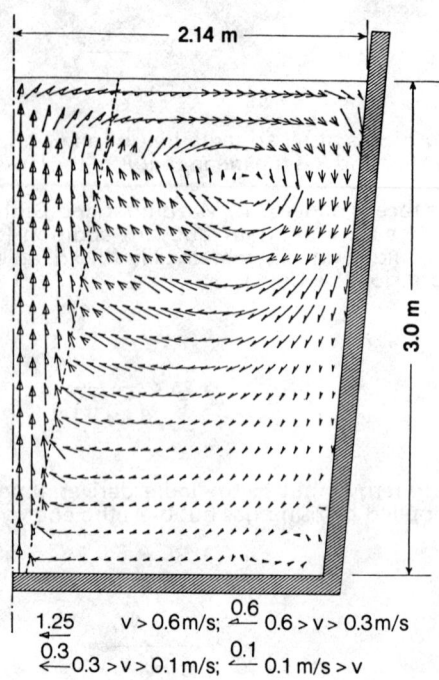

v > 0.6m/s; $\underset{\longleftarrow}{0.6}$ **0.6 > v > 0.3m/s**

$\underset{\longleftarrow}{1.25}$

$\underset{\longleftarrow}{0.3}$ **0.3 > v > 0.1m/s;** $\underset{\longleftarrow}{0.1}$ **0.1 m/s > v**

Figure 6   *Predicted flow fields generated by argon stirring in a typical 250 tonne cylindrical ladle by a flow of 0.25 Nm³/min. (9 SCFM) argon from a centrally located porous plug.*

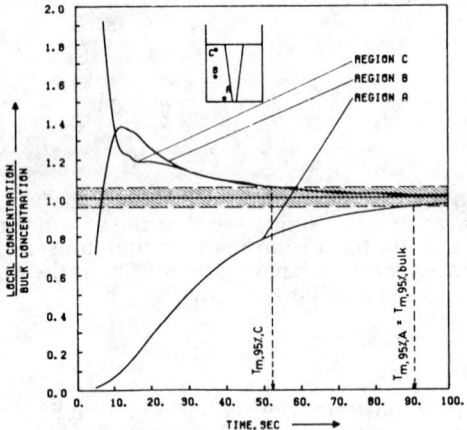

Figure 7   *Variation in local mixing time for three different regions, A, B, and C, in the water model ladle (L = 0.93 m, R = 0.56 m, β = 0.96, and Q = 6.67 x 10⁻⁴ m³/s.*

and clearly demonstrate that regions B & C approach well mixed conditions more rapidly than region A deep within the ladle and close to the rising plume.

That paper goes on to show that mixing times in ladles are a function of both turbulent diffusion and convective mixing processes. It is demonstrated that under typical ladle operating conditions, both should be weighted about equally:

"Universal relationships" such as that proposed by Nakanishi, Fujii and Szekely (31) (τ_m = 12.68 x 10³ $\dot{\epsilon}_m^{-0.4}$) ignore the effects of vessel shape, liquid convection, the presence of baffles or mixing zones, etc., and are therefore flawed. The application of such expressions should be limited to intensively stirred reactors, i.e. back-mix reactors.

Off-Centre Stirred Ladles

Figure 8 shows experimental data by S. Joo (32), using an aqueous one third scale model of an industrial 150 tonne ladle, in which 95% mixing times are plotted versus the radial position of the porous plug. As seen, mixing rates within the bulk of the liquid are increased, and mixing times shortened, as the plug is moved away from the centre towards the half radius. Beyond a minimum mixing time, reached at half radius, plugs set closer to the ladle sidewalls tend to give slightly longer mixing times for a give flow of gas.

These results were obtained using a one third scale model of a 150 ton steelworks in the absence of any slag layers. Fluid slags are known to absorb energy and lead to longer mixing times Also seen from Figure 8, is the fact that an increasing gas flow leads to decreases in mixing times, τ_m being proportional to $1/Q^{1/3}$, as suggested by the form of equations 35 and 36.

In some circumstances, it is necessary to bubble a ladle with two or more plugs so as to achieve rapid mixing, as well as to promote slag/metal intermixing, and to avoid explosive degassing effects under vacuum. A specific example is the tank degasser unit to be operated by Dofasco (Hamilton, Canada) in 1988, as an alternative to an RH degasser unit. Figure 9 shows that the optimum location for a second plug with respect to minimising mixing times, is to place the second plug at mid-radius, diametrically opposite the first.

Such smaller scale physical models, which are normally based on equivalence of Froude Numbers, need to be complemented by mathematical models. Thus, mixing times do not scale solely with Froude (convective/potential), but also with Reynolds (convective/viscous), Numbers, so that mathematical models and/or full scale physical models are needed for studying such matters.

Figures 10A, B, C, D, pictorially illustrate the characteristics of a single plume rising through water, from plugs set at centre, one third, half and two thirds radius respectively. One should note that the flow field can distort the plume from rising vertically, this being a function of cross flows within the ladle. Figures 10 E, F, G show plume interactions for two plug arrangements. Numerical procedures such as those adopted by Boysan and Johansen (25), can be used for predicting such behaviour.

Mathematical modelling of a plume set at two thirds radius in a 240 ton bulk degasser ladle is shown in Figure 11. Three vertical axial planes, set through

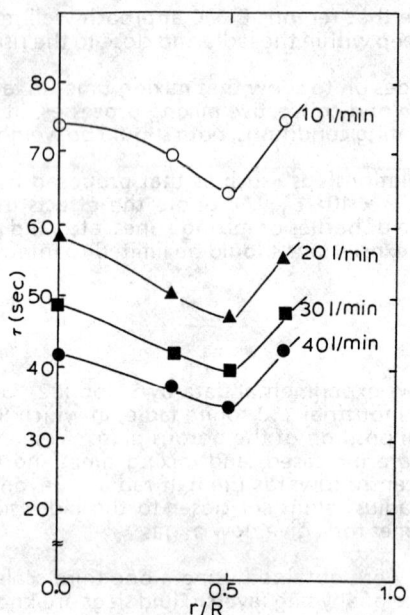

Figure 8   A plot of mixing time versus radial
             position for a single plug for various flow
             rates.

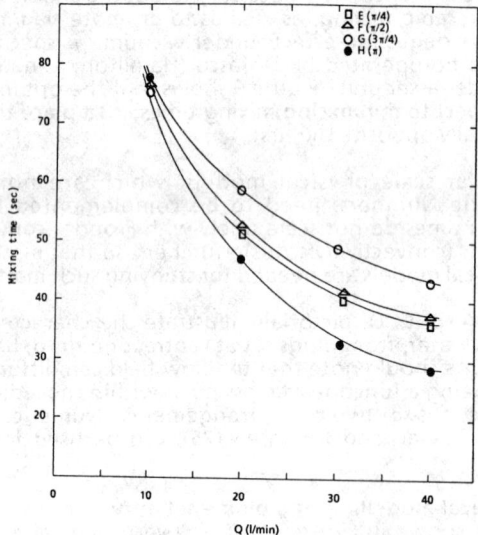

Figure 9   Variation in 95% mixing times with
             flowrate for double plug arrangements,
             placed at mid-radius. The effect of the
             angle, θ, subtended between the two
             plugs is illustrated.

A

B

Figure 10   Photographs of model ladle illustrating gas bubbling through single and double porous plug arrangements. A-centre,   B - third radius,   C - mid-radius,   D - two thirds radius,   E - one third radius,   F - mid radius,   G - two thirds radius.

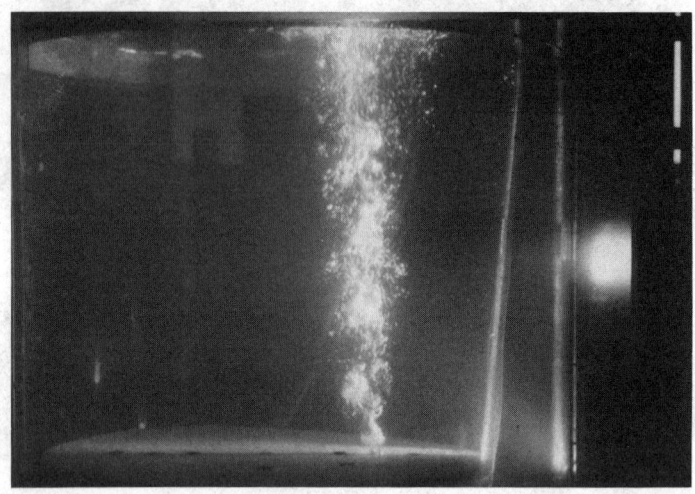

C

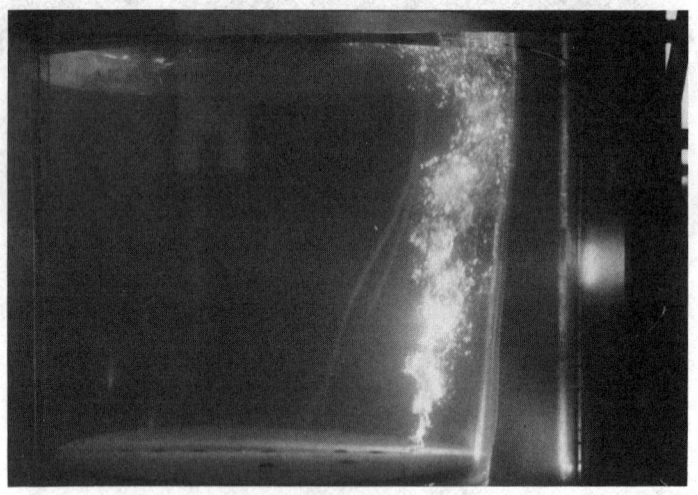

D

468

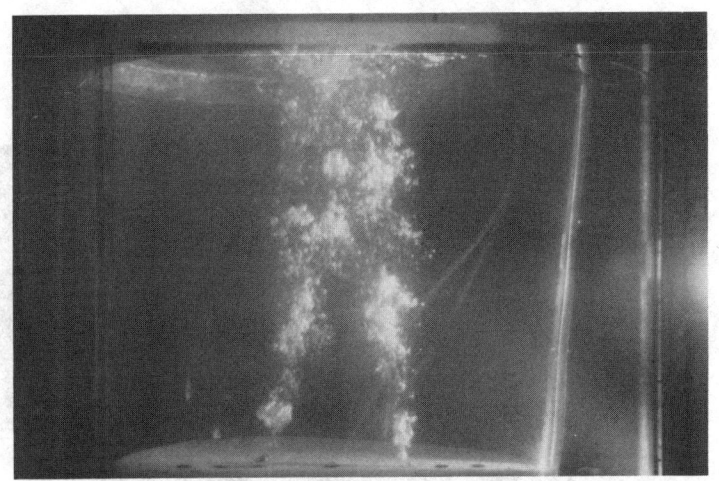

E

F

469

G

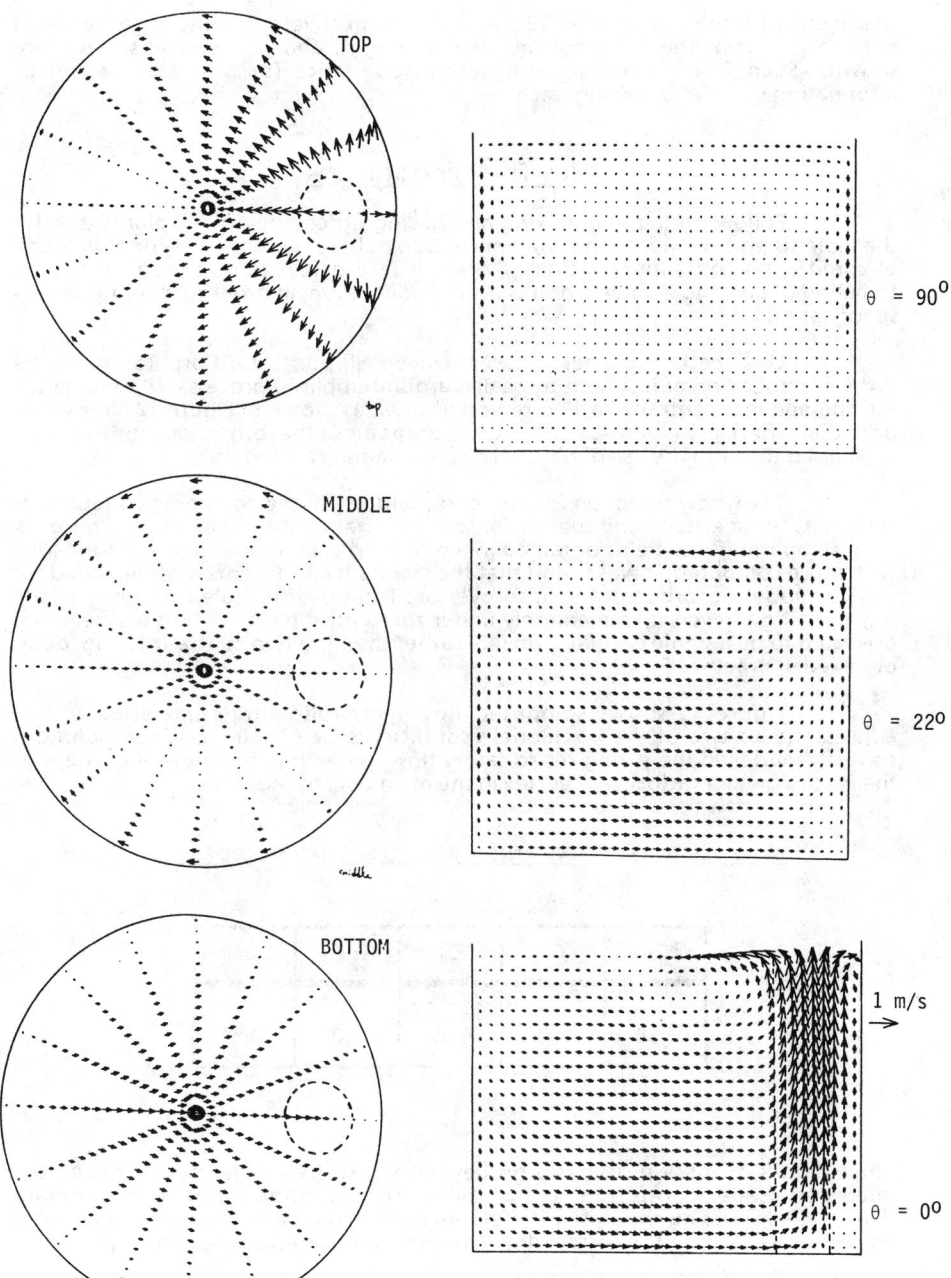

Figure 11   Computational analysis of flows predicted for a 300 tonne
ladle filled to 240 tonne level prior to tank degassing.
Specific conditions: Atmospheric Pressure, single plug at
two-thirds radius, Bath Temp. 1600°C   Gas Flow
500Nℓ/minute.

471

the plane of bubbling, and at 22°, and, 90° respectively to this, together with horizontal cuts at the ladle bottom, at mid-height, and at the steel surface, are shown. Such flow fields can provide an abundance (or over-abundance) of information on rates of mixing, etc.

Alloy Trimming Stations

Following tapping, mixing and rinsing procedures, many plants transfer the ladle to an alloy addition trimming station. There, wire feeding of aluminum, or cored wires containing calcium silicide, or rare earths can be used for precise trimming of alloying elements or inclusion modification, in order to meet the steel's specifications.

One method of merit, developed by Nippon Steel Corporation, is the C.A.S. (composition adjustment by sealed argon bubbling) process.[33] Here, alloys are dumped into centrally baffled region of steel, as shown in Figure 12. In view of potential interest by various U.S. steel companies, the proposed method was scrutinised through the use of physical and mathematical models.

The gridwork set up for numerical solution of the governing equations of continuity, momentum, and species conservation has already been shown in Figure 3. In comparing flow fields obtained in a one third scale model of a 150 ton ladle, with those computed, it was found that the standard coefficients recommended for the k-ε model of turbulence lead to over predictions of turbulent viscosity in the region of flow reversal, immediately under the central baffle. The flow field was predicted to be a single circular loop (18) rather than the two contra-rotating loops, observed in practise.

Figure 13 shows computed flowfields following somewhat ad-hoc adjustments to two of the k-ε model's coefficients (see Table 2). These solutions correctly modelled the strong recirculatory flow beneath the cylindrical baffle, and the much slower anticlockwise vortex in the main bulk of the ladle.

TABLE 2   VALUES OF CONSTANTS FOR K-ε TURBULENCE MODEL FOR C.A.S.

C_1	C_2	C_μ	σ_k	σ_ε
1.58	1.75	0.09	1.0	1.3

While details of these procedures go beyond the scope of the present paper, the interested reader is referred to the following publications for further details [17,18,19,20] In passing though, it is useful to note that an algebraic model by Sahai and Guthrie (11) gave the correct flow patterns, but was inferior to the adjusted k-ε

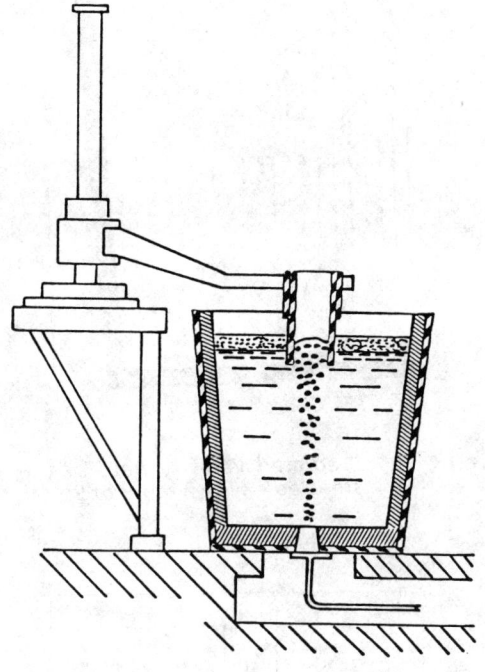

Figure 12   Schematic of C.A.S. operation

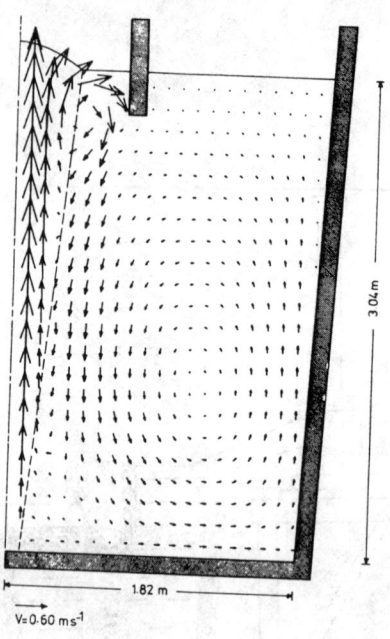

Figure 13   Predicted velocity field in the 150 ton steel ladle with a taper of 5°.

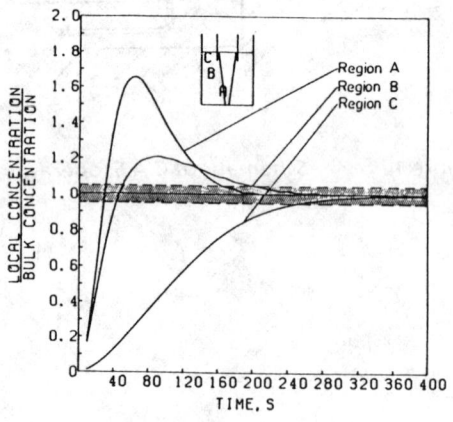

Figure 14   Predicted distribution of dimensionless mass fraction $(m_i/m_{i,b})$ as function of time in three different regions of 150 ton cylindrical ladle.

model, since a constant viscosity prescription was too crude an approximation for matching local rates of tracer dispersion within the ladle. Such features are very sensitive to correct turbulence prescriptions since, as one will note, the governing equations contain no overwhelming pressure gradient term on their right side equivalent to those appearing in the momentum equations for such systems

Figure 14 presents predictions of variations in local tracer concentration with time at positions A, B, and C within the ladle following its injection into the central baffled region at time zero. In this case, sampling the steel bath close to the suface would provide a satisfactory estimate of the time needed for all the ladle's contents to be fully mixed. The result is opposite for non-baffled ladles, as seen from Figure 7.

Figure 15 predicts the trajectory of ferro-alloy additions and aluminum after being projected into steel within the central baffled region of a 150 ton ladle. As seen, the light additions (Al and Fe-Si) will remain within this region, while more dense additions (γ)1) will sink to the bottom of the ladle, and then only gradually disperse. Consequently, additions such as ferro-niobium, which dissolves comparatively slowly compared to the release times of Class 1 (low melting point) additions (13,14) need to be finely crushed (d⟨2cm) if they are to dissolve completely within a reasonable length of time.

The final figure, Figure 16, shows predictions of the way in which concentrations of lighter (Al), neutrally buoyant (Fe-Mn), and heavier (Fe-Nb) additions would change with time at a surface sampling point just outside the baffle, following their initial projection into liquid steel within the baffled region.

Conclusions

A variety of ladle processing operations have been analysed through the use of physical and mathematical models. Experience has shown that physical and mathematical models need to be meshed in a complementary manner, when tackling new events, or phenomena. Given the power and speed of today's computers, many industrial problems such as those illustrated, are now amenable for study at reasonable cost. In the future, one can confidently predict that metallurgical processes will be designed more and more from the 'ground-up', so that new processing operations can be 'pre-optimised', and costly 'retrofits' become a thing of the past.

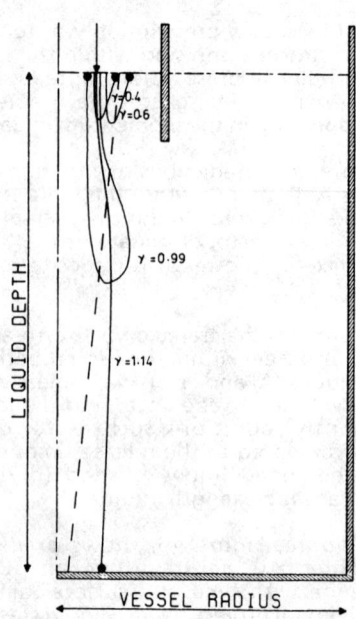

Figure 15   Predicted trajectories of four typical spherical alloy additions in the 150 ton ladle during C.A.S. alloy addition operation.

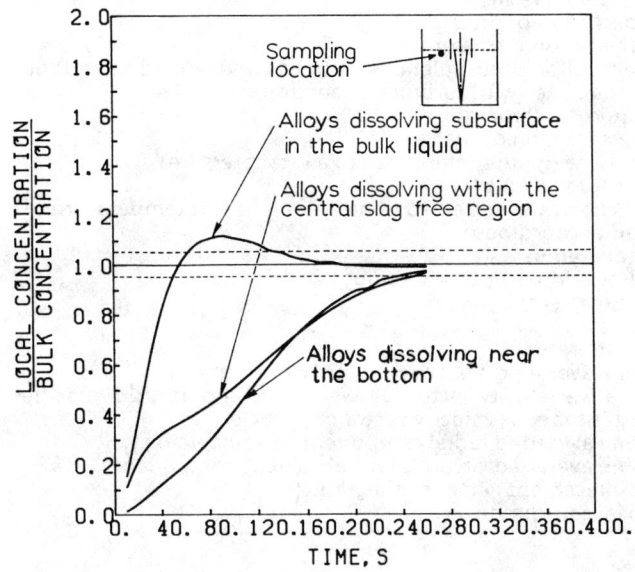

Figure 16   Rates of mixing-in of alloy additions when sampling steel bath
close to slag-metal interface for typical
A)   Low melting point, less dense additions melting within the
central baffle and then dispersing
B)   Alloy additions of neutral buoyancy melting within the
bath (e.g. Fe-Mu)
C)   Alloy additions dissolving into steel from bottom of ladle
(e.g. ferro-niobium)

LIST OF SYMBOLS

C_A	Added Mass Coefficient (Eq. 22)
C_1	Constant (Table 1)
C_2	Constant (Table 1)
C_u	Constant (Table 1)
C_D	Coefficient of drag
g	Gravitational acceleration
G	Generation of turbulence kinetic energy defined by Equation 7
k	Kinetic energy of turbulence per unit mass
L, H	Liquid depth/height
m_i	Mass fraction of species 'i'
P	Pressure (gauge referenced to a local pressure)
P	Pressure (absolute)
Q	Gas flowrate, corrected to mean height and temperature of liquid
r	Radial coordinate
R	Ladle radius
R_p	Alloy addition/particle radius
S	Source Term
t	Time
T	Temperature
u	Time averaged axial component of velocity
U_r	Relative velocity vector (between fluid and particle velocities)
U_p	Instantaneous velocity vector of particle
v	Time averaged radial component of liquid velocity
w	Time averaged circumferential velocity component
x	Displacement vector of the particle
z	Axial coordinate

GREEK SYMBOLS

α	Volume fraction of gas in the gas liquid plume
β	Fractional depth of lance submergence
$\dot{\epsilon}$	Rate of turbulence energy dissipation
$\dot{\epsilon}_m$	Specific energy input per unit mass of liquid
λ	Geometrical scaling factor ($= L_m/L_{f.s}$)
μ_{eff}	Effective viscosity
μ_ℓ	Dynamic viscosity
μ_t	Turbulent viscosity
ρ	Density
ρ_G	Density of gas
ρ_L	Density of liquid
ρ_p	Density of particle
σ_k	Schmidt Number for k (Table 1)
σ_ℓ	Laminar Schmidt number
σ_t	Turbulent Schmidt number
σ	Schmidt Number for (Table 1)
τ_m	95% Mixing Time
γ	Particle-liquid density ratio
θ	Circumferential co-ordinate
Γ	Exchange coefficient (diffusivity)

SUBSCRIPTS

eff	effective
G	gas
ℓ	laminar
L	liquid
t	turbulent

REFERENCES

1) A.W.D. Hills, ed., Heat and Mass Transfer in Process Metallurgy (London, U.K.: The Institute of Mining and Metallurgy, 1967).

2) Y. Sahai, R.I.L. Guthrie "Recent Advances in the Hydrodynamics of Metallurgical Processing" Advances in Transport Processes, Vol. IV.

   ed: Arun. S. Majumdar and R.A. Mashelkar (Wiley Eastern Limited, New Delhi, India, 1986) pp. 1-48.

3) H. Nakajima "On the detection and behaviour of second phase particles in steel melts", (Ph.D. thesis, McGillUniversity, 1986), 18-20.

4) R.I.L. Guthrie, R.Clift, H.Henein, 1975, 'Contacting Problems Associated with Aluminumand Ferro-Alloy Additions in Steelmaking - Hydrodynamic Aspects', Met. Trans., 6B, pp. 321, 329.

5) M. Salcudean, R.I.L. Guthrie, 1978, 'Turbulent Flow in Filling Ladles', Met. Trans., 9B, pp. 673-680.

6) M. Salcudean, R.I.L. Guthrie, 1978, 'Fluid Flow in Filling Ladles', Met. Trans., 9B, pp. 151-154.

7) R.I.L. Guthrie, 1982, 'The Role of Fluid Mechanics in Ladle Metallurgy', Iron and Steelmaker, Vol. 9, No. 1, pp. 41-45.

8) M. Salcudean, C.H. Low, A. Hurda, R.I.L. Guthrie, 1982, 'Computation of Three Dimensional Flow and Heat Transfer in Gas Agitated Reactors', Chem. Eng. Comm., Vol. 21, pp. 89-103.

9) Y. Sahai, R.I.L. Guthrie, 1982, 'Hydrodynamics of Gas Stirred Melts, Part I, Gas Liquid Coupling', Met. Trans. Vol. 13B, No. 2, pp. 193-202.

10) Y. Sahai and R.I.L. Guthrie, 1982, 'Hydrodynamics of Gas Stirred Melts, Part II, Axisymmetric Flows', Met. Trans. Vol. 13B, No. 2, pp. 203-211.

11) Y. Sahai, R.I.L. Guthrie, 1981, 'An Effective Viscosity Model for Gas Stirred Liquid Metal Reactors', Met. Trans. B., Vol. 13B pp. 125-127.

12) Y. Sahai, R.I.L. Guthrie, 1982, 'An Effective Viscosity Model for Gas Stirred Reactors', T.M.S.-A.I.M.E. Paper A82-11, 111th A.I.M.E. Annual Meeting, Dallas, pp. 1-9.

13) S.A. Argyropoulos, R.I.L. Guthrie, 'Dissolution Kinetics of Ferro-Alloys in Steelmaking', Steelmaking Proceedings, Vol. 65, Pittsburgh 1982, pp. 156-167.

14) L. Gourtsoyannis, R.I.L. Guthrie, G. Ratz, 'The Dissolution of Ferromolybdenum, Ferroniobium and Rare Earth (Lanthanide) Silicide in Cast Iron and Steel Melts', 1984, Proceedings of the Electric Furnace Conference, A.I.M.E., Vol. 42, Toronto, pp. 119-132.

15) D. Mazumdar, R.I.L. Guthrie, 1985, 'Hydrodynamic Modelling of some Gas Injection Procedures in Ladle Metallurgy Operations', Met. Trans. B., Vol. 16B, March 1985, pp. 83-90.

16) M. Salcudean, K.Y.M. Lai, R.I.L. Guthrie, 1985, 'Multi-dimensional Heat, Mass and Flow Phenomena in Gas Stirred Reactors', Canadian Journal of Chem. Eng., Vol. 63, February, 1985, pp. 51-61.

17) D. Mazumdar, R.I.L. Guthrie, 1985, "Hydrodynamics of the C.A.S. Method of Alloy Additions", Ironmaking and Steelmaking, Vol. 12, No. 6, pp. 256-264.

18) D. Mazumdar, R.I.L. Guthrie, 1985, "Numerical Computation of Flow and Mixing in Ladle Metallurgy Steelmaking Operations (C.A.S. Method)", Applied Mathematical Modelling, 1986, Vol. 10, pp. 25-32.

19) S. Tanaka, R.I.L. Guthrie, 1986; 'Concerning Slag Entrainment by Steel during Simulated Top and Bottom Blowing Operations in Steelmaking', Process Technology Proceedings, Volume 6, Washington, D.C., pp. 249-255.

20) D. Mazumdar, R.I.L. Guthrie, 'Alloying in Ladles with CAS', 1986; Process Technology Proceedings, Vol. 6, Vth Int. Iron and Steel Congress, I.S.S. of AIME, Washington, D.C. April 6-9th pp. 1147-1158.

21) D. Mazumdar, R.I.L. Guthrie, 1986, 'Mixing Models for Gas Stirred Metallurgical Reactors' Met. Trans. B. Vol. 17B, pp. 725-733.

22) M. Tanaka, D. Mazumdar and R.I.L. Guthrie, 1987, 'Alloy Additions Practices in Steelmaking Operations, Part I, Alloy Additions to Ladles during Furnace Tapping Operations', Met. Trans. B. Vol. 18B, (In press).

23) D. Mazumdar, R.I.L. Guthrie, 1987, Alloy Addition Practices in Steelmaking Operations, Part II, The CAS Alloy Addition Procedure, Met. Trans. B. vol. 18B, (In press).

24) S. V. Patankar, 'Numerical Heat Transfer and Fluid Flow, Hemisphere Publishing Corporation, New York, 1980.

25) F. Boysan and S.T. Johansen, 'Mathematical Modelling of Gas Stirred Reactor', Int'l Seminar on Refining and Alloying of Liquid Aluminium and Ferroalloys, Ed, T.A. Engh, S. Lyng, H.A. Oye, N.I.T. Trondheim, 1985, Aluminum-Verlag, Dusseldorf.

26) B.E. Launder and D.B. Spalding, Computer a Methods in Applied Mechanics and Engineering, Vol. 3, pp. 269, 1974.

27) D. Mazumdar, Ph.D. Thesis, Department of Mining and Metallurgical Egineering, McGill University, Montreal, Canada, 1985.

28) R.I.L. Guthrie, L. Gourtsoyannis, H. Henein, 1976, 'An Experimental and Mathematical Evaluation of Shooting Methods for Projecting Buoyant Additions into Liquid Steel Baths', Can. Met. Quarterly, 15, No. 2, pp. 145-153.

29)   S.V. Patankar and D.B. Spalding, International Journal of Heat and Mass Transfer, Vol. 15, pp. 1787, 1972.

30)   D.B. Spalding, Heat Transfer in Turbulent Buoyant Convection, Eds. D.B. Spalding and N. Afgan, Hemisphere Publishing Corporation, New York, 1977, pp. 569.

31)   K. Nakanishi, T. Fujii and J. Szekely, Ironmaking and Steelmaking, 1975, No. 3, pp. 193-197.

32)   S. Joo, Doctoral Candidate, McGill University, Metallurgical Eng. Unpublished research.

33)   K. Takashima, K. Arima, T. Shozi, and H. Mori: US Patent, 3971, 655, 1976.

34)   B. Kulunk, 'Agueous Modelling of Aluminum Wire Injection Procedures in Steelmaking'. (M.Eng. Thesis, McGill University, 1986).

MATHEMATICAL MODELLING OF EXOTHERMIC DISSOLUTION

Panagiotis G. Sismanis and Stavros A. Argyropoulos

Department of Metallurgy and Materials Science
University of Toronto
Toronto, Ontario, Canada M5S 1A4

Abstract

A mathematical model which describes heat and mass transfer events of exothermic dissolution of additions in liquid steel has been developed. Experimental results and model predictions will be shown on the dissolution of zirconium, titanium, niobium and tantalum in liquid steel. The good agreement between experimental results and model predictions indicates that the mathematical model describes well the complex coupled heat and mass transfer phenomena involved.

Introduction

Solid additives with melting temperatures above the melting point of steel, generally dissolve when they are immersed in liquid steel. Depending on the heat of mixing of the addition with the liquid steel, the dissolution may proceed with continuous supply of heat (i.e., endothermic dissolution; e.g.: Cr, Mo, W), or with heat generation (i.e., exothermic dissolution; e.g.: Ta, Nb, Ti, Zr). During the endothermic dissolution of high melting point metals in liquid steel, relatively small dissolution rates should be expected [1]. In addition to this, the temperature distribution inside the dissolving metal is smaller (or at most equal) in magnitude than the temperature of the surrounding steel bath. However, in the case of exothermic dissolution of high melting point metals in liquid steel the generated heat can give rise to temperature distributions inside the additive which can exceed the bath temperature. Furthermore, the heat and mass transfer events seem to influence each other so that the dissolution can even become self-accelerated. In this study, a mathematical model has been developed in order to predict and quantify these phenomena; predictions from this model have been tested against experimental results from dissolution studies of zirconium, titanium, and niobium cylinders in liquid steel.

Theoretical Considerations and Differential Equations

When a metallic addition at room temperature is immersed in liquid steel a layer of material from the melt solidifies around it. The thickness of this shell initially increases, reaches a maximum value, and finally diminishes because of the continuous supply of heat from the melt. This time duration is referred to as the steel shell period, after which follows the free dissolution period which ends with the complete assimilation of the additive in the melt.

Assuming that the metallic addition is of a cylindrical shape, then the temperature distribution inside it can be described by the heat conduction equation expressed in cylindrical coordinates:

$$\frac{1}{r} \frac{\partial}{\partial r} \left(kr \frac{\partial T}{\partial r} \right) + \frac{1}{r^2} \frac{\partial}{\partial \phi} \left(k \frac{\partial T}{\partial \phi} \right) + \frac{\partial}{\partial z} \left(k \frac{\partial T}{\partial z} \right)$$

$$+ \dot{q} = \rho c_p \frac{\partial T}{\partial t} \tag{1}$$

In the case of infinitely long cylinder, temperature variations along the vertical axial direction are ignored. The same is generally true for practical cases in which the length-over-diameter ratio of the cylinder is relatively large. Furthermore, if the temperatures within the main body of the cylinder are affected mainly by heat entering in the radial direction from the bath, ignoring any heat transfer by conduction up through the bottom surface, then it is inferred that there is no heat conduction in the axial direction and hence, $\partial T/\partial z = 0$. If in addition, the heat flow is symmetrical with respect to the angular component, (i.e., conditions at r and z, uniform and independent of ϕ), then the above equation reduces to:

$$\frac{1}{r} \frac{\partial}{\partial r} \left(kr \frac{\partial T}{\partial r} \right) + \dot{q} = \rho c_p \frac{\partial T}{\partial t} \tag{2}$$

Finally, without any heat generation ($\dot{q} = 0$), this general equation simplifies to that for transient radial conduction of heat in a cylinder:

$$\rho c_p \frac{\partial T}{\partial r} = \frac{1}{r} \frac{\partial}{\partial r} \left(kr \frac{\partial T}{\partial r} \right) \tag{3}$$

Equation (3) is solved to determine the temperature distribution inside the solid additive and the solidified steel shell. Figure 1 presents a typical

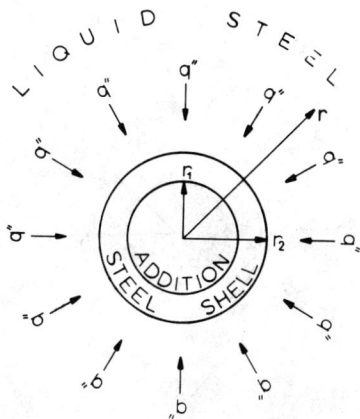

Figure 1 - Schematic representation of an addition with the solidified steel shell and the liquid steel; the various interfaces and the coordinate system used in the mathematical model are also illustrated.

set of events taking place during the steel shell period for a cylindrical additive immersed in liquid steel. The diagram represents a schematic cross-section perpendicular to the cylinder's axis. One can use equation (3) to write appropriate expressions for transient heat conduction in the solid additive and the steel shell:

Solid additive:

$$0 \leqslant r \leqslant r_1 , \qquad 0 \leqslant t \leqslant t_{tot}$$

$$\rho_s \, c_{p_s} \, \frac{\partial T}{\partial t} = \frac{1}{r} \frac{\partial}{\partial r} \left(k_s \, r \, \frac{\partial T}{\partial r} \right) \tag{4}$$

Solid steel shell:

$$r_1 \leqslant r \leqslant r_2 , \qquad 0 \leqslant t \leqslant t_{tot}$$

$$\rho_{Fe} \, c_{p_{Fe}} \, \frac{\partial T}{\partial t} = \frac{1}{r} \frac{\partial}{\partial r} \left(k_{Fe} \, r \, \frac{\partial T}{\partial r} \right) \tag{5}$$

Depending upon the type of the immersed metal, an exothermic reaction can start at the steel shell/additive interface once a certain threshold temperature is exceeded at this interface (2). This phenomenon has been approximated (in its simplest form) with a constant heat generation term at the inner steel shell interface, together with associated erosion of the steel shell. This erosion is caused by a flux of dissolved additive (i.e.,

485

zirconium or titanium) supplied through dissolution of its core. Figure 2 presents these phenomena in schematic form.

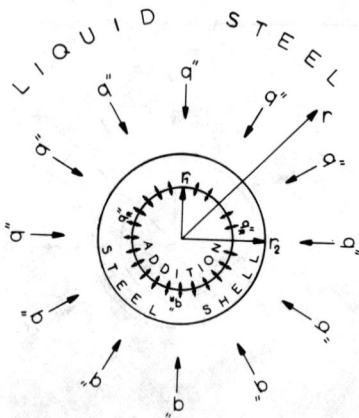

Figure 2 – Schematic representation of an addition reacting exothermically during the steel shell period; the various interfaces and the coordinate system used in the mathematical model are also illustrated.

A) Initial Conditions

Prior to immersion, the temperature of the cylinder can be taken to be uniform at T_o. Expressing the above condition mathematically,

(I.C.1) for $t = o$ and $o \leqslant r \leqslant r_1$ $T = T_o$ (6)

Similarly, bulk temperatures within the steel bath can be taken to be constant prior to immersion:

(I.C.2) for $t = o$ and $r > r_2$ $T = T_B$ (7)

B) Boundary Conditions

Boundary conditions have been listed in a systematic way starting at the cylinder's center and time zero, and proceeding radially outward towards the steel bath. Although the problem has been expressed in terms of five boundary conditions, those applying at the inside steel-shell boundary change once the exothermic reaction begins. The boundary conditions are:

(B.C.1) $o \leqslant t \leqslant t_{tot}$, $r = o$

$$\frac{\partial T}{\partial r} = 0$$ (8)

(B.C.2A) $o \leqslant t \leqslant t_R$, $r = r_1$

$$q'' = k_{Fe} \frac{\partial T}{\partial r} = k_s \frac{\partial T}{\partial r}$$

and $\quad q'' = \dfrac{T_{Fe}^* - T_s^*}{R_T}$ \hfill (9)

where T_{Fe}^* and T_s^* represent the temperature of the iron and solid additive interfaces in imperfect thermal contact.

(B.C.2B) $\quad t_R \leqslant t \leqslant t_{SSP}$, $\quad r = r_1$

$$- k_{Fe} \frac{\partial T}{\partial r} + N_s'' \Delta H_R = - k_s \frac{\partial T}{\partial r} \hfill (10)$$

where $R_T = 0$, and $N_s'' = K_s(C_s - C_{s/Fe})$.

The boundary condition (B.C.2B) suggests that when the exothermic reaction starts, the difference between heat fluxes into and out of the interface or reaction zone must be balanced by the heat flux generated by the exothermic reaction. Furthermore, when the reaction starts the contact resistance disappears (i.e., $R_T = 0$), and it is assumed that the reaction zone remains at the original iron/additive interface.

(B.C.3) $\quad o \leqslant t \leqslant t_{SSP}$, $\quad r = r_2$

$$T = T_{M.P.Fe} \hfill (11)$$

Equation (11) means that the temperature at the interface between the steel shell and the liquid steel is, in fact, the melting point of the steel bath.

(B.C.4) $\quad o < t < t_{SSP}$, $r = r_2$

$$\left(k_{Fe} \frac{\partial T}{\partial r}\right)_{Fe,shell} = \rho_{Fe} \lambda_{Fe} \frac{\partial r}{\partial t} + h \left(T_{BATH} - T_{M.P.Fe}\right) \hfill (12)$$

where λ_{Fe} is the latent heat of fusion of the melt. Equation (12) represents a heat balance for the moving steel solidification front at the steel shell/liquid steel interface. The heat transfer coefficient (h) from the bath to the enclosing steel shell surface, has been deduced from the following dimensionless correlation:

$$Nu_L = c \, Ra_L^{1/3} \hfill (13)$$

where c is a constant depending on the experimental conditions. Actually, c = 0.23 when there is some induction stirring to the melt (i.e., the power to the induction furnace is kept on during the test), and c = 0.17 when the induction furnace is turned off during the test. The validity of the equation (13) has been verified from another independent study [3,4], in which cylindrical specimens were immersed in similar steel baths to those used in the dissolution studies of zirconium, titanium, and niobium cylinders in liquid steel.

(B.C.5) $\quad o < t < t_{tot}$, $\quad r \to \infty$

$$T = T_{BATH} \hfill (14)$$

The last boundary condition demonstrates that the temperature of the steel bath far from the cylinder can be regarded as being constant.

487

In order to model the free dissolution, an effective heat transfer coefficient (h_{eff}) was introduced to take into account the convective heat transfer between the dissolving cylindrical additive and the steel bath. The heat flux that leaves the cylinder and goes into the melt during the free dissolution period, is called the 'outward flux', and is given by the formula:

$$q''_{out} = h_{eff} (T_I - T_B) \qquad (15)$$

The outward flux (q''_{out}) can be positive or negative depending on the sign of the difference ($T_I - T_B$). It can become negative at the very beginning of the free dissolution period, just after the shell has melted back, when the surface temperature of the cylinder is less than the bath temperature. The inward flux q''_{in}, that is the flux that enters the cylindrical additive at its dissolving front, is given by the following equation:

$$q''_{in} = k \left(\frac{\partial T}{\partial r}\right)_I \qquad (16)$$

where k is the thermal conductivity of the additive. The derivative of the temperature with respect to the radial distance ($\partial T/\partial r$), is taken at the interface (I).

A simple energy balance at the dissolving front demonstrates that the sum of the inward and outward fluxes is equal to the generated heat per unit surface per unit time due to the exothermic dissolution:

$$k \left(\frac{\partial T}{\partial r}\right)_I + h_{eff}(T_I - T_B) = N''_j \, \Delta\bar{H}_j^{\,o} \qquad (17)$$

The molar flux N''_j of the dissolving element j can also be written as:

$$N''_j = \frac{\rho}{M} \frac{dr}{dt} \qquad (18)$$

where $\dfrac{dr}{dt}$ = the reduction of the cylindrical additive's radius per time, or simply the dissolution speed.

One can group some constant parameters together:

$$Q_v = \frac{\rho}{M} \Delta\bar{H}_j^{\,o} \qquad (19)$$

where Q_v is the generated heat per unit volume. So, equation (17) now becomes:

$$k\left(\frac{\partial T}{\partial r}\right)_I + h_{eff} (T_I - T_B) = Q_v \frac{dr}{dt} \qquad (20)$$

The above formula (20) has been used to model heat transfer events during the free dissolution of cylindrical metallic additions (5), which exhibit exothermic dissolution in liquid steel. Actually, there is no information available on heat transfer characteristics for problems of this nature or, in other words, there is no correlation in the literature from which a value for the heat transfer coefficient (h_{eff}) could be obtained. However, there is a way that an approximation can be made.

When the dissolution takes place in an inductively stirred melt it is more likely that conditions of forced rather than natural convection prevail. For the cylinder in cross-flow, where the fluid flow is normal to the axis of the cylinder, the following empirical correlation is important from the standpoint of engineering calculations for heat transfer in forced convection environments [6]:

$$Nu_D = c_1 \; Re_D^{\;m} \; Pr^{1/3} \tag{21}$$

where Nu_D (= hD/k): the Nusselt number;

$\qquad Re_D$ (= $U_c D/\nu$): the Reynolds number;

$\qquad Pr$: the Prandtl number of the fluid;

$\qquad c_1, m$ are constants.

For a specific fluid and for a certain temperature range the Pr number is almost constant and hence, equation (21) can be written as:

$$Nu_D = c_2 \; Re_D^{\;m} \tag{22}$$

Recently, Churchill and Bernstein [7] have proposed a single comprehensive equation which covers the entire range of Re_D for which data are available, as well as a wide range of Pr. The equation is recommended for all $Re_D Pr > 0.2$ and has the form:

$$Nu_D = 0.3 + \frac{0.62 \; Re_D^{\;1/2} \; Pr^{1/3}}{[1 + (0.4/Pr)^{2/3}]^{1/4}} \left[1 + \left(\frac{Re_D}{282000}\right)^{5/8} \right]^{4/5} \tag{23}$$

For liquid steel where $Pr \simeq 0.12$ equation (23) can be used when $Re_D >$ $(0.2/0.12) = 1.7$. But for a 2.54 cm diameter cylinder, and a steel melt of a $7.6 \; 10^{-7}$ m^2/sec kinematic viscosity, it takes only a minimum cross-flow velocity of the order of 0.05 mm/sec for this constraint to be satisfied. Minimal natural convection driven forces can develop velocities of this order in liquid steel, although the velocity of the liquid steel in the vicinity of the dissolving cylinder should be of the order of a few centimeters per second. In any case, equation (23) can be applied to liquid steel, and the present authors used this correlation to deduce a relationship of the form (22), for engineering calculations in liquid steel (Pr $\simeq 0.12$) only:

$$Nu_D = 0.22 \; Re_D^{\;0.52} \tag{24}$$

The above correlation gives the same Nu_D numbers as equation (23) for liquid steel, for Reynolds numbers in the range of 300 to 15000; least-squares fitting was employed for the determination of its coefficients and the correlation coefficient was found to be 0.999. What is of great importance in equation (24) is the power to which the Reynolds number is raised or in other words, the functional relationship between the Nu_D number with the Re_D number. One has

$$Nu_D \propto Re_D^{\;0.52}$$

or

$$\frac{h_{eff} \; D}{k} \propto \left(\frac{U_c D}{\nu}\right)^{0.52} \tag{25}$$

It is clear from equation (25) that the heat transfer coefficient (h_{eff}) is related to the cross-flow velocity (U_c). During the free dissolution period however, the density difference between the dissolving metal and the liquid steel generates motion of the fluid along the cylinder axis and not across it, leaving the cross-flow velocity (U_c) almost constant. Therefore, from equation (25) one can derive that:

$$h_{eff} \quad \propto \quad D^{-0.48} \tag{26}$$

Equation (26) clearly states that the effective heat transfer coefficient increases as the free dissolution proceeds. This is a reasonable result since during the free dissolution period mass-transfer phenomena appear, and become equally important with the heat transfer ones, so that the overall convective heat transfer increases due to the cumulative effect of two events. Also, as the heat transfer coefficient (h_0) at the end of the steel shell period cannot be different from the heat transfer coefficient at the beginning of the free dissolution period, the following equation is finally derived:

$$h_{eff} \quad = \quad h_0 \left(\frac{D_0}{D}\right)^{0.48} \tag{27}$$

When the dissolution proceeds in a natural convection environment there will be some degree of turbulence associated with the exothermic dissolution, and hence, the Nusselt number (Nu_D) will be proportional to the Rayleigh number (Ra_D) raised to the power of 1/3. A correlation of this type suggests that the effective heat transfer coefficient will be independent of the diameter (D) of the dissolving cylindrical specimen. But as was mentioned previously, the heat transfer coefficient at the beginning of the free dissolution must be equal to the heat transfer coefficient (h_0) estimated at the end of the steel shell period using the correlation (13) and c = 0.17 (i.e., power off). So, for the free dissolution of a metallic cylinder under conditions of natural convection in a quiescent (i.e., no inductively stirred) steel bath, the following formula can be used:

$$h_{eff} \quad = \quad h_0 \tag{28}$$

It is interesting to note that equations (27) and (28) give almost the same results when the rate of change of the cylinder radius per unit time is relatively small, so that $D \simeq D_0$, and hence $h_{eff} \simeq h_0$.

Numerical Solution of the Differential Equations

As the set of partial differential equations (4) and (5) is too complex for analytical solutions with such boundary conditions, numerical procedures were employed to predict shell thickness, melting rates, etc. as a function of time. The explicit finite-difference numerical technique was employed [8] for solving the set of partial differential equations for the boundary conditions just described. Assumptions involved in describing the physical phenomena of a cylinder's dissolution/melting were:

(a) Conduction up through the bottom surface was ignored. Thus temperatures in the main body of the cylinder are affected only by heat entering in the radial direction from the bath.

(b) The heat transfer was taken to be symmetrical about the cylinder's axis (i.e., slight variations with respect to length are ignored).

(c) The immersion of the cylindrical specimen in the liquid steel is assumed to take place instantaneously.

(d) The cylinder maintains its outer physical dimensions during the freezing and melting process involved (i.e., variations of density were neglected).

(e) The material within each elemental volume or node was assumed to have uniform temperature and thermophysical properties.

(f) The temperature of the solid-liquid steel interface was constant during the freezing or melting process of the steel shell.

The numerical equivalent of differential equations (4) and (5) cast in fully-explicit finite difference form is given by the following formula:

$$T_N' = T_N (1 - 2M) + M \left(\frac{2N - 1}{2N - 2}\right) T_{N+1} + M \left(\frac{2N - 3}{2N - 2}\right) T_{N-1} \tag{29}$$

where $M = \dfrac{k \, \Delta t}{\rho \, c_p \, (\Delta r)^2}$ is the dimensionless Fourier Modulus:

Δr: distance between two adjacent nodal points;

Δt: iteration time;

N: is the corresponding (Nth) nodal point; the central nodal point located on the cylinder's axis was designated as number 1, and successive points as 2,3,4, etc.

T_N': the new temperature of nodal point N at time $t + \Delta t$. As seen, a knowledge of the nodal point temperature at time t allows new nodal point temperature (T') to be computed at time $t + \Delta t$ in any region of the cylinder (i.e., addition, steel shell).

The finite difference equation for the central nodal point 1, satisfying boundary condition (8) is:

$$T_1' = T_1 (1 - 4M) + 4 M T_2 \tag{30}$$

Referring to Figure 3, let us assume that the solid-liquid steel interface at time t is in position A. Let X_t be the distance between the last nodal point N + 1 and the liquid-solid steel interface. At time $t + \delta t$, the new interface position is given by a dashed line. Then the relevant boundary condition equation (12) can be written in finite difference form as follows [8]:

$$\rho_{Fe} \, \lambda_{Fe} \, \frac{\delta r}{\delta t} = k_{Fe} \left(\frac{T_{M.P.Fe} - T_{N+1}}{X_t}\right) - h \left(T_B - T_{M.P.Fe}\right) \tag{31}$$

Each time X_t exceeded Δr (radial nodal point distance), a new nodal point was assigned, labelled N+1. The remaining nodes were relabelled, becoming N, etc. Similarly, when δr became negative, the reverse procedure was applied. In the case where a new nodal point was assigned and labelled N+1, the temperature for this new nodal point was calculated, using linear interpolation between the temperature of nodal point N and the melting point of steel (temperature of solid-liquid steel interface). The following equation gives the temperature of the newly assigned nodal point, N+1:

$$T_{N+1}' = \frac{T_{M.P.Fe} \, \Delta r + X_t \, T_N'}{X_t + \Delta r} \tag{32}$$

where T_{N+1}: new temperature for new nodal point (N+1);

$T_N{}'$:   new temperature for the N nodal point;
$T_{M.P.Fe}$:   steel melting point.

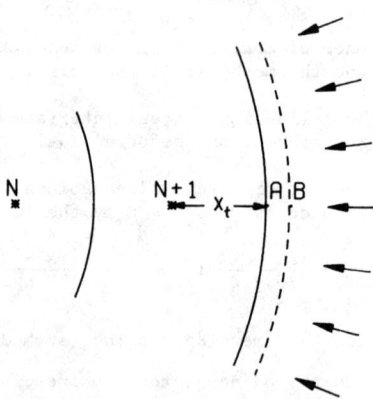

Figure 3 — Schematic representation of the movement of the steel shell/liquid interface. X_t is the distance of last nodal point from this interface at time t. The dashed line represents the position for this interface at time t + δt.

In the case where there is heat generation inside an elemental volume N, this is taken into account through proper modification of the corresponding finite difference equation. Let us assume that heat is generated in the nodal point, N, at a rate of $\dot{q}$ per unit volume. Writing a heat balance for nodal point N for unsteady-state heat conduction between N and two adjacent nodal points (N+1) and (N−1) yields the general equation:

$$T_N{}' = T_N (1 - 2M) + M \left(\frac{2N - 1}{2N - 2}\right) T_{N+1} +$$

$$+ M \left(\frac{2N - 3}{2N - 2}\right) T_{N-1} + \frac{\dot{q} \, \Delta t}{\rho \, c_p} \tag{33}$$

Equation (33) represents the generalized transformation of equation (2) into an equivalent finite difference form.

Any phase transformation that the metallic addition may suffer (i.e., α-Ti transforms to β-Ti endothermically at 1155 K or 882°C), is taken into account by the model as well.

The numerical computations were performed in the following way. The temperatures and the M modules for all nodal points were initially assigned with values corresponding to the initial temperatures and initial M modules for each material used. Then, over the next time step (Δt), the temperature for each nodal point was recalculated using the finite difference equations (29), (30), and (32). The shell thickness was computed using equation (31). After that, the corresponding M modules for each nodal point were assigned and the calculations for the new temperatures were repeated over successive time intervals.

492

This model was coded in FORTRAN and was compiled by a FORTRAN-H compiler residing in an AMDAHL-V7 computer, as well as by two FORTRAN-77 compilers which have been installed in two different systems: an HP-9000 minicomputer and an EXORmacs (MOTOROLA/M68000) microcomputer. Double precision arithmetic was employed, and although the results obtained by the three different systems were identical, execution speed was proven to be remarkably different. The AMDAHL-V7 computer was found to be 8 times faster than the HP-9000 minicomputer, and the latter was proven to be 10 times faster than the EXORmacs.

Application of the Mathematical Model

(i) Titanium

In Figure 4 a typical experimental result from the dissolution of a 3.81 cm (1.5 in) diameter titanium cylinder [9] in liquid steel is presented. Curve 1 shows the net downward force as registered by a weight sensor which was attached to the specimen, and in this way it monitored the apparent weight of the sample.

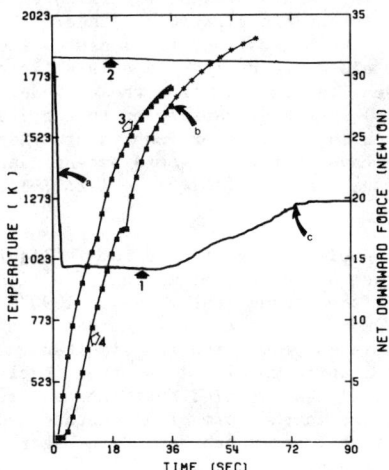

Figure 4 - Results from a titanium dissolution experiment in liquid steel. Cylinder diameter 3.81 cm and 21 cm in length: 1) Registered net downward force; 2) Measured steel bath temperature; 3) Measured and predicted temperatures at a point 1.35 cm from cylinder centerline; 4) Measured and predicted temperatures at the cylinder's centerline.

Actually, the following formula is true:

$$F_{NDF} = F_G - F_B \qquad (34)$$

where F_{NDF}: the net downward force;
  F_G: the weight of the specimen;
  F_B: the buoyancy force.

493

Equation (34) can be differentiated by time, and after a few manipulations it can be written in an equivalent form:

$$\frac{dF_{NDF}}{dt} = \frac{dV_{IMM}}{dt} (\rho_s - \rho_{Fe,\ell}) g \tag{35}$$

where V_{IMM}: the immersed portion of the cylinder.

Segment AB of Curve 1 (Figure 4) depicts the immersion period of the titanium cylinder in liquid steel. Segment BC corresponds to the steel shell period; as stated previously, there is a shell of steel which solidifies around the cylinder immediately after immersion, due to the large temperature difference between the melt and the addition. During this period there are no significant changes of the net downward force recorded, as the shell formation is in a manner compensated by the volumetric expansion of the addition. When the shell finally melts back, the cylinder is directly exposed to the melt and the main dissolution period commences (segment CD). As the density of titanium is less than the density of liquid steel, the factor $(\rho_s - \rho_{Fe,\ell})$ is negative; this negative factor is multiplied by (dV_{IMM}/dt) which is also negative, so according to equation (35) the net downward force increases during the free dissolution period [9]. After D, no significant changes were recorded since the addition had already been assimilated in liquid steel. Curve 2 shows the steel bath temperature during the test. Curves 3 and 4 present experimental and predicted temperatures in the interior of the titanium cylinder. The solid lines represent predictions while the points show the measured temperatures. Curve 4 shows the temperature along the cylinder's axis while curve 3 presents temperatures at a distance 1.35 cm away from the cylinder's axis. The centerline temperature (4) increases rapidly up to the 21st second of immersion. A reduction in the rate of temperature rise then follows up to the 23rd second. The cause for this slowed ascent can be explained in terms of the endothermic transformation [10] of α-Ti to β-Ti, which occurs at 882°C (1155 K).

During the steel shell period of titanium in liquid steel, an exothermic reaction takes place at the titanium/steel shell interface once the temperature at this interface reaches the value of 1090°C (1363 K). It has been determined experimentally that the formation of the FeTi and Fe$_2$Ti liquid intermetallics takes place with the simultaneous erosion of the inner part of the steel shell which results in shorter steel shell periods [9]. For modelling purposes, it was assumed that the heat released due to the exothermic reaction acted in the form of a constant heat flue (q_s") at the titanium/steel shell interface (noted as q_R" in Figure 2). From equation (10) (B.C.2B), one then has:

$$q_s" = N_s" \Delta H_R \tag{36}$$

q_s" was determined semi-empirically as the value which resulted in the best agreement between experimental and predicted data. For the case presented in Figure 4 q_s" was found to be 9.7 cal/cm^2/sec or 0.4 MW/m^2. The arrows a, b, and c signify the two periods: steel shell period (a,b), and free dissolution period (b,c).

(ii) Zirconium

In Figure 5 a typical experimental result from the dissolution of a 3.81 cm-diameter zirconium cylinder [5] in liquid steel is presented. Curve 1 shows the bath temperature, and curve 2 depicts the net downward force as registered by the weight sensor. During the period AB the cylinder was above the melt while the segment BC signifies the immersion period.

494

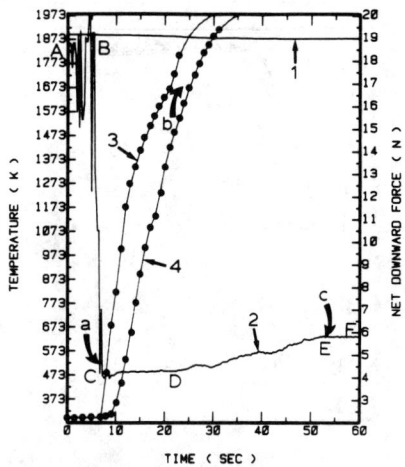

Figure 5 – Results from a typical zirconium dissolution experiment
in liquid steel. The cylinder diameter was 3.81 cm and its length
was 20.5 cm. 1) Measured steel bath temperature; 2) Registered net
downward force; 3) Measured and predicted temperature at a distance
1.4 cm apart from the center; 4) Measured and predicted centerline
temperature.

Segments CD and DE denote the steel shell and the free dissolution periods
while no significant changes are recorded after point E. Equation (35)
still applies and as the density of zirconium is smaller than the density of
liquid steel, the net downward force increases during the free dissolution
period, as in the case of titanium. Curves 3 and 4 present measured and
predicted temperatures, respectively, at a distance 1.4 cm apart from the
center of the cylinder and along the centerline. The solid lines represent
predictions, while the points show the measured temperatures. The centerline temperature (curve 4) increases rapidly up to the 17th second of
immersion. A reduction in the rate of the temperature rise then follows up
to the 19th second. The cause for this slowed ascent can be explained in
terms of the endothermic transformation of α-Zr to β-Zr. This occurs [10]
at a temperature of 863°C (1136 K).

As in the titanium case, zirconium also exhibits an exothermic reaction
during the steel shell period. This happens when the temperature at the
steel shell/zirconium interface exceeds the value of 947°C (1220 K). It was
experimentally determined that this reaction gives rise to the formation of
the liquid Fe_2Zr and $FeZr_2$ intermetallic compounds [5]. The value of q_s''
was semi-empirically deduced and for the test presented in Figure 5 was
found to be 11.2 cal/cm^2/sec or 0.47 MW/m^2.

(iii) Niobium

Figure 6 illustrates a typical experimental result of a 2.54 cm (1 in)
niobium cylinder [5] in liquid steel. Curve 1 shows the bath temperature
and Curve 2 depicts the net downard force. Segment BC denotes the immersion

495

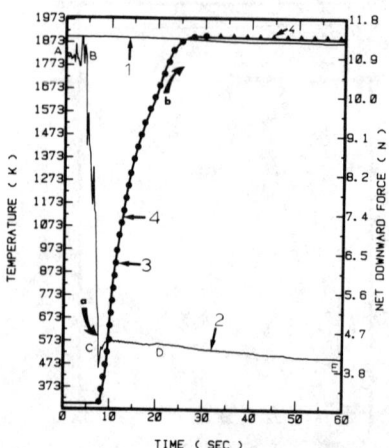

Figure 6 – Results from a typical niobium dissolution
experiment in liquid steel. The cylinder diameter was
2.54 cm and its length was 18 cm. 1) Measured steel
bath temperature; 2) Registered net downward force;
3) Measured and predicted (4) centerline temperature.

period; segment CD signifies the steel shell period. The free dissolution
period starts at D but, as niobium dissolves slowly in liquid steel, it does
not end at E. As the density of niobium is larger than the density of
liquid steel, the net downward force increases during the free dissolution
according to equation (35). Measured (3) and predicted (4) values of the
centerline temperature are also presented in the same figure. Niobium does
not suffer any phase transformation. It has been computationally found and
experimentally verified that a reaction may or may not take place during the
steel shell period of niobium in liquid steel, the phenomenon being exclusively controlled by heat transfer. In other words, at low superheats where
the steel shell periods are longer, there is more time for the temperature
at the steel shell/niobium interface to reach the value of 1370°C (1643 K),
at which the exothermic formation of the Fe_2Nb and Fe_2Nb_3 intermetallics
takes place. As the starting point for the reaction is relatively high
(1370°C), even when the reaction happens, this occurs at the final stages of
the steel shell period and therefore the semi-empirical determination of $q_s{}''$
is not possible.

Discussion

As is described for the titanium case (Figure 4), it is observed that
the large exothermic heat of mixing of Ti in liquid steel gives rise to an
increase in the temperature distribution in the interior of the cylinder,
which finally melts. In Figure 7, Curve 1 shows the cylinder radius for the
same test. During the steel shell period (region A) the radius of the
cylinder is larger than the initial one due to the existence of the shell.
In the free dissolution period (region B) however, the radius of the titanium cylinder decreases relatively slowly in the beginning, and faster in

496

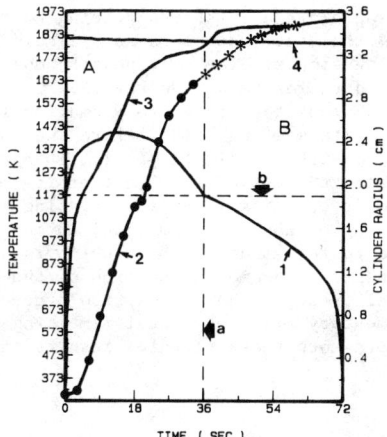

Figure 7 - Experimental and predicted results from a titanium
experiment in liquid steel (see Fig. 4). The cylinder
diameter was 3.81 cm and its length was 21 cm. 1) Predicted
cylinder radius; 2) Measured and predicted centerline temperature; 3) Predicted temperature at the titanium/steel interface;
4) Measured steel bath temperature.

the end when melting occurs. One can notice that both the temperatures, at
the centerline (Curve 2) and at the titanium/steel interface (Curve 3)
approach the melting point of Ti, which is far above the bath temperature
(Curve 4). Titanium has a relatively low melting point and exhibits a very
exothermic mixing and accelerated dissolution in liquid steel. The rate of
change of its radius by time (i.e., $-dr/dt$) increases rapidly during the
free dissolution period. In fact, computational results show that melting
follows once the dissolution speed has reached the value of 0.035 cm/sec,
for a 3.81 cm (1.5 in) initial-diameter cylinder. Another important aspect
of the accelerated exothermic dissolution is that the total generated heat
flux increases. This flux is given by the following formula:

$$q_{tot}^{"} = N_j^{"} \Delta \bar{H}_j^0 = Q_v \frac{dr}{dt} \qquad (37)$$

where j = the dissolving element (i.e., Ti, Zr, Nb).

It is obvious from equation (37) that as the dissolution speed
increases, so does the total generated heat flux. This heat is distributed
between the addition ($q_{in}^{"}$) and the melt ($q_{out}^{"}$), so that the following
equation holds true:

$$q_{tot}^{"} = q_{in}^{"} + q_{out}^{"} \qquad (38)$$

Equation (38) is identical to equation (20) presented earlier in this work.
The inwards directed heat flux ($q_{in}^{"}$) is critical to the computation of the
temperature distribution inside the dissolving cylinder. One should expect
that $q_{in}^{"}$ should be larger in the beginning of the free dissolution, just

after the shell has melted back, and when the cylindrical specimen is relatively cooler in comparison with the melt. The direct exposure of the titanium cylinder to the melt together with its exothermic fashion of
dissolution, heat it up rapidly and it starts getting hotter than the liquid
steel. At the same time, the steel bath acts as a liquid metal heat-
exchanger to the dissolving sample. The melt absorbs most of the exothermically generated heat, and therefore q_{in}'' decreases. However, although the
inwards directed heat flux drops to a relatively small value, it is enough
to heat the cylinder up to its melting point before the whole specimen has
been completely assimilated with the liquid steel. Then an interesting phenomenon takes place in which the cylinder is kept at its melting point with
the exothermic dissolution proceeding in an accelerating manner. In this
way there is partial melting of the titanium specimen and the inwards heat
flux increases again, due to the amount of heat which is absorbed in the
form of latent heat of fusion. The molten portion of the totally dissolving
material is initially small, but as the dissolution proceeds, it becomes
larger with the last elementary piece of titanium melting completely in
liquid steel. Figure 8 presents these computed results in graphical form.

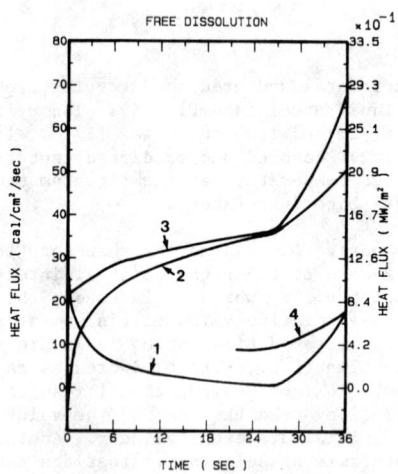

Figure 8 - Predicted heat fluxes during the free dissolution
period for a titanium dissolution test in liquid steel (see
Fig. 4). 1) The inwards directed heat flux; 2) The outwards
directed heat flux; 3) The generated (total) heat flux; 4)
Required flux for melting.

Curve 1 depicts the inwards directed heat flux q_{in}''. Curve 2 shows the outwards directed heat flux q_{out}'', and Curve 3 illustrates the generated (total)
heat flux q_{tot}''. Curve 4 depicts the heat flux attributed to melting. In
other words, this would be the necessary heat flux to melt the titanium
sample with the same rate as recorded experimentally. The average inwards
directed heat flux was estimated to be 5.3 cal/cm^2/sec or 0.22 MW/m^2, for
the test presented in Figure 4.

Very similar results have been observed for the dissolution of zirconium in liquid steel. Figure 9 shows the predicted cylinder radius (Curve

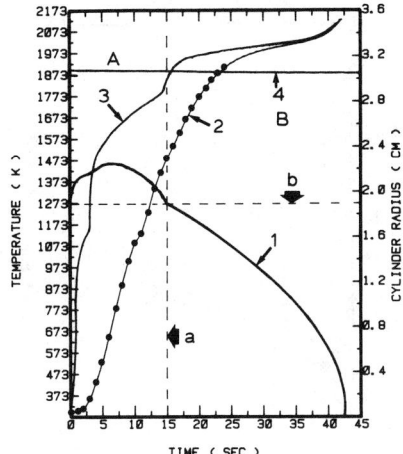

Figure 9 — Experimental and predicted results from a zirconium
experiment in liquid steel (see Fig. 5). The cylinder diameter
was 3.81 cm and its length was 20.5 cm. 1) Predicted cylinder
radius; 2) Measured and predicted centerline temperature;
3) Predicted temperature at the zirconium/steel interface;
4) Measured steel bath temperature.

1) for the test presented in Figure 5. Curves 2 and 3 illustrate the manner
with which the temperatures inside the cylinder approach the melting point
of zirconium in the final stages of the free dissolution. More time is
required for the zirconium melting point to be reached and therefore, more
vigorous dissolution phenomena and larger dissolution speeds than titanium
are recorded. Actually, computational results reveal that for the zirconium
case melting follows when the dissolution speed has reached the value of
0.165 cm/sec, for a 3.81 cm (1.5 in) initial-diameter cylinder. In other
words, as Zr has a higher melting point than Ti, the dissolution speed can
become up to 5 times larger than the one for Ti before melting begins.
Figure 10 illustrates the heat fluxes which were predicted for the zirconium
test presented in Figure 5. In average terms, the inwards directed heat
flux is larger in the case of zirconium than in the case of titanium. In
fact, for the zirconium test under consideration the average inwards
directed heat flux was found to be 6.2 cal/cm^2/sec or 0.26 MW/m^2. This can
be attributed to the fact that zirconium exhibits larger dissolution rates
with a larger value for the heat of mixing at infinite dilution. One may
notice (Figure 10) that the generated (total) heat flux can become up to 4
times larger in the dissolution of a 3.81 cm initial-diameter zirconium
cylinder than during the dissolution of a similar titanium cylinder. It
should be pointed out that the increase in the inward heat flux which was
observed in the case of titanium cannot happen for zirconium. The zirconium
melting point is almost approached in the end of its free dissolution
period.

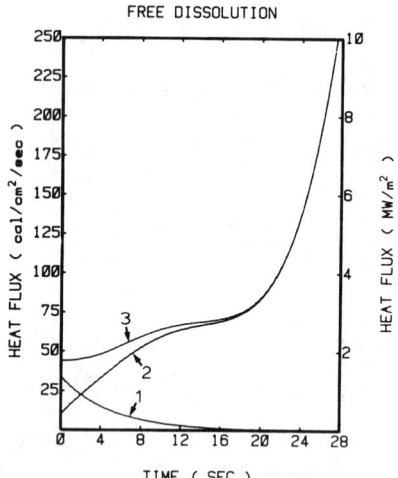

FREE DISSOLUTION

Figure 10 - Predicted heat fluxes during the free dissolution period
for a zirconium dissolution test in liquid steel (see Fig. 5).
1) The inwards directed heat flux; 2) The outwards directed heat
flux; 3) The generated (total) heat flux.

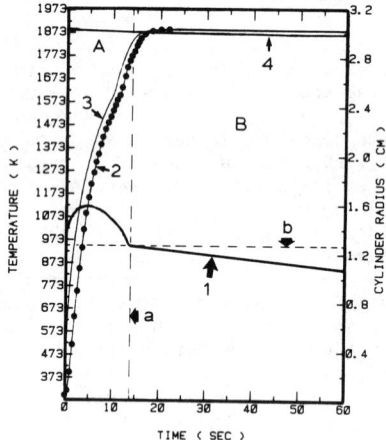

Figure 11 - Experimental and predicted results from a niobium experiment
in liquid steel (see Fig. 6). The cylinder diameter was 2.54 cm and its
length was 18 cm. 1) Predicted cylinder radius; 2) Measured and predicted centerline temperature; 3) Predicted temperature at the niobium/steel
interface; 4) Measured steel bath temperature.

500

Niobium is a high melting point metal which exhibits exothermic dissolution in liquid steel. From the experimental test described in Figure 6 it was shown that niobium dissolves slower than titanium and zirconium. Experimental results exist during the initial stages of the free dissolution and therefore, the analysis presented in this work will be restricted only to this period. It has been verified experimentally that the dissolution speed of a niobium cylinder remains almost constant during the early stages of the free dissolution period. Curve 1 in Figure 11 shows the predicted radius of the cylinder whose test was described in Figure 6. It is easily noticed that the radius decreases in an almost linear manner during the free dissolution period (region B). On the other hand it is observed that the temperature distribution inside the cylinder (Curves 2 and 3) ultimately exceeds the bath temperature (Curve 4) by an almost constant value. This can be explained by the fact that the dissolution speed of niobium in liquid steel is small, and also by the fact that the generated heat due to the exothermic dissolution is small. Furthermore, the niobium thermal conductivity is large (almost two times the thermal conductivity of zirconium, and three times the thermal conductivity of titanium at steelmaking temperatures), therefore the temperature distribution inside the cylinder becomes almost flat, or in mathematical terms:

$$k \frac{\partial T}{\partial r} \simeq 0 \qquad (39)$$

From equation (39) one then has (because of equation (20)):

$$h_{eff} (T_I - T_B) \simeq Q_v \frac{dr}{dt}$$

or

$$T_I - T_B \simeq \frac{Q_v}{h_{eff}} \frac{dr}{dt} \qquad (40)$$

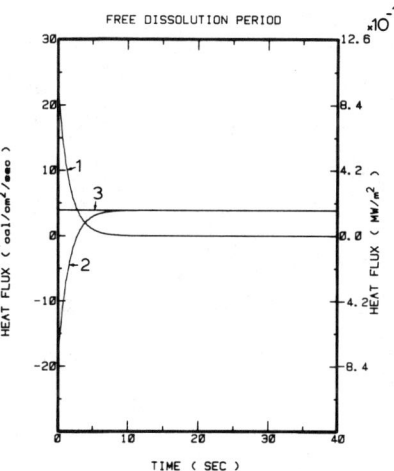

Figure 12 — Predicted heat fluxes during the free dissolution period for a niobium dissolution test in liquid steel (see Fig. 6). 1) The inwards directed heat flux; 2) The outwards directed heat flux; 3) The generated (total) heat flux.

For the niobium dissolution, the right-hand-side part of equation (40) is almost constant. So this formula suggests that the temperature of the dissolving front of a niobium cylinder in liquid steel exceeds the melt temperature by a constant value, at least for the initial stages of the free dissolution. In Figure 12, the estimated heat fluxes during the initial stages of the free dissolution period of niobium in liquid steel are presented.

Niobium may exhibit accelerated exothermic dissolution, similar to the ones observed for titanium and zirconium, in the final stages of the free dissolution when the cylinder radius approaches the value of zero. Experimental difficulties, however, leave this theoretical assumption untested at present [5].

Figure 13 illustrates the radii of cylindrical specimens with a 3.81 cm (1.5 in) initial diameter, during their immersion in liquid steel at 1600°C under static conditions (i.e., quiescent melt). The predictions were made for cylindrical specimens of pure titanium (Curve 2), zirconium (Curve 3), and niobium (Curve 4). Although zirconium exhibits a shorter steel shell period than titanium, their total dissolution times are almost equal. Niobium requires much more time than one hundred seconds (i.e., about 445 seconds) to dissolve completely in liquid steel. It should be pointed out that a cylindrical specimen of pure vanadium (3.81 cm initial diameter) should exhibit almost the same dissolution behaviour [8] as the one presented by Curve 4 of Figure 13.

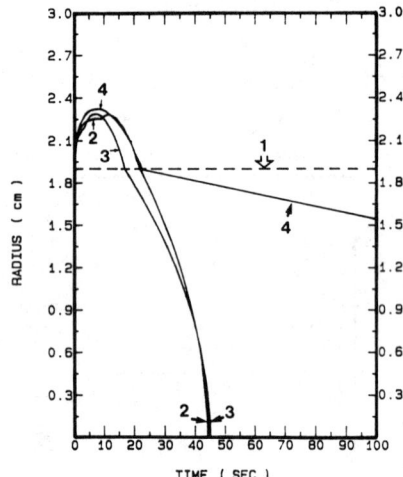

Figure 13 - Predicted radii as a function of time for the dissolution of 3.81 cm initial-diameter cylindrical specimens made of (2) pure titanium, (3) pure zirconium, and (4) pure niobium.

An attempt to predict the first sixty seconds of tantalum dissolution in liquid steel was undertaken as well. There is no experimental data available for its dissolution in liquid steel. However, it seems that similar dissolution behavior in liquid steel is exhibited by metals which belong

502

to specific groups of the periodic table of the elements. So far, experimental evidence exists for the metals of the groups IVA (Ti, Zr), VA (V, Nb), and VIA (Cr, Mo, W). Hence, it was assumed that tantalum exhibits similar dissolution behavior with niobium. Liquid phase mass-transfer was considered the rate-controlling step and a value for the dissolution speed was estimated (Appendix II) to be $2.5 \ 10^{-3}$ cm/sec. In Figure 14, Curve 1 shows the radius of the cylinder during the steel shell period (region A) and the free dissolution period (region B), for a tantalum cylinder of 2.54 cm initial diameter. Curve 2 is the predicted temperature along the centerline and Curve 3 is the temperature at the interface. Tantalum shows exothermic mixing in liquid iron and, as a result of this, the temperature at the interface must be somewhat greater than the bath temperature. Tantalum should exhibit small dissolution rates in liquid steel, and the total dissolution time for a 2.54 cm initial-diameter cylinder was estimated to be about 523 seconds.

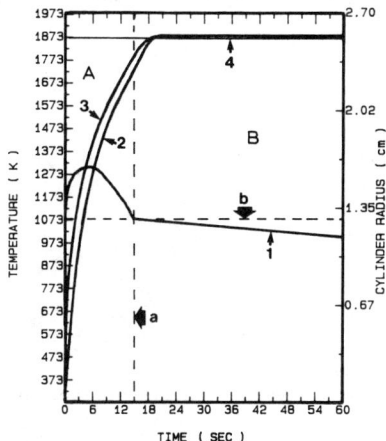

Figure 14 - Predictions for the dissolution of a 2.54 cm initial-diameter cylindrical specimen of tantalum in liquid steel, at 1600°C. 1) Cylinder radius; 2) Centerline temperature; 3) Interface temperature.

Conclusions

The mathematical model developed can successfully predict the complex heat and mass transfer events which take place during the exothermic dissolution of various metals in liquid steel. The predictions are in good agreement with experimental results. The dissolution rate in liquid steel increases in the order of Nb - Ti - Zr. For these elements, the heats of mixing with liquid iron as well as their melting points seem to be the most important factors affecting their dissolution speed.

Acknowledgements

The authors would like to express their sincere appreciation to the Natural Sciences and Engineering Research Council of Canada for providing support for this work.

Nomenclature

c_p	:	heat capacity at constant pressure
C	:	concentration
D_s	:	diameter
F	:	force
Gr_M	:	Grashof number for mass transfer
g	:	acceleration of gravity (9.81 m/s^2)
h	:	heat transfer coefficient
k	:	thermal conductivity
K_s	:	mass transfer coefficient during the steel shell period
L	:	length
M	:	molecular weight; Fourier modulus
N''	:	molar flux
Nu	:	Nusselt number
Pr	:	Prandtl number
q''	:	heat flux
$\dot{q}$	:	heat generation per unit volume per unit time inside the cylinder
Q_v	:	heat generation per unit volume
r	:	radius; radial distance
Ra	:	Rayleigh number
Re	:	Reynolds number
R_T	:	contact resistance
Sc	:	Schmidt number
Sh	:	Sherwood number
T	:	temperature
t	:	time
U_c	:	cross-flow velocity
V	:	volume
z	:	axial distance
ρ	:	density
ϕ	:	angle
ν	:	kinematic viscosity
λ	:	latent heat of fusion
ΔH_R :	:	heat generated at the steel shell/additive interface per mole of reacting additive
$\Delta \bar{H}_j^0$	:	partial molar heat of mixing at infinite dilution of element j in liquid iron

Subscripts

B	:	referring to the bath
eff	:	effective
IMM	:	immersed
I	:	referring to the interface
j	:	element
ℓ	:	liquid
L	:	length
M.P.	:	melting point
N	:	nodal point
SSP	:	steel shell period
in	:	inward
out	:	outward
tot	:	total

Thermophysical and Thermodynamic Data used for the Computations

Titanium

Transformation temperature [10]: 882°C or 1155 K
Melting point [11]: 1667°C or 1940 K

Heat capacity [8]	cal/g/°C	J/kg/K
α-Ti	0.1478	618.4
β-Ti	0.1600	669.4
liquid Ti	0.1774	742.2

Thermal conductivity [8]	cal/cm/sec/°C	W/m/K
α-Ti	0.0485	20.3
β-Ti	0.0578	24.2
liquid Ti	0.0580	24.3

Density at 20°C: 4.5 g/cm³
Latent heat of transformation [10]: 21.232 cal/g or 88.8 kJ/kg
Latent heat of fusion [10]: 77.0772 cal/g or 322.5 kJ/kg
Reaction starting point [8]: 1090°C or 1363 K
Partial molar heat of mixing at infinite dilution in liquid iron [12]:
    -15100 cal/gr-atom or -63.2 kJ/mol atoms

Zirconium

Transformation temperature [10]: 863°C or 1136 K
Melting poing [10]: 1852°C or 2125 K

Heat capacity [8]	cal/g/°C	J/kg/K
α-Zr	0.0794	332.2
β-Zr	0.0822	343.9
liquid Zr	0.0877	366.9

Thermal conductivity [13]	cal/cm/sec/°C	W/m/K
α-Zr	0.0550	23.0
β-Zr	0.0690	28.9
liquid Zr	0.0813	34.0

Density at 20°C: 6.49 g/cm³
Latent heat of transformation [10]: 10.3157 cal/g or 43.16 kJ/kg
Latent heat of fusion [10]: 44.2666 cal/g or 185.2 kJ/kg
Reaction starting point [11]: 947°C of 1220 K
Partial molar heat of mixing at infinite dilution in liquid iron [5]
    -19750 cal/gr-atom or -82.634 kJ/mol atoms

Tantalum

Melting point [11]: 3020°C or 3293 K

Heat capacity [14]	cal/g/°C	J/kg/K
20-1000°C	0.034	142.2
1000-3020°C	0.036	150.6
liquid Ta	0.041	171.5

Thermal conductivity [14]	cal/cm/sec/°K	W/m/K
20-1000°C	0.141	59.0
1000-3020°C	0.148	61.9
liquid Ta	0.163	68.2

Density at 20°C: 16.6 g/cm^3
Latent heat of fusion [10]: 41.45 cal/g or 173.4 kJ/kg
Reaction starting point [10]: 1440°C of 1713 K
Partial molar heat of mixing at infinite dilution in liquid iron [15]:
   -10300 cal/gr-atom or -43.1 kJ/mol atoms
Molecular weight : 180.948

Niobium

Melting poing [10]: 2467°C or 2740 K

Heat capacity [13]	cal/g/°C	J/kg/K
20-927°C	0.0691	289.1
927-2467°C	0.0786	328.8
liquid Nb	0.0861	360.2

Thermal conductivity [13]	cal/cm/sec/°C	W/m/K
20-927°C	0.1446	60.5
927-2467°C	0.1752	73.3
liquid Nb	0.1528	63.9

Density at 20°C: 8.6 g/cm^3
Latent heat of fusion [10]: 67.8317 cal/g or 283.8 kJ/kg
Reaction starting point [11]: 1370°C of 1643 K
Partial molar heat of mixing at infinite dilution in liquid iron
[5,15]: -9000 cal/gr-atom or -37.7 kJ/mol atoms

Liquid steel (ARMCO iron)

Melting point: 1520°C or 1793 K
Heat capacity: 0.1898 cal/g/°C or 794 J/kg/K
Thermal conductivity: 0.0828 cal/cm/sec/°C or 34.6 W/m/K
Density at the melting point: 7.0 g/cm^3
Coefficient of thermal volumetric expansion: $1.3 \ 10^{-4} \ K^{-1}$
Kinematic viscosity: $7.6 \ 10^{-7} \ m^2/sec$
Latent heat of fusion: 65.5 cal/g or 274 kJ/kg
Resistance (R_T): 8-9 cm^2 sec °C/cal or $1.9 \ 10^{-4} - 2.1 \ 10^{-4} \ m^2$ K/W

APPENDIX II

From the liquidus curve of the Fe-Ta phase diagram [11] at 1600°C, the concentration at the interface should be 45 wt % Ta and 55 wt % Fe. The density at the interface should be 10.995 g/cm^3 and the driving force 0.02734 mol Ta/cm^3. Assuming a density of liquid steel equal to 6.95 g/cm^3, the driving force with respect to density change is (10.995 - 6.95)/6.95 = 0.5820. The Gr_M and Sc numbers are estimated to be equal to $9.885 \ 10^{12} \ L^3$ and 131, respectively. The following correlation has been proven to give good predictions for the dissolution of pure metals in liquid metals when the rate-controlling step is liquid phase mass-transfer [16,5,17]:

$$Sh_L = 0.13 \ (Gr_M \ Sc)^{1/3} \qquad \text{(II-1)}$$

A value of 8.3 10^{-3} cm/sec was estimated for the mass-transfer coefficient of pure tantalum in liquid steel. An estimation of the dissolution speed can be made:

$$-(dr/dt) \; = \; 180.948 \; 8.3 \; 10^{-3} \; 0.02734/16.6$$

$$= \; 2.5 \; 10^{-3} \; cm/sec$$

References

1. K. Bungardt, K. Wiebking, H. Brandis, and H. Schmalzried, "The Dissolution of Molybdenum and Tungsten in Iron Melts as a Contribution to the Process of Dissolution of Solid in Liquid Materials. I. - Dissolution of Solid Metallic Bodies in Metal Melts Unaffected by Convection; II. - Dissolution of Solid Metallic Bodies in Metal Melts Influenced by Convection", DEW Techn. Ber., 9(3)(1969), 407-438; discussion 431-437.

2. S.A. Argyropoulos and R.I.L. Guthrie, "Titanium Additions in Steelmaking", Titanium Science and Technology, ed. G. Lütjering, U. Zwicker, W. Bunk, Proceedings of the Fifth International Conference on Titanium, Munich, FRG, September 1984, 2, 1135-1142.

3. P.G. Sismanis, "An Experimental Technique to Measure Convection in Liquid Metals" (M.Eng. thesis, McGill University, Montreal, 1985).

4. P.G. Sismanis and S.A. Argyropoulos, "Measuring Convection in Steel Metal Baths", Proceedings of the Fifth Technology Conference on the Measurement and Control Instrumentation in the Iron and Steel Industry, ISS-AIME, Detroit, Michigan, April 1985, 167-177.

5. P.G. Sismanis, "The Dissolution of Niobium and Zirconium in Liquid Steel", (Ph.D. thesis, McGill University, Montreal, 1987).

6. F.P. Incropera and D.P. DeWitt, Fundamentals of Heat Transfer (New York, NY: John Wiley & Sons, 1981).

7. S.W. Churchill, M. Bernstein, "A Correlating Equation for Forced Convection From gases and Liquids to a Circular Cylinder in Cross-flow", Journal of Heat Transfer, Transactions of the ASME, 99 (1977) 300-306.

8. S.A. Argyropoulos, "Dissolution of High Melting Point Additions in Liquid Steel" (Ph.D. thesis, McGill University, Montreal, 1981).

9. S.A. Argyropoulos and R.I.L. Guthrie, "The Dissolution of Titanium in Liquid Steel", Met. Trans. B, 15B (1984) 47-58.

10. R. Hultgren et al., Selected Values of the Thermodynamic Properties of the Elements (American Society for Metals, 1973).

11. O. Kubaschewski, IRON-Binary Phase Diagrams (Springer-Verlag, Berlin, 1982).

12. Yu. O. Esin, M.G. Valishev, A.F. Ermakov, P.V. Gel'd, and M.S. Petrushevskiy, "Partial and Integral Enthalpies of Formation of Liquid Alloys of Titanium with Iron", Russian Metallurgy, 3 (1981) 15-17.

13. Y.S. Touloukian, C.Y. Ho, eds., Thermophysical Properties of Matter, The TPRC Data Series, 13 volumes on Thermophysical Properties (New York, NY: Plenum, 1970-1977).

14. Handbook of Chemistry and Physics (CRC Press) 60th edition.

15. Y. Iguchi, S. Nobori, K. Saito, and T. Fuwa, "A Calorimetric Study of Heats of Mixing of Liquid Iron Alloys — Fe-Cr, Fe-Mo, Fe-W, Fe-V, Fe-Nb, Fe-Ta", Tetsu-to-Hagane, 68(6)(1982), 89-96.

16. E. Ravoo, J.W. Rotte, F.W. Sevenstern, "Theoretical and Electrochemical Investigation of Free Convection Mass Transfer at Vertical Cylinders", Chemical Eng. Science, 25 (1970), 1637-1652.

17. T. Ishida, "Rate of Dissolution of Solid Nickel in Liquid Tin under Static Conditions", Met. Trans., 17B (1986), 281-289.

ASSESSMENT OF IRON ORE INDURATION SYSTEMS

USING COMPUTER SIMULATION

by

M Cross[*] and D Englund[+]

*Centre for Numerical Modelling and Process Analysis
Thames Polytechnic, London, UK.

+MA Hanna Company, Research Center, Nashwauk,
Minnesota, USA.

ABSTRACT

Pelletising of iron ore is an important process in the preparation of
burden for the iron blast furnace. Over the last few years the need to
reduce energy costs and operate efficiently at other than optimum
production rates has provided a motivation to examine more carefully the
heat distribution within the system with a view to minimising energy costs
at all production levels. All induration systems are highly coupled with
respect to countercurrent gas and solid material flows. Analysis of such
systems is complex because each unit of the process affects all others
through the above coupling. Over the past decade a great deal of effort
has gone into the development of a detailed computer simulation of the
pellet induration process. The models are now able to reflect the
influence of bed dimensions, moisture, magnetite, limestone (or dolomite)
and carbon in the charged pellets, as well as a range of solid and gas flow
distributions (including leakage) and supplementary burners.

Although, the validation of the underlying mathematical models against
pot-grate laboratory data has been performed, comparison with plant
measurements is not straightforward. In this paper, detailed comparison
with plant data, under a variety of operating conditions and gas flow
configurations is presented. The comparison is impressive and permits the
reasons for changes in performance to be identified. This information
provides a powerful basis for optimising an operation at any given level of
production.

Introduction

The pelletising of iron ore is an important process in the preparation of burden for the iron blast furnace. Over the past decade a combination of the need to:

- reduce fuel costs
- operate more flexibly at different production rates, and
- produce high quality product with a range of flux and/or carbon additions

has provided a strong motivation to examine more carefully the heat distribution within induration systems with a view to process optimisation. Within this context process optimisation may be interpreted as:

- making a suitable quality fired pellet, which has adequate strength to survive each stage of the induration process and has the properties specified by the client market sector
- operating at minimum cost at any prescribed rate of production.

For more than 30 years the process analysis tools used by the pelletising industry have involved a combination of

- pot grate and pellet quality apparatus
- plant measurements
- heat and mass balance models.

The pot grate simulation provides a useful means to assess how a prescribed gas flow pattern (in terms of input temperature and flow rate per unit area of bed) will affect the heat exchange with the packed bed of pellets[1-3]. It will provide indications of when the moisture will be evaporated off the pellet bed together with oxidation rates, etc. Usually, any associated 'pot' kiln simulator is very approximate in the way it attempts to emulate the plant operation. Pot grate systems rarely have the facility to simulate the cooling stages of the induration process. Given these constraints, the 'pot' systems provide an adequate means of assessing how a prescribed gas flow configuration will affect the quality of the fired pellet quality.

All pellet induration systems are highly coupled with respect to the countercurrent gas and solid materials flows. It is the substantial degree of heat recuperation within the system that makes pellet induration an economically viable process. Unfortunately, the 'pot' systems only possess the ability to reflect the influence of a prescribed gas flow configuration where the entry temperatures at each zone are specified. It is, thus, not surprising that heat and mass balance models with varying degrees of sophistication, have been used extensively over the years to complement the experimental pot grate work. The limitation of the conventional heat and mass balance models is that they have to make assumptions as to how much magnetite oxidises in each zone, where the moisture evaporates, what the solids temperature is on exit from each zone, etc. By using a combination of pot grate and heat and mass balance models it is possible to eliminate most of the shortcomings of each approach. Draskovich et al[4] describe one such successful venture in optimising the National Steel Pellet Plant in Northern Minnesota.

Although the combined pot grate/heat-and-mass balance approach can be an effective means of process analysis, it is expensive and time-consuming because it requires a number of iterations between the experiments and the model to obtain a simulation of one operating condition. The elapsed time to obtain a reasonably reliable prediction of one operating condition is then an order of days. Moreover, the approach is inflexible and not entirely reliable. Certainly, a full analysis of a system can take many months with these analysis tools alone.

The acquisition of reliable plant data is a difficult task, not only because of the hostile environment, but also the important influence of leakages into and out of the system. However, it is usually possible to obtain reliable estimates of the gas flow rates through the fans, the heat input via the burners and the gas entry and exit temperatures from each zone. The latter are often influenced by leakages and so they may be in error of 10-20%. Fortunately, the input conditions of the pellet bed are usually known with a reasonable degree of precision.

The problems described above, therefore, render the development of reliable fully predictive mathematical models of pellet induration processes attractive, as a means of analysing such systems. The availability of such models would enable the evaluation of the heat distribution within the system, whilst the pot grate could be used to assess the pellet quality that would result from that distribution. These complementary tools would then provide a powerful and efficient framework for analysing pellet induration systems. Initial numerical studies by Young et al in 1963[5] have been followed by a range of model developments over the next twenty years or so[6-12]. Lebelle et al[6] developed a model of a hematite based straight grate system which was validated against the Hoogovens plant. Drugge[7,8] developed straight grate models to treat magnetite ores though there was no published validation. Batterham and co-workers[9] developed detailed models of the straight plant at Dampier processing hematite materials. Although this model was carefully validated against the plant and was used to help increase the efficiency of the process, it was ultimately to little avail as the plant was subsequently closed. A fate that has fallen to many others in North America also, in recent years.

Background to current modelling activities

In the early 1970's the British Steel Corporation initiated a project to develop models of the grate-kiln induration system to assist in the understanding and optimisation of the plant then under construction at Redcar. The grate and kiln models subsequently developed were validated against laboratory and plant data separately[10,11].

In the late 1970's work commenced in collaboration with MA Hanna to develop a general purpose software framework for modelling a wide range of iron ore pelletising systems. The component grate and kiln mathematical models were to be based upon those of Young et al[10,11], though the numerical solution procedures were to utilise control volume methods which have proved to be much faster and more accurate than the original solution techniques. Aside from careful validation of the models against reliable plant data, the project involved the development of a comprehensive simulation framework which will permit the analysis of a wide range of pellet induration systems[12]. During the course of validation the submodels describing the moisture evaporation and the oxidation of magnetite were both recast and refined.

The remainder of the paper focusses upon a description of INDSYS, its validation on a wide variety of systems and some indication of how it might be used in process analysis. INDSYS, therefore, provides a framework for the numerical modelling of iron ore pellet induration systems.

Indsys-induration system simulator

The objective of INDSYS is to provide a framework for the analysis of the straight grate, grate-kiln or circular grate type iron ore pellet induration systems. The structure of the INDSYS software is illustrated in Figure 1 and consists of essentially five components:

- DATINP - which permits the user to define the geometrical configuration and input operating conditions.

- SYSMAN - manages the simulation, performs the various heat and mass balances required, prepares reports on each zone and governs the results output.

- GRATE - simulates the packed bed processes of either the grate or cooler zones.

- KILN - simulates the heat and mass transfer processes in the kiln.

- PRINP/R- file printout facilities for both the problem definition data and simulation results.

The solution procedures of the grate and kiln models will be considered in detail elsewhere; here we will focus upon a description of the software as a modelling framework.

In defining a system it is helpful to recognise that INDSYS considers an induration system to be composed of a number of 'simulation' zones, which do not necessarily correspond to a physical process zone. The essential feature that defines a 'simulation' zone is that the gas flow into it may be considered to be well mixed and so at a uniform temperature. Also, although the off gas may be split in a number of ways the second essential feature that defines a single 'simulation' zone, is that before it is divided the gas flow is assumed to be well mixed. This means that depending on the gas flow configuration around a process zone it could be one or a number of 'simulation' zones. Figure 2 shows various ways in which gas flow through process zones may be represented in the simulation.

Figure 3 shows an idealisation of a five zone system, where zone III represents a kiln. Note that the exit gas from each zone is either redistributed as entry gas to other "earlier" zones (with respect to solids flow) or exhausted to a stack. Figure 4 shows how the simulator represents the zonal heat balance, at least, with respect to the gas flows and supplementary burners. The entry gas to a zone is composed of contributions of the exit gas from "later" zones together with an inflow source which represents the leakage into or out of the system. The inflow source also provides the means of representing the entry gas into the cooler zones. Both the entry and exit gas streams can represent the influence of supplementary burners. The gas inflow burner can be controlled to ensure a minimum or prescribed temperature.

The kernel components of the software are the numerical models of the grate and kiln processes. The mathematical formulation of the models is essentially that described in refs [10,11], with some refinements. The numerical solution procedures, however, have been completely reformulated and are now based upon the control volume methods as explained, for example, by Patankar[13]. The solution procedure of the whole system simulation is necessarily iterative because the gas inlet temperatures cannot be prescribed but arise as a consequence of the calculation.

Simply, since the gas flow is recuperative, then its inlet heat content to each zone receiving recuperative gas has to be evaluated. Because there usually exists contraints, such as a peak pellet temperature in the kiln, or specified hood temperatures in the straight grate process, the convergence occurs faster than it might otherwise do. Experience has shown that for a system with N simulation zones, N complete passes through the simulation is more than adequate to reach practical convergence.

The results of the simulation are:

- temperature profiles across and along the bed for the grate zones, and along the length of the kiln for both the gas and pellet streams, within each zone.
- heat and mass balances for each zone on each gaseous or solid constituent (e.g. moisture, magnetite, etc.)
- a global heat balance.

Thus, given a plant configuration, initial pellet properties, production rate and gas flow distribution, the simulation will predict the gas entry temperature into each zone, the heat and mass transfer between two gas and pellet streams plus the evaporation of moisture, oxidation of magnetite within each zone and the heat required by the burners to achieve the desired pellet (or gas) temperatures at prescribed locations in the system.

INDSYS is currently implemented on an IBM-AT or compatible system and a run takes from 15-30 minutes depending on the number of zones.

Software validation and process analysis

Validation of pellet induration models has been a thorny topic for many years. It is possible, by making careful laboratory measurements in a pot grate, to obtain some reliable data for comparison with the models. However, even here it is necessary to proceed with care because the heat capacity of the pot is large compared to the pellet bed sample and this can distort the basis for comparison. Certainly earlier versions of the models in INDSYS were successfully compared with laboratory data. Comparison with laboratory data, though useful, does not necessarily mean the simulation will provide an accurate reflection of a plant operation. In an earlier paper[12], it has been pointed out that there are substantial problems to be overcome in comparing induration models with plant data. One of those problems concerns the reliability and completeness of the plant data that can be acquired. It is, for example, notoriously difficult to obtain reliable estimates of the leakage into a system, though it is now accepted that leakage (and bleed in air) has a substantial effect on the efficient operation of a process. The second major problem is that the total heat present in an induration system can be over four times that which enters via the burners. Thus, it should be expected that the error in the prediction of the heat input via the burners could be large.

For the past five years or so the Hanna Research Laboratory[4] has been involved in a collaborative programme of work to increase both the efficiency and capacity of the National Steel Pellet Plant located at Keewatin, Minnesota. This Allis Chalmers grate-kiln plant commenced operation in 1969 with the induration system in use today coming on stream in 1976. Its rated capacity at that time was 4.06 million tonnes/year. The original system had a grate with one drying zone consisting of eight windboxes and a six windbox preheat zone. The cooler also had two zones and the gas flow pattern was similar to that shown in Figure 5. Notice that gas from the second cooling zone is recouped and used to provide the ongas into the drying zone on the grate. In terms of the simulation, then the cooler is perceived as having three zones - 1st stage cooling, 2nd stage cooling to recoup, with the remainder exhausted to a stack. Over a period of five or six years Hanna Research collaborated with the NSPC management in the development and implementation of process design revisions to the NSPCII induration system. These revisions were primarily associated with the effective creation of a third zone in the grate and the modification of the size of the cooler stages. In effect, the "recoup" cooler zone's hotter (and untempered) exit gas was used as the inflow gas to the new secondary drying zone inserted between the drying and preheat zones on the grate. Over the period of time a number of modifications to the basic concept were made, each with the aim of further increasing the thermal efficiency of the system.

Over the six year period Hanna Research performed a large number of air surveys covering a wide range of operating conditions. A selection of these operating conditions is shown in Table 1, which shows variations in the gas flow configuration, (i.e. tonnage rates and distribution), production rate of pellets and magnetite levels in the charged pellet bed. The gas flow distributions referred to in Table 1 are illustrated in Figure 5.

The measurements which can be relied upon include the gas inlet temperature to the primary drying zone, the gas temperature through the recoup ducting and the amount of burner fuel injected into the system. Figure 6 shows the comparison between measured and predicted gas inlet temperatures for Zone 1 for the seven different operating conditions observed on the NSPC plant over a six year period. Note that in all cases the comparison is good to within 30°C or better than 8% on a relative error basis. This result is encouraging because from the gas flow perspective, this is the tail-end of the process and shortcomings earlier in the representation of the heat and mass transfer between the gas and solids flow should reflect in the ongas temperature into Zone 1. Figure 7 shows the corresponding comparisons between model and experiment for the gas temperature through the recoup hood. Here the comparisons are also good; the predictions are within 10% of the measurements and show exactly the same trends. Again this is an encouraging result when the uncertainties of the leakage distribution in the system is taken into account.

Figure 8 shows the comparison between model prediction and measurements of the burner fuel heat demand in the system for the same range of operating conditions. The comparisons have show a high degree of correlation, accurately reflecting the changes in the operational conditions. The results of Figures 6,7 and 8 provide a great deal of encouragement in assessing the reliability with which INDSYS can be used to predict how a process will respond to a prescribed gas flow configuration and set of operating conditions.

At this stage, the software has essentially reached beyond that of other published induration models in that although it has only been compared against one process in detail, it has borne comparison under a wide variety of operating conditions. Recently, it has also proved possible to compare the performance of the model against a further grate-kiln operation and two straight-grate systems. The comparisons of between prediction and measurement of the fuel input via the burners are also shown in Figure 8 which provides further evidence that INDSYS is a reliable framework for the analysis and optimisation of pellet induration systems.

Discussion and Conclusion

The effort required to develop, validate and effectively utilise mathematical models of complex industrial operations which feature sophisticated physicochemical processes, is often measured in many man years. Modelling results at intermediate stages of a development process may well provide insight into the key features of a system, but it has to be admitted that the investment required to obtain a measurable payback is far beyond the mere production of a working code - crucial though this step might be. The numerical modelling of induration systems has proved to be no less of a challenge in this respect. Although the mathematical formulation of the models has been in the open literature for nearly 10 years, it is only now that the software encompassing those models can be regarded as a reliable tool for use by those who are primarily involved in the operation and process analysis of pellet induration systems. Notice that in Figure 8 the model predicts substantial savings on fuel with an operation at a significantly higher throughput rate on the NSPC operation. It should also be noted that a relatively small change to the gas flow configuration on one of the straight grate processes simulated indicates considerable potential for energy savings with operations down to the order of 150,000 Btu/tonne.

Finally, current developments to the models and associated INDSYS software include:

- the inclusion of limestone decomposition and carbon additions,

- automatic evaluation of gas flow rate, based on operating fan characteristics,

- the development of an operational advisor for the plant control room.

It is expected that INDSYS and its enhancements will find substantial utility in the iron ore pelletising industry.

References

1. C G Thomas and K N Clark "Aspects of simulating iron ore production", Aust. IMM, Pellets and Granules Symposium, Newcastle, October 1974.

2. J A Thurlby et al, "The role of mathematical models and pot grate simulators in studies of new and existing pelletising plants", Int Symposium of Agglomeration, Nurnberg, May 1981.

3. D F Ball et al, Agglomeration of Iron Ores, Heinneman Ed. Books (1973).

4. E F Draskovitch et al, "Design and performance of the National Steel Pellet Plant high temperature heat recuperation system", AIME-SME Annual Meeting, New Orleans, March 1986; SME Preprint No.86/128.

5. P A Young et al, "Packed beds in metallurgical operations", Inst Chem Engrs, London (1963).

6. P A M Lebelle et al, "The induration process of pellets on a moving strand", Mathematical Process Models in Iron and Steelmaking, pub. Metals Society, London, 6-16 (1975).

7. R Drugge, "Modelling of heat exchange phenomena during the induration of magnetite pellets on a travelling grate", AIME-SME Preprint

No.75-B-51 (1975).

8. R Drugge, "Optimisatiion of pelletising processes in sintering of iron ore concentrates at LKAB, Sweden", Int Symposium on Agglomeration, Nurnberg, May 1981.

9. J A Thurlby et al, "Development and validation of a mathematical model for the moving grate induration of iron ore pellets", Int J Mineral Processing, 6, 43 (1979).

10. M Cross and R W Young, "Mathematical model of rotary kilns used in the production of iron ore pellets", Ironmaking and Steelmaking, 3, 129 (1976).

11. R W Young et al, "Mathematical model of the grate-kiln-cooler process used for the induration of iron ore pellets", Ironmaking and Steelmaking, 6, 1 (1979).

12. M Cross et al, "Mathematical models of iron ore pellet induration-validation and application", in Applications of Mathematical and Physical Models in the Iron and Steel Industry, proceedings 3rd PTD Conference, pub ISS-AIME, 101 (1982).

13. S V Patankar, Numerical Fluid Flow and Heat Transfer, Hemisphere (1980).

Acknowledgements

It is a pleasure to acknowledge the contribution of John Wakeman, Robert Frans and Tom Wolke to this programme of research.

Run No.	Date	Throughput (tonnes/hr)	Feed magnetite level (% wt)	Gas flow through cooler (tonnes/hr)	Wind box distn in grate zones			Wind box distn in cooler zones			Gas Flow Config
					DD1	DD2	PH	C1	C2-Rec	C2-Exh	
1	8/80	534	89	700	8	0	6	10	4	5	a)
2	5/81	507	78	700	8	0	6	10	4	5	a)
3	9/84	543	76	751	5	3	6	10	4	5	b)
4	9/84	543	76	616	5	3	6	10	4	5	b)
5	7/85	577	78	768	5	4	5	8	6	5	b)
6	5/86	597	82	893	5	4	5	8	6	5	b)
7	10/86	574	81	813	5	4	5	9	5	5	b)

Table 1 A summary of the seven sets of operating conditions for comparison between the NSPC grate kiln system and the simulation.

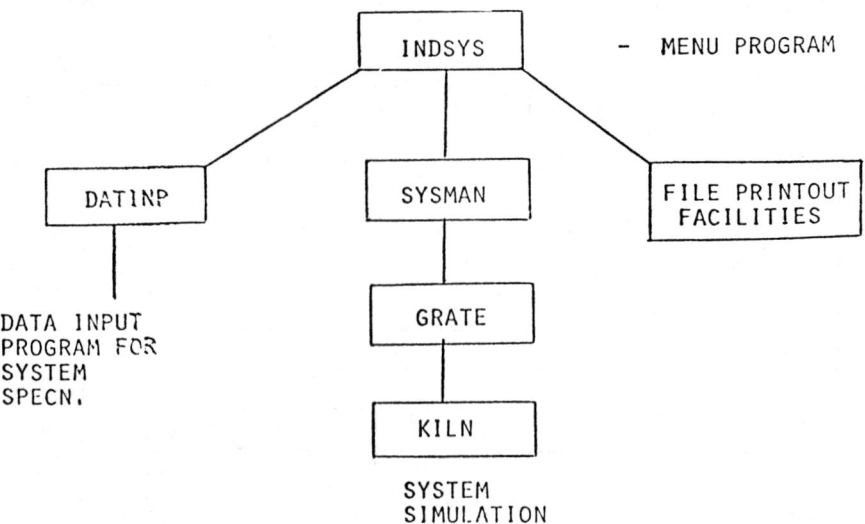

Figure 1   Structure of the INDSYS software

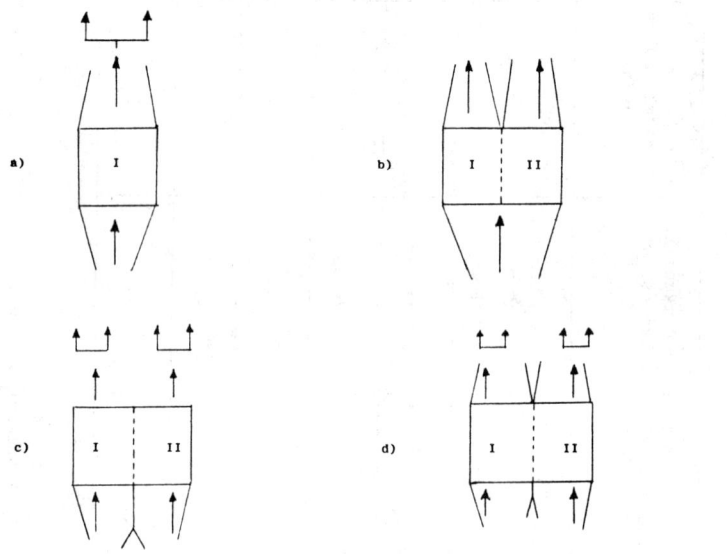

Figure 2   Ways to represent the gas flow through a zone and
its subsequent distribution to other zones (or
the exhaust stack)

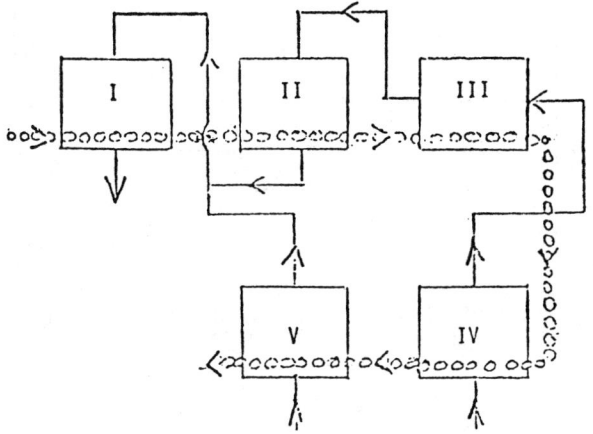

Figure 3   Idealisation of the gas flow distribution through
            a pellet induration system

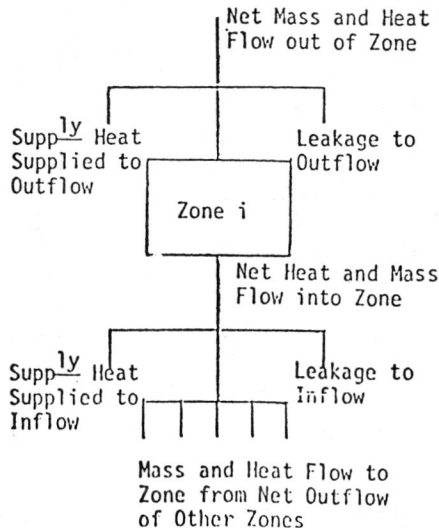

Figure 4   Structure of the gas flow and burner fuel on both
            entry and exit to a simulation zone in INDSYS

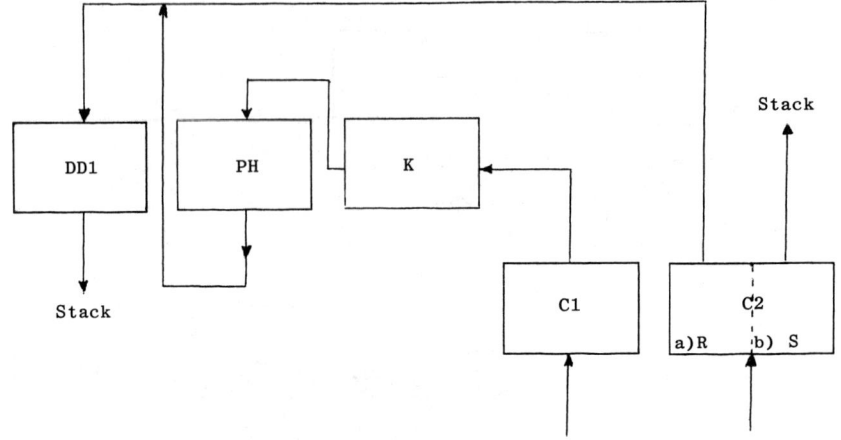

(a) Original flow concept

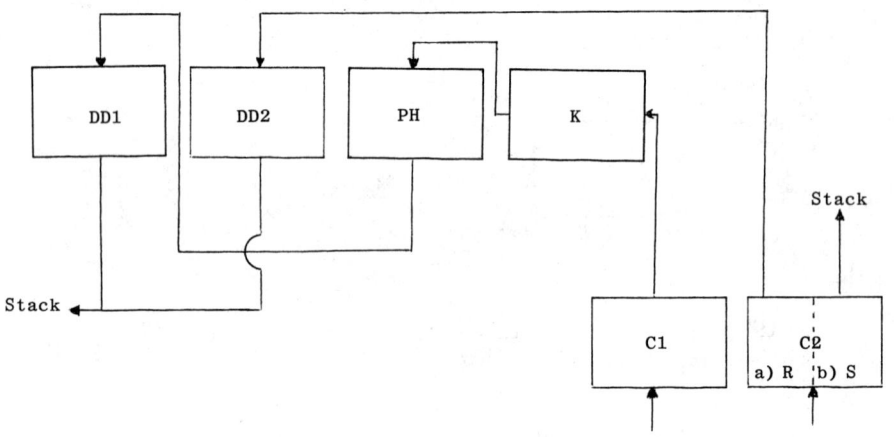

(b) High temperature recoup concept

Figure 5   The gas flow configurations used in the NSPC system

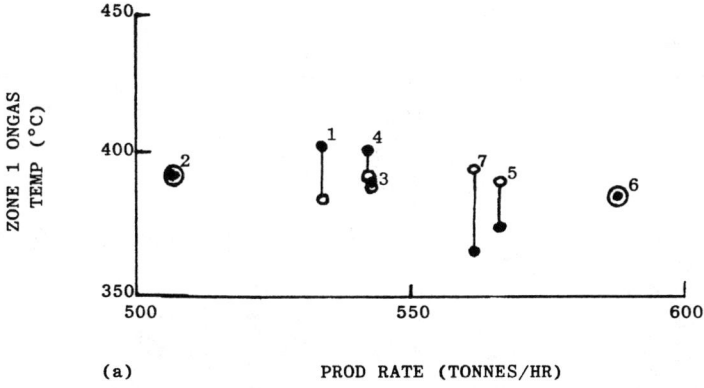

(a)   PROD RATE (TONNES/HR)

Figure 6   Comparison between measured and predicted primary
            drying zone on gas temperatures for the various
            NSPC operating conditions

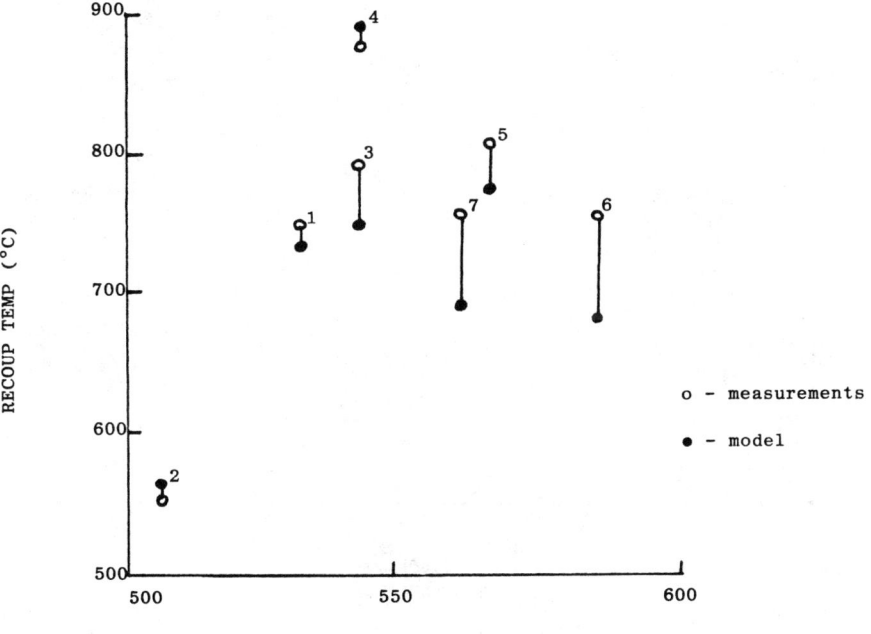

(b)   PROD RATE (TONNES/HR)

Figure 7   Comparison between predicted and measured gas temperature
            in the recoup ducting for the various NSPC operating
            conditions

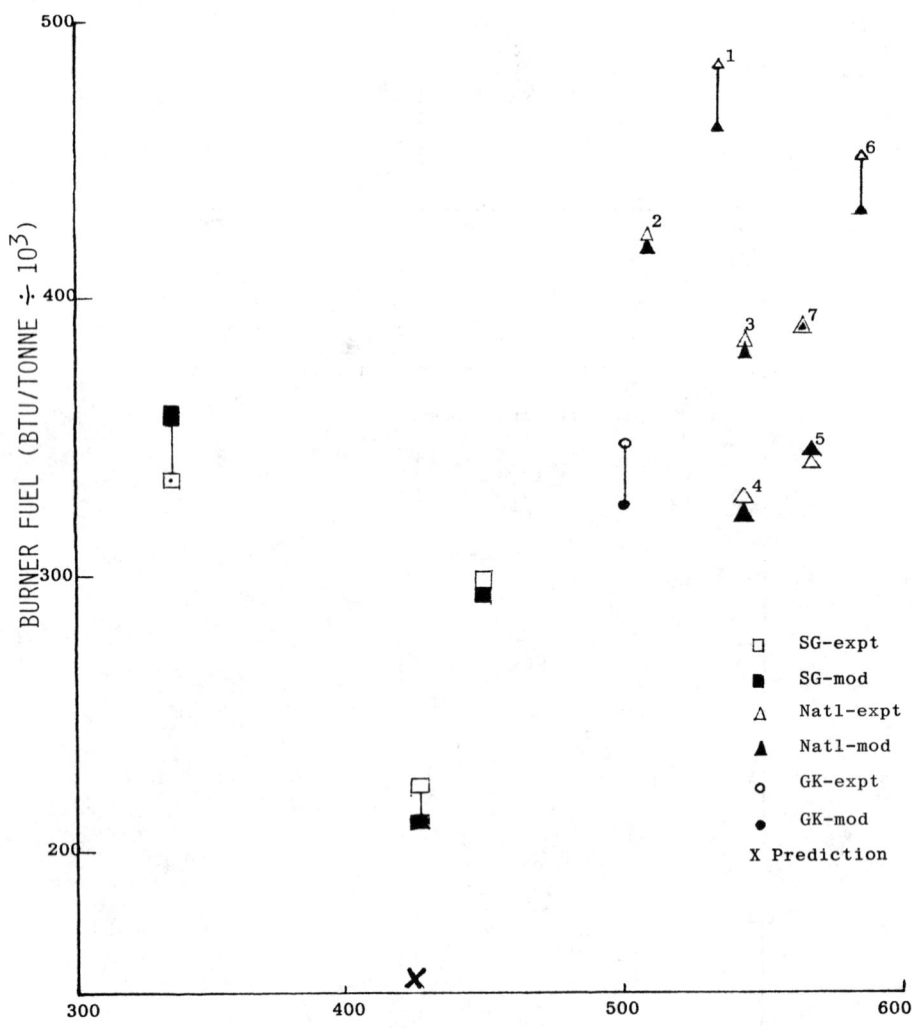

Figure 8   Comparison between the predicted and measured burner
heat input required to fire the pellets for all
induration systems referred to

A MATHEMATICAL MODEL FOR THE ANALYSIS OF THE REDUCTION OF IRON OXIDE

PELLETS IN THE FIXED BED WITH MULTICOMPONENT GAS MIXTURE

TAKEAKI MURAYAMA and YOICHI ONO

Department of Iron and Steel Metallurgy, Kyushu University
Hakozaki, Fukuoka 812, JAPAN.

Abstract

A mathematical model for the analysis of the reduction of iron oxide pellets in a fixed bed reactor with multi-component gas mixtures was developed based on the unreacted-core model. The rate parameters included in the model were modified according to the gas composition. The water-gas shift reaction and the oxygen diffusion in the product dense iron layer were also taken into account. The calculated results were in good agreement with the experimental data.

Introduction

The reduction of iron oxide pellets in the industrial scale such as a blast furnace and a shaft furnace for direct reduction is usually carried out in the flow of multi-component gas. Nevertheless, our knowledge on the reduction of iron oxide with a multi-component gas such as H_2-CO gas mixture (1-12) is still insufficient as compared with that on the reduction with a one-component gas (13,14).

In this work, a mathematical model for the analysis of the reduction of iron oxide pellets in an isothermal fixed bed reactor with multi-component gas mixtures was developed based on the unreacted-core model. The outline of the model is as follows.

In this work, the parameters included in the unreacted-core model were modified depending on the gas composition, because the applicability of the unreacted-core model to the reduction of porous pellets is only apparent (15) and they may therefore vary with the composition of the gas mixture (11,15,16).

In our model, the effect of high void fraction in the peripheral part of the fixed bed adjacent to the reaction tube wall on the gas flow was also taken into account, because the ratio of the reaction tube diameter against the pellet diameter is usually small in a laboratory experiment (11,13,17).

At higher reduction degree, the reduction rate is often controlled by the oxygen diffusion in the dense reduced iron layer (18-20). In our model, the effect was taken into account and the rate equation taking into account both the oxygen diffusion in the dense reduced iron layer and the reaction between reducing gas and oxygen in the reduced iron was developed.

The effect of the water-gas shift reaction in the reduced iron layer was also taken into account in our model.

Mathematical Model

It was assumed that the gaseous reduction of wustite pellets in an isothermal fixed bed reactor took place according to the following equation.

$$Fe_xO + A(CO \text{ or } H_2) = xFe + B(CO_2 \text{ or } H_2O) \tag{1}$$

In this model, the following effects were taken into account.

(1) the effect of high void fraction in the peripheral part of the fixed bed adjacent to the wall of the reaction tube on the gas flow. We call this the wall effect.
(2) the effect of the gas composition on the parameters included in the unreacted-core model.
(3) the effect of oxygen diffusion in the dense reduced iron layer on the reduction rate.
(4) the effect of the water-gas shift reaction on the gas composition.

These effects will be explained in the construction of the fundamental equations except item (2), which will be explained in the next section. Each symbol appearing in this paper has the meaning given in the nomenclature.

Effect of High Void Fraction in the Peripheral Part of the Fixed Bed (The Wall Effect)

When the ratio of reaction tube diameter against pellet diameter $(RTP=D_{tube}/dp)$ is small, the effect of high void fraction in the peripheral part of the fixed bed on the gas flow is significant. Therefore, in this work, the fixed bed in the reaction tube is separated into two regions in radial direction, which are the central part of the tube $(0<\zeta<\zeta_1$: region I) and the peripheral part of the bed $(\zeta_1<\zeta<1$: region II). In each region, the fundamental equations are solved independently and the calculated results are combined with each other (11,13).

In each region, void fraction (ε_I and ε_{II}), gas flow rate (Q_I and Q_{II}), and number of pellets per unit volume of the bed (Np_I and Np_{II}) are given as follows (11,13,21).

Region I ($0<\zeta<\zeta_1$)

$$\varepsilon_I = 1-3(1-\bar{\varepsilon}_b)/(\zeta_1^2+\zeta_1+1) \tag{2}$$

$$Q_I = S_I\varepsilon_I Q/(S\bar{\varepsilon}_b) \tag{3}$$

$$Np_I = 3(1-\varepsilon_I)/(4\pi r_o^3) \tag{4}$$

Region II ($\zeta_1<\zeta<1$)

$$\varepsilon_{II} = (\bar{\varepsilon}_b -\zeta_1^2\varepsilon_I)/(1 - \zeta_1^2) \tag{5}$$

$$Q_{II} = S_{II}\varepsilon_{II}Q/(S\bar{\varepsilon}_b) \tag{6}$$

$$Np_{II} = 3(1-\varepsilon_{II})/(4\pi r_o^3) \tag{7}$$

where:

$$\zeta_1 = 1 - 2\sigma/RTP \tag{8}$$

$$\sigma = (RTP -\sqrt{(RTP-1)^2-1}\ )/2 \tag{9}$$

ζ = dimensionless radial distance in the bed

The average values of fractional reduction F_{ave} and gas concentration Xj_{ave} over the cross section of the bed are given by

$$F_{ave} = (F_I S_I Np_I + F_{II}S_{II}Np_{II})/(SNp) \tag{10}$$

$$Xj_{ave} = (Xj_I Q_I + Xj_{II}Q_{II})/Q \tag{11}$$

where:

$$Np = 3(1-\bar{\varepsilon}_b)/(4\pi r_o^3) \tag{12}$$

Overall fractional reduction $\bar{F}$ in the bed is given by

$$\bar{F} = \frac{\int_0^L \int_0^1 F2\pi\zeta d\zeta dz}{\int_0^L \int_0^1 2\pi\zeta d\zeta dz} \tag{13}$$

Fundamental Equations for Fixed Bed Reactor

In each of the two regions, the fundamental equations are given by

$$(\varepsilon_b/RT)(\partial P_j/\partial t) + (u_G/RT)(\partial P_j/\partial z) + R_j + (-1)^n Rw = 0 \qquad (14)$$

$$Np(\partial Cs/\partial t) + \sum_j R_j = 0 \qquad (15)$$

where

$$j = H_2, \quad n=1$$

$$CO, \quad n=2$$

$$Cs=(4/3)\pi r_i^3 d_o = (4/3)\pi r_o^3 d_o(1-F) \qquad (16)$$

The symbol R_j denotes the consuming rate of species j per unit volume of the bed and one of the following two equations was used for R_j depending on the conditions.

1) $0 < F < F_1$ and $Xj > Xj^e + \Delta X$ (j= H_2, CO)

Under this condition, the reduction was assumed to proceed topochemically according to the unreacted-core model. The following equation was used (11).

$$R_j = Np \cdot \dot{n}_j \qquad (17)$$

$$\dot{n}_j = \frac{4\pi r_o^2 (C_j - C_{je})}{\dfrac{1}{kf^j} + \dfrac{r_o(r_o - r_i)}{r_i De^j} + \dfrac{(r_o/r_i)^2}{kc_j(1+1/Ke^j)}} \qquad (18)$$

where:

$$\frac{1}{kf^j} = \left(\frac{1}{kf_j} + \frac{1}{kf_k Ke^j}\right)\left(\frac{1}{1+1/Ke^j}\right) \qquad (19)$$

$$\frac{1}{De^j} = \left(\frac{1}{De_j} + \frac{1}{De_k Ke^j}\right)\left(\frac{1}{1+1/Ke^j}\right) \qquad (20)$$

2) $F_1 \leq F < 1$ and $X_j \leq X_j^e + \Delta X$ (j=H_2, CO)

Under this condition, the reaction was assumed to proceed according to the oxygen diffusion model in the reduced dense iron layer. The following equation was used.

$$R_j = Np \cdot \dot{n}_{oDj}(r_o/r_g)^3(1-\varepsilon_w) \qquad (21)$$

$$\dot{n}_{oDj} = \frac{4\pi r_g^2 (C_j - C_j^e)/(1+1/Ke^j)}{\dfrac{r_g(C_j + C_k ks^k/ks^j)((1-F)^{-1/3} - (1-\alpha F)^{-1/3})}{D_o C_o}} \qquad (22)$$

$$j = H_2, \ CO$$
$$k = CO, \ H_2$$

In this equation, the reaction between reducing gas and oxygen in reduced iron expressed by

$$j(H_2 \text{ or } CO) + \underline{O}(\text{in Fe}) = k(H_2O \text{ or } CO_2) \qquad (23)$$

was also taken into account. This reaction was not considered by previous investigators (19,20). In Eq.22, the oxygen concentration in reduced iron at the gas/iron interface was assumed to be in equilibrium with the reducing gas. As shown in Eq.22, the reduction rate varies with the concentration of the reducing gas.

The rate of the water-gas shift reaction Rw appearing in Eq.14 was assumed to be given by Eq.24 (22).

$$R_w = (4/3)\pi r_o^3 N pk_w (P_{CO}P_{H2O} - P_{CO2}P_{H2}/K_w) \tag{24}$$

Initial and Boundary Conditions

The fundamental equations were solved numerically under the following initial and boundary conditions.

$$R_i = r_o \quad (0 < z < L) \text{ at } t = \varepsilon_b z/u_G \tag{25}$$

$$P_j = P_{j \text{ inlet}} \quad \text{at } z=0 \tag{26}$$

Estimation of Physico-Chemical Constants

Equilibrium constants for the reactions were estimated from Eqs.27-29 (11,12,23-25).

$$Fe_xO + H_2 = xFe + H_2O: \quad Ke^{H2} = \exp(0.9733-1743.3/T) \tag{27}$$

$$Fe_xO + CO = xFe + CO_2: \quad Ke^{CO} = \exp(-3.127+2879/T) \tag{28}$$

$$CO + H_2O = CO_2 + H_2: \quad K_w = Ke^{CO}/Ke^{H2} \tag{29}$$

The mass transfer coefficient kf_j was estimated by Eq.30 (26).

$$\varepsilon_b kf_j 2r_o/Djm = \varepsilon_b Sh = 2 + 0.6Re^{1/2}Sc^{1/3} \tag{30}$$

where Djm is the effective binary diffusion coefficient in a multi-component mixture. The value of Djm was estimated by Eq.31 based on the STEFAN-MAXWELL's equations (27).

$$1/Djm = \sum_{\substack{k=1 \\ (k \neq j)}}^{4} (X_k - X_j N_k/N_j)/Djk \tag{31}$$
$$(k=1-4, \ 1:H_2, \ 2:H_2O, \ 3:CO, \ 4:CO_2)$$

where Djk is the binary diffusion coefficient of j-k pair in a binary mixture. Binary diffusion coefficient Djk was estimated by Fujita's equation (28) for the systems except for H_2-H_2O system and Andrussow's equation (29) for H_2-H_2O system.

Effective diffusion coefficients De_j in the product layer were estimated by Eq.32 (12).

$$De_j = Djm\delta \quad (j= H_2, H_2O, CO, CO_2) \tag{32}$$

where δ is the diffusibility of the product layer given by Eq.33.

$$\delta = \delta_{H2}(X_{H2}^b + X_{H2O}^b) + \delta_{CO}(X_{CO}^b + X_{CO2}^b) \tag{33}$$

527

where δ_{H2} and δ_{CO} are diffusibilities of the product layers in the reduction with H_2 gas and CO gas respectively and the following values were used.

$$\delta_{H2} = 0.086, \quad \delta_{CO} = 0.099$$

These values were determined from the reduction of single wustite pellets with H_2 and CO gas respectively at $900^{\circ}C$ (12).

The applicability of the unreacted-core model to the reduction of porous pellets is only apparent. According to the relation between parameters included in the unreacted-core model, Ishida-Wen's model (30) and the grain model (31), the chemical reaction rate constant kc for the unreacted-core model also depends on the diffusion coefficient of the gas taking part in reduction reaction (11,12,15,16). Then, the chemical reaction rate constants determined from the reduction with pure CO or H_2 were modified as follows (11,12,15,16).

$$kc_A = kc_A^{\,o} \sqrt{De^A/(D_{AB}\delta_A)} \quad (A=H_2,\ CO\ ;\ B=H_2O,\ CO_2) \tag{34}$$

The following values (12) were used as $kc_A^{\,o}$.

$$kc_{H2}^{\,o} = 6.275 cm/s, \quad kc_{CO}^{\,o} = 1.509 cm/s$$

These values were determined from the reduction of the single pellets with pure H_2 and CO at $900^{\circ}C$ respectively.

Since Djm varies depending on the gas composition as shown in Eq.31, the values of kf_j, De_j, and kc_j may also vary with the gas composition.

The values of $C_o D_o$ was estimated from Eq.35 obtained by Riecke et al.(18).

$$C_o D_o = 0.03 exp(-57000/RT) \quad (mol\ O/s\ cm) \tag{35}$$

As the value of the ratio of ks^{CO}/ks^{H2} was not available, the following value was used as an approximation referring to the ratio of $kc_{CO}^{\,o}/kc_{H2}^{\,o}$.

$$ks^{CO}/ks^{H2} = 1/4$$

The rate constant of the water-gas shift reaction kw was estimated by the following equation obtained by Ishigaki et al.(22).

$$kw = 93.32 exp(-30620/1.987T) \quad (mol/s\ cm^3\ atm^2) \tag{36}$$

Experimental

Samples

Commercial pellets were used as samples. The chemical composition of the pellets is shown in table I. The pellets were 2.6g in weight, 1.15cm in diameter, and 0.28 in porosity.

Table I. Chemical Composition of the Pellets
          Used in the Experiments.

Chemical Composition (wt%)				
T.Fe	FeO	SiO_2	CaO	Al_2O_3
61.9	0.5	3.7	4.3	1.3

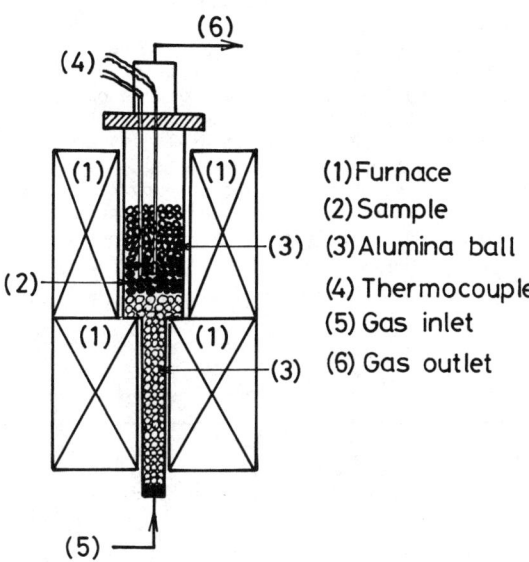

(1) Furnace
(2) Sample
(3) Alumina ball
(4) Thermocouple
(5) Gas inlet
(6) Gas outlet

Fig. 1. Schematic diagram of experimental apparatus.

Reduction in Fixed Bed Reactor(32)

The schematic diagram of the fixed bed reactor used in this work is shown in Fig. 1. The reaction tube was made of SUS304 stainless tube (I.D. 13cm). The reactor was equipped with a furnace for preheating the gas. The temperature in the bed was kept uniform at $900^{\circ}C\pm10^{\circ}C$. Gas flow rate was $1.1\times10^4 cm^3$(STP)/min. The total weight of the pellets in the bed was about 1200g and the height of the bed was 5.4cm.

At first, the bed was heated up to $900^{\circ}C$ in N_2 gas atmosphere and then reduced to wustite with $CO-CO_2$ (1:1) gas mixture. After that, the bed was reduced to iron with H_2-CO gas mixtures of desired gas composition. During the reduction, exit gas from the bed was cooled and dried and then analyzed by a gas chromatograph and an infrared analyzer. The utilization efficiency of H_2 was determined from the difference between the input gas flow rate and the exit gas flow rate after eliminating water.

Results and Discussion

The Wall Effect

The wall effect was first investigated. In the calculation, the value of RTP was changed by changing the diameter of the reaction tube (D_{tube}), while the diameter of pellets (dp) and the void fraction in the central part of the bed (ε_I) were kept constant. Therefore, the gas flow rate (Q) and the overall void fraction of the bed $(\bar{\varepsilon}_b)$ changed with the value of RTP. As shown in Fig. 2, when the gas flow rate Q is low $(1.1 \times 10^4$ cm^3(STP)/min at RTP=11), the results calculated with ignoring the effect do not agree with the results considering the effect even at high RTP of 20, which indicates significant wall effect. On the other hand, as shown in Fig. 3, when the gas flow rate is high $(4.0 \times 10^4$cm^3(STP)/min), the results calculated with and without the effect are getting close to each other even at low RTP of 10. Figure 4 shows the comparison of experimental results with calculated results. In this experiment, where RTP was 11 and Q was 1.1×10^4cm^3(STP)/min at RTP=11, experimental data agreed well with the results calculated with considering the wall effect. As shown in Fig.4, the model gives a good prediction for not only overall fractional reduction $\bar{F}$ in a bed, but also variations of H_2O and CO_2 concentrations in exit gas. Although the ratio of S_{II}/S is relatively small at RTP>10, high gas flow rate is required in order to ignore the wall effect even if the ratio of D_{tube}/dp is high.

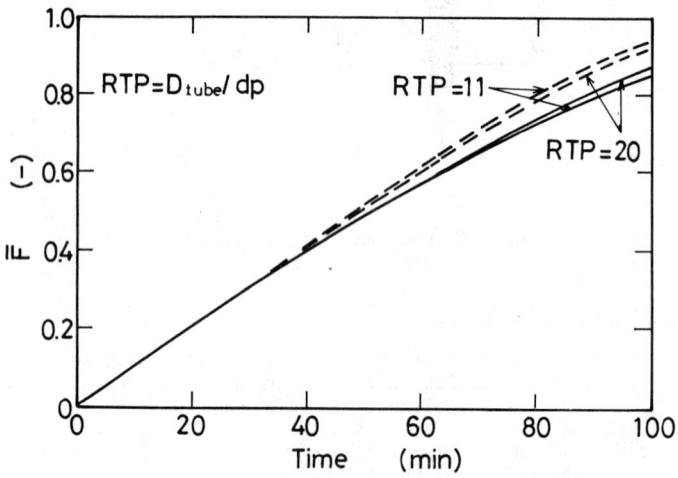

——— : the wall effect was considered.
----- : the wall effect was ignored.

Fig. 2. Effects of diameter ratio of reaction tube against pellet ,RTP, on the reduction curve of the wustite pellets in the fixed bed.
($900°$, Q=1.1×10^4cm^3(STP)/min at RTP=11,H_2=52%, CO=44.5%, CO_2=3.5%)

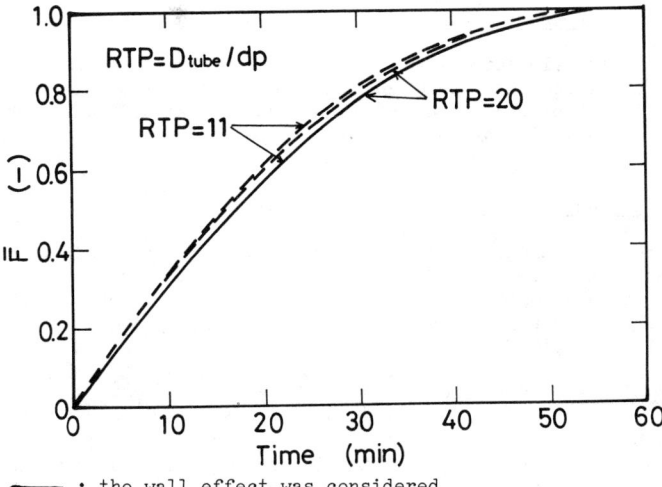

: the wall effect was considered.
: the wall effect was ignored.

Fig. 3. Effects of diameter ratio of reaction tube against pellet ,RTP, on the reduction curve of the wustite pellets in the fixed bed. ($900°$, $Q=4.0×10^4 cm^3$(STP)/min at RTP=11,H_2=52%, CO=44.5%, CO_2=3.5%)

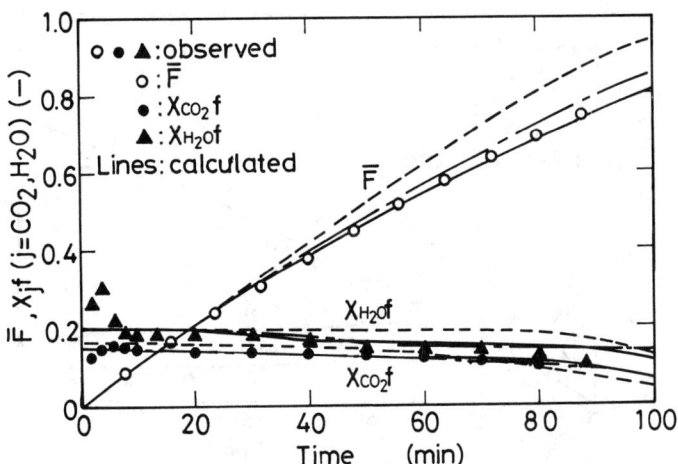

: the all effects were considered.
: the effect of oxygen diffusion was ignored.
: the wall effect and the effect of oxygen diffusion were ignored.

Fig. 4. Comparison of calculated curves of overall fractional reduction of the pellets over the whole bed ($\overline{F}$) and CO_2 and H_2O concentrations in exit gas ($X_{CO_2}f$, $X_{H_2O}f$) with experimental data for H_2-CO reduction of wustite pellets in the fixed bed. ($900°$, $Q=1.1×10^4 cm^3$(STP)/min, H_2=52%, CO=44.5%, CO_2=3.5%)

531

The Effect of the Water-Gas Shift Reaction

The results calculated with ignoring the effect of the water-gas shift reaction also agreed with the experimental data under the same condition of Fig. 4. In this case, the effect was not so significant, presumably because the deviation of the inlet gas composition from the equilibrium composition of the water-gas shift reaction was small.

The Effect of Gas Composition on the Parameters

The results calculated by using the parameters obtained by reduction experiments with pure H_2 and CO also agreed well with the experimental data under the same condition of Fig. 4. However, this effect should be reinvestigated, because this effect was significant in the reduction of single pellets (12,16).

The Effect of Oxygen Diffusion

If this effect is ignored, the calculated reduction curve is getting higher than the data especially at higher reduction degree as shown in Figs. 5 and 6. The cross section of pellets reduced partially in the bed showed that reaction interface was not observed clearly and the wustite grains covered with dense iron layer were observed. Therefore, this effect should be taken into account. In these calculations, the values of F_1 and ΔX were determined as $F_1 = 0.8$ and $\Delta X = 0.05 X_1^e$ on the basis of the microscopic observation of the cross section of pellets reduced partially.

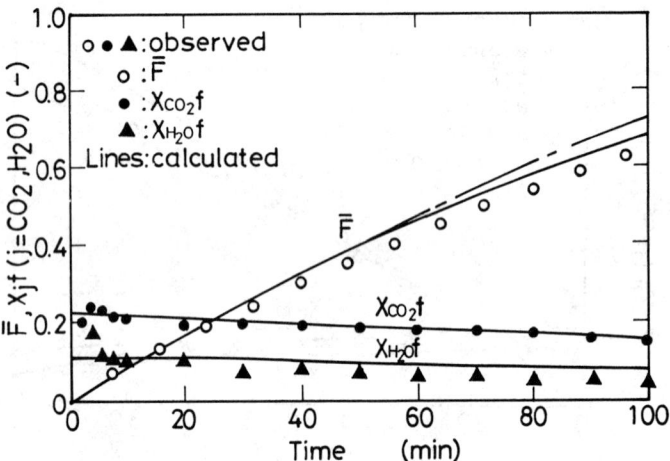

——— : the all effects were considered.
—·— : the effect of oxygen diffusion was ignored.

Fig. 5. Comparison of calculated curves of overall fractional reduction of the pellets over the whole bed ($\overline{F}$) and CO_2 and H_2O concentrations in exit gas (Xco_2f, $X_{H2O}f$) with experimental data for H_2-CO reduction of wustite pellets in the fixed bed.
(900°, $Q = 1.1 \times 10^4 cm^3$(STP)/min, $H_2 = 29.3\%$, CO=64.2\%, $CO_2 = 6.2\%$)

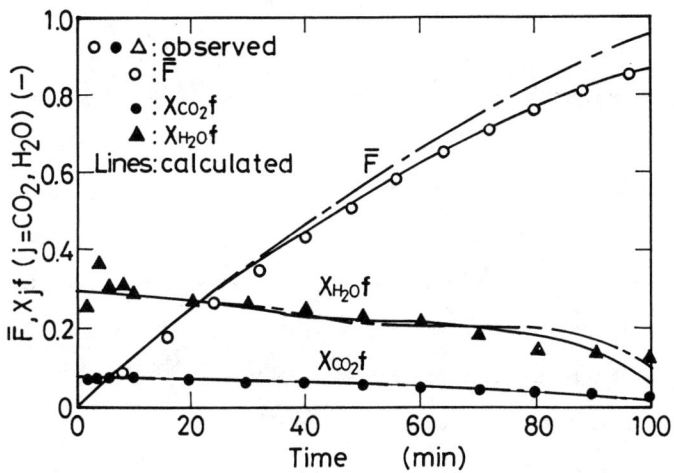

Fig. 6. Comparison of calculated curves of overall fractional reduction of the pellets over the whole bed ($\overline{F}$) and CO_2 and H_2O concentrations in exit gas ($X_{CO_2}f$, $X_{H_2O}f$) with experimental data for H_2-CO reduction of wustite pellets in the fixed bed. ($900°$, Q=1.1x10^{4}cm^3(STP)/min,H_2=75%, CO=23.4%, CO$_2$=1.6%)

———— : the all effects were considered.
—·— : the effect of oxygen diffusion was ignored.

Conclusion

A mathematical model for the analysis of the reduction of iron oxide pellets in a fixed bed reactor with multi-component gas mixtures was developed. In this model, the wall effect, the effect of the water-gas shift reaction, the effect of the gas composition on the parameters, and the effect of oxygen diffusion in the dense iron layer were taken into account. These effects were investigated in the comparison between the calculated results and experimental data. Following results were obtained:

(1) High gas flow rate is required in order to ignore the wall effect even if the diameter ratio of reaction tube to pellet is high.
(2) The effect of oxygen diffusion in the dense iron layer should be taken into account especially at higher reduction degree.
(3) In this experimental condition, the effects of the water-gas shift reaction was not so significant.
(4) In this experimental condition, the effect of gas composition on the parameters was not so significant. However, this effect should be reinvestigated, because this effect was significant in the reduction of single pellet.

Acknowledgement

The present authors would like to express their deep thanks to Mr. Chong Min Cho, formerly a student of Graduate School of Kyushu University, now at Research Insitute of Industrial Science and Technology in Korea, for his experimental work.

Nomenclature

C_j = molar concentration of species j

C_j^e = equilibrium molar concentration of species j

C_o = oxygen concentration dissolved in metallic iron

Cs = reducible oxygen in a pellet

Djk = binary diffusion coefficient of j-k pair in a binary mixture

Djm = effective binary diffusion coefficient in a multi-component mixture

D_o = diffusion coefficient of oxygen in metallic iron layer

D_{tube} = inside diameter of the reaction tube

d_o = concentration of reducible oxygen

dp = diameter of a pellet

F = fractional reduction of a pellet

$\bar{F}$ = overall fractional reduction of pellets over the whole bed

Ke^j = equilibrium constant of the reduction with j

Kw = equilibrium constant of the water-gas shift reaction

kc_j = chemical reaction rate constant of the reduction with j

kf_j = mass transfer coefficient in the gas film

ks^j = chemical reaction rate constant of the following reaction:

$$j(H_2 \text{ or } CO) + \underline{O}(\text{in Fe}) = k(H_2O \text{ or } CO_2)$$

kw = rate constant of the water-gas shift reaction

L = length of the bed

M_j = molecular weight of species j

N_k = molar flux of species k

Np = number of pellets per unit volume of bed

$\dot{n}_j$ = reaction rate of a pellet with reducing gas j

$\dot{n}_{oDj}$ = the rate of oxygen diffusion in the dense iron layer of a grain

$P_{j\ inlet}$ = partial pressure of gas species j at the inlet (i.e. bottom of the bed)

P_j = partial pressure of the gas component j

Q = gas flow rate as a whole

R = gas constant

Re = Reynolds number

R_j = consuming rate of gas j per unit bed volume

RTP = (D_{tube}/dp): diameter ratio of the reaction tube to pellet

Rw = the rate of water-gas shift reaction

r_g = radius of a grain within a pellet

r_i = radius of unreacted-core

r_o = radius of a pellet

S, S_I, S_{II} = cross sectional area of the bed as a whole, in region I,
and in region II, respectively

Sc = Schmidt number

Sh = Scherwood number

T   = temperature

t   = time

u_G = gas velocity

X_j = mole fraction of species j

X_j^e = equilibrium mole fraction of species j

X_jf = mole fraction of species j in exit gas

z   = distance from the bottom of the bed

Greek:

α = $1 - (\rho_{Fex0}/\rho_{Fe})[xM_{Fe}/(xM_{Fe} + M_0)]$; rate of volume reduction

δ = diffusibility of the reduced iron layer

ε_b = void fraction

ε_I, ε_{II} = void fraction in region I and II respectively

ε_w = porosity of the wustite pellet

ζ = dimensionless radial distance in the bed

ρ_j = density of j

Subscripts:

ave = average over the cross section of the bed
I   = region I (central part of the bed)
II = region II (peripheral part of the bed)

References

1. H.K.Kohl and B.Marincek,"On the Reaction Kinetics of Wustite with
   Carbon Monoxide and Hydrogen",Helv. Chim. Acta.,48(1965),1857-1867.

2. E.T.Turkdogan and J.V.Vinters,"Gaseous Reduction of Iron Oxide: Part III. Reduction-Oxidation of Porous and Dense Iron Oxides and Iron", Metall. Trans., 3(1972),1561-1574.

3. J.Szekely and Y.El-Tawil,"The Reduction of Hematite Pellets with Carbon Monoxide-Hydrogen Mixtures",Metall. Trans.,B,7(1976),490-492.

4. Q.T.Tsay, W.H.Ray and J.Szekely,"The Modelling of Hematite Reduction with Hydrogen Plus Carbon Monoxide Mixtures: Part I.The Behavior of Single Pellets",AIChE J.,22(1976),1064-1072;"Part II,The Direct Reduction Process in a Shaft Furnace Arrangement",AIChE J.,22(1976), 1072-1079.

5. J.Szekely M.Choudhary and Y.El-Tawil,"On the Reaction of Solids with Mixed Gases",Metall. Trans., B, 8(1977),639-643.

6. V.Croft,"Diffusion in Mixed-Gas Reduction",Metall. Trans., B,10(1979), 121-122.

7. R.Takahashi, S.Kurozu and Y.Takahashi,"Reduction Rates of Iron Oxide Pellet with Mixtures of Hydrogen and Carbon Monoxide at High Pressures", Tetsu-to-Hagané, 66(1980),336-345.

8. N.Towhidi and J.Szekely,"Reduction Kinetics of Commercial Low-Silica Hematite Pellets with CO-H_2 Mixtures over Temperature Range $600°$-$1234°$C", Ironmaking Steelmaking, 8(1981),237-249.

9. T.Chida, N.Sakai and T.Tadaki,"Approximate Analysis of Metallic Oxide Reduction with Mixtures of Hydrogen and Carbon Monoxide", Tetsu-to-Hagané, 67(1981),1485-1490.

10. A.A.El-Geassy and V. Rajakumar,"Gaseous Reduction of Wustite with H_2, CO and H_2-CO Mixtures", Trans. Iron Steel Inst. Jpn., 25(1985),449-458.

11. T.Murayama and Y.Ono,"Kinetics of Gaseous Reduction of Iron Oxide Pellets and Its Application to the Analysis of Bed Reactor Operations", Fifth International Iron & Steel Congress, 6th Process Technology Conference Proceedings, 6(1986),53-65.

12. C.M.Cho, T.Maeda, T.Murayama, and Y.Ono, "Reduction Rate of Wustite Pellets with CO-H_2 Gas Mixtures",Tetsu-to-Hagané, 73(1987),972-979; Trans. Iron Steel Inst. Jpn.,25(1985),B330.

13. T. Murayama, Y. Ono, and Y. Kawai,"Analysis of CO Reduction of Hematite Pellets in an Isothermal Fixed Bed by Multi-interface Model," Tetsu-to-Hagané, 64(1978),1509-1517.

14. M. A. Osman, F. S. Manning, and W. O. Philbrook, "Reduction of Single Particles and Packed Beds of Hematite with Carbon Monoxide," AIChE. J., 12(1966),685-692.

15. T. Murayama and Y. Ono, "Relationship between Parameters Included in Wen's Model and Unreacted-core Model," Tetsu-to-Hagané, 71(1985),S819; Trans. Iron Steel Inst. Jpn.,26(1986),B89.

16. T. Murayama, T. Yoshihara, and Y. Ono, "Analysis of Reduction Rate of an Iron Oxide(Wustite) Pellet with H_2-N_2 and CO-N_2 by Unreacted-Core Model," Tetsu-to-Hagané, 68(1982),2253-2262.

17. T. Murayama, Y. Ono, and Y. Kawai,"Analysis of CO Reduction of Hematite Pellets in an Isothermal Moving Bed by Multi-interface Model," Tetsu-to-Hagané, 64(1978),1518-1527.

18. E.Riecke, K.Bohnenkamp and H.-J.Engell,"On the Reduction of Wustite by Means of Hydrogen-Water Vapour and Carbon Monoxide-Carbon Dioxide Mixtures", Archiv Eisenhüttenwes.,38(1967),249-255.

19. Y.Iguchi and M.Inoue,"On the Rate of the Reduction of Wustite, Magnetite, and Hematite Containing Al_2O_3, CaO, and MgO", Tetsu-to-Hagané,65(1979),1692-1701.

20. M.Ohmi and T.Usui,"Improved Theory on the Rate of Reduction of Single Particles and Fixed Beds of Iron Oxide Pellets with Hydrogen", Trans. Iron Steel Inst. Jpn., 22(1982),66-74.

21. R. F. Benenati and C. B. Brosilow,"Void Fraction Distribution in Beds of Spheres," AIChE J., 8(1962),359-361.

22. M.Ishigaki, R.Takahashi and Y.Takahashi,"Kinetics of the Side Reaction in the Reduction of Iron Oxide by the Gas Mixtures", Tetsu-to-Hagané, 68(1982),S827.

23. L. Von Bogdandy and H.-J. Engell, Die Reduktion der Eisenerze, Verlag Stahleisen m.b.H., Dusseldorf,(1967),3-12.

24. M. Ohmi and T. Usui,"Study on the Rate Of Reduction of Single Iron Oxide Pellet with Hydrogen," Tetsu-to-Hagané, 59(1973),1888-1901; "On the Unreacted-core Shrinking Model for Reduction of a Single Hematite Pellet with Hydrogen," Trans. Iron Steel Inst. Jpn., 16(1976),77-84.

25. T. Murayama, Y. Ono, and Y. Kawai, "Step-wise Reduction of Hematite Pellets with CO-CO_2 Gas MIxtures," Tetsu-to-Hagané, 63(1977), 1099-1107; Trans. Iron Steel Inst. Jpn., 18(1978),579-587.

26. T. Shirotsuka, A. Hirata, and A. Murakami, Kagaku-Gijutsusha-no-tame-no-Idou-Sokudoron, in Japanese,Ohm-sha,Tokyo, (1965),240-243.

27. R. B. Bird, W. E. Stewart, and E.N. Lightfoot, Transport Phenomena, John Wiley and Sons, New York, (1960),570-571.

28. S. Fujita, "Comparison of Equations for Estimation of Diffusion Coefficient in the Gas Phase," Kagaku-kougaku, 28(1964),251-254.

29. L. Andrussow,"Uber die Diffusion in Gasen I," Z. Elektrochem., 54(1950), 556-571;"Uber die Diffusion in Gasen II," Z. Elektrochem., 55(1951),51-53.

30. M. Ishida and C. Y. Wen, "Comparison of Kinetic and Diffusional Models for Solid-Gas Reactions," AIChE. J., 14(1968),311-317.

31. H. Y. Sohn and J. Szekely, "A Structural Model for Gas-Solid Reactions with a Moving Boundary-III," Chem. Eng. Sci., 27(1972),763-778.

32. C.M.Cho, T.Murayama and Y.Ono, "Reduction of Iron Oxide Pellets in a Fixed Bed with H_2-CO Gas Mixtures",Tetsu-to-Hagané, 72(1986),S896; Trans. Iron Steel Inst. Jpn.,27(1987),B143.

COMPUTATION OF COMPLEX MULTICOMPONENT MULTIPHASE EQUILIBRIA

: A NEW ALGORITHM AND APPLICATION TO METALLURGICAL SYSTEMS

P.Sastri

Research and Development Division
Mukand Iron and Steel Works Limited
Belapur Road, Kalwe, District Thane 400605, India

Abstract

An algorithm is presented to calculate the equilibrium compositions
in a multicomponent multiphase system. The algorithm is a modification
of the element potential method and does not require the specification
of the condensed phases present initially. The minimisation is done
using the Newton Raphson method. The algorithm does not need any
initial guess values and the convergence is good in all cases. The
utility of the method is demonstrated by applying it to the problem
of silicon reduction in the blast furnace.

539

Introduction

There has been a surge of interest in the calculation of chemical equilibria of importance in metallurgical systems. Two classes of problems are dealt with in such calculations : i) those involving no condensed phase, and ii) those involving one or more condensed phases. Literature is full of examples of the use of such algorithms and a satisfactory one capable of being directly used for metallurgical problems seems an urgent necessity.

Rao (1) has recently presented a novel method of calculating gas phase equilibria in the C-O-H system. Another method presented by him suggests that calculation of the amount of different phases present at equilibrium is possible but only the composition of the gas phase is calculable (2) . Bale et al (3) have recently proposed a new algorithm for the calculation of gas phase equilibria with very low concentration of the components present. However, the utility of such calculations are limited as these are applicable to cases where no condensed phase is present and most metallurgical processes actively involve one or more condensed phases.

In the class of problems involving condensed phases, the program called SOLGASMIX developed by Eriksson seems to be the most popular one (4). It has been applied to the modelling of the silicon arc furnace and given useful results (5,6). However, its restricted availability seems to be a retardant in its widespread use. Other algorithms that can be applied to metallurgical systems are well documented and many have been reviewed by Van Zeggeren and Storey (7). These methods have been discussed by Lahiri (8) who also proposes an algorithm for solving multicomponent multiphase equilbria. As indicated by Lahiri (8) the second order free energy minimisation techniques have the following inherent limitations : i) the phases which are present at equilibrium must be known, and ii) the initial guess or starting solution must be close to the final solution otherwise the iterative scheme will not converge.

By modifying White's element potential method (9), Lahiri has been able to overcome the first limitation and apply the method to such complex systems such as Fe-O-C-Cl and Fe-O-C-H-Ca-N containing more than one condensed species. However the algorithm requires initial guess values which are chosen on a somewhat arbitary basis.

The object of this paper is to show that by a rational choice of initial guess values (computed within the algorithm itself) the algorithm proposed here converges very rapidly to the final solution. The utility of this new methodology is then demonstrated by applying it to the problem of silicon reduction mechanism in the blast furnace.

Theoretical Considerations

Consider the free energy of the system defined as

$$G = \sum_{i=1}^{N} \mu_i \, n_i \tag{1}$$

where μ_i is the partial molar free energy of species i, and

n_i the number of moles of species i in the system (see nomenclature

540

for complete definition of terms used). The free minimization problem then becomes that of finding a set of n_i which makes G minimum. However, two constraints must be considered before any solution can be attempted. First, the mole numbers, n_i, represent the amount of species i present in the system. Obviously all of them must satisfy the constraint shown in equation 2 :

$$n_i >= 0 \tag{2}$$

The second constraint refers to the mass balance constraint which can be expressed as

$$\sum_{i=1}^{N} b_{ie} n_i = B_e \quad , \ e=1,2,\ldots,M \tag{3}$$

where N is the number of species present, M the number of elements present (N>=M), b_{ie} number of atoms of element e present in species i and B_e the total number of gram atoms of element e present (also called elemental abundances).

Usually the first constraint is ignored until the actual iterative scheme is formulated. Using the method of Lagrangian multipliers(10) the objective function can then be written as

$$L= \sum_{i=1}^{N} /u_i n_i - \sum_{e=1}^{M} \lambda_e (\sum_{i=1}^{N} b_{ie} n_i - B_e) \tag{4}$$

where λ_e 's are the Lagrangian multipliers.

Equating

$$\frac{\partial L}{\partial n_i} = 0 \tag{5}$$

one obtains

$$/u_i = \sum_{i=1}^{N} \lambda_e b_{ie} \tag{6}$$

In the method of element potentials, Powell and Sarner(11) have shown that

$$\lambda_e = /u_e \tag{7}$$

where $/u_e$ is the element potential of element e, or as defined by White(9), it is the Gibb's free energy contribution per gram atom of element e.

Since for any species i,

$$/u_i = /u_i^{o} + RT \ln a_i \tag{8}$$

we can write

$$a_i = \exp(- \frac{/u_i^{o}}{RT} + \frac{1}{RT} \sum_{e=1}^{M} b_{ie} /u_e \tag{9}$$

or
$$a_i = \exp(-C_i + M\sum_{e=1} b_{ie} \lambda'_e) \quad (10)$$

where
$$C_i = {}^0\!/u_i / RT \quad (11)$$

and
$$\lambda'_e = /u_e / RT \quad (12)$$

For ideal gases, equation 10 can be written as

$$x_i = (1/p) \exp(-C_i + \sum_{e=1}^{M} \lambda'_e b_{ie}), i=1,2,\ldots,L \quad (13)$$

and for pure condensed phases, equation 10 takes the form

$$0 = C_j + \sum_{e=1}^{M} \lambda'_e b_{je}, \quad j = 1,2,\ldots,K \quad (14)$$

where $K + L = N$, the total number of species present.

The mass balance equation 3 can be then be written as

$$n\sum_{i=1}^{L} b_{ie} x_i + \sum_{j=1}^{K} b_{je} m_j - B_e = 0 \quad e=1,2,\ldots,M \quad (15)$$

and
$$\sum_{j=1}^{L} x_i - 1 = 0 \quad (16)$$

Equations 13 to 16 can be solved for λ_e, n and m_j by the Newton Raphson method to obtain the equilibrium compositions.

At any iteration step equations 13 are exactly satisfied as they represent the equilibrium equations for gaseous species but equations 14 to 16 are satisfied only when the final solution has been reached. However it is expected that the residues for these equations will decrease with each iteration. For inclusion and elimination of condensed phases Zeggeren and Storey suggested that if at the end of the iteration, for the k th. condensed phase which is not included in the calculation, the inequality

$$-C_k + \sum_{e=1}^{M} (\lambda'_e + \Delta\lambda'_e) b_{ke} > 0 \quad (17)$$

is found valid, the k th. phase is likely to be present at equilibrium. On the other hand the inequality

$$m_r + \Delta m_r < 0 \quad (18)$$

indicates that the r th. condensed phase which has been included in the calculation is likely to be absent at the final equilibrium.

542

The method of solution suggested by Lahiri consists of imposing the following restriction on the correction factor obtained by the Newton Raphson method to restrict the search to physically meaningful zones

$$-C_i + \sum_{e=1}^{M} (\lambda'_e + \Delta\lambda'_e) b_{ie} - \ln p \leq 0 \qquad (19)$$

$$0 \leq (n + \Delta n) \leq \sum_{e=1}^{M} B_e \qquad (20)$$

$$(m_j + \Delta m_j) \leq \min_e (B_e / b_{je}) \qquad (21)$$

Equations 19 ensure that the mole fraction of the i th. gaseous species is less than or equal to unity. Since this equation will not damp the $\Delta\lambda'_e$ which are less than zero, the restriction

$$\Delta\lambda'_e \geq - \beta \qquad (22)$$

was arbitrarily introduced by Lahiri. An arbitary scheme was used for choosing the initial values and reasonably good results obtained in such complicated systems as Fe-O-C-H-Ca-N at equilibrium. The number of iterations ranged from 20 to 30 for a variety of examples chosen.

In the present algorithm the initial values are calculated based on the following criteria

a) calculate $\Delta/u = \Delta G^0 /RT$ for all the species considered to be present in the system.

b) a total of M species are chosen to be present. These are selected on the basis of those having the maximum probability of existing at equilibrium i.e. those with most negative chemical potentials. It is necessary that all the elements be represented in the species so chosen. If it is not so, the next successive most negative ones which will ensure that all the the elements are present are included.

c) the initial value of n is calculated as

$$n = 0.5 \sum_e b_{ie} B_e \qquad e = \max(1 ,M) \qquad (23)$$

for all the gaseous species included above.

d) the amount of the condensed phase present m_j is calculated as

$$m_j = 0.5 \min_e (B_e / b_{je}) \qquad (24)$$

All these calculation are included in the algorithm itself.

A computer programme written in FORTRAN IV has been developed based on the above scheme. As a test case some of the examples used by Lahiri(8) have been recalculated and the results of the two algorithms compared in Table I. The new algorithm demonstrated above is clearly quicker in its approach to the final solution. The utility of the algorithm in metallurgical systems is shown in the following sections.

Table I : Comparison of Present Algorithm and Lahiri's Algorithm

S.No.	Elemental Abundances	Equilibrium Composition (final solution)	No. of Iterations Lahiri's / Present algorithm	
1.	Fe = 0.3	$FeCl_3$ = 0.0982	16	8
	O = 0.45	Fe_2Cl_6 = 0.0193		
	C = 0.7	CCl_4 = 0.0088		
	Cl = 4.0	$COCl_2$ = 0.00456		
	Temp. 1000 K	CO = 0.168 , CO_2 = 0.0162		
		Cl_2 = 0.684, Cl = .000317		
		O_2 = 0.332 x 10^{-23}		
		n = 2.193		
		C (s) = 0.266		
2.	Fe = 2.0	CO = 0.327 ,CO_2 = 0.206	26	12
	O = 5.82	CH_4 = 0.192 x 10-4		
	C = 3.2	H_2 = 0.0255, H_2O = 0.0122		
	H = 0.44	N_2 = 0.43, n = 5.82		
	Ca = 0.1	Fe (s) = 0.905		
	N = 5.0	$Fe_{0.95}O(s)$ = 1.152		
		$CaCO_3$ = 0.1		

For gaseous species, gas compositions in mole fractions and n is the total number of moles of gas.

The Silicon Reduction Mechanism in the Blast Furnace

Control of silicon level in hot metal has assumed great importance with the replacement of the OHFs with BOF units. A clear understanding of the mechanism through which silica present in the burden gets reduced to silicon and ends up in the hot metal is necessary to control its level in the hot metal.

A review of the work of many investigators has been recently by Batra(12). The investigations of Fulton and Chipman(13) and Rein and Chipman(14,15) are considered the most important of the early work. The possibility of silicon transfer occuring via gas-metal reaction was mentioned initially by Taylor(16), wherein the silicon content of the hot metal was determined for many blast furnaces by applying thermodynamic data on metal and slag and compared with actual values. It was concluded that the silicon content of the blast furnace metal tends towards the equilibrium value in most of the furnaces investigated. The equilibrium calculation was based on the equation

$$SiO_2 \; (\text{ in slag }) + 2 \, C = Si + 2 \, CO \text{ (gas)} \qquad (25)$$

Tsuchiya et al(17) have argued that Taylor's approach should yield a silicon content in pig iron between 0.06 to 0.12 % which is much smaller than the usual value of 0.4 to 1.0 %. In thier study they have critically reviewed the silica reduction through slag-metal reaction mechanism and investigated the kinetics of reaction between SiO containing gas and liquid iron.

Reaction Mechanism

Silica in slag or coke ash can be reduced to SiO at high temperatures and the strongly reducing atmosphere prevailing in the coke wall behind the raceway in the blast furnace according to the reactions

$$SiO_2 + C \text{ (s)} = SiO \text{ (g)} + CO \text{ (g)} \qquad (26)$$

$$SiO_2 + CO \text{ (g)} = SiO \text{ (g)} + CO_2 \text{ (g)} \qquad (27)$$

$$CO_2 \text{ (g)} + C \text{ (s)} = 2CO \text{ (g)} \qquad (28)$$

The pO_2 – T diagram for the Si-O system indicates that SiO is stable between the temperatures of 1750 - 1500 C and partial pressures of oxygen between 10^{-2} and 10^{-3}. The SiO generated may either be oxidised back to silica as it travels upwards from the tuyere region or it can be easily reduced to silicon in iron if it meets liquid iron containing sufficient carbon. From experimental studies Tsuchiya et al (17) concluded that SiO plays an important role in the transfer of silicon into iron in industrial blast furnaces.

The Problem Restated

The silicon reduction mechanism in this paper is presented as that of finding the amounts of silicon monoxide present in the blast furnace under operating conditions. The following assumptions are necessary to carry the calculations :

i) the operating conditions in the blast furnace are conditions which are close to that required for thermmodynamic equilirium to be achieved. This seems reasonable from Taylor's calculations as also that of Tsuchiya et al.

ii) For determining the activity of silicon in the hot metal only binary interaction coefficients have been considered. Though not strictly true, this facilitates calculations.

iii) The slag is assumed to be a quarternary system of Al_2O_3 - SiO_2 -CaO-M_xO_y and the activity of the fourth component is taken from the respective quarternary when available. Otherwise it is taken from the system CaO-SiO_2 -M_xO_y . Consequently the effect of other oxides are not accounted for.

Calculations and Results

Detailed data from blast furnaces are required for performing the calculations. The data have to be reduced to a form suitable for the algorithm. Since the object here is to prove the utility of the algorithm data only from one blast furnace, namely Bhilai Steel Plant(19) was used. The data were collated and reduced to the form required. Table II shows the data as is used here. Table III is the transformed data as required by the algorithm. Table IV lists the various gaseous and condensed species included in the calculations of equilibrium compositions. Thermodynamic data were taken from Barin and Knacke(20, 21). The calculations were performed over a range of temperatures from 800 C to 2100 C to determine the amounts of SiO present at equilibrium and results are presented in Table V.

Discussion

The results presented in Table V show the presence of SiO in the gaseous phase under the operating conditions, if equilibrium is achieved in the blast furnace. The amount of SiO in the gas phase is highest at 2100 C. If temperatures are higher the amount of SiO is expected to go up. With the fall in temperature the fall in SiO fraction in the gas phase is sharp. With a higher SiO content the metal trickling down to the hearth will pick up more silicon and therefore the silicon content of the metal is expected to increase with higher blast temperatures.

Other factors will however influence the final silicon content of the hot metal. The effect of various additives in the blast has not been examined here in detail. The calculations show the presence of SiO in the gaseous phase in a blast furnace under equilibrium conditions. A detailed examination of silicon levels in the blast furnace hot metal vis-a-vis the mechanism is not included here and may form the subject matter of a different publication.

Table II : Operating Conditions of Bhilai Blast Furnace used in the
Calculations

Comsumption Rates : Iron ore   686 kgs / thm
                    Sinter   587 kgs / thm
                    Coke   780 kgs / thm
                    Limestone   186 kgs / thm
                    Mn ore   85 kgs / thm

Blast Temperature 1100 C

Top Pressure 1.8 atmospheres

---

thm : tonne hot metal

Table III : Calculated Elemental Abundances

C   = 50.58 kg moles

Fe   = 165.30 kg moles

O_2 = 69.05 kg moles

H_2 =   9.22 kg moles

Si   =   3.75 kg moles

S   =   0.18 kg moles

P   =   0.09 kg moles

Ca   =   3.78 kg moles

Mn   =   0.23 kg moles

Al   =   1.90 kg moles

N_2 = 80.80 kg moles

Mg   =   1.57 kg moles

Table IV : List of Species Considered for Calculations

Gaseous Species

CO	CH_4	O_2	CH_3OH	SiO
CO_2	PH_3	N_2	P_4O_{10}	SiS
CS	H_2O	H_2	S_2	SiH_4
CS_2	NH_3	NO_2	SO_2	H_2O_2

Condensed Species

C	CaO	FeO	MnO	MgO
Fe	CaS	FeS	MnS	$MgSiO_3$
Si	$CaCO_3$	$FeSiO_3$	$MnSiO_3$	Mg_2SiO_4
Ca	$CaSiO_3$	$FeAl_2O_4$	$MnAl_2O_4$	$MgAl_2O_4$
Mn	Ca_2SiO_4	Fe_3O_4	Mn_2SiO_4	Al_2O_3
Al	$CaAl_2O_4$	Fe_2O_3	AlN	AlP
P	$Ca_3P_2O_8$	Fe_3C	Al_4C_3	$Al_2Si_2O_7$
Mg	$CaSi_2$	FeS_2	Al_2SiO_4	CaC_2
	$CaSi$	Fe_2SiO_4		
Si	Mn	C	S	P (in metal)

Table V : Amounts of SiO Present in the Gaseous Phase under
Operating Conditions in the Blast Furnace

S.No.	Temperature ($°C$)	Amount of SiO (mole fraction)	No. of iterations required for convergence
1.	2100	.015	48
2.	1900	.013	46
3.	1700	.0059	48
4.	1500	.0023	54
5.	1300	.00067	63
6.	1200	.00014	72
7.	1100	.000058	78
8.	1000	.0000164	85

Conclusions

The algorithm proposed to calculate multicomponent multiphase equilibria has been successfully utilised for a large system and is found to be useful for metallurgical calculations assuming equilibria exist in these systems.

Acknowledgements

The author would like to thank Dr.R.H.G.Rau for permission to publish this paper.

Nomenclature

G	:	free energy of the system
$/u_i$	:	chemical potential of i th. species
x_i	:	mole fraction of species i
b_{ie}	:	number of atoms of element e in species i
B_e	:	elemental abundances of element e
e	:	chemical potential of element e
a_i	:	activity of species i
$\check{}_e$	:	natural logarithm of the activity of element e
n_i	:	number of moles of species i
p	:	total pressure
m_j	:	mole number of j th. condensed species
	:	constant

References

1. Y.K.Rao and H.G.Lee,"Discussion on Calculational Method of Equilibrium Composition in the Carbon-Hydrogen-Oxygen System and its Application to Environments of a High Temperature Gas Cooled Reactor",Metall. Trans.,15B(1984)396-399.

2. Y.K.Rao,"The Analysis and Calculation of Equilibria in Complex Metallurgical Systems",Metall. Trans.,14B(1983)701-710.

3. J.Melancon, Y.Blanchette and C.W.Bale,"Computation of Gas Phase Equilibria - A General Algorithm Using the Newton Raphson Method", Metall. Trans.,16B(1985)793-799.

4. G.Eriksson,"Thermodynamic Studies of High Temperature Equilibria XII. SOLGASMIX, A Computer Programme for Calculation of Equilibrium Compositions in Multiphase Systems",Chem. Scr.,8(1975)100-103.

5. G.Eriksson and T.Johansson,"Chemical and Thermal Equilibrium Calculations for a Quantitative Description of a Non-isothermal Reactor, with Application to the Silicon Arc Furnace",Scand. J. Metallurgy, 7(1978)264-270.

6. T.Johansson and G.Eriksson,"Technical Aspects on the Silicon Arc Furnace Process Based on Chemical and Thermal Equilibrium Computations",Scand. J. Metallurgy,9(1980)283-291.

7. F.Van Zeggeren and S.H.Storey,The Computation of Chemical Equilibria,(Cambridge University Press,1970).

8. A.K.Lahiri,"Multicomponent Multiphase Equilibria",Fluid Phase Equilibria,3(1979)113-121.

9. W.B.White,"Numerical Determination of Chemical Equilibrium and the Partitioning of Free Energy",J. Chem. Phys.,46(1967)4171-4175.

10. T.L.Hill,An Introduction to Statistical Thermodynamics,(Addison Wesley Publishing Co.Inc., Reading, mass.,1960).

11. H.N.Powell ans S.F.Sarner,"The Use of Element Potential in the Analysis of Chemical Equilibrium",Vol 1.(General Electric Co. Report NO. R59/FDD796,1959).

12. N.K.Batra,"Silicon Reduction Mechanism in BF",Proc. International Symposium on Ironmaking, Bhilai 1982,p III-21.

13. J.C.Fulton and J.Chipman,"Slag-Metal-Graphite Reaction and the Activity of Silicon in Lime-Alumina-Silica Slags",Trans. AIME 200(1954) 1136-1146.

14. H.Rein and J.Chipman,"The Distribution of Silicon Between Fe-Si-C Alloy and SiO_2-CaO-MgO-Al_2O_3 Slag at 1600 C",Trans. AIME 227(1963) 1193-1203.

15. H.Rein and J.Chipman,"Activities in the Liquid Solution SiO_2-CaO-MgO-Al_2O_3 Slags",Trans. AIME 233(1965)415-424.

16. J.Taylor,"Silica Reduction in the Blast Furnace",J. Iron and Steel Institute 202(1964)420-423.

17. N.Tsuchiya, M.Tokuda and M.Ohtani,"The Transfer of Silicon from Gas Phase to Molten Iron in the Blast Furnace",Metall. Trans.,7B(1976)315-320.

18. D.J.Young and C.J.Cripps Clark,"Sulphur Partition in Blast Furnace Hearth",Ironmaking and Steelmaking,5(1980)209-214.

19. Annual Report for 1979 : Bhilai Steel Plant (SAIL, Bhilai Steel Plant,1980).

20. K.Barin and O.Knacke,Thermodynamic Properties of Inorganic Substances,(Springer Verlag,Berlin,1973).

21. K.Barin, O.Knacke and O.Kubaschweski,Thermodynamic Properties of Inorganic Substances - Supplement,(Springer Verlag,Berlin,1980).

MATHEMATICAL MODELLING OF THE COMBUSTION

CHAMBER IN A NEW METALLURGICAL FURNACE

H. Y. Gou and W-K. Lu

Department of Materials Science and Engineering
McMaster University, Hamilton, Ontario
Canada, L8S 4L7

Abstract

In the last three years, some experimental data and basic characteristics of a new metallurgical process which is based on a smelting reduction furnace, have become available. Preliminary results are very encouraging for ironmaking and other purposes. The gaseous flow, mixing, combustion and heat transfer in the combustion chamber of this furnace is considered to be one of the critical steps in the overall process. To further our understanding of phenomena observed and future development, attempts have been made at mathematical modelling for this kind of combustion chamber which is cylindrical in shape with reversed flow and strong recirculation. The turbulent Reynolds equations which govern the two-dimensional, axisymmetric elliptical flow are solved numerically under non-reacting condition to obtain the velocity fields and concentration profiles in this chamber. The computed results are compared with measurements in cold models by using pitot probe and CO_2 tracer. The influences of the air inlet position and gaseous fuel to air mass flow ratio, on the overall flow and mixing patterns, are discussed.

Introduction

A new metallurgical process called the LB furnace for ironmaking, cokemaking and other purposes is being developed at McMaster University. The experimental work in the LB furnace of an earlier design and the feasibility analysis of this process were presented elsewhere(1) and the detailed description of this process will not be repeated here. However, it was noticed in the earlier work that the combustion was incomplete, resulting in less efficient heat utilization. A better understanding of the gaseous flow, combustion and heat transfer in the combustion chamber is certainly a prerequisite for any progress to be made in overall process development.

Although significant progress has been reported in the literature on predictive modelling techniques for combustion furnaces, it is still rather difficult to predict the detailed flow pattern, turbulent properties, flame characteristics and heat flux distribution even in a system of very simple geometry. The difficulties experienced in modelling may account for the:

(1) defect in turbulent models.

(2) defect in combustion models.

(3) defect in energy transfer models.

The works of Beer(2), Gosman et al.(3), Elghobashi and Pun(4), Hutchinson et al.(5), Spalding(6), Libby and Williams(7), Khalil(8), and Vanka(9) are just a few examples of modelling of fluid flow, combustion and heat transfer.

It is generally accepted that in the turbulent, non-premixed diffusion flame, the overall chemical reaction rate is dominated by the mixing of the reagents. The present work mainly concerns the mathematical modelling of the non-reacting gaseous flow and mixing in this particular combustion chamber in order to free fluid dynamic phenomena from coupled chemical reactions and energy transfer effects. The purpose of this study is aimed at enhancing the understanding of the transport phenomena in this chamber and providing assistance for future process development.

The turbulent model used in mathematical modelling is the two-equation k, model. A computer program has been developed, which is based on TEACH code(10), for this study. Besides the original function of the code for calculating velocities and k and , an additional conservation equation was included to allow the computation of species concentration distribution.

The measurements of velocity field and tracer gas concentration were conducted in cold models to compare with computed results. Three different air inlet arrangements and their effects on flow and mixing patterns were evaluated by using this mathematical model.

Experimental Details

Characteristics of The Combustion Chamber

Fig.1 is the sketches of the combustion chamber under consideration. The chamber is a cylinder with one end closed. From the other end, a tube, called the reduction tube, extends inside the cylinder for a certain distance along its axis . The gaseous fuel, mainly consisting of CO and

H$_2$, generated from iron oxide reduction by corbon leaves this reduction tube concurrently with sponge iron(1). The location of the air inlet is an important parameter to be studied. In the mathematical model, the air inlet is in the shape of an annular ring maintaining its axial symmetry for computational purposes. For the physical model, annular air inlet is also adopted for effective comparison of results. This restriction is necessary in order to have a two-dimensional mathematical model. The cross-sections of three air inlet arrangements relevant to the existing furnace construction are shown in Figs.1-a, 1-b and 1-c. In all cases, the cross-sectional areas are chosen for predetermined fuel and air mass flow rates and linear velocities. The flue gas exit is located at the same end as the fuel inlet in the form of an annular gap resulting in a reversed flow and strong recirculation in the chamber.

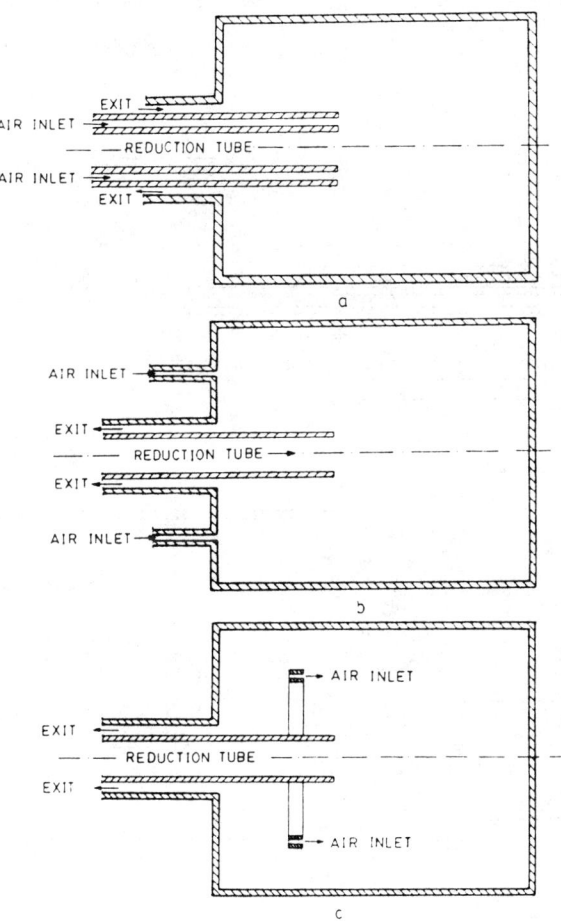

Fig. 1 Cross-sections of the combustion chamber of the LB furnace relevant to three air inlet positions: (a) Model A-1; (b) Model A-2; and, (c) Model A-3.

The Experimental Rigs

In order to achieve the best experimental measurements, different sizes of cold model were used. The larger size Model A-1 was used for tracer concentration determination and the smaller sizes Model B-1 and B-2 were used for velocity measurement. Model B-1 and B-2 have different chamber lengths obtained by adjusting the piston inserted in the cylinder. The actual laboratory furnace has a dimension similar to that of Model A-1. The sketches and dimensions of the models are shown in Fig.2.

Model A-1 has eight sampling holes located on the cylinder wall along the axial direction and distinguished by NA-1 to NA-8. In experiments, air as gaseous fuel was introduced through the reduction tube and air with 2% CO_2 as combustion air through the ring shaped inlet to the combustion chamber. Model B-1 has nine and Model B-2 has six holes labelled NB-1 to NB-9 and NB-1 to NB-6, respectively, which allow the access of the pitot probe for velocity measurement.

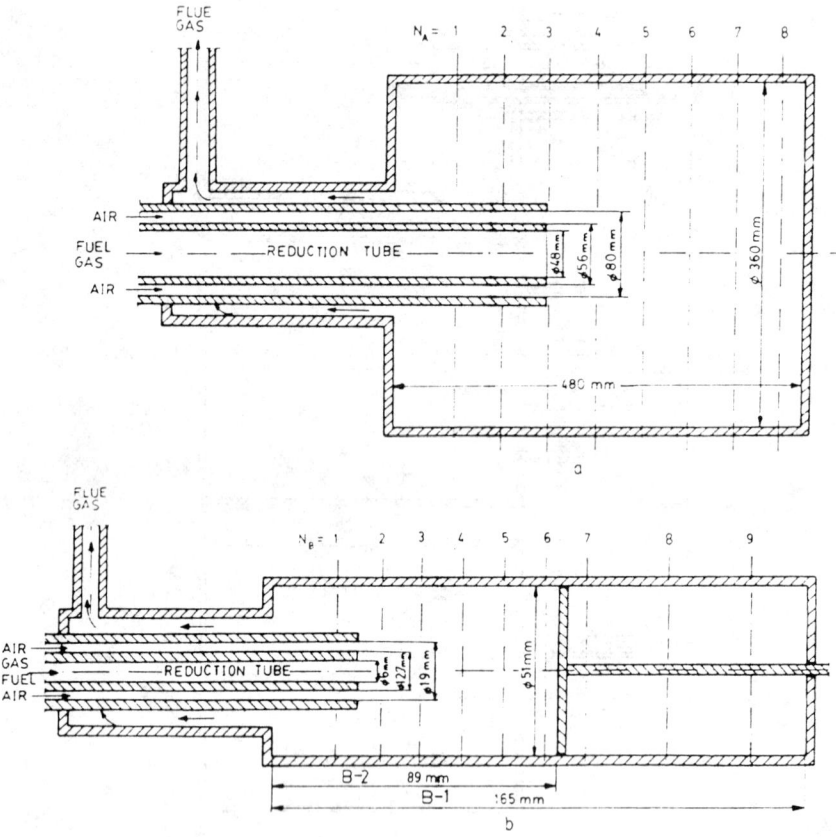

Fig. 2 Experimental rigs and the main dimensions. (a) Model A-1; and
(b) Model B-1 and B-2.

The CO_2 tracer concentration detecting equipment consists of a sampling probe which is made from a 1/16" ID and 3/32" OD steel tube, and an Horiba Infrared CO_2 Analyzer which continuously analyzes for CO_2. It was operated with full scale corresponding to 20% CO_2 in air. The accuracy is considered to be $\pm0.5\%$ full scale.

The axial and radial velocities were measured by using a pitot probe, a diaphragm-type pressure transducer and a demodulater. The normal predictive range of this pressure transducer is 0-0.086 kpa with the accuracy of $\pm0.5\%$ of full scale.

The information of velocity and CO_2 concentration obtained was analyzed and demonstrated graphically through a computer by specially designed computer programs.

The Experimental Conditions

Due to the fact that computed results are much more readily available than that measured data from cold models, only a few selected cases are compared and discussed to validate the mathematical model. Five combustion chamber configurations labelled Model A-1, A-2 ,A-3, B-1 and B-2 were chosen in this study. Model A-1, A-2 and A-3 have the same chamber dimension of the Fig.2-a and different air inlet arrangements corresponds to those shown in Fig.1-a, 1-b and 1-c. Only Model A-1 was used in tracer concentration experiment. The effects of different air inlet are discussed later. Model B-1 and B-2 differ only in chamber length as shown in Fig.2-b and were used in velocity field determination. The experimental conditions under consideration are listed in Table 1.

Table 1. The Experimental Conditions

Model	F/A	M_F(kg/min)	M_A(kg/min)	$CO_2\%_A$	U_F(m/s)	U_A(m/s)
A-1	0.317	0.324	1.023	2.0	4.0	8.4
A-1	0.158	0.162	1.023	2.0	2.0	8.4
A-2	0.317	0.324	1.023	---	4.0	8.4
A-2	0.158	0.162	1.023	---	2.0	8.4
A-3	0.317	0.324	1.023	---	4.0	8.4
A-3	0.158	0.162	1.023	---	2.0	8.4
B-1	0.250	0.028	0.112	---	12.5	10.0
B-2	0.250	0.028	0.112	---	12.5	10.0

The fuel/air mass flow ratio F/A=0.317 is equal to stoichiometric requirement in the actual furnace. In the Model A, a lower F/A ratio of one-half of the stoichiometric value was also used. This lower F/A ratio was used to reveal the extreme case in which the reoxidation of the reduced product might take place(1). In computation, both the initial velocities U_F and U_A for reduction tube flow and air inlet flow represent the uniform axial velocity based on the mass flow rate M_F and M_A, and cross section areas, respectively. $CO_2\%_A$ is the inlet CO_2 value in air flow.

The Mathematical Modelling

The Governing Differential Equations

The present work is designed to compute axial and radial velocities u and v, and species i concentration m_i in a two-dimension axisymmetric elliptical flow at steady state. In doing so, the additional variables of pressure P and two turbulent quantities k and ε have to be determined simultaneously. There are six differential equations; all of them can be generalized to a common form:

$$1/r[\partial/\partial x(\rho ur\Phi) + \partial/\partial r(\rho vr\Phi) - \partial/\partial x(r\Gamma\phi\ \partial\Phi/\partial x) - \partial/\partial r(r\Gamma\phi\ \partial\Phi/\partial r)] = S_\phi \quad (1)$$

where Φ = u, v, m_i, P, k and ε. The first two terms in square brackets are the convective term and the last two terms are the diffusion term. ρ is the so-called turbulent exchange coefficient and S_ϕ is a source term including all those terms which are not included in convective and diffusion terms. For different variables, S_ϕ and Γ_ϕ have different contents. The detailed differential equations and particular S_ϕ and Γ_ϕ expressions were given in several earlier works(5,8,11). The set of constants in turbulent model employed in this work was reported by Launder and Spalding(14).

The finite difference analog of equation (1) for Φ = u, v, m_i, P, k, and ε is then solved by so-called SIMPLE(Semi-Implicit Method for Pressure Linked Equation) algorithm(14). The iterative method of solution employed is a double sweep of TDMA(Tri-Diagonal Matrix Algorithm).

The Turbulent Model

The two-equation k, ε turbulent model(11) which involves the local kinetic energy k and the energy dissipation rate ε is employed to close the differential equations. The calculation of local values of k and ε are then related to the evaluation of a local turbulent viscosity:

$$\mu_t = \rho C_\mu k^2/\varepsilon \quad (2)$$

from which the turbulent shear stresses are calculated, where μ_t is turbulent viscosity, ρ is fluid density and C_μ is a constant.

This two-equation model was used and its validity was assessed in simple non-reacting flows for a wide range of flow configurations (3,11,12,13). The equations for reacting flows to a certain extent were also developed and extended to elliptical flows but much less modelling work has been conducted(8).

The Grid Arrangement

The calculation domain is one-half axisymmetric x-r plane. In the Model A-1, A-2 and A-3 computations, different 24X22 non-uniform grid nodes in x and r directions, respectively, were employed. The Model B-1 has 15X18 grid nodes and B2 corresponds to 9X18 grid nodes, both uniformly distributed.

The Comparisons of Computed And Measured Results

The Velocity Profile

The experiments for measuring velocity profile were carried out in Model B. In Figs.3,4 and Figs.5,6, two types of velocity vectors, measured and computed, in one-half of the x-r symmetrical plane are displayed for comparison. Figs.3 and 4 are the results from Model B-1 and Figs.5 and 6 from B-2. In Fig.3-a to 6-a, the arrows represent both magnitude and direction of velocity, while the arrows in Fig.3-b to 6-b all have the same length and represent flow direction only. Fig.7 is the radial profile of axial velocity comparison at different axial levels. The actual NB probe positions are shown in Fig.2-b.

In general, two velocity profiles, measured and computed, agree well in view of the errors in measurement and assumptions in computation. They exhibit a relatively large recirculation zone over the main body of the chamber. The two streams from the reduction tube and the air inlet move down to the end zone of the cylinder. When it reaches the location near the end wall, the flow turns sharply to the radial direction. Part of the flow goes backwards along the cylinder side wall with almost uniform reversed velocity until close to the annular exit. The remaining gas is entrained in the recirculation. In both measured and computed results, Model B-2 with a shorter chamber length has a similar profile as that of Model B-1. The influence of the end wall location under consideration on the general flow pattern seems not to be very important.

However, there are two main discrepancies in the comparison of velocity profiles. Firstly, there is an evident radial velocity component in the forward flow beginning from the inlet in the measured profiles, while the computed results show almost parallel axial forward flow from the inlet to the end zone. The computed radial velocity of any significant value appears near the recirculation zone and at a location very close to the end wall. Secondly, the mixing between the two streams is more intensive in measured profiles than that shown in computed results. The distinct two velocity peaks shown in Fig.7 represent the air and fuel streams, respectively, and remain over the distance down to the closed end in computed results. On the other hand, in measured profiles the peaks become diffused and disappeared at a short distance from the inlet. The first phenomena is probably directly related to the second one as the high radial velocity component causes the intensive mixing at the interface between two streams. Furthermore, one would notice that these discrepancies are restricted to the region near the opening of the coaxial tube. Far down stream and away from the axial, the measured and computed profiles agree very well.

The Tracer Concentration Profile

Experiments were conducted in Model A-1 with conditions listed in Table 1. The measured and computed tracer concentration profiles are compared and shown in contourlines in Fig.8. Fig.8-a is the result of F/A=0.317 and Fig.8-b shows the result for the case of F/A=0.158. These two sets of contourlines from measurement and computation resemble each other in general shape and tendency. The largest concentration gradient appears in the flow inlet region at the interface between two streams, where intensive mixing is expected. Decreasing fuel/air flow ratio brings the air inlet flow closer to cylinder axis. In the regions of reversed flow and recirculation, the tracer concentration is close to the average value

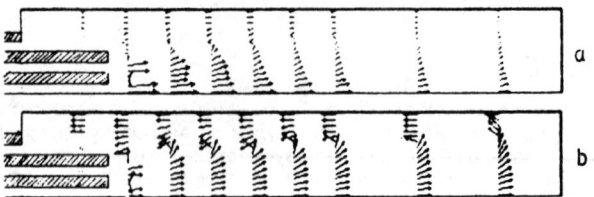

Fig. 3 Measured velocity vectors in Model B-1, F/A = 0.317 (a) magnitude and direction and (b) direction only.

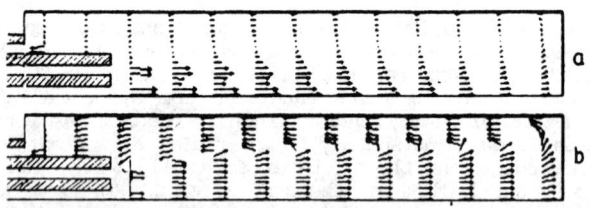

Fig. 4 Computed velocity vectors in Model B-1, F/A = 0.317 (a) magnitude and direction and (b) direction only.

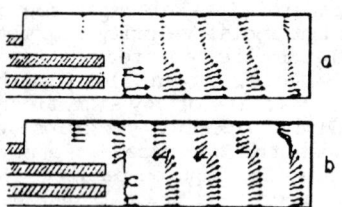

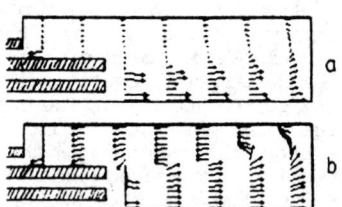

Fig. 5 Measured velocity vectors in Model B-2, F/A = 0.317 (a) magnitude and direction; (b) direction only.

Fig. 6 Computed velocity vectors in Model B-2, F/A = 0.317 (a) magnitude and direction; (b) direction only.

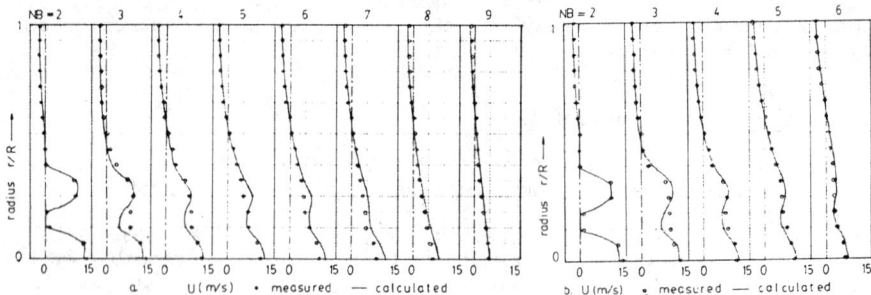

Fig. 7 Comparisons of radial profiles of axial velocity at different axial levels. F/A = 0.317 (a) Model B-1; and (b) Model B-2.

558

of the exit gas. The same discrepancies noticed in velocity profiles also occur in tracer concentration contours; i.e. the measured results show a faster mixing along the axial direction than that of computed results.

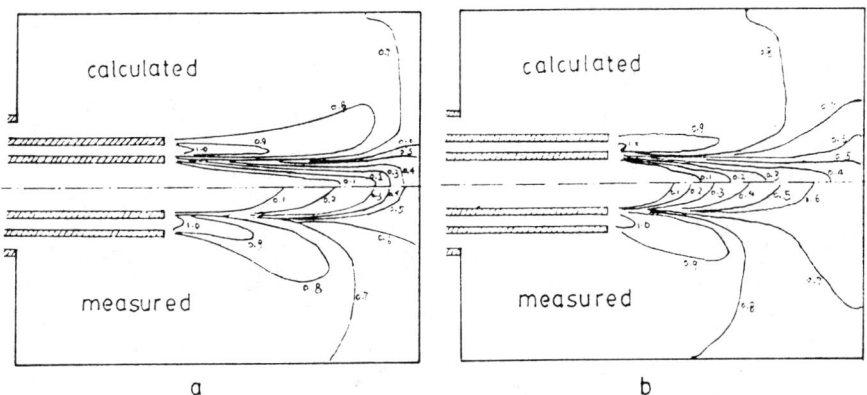

Fig. 8 CO_2 concentration countour lines in Model A-1:   (a) F/A = 0.317; and, (b) F/A = 0.158.

Discussion of Results

The general agreement between computed and measured velocity profiles as well as tracer concentration profiles illustrates that this mathematical model is capable, at least qualitatively, of giving a good picture of fluid flows and mixing in a two-dimensional reversed flow field. The turbulent model employed here predicts the turbulent properties of the flow which are generally acceptable except near the flow inlet region in our combustion chamber.

The discrepancies near the flow inlet may indicate that the turbulent viscosities were underestimated and the spatial distributions of the turbulent viscosity were not adequately accounted for at this part of the system.   Certain improvements may be acheived by using a finer grid mesh, employing a more realistic inlet velocity profile which might include radial velocity component, or adopting a more complex turbulent model which accounts for the non-isotropy of the turbulent shear stress and viscosity.

The Influences of Air Inlet Position And F/A Ratio

Three selected air inlet arrangements which are of interest to us illustrated in Fig.1 were evaluated and the sensitivities of their flow patterns against the F/A ratio change were also examined by using the above-mentioned mathematical model.   This set of experimental conditions is listed in Table 1 and the same for Model A-1, A-2 and A-3. Two F/A ratios of 0.317 and 0.158 were used in computations.

Figs.9 and 10 are the computed results in Model A-2 with the air inlet position shown in Fig.1-b. This model has an annular air inlet located about half way between the exit and the chamber side wall.   It can be seen

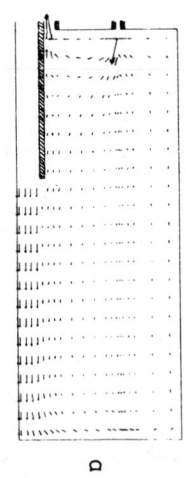
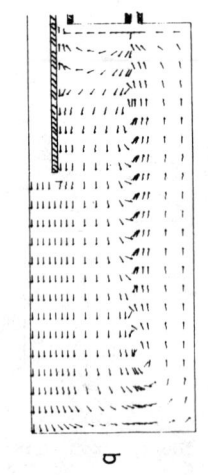
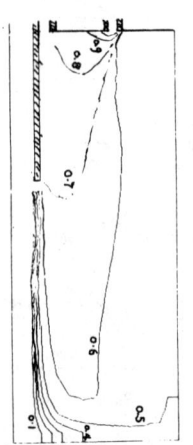

Fig. 9 Computed velocity vectors and air concentration contours in Model A-2, F/A = 0.317:
(a) velocity magnitude and direction;
(b) velocity direction only; and,
(c) air concentration contours.

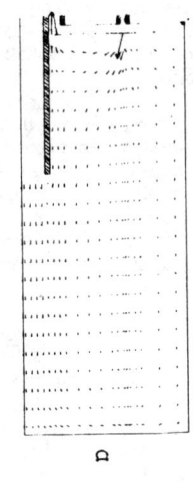
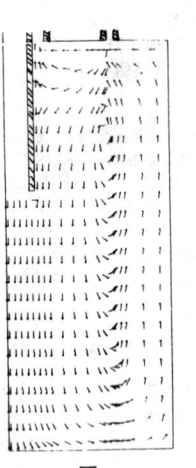
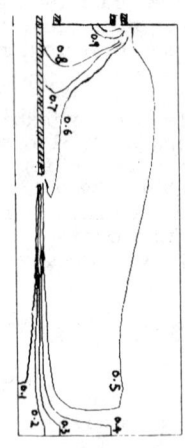

Fig. 10 Computed velocity vectors and air concentration contours in Model A-2, F/A = 0.158:
(a) velocity magnitude and direction;
(b) velocity direction only; and,
(c) air concentration contours.

560

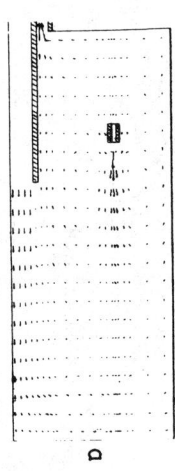

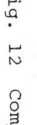

Fig. 11 Computed velocity vectors and air concentration contours in Model A-3, F/A = 0.317:
(a) velocity magnitude and direction;
(b) velocity direction only; and,
(c) air concentration contours.

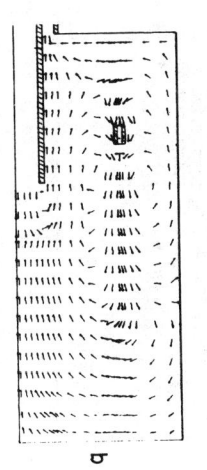

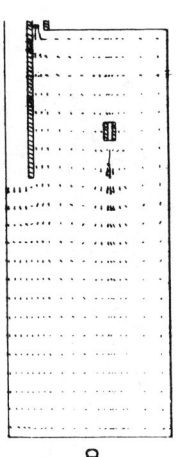

Fig. 12 Computed velocity vectors and air concentration contours in Model A-3, F/A = 0.158:
(a) velocity magnitude and direction;
(b) velocity direction only; and
(c) air concentration contours.

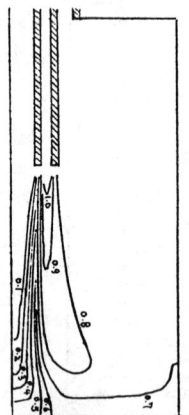

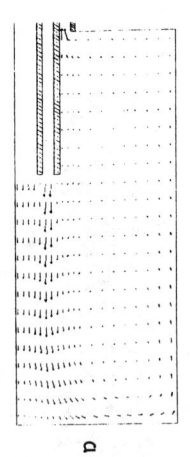

Fig. 13 Computed velocity vectors and air concentration contours in Model A-1, F/A = 0.158:
(a) velocity magnitude and direction;
(b) velocity direction only; and,
(c) air concentration contours.

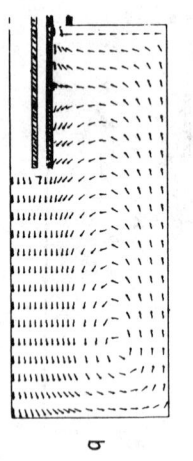

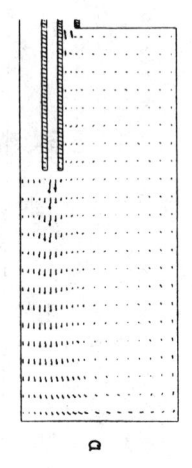

Fig. 14 Computed velocity vectors and air concentration contours in Model A-1, F/A = 0.158:
(a) velocity magnitude and direction;
(b) velocity direction only; and,
(c) air concentration contours.

that since the reversed flow goes backwards along cylinder wall, it has to make its way across the inlet air jet stream in order to reach the exit, so that there exists a strong interference between these streams. A short circuit of air flow can not be avoided.

In order to avoid the interference shown in the case of Figs.9 and 10, in Model A-3 a ring-shaped air inlet is extended inside the chamber as shown in Fig.1-c. Figs.11 and 12 show that the computed results of Model A-3 have very complicated flow patterns with more than one recirculation zone. Comparing these two profiles for the two F/A ratios, it is found that the flow fields generated from this arrangement are very unstable, i.e. sensitive to change of F/A ratio. When F/A=0.158, most of the mixing takes place very near the opening of the reduction tube where large gradients of velocity and concentration exist.

Model A-1 is similar to a coaxial burner with the air inlet form an annular tube outside the reduction tube shown in Fig.1-a. The computed results which give a relative stable and simple flow pattern, are displayed in Figs.13 and 14. In this model the mixing takes place along the interface between the two streams where highest gradients exist. By comparing Figs.13 and 14, it seems that the F/A ratio has a much less influence on flow pattern.

The positioning of the air inlet similar to the case of Model A-1 may be recommended in future furnace design, at least as the best among the three evaluated. But the actual adoption of this arrangement still has some technical problems to be solved, e.g. its influence on heat transfer. A hot model will be used to continue our investigation.

Concluding Remarks

A mathematical model for two-dimensional turbulent reversed flow has been established based on the existing computer code, to evaluate the non-reacting velocity profile and species concentration distribution in a cylindrical combustion chamber relevant to a new metallurgical process.

The cold model experiments were conducted to measure velocity and tracer concentration profiles. The generally good agreement between measured and computed results indicates that the turbulent aerodynamic model employed here is adequate in those regions where the non-isotropy of the turbulent shear stress is not very strong. The discrepancies observed near the flow inlet show that the inadequacy of this model because of the spatial variance of turbulent viscosity can not be totally ignored.

With the aid of this model the design of a new combustion chamber for the LB furnace is being carried out at McMaster University.

Acknowledge

The authors wish to thank Natural Science and Engineering Research Council of Canada and Stelco Inc. of Toronto for their financial support of this project. Advice offered by Dr. P. E. Wood and Dr. G. A. Irons in mathematical modelling and by Dr. J. H. T. Wade, DR. B. Latto and Dr. J. S. Chang in physical modelling is sincerely appreciated. Stimulating discussions with members of the "LB Furnace Team", particularly, Dr. C. Bryk, have been invaluable in the successful conclusion of the present work.

References

1. W-K. Lu, C. Bryk and H. Y. Gou, The LB Furnace for Smelting Reduction of Iron Ores, 5th International Iron and Steel Congress, Washington, D.C., U.S.A., B3, pp.1065-1075, 1986.

2. J.M.Beer, The Significance of Modelling, Proceedings of The 3rd Symposium on Flames and Industry, Institute of Fuel, pp.B1-B8, 1966.

3. A.D.Gosman, W.M.Pun, A.K.Renchal, D.B. Spalding and M.Wolfshtein, Heat and Mass Transfer in Recirculating Flows, Academic Press, London, 1969.

4. S. E. Elghobashi and W. M. Pun, A Theoretical and Experimental Study of Turbulent Diffusion Flames in Cylindrical Furnaces, Fifteenth Symposium (international) on Combustion, pp.1353-1365, 1974.

5. P. Hutchinson, E. E. Khalil, J.H. Whitelaw and G.Wigley, The Calculation of Furnace-Flow Properties and Their Experimental Verification, J. Heat Transfer, Vol.98, pp.276-283, May 1976.

6. D.B. Spalding, Development of the Eddy-break-up of Turbulent Combustion, Sixteenth Symposium (International) on Combustion, pp.1657-1663, 1977.

7. P. A. Libby and F. Williams, Turbulent Reacting Flows, Springer-Verlag, New York, 1980.

8. E.E. Khalil, Modelling of Furnaces and Combustors, Abacus Press, 1982.

9. S.P. Vanka, Calculation of Axisymmetrical, Turbulent, Confined Diffusion Flames, AIAA J., Vol.24, No.3, pp.462-469, May 1986.

10. A. D. Gosman and W. M. Pun, Calculation of Recircualating Flows, Rept. No. HTS/74/12, Mechanical Eng. Dept, Imperial College, London, 1974.

11. B. E. Launder and D.B. Spalding, The Numerical Computation of Turbulent Flows, Computer Methods in Applied Mechanics and Engineering, Vol.3, pp.269-289, 1974.

12. D.R.Boyle and M. W.Golay, Measurement of A Recirculating, Two-Dimensional, Turbulent Flow and Comparison to Turbulent Model Predictions, J. Fluids Engineering, Vol.105, pp.439-440, 1983.

13. S.E. Elghobashi, G.S. Samuelsen, J.E. Wuerer and J.C. LaRue, Prediction and Measurement of Mass, Heat, and Momentum Transport in A Nonreacting Turbulent Flow of A Jet in An Opposing Stream, J. Fluids Engineering, Vol.103, pp.127-132, 1981.

14. B.E.Launder and D. B. Spalding, Mathematical Models of Turbulence, Academic Press, London, 1972.

15. S. V. Patankar and D. B. Spalding, A Calculation Procedure for Heat, Mass and Momentum Transfer in Three-Dimensional Parabolic Flows, Int. J. Heat Mass Transfer, Vol.15, pp.1787-1806, 1972.

MEASUREMENT AND SIMULATION OF THE FLOW IN AN IRON BATH

STIRRED WITH BOTTOM-INJECTED NITROGEN

M. P. Schwarz, J. K. Wright and B. R. Baldock

CSIRO Division of Mineral Engineering
Clayton, Victoria, 3168   Australia

Abstract

    The general purpose flow simulation package, PHOENICS, is used to
model the flow in a 1 kg capacity experimental iron bath reactor which is
stirred by bottom-injected nitrogen. A two-fluid formulation is used to
treat the gas-liquid region, and turbulence is modelled using the
two-equation k-ε model. The liquid velocity in the plume region of the
bottom blown iron bath is measured at several gas flow rates between 0.5
and 6.0 Nl/min using an iron rod dissolution technique. Mass transfer
correlations developed for the dissolution of iron cylinders in Fe/C melts
under forced convection conditions are used to derive the plume velocities.
The predicted velocities agree with the experimentally inferred plume
velocities in order of magnitude although the computed velocity increases
more steeply with increasing flowrate than do the measured values. The
study has suggested improvements, such as more realistic descriptions of
interphase friction and turbulence, which could be made to the mathematical
model.

Introduction

The CSIRO Division of Mineral Engineering is undertaking a comprehensive investigation of gas-stirred iron-bath reactors as part of a programme of research into intensive smelting processes. The study includes experimental work on small scale iron baths, mathematical modelling of gas-stirred reactors, and cold air-water modelling.

In this paper we report the development of a mathematical model of the flow in a 1 kg capacity bottom-blown iron bath. The predicted liquid velocity in the gas plume is compared with the value measured using an iron rod dissolution technique.

The modelling of gas driven bath circulation has received considerable attention in recent years particularly in connection with steel ladle stirring. Momentum and energy balances on the system as a whole have been used (1) to obtain a theoretical relationship between plume velocity, U, and gas flow rate, Q, :

$$U \propto Q^{1/3}. \tag{1}$$

This relationship is consistent with the correlation between mixing time, τ, and energy dissipation rate, ε_m, first proposed by Nakanishi et al (2) on the basis of water model measurements:

$$\tau \propto \varepsilon_m^{-0.4}, \tag{2}$$

if it is assumed that the mixing time is inversely proportional to plume velocity, and dissipation rate proportional to gas flow rate.

More recently, detailed mathematical models have been constructed of gas-stirred baths so as to predict the flow velocities throughout the bath, e.g. (3,4,5,6,7). Some of these models have been tested against experimental measurements of velocities in air-water systems. In general, good agreement is obtained over most of the bath, although little data exists for the plume region itself because of the difficulty of measurement in gas-liquid mixtures.

Gas-stirred baths are two-phase systems and in order to overcome the computational difficulties involved in modelling such flows, most researchers (3,5) have assumed they know beforehand the width of the plume, and the distribution of void fraction across it. While these may be measured or estimated for an air-water system, such is not the case for an iron bath. We have therefore used a more sophisticated formulation which solves for the distribution of void fraction in the bath. Furthermore we solve for the velocity of each phase separately, thus employing the full "two-fluid" model. Almost all researchers (3,4,5,6) previously have solved the mixture equations, ie one phase, variable density equations representing the gas-liquid mixture; McKelliget et al (4) have suggested that it may be limitations in this approach which prevented them from obtaining realistic solutions in the case of horizontal injection.

Another difference between this work and that previously published is in the gas flow rate. Since we are interested in intensive smelting processes with much more violent mixing than is required in a steel ladle, the superficial gas velocity is much higher than in most studies to date of gas stirred baths. For example the models constructed by Grevet et al (3)

566

were at superficial velocities of 6.0×10^{-4} m/s and 12.0×10^{-4} m/s, whereas the maximum value for models calculated in this paper is 0.3 m/s.

Experimental Set-up and Results

A schematic diagram of the experimental apparatus is shown in Figure 1. Nitrogen was injected at rates up to 6 Nl/min through a single orifice of 0.5, 1.0 or 2.0 mm inside diameter situated in the bottom of a tapered conical porcelain crucible. The crucible contained 1 kg molten Fe/C charge (4% carbon) at 1400°C, occupying a depth of about 58 mm. In each case a steel rod of 12.1 mm diameter, after being preheated to within 20°C of the melt temperature, was lowered into the bath and positioned 10 mm directly above the inlet port. After 5 seconds immersion the rods were removed, cooled and the rod diameter measured with calipers at several positions along the immersed length. An average diameter was then calculated. The rod width was fairly uniform except for a neck near the nominal bath surface. The cylinders were also weighed before and after immersion: the diameter change so deduced was consistent with the average measured diameter.

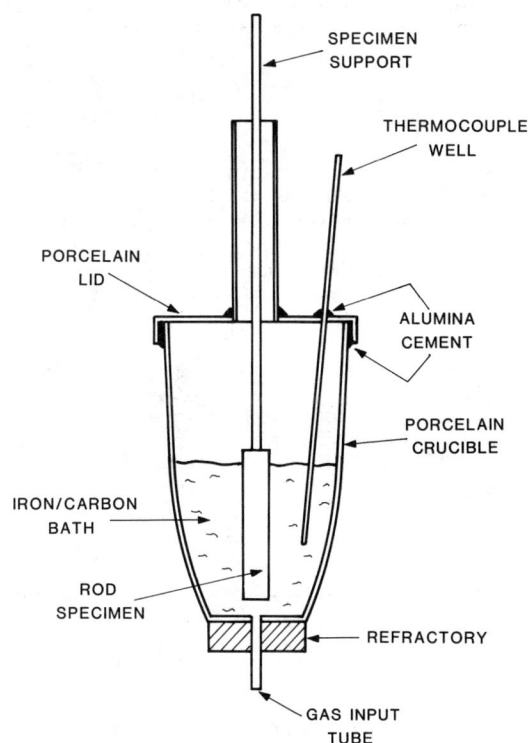

Figure 1. Schematic diagram of the bottom-blown crucible
and iron rod specimen

For each rod the procedure was repeated until a substantial amount of dissolution had occurred. The dissolution rate (-dr/dt) was then determined by plotting the average rod diameters against immersion time. The diameter was found to decrease linearly with immersion time.

In order to derive mass transfer coefficients, it is assumed that the dissolution rates are controlled by diffusion processes in a boundary layer. It is also assumed that the interface temperature is close to the bulk melt temperature and that the liquid and solid at the interface are in equilibrium. Given these assumptions the mass transport coefficients at each gas flow rate were derived using the Lommel and Chambers equation (8):

$$-\frac{dr}{dt} = k_m \ln \left[1 + \left(\frac{C_\ell - C_b}{C_o - C_b}\right) \right]. \tag{3}$$

The derived mass transport coefficients, as functions of gas flowrate and orifice size, are plotted in Figure 2. The mass transfer coefficients appear to be independent of the orifice size. This suggests that the kinetic power contribution is unimportant even in an experimental system in which the kinetic power input can be an appreciable proportion of the total mixing power. [Power contributions were calculated using formulae given in (9).] A regression analysis on the data returned the dependence:

$$k_m \propto Q_o^{0.21}. \tag{4}$$

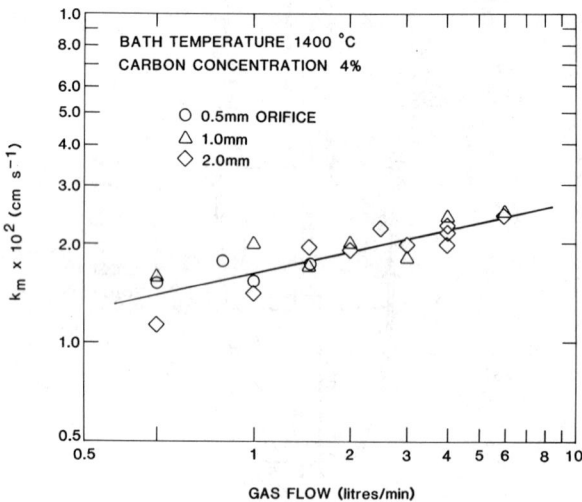

Figure 2. Plot of mass transfer coefficient as a function of gas flowrate (Nl/min) for three different orifice sizes. The line gives the least squares fit on all the data.

There have been a number of investigations in which correlations have been derived for the dissolution of steel cylinders in Fe/C melts under known hydrodynamic conditions. If it is assumed that these correlations are applicable to the present gas stirred system, then it is possible to derive indirectly the plume velocities. Kim and Pehlke (10) have summarized the results of several investigators by the relation:

$$j_d = b \; Re^{-n}, \qquad (5)$$

where j_d is a dimensionless mass transfer coefficient given by

$$j_d = \frac{k_m}{U} \; Sc^{0.644}. \qquad (6)$$

A regression analysis of the plume velocities obtained using this correlation gives:

$$U = Q_o^{0.3}, \qquad (7)$$

which agrees closely with other work discussed in the Introduction.

The Simulation Technique

To compute numerical models of the bath flow we have used the general purpose flow simulation package, PHOENICS, in the steady-state two-phase mode. The equations solved are the mass conservation equations for each phase,

$$div \left(R_i \, \rho_i \, \underline{V}_i \; - \; \rho_i \, D_t \; grad \; R_i \right) = m_i, \qquad (8)$$

and conservation equations for momentum and turbulence quantities all of which can be written in the general form

$$div \left(R_i \, \rho_i \, \underline{V}_i \; \phi_i \; - \; R_i \, \Gamma_{\phi_i} grad \; \phi_i \right) = R_i \, S_{\phi_i}, \qquad (9)$$

where ϕ_i is the general conserved property. The discretized form of the equations is obtained by integrating the above differential equations over control volumes of finite size (or 'cells'), hence the name 'finite-volume' formulation. The resulting scheme is fully conservative and implicit. Details of the solution procedure can be found elsewhere (11).

The simulation technique is similar to that used by Schwarz and Koh (12) and Schwarz and Turner (13) to model air-water systems. Good agreement with the data was obtained in those cases. The flow is assumed to be axisymmetric so that the problem can be solved on a two-dimensional polar grid. The variables, ϕ_i, solved for are then pressure, pressure correction, V_ℓ, V_g, W_ℓ, W_g, k, and ε.

Each phase is assumed to be incompressible. The gas density, ρ_g, is taken to be that at bath temperature and average bath pressure. For example for the 1 kg reactor, assuming the temperature is 1400°C, and the average pressure 1. x 10^5 Pa, gives a nitrogen density of 0.21 kg m^{-3}. The iron density is taken to be 6900 kg m^{-3}.

Turbulence Model

Turbulence is modelled by replacing the laminar viscosity by an effective eddy viscosity calculated using the k-ε model (14) applied to the liquid phase. The standard values of the empirical constants which enter into the source terms for k and ε are used.

Furthermore the turbulent diffusion of bubbles through the liquid is accounted for by a term,

$$- \text{div} \left(\rho_i \, D_t \, \text{grad} \, R_i \right) \qquad (10)$$

in the continuity equations (8), with the turbulent diffusivity, D_t, taken to be equal to the kinematic effective eddy viscosity. This phase diffusion term has been specially coded into the 81 version of PHOENICS. It is worth emphasizing that the velocities which are solved for and presented in this paper are the mean flow velocities. The fluctuating turbulent component is superimposed on this mean flow and its effect is accounted for by an enhanced effective viscosity and diffusion coefficient.

Boundary Conditions

To simplify the problem, we assume the bath surface to be flat and to coincide with the top boundary of the computational domain, and allow no flow of liquid across it, ie $W_\ell = 0$. This is clearly a gross simplification, especially at high injection rates when vigorous splashing occurs. Gas is allowed to leave the bath at the rate at which it arrives at the surface, ie

$$\frac{\partial W_g}{\partial z} = 0 . \qquad (11)$$

We also apply the condition

$$\frac{\partial R_g}{\partial z} = 0 \qquad (12)$$

which prevents an accumulation of gas which otherwise occurs near the surface. Though difficult to justify on physical grounds, this boundary condition allows the problem to be solved as steady state. Otherwise neither the bath surface nor the total volume of liquid is specified by the boundary conditions and the problem must be solved as an initial value problem, or transient.

A zero shear stress condition is applied at the bath surface, i.e.

$$\frac{\partial V_\ell}{\partial z} = 0 , \qquad (13)$$

and no sources or sinks of k or ε are introduced:

$$\frac{\partial k}{\partial z} = 0, \quad \frac{\partial \varepsilon}{\partial z} = 0 . \qquad (14)$$

We also fix the pressure to zero at one cell on the surface, so as to prevent a drift to large values.

On the bottom of the vessel and on the side of the rod, friction is applied according to the familiar logarithmic wall law, i.e. a sink is applied to the component of velocity parallel to the wall using the skin-friction factor, s, deduced from

$$s = \left[\kappa / \ln \left(1.01 + 9.0 \, Re_w \, s^{1/2} \right) \right]^2 , \quad Re_w > 132.5$$

$$s = 1 / Re_w \qquad\qquad , \quad Re_w < 132.5 \qquad (15)$$

where

$$Re_w = \frac{\rho_\ell \, \delta \, U_{\shortparallel}}{\mu} . \qquad (16)$$

The values of k and ε in the cells closest to the wall are then set to

$$k = U_\tau^2 / C_\mu^{1/2}, \quad \varepsilon = U_\tau^3 / (\kappa \delta), \qquad (17)$$

where
$$U_\tau = s^{1/2} |U_{||}|, \tag{18}$$
and δ is the normal distance to the wall.

The sloping wall of the reactor is approximated by a section of a right cone, and simulated by 'blocking off' the appropriate cells using "partial porosities" in the way described by Patankar (15).

The source of gas mass applied to the cell adjacent to the orifice is
$$\rho_o \, Q_o, \tag{19}$$
where ρ_o is the gas density at NTP. The vertical gas velocity which is convected into this cell is the cold velocity,
$$Q/(\pi \, r_o^2). \tag{20}$$
That is we assume that the gas expands to ρ_g (corresponding to bath temperature) immediately after entering the bath. Turbulence energy convected into the bath with the gas is neglected.

To account for gravity, a buoyancy force equal to $R_g \, (\rho_\ell - \rho_g)g$ per unit volume acts on the gas phase.

Interphase Friction

Interphase friction is modelled by the formula built into PHOENICS: the force per unit volume is
$$CFIPS \times R_\ell \times R_g \times V_{slip}, \tag{21}$$
with $CFIPS = 2.0 \times 10^5 \ kg \ m^{-3} s^{-1}$. This is equivalent to assuming that the equilibrium gas-liquid slip velocity, i.e. the value at which buoyancy and drag balance, is 0.35 m/s. This slip velocity in turn is the value calculated for an isolated spherical bubble of radius 4 mm using the following expression for drag:
$$\frac{\pi}{8} \, c_d \, \rho_\ell d_b^2 \, V_{slip}^2, \tag{22}$$
with the drag coefficient given by (16):
$$C_D = 24 \left(1 + 0.15 \, Re_b^{0.687}\right)/Re_b + 0.42/\left(1 + 4.25 \times 10^4/Re_b^{1.16}\right) \tag{23}$$

In regions of high void fraction, equation (21) automatically simulates the drag on spherical iron droplets of diameter 0.07 mm., and in regions of intermediate void fraction it provides a smooth transition between the bubble and droplet regimes.

Model Validation

Extensive validation of this simulation procedure has been carried out using data from air-water systems. One test is described here; full details will be published elsewhere.

The experiment which we simulate was performed by Grevet et al (3), and consisted of injection of air into the centre of the bottom of a cylindrical tank of water. A laser anemometer was used to measure the mean flow velocities and turbulence intensities at many points in the tank. The diameter of the vessel was 0.6 m, the water depth 0.6 m, the orifice diameter 0.0127 m and the air velocity at the nozzle 3.2 m/s.

571

This experiment has been simulated by several other researchers (e.g. 3,4,5,7). PHOENICS was used by Cross et al (7), but they did not include turbulent diffusion of the gas phase into the liquid (or vice versa), and they employed a different boundary condition at the surface.

The simulation procedure used here differs from that applied to the iron bath only in the values of constants used: gas density 1.2 kg m^{-3}, liquid density 1000 kg m^{-3}, liquid laminar viscosity 1.0 x 10^{-3} kg m^{-1} s^{-1}, and CFIPS 5.0 x 10^4 kg m^{-3} s^{-1}. The value of CFIPS is based on a gas-liquid slip velocity of about 0.2 m/s, which is close to the measured value over a wide range of bubble sizes.

Measured and computed velocity magnitudes are compared at two different heights in the bath in Figure 3. There is excellent agreement over most of the bath, with the maximum error being about 0.02 m/s. This compares favourably with other authors' simulations of the same experiment. The width of the model plume is similar to that indicated by Grevet et al (3), though void fraction distributions were not measured.

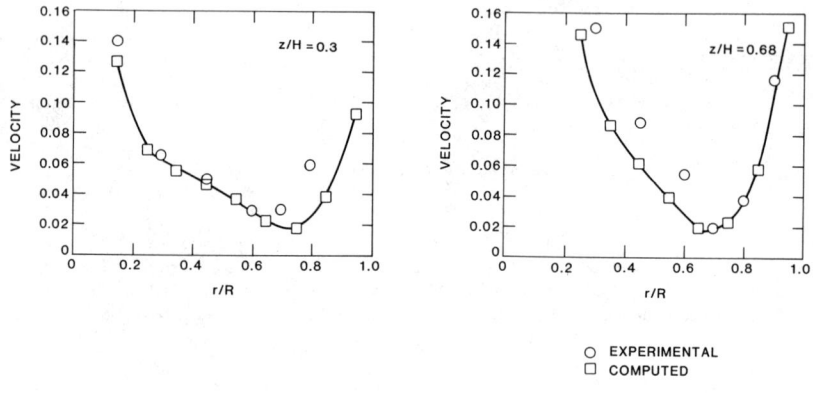

Figure 3. Comparison between measured and computed velocity magnitudes for air-water experiments of Grevet et al. Values are plotted at two different heights, z, in the bath.

Model Results and Comparison with Experiment

Numerical models were constructed for flow rates between 0.1 and 10.0 Nl/min for orifice diameter 2 mm, both with and without the rod present.

Models for Flowrate 2 Nl/min

Figure 4 shows the computed liquid flow field for gas rate 2 Nl/min. The expected recirculation is seen, with a mean local velocity of more than 0.1 m/s over most of the bath. The void fraction distribution is shown in Figure 5 together with the liquid velocity in the plume region. The predicted gas volume fraction is very high (> 0.7) along the centreline, and the plume widens with height as expected, because of the outward turbulent diffusion of bubbles into the melt. The local turbulence level is determined from k and ε for which the distributions are plotted in Figure 6. The predicted rms turbulence velocity in the centre of the plume is about 0.5 m/s, not much smaller than the bulk velocities which are of order 1.0-2.0 m/s.

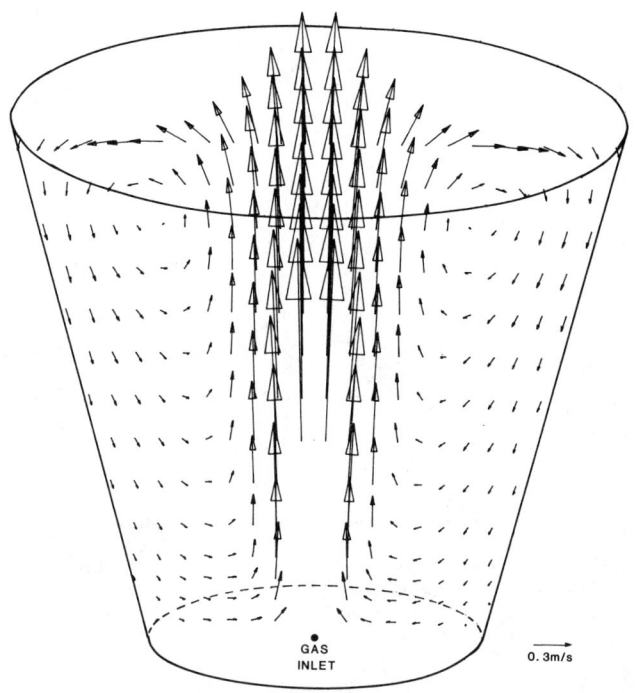

Figure 4. Computed liquid flow field for bath without rod
at gas flow rate 2 Nl/min. The vectors are plotted on
a vertical plane through the axis of the reactor.
Large velocities near the tuyere are not plotted.

It is interesting to note that the effective eddy viscosity predicted
by the k-ε model is of the same order as that given by the formula of Sahai
and Guthrie (1) for the spatial average:

$$\mu_{eff} = 5.5 \times 10^{-3} \rho_\ell H \left[(1 - \alpha)g \ Q/W \right]^{1/3} \qquad (24)$$

For the present case this yields roughly 0.5 kg m^{-1} s^{-1}, compared
with a value predicted by the k-ε model of about 0.6 kg m^{-1} s^{-1}
over much of the bath. The effective viscosity is higher in the plume
(2.0-3.0 kg m^{-1} s^{-1}). A model has been run with a uniform effective
viscosity of 0.5 kg m^{-1} s^{-1}: the velocities in the plume and in the
descending stream of liquid at the outer wall are different by as much
as a factor of two from the values derived using the k-ε model.

Figure 7 shows that the flow field is significantly disturbed by the
presence of the rod: the plume velocity is suddenly reduced at the bottom
end of the rod, then increases steadily under the influence of buoyancy
from 0.5 m/s to 1.1 m/s just below the surface. With the rod absent the
liquid plume velocity decreases from 2 m/s near the tuyere to 1.3 m/s just
below the surface. The decrease is caused by entrainment of liquid from
the bath into the plume.

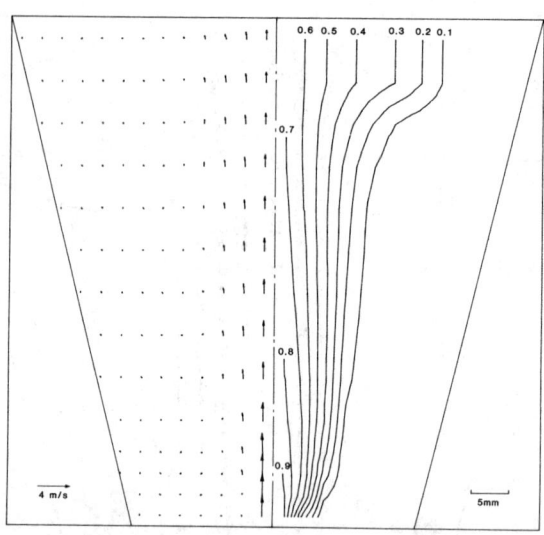

Figure 5. Computed liquid flow field and void fraction map for bath
without rod at gas flow rates 2 Nl/min. The quantities are plotted
on a vertical plane through the axis of the reactor.

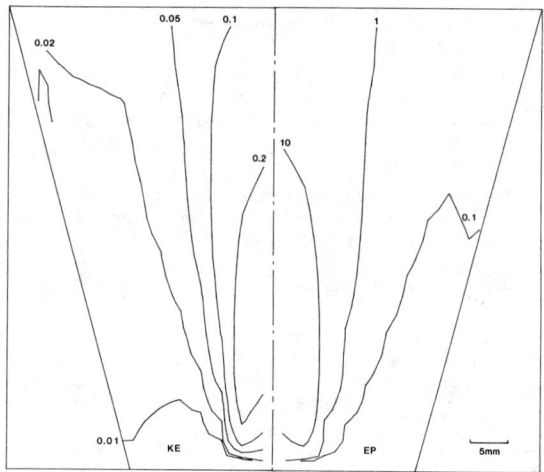

Figure 6. Computed distributions of k and ε for bath without rod
at gas flow rate 2 Nl/min. The quantities are plotted
on a vertical plane through the axis of the reactor.

The corresponding distributions of k and ε are plotted in Figure 8.
The effect of the rod is to damp the turbulence in the plume, as can be
seen by comparing Figures 8 and 6. For this reason the plume is 1.5 to 2
times wider in the absence of the rod.

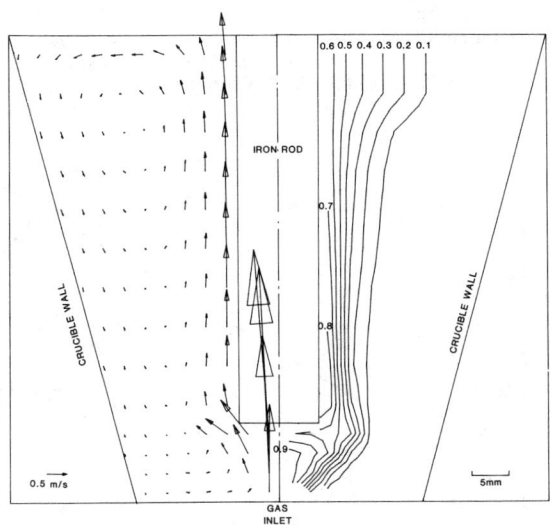

Figure 7. Computed liquid flow field and void fraction map for bath with rod at gas flow rate of 2 Nl/min. The quantities are plotted on a vertical plane through the axis of the reactor.

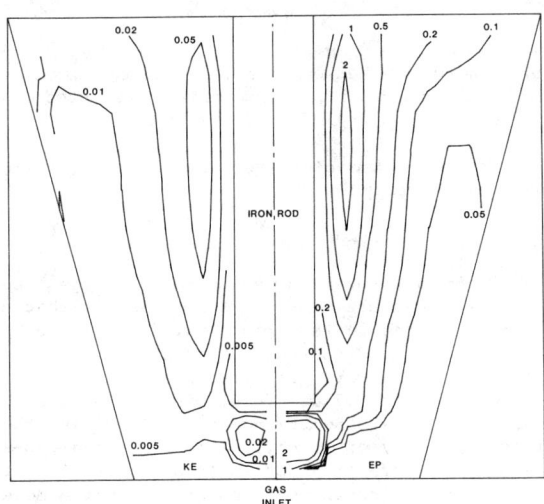

Figure 8. Computed distributions of k and ε for bath with rod at gas flow rate 2 Nl/min. The quantities are plotted on a vertical plane through the axis of the reactor.

Dependence on Gas Flowrate

The computed and experimentally derived plume velocities are plotted against gas flow rate in Figure 9. The computed velocity is the average along the length of the rod, or in the 'no rod' cases, the average along the centreline over the same range of depths. Although the velocities computed in the absence of the rod are much greater than those obtained experimentally, the presence of the rod reduces the values so that they are closer to those derived from the dissolution rates.

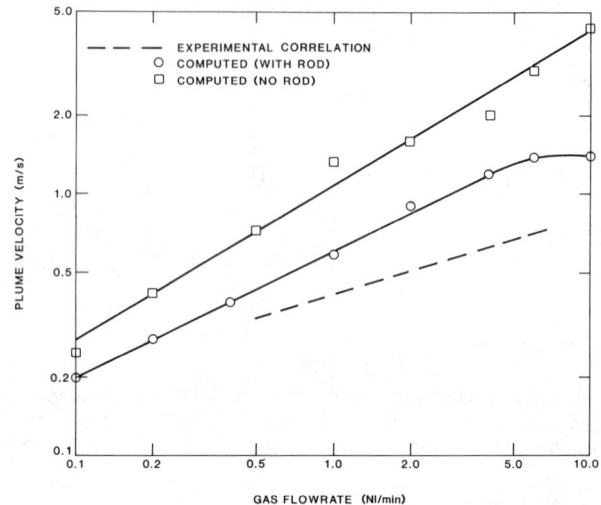

Figure 9. Comparison between computed and experimental plume velocity as a function of gas flow rate. Predicted velocities are shown for bath with and without rod.

The numerically determined plume velocity in the presence of the rod can be fitted well by the relationship:

$$U \propto Q_o^{0.48} \qquad \qquad (25)$$

up to gas rate 4 Nl/min. This is steeper than the dependence found experimentally. One possible reason for this is that as the gas rate increases, the void fraction in the plume will increase, leading to a reduction in the time liquid contacts the rod. For example the numerical models predict that the mean void fraction near the rod increases from about 0.3 at 0.5 Nl/min to 0.7 at 2.0 Nl/min. Other possible reasons for the difference in slopes are simplifying assumptions in the mathematical model, and errors caused by applying equation (5) to a gas-stirred system. It is unlikely that the expression for interphase friction will be valid up to the very high void fractions encountered in the models, and we are developing more realistic formulas for future work. The k-ε model may also be inaccurate in regions of high void fraction: it is after all just a one phase model. That the predicted plume velocity is higher than the measured value may imply that the k-ε model underestimates the turbulence level in a bubbly flow.

Conclusions

Flow velocities in a bottom blown iron bath have been studied by a combination of experimental measurement and mathematical modelling. An iron rod dissolution technique was used to derive the liquid velocity in the plume at several gas flow rates between 0.5 and 6.0 Nl/min. The velocities, inferred from mass transfer coefficients using correlations developed for the dissolution of iron cylinders in Fe/C melts under forced convection, increased with gas injection rate as

$$U \propto Q_o^{0.3} \qquad (26)$$

This is similar to relationships obtained by other authors at lower superficial gas velocity.

The flow simulation package, PHOENICS, was used to compute the flow field and void fraction distribution at each gas rate. The presence of the rod was found to make a significant change to the bath flow. The predicted plume velocities were of similar magnitude to the experimental values though larger, and the dependence on gas rate was steeper:

$$U \propto Q_o^{0.48} \qquad (27)$$

The study has suggested improvements to the mathematical model, and also further experimental work to test the applicability of the Kim and Pehlke correlation to gas stirred baths, particularly with void fraction as high as 0.7. That the predicted plume velocity was higher than the measured values suggests that the k-ε model may underestimate the turbulence level in bubbly flow.

Acknowlegment

The authors would like to ackowledge the support of CRA.

Nomenclature

b	Empirical constant in equation (5) (0.064)
C_ℓ	Liquidus carbon concentration at bath temperature
C_b	Carbon concentration in the bulk melt
C_o	Initial uniform carbon concentration in the rod
C_μ	Standard k-ε model constant, 0.09
CFIPS	Interphase friction coefficient
d	Rod diameter
d_b	Bubble diameter
D	Diffusivity
D_t	Turbulent diffusivity
g	Acceleration due to gravity
H	Bath depth
j_d	"j-factor" for mass transfer
k	Kinetic energy of turbulence
k_m	Mass transfer coefficient
m_i	Mass source per unit volume of phase i
n	Empirical constant in equation (5) (0.25)
Q	Volumetric gas flow rate
Q_o	Gas flow rate at NTP

r	Rod radius
r_o	Orifice radius
R_g, R_ℓ	Gas, liquid volume fraction
Re	Rod Reynolds number $(\rho_\ell\, d\, U/\mu)$
Re_b	Bubble Reynolds number $(\rho_\ell\, d_b\, V_{slip}/\mu)$
Re_w	Wall Reynolds number $(\rho_\ell\, \delta\, U\, /\mu)$
s	Skin-friction factor
S_{ϕ_i}	Source of ϕ_i per unit phase volume
Sc	Schmidt number $(\mu/\rho_\ell\, D)$
t	Time
U	Liquid velocity in the plume
$U_{\shortparallel}$	Velocity component parallel to wall
U_τ	Friction velocity
$\underline{V}_i$	Vector velocity of phase
V_g, V_ℓ	Gas, liquid radial velocity component
V_{slip}	Gas-liquid slip velocity
W	Bath diameter
W_g, W_ℓ	Gas, liquid vertical velocity component
z	Vertical coordinate
α	Gas holdup
δ	Normal distance of cell centre to wall
ε	Dissipation rate of turbulence kinetic energy
ε_m	Mixing power
Γ_{ϕ_i}	Exchange coefficient for ϕ_i
κ	von Karman constant, 0.435
μ	Liquid viscosity
μ_{eff}	Effective eddy viscosity
ϕ_i	General conserved variable
ρ_g, ρ_ℓ	Gas, liquid density
ρ_o	Gas density at NTP
τ	Mixing time

Subscripts
g	gas phase
i	phase i (gas or liquid)
ℓ	liquid phase

References

1. Y. Sahai and R.I.L. Guthrie, "Hydrodynamics of Gas Stirred Melts: Part I. Gas/Liquid Coupling," Metallurgical Transactions B, 13B (1982) 193-202.

2. K. Nakanishi, T. Fujii, and J. Szekely, "Possible Relationship between Energy Dissipation and Agitation in Steel Processing Operations," Ironmaking and Steelmaking, 1975, no. 3:193-197.

3. J.H. Grevet, J. Szekely, and N. El-Kaddah, "An Experimental and Theoretical Study of Gas Bubble Driven Circulation Systems," Int. J. Heat Mass Transfer, 25 (1982) 487-497.

4. J.W. McKelliget, M. Cross, and R.D. Gibson. "A Turbulent Fluid Flow Model of Gas Agitated Reactors," Applied Mathematical Modelling, 6 (1982) 469-480.

5. M. Salcudean, K.Y.M. Lai, and R.I.L Guthrie, "Multi-dimensional Heat, Mass and Flow Phenomena in Gas Stirred Reactors," Can. J. Chemical Engineering, 63 (1985) 51-61.

6. N. Bessho, S. Taniguchi, and A. Kikuchi, "Fluid Flow in a Gas-stirred Vessel," Tetsu-to-Hagane, 71 (1985) 1117-1124.

7. M. Cross, N.C. Markatos, and C. Aldham, "Gas Injection in Ladle Processing," Control '84: Proc. 1st Inter Symp. on Automatic Control in Mineral Processing and Process Metallurgy, ed. J.A. Herbst (New York, NY: AIME-SME, 1984), 291-297.

8. J.M. Lommel and B. Chambers, "The Isothermal Transfer from Solid to Liquid in Metal Systems", Ironmaking and Steelmaking, 1975, no. 3:193-197.

9. N.J. Themelis and P. Goyal, "Gas Injection in Steelmaking: Mechanism and Effects," Canadian Metallurgical Quarterly, 22 (1983) 313-320.

10. Y.-U. Kim and R.D. Pehlke, "Mass Transfer during Dissolution of a Solid in the Iron-Carbon System," Metallurgical Transactions, 5 (1974) 2527-2532.

11. D.B. Spalding, "Four Lectures on the PHOENICS Computer Code" (Report CFD/82/5, Imperial College, 1982).

12. M.P. Schwarz and P.T.L. Koh, "Numerical Modelling of Bath Mixing by Swirled Gas Injection," SCANINJECT IV: Proc. 4th Inter. Conf. on Injection Metallurgy, ed. G. Carlsson (Lulea, Sweden: MEFOS, 1986), 6:1-6:17.

13. M.P. Schwarz and W.J. Turner, "On the Applicability of the Standard k-ε Model to Two-Phase Flows," Submitted to Applied Mathematical Modelling.

14. B.E. Launder and D.B. Spalding, Mathematical Models of Turbulence, (New York, NY: Academic Press Inc., 1972).

15 S.V. Patankar, Numerical Heat Transfer and Fluid Flow (Washington, DC: Hemisphere, 1980).

16. R. Clift, J.R. Grace, and M.E. Weber, Bubbles, Drops and Particles, (New York, NY: Academic Press Inc., 1978).

579

MULTI-PHASE MODELLING OF POWDER INJECTION REFINING

G.A. Irons, L.-K. Chiang and W.-K. Lu

Department of Materials Science and Engineering
McMaster University
Hamilton, Ontario, Canada

Abstract

A "two-fluid" model was developed for momentum, heat and mass transfer in the gas-liquid-solid three-phase plume which forms when reactive powders are pneumatically conveyed into liquid metals, such as iron or steel. Separate equations are written which explicitly account for coupling or transport between phases. The model is steady-state and one-dimensional; quantities are averaged across the plume. The volume fractions, velocities, temperatures and utilizations of the solid phase, as well as the extent of refining are calculated as the phases rise. The position of the particles (in the gas or on the carrier gas bubbles) was an important parameter. Refining rates for particles in the liquid were strongly dependent on the solid flow rate and virtually indepedent of the gas flow rate, whereas the converse was true for particles on the carrier gas bubbles. Comparison of the model results with experimental first-order rate constants for calcium carbide desulfurization of three tonne heats of iron suggests that approximately 30% of the particles enter the liquid.

Introduction

Most fluid dynamic models of gas-liquid plumes in metallurgical applications (1–3) employ "mixture models" for the two-phase region, that is the mixture density is calculated from the void fraction at each point (4) Thus the two-phase region is treated as a single-phase region of intermediate density. It is very important to obtain the correct void fraction because it determines the buoyancy forces which in turn control the amount of liquid entrainment and the liquid recirculation rate. These earlier models have generally used constant void fractions in the two-phase region.

More recently, Boysan and Johansen (5) have used a Lagrangian framework to calculate the void fraction. In such schemes thousands of trajectories of individual bubbles are tracked under the influence of the turbulent liquid phase (6). In this way a stochastic void fraction distribution is established which is in reasonable agreement with experiment This new void fraction distribution is used to recalculate liquid velocities until some convergence criterion is satisfied. The computational demands of such schemes quickly become out of hand when one takes into account additional factors such as the effect of the gas phase on turbulence, the introduction of more than one bubble size or a third phase or to consider transport between phases. Nevertheless, they are fundamentally attractive.

The present modelling effort was driven by a more pragmatic need to describe the inter-phase momentum, heat and mass transport in three-phase plumes characteristic of powder injection processes in the Iron and Steel Industry. The "two-fluid" models developed extensively in the thermohydralics field have been adapted for this purpose (4). In such models separate equations are written for each phase, explicitly allowing for transport between phases; thus the averaging inherent in the mixture models is avoided. The present model extends the work of Farias and Irons (7) on momentum transfer in the plume to include heat transfer, mass transfer and chemical reaction. The model yields the volume fraction, velocity, and temperature profiles of the gas, liquid, and solid phases, the utilization of the solid phase and the extent of refining in the rising plume. While the model is quite general for powder injection processes, the particular case of calcium carbide injection into 3 Tonne heats of hot metal is used so that the computational results can be compared with experiment. This type of model is also considered to be particularly useful in situations where there are large rates of mass or heat transfer between phases which results in strong coupling between the momentum, heat and mass transfer equations, such as submerged converting processes or the injection of volatile reagents such as magnesium or calcium.

Development of the Model

The physical situation in the plume is shown schematically in Figure 1. The gas and powder penetrate into the liquid as a gas-particle jet until their momentum is dissipated. A model of the jet is used to calculate the maximum penetration (8). At the bottom of the jet, a plume is created where the gas forms bubbles, usually spherical-cap bubbles, 20 to 100 mm in diameter. The particles may position themselves on the bubble interfaces or reside in the liquid. The gas and particles are buoyant, and start to rise. The drag forces on the liquid cause it to accelerate as well, and consequently liquid is entrained into the plume. As the gas and particles rise they are heated, and the particles react. The rate of mass transfer is considered to be governed by three consecutive steps:

1. The rate at which liquid is pumped into the plume by entrainment; this is termed pumping control.

2. The rate at which the reactive solute diffuses through the boundary layers to the particles. This rate depends on whether the particles are located on the bubble interfaces or in the liquid, thus it is called contact control.

3. The rate at which reactants diffuse through the product layers; this is called product layer control. Based on an analysis of previous experimental work for the case of calcium carbide desulphurization (9), it was concluded that this rate is fast enough not to control this particular process.

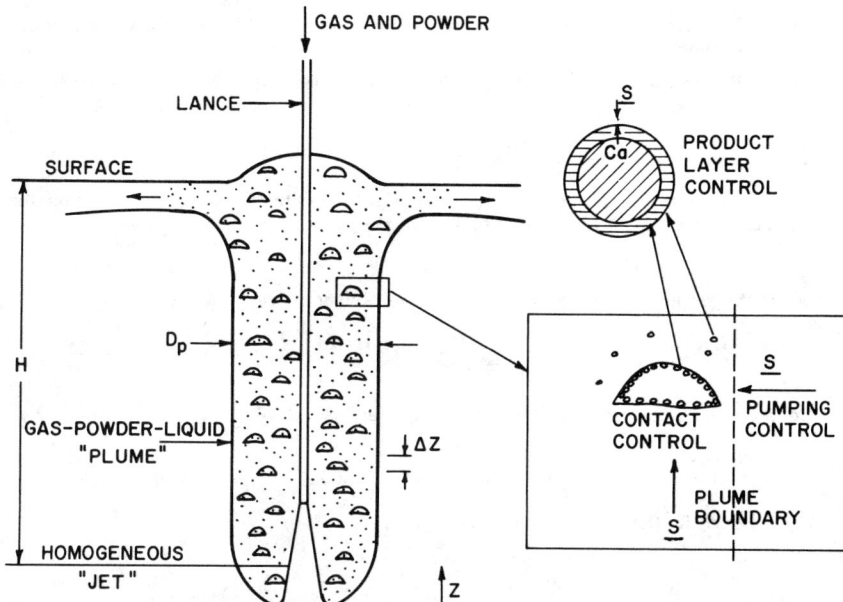

Fig. 1   Schematic representation of the physical and chemical phenomena in the rising plume. The carrier gas and injected particles rise and react through the plume The inset indicates that particles may be on the bubble interfaces or in the liquid. At individual particles, calcium vapour diffuses through the product layers

This model is an extension of the model by Farias and Irons (7) to describe the fluid dynamics of a three-phase plume. The model was steady-state, three-phase, and one-dimensional; quantities were averaged over the plume cross-section. The present model includes heat transfer and mass transfer.

The principal assumptions are:

(a)   All variables are averaged across the plume.

(b)   The plume is cylindrical in shape and its diameter is taken from photographs of the "break-through" area on the melt surface.

(c)   Once the particles reach the top of the plume, they are incorporated into the slag.

(d)   Gas and particle concentrations in the plume are low enough that single-bubble drag, heat and mass transfer coefficients can be used, and that bubble-bubble and particle-particle interactions can be ignored.

(e)   The average bubble equivalent diameter is 25 mm and constant during rise.

(f)   A fraction of the particles, f, is assumed to rise on the bubble interfaces, and the remainder, 1–f, is dispersed in the liquid.

(g)   The particles are small enough that internal temperature gradients can be ignored.

(h)   Convection inside the gas bubbles eliminates any thermal gradients except in the boundary layer.

(i)   The heat of desulfurization is negligible.

(j)   The liquid entrained into the plume is at a constant temperature maintained by external heating, and has the average concentration in the ladle.

583

(k)   The particles on the bubble surfaces are heated as quickly as if they were in the liquid.

(l)   The particles on the bubble interface do not influence the mass transfer coefficients to the bubbles.

(m)   Due to the low solubility of calcium in the iron and the relatively large sulfur concentration in the iron, desulfurization occurs at the particle-iron interface.

Continuity and Momentum Equations

Following Farias and Irons (7), the bubbles are assumed to be composed of a mixture of gas and a fraction of the particles, so that the mixture density can be written as:

$$\rho_m = \theta_g \rho_g + f\theta_p \rho_p \tag{1}$$

where the gas density is a function of temperature and pressure:

$$\rho_g = \frac{P_{atm} + \rho_\ell g(H-Z)}{RT_g} \tag{2}$$

Conservation of mass in the bubbles can be expressed as.

$$\frac{d(\rho_m U_g A_{pl})}{dZ} = 0 \tag{3}$$

and similarly for the powder in the liquid:

$$\frac{d((1-f)\theta_p \rho_p U_p A_{pl})}{dZ} = 0 \tag{4}$$

The momentum of bubble and particle phases is increased by the buoyancy forces, while the drag forces from the liquid phase resist this acceleration:

$$\frac{d(\rho_m U_g^2 A_{pl})}{dZ} = F_{gp}^b - F_{gp-\ell}^d \tag{5}$$

$$\frac{d(1-f)\theta_p \rho_p U_p^2 A_{pl}}{dZ} = F_p^b - F_{p-\ell}^d \tag{6}$$

The liquid experiences the reactive component of the drag forces:

$$\frac{d(\theta_\ell \rho_\ell U_\ell^2 A_{pl})}{dZ} = F_{gp-\ell}^d + F_{p-\ell}^d \tag{7}$$

The buoyancy forces for the gas and powder mixture in the bubbles and for the powder in the liquid are:

$$F_{gp}^b = (\rho_\ell(\theta_g + f\theta_p) - \rho_m)g A_{pl} \tag{8}$$

$$F_p^b = (1-f)(\rho_\ell - \rho_p)\theta_p g A_{pl} \tag{9}$$

The interphase drag function for the bubbles in the liquid is:

$$F_{gp-\ell}^d = \frac{0.75 \rho_\ell C_{dg}(U_g - U_\ell)^2(\theta_g + f\theta_p)A_{pl}}{d_b} \tag{10}$$

and for the particles rising in the liquid is:

$$F^d_{p-\ell} = \frac{0.75 \rho_\ell C_{dp} (U_p - U_\ell)^2 (1-f) \theta_p A_{pl}}{d_p} \quad (11)$$

Heat Transfer Equations

The heat balance is written for each phase:

Liquid·

$$\theta_\ell \rho_\ell U_\ell C_{p\ell} \frac{dT_\ell}{dZ} = -G_1 (T_\ell - T_{p\ell}) - G_2 (T_\ell - T_g) + \rho_\ell C_{p\ell} (T_{b\ell} - T_\ell) \frac{d\theta_\ell U_\ell}{dZ} \quad (12)$$

Particles in the liquid:

$$(1-f) \theta_p \rho_p U_p C_{pp} \frac{dT_{p\ell}}{dZ} = G_1 (T_\ell - T_{p\ell}) \quad (13)$$

Gas phase·

$$\theta_g \rho_g U_g C_{pg} \frac{dT_g}{dZ} = G_2 (T_\ell - T_g) - G_3 (T_g - T_{gp}) \quad (14)$$

For convection to the particles in the liquid

$$G_1 = \frac{6 h_p (1-f) \theta_p}{d_p} \quad (15)$$

where the heat transfer coefficient, h_p, was obtained from turbulent mass transfer correlations for spheres in liquids (10), adapted to heat transfer:

$$h_p = \frac{Nu \, k_\ell}{d_p} \quad (16)$$

$$Nu = 2 + 0.4 \left(\frac{\varepsilon d_p^4}{\nu^3} \right)^{1/4} Pr^{1/3} \quad (17)$$

where the energy dissipation is defined as:

$$\varepsilon = (U_g - U_p) g \quad (18)$$

For convection to the bubbles:

$$G_2 = \frac{6 h_b (\theta_g + f \theta_p)}{d_b} \quad (19)$$

where the heat transfer coefficient, h_b, was obtained from penetration theory applied to the gas phase resistance:

$$h_p = 2 \left(\frac{k_g \rho_g C_{pg} (U_g - U_\ell)}{\pi d_b} \right)^{1/2} \quad (20)$$

585

Mass Transfer Governing Equations

The sulfur balance in the plume can be expressed as:

$$\frac{1}{A_{pl}} \frac{d\, C_S^{p\ell}\, U_\ell A_{pl}\, \theta_\ell}{d\, Z} = \frac{C_S^B}{A_{pl}} \frac{d\, \theta_\ell\, U_\ell A_{pl}}{d\, Z} - J_1 (C_S^{p\ell} - C_S^{ip}) - J_2 (C_S^{p\ell} - C_S^{ib}) \tag{21}$$

The first term on the right side of the equation represents the influx of sulfur due to entrainment, while the second and third terms represent desulfurization by particles in the melt and by particles on the bubble surfaces, respectively. The extent of reaction of the calcium carbide particles in the liquid, α (utilization) is given by:

$$\frac{d\,(1-f)\,\theta_p\, U_p\, C_p'\, \alpha_{p\ell}}{d\, Z} = J_1 (C_S^{p\ell} - C_S^{ip}) \tag{22}$$

and for the particles in the bubbles by:

$$\frac{1}{A_{pl}} \frac{d\,\theta_p\, f f_R\, U_p\, C_p'\, \alpha_{pg}\, A_{pl}}{d\, Z} = J_2 (C_S^{p\ell} - C_S^{ib}) \tag{23}$$

The mass transfer functions, J_1 and J_2 are given by:

$$J_1 = \frac{6\theta_p\, k_p\, (1-f)}{d_p} \tag{24}$$

and

$$J_2 = \frac{6\,\theta_g\, k_b}{d_b} \tag{25}$$

Turbulent mass transfer coefficients to spheres in liquids may be obtained from the following formula (10):

$$Sh = \frac{k_p\, d_p}{D_S} = 2 + 0.4 \left(\frac{\varepsilon d_p^4}{\nu^3} \right)^{1/4} Sc^{1/3} \tag{26}$$

The mass transfer coefficient to spherical-cap bubbles was taken from the the Baird and Davidson (11) formula:

$$k_b = 0.951\, g^{0.25}\, d_b^{-0.25}\, D_S^{0.50} \tag{27}$$

Boundary Conditions

The volume fractions, velocities, and temperatures of the three phases as well as the sulfur concentration in the iron must be specified at the bottom of the plume ($Z = 0$). The jet model used to calculate the jet penetration (9) also calculates the volume fraction of liquid entrained. Typically, it is less than 0.1, thus the volume fractions of gas and particles at the plume bottom are diluted correspondingly from conditions at the lance tip. From these volume fractions of gas and particles at the bottom, the phase velocities are calculated:

$$U_p = \frac{W_p}{\theta_p \rho_p A_{pl}} \tag{28}$$

$$U_g = \frac{Q_g}{\theta_g A_{pl}} \tag{29}$$

The liquid velocity is set equal to the gas velocity. As was previously demonstrated (7), the resulting velocities and phase fractions higher in the plume are insensitive to changes in the initial liquid fraction and velocity.

Irons has shown that gas and particles travelling down a lance are not heated very quickly because the residence time is short (12). The same can be said for the descending jet, so that the gas and particle temperatures at $Z = 0$ are assumed to be at 25°C.

Consistent with assumption (1), the liquid at that point is assumed to be at the bulk liquid temperature, and the sulfur content of the iron at $Z = 0$ is assumed to be at the bulk composition.

Method of Solution

The governing differential equations were manipulated into ten coupled, first-order differential equations for the three phase velocities, the temperatures of the liquid, gas and particles in the liquid and on the bubbles, the sulfur concentration in the plume, and the utilization of the particles in the liquid and on the bubbles. These equations were solved simultaneously using Gear's Method (IMSL routine, DGEAR (13)).

From these solutions other parameters can be calculated. The fraction of control exerted due to pumping control is simply the ratio of concentration driving force across the plume compared to the total:

$$R_{pump} = \frac{C_S^B - C_S^{PL}}{C_S^B - C_S^{eq}} \tag{30}$$

The rate of desulfurization with respect to time can be examined with first-order rate constants which are usually defined by:

$$\frac{dC_S^B}{dt} = -K_1(C_S^B - C_S^{eq}) \tag{31}$$

In the present model the rate of desulfurization in the plume can be determined by the reduction of sulfur content and the flow rate of liquid at the top of the plume:

$$\frac{dC_S^B}{dt} = \frac{\theta_\ell U_\ell A_{p\ell}}{V}(C_S^{p\ell} - C_S^B) \tag{32}$$

Equating these last two equations yields the first order rate constant:

$$K_1 = \frac{\theta_\ell U_\ell A_{pl}}{V} \frac{(C_S^B - C_S^{p\ell})}{(C_S^B - C_S^{eq})} \tag{33}$$

which will be compared with experimental results.

The numerical values used throughout the model are summarized in Table I.

587

Table I. Numerical Values Used for Calculation

Phase	Density (kg/m³)	Viscocity (kg/m/s)	Thermal Conductivity (W/m/K)	Heat Capacity (J/kg/K)
Hot Metal	7000	0.006	33	824
Calcium Carbide	2220	–	–	1005. + 0.13T
Argon	Ideal	2.15×10^{-5} $+ 4.25 \times 10^{-9}$ T	0.0193 $+ 1.71 \times 10^{-5}$ T	520
Sulphur Diffusivity		1.0×10^{-9} m²/s		
Metal Weight		2500 kg		
Furnace Diameter		0.76 m		
Lance Immersion		0.65 m		
Lance Inner Diameter		0.01 m		

Computed Results

Typical computed results are presented in Figure 2. Figure 2A shows that at the bottom of the plume the volume fraction of gas is large, however liquid is quickly entrained into the rising plume. The volume fractions of gas and solid for most of the plume are less than 5% and 0.1%, respectively. This supports assumption (d), regarding the use of single bubble and particle transport coefficients. Figure 2B shows that the velocities of the phases quickly rearrange themselves as well. The slip velocity between the gas and liquid is the single bubble rising velocity because single-bubble drag coefficients were used. The particles rise with Stokes velocity with respect to the liquid which is indistinguishable from the liquid velocity. The particle temperatures shown in Figure 2C quickly rise to the liquid temperature. This causes a slight drop in the liquid temperature which recovers as hot liquid is entrained into the plume. The gas temperatures rise more slowly. Since the volumetric gas flow rate is the most important factor in determining the phase velocities (7), the gas temperature is a significant, coupled variable between the momentum and thermal equations. The utilizations, a, are shown in Figure 2D along with the sulfur content in the plume. The utilization of the particles in the gas is low because it is limited by the bubble area. The particles in the liquid undergo much more reaction, however it is far from complete at the bath surface. The curvature of the curve for the utilization of the particles in the liquid reflects the longer residence times near the bottom of the plume due to lower liquid velocity. The average utilization is weighted according to the fractions in the bubbles and in the liquid:

$$\bar{a} = a_{p\ell}(1 - f) + a_{pg} f f_R \tag{34}$$

In this case f = 0.7 and f_R = 1.0.

The effect of carrier gas flow rate at constant solids flow rate is shown in Figure 3. The average gas and liquid velocities in the plume are presented in Figure 3A. The average liquid and gas velocities from the iso-thermal predecessor of the present model (7) have been shown to be in good agreement with experimental water model results and the mathematical models of Sahai and Guthrie (2) and Sano and Mori (14). The utilization of the particles in the gas bubbles is a strong function of gas flow rate because as flow rate increases the particles are spread over greater bubble surface presenting more area for reaction. The utilization of the particles in the liquid first decreases with increasing gas flow rate because the liquid velocity

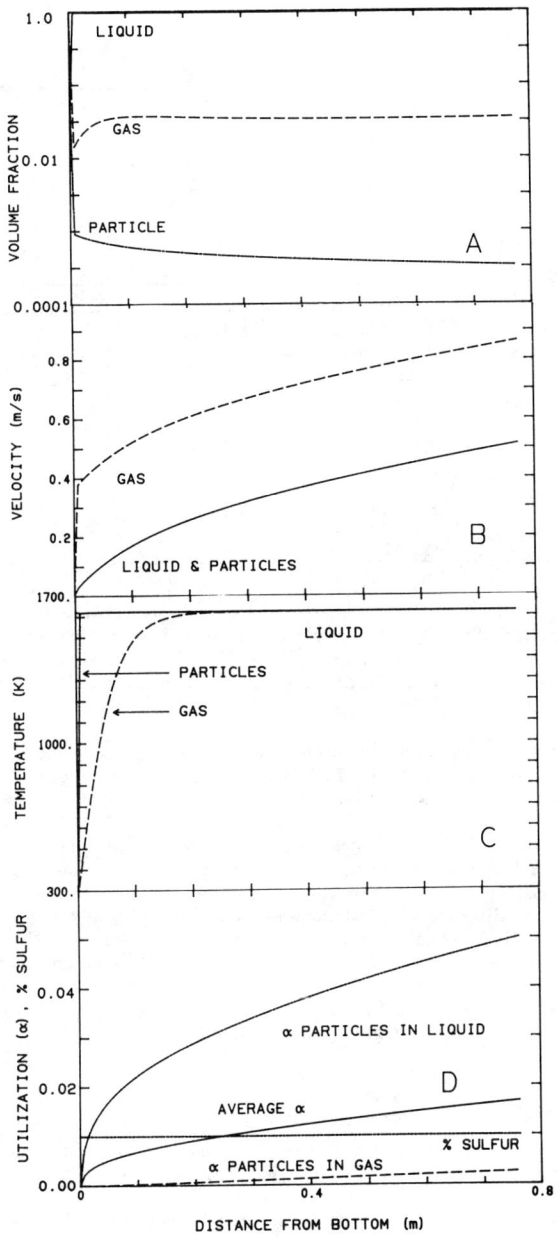

Fig. 2   Typical computed results for injection rates of 6 kg CaC₂/min and 0.16 Nm3/min of argon for Case 2 (30% of the particles in the liquid) from the bottom of the plume to the top. A: phase volume fraction, B: phase velocities, C: temperatures, D: particle utilizations and sulfur content in the plume.

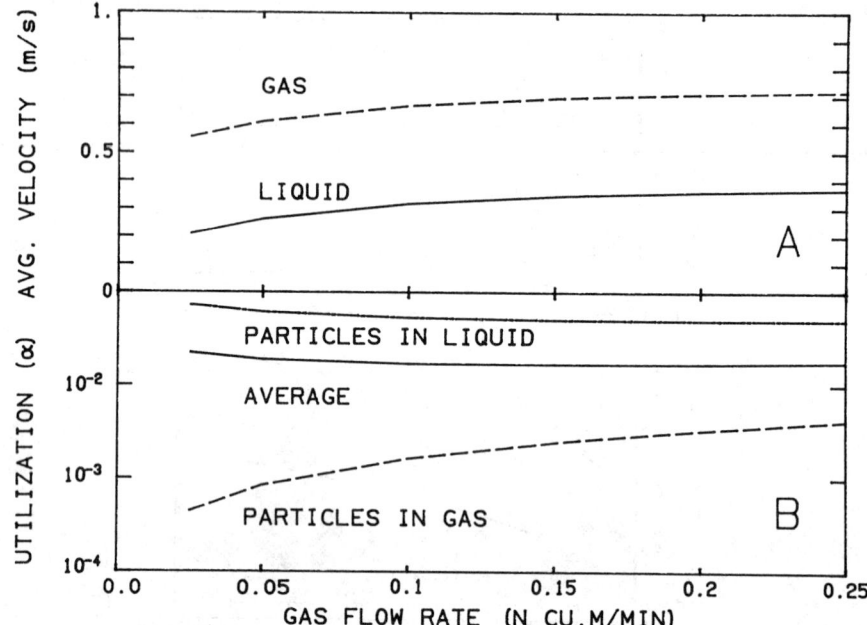

Fig. 3   The effect of gas flow rate (at a constant solids flow rate of 6 kg/min) on the average gas and liquid velocities in the plume (A) and the utilization of the particles in the liquid, on the gas bubbles and the average (B) for Case 2 (30% of the particles in the liquid).

increases, and then it increases slightly because at increasing gas flow rate the jet goes deeper, thereby increasing residence time. In this particular case the average utilization passes through a shallow minimum at 0.15 Nm^3/min.

In Figure 4 the effects of increased solids flow rate at constant gas flow rate are examined. The gas and liquid velocities are almost independent of the solids flow rate because the buoyancy of the particles is small compared to the gas under typical injection conditions. Consequently the utilization of the particles in the liquid is virtually independent of the solids flow rate as well. However as the solids flow rate is increased the utilization of the particles in the gas is dramatically decreased because of the greater loading on each bubble.

In all cases the fraction of resistance due to pumping, R_{pump}, was less than 0.01. The carrier gas always provides enough liquid recirculation to keep pace with the contact control step.

Sensitivity Analyses

There are a number of model factors and parameters which have been assigned particular values; the impact of these choices on the computed results is assessed and discussed.

Particle Position. The position of the particles with respect to the bubbles strongly influences the calcium carbide utilization and the desulfurization rate. To examine this effect in more detail, several different particle-liquid contact patterns were simulated (Table II).

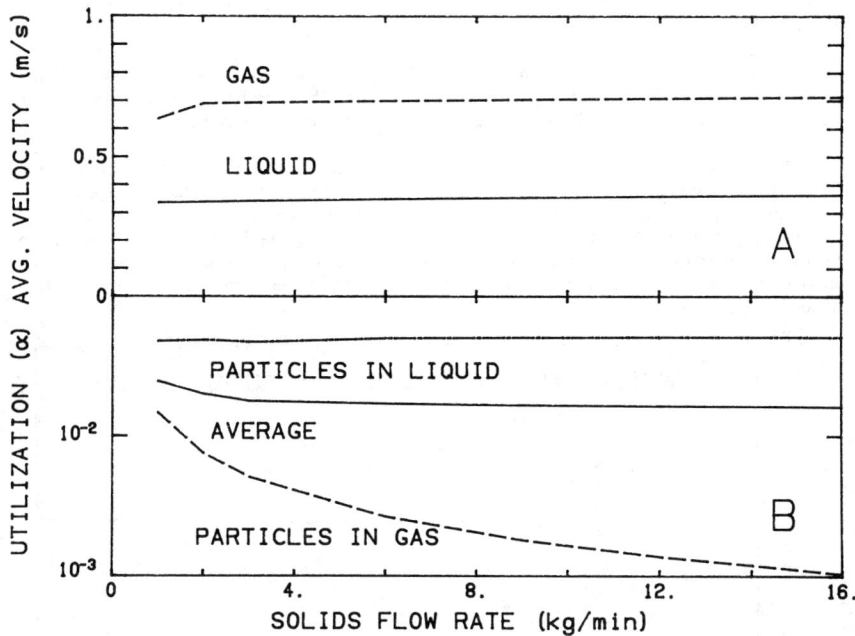

Fig. 4   The effect of solids flow rate (at a constant gas flow rate of 0.16 Nm³/min) on the
average gas and liquid velocities in the plume (A) and the utilization of the particles
in the liquid, on the gas bubbles and the average (B) for Case 2 (30% of the particles
in the liquid).

Table II. Particle Contact Cases

Case	Description	f	f_R
1	All Particles on Bubble Surfaces, Free to Circulate	1.0	1.0
1A	Same as 1, Except Only Monolayer Reacts	1.0	f_m
2	All Particles in Liquid	0.0	–
3	30% in Liquid, 70% on Bubbles	0.7	1.0

For Case 1 it is assumed that all the particles are trapped on the bubble surfaces. This
is a plausible situation because calcium carbide is not wetted by liquid iron. Furthermore, it is
assumed that the particles are free to circulate on the bubble surfaces so that each particle has
an equal chance to contact the liquid and react. Hence the rate of reaction is governed by
diffusion through the boundary layer around the bubbles.

Case 1A is the same as Case 1, except that the particles on the bubble surface are
stagnant, thus only a monolayer can react. This case was chosen to show the effects of limited
particle recirculation which may occur in thick layers or if the particles sinter together. This
represents the slowest possible desulfurization rate.

In Case 2, it is assumed that all the particles enter the liquid which represents the best particle-liquid contact and therefore the fastest reaction rate.

For Case 3, it was assumed that only 30% of the particles enter the liquid. Thirty percent was chosen because Irons and Farias (15) found that the bath cooling rate for silica injection into lead was only 30% of that expected if all the particles were in the liquid

The average utilizations for these are presented as a function of gas flow rate in Figure 5; the trends are similar to Figure 3B. Starting with the highest utilizations, Case 2, corresponding to all particles in the liquid, one sees that there is a weak dependency on gas flow rate. Increasing gas flow rate increases jet penetration, lengthens the distance through which the particles rise and thereby increases residence time. However, increasing gas flow rate also increases the liquid velocity which tends to decrease residence time; the net effect is little dependency on gas flow rate.

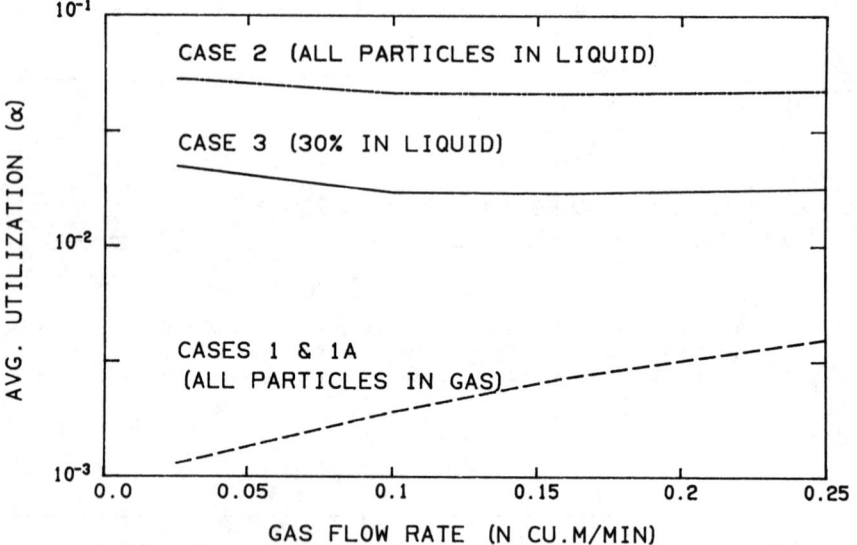

Fig. 5   The dependencies of the average utilizations of the particles on gas flow rate for the various cases at 6 kg CaC$_2$/min.

The lowest utilizations in Figure 5 occur when all the particles are positioned on the bubble interfaces, Cases 1 and 1A. In these cases the utilization increase rapidly with gas flow rate because the greater number of bubbles increases the interfacial area for reaction. There is little difference between Cases 1 and 1A which assume circulating and non-circulating particles, respectively. This is a result of the low utilization of the particles; even for the particles in the stagnant monolayer the utilization only reached 0.1 in the highest case. A stagnant monolayer would only affect reaction rates and utilization if the utilization reached 1.0 before the particles reached the bath surface. The average utilization for Case 3 has been discussed in connection with Figure 3B.

Bubble Size. In the present model the bubble diameter was chosen as 25 mm. However it depends on liquid properties and gas flow rates; for iron systems the bubble diameter is likely to be between 20 and 100 mm as discussed by Sano and Mori (16). If the bubble diameter in the model is increased by 100%, this results in a 31% increase in utilization of the particles in the liquid because the liquid velocity is decreased which increases residence time. Conversely there is a 70% decrease in the utilization of particles in the liquid for a 100% increase in bubble diameter which is primarily attributable to decreases in surface area to volume ratio and mass transfer coefficients for the bubbles. A constant bubble size was assumed (assumption (e)) despite the decreasing ferrostatic head during rise, which implies bubble break-up. An alternate assumption, that the initial number of bubbles simply increase in volume, would produce similar results (7).

Particle Size. The mean particle size in the experimental work was 24 μm. If the particle size in the model is reduced by a factor of two, the utilization of the particles in the liquid is increased by a factor of 3.97. This result is expected because the Sherwood Numbers for the particles are always close to two, so that the utilization is inversely proportional to the square of the diameter. The utilization of the particles in the gas was unchanged by reducing particle diameter because control lies at the bubble interfaces.

Heat Transfer Coefficients. There is some uncertainty in the heat transfer coefficients for the particles and the gas bubbles. Figure 2C shows that the particles in the liquid are heated very rapidly to the bath temperature; the particles are heated to 1600 K after rising 7 mm. Even if the heat transfer coefficients are considerably over-estimated, the particles will be very close to the bath temperature for most of their rise and will therefore be capable of desulfurization. It was assumed that the particles on the bubble interfaces were heated as rapidly as particles completely immersed in the liquid, (assumption (k)), principally because heat transfer coefficients to particles on bubble surfaces have not been measured. The monolayer on the surface will be heated quickly, and as shown earlier, only monolayer of particles at the bubble surface need react to produce the rates of desulfurization for Case 1 as indicated by the similarity between cases 1 and 1A. (The particles not in the monolayer in Case 1A could have been considered cold and unreactive).

The bubbles are heated more slowly as seen in Figure 2C. If the heat transfer coefficient to the bubbles is reduced by 50%, this results in a 7% increase in the utilization of particles in the liquid because the smaller bubble volume at lower temperature results in lower liquid velocity and longer residence times.

Mass Transfer Coefficients. The mass transfer coefficient for the bubbles in Equation 27 may over-predict the rate if the bubble surface is immobilized by surface-active agents or by the particles on the surface. Reduction of the bubble mass transfer coefficient by 50% results a 50% reduction utilization of the particles in the liquid.

In the cases where particles are in the liquid, Sherwood Numbers, according to Equation 26 were typically 2.01, just slightly larger than for the limiting case of 2.0 for molecular diffusion. This is due to the fact that the particles are much smaller than the scale of turbulence making the mass transfer coefficients relatively insensitive to the level of turbulence.

Comparison With Experiment

The experimental apparatus and results have been previously reported (17). Briefly, 2.5 Tonne heats of hot metal (3.5–4.0% C, 0.1–0.01% S, 0.2–1.5% Si, 0.5% Mn and 0.04% P) were melted and held in an induction furnace at 1350°C. Calcium carbide was injected into the iron at a wide variety of rates and solid-to-gas loadings. Samples were taken frequently during and immediately after injection. After injection, with carrier gas alone, the rate of desulfurization was found to be first-order with respect to the sulfur content in the iron. When this rate was subtracted from the total rate of desulfurization during injection, the rate of desulfurization for the plume reactions were also found to be first-order with respect to sulfur in the iron.

The first order rate constants as a function of solid and gas flow rates are shown in Figures 6 and 7. All the data falls between the two limiting cases, 1 and 2. For Case 2 (all the particles in the liquid), the first-order rate constant has a very strong dependency on solids flow rate because more particles are in the melt at higher flow rates. The gas flow rate has virtually no effect on the rate constant for Case 2, again because of the compensating effects of increased jet penetration and increased liquid velocity at high flow rates. Conversely for Case 1 (all the particles on the bubble interfaces), the rate constant has a weak dependency on solids flow rate and a strong dependency on gas flow rate because the rate depends on the bubble area available for reaction.

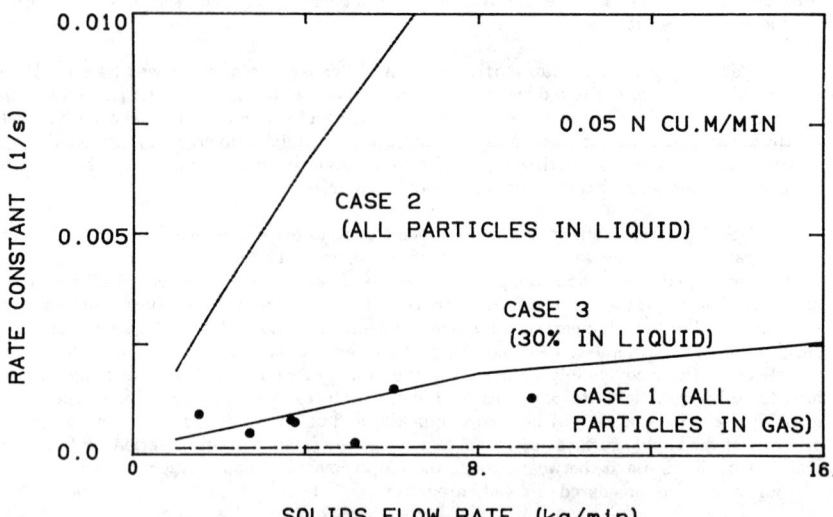

Fig. 6   The experimental first order rate constants for desulfurization in the plume as a function of CaC$_2$ injection rate at 0.05 Nm^3/min of argon compared with the model cases.

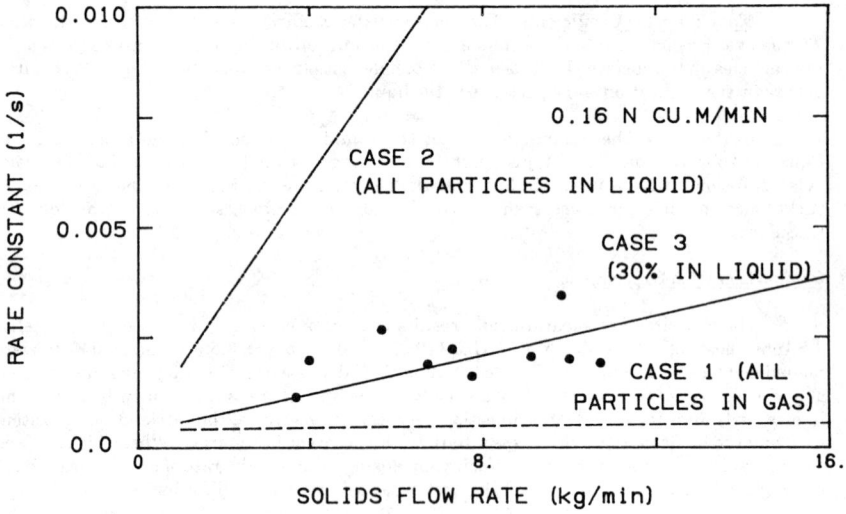

Fig. 7   The experimental first order rate constants for desulfurization in the plume as a function of CaC$_2$ injection rate at 0.16 Nm^3/min of argon compared with the model cases.

The experimental first order rate constants agree reasonably well with those generated by Case 3 (30% of the particles in the liquid) in terms of dependency on gas flow rate, solids flow rate as well as the absolute value.

Discussion

The first order rate constants for desulfurization are consistent with the model Case 3 in which 30 % of the particles enter the liquid and the remainder are positioned on the carrier gas bubble interfaces. There is no way the predicted rates of Case 2 could be reduced to match the data in Figures 6 and 7 because the mass transfer coefficients to the particles in the liquid are already at their lowest values (Sh = 2). Case 1 could be fitted to the data by decreasing the bubble diameter by approximately an order of magnitude to 2.5 mm, however with this reduced bubble diameter the dependence of the rate constant on gas flow rate would be stronger and the dependence on the solids flow rate would be weaker than the experimental data. To further account for this, one would have to postulate that these smaller bubble diameters decrease with increasing solid-to-gas loading. Small bubbles may be produced at the bottom of the plume, but it is likely that these will coalesce into larger ones, based on bubbles observed in water and mercury models (16). This small bubble hypothesis is also unlikely based on recent work by Irons and Farias (15) who obtained evidence of incomplete particle-liquid contact in experiments in which they measured bath cooling rates during silica injection into lead for vertically downward injection. The bath cooling rates were only 30% of that expected if all the particles had contacted the melt. The flow rates and loadings were similar to the present experiment, however the liquid head was much smaller (0.11 m compared to 0.65 m in the present study) Therefore this work in lead reflects the conditions prevailing at the lance tip and bottom of the plume more than the present work. Even in this situation the particle-liquid contact was not dependent on solid-to gas loading as the small bubble hypothesis would have to be to be consistent with the present experimental rate constants. Therefore Case 3 is the most likely situation. These lead studies and the present work all indicate that there can be significant problems in particle-liquid contact, and moreover they are likely to be system specific. It is probably coincidental that the lead work and the present work indicate that 30% of the particles are in the liquid. In fact, Irons and Farias (15) found that particle liquid-contact could be significantly improved by angled injection or by the injection of gas evolving compounds.

There have been other studies in which more direct evidence of particle attachment to bubbles has been obtained. Coppus observed layers of aluminum particles attached to the backsides of bubbles in water (18). Robertson and Conochie (19,20) also observed particles attached to their "open-top bubbles" in water and liquid lead. In work with the injection of low-melting point fluxes, Engh (21) found that the refining rate was consistent with sulfur diffusion to bubbles coated with a liquid layer of the flux.

In several injections, both the powder and the gas were stopped simultaneously, for example, a lance blockage. In these cases there was no further desulfurization, indicating that either significant numbers of particles were not entrained into the bulk of the iron, or that the rate of desulfurization with these particles was slow. This corroborates assumption (c), that the particles rise through the plume, and are immediately incorporated in the top slag.

Conclusions

1.   A mathematical model to describe the transport phenomena and chemical reactions occurring in the plume during powder injection was developed. The particles could be positioned either in the liquid or on the carrier gas bubbles. For particles in the liquid, the rate of refining was strongly dependent on the solid flow rate and virtually independent of gas flow rate. Conversely, reaction rates for particles on the bubbles were strongly dependent on gas flow rate and weakly dependent on solids flow rate.

2.   The rate at which liquid is entrained into the plume exerts little control over the refining at moderate reaction rates.

3.   The first-order rate constants for calcium carbide desulfurization with respect to sulfur in the iron were found to be consistent with those expected if 30% of the particles enter the liquid.

<u>Nomenclature</u>

A_{pl} — area of plume, m^2

C_S^b — sulphur concentration in bulk phase, $mole/m^3$

$C_S^{p\ell}$ — sulphur concentration in plume, $mole/m^3$

C_d — drag coefficient, $(-)$

C_p — heat capacity at constant pressure, $J/kg/K$

C' — molar density, $mole/m^3$

D_S — diffusivity of sulphur in molten metal, m^2/s

D_{eff} — effective diffusivity of calcium vapour through the reaction product layer, m^2/s

d — diameter, m

f — fraction of powder on the bubble surfaces

f_m — fraction of powder on bubble surface which is in monolayer on bubble surface

f_R — fraction of powder on the bubble surface which is allowed to react

F_{gp}^b — buoyancy force on the gas-powder mixture, N/m

F_p^b — buoyancy force on the powder in liquid, N/m

$F_{gp-\ell}^d$ — drag force on the gas-powder mixture, N/m

$F_{p-\ell}^d$ — drag force on the powder in liquid, N/m

G_1, G_2, G_3 — heat transfer functions, $W/m^3/K$

g — gravitational acceleration, m/s^2

H — depth of the melt, m

h — heat transfer coefficient, $W/m^2/K$

J_1, J_2 — mass transfer function, s^{-1}

k — thermal conductivity, $W/m/K$

k_p — mass transfer coefficient to particles, m/s

k_b — mass transfer coefficient to bubbles, m/s

K_1 — first order rate constant, s^{-1}

Nu — Nusselt Number, hd/k

Pr — Prandtl Number

P — pressure, Pa

Q — gas flow rate, Nm^3/s

R — Engineering Gas Constant, $J/kg/K$

T — temperature, K

U — velocity, m/s

V — volume of iron, m^3

Wp — solids mass flow rate, kg/s

Z — vertical distance from bottom of plume, m

α — molar fraction of particle phase reacted

ε — energy dissipation per unit mass, m^2/s^3

σ — phase volume concentration, $-$

v	kinematic viscosity, m^2/s
ρ	density, kg/m^3

Subscripts

atm	atmospheric
b	bubble
g	gas
ℓ	liquid
m	gas-particle mixture in bubbles
p	particle
pl	plume
S	Sulphur

Superscripts

b	bulk or buoyancy
d	drag force
ib	bubble-liquid interface
ip	particle-liquid interface
L	product layer

References

1. N. El-Kaddah and J. Szekely: Ironmaking and Steelmaking, 1981, no. 6, p 269-78.

2. Y. Sahai and R.I.L. Guthrie: Metall. Trans. B, 1982, vol. 13B, pp.193-202

3. J.W. McKelliget, M. Cross and R.D. Gibson: Appl. Math. Modelling, 1982, vol. 6, pp.469-80.

4. J.A. Boure and J.M. Delhaye: in Handbook of Multiphase Systems, Hemisphere, Washington, G. Hetsroni, ed. 1982, pp. 1-36 to 1-95.

5. F. Boysan and S.T. Johansen: Intern. Seminar on Refining and Alloying of Liquid Aluminum and Ferro-Alloys, Aug. 26-28, 1985, Trondheim, Norway.

6. C.T. Crowe: J. Fluid Eng., 1982, vol. 104, p. 297.

7. L.R. Farias and G.A. Irons: Metall. Trans. B, 1986, vol. 17B, pp. 77-85.

8. L.R. Farias and G.A. Irons: Metall. Trans. B, 1985, vol 16B, pp. 211-25.

9. M. Talballa et al.: AFS Transactions, 76-122, pp. 775-786.

10. Y. Sano et al.: J. of Chemical Engineering of Japan, Vol. 7, No. 4(1974), pp.255-261.

11. M.H.I. Baird and J.F. Davidson: Chem. Engng Sci., 1962, vol. 17, pp.87-93

12 G.A. Irons: Metall. Trans. B, 1987, vol 18B, pp. 105-117.

13. C.W. Gear, Numerical Initial Value Problem in Ordinary Differential Equations, Prentice-Hall, Englewood Cliffs, NJ, 1971.

14. M. Sano and K. Mori: Proceedings of Scaninject III Conference, 1983, Lulea, Sweden, June 15-17, Mefos and Jernkontoret, pp. 6:1-6:17.

15. G.A. Irons and L.R. Farias: in press Can. Metall. Quart.

16. M. Sano and K. Mori: Trans. ISIJ, 1980, vol. 20, p. 675.

17. L.K. Chiang, I.A. Cameron, G.A. Irons and W.K. Lu: Proceedings of the Fifth International Iron and Steel Congress, Washington, DC, Apr. 6-9, 1986, ISS of AIME, pp. 441-451.

18. J.H. Coppus, PhD. Thesis, Eindhoven University of Technology, Holland, 1977.

19. D.G.C. Robertson, D.S. Conochie and A.H. Castillejos, Proceedings of Scaninject II Conference, 1980, Lulea, Sweden, June 12-13, Mefos and Jernkontoret, pp 4:1-4:36.

20. D.G.C. Robertson, University of Missouri-Rolla, private communication, 1986.

21. T.A. Engh et al.: Scan. J. Metallurgy, 1972, vol. 1, p.103.

NON FERROUS

Session Chairmen
H. Y. Sohn, University of Utah
Alfred F. LaCamera, Alcoa Technical Center

One of the two sessions in this broad topical area examines aluminum production from computer models to evaluate interface shape changes associated with specific Hall cell designs and to illustrate how fluid flow simulations assisted in commercial development of a bipolar cell. Papers on carbothermic reduction analyze the particle behavior in plasma and study the prevailing chemical equilibria at various stages of the process.

Presentations on other metals include the application of a diffusion model sulfide reduction by CO assisted by S scavenging using excess lime and a Monte Carlo algorithm to help understand the gas-solid, diffusion controlled reactions occurring in the interior of an irregularly shaped partical. Several conventional smelting processes are simulated and phenomenalogical models of flash smelting and flash converting are also highlighted.

ANALYTICALLY MODELING THE CARBOTHERMIC REDUCTION

OF ALUMINA IN A THERMAL PLASMA

D. J. Varacalle, Jr., W. C. Schutte, C. D. VanSiclen,
R. W. Bartlett

Idaho National Engineering Laboratory
EG&G Idaho, Inc.
P.O. Box 1625
Idaho Falls, Idaho 83415

Abstract

The in-flight carbothermic reduction of vaporized alumina (Al_2O_3) particles to aluminum was analytically investigated in an argon nontransferred plasma torch. The computational procedure for approximating the plasma physics, particle dynamics, and plasma chemistry of the process is presented. The equations that govern the energy and mass transport in the plasma column and plume are solved numerically to determine the plasma temperature and velocity. The plasma/particle interaction analysis procedure then approximates the trajectories and vaporization of injected particulates in the plasma jet. The results obtained to date indicate the potential of vaporizing a significant amount of alumina and carbon in a plasma, if the particles are small. The thermochemistry analysis indicates that the carbothermic reduction of alumina is possible for in-flight chemical reactions in a dc nontransferred arc configuration. The economics of the process using a nontransferred configuration is discussed on the assumption that quenching to liquid metal can be achieved.

601

Introduction

The use of electrically produced plasmas for reduction of ores to metals has been proposed for many years to replace conventional electrolysis methods. Potential advantages of using plasma processing instead of conventional methods include better control, more complete reactions, lower energy consumption, and reduced capital and labor costs. However, evaluation of the technical and economic viability of a proposed process requires knowledge of what actually occurs within the plasma.

The Hall-Heroult process has undergone impressive progress over the hundred years since its invention, with the electrical energy required for the best cells reduced from 17 to 13 kWh/kg in the last two decades. Nevertheless, it remains a very electrical energy intensive process, and because it is a monopolar electrolytic process, large amounts of capital are required.

The use of shaft furnaces and electric arc furnaces for carbothermic reduction of alumina was investigated by Alcoa (1, 2). Mixtures of carbon and ore, made into pellets, were heated in a shaft furnace. Reduction reactions occurred to Al_4O_4C, Al_4C_3, and a liquid mixture of these carbides and aluminum oxide. This oxycarbide liquid causes pellets to bridge and stick together, plugging the bed under all feasible temperatures and CO pressures. Sticking is furthered by vapor transport, particularly Al_2O, and recondensation of aluminum-bearing vapor species as the gases cool on rising in the shaft. This phenomenon defeats carbothermic reduction of alumina using conventional technology. Two options remain.

The first option is to operate at limited temperatures to avoid oxycarbide liquid while using low back pressures of carbon monoxide and a very dilute aluminum alloy product to drive the reaction forward against adverse thermochemistry (3). The dilute alloy must be separated or refined to produce commercially pure aluminum. The second option is to operate at extremely high temperatures to avoid the oxycarbide liquid problem. An entrained flow system using dc arc plasma heating is one possible way to achieve these high temperatures. The use of an RF plasma for reduction of alumina has also been investigated (4). The purpose of the present study was to model carbothermic reduction of metal oxide particles entrained in a dc nontransferred arc plasma; alumina was selected as the first system to be investigated.

There is an extensive effort at the Idaho National Engineering Laboratory (INEL) to develop models that accurately describe thermal plasma physics and thermochemistry. The intent of the analytical effort described here was to establish a mathematical process model that fully describes the interacting electrical, magnetic, fluid, and thermal effects of a plasma in conjunction with the dynamics and thermochemistry of injected species. This method uses existing computer models for plasma physics, plasma/particle dynamics, and thermochemistry which yield predictions of temperature, velocity, particle trajectory and vaporization, and species composition in a commercially available arc torch. The models characterize the gas species and develop the flow field from the cathode through the free plume. In addition, the particle dynamics of the process was quantified, and the thermochemistry of the carbothermic reduction of alumina was approximated. Finally, the total process economics was estimated.

Plasma Physics Computer Modeling and Results

Evaluating the technical and economic viability of any proposed plasma process requires knowledge of the dynamics within the arc column and free plume, which can also lead to development of new or improved processing techniques. Over the last several years, increasing attention has been devoted to the mathematical modeling of plasma systems. The first detailed arc column code can be traced back to Watson and Pegot (5). Further work in this area was carried out by Pfender et al. (6-13), ranging from anode/cathode studies to the computation of the flow and temperature fields in plasma reactors. Attempts to model the flow and temperature fields in a dc plasma jet can be traced to the work of Donaldson (14) and Boulos and Gauvin (15). The problem of modeling the flow and temperature fields in a turbulent dc plasma jet has been addressed by Correa (16), McKelliget (17), and Varacalle (18).

An arc plasma torch for use in materials processing was numerically modeled, as described in this paper, using state-of-the-art computer codes. The torch (Figure 1) is of the free arc length geometry and can operate at power levels up to 30 kW. Typical operating conditions are 21.5 V, 500 A, 10.8 kW, 38 scfh argon gas flow, and an argon chamber pressure of 1.0 atm. A 50% electrical to thermal conversion efficiency was used for this computational study, based on an experimental enthalpy balance of the torch cooling water. The thermodynamic and transport properties of argon at one atmosphere pressure and temperatures from 1000 to 20,000 K were taken from tabulated sources (19). These properties, along with operational parameters, were used as input to a model of the plasma column (20), producing output of predicted temperature, enthalpy, velocity, and other properties as a function of radial and axial position. In turn, this output was used as input to the free plume expansion model (21).

Modeling energy transport in a current-carrying thermal plasma requires solving the equations for mass, momentum, and energy conservation. Over the last several years, increasing attention at the INEL has been devoted to the mathematical modeling of the specific plasma system of this study (22, 23). In the plasma column model, the electric arc is assumed to originate at the cathode tip and to attach at the start of the anode (length = 0.03175 m). The following assumptions were made for the plasma column analysis: (1) the flow in the nozzle and in the jet is axial (no swirl component), steady state, and rotationally symmetric; (2) turbulence is represented by the K-ε model (21); (3) local thermodynamic equilibrium prevails over the entire calculation domain and a continuum is assumed; (4) axial thermal diffusion effects are neglected; (5) electric and magnetic forces are negligible; (6) viscous heat dissipation and gravity forces are negligible; (7) the plasma is optically thin, and emitted radiant energy is taken from a table of energy emitted as a function of temperature (20); (8) pressure is constant and equal to atmospheric pressure around the plasma plume; (9) the boundary layer approximation is used, i.e., axial gradients are considered negligible with respect to radial gradients and axial velocities are assumed much larger than radial velocities; (10) the electric discharge is stationary; (11) the plasma is treated as an ideal gas, therefore dh = C_p dT; (12) since the copper anode is water cooled, the wall temperature is assumed to be 700 K for the entire column.

Based on these assumptions, the conservation equations expressed in cylindrical coordinates, where r is the radial and the x the axial distance, can be written for the arc column as shown in Table I. The boundary conditions for the differential equations are those of symmetry at the centerline, and a given fixed temperature and velocity at the outer boundary. The assumptions used for the column analysis (excluding

603

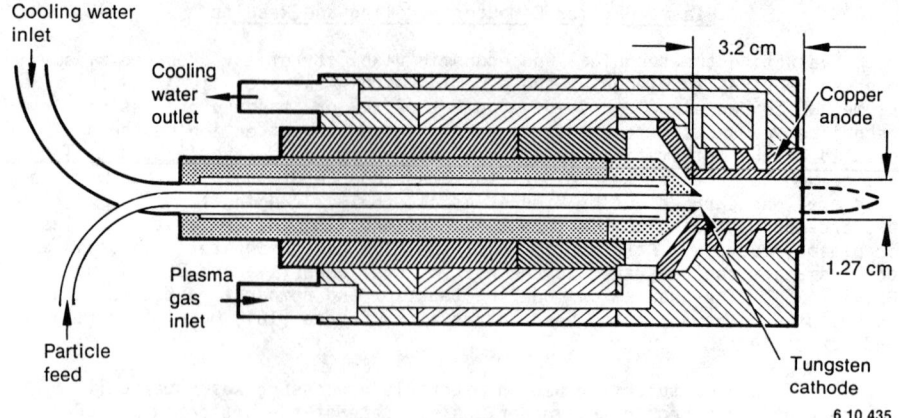

Figure 1 - Schematic of 11 kW plasma torch.

assumptions 5 and 10 above) also apply to the plume analysis. The notations used in the equations are given in the Notations and Subscripts sections toward the end of this paper.

Figures 2 through 5 illustrate predictions of temperature and velocity in the plasma column and free plume. As shown in Figure 2, the centerline temperature decreases over the length of the column, with a maximum value of 13,700 K at the arc attachment point (0.0 m) and decreasing to 11,550 K at the nozzle exit (0.03175 m). The decrease in the axial profile is mostly attributed to radiation losses, while the radial distribution is largely due to heat conduction to the cold wall (boundary condition of 700 K on the water cooled copper anode). The 50% loss in energy from the plasma is dominated by radiation heat transfer. The average temperature of the plasma column is 10,500 K, and the average radiation loss in the column is 8.11×10^8 W/m^3, indicating that approximately 60% of the losses from the plasma are attributed to radiation heat transfer.

The centerline velocity decreases from 300 m/s at the arc initiation point to 200 m/s at the nozzle exit (Figure 3). The wall effects (shear and heat transfer to the cold wall) can be clearly seen in the sharp radial drops in the boundary region. The radial velocity profile is dually dominated by the viscous force contribution and the density gradient (inertia forces).

The plasma plume program models a free expanding plasma jet, which is the main processing zone in applications such as ore reduction, plasma spraying, or chemical synthesis. Only field-free plasma conditions are considered in this model, characterized by one-way or parabolic behavior, due to the very strong convective flows in one predominant direction. The plasma column code's predicted temperature and velocity at the nozzle exit are used as boundary conditions in the free plume code.

The plume centerline temperature at the nozzle exit (0.0 m) is 11,550 K and decreases to 2100 K at 0.2 m from the nozzle exit (Figure 4). The calculated boundary growth radially into the ambient atmosphere is determined by monitoring the edge gradients within prescribed bounds. As illustrated in Figure 5, the corresponding centerline velocity decreases from 200 m/s at the nozzle exit to 16 m/s at 0.2 m into the plume. The sharp drop in temperature and velocity from the exit is due to mixing with the ambient argon atmosphere. The plasma jet radius increases by a factor of 5 (r_{exit}/r_{plume}) in this axial distance.

Table I. Plasma Differential Equations

Mass:

$$\frac{\partial(\rho u)}{\partial x} + \frac{1}{r} \frac{\partial(r\rho v)}{\partial r} = 0$$

Axial momentum:

$$\rho u \frac{\partial u}{\partial x} + \rho v \frac{\partial u}{\partial r} = -\frac{\partial P}{\partial x} + \frac{1}{r} \frac{\partial}{\partial r} \left(\mu \sqrt{r} \frac{\partial u}{\partial r} \right)$$

where $\mu = \mu_1 + \mu_t$

Energy (in column):

$$\rho u \frac{\partial h}{\partial x} + \rho v \frac{\partial h}{\partial r} = \frac{1}{r} \frac{\partial}{\partial r} \left(r\Gamma \frac{\partial h}{\partial r} \right) + \sigma E_x^2 - R$$

where $E_x = \frac{I_x}{2\pi} \frac{1}{\int_0^R r\sigma dr}$ (Ohm's Law) and $\Gamma = \frac{k}{C_p} + \frac{\mu_t}{Pr_t}$

Energy (in plume):

$$\rho u \frac{\partial T}{\partial x} + \rho v \frac{\partial T}{\partial r} = \frac{1}{r} \frac{\partial}{\partial r} \left(r\Gamma \frac{\partial T}{\partial r} \right) - \Gamma \quad C_p \frac{\partial T}{\partial r} \frac{\partial}{\partial r} \frac{1}{C_p}$$

$$+ \frac{1}{rC_p} \frac{\partial}{\partial r} (r\Gamma) \ (h_{argon} - h_{air}) \frac{\partial \omega}{\partial r} - \frac{\rho u}{C_p} (h_{argon} - h_{air}) \frac{\partial \omega}{\partial x} - R$$

Argon concentration:

$$\rho u \frac{\partial \omega}{\partial x} + \rho v \frac{\partial \omega}{\partial r} = \frac{1}{r} \frac{\partial}{\partial r} \ r(\rho D + \mu_t/Pr_t) \frac{\partial \omega}{\partial r}$$

Turbulence:

$$\rho u \frac{\partial K}{\partial x} + \rho v \frac{\partial K}{\partial r} = \frac{1}{r} \frac{\partial}{\partial r} \left(r(\mu_1 + \mu_t) \frac{\partial K}{\partial r} \right) + \mu_t \left(\frac{\partial u}{\partial r} \right)^2 - \rho\varepsilon \quad \text{where } K_i = 0.005 \ u^2$$

$$\rho u \frac{\partial \varepsilon}{\partial x} + \rho v \frac{\partial \varepsilon}{\partial r} = \frac{1}{r} \frac{\partial}{\partial r} \left(r(\mu_1 + \mu_t/1.3) \frac{\partial \varepsilon}{\partial r} \right) + 1.44 \frac{\varepsilon}{K} \mu_t \left(\frac{\partial u}{\partial r} \right)^2 - 1.92 \ \rho \ \frac{\varepsilon^2}{K}$$

where $\varepsilon_i = 0.09 \frac{\rho K^2}{\mu}$ and $\mu_t = 0.09 \frac{\rho K^2}{\varepsilon}$

Temperature fluctuations:

$$\rho u \frac{\partial L}{\partial x} + \rho v \frac{\partial L}{\partial r} = \frac{1}{r} \frac{\partial}{\partial r} \left(r\Gamma \frac{\partial L}{\partial r} \right) + \frac{\mu_t}{Pr_t} \left(\frac{\partial T}{\partial r} \right)^2 - 2 \ \rho\eta$$

$$\rho u \frac{\partial \eta}{\partial x} + \rho v \frac{\partial \eta}{\partial r} = \frac{1}{r} \frac{\partial}{\partial r} \left(r\Gamma \ \frac{\partial \eta}{\partial r} \right) + \frac{\eta}{L}\left(\frac{\mu_t}{Pr_t}\right)\left(\frac{\partial T}{\partial r}\right)^2 - 2.2 \frac{\eta}{L} \ \rho\eta - 0.001 \ \rho \ \frac{\varepsilon\eta}{K} \quad \text{where}$$

$$Pr_t = \left(-0.35 \ \frac{\partial u}{\partial r} \sqrt{\frac{K \ L}{\varepsilon \ \eta}} \right)^{-1}$$

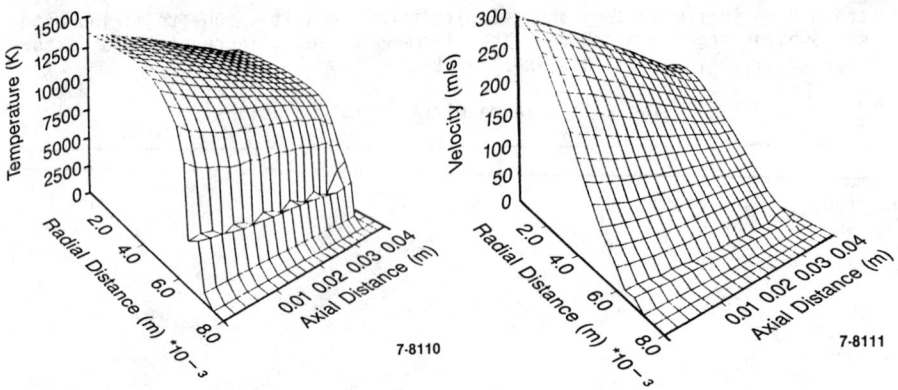

Figure 2 - Prediction of temperature versus location in the plasma column.

Figure 3 - Prediction of velocity versus location in the plasma column.

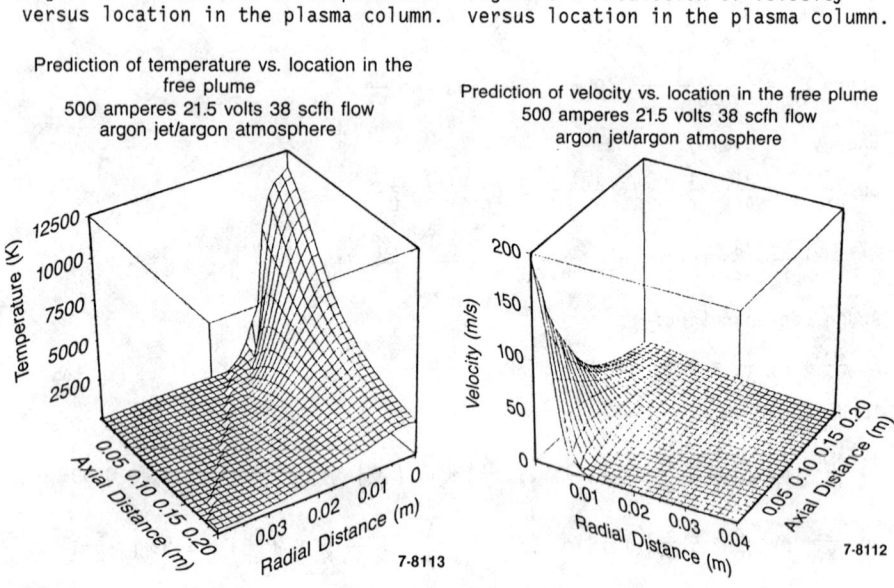

Figure 4 - Prediction of temperature versus location in the free plume.

Figure 5 - Prediction of velocity versus location in the free plume.

Plasma/Particle Computer Modeling and Results

Since the thermal treatment of ores and powders in plasma torches and furnaces represents a promising application of plasma technology, considerable attention has been given to the important problem of plasma/particle heat transfer. A number of mathematical models have been developed for the thermal treatment of particulates in dc plasma torches (24-31).

The plasma/particle computer program (32) used in this study uses the temperature and velocity fields generated by the plasma plume program to calculate the dynamics of particles injected into the plasma. The particles

606

are assumed to be injected radially at the nozzle exit. The primary result
of the plasma/particle code is a description of the injected material
vaporization rate (vapor mass per volume per time) as a function of position
in the plasma. This information is necessary for calculation of the
vaporized chemical reaction rates. In addition, the code determines the
amount of mass, momentum, and energy transferred from the particles to the
plasma.

The code calculates the histories of injected particles one at a time
and accumulates the results for as many particles as stipulated. This
process can thus give results representing the total impact of a number of
particles having specified distributions of particle size, sphericity, etc.
After the effects of the individual particles have been accumulated, the
total effects are related to the specified particle mass injection rate.

For each injected particle, the specified initial conditions are
injection location, velocity, diameter, sphericity, and temperature. Also
specified are the particle material properties--density, specific heat
(versus temperature), melting temperature, boiling temperature, heat of
fusion, heat of vaporization, and emissivity.

The force on the particle is calculated by

$$F = \frac{1}{2} C_D \, \rho |v_e - v_p| (v_e - v_p) \pi D_p^2 \tag{1}$$

where C_D is the drag coefficient corrected for the variable property
effect and the noncontinuum (Knudsen) effect, and v is the mean velocity.
The calculation of C_D and the derivation of the correction factor for the
Knudsen effect are detailed in Reference 22. Force corrections for
thermophoresis, vaporization, and Basset history effects are assumed to be
minimal and are not included. The model does account for the effects of
turbulence on the particle trajectory.

The rate of heat transfer from the plasma to the particle is given by

$$Q = \left[2 + 0.6 \, Re^{1/2} \, Pr^{1/3}\right] \left[\frac{C_{p_e}}{C_{p_p}}\right]^{0.38} \left[\frac{\rho_e \cdot \mu_e}{\rho_p \cdot \mu_p}\right]^{0.60} \frac{k(T_f)}{D_p} \frac{\pi D_p^2}{\psi} (T_e - T_p) \tag{2}$$

where Re is the Reynolds number, and Pr is the Prandtl number based on film
properties. This expression for Q includes a correction for the variable
property effect but does not include vaporization, noncontinuum, or particle
charging effects since they are estimated to be small. The model includes a
standard term describing radiative heat loss by the particle, but it does
not include radiative transfer from the plasma to the particle. The entire
particle is assumed to be at uniform temperature at all times; thus, the
time constant for heat diffusion in the particle is assumed to be negligibly
small. A simple treatment of particle vaporization is used: no
vaporization occurs while the particle temperature is less than its boiling
temperature, and at the boiling point temperature all the heat transferred
into the particle is used to vaporize material from the particle.

Figure 6 gives the calculated isotherms in the free plume. Overlaid on
these isotherms are typical trajectories of 5, 10, and 20 μm alumina
particles. Calculations for many such particle paths yield a map of alumina
vapor generation rate versus position. These data, in conjunction with
velocity and temperature field data shown in Figures 2 through 5, are
required for vapor phase chemical reaction calculations. Simplified forms
of the data, obtained by summing (integrating) and averaging over the cross

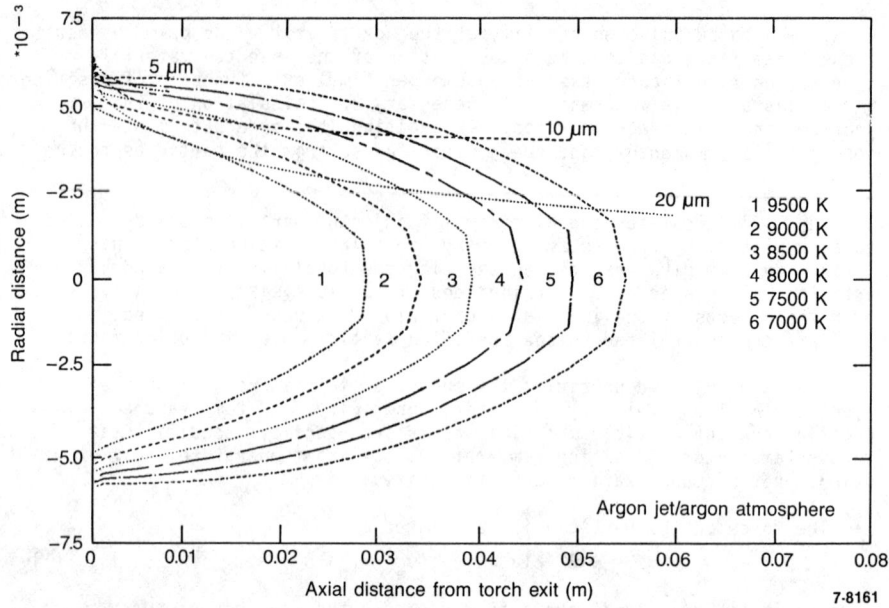

Figure 6 - Calculated argon isotherms and trajectories for 5, 10, and 20 μm alumina particles.

section of the plume for each axial location, can be used as input for the thermochemistry calculations. Examples of these data are shown in Table II for 5 μm alumina particles.

The data in Table II are based on a particle injection rate of 0.00534 g/s (carbon + alumina) with a radial injection velocity of 5 m/s. The "average radial location" value is the average radial coordinate of the vapor generation. The last two columns give the gas velocity and temperature at this average radius. The optimum size for achieving total evaporation of alumina particles for the torch of this study is between 5 and 10 μm (22).

Thermochemistry

ACE81 Computer Code - Equilibrium Thermochemistry

The ACE81 (33) equilibrium thermochemistry code for high temperature gas applications is an extremely versatile program for theoretical determination of transport and thermodynamic properties relevant to most thermochemical processes. A variety of phenomena may be treated by this code, including the thermodynamic and chemical state in a reaction, shock state chemistry, and boundary layer composition. The basic solution technique in ACE81 is to define the problem with the conservation equations, the total pressure equation, and one heterogeneous vapor pressure relation. The solution of these simultaneous nonlinear algebraic equations is based on a Newton-Raphson iteration. With input of the elemental data and thermodynamic information [H_f at 298 K, ΔH from 298 to 3000 K, $C_p(T)$, and S at 3000 K], the code selects base species, determines stoichiometric coefficients of formation for reactions of all base species, and evaluates molecular thermodynamic properties. The thermodynamic and transport properties of the homogeneous control volume are then evaluated. The

608

Table II. Calculated Axial Dependence of Vaporization for 5 μm
Alumina Particles as a Function of Location in the Free Plume

Axial Location[a] (mm)	Alumina Vaporization (g/s)	Average Radial Location (mm)	Plasma Gas Velocity (m/s)	Plasma Gas Temperature (K)
26.2	0.0	--	--	--
28.7	0.00066	4.86	71	8279
31.2	0.00117	4.45	86	9134
33.7	0.00084	4.25	92	9338
36.2	0.00049	4.29	89	9063
38.7	0.00026	4.40	84	8691
41.2	0.00014	4.51	79	8328
43.7	0.00008	4.77	71	7770
46.2	0.00005	5.08	63	7038
48.7	0.00004	5.21	59	6709
51.2	0.00003	5.39	55	6343
53.7	0.00002	5.36	55	6308
56.2	0.00002	5.25	57	6368
58.7	0.00001	5.21	57	6319
61.2	0.00001	5.21	56	6197
63.7	0.00001	5.52	51	5721
66.2	0.00001	5.72	48	5416

a. Axial location of particle injection is 25 mm; axial location of nozzle
   exit is 31.75 mm.

transport properties are calculated from kinetic theory. The thermodynamic
computational procedure includes the Debye-Huckel correction and is used to
calculate the mixture density, enthalpy, and species mole fraction from
first principles.

The ACE81 program (33) uses standard relations and relatively
conventional procedures to solve for the equilibrium thermodynamic state.
Curve fit relations are used to represent the thermodynamic data for each
species, given by a set of six constants for two allowable temperature
ranges. These constants are determined by a least-squares curve fit of
tabulated data. The specific heat is then used to calculate the enthalpy
and entropy of each species.

For the equilibrium calculation, it can be shown (34) that the
equilibrium constant, K, for each reaction is

$$\ln K = - \frac{\Delta G^{\circ}}{RT} \ . \tag{3}$$

The standard state Gibbs free energy is a function of temperature only,
and is obtained for each molecular species from

$$G = H - TS \tag{4}$$

where enthalpies are obtained relative to some chemical base state, often
the elements in their most natural form at 298 K and one atmosphere (JANAF
base state). Closure of the above formulation is achieved with the
elemental conservation equations and the total system pressure equation.

The stationary condition of the Gibbs free energy at equilibrium expressed in the above equations is consistent with the minimum Gibbs free energy statement often utilized in seeking the equilibrium state.

Using the temperatures calculated by the plasma physics code, the ACE81 code can predict species concentrations in the Al/O/C system as a function of location. Several simplifying assumptions were made for these calculations; dispersal of alumina and carbon was assumed to be homogeneous at the selected location, mass injection rate was very low (0.5% of the argon base flow), and energy transfer to the particles was instantaneous. Various stoichiometries were assumed in the process of converting alumina completely to Al and CO. Figure 7 shows the relationship between concentration and temperature for the six dominant species of the 37 candidates, as calculated by ACE81 for a 3 carbon to 1 alumina stoichiometry (Case 1). Using the data in Figure 7 and Table I, the yield of total gaseous aluminum (neutrals plus ions) as function of location in the plume can be approximated. As illustrated in Figure 7, the mole fraction of Al^+ is insignificant at temperatures below 4000 K. The highest mole fraction of gaseous aluminum is found to be 0.002 at 3000 K (axial location = 65 mm). Since the injected Al_2O_3 amounted to a mole fraction of 0.0011, the maximum gaseous yield in the plume would be 98%. Once free gaseous aluminum is generated, it must be removed from the plasma before it is reoxidized downstream in the cooler portion of the plume. This is done by condensing the aluminum onto a water-cooled surface (e.g., containment wall or probe). The concentration and temperature gradients are very steep in the quench region between the plasma and the cool surface. Thus, a condensable species entering this region is condensed with extreme rapidity. If the rate of condensation is greater than the rate of oxidation, free aluminum can be obtained.

The gaseous aluminum suboxides are of interest since they are expected to occur either during the dissociation of alumina or by a recombination of aluminum and oxygen atoms. It is concluded from this study that AlO and Al_2O are the only two important gaseous oxides of aluminum. The former is the principal gaseous suboxide for the thermal degradation of alumina in an argon thermal plasma, and the latter predominates under reducing conditions (alumina/carbon/argon system). There has been no conclusive evidence that solid suboxides can exist at room temperature (4), but solid AlO and Al_2O have been identified between 1400 and 1900 K. However, this does not preclude the possibility of a solid suboxide forming from the rapid quench of the gaseous oxide.

The addition of carbon to the plasma aids the recovery of aluminum. The carbon condenses along with aluminum in the quench region. Thermodynamically, oxygen reacts first with carbon and then aluminum, as seen by a comparison of the free energies of formation of CO, Al_2O, and AlO. Therefore, carbon removes oxygen during cooling, lowering the amount of aluminum oxide formation.

At any quench location, the aluminum recombination reactions should be inhibited to enhance condensation of liquid aluminum. As is evident, quenching should occur rapidly, and at temperatures no lower than 3000 K to alleviate this recombination problem.

Figures 8 through 10 illustrate the problem of recombination reactions, showing the yield of gaseous aluminum for stoichiometries of 3 carbon to 1 alumina (Figure 8), 2 carbon to 1 alumina (Figure 9), and 4 carbon to 1 alumina (Figure 10). As previously noted, the yield of gaseous Al plus Al^+ for the 3:1 stoichiometry is extremely high (>98%) for temperatures exceeding 3000 K. As the temperature decreases, aluminum recombination

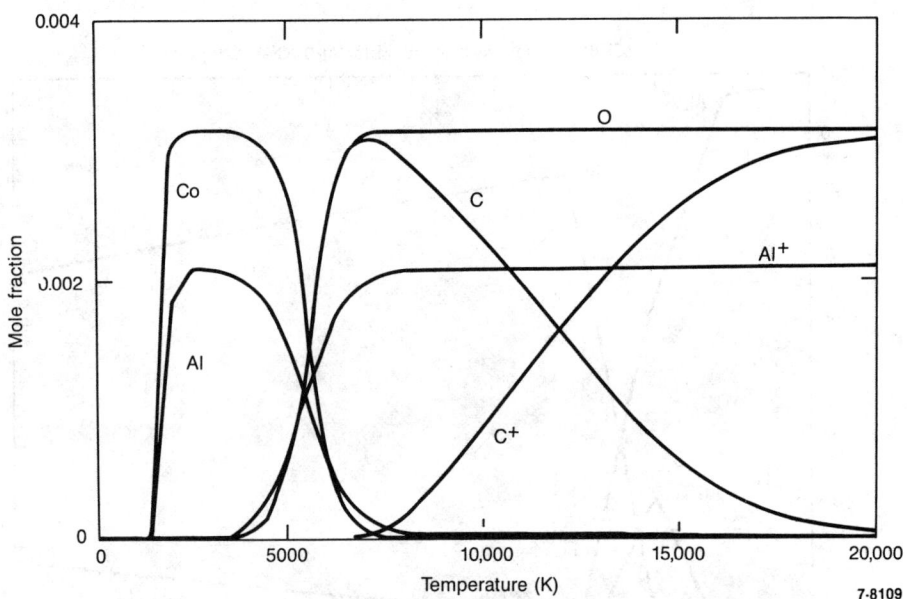

Figure 7 - ACE81 prediction of aluminum yield for 3 to 1 stoichiometry.

ACE81 equilibrium calculation for Al_2O_3/C/Ar (3-1)

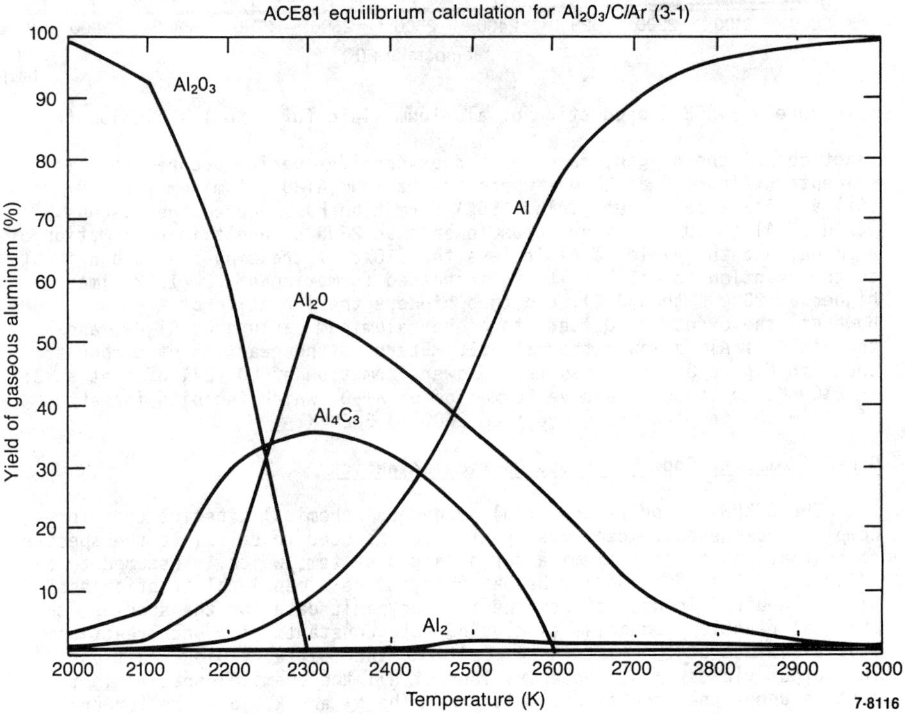

Figure 8 - ACE81 prediction of aluminum yield for 3 to 1 stoichiometry.

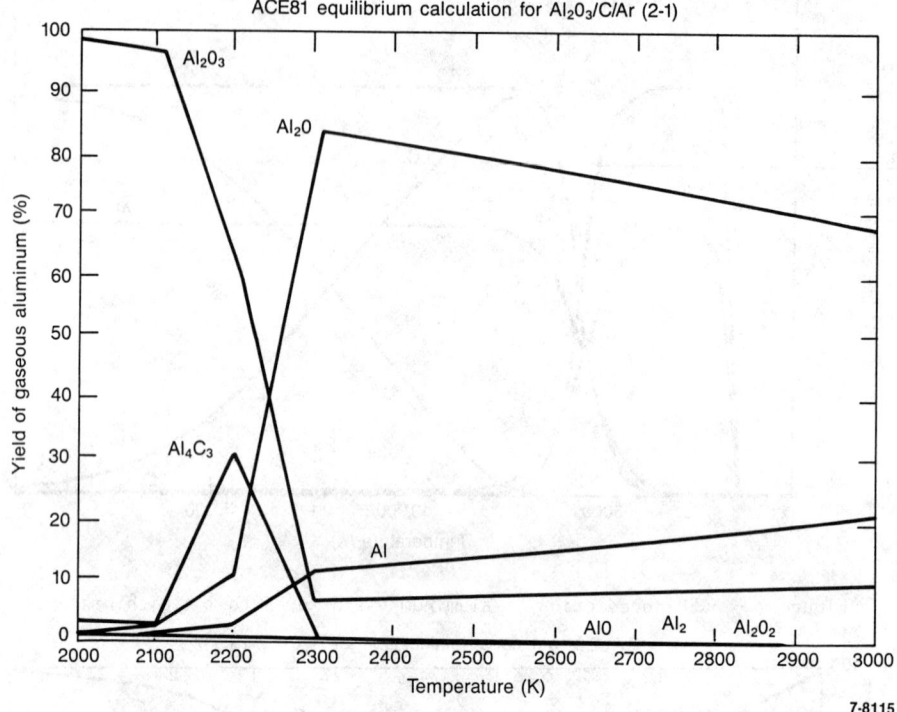

7-8115

Figure 9 - ACE81 prediction of aluminum yield for 2 to 1 stoichiometry.

reactions of the oxygen, carbide, and oxycarbide species occur. This is evident in Figure 8 at 2300 K where the maximum Al_2O (aluminum yield = 55%) and Al_4C_3 (aluminum yield = 35%) formation has lowered the gaseous yield of Al to 10%. Temperatures lower than 2300 K result in reformation of alumina, and the yield of Al is less than 10%. Increasing the carbon input to the reaction to 4:1 results in increased formation of CO (1.12 times higher at 2300 K than 3:1), and thus hinders the formation of Al_2O. However, the excess of C leads to higher aluminum capture by Al_4C_3, and the yield of Al is lower than at 3:1. Starving the reaction of carbon as shown in Figure 8 (2:1) results in lower formation of CO (84% of that at 3:1 at 2300 K) and thus excessive formation of Al_2O, which inhibits formation of Al in the temperature range from 2000 to 3000 K.

GCKP84 Computer Code - Gaseous Chemical Kinetics

The GCKP84 computer code (35), a general chemical kinetics code for complex, homogeneous ideal gas reactions, was used to calculate the species concentrations for the alumina/carbon/argon system, which is assumed to be adiabatic. The reaction may be batch or in one-dimensional frictionless flow. Required input includes the thermodynamic data for the species, the chemical reaction mechanism (including rate constants) for each reaction, initial values of the mixture composition, and initial fluid conditions. The output yields final concentrations of all the chemical species in the mixture under the specified conditions. The k_j and k_{-j} are the forward and reverse rate constants for the reaction. Each k_j is a function of temperature and usually given by the modified Arrhenius expression

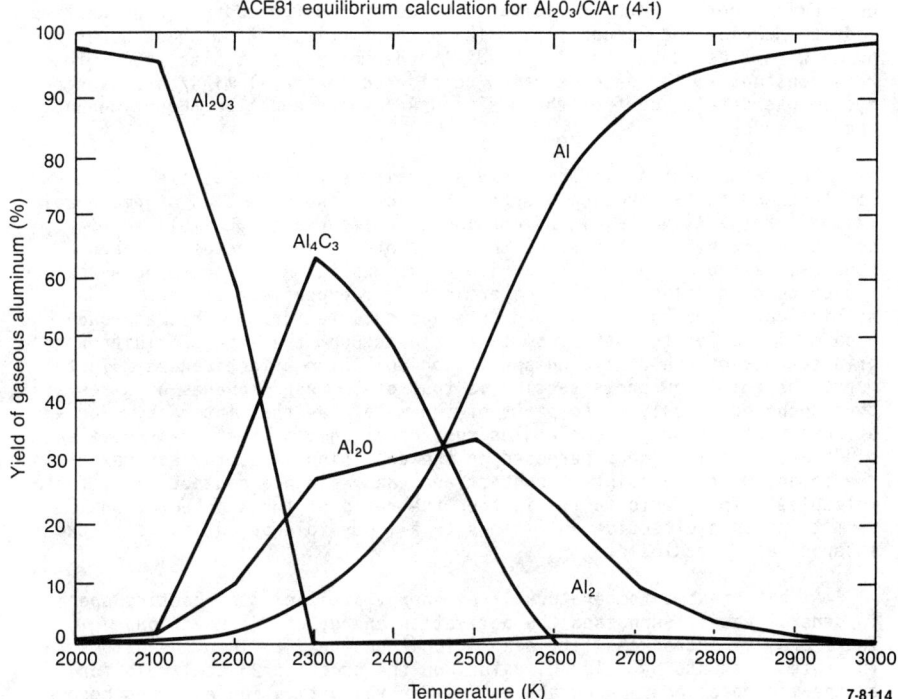

Figure 10 - ACE81 prediction of aluminum yield for 4 to 1 stoichiometry.

$$k_j = A_j\, T^n\, e^{-E/RT} . \tag{5}$$

Most reactions are considered to be reversible, and the reverse-direction rate constant k_{-j} is calculated from k_j and the equilibrium constant K_j (in concentration units) by the principle of microscopic reversibility.

The GCKP84 code allows the consideration of many different types of chemical reactions. In addition, any reaction may be treated as either reversible or irreversible, at the user's option. Any reaction of the general type

$$n_1A_1 + n_2B_2 = n_3C_3 + n_4D_4 \tag{6}$$

may be used. GCKP84 uses a new implicit numerical integration procedure for sets of "stiff" differential equations developed by Zeleznik (36). This property of stiffness often arises in the differential equations of chemical kinetics because of widely varying rates of relaxation of the fluid dynamic and chemical processes toward the equilibrium conditions. Reference 36 develops a generalized theory of implicit numerical integration that includes the methods of Gear (37) and Nordsieck (38) as special cases. This variable-order, predictor-corrector method selects the largest step size consistent with the gradients of the unknowns and with the accuracy requirements.

Thermodynamic functions are computed by using polynomial equations for $(C_p)_i$, H_i, S_i, and G_i as a function of temperature. The polynomial coefficients are computed by using the thermodynamic properties code of McBride and Gordon (39). The data calculated from these equations agree with those tabulated in the JANAF thermochemical tables (40). The rate constant k_j for each reaction considered in the alumina/carbon/argon system was determined from the modified Arrhenius equation (Equation 5), where n = 0.5.

The rate constant is obviously a sensitive function of the Arrhenius constant and the activation energy; therefore, the selection of reasonable values for the A_j and E_j is of considerable importance. Simple collision theory, which assumes that the interacting species are hard spheres, is typically used to arrive at values for A. Discrepancies between the calculated value and that determined from experiment or more sophisticated models is absorbed into a steric factor p. This approach has been modified for this study to take into account the effect of different orientations of the colliding species on collision effectiveness. In this work, the collision cross-section is that of two hard spheres of sizes corresponding to only those parts of the molecules that interact. For an effective collision, the molecules must be aligned so that their covalent bonds are parallel and superposed on the collision axis, thereby maximizing the momentum at the point of contact and the residence contact time of the molecules. The steric factor is then the ratio of those orientations permitting such effective collisions to all possible orientations of the interacting molecules.

At high plasma temperatures, the energy available to reacting species in general easily surpasses the activation energy of the reaction. Therefore, it introduces little additional uncertainty into the values for the rate constants by using E_j values on the order of 25 kcal/mole for exothermic reactions. Accurate values for the activation energies become more important in considering reactions in the outer regions of the plume, where the temperatures are considerably lower. Activation energies, however, are unavailable in the literature for these lower temperatures.

To use the GCKP84 code to determine the final concentrations of the various products, it was necessary to develop a procedure to track an alumina pellet through the plasma plume. The crucial assumption was made that the code could be used in such a dynamic fashion, i.e., close to kinetic equilibrium, the code mathematically approaches this equilibrium and approximates the actual, kinetic equilibration process.

Earlier theoretical studies (22) of the transport and flow properties of small, solid particles through the plasma plume indicate that the spread in velocity (speed and direction) of a collection of such particles released into the plume is very small. This suggests that a small volume, with density equal to the average density of the plasma and containing a specified mass of alumina, can be followed through the plasma. It will initially contain a 5 μm diameter alumina pellet assumed to possess initial conditions of temperature and velocity appropriate to its position in the plasma plume. As the volume moves along a typical path through the plasma (computer simulations show this to be essentially parallel to the plume axis), the pellet will evaporate at a rate consistent with the conditions at that volume location and thereby provide alumina molecules that will decompose and interact with other species in the volume. The gas-phase reactions inside the volume are also assumed to take place under the plasma conditions characterizing the volume location.

To calculate the changes in concentrations of the species within the volume, the path through the plume was divided into 16 nodes. The volume was considered to move from one cell to the adjacent "downstream" cell after

614

a time equal to the linear dimension of the cell divided by the average velocity of the material in the cell as determined from computer-generated velocity profiles. The GCKP84 code calculated final species concentrations for each cell, using as initial input concentrations the final concentrations from the adjacent "upstream" cell, and letting the system evolve during a time period determined in the manner just described.

Sixteen cells were considered sufficient to sample the plume. The thermodynamic data and the characterization of the plasma plume are too crude to justify significantly increasing the number of cells and decreasing their size in order to make each cell reflect local conditions more accurately. As more is learned about the plasma conditions in the fringe of the plume (where changes occur most abruptly as a volume moves radially or axially), more, and smaller, cells may profitably be used.

From thermodynamics, we say that a system is in equilibrium with its surroundings of given temperature and pressure if the Gibbs free energy of the system is at its lowest possible value. For a given reaction, this assumes infinite reaction time. From a chemical kinetics viewpoint, the system is at equilibrium if the rates of change of all the forward and reverse elementary reactions are equal.

Experimental values for rates of reaction are generally either on the order of magnitude of, or are below, those predicted by collision theory. Thus, collision theory may be used to estimate the upper bound to the expected rate of reaction.

Table III presents the chemical reactions utilized for the carbothermic reduction of alumina problem. Thirty-four reactions, which involved 19 species, were considered. The plasma temperature and velocity outputs were combined with the trajectory analysis for 5 μm alumina particles injected into the argon plume for this model. Thus, the vaporization rates, velocity, and temperature profiles of Table I were used as input to the code. A 1% loading of alumina + carbon was assumed with a 3 carbon to 1 alumina stoichiometry. The carbon particulates were assumed to follow the alumina trajectory. Table IV presents the mole fractions of the possible species of the carbothermic reduction of alumina at 3000 K. As shown, aluminum is predominantly bound in Al_2, with lesser amounts bound in Al^+, AlC, and Al. Since Al_2 would be thermodynamically expected to decompose to Al, the yield of gaseous aluminum (i.e., Al, Al^+, Al_2) is very high (92%).

It is interesting to compare the output from the equilibrium calculations to that of the kinetic calculations. As shown in Table V, at 3000 K, the ACE81 code predicts atomic carbon and oxygen to be tied up predominantly in CO, and predicts atomic aluminum to be predominantly in the form of Al, whereas the kinetic calculation predicts C, O, and Al_2 respectively. Examination of the GCKP84 equilibration factors for reactions 3 and 17 of Table III indicates that because of the small residence time in this particular cell (0.0002 s), both reactions are far from an equilibrium state. This comparison, incidently, suggests a way to evaluate the many possible sets of chemical reactions for their suitability for inclusion in the kinetics calculations. One could chose a particular group of reactions, run the GCKP84 code all the way to equilibrium to determine their equilibrium species concentrations, and compare these results with those of the ACE81 code at the same temperature. Reactions can then be added, deleted, or changed to bring the results of both calculations into reasonable agreement. Reactions that survive this iterative process over the range of temperatures characteristic of the plasma plume would then

Reaction No.		Reaction No.	
1	$2\ AlO = 1\ Al_2O_2$	18	$C+ + e- = C$
2	$Al + O = AlO$	19	$AlC + C = Al + C_2$
3	$C + O = CO$	20	$C_2 + O = CO + C$
4	$Al_2O_2 + O = Al_2O_3$	21	$2\ C = C_2$
5	$Al+ + E- = Al$	22	$C_2 + e- = C_2-$
6	$AlO + C = Al + CO$	23	$AlC + C = Al + C_2$
7	$AlC + O = AlO + C$	24	$CO + O = CO_2$
8	$AlC + O = Al + CO$	25	$C + O_2 = CO + O$
9	$AlC + O = CO + Al$	26	$AlO + CO = CO_2 +$
10	$AlO + Al = Al_2O$	27	$CO_2 + C = CO$
11	$Al_2O + O = Al_2O_2$	28	$AlO + O = Al + O_2$
12	$Al_2O + AlO = Al_2O_2 + 1\ Al$	29	$Al + C = AlC$
13	$Al_2O + C = Al + CO$	30	$2\ O = O_2$
14	$Al_2 + AlO = Al_2O + Al$	31	$C + e- = C-$
15	$Al_2 + O = AlO + Al$	32	$O + e- = O-$
16	$Al_2 + C = AlC + Al$	33	$2\ AlO = Al_2O + O$
17	$2\ Al = Al_2$	34	$AlO + AlC = Al_2O$

Table IV. GCKP84 Predicted Mole Fractions of Possible Species for the
Carbothermic Reduction of Alumina (3 to 1 Stoichiometry) at 3000 K

Species	Mole Fraction
AlO	0.944 E-9
Al_2O_2	0.235 E-6
Al	0.0046
O	0.4681
C	0.3366
CO	0.0114
Al^+	0.151 E-6
Al_2O_3	0.0255
AlC	0.0152
Al_2O	0.011 E-15
Al_2	0.1374
$C+$	0.839 E-3
C_2	0.075 E-6
C_2-	0.759 E-6
CO_2	0.008 E3
O_2	0.199 E-3
$C-$	0.045 E-3
$O-$	0.104 E-3

Table V. Comparison of Dominant Predicted Species at 3000 K for the
Carbothermic Reduction of Alumina (ACE81 versus GCKP84).

Species	ACE81 Mole Fraction	GCKP84 Mole Fraction
C	0.247×10^{-6}	0.3366
O	0.001	0.4681
Al	0.396	0.0046
Al^+	0.835×10^{-3}	0.0255
CO	0.599	0.0114
AlO	0.0016	0.944×10^{-9}
Al_2O	0.0017	0.011×10^{-15}
Al_2	0.784×10^{-5}	0.1374

be used in the tracking procedure described above. Although the computer
time required (and hence the expense) will be increased considerably, we
hope to incorporate this selection process into our future work.

Since the average plume residence time of a particle in this study is
on the order of 0.004 s, the prediction of the proper reaction rate is of
paramount importance. The rate expression involves two factors, the
temperature-dependent and the concentration-dependent factors. If a
reaction has available a number of competing paths, as shown in Table III,
it will in fact proceed by all of these paths, although primarily by the one
of least resistance. Only a knowledge of the energies of all possible
intermediates will allow prediction of the dominant path and its
corresponding rate expression (41). Since such information cannot be found
with the present state of knowledge, a priori prediction of the form of the
concentration term is not possible. Assuming that we can use our predicted
mechanisms of reaction of Table III, we may then proceed to determine the
frequency factor and activation energy terms of the rate constant.
Frequency factor predictions from either collision or transition-state
theory may come within a factor of 100 of the correct value; however, in
specific cases predictions may be much further off. In this study, the
methodology used to calculate this factor was compared to experimental
values for the H-N-O-C systems with good correlation. Although activation
energies can be estimated from transition-state theory, reliability is poor,
and it is probably best to estimate them from the experimental findings for
reactions of similar compounds. For this study, the activation energies,
estimated from low temperature reactions of the aluminum family, ranged from
65 to 130 kcal/mole per reaction. Thus, this is the most obvious source of
error in the kinetic calculations. The exponential expression ($e^{-E/RT}$) in
Equation 5 is capable of modifying the reaction rate constant, k_j, by
three orders of magnitude with this range of activation energy for any one
reaction. This level of perturbation of the reaction rate constant is
capable of modifying the reaction rate and thus the yield for any one
species by ±20%.

Economic Considerations

A preliminary economic assessment was made for plasma in-flight
carbothermic reduction of vaporized alumina by the process outlined in
Figure 11. Alumina is pulverized and classified in the argon carrier gas to
obtain a particle distribution below a critical maximum size consistent with
the limited plasma plume residence time. Lampblack, used as a source of
carbon to obtain fine particle size, is dispersed in the argon carrier gas
and classified. The alumina-lampblack mixture with recycled feed material
is injected into the plasma plume. The resulting aluminum vapor is quenched

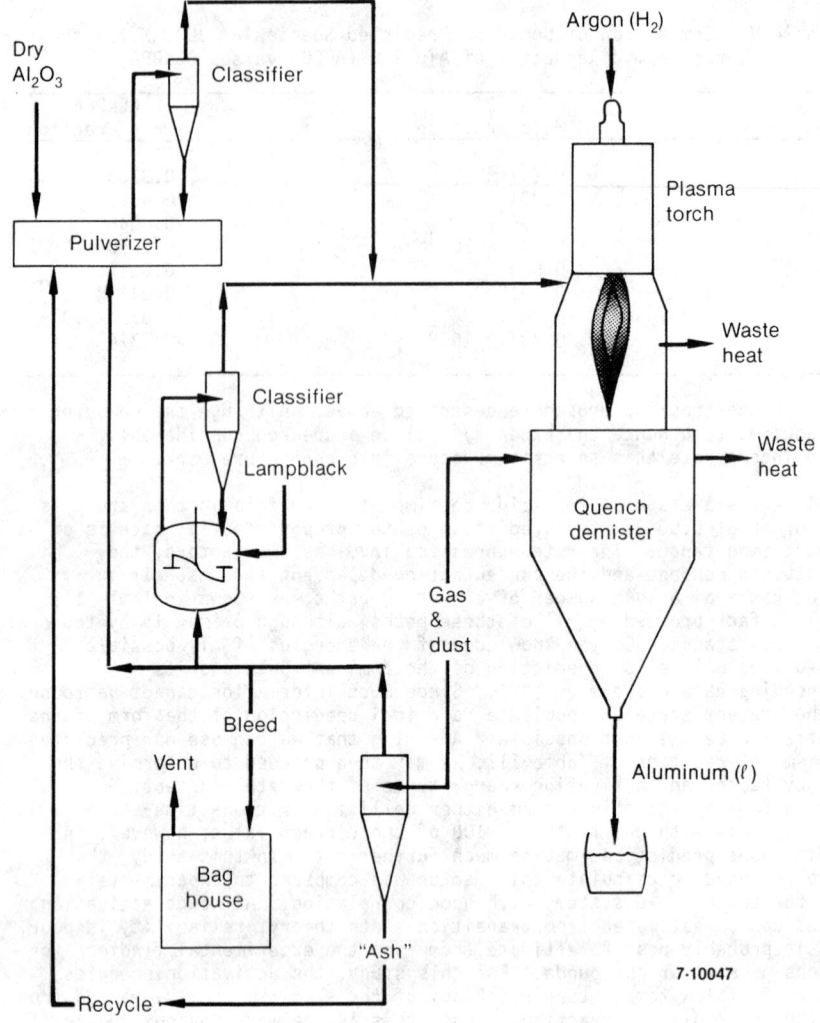

Figure 11 - Plasma carbothermic reduction of vaporized Al_2O_3 schematic process flow sheet.

to mist and eliminated by a demister that separates liquid aluminum from unreacted particulates, which are recycled. The technological basis for the plasma quench/demister has not been established. Argon, unreacted alumina, and carbon are recycled.

Because electrical energy consumption is a dominant cost factor in both the plasma carbothermic reduction and the commercial Hall-Heroult process, the economic assessment was limited to an estimate of electrical power consumption per unit of aluminum metal produced (kWh/lb-Al).

Computer process code analysis was made for torches at 10 kW, 160 kW, and 6 MW (commercial) using the codes discussed above. Because of radiation losses and gas velocity effects, available residence time for particle heating and vaporization scales less than linearly with torch size. The computer runs provide solutions of the plasma physics and plasma-particle dynamics for alumina and carbon particles in an argon plasma and estimate the amount of vaporization that will occur for selected trajectories in the free plume. To obtain a statistically valid vapor conversion, 100 individual particle trajectories through the plume were selected and averaged for each run. The thermochemistry and kinetic calculations then provide the amount of gaseous aluminum that is produced from vaporized alumina and available to be quenched to metallic aluminum. For the conditions evaluated, conversion of vaporized alumina, based on thermochemical and gaseous kinetic considerations, is above 94%. A 90% combination metal yield was assumed, based on thermochemical reduction of the alumina vapor and quench-condensation of the resulting aluminum metal vapor to liquid and removal from the system. Lower yields will require greater power consumption for producing aluminum by this method.

Some general observations obtained from computer runs using various discrete particle sizes for alumina and carbon are worth discussing before proceeding to the electric power consumption estimates. Very small particles are fully vaporized. However, large particles are not completely vaporized and absorb considerable heat in material that does not vaporize. Data showing the mass percent vaporized as a function of particle diameter are shown in Figure 12, based on several computer runs for the 6 MW torch at various discrete particle sizes. Five-micron particles of both carbon alumina are completely vaporized. Ten-micron particles are substantially vaporized for alumina, but only about 42% vaporized for carbon. Twenty-micron particles for both feed materials are only about 10% vaporized. Clearly, in large particles (in this context, a 325 mesh particle is large) there will be a very high circulating load of incompletely reacted particles that must be cooled during the quench process and then reheated during subsequent passes through the plasma torch.

As expected, the total specific energy absorbed in heating and partially vaporizing the particle increases as the extent of vaporization increases. This increase is substantial, due to the high heat of vaporization of carbon and alumina. The case for alumina is shown in Figure 13. Without vaporization, the heat absorbed in alumina is approximately 10 MJ/kg and rises to approximately 45 MJ/kg upon complete vaporization. These results will change slightly with different plasma run conditions such as average plasma temperature, but the variation under different run conditions is not substantial. All of the data presented in these figures are averages for 100 trajectories for each run.

Although specific energy consumption increases with increasing vaporization, energy efficiency decreases if the extent of vaporization decreases. This occurs because a proportionally higher amount of energy is used for heating and melting material which never vaporizes, which represents a heating loss. The specific energy input for production of alumina vapor thus increases as the mass percent of vaporized material decreases, i.e., as particle size increases. This effect is shown in Figure 14, where specific energy input for alumina vapor production is shown on a normalized basis. The energy requirement for alumina vaporization increases to 145% of that required for full vaporization when only 20% is vaporized and the energy requirement doubles when the vaporized mass percent drops to 10%. This is shown in Figure 14.

In summary, repetitive heating of large particles that are only partially vaporized is energy inefficient. This dictates that the particles fed to a commercial carbothermic alumina reduction unit must be extremely small and their size distribution must be carefully controlled to limit them to a small top size. Ideally, particles not much larger than 5 μm should be used and the loading density of these particles in the plasma gas should

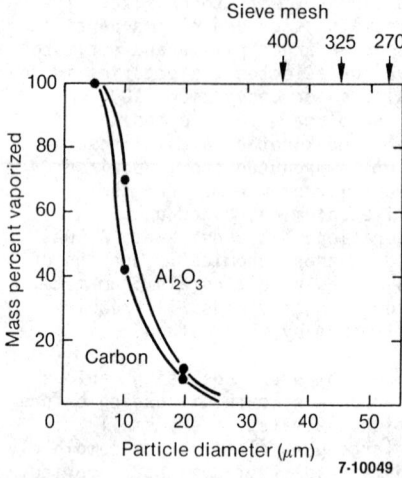
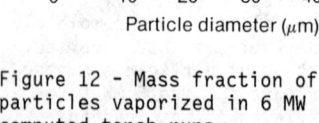

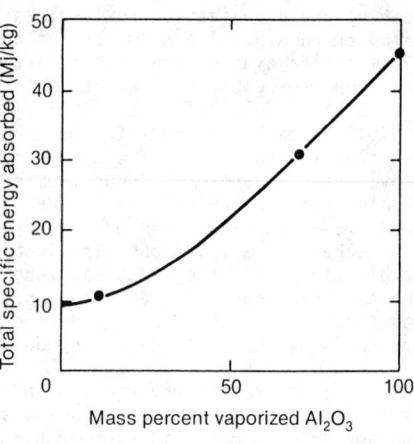

Figure 12 - Mass fraction of particles vaporized in 6 MW computed torch runs.

Figure 13 - Energy to heat and partially vaporize Al_2O_3 particles, for 6 MW computed torch runs.

be very high to minimize electric power consumption. Grinding and dispersing small particles at high mass loadings in the carrier gas is extremely difficult, if not impossible to achieve.

Power consumption was estimated for the hypothetical 6-MW plasma torch processing unit using four cases: 5-micron alumina particles and three size distributions of aluminum particles with top sizes of 25 μm, 44 μm (400 mesh), and 53 μm (270 mesh). The particle size distributions are shown in a log-log plot in Figure 15. These distributions are arbitrary, but they follow a Gates-Gaudin-Schuhmann distribution with a coefficient of 0.6, which is fairly typical of ground and classified brittle solids such as minerals. Each particle size distribution was discretized into a histogram of short size ranges. Using mass percent vaporized data shown in Figure 12, derived from the computer runs, materials balances were generated for each of these four cases and results are shown in Table VI. The basic assumptions and input parameters used to calculate the materials balances shown in Table VI are listed in Table VII. Different input parameters and assumptions will lead to somewhat different results than in Table VI. However, the parameters selected, which were derived from the computer model, and the assumptions made are believed to be reasonable with respect to a potential commercial application. The results shown in Table VI show that 100% of the alumina is vaporized at a 5 μm top size, but the percent of alumina vaporized drops to 60% at a 25-micron top size, to 45% at a top size of 400 mesh, and to 33% with a top size of 270 mesh, using the discretized particle size distributions of Figure 15.

Similarly, using the discretized particle size distributions and their associated fraction vaporized and energy absorption, it is possible to generate (1) energy inputs to the alumina particles, (2) the amount of carbon required, (3) the energy inputs to the carbon particles, (4) the energy required for the residual plasma gas, and (5) radiation/convection heating of the surroundings based on a 20 weight percent feed particle loading of the total plasma and carrier gas. These data, expressed in megajoules and matched to the materials balance mass numbers (kilograms) shown in Table VI, are presented in Table VIII. The energy requirements can be divided by the corresponding production of reduced and quenched aluminum

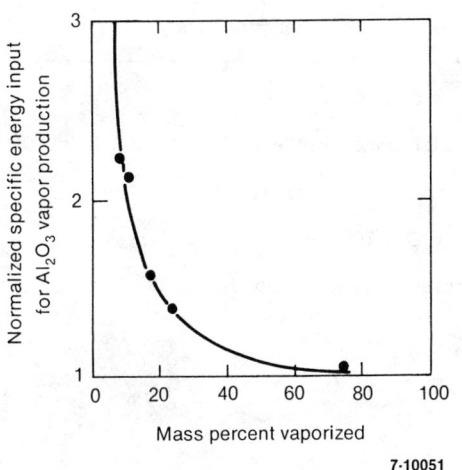

7-10051

Figure 14 - Increasing specific
energy input for vaporization with
reduced fraction vaporized
(larger particles).

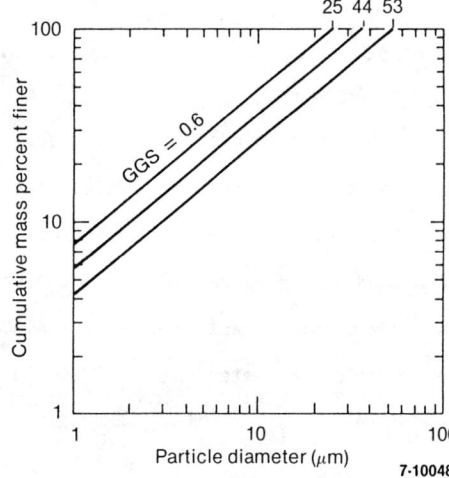

7-10048

Figure 15 - Arbitrary classified
particle size distribution (typical
at GGS = 0.6) with top sizes of
25 μm, 400 mesh (44 μm), and
270 mesh (53 μm) shown.

Table VI. Materials Balance

	Al_2O_3 Feed Particle Top Size			
	5 μm	25 μm	44 μm 400 mesh	53 μm 270 mesh
Al_2O_3 Feed[a] (kg)	1.0	1.0	1.0	1.0
Carbon (Lampblack) Feed[a] (kg)	0.59	0.35	0.265	0.195
Argon (H_2)[b] (kg)	6.4	5.4	5.1	4.8
Al_2O_3 Vaporized[c] (kg) (% Al_2O_3 Vaporized)	1.0 (100%)	0.60 (60%)	0.45 (45%)	0.33 (33%)
Al reduced and quenched (90% Yield) (kg)	0.476	0.285	0.215	0.157

a. Includes recycled "ash" (C + Al_2O_3).

b. Includes recycled gas and both the primary plasma gas and particle
carrier gas. Solid feed loading is 20 wt.%.

c. Estimated from Al_2O_3 particle size distribution and fraction
vaporized as a function of particle size from 100 computer code trajectories
for each particle size.

Table VII. Technology Assumptions for Economic Study

o   6 MW power

o   Primary argon gas flow discharges to argon atmosphere.

o   Torch efficiency of 85% (electrical to plasma energy)

o   Average enthalpy of the argon plume = 2×10^6 J/kg

o   Combined plume radiation and convection losses of 2×10^6 J/kg

o   20% particle mass loading (C + Al_2O_3)

o   Particle diameter range of 5 to 50 μm

o $M_C/M_{Al_2O_3}$ (g) = 5/1

o   Alumina vaporized to condensed liquid aluminum yield is 90%.

Table VIII. Energy Balance (MJ) Matched to Materials Balance of Table VI

| | Al_2O_3 Feed Particle Top Size | | | |
	5 μm	25 μm	44 μm 400 mesh	53 μm 270 mesh
Heating and vaporizing alumina[a]	45.3	29.2	24.2	20.4
Heating and vaporizing carbon	45.5	27	21	15
Heating plasma gas (net)	25.6	21.6	20	19.2
TOTAL	116.4	77.8	65.6	54.6
Electrical requirement at 85% efficiency of conversion to plasma	136.9	91.5	77.2	64.2
Al net production electrical energy (MJ/kg-Al)	287.4	321.1	359.1	409.1
(kWh/lb-Al)	36.2	40.4	45.2	51.5

a. Based on : (1) Particle size distribution, (2) Fraction vaporized by particle size, and (3) energy required for particle heating and particle vaporization by particle size.

shown as the last row in Table VI for each of the 4 cases to generate a specific power requirement. The results of these calculations are also presented at the bottom of Table VIII.

The results show that for complete vaporization of alumina using particles no larger than about 5 μm, 36.2 kWh hours of electricity are required per lb of aluminum. As the top size of the alumina particle size distribution increases, the power requirement increases gradually up to 51.5 kWh/lb-Al at a particle top size feed of 53 μm (270 mesh). These power requirements are substantially higher than the power required for Hall-Heroult fused salt electrochemical production of aluminum metal, at about 7 kWh/lb-Al. The assumptions used in these calculations are believed to be reasonable, but if in error, probably somewhat optimistic. Therefore, it is concluded that carbothermic reduction of plasma vaporized alumina in a nontransferred axial flow torch unit operation is not economically promising. Although the excess energy can be recovered as waste heat, converted to steam, and used to generate electrical power, the power credit at best would only be about 1/3 of the power consumption, and the net power requirement would still make the process uneconomic as presently seen.

Summary and Conclusions

A computational procedure for approximating the plasma physics, particle dynamics, and plasma chemistry of a dc nontransferred plasma torch has been presented. The process presented is the carbothermic reduction of alumina. The equations that govern the energy and mass transport in the plasma column and plume have been solved numerically to determine the plasma temperature, velocity, and other variables. Modeling the plasma physics phenomena in the arc column and plume is reasonable. Prior numerical calculations (19) for an argon jet discharging into an air environment agreed closely with the experimental measurements and illustrated the qualitative trends of the measurements.

The plasma/particle interaction analysis procedure presented in this study can approximate the trajectories and vaporization of injected particulates in a plasma jet. The results obtained to date are encouraging, and indicate the potential of vaporizing a significant amount of alumina and carbon in a nontransferred plasma torch, using small particles. For plasma/particle interaction modeling for a given application, there are optimum particle size and injection velocity combinations for achieving maximum vaporization. For this study, the optimum particle size is between 5 and 10 μm with a particle injection velocity between 5 and 10 m/s.

The thermochemistry analysis procedure presented in this study indicates that the carbothermic reduction of alumina is possible for in-flight chemical reactions in a dc nontransferred arc configuration. However, this study does not conclusively demonstrate the feasibility of using plasmas for alumina reduction. There are other important questions yet to be answered such as: will carbon be vaporized in the plasma, in the same regions as the alumina is vaporized, so that all the necessary reactants are present in the same place? Are the reaction rates fast enough so that the reactions can take place before the materials exit the hot plasma? Can the desired product (metallic aluminum) be extracted before reoxidation or some other competing reaction destroys it? For the equilibrium and kinetic thermochemistry calculations, maximum yields at 3000 K for gaseous aluminum species before quench were 98% for the equilibrium calculations and 92% for the kinetic calculations.

The computational procedure described in this paper will provide substantial assistance in selecting and optimizing the operational parameters for future materials processing experiments and applications.

The following conclusion can be made relative to process development for the in-flight carbothermic reduction of alumina in an argon thermal plasma: the carbothermic reduction of alumina in a nontransferred configuration is not econonically feasible. Any optimization of the process should involve a transferred arc configuration in conjunction with a plasma reactor for increased residence time for complete dissociation and enhanced energy transfer at high loading rates (>20%).

Acknowledgments

The consulting services of Professor Emil Pfender of the University of Minnesota are gratefully acknowledged. The authors are also grateful for many stimulating and fruitful technical discussions with G. D. Lassahn, INEL. The work described in this paper was supported by the Interior Department's Bureau of Mines under Contract No. J0134035 through Department of Energy Contract No. DE-AC07-76ID01570.

Notations

A_j	Arrhenius constant for reaction j	r	radial coordinate
C_D	drag coefficient	T	mass-weighted average temperature
C_p	specific heat	S	entropy
D	species diffusivity	u	axial velocity
D_p	particle diameter	v	radial velocity
E	activation energy	x	axial coordinate
E_x	axial electric field strength	Γ	thermal diffusivity =
G	Gibbs free energy		$k/C_p + \mu_t/Pr_t$
H_f	heat of formation (enthalpy of formation)	ε	dissipation rate for K
		η	dissipation rate for L
ΔH	enthalpy difference	μ	viscosity
H	enthalpy	ρ	density
h	specific enthalpy	σ	electrical conductivity
I_x	axial current	ω	plasma gas concentration
K	equilibrium constant, turbulent kinetic energy, Kelvins		
			Subscripts
k	thermal conductivity	e	plasma, away from the particle
L	temperature variance	ef	effective
n	constant or moles of species	f	film conditions, in the plasma near the particle
P	pressure		
Pr	Prandtl number	i	species or initial condition
R	radiant energy emitted per unit time and volume or gas constant	j	forward reaction j
		-j	reverse reaction j
		l	laminar
Re	Reynold's number	n	constant
		p	particle
		t	turbulent

References

1. A. S. Russell, Metallurgical Trans., vol. 12B (1981), 201-215.

2. M. J. Bruno, "Production of Al-Si Alloy, and Fe Si and Commercial Purity Al by the Direct Reduction Process," [DOE Conservation Report CONS-5089-16 (DE 83009361) Fuel Report, February 1983, by Aluminum Company of American ALCOA Laboratories, ALCOA Center, PA].

3. Professor John Elliott, private communication, MIT, Cambridge, MA, 19 May 1986.

4. R. R. Rains and R. H. Kadlec, Metallurgical Trans., vol. I (1970), 1501-1506.

5. V. R. Watson and E. B. Pegot, "Numerical Calculations for the Characteristics of a Gas Flowing Axially Through a Constricted Arc," D-4042, NASA, TN, (1967).

6. E. Pfender, "Electric Arcs and Arc Gas Heaters," Gaseous Electronics, M. N. Hirsh and H. J. Oskam, eds., (Academic Press, 1978) 291-398 Vol. 1.

7. E. Pfender, Pure and Applied Chemistry, vol. 52 (1980), 1773.

8. H. A. Dinuleseu and E. Pfender, J. Appl. Phys., vol. 51 (1980) 3149.

9. D. M. Chen and E. Pfender, IEEE, Trans. Plasma Sci., vol. PS-8 (1980), 252.

10. D. M. Chen and E. Pfender, IEEE, Trans. Plasma Sci., vol. PS-9 (1981), 265.

11. M. Chen, K. C. Hsu, and E. Pfender, Plasma Chemistry and Plasma Processing, vol. 1 (1981), 295.

12. A. Mazza and E. Pfender, ISPC-6, Montreal, vol. 1 (1983), 41.

13. R. M. Young et al., Ibid., vol. 1 (1983), 211.

14. C. P. Donaldson and K. E. Gray, AIAAJ, vol. 4 (1966), 2017.

15. M. I. Boulos and W. H. Gauvin, C. J. Ch. E., vol. 52 (1974), 355.

16. S. M. Correa, ISPC-6, vol. 1, (Montreal, 1983), 77.

17. J. McKellizet et al., Plasma Chemistry and Plasma Processing, vol. 2 (1982), 317.

18. D. J. Varacalle, Jr. et al., "Modeling Thermal Plasma Material Processing Experiments," Materials Research Society Proceedings, vol. 38 (1985).

19. Y. P. Chyou, "Modelling of a Convection-Stabilized Arc with Particle Injection" (M.S. thesis, University of Minnesota, March 1984).

20. A. Mazza, "Studies of an Arc Plasma Reactor for Thermal Plasma Synthesis" (Ph.D. thesis, University of Minnesota, 1983).

21. Y. C. Lee, "Modeling Work in Thermal Plasma Processing" (Ph.D. thesis, University of Minnesota, 1984).

22. D. J. Varacalle, Jr. et al., "Modeling Particulate Injection in Thermal Plasma Material Processing Experiments" (Paper presented at 1986 Metallurgical Society Extractive and Process Metallurgy Fall Meeting, Colorado, Springs, Colorado, November 1986).

23. D. J. Varacalle, Jr. et al., "Modeling Turbulence in a Thermal Plasma Reactor" (Paper presented 1987 Spring Meeting of the Materials Research Society, 21-25 April 1987).

24. M. I. Boulos and W. H. Gauvin, C. J. Ch. E., vol. 52 (1974), 355.

25. D. Bhattacharya and W. H. Gauvin, AIChEJ, vol. 21 (1975), 879.

26. B. Gal-Or, J. of Eng. for Power, vol. 102 (1980), 589.

27. J. K. Fiszdon, Int. Heat & Mass Transfer, vol. 22 (1979), 749.

28. M. Vardelle et al., AiChEJ, vol. 29 (1983), 236.

29. D. Wei et al., ISPC-6, vol. 1 (1983), 83.

30. E. Pfender and Y. C. Lee, "Particle Dynamics and Particle Heat and Mass Transfer in Thermal Plasmas. Part I. The Motion of a Single Particle without Thermal Effects," Plasma Chemistry and Plasma Processing, 5(3) (September 1985), 211-237.

31. Y. C. Lee, Y. P. Chyou, and E. Pfender, "Particle Dynamics and Particle Heat and Mass Transfer in Thermal Plasmas. Part II. Particle Heat and Mass Transfer in Thermal Plasmas," Plasma Chemistry and Plasma Processing, 5(4) (December 1985), 391-414.

32. Y. C. Lee, "Trajectories and Heating of Particles Injected into a Thermal Plasma" (Masters thesis, Department of Mechanical Engineering, University of Minnesota, 1982).

33. Acurex Corporation, "Users Manual Aerotherm Chemical Equilibrium Computer Program (ACE81)" (Acurex Corporation, Aerotherm Division, Mountain View, California, August 1981).

34. R. M. Kendall, "An Analysis of the Coupled Chemically Reacting Boundary Layer and Charring Ablator, Part V. A General Approach to the Thermochemical Solution of Mixed Equilibrium-Nonequilibrium, Homogeneous or Heterogeneous Systems" (Final Report No. 66-7, Part V, NASA Contract NASA9-4599, Aerotherm Corporation, Palo Alto, California, 14 March 1967).

35. D. A. Bittker and V. J. Scullin, "GCKP84-General Chemical Kinetics Code for Gas Phase Flow and Batch Processes Including Heat Transfer Effects," NASA Technical Paper 2320 (1984).

36. F. J. Zeleznik and J. B. McBride, "Modeling the Internal Combustion Engine (NASA RP-1094, 1984).

37. C. W. Gear, Numerical Initial Value Problems in Ordinary Differential Equations (Prentice Hall, 1971).

38. "Differential Equations: Method Formulations, Stability, and the Methods of Nordsieck and Gear" (UCRL-51186-Rev. 1, Lawrence Livermore Lab., 1972).

39. B. J. McBride and S. Gordon, "FORTRAN IV Program for Calculation of Thermodynamic Data" (NASA TN D-4097, 1967.)

40. D. R. Stull and H. Prophet, "JANAF Thermochemical Tables," Second ed., Report NSRDS-NBS-37, National Bureau of Standards, 1971.

41. O. Levenspiel, Chemical Reaction Engineering (Wiley 1962).

MATHEMATICAL MODELS FOR THE STUDY OF

CARBOTHERMIC REDUCTION OF ALUMINA

R. M. Kibby & A. F. Saavedra

Manufacturing Technology Laboratory
Reynolds Metals Company
P. O. Box 1200
Sheffield, Alabama   35660

Abstract

Two mathematical models, which have been developed to facilitate studies in carbothermic reduction of alumina, are described and discussed. One of these models is a closed material and energy balance of the process as expressed in eight stages. The other simulates the behavior of the hearth of a primary reduction furnace. Both models employ the same set of thermodynamic properties and simple solution models for the slag and metal phases. Calculations of the compositions of slag, metal and gas phases, where they exist in each stage are based on the assumption that certain chemical reactions achieve and dominate the equilibrium in each stage.

Introduction

In an era when the art of mathematical modeling appears to have become a matter of articulating software packages and even data sets previously prepared by others, the models to be presented may be the last of a kind. They were scratch-built to meet an immediate need with very little computing power at hand.

The field of technology involved is carbothermic reduction of alumina to produce commercially pure aluminum. For the past century, aluminum has been produced commercially by the Hall-Heroult electrolytic process. Over this period, major efforts have been made by many producers to replace this electrolytic process with a carbothermic process. The underlying reason for this abiding interest lies in the inherent advantage of thermal reduction over electrolytic: one large thermal furnace would be expected to produce 50 times as much as a modern electrolytic cell.

Silicon has been produced for many years by a carbothermic process. There are many similarities between the silicon process and hypothetical carbothermic processes for producing aluminum. But there are two important differences: in contrast to the silicon process, (a) aluminum is produced carbothermically at a temperature where the solubility of carbide in the metal is about 35%, and (b) liquids are formed in the charge column or burden of an aluminum furnace as this bed of materials descends to the reaction zone. A successful carbothermic process to produce aluminum will undoubtedly have method and apparatus to address these differences.

The first of the models to be presented was designed to answer the question: If certain chemical reactions could be brought to thermodynamic equilibrium, what would the process energy demand be for various process configurations? This model is referred to as the Closed Material and Energy Balance Model. The second model to be presented uses the same thermodynamic property set and assumptions as the first but is limited to forecasting the state of the hearth system in experimental and commercial furnaces as a function of the material charge and power delivery history of a furnace run. This model is called The Control Model.

The foundation of both models is a thermochemical model comprising: a set of thermodynamic properties for the chemical compounds involved, a slag solution model, and routines for calculating the thermodynamic activities of the species in solution in the furnace slag.

In a previous work (1), the authors described the results of a program named MODEL3 used to predict the carbon partition between carbothermically reduced Al and $Al_2O_3-Al_4C_3$ melts. MODEL3 functions were used in the Closed Material and Energy Balance Model.

The Thermochemical Model

Where available, the thermochemical properties of the compounds involved were taken from the JANAF Tables, Second Edition (2). An exception to this rule was the properties of solid Al_4C_3, which along with the properties for solid Al_4O_4C were taken from an assessment by Belton (3). The best fit we found for the thermodynamic properties selected and the liquidus lines reported in the literature (4) was with the assumption that the metal phase behaves as an ideal solution and the slag phase behaves as a regular solution in the form:

$$\ln \left(\frac{a_i}{n}\right) = \frac{\alpha}{RT} (1 - n)^2$$

where:

a_i = the activity of the constituents in the slag $Al_2O_3(l)$, $Al_4C_3(l)$, $Al_4O_4C(l)$.

n = the mole fraction of the constituent i in the slag.

α = a solution model parameter. The best fit was obtained with a value of -6150 cal/mol².

These properties were supplemented by those in Table I derived by the authors for the best fit to the phase equilibria reported by Gjerstad (4).

Table I
Thermodynamic Properties Other Than
JANAF Second Edition

°K	Al_2O	C (l)	Al_4O_4C (s)	Al_4O_4C (l)	Al_4C_3 (s)	Al_4C_3 (l)
Free Energy Functions, $-(F°-H°_{298})/T$, cal/deg						
1900	75.41	11.98	69.03	89.30	60.45	79.55
2100	76.48	12.40	73.54	93.81	63.93	83.03
2300	77.48	12.80	77.77	98.04	67.23	86.33
2500	78.49	13.17	81.76	102.03	70.37	89.47
Heat of Formation, $\Delta H°_f$, kcal/mole						
298	-32.4	+26.8	-557.8	-503.8	-49.5	+7.5
Enthalpy Above 298°K, $(H° -H°_{298})$, kcal/mole						
1900	21.15		85.18	85.18	67.98	67.98
2100	23.91		97.36	97.36	78.13	78.13
2300	26.67		109.76	109.76	88.56	88.56
2500	29.43		122.36	122.36	99.28	99.28

Figure 1 expresses the thermochemical model for the pseudobinary Al_2O_3-Al_4C_3 system showing equilibrium lines calculated from the thermodynamic property set and slag solution model adopted.

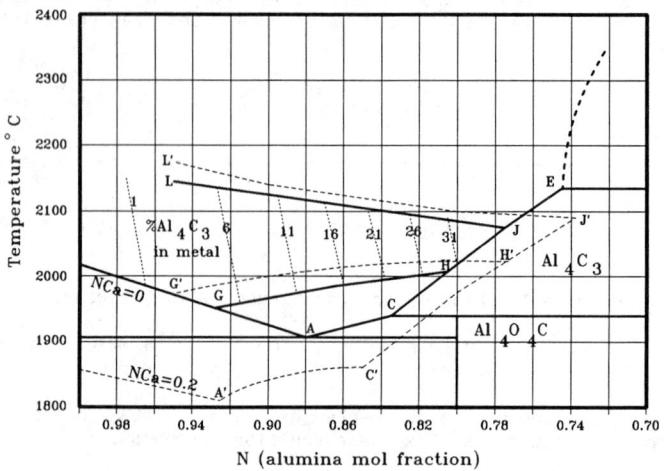

Figure 1 - Al_2O_3-Al_4C_3 Pseudo-binary phase
and reaction equilibrium diagram at 1 atm.

In Figure 1, N is the equivalent mole fraction of alumina in an
Al_2O_3-Al_4C_3 mixture, defined as:

$$N = \frac{\text{moles of } Al_2O_3}{\text{moles } Al_2O_3 + \text{moles } Al_4C_3}, \text{ and}$$

NCA is the equivalent mole fraction of CaO in an Al_2O_3-Al_4C_3-CaO system.

$$NCA = \frac{\text{moles CaO}}{\text{moles CaO} + \text{moles } Al_2O_3 + \text{moles } Al_4C_3}$$

In the above definitions for N and NCA, the moles of Al_2O_3 and moles of
Al_4C_3 include those derivable from Al_4O_4C according to the reaction:

$$3Al_4O_4C \longleftrightarrow 4Al_2O_3 + Al_4C_3$$

The liquidus line A-C in the pseudobinary diagram expresses the
temperature in degrees C at which a liquid having composition N is in
equilibrium with solid Al_4O_4C.

The liquidus line C-E expresses the temperature at which liquid having
composition N is in equilibrium with solid Al_4C_3.

The line G-H expresses the temperature at which liquid of composition N
reacts with solid carbon to produce Al_4C_3 in solution and Al_4O_4C in solution
at 1 atmosphere CO pressure.

The liquid thus far cited is called the "slag" herein.

630

Table II.
Relevant Reactions in Carbothermic
Reduction of Alumina

$2Al_2O_3(s) + 3C \leftrightarrow Al_4O_4C(s) + 2CO$ R1

$Al_2O_3(slag) + Al_4C_3(s) \leftrightarrow 6Al(l) + 3CO$ R2

$Al_2O_3(slag) + Al_4C_3(slag) \leftrightarrow 6Al(l) + 3CO$ R3

$2Al_2O_3(slag) + 9C \leftrightarrow Al_4C_3(s) + 6CO$ R4

$Al(l) \leftrightarrow Al(g)$ R5

$Al(g) + Al_2O_3(slag) \leftrightarrow 3Al_2O$ R6

$Al_2O_3(l) + 2C \leftrightarrow Al_2O + 2CO$ R7

$Al_4O_4C(s) + C \leftrightarrow 2Al_2O + 2CO$ R8

$Al_2O_3(slag) + Al_4C_3(s) \leftrightarrow 3Al_2O + 3C$ R9

$4Al(g) + 3C \leftrightarrow Al_4C_3(s)$ R10

Stages 0 to 3 take place in the upper regions of the furnace (charge column) and are concurrent with stages 4 to 7, which take place in the hearth.

Stage 0 is the fume collection stage, where CO and Al_2O leaving the reduction furnace are burned with air and the resulting alumina is separated and returned to the charge column of the reduction furnace.

Stage 1 is a charge preheating stage. It is assumed that aluminum vapor and recycled aluminum entering this stage are converted to Al_4C_3.

Stage 2 is the first "pre-reduction" stage in which solid alumina and carbon react to produce Al_4O_4C, which then converts to equilibrium proportions of slag and solid Al_4C_3. Al_2O gas entering stage 2 reacts according to R8 forming solid Al_4O_4C. Aluminum vapor entering stage 2 reacts with carbon to form Al_4C_3 according to R10.

Stage 3 is the second "pre-reduction" stage in which all remaining Al_2O_3 or carbon (remains) reacts to form solid Al_4C_3. Al_2O gas entering stage 3 reacts according to R8. Aluminum vapor entering stage 3 condenses to liquid aluminum and reports to stage 4.

Stage 4 is primarily a mixing stage where alumina is added or slag is recycled from subsequent stages to adjust the slag composition for the metal producing steps. When unreacted carbon remains after stage 3, reaction R4 occurs in this stage.

Stage 5 carries out reaction R2 until solid Al_4C_3 disappears from the slag. Vaporization losses in this stage are by reactions R5 and R6.

Stage 6 continues the production of metal by slag decomposition reaction R3 until N becomes the final value set for this stage. Vaporization losses in this stage are by reactions R5 and R6. In changing from the slag composition at the end of stage 5 to the higher values of stage 6, the metal made in stage 5 is decarbonized according to the "reduction" decarbonizing mode defined in U.S. Patent No. 4,216,010 (5).

Stage 7 is another decarbonizing stage in which, to the extent that alumina is consumed in the stage, decarbonization is achieved in the "extraction" mode as defined in the same U.S. patent.

Stage 8 is a conventional holding furnace operation in which the decarbonization is accomplished by gas fluxing of the molten product of the previous stage. Tri-Gas is used for fluxing, which yields two fractions: one is the product metal and the other is a dross containing aluminum, aluminum carbide and slag. The dross is recycled to Stage 1 of the process.

The closed balance program computes a mass balance in the elements Al, C, O and the compound CaO for the process stages outlined above. It then computes an energy balance for each stage and an overall process energy demand. As presently written, the program adds up the positive enthalpy changes (heat absorbed) for the stages and disregards the negative enthalpy changes (heat released) to compute the overall process heat demand. Gases produced in certain stages are recycled to previous stages where they back react to equilibrium, releasing heat for pre-heat and pre-reduction reactions.

The program computes the equivalent mole fraction of alumina N and the mole fraction NCA and weight percent of CaO (PCA) in the slag of stages 4, 5, 6 and 7. It also computes the liquids/solids ratio in the charge column, stages 2 and 3.

The following input data must be specified:

1. The operating temperature of each stage,

2. The liquidus temperature of stages 4, 5, 6 and 7,

3. The percent aluminum carbide in the metal product of stages 6 and 7,

4. The weight and composition of slag inventory at the start of stage 4,

5. The weight percent of "inerts" (i.e., Al_2O_3) in the metal product of stage 6,

6. The weight of alumina charged to stages 1, 4, 5, 6, and 7,

7. The carbon charged to stages 2, 4, 5, and 6, and

8. The metal recovery in stages 7 and 8.

Items 1 and 2 are derivable from Figure 1 or, to the initiated, from the computer subroutines yielding the data for Figure 1. Item 3 may be (and is) set on the basis of experimental evidence. Item 4 is set at the process designer's option. Items 5 and 8 are based on experimental evidence. Items 6 and 7 are set at the process designer's option.

Line J-L expresses the temperature at which slag of composition N decomposes to produce liquid aluminum and CO at 1 atm. pressure.

Corresponding lines are drawn for selected values of NCA. The lines corresponding to points A', C', G', H', J' and L' in Figure 1 were calculated for NCA value of 0.2. In this model the assumption is made that calcium oxide behaves as an inert diluent with respect to its effect upon the activities of alumina, aluminum tetraoxycarbide, and aluminum carbide in the slag.

The Closed Material and Energy Balance Model

The block flow diagram in Figure 2 shows the eight stages involved in the carbothermic reduction of alumina as practiced in the RMC process. Various process schemes can be represented by deleting stages, deleting recycle streams and specifying slag conditions.

This model is based on the assumption that certain selected chemical reactions go to thermodynamic equilibrium and determine the compositions of exit streams from each stage. The relevant reactions are listed in Table II.

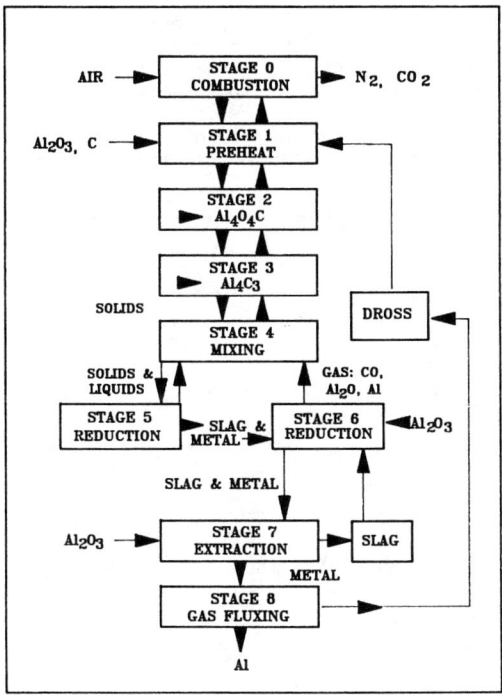

Figure 2 - Flow diagram of the RMC
Carbothermic Reduction Process

This program achieves closure of the material balance by first computing the overall results of CO produced by the amount of carbon charged to produce a stated amount (i.e. 100 lb) of final aluminum product. CO is the mass carrier in every stage, which permits the calculation of the mass of other gases, such as Al_2O and aluminum vapor leaving a process stage. The RUN program of this model calculates down from stage 0 to stage 4 and up from stage 8 to stage 4. The criterion of balance is that the elemental mass input to stage 4 equals the elemental mass output from stage 4. The RUN program provides the facility to repeat successive iterations at the users option until the criterion of closure is achieved

Obviously, not all selected sets of conditions choose to close the balance at stage 4. For example, closure is not achieved when the amount of alumina charged to stage 7 is not sufficient to decarbonize the product of stage 6 to the target selected by the user. Other similar situations abound. The user is often skirting limits imposed by the real world situations.

A typical output print for the closed material and energy balance is shown on Table III. A subordinate output print covering stage 3 of this case is shown on Table IV.

Many hypothetical conditions have been subjected to these balance calculations. Some of the results were reported in a previous work by the authors (6). It presented conclusions drawn from the closed material and energy balance model with respect to the effect of alumina charge strategies on L/S ratios in the charge column, the effect of conversion in the charge to Al_4C_3, and the ability of the charge column to recover usefully the heat expended in producing Al_2O and Al vapors in stages 5 and 6 of the process.

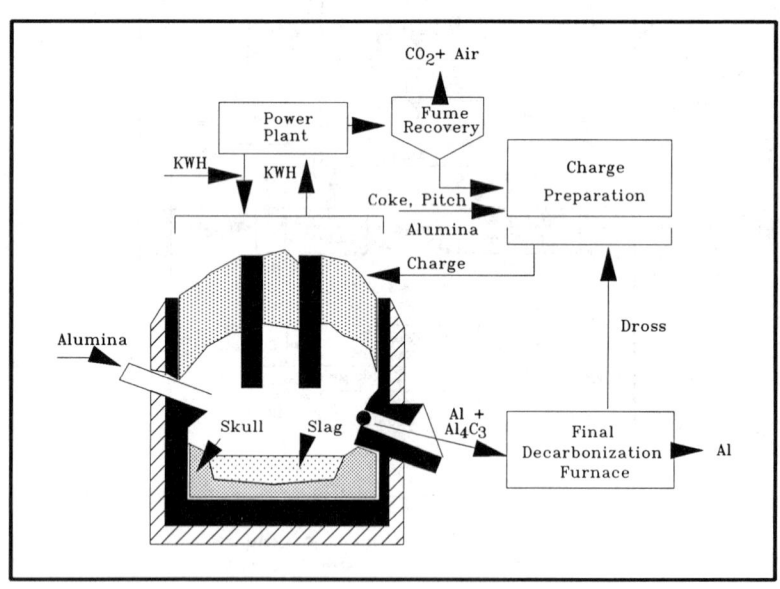

Figure 3 - Furnace Configuration for Carbothermic
Reduction of Alumina

Table III.
Summary of Material and Energy Balance

STAGE		0	1	2	3	4	5	6	7	8
CHARGE:	Al_2O_3	.0	.0			.0	.0	3.0	185.9	
	C	.0	71.9							
RECYCL:	Al_2O_3	.0	44.0			.0		285.9		
	Al_4C_3	.0	2.6			.0		33.9		
	Al	.0	25.0			.0		22.1		
	CaO					.0		57.2		
I SLAG:	Al_2O_3					183.0			100.0	
	Al_4C_3					42.0			5.9	
	CaO					104.6			20.0	

OUTPUT SLAG/SOLID:

		0	1	2	3	4	5	6	7	8
	Al_2O_3	22.1	44.0	51.1	.0	133.8	105.3	283.0	385.9	
	C	.0	62.2	55.6	18.9	.0				
	Al_4C_3	.0	41.3	62.4	121.5	165.2	104.2	65.0	39.8	
	CaO	.0	.0	.0	.0	104.6	104.6	161.8	77.2	
	Al								22.1	

OUTPUT METAL:

		0	1	2	3	4	5	6	7	8
	Al_4C_3					16.1	42.7	13.5	2.6	
	Al_2O_3						.0	21.9	21.9	
	Al				34.7	34.7	65.8	147.1	125.0	100.0

OUTPUT GAS:

		0	1	2	3	4	5	6	7	8
	CO_2	263.5								
	CO	.0	167.7	167.7	164.6	113.6	20.0	59.7		
	Al_2O	.0	15.2	15.2	37.5	59.9	8.5	34.9		
	Al	.0	.0	4.1	6.4	41.1	3.1	38.0		
N					.000	.534	.588	.860	.960	
NCA						.525	.515	.472	.252	
L/S				38/62	0/100	63/37				
% CaO in Slag				0	0	36	33	32	15	
Temp °C		250	235	1970	2010	2133	2133	2300	1795	1000
LIQ Temp. °C						2118	2118	1795	1795	
Kwhr/lb PROD		⁻2.602	.007	⁻.138	.161	.700	.683	2.085	.557	⁻.287

Process Heat Demand: 4.195 kwhr/lb Al
Slag Tapped From Stage 6:   180

635

Table IV.
Detail of Stage 3 Material and Energy Balance

Stage 3 Temp. C: 2010   Second Pre-reduction

	Al	C	O	Total	Kwhr
Input Solid					
Al_4C_3	37.27	12.42	0.00	49.70	6.56
C	0.00	55.57	0.00	55.57	24.12
Input Liquid					
Al_2O_3	27.04	0.00	24.03	51.07	-82.89
Al_4C_3	9.54	3.18	0.00	12.72	4.34
Input Gas					
CO	0.00	48.67	64.90	113.57	-22.23
Al_2O_3	46.23	0.00	13.70	59.92	-2.24
Al(g)	41.10	0.00	0.00	41.10	70.57
Total Input	161.18	119.85	102.63	383.65	-1.78
Output Solid					
Al_4C_3	91.11	30.37	0.00	121.48	16.97
C	0.00	18.94	0.00	18.94	8.42
Output Liquid					
Al	34.72	0.00	0.00	34.72	11.62
Al_2O_3	0.00	0.00	0.00	0.00	0.00
Al_3C_3	0.00	0.00	0.00	0.00	0.00
Output Gas					
CO	0.00	70.53	94.05	164.58	-32.21
Al_2O_3	28.96	0.00	8.58	37.54	-1.41
Al(g)	6.38	0.00	0.00	6.38	10.96
Total Output	161.17	119.85	102.63	383.65	14.35

N3: .000   L/S: 0/100   kwhr/lb heat demand: .161

The Hearth Behavior Simulation Model

Several carbothermic furnace configurations have been described in the patent literature (7). The one providing what appears to be the best control over process conditions involves two furnaces for the carbothermic process: a "primary" furnace for stages 2 through 6 of the flow diagram of Figure 2 and a "decarbonization" furnace for stage 7 of the flow diagram. Figure 3 shows a typical furnace configuration.

The charge column of the primary furnace is where stages 1, 2 and 3 of the flow diagram are conducted. The hearth system of the primary furnace is where stages 4, 5 and 6 are conducted.

A requirement for the control of the hearth system is the capacity to predict the effects of proposed control actions upon future states of the system. For this treatment, the state of the hearth system is defined in terms of the quantities, composition and certain dimensions of the liquid and solid zones within the hearth.

The state of the system is computed at the end of every five-minute period of a production cycle based on the state at the beginning of the period and the charging, tapping and power delivery activities during the period. Values for the variables identifying and defining the state, and other variables important in the calculations are catenated into a vector SV

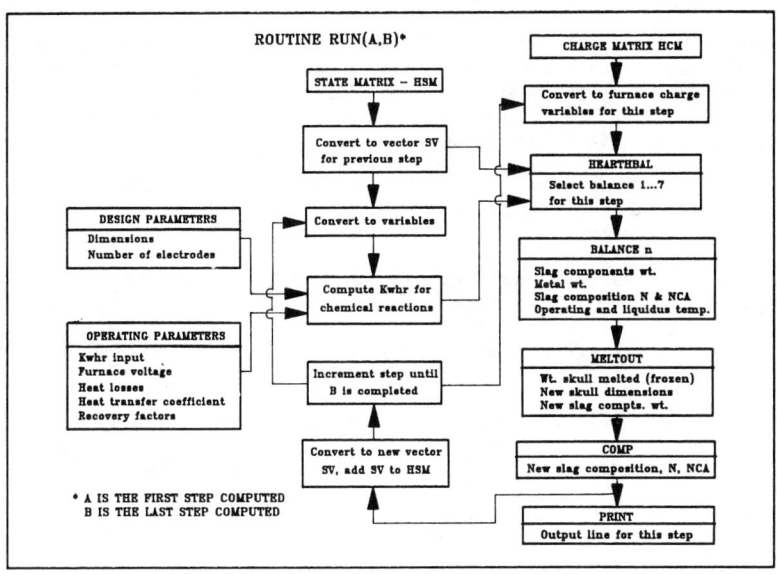

Figure 4 - The hearth behavior simulation
model RUN program scheme

637

of 76 elements. A hearth state matrix HSM is generated comprising a vector SV for each step that has been completed. A charge matrix HCM is generated from input data. This provides information for the weights of liquid and solid materials charged to the hearth during selected steps.

A block flow diagram for the RUN routine is shown on Figure 4. The RUN routine has two explicit arguments: the first step to be computed and the last. It operates successively from the first step specified through the last. Data for kwhr furnace input and furnace voltage must be entered before RUN is invoked.

RUN starts by taking as a starting state vector the line of HSM for the step before the first step to be calculated. It converts the starting state vector to variables. RUN then determines the heat available for reaction after deducting from the furnace input the losses to roof and sidewall, cooling water and the heat transferred from slag to skull. For the step being calculated, RUN enters data from the charge matrix HCM.

Then a routine called HEARTHBAL is invoked. HEARTHBAL decides which of seven equilibrium controlled mass and energy balances will be computed for the five minute period of the step. The decision is based on the composition of the material being charged, the unreacted materials in the hearth pool and the composition of the slag in the pool at the beginning of the step. Each of these balances is based on finding how far a selected set of reactions proceeds toward completion with the reaction heat available after allowing for roof losses, cooling water losses and the heat transferred during the previous step between the slag and the frozen skull.

After HEARTHBAL a routine called MELTOUT is invoked. MELTOUT computes, the heat transferred between the slag and skull, the amount of slag melted or frozen during the step and the new skull dimensions at the end of the step. The composition of the liquid slag is adjusted to account for the skull melted or frozen during the step.

Finally, RUN creates a vector SV for the step just calculated, adds it to the matrix HSM, increases the step number by one and repeats until the final step in the argument for RUN has been calculated.

A routine called POWER is provided to treat open arc situations by dividing open arc heat into three parts:

A. A part delivered by the arc spot to the liquid metal-slag arc target area,

B. A part radiating heat to the slag for chemical reactions, and

C. A part radiating heat to the furnace walls and roof.

The part delivered by the arc spot (Part A) is assumed to be used entirely in the production of CO, Al_2O and $Al(g)$. The computer program decides, based on furnace voltage, how many electrodes are on open arc, if any.

A routine called RESET is provided to facilitate the adjustment of the hearth state matrix HSM to reflect dimensional readings, such as skull thickness, slag depth, metal depth, etc. It is also useful in setting up a new matrix HSM based on pool and skull dimensions and on slag composition expressed as N and NCA. A routine called TAP is provided to make HSM adjustments on the basis of the weight of metal and slag tapped.

The existence of the state matrix HSM and the charge matrix HCM facilitates the use of this model to forecast the consequences of a proposed control action. If the program is on-line with a furnace operation, it can be taken off line, proposed changes can be entered and the state at some time in the future of the production cycle can be compared with what it would have been without the change.

Not all of the information required to forecast future states of the hearth system is retrievable from the state matrix HSM and the charge matrix HCM. The power input to the furnace and the furnace voltage during a step are specified independently and remain the same for all subsequent steps until respecified. There is also a set of design and operating parameters which are specified independently and remain the same until respecified.

The model is "calibrated" for a particular situation by finding values for the design and operating parameters, which cause the model to account for certain results of furnace operations. The results for which accountability is sought are skull thickness, slag depth, slag composition, metal weight and carbon content of the metal. This procedure establishes the heat transfer coefficients of side and bottom linings and of the slag-skull interfaces at the sides and bottom of the hearth zone.

Table V.
Calibration Parameters

Campaign	Symbol	A	B	C
Bottom Insulating Thickness – in	BID	5	5	24
Bottom Lining Thickness – in	BLD	6	6	48
Side Lining Thickness – in	SCD	9	9	24
Center of Rotation Spacing – in for Endwall Radius	CCD	16	16	96
Radius – Endwall – in	EWR	15	15	96
Electrode Diameter – in	ELD	8	8	40
Insulation – Sides – Btu/hr sq ft°F	HIN	.11	.11	.11
Insulation – Bottom "	HIB	.21	.21	.21
Slag-Skull Interface, " Side	HSN	9	9	9
Slag-Skull Interface, " Bottom	HSB	7	7	7
Heat Loss, Baseline				
Submerged Arc – kwhr/5 Min Step	KWHATER	1.8	1.8	45
Submergeg Arc – kwhr/5 Min Step	KWHSIDE	3.1	3.1	42.2
Submerged Arc – kwhr/5 Min Step	KWHROOF	0.4	0.9	21.1
Decimal Part of Al Converted to – Al_4C_3 When Carbon is First Charged	ALCF	.1	.1	.1
Decimal Part of Aluminum Vapor – Returned as Condensate	ALVCF	.8	.8	.8

Table VI.
Comparison of Results

Case	Closed Balance	Small Furnace	Large Furnace Weights x 100			
	1	2	3	4	5	6
Open Arc	No	No	No	No	No	Yes
Initial Slag - lb	330	295	301	301	301	301
N Initial Slag	.86	.9	.9	.9	.9	.9
Carbon Rich Charge:						
o Weight	140	70	150	150	120	120
o Composition	A	B	B	B	A	A
Alumina Rich Charge:						
o Weight	402	120	360	383	383	383
o Composition	C	D	D	C	C	C
Primary Furnace Reaction Heat Demand						
o kwhr/lb Al in PFP	2.36	9.6	9.6	3.5	2.3	2.6
Primary Furnace Total Heat Demand						
o kwhr/lb Al in PFP	-	21.7	10.8	4.0	2.8	3.9

* * * * * * * * * * * * * * * * *

		CHARGES - WT %			
		A	B	C	D
Carbon	(s)	13.4	47.4	0	0
Alumina	(l)		30.0	71.1	0
Alumina	(s)		-	.7	100.0
Al_4C_3	(l)		19.0	8.4	
Al_4C_3	(s)	86.6	3.1	0	
CaO	(l)		0	14.2	
Al	(l)		0.4	5.5	

Table V shows in columns A and B "calibration" values derived from two experimental runs in a top entry twin electrode experimental furnace rated at 100 kw. Column C shows the corresponding values selected to represent a commercial scale furnace rated at 10,000 kw.

The hearth behavior model has been applied to estimate the performance of a commercial scale furnace. A comparison of calculated results is shown in Table VI. It is seen that, when the charge strategy for the large furnace (case 3) is the same as for the small furnace (case 2), the large furnace has the same primary furnace reaction heat demand as calculated for the small furnace but the large furnace has a lower overall power consumption. When the charge strategy is the same as for the closed balance of case 1, which means that solid Al_4C_3 produced in the column is charged to the hearth, then case 5 results show the same process energy demand as the closed balance of case 1. Under a 450 volt open arc, and the same charge strategy as in the closed balance, case 6 shows an increase in reaction heat demand from 2.3 to 2.6 kwhr/lb Al in the primary furnace product (PFP) and an increase in total heat demand, including roof and sidewall losses, of from 2.8 to 3.9 kwhr/lb Al in PFP. According to the closed balance, it takes 147 lb of PFP to yield 100 lb of final product.

Discussion

The basic assumption of both models described in this work is that selected chemical reactions go to thermodynamic equilibrium and determine the composition of the products of a process stage. It is recognized that some of the possible reactions may not dominate, even though thermodynamically favored, because of mass transport impediments. For example, in the charge column, reactions R8 and R9 are very close together in their driving forces, but R9 involves a reaction between gaseous components while reaction R8 requires access to carbon. In view of the liquid forming tendencies in the charge column, the availability of carbon to reaction R9 is not promising. The judgement call made for this work was that reaction R8 determines the composition to the products of stages 2 and 3. It is hoped that the importance of carbothermic reduction of alumina will bring forth the reaction rate studies needed to make better selections.

Another area deserving basic research is the determination of the thermodynamic properties of liquid Al_4O_4C and Al_4C_3, and appropriate solution models for the slag and metal phases. In the present work, the necessary values were selected to give the best fit to published phase diagrams. These mathematical models were most helpful in verifying and modifying operation of successful experimental runs for the production of carbothermic aluminum.

References

1. L. M. Ruch, A. F. Saavedra, and R. M. Kibby, "Carbon Partition Between Carbothermically Reduced Al and Al_2O_3-Al_4O_3 Melts," Light Metals 1984, ed. J. P. McGeer (Warrendale, PA: The Metallurgical Society, 1984), 589-599.

2. D. R. Stull and H. Prophet, eds., JANAF Thermochemical Tables, (U. S. Department of Commerce, National Bureau of Standards, 1971).

3. G. R. Belton, private communication with authors, University of Strathclyde, Glasgow, December 1974.

4. S. Gjerstad, "Kjemish-Metallurgiske undersokelser verdrorende karbotermisk reduksjon av aluminumoksyd og silisiumoksid," (Thesis, Norges tekniske hogskole, Trondheim, Norway, 1968).

5. R. M. Kibby, "Aluminum Purification System," U. S. Patent 4,216,010, August 5, 1980.

6. R. M. Kibby and A. F. Saavedra, "Model Studies in Carbothermic Reduction of Alumina," Light Metals 1987, ed. R. D. Zabreznik (Warrendale, PA: The Metallurgical Society, 1987), 263-268.

7. R. M. Kibby, "Minimum-Energy Process for Carbothermic Reduction of Alumina," U. S. Patent 4,388,107, June 14, 1983.

HYDRODYNAMIC MODELING OF THE P-155 HALL CELL

Walter E. Wahnsiedler
Aluminum Company of America
Alcoa Laboratories
New Kensington, PA 15068

Abstract

A hydrodynamic model of the two-fluid system in the Hall cell was constructed. This model is fully three-dimensional and includes calculation of the metal/bath interface contour from the pressure field, calculation of the cell current distribution from the metal/bath interface shape, calculation of the Lorentz force from the calculated current distribution and input magnetic fields, magnetic field redistribution due to current distribution variations, and allowance for the presence of frozen bath (ledge) in the cell. The model has been used to study the P-155 and other Hall cell designs in both steady-state and transient time dependence (the steady-state results can be viewed as approximately representing the average behavior in an operating cell). A variety of flow patterns and interface shapes are predicted for different cell designs and modes of operation, with the interface shape most dependent on the divergence of the Lorentz force. Transient simulations including magnetic field redistribution have shown considerable instability. The wave motions in operating cells may be due to interaction between the current and the variable magnetic field. The inclusion of current induced by the fluid motion in the calculation partially damps this effect.

Introduction

Previous efforts (1-4) to model the hydrodynamics of Hall cells have been hampered by such difficulties as confinement to two dimensions, coarse calculation meshes, simplifications in the calculation of current densities and/or magnetic fields, complex usage procedures, and lack of sufficiently powerful computational facilities to fully consider all the interacting phenomena. Due to calculational constraints, most previous work has been confined to steady-state flow simulations and has not been able to predict the time-dependent behavior of the fluid flow and metal/bath interface shape. Owing to the great economic impact of shorting's tendency to limit efforts to improve power efficiency by reducing anode/cathode distance, and to current efficiency losses due to shorting, it was felt that more complete modeling of the behavior of the metal/bath interface in commercial cells was justified. The availability of significantly more capable computational hardware, at reduced cost, promises the possibility of refining calculational techniques and model detail to a level previously unattainable. Simulations of time-dependent behavior of the interface are now possible in reasonable turnaround time. The approach taken in this work was to build on a previously developed model specialized for Hall cell simulation. With the additions to be described and the greater computational capability now available, a much more complete model of the hydrodynamics of Hall cells has been generated and utilized to produce detailed predictions of the steady-state and transient fluid behavior.

GA 10066

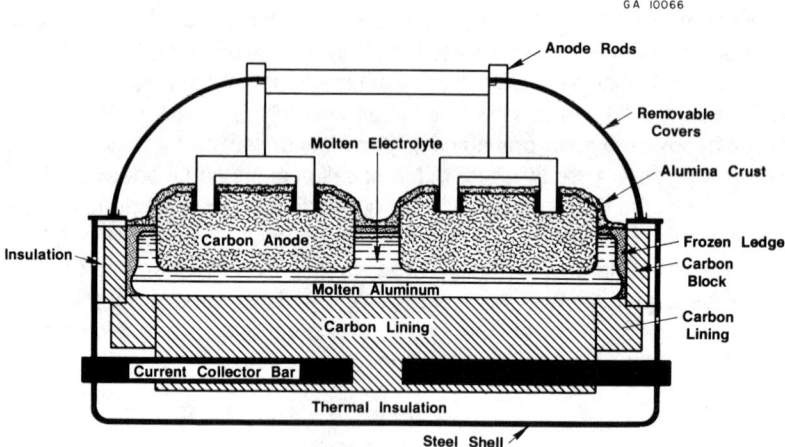

Figure 1. Cross Section of a Hall Cell

Figure 1 depicts a transverse vertical section of a typical prebake Hall cell. Two carbon anodes are shown suspended in the electrolyte from copper-aluminum anode rods. The electrolyte is a mixture of cryolite, aluminum oxide, and other additives. A total of 20-32 anodes is used in a typical cell. The anodes are covered by a "crust" of frozen cryolite and alumina. The fluids are contained in a carbonaceous lining, shaped more or less as a rectangular box, which serves as a cathodic current carrier. A pool of molten aluminum collects on the bottom of the cavity, due to its slightly higher density (the density of the electrolyte is approximately 2.1 g/cm^3, that of the aluminum approximately 2.3 g/cm^3), and this pool serves as the cell cathode. The carbonaceous lining is surrounded by refractory insulation and an iron shell, and penetrated by iron "collector rods", which serve to conduct the current out of the cell into buswork which carries the current to the next cell, which is wired in series. A "ledge" of frozen cryolite forms on the walls of the cavity and serves to protect the lining from carbiding and erosion. This ledge is essentially non-conductive. Alumina is periodically fed to the cell by pneumatic systems to compensate for its consumption.

Anodic current densities of about 1 Amp/cm^2 are passed through the cell. The magnetic field produced by the cell current, buswork currents, and adjacent cell currents interacts with the cell current to drive fluid flow. Imbalances in cell current, which may be transient in nature, can set up oscillatory behavior in the fields and flow which result in wave-like deformations of the aluminum/electrolyte (metal/bath) interface. Under some conditions, this interface can touch the undersides of the anodes, thus shorting the cell and reducing the production of aluminum. This limits the nominal distance between the anode bottoms and metal pool to a distance of at least 4 cm in most commercial cells. Since the most electrically resistive component is the electrolyte, electrical losses are almost proportional to this distance. If the wave motions of the aluminum pool can be quieted, closer approach of the anodes to the pool can be made and power savings thus attained.

Mathematics Describing the System

The fundamental physical laws applicable to this system are the Navier-Stokes equations which describe the fluid motion (underlines denote vectors):

$$\rho \, \partial \underline{u}/\partial t + \rho \, \underline{u} \cdot \underline{\nabla}\underline{u} = -\underline{\nabla}p + \langle \epsilon \rangle + \rho \, \underline{g} + \underline{L} \tag{1}$$

$$\rho \, c_p \, (\partial T/\partial t + \underline{u} \cdot \underline{\nabla}T) = \underline{\nabla} \cdot (k \, \underline{\nabla}T) + \mu \, \phi + q_s \tag{2}$$

the continuity equation which expresses conservation of mass:

$$\underline{\nabla} \cdot (\rho \, \underline{u}) = m' \tag{3}$$

Ohm's and Coulomb's Laws which govern the electric current distribution:

$$\underline{J} = \sigma \, \underline{E} \tag{4}$$

$$\underline{\nabla} \cdot \underline{E} = 0 \tag{5}$$

Ampere's Law and the magnetic permeability definition which govern the magnetic field distribution:

$$\underline{\nabla} \times \underline{H} = C_a \, \underline{J} \tag{6}$$

$$\underline{B} = \mu \, \underline{H} \tag{7}$$

and the Lorentz force definition which defines the electromagnetic force which enters the Navier-Stokes equations:

$$\underline{L} = C_L \, (\underline{J} \times \underline{B}) \tag{8}$$

The behavior of the interface can be deduced from application of the Navier-Stokes equations to the two fluids near the interface (possibly including a surface tension term), but the equation used in the present model to define the interface height:

$$p + \rho_M \, g \, h = C_i \tag{9}$$

is shown here because it is treated separately in the model. The present model neglects thermal phenomena as much smaller than electromagnetic and gas-driven phenomena (for that matter, in the results to be shown here, the effect of gas is also neglected, although the model can in principle consider it).

It has been observed experimentally, however, that the main effect of gas on the interface (at least in water models) is to distort it only locally near the anode edges (5). Thus, ignoring the gas effect in the results presented here may somewhat underestimate the potential for shorting in the system, but should not significantly alter the overall shape of the predicted interface.

The method used in the model to predict the interface shape is an "ad hoc" technique which has been found to perform well in all cases so far investigated (including a test of a "wave-in-a-tank" model, for which analytic solutions are known). The reduced pressure is set equal to a constant at all points on the interface. The interface height equation which results from a rigorous application of the Navier-Stokes equations to the fluids contains more terms, but they are small in the situations considered here.

Modeling Methodology

The fluid flow in the modeled domain is calculated using PHOENICS (6), a fluid dynamics code commercially available from Concentration, Heat and Momentum, Ltd. (CHAM). A "satellite" appendage program, named ESTER (Electrolytic SmelTER) (7), is used to specialize the general PHOENICS package to Hall cell simulation. ESTER contains an algorithm to set up appropriate geometric quantities representing the ledge and anodes, define the solution mesh, define the fluid properties, manage the solution of the current distribution and interface height equations.

The original CHAM ESTER model simulated ledge using flow blockages which were specified as a thickness from the bottom of the cell. This may be more appropriate for Soderberg cells in which frozen material collects along the bottom edges from side feeding. For Alcoa's prebake cells, simulation of the ledge as a frozen material on the cell walls was more appropriate. Furthermore, the original model could not handle partial blockage of a "domain" (the smallest geometrical unit with which PHOENICS works). Therefore, the blockage determination subroutine was completely recoded to allow the ledge to be specified as a thickness on the cell walls and to handle partial domain blockages.

The original CHAM program had two unimplemented provisions for user code to adjust the currents through each anode and to recalculate the magnetic field in response to current density changes. A set of subroutines was written to solve the DC electrical network constituted by the fluids and the anodes. The system modeled is a parallel connection from the anode bus to the cathode. At the end of each iteration in which anode current adjustment is to be done, the voltages and current densities determined by ESTER for each anode are retrieved from ESTER data storage and used to calculate the effective cell resistance seen by each anode using Ohm's law. This effective resistance varies with the interface position. Then the electrical network is solved using standard parallel connection methodology. An iteration is involved because the overvoltage on each anode is a function of the anode's current density. The anode overvoltages were computed using correlations due primarily to Haupin (8).

The magnetic fields are updated when the current distribution changes. The method used was to provide the program initially with both a magnetic field distribution and current distribution, which are assumed to be consistent with each other. Whenever a new current distribution is calculated, the differences between the initial and new current distributions are calculated and used to calculate the differences between the initial and new magnetic field. This is done using the Law of Biot and Savart (9). Symmetry considerations allow the effect of a domain on itself to be ignored. The effect of any current redistribution in the surrounding buswork is presently ignored.

These changes provided a tool very useful for simulating steady state interface behavior. The predicted interface shapes in steady state can be viewed as a prediction of how the "time averaged" interface shape should appear in an operating cell. However, since interface motion is the primary variable of interest, it was desired to be able to predict the time response of the interface to various stimuli. To test the dynamic interface model, a test problem involving a "wave-in-a-tank" was constructed. This problem involved a single fluid in a rectangular tank which initially had a perturbed surface shape (a sine function was chosen for the initial surface pertubation). The solution of this system is known analytically, and the ESTER results were compared with the analytic values. ESTER predicted the frequency of the resulting surface wave extremely accurately, but exhibited a phenomenon called "numerical damping" in which the predicted solutions exhibit more dissipation than is actually present. This test problem was used to study the numerical damping phenomenon in detail, and it was found that reduction of time step size allowed the numerical damping to be reduced to any desired level (at the expense of extra calculation time, of course). The size of the spatial mesh was found to have little effect on the numerical damping.

Some discussion of the methodology used by PHOENICS and ESTER to actually solve the equations shown above is appropriate at this point. Since the entire system is built on the capabilities of PHOENICS, which provides the ability to solve generalized transport equations (mass, heat, electrical transport), a discussion of PHOENICS will be given first, followed by a discussion of the way in which these general techniques are used by ESTER.

PHOENICS solves the Navier-Stokes equations defining the fluid flow using an algorithm derived from the SIMPLE (10) method. SIMPLE is an acronym for Semi-Implicit Method for Pressure-Linked Equations (although the current version of PHOENICS uses a fully implicit time integration scheme). The coordinate system used is a rectangular grid of "domains", with pressures stored at the domain centers and velocity components stored on the domain faces. This is depicted in Figure 2. Irregular shapes may be modeled using a combination of the rectangular grid with "porosities", specified for each domain volume and face. In addition to the continuity equation and the momentum conservation and energy conservation equations which define the Navier-Stokes equation set, PHOENICS can also simultaneously solve mass conservation equations for up to four individual components, thus allowing solute tracking and reaction to be considered. Turbulence is handled in PHOENICS with the k-epsilon (11) turbulence model, which was not activated for the results described here. Instead an artifically augmented effective viscosity model was used to account for turbulence. PHOENICS has a "grid-stretching" feature, available in one direction only, which allows the user to adaptively alter the grid dimensions to model varying simulation volume (as in an engine cylinder) or wave motions.

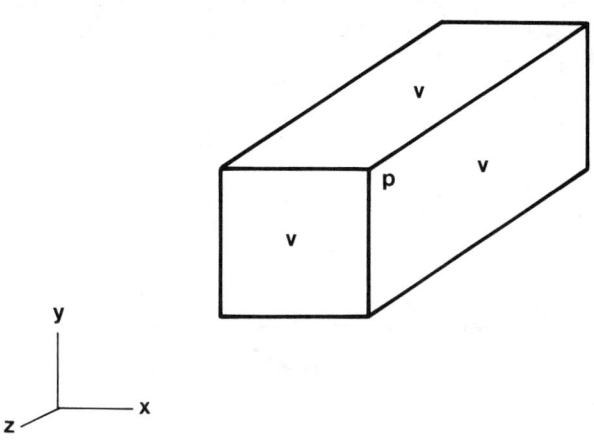

v = Velocity component storage locations

p = Pressure storage location (center of domain)

Figure 2. Field Storage Locations for a Single Domain

The heart of the PHOENICS flow algorithm is the system of pressure correction computation. At each step in the solution, a pressure field is assumed. This field is provided as an initial guess during the first calculation sweep, and is updated as the calculation proceeds. Typically, this assumed pressure field will not allow the momentum conservation and continuity equations to be simultaneously satisfied. The process of finding a solution reduces to finding a pressure field (and corresponding velocities) which satisfy all governing equations simultaneously. The method used to do this involves calculating the velocities corresponding to the assumed pressure field (this is done from the momentum conservation equations, in discretized form, directly), and then calculating a set of "pressure corrections" which will improve the level at which the continuity equation is satisfied. In the process of calculating these pressure corrections, the relationship between pressure gradient and velocity is linearized and for this reason the pressure corrections calculated do not completely solve the continuity equation in one sweep. However, as the iterative procedure progresses, the approximate pressure corrections eventually drive the system toward a stable solution if all goes well. The pressure correction equation which results from this analysis is analogous to the Poisson equation, and is sometimes called the "Poisson equation for pressure":

$$\rho \, (\partial \underline{u}/\partial p \cdot \nabla^2 p) + m' = 0 \tag{10}$$

in which the unusual convention of taking a dot product between a vector and a Laplacian is intended to imply that the corresponding coordinate direction terms should multiply each other and sum.

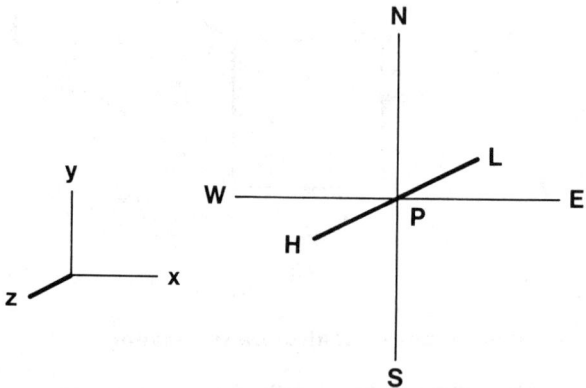

Velocities are stored between nodes, all other quantities at nodes.

Figure 3. Coordinate System Used

These equations are converted to discretized form by applying them to each domain in turn. The coordinate system used to represent the relevant values and coefficients is shown in Figure 3. The conservation equations in discretized form take the general form:

$$\phi_P = A_N{}^\phi \, \phi_N + A_S{}^\phi \, \phi_S + A_E{}^\phi \, \phi_E + A_W{}^\phi \, \phi_W + A_H{}^\phi \, \phi_H + A_L{}^\phi \, \phi_L + B^\phi \tag{11}$$

Solution of the discretized forms of the equations is carried out in PHOENICS using successive-approximation algorithms rather than direct matrix inversion. The details of the Tri-Diagonal Matrix Algorithm (TDMA) can be found in Reference 10.

In the results to be described, ESTER has been used to solve the Poisson equation for the current distribution using the pressure correction solver built into PHOENICS. Current boundary conditions on the anode bottom faces and across the cell bottom were provided. Note that this considers only resistive components of the cell voltage, and thus produces a primary current distribution only. The anode current distribution was either assumed uniform or specified using anode current adjustment as explained above, and the cathode current distribution was assumed uniform. A feature is also included in ESTER to calculate the "induced" current density $\sigma(\underline{u} \times \underline{B})$ and include it in the current distribution. This feature was not activated in the steady state simulations reported here, but was used in time-dependent simulations as explained below. Steady state simulations including the feature produced interface shapes similar to those not including it.

In the results quoted here, measured magnetic fields were specified at the normal measuring locations (on the upstream and downstream sides of the cell, between each set of anodes) and were linearly interpolated by ESTER in the horizontal plane. The magnetic fields were assumed constant in the vertical direction. For many cells, the area in which most error is created by this interpolation technique is near the ends of the cells.

The fluid-filled volume of the cell is modeled by ESTER as a single fluid with different density above and below the metal/electrolyte interface. This effectively represents the interface itself as a non-slip boundary. The PHOENICS grid-stretching provisions are used by ESTER to deform the grid in the vertical direction to follow the calculated height of the metal/electrolyte interface.

The effect of walls in the modeling domain is included using wall functions available in the main PHOENICS code. These use a standard (6) power-law representation of the logarithmic velocity in the vicinity of the wall. The Reynolds number in the vicinity of the wall is calculated using a characteristic length equal to half the thickness of the domains which border the wall.

This approach has produced a model which is believed to be a relatively realistic representation of the complex and interacting phenomena occurring in a Hall cell. The only major phenomenon not included in the actual iterative calculation is the formation and thawing of ledge. The ledge thickness was chosen as an input because ledge typically varies slowly in comparison to the other phenomena considered here and can to some extent be calculated independently of this model.

Results

The P-155 is a 24-anode cell, originally designed for a current of 155,000 amperes. This cell was used as the primary test case throughout most of the model development, and therefore the results quoted here will be for the P-155. Figure 4 depicts the horizontal mesh used in the P-155 models. This mesh consists of 25 "domains" in the longitudinal ("x") direction and 10 domains in the transverse ("y") direction. In the longitudinal direction, there is one domain per anode and one domain per anode gap. In the transverse direction, there are three domains per anode and gap and two rows of domains define the "center aisle" of the cell. A row of domains around the perimeter of the cell defines the anode-to-wall gap. Figure 5 depicts the vertical mesh used in the P-155 models. The bottom two layers of domains represent the resting metal pool, the next two layers represent the electrolyte and the top layer of domains represents the anodes and the anode gaps.

GA-19457.7

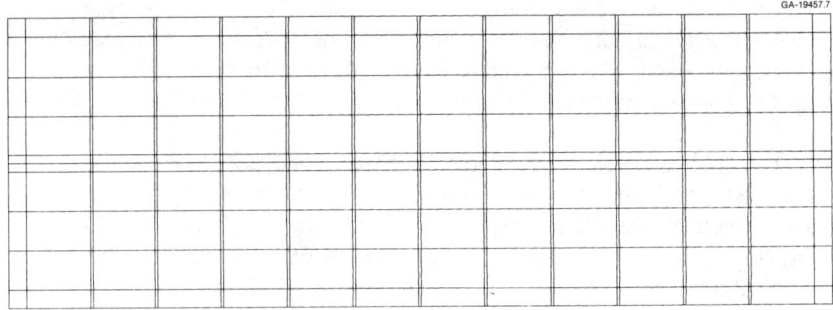

Figure 4. Horizontal Calculation Mesh

GA-19457.8

Figure 5. Vertical Calculation Mesh

Figure 6 depicts the metal circulation at a height halfway between the cell lining and the resting metal/electrolyte interface for the "base case"

P-155 in steady state. This is a model which assumes uniform anode and cathode current densities, nominal measured magnetic fields, a flat cell bottom and flat setting positions for the anodes, no gas drag, and a "tailored ledge" whose shape is intended to simulate a typical cell ledge profile. This profile is shown in Figure 7 and varies from a thickness of zero near the metal/electrolyte interface to a maximum of 0.2 m at the bottom of the metal layer. Above the metal layer, the profile has a uniform thickness of 0.1 m. Values for the rest of the simulation parameters for this base case are shown in Table 1.

Downstream

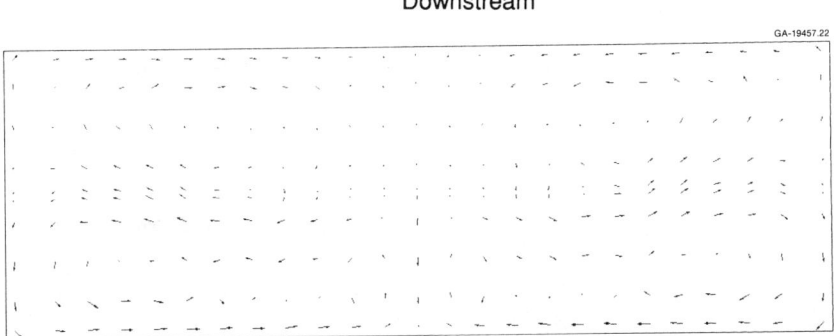

Upstream

Figure 6. Base Case Metal Circulation Pattern

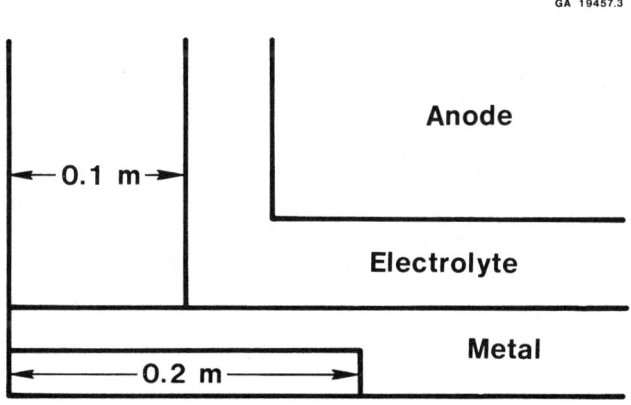

Figure 7. "Tailored Ledge" Shape

The results in Figure 6 show a typical pattern for the P-155 of four interconnected flow loops, two on the upstream side of the cell (bottom of the figure) and two on the downstream side of the cell (top of the figure). The loops meet along the longitudinal centerline of the cell. The influence of the current risers (which are located at the "quarter points" of the cell) can be discerned from the turning of the flow vectors which occurs near the quarter points on the upstream side. The maximum velocity in Figure 6 is 0.0956 m/sec. The reader should note that all results figures have been automatically scaled by the software with which they were generated, and therefore direct comparison of magnitudes between the figures is not appropriate. In all cases, the magnitude of variables plotted will be given in the text.

Table 1. Base Case P-155 Simulation Parameters

Name	Value	Units	Significance
SLOPE	1.0	degrees	Slope of anode bottoms toward center aisle
GMDOT	0.0	kg/m2/s	Gas generation rate on anode bottoms
RHO1	2250	kg/m3	Aluminum density
CRHO1	2050	kg/m3	Electrolyte density
EMU1	1.0	Pa-s	Turbulent effective viscosity
EMULAM	0.001	Pa-s	Laminar (wall) viscosity
SIGMA(14)	5000000	mho/m	Aluminum electrical conductivity
SIGMA(15)	200	mho/m	Electrolyte electrical conductivity
LITER(2)	15	-	Pressure iteration limit
LITER(14)	100	-	Electrical iteration limit
LSWEEP	200	-	Flow iteration limit
ENDIT(2)	0.001	Pa	Pressure convergence criterion
ENDIT(14)	0.000001	V	Electrical convergence criterion
DTFALS(3)	10	s	Relaxation "time step" for x velocity
DTFALS(5)	10	s	Relaxation "time step" for y velocity
DTFALS(7)	0.1	s	Relaxation "time step" for z velocity
ISPCSO(23)	1(on)	-	Flag to turn on interface height calculation
MODE	2	-	Flag to control Lorentz force calculation
NEP	5	-	Number of flow "sweeps" between electrical iterations
NIH	5	-	Number of flow "sweeps" between interface height calculations
SLOH	0.5	-	Relaxation factor for interface height
INDCUR	0(off)	-	Flag to turn on induced current calculations

Table 1. (Continued)

Name	Value	Units	Significance
NZ	5	-	Number of "domains" in vertical direction
NY	10	-	Number of "domains" in transverse direction
NANODT	2	-	Number of anodes in transverse direction
CLENGT	9	m	Cell internal cavity length
CWIDTH	3.05	m	Cell internal cavity width
ALONG	0.69	m	Anode dimension in longitudinal direction
ATRAN	1.24	m	Anode dimension in transverse direction
GAPLNG	0.2	m	Anode to wall gap in longitudinal direction (ledge included)
GAPTRS	0.2	m	Anode to wall gap in transverse direction
NANODL	12	-	Number of anodes in longitudinal direction
DPMETL	0.1	m	Quiescent aluminum (metal) depth
DPBATH	0.2	m	Quiescent electrolyte depth
DPANOD	0.15	m	Height of outside edges of anode bottoms above cavity bottom
TOTCUR	170000	Amp	Total cell current
NX	25	-	Number of "domains" in longitudinal direction

Figure 8 shows the flow pattern in the electrolyte for the base case P-155 model, halfway between the resting metal/electrolyte interface and the undersides of the anodes. There is relatively little difference between the electrolyte and metal circulation patterns, but careful study of the electrolyte pattern can discern a slight turning of every other flow vector along both the upstream and downstream sides of the cell (more discernable along the upstream side, bottom of the figure). This is due to the requirement that, in the vicinity of the anodes, the electrolyte must flow only through the anode gaps. This tendency to "channel" through the anode gaps is noticeable even in the interelectrode space, here about one inch (2 cm) below the anode bottoms. Another phenomenon which may be partly responsible for this "channeling" is the difference in current density beneath the anodes as opposed to beneath the channels between them. The maximum velocity in Figure 8 is 0.169 m/sec.

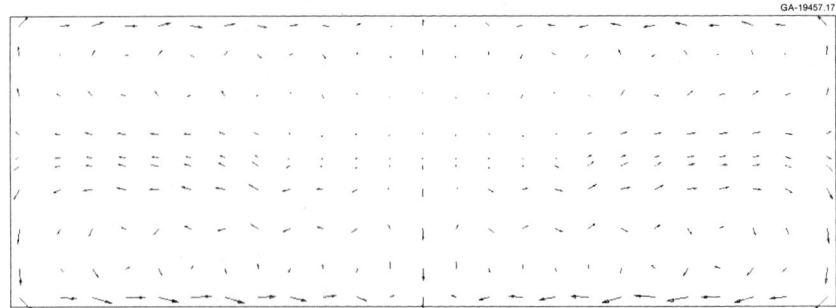

GA-19457.17

Figure 8. Base Case Bath Circulation Pattern

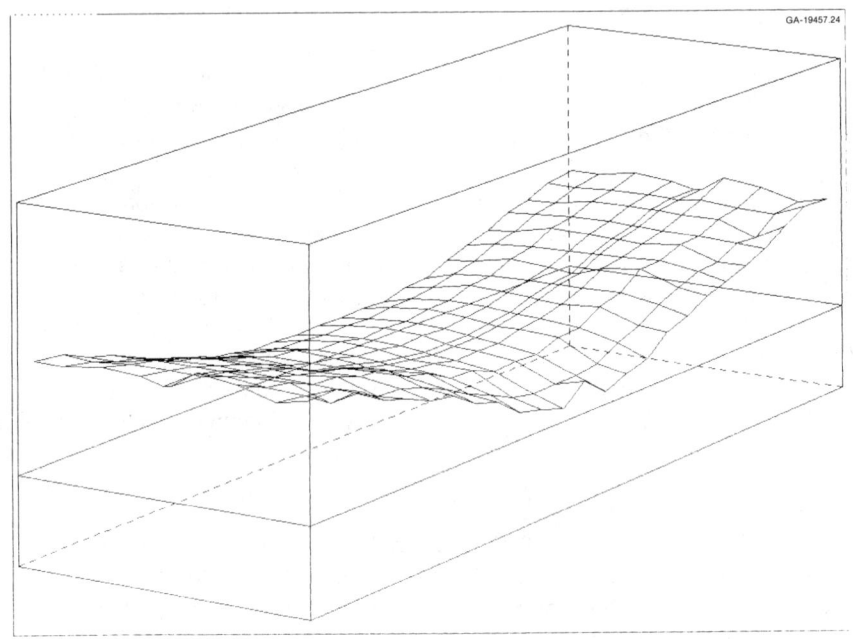

GA-19457.24

Figure 9. Base Case Interface Shape

Figure 9 depicts the predicted shape of the metal/electrolyte interface for the P-155 base case model. The interface plot is expanded greatly in the vertical direction for ease of viewing. The basically "concave" shape of the interface is due primarily to the divergent nature of the Lorentz force for this cell, which is plotted in the horizontal plane in Figure 10. Particularly near the ends of the cell, the interaction of current density and magnetic field produces a horizontal force which directs the flow toward the ends of the cell. This force is opposed in steady state by the hydrostatic head of the metal collected at the cell ends. A small ripple in the interface near the center of the upstream side of the cell (near the viewer) is due to the confluence of two flow loops there. The amount of horizontal divergence in the Lorentz force is determined primarily by the sign of the longitudinal component of the Lorentz force. This component is contributed to by the product of the vertical current and the transverse field and by the product of the transverse current and the vertical field. Of these, probably the first contribution is the larger. With the exception of the vertical current, all the components which contribute to the longitudinal Lorentz force are minor and not easily predicted (in magnitude or direction) by simple arguments. Further, the horizontal current components are largely influenced by the ledge shape. These observations point up the need to use detailed numerical models to investigate the behavior of these cells. Simple analyses often yield incorrect results, especially as regards the interface shape. The minimum metal depth in Figure 9 is 0.082 m and the maximum metal depth is 0.115 m.

GA-19457.5

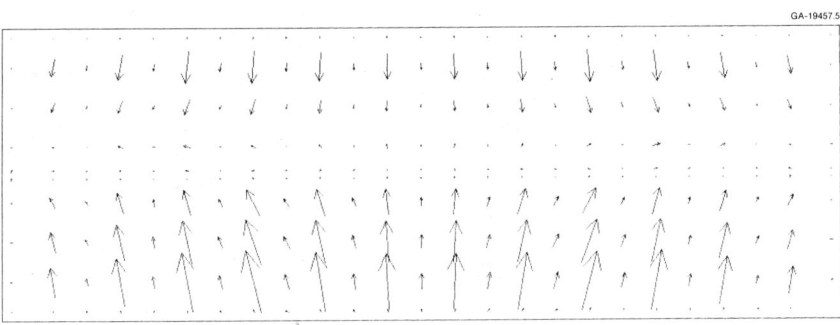

Figure 10. Base Case Lorentz Force Pattern

Figure 11 depicts the metal flow pattern in a P-155 simulation identical to the base case above except that ledge was omitted. Two flow loops are notable near the center of the upstream side of the cell which were nearly missing from Figure 6. The flow of metal across the longitudinal centerline of the cell is reversed. The maximum velocity in Figure 11 is 0.211 m/sec, more than twice as high as the maximum velocity of Figure 6. This is probably due to the combination of reduced flow resistance due to wider channels when ledge is missing and increased horizontal currents producing increased Lorentz force. It should be noted that the model in its present form does not allow current to flow to the walls of the cell (even when no ledge is present) nor does it allow current to flow from the sides of the anodes.

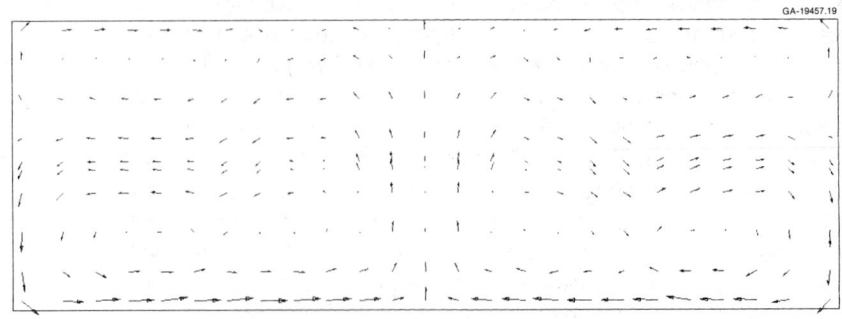

Figure 11. No Ledge Metal Flow Pattern

Figure 12 depicts the electrolyte flow for the no-ledge P-155 case. As before, the electrolyte flow is similar to the metal flow except for flows which are turned near the walls by anode channeling. The maximum velocity in Figure 12 is 0.161 m/sec.

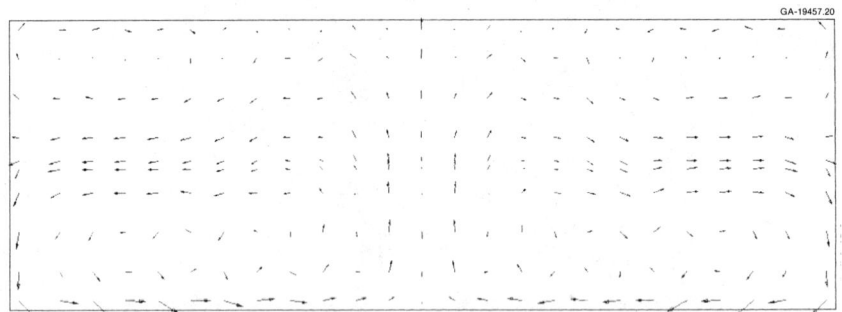

Figure 12. No Ledge Bath Flow Pattern

Figure 13 depicts the interface shape for the no-ledge P-155 case. The increased flow velocities in this case cause greater fine structure near the walls resulting from wall impingement and result in a greater peak-to-valley distance for this interface. However, the behavior of the interface away from the cell walls is similar to that for the P-155 base case. The maximum interface height in Figure 13 is 0.127 m and the minimum height is 0.078 m.

Figure 14 depicts the anode current densities in side view for a case in which a P-155 was simulated with the "tailored ledge" mentioned above and with anode current adjustment as described. The variation in current density

with anode current adjustment as described. The variation in current density brought about by the interface shape variation is small (about 10 percent). The maximum current density in Figure 14 is 9310 amp/sq m.

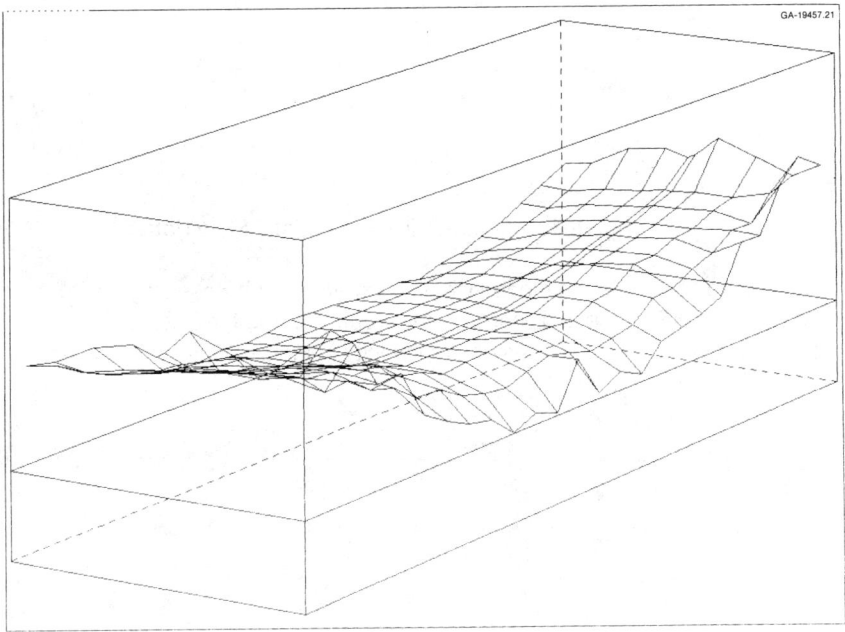

Figure 13. No Ledge Interface Shape

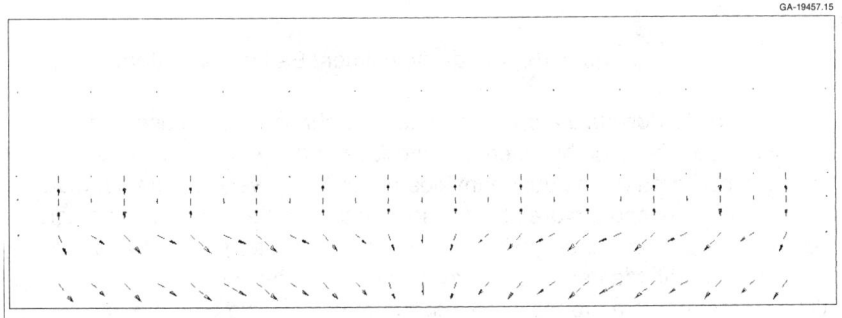

Figure 14. Anode Adjustment Current Density (Side View)

Figure 15 depicts the metal flow pattern in this case. There is little difference from the base case P-155. The maximum velocity in Figure 15 is 0.098 m/sec.

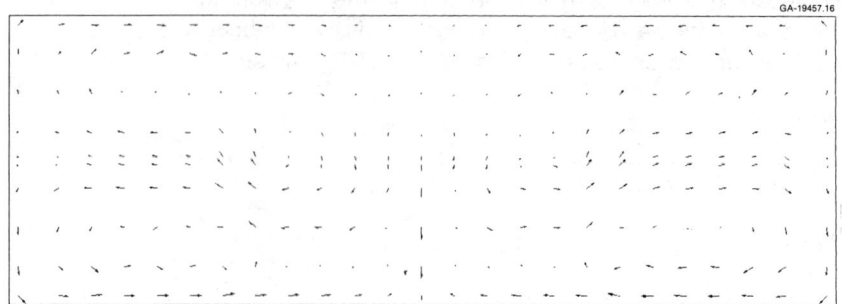

Figure 15. Anode Adjustment Metal Flow Pattern

Figure 16 shows the electrolyte flow pattern for this case. Again, relatively little difference is discernable between this and the base case electrolyte flow pattern. The maximum velocity in Figure 16 is 0.166 m/sec.

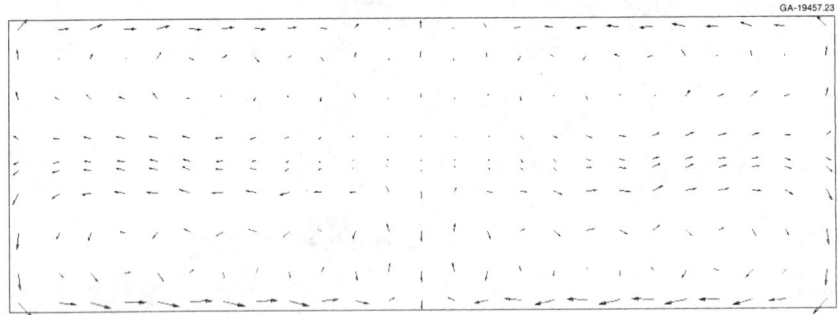

Figure 16. Anode Adjustment Bath Flow Pattern

Figure 17 depicts the predicted interface shape for this case. There is relatively little difference between this interface and that of the base case, but the center area on the upstream side (near the viewer) is more flat than in the base case, and a more abrupt rise is noted on the upstream side near the quarter points of the cell (where the risers are located). Also two small "dimples" are notable near the quarter points along the centerline of the cell. All these features are probably the result of magnetic field structure near the risers. The minimum height in Figure 17 is 0.081 m and the maximum height is 0.116 m.

Several simulations were made of the time-dependent response of the P-155 to current distribution variations. The first of these, which involved simulating the lowering of an anode (by one inch) to approximately double its current density, resulted only in a slow relaxation to a new steady state. However, these simulations were performed with time steps which were subsequently found to be much too large from the standpoint of numerical damping. It is, however, worth noting that when these simulations were performed with anode current adjustment, the final anode current was considerably less than double the current on the rest of the anodes. This was due to a tendency for the interface to draw away from the anode which was lowered.

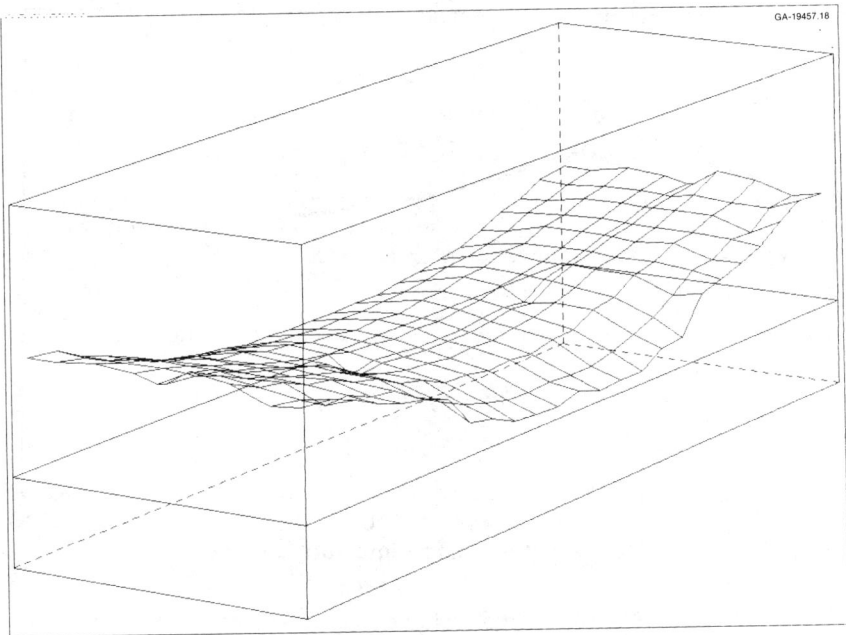

GA-19457.18

Figure 17. Anode Adjustment Interface Shape

It was then decided to simulate the most drastic upset which cells usually experience, in an effort to make the calculation produce a wave (i.e., periodic interface behavior). This was chosen to be an anode change. The simulation selected involved starting from the converged P-155 steady state and interrupting the current on one anode (near an upstream corner) at time zero. The anode was not removed from the cell. The simulation was carried out with time steps of 0.05 sec and 0.02 sec, and the results were found to be comparable (although the 0.02 sec results showed a slightly greater tendency to oscillate). Thus, 0.05 sec was identified as a small enough time step for this simulation.

The response observed was a tendency for the metal to "well up" against the wall adjacent to the location where the current interruption occurred. Figure 18 depicts the height of this interface peak from the simulation as a function of time. It can be seen that the peak reaches a maximum height in about 8.5 sec of simulated time and then gradually falls off to a steady state. When this identical simulation was run as a steady state calculation, the result was similar to the final result found here.

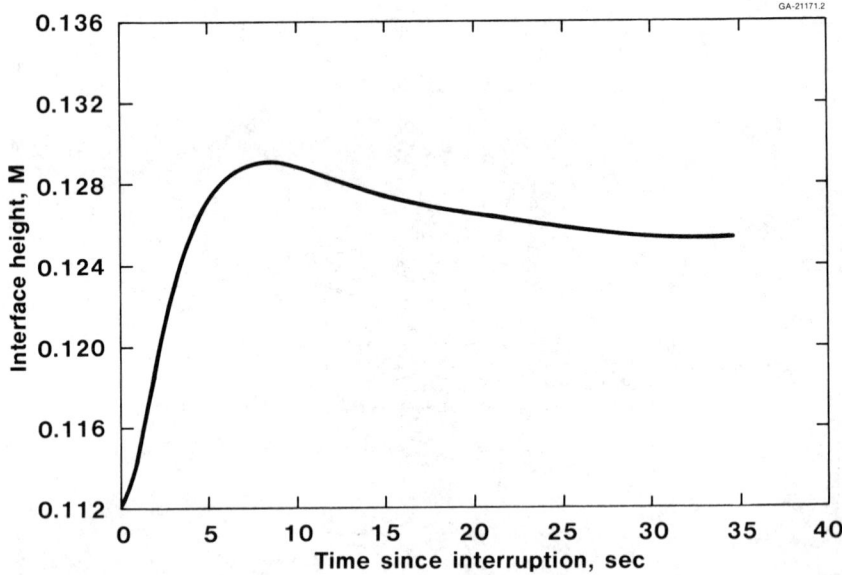

Figure 18. Interrupted Anode Interface Peak Height

Figure 19 shows the predicted metal flow pattern at time 8.5 sec (the maximum peak height). This is similar to the steady state P-155 base case pattern, but the flow loop under the anode whose current was interrupted (at the bottom left corner of the figure) is stronger in velocity than the corresponding one in the bottom right corner. The maximum velocity in Figure 19 is 0.123 m/sec.

Figure 20 depicts the corresponding electrolyte flow pattern. Again, the only discernable difference is the increase of velocity of the flow loop under the affected anode. The maximum velocity in Figure 20 is 0.182 m/sec.

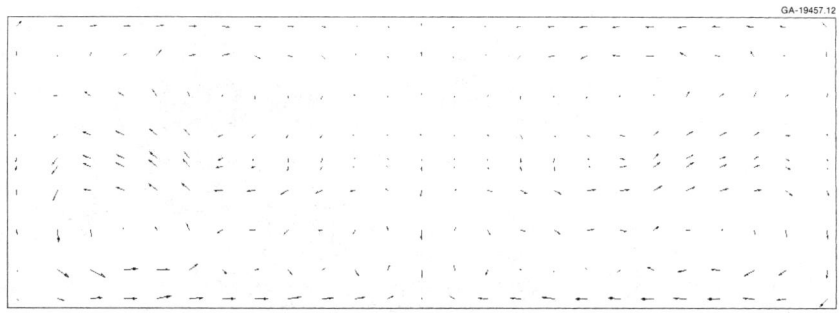

GA-19457.12

Figure 19. Metal Flow Pattern, 8.5 sec

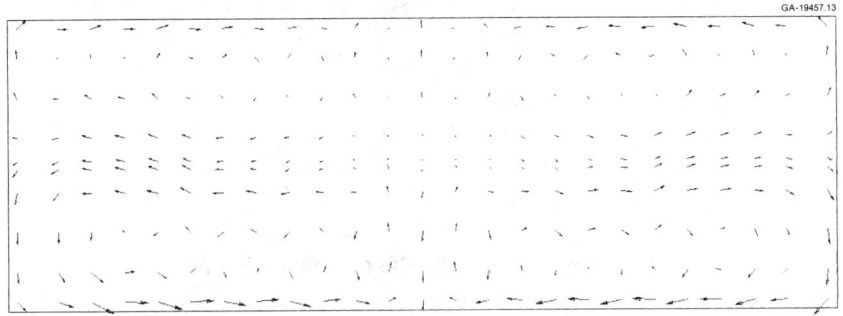

GA-19457.13

Figure 20. Bath Flow Pattern, 8.5 sec

Figure 21 depicts the metal/electrolyte interface for the P-155 steady state base case. This has been presented before, but is shown here in a view corresponding to the view used for the interface depictions to follow. This should permit easier comparison of the figures. The anode on which the current interruption was simulated is at the corner opposite the viewer in the figure.

Figure 22 shows the predicted metal/electrolyte interface at time 8.5 sec. The large peak opposite the viewer is adjacent to the anode on which the current interruption occurred, and is probably due to wall impingement of the increased flow under that anode. A small ripple in the interface is also notable opposite the viewer near the quarter-point of the cell. Again, this is probably due to the interaction of the riser at that location with the flow. The minimum height in Figure 22 is 0.082 m and the maximum height is 0.129 m.

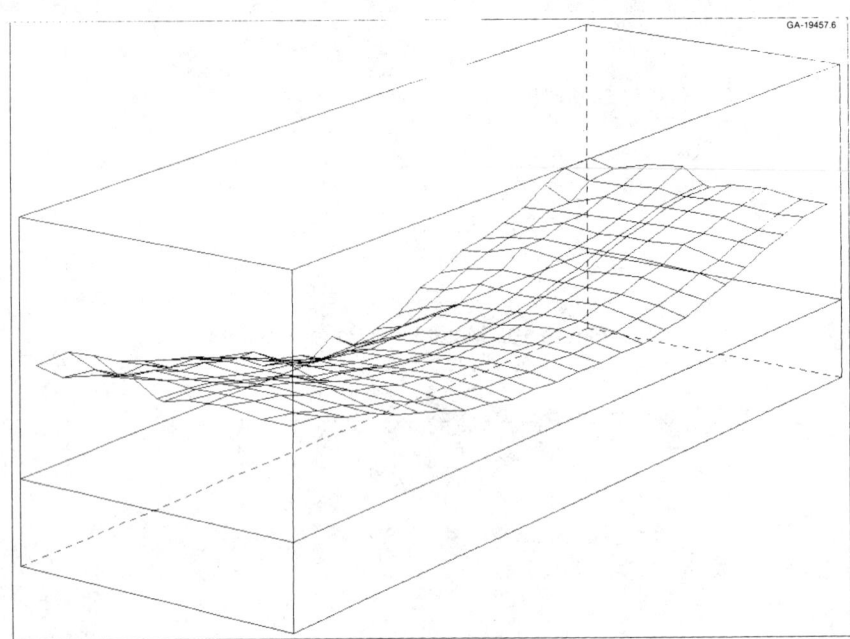

Figure 21. Base Case Interface Shape

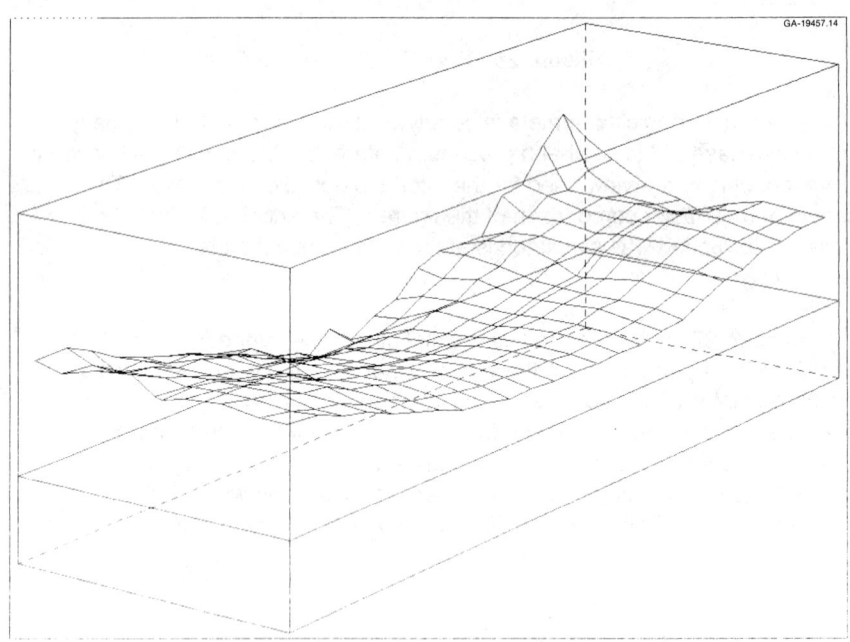

Figure 22. Interface Shape, 8.5 sec

Figure 23 depicts the metal flow pattern at time 35 sec (i.e., after achieving steady state). The major difference from the base case flow pattern is the increased velocity of the flow loop between the interrupted anode and the first riser (lower left again in the figure). One can also see that the flow near the wall adjacent to the anode whose current was interrupted tends to turn more gradually along the wall. The maximum velocity in Figure 23 is 0.141 m/sec.

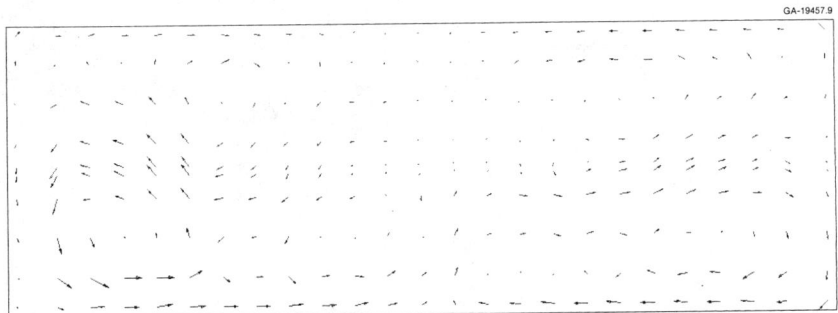

Figure 23. Metal Flow Pattern, 35 sec

Figure 24 shows the electrolyte flow pattern at time 35 sec. Again, the flow loop at the lower left corner is the only notable difference. The maximum velocity in Figure 24 is 0.191 m/sec.

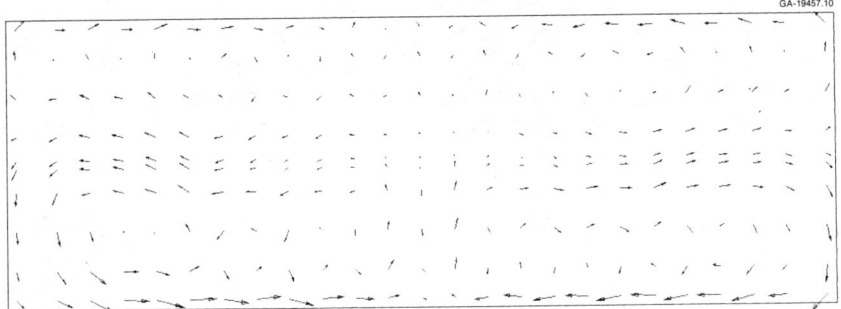

Figure 24. Bath Flow Pattern, 35 sec

Figure 25 depicts the metal/electrolyte interface at time 35 sec. The peak adjacent to the interrupted anode can be seen to have receded somewhat, and the "dimple" near the riser is slightly deeper. The minimum height in Figure 25 is 0.083 m and the maximum height is 0.125 m.

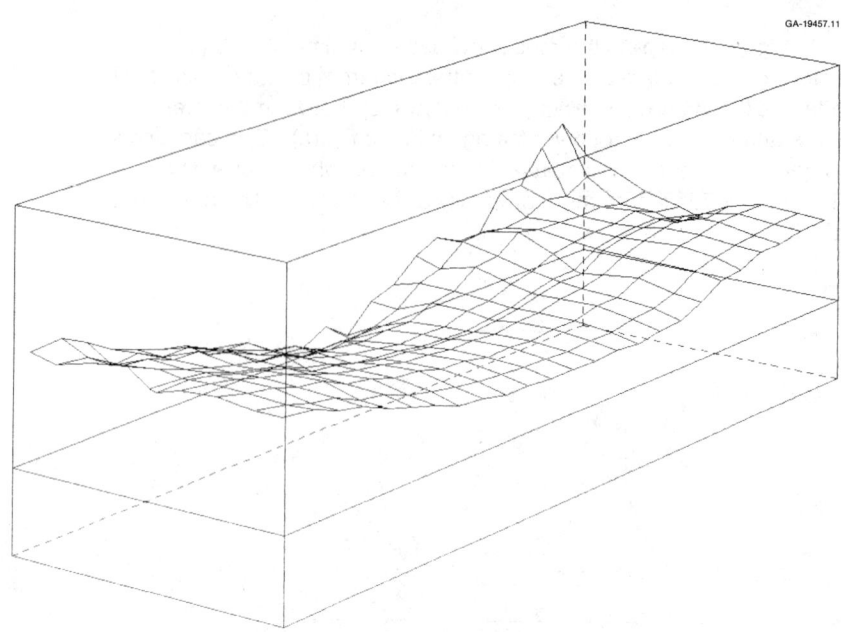

Figure 25. Interface Shape, 35 sec

  Calculations of time-dependent behavior using magnetic field updating as described above have shown a marked tendency toward instability in the predicted interface shapes. Within 20 seconds of simulated time (even without any disturbing factors) the predicted interface shapes are quite erratic. This may indicate that the coupling between the current distribution and the magnetic field is the mechanism that excites wave motions in the fluid interface. The introduction of induced current effects (which should act to damp the oscillations) produces a partial smoothing of the interface distortion. However, even with induced current effects included, some distortion of the interface shape remains.

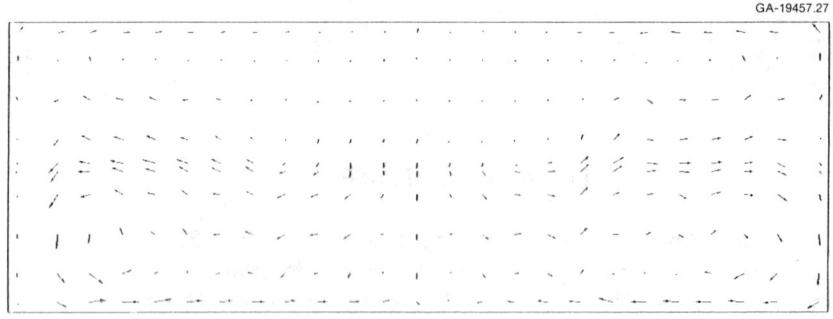

Figure 26. Metal Flow Pattern, 8 sec

Figure 26 shows the metal flow pattern for a case identical to the one discussed above, but with magnetic field updating and induced current effects included. The figure is plotted for a time 8 seconds after anode current interruption, essentially the same time as for Figures 19, 20 and 22. Again the tendency for flow to avoid the end of the cell where the anode without current resides is notable. Figure 27 depicts the corresponding electrolyte flow pattern. The "anode channeling" effect is more pronounced, and the flows, particularly along the cell centerline and the downstream wall, are more concentrated.

GA-19457.26

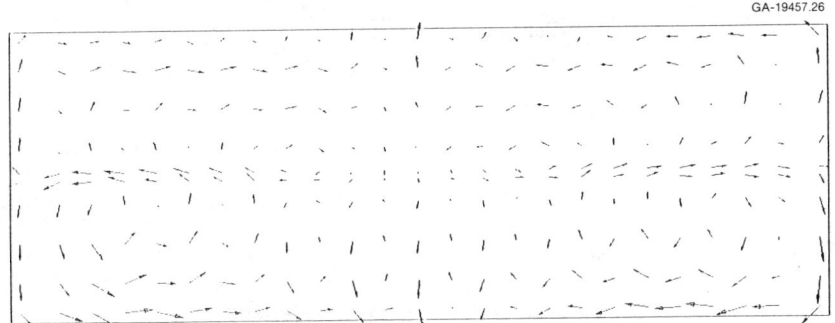

Figure 27. Bath Flow Pattern, 8 sec

Figure 28 shows the predicted metal/electrolyte interface shape for this case 8 seconds after current interruption. The result is similar to Figure 22, but there is an increased tendency for the interface to rise at the ends of the cell. As the simulation evolves from the 8-second point, this effect becomes more pronounced. When a fixed voltage boundary condition is applied to the cathode instead of a uniform current distribution, this redistribution of metal to the ends of the cell is markedly reduced. This is probably due to reduction of horizontal currents in the metal, because the voltage boundary condition on the cathode allows currents to flow almost vertically through the metal.

Qualitative verification of many of the results presented here (and of results of sensitivity analyses which have not yet been published) has been obtained from plants. A program is underway to quantitatively verify the results presented here.

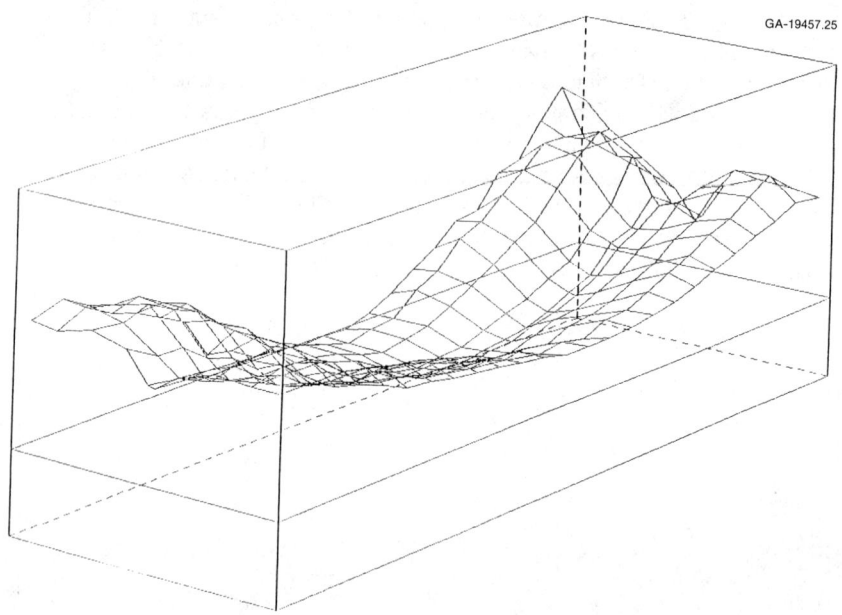

GA-19457.25

Figure 28. Metal/Bath Interface

Conclusions

The capability of the present model to simulate details of the flow field of Hall cells, both in steady state and transient modes, has been demonstrated. The need for a model of the Hall process of this level of detail (in fact, an even greater level is probably justified) is shown by the influence of relatively minor changes (ledge profile, fluid depth, cell shape) on the predictions. Insights gained in manipulating the model should be helpful in overall understanding of the process, and the tools developed in this effort should prove useful in the design of process improvements and new cells.

Acknowledgments

The author wishes to acknowledge the contributions of R. F. Robl, whose earlier efforts provided the basis for the current model, and J. Reichenbaugh, who performed much of the work reported here.

References

1. N. Urata, Y. Arita and H. Ikeuchi, Light Metals, 233(1975).

2. E. D. Tarapore, Light Metals, 341(1981).

3. R. F. Robl, Light Metals, 449(1983).

4. S. D. Lympany, D. P. Ziegler and J. W. Evans, Light Metals, 507(1983).

5. D. C. Chesonis, private communication.

6. M. C. Gunton, H. I. Rosten, D. B. Spalding and D. G. Tatchell, "PHOENICS -An Instruction Manual", CHAM Report TR/75, 1983, Sec. 3.2.

7 H. I. Rosten, "The Mathematical Foundation of the ESTER Computer Code", CHAM Report TR/84, 1982. See also H. I. Rosten, "The Instruction Manual for the ESTER Computer Code", CHAM Report 668/2, 1981.

8. W. E. Haupin, J. Metals, 23(10), 46(1971). See also W. E. Haupin and W. B. Frank, Electrometallurgy of Aluminium, in Comprehensive Treatise of Electrochemistry, Volume 2, Electrochemical Processing, Plenum Press, NY 1981.

9. J. D. Jackson, "Classical Electrodynamics", J. Wiley, New York, 1966, Sect. 5.2.

10. S. V. Patankar and D. B. Spalding, Int. J. of Heat and Mass Transfer, 15, 1787(1972).

11. W. Rodi and D. B. Spalding, Warme- und Stoffubertragung, 3, 85(1970).

Nomenclature

A^ϕ	Constant in PHOENICS finite difference equations
B^ϕ	Source term in PHOENICS finite difference equations
$\underline{B}$	Magnetic induction
C_a	Ampere's Law constant (dependent on units used)
C_i	Constant in interface height equation
C_L	Lorentz force constant (dependent on units used)
c_p	Heat capacity
$\underline{E}$	Electric field

E,W,N,S,H,L   PHOENICS face position designators
$\underline{g}$ Gravitational acceleration
h   Metal/bath interface height
$\underline{H}$ Magnetic field
$\underline{J}$ Current density
k   Thermal conductivity
$\underline{L}$ Lorentz force
m'   Mass source term (zero if gas not included)
p   Pressure
p'   Pressure correction
q_S Heat source
t   Time
T   Temperature
$\underline{u}$ Velocity

$\langle \epsilon \rangle$ Fluid stress tensor
ρ Fluid density
ρ_M Metal density
ρ_E Electrolyte (bath) density
μ Viscosity; magnetic permeability
ϕ General PHOENICS variable; viscous dissipation function
σ Electrical conductivity

FLUID DYNAMICS, A KEY FACTOR IN THE DESIGN OF A BIPOLAR ELECTROCHEMICAL CELL FOR ALUMINUM PRODUCTION

Alfred F. LaCamera
Chemical Systems Division
Alcoa Laboratories
New Kensington, PA

Abstract

This paper describes a vertically propagated bipolar electrochemical cell used for the production of aluminum from a molten salt containing aluminum chloride. It demonstrates the key role understanding the fluid dynamics played in the development and performance of the cell. The use of physical modeling and bench scale cell design are demonstrated as tools in the development of a commercial bipolar cell.

Introduction

Electrochemical cells using bipolar electrodes have been used in many applications in the general areas of energy conversion and chemical synthesis. During the early 1980's Alcoa developed the first commercial scale molten salt bipolar cell for the production of aluminum. The cell had a demonstrated production capacity of 32,000 pounds of aluminum a day and an energy usage of 4.2 kWh/lb, compared to 2000-4000 lb/day and 6-7 kWh/lb for Hall Cells. The evolution of this cell from its initial commercial installation to its final standard design involved both the development of materials and process understanding. This paper deals with the evolution of the cell as process understanding related to the fluid dynamics was uncovered. Two main areas of development are described: the application of physical modeling to the design of commercial cells, and the determination of the key phenomena associated with shorting using a uniquely designed bench scale cell. In this paper the term bipolar will be used to denote an single electrode for which the top surface is a cathode and the bottom is an anode.

Description of the Bipolar Cell

Alcoa has developed molten salt bipolar cell technology to produce aluminum from aluminum chloride. A typical industrial cell for this purpose is illustrated in Figure 1.

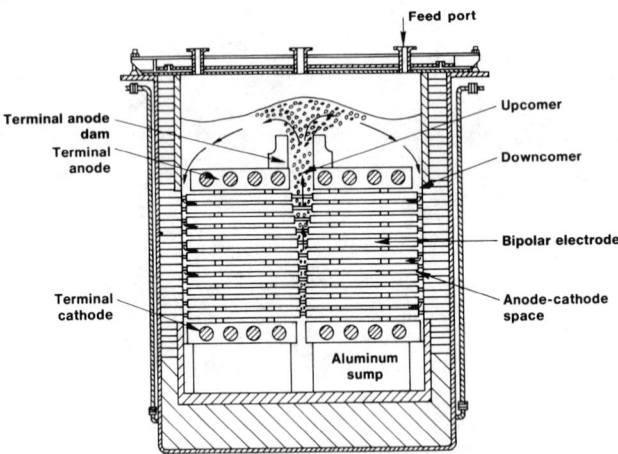

Figure 1. Industrial Bipolar Cell

The electrolyte used is composed primarily of sodium, lithium and aluminum chloride. Aluminum chloride for the cell is produced by the carbo-chlorination of alumina at 1560°F. The cell is operated at 1340°F. A unique aspect of the cell design is the use of a gas lift to create the major driving force for electrolyte circulation. The cell is composed of three vertical flow channels. The center is a gas lift (upcomer), and the two side channels (downcomers) act as fluid distributors for the bipolar stack. The geometry of the upcomer and downcomer in these cells is illustrated in Figure1. Electrical current passes vertically down through the bipolar stack from the terminal anode to the terminal cathode. All electrodes are fabricated of graphite. The primary electrochemical reaction is the decomposition of aluminum chloride to molten aluminum and chlorine gas. Chlorine is formed at the anode and aluminum at the cathode of vertically stacked bipolar electrodes. The chlorine gas generated is geometrically directed to the upcomer providing electrolyte circulation by means of a gas lift. The anode is designed with a land-slot geometry to augment the removal of gas from the land area and to provide a gravity driving force to direct the chlorine to the upcomer region of the cell. Various anode designs that provide this function are illustrated in Figure 2. Aluminum chloride is charged to the cell through a feed port in the lid and distributed throughout the cell by virtue of the gas-driven circulation. Aluminum entrained in the flowing electrolyte, circulates throughout the cell settling from slower flowing electrolyte in the sump region, from which it is periodically tapped.

The cell is built into a refractory lined, water cooled steel vessel. The inside lining dimensions are ten feet wide, nineteen feet long and twelve feet high. The upcomer plus downcomer averages 5.0% of the cell width. The total metal production surface area is 1740 square feet.

The flow of electrolyte through the anode-cathode spaces in the cell has two main functions. The first is to supply aluminum chloride dissolved in the electrolyte at a sufficient rate so that adequate concentration is maintained throughout each anode-cathode space, avoiding electrolysis of the supporting electrolyte. Decomposition of sodium chloride with the formation and subsequent intercalation of sodium at the graphite cathode is catastrophic. The control point for aluminum chloride in these cells was 7 wt%. The second function of the electrolyte flow is the entrainment of aluminum droplets as they form on the cathode surface, preventing metal droplets from growing and contacting the anode surface. A metal connection from anode to cathode creates an electronic short, virtually eliminating metal production in the anode-cathode space, significantly reducing the efficiency of the cell. Anode-cathode distances of 0.25 inch are possible using this approach for metal removal. Other phenomena affected by the electrolyte flow are the ohmic resistance of the chlorine gas layer on the anode and the reaction of aluminum on the cathode with dissolved chlorine in the electrolyte.

The upcomer and downcomer determine the fluid circulation in the cell but they also cause a loss of efficiency. This loss is due to a fraction of the current bypassing the bipolar stack through these flow channels reducing the production capacity of the cell. Consequently the design of the upcomer and downcomer had to consider not only the electrolyte circulation but also the loss due to bypass current.

Side View of Bipolar

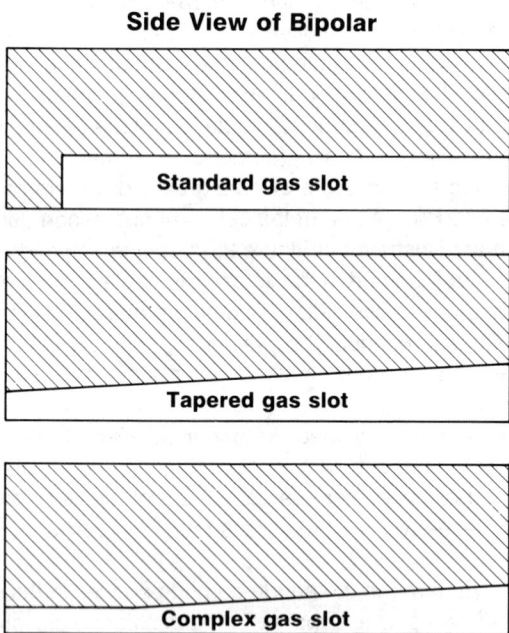

Upcomer View of Bipolar

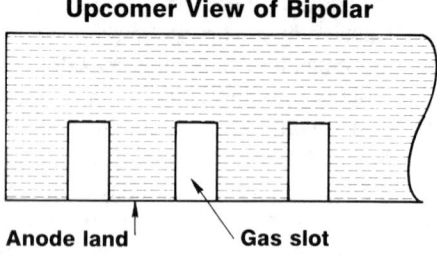

Figure 2. Anode Gas Slot Design

1976 Commercial Cell

The 1976 commercial cell goals were a production capacity of 20,000 lb/day and a energy consumption of 4.5 kWh/lb. The design employed twelve compartments, composed of a terminal anode, eleven bipolar electrodes, and a terminal cathode. A compartment in the bipolar cell is defined as the volume between the bottom of one electrode (an anode) and the top of the next electrode (a cathode). The cell life, at this time, was limited by two factors: shorting and materials of construction. In the early cell designs, shorting occurred within weeks of start-up; the lifetime of the construction materials was beyond three months. Shorting was the crucial problem blocking the commercialization of the cell. As indicated, the velocity of the electrolyte through the anode-cathode space was the major factor influencing the entrainment of metal from the cathode surface and therefore the tendency for a short to form. Autopsy of the cells after failures due to shorting generally showed that less metal had accumulated on the cathode of the top compartment than those at the bottom. This pointed to a maldistribution of velocity in the cell. A design tool was needed to determine the geometry of the upcomer and downcomer to provide adequate circulation to each of the compartments in the cell.

The multiplicity and geometric complexity of the flow channels, the effect of flow splitting and converging, and the coupling of the single phase and two phase flow regions made a computational solution to the problem uncertain. The time required to develop a effective design tool was also a factor because the process was in the initial stages of commercialization. For these reasons, physical modeling was investigated as a means of providing the technology needed. Also, as later discovered, there were important effects occurring in the electrolyte region over the bipolar stack that would have been difficult to address by a computational scheme.

Physical Modeling Justification

In establishing the physical model, the key concern was to provide similarity between the gas driven buoyancy forces of the physical model and the cell. A similarity analysis was conducted by flow region in the cell. These regions included: the evolution of gas bubbles from the anode slots into the gas lift, the two phase flow in the upcomer, and the single phase regions in the cell.

The process fluids were chlorine gas and a mixed chloride molten salt at 1340°F. An air-water model was evaluated for similarity, because of its convenience and the large scale of the model being considered. Table I shows the properties considered for the process and water model. The water properties are given at two temperatures to show their sensitivity.

Table I. Comparison of Properties
Molten Salt versus Water

Properties	Molten salt	Water	Temp. °F
Surface tension	8.4E-3[8, 9]	5.0E-3	68.0
		4.8E-3	99.0
Viscosity	2.66[5, 6]	2.42	68.0
		1.67	99.0
Density	93.6[7]	62.4	68.0
		61.8	99.0

Molten salt data at 1340°F

Surface tension - σ - lb$_f$/ft

Viscosity - μ - lb$_m$/(ft hr)

Density - ρ - lb/ft^3

Table II. Dimensionless Groups

Reynolds No. (RE)	$\rho v D_e/\mu$	Inertial/viscous
Bubble RE No. (RE$_b$)	$\rho v D_b/\mu$	Inertial/viscous
Weber No. (We)	$\rho v^2 D_b/\sigma$	Inertial/surface
Froude No. (Fr)	$v^2/D_e g$	Inertial/gravity
Bond No. (Bo)	$g D_e^2 \rho/\sigma$	Gravity/surface
Morton No. (Mo)	$g\mu^4/\rho\sigma^3$	We^3/(Re$_b^4$ Fr)

D$_e$ - Equivalent diameter

D$_b$ - Bubble diameter

v - Velocity

Significant property differences exist between the fluid densities and gas-fluid surface tensions. The forces involved in the defined regions are as follows. In the two phase regions, surface tension, inertial, viscous and buoyancy forces are considered. In the single phase region, the key forces are viscous and inertial. The dimensionless numbers that were considered to establish similarity included Reynolds number, Bubble Reynolds number, Weber number, Bond or Eotvos number, Froude number and Morton number, Table II. The rederivation of the importance of these groups was not considered necessary. The literature adequately covers this. Instead, based on literature correlations for the regions defined, fluid and flow regime similarity was examined.

A review of the two phase flow literature by R.S. Brodkey[1] shows that the formation of gas bubbles issuing from an orifice submerged in a liquid can be characterized into three regimes of bubble growth: a regime of constant volume, a constant frequency regime and a regime of coalescence or disruption. Similarity for the evolution of gas from the anode slots was established using these bubble growth phenomena. The dimensionless group that was used to identify the transition from one regime to the next was the ratio of $We/Fr^{.5}$. The constant volume regime is characterized by values of this group less than 0.1, the constant frequency regime by values greater than or equal to 0.1 and less than 18, and the coalescence regime by values greater than 18. The $We/Fr^{.5}$ ratio for the molten salt and water model are 24 and 27. This illustrates that the two systems are in the same bubble growth regime. The gas velocity used in this dimensionless group is estimated from the current density, electrode geometry, and the cross sectional area of the gas slot. The property ratio present in this dimensionless group is the ratio of density to surface tension. The relative difference between the process and the model is approximately 12%. Based on this analysis gas flow from the anode slot to the upcomer was expected to be similar.

The similarity of the process and the model in the gas lift region was established by comparing the bubble shape regimes, and the two phase flow regimes (bubble, slug, churn flow or annular flow). J.R. Grace[2] established a map of the shapes and velocities of bubbles rising through a quiescent fluid which is shown in Figure 3[2].

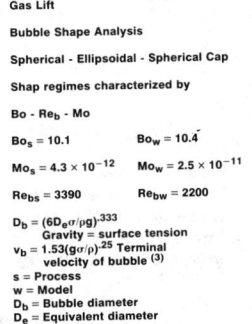

Gas Lift

Bubble Shape Analysis

Spherical - Ellipsoidal - Spherical Cap

Shap regimes characterized by

Bo - Re$_b$ - Mo

$Bo_s = 10.1$ $Bo_w = 10.4$

$Mo_s = 4.3 \times 10^{-12}$ $Mo_w = 2.5 \times 10^{-11}$

$Re_{bs} = 3390$ $Re_{bw} = 2200$

$D_b = (6D_e\sigma/\rho g)^{.333}$
Gravity = surface tension
$v_b = 1.53(g\sigma/\rho)^{.25}$ Terminal
velocity of bubble [3]
s = Process
w = Model
D_b = Bubble diameter
D_e = Equivalent diameter

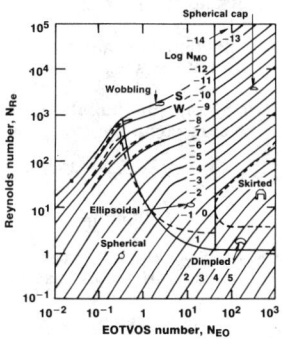

Figure 3a. Bubble Shape Analysis   Figure 3b. Shapes of Bubbles Rising in Liquids by J. R. Grace [2]

These conditions are appropriate for the upcomer region near the terminal cathode. The correlation involves the Bubble Reynolds number, the Bond number (Eotvos number) and the Morton number. The main regions of bubble shape are spherical, ellipsoidal and spherical cap. As indicated by the points on Figure 3, the salt-Cl_2 (S) and water-air (W) systems react similarly and are in the ellipsoidal region close to the transition to spherical cap. The terminal velocity of the bubble was used in the estimation of the Re_b. The bubble diameter was established by a balance between gravity and surface tension forces. A comparison of the dimensionless groups and the equations used to estimate the terminal velocity and bubble diameter is also given in Figure 3.

The vertical two phase flow regime most likely for this molten salt-chlorine system in the middle and upper sections of the gas lift was either bubble or slug flow. In the bubble regime, the gas-liquid mixture is homogeneous. The bubbles are relatively small compared to the size of the flow channel. In the slug flow regime, the gas phase extends to the walls of the channel; vertically there are alternating regions of gas and liquid. The transition from the bubble to slug flow regime has been shown to occur at gas fraction between 25-30%[4]. A method of computing the superficial liquid velocity, at transition, as a function of superficial gas velocity and fluid properties has been developed by A. E. Dukler[4]. This relationship is shown in Figure 4.

Gas Lift
Two Phase Flow Regime

Bubble-Slug-Churn-Annular

Most likely flow regimes
bubble or slug

Transition from bubble
to slug given by

$$U_{ls} = 3.0U_{gs} - 1.15(g\sigma/\rho)^{.25(4)}$$

Gas fraction for transition = 25%

$(g\sigma/\rho)^{.25}_{Salt} = .55$ $U_{ls} = 3.0$ ft/sec
$(g\sigma/\rho)^{.25}_{Water} = .54$ $U_{ls} = 3.1$ ft/sec

U_{gs} - Superficial gas velocity
in upcomer

U_{ls} - Superficial liquid velocity
in upcomer

g - Acceleration of gravity

Figure 4a. Two Phase Flow
Transition

Single Phase Flow Region

Anode Cathode Space

$$Re_{AC} = \rho v D_e / \mu$$

$$Re_{Salt} = 7870$$

$$Re_{Water} = 5790$$

Electrolyte velocity
in the AC space based
on saltation velocity
(inertial = gravity) for .125
inch aluminum droplet
is .75 ft/sec.

D_e = Equivalent diameter
$v = (6.07Rg(\rho_d - \rho_f)/\rho_f)^{.5}$
R = Radius of droplet
g = Acceleration of gravity
d = Droplet
f = Fluid

Figure 4b. Reynolds Analogy

The key fluid properties are surface tension and density. The predicted transition velocity for the process and the model was 3.0 and 3.1 ft/sec, excellent agreement. The estimate was made for the top compartment of a twelve compartment cell operating at 6 amp/in^2, assuming geometric similarity.

The last area of dimensional analysis is that of flow in the single phase regions of the cell. Flow in the anode-cathode gap, and at entrances and exits is characterized by the Reynolds number (Re). Determination of Re requires an estimation of fluid velocity. The rationale for establishing a fluid velocity was to determine the minimum velocity required to suspend aluminum droplets in the flow through a compartment. This is known as the minimum saltation velocity and is based on the balance between buoyancy and drag forces on a spherical droplet in the turbulent drag regime. The velocity estimated for a 0.125 inch diameter droplet was 0.75 ft/sec. As the equation in Figure 4 shows the saltation velocity increases as the square root of the droplet diameter, making entrainment of larger droplets more demanding. A 0.125 inch droplet was chosen, somewhat arbitrarily, at 25% of the anode-cathode distance in the cell. Values for the Re of 7870 for salt and 5790 for water were estimated using the saltation velocity. The difference in Re was important because it reflected force differences between the process and the model in an area of critical importance. Geometric ratios could have been adjusted to match Reynolds number; however, it was decided to maintain geometric similarity throughout the model and compensate by requiring the design velocities in the model to be increased by the Re ratio. Given the same driving force, the molten salt velocity would be 74% of the velocity achieved in the water model. This is experimentally verified in the section of this paper entitled Molten Salt Studies. With this consideration the minimum design velocity for the water model was set to 1.0 ft/sec.

The downcomer region of the cell is analogous to the single phase flow in the anode-cathode space and therefore the same reasoning holds.

Generally this analysis showed that water-air was a good choice for the physical model.

Model Verification

As a means of developing confidence in the flow similarity between the physical model and a cell prior to its use for commercial design, the model was used to design the upcomer and downcomer of a pilot cell. The cell was a nominal 5000 ampere, six compartment bipolar cell operated at the laboratory for development purposes. The pilot cell previously could not be operated at an anode-cathode distance smaller than 0.75 inch without shorting. The goal was to develop a design that would operate without shorting at a 0.50 inch anode-cathode space. This would be a significant demonstration of the effectiveness of the model. Although an estimation of the minimum saltation velocity for 0.125 inch aluminum droplets had been made, the velocity required to disengage metal from the cathode could not be determined. The force holding the droplet to the cathode was dependent on the interfacial tension among the bath, metal, and graphite cathode under conditions of electrolysis in the cell. This interfacial property was not known. In lieu of actual measurements an arbitrary design factor of 20% was added to the minimum design velocity to compensate for this unknown and the possibility of growth of larger aluminum droplets than assumed. The minimum design velocity for the water model was increased to 1.2 ft/sec.

The flow design of the pilot cell was accomplished using a full-scale water model. Air was used to simulate the chlorine generation in the cell. The rates were based on the current density and the difference in operating temperatures. The velocity in each compartment was measured by timing the movement of a nylon ball, having a density 5% greater than water, over a specified distance across the simulated cathodes. The velocity in each compartment was achieved by adjusting upcomer spacing. The cell is shown in Figure 5.

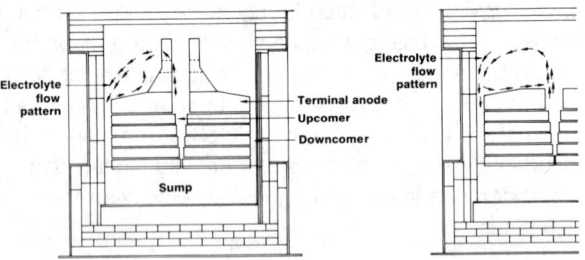

Figure 5a. Pilot Cell with
Upcomer Dams

Figure 5b. Pilot Cell with
No Upcomer Dams

The maximum velocity was 1.8 ft/sec and the minimum was 1.2 ft/sec, which satisfied the design criteria, Figure 6. Figure 5 shows the cell with and without dams at the upcomer end of the terminal anode. This dam was added during the design phase and was found to have a significant influence on circulation. Without the dam in place, a low pressure area existed in the gas lift at the top of the terminal anode. The electrolyte flow towards the downcomer was primarily at the salt-gas interface, and there was a significant reverse flow towards the upcomer directly over the terminal anode. This was of concern because metal carried in the upcomer could drop out in the low velocity region near the surface of the electrolyte and recirculate back to the chlorine lift for possible back reaction. Also, metal accumulated on the anode could be transported electrochemically to lower electrodes consuming a portion of the current for aluminum chloride electrolysis. The flow dam shown in Figure 5 was used to minimize these effects. Figure 6 shows the influence the dam had on the velocities in the cell. The minimum, without the dam, changes to 0.8 ft/sec and is now located on the top compartment as opposed to the bottom. The design, including the dam, resulted in the first successful operation of a pilot cell at a 0.50 inch anode-cathode space, validating the physical model.

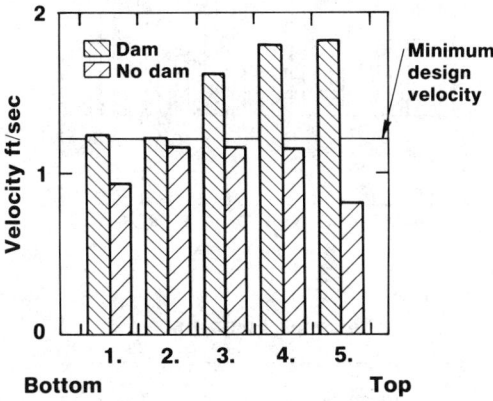

Figure 6. Electrolyte Velocity in Anode-Cathode Space of the Pilot Cell

Commercial Cell Design

Having verified the effectiveness of the physical model in establishing the design of the upcomer and downcomer, a program was initiated to use the technology for commercial cell design. During the next few years, three major accomplishments in cell design were achieved through the use of the model. First, the physical model made a significant contribution to the achievement of the initial performance goals set for the cell; second, the model was used to optimize the productivity of the existing cell; and, last, it was used to demonstrate the potential of the basic design concept.

Initially, the evaluation of the design of the planned commercial cell was investigated to determine the magnitude and uniformity of the velocity in the compartments of the cell. A full scale model (height and width) of the cell was constructed. A section sixteen inches in length was considered adequate to eliminate edge effects. The anode-cathode distance was 0.50 inch, and the cell contained twelve bipolar compartments. It was designed to operate at 120,000 amperes and produce 20,000 lb/day. The expected energy usage was 4.5 kWh/lb. The velocities determined using the physical model for this design indicated why the life was limited. Figure 7 gives the results obtained. The flow was reversed and very low in compartments nine and ten, and only two compartments were above the minimum velocity desired. The bottom compartment was stagnant. The spacing in the upcomer and downcomer from the terminal cathode to the terminal anode in the cell is also given in Figure 7.

12 Compartment Cell

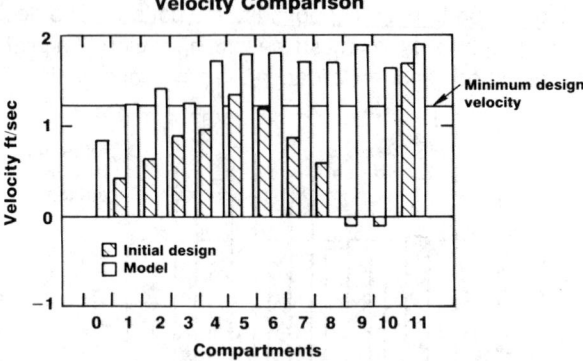

Velocity Comparison

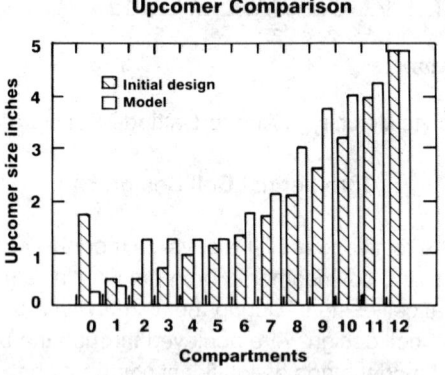

Upcomer Comparison

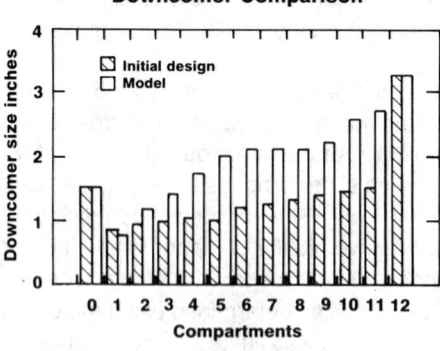

Downcomer Comparison

Figure 7. Twelve Compartment Commercial Cell Analysis

Using the physical model and the understanding developed in the design of the pilot cell, the cell upcomer and downcomer were redesigned to achieve the 1.2 ft/sec minimum velocity desired; this included the addition of upcomer dams. These velocities are reported on the same bar graph as the post water model design, Figure 7. As seen, the design criteria is met in all compartments except the bottom. The anode-cathode space in the terminal cathode compartment was 0.75 inch. This was the compartment most prone to shorting and was, therefore, allowed additional height as a safety factor. This water model design changed the average upcomer-downcomer opening from 6 inches to 9 inches, increasing the potential for efficiency losses due to bypass current. In fact, the current efficiency actually increased due to reduced shorting. This design now became limiting because of material life, and not because of failures due to shorting. This cell operated for 150 days with an average production rate of 21,400 pounds of aluminum per day at 4.34 kWh/lb; the current efficiency averaged 84.3%. Table III compares the performance of the design before and after modeling. Expected performances were achieved in all areas and cell life was increased by a factor of three. Shorting was still an issue but no longer a major limitation .

Table III. Comparison of Cell Performance, Post Modelling through Cell Optimization

	Commercial Specifications	1st Plant Cell	Modelled Cell	14Compartment Cell
Pounds/day	20,000	16,600	21,400	30,200
kWh/lb	4.5	4.62	4.34	4.2
Bipolar days	365	45	150	233
Kamps	120	109	119	145
C.E.%	85	72	84	83
Compartments	12	12	12	14
Electrode spacing (Inch)	.50	.50	.50	.25

CELL OPTIMIZATION

Recognizing the potential of the physical model as a tool for cell design, the course was set to increase the productivity of the cell. This could be accomplished by reducing anode-cathode distance and increasing the number of bipolar compartments within the confines of the present cell dimensions. The economic payoff would be brought about by reduced capital and reduced labor because fewer cell stations would be required.

The pilot cell was used, as previously, to establish feasibility for operation at a lower anode-cathode distance. A six compartment, 0.25 inch anode-cathode space cell was designed using the physical model. The design was similar to the previous cell except the location of the upcomer and downcomer were reversed. This design used two side upcomers and a central downcomer. This change was found to reduce the height of the gas-liquid plume at the top of the upcomer and the subsequent wave formation at the surface of the liquid. Higher velocities were also achieved using side upcomers versus center upcomer of the same overall dimensions. This design was successfully operated with 0.25 inch anode-cathode distance.

While the pilot cell was being operated, the design of a fourteen compartment, 0.25 inch anode-cathode distance commercial cell was being developed. The upcomer-downcomer design established through physical modeling is shown in Figure 8. The velocities achieved in 0.25 inch compartments were above the minimum desired. The performance of this cell is compared to the 12 compartment, 0.50 inch anode-cathode distance cell in Table 3. The capacity was increased 55% from 21,400 to 33,300 pounds per day. The current efficiency remained constant at 84%, and the energy utilization improved slightly to 4.18 kWh/lb. This reduced the number of cells required by 35% for a constant capacity plant. Another indicator of the improved circulation in the cell was the ability to operate at aluminum chloride concentrations of 2 wt% as compared to the 7 wt% necessary in early cell designs.

Velocity measurements were made by the technique previously reported, timed floating balls, and also by laser anemometry. The laser permitted a more detailed measure of the flow in the small anode-cathode gaps without creating disturbances in the flow field. The laser system was composed of a 15 mw helium-neon laser, frequency shifting for low velocity and component measurements, accompanying optics, a photodetector, and a counter-type signal processor. The digital output of the signal processor was interfaced to a HP-9825 micro-processor. The HP-9825 was programmed to compute mean velocity and turbulence intensity, plot velocity frequency distributions, and compute velocity components.

A comparison of velocities measured by the floating ball and laser is shown in Figure 8. The laser measurements were made at nine positions uniformly spaced over the length and width of the electrode, mid-height in the anode-cathode space. The velocities are in the same range. One compartment is shown by laser measurement to be .1 ft/sec below the minimum desired. Figure 8 also shows the composite average of velocities in all compartments at three positions, the downcomer, centerline, and upcomer. The velocity at the downcomer end is .4 ft/sec higher than the upcomer end of the compartment. Figure 2 illustrates the various gas slot geometries employed. This cell used the complex geometry. The depth of the gas slot was constant at one inch deep for the first twenty-one inches from the downcomer and then increased to 1.5 inches at the upcomer end. The lower velocity at the upcomer is attributed to an excess gas slot depth increasing the cross sectional flow area. Optimizing the depth of the gas slot geometry was another means of maximizing the velocity in the anode-cathode space for a given upcomer-downcomer design.

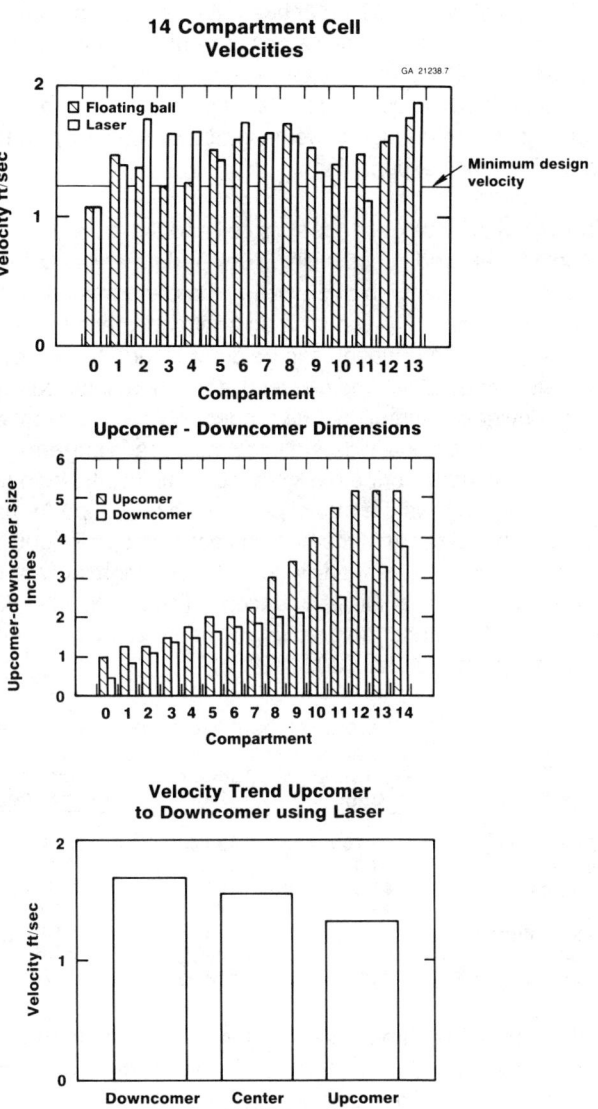

Figure 8. Commercial Cell Optimization

CELL POTENTIAL

The last commercial cell designed using the water model was aimed at establishing the potential of the basic design concept. A twenty-one compartment cell, using a 0.25 inch anode-cathode distance, with approximately double the height between the terminal anode and terminal cathode was designed. A prototype cell was operated full height and width, approximately one third the commercial length. The cell had a structural failure but operated long enough to demonstrate its feasibility from a flow design and performance expectation.

Table IV shows the operating data for this cell and the predictions of a semi-empirical model used in establishing cell performance. As illustrated the model does an excellent job of predicting the performance of the prototype cell. The heart of the model is an energy balance for the cell from which key performance information can be determined. It used basic cell design information to establish the cell resistance, heat loss, current and voltage rating, power consumption, and productivity. The current efficiency was established by theoretically determining the bypass current and estimating the aluminum-chlorine back reaction empirically, from historical data. The table also shows the model predictions for the performance of a commercial cell. The capacity of the cell was estimated at 71,000 lb/day at 4.19 kWh/lb with an 87% current efficiency. The commercial data demonstrates the potential of this cell concept. The electrical energy required is reduced 35%, and the productivity of the cell was potential twenty times greater than a Hall Cell.

Table IV. 21 Compartment Prototype Performance

	Operating Data	Computer Model	Commercial Prediction
Pounds/day	29,700	30,800	71,000
kWh/lb	4.03	3.94	4.19
Kamps	93.0	94.2	220
C.E.%	86.0	87.9	87.1
10^{-6} Ohms	137	136	71.6
Volts	53.7	53.6	56.6
Heat loss kW	461	412	1678

The development of the physical water model as a design tool had a major impact on the reduction of shorting failures and the ability to capture the potential of the cell in a relatively short period of time.

Molten Salt Studies

The physical model provided a means of designing the upcomer and downcomer to achieve minimum velocities in the anode-cathode space to avoid shorting. However the velocity in the molten salt at which the initiation of shorting occurred was not known. As previously indicated, the other factors that impact on shorting include the surface of the cathode and the electrolyte itself as they modify the interfacial tension of the three phase system.

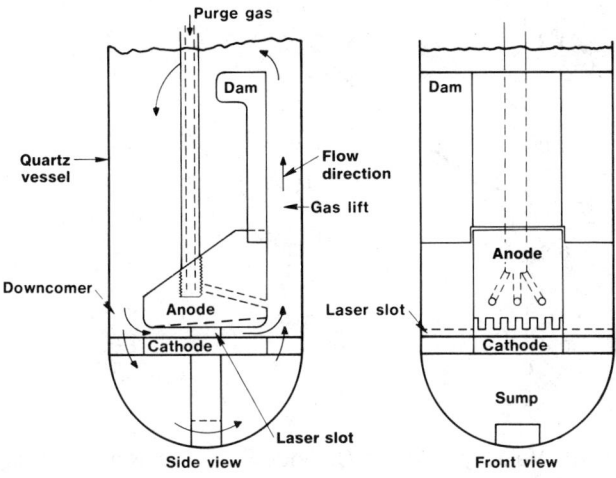

Figure 9. Bench Scale Cell

A novel approach was used to study these effects. The bench scale molten salt cell was designed using the same physical modeling approach as for commercial cell. The design of the cell is illustrated in Figure 9. It was designed to fit into a quartz vessel for viewing during operation and to permit velocity measurements using the laser. A unique feature of this cell was the ability to control the velocity in the anode-cathode distance by a purge of nitrogen into the gas lift. The anode current collector was a tube through which nitrogen was purged into holes in the anode that evolved into the upcomer just above the gas slot. Laser velocity measurements were made in both the water model and the molten salt cell. This provided the first means of establishing the difference between velocities measured in a water model and those achieved in a molten salt cell of the same design.

Figure 10 shows the velocities measured in the molten salt as a function of nitrogen purge rate at standard cell operating conditions. As shown, these are in the range of velocities achieved in the physical models for commercial cells.

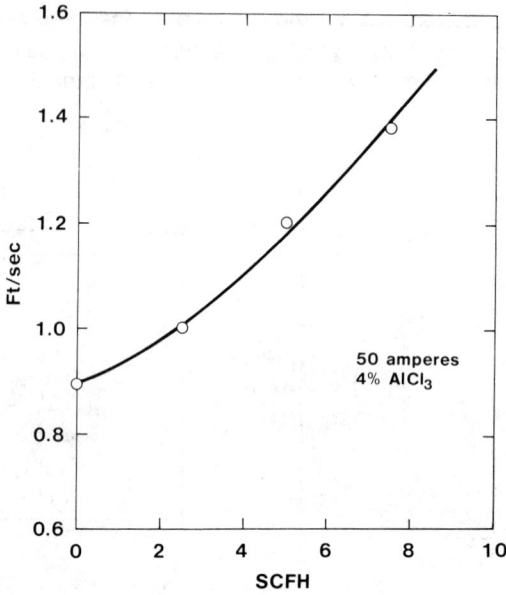

Figure 10. Velocity in Anode Cathode Space vs. Nitrogen Purge Rate

Figure 11 compares the molten salt versus water model velocities. The data are for one water model design and three different molten salt cells of the same design.

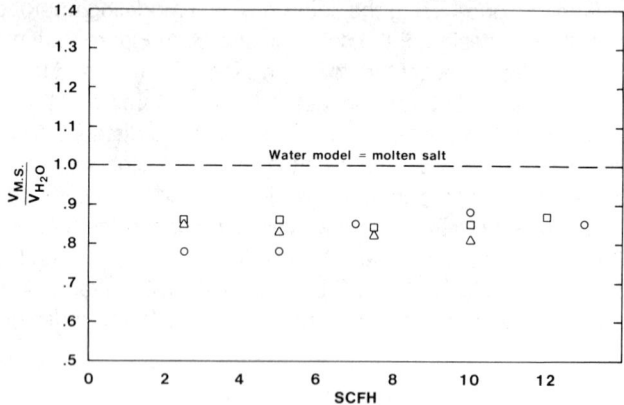

Figure 11. Comparison Data - Molten Salt Veloctiy/Water Model Velocity

The ratio of molten salt to water model velocity averages 0.85. The expected value was .74 based on the Re ratio. This results in a slightly larger safety factor in the water model design than expected and could be a result of differences in the two phase similarity.

Many phenomena that could affect shorting were investigated with this cell. Two significant effects reported here are the influence of cathode surface and the importance of electrolyte chemistry.

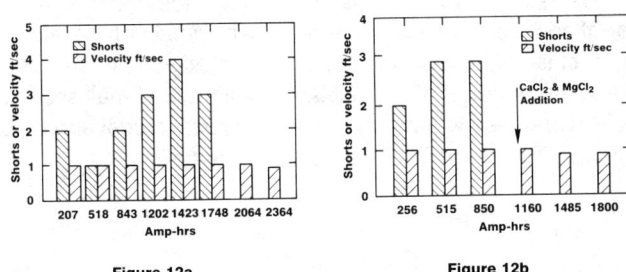

Figure 12a   Figure 12b

Figure 12. Bench Scale Tests for Surface Effects
Temperature 1340°F;
Electrolyte 60% NaCl-37%LiCl-3%AlCl$_3$

Shorting could be easily determined during electrolysis by a sharp change in voltage from approximately 2.5 volts to less than .5 volts. Figure 12 shows a record of cell shorting as a function of time for a cathode current density of 5.5 amps/inch^2 at the condition shown. Shorting is occurring on a regular basis for the first 1750 ampere-hrs and then abruptly stops. Even if the velocity is lowered from 1.0 to 0.9 ft/sec, the shorting does not reoccur. Analysis of the cathode, as a function of thickness, at the completion of the cell test program (6800 ampere-hrs) shows that a slice 1mm thick at the cathode surface contains 49,000 ppm aluminum carbide. A slice at the 3m to 4 mm thickness contains 320 ppm. The change in minimum velocity required to avert shorting was felt to be due to the formation of the aluminum carbide layer at the graphite cathode-aluminum interface.

Similar tests were conducted to determine the effect of electrolyte chemistry. Magnesium chloride and calcium chloride were two of the additives evaluated. The identical start-up procedure was used and after 1000 ampere-hrs, an addition of magnesium and calcium chloride (10 wt%) was made to the electrolyte. The result of this experiment is shown in Figure 12. Again an abrupt change in the occurrence of shorting was observed, and was sustained when the velocity was changed from 1.0 to 0.9 ft/sec. The effect of the calcium and magnesium chloride was to arrest shorting at a time when the cathode was most sensitive to short formation. These examples serve to illustrate the power of this small scale experiment in establishing key phenomena associated with shorting.

Conclusions

Shorting was a major issue in the operation of Alcoa's bipolar cell for the electrolysis of aluminum chloride. This was particularly true because of the 0.25 inch anode-cathode distance used to achieve productivity and power efficiency. The use of physical modeling to design the upcomer and downcomer of commercial cells had a significant impact on developing process understanding, enhancing the implementation time, and demonstrating the potential of the cell design concept. Small scale, 50 ampere, cells were used to further verify the physical model and to refine the understanding of shorting in commercial scale cells.

References

1. R. S. Brodkey, "The Phenomena of Fluid Motion", edited By H. Hoelscher, Addison-Wesley Publishing Co., Reading, Mass., 1976, pp. 543-545.

2. J. R. Grace, "Shapes and Velocities of Bubbles Rising in Infinite Liquids", Trans. Instn. Chem. Engrs., vol. 51, 1973, pp. 116-120.

3. G. B. Wallis, "One-Dimensional Two-Phase Flow", McGraw-Hill Book Co., New York, N.Y., 1969, pp. 243-253.

4. A. E. Dukler, "Modeling Flow Pattern Transitions for Steady Upward Gas-Liquid Flow in Vertical Tubes", AICHE Journal, (Vol. 26, No.3), May 1980, pp. 345-354.

5. W. Brockner, K. Torklep and H. Oye, "Viscosity of Molten Alkali Chlorides", J. Chem. Eng. Data, 1981, 26, pp 250-253.

6. T. Ejima and Y. Sato, "Viscosity of $LiCl-NaCl-AlCl_3$ Ternary Melt", Molten Salt Chemistry and Technology, Molten Salt Committee of The Electrochemical Society of Japan, Kyoto, Japan, 1983, pp 315-318.

7. G. J. Janz, "Molten Salt Handbook", Academic Press Inc., New York, N.Y. 1967, p 39.

8. G. J. Janz, ibid, p 80.

9. D. A. Nissen and R.W. Carlsten, "The Surface Tension of the Molten Binary System LiCl-KCl", J. Electrochem. Soc., April 1974, pp 500-505.

REACTION OF AN IRREGULAR PARTICLE WITH A GAS

K. Rajamani

Department of Metallurgy and Metallurgical Engineering
University of Utah
Salt Lake City, Utah 84112

Abstract

A number of industrial metal extraction processes involve the reaction of solid particles with a gas phase, and therefore several gas/solid reaction models have been proposed for process analysis. Though in most applications, the particles are irregular in shape, the assumption that particles are of a regular shape, i.e., spherical, cylindrical, or cubical, is made to simplify the solution of the model equations.

In this work, the solution of the pellet-grain model for an arbitrary irregular particle is shown. A Monte Carlo algorithm enables the solution of diffusion and chemical reaction equations in the interior domain bounded by the surface of an irregular particle. Comparison of the numerical solution with known analytical solution is shown followed by computational results for irregular particles. Further, this computationally simple technique is also applicable in similar problems involving heat, momentum, or mass transfer in an irregular domain.

Introduction

The description of gas-solid or gas-liquid reaction is of considerable interest since many industrial processes fall in this category. Starting with the shrinking-core model, several other detailed models (1) have been proposed for the prediction and interpretation of experimental data. A common feature of all these models is that the solid particles are assumed to be made up of regular geometrical shapes, such as spheres, rectangular solids, or flat plates to facilitate analytical or numerical solution of the model equations.

The purpose of this paper is to show the numerical solution of one of the gas-solid reaction models, i.e, the pellet-grain model, for irregular or arbitrary shape of the pellet. This solution is of interest for instance in coal-char burning, oil-shale retorting, and industrial roasting processes.

Model Formulation

Solid-gas reactions may be represented by the general reaction scheme

$$A(gas) + bB(solid) = cC(gas) + dD(solid) \tag{1}$$

We will follow the model proposed and solved by Szekely and Evans (2,3). This model deserves special interest since it has been extended to several special reaction applications. Szekely et al. (4) verified the model expressions in an experimental study of the reduction of porous nickel-oxide pellets with hydrogen. The model has been extended to describe successive gas-solid reactions occurring in the reduction of metal sulfides in the presence of lime (5). The case of a mixture of metal sulfide and lime surrounded by a lime coating to minimize sulfur-bearing gases escaping to the surroundings was analyzed by Rajamani and Sohn (6). In this model, a porous pellet made up of individual grains constitutes the solid phase (Fig. 1). The shape of the grains can be spheres, long cylinders slabs, and the shape of the pellet is spheres in most real cases. Gas A diffuses through the pellet reacting with solid grains of B. The resulting product is partially reacted grains of B and gas C which diffuses out. We follow the model development given in Sohn and Szekely (7) except that we depart from the fact that the pellet is spherical and instead take it to be of irregular shape. The interested reader is asked to refer to the original work on model development (7), since we present the final model equations only for the sake of brevity. The problem of conservation of the gaseous reactant with a mass balance for the reacting solid yields

$$D_e \nabla^2 C_A - v_A = 0 \tag{2}$$

The Laplacian operator is used here to indicate that the geometry of the solid is contained in the three-dimensional Cartesian coordinate system. The local rate of reaction, v_A, is given as

$$v_A = (1 - \varepsilon)\, k\, C_A\, \frac{A_g}{V_g} \left(\frac{A_g\, r_g}{V_g\, F_g}\right)^{F_g - 1} \tag{3}$$

The local rate of reaction of the grains may be expressed as

$$\rho_g \frac{dr}{dt} = -\, bkC_A \tag{4}$$

694

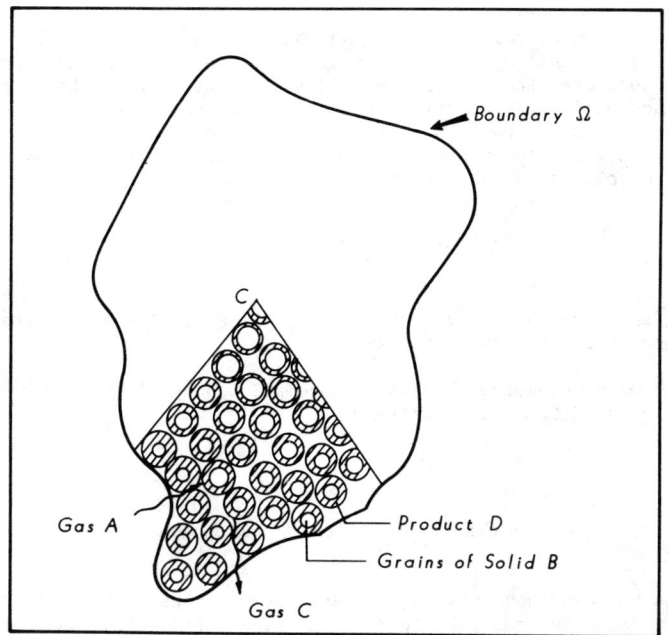

Figure 1. An irregular pellet made up of spherical grains.

After introducing dimensionless variables, Equations (2) and (3) can be written as

$$\nabla^2 \psi - \sigma^2 \psi \xi^{F_g - 1} = 0 \qquad (5)$$

and

$$\frac{d\xi}{dt^*} = -\psi \qquad (6)$$

The boundary conditions for Equations (5) and (6) are readily written as follows:

$$\xi = 1 \text{ for all } x,y,z \ \epsilon \ D \text{ for } t^* = 0 \qquad (7)$$

$$\psi = 1 \text{ at the boundary } \Omega \ t^* \geq 0$$

where D denotes the interior domain bounded by the surface of the particle. In general, the governing equations must be solved for ψ and ξ as a function of t^* and coordinate position within the pellet. Finally, knowing the grain radii in all locations of the pellet, the overall conversion can be computed by integrating across the pellet the individual conversion of each grain. For a two-dimensional irregular pellet, the overall conversion is given by

$$X = \frac{\underset{xyz}{\iiint} (1 - \xi^{F_g}) \ dx \ dy \ dz}{\underset{xyz}{\iiint} dx \ dy \ dz} \qquad (8)$$

695

Monte Carlo solution

The solution of differential equations for which boundary conditions are defined on an irregular domain generally requires numerical methods. Kakutani (8) first suggested the use of the random walk method or Monte Carlo method, that had arisen for the solution of Brownian motion, to the solution of Dirichlet's problem. This led to the development of Monte Carlo solutions (9-11) of parabolic and elliptic partial differential equations. Here, we present the Monte Carlo solution of the differential equation

$$\nabla^2 u(x,y) + g(x,y)\, u(x,y) = 0 \qquad (9)$$

following the development by Wasow (9). The essential steps of the algorithm only are shown for the sake of brevity.

Solution of Equation (8) can be accomplished by finite-difference schemes which yield the difference equation

$$\frac{1}{h^2}\, [u(x+h,y) + u(x-h,y) + u(x,y+h)$$
$$+ u(x,y-h) - 4u(x,y)] + g(x,y)u(x,y) = 0 \qquad (10)$$

where h is the distance between adjacent nodes in the x and y directions. Consider a random walk from point $P(x,y)$ in which a fictitious particle of mass unity starts from the point P and follows a random path. The walk adheres to the rule that the probability of moving to any of the four neighboring points P_j (j = 1, 2, 3, or 4) is equal to 1/4. Thus, the particle is equally likely to reach any of the four neighboring points. Let the mass of the particle be multiplied by the unknown function $k(P)$ at each point during its sojourn. The value of the function $k(P)$ depends on the point $P(x,y)$. After a number of steps, the particle arrives at the boundary R, at which time the mass of the particle is multiplied by the value of $u(x,y)$ on the boundary (i.e., known boundary condition) and the random walk is terminated.

Following Wasow's notation, let $L_n(P,R)$ be the set of all possible paths starting at P and ending on the boundary at R on exactly the nth step. If $m_n(P,R)$ is the mass upon arrival at the boundary for a given path then the expected value of the mass of the particle upon arrival at the boundary is given by

$$e_n(P,R) = \frac{1}{4^n} \sum_{L_n(P,R)} m_n(P,R) \qquad (11)$$

where $1/4^n$ stands for the probability that the particle follows any single n-step path. Define

$$E_n(P,R) = \sum_{i=0}^{n} e_i(P,R) \qquad (12)$$

and

$$E(P,R) = \lim_{n \to \infty} E_n(P,R) \qquad (13)$$

Consider a single n-step path for which the following holds:

$$m_n(P,R) = k(P) \ m_{n-1}(P_j,R) \tag{14}$$

since the particle reaches one of the neighboring points in the first step. Therefore Equation (11) can also be written as

$$e_n(P,R) = \frac{1}{4^n} \sum_{j=1}^{4} \sum_{L_{n-1}(P_j,R)} k(P) \ m_{n-1}(P_j,R) \tag{15}$$

and

$$e_n(P,R) = \frac{k(P)}{4} \sum_{j=1}^{4} e_{n-1}(P_j,R) \tag{16}$$

Using the definitions (12) and (13), Equation (16) can be written as

$$E_n(P,R) = \frac{k(P)}{4} \sum_{j=1}^{4} E_{n-1}(P_j,R) \tag{17}$$

The Laplacian of the continuous function E(P,R) may be written in finite difference notation as

$$\nabla^2 E(P,R) = \{ \sum_{j=1}^{4} E(P_j,R) - 4E(P,R) \} \frac{1}{h^2} \tag{18}$$

Substituting (18) in (17) and eliminating $E(P_j,R)$ yields

$$\nabla^2 E(P,R) + \frac{4[k(P) - 1]}{h^2 \ k(P)} E(P,R) = 0 \tag{19}$$

Comparing Equation (19) with (9), we get

$$g(P) = \frac{4[k(P) - 1]}{h^2 \ k(P)} \quad \text{or} \quad k(P) = [1 + \frac{h^2 g(P)}{4}]^{-1} \tag{20}$$

Therefore, the numerical value of u(x,y) at some interior point P is equal to the value of the function E(P,R), which is obtained by generating a large number of random walks starting at P(x,y) and terminating at any point on the boundary and finding the mean value of the mass of the particle given as

$$u(P) = \frac{1}{N} \sum_{i=1}^{N} m_i(P,R) \tag{21}$$

where N is the total number of random walks carried out.

Numerical solution

Numerical solution involves defining the boundary and interior of the particle and solving Equations (5) and (6) in the domain D bounded by the particle. One-dimensional problems are solved by a random walk on a straight line (x-axis), two-dimensional problems by a random walk on a plane (x and y axes), and three-dimensional problems by a random walk in the three dimensions (x, y, and z axes).

For example, given a two-dimensional particle, a square grid of mesh-width h is appended to the particle domain. If the edges of the particle do not fall on the nodes, a smaller mesh width is suggested. While Equation (5) is solved by the Monte Carlo algorithm, Equation (6) is solved simultaneously with a third-order Runge-Kutta algorithm. The latter algorithm requires knowing values of ψ at each x,y-coordinate position for intermediate times between successive integration intervals; therefore, the Monte Carlo algorithm is executed three times in between each integration interval. The random walk is initiated at a given node-x,y and a value of unity is assigned to the fictitious particle. When a move occurs toward one of the neighboring nodes, the mass of the particle is multiplied by the value of the function k(P) given as

$$k(P) = (1 + \frac{h^2}{4} \sigma^2 \xi^{F} g^{-1})^{-1}$$ (22)

After simulating a total of N random walks, the value of the function $\psi(x,y)$ is given by

$$\psi(x,y) = \frac{1}{N} \sum_{i=1}^{N} m_i(P,R)$$ (23)

where $m_i(P,R)$ is the mass of the particle upon arrival at the boundary R. Typically a 20x20 grid and 40 repetitions of the random walk are sufficient to ensure numerical accuracy. Since a large number of random numbers are used in the solution, caution must be exercised in the generation of uniformly distributed random numbers. The URAND function (12) which exhibits a large period is found to be adequate.

Computational Results

In this section we present numerical solutions for regular pellets first, compare them with known analytical solution, and having established the numerical accuracy we then present the results for a three-dimensional irregular pellet.

When σ approaches zero, the gas-solid reaction is controlled by chemical kinetics, and intrapellet diffusion is rapid compared with the rate of chemical reaction. Hence, the concentration of gas is uniform throughout the pellet ($\psi = 1$). Under this condition, the conversion versus time can be solved analytically to give

$$t^* = 1 - (1 - X)^{1/F}g$$ (24)

On the other hand, when σ approaches infinity, the gas-solid reaction rate is controlled solely by intrapellet diffusion of the gaseous reactant. This case corresponds to the familiar shrinking-core model, and the analytical solution for a flat square pellet is given as

$$t^* = \sigma^2 X^2/2F_g F_p \qquad \text{for } F_p = 1 \quad (25)$$

$$t^* = [X + (1 - X) \ln (1 - X)]\sigma^2/2F_g F_p \qquad \text{for } F_p = 2 \quad (26)$$

$$t^* = [1 - 3(1 - X)^{2/3} + 2(1 - X)]\sigma^2/2F_g F_p \qquad \text{for } F_p = 3 \quad (27)$$

698

The progress of reaction in the solid is solved using a 23x1 grid for flat pellets, a 23x23 grid for cylindrical pellets, and a 13x13x13 grid for spherical pellets; a selected set of results is shown in Figs. 2 and 3.

The straight-line relationship predicted by asymptotic solutions are closely followed by the Monte Carlo solution. Sohn and Szekely (7) show a similar trend in their computational results obtained with a finite-difference numerical solution. Interested readers may refer to Sohn and Szekely (7) for the practical implications. These results attest to the accuracy of the numerical results obtained with the Monte Carlo algorithm.

The progress of reaction in an irregular pellet is shown in Fig. 4. The pellet is generated using the particle-shape algorithm. A 50x40-mesh point is used in dividing the pellet, and the calculations were performed on a UNIVAC-1160 mainframe computer. A hole is intentionally introduced in the pellet to test the algorithm. The hole is exposed to the surrounding gas and so the concentration of gas A within the hole is the same as that surrounding the periphery of the pellet. The equigrain-conversion and equiconcentration lines are shown in Figs. 4 and 5 respectively. The value of the dimensionless constant for this case is 10^4. A noteworthy feature of the Monte Carlo method is that the boundary can be as irregular as possible. Normal numerical procedures for solving partial differential equations encounter severe complications especially near ragged edges and holes.

Finally, the conversion of three-dimensional pellets is shown in Fig. 6. A cylinder, a pentagonal prism, and a cube were chosen for comparative study. In each case, the base area and height were so chosen that the dimensionless volume is close to unity. The surface area is greatest for the cube and smallest for the cylinder, and the average diffusion path is approximately the same for all three objects. Since the reaction is initially chemical-reaction controlled, the conversion is highest for the cube, followed by the pentagonal prism and cylinder at any given time.

Conclusions

The diffusion/chemical-reaction equation can be readily solved with finite-difference or other numerical schemes for regular geometries; however, the solution procedure becomes cumbersome for irregular geometries. The latter can be solved quite simply with the Monte Carlo algorithm. It is shown that the results of the Monte Carlo solution agrees closely with known asymptotic solutions for regular pellets. Later, the results for an irregular pellet reacting with a gas is shown.

This procedure is attractive from the point of view of numerical solution on a computer since the algorithm is extremely simple. The gas-solid reaction results shown here are useful for industrial roasting and fluidized-bed reaction analysis. Also, the algorithm can be used for the solution of elliptic and parabolic partial differential equation encountered in mass-, heat-, and momentum-transfer problems which must be solved in an irregular domain.

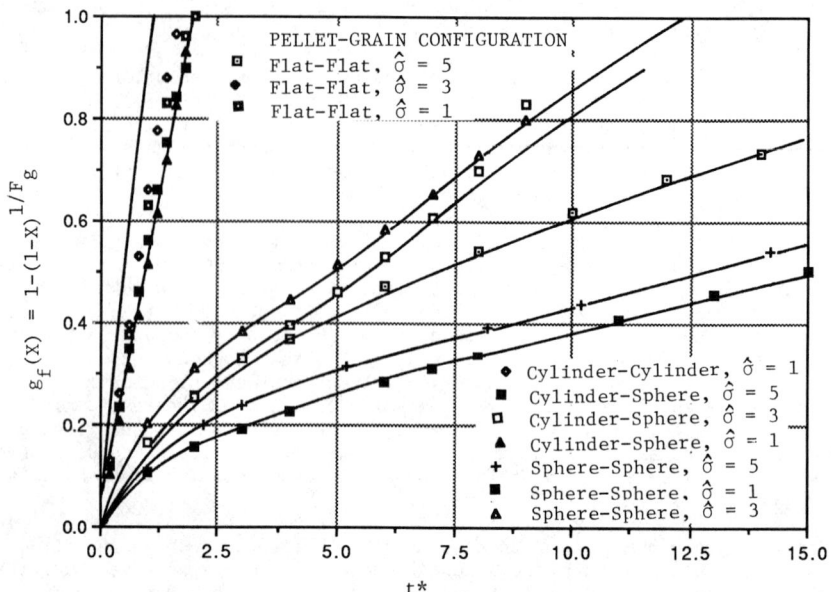

Figure 2. Conversion function $g_f(X)$ versus dimensionless
time

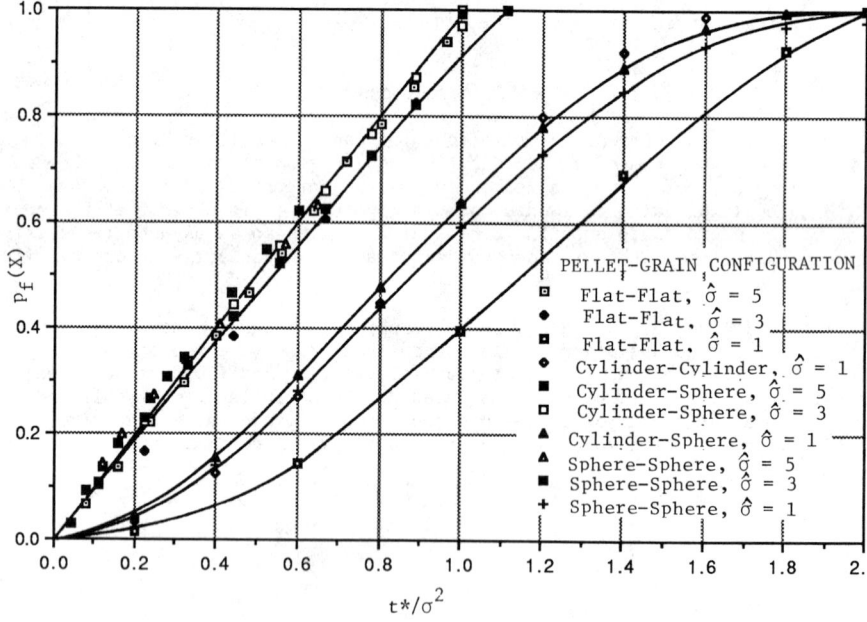

Figure 3. Conversion function $p_f(X)$ versus reduced
time

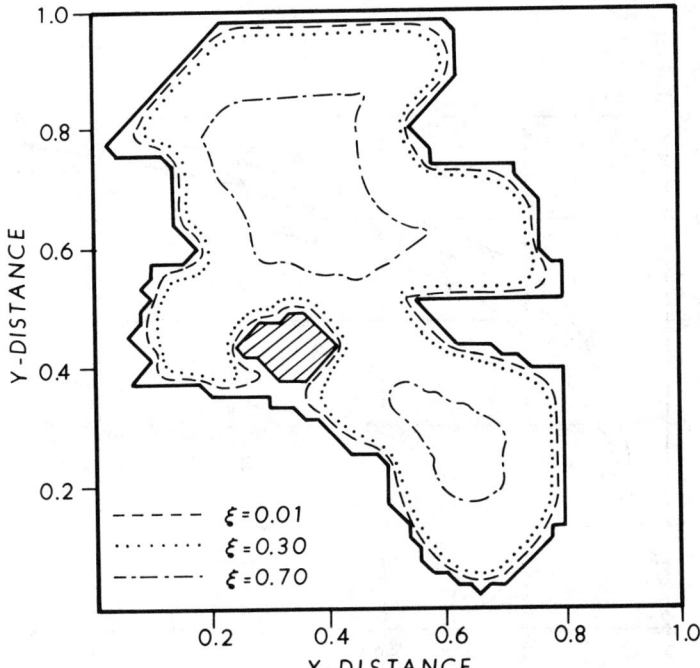

Figure 4. Progress of reaction in an irregular pellet with a hole (shaded region). Dotted lines indicate loci of equigrain conversion.

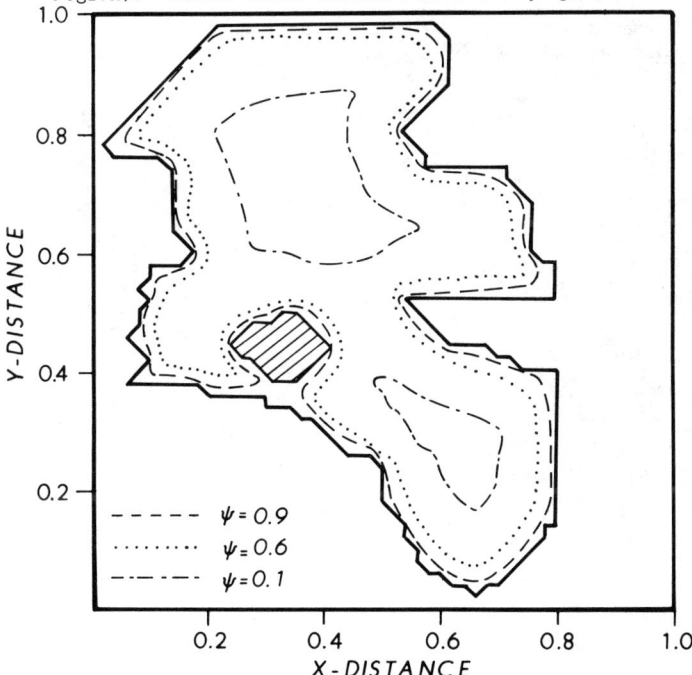

Figure 5. Dimensionless concentration of species A in the irregular pellet. Dotted lines show loci of equal concentration.

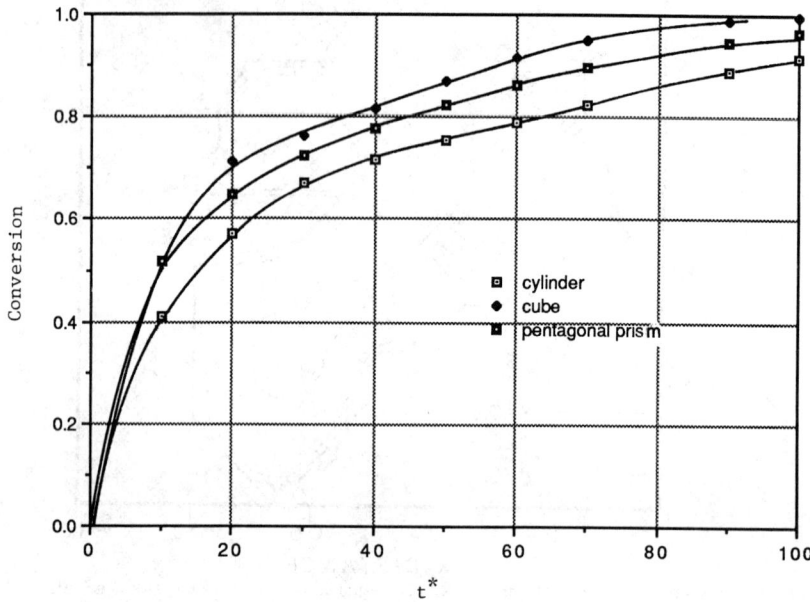

Figure 6. Comparison of conversion of cylindrical, cubrical and prismatic pellets.

NOTATION

A_g external surface area of the individual grains, cm^2

b   stoichiometric coefficients in Eq. (1)

C_A, C_{AO} gaseous reactant concentration and concentration surrounding the pellet

d_p principal dimension of the particle ($= F_p V_p / A_p$)

D   interior domain of the particle

D_e effective diffusivity of A in the porous pellet

$e_n(P,R)$ expected mass of particle starting at P upon arrival at R

$E_n(P,R)$ as defined in Eq. (12)

$E(P,R)$ as defined in Eq. (13)

F_g shape factor for the grain ($= 1$, 2, and 3 for flat plates, cylinders, and spheres, respectively)

F_p shape factor for the pellet ($= 1$, 2, and 3 for flat plates, cylinders, and spheres, respectively)

$g(x,y)$ known function in Eq. (9).

h   distance between adjacent mesh points

k   reaction-rate constant

$k(P)$ as defined in Eq. (18)

$L_n(P,R)$ set of all possible paths from P to R

$m_n(P,R)$ mass of fictitious particles starting at R upon arrival at R

N   total number of random walks used in the computations

P   a general point within the irregular domain bounded by the particle

P_j the four adjacent points to P lying on the mesh

r   distance coordinate in the grain

r_g radius of the grain

t   time variable

t^* dimensionless time ($= \dfrac{bkC_{AO}A_g t}{\rho_m F_g V_g}$)

$u(x,y)$ unknown function in Eq. (9)

v_A reaction rate for A

V_g volume of the individual grain

x,y,z x, y, and z coordinate directions

REDUCTION OF $Cu_2S(s)$ WITH $CO-CO_2$-He GAS MIXTURES

IN THE PRESENCE OF LIME: MATHEMATICAL MODELING

Y. K. Rao and M. Moinpour

Department of Materials Science and Engineering
University of Washington
Seattle, Washington 98195

Abstract

The kinetics of reduction of $Cu_2S(s)$ with $CO-CO_2$-He gas mixtures, using an excess of $CaO(s)$ as a sulfur-scavenging substance were determined in the range 850-960°C by a thermogravimetric method. A mathematical model developed to correlate the experimental results is presented here. The reaction mechanism involves the reaction of $CO(g)$ with $Cu_2S(s)$ to produce $COS(g)$ and $Cu(s)$; the product $COS(g)$ then reacts with $CaO(s)$-grains forming $CO_2(g)$ and $CaS(s)$. As the reduction progresses, a layer of $CaS(s)$ is built up around CaO-core. The model assumes that the pore-diffusion of $CO_2(g)$ through $CaS(s)$-product layer is rate-limiting and the reaction steps are relatively fast. The effective pore-diffusivities for $CO_2(g)$, under different conditions, were derived from the experimental weight loss data by the application of the model.

Introduction

The extraction of copper from sulfide concentrates without entailing simultaneous release of copious amounts of SO_2-bearing gases has attracted considerable attention in recent years. The reduction of sulfide with a suitable reducing agent (H_2, CO, or carbon), in the presence of a sulfur-scavenging agent (CaO, $CaCO_3$, or Na_2CO_3), produces metallic copper and the sulfur in the charge is "fixed" as sulfide of calcium (or sodium).

In a recent paper, the authors (1) reported the results of their investigation of the direct reduction of copper sulfide with carbon using excess lime to ensure complete retention of sulfur in the solid residue CaS(s). Potassium carbonate, sodium fluoride, and ternary (K,Li,Na)$_2CO_3$ eutectic exhibited (1) significant catalytic activities in the reduction of Cu_2S:4CaO:4C mixtures at temperatures in the range 858-996°C.

To elucidate the mechanism of the direct reduction process, a detailed thermogravimetric study was carried out to determine the rates of reaction between uncatalyzed Cu_2S/4CaO samples and $CO-CO_2$-He gas mixtures. Experiments were conducted at three different temperatures--855,904, and 950°C; several different combinations of partial pressures of CO, CO_2, and He were used. The materials Cu_2S(s) and CaO(s) were of high-purity (1) with average particle sizes of 8.2 μm for Cu_2S(s) and 5.4 μm for CaO(s), respectively. The overall reaction may be represented as follows:

$$Cu_2S(s) + CaO(s) + CO = 2 Cu(s) + CaS(s) + CO_2 \qquad (1)$$

The weight loss registered by the Cu_2S/4CaO sample is a direct measure of the extent of reaction. The fractional reduction α, is given by

$$\alpha = \Delta W/\Delta W_{max} \qquad (2)$$

where ΔW is the corrected weight loss at time t and ΔW_{max} (= $0.415 \, M_0 W_T^\circ/M_{Cu_2S}$) is the maximum weight loss corresponding to complete reduction. Note that W_T° is the initial weight of the Cu_2S/4CaO sample, M_0 and M_{Cu_2S} are the molecular weights of atomic oxygen and cuprous sulfide, respectively.

The details of the experimental work and the results are reported elsewhere (2). A typical set of α-t plots for the reduction of Cu_2S/4CaO with $CO-CO_2$-He gas mixtures is given in Figure 1. Normally it is expected that the rate of reduction increases with increasing temperature. However, as shown in Figure 1, the rate decreases with increasing temperature; a similar behavior was noted in the majority of the experiments with $CO-CO_2$-He gas mixtures. This apparently paradoxical behavior is due to equilibrium limitations that constrain the reduction process as well as the formation of a relatively impervious product layer CaS(s) around an unreacted CaO(s)-core. X-ray and SEM/EDAX analyses of partially-reacted samples confirmed the growth of CaS(s)-shell on CaO(s)-grains.

Figure 2 shows a schematic diagram of the partly-reacted mixture contained in a graphite crucible. Also shown is an amplified view of the Cu_2S/CaO particulates; this indicates the partial pressures of gas species CO, CO_2, He, and COS at a specified position (Z = Z) in the crucible and also at the CaS/CaO interface.

The overall reaction (1) consists of the following consecutive gas-solid reactions:

$$Cu_2S(s) + CO = 2 Cu(s) + COS \tag{A1}$$

$$CaO(s) + COS = CaS(s) + CO_2 \tag{A2}$$

The standard free energy changes for these reactions are drawn from published sources (2,3):

$$\Delta G_1^\circ = 35,865 - 12.385 \ T \log T + 89.747 \ T, \ J$$

$$\Delta G_2^\circ = -95,065 + 3.682 \ T \log T - 10.544 \ T, \ J$$

As shown in Figure 2, at level Z = Z in the crucible, the gas phase around the partially-reduced cuprous sulphide and partly-reacted lime grains consists of CO, CO_2, He, and COS at partial pressures P_{CO}, P_{CO_2}, P_{He}, and P_{COS}. The total pressure within the $Cu_2S/4CaO$ mixture in the crucible is a uniform 1 atm. Therefore,

$$P_{CO} + P_{CO_2} + P_{He} + P_{COS} = 1.0 \tag{3}$$

The intermediate COS(g) is produced by reaction (A1) and is consumed by reaction (A2). Within the CaS(s)-product layer that builds up around the CaO(s)-core, there are micropores that permit the diffusion of COS(g) inwards and that of $CO_2(g)$ outwards. At the CaS(s)/CaO(s) interface, it is reasonable to expect the sulfidizing reaction (A2) to be in a state of "virtual equilibrium." The published kinetic studies (4,5) reveal that reaction (A2) occurs rapidly at temperatures as low as 600°C. Thus, in Figure 2, the partial pressures of the gas species at the CaS(s)/CaO(s) interface are: $P_{CO_2}^e$, P_{COS}^e, P_{CO}, and P_{He}; the last two are to be regarded as non-diffusing species and their partial pressure gradients across the CaS(s)-layer are taken as nil. It will also be noted that $P_{CO_2}^e > P_{CO_2}$ and $P_{COS}^e < P_{COS}$ at level Z = Z.

$$P_{CO_2}^e + P_{COS}^e + P_{CO} + P_{He} = 1.0 \tag{4}$$

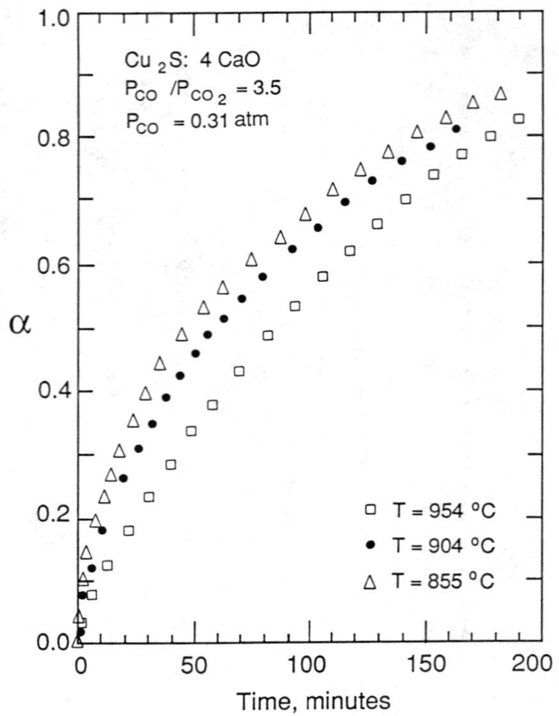

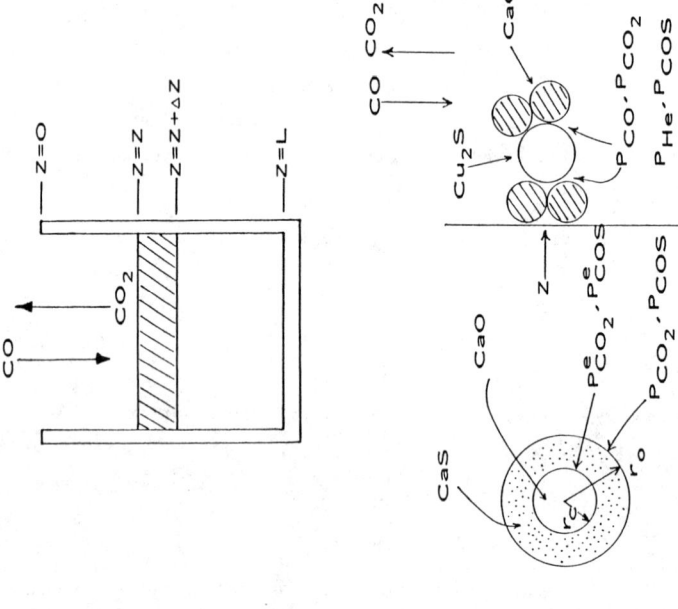

At Z=Z
in Crucible

Figure 2 – Schematic diagram of the $Cu_2S/4CaO$ mixture contained in a graphite crucible and an expanded view of reacted grains.

Figure 1 – Uncatalyzed reduction of $Cu_2S(s)$ with $CO-CO_2$-He gas mixtures in the presence of excess CaO(s); the effect of temperature is shown.

Noting that $K_2 = P^e_{CO_2}/P^e_{COS}$, where K_2 is the equilibrium constant for reaction (A2), we can obtain

$$P^e_{CO_2} = (1.0 - P_{CO} - P_{He}) [1 + (K_2)^{-1}]^{-1} \tag{5}$$

From equations (3) and (5) it can be shown that

$$P^e_{CO_2} - P_{CO_2} = \frac{P_{COS} - (P_{CO_2}/K_2)}{1 + (K_2)^{-1}} \tag{6}$$

The partial pressure P_{COS} of COS(g) in the intergranular pore-space can be found by assuming that reaction (A1) is in a state of virtual equilibrium. That is, $P_{COS} = K_1 P_{CO}$, where K_1 is the equilibrium constant for reaction (A1). Substitution gives

$$P^e_{CO_2} - P_{CO_2} = \frac{K_1 P_{CO} - (P_{CO_2}/K_2)}{1 + (K_2)^{-1}} \tag{7}$$

The difference $(P^e_{CO_2} - P_{CO_2})$ is the driving force for pore-diffusion through the CaS(s)-product shell (Figure 2). Thus, reduction is possible so long as the numerator on the right-side of equation (7) is a positive quantity. Calculations showed that the partial pressure difference is greater than zero for the 37 gas-compositions used in the present work (2); it ranged from 3.5845×10^{-6} atm for a 29.7% CO-14.9% CO_2-55.4% He mixture at 947°C to 2.6388×10^{-5} atm for the 49.8% CO-10% CO_2-40.2% He mixture at 951°C. It was found that for most gas mixtures, the driving force decreased with rising temperature, at a specified gas composition. This may account, at least in part, for the observed rate-decrease with increasing temperature.

In general, the heterogeneous reduction of $Cu_2S/4CaO$ Particulate samples by CO-CO_2-He gas mixtures involves a number of elementary process steps; the slowest step or a combination of slower steps in effect determines the rate of the overall reduction process. The following are considered important: (1) Film mass-transfer of the reactant and product gases across the stagnant boundary layer around the sample-filled crucible. (2) Diffusion of these gases through the pores around the grains Cu_2S and CaO. (3) In partially-reduced samples, with product layer formed on unreacted-core, the gases must diffuse through micropores of the product shell. (4) Chemical reactions between gaseous species, cuprous sulfide, and lime grains at the respective reaction interfaces: reaction (A1) at Cu/Cu_2S and reaction (A2) at CaS/CaO interfaces, respectively. The first of these two reactions establishes the COS-pressure in the pore-space between the grains in the sample.

709

The film mass-transfer effects were made negligible by employing relatively high flow-rates (1900-2300 ml/min at STP) for the $CO-CO_2$-He gas mixtures. The experimental results (2) suggest that reactions (Al) and (A2) are not rate-controlling; evidence from the literature (4,5) shows that the kinetics of these reactions are quite rapid at 800°C and above. Consequently, our focus will be on the pore-diffusion of gaseous species through the porous particulate mixture; especially, the diffusion of gases through the micropores of the product shell will receive a careful scrutiny.

The conversion of $Cu_2S(s)$ to $Cu(s)$, as per reaction (Al), involves a net transfer of matter from the solid to the gas phase. Thus, the metallic product layer surrounding the unreduced cuprous sulfide core is expected to be porous; it is readily seen that

$$\theta_{Cu} = \frac{v^m_{Cu_2S} - 2\,v^m_{Cu}}{v^m_{Cu_2S}} = \frac{28.42 - 2(7.12)}{28.42} = 0.5$$

where θ_{Cu} is the fractional porosity of the copper shell, and $v^m_{Cu_2S}$ and v^m_{Cu} are the molar volumes (in cm^3/mole) of $Cu_2S(s)$ and $Cu(s)$, respectively (6). This porosity is distributed in the form of pore-capillaries that traverse the product layer of the cuprous sulfide grain.

The sulfidization of $CaO(s)$ according to reaction (A2) produces a $CaS(s)$-shell. The density of calcium sulfide is only about three-fourths that of calcium oxide (6); hence, there is a volume-increase on forming $CaS(s)$. The fractional increase in volume, ε_{CaS}, is given by

$$\varepsilon_{CaS} = \frac{v^m_{CaS} - v^m_{CaO}}{v^m_{CaO}} = \frac{28.86 - 17.00}{17.00} = 0.7$$

where v^m_{CaO} and v^m_{Cas} are the molar volumes (in cm^3/mole) of $CaO(s)$ and $CaS(s)$, respectively. The substantial difference between these quantities and the large value for ε_{CaS} make it unlikely that significant porosity will develop within the $CaS(s)$-product shell. The substitution of $CaS(s)$ for $CaO(s)$ introduces mechanical strains which result in microcracks in the $CaS(s)$-product shell. Thus, a small fractional microporosity does come into existence as the product layer builds up around the unreacted $CaO(s)$-core.

In view of the foregoing, it is clear that the preponderant resistance to the diffusion of gaseous species originates with the $CaS(s)$ product layer. The equimolar counterdiffusion of $COS(g)$ and $CO_2(g)$ species through the $CaS(s)$-shell is regarded as the rate-controlling elementary process.

Mass-Transfer

The flow of gas species through porous media is frequently expressed by the well-known Kozeny-Carman equation (7,8). For the flux of a gas through the porous aggregate, Carman (7,8) provides,

$$J = \left[\frac{P\theta^3}{KS_v^2 \eta(1-\theta)^2} + \frac{8\delta\theta^2}{3 \ K'S_v(1-\theta)} (\frac{2 \ RT}{\pi M})^{1/2} \right] \frac{\Delta P}{RTL} \tag{8}$$

where P is pressure, S_v, specific pore-surface area, M, molecular weight of gas, θ, fractional porosity of aggregate, L, thickness of the porous mixture, ΔP, pressure drop, R, gas constant, T, temperature, and η, viscosity of gas. Note that K, K', and δ are constants. In this equation, the first term denotes the viscous flow and the second term represents the "slip flow" contribution. Rao (9,10) showed that viscous flow becomes important when the average pore diameter is large. For small pore-size, the slip flow term predominates. In the experimental work (2) considered here, the particle sizes are such that the viscous flow can be neglected without introducing any appreciable error. Rao (9) showed that the slip-flow term can be written as follows:

$$J_i = \frac{D_{ie}}{RT} (\frac{dP_i}{dZ}) \tag{9}$$

where D_{ie} is the effective diffusivity of the ith gas species and (dP_i/dZ) is its partial-pressure gradient in the volume element at Z = Z in Figure 2. The fluxes of $CO_2(g)$ and $COS(g)$ through the micropores of CaS(s)-product layer can be represented by expressions similar to equation (9). The calculation of effective diffusivities in multi-component gas mixtures is presented in the Appendix.

Mechanism

The mechanism of reduction of a porous bed composed of dense $Cu_2S(s)$ and CaO(s) grains with $CO-CO_2$-He gas mixtures consists of reactions (A1) and (A2), given earlier. As shown in Figure 2, the reactant gas CO diffuses inwards through pore-passages around the grains, then through porous Cu(s)-layer, and reacts with $Cu_2S(s)$; this produces COS(g) which difuses outwards into the intergranular pore-space. Thereafter it diffuses through micropores in the CaS(s)-product layer to the CaS/CaO interface where reaction (A2) occurs. The product of this reaction, $CO_2(g)$, diffuses outwards through CaS(s) and then upwards through pore-passages, as shown in Figure 2.

711

The lack of significant temperature-influence on the rate of reduction indicates that the interfacial reaction steps (A1) and (A2) are at near (or virtual) equilibrium. The intergranular effective diffusivities of the porous aggregate are much larger than those for the CaS(s)-product shell (Appendix). Thus, of the various diffusion processes, the diffusion of $CO_2(g)$, produced by reaction (A2), through the relatively impervious layer of CaS(s) which forms around CaO(s)-grain appears to be the rate-determining step.

Development of the Model

The porous sample is seen to fill the graphite crucible up to about 1.5 mm from the top (Figure 2). This position is designated $Z = 0$. For the purposes of developing the mass-conservation relations, a reference volume element ΔZ thick is chosen at a depth of $Z = Z$; only the upper surface of the porous $Cu_2S/4CaO$ mixture is exposed to the incoming $CO-CO_2-He$ reducing gas mixture. The average particle diameter, d_o, for the mixture is given by

$$d_o = [(\nu_1/d_1) + (\nu_2/d_2)]^{-1} \tag{10}$$

where ν_1 and ν_2 are the volume-fractions and d_1 and d_2 are the average particle sizes for the $Cu_2S(s)$ and CaO(s) components, respectively. The value of d_o was found to be 6.01 μm. The average pore-diameter $\bar{d}_p$ for a close-packed bed of spherical particles is approximately $\bar{d}_p/3$; hence $\bar{d}_p$ is 2.003 μm (11). For this pore-size, both Knudsen and molecular diffusion mechanisms are important. To facilitate the development of the model, the following assumptions are introduced: (1) the resistance to reduction process due to film mass-transfer is negligible; (2) the porous aggregate remains isothermal throughout the course of reduction; (3) the porous packed-bed maintains its structural integrity during reduction; (4) the grains are assumed to be uniform-sized spheres which undergo no significant changes during the process; (5) only diffusion in the Z-direction need be considered; and (5) the quasi-steady-state approximation applies (12-14).

Diffusion through CaS(s)-shell

Figure 2 shows the cross-section of a partially-sulfidized lime grain. Let r_c be the radius of the unreacted CaO(s)-core with r_o denoting the initial radius of the grain. At a depth of $Z = Z$ in the porous sample, let P_{CO_2} and $P_{CO_2}^e$ be the partial pressures of CO_2 at the grain-surface and at the CaS/CaO interface, respectively. Note that $P_{CO_2}^e$ is the equilibrium partial pressure of CO_2 for reaction (A2), given by equation (5). The partial pressure profile for $CO_2(g)$ across the CaS(s)-layer can be derived by the integration of Fick's second law,

$$\frac{\partial}{\partial r} (r^2 D_{es} \frac{\partial \overset{\circ}{P}_{CO_2}}{\partial r}) = 0 \qquad (11)$$

using the following boundary conditions:

$$(i) \quad \overset{\circ}{P}_{CO_2} = P_{CO_2} \quad at \quad r = r_o \text{ (outer surface)} \qquad (12.a)$$

$$(ii) \quad \overset{\circ}{P}_{CO_2} = P^e_{CO_2} \quad at \quad r = r_c \text{ (CaS/CaO interface)} \qquad (12.b)$$

Solution to equation (11) subject to these yields

$$\overset{\circ}{P}_{CO_2} = P_{CO_2} - \frac{P^e_{CO_2} - P_{CO_2}}{(\frac{1}{r_o} - \frac{1}{r_c})} (\frac{1}{r} - \frac{1}{r_o}) \qquad (13)$$

where $\overset{\circ}{P}_{CO_2}$ is the partial pressure of CO_2 at any position $r = r$ within the CaS-product layer $(r_c < r < r_o)$. The flux of $CO_2(g)$ out of the CaO/CaS grain is given by

$$\overset{\circ}{J}_{CO_2}\Big|_r = - (D_{es}/RT) \cdot \frac{\partial \overset{\circ}{P}_{CO_2}}{\partial r}\Big|_r , \quad mole\ CO_2/cm^2 \cdot s \qquad (14)$$

where D_{es} is the effective diffusivity (in cm^2/s) of CO_2 in the CaS(s)-layer. The steady-state outward diffusion-rate of $CO_2(g)$, from the CaS/CaO grain, is found by multiplying the flux at $r = r$ with the surface area. As a result, for a single spherical grain, we have,

$$\pi^{CaO}_g = \frac{4 \pi D_{es}}{RT} \cdot \frac{P^e_{CO_2} - P_{CO_2}}{(\frac{1}{r_c} - \frac{1}{r_o})} , \quad mole\ CO_2/s \qquad (15)$$

This is the rate of formation of $CO_2(g)$ for one CaO-grain.

The number $N^{\circ}_{Cu_2S}$ of $Cu_2S(s)$ grains per unit volume of the unreacted $Cu_2S/4CaO$ porous aggregate can be found using

$$\frac{4}{3} \pi (r^{\circ}_{Cu_2S})^3 \rho^t_{Cu_2S} N^{\circ}_{Cu_2S} V_o = W_o y_{Cu_2S} \qquad (16)$$

Similarly, for the number N°_{CaO} of CaO(s) grains, we have,

$$\frac{4}{3} \pi (r^{\circ}_{CaO})^3 \rho^t_{CaO} N^{\circ}_{CaO} V_o = W_o y_{CaO} \qquad (17)$$

where $r^{\circ}_{Cu_2S}(= 4.1\ \mu m)$ and $r^{\circ}_{CaO}(= 2.7\ \mu m)$ are the average grain radii, $\rho^t_{Cu_2S}$ and ρ^t_{CaO} are true densities, and y_{Cu_2S} $(= 0.41503)$ and $y_{CaO}(= 0.58497)$ are weight-fractions, for $Cu_2S(s)$ and CaO(s) respectively. Note that

V_o (=0.5081 cm^3) is the volume of the porous mixture having an initial weight of W_o (=0.6356 gm). Rearrangement of terms yields

$$N^\circ_{Cu_2S} = 3 \rho_b y_{Cu_2S}/4 \pi (r^\circ_{Cu_2S})^3 \rho^t_{Cu_2S} \quad , \quad cm^{-3} \tag{18.a}$$

$$N^\circ_{CaO} = 3 \rho_b y_{CaO}/4 \pi (r^\circ_{CaO})^3 \rho^t_{CaO} \quad , \quad cm^{-3} \tag{18.b}$$

where ρ_b (=1.2509 gm/cm^3) is the bulk density of the porous mixture.

We can now calculate the rate of formation of $CO_2(g)$ per unit volume of the porous sample. Because there is 300 mol% excess lime in the mixture, stoichiometric considerations dictate that, on the average, only one-fourth of the CaO(s) grains will be sulfidized. Hence, the rate of formation of $CO_2(g)$ per unit volume is given by

$$R^{CaO}_g = \frac{\pi D_{es} N^\circ_{CaO}}{RT} \frac{P^e_{CO_2} - P_{CO_2}}{(\frac{1}{r_c} - \frac{1}{r_o})} \quad , \quad mole \ CO_2/cm^3 \cdot s \tag{19}$$

Note that each mole of $CO_2(g)$ produced corresponds to one mole of $Cu_2S(s)$ reduced to Cu(s) (or one mole of CaO sulfidized to CaS). Thus, the rate of production of $CO_2(g)$ is equal to the rate of reduction of $Cu_2S(s)$, in conformity with the stoichiometric equation (1).

Mass-transfer through Porous Packed-bed

As discussed earlier, the flow through porous media can be expressed by equation (8). For the experimental conditions used in the present work, the flux equation (9) is more apt. There is a net inward diffusion of CO(g) and an equivalent outward diffusion of $CO_2(g)$ through the particulate mixture. We shall now develop the mass-conservation relationships for CO and CO_2. The flux of CO_2 outwards at $Z = Z$ in Figure 2 is given by

$$J_{CO_2}\Big|_Z = - (D'_b/RT)\frac{\partial P_{CO_2}}{\partial Z}\Big|_Z \quad , \quad mole/cm^2 \cdot s \tag{20}$$

Performing mass-balance over the volume element ΔZ thick, we obtain,

$$\frac{\partial J_{CO_2}}{\partial Z} \cdot A_c = R^{CaO}_g \cdot A_c \cdot \Delta Z$$

where A_c (=0.4537 cm^2) is the area of cross-section of the graphite crucible holding the porous aggregate. Simplifying,

$$\frac{\partial^2 P_{CO_2}}{\partial Z^2} = - \frac{\pi D_{es} N^\circ_{CaO}}{D'_b} \cdot \frac{P^e_{CO_2} - P_{CO_2}}{(\frac{1}{r_c} - \frac{1}{r_o})} \tag{21}$$

714

Similarly, for the reactant species $CO(g)$, we can derive

$$\frac{\partial^2 P_{CO}}{\partial Z^2} = \frac{\pi D_{es} N^{\circ}_{CaO}}{D_b} \cdot \frac{P^e_{CO_2} - P_{CO_2}}{(\frac{1}{r_c} - \frac{1}{r_o})} \tag{22}$$

In the foregoing D_b and D'_b are the effective diffusivities of CO and CO_2 through the porous solid mixture (Appendix).

By comparing the non-linear second-order differential equations (21) and (22), we obtain,

$$D'_b(\partial^2 P_{CO_2}/\partial Z^2) + D_b(\partial^2 P_{CO}/\partial Z^2) = 0$$

Rearranging terms,

$$\frac{\partial^2[(D_b/D'_b)P_{CO} + P_{CO_2}]}{\partial Z^2} = 0 \tag{23}$$

The boundary conditions are as follows:

(i) $P_{CO} = P^b_{CO}$ and $P_{CO_2} = P^b_{CO_2}$ at $Z = 0$ (top) (24.a)

(ii) $\dfrac{\partial P_{CO}}{\partial Z} = \dfrac{\partial P_{CO_2}}{\partial Z} = 0$ at $Z = L$ (bottom) (24.b)

On integrating equation (23) subject to equations (24.a,b),

$$(D_b/D'_b)P_{CO} + P_{CO_2} = (D_b/D'_b)P^b_{CO} + P^b_{CO_2}$$

Solving for P_{CO}, we have, at position $Z = Z$ in the crucible,

$$P_{CO} = P^b_{CO} - (D'_b/D_b)(P_{CO_2} - P^b_{CO_2}) \tag{25}$$

where P^b_{CO} and $P^b_{CO_2}$ are the partial pressures of CO and CO_2, respectively, in the bulk gas phase. Thus, only one of the two equations (21) and (22) is truly independent. On combining equations (7) and (25) and simplifying, we obtain for the driving force,

$$P^e_{CO_2} - P_{CO_2} = u - vP_{CO_2} \tag{26}$$

where

$$u = \frac{K_1 P^b_{CO} + K_1(D'_b/D_b)P^b_{CO_2}}{1 + (K_2)^{-1}} \tag{27.a}$$

$$v = \frac{K_1(D'_b/D_b) + (K_2)^{-1}}{1 + (K_2)^{-1}} \tag{27.b}$$

715

On substituting equation (26) into equations (15) and (21), we find

$$\pi_g^{CaO} = 4 \pi D_{es} \frac{u - vP_{CO_2}}{RT(1/r_c - 1/r_o)} \quad , \quad \text{mole } CO_2/s \tag{28}$$

$$-\frac{\partial^2 P_{CO_2}}{\partial z^2} = \frac{\pi D_{es} N_{CaO}^o (u - vP_{CO_2})}{D_b'(1/r_c - 1/r_o)} \tag{29}$$

Formulation of the Model: First Stage

During this stage, the topmost layer of the particulate mixture still contains some unreduced $Cu_2S(s)$, distributed among numerous grains in the form of a core enveloped by a shell of metallic copper. As the reduction progresses, the $COS(g)$ produced by reaction (A1) is instantaneously consumed by reaction (A2) to form $CaS(s)$; with passage of time, the core radius r_c of CaS/CaO grain decreases. Mass balance for a single grain gives

$$-\frac{\partial n_{CaO}}{\partial t} = \pi_g^{CaO} \tag{30}$$

The number of moles n_{CaO} of CaO(s)-core of the grain at t = t is given by

$$n_{CaO} = \frac{4}{3} \pi r_c^3 \rho_{CaO}^t / M_{CaO} \tag{31}$$

On combining equations (30) and (31) with (28) and simplifying, we obtain,

$$-\frac{\partial r_c}{\partial t} = \frac{(u - vP_{CO_2}) D_{es} M_{CaO}}{RTr_c^2 (1/r_c - 1/r_o) \rho_{CaO}^t} \tag{32}$$

where M_{CaO} is the molecular weight of CaO. Equations (29) and (32) constitute the essence of the mathematical model the solution of which provides information on the reduction process. It is well to note that in these equations the value of D_{es}, the effective diffusivity of $CO_2(g)$ through CaS-product layer, is not known a priori; it is calculated by the application of the model. The anticipated partial pressure profiles of $CO(g)$ and $CO_2(g)$ in the reduction system are depicted schematically in Figure 3. The initial and boundary conditions are as follows:

(i) At $t = 0$, $r_c = r_o$ (33.a)

(ii) At $Z = 0$, $P_{CO_2} = P_{CO_2}^s = P_{CO_2}^b$ (top)   (33.b)

(iii) At $Z = L$, $\left.\frac{\partial P_{CO_2}}{\partial Z}\right|_{Z=L} = 0$ (bottom)   (33.c)

The model equations (29) and (32) are transformed into a dimensionless form:

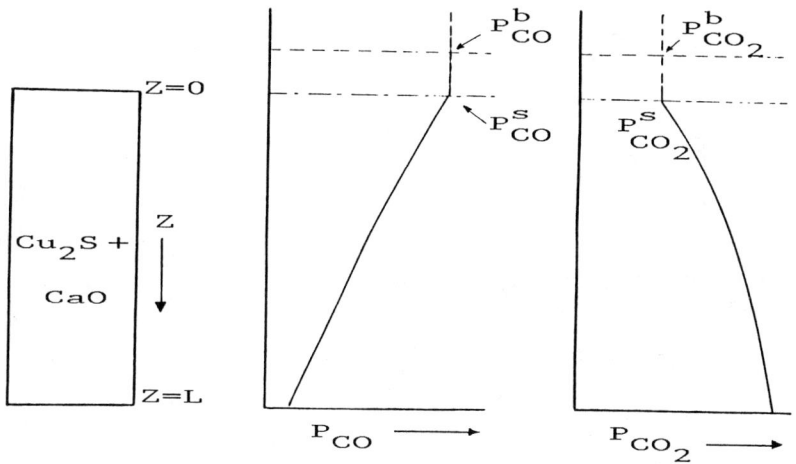

Figure 3 - Partial pressure profiles for $CO(g)$ and $CO_2(g)$ during the
first stage of reaction.

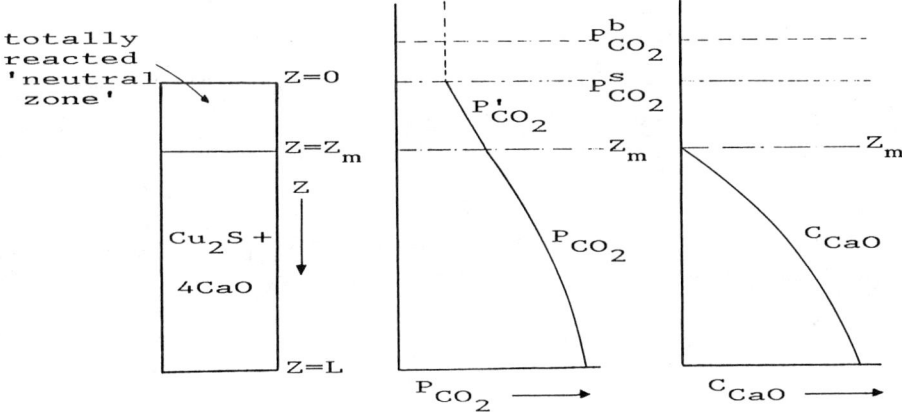

Figure 4 - Profiles of CO_2-pressure and CaO-concentration during the
second stage; the "neutral zone" is shown.

717

$$\frac{\partial^2 \gamma}{\partial \sigma^2} - \phi^2 \frac{\gamma \varepsilon}{(1 - \varepsilon)} = 0 \qquad (34)$$

$$\frac{\partial \varepsilon}{\partial \tau} + \frac{\gamma}{\varepsilon(1-\varepsilon)} = 0 \qquad (35)$$

where

$$\gamma = \frac{u - vP_{CO_2}}{u - vP_{CO_2}^b} \quad ; \quad \varepsilon = r_c/r_o \quad ; \quad \sigma = Z/L \qquad (36.a\text{-}c)$$

$$\tau = [(u - vP_{CO_2}^b)D_{es}M_{CaO}/RT \, r_o^2 \rho_{CaO}^t]t \qquad (37)$$

$$\phi = [\pi D_{es} vL^2 r_o N_{CaO}^\circ/D_b']^{1/2} \qquad (38)$$

In the foregoing, the functions u and v have the same meaning as before and are defined by equations (27.a) and (27.b) respectively. The initial and boundary conditions now become

(i) At $\tau = 0$, $\varepsilon = 1$ $\qquad\qquad\qquad\qquad\qquad$ (39.a)

(ii) At $\sigma = 0$, $\gamma = 1$ $\quad$ (top) $\qquad\qquad\qquad$ (39.b)

(iii) At $\sigma = 1$, $\frac{\partial \gamma}{\partial \sigma} = 0$ $\quad$ (bottom) $\qquad\qquad$ (39.c)

It is well to note that γ, the dimensionless CO_2-pressure in the porous aggregate, is a function of both the position-coordinate σ and the reaction time τ; so is the dimensionless core radius ε. By solving equations (34) and (35), we can obtain γ vs. σ and ε vs. σ profiles at specified τ; and the partial pressure profiles of CO_2 and CO can be determined therefrom using equations (36.a) and (25), respectively.

Conversion of CaO to CaS

At position $Z = Z$ in the crucible (Figure 2), at time t, in a partially-reacted grain of CaO(s), the core-radius is r_c and a layer of CaS(s) is built up around the CaO-core. In the grain,,

$$n_{CaO}^\circ = \frac{4}{3} \pi r_o^3 \rho_{CaO}^t/M_{CaO} \qquad \text{at } t = 0$$

$$n_{CaO} = \frac{4}{3} \pi r_c^3 \rho_{CaO}^t/M_{CaO} \qquad \text{at } t = t$$

The number of moles of CaO(s) converted to CaS(s) in a single grain is given by

$$\Delta n_{CaS}^g = [4 \pi \rho_{CaO}^t (r_o^3 - r_c^3)/3 \, M_{CaO}] \qquad (40)$$

In the elemental volume ΔZ thick at $Z = Z$, there are $(N_{CaO}^\circ A_c \Delta Z/4)$ grains of

718

partially-sulfidized CaO(s). Hence, the number of moles of CaO(s) converted to CaS(s) in this volume element are

$$\Delta n_{CaS} = [\pi \rho_{CaO}^t N_{CaO}^\circ A_c r_o^3 (1-\varepsilon^3) \Delta Z/3 \; M_{CaO}] \tag{41}$$

It can be shown that for the entire volume of the porous sample in crucible, the number of moles of CaO(s) transformed to CaS(s) is:

$$n_{CaS} = [\pi \rho_{CaO}^t N_{CaO}^\circ A_c r_o^3 L/3 \; M_{CaO}] \int_0^1 (1-\varepsilon^3) d\sigma \tag{42}$$

Each mole of CaO(s) converted results in the transfer of one gram-atom (or 16 gms) of oxygen to the gas-phase. Hence the total weight loss accompanying reduction of $Cu_2S/4CaO$ mixture at $t = t$ is found to be

$$\Delta W_{cal} = [16\pi \rho_{CaO}^t N_{CaO}^\circ A_c r_o^3 L/3 \; M_{CaO}] \int_0^1 (1-\varepsilon^3) d\sigma \tag{43}$$

The objective is to calculate ΔW_{cal}, from the model, at $t = t$ and compare it with experimental result ΔW_{obs}. If the agreement is good, it can be concluded that the particular value of D_{es} used in equations (37) and (38) is the correct effective diffusivity of $CO_2(g)$ through CaS(s)-product layer. To apply the model, equations (34) and (35) have to be solved numerically.

Application of the Model
Numerical Solution

The solution to the partial differential equations (34) and (35) is obtained by the finite-difference method (15). A rectangular mesh of size $(\Delta\sigma, \Delta\tau)$ was constructed to represent the $Cu_2S/4CaO$ in space and time. The finite-difference form of the second-differential of the function $\gamma(\sigma,\tau)$ may be represented by the central-difference formula (15):

$$\frac{\partial^2 \gamma}{\partial \sigma^2} = \frac{\gamma_{i+1,j} - 2\gamma_{ij} + \gamma_{i-1,j}}{(\Delta\sigma)^2} \tag{44}$$

Similarly, for the first-differential of the function $\varepsilon(\sigma,\tau)$ with respect to τ, at any given node, we can use the forward-difference formula (15):

$$\frac{\partial \varepsilon}{\partial \tau} = \frac{\varepsilon_{i,j+1} - \varepsilon_{ij}}{(\Delta\tau)} \tag{45}$$

Substitution of equation (45) into equation (35) yields the following expression:

$$\varepsilon_{i,j+1} = \varepsilon_{ij} - \frac{\gamma_{ij}}{\varepsilon_{ij}(1-\varepsilon_{ij})} \cdot \Delta\tau \tag{46}$$

On combining equations (34) and (44), we obtain,

$$\frac{\gamma_{i+1,j} - 2\gamma_{ij} + \gamma_{i-1,j}}{(\Delta\sigma)^2} - \phi^2 \frac{\gamma_{ij}\,\varepsilon_{ij}}{(1-\varepsilon_{ij})} = 0 \tag{47}$$

At the closed end, the bottom surface of the crucible, which corresponds to i = 0, the boundary condition (39.c) can be written as

$$\frac{\partial\gamma}{\partial\sigma} = (\gamma_{1,j} - \gamma_{-1,j})/(2\,\Delta\sigma) = 0$$

Thus, at the bottom, at i = 0, we have

$$\gamma_{-1,j} = \gamma_{1,j} \qquad \text{(mirror-image)} \tag{48}$$

Equations (47) and (48) can now be combined to obtain the following expressions for γ, the dimensionless CO_2-pressure:

$$\gamma_{1,j} = \gamma_{0,j} + 0.5(\Delta\sigma)^2\phi^2 \frac{\gamma_{0,j}\,\varepsilon_{0,j}}{(1-\varepsilon_{0,j})} \qquad \text{at } i = 0 \tag{49.a}$$

$$\gamma_{i+1,j} = 2\gamma_{ij} - \gamma_{i-1,j} + (\Delta\sigma)^2\phi^2 \frac{\gamma_{ij}\,\varepsilon_{ij}}{(1-\varepsilon_{ij})} \qquad \text{at } i > 0 \tag{49.b}$$

To perform numerical calculations, the sample-height is subdivided into 100 equi-sized disc-like sections which are numbered sequentially starting at the bottom-surface (i = 0, or actually I = 1 in the program) and concluding at the top surface (i = 100, or actually I = 101 in the program); hence, in equations (49.a,b), $\Delta\sigma$ = 0.01. Also a time-interval of 10 minutes is selected; the corresponding dimensionless value $\Delta\tau$ is found using equation (37). It will be noted that the parameters P_{CO}^b, $P_{CO_2}^b$, D_b, D_b', K_1, K_2, r_o and N_{CaO}^o are calculated prior to the start of numerical integration.

An iterative procedure is adopted to solve the finite-difference equations given above. In order to carry out the computation of γ vs. σ and ε vs. σ at any given τ, it is necessary to assign an arbitrary initial value to D_{es} and determine therefrom the diffusion modulus ϕ. Based on the experimental results for the sulfidation of CaO(s) with COS(g) reported by Yang and Chen (4), D_{es} is estimated and used as starting value in the numerical calculations. Although the initial condition (39.a) implies that $\varepsilon_{0,0}$ = 1.0, insertion of this value into equation (46) makes the right-side go to infinity. To overcome this, an initial value of $\varepsilon_{0,0}$ = 0.975 is made use of in equation (46) at the start of the calculations. The corresponding time of reduction is determined by comparing $\Delta W_{cal}^{0,0}$, computed using equation (43), with experimental weight loss data. In this manner, the zero time of reduction is adjusted.

To begin the calculation, an arbitrary value is assigned to $\gamma_{0,0}$, at the closed end of the crucible at zero time. The successive γ-values (at i = 1 to i = 100) are found by means of equations (49.a,b); in these calculations, use is made of $\varepsilon_{i,0}$ = 0.975 for all values of i. The γ-value at i = 100 (or I = 101) is compared to 1.0, the expected value at the top-surface of the sample, as stipulated by the boundary condition (39.b). In general, since $\gamma_{0,0}$ is chosen arbitrarily, equation (39.b) is not satisfied in the first iteration. Then $\gamma_{0,0}$ is modified and the procedure repeated until convergence is achieved whence $\gamma_{100,0}$ approaches 1.0 within 1% accuracy limit. Then the value of $\varepsilon_{0,1}$ is found with equation (46) using the chosen time interval $\Delta\tau$. The calculations are repeated until $\gamma(\sigma,\tau)$ and $\varepsilon(\sigma,\tau)$ are found at all (σ,τ) nodes of interest. In this manner, the γ vs. σ and ε vs. σ profiles can be calculated for any given set of P_{CO}^b, $P_{CO_2}^b$, T, and D_{es}.

The integrand in equation (43) is found by trapezoidal numerical integration at each time interval considered and the corresponding ΔW_{cal} is determined. The experimental ΔW_{obs} values are compared to the calculated ΔW_{cal} values. The differences between the two sets are lessened by changing the values assigned to D_{es}. The procedure is continued until D_{es}-value that corresponds to the smallest sum of squared differences between the sets ΔW_{obs} and ΔW_{cal} is obtained for each experiment. Preliminary calculations showed that better fits are achieved if, instead of assigning constant values to D_{es}, a time-dependent function is employed for the diffusivity of CO_2.

$$D_{es} = D_o - E_o t^{1/n} \, , \, cm^2/s \qquad (50)$$

In this equation, D_o is the initial effective diffusivity for CO_2(g) at t = 0 and E_o is a constant. This means that D_{es} is a decreasing function of time. This is consistent with the earlier observation that the thickness of CaS-product layer grows with elapse of time becoming more impervious thus making the diffusion of CO_2(g) slower. The increase in the thickness of the product layer affects both the CO_2-concentration gradient across the CaS-shell and CO_2-diffusivity, D_{es}. It also appears that the temperature of reduction affects the stress-relaxation characteristics of the CaS(s)-product layer and thereby influences the micro-crack development. In equation (50), the best value for exponent 1/n is found to be 1/3. This equation is incorporated into the model with arbitrary values assigned to D_o and E_o at the start of calculation.

From equation (37), the following relationship can be written for the dimensionless time interval, $\Delta\tau$.

$$\Delta\tau = [(u - vP_{CO_2}^b)D_{es}M_{CaO}/RT \, r_o^2\rho_{CaO}^t]\Delta t \qquad (51)$$

Furthermore

$$t = j \cdot \Delta t \quad \text{and} \quad \tau = j \cdot \Delta \tau$$

As the reduction progresses, at a particular time $t_c(=j \cdot \Delta t)$ or $\tau_c(=j \cdot \Delta \tau)$, the sulfidization of reacting CaO(s)-grains in the exposed upper layer in the crucible is complete; hence $r_c/r_o = 0$ for these lime particles. In other words,

$$(\varepsilon_{i,j})_{t=t_c} = 0 \tag{52}$$

Therefore, all subsequent values $\varepsilon_{i,j+1}$ at $t > t_c$ have to be zero. This gives, from equation (49.b),

$$\gamma_{i+1,j} = 2 \gamma_{ij} - \gamma_{i-1,j} \tag{53}$$

The mathematical model must be modified to take into consideration the presence of the inert layer of spent solid reactants in the crucible.

Formation of Neutral Zone: Second Stage

Continued reaction will result in the build-up of a totally reacted layer atop the partially-reacted mixture; usually, this occurs during final stages of reduction. The effective diffusivities of CO(g) and $CO_2(g)$ through this "neutral zone" (Figure 4) may be regarded as equal to those in the partially-reacted $Cu_2S/4CaO$ mixture because the presence of 300% excess lime in the porous bed ensures that the pore-structure does not change significantly. However, the partial-pressure profile of $CO_2(g)$ and the concentration profile of CaO(s), shown schematically in Figure 4, are different from those during first-stage. The material balance relations for $CO_2(g)$ in the two regions can be expressed as follows.

Within the neutral zone: $0 < Z < Z_m$

$$- (D_b'/RT)(\partial^2 P_{CO_2}'/\partial Z^2) = 0 \tag{54}$$

(i) At $Z = Z_m$, $P_{CO_2}' = P_{CO_2}^m$ $\tag{55.a}$

(ii) At $Z = 0$ (top), $P_{CO_2}' = P_{CO_2}^b$ $\tag{55.b}$

Within the partially-reacted zone: $Z_m < Z < L$

$$- (D_b'/RT)(\partial^2 P_{CO_2}/\partial Z^2) = R_g^{CaO} \tag{56}$$

(iii) At $Z = L$ (bottom), $(\partial P_{CO_2}/\partial Z) = 0$ $\tag{55.c}$

At the boundary between the two zones, $Z = Z_m$:

722

$$P_{CO_2} = P'_{CO_2} = P^m_{CO_2} \tag{57.a}$$

$$\left.\frac{\partial P_{CO_2}}{\partial Z}\right|_{Z=Z_m} = \left.\frac{\partial P'_{CO_2}}{\partial Z}\right|_{Z=Z_m} \tag{57.b}$$

The conditions (57.a,b) simply ensure that there is continuity in the $CO_2(g)$-profile along the height of the porous aggregate. The solution to equation (54) subject to boundary conditions (55.a,b) is as follows:

$$P'_{CO_2} = P^b_{CO_2} - (P^b_{CO_2} - P^m_{CO_2})(Z/Z_m) \tag{58}$$

The solution to equation (56) has already been obtained by numerical (finite-difference) method. Thus, the values of $P^m_{CO_2}$ and Z_m, both functions of time t, are found through numerical solution of equation (56); these correspond to $(\varepsilon_{i,j})_{t=t_c}$. A careful examination of equation (58) shows that within the "neutral zone" $(0 < Z < Z_m)$, the CO_2-partial pressure profile is linear with Z and the slope is

$$\frac{\partial P'_{CO_2}}{\partial Z} = \frac{P^m_{CO_2} - P^b_{CO_2}}{Z_m} \tag{59}$$

This equation is transformed into the following dimensionless form:

$$\frac{\partial \gamma'}{\partial \sigma} = \mu \tag{60}$$

where

$$\gamma' = \frac{u - vP'_{CO_2}}{u - vP^b_{CO_2}} \quad ; \quad \sigma = Z/L \tag{61.a,b}$$

$$\mu = [(P^m_{CO_2} - P^b_{CO_2})vL]/[(u - vP^b_{CO_2})Z_m] \tag{62}$$

Using the finite-difference method (forward-difference formula) to solve equation (60), we obtain,

$$\gamma'_{i+1,j} = \gamma'_{i,j} + \mu\Delta\sigma \tag{63}$$

It will be noted that after the "neutral zone" has formed $(t > t_c)$, the γ vs. σ profile (as well as the P_{CO_2} vs. Z profile) for the remaining grid points is calculated using equation (63) instead of equation (49.b). Accordingly,

At $i \geqslant i_c$ and $j \geqslant j_c$, $\tau = \tau_c$

$$\gamma(\sigma,\tau) = \gamma'(i\Delta\sigma, j\Delta\tau) \equiv \gamma'_{i,j} \tag{64}$$

and

$$\epsilon(\sigma,\tau) = (\epsilon_{i,j})_{\tau \geqslant \tau_c} = 0 \tag{65}$$

In order to solve equation (63) numerically, the values of μ and $P^m_{CO_2}$ are calculated first; this is done by considering the boundary condition[2](57.b).

$$\frac{\partial P_{CO_2}}{\partial Z} = \frac{\partial P'_{CO_2}}{\partial Z} = \frac{P^m_{CO_2} - P^b_{CO_2}}{\sigma_m L} \tag{66}$$

In the absence of an analytical solution to equation (56), the partial-differential in the above equation is found numerically.

At $Z = Z_m = \sigma_m L$; $i = m$

$$\text{Slope (i)} = \left.\frac{\partial P_{CO_2}}{\partial Z}\right|_{i=m} = \hat{\alpha} \cdot \frac{(\gamma_{i+1,j} - \gamma_{i-1,j})}{(\sigma_{i-1} - \sigma_{i+1})} \tag{67}$$

where

$$\hat{\alpha} = (u - vP^b_{CO_2})/vL \tag{68}$$

On combining equations (66) and (67), we obtain,

$$P^m_{CO_2} = P^b_{CO_2} + \sigma_m L \text{ Slope(i)} \tag{69}$$

By means of equations (58) and (69), the P'_{CO_2} vs. Z profile in the neutral zone ($0 < Z < Z_m$) can be calculated. The remaining steps in the numerical integration are identical to those described previously.

Results and Discussion

The numerical calculations were performed on the CDC Cyber 180-855 digital computer; the complete Fortran Program used is given elsewhere (2). The input data to the program consisted of experimental partial pressures of CO and CO_2, reaction temperatures, experimental weight loss data (ΔW_{obs}), calculated effective diffusivities for CO and CO_2, equilibrium constants K_1 and K_2 for reactions (A1) and (A2), apparent density of the mixture, and the first estimates of D_o and E_o in equation (50). The output of the program provided values of ΔW_{cal}, fractional reduction ($\bar{F}$), and both actual and dimensionless CO and CO_2 partial pressure profiles.

The calculated weight loss data ΔW_{cal} are compared with the measurements ΔW_{obs} in Figures 5 through 8. In general, there is good fit between the experimental results and the numerical solution. In Figure 6, for run #E203, the standard deviation between the calculated and experimental sets (a total of 19 data points) is ±1.2 mg; for this run, the initial $Cu_2S/4CaO$ mixture weighed 524.17 mg which means the stoichiometric maximum weight loss for complete reduction amounts to 21.87 mg. The poorest fit obtained is shown

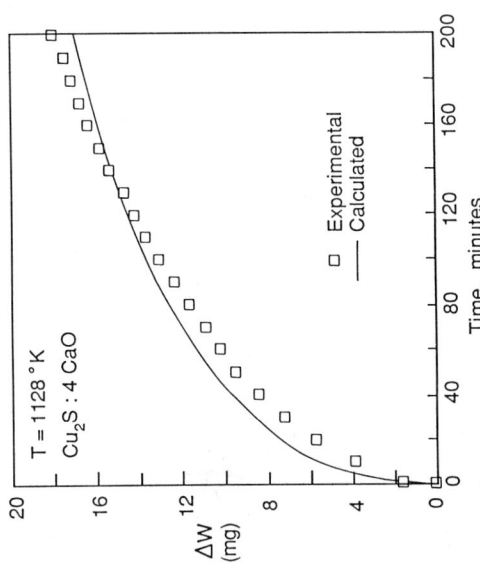

Figure 6 - Correlation of experimental ΔW vs. t data for $Cu_2S/4CaO$ samples reduced with $CO-CO_2$-He mixtures. Run No. E203; CO, 0.165 atm; CO_2, 0.047 atm.

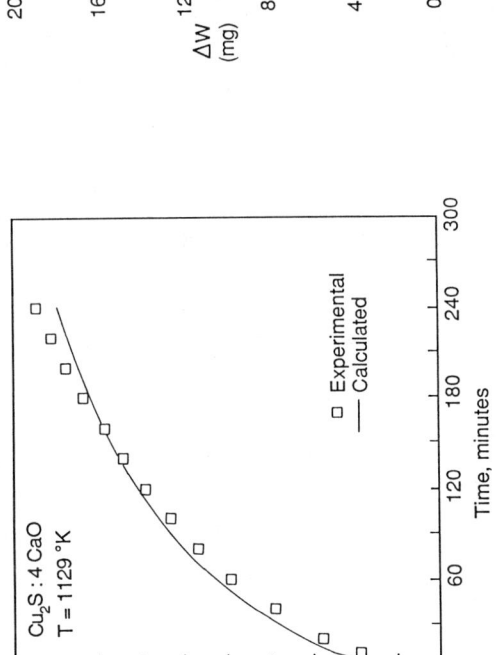

Figure 5 - Correlation of experimental ΔW vs. t data for $Cu_2S/4CaO$ samples reduced with $CO-CO_2$-He mixtures. Run No. E198: CO, 0.654 atm; CO_2, 0.262 atm.

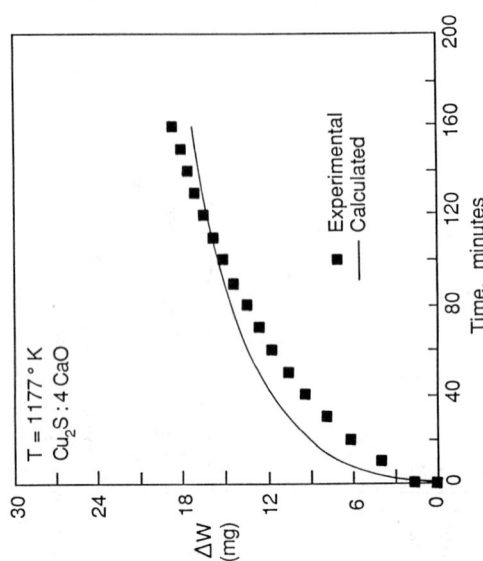

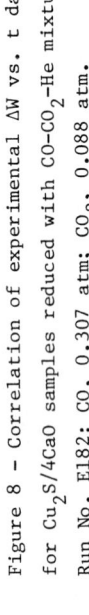

Figure 8 - Correlation of experimental ΔW vs. t data for Cu₂S/4CaO samples reduced with CO-CO₂-He mixtures. Run No. E182: CO, 0.307 atm; CO₂, 0.088 atm.

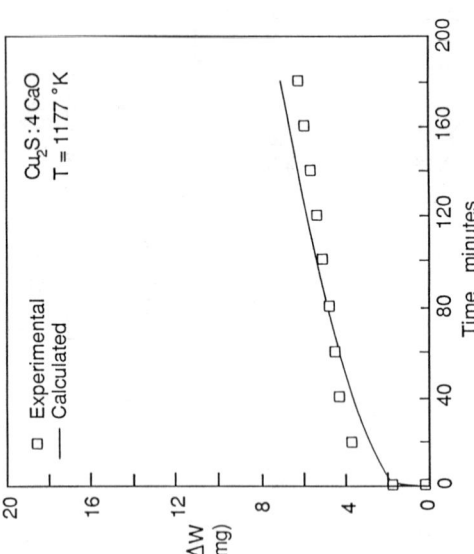

Figure 7 - Correlation of experimental ΔW vs. t data for Cu₂S/4CaO samples reduced with CO-CO₂-He mixtures. Run No. E178: CO, 0.298 atm; CO₂, 0.149 atm.

in Figure 8 for run #E182. It should be noted that this kind of discrepancy was found only in a small number of experiments. As described earlier, the best values of the micropore diffusion coefficients for $CO_2(g)$ in $CaS(s)$-product layer were obtained from the numerical solution. These are expressed in the form of the following two-term equation:

$$D_{es} = D_o - E_o t^{1/3} \quad , \quad cm^2/s \tag{70}$$

where t is in minutes. The values of D_o and E_o are summarized in Table I for the thirty-three experiments considered; the corresponding bulk-gas compositions are given in the Appendix.

The results of the present study are compared to those of Yang and Chen (4) who determined the kinetics of the reaction between $CaO(s)$ grains and COS-bearing $CO-N_2$ gas mixtures in the temperature range 500-900°C. The granular lime sample (1 μm-5 μm size range), about 28 mg in weight, was placed in a porous alumina pan and contacted with a $CO-N_2$ gas stream containing a small concentration of COS. The weight gain was monitored continuously. The sulfidation can be represented by

$$CaO(s) + COS = CaS(s) + CO_2 \tag{A2}$$

These authors reported (4) that the $CaO(s)$ grains were shaped like flakes (or flat plates); they used unreacted core model to correlate the experimental data. For the case of $CaS(s)$-product layer diffusion control, the fractional reaction, α, is given by

$$\alpha^2 = t/t_o \tag{71}$$

where t_o, the time required for completion of sulfidation, is

$$t_o = [R_o^2 C_{CaO}^\circ / 2 \, D_{es}' (C_{COS}^b - C_{COS}^e)] , \text{ sec} \tag{72}$$

where R_o (=1.5 μm) is the half-thickness of the $CaO(s)$-grain, C_{CaO}° (=0.058845 $mole/cm^3$) the molar density of lime, D_{es}' the effective diffusivity of COS(g) through the $CaS(s)$-product layer, C_{COS}^b the molar concentration of COS(g) in the bulk gas and C_{COS}^e the equilibrium concentration of COS(g) at the CaS/CaO interface, respectively. Because the gas phase used in the experiments reported by Yang and Chen (4) had no $CO_2(g)$, the corresponding C_{COS}^e at the CaS/CaO interface is virtually zero. The use of C_{COS}^b in equation (72) implicitly assumes that the effectiveness factor for the porous sample in the alumina pan is unity; the authors (4) provide no verification of this assumption. Table II summarizes the effective diffusivity data for COS(s) through the $CaS(s)$-product layer; these are derived from the results of Yang and Chen (4) using equations (71) and (72). The COS-content of the $CO-N_2$ gas stream ranged from 0.02% to 0.2%, by volume.

Table I. Values of the Effective Diffusivity (D_{es}) of $CO_2(g)$ through the CaS(s)-product layer, calculated by the numerical model. $D_{es} = D_o - E_o t^{1/3}$, cm^2/s with t in minutes.

Run No.	Temp. $^\circ K$	D_o, $cm^2.s^{-1}$	E_o, $cm^2.s^{-1}.min^{-1/3}$
E124	1220	4.2143×10^{-5}	5.7509×10^{-6}
E127	1224	3.2040×10^{-5}	4.1946×10^{-6}
E128	1219	2.8355×10^{-4}	4.3922×10^{-5}
E129	1225	3.4834×10^{-4}	3.9368×10^{-5}
E130	1224	3.9261×10^{-4}	4.4146×10^{-5}
E131	1228	3.6536×10^{-4}	3.7397×10^{-5}
E132	1225	4.1044×10^{-4}	5.1842×10^{-5}
E133	1227	3.7046×10^{-4}	3.5613×10^{-5}
E134	1222	5.4154×10^{-4}	3.4593×10^{-5}
E135	1222	7.2264×10^{-4}	9.0140×10^{-5}
E136	1221	7.9125×10^{-4}	6.0565×10^{-5}
E171	1176	4.8345×10^{-4}	5.4738×10^{-5}
E172	1177	1.1638×10^{-4}	9.2286×10^{-5}
E173	1177	5.2864×10^{-4}	5.0907×10^{-5}
E174	1177	3.8945×10^{-4}	5.7330×10^{-5}
E175	1177	5.0539×10^{-4}	6.4813×10^{-5}
E177	1177	1.7541×10^{-4}	1.5997×10^{-5}
E178	1177	6.7042×10^{-5}	6.5102×10^{-6}
E181	1177	7.7434×10^{-4}	1.2384×10^{-4}
E182	1177	6.1140×10^{-4}	9.2080×10^{-5}
E183	1177	9.8734×10^{-4}	6.0841×10^{-5}
E184	1177	8.2734×10^{-4}	6.4721×10^{-5}
E193	1127	1.4502×10^{-5}	1.6221×10^{-6}
E194	1128	5.3780×10^{-4}	6.5642×10^{-5}
E195	1129	2.6035×10^{-4}	2.9845×10^{-5}
E196	1130	3.8236×10^{-4}	5.2730×10^{-5}
E197	1129	2.6864×10^{-4}	3.0040×10^{-5}
E198	1129	1.5436×10^{-4}	1.3144×10^{-5}
E199	1129	9.2143×10^{-5}	3.6834×10^{-6}
E200	1128	6.3854×10^{-4}	8.2021×10^{-5}
E201	1128	5.4842×10^{-4}	6.5970×10^{-5}
E202	1128	9.4139×10^{-4}	6.6341×10^{-5}
E203	1128	6.8236×10^{-4}	5.9374×10^{-5}

Table II. Values of the Effective Diffusivity (D'_{es}) of COS(g) through CaS(s)-product layer derived from the published Results of Yang and Chen (4).

Run No.	Temp. °C	%COS	t_o (mins.)	c_{COS}^b mole/cm^3	D'_{es} cm^2/s
1	800	0.02	7,000	2.271×10^{-9}	0.694×10^{-6}
2	800	0.05	560	5.679×10^{-9}	3.469×10^{-6}
3	800	0.10	155	1.136×10^{-8}	6.262×10^{-6}
4	800	0.15	140	1.704×10^{-8}	4.626×10^{-6}
5	800	0.20	94	2.271×10^{-8}	5.168×10^{-6}
6	900	0.10	128	1.039×10^{-8}	8.297×10^{-6}

The effective diffusivity D'_{es} exhibits a strong-dependence on COS-concentration; it increases with COS-content at first and then levels off. The temperature has a positive effect on the value of D'_{es}.

In the present work, the COS-concentration in the intergranular pore space is set by reaction (A1):

$$Cu_2S(s) + CO = 2 Cu(s) + COS \tag{A1}$$

The COS-content depends on temperature and the CO-partial pressure. The variation of the effective diffusivity D_{es} of CO_2(g) through the CaS(s)-product layer is displayed in Table III which lists D_{es}^{60} and D_{es}^{150}, the values of D_{es} at t = 60 min. and t = 150 min., respectively.

Table III. Effects of Gas Composition and Temperature on D_{es}; values computed from Table I using equation (70).

Run No.	Temp. (K)	P_{CO}^b (atm)	K_1	P_{COS} (atm)	D_{es}, cm^2/s t = 60 min	t = 150 min
E203	1128	0.165	4.308×10^{-5}	0.711×10^{-5}	4.499×10^{-4}	3.669×10^{-4}
E201	1128	0.307	4.308×10^{-5}	1.322×10^{-5}	2.902×10^{-4}	1.979×10^{-4}
E194	1128	0.496	4.308×10^{-5}	2.137×10^{-5}	2.808×10^{-4}	1.890×10^{-4}
E198	1129	0.654	4.308×10^{-5}	2.817×10^{-5}	1.029×10^{-4}	0.845×10^{-4}
E173	1177	0.495	5.093×10^{-5}	2.521×10^{-5}	3.294×10^{-4}	2.582×10^{-4}
E131	1228	0.495	6.096×10^{-5}	3.017×10^{-5}	2.190×10^{-4}	1.167×10^{-4}

It is seen that at 1128 K, with increasing CO-pressure (hence, the COS-content) there is a definite decrease in D_{es}; the temperature-influence is much harder to determine. The difference between the D_{es} and D'_{es} data can be attributed to the influence of intergranular pore-diffusion which was not corrected for in the deduction of the latter from the results of Yang and Chen (4).

Figures 9 and 10 show that γ vs. σ plots for $Cu_2S/4CaO$ samples reduced with CO-CO_2-He gas mixtures at 1128 K and 1177 K, respectively. It is seen that at any given position inside the crucible ($\sigma = \sigma$), the value of γ increases with the fractional reduction; this means less $CO_2(g)$ is produced per unit time, as per reaction (A2). As the reduction progresses, the solid reactants Cu_2S and CaO, being enclosed in product shells, become less accessible to reactant gases CO and COS; hence the lower $CO_2(g)$ product at reaction interfaces. This trend is depicted in Figure 11 which shows the partial pressure profiles for $CO(g)$ and $CO_2(g)$ at $\alpha = 0.26$ and $\alpha = 0.77$, respectively. At higher fractional reduction, there is more $CO(g)$ available at the closed end ($\sigma = 1.0$) implying that there exist less number of $Cu_2S(s)$ particles to be converted to metallic copper. The variations of γ and ε with time at two different positions ($\sigma = 0.5$ and $\sigma = 1.0$) in the partially-reduced sample are shown in Figure 12. Thus, with increasing fractional reaction, the dimensionless CaO(s)-core radius $\varepsilon(=r_c/r_o)$ is seen to decrease, more rapidly in the top layers of the sample. It was found that the neutral zone of completely reacted sample did not develop until after 80% or greater reduction was achieved.

Concluding Remarks

The experimental data for the uncatalyzed reduction of Cu_2S:4CaO samples contained in graphite crucibles with CO-CO_2-He gas mixtures which are reported elsewhere (2) are interpreted by means of a numerical model. The model provides a good fit with the experimental results for the entire range of ΔW vs. t plots. In developing the model, it was assumed that the diffusion of $CO_2(g)$ through micropores of CaS(s)-product layer which forms around the CaO(s)-core is the rate-limiting step. The values of the micropore diffusion coefficient for $CO_2(g)$, D_{es}, were calculated for the thirty-three experiments. It was found that D_{es} varies with both temperature and bulk-gas compositions, apart from its dependence on time.

Acknowledgments

The financial support of this work by the National Science Foundation, Washington, D.C., through Grant No. CPE-8202588 is gratefully acknowledged.

730

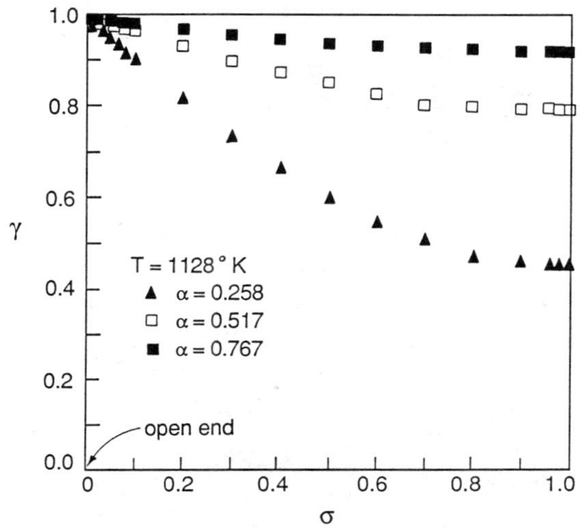

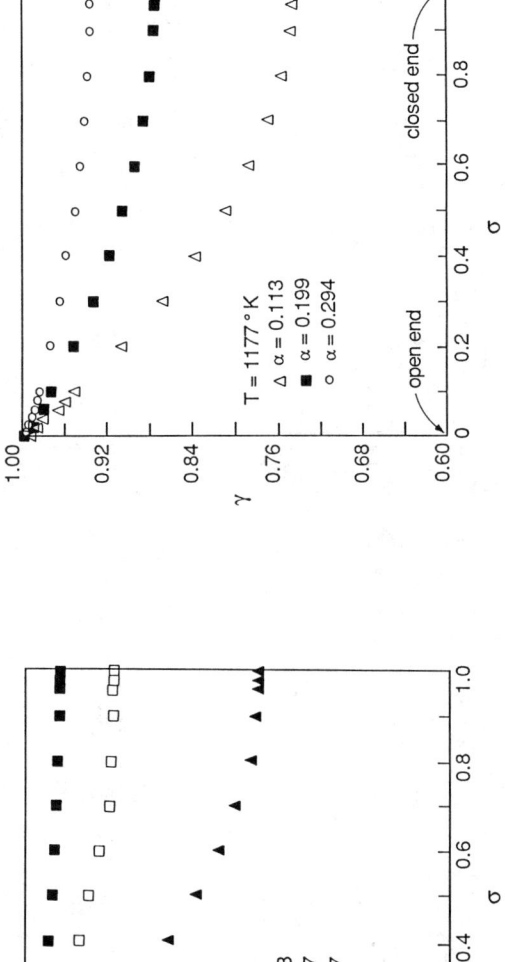

Figure 9 - Profiles of dimensionless CO_2-pressure (γ) in $Cu_2S/4CaO$ samples at various degrees of reduction. Run No. E203: CO, 0.165 atm; CO_2, 0.047 atm.

Figure 10 - Profiles of dimensionless CO_2-pressure (γ) in $Cu_2S/4CaO$ samples at various degrees of reduction. Run No. E178: CO, 0.298 atm; CO_2, 0.149 atm.

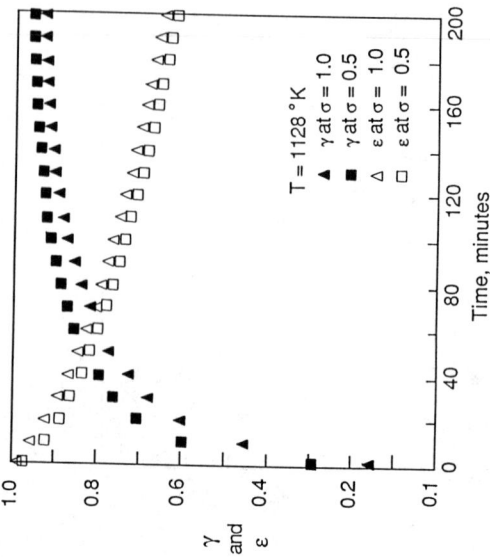

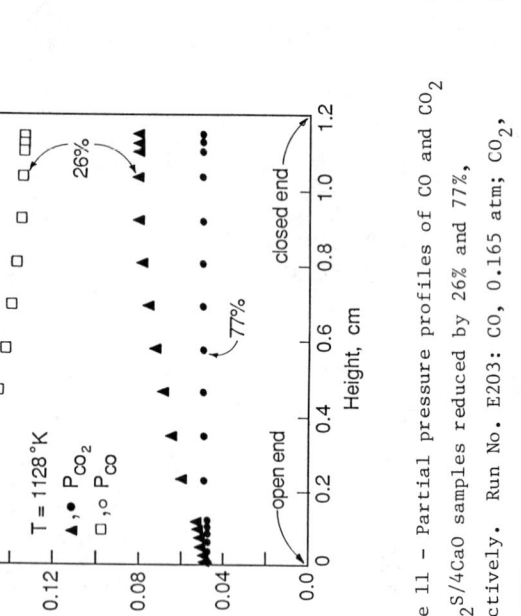

Figure 12 – Variation of γ and ϵ with time in $Cu_2S/4CaO$ samples reduced with $CO-CO_2$–He mixtures. Run No. E203: CO, 0.165 atm; CO_2, 0.047 atm.

Figure 11 – Partial pressure profiles of CO and CO_2 in $Cu_2S/4CaO$ samples reduced by 26% and 77%, respectively. Run No. E203: CO, 0.165 atm; CO_2, 0.047 atm.

Thanks are also due to the Washington State Mining and Mineral Resources Research Institute, University of Washington, Seattle, Washington.

References

1. M. Moinpour and Y. K. Rao, "Direct Reduction of Copper Sulphide with Carbon in the Presence of Lime," Canadian Metallurgical Quarterly, 24 (1985), 69-81.

2. M. Moinpour, Reduction of Copper Sulfide with Carbon and CO as Reducing Agents in the Presence of CaO; Catalytic Effects, Ph. D. Thesis, University of Washington, Seattle, Washington (1987).

3. D. R. Stull and H. Prophet, JANAF Thermochemical Tables, 2nd ed., NSRDS-NBS37 (Washington, D.C.: U.S. Department of Commerce, 1971), 1141 pp.

4. R. T. Yang and J. M. Chen, "Kinetics of Desulfurization of Hot Fuel Gas with Calcium Oxide. Reaction between Carbonyl Sulfide and Calcium Oxide," Environmental Science & Technology, 13(1979), 549-553.

5. V. S. Kamath and T. W. Petrie, "Rate of Reaction of Hydrogen Sulfide-Carbonyl Sulfide Mixtures with Fully Calcined Dolomite," Environmental Science & Technology, 15(1981), 966-968.

6. R. C. Weast and M. J. Astle, ed., CRC Handbook of Chemistry and Physics, 60th edn. (Boca Raton, Florida: CRC Press, Inc., 1979), B50-B144.

7. P. C. Carman, "Diffusion and Flow of Gases and Vapors through Micropores. I. Slip flow and Molecular Streaming," Proceedings of Royal Society, A203(1950), 55-74.

8. P. C. Carman, Flow of Gases through Porous Media (London, U.K.: Butterworths, Ltd., 1956), 182.

9. Y. K. Rao, "A Physico-Chemical Model for Reactions between Particulate Solids occurring through Gaseous Intermediates-I. Reduction of Hematite by Carbon," Chemical Engineering Science, 29(1974), 1435-1445.

10. I. J. Lin and Y. K. Rao, "Reduction of Lead oxide by Carbon," Transactions of the Institution of Mining and Metallurgy, Section C, 84(1975), 76-82.

11. L. H. Van Vlack, Elements of Materials Science and Engineering, 4th edn. (Reading, Massachusetts: Addison-Wesley Publishing Company, Inc., 1980), 74-80.

12. J. R. Bowen, "Comments on Pseudo-steady-state Approximation for Moving Boundary Problems," Chemical Engineering Science, 20(1965), 712-713.

13. K. B. Bischoff, "Further Comments on the Pseudo-Steady-State Approximation for Moving Boundary Diffusion Problems," Chemical Engineering Science, 20(1965), 783-784.

14. T. G. Theofanous and H. C. Lim, "An approximate Analytical Solution for Non-Planar Moving Boundary Problems," Chemical Engineering Science, 26(1971), 1297-1300.

15. C. F. Gerald, Applied Numerical Analysis, 2nd edn. (Reading, Massachusetts: Addison-Wesley Publishing Company, Inc., 1978), 190-219.

16. H-W. Hsu and R. B. Bird, "Multicomponent Diffusion Problems," A.I.Ch.E., Journal, 6(1960), 516-524.

17. Y. K. Rao, "Diffusion-limited Heterogeneous Processes," Canadian Metallurgical Quarterly, 18(1979), 379-381.

18. E. N. Fuller, P. D. Schettler, and J. C. Giddings," A New Method for Prediction of Binary Gas-phase Diffusion Coefficients," Industrial and Engineering Chemistry, 58(5)(1966), 18-27.

19. P. B. Weisz and A. B. Schwartz, "Diffusivity of Porous-Oxide-Gel-Derived Catalyst Particles," Journal of Catalysis, 1(1962), 399-406.

Appendix

Calculation of Diffusivities

The effective diffusivities for $CO(g)$ and $CO_2(g)$ in the porous $Cu_2S/4CaO$ mixture are calculated from the structural properties of the aggregate. The average pore-diameter, d_p, for the granular mixture is estimated to be 2.003 μm. For this pore-size, both Knudsen and molecular diffusion mechanisms must be considered (2). The calculation of molecular diffusivity D_i^V of the ith species in a multicomponent gas mixture is greatly facilitated by the Stefan-Maxwell formalism proposed by the Hsu and Bird (16) and developed further by Rao (17). Consider a system in which the following heterogeneous reaction occurs:

$$\text{Solid I} + aA(g) + bB(g) = pP(g) + qQ(g) + \text{Solid II} \tag{1.A}$$

In a quasi-steady-state process, the diffusion fluxes are related as follows:

$$N_A/a = N_B/b = -N_P/p = -N_q/q \tag{2.A}$$

The inert species I which is present in the mixture can be regarded as stagnant with the result $N_I = 0$. Defining

$$\psi_A = (N_A + N_B + N_P + N_Q)/N_A \tag{3.A}$$

we find, on substitution,

$$\psi_A = 1 + (b/a) - (p/a) - (q/a) \tag{4.A}$$

Following the method reported by Hsu and Bird (16), the equivalent binary diffusivity D_{AV} of species A in the multicomponent gas mixture is found to be given by

$$D_{AV} = \frac{(1 - \psi_A x_A)}{(\dfrac{x_B}{D_{AB}} + \dfrac{x_P}{D_{AP}} + \dfrac{x_Q}{D_{AQ}} + \dfrac{x_I}{D_{AI}}) - \dfrac{x_A}{a}(\dfrac{b}{D_{AB}} - \dfrac{p}{D_{AP}} - \dfrac{q}{D_{AQ}})} \tag{5.A}$$

For the system under consideration, the overall reaction can be represented by

$$Cu_2S(s) + CaO(s) + CO(g) = 2\ Cu(s) + CaS(s) + CO_2(g)$$

The helium component of the ternary gas mixture is virtually stagnant. Thus,

$$A : CO \ ; \ P : CO_2 \ ; \ I : He$$

$$a = P = 1 \ ; \ b = q = 0$$

Substitutions into equation (5.A) provide

$$D_{CO}^{V} = \frac{(1 - \psi_A x_{CO})}{(\dfrac{x_{CO_2}}{D_{CO-CO_2}} + \dfrac{x_{He}}{D_{CO-He}}) - x_{CO}(-\dfrac{1}{D_{CO-CO_2}})}$$ (6.A)

$$D_{CO_2}^{V} = \frac{(1 - \psi_B x_{CO_2})}{(\dfrac{x_{CO}}{D_{CO-CO_2}} + \dfrac{x_{He}}{D_{CO_2-He}}) - x_{CO_2}(-\dfrac{1}{D_{CO-CO_2}})}$$ (7.A)

where

$$\psi_A = 1 + (-p/a) = 0 \; ; \; \psi_B = 1 + (-a/p) = 0$$

In these equations, x_i denotes the mole fraction of the ith species in the gas-mixture; the bulk gas composition was used to ascertain the diffusivities by equations (6.A) and (7.A). The binary molecular diffusivities D_{ij} for i-j pairs are calculated using the Fuller-Schettler-Giddings formula [18]. The Knudsen diffusivity for the ith species is given by

$$D_{Ki} = \frac{1}{3} d_p (8 \, RT/\pi M_i)^{1/2}$$ (8.A)

where M_i is the molecular weight of the ith species. The total diffusivity D_i^{tot} for the transition-regime is found using

$$D_i^{tot} = [(D_i^{V})^{-1} + (D_{Ki})^{-1}]^{-1}$$ (9.A)

The values of D_{CO}^{V}, $D_{CO_2}^{V}$, D_{CO}^{tot}, and $D_{CO_2}^{tot}$ are listed in Tables IV and V, respectively.

The effective diffusivity D_i^{b} of the ith species in the porous $Cu_2S/4CaO$ aggregate was calculated using the Weisz-Schwartz correlation [19]:

$$D_i^{b} = (\theta^2/\sqrt{3}) \, D_i^{tot}$$ (10.A)

where θ is the fractional porosity of the mixture. The effective diffusivities for CO and CO_2 computed with the formula (10.A) are given in Table V. It will be noted that these values are used in the numerical model as D_b and D_b' for CO(g) and CO_2(g), respectively.

736

Table IV. Calculated equivalent binary diffusivities for CO and CO_2 in ternary gas mixture $CO-CO_2-He$.

Run No.	Temp. $^{\circ}K$	Mole Fractions, x_i			D_{CO}^V $cm^2 \cdot s^{-1}$	$D_{CO_2}^V$ $cm^2 \cdot s^{-1}$
		CO_2	CO	He		
E124	1220	0.297	0.149	0.554	3.2953	3.1546
E127	1224	0.498	0.249	0.253	2.3691	2.3351
E128	1219	0.651	0.130	0.219	2.2788	2.2513
E129	1225	0.649	0.260	0.091	2.0569	2.0476
E130	1224	0.498	0.100	0.402	2.7585	2.6862
E131	1228	0.495	0.141	0.364	2.6627	2.6018
E132	1225	0.647	0.185	0.168	2.1957	2.1762
E133	1227	0.307	0.088	0.605	3.5697	3.3925
E134	1222	0.295	0.059	0.646	3.7636	3.5531
E135	1222	0.165	0.047	0.788	4.7902	4.3868
E136	1221	0.157	0.031	0.812	5.0146	4.5614
E171	1176	0.499	0.100	0.401	2.5692	2.5020
E172	1177	0.491	0.246	0.263	2.2334	2.1998
E173	1177	0.495	0.141	0.364	2.4722	2.4157
E174	1177	0.653	0.131	0.216	2.1373	2.1120
E175	1177	0.647	0.185	0.168	2.0474	2.0292
E177	1177	0.654	0.262	0.084	1.9069	1.8990
E178	1177	0.298	0.149	0.553	3.0906	2.9590
E180	1178	0.295	0.059	0.646	3.5297	3.3323
E181	1177	0.295	0.059	0.646	3.5244	3.3273
E182	1177	0.307	0.088	0.605	3.3190	3.1542
E183	1177	0.157	0.031	0.812	4.7026	4.2776
E184	1177	0.165	0.047	0.788	4.4858	4.1080
E189	1126	0.499	0.100	0.401	2.3811	2.3188
E192	1126	0.499	0.100	0.401	2.3811	2.3188
E193	1127	0.491	0.246	0.263	2.0700	2.0389
E194	1128	0.496	0.142	0.362	2.2901	2.2381
E195	1129	0.499	0.100	0.401	2.3922	2.3297
E196	1130	0.653	0.131	0.216	1.9902	1.9666
E197	1129	0.647	0.185	0.168	1.9035	1.8866
E198	1129	0.654	0.262	0.084	1.7729	1.7656
E199	1129	0.298	0.149	0.553	2.8734	2.7511
E200	1128	0.295	0.059	0.646	3.2717	3.0887
E201	1128	0.307	0.088	0.605	3.0810	2.9280
E202	1128	0.157	0.031	0.812	4.3654	3.9709
E203	1128	0.165	0.047	0.788	4.1641	3.8134

Table V. Calculated total and effective diffusivities for CO and CO_2.

Run No.	Temp. $^\circ K$	θ	D_{CO}^{tot}	$D_{CO_2}^{tot}$	D_{CO}^b	$D_{CO_2}^b$
E124	1220	0.706	2.1769	1.9515	0.6265	0.6516
E127	1224	0.685	1.7309	1.6043	0.4689	0.4346
E128	1219	0.686	1.6813	1.5633	0.4568	0.4247
E129	1225	0.685	1.5582	1.4633	0.4221	0.3964
E130	1224	0.690	1.9299	1.7625	0.5305	0.4845
E131	1228	0.685	1.8834	1.7267	0.5102	0.4678
E132	1225	0.681	1.6366	1.5278	0.4382	0.4091
E133	1227	0.680	2.2957	2.0423	0.6129	0.5452
E134	1222	0.683	2.3726	2.0977	0.6390	0.5650
E135	1222	0.675	2.7432	2.3628	0.7216	0.6216
E136	1221	0.683	2.8148	2.4121	0.7581	0.6496
E171	1176	0.705	1.8247	1.6702	0.5236	0.4793
E172	1177	0.700	1.6489	1.5301	0.4665	0.4329
E173	1177	0.686	1.7755	1.6315	0.4824	0.4433
E174	1177	0.687	1.5959	1.4871	0.4349	0.4052
E175	1177	0.687	1.5452	1.4456	0.4211	0.3939
E177	1177	0.687	1.4639	1.3783	0.3989	0.3756
E178	1177	0.721	2.0735	1.8625	0.6223	0.5590
E180	1178	0.721	2.2626	2.0041	0.6791	0.6015
E181	1177	0.725	2.2601	2.0020	0.6859	0.6075
E182	1177	0.727	2.1738	1.9380	0.6633	0.5914
E183	1177	0.733	2.6927	2.3109	0.8353	0.7168
E184	1177	0.743	2.6202	2.2605	0.8351	0.7205
E189	1126	0.742	1.7174	1.5756	0.5459	0.5008
E192	1126	0.743	1.7174	1.5756	0.5474	0.5022
E193	1127	0.750	1.5497	1.4413	0.5033	0.4681
E194	1128	0.744	1.6700	1.5383	0.5337	0.4916
E195	1129	0.740	1.7239	1.5813	0.5450	0.4999
E196	1130	0.736	1.5050	1.4054	0.4707	0.4395
E197	1129	0.741	1.4547	1.3639	0.4612	0.4324
E198	1129	0.737	1.3772	1.2995	0.4319	0.4075
E199	1129	0.739	1.9605	1.7648	0.6181	0.5564
E200	1128	0.744	2.1377	1.8975	0.6832	0.6064
E201	1128	0.748	2.0546	1.8357	0.6531	0.5835
E202	1128	0.744	2.5562	2.1975	0.8169	0.7023
E203	1128	0.748	2.4858	2.1485	0.8030	0.6940

PHENOMENOLOGICAL MODELING OF GREEN CHARGE SMELTING

OF COPPER CONCENTRATE IN REVERBERATORY FURNACE

*Alfonso Otero M., **Igor Wilkomirsky F., ***Antonio Luraschi G.

*, *** Centro de Investigación Minera y Metalúrgica (CIMM)
P.O. Box 170, Santiago 10, Chile

** Department of Metallurgical Engineering, Concepción
University, P.O. Box 53-C, Concepción, Chile

Abstract

A mathematical representation that considers mainly a dynamic phenomeno
logical model of green charge smelting in reverberatory furnace has been developed. The phenomena which are treated in the model are heat transfer,
fluid-dynamics, smelting and combustion kinetics, and mass and heat balance.
As a result, temperatures, smelting rate, gas composition and velocity profiles as functions of time and/or position along the furnace are determined.
The theoretical predictions appear to be consistent with industrial data for
two operating reverberatory furnaces, measured as part of this study, one
using coal and the other fuel oil for combustion.

The model was used to analyse the effect of operational variables on
production indexes through simulation. A discussion is presented for the
convenience and potential of automatic control of reverberatory combustion
and the incorporation of the mathematical model developed into optimization
schemes, considering the results obtained in the simulation of a reverb under
automatic control, using the dynamic model with a PID algorithm.

In addition to the dynamic model, steady state models were formulated
to study: the tridimensional heat transfer interactions among internal zones,
the combustion of coal and oil, and the flow gas regime inside the furnace.
Results indicate the existence of an optimum furnace roof height and air/
fuel ratio (RAC) to obtain a maximum furnace productivity.

An optimized strategy to recharge reverbs was also developed. Increasing fuel rate for a given RAC does not imply necessarily an increase in smelt
ing rate; this is due to the axial gas temperature profile displacement
towards the furnace uptake which creates an unsmelted bank zone.

The mechanism of melting by ablation of the material on the banks is
described and utilized in the model. The most important parameters in determining the smelting rate according to this mechanism are: the viscosity of
molten material flowing down the bank, bulk thermal conductivity of particulate solids which is mainly controlled by particle size and the absorptivity of the liquid surface.

I. Introduction

Reverberatory furnaces are still the main units used in Chile for smelting copper concentrates. Although the high cost of energy has created big incentives to adopt other smelting technologies which make better use of the fuel value of concentrates, the scarcity and high cost of capital has meant a very slow replacement of traditional reverbs. They have thus survived either as conventional units or in the oxygen smelting process where reverberatory furnaces are furnished with oxy-fuel roof burners, coupled with the Teniente Reactor.

Traditionally, smelting in reverbs has been an inefficient and expensive stage in the pyrometallurgical extraction of copper from concentrates to anodes, consuming 90% of the total energy requirement, with an expenditure of 5200 Mcal/ton Cu, thermal efficiency of 29% and 40% of direct operation costs, the main items being fuel and refractory consumption and labor with 35%, 15% and 30%, respectively (1).

In the early 1960's (2) the first known attempts to correlate smelting behaviour of reverbs with operational variables were published. Because smelting is the main function of the reverberatory furnace, studies have centered on aspects of furnace design, smelting, heat transfer and combustion.

Later contributions have been oriented to decrease fuel consumption (19, 9, 10), increase furnace capacity (3,4), increase thermal efficiency (5,6), by developing empirical and hybrid models which permit the optimization of furnace operation and design. Other studies have considered chemical composition of the charge (7, 8) and smelting rate of concentrate on charge banks (4,8), steady state heat transfer for green charge smelting (16), as well as unsteady for calcine smelting (11). Also, burner position schemes, related combustion conditions and automatic control strategies (11), have been analyzed. From the literature available there is no established experimental evidence on the mechanism for charge bank smelting, considering various possibilities such as ablative smelting on the banks, digestion of material from the solid charge by the liquids and carryover of solids by the molten material flowing down the banks. This represents a great challenge to the development of a phenomenological model of the green charge reverberatory smelting of concentrates.

While it is apparent that from operational experience of reverberatory furnaces has evolved a technology that is reasonably efficient, this has been achieved almost completely empirically in the absence of any detailed knowledge concerning the fundamentals of the melting kinetics and mechanism for concentrate smelting on a bank.

The purpose of this work, developed as the thesis of one of the authors (12), was therefore to analyze ways to increase productivity and thermal efficiency by making use of mathematical models that include all aspects such as the kinetics of the melting mechanism (the heating, melting and flow of molten materials from charge banks), and the heat transfer among different surfaces and volumes (gas, banks, liquid channel and roof). This would allow determination of several variables in order to quantify different aspects of reverberatory operation and design. In turn, this makes it possible to evaluate the ultimate potential of this furnace and to determine strategies for automatic control of its operation.

II. Mathematical Models

In this section the dynamic model and the complementary ones are described.

A. Dynamic Phenomenological Model

In the main mathematical model the furnace is divided into a number of cross sectional elements (see Figure 1). This model calculates the differential longitudinal and transversal heat transfer interactions between a given surface zone inside the furnace (charge bank, roof or liquid channel) and a volume zone (gas) for each furnace element. The model is tri-dimensional (two space coordinates and time) and takes into account mainly the combustion kinetics, type of combustion, thermal efficiency, charge bank wear through ablative smelting, internal zone heat exchange, gas residence time and others. The model is flexible, allowing variations of: i) burner location, to simulate furnaces with end-wall and roof burners and ii) charge bank angle to simulate green or calcine charge smelting.

The basic considerations of the model are that a net heat flux to the surface of the bank, $q_i(x,t)$ at any given element along the furnace causes the material on the bank to be heated until it reaches the melting temperature of the charge.

During smelting is formed a liquid layer of molten material flowing down the bank, whose thickness and continuity determine the conduction of heat across the layer and thus the temperature difference between the liquid surface of the bank and the liquid/solid interface, which is supposed to remain at T_m.

The thickness of the liquid layer which in fact represents a resistance to the heat transfer from the roof and flames to the bank, is mainly determined by the viscosity of the liquid and the bank slope. For a given flame temperature, an increase in this resistance will produce a higher temperature of the liquid surface. This in turn will determine a lower net heat conducted into the bank, since this depends on T^4 difference in temperature between gas and bank surface. Finally, it is assumed that bank slope remains practically constant during smelting; therefore the viscosity of the liquid flowing on the bank will be the most important variable in terms of heat transfer to the bank and smelting rate.

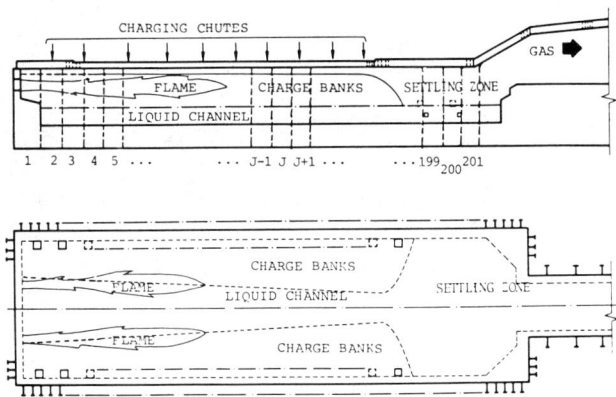

FIGURE 1 : Reverbs Longitudinal Cross Section and Plant View

The following assumptions have been made in the model formulation:

a) For end-wall burners the main heat transfer mechanism to charge bank is radiation, while for roof burners is distributed between radiation and convection.

b) Conduction heat transfer is not considered for the interaction between different solid, liquid or gas elements inside the furnace.

c) Combustion process is approximated by a first order kinetics (11) of the type, fuel burned = 1 - exp($-k_c t$), and starts at the burner nozzle.

d) The reverb gas is an absorbing medium, uniform through any furnace element, and is the main heat source.

e) Refractory, liquid channel and liquid over charge banks are considered black surfaces.

f) Variations in heat supplied to furnace roof are negligible because of its low heat capacity.

g) Infiltrated air increases linearly along the furnace.

h) Molten material fed to the furnace and tapped from it is in thermal equilibrium with material inside the furnace.

i) Feeding time to the furnace is assumed to be nil.

By performing a dynamic heat balance for each furnace element, the surface temperature of the bank as a function of the heat supplied from the combustion gases can be calculated.

Heat inputs to the balance are heats of combustion and chemical reactions of the charge, while outputs are radiation heat losses to neighbour furnace elements, infiltrated air, and heats to bank and liquid channel surface. Therefore, the rate of change with time of the sensible heat of the gas along the furnace is given by:

$$(m_G c p_G) \frac{\partial T_G(x,t)}{\partial t} = \sum_i Q^i_{COMB} (k_c,t) + Q_{REAC}(\Delta H(T_s), \dot{m}(C_s)) -$$

$$- [Q_{RAD}(T_G(x,t), F, \varepsilon, \sigma, K_1) + Q_{INFIL}(T_G(x,t), T_o, \dot{v}_{ia}(t),K_2)$$

$$+ Q_{BANK}(T_G(x,t), T_{SB}(x,t), T_{LB}(x,t), K_3) +$$

$$+ Q_{CHANN.}(T_G(x,t), T_{LC}(x,t), T_{LS}(x,t), K_4) \qquad (1)$$

Where element position along the furnace is x = J·L/N, subscript i denotes several fuels to be burnt alternatively.

Rigurously $T_{SB}(x,t)$ which represents bank surface temperatures in equation 1 should be written as $T_{SB}(x,y,t)$ (Figure 2) but for simplicity the solid internal temperature profile has been treated separetely.

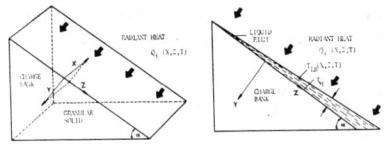

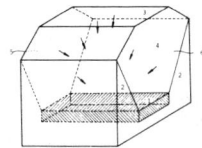

FIGURE 2 : Sectional View of Reverbs Charge Bank

  a) Scheme of Charge Bank Heating Stage
  b) Scheme of Charge Bank Smelting Stage
  c) Scheme of Zones in Furnace Element

The bank smelting process consists of two sequential stages: a) unsteady state heating, b) unsteady and steady state melting.

The phenomena occuring during charge bank heating and melting can be included considering a semi-infinite solid initially at T_o extending from $y = 0$ to $y = + \infty$[15]. At $t = 0$, the external surface, $y = 0$, receives a net radiant heat flux, q_i, which will produce an increase of solid temperatures until $t = t_m$. At this time, the charge bank surface will reach melting temperature. Molten material flows to liquid channel and the solid melting surface receads to $y = + \infty$ according to the smelting rate.

Calculations have shown that unsteady state melting time is negligible compared with steady state melting time, so only steady state is considered in the model. Bank surface receives a variable net radiation heat, $q_i(x,t)$ during unsteady state heating. The solution of the unsteady state heat conduction equation presented later is very difficult. However, adapting Duhamel Theorem (13) to this case and using the recursive form for the final equation the surface bank temperature takes the form:

$$T_{SB}(x,t_n) = T_o + 2 \Pi^{-\frac{1}{2}} \sum_{k=1}^{n} q_i(x,t_k)[(t_n-t_{k-1})^{\frac{1}{2}} - (t_n-t_k)^{\frac{1}{2}}] \quad t_n \leq t_m \quad (2)$$

where the net radiation heat on bank surface is given by:

$$q_i(x,t) = \sigma \xi [T_G^4(x,t) - T_{SB}^4(x,t)] \quad (3)$$

and

$$\frac{1}{\xi_{g \to i}} = \frac{1 - \varepsilon_g}{\varepsilon_g} + \frac{1}{\varepsilon_g (F_{g \to i} + \frac{1}{2})} + \frac{1 - \varepsilon_i}{\varepsilon_i} \quad (4)$$

Equation 2 is valid, as expressed by the condition $t_n \leq t_m$, before melting starts on the bank.

During steady state melting, liquid film flowing over the banks constitutes an important resistance to heat exchange between gas and solids, therefore the bank liquid surface temperature can be obtained considering the following analysis:

For liquid flowing under laminar regime (14) over the bank, the mass
flowrate of the liquid per unit width of the bank is:

$$G = \rho_\ell a(z) \, \bar{u}_\ell = \rho_\ell \left(\frac{ds}{dt}\right)_E z \tag{5}$$

where $\bar{u}_\ell = \rho_\ell \, g \, sen \, \alpha \, a^2(z)/3\mu$ (6)

$$\left(\frac{ds}{dt}\right)_E = q_i(x,t)/Q_m \tag{7}$$

As melting starts $q_i(x,t)$ is expressed by equation 3, after it is corrected for liquid layer formation.

$$Q_m = \rho_s[\Delta H_m + Cp_s(T_m - T_o)] \tag{8}$$

The liquid film thickness as a function of distance down the bank, z,
becomes:

$$a(z) = [3\mu \left(\frac{ds}{dt}\right)_E z/(\rho_\ell g \, sen\alpha)]^{1/3} \tag{9}$$

By performing a heat balance in the liquid film and considering that
heat transported with the flowing film is negligible compared with heat conducted through the liquid layer, the liquid film superficial temperature becomes:

$$T_{LB}(x,z,t) = T_m + \frac{\sigma \, \epsilon \, F \, a(z)}{k_\ell} [T_G^4(x,t) - T_{LB}^4(x,z,t)] \tag{10}$$

Corrected net radiation heat becomes:

$$q_i(x,z,t) = K_5 [T_G^4(x,t) - T_{LB}^4(x,z,t)] \tag{11}$$

Then melting velocity and smelting rate take the forms:

$$\left(\frac{ds}{dt}\right)_E = q_i(x,z,t)/Q_m \tag{12}$$

and

$$\left(\frac{dsR}{dt}\right)_E = K_p [L_z - 2 \left(\frac{ds}{dt}\right)_E tg \, \alpha \cdot t]\left(\frac{ds}{dt}\right)_E \rho_\ell \tag{13}$$

Where K_p is the partition coefficient which corresponds to the mass
ratio of liquid material obtained to starting solid material supplied.

To obtain internal solid temperature profiles during bank heating and
melting, the transient solid conduction is considered, whose principal assumptions are:

744

a) Unidirectional heat conduction.
b) Bank thermal conductivity is independent of temperature.
c) Different net radiant heat to the banks during heating and melting.
d) Bank consumption is mainly by ablative melting.
e) Bank inclination to horizontal (α) is constant during melting.
f) Steady state melting is reached instantaneously after melting starts.

$$\frac{\partial T_s}{\partial t} = \alpha_s \frac{\partial^2 T_s}{\partial y^2} \tag{14}$$

The relevants boundary conditions are:

i) <u>Heating stage</u>

$$q_i = -k_s \frac{\partial T_s}{\partial y} + q_e, \qquad \forall y, \ t < t_m \tag{14a}$$

where subscript e is the evaporation and thermal decomposition of solid depth "y".

ii) <u>Smelting stage</u>

$$q_i = -k_s \frac{\partial T_s}{\partial y} + \rho_s \Delta H_m \frac{dY}{dt} + q_r, \ y = Y, \ t \geq t_m \tag{14b}$$

Where subscript r is the FeS to FeO and fayalite slag formation reaction.

The analytical solution for heating stage, equation 14 with boundary condition (14a) is:

$$T_s(y,t) = T_o + 2 q_i (t/k_s \rho_s Cp_s)^{\frac{1}{2}} \ \text{ierfc}(y/2(\alpha_s t)^{\frac{1}{2}}) \tag{15}$$

To obtain more flexibility for simulation of this stage, a finite difference explicit formulation was applied. Its stability is guaranteed if dimensionless parameter $\alpha_s (\Delta t/\Delta y^2)$ remains constant with a value of $\leq \frac{1}{2}$, with proper values for density, specific heat and thermal conductivity.

It is to be noted that steady state solution of equation 14 with boundary condition 14b will correspond to equation 7 presented earlier.

Other equations derived were written to calculate:

a) Superficial liquid channel temperature along the furnace.

This temperature depends mainly on the melted bank mass, the radiative heat to liquid channel and the convective heat transfered between the surface and the bulk of the liquid channel. Its final expression is:

$$\frac{\partial T_{LS}(x,t)}{\partial t} = \frac{[K_6(T_G^4(x,t) - T_{LS}^4(x,t)) - K_7(T_{LS}(x,t) - T_{LC}(x,t))]}{(\rho_{LS} \, Cp_{LS} \, D_{LS}(x))} \tag{16}$$

Where $T_{LC}(x,t)$, liquid bulk channel temperature, is determined through melted mass as follows:

$$\frac{\partial T_{LC}(x,t)}{\partial t} = \frac{[K_8(T_G^4(x,t) - T_{LS}^4(x,t)) + K_9(T_{LB}(x,t) - T_{LS}(x,t))]}{(\rho_{LC} \, Cp_{LC} \, D_{LC}(x))} \tag{17}$$

b) Liquid channel mass fed to settling zone.

By performing a heat balance for a liquid channel element "dx" assuming that heat transfer by convection is larger than the conductive heat transfered along "dx" and taking into account radiation and convection heat from the gas, with constant thermal conductivity, heat capacity and density for the liquid, the equation becomes:

$$\frac{\partial T_{LM}(x,t)}{\partial x} = \frac{K_{10} A_C(x) \ \bar{h}}{A_{TC}(x) \ \rho_\ell \ Cp_\ell \ u(x)} \ [T_G(x,t) \ - \ T_{LM}(x,t)] \tag{18}$$

From equation 18 the liquid channel temperature arriving to the settling zone can be calculated. Using this information, liquid temperature variation of matte and slag can be obtained.

c) Heat obtained through complete or incomplete fuel combustion.

For example, if oxygen mass is enough for H_2O but is not enough for CO formation, then net heat available from combustion takes the following form:

$$H_{Net} = \ m_F[CFV \ - \ \Delta \bar{m}_{O_2,i} \ \Delta \bar{H}_i \ - \ (m_{O_2,j} \ \bar{H}_j)/m_F] \tag{19}$$

Where $\Delta \bar{m}_{O_2,i}$; $\overline{\Delta H}_i$: specific oxygen mass difference and specific combustion heat difference between that required for CO_2 and CO formation starting from C.

$\bar{H}_j$: specific combustion heat from C to CO.

$\bar{m}_{O_2,j}$: oxygen enough for $H_2 \rightarrow H_2O$ and not enough for C to CO.

d) Fuel combustion thermal efficiency.

This was obtained by considering maximum heat generation from complete combustion, sensible heat from combustion gas, heat losses through refractories and furnace openings.

$$\frac{d \ \eta(t)}{dt} = [\Sigma_i \ (H_i exp(k_{ci} t_i)) \ - \ \Sigma_i \ m_i Cp_i(T_G(L_x,t) \ - \ T_G(o,t)) \ -$$

$$- \ K_{11} \ \Sigma_i \ Q_{LO} \ ]100/\Sigma_i \ H_i \tag{20}$$

e) Gas residence time along the furnace.

For each furnace element gas residence time is calculated considering all sources of gas generated from combustion and bank reactions, residual bank surface and furnace cross sectional area.

$$t_G = \Sigma_i \ (x_i(J,L,N)/v_{Gi}(x,t)) \tag{21}$$

where $\quad v_{Gi}(x,t) = T_G(x,t) \ \Sigma_i \ \dot{v}_{ai}/(T_0 A_{TF}(x,t)) \tag{22}$

$$A_{TF}(x,t) = f(\left(\frac{ds}{dt}\right)_E, \ L, \ tg \ \alpha, \ J, \ N) \tag{23}$$

746

This set of equations is solved step by step, but previously it is necessary to define initial conditions just after furnace is charged and heat generated from fuel combustion. Therefore, the following conditions must be defined: a) steady state profiles for bank and liquid channel temperatures; superficial bank area and cross section furnace area; roof, banks and liquid channel emisivities and view factors; b) calorific value of fuel and kinetic constant for combustion; temperature of liquid flowing along the banks; fuel combustion conditions; inner furnace geometry charge mineralogical and chemical composition, heat requirement for charge smelting; and physico-chemical data for solids liquids and gases: densities, specific heat, thermal conductivity and viscosity.

The influence of the following variables on combustion were simulated in the model:

1) Fuel feed rate.
2) Combustion air temperature.
3) Excess combustion air.
4) Combustion constant.
5) Longitudinal bank surface area.
6) Furnace cross section area.

By changing type of fuel and burner position it was possible to calculate thermal conditions prevailing in end wall and roof burners furnace reverbs.

B. Complementary Models

Three Dimensional Steady State Heat Transfer Zonal Model

This model describes more rigorously the heat transfer between zones and sections inside the furnace. The furnace was divided into eight sections, each containing four zones, plus two additional zones for the end and front walls. Axial temperature profiles of gas, banks, liquid channel and roof were obtained by solving the 34 simultaneous algebraic equations derived for radiation, convection, sensible heat, melting heat, charge reaction heats and heat losses.

The net incident radiant heat to zone i in section k is given by:

$$q_{ik} = A_{ik} \sigma \alpha_{ik} T_{ik}^4 - \sum_{jk}^{i \neq j} A_{jk} \alpha \beta_{jk} F_{jk} T_{jk}^4 \qquad 1 \leq k \leq 8 \qquad (24)$$

The set of equations can be written as follows:

$$\sum_{j=1}^{34} a_{ij} T_j^4 + \sum_{j=1}^{34} b_{ij} T_j + c_i = 0 \qquad (i = 1, 2, \ldots, 34) \qquad (25)$$

where a_{ij}, radiant heat coefficient between zone i and j.

b_{ij}, convection heat coefficient between zone i and j.

c_i , internal and external heat flowrate.

T_j , temperatures.

To solve these equations the generalized Newton Raphson Iteration Method was used.

Gas Phase Fluid Dynamics Model

A fluid dynamics simplified model was constructed to obtain the gas velocity profiles inside the furnace and to evaluate associated dust losses under laminar and turbulent regimes.

For forced convection and assuming constant gas viscosity, heat and momentum transfer equations can be formulated independently without serious loss of accuracy, avoiding a coupled transport phenomena treatment. The equations developed assumed Newtonian fluid and Prandtl laminar and turbulent regimes.

For unidimensional laminar flow the momentum equation becomes:

$$\frac{\partial v_x}{\partial t} + v \cdot \nabla v_x = -\frac{1}{\rho} \frac{\partial p_x}{\partial x_s} + \frac{\mu}{\rho} \nabla^2 v_x + g_x \qquad (26)$$

Neglecting g_x and $\frac{\partial^2 v_x}{\partial y^2}$, dimentionless equation becomes:

$$A \, V \, \frac{\partial V}{\partial X} + B - C \, \frac{\partial^2 V}{\partial Y^2} = 0 \qquad (27)$$

where: $A = \dfrac{\rho v_\infty^2}{x_T}$; $B = \dfrac{2\mu\bar{v}}{R_h^2}$; $C = \dfrac{\mu v_\infty}{y_T^2}$

with boundary conditions given by:

$\dfrac{\partial V}{\partial Y} = 0$ at $Y = 0$; $\qquad V = 0$ at $Y = \pm 1$; $\qquad V = 1$ at $X = 0$

For two dimensional turbulent flow the dimensionless equation for momentum and continuity can be written as:

$$E \, V \, \frac{\partial V}{\partial X} + F \, W \, \frac{\partial V}{\partial Y} = G + D \, \frac{\partial V}{\partial Y} \frac{\partial^2 V}{\partial Y^2} \qquad (28)$$

$$H \, V \, \frac{\partial \rho}{\partial X} + I \, \frac{\partial V}{\partial X} + J \, \frac{\partial W}{\partial Y} = 0 \qquad (29)$$

where: $E = \dfrac{\rho v_\infty^2}{x_T}$: $F = \dfrac{\rho w_\infty^2}{y_T}$; $G = -\dfrac{\Delta P}{\Delta L}$; $D = \dfrac{2 \rho L^2 w_\infty^2}{y_T^3}$

$H = v_\infty$; $\quad I = \dfrac{\rho v_\infty}{x_T}$; $\quad J = \dfrac{\rho w_\infty}{y_T}$

The boundary conditions are :

$\dfrac{\partial V}{\partial Y}$; $\dfrac{\partial W}{\partial Y} = 0$ at $Y = 0$

$V, W = 0$ at $Y = \pm 1$

$V, W = 1$ at $X = 0$

Equations 28 and 29 must be solved simultaneously. For simplicity only one direction has been considered (W = 0). Laminar and turbulent equations have been solved by the finite difference numerical method.

Fuel Combustion Model

This model predicts gas temperature and chemical composition profiles for coal and oil combustion performing for the gas phase:mass, heat and momentum balance, conversion and reaction kinetics,while for particulate phase: population and momentum balance.

This phenomenological model is composed by eight ordinary differential equations and several algebraic equations, which must be solved simultaneously. This system can be represented as a function of flame length x_1, in the following vectorial form:

$$\vec{Y}'(x_1) = f(x_1, \vec{Y}(x_1))$$

$$\vec{Y}(o) = \vec{Y}_o \tag{30}$$

and

$$Y_i(x_1) = g(T_i, X_{O_2}^j, n_i, \dot{v}_i, p_i, \Delta H_i, \bar{h}, \rho_s, K_{11}) \tag{31}$$

where

$$f(x_1, \vec{Y}(x_1)) = \begin{bmatrix} f_1(x_1,\vec{Y}) \\ \cdot \\ \cdot \\ f_8(x_1,\vec{Y}) \end{bmatrix}, \quad \vec{Y}_o = \begin{bmatrix} Y_1 \\ \cdot \\ \cdot \\ Y_8 \end{bmatrix}$$

$$Y_1, \ldots Y_8 \text{ are } X_{O_2}^j (j = 1,3), T, v, u, \tau_C, \frac{\partial u}{\partial x_1}$$

The set of simultaneous equations 30 can be solved numerically by the Runge Kutta method, using previous results of system formed by equations 31.

PID Control Algorithm

The objective of the control system is to regulate the process variable so that it remains as close to a constant (set point) as possible in a feedback fashion.

The PID controller parameters (K_C, T_I, T_D) have been obtained through Ziegler-Nichols relationships (17), by analyzing the step response curve and optimized through ISE technique. Basically the curve has to be identified with a first order plant plus delay to obtain K_p, σ and τ.

Then, $\Delta CV = [PLANT] \Delta M$ (32)

$$[PLANT] = G_p(s) = K_p e^{-\theta s}/(\tau s + 1) \tag{33}$$

For simulation of PID behaviour it is necessary to discretize the following function:

$$M(t) = CV(e(t)) \tag{34}$$

whose final form is:

749

$$M_{k+1} = M_k + K_C[(e_k - e_{k-1}) + \frac{T_D}{T_M}(e_k - 2e_{k-1} + e_{k-2}) + \frac{T_M}{T_I}e_k] \qquad (35)$$

$$k = 0, \ldots, ITOP$$

III. Results

1. Accuracy of the Model

As said previously the reverberatory furnace was divided into a number of transversal elements. By analysing gas temperatures profiles which are the most sensitive results of the model, and applying Lothar's method to obtain an upper limit for the error (18) it was found that for 200 elements the mean percent error (MPE) is of 6%, while for 110 and 550 elements MPE could be 9% and 2%, respectively. Therefore, 200 elements were selected considering the error produced and the computing time involved.

2. Application of the Dynamic Model to Industrial Reverbs

The validation of phenomenological models for high temperature metallurgical processes is difficult, due to the lack of reliable sensors to operate under extreme conditions of temperature and corrosive environment, as well as the impossibility to decouple process and operational fluctuations. However, conventional and non-conventional measurements have been made in two reverbs, one operating with coal and the other with oil.

Conventional measurements were: longitudinal temperature, velocity and composition profiles in smelting zones; gas composition and gas flowrate determinations in the settling zone. Non-conventional measurements were: bank smelting rate; charge bank internal temperature; color photographs of burner, charge banks and furnace settling zone; and charge feed to each bin along the furnace.

The measurements performed were not as precise as desired. Nevertheless, there is a reasonable agreement between the predictions of the model and corresponding measured values that appears to validate the dynamic model. One advantage of the model is that it predicts some values that experimentally would be extremely difficult to obtain. Some of these results will be discussed in the following paragraphs.

It was found that during operation the longitudinal gas temperature profiles move towards the settling zone as shown in Figure 3 for oil burners simulation.

This effect is due to the heat balance between gas and charge banks. As melting progresses and the bank surface decreases, preferentially in the highest temperature region, heat transfer also decreases and the gas takes the heat further along the furnace, thus moving the maximum in the temperature profile towards the settling zone of the furnace.

For oil burners, the maximum gas temperature was lower than for coal burners and for both the maximum occurs at about 50% of the furnace length.

Results of the individual charge bank smelting rate versus time are shown in Figure 4 and longitudinal smelting rate profiles in Figure 5.

The information derived from the curves of Figure 4 can be used to optimize the recharging schedule of the furnace in order to obtain a maximum overall smelting rate.

750

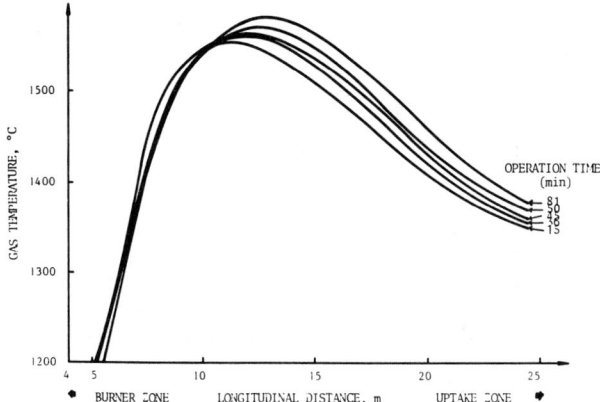

FIGURE 3 : Longitudinal Gas Temperature Profiles as a Function of the Furnace Length for Different Operation Times. Oil Burners Operation.

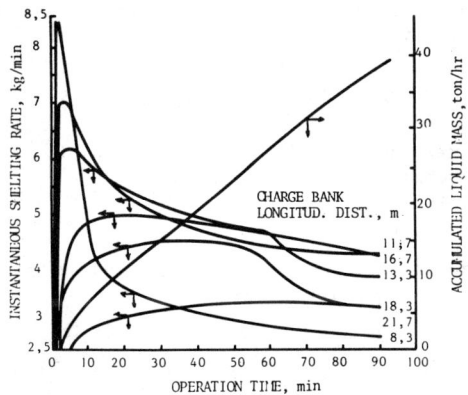

FIGURE 4 : Charge Bank Smelting Rate for Different Charge Banks and Accumulated Liquid Material as Functions of Operation Time. Oil Burners Operation.

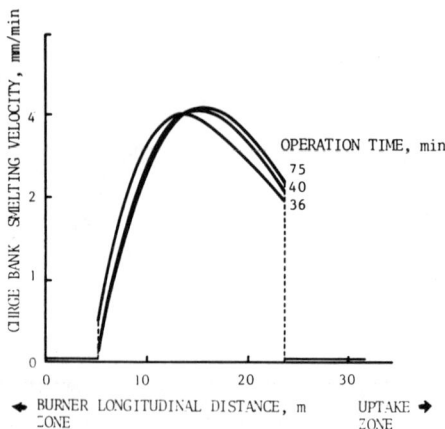

FIGURE 5 : Charge Bank Smelting Velocity for Different Operation Time as a Function of Furnace Longitudinal Distance. Oil Burners Operation.

If it is assumed that the bank is replenished at fixed intervals τ and that the smelting curve $sR(J,t)$ (Fig.4) has a maximum within the interval , the condition to maximize the smelting rate at that position, as found by Del Campo and Luraschi (20), is expressed by:

$$\tau \cdot sR(J,t) \;=\; \int_{0}^{\tau} sR(J,t)\,dt \tag{36}$$

To obtain the maximum overall smelting rate in all the length of the banks different recharging intervals would have to be used at different charging pipes, utilizing the analysis of Eqn. 36 for each curve of Figure 4 to find its optimum charging frequency.

Accumulated liquid material can be represented by the expression:

$$\int_{t}\left[\int_{x} \frac{sR(x,t)}{\ell}\,dx\right]dt \tag{37}$$

Initially, liquid is formed at a fast rate due to the high heating rate, to decrease later when available surface bank decreases. Smelting rate remains approximately constant during most of the operation, which validates the assumption that reverbs operate under steady state condition or close to it.

From Figure 5, the longitudinal velocity distribution can be verified, with a maximum velocity which moves towards the settling zone with time. By comparison with Figure 3, this maximum nearly coincides with the maximum gas temperature.

To validate the model it is suggested to perform further conventional and non-conventional measurements, to develop appropiate sensors in order to determine smelting mechanism, as well as physical and chemical properties related to heat transfer phenomena. It is also important to determine parameters through experimentation such as bank thermal conductivity and liquid viscosity, and to reach a more fundamental knowledge about other phenomena involved in the process, like the digestion of charge in liquid slags and the erosion and digestion of liquid flowing down the bank, combined with bubbling of the gases produced at the solid/liquid interface through the molten flowing material.

3.   Simulation Results

Simulation was performed for oil and coal combustion.

Coal Operation. Optimum air/coal ratio (RAC) must generate maximum heat without an excesive heat loss to the exit gases. This condition is described by curve 0,2 in Figure 6. Above 15% excess air, although a complete combustion takes place, heat transfered to banks decreases as a result of high gas velocity which permits a high heat content in combustion gases. Below 15% excess air, heat transfered to banks decreases also as a resultant of incomplete combustion and heat potential inability to compensate other heat losses.

Also, it can be observed that decreasing coal rate which increase RAC for constant air supply, or decreasing air flow rate which decreases RAC for constant coal supply, will result in an increase in smelting rate. For process control, it is simpler to maintain constant fuel rate and to vary RAC through air supply.

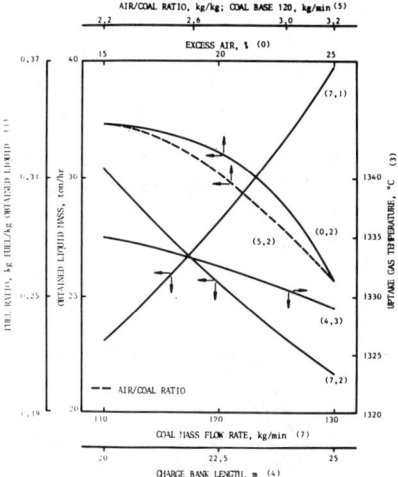

FIGURE 6 : Effect of Coal Mass Flowrate, Charge Bank Longitudinal Zone, Excess Air and Air/Coal Ratio in Fuel Ratio, Obtained Liquid Mass and Uptake Gas Temperature. Operation Time: 72 min

Decreasing banks length, exit gas temperature increases slightly which confirms the highly efficient heat transfer process found previously (1). Thermal inefficiency is due only to the sensible heat content of the large gas volume generated.

A lower combustion excess air produces a displacement of the axial gas temperature profile to the burner zone and thus a lower exit gas temperature, greater thermal efficiency and larger amount of molten material (Figure 7).

Greater bank surface area exposed to radiant heat allows a larger smelting rate; its limit is the condition that avoids cooling the gas phase and the roof below the desired minimum temperature. The effect of bank surface area and its dependence on roof height are seen in Figure 8. Thus, there must exist an optimal roof height to produce maximum smelting capacity.

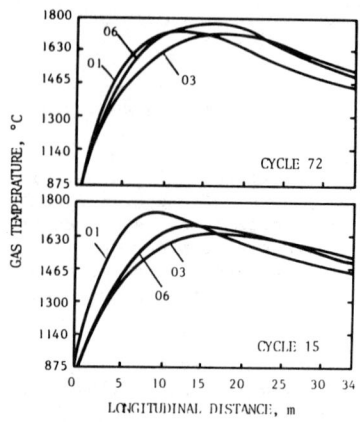

FIGURE 7 : Effect of Oil Flowrate and Excess Air in
Gas Temperature Profiles

01 : Oil 80 (kg/min); 17% excess air
06 : Oil 80 (kg/min); 10% excess air
03 : Oil 90 (kg/min); 17% excess air

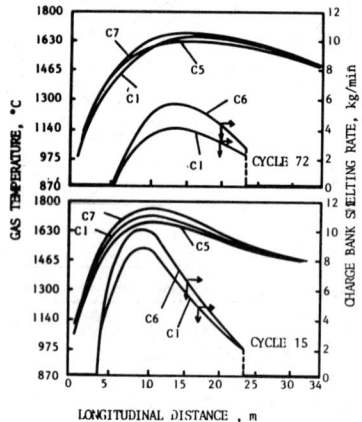

FIGURE 8 : Effect of Charge Bank and Transversal Furnace
Area in Longitudinal Gas Temperature Profiles.
Cycles 15 and 72 mean 15 and 72 minutes

C1 : Charge Bank Area and Charge Supply. Base Case
C5 : Increased Charge Bank Area: more charge supply
without furnace roof elevation
C7 : Decreased Charge Bank Area: less charge supply
without furnace roof elevation
C6 : Increased Charge Bank Area: more charge supply
with furnace roof elevation

754

Oil Operation. Operating only with the stoichiometric air, it is theoretically possible to reach high temperatures (Figure 9), obtaining high
smelting rates. Three practical problems arise from this situation: a) excessive refractory wear, b) short flame that leaves a large unsmelted bank
zone and c) inefficient fuel-air mixing. These factors force to operate the
furnace with some level of excess air. Combustion rate constant, combustion
air temperature and oil mass flowrate are less relevant to the amount of melt
ed material.

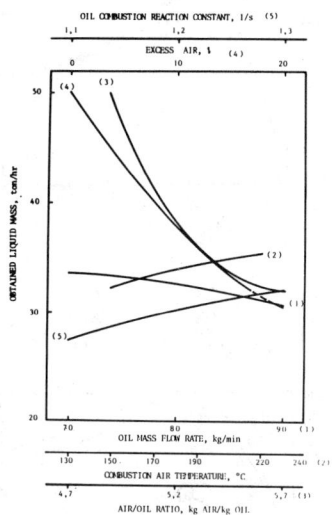

FIGURE 9   :   Effect of Oil Mass Flowrate, Combustion Air
              Temperature, Air/Oil Ratio, Excess Air and Oil
              Combustion Reaction Constant in Obtained Liquid
              Mass. Operation Time: 72 min

Increasing oil mass flowrate allows to move the axial gas temperature
profile to the settling zone, which has a negative effect in terms of smelted
concentrate (Figure 10). Decreasing excess air permits to increase flame
temperature over the maximum value which produces an excessive refractory
wear and increasing also exit gas temperature, which on the other hand can
minimize magnetite precipitation in the settling zone. Less excess air would
also hamper extra heat generation by burning the labile sulphur generated
from concentrate by thermal decomposition and makes gas temperature profile
displacement less flexible. If secondary air temperature is increased from
150 °C to 220 °C, the smelting rate would increase accordingly and an axial
gas temperature profile will be distorted as it is seen in Figure 11.

4.   Application of the Model to Automatic Control

As a first approach, it can be assumed that a dynamic model coupled with
a control algorithm is adequate to determine automatic control strategies,
when the algorithm considers two interactive feedback control loops: one for
the gas temperature and the other for gas composition. With a PID algorithm
it was not possible to reach a constant final value (steady state) for control
led variables. It is proposed that this can be achieved by considering K_p and
τ as functions of temperature and K_C only as a function of time.

755

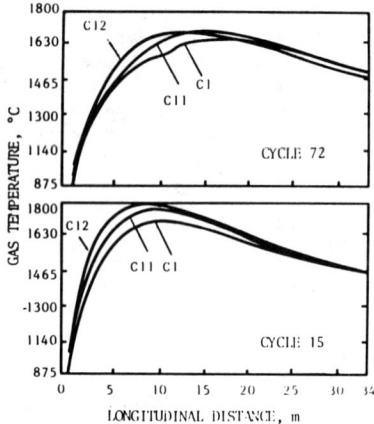

FIGURE 10 : Effect of Excess Air in Gas Temperature Longitudinal Profiles

C1 : 25% Excess Air
C11 : 20% Excess Air
C12 : 15% Excess Air

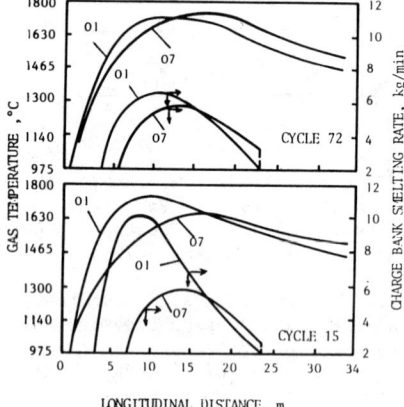

FIGURE 11 : Effect of Secondary Air Temperature in Gas Temperature
Longitudinal Profiles

01 : Secondary Air Temperature, 220 °C
07 : Secondary Air Temperature, 150 °C

In order to show the dynamic behaviour of the controlled variable (coal mass flowrate), when it is imposed to the furnace to reach 0% of O_2 and CO in exhaust gases and to reach 1400 °C at 16,7 m (middle length), Figure 12 shows the relationship among these variables. It can be seen that initially, the controlled variable oscillates further reaching an apparent steady state for the desired gas temperature after 100 min of operation with 100 (kg/min). This time should be smaller for an adaptive control to be used. Also, composition of exit gas reaches steady state for oxygen and carbon monoxide after 25 min. Figure 13 shows how axial gas temperature profile reaches the value of 1400 °C with time.

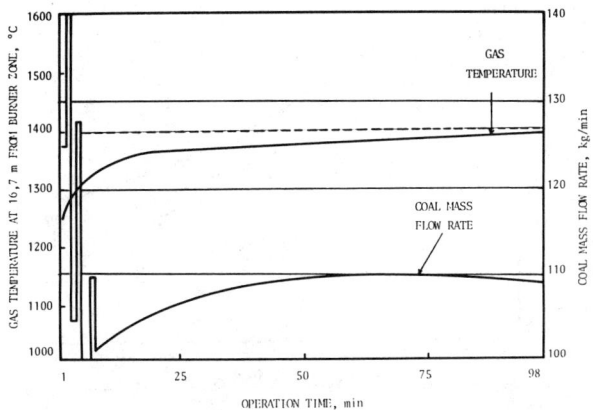

FIGURE 12 : Dynamic Coal Mass Flowrate and Gas Temperature
Localized at 16,7 m from Burner Zone

Condition: 0 $\%O_2$ and 0 $\%CO$ in exhaust gas

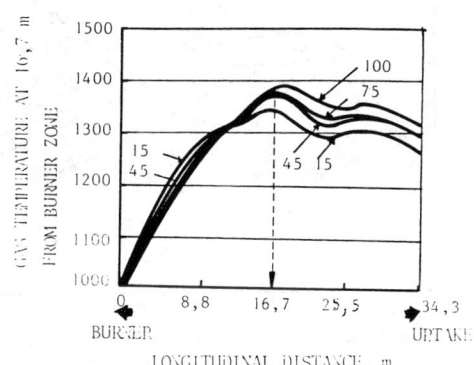

FIGURE 13 : Dynamic Gas Temperature Longitudinal Profiles.
Condition: 0 $\%O_2$ and 0 $\%CO$ in evacuated gas and
gas temperature control at 16,7 m
from burner zone

5. Complementary Models Results

Three Dimensional Heat Transfer Zone Model. Figure 14 shows steady
state longitudinal gas, roof, liquid channel and charge bank temperature profiles for a furnace operating with oil end-wall burners. It can be observed
that its tendency is similar to the one obtained with the dynamic model.
Similarly, Figure 15 shows temperature profiles for a furnace operating with
oil roof burners. Here a more homogeneous profile is obtained along the furnace which allows a larger amount of smelted material being obtained for the
same thermal intensity from the gas to bank surface.

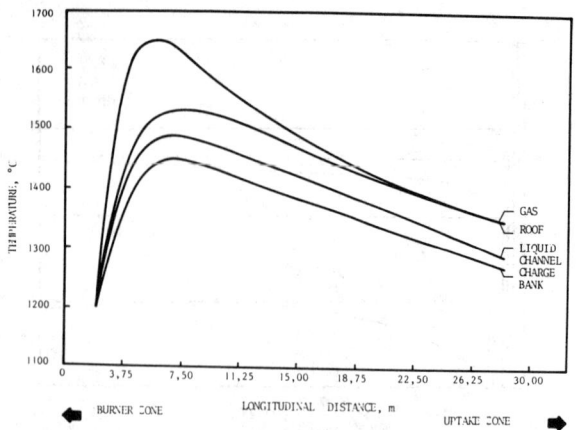

FIGURE 14 : Temperatures Longitudinal Profiles for Different
Zones Inside the Furnaces

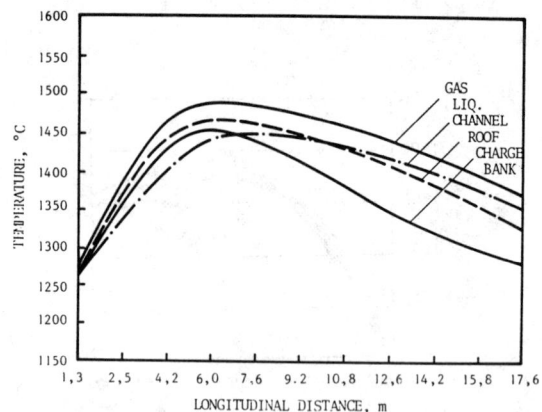

FIGURE 15 : Charge Bank, Liquid Channel, Roof and Gas Temperature
Longitudinal Profiles

Gas Phase Fluid-Dynamic Model. Figure 16 shows steady state axial and
transversal gas velocity profiles for a section of the smelting zone of a
furnace operating with oil. It can be observed that the experimental data
agrees well with the turbulent regime profile proposed. To have a more realistic description, it is suggested to incorporate recirculation gas flow
patterns, particularly near the burner zone, as well as transversal turbulence.

IV. Discussion. Implications on Smelting Practice

The most important variables which determine the efficiency and economic
results of reverberatory smelting are the fuel rate and concentrate charging
practice. These variables can only be controlled under some restrictions
imposed mainly by operational reasons and the need to keep refractory consump
tion low.

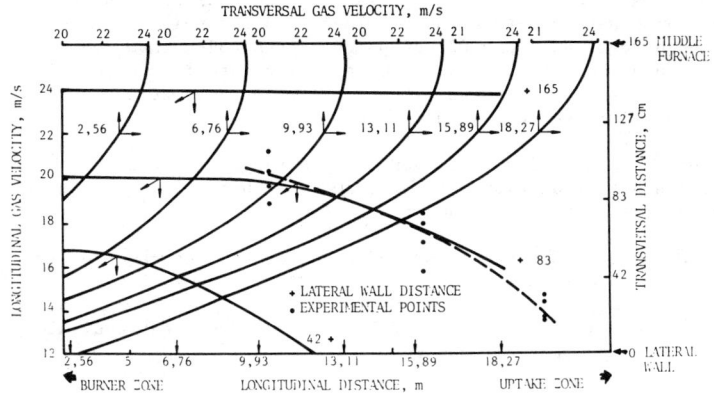

FIGURE 16 : Axial and Gas Temperatures Longitudinal Profiles.
Turbulent Regime in Oil Operating Furnace.
Experimental Points at 1 m of Lateral Wall

From a practical view point the dynamic model may be useful: a) as a basis for decisions on production rates and how to achieve them, b) as a tool for operational control of the furnace, c) as a subroutine of a simulator of the reverberatory-boiler system to optimize individual process or joint efficiency, d) subroutine of an algorithm of adaptive optimizing control of the combustion process, and e) as a tool for training of operating and supervisory personnel.

Model validation was only partially possible specially with respect to precision of model predictions, due to problems with measurements deriving from the lack of sensor for some of the variables affected by agressive conditions in the furnace. According to this experience, the following is suggested for further progress: i) to develop sensors that can work in the furnace environment and to perform continuous measurements in plant, ii) to conduct experimental studies on the smelting phenomena at the lab or pilot plant scale in order to clarify the mechanisms involved.

The results of simulations using the dynamic model show that: i) there is an optimum in the air to fuel ratio for each value of fuel rate, under a given set of operational conditions and restrictions, ii) confirmation that the low thermal efficiency of the reverberatory furnace is due to the high heat content of exhaust gases rather than heat transfer from gases to roof and banks, iii) it is possible to determine an optimum roof height of a reverb with end-wall burners to maximize smelting rate without deteriorating the settling function of the furnace, iv) the recharging schedule of the furnace is subject to optimization through the analysis of the smelting versus time function, as determined in this work, v) under furnace conditions studied, it is not possible to increase the smelting rate by just increasing the fuel rate, since this will cause the flame profile to move along the furnace, thus decreasing the smelting rate in the back zone of the furnace.

Other conditions under which the smelting rate may be increased are: a) to alter the flame shape and length, b) to incorporate roof burners (oxy-fuel, c) to preheat the secondary combustion air.

The simulation results of combustion automatic control, considering as an example a case of 18% base excess air with coal burners, showed that by establishing a temperature control loop with set point of 1400 C at 16,5 m from the burner end and a 0% set point oxygen and carbon monoxide control loop through the air/fuel ratio, a large increase in the smelting rate is obtained with respect to the situation without automatic control, with only a marginal increase in the fuel rate.

Other results indicate that, in the ranges of fuel ratios from 100 to 150 kg/ton and production rate from 500 to 800 ton/day green charge, the following are valid:

i)   Increasing 1% the moisture of the green charge requires from 1.75 to 2.5 tpd additional fuel.

ii)   One ton nitrogen in the furnace air consumes heat equivalent to smelt from 1 to 2 tons of charge.

iii) Each ton of oxygen added to the furnace air allows increasing the charge smelted by 4.5 ton or decreasing fuel oil used by 0.25 ton.

Through the three zone heat transfer model it was found that a lower roof height is required for a roof burner furnace than for an end-wall burner furnace, in the first case the smelting rate being more sensitive to this variable, other conditions remaining constant.

The analysis of fluid dynamics conditions of gas flow determines that turbulent flow prevails in the furnace. On the other hand it is apparent that smelting on the bank is of the ablative type, which means that flowing molten liquid along the bank carries the great majority of heat received from the gases, thus permiting only a small penetration of the temperature gradient inside the solid. Since the former is determined mainly by the thermal conductivity of the charge, and this in turn by particle size, and liquid film viscosity over bank surface, this means that greater control of furnace operation and smelting rate could be achieved through these variables.

Finally, it is considered possible to determine a recharging schedule, through use of the model, to optimize the smelting rate.

V. Summary

A dynamic phenomenological model was developed to predict axial profiles for gas temperature, composition and velocity, banks temperature, smelting rate and melting velocity, distribution of heat supply to the furnace and total mass smelted. Factors such as different fuel, and combustion conditions, as well as, smelting conditions, burner location, bank surface area were studied. From the results obtained the convenience to establish some automatic control strategy was defined. It was also attempted to correlate smelting practice to operating problems.

In general, the model agrees reasonably well with tendencies of measurements performed in industrial reverbs. Evaluation of the model sensibility shows that internal reverbs profiles can change over a wide range, depending upon the bank surface area, furnace transversal area, combustion kinetic constant, fuel type and fuel mass rate, combustion air temperature, excess combustion air, charge chemical composition, heat losses, burners location and internal furnace geometry.

Additional to the dynamic model and in order to make it more practical it has been developed a steady state models which are a: tridimensional
heat transfer zone model for internal furnace heat distribution; fuel combustion model for temperature and flame profiles, momemtum and heat transfer model for gas velocity profiles.

Main model prediction indicated the following results:

1.   During banks heating and melting, internal banks temperature profiles
changes abruptly from high surface temperatures to ambient temperatures in a small bank depth which indicates that its behaviour is
similar to an insulating material. Bank surface melting starts between the first 2 and 8 min, depending on the bank thermal conductivity. Viscosity and thickness of the liquid film flowing down over
bank surface area, are the main variables that could affect melting
velocity and smelting mass rate.

2.   From a heat transfer point of view, there is an optimum furnace roof
height, when main factor considered are radiation and convection heat
transfer mechanism from flame to bank surface and the residence time
of the heating gas along the furnace for a constant smelting productivity. Calculation shows that for wall-end burners furnace optimum
roof height would be higher than for roof burners furnace.

3.   It is confirmed that thermal inefficiency in reverbs is due to the
larger amount of heat transported by exit gases, which represent a
40% to 50% of the total heat output, while heat transfered from roof
and gas to the banks is an efficient heat transfer process.

4.   From a combustion point of view, it exists an optimum air/fuel ratio
for a given fuel mass flowrate and operational variables restriction.

5.   By analysing the predicted axial smelting rate functions, it can be
derived an optimized strategy to recharge reverbs.

6.   Furnace productivity cannot be increased solely by increasing the fuel
mass rate, bank will not be smelted due to the axial gas temperatures
profile displacement towards the furnace uptake. To overcome this
limitation, others solutions can be implemented such as using additional burner supply at the end wall or in the roof: use secondary air
preheat; or modify flame length and shape.

7.   To increase reverbs furnace productivity with particular restrictions,
it has been found convenient to establish an automatic combustion
control strategy considering a control loop for gas temperature localized at the middle of the furnace and a control loop for CO and O_2
composition simultaneously.

Finally, the model could be used: to optimize bank smelting; to
implement modifications such as type and location of burners, frequency of
bank recharge; to determine strategies for automatic combustion control or
to apply direct digital control (D.D.C.) with on-line computer; to develope
metallurgical process control to have a proper decision tool for improving
operational indexes; to select type and location for sensors implementation
in reverbs instrumentation; and to improve furnace behaviour through operators training.

761

Nomenclature

A	Longitudinal surface area
$A_{i \text{ or } j,k}$	Internal surface area of zone i or j in section k
a	Liquid film thickness over bank surface
C	Chemical composition
CFV	Calorific fuel value
Cp	Heat capacity
CV	Controlled variable
ΔCV	Change in controlled variable
D	Depth
e	Error
F	View factor
g	Gravity acceleration constant
$\bar{h}$	Radiation-convection heat transfer coefficient
H	Combustion heat
$\bar{H}$	Combustion heat per unit of comburent mass
ΔH	Entalphy change
J	Assigned number for each furnace element
K_i	Constants $i = 1, \ldots, 11$
K_p, K_C	Plant and controller gain
k	Thermal conductivity
k_c	Combustion kinetic constant
L	Furnace length
L_z	Bank generatrix
ℓ	Unit of bank surface length
m	Mass
$\bar{m}$	Mass per unit of fuel
Δm	Change in mass
$\dot{m}$	Mass flowrate
M	Manipulated variable
ΔM	Change in manipulated variable
N	Number of furnace elements
n	Mol amount
p	Pressure
Q	Heat component
q	Heat transfer rate
R_h	Hidraulic radio
$\left(\dfrac{ds}{dt}\right)_{\bar{E}} \left(\dfrac{dsR}{dt}\right)_E$	Steady state melting velocity and smelting rate

762

s	Time conversion in Laplace Transform
T	Temperature
T_I, T_D, T_M	Integration, derivative and sampling time
t	Time
V, W	Dimensionless velocity
u, v or w	Particle and gas velocity
$\dot{v}$	Volumetric flowrate
X, Y	Dimensionless coordinates
$x_{O_2}^j$	Oxygen conversion by chemical reaction j
x, y and z	Spatial coordinate in furnace length, cross section inside bank and length down surface bank (generatrix)
x_l	Variable flame length
α_s	Thermal diffusivity
α_{ik} and β_{jk}	Coefficient dependent on emisivities and view factors
α	Bank angle
ε	Emisivity
μ	Viscosity
η	Thermal efficiency
ρ	Density
σ	Stefan Boltzmann constant
τ, Θ	Time constant and death time
τ_c	Combustion time

Subscripts

a, ai	Air and infiltration air
B	Charge bank
C	Liquid channel
Comb	Fuel combustion
F	Fuel
G, g	Gas
Infil	Infiltration
LB	Liquid bank surface
LS	Liquid channel surface
LC	Liquid channel bulk
LM	Liquid mass in settling zone
LO	Heat loss
ℓ	Liquid
m	Melting
n	Time interval
o	Initial condition

Rad,Reac   Radiation and reaction

SB   Solid bank surface

TC   Liquid channel transversal area and inner furnace cross section

Acknowledgements

The work has been possible through Dr. Otero's finantial support from the
Mining and Metallurgical Research Center (CIMM), the University of Concepción
and Dow Chemical (Chile).

Also thanks are extended to Dr. Carlos Díaz from INCO and Professor
Charles Cooper from Queen's University (Canada) for their valuable comments.

References

1. A.K. Biswas and W.G. Davenport, Extractive Metallurgy of Copper (Perga-
    mon Press, Oxford, 1976) 80-160.

2. G.J. Allen et al., "Empirical Models of a Copper Reverberatory Furnace-
    Preliminary Results", Quarterly of the Colorado School of Mines, 10
    (1964) 385-414.

3. M.G. Burcher, J.G. Eacott and M.A.T. Cocquerel, "A Study of Copper
    Smelting Reverberatory Furnace Design and Performance, and Methods
    Available for Increasing Throughput", CIM Bulletin, 73 (1980) 137-145.

4. I.A. Onaev and K. Tungushbaev, "Effect of Reverberatory Furnace Basic
    Dimensions Upon Output", Tsvetnye Metally/Non Ferrous Metals, 49 (1976)
    37-38.

5. Yu. A. Zhuravlev, "Application of a Mathematical Model for Optimization
    of Heat Efficiency in a Reverberatory Furnace", Tsvetnye Metally/Non
    Ferrous Metals, 9 (1979) 21-24.

6. D.J. Milne, G.E. Casley and G.S. Stacey, "The Efficiency of Reverbera-
    tory Furnace Smelting", The Journal of the Australian Institute of
    Metals, 16 (1971) 42-62.

7. L.R. Verney, "Factors Affecting Copper Reverberatory Furnace Performance
    and Their Influence on the Choice of Various Smelting Methods", Trans.
    Instn. Min. Metall., 69 (1960) 221-236.

8. J.N. Mashurian et al., "Effect of Charge Composition in the Smelting
    Rate Through Reverberatory Furnace", Tsvetnye Metally/Non Ferrous Metals,
    11 (1978) 34-36.

9. J.G. Eacott, "The Role of Oxygen Potential and Use of Tonnage Oxygen in
    Copper Smelting", Extraction Metallurgy'85, ed. The Institution of Mining
    and Metallurgy (London 9 to 12 September, 1985).

10. P. Wrampe and E.C. Nollmann, "Oxygen Utilization in the Copper Reverbera-
    tory Furnace: Theory and Practice", TMS Paper A74-25B.

11. I.J. Harris, "Development of a Mathematical Model for Reverberatory
    Furnace Heat Transfer", Institution of Mining and Metallurgy, Section C,
    8 (1972) 104-109.

12. A. Otero, "Modelo para la Fusión de Concentrados de Cobre y Simulación
    de Perfiles para Control de la Operación en Hornos de Reverbero",
    (D.Sc. Thesis, Concepción University, 1986) 677.

13. H.S. Carslaw and J.C. Jaeger, "Conduction of Heat in Solids", (Clarendon
    Press, Oxford, 1959) 255-320.

14. J. Szekely and N.J. Themelis, "Rate Phenomena in Process Metallurgy", (John Wiley & Sons, Inc. 1st Ed., 1971) 346-347.

15. H.G. Landau, "Heat Conduction in a Melting Solid", Quarterly of Applied Mathematics, 8 (1951) 81-94.

16. Yu. A. Zhuravlev, "Three Dimensional Zone Model and Calculation of the Heat Exchange in a Copper Reverberatory Furnace", Soviet Non Ferrous Met. Res., 3 (1975) 62-65.

17. D.R. Coughanowr and B.L. Koppel, "Process System Analysis and Control", (Mc Graw-Hill, N.Y. 1965).

18. C. Lothar, "The Numerical Treatment of Differential Equations" (Springer-Verlog, 3rd Ed., Berlin, 1966).

19. L.R. Verney, "Factors Affecting Copper Reverberatory Furnace Performance and Their Influence on the Choice of Various Smelting Methods", Trans. Instn. Min. Metall., Vol. 69, pp 211-236, (1966).

20. A.O. del Campo and A.A. Luraschi, "Strategies for Application of Automatic Control to Copper Reverberatory Furnaces", Control'84, ed. Herbst, (AIME-SME/TMS 1984).

EXPERIMENTAL VERIFICATION OF THE USE OF FREE-ENERGY MINIMIZATION

TECHNIQUES FOR MODELLING COMPLEX SULFIDE SMELTING

Harry E. Flynn* and Arthur E. Morris**

*Harry Flynn formerly a Graduate Student, is now Project Engineer,
Kerr-McGee Corporation, Oklahoma City, Oklahoma
**Arthur Morris is a professor, Department of Metallurgical Engineering
University of Missouri-Rolla, Rolla, Missouri

Abstract

The use of free-energy minimization techniques was examined as a method for modelling complex sulfide smelting. Model results were verified experimentally by smelting a variety of sulfide concentrates at 1523 K using a submerged lance technique, where the matte, slag, and gas phases are in virtual equilibrium. It was found that a combination of the submerged lance technique with an equilibrium computer program was a useful way for estimating activity coefficients in systems where the volatility of species such as lead and zinc and their compounds make it difficult to determine activities using conventional experimental techniques. In addition, it was found that a chemical equilibrium program using activity coefficient expressions derived from simpler systems can be applied to complex sulfide smelting to determine the distribution of elements between the matte, slag, and gas phases.

This research has been supported by the Department of the Interior's Mineral Institutes program administered by the Bureau of Mines through the Generic Mineral Technology Center for Pyrometallurgy under allotment grant number G1125129.

Introduction

In any chemical process the thermodynamics and mass balance of the system limits the extent to which the reactions proceed. For relatively simple smelting systems (e.g. smelting of a simple copper sulfide concentrate) the system can be adequately described by a few reactions involving a small number of species. Thus, calculations involving a few major reactions are sufficient to determine the composition of the matte, slag, and gas phases.

For slightly more complex systems where the total number of moles of minor elements is much less than the total number of moles of the major elements, the use of distribution ratios and Henrian activity coefficients allows calculation of the distribution of the minor species between slag, matte and gas phase. The extent of the major reactions is only slightly affected by the presence of the minor species.

For more complex system the concentration of "minor" species can be large enough that the simplifying assumptions made in the above systems are no longer valid. In this case no single reaction or small number of reactions dominate the system, so reactions involving all of the species in the system must be considered. Also, thermodynamic models for solution phases are very complex. As the number of species and phases increases, the ease of solving a system of nonlinear simultaneous equations diminishes rapidly, and the equilibrium composition becomes virtually impossible to solve by hand.

An alternative method for calculating equilibrium is by free energy minimization. By definition, the equilibrium state of a closed system is that in which the free energy of the system is a minimum, subject to the mass balance constraints. Stated in this form the criteria can be used as the basis of a nonlinear optimization problem; optimize (minimize) the free energy of the system subject to the number of moles of the elements in the system.

The determination of the equilibrium composition of a complex system using a chemical equilibrium program is not difficult in theory if the thermodynamic data for each specie and phase are available. However, in practice even with a computer program it takes considerable programming skill to take a general purpose equilibrium program and use it to calculate specific equilibrium problems. There is an additional complication in that existing programs are written for a closed system; i.e. one where there is no exchange of elements between the outside world and the reactor. This is not the case in many industrial processes, so modifications are necessary to account for loss of elements via the gas phase. Additionally, in multicomponent systems such as exist in complex sulfide smelting there is little or no activity data for many substances (species) in the matte and slag.

Because of the interest in complex sulfides as economically viable ore-bodies and possible domestic sources of elements such as cobalt and nickel, the objectives of this study were to:

a) Determine whether an equilibrium model could be used to calculate the distribution of elements between phases as smelting proceeds. The model should use existing activity data as much as possible. Missing data must be derived or estimated to describe the system. Such a model would provide a way of examining various smelting options for processing complex sulfides.

b) Determine if a submerged lance blowing technique can be used to create turbulent conditions within a crucible such that the matte, slag, and gas phases are in virtual equilibrium. The validity of the assumption of equilibrium in the crucible would be checked by comparing the equilibrium calculations on simple systems with the experimental data.

c) Determine whether the submerged lance blowing technique could be used to generate activity data for species in multicomponent melts containing species with a significant vapor pressure, such as lead, zinc, and their compounds.

Review of Literature

Articles dealing with the subject are of two types: 1. modelling of sulfide smelting, and 2. methods for calculating complex chemical equilibrium. Models for copper and complex sulfide smelting will be discussed first, followed by a discussion of methods for calculating complex chemical equilibrium.

Copper Sulfide Smelting

The last twenty years have seen a dramatic growth in the understanding of the physical chemistry of sulfide smelting and the use of computer programs models to predict the behavior of elements in smelting systems. Nagamori et al (1-9) have published a series of papers describing a thermodynamic model of the Noranda reactor for determining the distribution of elements between the blister copper, matte, slag, and gas phases. The Noranda reactor typically operates on a feed material containing relatively low amounts of impurities, so the oxidation of iron and sulfur can be assumed to dominate the system. By assuming certain activities of FeO and Fe_3O_4 in the slag and a calculated partial pressure of SO_2 in the gas phase, algebraic equations can be set up to predict the distribution of species between the various phases. Minor elements were assumed to be present at such low levels that Henrian behavior holds, and thus their distribution can be predicted using distribution coefficients. In addition, "suspension indices" were also used to assess the mechanical entrainment of matte in the slag phase to modify the thermodynamic model to agree with industrial practice.

Goto (10-13) has also approached the problem of predicting the distribution of species between the matte, slag and gas phases in copper smelting, using a somewhat different technique. His group has written a series of computer programs based on the work of Brinkley et al (14-19) for calculating the equilibrium state of a complex system. The programs have been compared with measurements taken in a smelter in Japan (20,21) with some success, although the nonequilibrium formation of magnetite in the furnace points out the difficulty of assuming equilibrium conditions in a reactor in which equilibrium is not obtained.

Eriksson and Bjorkman (22) have applied the computer program Solgasmix to copper smelting and converting of a chalcopyrite concentrate. Observed copper losses to slag were compared with calculated values, but no other experimental or industrial data were presented. The smelting, slag blowing and copper blowing stages were each assumed to take place in a single step.

Complex Sulfide Smelting

Complex sulfide orebodies have received a great deal of attention in recent years. Typically an ore is termed "complex" if it contains a variety of metals of which two or more have economic value. However, the term also refers to the mineralogy of an orebody where the minerals are finely disseminated and interlocked. The processing of these orebodies and concentrates from them have been recently reviewed by a number of authors (23-30). The mineralogy of Missouri complex concentrates has been discussed by Cornell et al (31) and Pignolet-Brandom and Hagni (32). Generally the consensus is that each orebody must be considered on a individual basis, because the best processing option or combination of options will depend on the particular circumstance.

Various pyrometallurgical techniques for processing complex ores or concentrates have been developed. In conjunction with these processes, efforts have been made to model the particular smelting process in order to predict the behavior of changes in feed composition, temperature, oxidation potential, etc. Fontainas et al (33) have developed a computer program based on algebraic relationships for the ratio of metals in matte and slag found in industrial practice. The user is required to make an assumption for the activity of copper in the system, and the program calculates the composition of the matte and slag based on this assumption.

Krug and Schwartz (34) have also developed a computer program for modelling the distribution of elements in a copper smelter based on industrial distribution ratios and mass balances, but by their own admission the program is said to be unreliable.

Barin and Sauert (35) have examined the cyclone smelting of copper concentrates containing approximately 10% zinc. Their calculation scheme is based on a system of nonlinear equations describing the reaction equilibria of the species considered in the system. The user specifies the temperature, partial pressure of SO_2, matte grade and the activity of FeO, and all other variables are calculated. Agreement between pilot plant studies and the model appear to be good.

Other papers concerning sulfide smelting can be found in various conference proceedings (36-38).

Chemical Equilibrium

The calculation of complex chemical equilibria has been studied by many researchers. As noted earlier, any time more than a few chemical reactions must be considered, the calculations become difficult to solve by hand. The problem is essentially one of solving a system of simultaneous nonlinear equations constrained by the mass balance of the system.

Over the years a number of techniques and programs have been written to solve complex equilibrium problems. The original work was done by Brinkley et al (14-19), who formulated a systematic method of hand calculation of equilibrium problems associated with the combustion of rocket fuel. With the advent of computers, these methods were adapted to computer programs.

A second method of solving the equations is by the Newton-Raphson method; in each successive iteration the number of moles of independent components is calculated from the number of moles of independent components in the previous iteration and a correction factor calculated from the stoichiometric factors for the chemical reactions describing the formation of the dependent components. The iterations continue until the difference between successive iterations is less than the required tolerance level. Other methods have also been implemented in computer programs. Dantzig et al (39,40) proposed the use of the method of Langrangian multipliers to solve the constrained optimization problem. This method was extended by Stephanou et al (41,42) to include condensed phases in the equilibrium. This is also the method adopted by Eriksson (43,44) in developing the program Solgasmix.

Napthali (45-47) used a reaction-adjustment method to solve systems of equations. Villars, (48-49) Cruise, (50), Meissner et al (51), Ma et al (52), and Smith and Missen (53) have proposed similar methods. In addition, several authors (54-56) have described small programs written specifically for a particular equilibrium problem using various algorithms. The advantages and disadvantages of the various methods have been discussed by a number of authors (57-61).

Of all the programs mentioned above, probably the most widely used is Solgasmix, written by Eriksson (43,44). This program forms the basis for FACT's Equilib program, (62) as well as CSIRO's Chemix program (63). It is available through sources such as NTIS (64) for installation on a mainframe computer.

The versatility of the Solgasmix program has been demonstrated by its application to many diverse problems, including solar-gas compositions (65,66), the composition of the earth's mantle (67,68), chemical vapor deposition (69,70), numerous gas-solid reaction (71-76), aqueous (77), and alloy (78,79) systems. There appears to have been only limited applications to smelting systems, and these were based on the simple Cu-Fe-S system (22). No work was found on the smelting of complex sulfides using a chemical equilibrium program.

The mainframe version of the program (64) is not particularly easy to use in the as-received state, in that it does not include a database and features such as parsing of the chemical species is not included. If these and other shortcomings could be rectified, the Solgasmix program would be convenient and useful for solving complex chemical equilibrium problems.

A number of users have worked on versions of Solgasmix for personal computers such as the IBM PC. The program can be compiled fairly easily, and can take advantage of the 8087 math coprocessor if it is installed. The main differences in the various PC versions are in the ease in which the data is entered, and the option to do heat balances (80,81).

Most recently, CSIRO has announced that version V of their comprehensive THERMODATA software package is also available for use on an IBM PC or equivalent (82). The Chemix equilibrium calculation program is capable of handling up to 15 elements, 100 species, and 30 solution phases, and contains 24 different solution phase activity coefficient models. Their database contains information on over 8000 compounds and activity coefficient parameters for dilute alloy and concentrated aqueous solutions.

A specialized version of Solgasmix called ITSOL was developed at UMR during the course of this investigation for the purpose of making equilibrium calculations in complex systems. A text editor was incorporated into the data entry portion of the program to prompt the user for input at each step, and uses error trapping to prevent the program from crashing. The text editor permits backpaging to modify entries, and parses the compound names to fill the element coefficient matrix. A permanent database of free energy of formation of compounds can be established. A routine for the entry of activity coefficients of solution species is provided as a function of temperature and mol fraction of one or more solutes. A main characteristic of ITSOL is its ability to calculate the change in composition of phases in a batch reactor during the continuous addition and removal of a reactant gas phase. A more complete discussion of ITSOL will be presented in a later section.

Thermodynamic Data

Thermodynamic data for the model was obtained from a variety of sources, which are listed for each specie in the Appendix. The data came primarily from JANAF (84), although some newer data for certain species was also taken from the USGS Bulletin 1452 (85). Data for Co_8S_9 and $ZnS_{(v)}$ were taken from Barin and Knacke's (86) compilation.

Activity coefficients were also taken from a variety of literature sources, as referenced in the Appendix. Activity coefficients were mainly taken from either Nagamori and Mackey (2), or Goto (10,12). Oxygen was assumed to be associated with FeO and Fe_3O_4 in matte, using the activity coefficient expressions formulated by Goto (10).

Experimental Procedure

Melt Tests

Experiments were conducted using a submerged lance injection technique. This technique uses the turbulence created by the jet to create conditions in the crucible such that the matte, slag, and gas phases are well-mixed, and should be in virtual equilibrium. If equilibrium is closely attained, then an equilibrium model should be suitable to describe the distribution of elements between the phases.

The experiments were made using the assembly shown in Figure 1. It is somewhat like CSIRO's Sirosmelt vessel (87-94). It consisted of a premium grade 20 gram fireclay crucible (Denver Fireclay Co.) that had been glazed with a Cone 6 - 8 porcelain glaze to reduce oxygen infiltration. The crucible held up to about 300 grams of concentrates. A scorifier dish with a 25 mm hole bored through the center was cemented onto the crucible. The scorifier was inverted to give a domed lid. A 40 mm diameter mullite chimney was cemented over the hole in the crucible lid to minimize splattering inside the furnace and to direct the fume and off-gas to the ventilation system.

The lance consisted of an upper section made of 8 mm i.d. by 200 mm McDanel MV-20 mullite or AP-35 alumina tubing and a lower section made of two-bore AP-35 alumina thermocouple sheathing with 0.8 mm holes. The lower section was inserted into the upper section approximately 40 mm and cemented with alumina cement. The upper section served as a pre-heat exchange zone for the incoming cold air to minimize the possibility of freezing at the lance tip. A 1/4" NPT brass fitting was cemented onto the top of the upper section of the lance with an epoxy resin to provide a connection to the air supply tube. A protected thermocouple was positioned approximately 25 mm above the top of the crucible lid to monitor the temperature of the experiment.

Experiments were started by combining concentrate and the appropriate flux as required. Two different copper concentrates were used with compositions as noted in Table I. The Ajo concentrate was used for a number of preliminary runs in which the design of the lance and crucible were tested. A copper concentrate from the Chino mine were used for all of the copper-based smelting experiments. A complex sulfide concentrate obtained from a Missouri lead ore was used for complex sulfide smelting experiments.

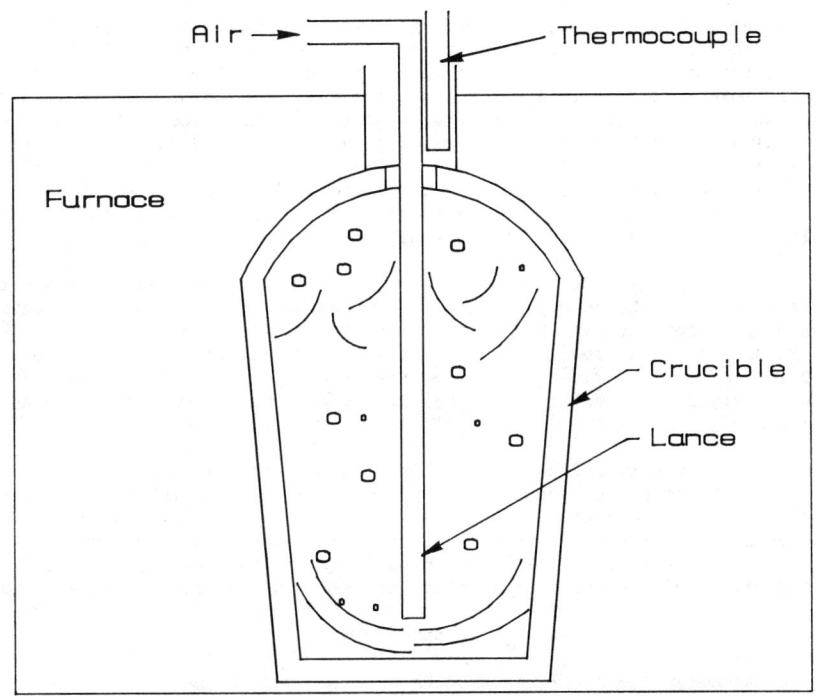

Figure 1. Crucible Assembly showing submerged lance injection of air into matte - slag bath.

Table I. Compositions of Copper and Complex Concentrates (wt pct)

Concentrate	Cu	Fe	S	Pb	Zn	Co	Ni	O_2	C	Mg	Ca	As	SiO_2	Total
Ajo	35.4	34.0	28.9	0.02	0.03	-	-	-	-	-	-	-	1.0	99.4
Chino	25.6	30.2	35.2	-	-	-	-	-	-	-	-	-	4.6	95.6
Complex	7.5	13.1	24.8	24.1	9.4	2.7	4.1	6.8	1.0	0.7	1.2	0.2	4.8	100.8

Although elements were being oxidized and transferred to the slag layer throughout the experiment, fluxes were added at periodic intervals to the crucible. This created periods of slight over fluxing and under fluxing of the slag, but on the average gave a slag typical of that from a reverberatory or similar bath smelting vessel used in a copper smelter*.

The crucible assembly was placed in the furnace and the lance was positioned such that the tip was inside the crucible but above the charge level. The furnace was heated at a rate of approximately ten degrees per minute to 1498 K. The crucible was purged with nitrogen at 20 ml per minute throughout heat up.

---

*Bob Johnson, Superintendent of Technical Services, Phelps-Dodge Corporation reports a typical Phelps-Dodge (Morenci, Az) reverberatory slag as being 37% SiO_2, 43% $FeO_{1.15}$, 7% Al_2O_3, 5% CaO, 1% Cu.

After the crucible had reached the set point temperature, the nitrogen was turned off, air was started and the lance was submerged such that the tip was approximately 12 mm above the bottom of the crucible. The lance could not be lowered beyond this depth because severe erosion of the crucible bottom would occur. Air flow was controlled by a Datametrics 1511 Mass Flow Controller and totalized using a Teledyne Hastings - Radist TR-1 Mass Flow Totalizer. Air flow was normally 1.5 liters per minute (STP). At various times accretions would build up on the tip of the lance causing the pressure to rise, but in general the pressure drop across the lance was about 48 KPa. Velocity at the lance tip was about 25 meters per second (STP).

Temperature was monitored using an Omega 660 Thermocouple Thermometer with a type "S" thermocouple. Due to the exothermic nature of the reactions, the bath temperature was hotter than the offgas temperature measured as shown in Figure 1. In a separate experiment a furnace set point temperature of 1498 K was found to give the desired bath temperature of 1523 K used in all runs. Temperature variation throughout the test series was plus or minus 15 degrees.

At the times specified by the flux addition schedule the lance was removed, the mass flow totalizer turned off, and a pre-weighed flux addition made. The lance was then re-inserted, and the totalizer started again. At the conclusion of the test the lance was removed, the furnace shut off, and the matte and slag layers allowed to solidify in the furnace. The crucible was removed and the phases separated, cleaned, and pulverized for assaying.

Modelling

A thermodynamic model was used to fit and compare the experimental results. A version of the Solgasmix free energy minimization program was modified to incorporate a feature called "iterative gas phase removal" (ITSOL), and adapted for use on an IBM PC. The iterative gas phase removal feature is useful in simulating processes where gases enter a system, react with solid and/or liquid phases, and then leave the system. Such processes are important for example in fluidized bed roasting or bath smelting.

This type of process cannot be accurately simulated by assuming that the reactor operates as a closed vessel over the entire processing cycle. In reality, the moles of gas in the reactor at any time are small in relationship to the total number of moles of gas evolved and the number of moles of elements in the condensed phases.

After equilibrium is calculated for a step, ITSOL calculates the number of moles of elements in the gas phase and removes all or any user-specified fraction of them from the system before the next calculation is made using the equation:

$$BG(I) = \sum_{J=1}^{N} TM(I,J) * BA(J,I) \qquad (1)$$

Where:

        $BG(I)$ = moles of element I in the gas phase.
        $TM(I,J)$ = matrix of elemental subscripts I for the J species in the system.
        $BA(J,1)$ = matrix of the moles of species J in phase 1 (gas phase).
        N = number of species in gas phase

Another increment of incoming gas (and if desired a specified amount of any other material) is added, and the next calculation is performed. In this manner a continuous process can be simulated as a series of steps. Since the step size can be varied at the user's discretion, one can achieve the desired degree of accuracy needed for a given simulation by increasing the number (and decreasing the increment) of steps until added steps give no significant difference. For this research a 15 liter air increment was deemed satisfactory. The current version of ITSOL assumes that the condensed phases remain in the system, although the program could be modified to remove any other phase as well by using the same methodology.

Thermodynamic data used for the free energy of formation for the various compounds is given in the Appendix. Activity coefficients for all species except PbS and ZnS in the matte phase were taken from literature sources, and are also listed in the Appendix. For the Cu - Fe - Pb - S Series 2 runs the activity coefficient expressions for Cu_2S and FeS were derived from phase diagrams that showed isoactivity lines for Cu_2S and FeS. In this case activity coefficient expressions were derived using a linear regression software package. Other investigators have used $CuS_{.5}$ rather than Cu_2S as a basis for modelling copper sulfide melts, but the activity coefficient expression for copper sulfide in matte was in a form that could not be readily converted to $CuS_{.5}$.

The activity coefficient expression for ZnS in the matte phase was determined by using published activity coefficients for all other species in the system, then varying the numerical value of the activity coefficient for ZnS in the model until the best fit between the experimental value for zinc in the matte in the Cu - Fe - Zn - S Series and the model prediction for zinc in the matte was attained. When the best fit for zinc was found, it was also found that the deviations for all the other elements in the matte phase were minimized. The derived activity coefficients of ZnS were fitted to a polynomial equation as a function of mole fraction of Cu_2S, as Cu_2S was the specie that represented the largest mole fraction in the system. The mole fraction of Cu_2S also changed considerably over the course of the experimental series, thus giving activity coefficient sensitivity to changes in the system as the oxidation occurred. The activity coefficient for PbS was derived using the same method on Cu - Fe - Pb - S Series 1 and 2 runs. The derived activity coefficient expressions for PbS and ZnS were used with the published values for the remainder of the species to model the Cu - Fe - Zn - S, Cu - Fe - Pb - S Series 1 and 2, Cu - Fe - Pb - Zn - S and Complex Sulfide Series runs.

Model predictions and experimental results for the matte phase were compared on the basis of chemical analyses that had been corrected for the small amount of insoluble portion in the matte. The samples were corrected for insol using the following formula:

$$TW = W*(M+I)/(M-I) \qquad (2)$$

Where:

TW = True weight % of element in matte
W = Assayed weight % of element in matte
M+I = Total weight percent accounted for by chemical analysis including insol.
M-I = Total weight percent accounted for by chemical analysis without insol.

This was justified because any insol was most likely slag entrained in the matte.

Results and Discussion

Results

Cu – Fe – S Series 1 and 2. The results of the Cu – Fe – S Series runs are shown in Figures 2 and 3. Two test series were run on the Cu – Fe – S system. The Ajo concentrate was used for preliminary tests and the Chino concentrate was used for all the other copper-based smelting tests. As can be seen, the results of the tests and the model predictions for the matte phase agree quite well for both test series. The agreement between equilibrium model predictions and the experimental data was a good indication mixing was intense and the system was operating at or at least very close to equilibrium.

As can be seen from the tests at high air volumes in the second Cu – Fe – S series, deviations between the model and experimental results begin to occur as the volume of matte decreases in the crucible. At 120 liters of air blown the volume of matte in the crucible is about 10 mls and the volume of slag is about 40 mls.

Cu – Fe – Zn – S Series. The results of the Cu – Fe – Zn – S Series are shown in Figure 4. As can be seen, the model predictions and experimental results for zinc agree quite well for the entire test series. The activity coefficient for ZnS, which controls the model prediction for the amount of zinc in the matte, was calculated as an expression of the type $a + bX + cX^2 + dX^3$, where X is the mole fraction of Cu_2S in the matte. It was the expression derived using the procedure given in the Experimental section. The model and experiments agree very well for copper, iron and sulfur over the entire test series.

The calculated activity coefficient of ZnS as a function of mole fraction of Cu_2S in the matte phase is shown in Figure 5. The points show the activity coefficient for ZnS as determined from the model by the method described in the experimental procedure. The curved line shows the data fitted to a third order polynomial equation. One data point at the highest matte grade (59% copper) was omitted because the activity coefficient was not very sensitive to matte grade at that composition (a change in gamma of ZnS from 2 to 4 caused a 2 percentage point change in copper and zinc in the matte, and less than a 1 percentage point change in iron and sulfur). The activity coefficient of ZnS in the matte can be adequately described by the equation:

$$\text{gamma ZnS} = 14.21 - 136.6X + 437.3X^2 - 405X^3 \tag{3}$$

where: X = mole fraction of Cu_2S in the matte.

Cu – Fe – Pb – S Series I and II. The Cu – Fe – Pb – S matte system was studied for mattes obtained from concentrates which initially contained 10% and 20% lead. Results of the 10% lead test (Series I) are shown in Figure 6. As can be seen, the model predictions and the experimental results for lead and sulfur agree quite well throughout the test series. However, there is an ever-increasing scatter between the copper and iron model data as the amount of air blown is increased. This scatter could be for two reasons; 1. nonequilibrium between the matte, slag, and gas phases in the crucible, or 2. invalid activity coefficient expressions. As will be shown in a later section, the deviation is probably due to nonequilibrium in the crucible. Since the test is a batch-type experiment, as the matte is oxidized the amount of matte in the crucible continually decreases, and the amount of slag continually increases. As the relative amounts of the two phases diverge, it becomes increasingly difficult to mix the matte and slag phases, leading to nonequilibrium conditions.

776

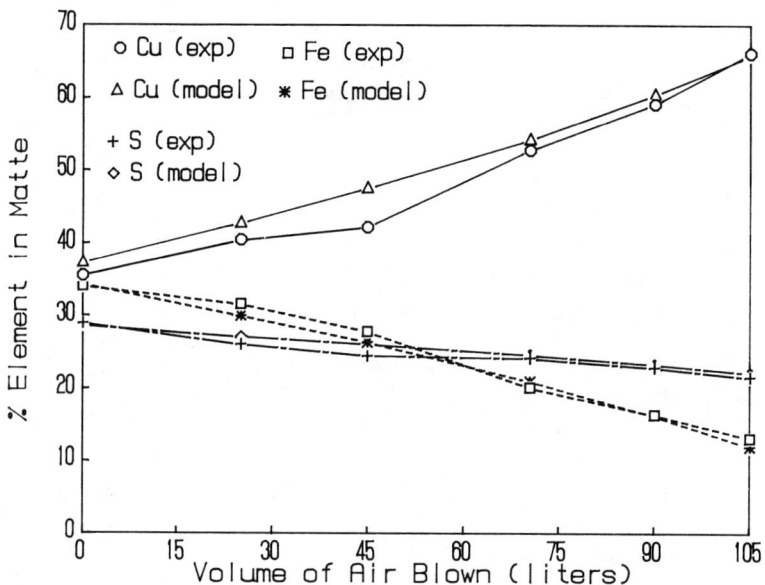

Figure 2. Cu - Fe - S Series 1 - Matte Phase. Experimental results and model predictions for copper, iron and sulfur at 1523 K. Ajo concentrate.

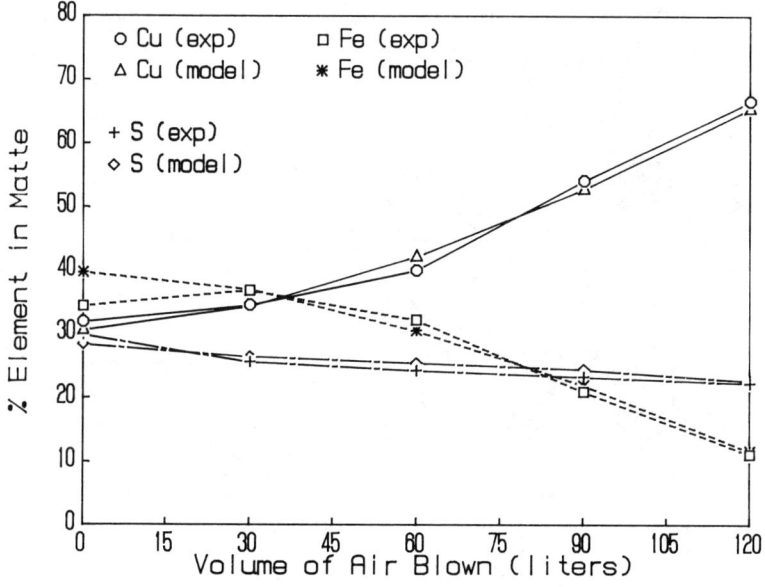

Figure 3. Cu - Fe - S Series 2 - Matte Phase. Experimental results and model predictions for copper, iron and sulfur at 1523 K. Chino concentrate.

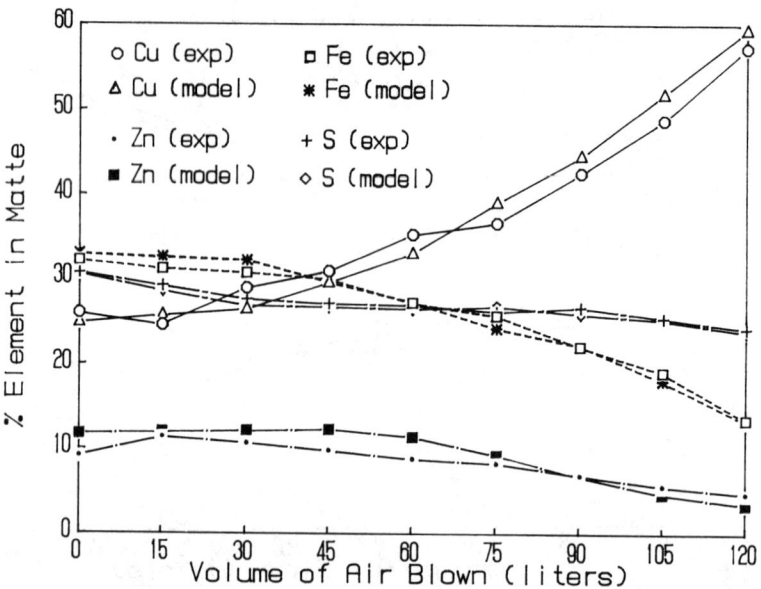

Figure 4. Cu - Fe - Zn - S Series - Matte Phase. Experimental results and model predictions for copper, iron, zinc and sulfur at 1523 K. Starting concentrate contained 10% zinc.

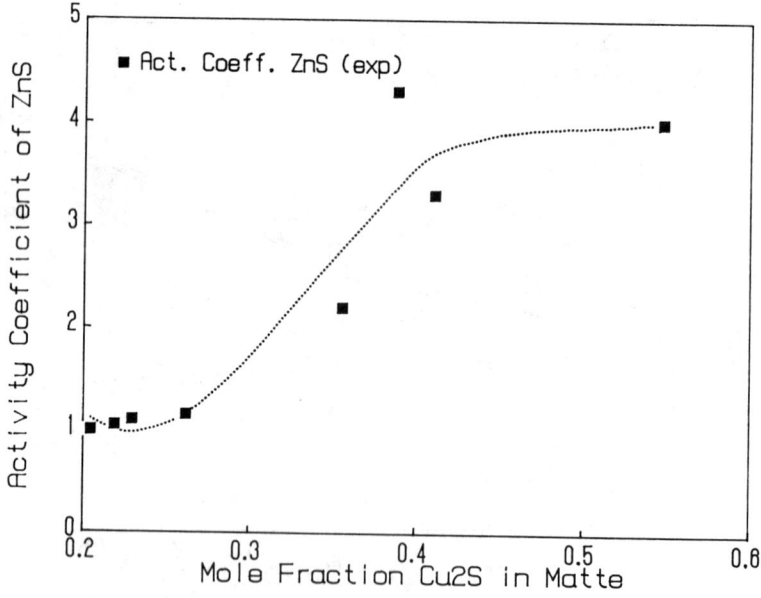

Figure 5. Activity Coefficient of ZnS in Matte. Points with square symbols are from experimental results. Dotted line represents values from third-order polynomial equation. T = 1523 K.

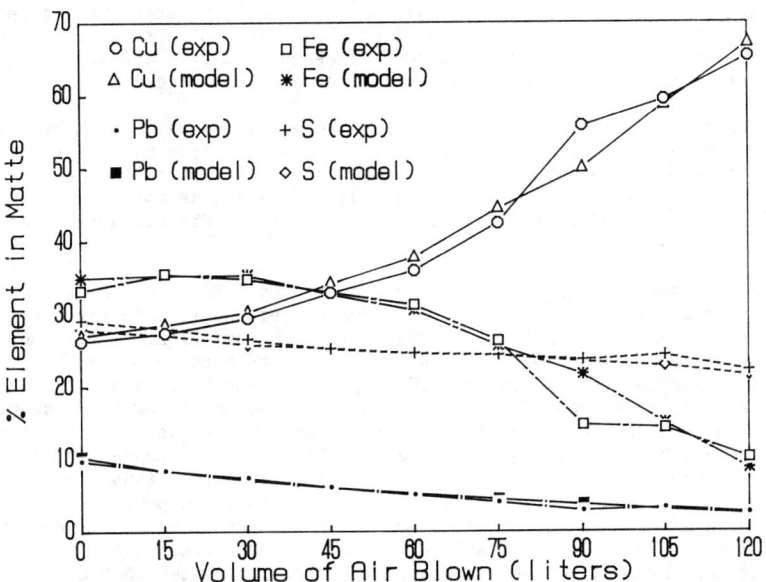

Figure 6. Cu – Fe – Pb – S Series 1 – Matte Phase. **Experimental results and model predictions for copper, iron, lead and sulfur at 1523 K. Starting concentrate contained 10% lead.**

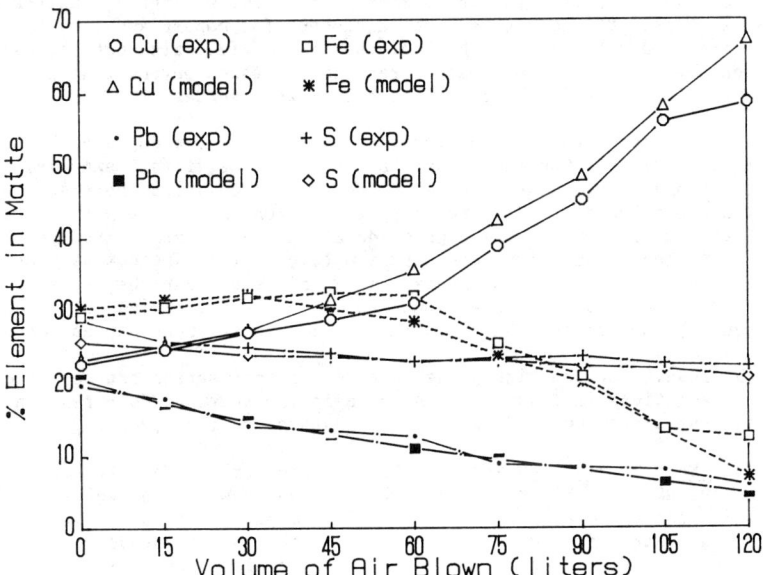

Figure 7. Cu – Fe – Pb – S Series 2 – Matte Phase. **Experimental results and model predictions for copper, iron, lead and sulfur at 1523 K. Starting concentrate contained 20% lead.**

The results of the high lead series (20% lead) are shown in Figure 7. The results for the weight percent elements in the matte show similar behavior to the low lead series, with good agreement for lead and sulfur, but increasing scatter between model predictions for copper and iron.

The activity coefficient for PbS in the matte was calculated by the same means as the activity coefficient for ZnS. After a number of different activity coefficient expressions were tested, it was found that a constant value of the activity coefficient gave as good a fit as more elaborate expressions, therefore the activity coefficient for PbS was determined to be: gamma PbS = 1.5.

High Grade Matte Series. As seen in the Cu - Fe - Pb - S Series 1 and 2, the model and experimental results for copper and iron in the matte, and to some extent lead for series 2, began to show variations as the amount of air blown into the matte increased. This could be because of invalid activity coefficients for the species considered, or because of nonequilibrium experimental conditions caused by a decrease in the volume of matte present. In order to test the validity of the matte volume hypothesis, a large amount of Cu - Fe - S matte assaying approximately 50% copper was doped with PbS in order to continue the Cu - Fe - Pb - S series at compositions near the compositions where the deviation between the model and the experimental results became large. Since the starting amount of high grade matte in the crucible was about the same as the starting amount of concentrate in the prior test series and the slag volume was small, the mixing problems should have been solved and any deviation between experiment and model could be attributed to invalid activity coefficients. Although the Cu - Fe - S and Cu - Fe - Zn - S series showed good agreement, it was decided to run high grade matte tests for these systems as well.

The results of the high grade matte series for Cu - Fe - S mattes are shown in Figure 8. As can be seen, agreement for copper and iron in the matte are good for the first half of the test series, although the deviation between the model and experimental results increases beyond that point. The agreement for sulfur is good for the entire test series.

The results of the test series for high grade matte doped with PbS are shown in Figure 9. The model uses the activity coefficient expression for PbS calculated previously. As can be seen, the deviations between the model and the experimental results for copper and iron in Cu - Fe - Pb - S Series 1 and 2 are not present in the high grade matte + PbS tests, although the copper, iron, lead, and sulfur are approximately in the same composition range in both sets of experiments. This likely indicates that the deviations in Cu -Fe - Pb - S Series 1 and 2 are because of poor phase mixing and consequent non-equilibrium, rather than invalid activity coefficient expressions.

The results of the high grade matte + ZnS test series are compared with model predictions in Figure 10. As seen in the previous Cu - Fe - Zn - S series, the agreement between model and assays are very good.

Cu - Fe - Pb - Zn - S Series. The results for the Cu - Fe - Pb - Zn - S Series are shown in Figures 11 and 12. As can be seen, the agreement between the model and the experimental results for copper, iron, zinc, lead, and sulfur in the matte is good, being generally within 5 percentage points over the entire set of experiments. The activity coefficient expressions calculated in the prior test series for ZnS and PbS were used in the model, indicating that the effect of zinc and lead on each other's activity coefficient in the matte is not strong.

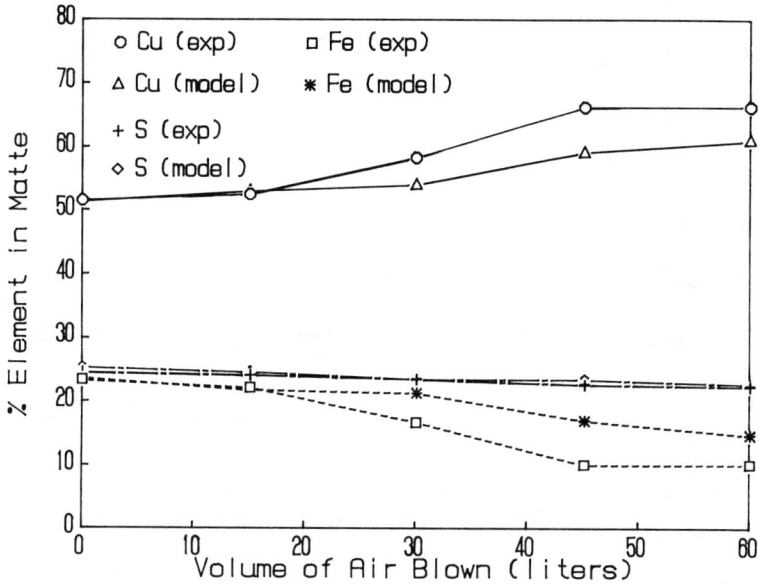

Figure 8. High Grade Matte Series - Matte Phase. Experimental results
and model predictions for copper, iron and sulfur at 1523 K. Starting
matte contained 49% copper.

Complex Sulfide Series. The results of the complex sulfide series are
shown in Figures 13 and 14. These results should be viewed as preliminary
only, as the mass balances of the assays in some of the tests are less than
95 percent, so differences between the model and experimental results may be
caused by analytical problems rather than poor fit. Generally the agreement
for all elements except copper and iron are within 5 percentage points.

Discussion

The accuracy with which an equilibrium model can predict the distribution
of elements between the various phases in a reaction system depends on two
factors: the accuracy of the thermodynamic data and models, and the degree to
which the system is at equilibrium. As stated in the introduction, one of
the problems in modelling of complex sulfide smelting is the lack of activity
coefficient expressions for species as a function of overall melt composition. However, these results show that if one uses activity coefficients
from simpler copper smelting systems and existing thermodynamic data, the
data can be used with reasonable accuracy to model complex sulfide smelting
systems. Furthermore, if an experimental set-up can be constructed such that
equilibrium conditions are created in the melts, the experimental data
obtained can be used to generate activity coefficient expressions for some
species if it is assumed that addition of the specie for which the activity
coefficient expression is to be determined has little or no effect on the
activity coefficients of the other species.

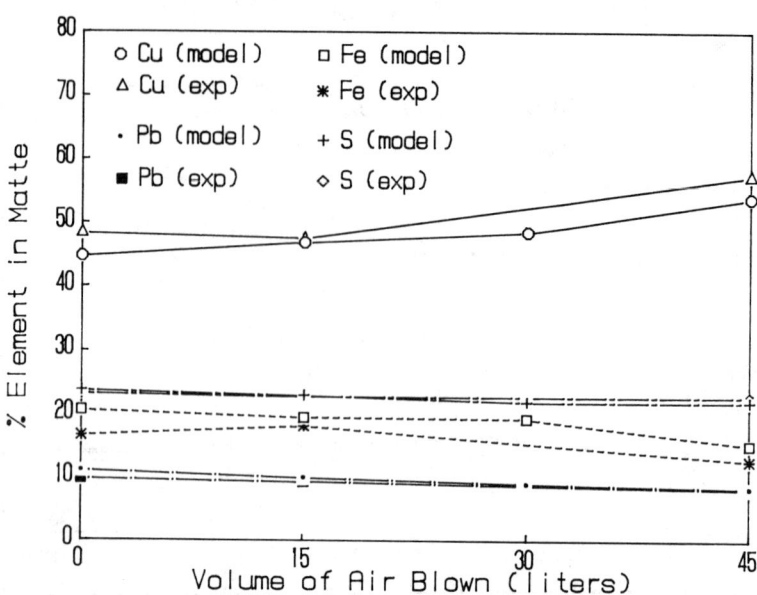

Figure 9. High Grade Matte + PbS Series - Matte Phase. Experimental results
and model predictions for copper, iron, lead and sulfur at 1523 K.

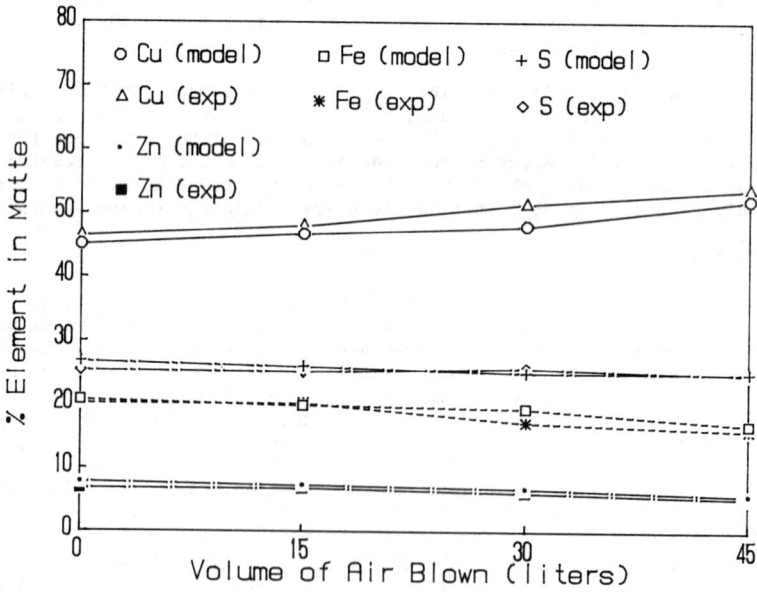

Figure 10. High Grade Matte + ZnS Series - Matte Phase. Experimental results
and model predictions for copper, iron, zinc and sulfur at 1523 K.

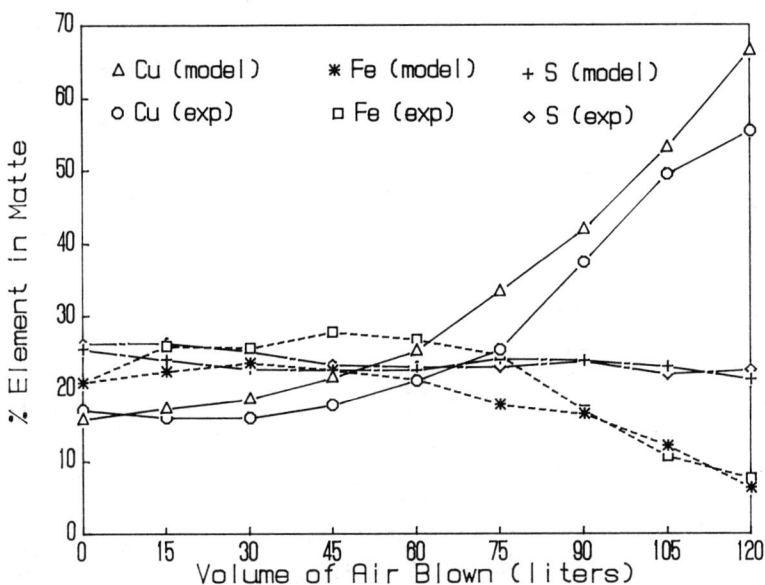

Figure 11. Cu – Fe – Pb – Zn – S Series – Matte Phase. Experimental results and model predictions for copper, iron, and sulfur at 1523 K.

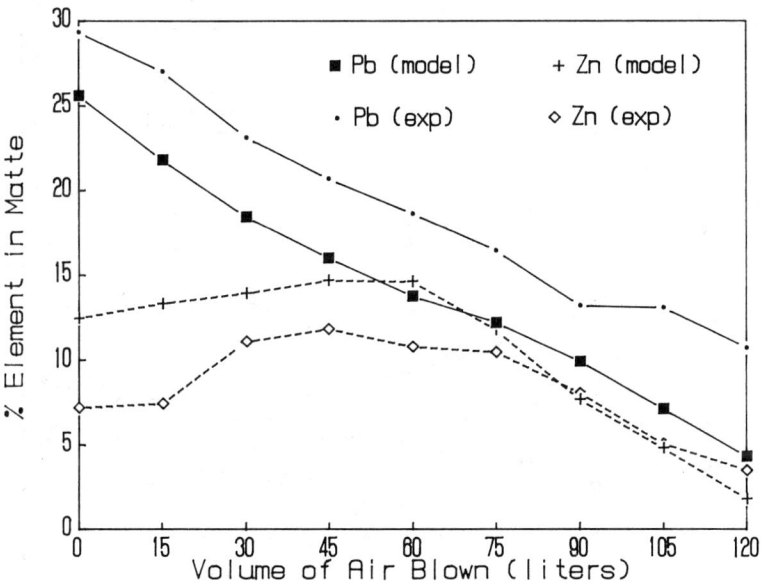

Figure 12. Cu – Fe – Pb – Zn – S Series – Matte Phase. Experimental results and model predictions for lead and zinc at 1523 K.

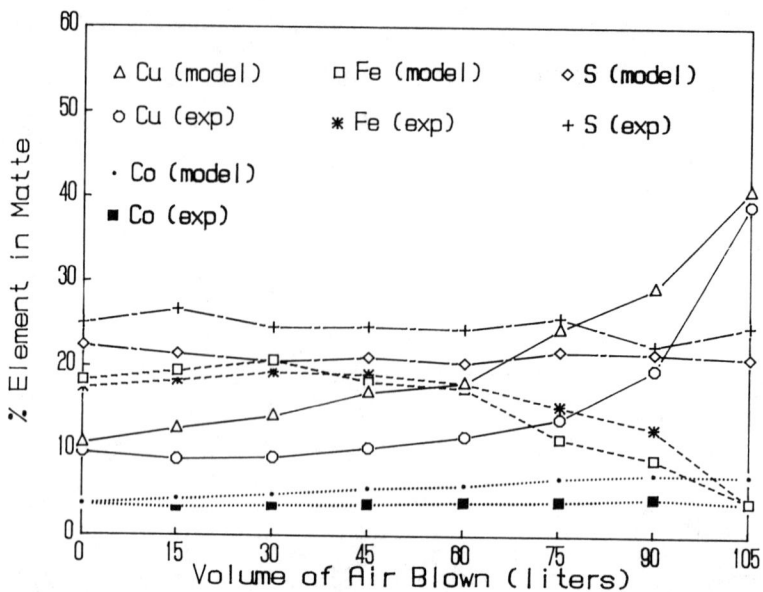

Figure 13. Complex Sulfide Series - Matte Phase. Experimental results and model predictions for copper, iron, cobalt and sulfur at 1523 K.

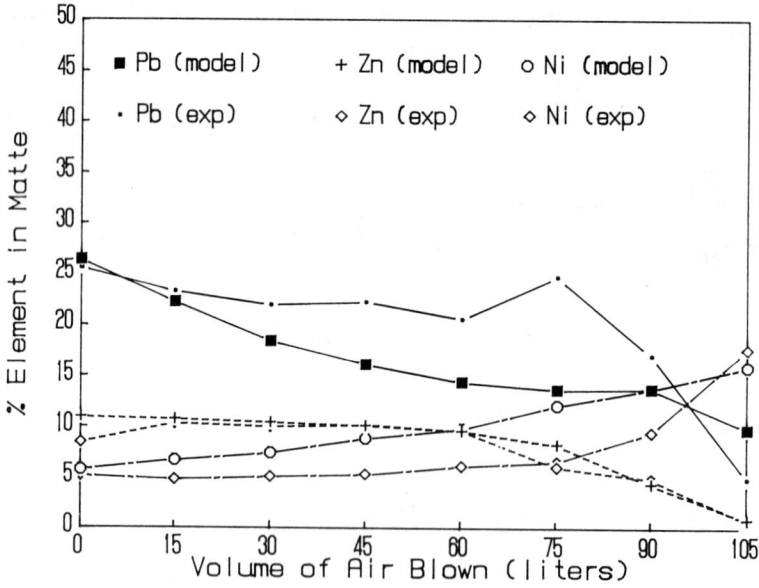

Figure 14. Complex Sulfide Series - Matte Phase. Experimental results and model predictions for lead, zinc and nickel at 1523 K.

The variety of systems examined shows the versatility of using programs such as Solgasmix. Once a database has been created, the addition or deletion of any new elements, species, or phases encountered is easy. Once a consistent model has been generated, it is easy to examine various processing options for a system, such as varying oxidizing or reducing conditions, preoxidation, removal of intermediate slags, or addition of fluxes. The modification of the program to include iterative gas phase removal provides an effective method for modelling open systems where the gas phase species are continuously being evolved and removed from the system.

As mentioned earlier, the ability of an equilibrium model to predict the distribution of elements depends on the experimental system being at or very close to equilibrium. In the submerged lance smelting method used in this study, the agreement of the model and the tests were generally very good in the portion of the experiment where the matte to slag ratio was high. As this ratio decreased, it may have become more difficult to create effective mixing of the phases. However, it is apparent that the problem is somewhat system-dependent; the Cu – Fe – S and Cu – Fe – Zn – S Series experiments agree well with model predictions through most of the experimental series, while deviations are seen in the Cu – Fe – Pb – S Series at a much lower degree of oxidation. This may be because of the high volatility of lead and its compounds, which would cause the volume of matte to decrease more rapidly than in the other systems. As mentioned earlier, lowering the lance was not an option for improving mixing because of crucible erosion.

Distribution of Elements

The distribution of elements for the Cu – Fe – Pb – Zn – S and complex sulfide systems are shown in Tables II and III. Distributions are shown based on both model predictions and experimental findings. Distributions for model predictions were calculated directly from the model, while the distributions for the experimental results were based on the following assumptions:

a) Elements such as copper, cobalt and nickel and their sulfide or oxide compounds have a low vapor pressure, so they will distribute between matte and slag only. Lead and zinc and their compounds have appreciable vapor pressures, so they will distribute between the matte, slag, and gas phases.

b) The mass balance obtained for a smelting experiment did not close because a portion of the matte and slag adhering to the inside of the crucible could not be recovered. For example, typical overall copper recoveries for the Cu – Fe – Pb – Zn – S series were 75 to 80 percent. Mass balances for copper in other test series were better than this. The distribution of copper, cobalt and nickel were determined by normalizing the relative amounts in the matte and slag to 100%.

c) The amount of lead and zinc distributed to the gas phase was determined by subtracting the amounts in the matte and slag phase from the initial input of lead and zinc.

785

Table II. Distribution of Elements in Cu - Fe - Pb - S Series

| | Model (wt pct) | | | Experiment (wt pct) | | |
	Matte	Slag	Gas	Matte	Slag	Gas
60 liters						
Cu	99.31	0.69	0.0	99.57	0.43	0.0
Pb	36.51	0.73	53.14	40.70	1.11	58.20
Zn	83.59	10.41	5.14	53.47	6.76	39.77
120 liters						
Cu	98.96	1.04	0.0	95.70	4.30	0.0
Pb	13.33	2.36	81.07	7.69	9.14	83.17
Zn	14.92	72.86	11.59	5.68	51.22	43.10

Table III. Distribution of Elements in Complex Sulfide Series

| | Model (wt pct) | | | Experiment (wt pct) | | |
	Matte	Slag	Gas	Matte	Slag	Gas
60 liters						
Cu	96.45	3.61	0.0	99.32	0.68	0.0
Pb	25.85	2.37	60.42	78.60	5.17	16.23
Zn	38.42	49.52	10.15	77.31	22.69	0.0
Co	85.74	14.26	0.0	93.83	6.17	0.0
Ni	92.65	6.62	0.0	98.89	1.05	0.0
105 liters						
Cu	96.39	3.87	0.0	98.28	1.72	0.0
Pb	10.97	6.11	80.30	43.30	10.10	46.50
Zn	1.06	84.59	13.87	29.21	70.79	0.0
Co	42.59	57.41	0.0	78.93	21.07	0.0
Ni	63.11	36.15	0.0	98.60	1.40	0.0

The distributions for the Cu - Fe - Pb - Zn - S series show the same trends for both the model and the experiments. At both 60 and 120 liters of air, copper remains almost entirely in the matte phase as expected. At 60 liters of air approximately 55% of the lead present initially has volatilized. By 120 liters of air, about 82% of the lead has volatilized. The model predictions and the experimental results for zinc show a wider variation, with the experiments showing zinc to be more volatile than the model indicates.

The complex sulfide series shows similar trends as the Cu - Fe - Pb - Zn - S series, although the variation between model and experiment is larger. The experimental distributions are based on the assays obtained, which in general show a poorer mass balance than the other systems examined. The model predicts that copper, cobalt, and nickel tend to remain in the matte at 60 liters of air, but by 105 liters of air predicts that about 57% of the cobalt and 36% of the nickel have been oxidized. The experiments show that copper, nickel, and to a large extent cobalt tend to remain in the matte both at 60 and 105 liters of air blown. The model predicts a much higher volatility of lead and a greater tendency for zinc to oxidize to the slag than the experiments show.

Model predictions are that of the zinc expelled to the gas phase, about 90% is in the form of elemental zinc, with the remainder being $ZnS_{(v)}$. Model predictions for lead volatilization are that the dominate vapor specie is $PbS_{(v)}$. At present, the reason for the difference in zinc and lead volatility between the model and experiments is unknown.

The model should be useful in predicting the effect of different smelting conditions on the matte composition. As an example of the possible effect of temperature, the model was used to simulate the smelting of complex sulfide concentrates at both 1523 and 1623 K. Calculations were made using the same activity coefficient expressions at both temperatures, so the results should be viewed as preliminary only. As can be seen from Figures 15 and 16, the model predicts that raising the smelting temperature to 1623 K would result in a higher matte grade for a given volume of air blown due to the increased volatilization of lead and zinc compounds. This might allow use of a process option to produce a high-grade matte containing most of the copper, nickel, and cobalt. The lead and zinc would be recovered as fume from the off-gas and from the slag. The high-grade matte could be treated hydrometallurgically by the Sheritt process (ammoniacal leach of copper, cobalt and nickel) to recover the metal values. Other possibilities would be to oxidize the melt with a gas containing a higher percentage of inert gas (for example 10% oxygen, 90% nitrogen). This would increase the volume of offgas produced and the amount of lead and zinc volatilized without over-oxidizing the bath. This type of simulation can be carried out quite easily using the model.

Conclusions

1. An iterative gas-phase removal version of Solgasmix was developed and shown to be effective for modelling of sulfide smelting systems which operate close to equilibrium. The main drawback is the lack of data and algorithmic expressions to calculate the activity coefficients of species in the matte and slag when smelting complex (multicomponent) sulfides.

2. The submerged lance injection technique creates sufficient turbulence within a small crucible such that for a considerable period of time equilibrium conditions are fairly closely met. As the volume of matte decreases, deviations from equilibrium and scatter in data increase.

3. The combination of an equilibrium program and the submerged lance blowing technique allows the development of activity coefficient expressions for species where no data exist and for which conventional equilibrium measurement techniques are inapplicable. In phases produced from complex sulfide smelting for example, the vapor pressure of lead and zinc species are too high for activities to be determined by simple gas-liquid equilibrium methods.

4. Submerged lance smelting results on a complex sulfide concentrate obtained from a Missouri lead belt show that up to 57% of the lead and 70% of the zinc can be volatilized or slagged while only 1.7% of the copper, 1.4% of the nickel, and 21% of the cobalt are lost to the slag phases. This may allow a processing option where the copper, cobalt, and nickel are concentrated in the matte phase for subsequent treatment, with the lead and zinc concentrated in the slag and gas phases for separate treatment.

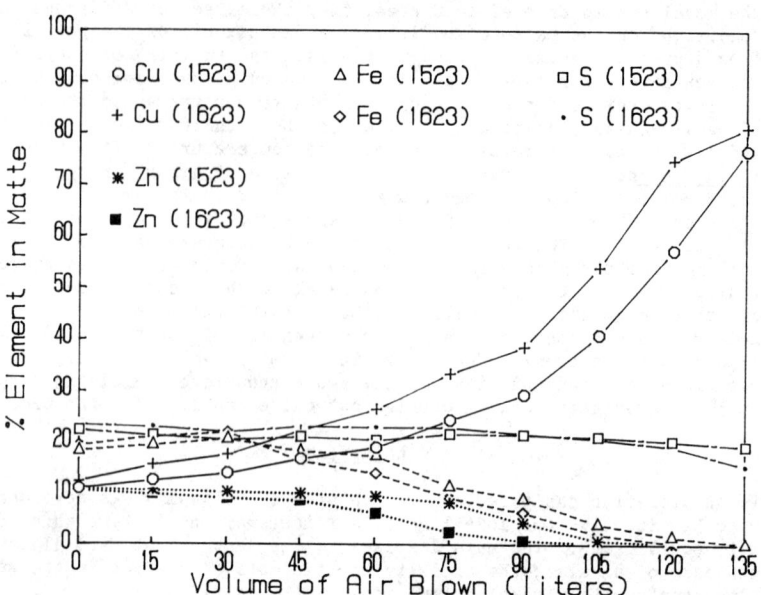

Figure 15. Complex Sulfide Smelting - Matte Phase. Model predictions for copper, iron, zinc and sulfur at 1523 and 1623 K.

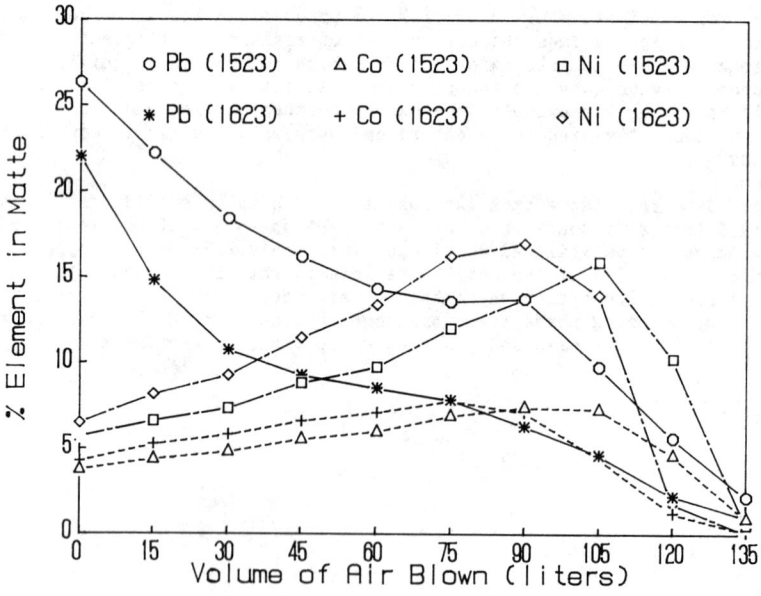

Figure 16. Complex Sulfide Smelting - Matte Phase. Model predictions for lead, cobalt and nickel at 1523 and 1623 K.

References

1. M. Nagamori and P.J. Mackey, "Thermodynamics of Copper Matte Converting: Part 1. Fundamentals of the Noranda Process," Metallurgical Transactions B, 9B (1978) 255 - 265.

2. M. Nagamori and P.J. Mackey, "Thermodynamics of Copper Matte Converting: Part 2. Distribution of Au, Ag, Pb, Zn, Ni, Se, Te, Bi, Sb and As between Copper, Matte and Slag in the Noranda Process," Metallurgical Transactions B, 9B (1978) 567 - 579.

3. M. Nagamori and P.C. Chaubal, "Thermodynamics of Copper Matte Converting: Part 3. Steady - State Volatilization of Au, Ag, Pb, Zn, Ni, Se, Te, Bi, Sb, and As from Slag, Matte, and Metallic Copper," Metallurgical Transactions B, 13B (1982) 319 - 329.

4. M. Nagamori, and P.C. Chaubal, "Thermodynamics of Copper Matte Converting: Part 4. A Priori Predictions of the Behavior of Au, Ag, Pb, Zn, Ni, Se, Te, Bi, Sb, and As in the Noranda Process Reactor," Metallurgical Transactions B, 13B (1982) 331 - 338.

5. M. Nagamori, "Metal Loss to Slag: Part 1. Sulfidic and Oxidic Dissolution of Copper in Fayalite Slag From Low Grade Matte," Metallurgical Transactions, 5 (1974) 531 - 538.

6. M. Nagmori, "Metal Loss to Slag: Part 2. Oxidic Dissolution of Nickel in Fayalite Slag and Thermodynamics of Continuous Converting of Nickel -Copper Matte," Metallurgical Transactions, 5 (1974) 539 - 547.

7. P.C. Chaubal and M. Nagamori, "Volatilization of Bismuth in Copper Matte Converting - Computer Simulation," Metallurgical Transactions B, 13B (1982) 339 - 348.

8. P.C. Chaubal and M. Nagamori, "Volatilization of Arsenic and Antimony in Copper Matte Converting," Metallurgical Transactions B, 14B (1983) 303 - 306.

9. P.C. Chaubal, M. Nagamori, and H.Y. Sohn, "Volatilization and Slagging of Lead in Copper Matte Converting: Computer Simulation," Canadian Metallurgy Quarterly, 23, (1984) 405 - 411.

10. R. Shimpo et al, "An Application of Equilibrium Calculations to the Copper Smelting Operation," Advances in Sulfide Smelting, Volume 1, Proceedings of the 1983 International Sulfide Smelting Symposium, San Francisco, CA, Nov. 6 - 9, 1983, (Warrendale, PA: TMS-AIME, 1983), 295 - 316.

11. Sakichi Goto, "The Application of Thermodynamic Calculations To Converter Practice," Copper and Nickel Converters, Proceedings of the Symposium on Converter Operating Practices, New Orleans, LA, Feb. 19-21, 1979, (Warrendale, PA: TMS-AIME, 1979), 33 - 55.

12. Sakichi Goto, "Equilibrium Calculations Between Matte, Slag, and Gaseous Phases in Copper Smelting," Copper Metallurgy, Practice and Theory (London: Institution of Mining and Metallurgy, 1974), 23 - 34.

13. Sakichi Goto, "Thermodynamic Consideration and Basic Tests for New Smelting Process of Zinc Calcine", TMS paper A84 - 12, AIME, (1984).

14. Harold J. Kandiner and Stuart R. Brinkley, "Calculation of Complex Equilibrium Relation," *Industrial and Engineering Chemistry*, 42 (1949) 850 - 855.

15. Stuart Brinkley, "Evaluation of Performance Factors of Fuel - Oxidant Mixtures," *Industrial and Engineering Chemistry*, 43 (1951) 2471 - 2475.

16. Stuart R. Brinkley and Bernard Lewis, "Combustion Gases... Equilibrium Composition and Thermodynamic Properties," *Chemical and Engineering News*, 27 (1949) 2540 - 2541.

17. Stuart R. Brinkley, "Calculation of the Equilibrium for Systems of Many Constituents," *The Journal of Chemical Physics*, 15 (1947) 107 - 110.

18. Stuart R. Brinkley, "Note on the Condition of Equilibrium for Systems of Many Constituents," *The Journal of Chemical Physics*, 14 (1946) 563 - 564, 686.

19. Stuart R. Brinkley, "Calculation of the Thermodynamic Properties of Multicomponent Systems and Evaluation of Propellant Performance Parameters," *Kinetics, Equilibria and Performance of High Temperature Systems, Proceedings of the First Conference*, (1960) 74 - 81.

20. N. Kemori et al, "Measurements of Oxygen Pressure in a Copper Flash Smelting Furnace by an EMF Method," *Metallurgical Transactions B*, 17B (1986) 111 -117.

21. N. Kemori, Y. Shibata, and K. Fukushia, "Thermodynamic Consideration for Oxygen Pressure in a Copper Flash Smelting Furnace at Toyko Smelter" (TMS Paper , 1985) A85 - 30.

22. Bo Bjorkman and Gunnar Eriksson, "Quantitative Equilibrium Calculations on Conventional Copper Smelting and Converting," *Canadian Metallurgical Quarterly*, 21 (1982) 329 - 337.

23. Rolf J. Wesely, "Complex Sulfides: New Technology and Opportunities," *Complex Sulfides*, Proceedings of the International Symposium on Complex Sulfides, San Diego, CA, Nov. 10 - 13, 1985, TMS-AIME, (Warrendale, PA: TMS-AIME, 1985), 5 - 32.

24. Jan H. Reimers, "Processing of Complex Sulfides. An Overview of Current and Proposed Processes," *Complex Sulfides*, Proceedings of the International Symposium on Complex Sulfides, San Diego, CA, Nov. 10 -13, 1985, TMS-AIME, (Warrendale, PA: TMS-AIME, 1985), 747 - 758.

25. H. Maczek, "The Economics of Processing Complex Sulfide Concentrates in the ISF," *Complex Sulfides*, Proceedings of the International Symposium on Complex Sulfides, San Diego, CA, Nov. 10 - 13, 1985, (Warrendale, PA: TMS-AIME, 1985), 773 - 782.

26. Robert W. Bartlett, "Economic Trade-offs in Increasing Revenue From a Massive Complex Sulfide Deposit - Caribou," *Complex Sulfides*, Proceedings of the International Symposium on Complex Sulfides, San Diego, CA, Nov. 10 - 13, 1985, (Warrendale, PA: TMS-AIME, 1985), 783 - 799.

27. N.A. Warner, "Towards Polymetallic Sulfide Smelting," *Complex Sulfides*, Proceedings of the International Symposium on Complex Sulfides, San Diego, CA, Nov. 10 - 13, 1985, (Warrendale, PA: TMS-AIME, 1985), 847 - 865.

28. J. Mehlbeer and G. Melcher, "A Combined Method for the Treatment and Smelting of Complex Non-ferrous Metal Ores," Complex Metallurgy 78, Proceedings of an International Symposium, Bad Harzburg, West Germany, Sept. 20 -22, 1978, (London: Institution of Mining and Metallurgy, 1978), 96 - 100.

29. J.M. Cases, "Finely Disseminated Complex Sulfide Ores," Complex Metallurgy 78, Proceedings of an International Symposium, Bad Harzburg, West Germany, Sept. 20 - 22, 1978, (London: Institution of Mining and Metallurgy, 1978), 234 - 247.

30. G. Barbery, A.W. Fletcher, and L.L. Sirais, "Exploitation of Complex Sulfide Deposits: A Review of Processing Options from Ore to Metals," Complex Metallurgy 78, Proceedings of an International Symposium, Bad Harzburg, West Germany, Sept. 20 - 22, 1978, (London: Institution of Mining and Metallurgy, 1978), 135 - 160.

31. W.L. Cornell, D.C. Holtgrefe, and F.H. Sharp, "Concentrating a Complex Cobalt and Nickel Sulfide from Missouri Lead Ores by Continuous Froth Flotation," Complex Sulfides, Proceedings of the International Symposium on Complex Sulfides, San Diego, CA, Nov. 10 - 13, 1985, (Warrendale, PA: TMS-AIME, 1985), 181 - 192.

32. Susanne Pignolet - Brandom and Richard D. Hagni, "Applied Minerology of the Complex Lead - Zinc -Copper - Cobalt - Nickel Ores and Concentrates from the Southeast Missouri Lead District," Complex Sulfides, Proceedings of the International Symposium on Complex Sulfides, San Diego, CA, Nov. 10 - 13, 1985, (Warrendale, PA: TMS-AIME, 1985), 831 - 842.

33. L. Fontainas, M. Coussement, and R. Maes,"Some Metallurgical Principles in the Smelting of Complex Materials," Complex Metallurgy 78, Proceedings of an International Symposium, Bad Harzburg, West Germany, Sept. 20 - 22, 1978, (London, 1978), 13 - 23.

34. H.P. Krug and W.H. Schwartz, "Effect of Impurities in Copper Smelting Input on Material Flow and Smelter and Refining Consumption Figures," Complex Metallurgy 78, Proceedings of an International Symposium, Bad Harzburg, West Germany, Sept. 20 - 22, 1978, (London, 1978), 24 - 32.

35. I. Barin and F. Sauert, "Thermodynamics and Kinetics of Cyclone Smelting of Sulphidic Copper Concentrates," Complex Metallurgy 78, Proceedings of an International Symposium, Bad Harzburg, West Germany, Sept. 20 - 22, 1978, (London, 1978), 193 - 198.

36. J.C. Yannopoulos and J.C. Agarwal, eds., Extractive Metallurgy of Copper, Proceedings of the International Symposium on Copper Smelting and Refining, Las Vegas, NV, Feb. 22 - 26, 1976, (Baltimore, MD: TMS-AIME, 1976).

37. A.D. Zunkel et al, eds., Complex Sulfides, Proceedings of the International Symposium on Complex Sulfides, San Diego, CA, Nov. 10 - 13, 1985, (Warrendale, PA: TMS-AIME, 1985).

38. H.Y. Sohn, D.B. George, and A.D. Zunkel, Advances in Sulfide Smelting, Proceedings of the 1983 International Sulfide Smelting Symposium, San Francisco, CA, Nov. 6 - 9, 1983, (Warrendale, PA: TMS-AIME, 1983).

39. W.B. White, S.M. Johnson, and G.B. Dantzig, "Chemical Equilibrium in Complex Mixtures," The Journal of Chemical Physics, 28 (1958) 751 - 755.

40. W.B. White, "Numerical Determination of Chemical Equilibrium and the Partitioning of Free Energy," The Journal of Chemical Physics, 46 (1967) 4171 - 4175.

41. R.C. Oliver, S.E. Stephanou, and R.W. Baier, R.W., "Calculating Free Energy Minimization", Chemical Engineering, 69 (1962) 121 - 128.

42. B.R. Kubert and S.E. Stephanou, "Extension to Multiphase Systems of the RAND Method for Determining Equilibrium Compositions," Kinetics, Equilibria and Performance, Proceedings of the First Conference, (1960) 166 - 170.

43. Gunnar Eriksson, "Thermodynamic Studies of High Temperature Equilibria. SOLGAS, a Computer Program for Calculating the Composition and Heat Condition of an Equilibrium Mixture", Acta Chem Scandanavia, 25 (1971) 11 - 18.

44. Gunnar Eriksson and Erik Rosen, "Thermodynamic Studies of High Temperature Equilibria 8. General Equations for the Calculation of Equilibria in Multiphase Systems," Chemica Scripta, (1973) 193 - 194.

45. Leonard M. Napthali, "Calculate Complex Chemical Equilibria," Industrial and Engineering Chemistry, 53 (1961) 387 - 388.

46. Leonard M. Naphthali, "Computing Complex Chemical Equilibria by Minimizing Free Energy," Kinetics, Equilibria, and Performance, Proceedings of the First Conference, (1960) 181 - 183.

47. Leonard M. Napthali, "Complex Chemical Equilibria by Minimizing Free Energy", Journal of Chemical Physics, 31 (1959) 263 - 264.

48. D.S. Villars, "A Method of Successive Approximations for Computing Combustion Equilibria on a High Speed Digital Computer," The Journal of Physical Chemistry, 63 (1959) 521 - 525.

49. D.S. Villars, "Computation of Complicated Combustion Equilibria on a High Speed Computer," Kinetics, Equilibria and Performance, Proceedings of the First Conference, (1960) 141 - 151.

50. D.R. Cruise, "Notes on the Rapid Computation of Chemical Equilibria," The Journal of Physical Chemistry, 68 (1964) 3797 - 3802.

51. H.P. Meissner, C.L. Kusik, and W.H. Dalzell, "Equilibrium Composition with Multiple Reactions," Industrial and Engineering Chemistry Fundamentals, 4 (1969) 659 - 665.

52. Y.H. Ma and C.W. Shipman, "On the Computation of Complex Equilibria," AIChE Journal, 18 (1972) 299 - 304.

53. W.R. Smith and R.W. Missen, "Calculating Complex Chemical Equilibria by an Improved Reaction - Adjustment Method," The Canadian Journal of Chemical Engineers, 46 (1968) 269 - 272.

54. Y.K. Rao, "The Analysis and Calculation of Equilibria in Complex Metallurgical Systems," Metallurgical Transactions B, 14B (1983) 701 - 710.

55. N. Kishimoto and H. Yoshida, "Calculation Method of Equilibrium Composition in the Carbon - Hydrogen - Oxygen System and Its Application to Environments of a High - Temperature Gas - Cooled Reactor," _Metallurgical Transactions B_, 14B (1983) 465 - 471.

56. Jacques Melancon, Yves Blanchette, and Christopher W. Bale, "Computation of Gas - Phase Equilibria - A General Algorithm Using the Newton - Raphson Method," _Metallurgical Transactions B_, 16B (1985) 793 - 799.

57. H.E. Brandmaier and J.J. Harnett, "A Brief History of Past and Current Methods of Solution for Equilibrium Composition," _Kinetics, Equilibria and Performance, Proceedings of the First Conference_, (1960) 69 - 73.

58. Warren D. Seider and Rajeev Gautum, "Computation of Phase and Chemical Equilibrium," _AIChE Journal_, 25 (1979) 991 - 1015.

59. Frank J. Zeleznik and Sanford Gordon, "Calculation of Complex Chemical Equilibria", _Industrial and Engineering Chemistry_, 60 (1968) 27 - 57.

60. William R. Smith, "The Computation of Chemical Equilibria in Complex Systems," _Industrial Engineering Chemistry Fundamentals_, 19 (1980) 1 - 10.

61. W.R. Smith, and R.W. Missen, _Chemical Reaction Equilibrium Analysis_, (New York: John Wiley, 1982).

62. W.T. Thompson, A.D. Pelton, and C.W. Bale, "Extension to Solgasmix for Interactive Calculations," _Calphad_, 7 (1983) 113 - 123.

63. A.G. Turnbull and M.W. Wadsley, "Thermodynamic Modelling of Metallurgical Processes by the CSIRO - SGTE Thermodata System," _The Australian Institute of Mining and Metallurgy Symposium on "Extractive Metallurgy_," (1984) 79 - 85.

64. Theodore M. Besmann, _SOLGASMIX - PV, a Computer Program to Calculate Equilibrium Relationships in Complex Chemical Systems_, (Report ORNL/TM-5775, Oak Ridge National Laboratory, 1977).

65. S.K. Saxena and Gunnar Eriksson, "High Temperature Phase Equilibria in a Solar - Composition Gas," _Geochimica et Cosmochimica Acta_, 47 (1983) 1865 - 1874.

66. S.K. Saxena and G. Eriksson, "Low to Medium Temperature Phase Equilibria in a Gas of Solar Composition," _Earth and Planetary Science Letters_, 65 (1983) 7 - 16.

67. S.K. Saxena and G. Eriksson, "Theoretical Computation of Mineral Assemblages in Pyrolite and Cherzolite," _Journal of Petrology_, 24 (1981) 538 - 555.

68. S.K. Saxena and G. Eriksson, "Anhydrous Phase Equilibria in Earth's Upper Mantle," _Journal of Petrology_, 26 (1985) 378 - 390.

69. K.E. Spear and C.F. Wan, "CVD of Niobium Germanides from Partially Reacted Input Gases," _Calphad_, 7 (1983) 149 - 155.

70. Angus I. Kingon et al, "Thermodynamic Calculations for the Chemical Vapor Deposition of Silicon Carbide," _Journal of the American Ceramic Society_, 66 (1983) 558 - 566.

71. Gunnar Eriksson and Erik Rosen, "Thermodynamic Studies of High Temperature Equilibria 12. Calculation of Equilibrium Compositions for the Reaction Between Chalcopyrite (CuFeS$_2$) and Air in the Temperature Interval 800 K - 1300 K," _Scandanavian Journal of Metallurgy_, 2 (1973) 95 - 99.

72. Gunnar Eriksson, and Erik Rosen, "Thermodynamic Studies of High Temperature Equilibria 9. Experimental Determination of the Stable Solid (Cu, Fe, S, O) - Phases in Equilibrium with Gas Mixtures of SO$_2$, O$_2$, and N$_2$ at Temperatures 1000 - 1300 K," _Scandanavian Journal of Metallurgy_, 3 (1974) 94 - 96.

73. J.A. Sell, "Chemical Equilibrium Calculations of Tungsten - Halogen Systems," _Journal of Applied Physics_, 54 (1983) 4605 - 4613.

74. R.J. Wesely, A.E. Morris, and H. Flynn, "Recovery of Elemental Sulfur from Massive Sulfides," _Complex Sulfides_, Proceedings of the International Symposium on Complex Sulfides, San Diego, CA, Nov. 10 - 13, 1985, (Warrendale, PA: TMS-AIME, 1985), 703.

75. A.E. Morris and H. Flynn, "Discussion of Lime - Enhanced Reduction of Sulfide Concentrates: A Thermodynamic Discussion," _Metallurgical Transactions B_, (1986) 914 - 917.

76. J. Lorentz et al, "The Stability of SiC - Based Ceramics Containing ZrO$_2$ and Other Oxides," _Calphad_, 7 (1983) 125 - 135.

77. Gunnar Eriksson, "An Algorithm for the Computation of Aqueous Multicomponent, Multiphase Equilibria," _Analytic Chim Acta_, 112 (1979).

78. R. Gallagher and P.J. Spencer, "Calculation of Precipitation Behavior in an Alloy Steel Using a Thermochemical Data Bank and the Program 'Solgasmix'," _Calphad_, 7 (1983) 157 - 163.

79. G. Eriksson and K. Hack, "Calculation of Phase Equilibria in Multicomponent Alloy Systems Using a Specially Adapted Version of the Program 'Solgasmix'," _Calphad_, 8 (1984) 15 - 24.

80. Karl Spear, _Penn State Solgasmix_ (The Pennsylvania State University, 1984).

81. Chum - Sien Lin, _Reference Manual of Solgasmix - PV, a Computer Program to Calculate Equilibrium Relationships in Complex Chemical Systems OSU Version 1.0_, (The Ohio State University, 1985).

82. Alan G. Turnbull, "New Thermodata Package," _Thermodata Update_, (Newsletter from CSIRO, P.O. Box 124, Port Melbourne, Vic. 3207, Australia, issue #2, January 1987).

83. H. Flynn, A.E. Morris, and D. Carter, _An Iterative Gas - Phase Removal Version of Solgasmix_, Canadian Institute of Mining and Metallurgy, Paper 6071, (1986).

84. D.R. Stull and H. Prophet, _JANAF Thermochemical Tables_, (Washington D.C.: U.S. Government Printing Office, 1971), and Supplements.

85. Richard A. Robie, Bruce S. Hemingway, and James R. Fisher, _Thermodynamic Properties of Minerals and Related Substances at 289.15 K and 1 Bar Pressure and at Higher Temperatures_, (Washington D.C.: U.S. Government Printing Office, 1979).

86. I. Barin, and O. Knacke, Thermochemical Properties of Inorganic Sub-stances, (New York: Springer - Verlag, 1973), and Supplement (1977).

87. J.M. Floyd and D.S. Conochie, "Sirosmelt - The First Ten Years," The Australian Institute of Mining and Metallurgy Symposium on "Extractive Metallurgy", (1984) 1 - 8.

88. R.A. McClelland and W.T. Denholm, "A Mathematical Model of the Sirosmelt Process," The Australian Institute of Mining and Metallurgy Symposium on "Extractive Metallurgy", (1984) 95 - 102.

89. J.M. Floyd et al, "Large Scale Development of Submerged Lancing Sirosmelt Tin Processes at Associated Tin Smelters," The Australian Institute of Mining and Metallurgy Symposium on "Extractive Metallurgy", (1984) 25 - 33.

90. L.E. Anderson et al, "Smelting of Olympic Dam Copper Concentrates using Sirosmelt Technology," Complex Sulfides, Proceedings of the International Symposium on Complex Sulfides, San Diego, CA, Nov. 10 - 13, 1985, (War-rendale, PA: TMS-AIME, 1985), 623 - 634.

91. W.T. Denholm, D. Chutinura, and R.N. Taylor, "A Theoretical and Exper-imental Study of Sulfide Tin Fuming from Low - Grade Mill Products," The Reinhardt Schuhmann International Symposium on Innovative Technology and Reactor Design in Extractive Metallurgy, Colorado Springs, CO, Nov. 9 - 12, 1986, (Warrendale, PA: TMS-AIME, 1986), 113 - 130.

92. N.B. Gray, M. Nilmani, and C.R. Fountain, "Investigation and Modelling of Gas Injection and Mixing in Molten Liquid Processes," The Australian Institute of Mining and Metallurgy Symposium on "Extractive Metallurgy", (1984) 269 - 277.

93. R.L. Biehl et al, "Design, Construction and Commissioning of a 4 Tonne/Hour Matte Fuming Pilot Plant," The Australian Institute of Mining and Metallurgy Symposium on "Extractive Metallurgy", (1984) 9 - 15.

94. M.C. Walton, "Refractories Performance in the Matte Fuming Process," The Australian Institute of Mining and Metallurgy Symposium on "Extractive Metallurgy", (1984) 17 - 23.

95. E.G. King, Alla D. Mah, and L.B. Pankratz, Thermodynamic Properties of Copper and Its Inorganic Compounds, (1973), INCRA Monograph II.

96. S.S. Wang, N.H. Santander, and J.M. Toguri, "The Solubility of Nickel and Cobalt in Iron Silicate Slags," Metallurgical Transactions, 5 (1974) 261 - 265.

97. H. Eric and M. Timucin, "Activities in Cu_2S - FeS - PbS Melts at 1200°C," Metallurgical Transactions B, 12B (1981) 493 - 500.

795

Free Energies of Formation for Species

Specie	Free energy, cal/mole	Temp. Range, K	Source
O_2	0.0		88
N_2	0.0		88
$S_{2(v)}$	0.0	718 and above	88
$0.5S_{2(v)} + 0.5O_2 ==> SO_{(v)}$	$-13830 - 1.22T$	1400 - 1600	95
$0.5S_{2(v)} + O_2 ===> SO_{2(v)}$	$-86570 + 17.3T$	1323 - 1573	88
$Zn_{(v)}$	0.0	1181 and above	88
$Zn_{(v)} + 0.5S_2 ===> ZnS_{(v)}$	$1189 + 7.25T$	1400 - 1600	90
$Pb_{(1)} ==> Pb_{(v)}$	$43120 - 21.35T$	1500 - 1700	88
$Pb_{(1)} + 0.5O_2 ===> PbO_{(v)}$	$7375 - 11.66T$	1500 - 1700	88
$Pb_{(1)} + 0.5S_{2(v)} ===> PbS_{(v)}$	$11829 - 11.47T$	1500 - 1700	88
$Cu_{(1)}$	0.0	1356 and above	88
$2Cu_{(1)} + 0.5S_{2(v)} ===> Cu_2S$	$-25460 + 3.0T$	1323 - 1573	88
$Cu(1) + 0.25O_2 ===> CuO._{5(1)}$	$-21679 + 9.62T$	1517 - 1700	88
$Fe + 0.5S_{2(v)} ===> FeS$	$-28476 + 9.14T$	1323 - 1573	88
$Zn_{(v)} + 0.5S_{2(v)} ==> ZnS_{(1)}$	$-89160 + 47.5T$	1323 - 1523	88
$Pb_{(1)} + 0.5S_{2(v)} ==> PbS_{(1)}$	$-32636 + 16.43T$	1500 - 1700	88
$Pb_{(1)}$	0.0	601 and above	88
$Fe + 0.5O_2 ===> FeO_{(s)}$	$-54890 + 10.55T$	1323 - 1573	88
$3Fe + 2O_2 ===> Fe_3O_{4(s)}$	$-260870 + 72.24T$	1323 - 1573	88
$Zn_{(v)} + 0.5O_2 ===> ZnO_{(s)}$	$-110140 + 47.35T$	1323 - 1523	88
$Pb_{(1)} + 0.5O_2 ===> PbO_{(1)}$	$-43616 + 16.84T$	1500 - 1700	89
$Si + O_2 ===> SiO_{2(s)}$	$-214000 + 40.21T$	1500 - 1700	89
$Ca + 0.5O_2 ===> CaO_{(s)}$	$-152953 + 26.0T$	1500 - 1700	89
$2Al + 1.5O_2 ===> Al_2O_{3(s)}$	$-403415 + 78.33T$	1500 - 1700	89
$2Ni_{(s)} + 0.5S_{2(v)} ==> Ni_2S_{(1)}$	$-28090 + 6.17T$	1500 - 1573	2
$Co + .445S_{2(v)} ==> Co_8S_{9(s)}$	$-33491 + 16.5T$	1500 - 1600	90
$Co + 0.5O_2 ===> CoO_{(1)}$	$-50428 + 14.28T$	1523 - 1623	96
$Ni + 0.5O_2 ===> NiO_{(s)}$	$-56075 + 20.46T$	1473 - 1523	2

Activity Coefficients

Specie	Activity Coefficient	Source
Matte Phase		
$Cu_{(1)}$	14.0	10
$Cu_2S_{(1)}$ (high grade tests)	1.0	10
$Cu_2S_{(1)}$ (complex sulfide)	$1.028 - 1.17*N_{Fes} + 0.0373*N_{Cu_2s}$ $-.3065*N_{Pbs}$	97
$FeS_{(1)}$ (high grade tests)	$0.925/(N_{Cu_2s} + 1)$	10
$FeS_{(1)}$ (complex sulfide)	$0.0805 + 1.021*N_{Fes} - 0.0412*N_{Cu_2s}$ $+ 0.2299*N_{Pbs}$	97
$ZnS_{(1)}$	$23.09 + N_{Cu_2s} * (-253.67 + N_{Cu_2s} * (992.3$ $+ N_{Cu_2s} * (-1521 + N_{Cu_2s} + 801.3)))$	PW*
$PbS_{(1)}$	1.5	PW*
$FeO_{(s)}$ (high grade tests)	$\exp(5.1 + 6.2 * (\ln(N_{Cu_2s})) + 6.41 *$ $(\ln(N_{Cu_2s}))^2 + 2.8 * (\ln(n_{Cu_2s}))^3$	
Slag Phase		
$FeO_{(s)}$	$1.42 * N_{Feo} - 0.44$	10
$Fe_3O_{4(s)}$	$0.68 + 56.8 * N_{Fe_3O_4} + 5.45 * N_{S_1O_2}$	10
$ZnO_{(s)}$	$\exp(400/T)$	2
$PbO_{(1)}$	$\exp(-3926/T)$	10
$SiO_{2(s)}$	2.1	10
$Cu_2S_{(1)}$	$\exp(2.46 + 6.22 * N_{Cu_2s \text{ in matte}})$	10
$FeS_{(1)}$	70	10
$NiO_{(s)}$	$\exp(3050/T - 1.31)$	2
$CoO_{(1)}$	1.16	2
$CuO_{.5(1)}$	3	2

*PW = Present Work

MATHEMATICAL MODELING OF THE COMBINED TURBULENT TRANSPORT PHENOMENA,

CHEMICAL REACTIONS, AND THERMAL RADIATION IN A FLASH-FURNACE SHAFT

Y. B. Hahn and H. Y. Sohn

Department of Metallurgy and Metallurgical Engineering
University of Utah
Salt Lake City, Utah   84112-1183

Abstract

A mathematical model has been developed to describe the processes occurring in an axisymmetric flash furnace shaft. The model incorporates turbulent
fluid dynamics, chemical reaction kinetics, and heat and mass transfer. The
key features include the use of the k-ε turbulence model, incorporating the
effect of particles on the turbulence, and the four-flux model for the radiative heat transfer. The model predictions were compared with experimental
data obtained from an Outokumpu pilot flash furnace. Good agreement was obtained between the predicted and measured data in terms of gas-phase temperature and SO_2 and O_2 concentrations. The model predictions show that reactions
of sulfide particles are almost completed in the upper zone of the furnace
within about 1 m of the burner, and the double-entry burner system with radial
feeding of the concentrate-laden distribution air gives better performance
than the single-entry burner system. Model predictions also showed that the
radiation between particles and their surroundings is the dominant mode of
heat transfer in the flash-smelting furnace.

Introduction

In the flash-smelting process, fine particles of dry sulfide concentrates and flux are injected into the furnace with oxygen-enriched air, forming a particle-laden turbulent jet. The concentrate particles are quickly dispersed within the furnace shaft and heated to ignition near the oxygen burner. As particles travel down the reaction shaft within the turbulent flow, they and the surrounding gas exchange momentum, mass, and energy. Eventually, the molten particles settle to the furnace bottom and are divided into molten slag and matte.

In spite of the increasing industrial stature of the process, the design of a flash-smelting furnace remains largely an art. This is mainly due to the difficulty of understanding the complex interactions of the individual subprocesses taking place in a flash furnace. In order to enhance systematic understanding of the overall process, a reliable mathematical model would be very helpful. Furthermore, the behavior of the complex reacting particle-laden turbulent gas flows can be predicted with a minimum amount of experimental work. It is only in very recent times that attention has been directed to mathematical modeling of the flash-smelting process.

The previous studies on mathematical modeling of the flash-smelting process have assumed the flame to be a one-dimensional stream (1,2), a two-dimensional free jet (3), or a two-dimensional confined jet (4).

Themelis and coworkers (1,2) developed a one-dimensional mathematical model to describe the transport phenomena occurring in the flash-smelting process. They assumed that the reaction of sulfide droplets was controlled by the combination of mass transfer of oxygen to the surface, diffusion through the reacted layer, and interfacial chemical reaction. The effects of phase transformation and particle fragmentation on the rate of flash oxidation of sulfides were incorporated. These authors divided the rate of sulfur removal into five stages according to subdivided temperature regions: induction region, flame-front region, flame region, and post-flame region. To describe the gas-phase flow phenomena, they simplified the flow system and considered only one-dimensional axial flow. For the particle phase, they assumed one-dimensional motion of particles only in the axial direction of gas flow by neglecting the dispersion of particles in the radial direction (2). Due to the one-dimensional nature of the model, the radial dispersion of fluid properties, the contribution of turbulent diffusion to the gas-phase momentum, the turbulence effect on flow field, the effect of particles on turbulence, and the dispersion of particles due to turbulent fluctuation could not be described.

Fukunaka et al. (3) carried out a modeling study on the flash-smelting of pyrite particles based on the previously studied behavior of gaseous free jets with the aid of correlation equations. Their model was limited only to the fully developed downstream zone, in spite of the importance of the upstream region in which particles are ignited and undergo thermal decomposition. They further did not consider the oxidation reaction of pyrite particles, the effect of the presence of particles on turbulence, dispersion of particles due to turbulent fluctuations, and the gravitational force acting on the particle.

As a two-dimensional model of the flash-smelting process in a confined furnace shaft, the only effort known to the authors is that by Ruottu (4). He considered that chalcopyrite and pyrite were first thermally decomposed to produce Cu_2S, FeS, and gaseous S_2, and then combustion took place. To describe the turbulence, he used the one-equation model under the assumption of constant mixing length, in spite of its strong dependency on location within a flow system. He did not account for the effect of the presence of particles on turbulence. He also assumed that the gas phase had the same temperature as the particle phase. These assumptions are not reasonable for a flash-smelting system in which, in the upstream zone, cold sulfide particles are first heated to the ignition temperature by convection from the surrounding hot gas and radiation from the furnace wall, and then particles undergo ignition, decomposition, and reaction with oxygen adding a large amount of heat to the gas phase as they travel along the furnace shaft. His assumption may be valid in the far downstream region from the burner where the heat exchange between the two phases is not significant. Furthermore, he neglected the scattering contribution to the calculation of the radiation field, only considering the absorbing and emitting phenomena, in spite of the fact that radiation by scattering is a very important subprocess in a particle-laden flow system.

The overall goal of this research was to develop a comprehensive mathematical model for the flash-smelting process incorporating most of the important subprocesses based on first principles. The flash smelting of chalcopyrite concentrate was used as an example in this work. In order to predict the overall phenomena occurring in the flash-smelting furnace, the comprehensive model must combine the turbulent fluid dynamics of a particle-laden gas jet, chemical kinetics, and the transfer of heat and mass.

In this study, the following aspects have been included:

(1) Use the two-equation (k-ε) turbulence model which accounts for the dependency of the mixing length on the location in a flow system.
(2) Incorporation of the effect of the presence of particles on the turbulence.
(3) Consideration of the effect of the turbulence on the particle phase, i.e., the dispersion of particles due to the turbulent fluctuations, and
(4) Use of the four-flux model by combining the absorbing, emitting, and scattering phenomena for radiative heat transfer.

Model Equations

The flash-furnace shaft can be schematically represented as shown in Fig. 1. The primary particle-laden gas stream with or without an oxygen-enriched secondary (or process) air stream enters the system through the burner nozzle and expands radially. The modeling equations to describe the overall phenomena of a reacting particle-laden turbulent gas jet in a confined system such as the flash-smelting furnace can be expressed by the general conservation equations of continuity, momentum, and energy for each phase.

Gas-Phase Equations

The gas phase, which can be considered as a turbulent continuous phase, is viewed from the Eulerian framework. The continuity and momentum equations combined with the effect of the presence of the reacting particles can, respectively, be expressed as:

Continuity:
$$\frac{\partial \rho}{\partial t} + \vec{\nabla} \cdot (\rho \vec{V}) = S_p^m \tag{1}$$

Momentum:
$$\frac{\partial}{\partial t}(\rho \vec{V}) + \vec{\nabla} \cdot (\rho \vec{V}\vec{V}) = - \vec{\nabla}p - \vec{\nabla} \cdot \overset{=}{\tau} + \rho \vec{g} + \vec{S}_p^V + \vec{V}S_p^m \tag{2}$$

where the arrow and the double bar represent a vector and a second-order tensor, respectively. In Eq. (1), the term S_p^m shows the net rate of mass addi-

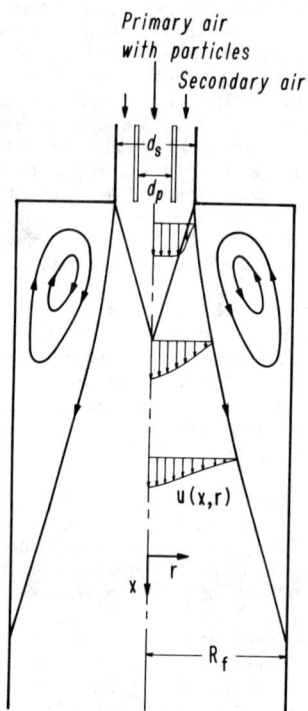

Fig. 1.   Particle-laden gas jet in a flash-smelting furnace.

tion to the gas phase per unit volume due to the reaction of solid particles.
In Eq. (2), the third and last terms represent momentum sources to the gas
phase due to, respectively, the presence of particles and the addition of mass
to the gas phase per unit volume.   The last term becomes zero for a nonreacting particle-laden gas jet.   The method for computing these terms is described
in the section on Gas-Particle Coupling.

      In order to reduce the above equations to simpler forms, the following
assumptions are made:

(1) The system is at steady state and axisymmetric in cylindrical coordinate,
(2) Body forces can be neglected,
(3) The contribution of the gas-phase momentum due to the dilation of gas,
     which describes the volume change of a fluid element due to expansion, can
     be neglected, and
(4) Gaseous properties are at local instantaneous equilibrium.

      Based on these assumptions, the time-averaged equations for the two-
dimensional elliptic flows without swirl can be written in some detail as
follows:

Continuity:

$$\frac{\partial}{\partial x}(\overline{\rho}\overline{u}) + \frac{1}{r}\frac{\partial}{\partial r}(r\overline{\rho}\overline{v}) = S_p^m \tag{3}$$

Axial momentum:

$$\frac{\partial}{\partial x} (\bar{\rho}\bar{u}\bar{u}) + \frac{1}{r} \frac{\partial}{\partial r} (r\bar{\rho}\bar{u}\bar{v}) - \frac{\partial}{\partial x} (\mu_e \frac{\partial \bar{u}}{\partial x}) - \frac{1}{r} \frac{\partial}{\partial r} (r\mu_e \frac{\partial \bar{u}}{\partial r})$$

$$= - \frac{\partial \bar{p}}{\partial x} + \frac{\partial}{\partial x} (\mu_e \frac{\partial \bar{u}}{\partial x}) + \frac{1}{r} \frac{\partial}{\partial r} (r\mu_e \frac{\partial \bar{v}}{\partial x}) + S_p^u + \bar{u} S_p^m \qquad (4)$$

Radial momentum:

$$\frac{\partial}{\partial x} (\bar{\rho}\bar{u}\bar{v}) + \frac{1}{r} \frac{\partial}{\partial r} (r\bar{\rho}\bar{v}\bar{v}) - \frac{\partial}{\partial x} (\mu_e \frac{\partial \bar{v}}{\partial x}) - \frac{1}{r} \frac{\partial}{\partial r} (r\mu_e \frac{\partial \bar{v}}{\partial r})$$

$$= - \frac{\partial \bar{p}}{\partial r} + \frac{\partial}{\partial x} (\mu_e \frac{\partial \bar{u}}{\partial r}) + \frac{1}{r} \frac{\partial}{\partial r} (r\mu_e \frac{\partial \bar{v}}{\partial r}) - 2\mu_e \bar{v}/r^2 + S_p^v + \bar{v} S_p^m \qquad (5)$$

where μ_e represents the effective viscosity given by

$$\mu_e = \mu_t + \mu_\ell \qquad (6)$$

Turbulence Model. Since it is difficult to completely understand the mechanism of turbulence because of its extremely complicated nature, so far, no rational theory which would enable us to quantitatively determine turbulence has been formulated. For this reason, many attempts have been made to create a mathematical basis for the study of turbulent motion with the aid of semi-empirical hypotheses.

The approach to the turbulence closure problem is credited to Boussinesq's analogy (5-8):

$$\overline{u'v'} = -\nu_g^t (\frac{\partial \bar{u}}{\partial r} + \frac{\partial \bar{v}}{\partial x}) \qquad (7)$$

where the over-bar denotes "time-averaged" and the prime represents a fluctuating velocity component. ν_g^t is the turbulent kinematic viscosity. The Reynolds stresses which designate the momentum transfer due to the turbulent fluctuations are thus expressed in terms of the local mean-velocity gradient coupled with the concept of eddy viscosity.

The eddy viscosity can be related to the turbulent kinetic energy (k) and its dissipation rate (ϵ) by the Prandtl-Kolmogorov relationship (5,8-10):

$$\nu_g^t = C_\mu k^2/\epsilon \qquad (8)$$

where C_μ is an empirical constant, and the turbulent kinetic energy is defined as

$$k = \frac{1}{2} (\overline{u'^2} + \overline{v'^2} + \overline{w'^2}) \qquad (9)$$

The turbulent kinetic energy and its dissipation rate can be expressed in terms of time-averaged variables by the following equations (5):

$$\frac{1}{r} \frac{\partial}{\partial r} [r(\bar{\rho}\bar{v}k - \frac{\mu_e}{\sigma_k} \frac{\partial k}{\partial r})] + \frac{\partial}{\partial x} [\bar{\rho}\bar{u}k - \frac{\mu_e}{\sigma_k} \frac{\partial k}{\partial x}] = G - \bar{\rho}\epsilon \qquad (10)$$

$$\frac{1}{r}\frac{\partial}{\partial r}\left[r\left(\bar{\rho}\bar{v}\varepsilon - \frac{\mu_e}{\sigma_\varepsilon}\frac{\partial\varepsilon}{\partial r}\right)\right] + \frac{\partial}{\partial x}\left[\bar{\rho}\bar{u}\varepsilon - \frac{\mu_e}{\sigma_\varepsilon}\frac{\partial\varepsilon}{\partial x}\right] = C_1 G\varepsilon/k - C_2\ \bar{\rho}\varepsilon^2/k \tag{11}$$

where
$$G = \mu_e\left\{2\left[\left(\frac{\partial\bar{u}}{\partial x}\right)^2 + \left(\frac{\partial\bar{v}}{\partial r}\right)^2 + \left(\frac{\bar{v}}{r}\right)^2\right] + \left(\frac{\partial\bar{u}}{\partial r} + \frac{\partial\bar{v}}{\partial x}\right)^2\right\} \tag{12}$$

σ_k and σ_ε and the turbulent Schmidt numbers for k and ε, respectively, which are simply empirical constants along with the constants C_1, C_2, and C_μ. These constants are known as "universal constants." Their values are given in Table I (8-10).

The turbulence closure model described above is known as the two-equation (k-ε) model. In spite of some shortcomings, the two-equation (k-ε) turbulence model has the widest application for confined turbulent jet systems. It has been tested in several nonreacting and reacting turbulent flow systems by several investigators over the last 15 years and resulted in satisfactory agreement with experimental data.

In flash-smelting processes, the turbulent flow confined in a flash furnace has added complexity due to the presence of particles. This makes it extremely difficult to use fundamental theories to describe the motion of the particle-laden turbulent gas jet. Hence, a great deal of empiricism must be included in describing turbulence.

The presence of solid or molten particles will affect the gas-phase turbulent field. The kinetic energy spectrum of the turbulent gas field is damped by the presence of particles. In addition, the gas phase affects the motion of particles through the turbulent fluctuations of flow properties. This latter effect is described in some detail in the section on Particle-Phase Equations.

Melville and Bray (11) proposed a semi-empirical correlation to account for the effect of particle phase on the gas turbulence. They used the ratio of the mean particle bulk density (i.e., mass of particles per unit volume) to the mean gas density as follows:

$$(\nu_g^t)_{particles} = (\nu_g^t)_{no\ particles}\ [1 + (\bar{\rho}_{bp}/\bar{\rho}_g)]^{-0.5} \tag{13}$$

The above equation suggests that the amount of gas-phase turbulence decreases as the particle number density increases. This equation was used throughout for the incorporation of the particulate influence on turbulence in this study, because no fundamental models are available at the present time.

Table I. Turbulence Model Constants

Constant	Value
C_μ	0.09
C_1	1.44
C_2	1.92
σ_k	0.9
σ_ε	1.22

Conservation Equation of Gaseous Species. The steady-state continuity equation for gaseous species j in a turbulent flow field can be expressed as:

$$\frac{\partial}{\partial x}(\overline{\rho u}\,\overline{m}_j) + \frac{1}{r}\frac{\partial}{\partial r}(r\overline{\rho v}\,\overline{m}_j) - \frac{\partial}{\partial x}(\Gamma_m \frac{\partial \overline{m}_j}{\partial x}) - \frac{1}{r}\frac{\partial}{\partial r}(r\Gamma_m \frac{\partial \overline{m}_j}{\partial x}) = \pm(\overline{S_p^m})_j \quad (14)$$

where $\overline{m}_j$ is the time-mean mass fraction of species j. The term on the right-hand side of Eq. (14) represents the time-averaged source (+) or sink (-) per unit volume due to the reactions of particles with gaseous reactants. This time-mean source or sink term is completely governed by the mean reaction rate of the individual sulfide particles.

It is worthwhile to note that the reaction rate of sulfide particles is assumed to be slow compared to the turbulent time scale but fast compared with the mean gas velocity. This assumption allows the kinetics of sulfide particles to be expressed in terms of the mean values of fluid properties.

The transport exchange coefficient Γ_m can be expressed as:

$$\Gamma_m = \mu_e/\sigma_m \quad (15)$$

where σ_m is the Schmidt number, having a value of about unity.

Gas-Phase Energy Equation. The steady-state energy equation for the gas phase can be expressed with the Eulerian framework by neglecting the energy change due to viscous dissipation, as follows:

$$\frac{\partial}{\partial x}(\overline{\rho u}\,\overline{h}_g) + \frac{1}{r}\frac{\partial}{\partial r}(r\overline{\rho v}\,\overline{h}_g) - \frac{\partial}{\partial x}(\Gamma_h \frac{\partial \overline{h}_g}{\partial x}) - \frac{1}{r}\frac{\partial}{\partial r}(r\Gamma_h \frac{\partial \overline{h}_g}{\partial r})$$

$$= Q_{rg} + \overline{u}\frac{\partial \overline{p}}{\partial x} + \overline{v}\frac{\partial \overline{p}}{\partial x} + S_p^h \quad (16)$$

On the righthand side of Eq. (16), the first and the last terms represent the net volumetric heat-transfer rate due to the gas-phase radiation and the heat addition to the gas phase due to the reaction of sulfide particles per unit volume. The former, Q_{rg}, is described in detail in the section on Radiative Heat Transfer.

Γ_h in Eq. (16) represents the transport coefficient for energy which is defined as

$$\Gamma_h = \mu_e/\sigma_h \quad (17)$$

where σ_h is the Prandtl number for the gas phase.

Particle-Phase Equations

The particle phase is described using the Lagrangian framework under the following assumptions:

(1) The interaction between particles is neglected because the particle concentration in the flash-smelting furnace is very low.

(2) The pressure gradient, virtual mass effect, and Basset force are negligibly small compared with the aerodynamic drag force (9,12).

Momentum Equation. The Lagrangian momentum equation for a single particle can be written as:

$$m_p \frac{d\vec{V}_p}{dt} = \frac{1}{2} C_D \rho_g A_p \; |\vec{V}_g - \vec{V}_p| (\vec{V}_g - \vec{V}_p) + m_p \vec{g} \tag{18}$$

Equation (18) indicates that the rate of change of the particle momentum is equal to the sum of the aerodynamic drag force and the gravitational forces acting on the particle.

The drag coefficient C_D can be calculated by the following correlation (9,14-16):

$$C_D = \frac{24}{Re} (1 + 0.15 \; Re^{0.687}) \tag{19}$$

where
$$Re = \rho_g \; |\vec{V}_g - \vec{V}_p| \; d_p / \mu_g \tag{20}$$

One of the major factors involved in the flash-smelting process is the turbulent dispersion of sulfide particles. Particles tracked through the flow field are considered to have different rates of dispersion from the gas phase. Turbulent particle-laden flows present many challenging problems, not only because they are difficult to describe mathematically but also due to the fact that some of the physical processes involved are not well understood.

In order to account for the dispersion of particles due to turbulent fluctuations, the particle velocity is broken down into a convective and a turbulent diffusive component (13,15,17):

$$\vec{V}_p = \vec{V}_{pc} + \vec{V}_{pd} \tag{21}$$

where subscripts pc and pd denote the convective and diffusive velocities, respectively. The convective velocity of the particle can be defined as the velocity that would result in the absence of turbulence or that based on the mean gas velocity. This velocity can then be obtained from Eq. (18) along a trajectory by numerical integration. The turbulent diffusive velocity accounting for the turbulent fluctuations can be modeled by assuming the turbulent diffusion of particles to be proportional to the gradient of mean particle number density:

$$\vec{V}_{pd} \bar{n}_p = D_p^t \; \vec{\nabla} \; \bar{n}_p \tag{22}$$

The transport coefficient D_p^t is defined as the turbulent-particle mass diffusivity and can be expressed as (11,13,15,17,18):

$$D_p^t = \nu_p^t / \sigma_p^t \tag{23}$$

where ν_p^t and σ_p^t are the turbulent-particle eddy viscosity and the Schmidt number of particles, respectively.

The turbulent kinematic viscosity of the particle must account for the degree of turbulence and particle size. Much work is currently being performed by many investigators on how to obtain ν_p^t (8,11,15,19-21). Although no reliable model based on a sound theory has been developed yet, a model by Melville and Bray (11) gave satisfactory results for the pulverized-coal combustion process (8,13,17), which is a system closely related to the flash-smelting process. For the present work, the following relationship by Melville and Bray (11) was selected for the expression of ν_p^t, because it is simpler than other models and is applicable to the size range of interest in the flash-smelting process:

$$\nu_p^t = \nu_g^t / [1 + (\tau_p / t_t)] \tag{24}$$

where τ_p and t_t are the particle relaxation time and the time scale of turbulence, respectively. The particle relaxation time is related to the Stokes particle drag by (11,15-17,21):

$$\tau_p = m_p / (3\pi\mu_g d_p) \tag{25}$$

By assuming an isotropic turbulence, the time scale of turbulence can be expressed as (11,13,15):

$$t_t = 1.5 \, C_\mu k / \varepsilon \tag{26}$$

For very small particles completely following the turbulent fluctuations, the turbulent-particle Schmidt number can be approximated to unity; for large particles failing to follow the turbulent fluctuations, it is less than unity. For the present work, the value of 0.35 was used for σ_p^t, as recommended by Smoot and Smith (13).

Particle Number Density. Since the local particle number density cannot be calculated from the Lagrangian particle-phase information, it can only be approximated using the Eulerian gas-phase information. The continuity equation for the particle number density of the jth size particle in a turbulent flow at steady state can be expressed as:

$$\frac{\partial}{\partial x} (\overline{u}\overline{n}_j) + \frac{1}{r} \frac{\partial}{\partial r} (r\overline{v}\overline{n}_j) - \frac{\partial}{\partial x} (D_j^t \frac{\partial \overline{n}_j}{\partial x}) - \frac{1}{r} \frac{\partial}{\partial r} (rD_j^t \frac{\partial \overline{n}_j}{\partial r}) = 0 \tag{27}$$

The diffusion coefficient D_j^t in Eq. (27) is the same as in Eq. (22).

Particle-Phase Energy Equation. The change of particle temperature is expressed by the following equation written in the Lagrangian framework:

$$\frac{d}{dt} (m_p h_p) = H_r + q_{rp} - Q_p - H_v - H_m \tag{28}$$

On the righthand side of Eq. (28), H_r represents the heat of reaction of sulfide particles, q_{rp} is the radiative heat transfer between the particles and the surrounding, Q_p is the heat loss to the gas phase by convection, H_v is the heat loss due to the volatilization of metal species, and H_m represents the heat loss due to the thermal decomposition or melting of the particle.

The heat of reaction of particles is a result of the oxidation reactions

807

of sulfide particles in the flash-smelting furnace which are described in detail in the section on Reaction of Chalcopyrite Particles. The radiation contribution q_{rp} is discussed in detail in the section on Radiative Heat Transfer.

For each particle size, the heat loss due to gas-phase convection is obtained by the following equations:

$$Q_p = Nu_j \pi d_j k_g (T_{pj} - T_g) \tag{29}$$

$$Nu_j = 2 + 0.65 \, Re_j^{1/2} Pr_g^{1/3} \tag{30}$$

$$Re_j = d_j \, |\vec{V}_g - \vec{V}_p| \, \rho_g/\mu_g \tag{31}$$

and
$$Pr_g = C_{pg}\mu_g/k_g \tag{32}$$

The heat loss due to volatilization of copper species is calculated by

$$H_v = \sum_i r_{vi} h_{vi} \tag{33}$$

where subscript i denotes the volatile copper species, r_{vi} is the vaporization rate of species i, and h_{vi} represents the enthalpy required for vaporization. The vaporization rate is described in the section on Reaction of Chalcopyrite Particles. To account for the heat loss due to melting of the particle, Richards' law was used:

$$H_m = 9.24 \, (\text{number of atoms}) \, (T_{mp}/M_p) \quad [J/kg] \tag{34}$$

where T_{mp} and M_p are the melting point of the particle and its molecular weight, respectively.

Reactions of Chalcopyrite Particles

In order to develop a comprehensive mathematical model of the flash-smelting process, chemical kinetics must be incorporated with the fluid dynamics of two-phase turbulent flow and the transfer of heat and mass. The reaction of chalcopyrite particles under flash-smelting conditions was studied by Chaubal and Sohn (22, 23). This section summarizes their results on the kinetics of oxidation of chalcopyrite particles.

Chemical Reactions

Copper concentrate charged to the flash furnace generally contains three major phases of chalcopyrite, pyrite, and silica. In the furnace shaft, chalcopyrite and pyrite will undergo oxidation reactions, but silica is essentially inert. The reactions include the oxidation of solid and molten particles, the vaporization of sulfur, and the volatilization of copper species. Concentrate particles are assumed to melt at 1153 K, which is the reported melting point of chalcopyrite (24).

808

When the concentrate particles are in solid state and below 873 K, the following reactions are assumed to occur (23):

$$CuFeS_2 + \frac{15}{4} O_2(g) \rightarrow CuSO_4(s) + \frac{1}{2} Fe_2O_3(s) + SO_2(g) \tag{35}$$

$$FeS_x(s) + (x + \frac{3}{4}) O_2(g) \rightarrow \frac{1}{2} Fe_2O_3(s) + xSO_2(g) \tag{36}$$

At temperatures between 873 K and 1153 K, the oxidation of solid particles proceeds primarily through the decomposition of sulfides. The overall reactions of solid particles in this temperature range are then expressed as (23):

$$CuFeS_2(s) + \frac{1}{2} O_2(g) \rightarrow \frac{1}{2} Cu_2S(s) + \frac{1}{2} SO_2(g) \tag{37}$$

$$FeS_x(s) + (x-1)O_2(g) \rightarrow FeS(s) + (x-1)SO_2(g) \tag{38}$$

Once the particles become molten, they are considered to be made up of Cu_2S and FeS (23). The overall reaction of the molten particle can be expressed as:

$$3FeS(\ell) + 5 O_2(g) \rightarrow Fe_3O_4(s,\ell) + 3SO_2(g) \tag{39}$$

$$Cu_2S(\ell) + O_2(g) \rightarrow 2Cu(\ell) + SO_2(g) \tag{40}$$

Although some FeS may react with silica in the particle, producing fayalite, the major oxidation product is magnetite. The produced magnetite and the initial silica in the particle are assumed to melt at 1873 K which is close to the melting point of magnetite and silica (23,25). For the copper-making mode, Cu_2S reacts with oxygen to produce liquid copper.

Reaction Kinetics

The data for the intrinsic kinetics of the oxidation of chalcopyrite were obtained by Chaubal and Sohn (22,23). Since no data on the intrinsic kinetics of the oxidation of pyrite are available, they assumed that the initial sulfur removal of pyrite was a result of decomposition and used the expression suggested by Fukunaka et al. (3) for fine particles, because some earlier work (26) suggested that very little oxidation of pyrite occurred before the onset of decomposition under a heating rate of 10 K/min.

The sulfur removal rate is related to the oxygen consumption by the stoichiometric coefficients of the oxidation reactions. The following equation was suggested for the rate of oxygen consumption (22,23):

$$N_{O_2,i} (\frac{dX_i}{dt}) = N_{O_2,i} k_0 \exp(-E/RT_p) f_1(p_{O_2}) f_2(X) f_3(d_p) \tag{41}$$

where i is either chalcopyrite or pyrite, and $N_{O_2,i}$ is the moles of oxygen required to react completely with either phase in the particle. X refers to the overall fractional degree of sulfur removed at a certain time.

The values of parameters in Eq. (41) were determined by Chaubal and Sohn (22,23) and are given in Table II. Since the kinetic information on the oxidation of pyrite is not available, it is assumed that pyrite oxidation occurs at the same rate as that of chalcopyrite (23).

Once the particles become molten, the rate of oxidation of sulfur is considered to be equal to the rate of external mass transfer of oxygen from the bulk. Assuming this mass-transfer control, the reaction rate of the molten particle can be obtained by

$$N_{O_2} \frac{dX}{dt} = k_m C_{O_2} A_p f_s \qquad (42)$$

where k_m and f_s are, respectively, the mass-transfer coefficient of oxygen and the fraction of the external surface area occupied by sulfides, which accounts for the fact that the produced oxide phases reduce the available surface area for oxygen transfer from the bulk. The above equation implies that the reaction between the oxide and the molten sulfides in the particle is negligible compared to the oxidation reaction of sulfides.

Volatilization of Copper

Copper loss by volatilization may be substantial at conditions of high oxygen content and temperature (27). Jorgensen (27) suggested that copper volatilization could play an important role in controlling the particle temperature. Chaubal and Sohn (23) also incorporated copper loss by volatilization using the following equation:

$$\frac{dm}{dt} = k_{m,Cu} \frac{(P_{Cu,e} - P_{Cu,b})}{RT} A_p f_s \qquad (43)$$

Table II. List of Chemical Reaction Kinetics[*]

	k_o	E	$f_1(P_{O_2})$	$f_2(X)$	$f_3(d_p)$
1. Chalcopyrite					
$T_p < 754$ K	2.4×10^8	215	P_{O_2}	$\dfrac{0.07}{\exp(X/0.07)}$	$1/d_p^2$
$754 < T_p < 873$ K	0.026	71.4	"	"	"
Sulfur vaporization:	2.72×10^9	208	--	$(1 - X)^2$	"
2. Pyrite					
Sulfur vaporization:	4.5×10^{10}	279	--	$(1 - X)^{2/3}$	"

[*]Adapted from Chaubal and Sohn (22,23).
Note: k_o in $cm^2/(s \cdot kPa)$, P_{O_2} in kPa, E in kJ/mol, and d_p in cm.

In the above equation, the partial pressure at the particle surface is assumed to be equal to the equilibrium partial pressure, which is obtained by multiplying the vapor pressure values given in Table III by the atomic fraction of copper in the molten sulfide droplets.

The conditions in Jorgensen's experiments were much more oxidizing and the temperature much higher than in an actual flash-smelting process. Therefore, in the latter the importance of copper volatilization is expected to be considerably less.

Physical Properties

This section summarizes the physical properties necessary for solving the gas-phase and particle-phase modeling equations. The proposed equations (9, 28,29) for these properties are summarized in Table IV.

Table III. Vapor Pressure of Copper Species[*]

| | | Vapor Pressure (N/m^2) | | |
Reaction	Species	1500 K	1800 K	2000 K
$Cu(\ell) \rightarrow Cu(g)$	$Cu(g)$	0.973	56.7	436

[*]Adapted from Chaubal and Sohn (23).

Table IV. Physical Properties

Gas species conductivity	$k_i = (C_{pi} + \frac{5R}{4M_i})\mu_i$	(44)
Gas mixture conductivity	$k_g = \sum_i (X_i k_i / \sum_j X_j \phi_{ij})$	(45)
Gas species viscosity	$\mu_i = 2.67 \times 10^{-6}(M_i T_g)^{1/2}/\sigma_i^2 \Omega_\mu$	(46)
Gas mixture viscosity	$\mu_g = \sum_i (X_i \mu_i / \sum_j X_j \phi_{ij})$	(47)
Interaction parameter in Eqs. (47) and (49)	$\phi_{ij} = (\frac{1}{8})^{1/2}[1 + \frac{M_i}{M_j}]^{-1/2}[1 + (\frac{\mu_i}{\mu_j})^{1/2}(\frac{M_j}{M_i})^{1/4}]^2$	(48)
Gas species diffusivity	$D_{ij} = 1.83\times10^{-12}T_g^{3/2}[\frac{1}{M_i} + \frac{1}{M_j}]^{1/2}/P\sigma_{ij}^2 \Omega_d$	(49)
Mixture diffusivity	$D_{im} = (1 - x_i)/\sum_{i \neq j}(x_j/D_{ij})$	(50)
Particle heat capacity	$C_{pj} = \sum_k (w_k C_{pk})_j$	(51)

The mass-transfer coefficient (k_m) is calculated with the following empirical correlation for the Sherwood number (9,28,29):

$$Sh = 2 + 0.6 \, Re_j^{1/2} \, Sc^{1/3} \tag{52}$$

where Sc represents the Schmidt number for gas species.

Since the size degradation of the particle is experimentally observed (2,27), the particle size is allowed to decrease linearly with the extent of overall conversion of the sulfide particle down to a final size. The size degradation is calculated with the equation:

$$d_j = (1 - X)d_{jo} + d_f X \tag{53}$$

where d_{jo} and d_j are, respectively, the initial size and the final particle size experimentally determined as 25 μm (2,27).

Radiative Heat Transfer

Due to the highly emitting, absorbing, and scattering nature of the components in a flash-smelting flame, thermal radiation may be a significant mode of heat transfer in a flash-smelting furnace. Radiative heat transfer in a furnace can be divided into nonluminous gaseous emission, luminous particulate emission, and radiation from the furnace wall.

Several investigators have studied and reviewed the radiative heat transfer occurring in pulverized-coal combustion systems (8,9,32-35). Fields (35) discussed the radiative heat-transfer rate in pulverized coal flames in detail. He used a four-flux model based on the work by Chu and Churchill (36). Very little work has been done on the incorporation of radiative heat transfer in describing a flash-smelting process except by Ruottu (4), who considered the emitting and absorbing phenomena. However, he neglected the scattering process which is an important subprocess in a particle-laden flow system. He further assumed the same temperatures between the gas and the particle.

In the flash-smelting process, the radiation undergoes attenuation and augmentation by various gas and particle sources which are continuously changing throughout the process. A flash-smelting flame is a multicomponent, nonuniform, emitting, absorbing, and scattering gas-solid cloud which generates heat by reaction and is surrounded by nonuniform, emitting, absorbing, and reflecting solid surfaces (i.e., furnace walls). The nonuniformity of the flash-smelting system and the lack of information on the radiative properties of the system components make the radiation problem extremely complex unless a number of approximations are made.

In this work, the following assumptions are made in describing the radiation field:

(1) The particles retain spherical shape,
(2) The temperature field is axisymmetric,
(3) The gas/particle medium is gray, and
(4) The control volume is at local equilibrium between components of the flame.

Based on these assumptions, the four-flux model reduced from the six-flux model of Chu and Churchill (36) has been used by considering absorbing, emitting, and anisotropic scattering phenomena, as follows:

$$\frac{1}{K_t} \frac{dI_x^+}{dX} = C_1 I_x^+ + C_2 I_x^- + C_3(I_r^+ + I_r^-) + C_4 I_b \tag{54}$$

$$-\frac{1}{K_t} \frac{dI_x^+}{dX} = C_1 I_x^- + C_2 I_x^+ + C_3(I_r^+ + I_r^-) + C_4 I_b \tag{55}$$

$$\frac{1}{K_t} \frac{1}{r} \frac{d(rI_r^+)}{dr} = C_1 I_r^+ + C_2 I_r^- + C_3(I_x^+ + I_x^-) + C_4 I_b \tag{56}$$

$$-\frac{1}{K_t} \frac{1}{r} \frac{d(rI_r^-)}{dr} = C_1 I_r^- + C_2 I_r^+ + C_3(I_x^+ + I_x^-) + C_4 I_b \tag{57}$$

where

$$C_1 = -(1 - \omega_0 f) - 2(\omega_0 s)^2/W \tag{58}$$

$$C_2 = \omega_0 b - 2(\omega_0 s)^2/W \tag{59}$$

$$C_3 = \omega_0 s - 2(\omega_0 s)^2/W \tag{60}$$

$$C_4 = (1 - \omega_0)[1 - 2(\omega_0 s)^2/W]/6 \tag{61}$$

$$W = \omega_0 b - (1 - \omega_0 f) \tag{62}$$

f, b, and s are the forward, backward, and sidewise scattering components, respectively, which are expressed as (9,35-37):

$$f = 2\pi \int_0^{\pi/2} p(\beta)\sin \beta \cos^2 \beta \, d\beta \tag{63}$$

$$b = 2\pi \int_{\pi/2}^{\pi} p(\beta)\sin \beta \, d\beta \tag{64}$$

$$s = (1 - b - f)/4 \tag{65}$$

where β and $p(\beta)$ represent, respectively, the scattering angle and the phase function for anisotropic particle scattering which describes the angular distribution of the scattered energy. It is noted that for isotropic scattering, the phase function has the value of 1, and $f = b = s = 1/6$.

In each equation, I represents a radiative flux, I_b is the black body emissive flux, and ω_0 is called the albedo for scatter, which is expressed as

$$\omega_0 = K_s/(K_a + K_s) = K_s/K_t \tag{66}$$

where K_a, K_s, and K_t are the absorption, scattering, and extinction coeffici-

813

ents, respectively, which depend on wavelength, particle size, and number density, as described by Eqs. (80) and (81).

The net radiative heat flux and the flux sum are, respectively, defined as (6,34,35):

$$\vec{Q} = \vec{I}^+ - \vec{I}^-$$ (67)

$$\vec{F} = \vec{I}^+ + \vec{I}^-$$ (68)

The addition of each pair of radiative flux equations then results in a set of mutually coupled second-order differential equations:

$$- \frac{d}{dx}\left(\Gamma_F \frac{dF_x}{dx}\right) = (C_1 + C_2) F_x + 2C_3 F_r + 2C_4 I_b$$ (69)

$$\frac{1}{r} \frac{d}{dr}\left[\Gamma_F \frac{d(rF_r)}{dr}\right] = (C_1 + C_2) F_r + 2C_3 F_x + 2C_4 I_b$$ (70)

where
$$\Gamma_F = K_t[1 - \omega_0(f - b)]^{-1}$$ (71)

The total volumetric radiative heat-transfer rate is then defined as (8,9,13,35):

$$Q_r = K_a(F_x + F_r - I_b)$$ (72)

Since the total absorption coefficient (K_a) can be partitioned into components for the gas and the particle, i.e., (8,9,13,35):

$$K_a = K_{ag} + K_{ap}$$ (73)

The radiative heat-transfer rates for the gas and the particle are then obtained, respectively, as

$$Q_{rg} = K_{ag}(F_x + F_r - I_b)$$ (74)

$$Q_{rp} = K_{ap}(F_x + F_r - I_b)$$ (75)

The Lagrangian form of the radiative heat-transfer rate to a single particle (i.e., energy per time) is then obtained by the following relationship:

$$q_{rp} = Q_{rp}/n_p$$ (76)

Equation (76) is used in the particle-phase energy equation, i.e., Eq. (28).

Phase Function

The effect of anisotropic scattering due to the presence of solid particles is characterized by the distribution of scattered intensity in the solid angle of 4π steradians surrounding a particle. The effect of scattering is incorporated by defining an appropriate form of the phase function. Due to the lack of information on the optical properties of the sulfide mineral particles, it is assumed that the sulfide particles can be considered opaque and diffuse. Based on this assumption, the phase function for such particles expressed by the following equation (31,35,37) can be used to determine the scattering components:

$$p(\beta) = \frac{8}{3\pi} (\sin \beta - \beta \cos \beta) \tag{77}$$

The above equation is used to determine the forward, backward, and sidewise scattering components given by Eqs. (63) to (65).

Absorption Coefficients of Gases

In general, gases emit radiation in rather narrow bands of wavelengths and solid particles emit radiation over a much broader range of wavelength.

Within the furnace, gaseous species O_2, N_2, SO, SO_2, and SO_3 are present. The absorbing heteropolar gases such as SO, SO_2, and SO_3 are important in the radiation calculation, while the simple symmetrical molecules such as O_2 and N_2 are essentially transparent to radiation (30,31). SO_2 is assumed to be the only gaseous absorbing component, because the amounts of SO and SO_3 are relatively small in a flash furnace.

The absorption coefficient is related to the emissivity of a gaseous component by (9,31,35,37):

$$K_{ag} = -[\ln (1 - \varepsilon_g)]/\ell_m \tag{78}$$

where ε_g and ℓ_m are the emissivity of the gas and the mean beam length, respectively. Equation (78) indicates that the absorption coefficient of the gas can be calculated by knowing the emissivity of SO_2. Detailed information on SO_2 emissivities at different temperatures and pressures is available in Hottel and Sarofim (37) and Chan and Tien (38). In this work, data obtained by Chan and Tien, which are more recent, are used. The mean beam length depends on the shape of the furnace and the size of the radiating medium and is calculated with (8,31,37)

$$\ell_m = 3.5 \ (V/A) \tag{79}$$

where V and A denote the volume and cross-sectional area of the furnace.

Absorption and Scattering Coefficients of Particles

The radiative scatter in a particle-laden reacting flow system is a complex topic, and a full discussion is beyond the scope of this paper. In this section, approximation methods are presented to express the absorption and scattering coefficients of particles in the flash-smelting furnace.

By assuming no interaction between particles, the absorption and scattering coefficients for particles are simply expressed with sums of the individual particle contributions in a unit volume, as follows (31,35,37):

$$K_a = \sum_j K_{aj} = \frac{\pi}{4} \sum_j \eta_a n_j d_j^2 \qquad (80)$$

$$K_s = \sum_j K_{sj} = \frac{\pi}{4} \sum_j \eta_s n_j d_j^2 \qquad (81)$$

where η_a and η_s are, respectively, the absorption and scattering efficiencies of particles, and n_j and d_j represent the particle number density and the particle diameter of the jth size particle, respectively.

The absorption and scattering efficiencies are dependent on the particle size parameter, which is defined as (8,35,37):

$$p_s = \pi d_j / \lambda \qquad (82)$$

where λ is the wavelength of radiation. The efficiencies may be calculated by the Mie theory (37,39) if the exact optical properties of sulfide particles are known. In the absence of the required data, the following approximations are introduced.

For a cloud of gray particles of which size parameter p_s is greater than 5, the following relationships can be used to calculate the absorption and scattering efficiencies (8,31,32):

$$\eta_a = \varepsilon_p \qquad (83)$$

$$\eta_s = C_s / \pi r_j^2 \qquad (84)$$

where ε_p is the emissivity of the particles and C_s represents the scattering cross-section, which is approximated to be equal to two times the cross-sectional area of the particle for $\lambda \ll d_j$ (8,31,32).

In the temperature range of coal combustion and gasification processes as well as the copper flash-smelting process, the most significant contributions to thermal energy are obtained by the radiation having wavelengths between 0.5 and 10 μm (8). Sulfide particles in the flash-smelting process are in the size range of 25 to 100 μm. By assuming that in the flash furnace the radiation wavelength is between 0.5 and 10 μm, size parameter p_s then becomes much greater than 5. Hence, the above approximation methods were used throughout for the calculation of radiative heat transfer occurring in this work.

Boundary Conditions

A complete specification of boundary conditions is necessary for solving the governing equations. The previous work done by Hahn and Sohn (40) examined extensively the effects of boundary conditions, especially those at the inlet, on the numerical results. They also tested various correlations for the dissipation rate of turbulent kinetic energy at the inlet and obtained the relation yielding the best results as

$$\varepsilon = C_\mu k^{1.5} / (0.015 \, d_e) \qquad (85)$$

where d_e = 4 x hydraulic radius. The detailed descriptions of boundary conditions of fluid properties except those for gas-phase energy and the radiation field can be found in References 40 and 41.

In the inlet, the gas-phase enthalpy is specified by inlet temperature. At the centerline, $\partial h_g/\partial r = 0$, and at the outlet, $\partial h_g/\partial x = 0$. For a nonadiabatic reactor, the heat losses by convection and conduction through the furnace wall must be considered. The following wall function derived by Launder and Spalding (5,9,10,42) was used for the wall-boundary condition of the gas-phase energy:

$$
q_w = \frac{(T_w - T)C_{p,mix}\,\bar{\rho}C_\mu^{1/4}k^{1/2}}{\frac{\sigma_{h,t}}{K}\ln[EC_\mu^{1/4}k^{1/2}y_n\bar{\rho}/\mu] + 9.24\sigma_h[\frac{\sigma_h}{\sigma_{h,t}} - 1][\frac{\sigma_{h,t}}{\sigma_h}]^{1/4}}
\tag{86}
$$

The values of σ_h and $\sigma_{h,t}$ are given as 0.8 and 0.9, respectively (8,42,43). The wall temperature also needs to be specified as a function of position.

The boundary conditions for the radiation equations are similar to those used for the gas-phase equations. At the centerline, $dF_r/dr = 0$, and at the outlet, $dF_x/dx = 0$. To define wall-boundary conditions, the following assumptions are made (9,13,35):

(1) The boundary is opaque to radiation,
(2) The boundary is gray, and
(3) The boundary surface is a diffuse emitter/reflecter (a Lambert surface).

Since the boundary surface is assumed to be opaque, the transmissivity of the walls is zero. Then it follows that

$$
\varepsilon_w + \rho_w = 1
\tag{87}
$$

where ε_w and ρ_w are the emissivity and the reflectivity of the wall, respectively.

According to the third assumption, all the radiation flux coming from the walls is due to emission or reflection. Hence, the following boundary conditions can be used for the ceiling and side walls, respectively (9,35):

$$
I_x^+ = \varepsilon_w\sigma T_w^4/\pi + (1 - \varepsilon_w)F_x^-
\tag{88}
$$

$$
I_r^- = \varepsilon_w\sigma T_w^4/\pi + (1 - \varepsilon_w)F_r^+
\tag{89}
$$

where I and F represent the radiative flux from the walls and the flux incident on the surface, respectively.

Numerical Methods

Modeling equations, except for the momentum equation of particle phase, can be expressed in the following form:

$$
\frac{\partial}{\partial x}(\bar{\rho}\bar{u}\phi) + \frac{1}{r}\frac{\partial}{\partial r}(r\bar{\rho}\bar{u}\phi) - \frac{\partial}{\partial x}(\Gamma_e\frac{\partial\phi}{\partial x}) - \frac{1}{r}\frac{\partial}{\partial r}(r\Gamma_e\frac{\partial\phi}{\partial r}) = S_\phi
\tag{90}
$$

where ϕ represents a dependent variable and S_ϕ is the "source term" which includes all other terms in the governing equations not embodied in the first four terms in Eq. (90). Γ_e denotes an effective transport exchange coefficient for individual variables in a turbulent flow. Equation (90) can be cast into finite difference equations to be solved by line-by-line or tri-diagonal matrix algorithm.

The TEACH code developed by Gosman and Pun (44) was used to solve the gas-phase equations. The SIMPLER algorithm devised by Patankar (45) was used to calculate the pressure field.

The Lagrangian equations for the motion of particles, which are easier to solve than the Eulerian gas-phase equations, were solved by the particle-source-in-cell (PSI-CELL) technique developed by Crowe and coworkers (12). The PSI-CELL technique has been followed directly to take account of the particle field in the Eulerian gas field. The procedure is described as follows:

(1) Solve the Eulerian gas field without particles.
(2) Solve the radiation field using the four-flux model.
(3) Calculate the Lagrangian particle field, i.e., particle trajectories are calculated.
(4) Evaluate the particle source terms.
(5) Solve the gas field with the updated particle source terms.
(6) Return to Step (2) and repeat until convergence is achieved.

The PSI-CELL technique is very efficient in terms of computer storage and computational time. Convergence is achieved when the gas field does not change between particle iterations. Steps (2) through (5) are termed a "particle-iteration." Three particle-iterations are usually sufficient to achieve convergence.

Particle/Gas Coupling

In order to include the effect of the presence of reacting particles, the particle-phase equations must be coupled to the Eulerian gas field through the source terms to fluid properties.

For particles of the ith size fraction injected at the jth starting location at the inlet, the particle mass source term S_p^m to the gas phase in a cell (or a control volume) can be obtained by:

$$(S_p^m)_{cell} = \frac{1}{V_{cell}} \{ \sum_i \sum_j \dot{n}_{ij} [(m_{pij})_{out} - (m_{pij})_{in}] \}_{cell} \qquad (91)$$

where V_{cell}, $\dot{n}_{ij}$ and m_{pij} represent the volume of the cell, the particle number flow rate, and the mass of particles of the ith size fraction injected through the jth starting location, respectively.

The momentum source terms of the u- and v-components of the gas-phase momentum are similarly described by:

$$(S_p^u)_{cell} = \frac{1}{V_{cell}} \{ \sum_i \sum_j \dot{n}_{ij} [(u_{pij}m_{pij})_{out} - (u_{pij}m_{pij})_{in}] \}_{cell} \qquad (92)$$

818

$$(S_p^v)_{cell} = \frac{1}{V_{cell}} \{ \sum_i \sum_j \dot{n}_{ij} [(v_{pij}m_{pij})_{out} - (v_{pij}m_{pij})_{in}] \}_{cell} \qquad (93)$$

The source or sink term of the gaseous species k due to the reactions of particles can be obtained by:

$$(S_p^m)_k = \frac{1}{V_{cell}} \{ \sum_i \sum_j \dot{n}_{ij} [(w_k)_{out} - (w_k)_{in}] \}_{cell} \qquad (94)$$

where w_k is the amount of the consumed or produced gaseous species k resulting from the reactions of particles. Equation (94) implies that particles are uniformly distributed in a given cell.

The energy source term to the gas phase is given by (8,13):

$$(S_p^h)_{cell} = \frac{1}{V_{cell}} \{ \sum_i \sum_j \dot{n}_{ij} [(h_{pij}m_{pij} - tq_{rp})_{out}$$

$$- (h_{pij}m_{pij} - tq_{rp})_{in}] \}_{cell} \qquad (95)$$

where q_{rp} accounts for radiation between the gas and the particle.

Results and Discussion

The present authors conducted grid-dependence tests in their previous work (40) and noted that virtual grid-independence was achieved with grid points more than 27 x 27 in a confined system they tested. For the present work, grid points of 32 x 32 were used throughout from the standpoint of overall convergence and computational economy.

In this section, in order to verify the two-dimensional mathematical model developed in this work, the model predictions were first compared with experimental data obtained from the Outokumpu pilot flash furnace (47). Finally, the model predictions for the overall phenomena occurring in a commercial-scale flash-smelting furnace are discussed.

In flash-smelting processes, there are two types of burner configurations: a single-entry burner system which allows the oxygen-enriched process air and concentrate particles to enter the furnace through a single entry nozzle without a distribution air stream, and a double-entry burner system in which the distribution air together with concentrate particles enter the furnace separately from the oxygen-enriched process air. The former type of burner is typically applicable to a horizontal flash-smelting furnace. For the latter type of burner, two modes of feeding the distribution (or primary) air were considered (41): (1) the axial flow of this air with zero radial-velocity component and (2) the radial flow with zero axial-velocity component at the inlet. The latter feeding mode (i.e., the radial feeding mode) represents a simplified approximation of an industrial flash-smelting burner system, in particular the Outokumpu flash furnace. A considerable amount of work has been done by Hahn and Sohn (41) to describe the behavior of a particle-laden gas jet under flash-smelting conditions but without reactions. They tested effects of burner configurations of the axial and radial feeding modes of the distribution air with varying inlet conditions such as inlet velocities, particle loading ratio, oxygen content, and particle size. They obtained good agreement between the predicted results and measured data taken from the literature in terms of velocity profiles and particle mass flux.

They concluded that, for a flash-smelting furnace, more uniform distribution of particles could be obtained by radially feeding the distribution air with particles. Detailed results can be found in Reference 41.

Comparison between Predictions and Pilot Flash Furnace Data

The mathematical model developed in this work was used to predict the performance of the Outokumpu pilot flash furnace whose dimensions and experimental conditions are described in Table V. The values corresponding to Tests 1-3 represent the experimental conditions for the Outokumpu double-entry burner system in which the distribution air is fed radially into the furnace, while Tests 4 and 5 represent those for a single-entry burner system. The

Table V. Boundary Conditions for the Outokumpu Pilot Flash Furnace

Inlet Geometry:

	Test No.	
	1-3	4-5
d_p (m)	0.09	0.13
d_s (m)	0.2	-
d_f (m)	1.2	1.2
L_f (m)	6.	5.

Particle Size: 50 μm (mass mean diameter)

Particle Density: 4300 kg/m^3

Turbulence Intensities: I_p = 0.02* (0.15**), I_s = 0.08*

	Test No.				
	1	2	3	4	5
Primary Stream:					
Gas flow (Nm^3/h)	30	30	30	1326	882
O_2 content (%)	21	21	21	28	28
Temperature (K)	298	298	298	298	463
Concentrate feed rate (kg/h)	470	500	500	1390	960
Cu (%)	18.1	17.7	19.2	17.7	18.1
Fe (%)	37.6	38.1	29.1	35.3	35.8
S (%)	35.9	36.1	35.4	36.3	35.5
SiO_2 (%)	3.55	3.65	2.61	5.6	5.0
Secondary Stream:					
Gas flow (Nm^3/h)	555	570	560	-	-
(axial velocity m/s)	(6.2)	(6.2)	6.3)	-	-
O_2 content (%)	27	28	28	-	-
Temperature (K)	298	298	298	-	-
Matte Grade (% Cu):	76.4	71.4	75.2	56	59

*Estimated values for Tests 1-3
**Estimated values for Tests 4-5

temperature at the furnace wall was assumed as 1473 K (48) for the purpose of computation.

To predict the performance of the flash-smelting furnace with the double-entry burner system, the radial feeding mode of the distribution air was used with u_p = 0 and v_p = 118 m/s (equivalent slit width = 0.25 mm). Figure 2 shows the results of the comparison between predictions and measurements for Tests 1-3. In these figures, the profiles along the centerline of the SO_2 and O_2 concentrations and the amount of oxygen consumed to produce oxide phases are plotted. Both the predicted and measured results for the three tests show almost the same profiles because the inlet conditions were similar. Although

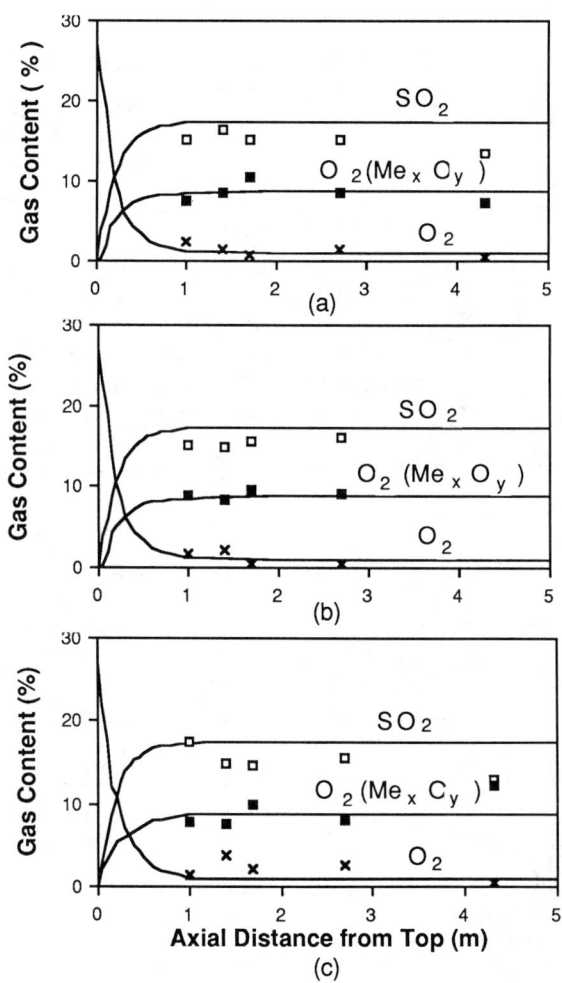

Fig. 2.   Comparison between the predicted and measured results.
(a) Test 1, (b) Test 2, (c) Test 3.
□■x measurements by Outokumpu (47)
──── prediction of this work.

the measured data are somewhat scattered, the overall agreement is satisfactory. Model predictions indicate that the reaction of chalcopyrite particles mostly occurs in the upper zone of the furnace shaft, i.e., between the burner exit and the location at x = 1 m. Hence we can conclude that the upper part of the flash furnace up to about 1 m from the top is a very important region in determining the extent of sulfide particles reacted. It is worthwhile to note that all of the measured data were obtained beyond this upper region. Hence, although the overall agreement is satisfactory, comparison could not be made in the upper region of the furnace due to the lack of experimental data at the present time.

More complete predictions for chalcopyrite concentrate combustion are shown in Fig. 3 in terms of the contours of the SO_2 and O_2 concentration and

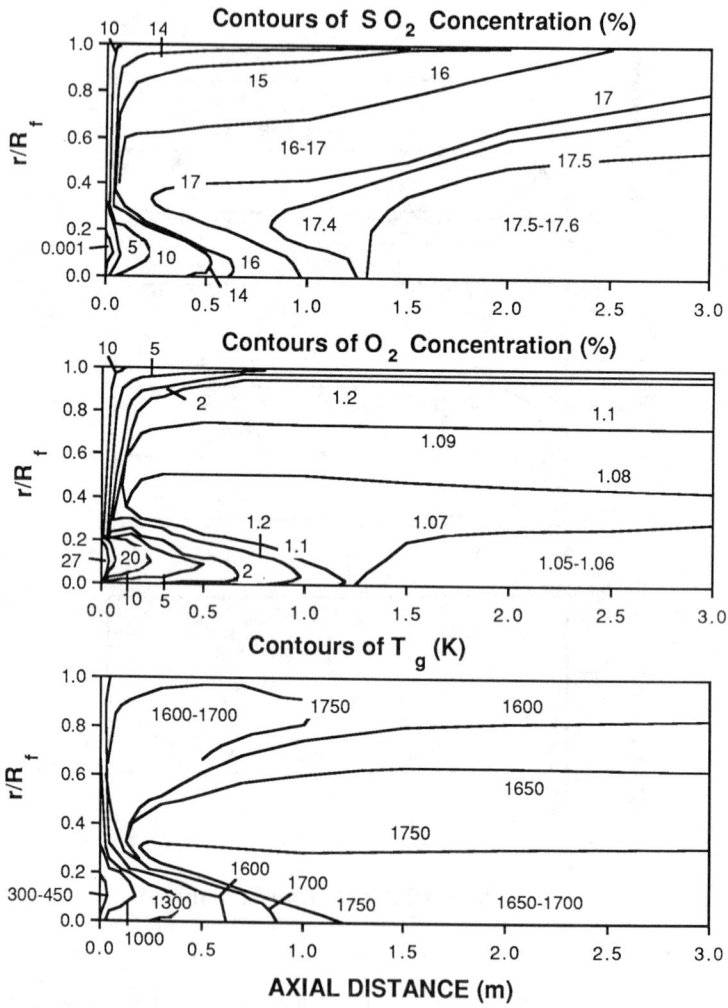

Fig. 3. Predicted contours of the SO_2 and O_2 concentrations and the gas-phase temperature for Test 1.

822

the gas-phase temperature for Test 1.   These contours again indicate that the
reaction of sulfide particles with oxygen occurs mostly in the upper zone of
the furnace up to about 1 m from the top, and little further reaction occurs
after that region.   The figures also show that reactions occur widely in the
radial direction because the concentrate particles are dispersed radially into
the furnace by the radial feeding of the distribution air.   The SO_2 and O_2
concentrations and gas temperature become relatively uniform below about 1.3 m
from the top.

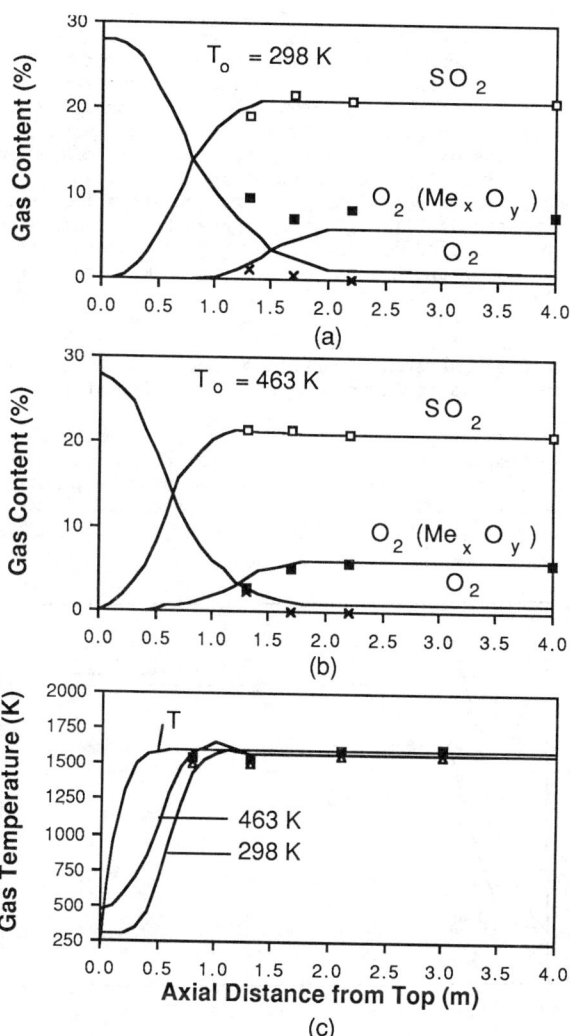

Fig. 4.   Comparison between the predicted and measured
results for various preheating temperatures.
Gas concentrations: (a) Test 4, (b) Test 5
□■x measurements by Outokumpu (47)
Gas temperature: T_o = 298 K   Δ ⎫ Measurements by
T_o = 463 K   ■ ⎭ Outokumpu (47)
—— predictions of this work

For the single-entry burner system, the predicted and measured results of the SO_2 and O_2 concentrations, the amount of oxygen consumed to produce oxides, and the gas temperature along the centerline are plotted in Fig. 4. Figures 4(a) and (b) show the results for inlet gas temperatures of 298 and 463 K, respectively. Predicted results show that reactions occur somewhat faster with a higher inlet temperature (Fig. 4(b)) than with a lower temperature. Comparing these results with those for the double-entry burner system (Fig. 3), it is seen that complete reaction of sulfide particles is attained further from the burner in the single-entry system. This can be explained by the fact that the wide dispersion of particles in the radial direction by the radial feeding of the distribution air facilitates the effective utilization of the furnace volume and, consequently, enhances the smelting rate of the sulfide mineral compared to the single-entry burner system. Again, the model predictions in the upper region of the furnace could not be compared with measurements due to the lack of data. Figure 4(c) shows the comparison of the predicted and measured gas-phase temperatures obtained by varying the preheating temperature of the inlet gas. This figure also shows the predicted result by Themelis et al. (curve T) (1). It is seen that their prediction shows the same profile for both cases of preheating and higher temperatures in the upper region of the furnace than those obtained by this work. According to their temperature predictions, the ignition of chalcopyrite particles occurs much nearer to the burner, and the oxidation reactions proceed faster than obtained in this work. Their higher temperature prediction in the upper zone is due to the fact that their one-dimensional model does not account for the turbulent dispersion of gases and heat transfer in the radial direction.

Figure 5 shows the comparison of the gas-phase temperature and the averaged particle-phase temperature along the centerline for the experimental conditions of Test 4. The solid squares are the gas-phase temperature values measured by Outokumpu (47). It is seen that the predicted temperature of the particle phase is much higher than the gas phase, especially in the upper zone of the furnace.

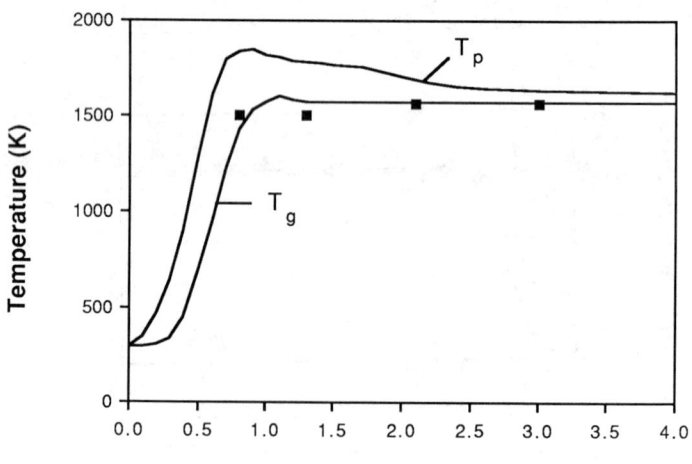

Fig. 5. Centerline profiles of the gas (T_g) and particle (T_p) temperatures for Test 4.
■ Measurements by Outokumpu (47),
—— prediction of this work.

More complete predictions for the chalcopyrite concentrate combustion in a single-entry system are shown in Fig. 6 in terms of the contours of the SO_2 and O_2 concentrations and the gas-phase temperature for Test 4. It is seen that the SO_2 and O_2 concentrations and the gas temperature near the ceiling have values similar to those in the downstream zone due to the recirculating flow, unlikely in the double-entry system with radial feeding of the distribution air. These contours indicate that complete reaction of sulfide particles is attained farther from the burner compared to that in the double-entry system (Fig. 3).

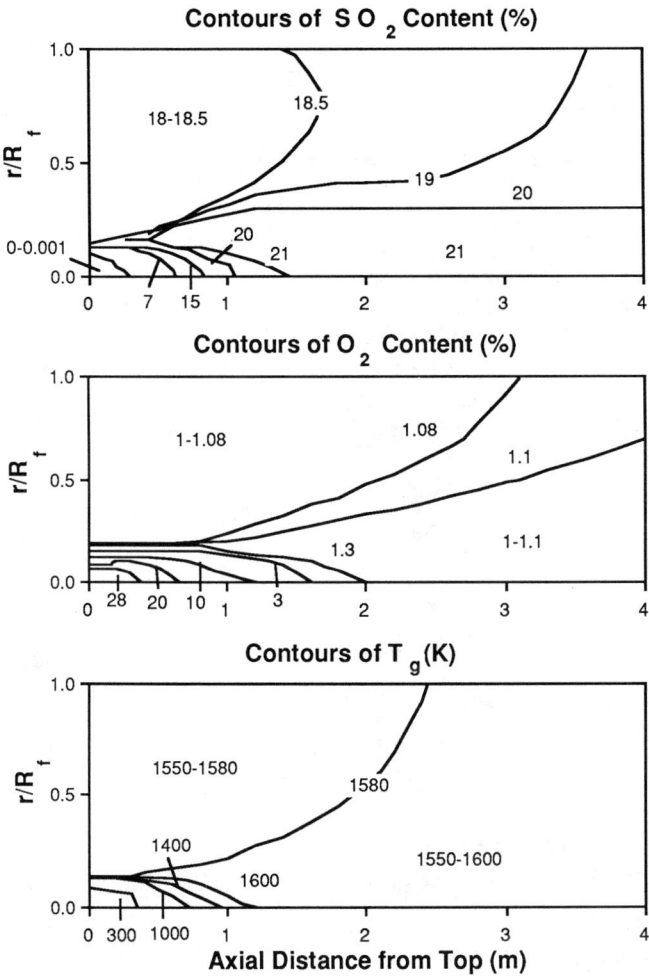

Fig. 6. Predicted contours of the SO_2 and O_2 concentrations and the gas-phase temperature in a single-entry system (Test 4).

825

Prediction of a Commercial Flash Furnace Operation

In order to predict the performance of a commercial flash-smelting furnace, the system described in Table VI was used. The dimensions of the system
were chosen to be close to a commercial-scale flash furnace. The primary air
(or the distribution air) laden with concentrate particles and the secondary
air (or the oxygen-enriched process air) enter the furnace radially and axially, respectively. The distribution air stream had velocity components of u_p
$= 0$ and $v_p = 120$ m/s (equivalent slit width $= 0.62$ mm). The oxygen-enriched
process air (50 vol. % O_2) preheated to 473 K enters the furnace axially with
a linear velocity of $u_s = 15$ m/s.

Figure 7 shows the contours of SO_2 and O_2 concentrations and gas-phase
temperature. These contours indicate that, in a commercial flash-smelting
furnace, the reactions of sulfide particles occur mostly within about 0.5 m of
the top. Relatively uniform profiles of the SO_2 and O_2 concentrations and the
gas temperature are observed within 0.5 m from the top. It is seen that the
overall trends are similar to those for the pilot furnace (see Fig. 3), except
that reactions in the commercial furnace take place nearer to the burner compared to those in the pilot furnace. These results may be explained by the
differences between the two systems in terms of furnace diameter, preheating
temperature, and oxygen content in the inlet gas: 1.2 m, 298 K, and 27 to 28%
O_2 for the pilot furnace (Table V), and 4 m, 463 K, and 50% O_2 for the commercial furnace (Table VI). For the same radial injection velocity of the dis-

Table VI. Boundary Conditions for Flash Furnace

Geometry:

$d_p = 0.36$ m
$d_s = 1$ m
$d_f = 4$ m
$L_f = 7$ m

Primary Stream (Distribution Air):

Gas flow = 307 m^3/h
O_2 content = 21%
Temperature = 300 K

Secondary Stream (Process Air):

Gas flow = 33,420 m^3/h (15 m/s linear velocity)
O_2 content = 50%
Temperature = 473 K

Concentrates in the primary stream:

Concentrate feed rate = 46,080 kg/h
Cu   (%)   28.85
Fe   (%)   28.91
S   (%)   33.48
SiO_2 (%)   5.93
Particle density = 4300 kg/m^3
Particle mean diameter = 50 μm

Matte grade: 70% Cu

Turbulent intensities: $I_p = 0.02$; $I_s = 0.1$

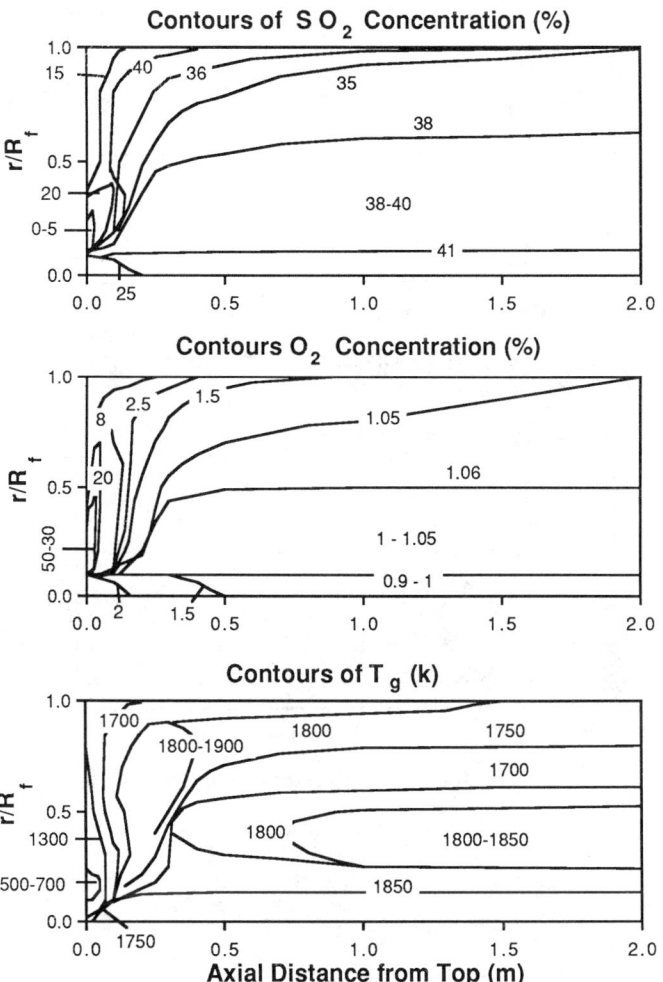

Fig. 7. Predicted contours of the SO_2 and O_2 concentrations and the gas-phase temperature in a commercial flash furnace described in Table VI.

tribution air, the larger diameter of the furnace results in a longer residence time of particles in the radial direction. Hence, particles in the larger furnace have more time to be heated up and ignite near the burner compared to those in a smaller furnace. As explained previously in conjunction with Fig. 4, reactions also occur somewhat faster with higher preheating temperature. The oxygen content in the inlet gas affects the smelting rate of the sulfide particles. To produce a given matte grade, the smelting capacity for treatment of concentrate particles increases with increasing oxygen enrichment (46); that is one of the main benefits of using oxygen-enriched process air. Figure 8 shows the effect of O_2 content in the inlet gas with the same inlet conditions listed in Table VI and for a fixed matte grade of 70%. Therefore,

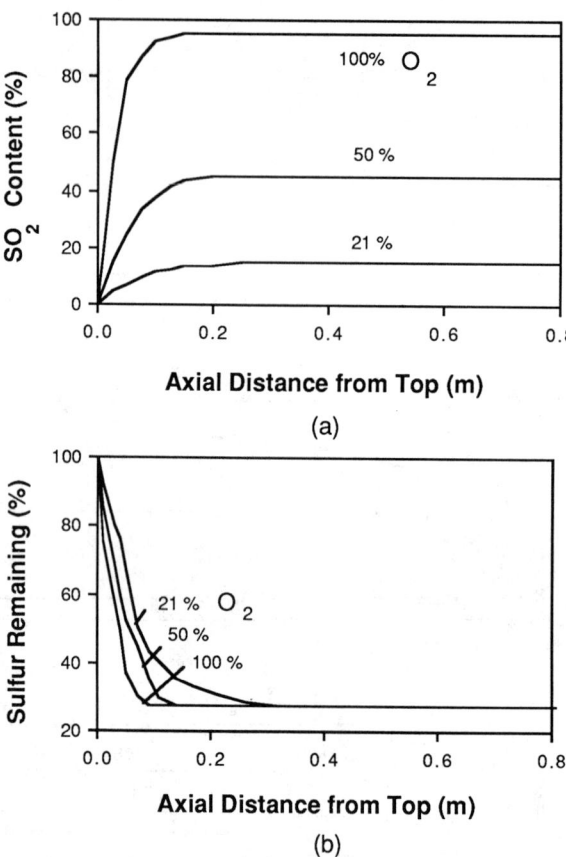

Fig. 8. Effect of O_2 content in the inlet gas:
(a) SO_2 concentration profile.
(b) Profile of sulfur remaining.

the concentrate feed rate would be 19,600 kg/h for 21% O_2 content and 92,000 kg/h for 100% O_2 content. In this figure, the concentration of SO_2 and the amount of sulfur remaining along the centerline are plotted aggainst the axial distance from the top. It is seen that the reaction rate of sulfide particles increases as the oxygen content increases. Ignition takes place nearer to the burner with higher oxygen content.

Although not illustrated, the computed results obtained by varying the matte grade from 50 to 70% Cu showed that ignition takes place somewhat nearer to the burner with higher matte grade, because particles are heated up faster for the higher matte grade due to the lower solids loading.

Conclusions

The numerical predictions for the flash-smelting furnace show overall satisfactory agreement with experimental data in spite of the complex nature

828

of the system as well as the mathematical model. This leads to the conclusion that the two-equation (k-ε) turbulence model combining the effect of particles on turbulence is adequate for the modeling of the flash-smelting furnace.

Model predictions of the performance of the flash-smelting furnace point to the following conclusions:

(1) Reactions of sulfide particles are almost completed in the upper zone of the flash-furnace shaft.
(2) The double-entry burner system with radial feeding of the concentrate-laden distribution air gives better performance than the single-entry burner system.
(3) Radiation between particles and their surroundings is a dominant mode of the heat-transfer process occuring in the flash-smelting furnace.
(4) Model predictions indicate that the overall performance of the flash-smelting furnace is affected by inlet conditions such as preheating temperature, oxygen content, solids loading, and furnace diameter.

Acknowledgment

The authors wish to express their appreciation to Dr. J. Asteljoki, Outokumpu Metallurgical Research Center, Pori, Finland for providing the pilot-plant data used in this work.

This work was supported in part by the National Science Foundation under Grant No. CPE-8204280 and by the Department of the Interior's Mineral Institute program administered by the Bureau of Mines through the Generic Mineral Technology Center for Pyrometallurgy under allotment grant No. G1125129.

Nomenclature

A_p	projected area of a particle
b	backward scattering component, defined in Eq. (63)
C_1, C_2, C_3	constants in turbulence model, Eqs. (8) and (11)
C_D	drag coefficient, defined in Eq. (19)
C_p	specific heat capacity
d_f	diameter of the furnace
d_p	diameter of the primary stream or particle diameter
d_s	diameter of the secondary stream
D	diffusivity
D_p^t	turbulent particle diffusivity
E	activation energy
f	forward scattering component, defined in Eq. (64)
$\vec{F}$	radiative flux sum vector, Eq. (68)
g	gravitational acceleration
h	enthalpy
H_m	rate of heat loss due to melting
H_r	rate of heat production by reaction
H_v	rate of heat loss due to volatilization

I	turbulent intensity or radiative flux
k	turbulent kinetic energy
k_g	gas-phase heat conductivity
k_m	mass-transfer coefficient
k_o	pre-exponential factor
K	radiative coefficient
ℓ_m	mean beam length
L_f	furnace length
m_j	mass fraction of gas species j
m_p	mass of a particle
n	particle number density
Nu	Nusselt number, defined in Eq. (30)
p	pressure
Pr	Prandtl number, defined in Eq. (31)
q_{rp}	radiative heat-transfer rate for the particle phase
$\vec{Q}$	net radiative heat flux, Eq. (87)
Q_p	rate of heat loss due to gas-phase convection
Q_{rg}	volumetric heat-transfer rate by gas-phase radiation
r	radial distance from the axis of symmetry
R	universal gas constant
Re	Reynolds number
R_f	furnace radius
s	sidewise scattering component, defined in Eq. (65)
S	source or sink term in conservation equations
Sc	Schmidt number
Sh	Sherwood number
t	time
t_t	time scale of turbulence
T	temperature
u	axial velocity
v	radial velocity
$\vec{V}$	velocity vector
w	tangential velocity, Eq. (9)
w_k	mass of gas species k
x	axial distance from the burner exit
X	overall fraction of sulfur removed

Greek Symbols

β	scattering angle
Γ_e	effective transport exchange coefficient

ε	dissipation rate of turbulent kinetic energy or emissivity
η_a, η_s	absorption and scattering efficiencies, respectively
λ	wavelength of radiation
μ	viscosity
ν	kinematic viscosity
ρ	density or reflectivity
ρ_{bp}	bulk particle density (mass of particles per volume)
σ_ϕ	Prandtl-Schmidt numbers for ϕ
τ	shear stress
τ_p	particle relaxation time
ϕ	general dependent variable
ω_o	albedo, defined in Eq. (66)

Subscripts

a	absorption
c	centerline
e	effective value
f	furnace
g	gas
l	laminar
o	inlet
p	primary stream or particle
s	secondary stream or scattering
t	turbulent
w	wall

Superscripts

t	turbulent
'	fluctuation component
+	positive direction on a major axis
−	negative direction on a major axis

Overlines

| − | time-averaged |
| = | second-order tensor |

References

1. N. J. Themelis, J. K. Makinen, and N. D. H. Munroe, "Rate Phenomena in the Outokumpu Flash Smelting Reaction Shaft," The Symposium on "Physical Chemistry of Extractive Metallurgy," eds. V. Kudryk and U. K. Rao (Warrendale, PA: TMS, 1985), 289-309.

831

2. Y. H. Kim and N. J. Themelis, "Effect of Phase Transformation and Parti-
   cle Fragmentation on the Flash Reaction of Complex Metal Sulfides," The
   Reinhardt Schuhmann International Symposium on Innovative Technology and
   Reactor Design in Extraction Metallurgy, eds. D. R. Gaskell, J. P. Hager,
   J. E. Hoffmann, and P. J. Mackey (Warrendale, PA: TMS, 1986), 349-369.

3. Y. Fukunaka, S. Nakashita, Z. Asaki, and Y. Kondo, "A Modeling Study on
   the Pyrite Smelting Process," World Mining and Metals Technology, vol. 1,
   ed. A. Weiss (New York, AIME, 1976), 481-504.

4. S. Ruottu, "The Description of a Mathematical Model for the Flash Smelting of Cu Concentrate," Combustion and Flame, 34 (1979), 1-11.

5. B. E. Launder and D. B. Spalding, Mathematical Models of Turbulence (Academic Press, London, 1972).

6. S. E. Elghobashi and W. M. Pun, "A Theoretical and Experimental Study of
   Turbulent Diffusion Flames in Cylindrical Furnaces," Fifteenth Symposium
   (International) on Combustion (The Combustion Institute, 1974), 1353-
   1364.

7. H. Schlichting, Boundary-Layer Theory (McGraw-Hill Inc., New York, 1979),
   473-480, 578-595.

8. L. D. Smoot and P. J. Smith, Coal Combustion and Gasification (Plenum
   Press, New York, 1985), 245-264, 349-371.

9. L. D. Smoot and D. T. Pratt, Pulverized Coal Combustion and Gasification
   (Plenum Press, New York, 1979), 57-64, 83-104, 217-231.

10. B. E. Launder and D. B. Spalding, "The Numerical Computation of Turbulent
    Flows," Computer Methods in Applied Mechanics and Engineering, 3 (1974),
    269-289.

11. E. K. Melville and N. C. Bray, "A Model of the Two-Phase Turbulent Jet,"
    Int. J. Heat Mass Transfer, 22 (1979), 647-656.

12. C. T. Crowe, M. P. Sharma, and D. E. Stock, "The Particle-Source-in-Cell
    (PSI-CELL) Model for Gas-Droplet Flows," Journal of Fluid Engineering,
    Trans. (ASME, 1977), 325-332.

13. L. D. Smoot and P. J. Smith, User's Manual for a Computer Program for
    2-Dimensional Coal Gasification or Combustion (PCGC-2) (Combustion Labor-
    atory, Brigham Young University, 1983).

14. P. J. Smith and L. D. Smoot, "One-Dimensional Model for Pulverized Coal
    Combustion and Gasification," Combustion and Flame, 23 (1980), 17-31.

15. A. S. Abbas, S. S. Koussa, and F. C. Lockwood, "The Prediction of the
    Particle Laden Gas Flows," Eighteenth Symposium (International) on Com-
    bustion (The Combustion Institute, 1980), 1427-1437.

16. F. C. Lockwood, A. P. Salooja, and S. A. Syed, "A Prediction Method for
    Coal-Fired Furnace," Combustion and Flame, 1980, 38 (1980), 1-15.

17. P. J. Smith, T. H. Fletcher, and L. D. Smoot, "Model for Pulverized Coal-
    Fired Reactors," Eighteenth Symposium (International) on Combustion (The
    Combustion Institute, 1980), 1285-1293.

18. T. H. Fletcher, "A Two-Dimensional Model for Coal Gasification and Combustion" (Ph.D. Dissertation, Brigham Young University, 1983).

19. S. E. Elghobashi and T. W. Abou-Arab, "A Two-Equation Turbulence Model for Two-Phase Flows," Phys. Fluids, 26 (1983), 931-938.

20. S. E. Elghobashi, T. W. Abou-Arab, M. Rizk, and A. Mostafa, "Prediction of the particle-Laden Jet with a Two-Equation Turbulence Model," Int. J. Multiphase F, 10 (1984), 697-710.

21. G. P. Lilly, "Effect of Particle Size on Particle Eddy Diffusivity," Ind. Eng. Chem. Fundam., 12 (1983), 268-275.

22. P. C. Chaubal and H. Y. Sohn, "Intrinsic Kinetics of the Oxidation of Chalcopyrite Partaicles under Isothermal and Nonisothermal Conditions," Metall. Trans. B, 17B (1986), 51-60.

23. P. C. Chaubal and H. Y. Sohn, "Combustion and Ignition of Chalcopyrite Particles under Suspension Smelting Conditions," Metall. Trans. B (submitted, 1987).

24. J. E. Dutrizac, "Reactions in Cubanite and Chalcopyrite," Can. Mineral., 14 (1976), 172-181.

25. R. C. Weast, Handbook of Physics and Chemistry, 60th edition (CRC Press, Boca Raton, 1980), B51-B144.

26. M. Nagamori and P. J. Mackey, "Thermodynamics of Copper Matte Converting, Part I: Fundamentals of Noranda Process," Metall. Trans. B, 9B (1978), 255-265.

27. F. R. A. Jorgensen, "Single Particle Combustion of Chalcopyrite," Proc. Australas Inst. Min. Metall., 288 (1983). 37-46.

28. R. B. Bird, W. E. Stewart, and E. N. Lightfoot, Transport Phenomena (John Wiley & Sons, Inc., N.Y., 1960), 19-25, 249-260, 504-513.

29. J. Szekely, J. W. Evans, and H. Y. Sohn, Gas-Solid Reactions (Academic Press, New York, 1976), 10-20.

30. D. K. Edwards, Radiation Heat Transfer Notes (Hemisphere Publishing Co., N.Y., 1981).

31. R. Siegel and J. R. Howell, Thermal Radiation Heat Transfer, 2nd edition (Hemisphere Publishing Co., N.Y., 1981).

32. A. F. Sarofim and H. C. Hottel, "Radiative Transfer in Combustion Chambers: Influence of Alternative Fuels" (paper presented at the Sixth International Heat Transfer Conference, Toronto, Canada, 1978).

33. A. M. Godridge and A. W. Read, "Combustion and Heat Transfer in Large Boiler Furnaces," Prog. Energy Combustion Sci., 2 (1976), 83-95.

34. A. D. Gosman and F. C. Lockwood, "Incorporation of a Flux Model for Radiation into a Finite-Difference Procedure for Surface Calculations," Fourteenth Symposium (International) on Combustion (The Combustion Institute, 1972), 661-671.

35. S. A. V. Fields, "A Mathematical Model for Radiation Heat Transfer during Combustion of Pulverized Coal in an Absorbing, Emitting and Anisotropic Scattering Medium" (Ph.D. Disseertation, University of Utah, 1981).

36. C.-M. Chu and S. W. Churchill, "Numerical Solution of Problems in Multiple Scattering of Electromagnetic Radiation," Journal of Physical Chemistry, 59 (1955), 855-863.

37. H. C. Hottel and A. F. Sarofim, Radiative Heat Transfer (McGraw-Hill Book Co., N.Y., 1967), 378-437.

38. S. H. Chan and C. L. Tien, "Infrared Radiation Properties of Sulfur Dioxide," Transactions of the ASME, May 1971, 172-178.

39. G. Mie, "Optics of Turbid Media," Ann. Phys., 25 (1908), 377-445.

40. Y. B. Hahn and H. Y. Sohn, "Computation of the Flow Field in a Recirculating Turbulent Gas Jet," Chem. Eng. Commun. (accepted, 1986).

41. Y. B. Hahn and H. Y. Sohn, "Prediction of the Behavior of a Particle-Laden Gas Jet as Related to the Flash-Smelting Process," The Reinhardt Schuhmann International Symposium on Innovative Technology and Reactor Design in Extractive Metallurgy, eds. D. R. Gaskell, J. P. Hager, J. E. Hoffmann, and P. J. Mackey (Warrendale, PA: TMS, 1986), 469-499.

42. E. E. Khalil, D. B. Spalding and J. H. Whitelaw, "The Calculation of Local Flow Properties in Two-Dimensional Furnaces," Int. J. Heat Mass Transfer, 18 (1975), 775-790.

43. P. J. Smith, "Theoretical Modeling of Coal and Gas Fired Turbulent Combustion or Gasification" (Ph.D. Dissertation, Brigham Young University, 1979).

44. A. D. Gosman and W. M. Pun, Lecture Notes for Course Entitled "Calculation of Recirculating Flows" (Imperial College, London, 1973).

45. S. V. Patankar, "Numerical Prediction of Three-Dimensional Flows," in Studies in Convection, vol. 1, ed. B. E. Launder (Academic Press, London, 1975), 1-78.

46. D. W. Rodolff, Y. E. Anjala, and P. T. Hanniala, "Review of Flash Smelting and Flash Converting Technology" (paper presented at the 115th AIME Annual Meeting, New Orleans, Louisiana, 1986).

47. J. Asteljoki, personal communication, Outokumpu Metallurgical Research Center, Pori, Finland, February 1987.

48. J. Makinen, personal communication, Outokumpu Oy Metallurgical Division, Harjavalta, Finland, February, 1987.

MATHEMATICAL MODELING OF FLASH CONVERTING OF COPPER MATTE

Q. Jiao, L. Wu and N.J. Themelis

Henry Krumb School of Mines
Columbia University
New York NY   10027

ABSTRACT

A two-dimensional mathematical model was developed to represent the reaction and transport phenomena in the flash converting of finely ground matte particles. In the calculations involving gas phase and particle phases, two basically distinctive methods, Eulerian and Lagrangian, were employed, respectively. The turbulence effect was accounted for by means of the k-ε model. The present model incorporated the information on the kinetics and reaction rate of the oxidation of suspended single matte particles and of entrained copper concentrate and matte particulates in oxygen enriched air streams, developed from the earlier work conducted in the pyrometallurgical laboratory of Columbia University and from the work of others. The equations involved in this model were solved numerically, on the basis of the dimensions and the parameters encountered in the operation of the Outokumpu flash smelting furnace.

Introduction

The flash converting process for producing blister copper from copper matte particles, under development by Kennecott Minerals and Outokumpu Oy, is a potential alternative to the conventional Peirce-Smith bath converting process and has been claimed to offer better environmental control and other advantages (1,2). In flash converting, the finely divided copper matte and flux particles are dispersed into an oxygen-enriched air stream which is injected into the reaction shaft of the furnace. As the gas-solid dispersion enters the high temperature enclosure, the particles are heated rapidly to the ignition temperature and are oxidized, first by gas-solid reaction and later, after melting, by gas-liquid phase reaction. Under these intense conditions of oxidation, nearly total sulfur removal can be effected in less than one second.

Evidently, the rate of oxidation of copper matte particles in a flash converting furnace is strongly dependent on the gas and particle velocity distributions, the gas and particle temperature profiles, and the oxygen concentration through the reaction shaft; in short, on the transport phenomena occurring within this space. Therefore, mathematical simulation of the fluid flow, and heat and mass transfer phenomena occurring in the shaft can provide an insight on the flash converting system and be a useful tool in the design of industrial flash converting furnaces.

Although there have been numerous studies of the kinetics and rate phenomena on the oxidation of synthetic copper and iron sulfides (3,4,5,6), very little work has been done on the oxidation of copper matte particles. Recently, experiments have been conducted at Columbia University on the oxidation of compacted matte pellets suspended in an oxygen-enriched air stream (7,8) and of fine matte particles entrained in an oxygen-nitrogen stream (9). In the first case, matte pellets were found to oxidize at rates in between sulfides (8). Also, the experimental flash converting tests have shown that matte particles oxidize at rates lower than, but in the same range as, copper concentrate particles (9). These results indicate that copper matte particles oxidize similarly to copper concentrates, under similar reaction conditions; and that information developed in investigations of the flash smelting process can be utilized to provide an insight in the flash converting process.

Chaubal and Sohn (10), Rao and Abraham (4), and Lenchev (11) studied experimentally the kinetics of oxidation of chalcopyrite and other copper sulfides and determined the activation energy of the oxidation reactions at different regions.

Jorgensen and Segnit (12), and Jorgensen (13,14) investigated extensively the thermal and reaction behavior of copper and other metal sulfides
in a laboratory flash furnace. These authors have provided some very useful information on ignition temperatures and on the degree of sulfur
removal as a function of residence time of the particles in the reaction
zone.

The information provided by the above authors and by others on the
thermodynamics and kinetics of the oxidation of copper and iron sulfides
has made possible the construction of mathematical models which attempt to
simulate conditions and phenomena in flash reactors.

The first model of flash smelting was published by Ruottu (15), who
developed a two-dimensional model to account for velocity, temperature and
concentration profiles in the reaction shaft. However, the assumptions
made and the form of presenting the results of the simulation have not lent
themselves to adoption and use of this model in the design and operation
of flash smelting furnaces.

Kim and Themelis (16,17) followed a simpler approach and developed a
uni-dimensional mathematical model to simulate industrial copper flash
smelting furnaces. This model incorporated the effect of particle fragmentation and assumed average gas and particles temperature and
compositions along the furnace axial direction. However, the omission of
the effect of turbulence and of the change of parameters in the radial
direction leaves room for further improvement of the model.

Thus, it is the objective of the present work to expand on earlier
work by Kim and Themelis (16,17), and Munroe and Themelis (9) and on
relevant publications by others, to describe by means of a two-dimensional
model the flash converting of matte particles to blister copper.

In the model developed in this work, the turbulence effect is accounted
for by means of the k-ε model proposed by Launder and Spalding (18,19), and
by Patankar (20), while the motion of the particles in the stream is
treated by a method developed by Crowe et al (21). The two-dimensional
model developed is based on the dimensions and principal parameters
encountered in the operation of the Outokumpu flash smelting furnace.

MATHEMATICAL FORMULATION

a) Assumptions made

Figure 1 is the schematic diagram of the reaction shaft of a flash
converting furnace, of similar cylindrical shape and size as an Outokumpu
flash smelting furnace. A burner is located at the center of the furnace
and is used for dispersing the matte particles into the oxidizing gas and
injecting the gas-solid stream in the reaction shaft.

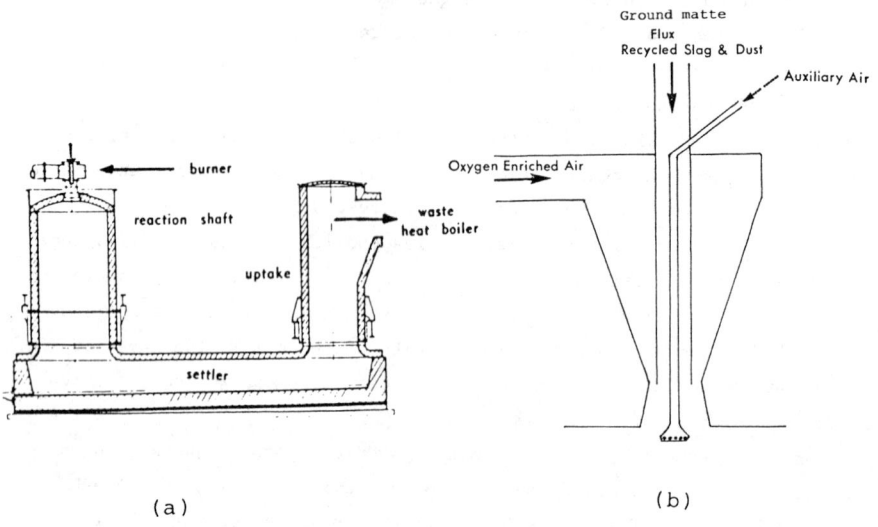

Ground matte
Flux
Recycled Slag & Dust

Auxiliary Air

Oxygen Enriched Air

burner

reaction shaft

waste
heat boiler

uptake

settler

(a)   (b)

Figure 1.   Schematic diagram of the reaction shaft of an Outokumpu flash
            converting furnace (a) and of burner (b).

The following assumptions were made in order to simplify the mathematical formulation of the flash converting process:

i) The furnace is axisymmetric and the tangential gas and particle
velocities are assumed to be negligible; i.e., it is assumed that the flow
is irrotational and that a two-dimensional model can be used to represent
adequately the two-phase gas-particle flow.

ii) The solid feed to the reactor consists of a number of discrete sets
of particle sizes each of which has uniform physical and chemical properties.

iii) Interactions or collisions among particles are negligible except
for radiation heat transfer.

iv) The temperature within a particle is uniform at any given time.

v) The particles are assumed to be spherical and of constant size
during their residence time in the furnace shaft.

vi) The gas phase is in local thermodynamic equilibrium so that local
gas composition can be obtained from equilibrium considerations.

With regard to assumption (ii), selection of a large number of sets
of particle sizes will result in a more accurate calculation of particle
trajectories and degree of reaction but will require much computing time
and is not practical. Three distinct particle sizes were used in this model.

Collisions between particles are not considered to be important (assumption iii) because the distance between adjacent particles has been calculated to be one order of magnitude larger than the particle size (22).

Assumption (iv) can be easily proven by a calculation of the rate of unsteady-state conduction through the fine particles, as compared to the rates of heat generation and heat loss by convection and radiation. It has been shown by some authors (23) that the temperature difference between the center and the surface of the particle is small enough to be negligible.

Assumption (v) would not be justified in the case of chalcopyrite oxidation where large pressures are built up within the reacting particle and result in particle fragmentation; this has not as yet been observed in the oxidation of matte particles but the model can be readily modified when quantitative information on particle fragmentation is obtained.

Assumption (vi) appears to be justified because, at flash converting temperatures, equilibrium among gas species is achieved in a few milliseconds.

b) Components of the model

During flash converting, there are many simultaneous physical and chemical phenomena, such as the motion of the gas and particles, mass transfer between gas and particles, heat transfer to and from the particles, melting and volatilization, oxidation of matte, chemical reaction among gas species, and so forth. A successful mathematical model must describe all the important phenomena, and usually consists of a number of submodels for accounting different physical and chemical phenomena. The present model encompasses four submodels, termed GFLOW, PREACT, GTHERM AND HTRAD; these account respectively for the gas mean flow (MFLOW) and turbulence (TURBU), the particle trajectory (PTRAJ) and reaction rate (PCHEM), the gas thermodynamic equilibria (GTHERM), and the radiation heat transfer (HTRAD). In solving the constituent equations in these submodels, two basically distinctive approaches were employed (24):

In calculations involving gas phase, the Eulerian method was used, whereby the variation of the fluid field with time was determined at fixed coordinates in space. On the other hand, for the computation of particle trajectories and other phenomena, the Lagrangian method was more convenient and was adopted; individual particles were followed through space and their velocities, temperatures, positions, etc. were recorded with time. Finally, the interactions between particles and gas were incorporated by considering the contributions of mass, momentum and heat transfer to or from the particles as source or sink terms in the corresponding equations for the gas species.

The overall organization is shown in Figure 2, which illustrates the output for each submodel and the linkage between the Eulerian and the Lagrangian approaches.

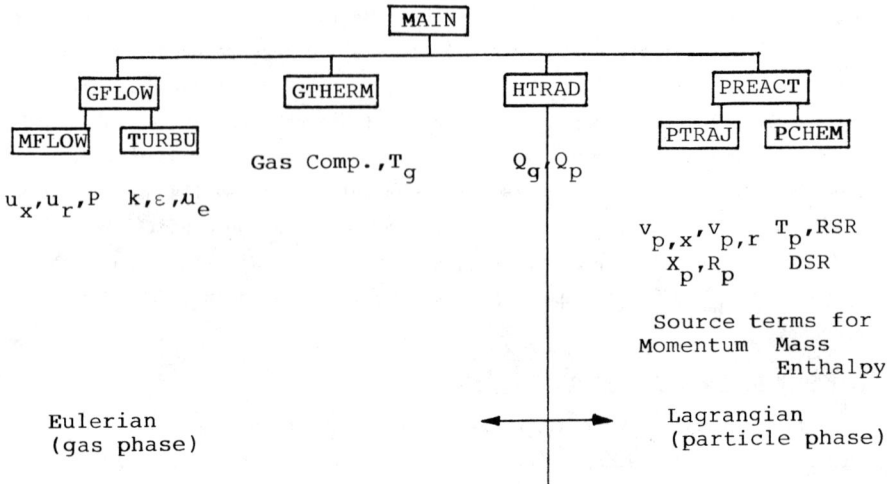

Figure 2. Organization of model

c) Gas phase fluid flow and turbulence

The conservation equations for mass, and for axial and radial momentum in two-dimensional compressible fluid flow can be expressed in cylindrical coordinates as follows (21):

1) Equation of continuity

$$\frac{\partial(\rho U_x)}{\partial x} + \frac{1}{r}\frac{\partial(\rho_r U_r r)}{\partial r} = S_p^m \tag{1}$$

2) Equation of momentum in axial direction

$$\frac{\partial}{\partial x}(\rho U_x U_x) + \frac{1}{r}\frac{\partial}{\partial r}(\rho U_r U_x) = -\frac{\partial P}{\partial x} + \frac{2\partial}{\partial x}(\mu_e\frac{\partial U_x}{\partial x}) + \frac{1}{r}\frac{\partial}{\partial r}\left[r\mu_e(\frac{\partial U_x}{\partial r}+\frac{\partial U_r}{\partial x})\right] + S_p^{U_x} \tag{2}$$

3) Equation of momentum in radial direction

$$\frac{\partial}{\partial x}(\rho U_x U_r) + \frac{\partial}{r\partial r}(\rho U_r U_r) = -\frac{\partial P}{\partial r} + \frac{2\partial}{r\partial r}(r\mu_e\frac{\partial U_r}{\partial r}) + \frac{\partial}{\partial x}\left[\mu_e(\frac{\partial U_r}{\partial x}+\frac{\partial U_x}{\partial r})\right] - \mu_e\frac{2U_r}{r^2} + S_p^{U_r} \tag{3}$$

where μ_e stands for the effective viscosity and is equal to the sum of the molecular and turbulent viscosities. The turbulent viscosity, μ_t, is expressed as a function of the turbulent energy, k, and its dissipation rate,

ϵ (18):

$$\mu_t = C_\mu \rho k^2 / \epsilon \qquad (4)$$

where C_μ is a constant and ϵ is dissipation rate of k.

The source term, S_p^m, in the mass conservation equation (equation 1) represents the effect of the chemical reaction which consumes oxygen and produces sulfur oxides.

The momentum source terms, S_p^{Ux}, S_p^{Ur}, account for momentum transfer to and from the particles and are expressed as follows:

$$S_p^{Ux} = \frac{1}{V_k} \sum_i \sum_j \{ (\dot{m}_{ij} v_{x,ij})_{out} - (\dot{m}_{ij} v_{x,ij})_{in} \} k \qquad (5)$$

$$S_p^{Ur} = \frac{1}{V_k} \sum_i \sum_j \{ (\dot{m}_{ij} v_{r,ij})_{out} - (\dot{m}_{ij} v_{r,ij})_{in} \} k \qquad (6)$$

where V_k, v, m and n represent, respectively, the volume of the control volume k, the particle velocity, the mass flow rate, and the number of particles of a certain size entering into the control volume from a particular entry port; the subscripts i, j, denote the different particle sizes and entry ports.

In its present form, the model encompasses 3 ports, evenly distributed at the entrance to the reaction shaft so that the first port is located at the burner outlet and the other two at 0.1 meters apart along the axis of the shaft. Also, the model considers three particle sizes, namely 80, 50 and 30 microns in diameter.

The kinetic energy, k, and the dissipation rate, ϵ, were calculated by the following two equations, derived from the Navier-Stokes equations for turbulent flow (18, 19):

$$\frac{\partial}{\partial x}(\rho U_x k) + \frac{1}{r}\frac{\partial}{\partial r}(\rho r U_r k) = \frac{\partial}{\partial x}(\frac{\mu_e}{\sigma_k}\frac{\partial k}{\partial x}) + \frac{1}{r}\frac{\partial}{\partial r}(r\frac{\mu_e}{\sigma_k}\frac{\partial k}{\partial r}) + G - \rho\epsilon \qquad (7)$$

$$\frac{\partial}{\partial x}(\rho U_x \epsilon) + \frac{1}{r}\frac{\partial}{\partial r}(r\rho U_r \epsilon) = \frac{\partial}{\partial x}(\frac{\mu_e}{\sigma_\epsilon}\frac{\partial\epsilon}{\partial x}) + \frac{1}{r}\frac{\partial}{\partial r}(r\frac{\mu_e}{\sigma_\epsilon}\frac{\partial\epsilon}{\partial r}) + \frac{\epsilon}{k}(C_1 G - C_2 \rho\epsilon) \qquad (8)$$

Where $G = \mu_t \{2[(\frac{\partial U_x}{\partial x})^2 + (\frac{\partial U_r}{\partial r})^2 + (\frac{U_r}{r})^2] + (\frac{\partial U_x}{\partial x} + \frac{\partial U_r}{\partial r})^2\}$; and

σ_k and σ_ϵ are termed Prandtl numbers for k and ϵ and their values are shown in Table 1, together with those of constants C_1, C_2, and C_μ.

Table 1 Values of various constants (18,19)

σ_k	σ_ϵ	C_μ	C_1	C_2
1.00	1.30	0.09	1.44	1.92

It can be seen in the above formulation that in order to compute the axial and radial mean velocities, two additional equations, for k and ϵ, were introduced. This method of computing turbulent viscosity is called the k-ε model and was developed by Launder and Spalding (18,19).

d) Particle trajectory, temperature and rate of oxidation

The equation of motion of a single particle in the Lagrangian frame can be expressed, on the basis of Newton's second law, as follows:

$$m_p \frac{dV}{dt} = \frac{1}{2} C_D \rho A_p \mid \vec{U} - \vec{v} \mid (\vec{U} - \vec{v}) + m_p \vec{g} \tag{9}$$

The drag coefficient, C_D, can be calculated from the following equation (21):

$$C_D = (24/Re) \ (1+0.15Re^{0.687}) \tag{10}$$

where the Reynolds number, Re, is expressed as a function of the slip velocity:

$$Re = \rho \mid \vec{U}-\vec{v} \mid d_p/\mu_g \tag{11}$$

Equation B is applicable for Reynolds numbers up to 1000 (21).

In the vicinity of the furnace inlet, the gas velocity is much higher than that of the particles and the Reynolds number is also high; e.g., a calculation by Kim (16) showed that under these conditions the value of Re can be as high as 70. Therefore, the Stokes law does not apply in this region. However, the rapid decrease of gas velocity in the entrance region, due to expansion, and the increasing particle velocity, due to the effect of both drag and gravity forces, rapidly reduce the slip velocity to the region in which the Stokes law is applicable. Apparently, equation 8 can be approximated to the Stokes equation as Re is very small.

The particle average temperature can be calculated from a heat balance equation as follows:

$$\frac{\rho_p C_{p,p}}{A_p} \frac{dT_p}{dt} = q_{cr} - h(T_p - T_g) - q_{p-w} \tag{12}$$

where ρ_p, $C_{p,p}$, A_p: particle density, heat capacity and surface area, respectively; T_p and T_g: particle and gas temperatures; q_{cr} and q_{p-w}: heat flux of chemical reaction and of radiation from the particles; h: heat transfer coefficient of convection.

The heat flux, q_{cr}, in the above equation is directly related to the rate of oxidation and the heat of reaction; the radiation heat flux, q_{p-w}, is calculated from the submodel, HTRAD. The calculation of these fluxes will be discussed later.

The reaction products were assumed, for the sake of simplicity, to be Cu_2O, Fe_3O_4, PbO, ZnO, and SO_2, regardless of temperature and oxygen potential. Also, the heat of oxidation of copper matte was taken to be the weighted average of the heats of reaction for each constituent at a mean temperature. It should be noted that the assumption of SO_2 being the only gas product formed by the oxidation of copper matte particles does not mean that there

842

will not be other sulfur oxides in the gas stream because the final gas composition is determined by the equilibrium calculation among gas species, which is the calculation task of the next submodel.

The oxidation of the copper matte particles was assumed to occur in the following stages, as shown in Figure 3: gas solid reaction stage, when particle temperature is less than 1250 K, the melting point of the matte particles (25); gas-solid liquid stage, when T_p is between 1250 and 1850 K, the melting point of the reacted oxides; and the gas liquid stage, when $T_p > 1850$ K.

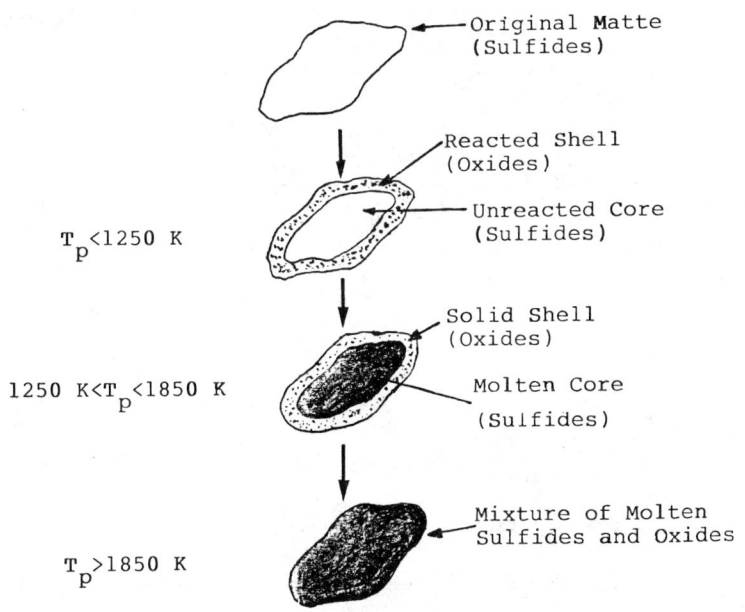

Figure 3. Proposed flash-reaction scheme of a copper matte particle.

In the gas-solid reaction stage, the oxidation rate of particles is represented by the topochemical model, i.e., the reaction is assumed to take place at a narrow interface between the reacted shell and the unreacted core of the particle; the overall reaction rate is determined by the resistances due to the boundary layer mass transfer of reactant and product gases, the reacted shell diffusion of the reactant and product gases, and the interface chemical reaction (26).

By considering the oxidation of the iron and copper sulfides in the matte particles is practically thermodynamically irreversible at converting temperatures, the overall reaction rate (i.e., the rate of sulfur removal) and the variation of the radius of the particle unreacted core can be computed from the following two equations (7):

$$N_S = \frac{A_p \, k_{ov} C_{O_2}}{3/2 \, X_{s,\,Cu_2S} + 5/2 \, X_{s,\,FeS} + 3/2 \, X_{s,\,PbS} + 3/2 \, X_{s,\,ZnS}} \tag{13}$$

$$\frac{dr}{dt} = \frac{M_s \, r_o^2}{r^2} \cdot \frac{k_{ov} \, C_{O_2}}{3/2 \, (X_{s,\,Cu_2S} + X_{s,\,PbS} + X_{s,\,ZnS}) + 5/2 \, X_{s,FeS}} \tag{14}$$

where $k_{ov} = 1/(\dfrac{1}{k_d} + \dfrac{r_o(r_o-r)}{r \, D_e} + \dfrac{r_o^2}{r^2 k_r})$

N_s: the rate of sulfur removal, mole/sec;

$X_{s,Cu_2S}, X_{s,FeS}$: the mole fraction of sulfur in Cu_2S and FeS, etc.

C_{O_2}: concentration of oxygen, mole per cubic cm;

r, r_o: radius of the unreacted core of particle and of the original particle, respectively;

M_s: molecular weight of sulfur;

k_d, D_e, k_r: respectively mass transfer coefficient, effective diffusivity of oxygen in porous reacted shell of particle, and rate constant of the interface chemical reaction.

The degree of sulfur removal (DSR) at any time can be obtained by integrating equation 14 numerically and the resulting DSR can be expressed as follows:

$$DSR = (r_o^3 - r^3)/r_o^3 \tag{15}$$

The value of the activation energy for the oxidation of the copper matte particles was taken as 65 kJ/mole (4,7,10) and the respective value for the Arrhenius constant are 20000.

The rate of the boundary layer mass transfer was calculated from the Ranz Marshall equation:

$$Sh = 2 + 0.6Re^{0.5}Sc^{0.33} \tag{16}$$

where Sh, Re, Sc are the Sherwood, Reynolds and Schmidt numbers, respectively.

The effective diffusivity of oxygen through the reacted shell is defined as a fraction of the molecular diffusivity of oxygen in the gas mixture:

$$D_e = f_p D_o \tag{17}$$

The proportionality, f_p, represents the combined effect of the shape, size, interconnection, and volumetric fraction of the pores, and is called the pore factor. Some authors define the pore factor as the ratio of particle porosity to tortuosity but unless the latter term can be defined on the basis of other known properties of the solid, equation (17) is more realistic.

When the particle unreacted core becomes molten and the outside reacted shall maintains solid state, the reaction mechanism is not substantially different from the gas-solid reaction, thus, the same equations as above were used to account for reaction rate in this stage.

The mechanism of gas-molten particle reaction is different from that of gas-solid reaction. The rate of chemical reaction at the gas-liquid interface is very fast, and again it is not the rate determining step. Hence, the overall reaction rate is controlled either by oxygen diffusion through the boundary layer or by sulfur transfer from the bulk to the surface of the droplet.

There are two ways to transfer sulfur from the core to the surface of the droplet: by molecular diffusion and by fluid circulation. This circulation is caused by the shear force imposed by the enveloping gas flow. According to Levich (27,28), the circulation velocity is proportional to the gas/liquid viscosity ratio and when the two viscosities are of the same order of magnitude, the circulation velocity will be of the same order as the bulk flow of the gas.

Since, in this case, the gas viscosity is two to three orders of magnitude smaller than that of the molten particle, the circulation velocity may be small. Therefore, diffusion of sulfur atoms from the core to the surface of the droplet could be rate controlling. However, it should be noted that Levich's calculation was based on the assumption that the Stokes law is applicable and that this assumption is not correct at the entry region to the reaction shaft, where the Reynolds number is much higher than that required for application of the Stokes law. Therefore, pending further experimental evidence, it will be assumed, as has been done by some other authors (25), that there is an abundance of sulfur at the droplet surface and that the rate controlling step is the transfer of oxygen through the boundary layer.

At higher particle temperature (T_p > 1800 K), the effect of volatilization of copper and copper compounds on particle temperature and reaction rate may be significant (14,25) and must be taken into account. By assuming that the particle boundary layer mass transfer is the rate controlling step (because of the high temperature), the rate of mass transfer of CuO (s) from the particle to the surrounding gas media can be calculated by the following equation:

$$dm/dt = A_p k_d (\rho_e - \rho_\infty)$$ (18)

where k_d is mass transfer coefficient and can be obtained from Eq. 16,
in which the diffusivity of copper monoxide gas in gas mixture can be calculated by the modified correlation of Gilliland, proposed by Andrussow (26);

ρ_e, ρ_∞: mass concentration of CuS (g) at equilibrium and in the bulk gas, respectively.

e) <u>Thermochemical calculations in gas phase</u>

The thermochemical calculation of species existing in the gas phase is done by means of the submodel GTHERM, which consists of two parts: the computation of the equilibrium gas composition and that of gas temperature distribution.

The concentration of each gas species at a particular location is obtained from the mass balance equation for that species:

$$\frac{\partial}{\partial x}(\rho U_x X_i) + \frac{1}{r}\frac{\partial}{\partial r}(r\rho U_r X_i) = \frac{\partial}{\partial x}(\frac{\mu e}{\sigma_{x_i}}\frac{\partial X_i}{\partial x}) + \frac{1}{r}\frac{\partial}{\partial r}(r\frac{\mu e}{\sigma_{x_i}}\frac{\partial X_i}{\partial r}) + S_p^{X_i} \quad (19)$$

where σ_{x_i} is the Prandtl number and $S_p^{X_i}$ the mass source term, for species i.

The above equation expresses the fact that the rate of production or depletion of a species in the control volume is equal to the rate of convection and diffusion transfer of that species through the control volume.

In the case of the flash converting system, where ten or more gas species can co-exist, simultaneous computation of the complex partial differential equations for all the gas species would require a great amount of computing time and is not considered to be essential.

Hence, only the equations for oxygen, nitrogen, sulfur dioxide and sulfur trioxide were set up in the model (sulfur monoxide was assumed to be a minor species because of the high oxygen potential prevailing in flash converting process) and the axial and radial concentration profiles for these species were computed. The concentrations of other species were assumed to be very small and can be obtained by setting up and solving the thermodynamic equilibrium equations of the respective species at the prevailing temperature and pressure.

The average gas temperature in a control volume of the converting furnace was determined by the following thermal energy balance equation:

$$\frac{\partial}{\partial x}(\rho U_x h_g) + \frac{1}{r}\frac{\partial}{\partial r}(r\rho U_r h_g) = \frac{\partial}{\partial x}(\frac{\mu e}{\sigma_h}\frac{\partial h_g}{\partial x}) + \frac{1}{r}\frac{\partial}{\partial r}(\frac{\mu e \partial h_g}{\sigma_h \partial r}) + q_{g-w} + S_p^h \quad (20)$$

where σ_h is the Prandtl number for enthalpy h_g, and is of the order of unity; the heat flux, q_{g-w} represents the heat losses by radiation to the wall. The source term, S_p^h, represents the heat contributed by the particles.

The gas temperature in each control volume is then calculated from the enthalpy of the gas mixture.

f) Radiation heat transfer

In a flash converting furnace, radiation heat transfer plays an important role in heating particles to their ignition temperature at the inlet, and is a main cause of the heat losses of the process. Radiation takes place among gas, particles and the wall. Since the particle concentration per unit volume of gas is very low, i.e., the distance between two particles is much larger than the particle size, the radiative heat among particles is assumed to be negligible. For the same reason, plus the small size of the matte particles, the emissivity of the cloud of particles can be very small (26), hence, the radiation heat transfer between particles and the enveloping gas will not be taken into account. Therefore, only the radiation between particles and the furnace wall (enclosure) and between the gas and the wall were included in this model. In addition, the gas, particles and the wall were all assumed to be grey bodies.

The radiation heat flux between a particle and the wall enclosure can be expressed by the following equation (26):

$$q_{p-w} = \varepsilon \sigma (T_p^4 - T_w^4) \tag{21}$$

where ε is emissivity of particles having a value of 0.6 (7) and σ Stefan-Boltzman constant.

The net heat flux from the wall to the gas by radiation can be obtained from the following equation (26):

$$q_{g-w} = \varepsilon_s \sigma (\varepsilon_g T_g^4 - \alpha_{g,w} T_w^4) \tag{22}$$

where ε_s is defined as the effective emissivity of the particles; ε_g effective emissivity of the enveloping gas; and $\alpha_{g,w}$ effective absorptivity of gas. The calculation of these properties was based on the method suggested by Szekely and Themelis (26).

g) Numerical method

The equations for describing particle trajectory, temperature and reaction rate are first order, ordinary differential equations and a backward difference scheme was used to discretize them to the corresponding difference equations.

The equations of momentum, kinetic energy, energy dissipation, thermal energy, and mass for the gas phase are second order, partial differential equations and much more complex than those for particles. To simplify the conputing process and to reduce the computing time, these equations were standardized and the generalized equation can be expressed as (25):

$$\frac{\partial}{\partial x}(\rho U_x \phi) + \frac{1}{r}\frac{\partial}{\partial r}(r\rho U_r \phi) - \frac{\partial}{\partial x}(\Gamma \frac{\partial \phi}{\partial x}) - \frac{1}{r}\frac{\partial}{\partial r}(r\Gamma \frac{\partial \phi}{\partial r}) = S_p^\phi \tag{23}$$

where ϕ is the dependent variable, such as u_x, U_r, h, or x_i, and r the transport coefficient, e.g., μ_e, μ_e/σ_k, etc.

The term, S^ϕ, denotes the source of the momentum, k, etc., produced by chemical reaction, interactions with particles and/or the wall.

The expressions of the dependent variable, the transport coefficient and the source term for each of the conservation equations are given in Table 2 (25).

Table 2

Equation	Φ	Γ	S^{Φ}
Axial momentum	U_x	μ_e	$\frac{-\partial p}{\partial x} + \frac{\partial}{\partial x}(\mu_e\frac{\partial U_x}{\partial x}) + \frac{1\partial}{r\partial r}(r\mu_e\frac{\partial U_r}{\partial x}) + S_p^{U_x}$
Radial momentum	U_r	μ_e	$\frac{-\partial p}{\partial r} + \frac{\partial}{\partial x}(\mu_e\frac{\partial U_x}{\partial r}) + \frac{1\partial}{r\partial r}(r\mu_e\frac{\partial U_r}{\partial r})-2\mu_e\frac{U_r}{r^2} + S_p^{U_r}$
Turbulent kinetic energy	k	μ_e/σ_k	$G-\rho\varepsilon$, where $G = 2\mu\{(\frac{\partial U_x}{\partial x})^2 + (\frac{\partial U_r}{\partial r}^2) + (\frac{U_r}{r})^2$ $+ \frac{1}{2}(\frac{\partial U_x}{\partial x} + \frac{\partial U_r}{\partial r})^2\}$
Dissipation rate of k	ε	μ_e/σ_ε	$\frac{\varepsilon}{k}(C_1 G - C_2\rho\varepsilon)$
Thermal energy	h_g	μ_e/σ_h	$q_{g-w} + s_p^h$
Concentration of species i	X_1	μ_e/σ_{x1}	$s_p^{X_i}$

RESULTS AND DISCUSSION

Table 3 shows the assumed operating parameters and initial conditions used in the model calculations. These values were based on information provided by Outokumpu Research for the envisaged flash converting operation. The value of the initial radial velocity was calculated on the basis of the volumetric gas flowrate through the distributor at the burner end (Fig.1).

Table 3. Boundary and initial conditions used in the model calculations

Geometry:	reactor diameter:	4.4 m
	burner outer diameter:	0.6 m
	burner inner diameter:	4.4 m
	reactor length:	8.0 m
Gas:	axial velocity:	17.09 m/s
	radial velocity	8.51 m/s
	mass flowrate:	3.94 kg/s
	oxygen enrichment:	90%
	turbulence intensity:	.15
	inlet temperature:	473 K
Particle:	axial velocity:	7.76 m/s
	radial velocity:	33.3 m/s
	matte feed rate:	10.4 kg/s
	flux feed rate:	2.1 kg/s
	inlet temperature:	323 K

The above boundary and initial conditions were used in solving the system of equations presented earlier. The model results were as follows:

a) Gas velocity and turbulence kinetic energy

Figure 4 shows the change in axial gas velocity, U_x, with distance from entry point. It can be seen that, there is an initial acceleration in gas velocity near the entry point; this is attributed to the high heating rate of the inlet gas stream by mixing with the enveloping hot gas and by radiation from the reactor wall.

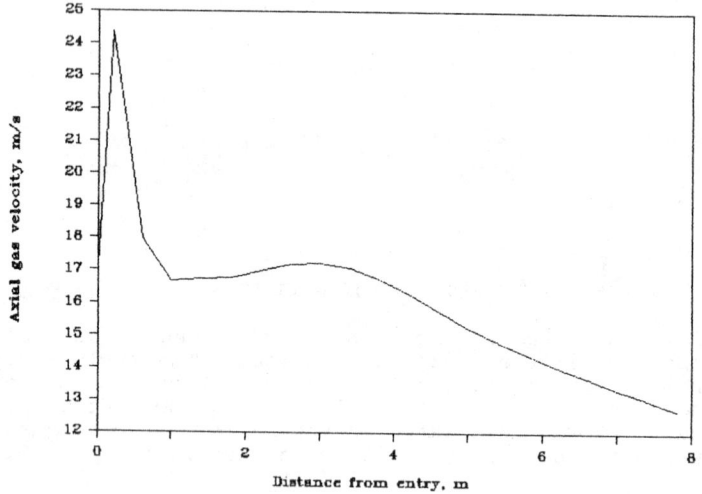

Figure 4. Axial gas velocity profile (r = 0

The gas radial velocity profiles (Figure 5) are similar to the flow development of an isothermal jet except fot the initial increase in gas velocity due to the rapid heat transfer to the gas-particles stream from the surroundings.

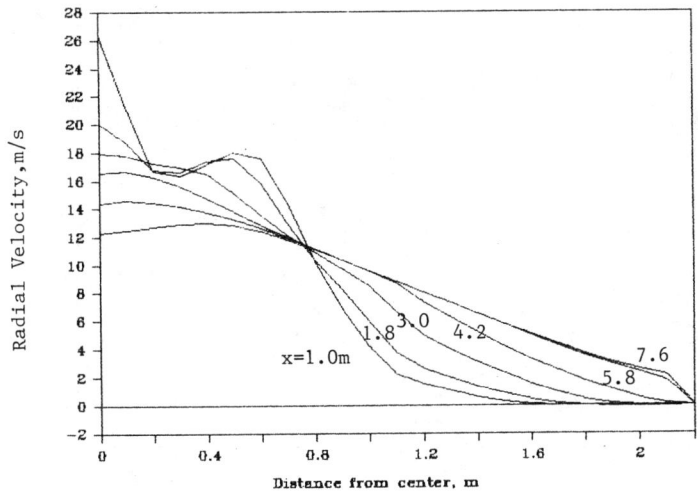

Figure 5. Radial gas velocity profiles at various locations from entry point.

Figures 6 and 7 show the axial and radial profiles of the turbulence kinetic energy, k, in the gas stream. Since turbulence is produced during the exchange of kinetic energy between adjacent fluid arrays, a high velocity difference across the adjacent fluids will result in a high turbulence. Therefore, higher values of k are attained at locations where high rates of momentum transfer prevail (Figure 7). The turbulence kinetic energy is also a function of the mean gas velocity and, as illustrated by Figure 6, it decreases with declining velocity.

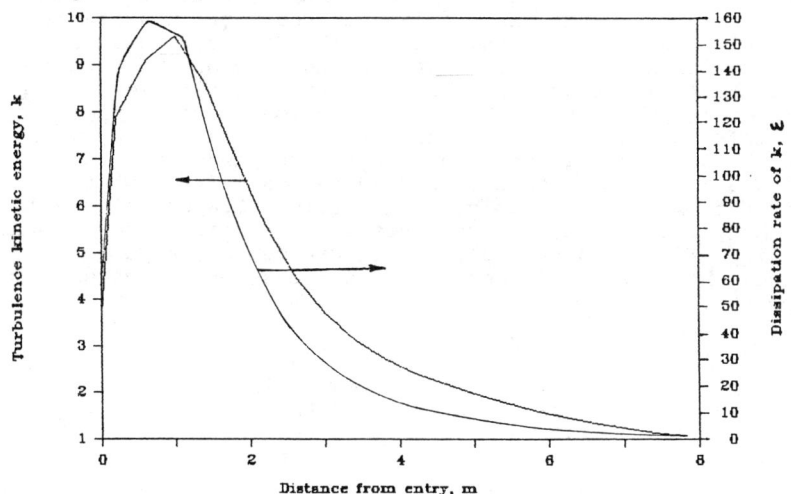

Figure 6. Axial profiles of k and ε

851

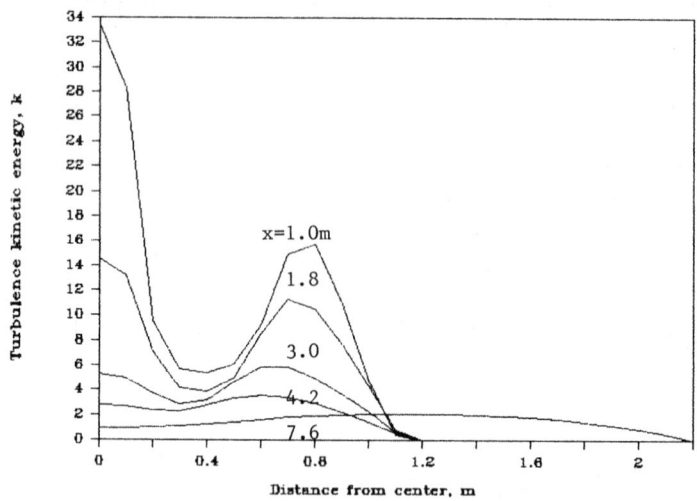

Figure 7. Radial profiles of k, at various locations from entry point.

The dissipation rate of the turbulence kinetic energy, which is also shown in Figure 6, behaves similarly to the turbulence kinetic energy: it is very high at the entrance to the reactor and declines with distance from the entry point.

b) Particle velocity and trajectory

The trajectories of particles of different sizes are shown in Figure 8, for particles entering the reactor through two entry ports: at 0.1 and 0.3 meters from center. All curves show a similar trend: initially they veer out in a radial direction, due to both the radial velocity component of the gas issuing from the distributor at the tip of the burner and to thermal expansion, and then tend to move parallel to the axis of the reactor. Under the assumption of the uniform particle initial velocity, the larger size particles, which have higher inertia, tend to move further away from the center (Figure 8). It is interesting to note that after two to three meters from the entry point, the particle trajectories move slightly toward the furnace center, due partially to the wall cooling effect and partially to oxygen depletion to form oxides, both of which result in a decrease in gas volumetric flowrate.

The fact that the matte particles accelerate to the enveloping gas velocity within a few milliseconds indicates that the particle trajectories are determined mainly by the gas axial and radial velocities, and that the distribution gas velocity does not have much influence on them.

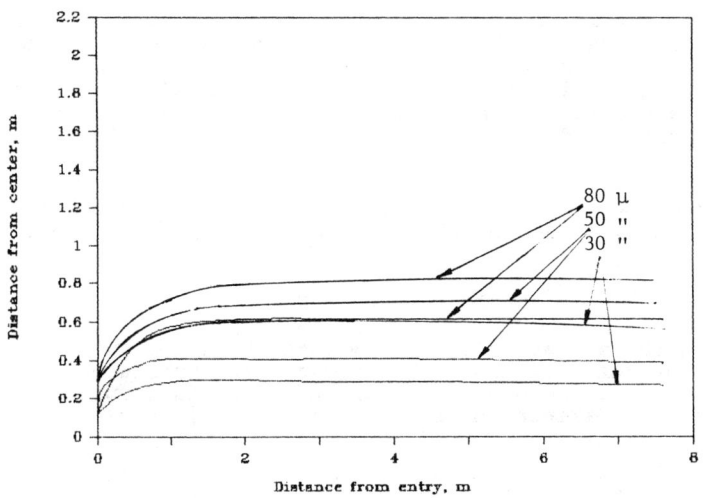

Figure 8. Particle trajectory as a function of particle size,
for particles entering from two entry ports (r=.1, .3m).

c)Particle temperature and oxidation rate

Figure 9 shows the particle temperature profile as a
function of particle size and of distance from the entry point
. As observed by other authors (13,14), the temperature of
fine particles increases faster than that of larger ones,
reaches a higher peak value, and then declines more sharply.
The corresponding fractional sulfur removal for these
particles is shown in Figure 10. The incomplete removal of
sulfur from the 0-micron particles is attributed to the
provision of inadequate excess of oxygen in the feed stream
(oxygen loading is defined as grams of oxygen input per gram
of sulfides in the feed to the reactor); the finer particles
consume most of the available oxygen, and the resulting oxygen
depletion prevents the complete converting of the coarser
particles.

CONCLUSIONS

A mathematical model was developed to represent the rate
phenomena associated with the industrial flash converting
process. This model provides more detailed information on the
transport phenomena in the reaction shaft of a flash furnace
than earlier models but the validity of its projections must
be judged by comparison with industrial or pilot furnace data
which, unfortunately are not available at the time of writing
this report. Therefore, the results presented are preliminary.

The projected particle trajectories indicate that the
distribution velocity does not play a role as important as
that of the main gas stream. Since the radial component of the
gas velocity declines very rapidly after entering the furnace

shaft, the particles traverse, under the conditions
considered, less than half of the furnace radius, i.e., the
reacting stream occupies only a fraction of the furnace
volume. It is believed that the model developed in this work
can be used to assess the effect of different burner
configurations and gas entry conditions on the performance of
flash reactors.

Work on this model and on its ramifications is continuing.
The results will be the subject of a future presentation.

ACKNOWLEDGEMENTS

The authors gratefully acknowledge the support of this
work by Outokumpu Oy and in particular the continuing
interaction and technical discussions with Drs. Juusela,
Asteljoki and Makinen and other Outokumpu personnel.

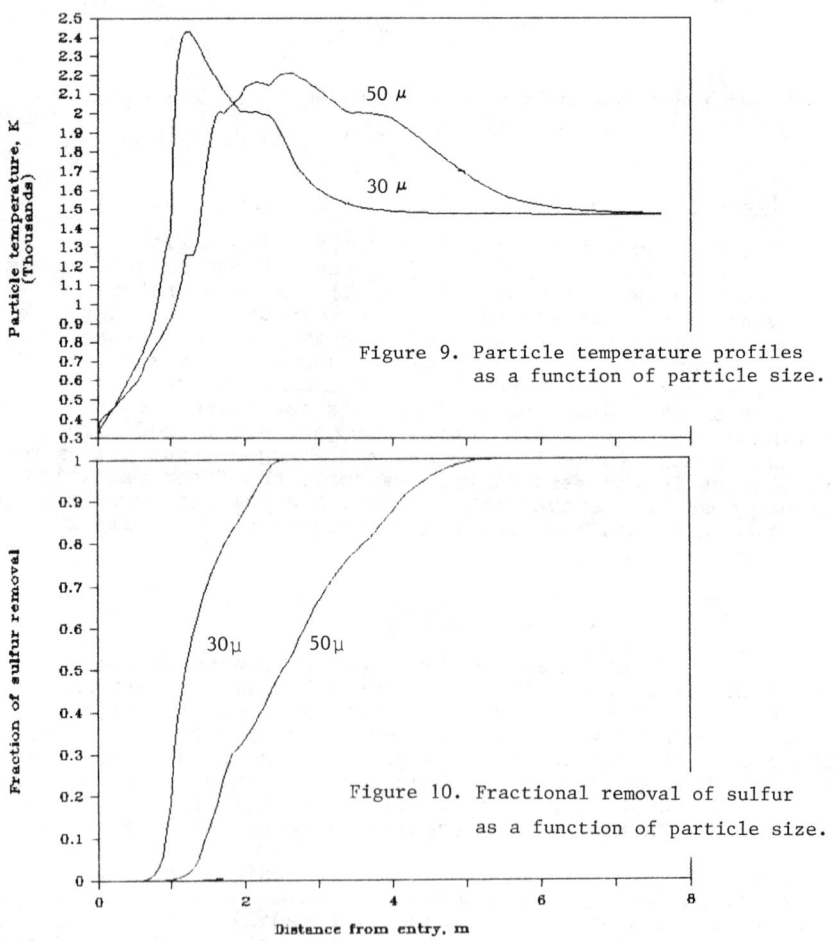

Figure 9. Particle temperature profiles
as a function of particle size.

Figure 10. Fractional removal of sulfur
as a function of particle size.

854

NOMENCLATURE

a_E, a_W, a_N, a_S: coefficients (equations 24-27)

a_p: coefficient at point P of a control volume

A_p: particle surface area

b: source term, eq.(23)

C_D: drag coefficient

$C_E, C_{W,N}, C_S$: convection coefficients

C_{O2}:   oxygen concentration

$C_{p,p}$:   average heat capacity of particles

C_1', C_2:   constants in eq.(8)

C: constant in eq. (4)

D_E, D_W, D_N, D_S: diffusion coefficients

D_e: effective diffusivity of oxygen through reacted shell of particles

D_{O2}:   molecular diffusivity of oxygen in gas mixture

d_p: particle diameter

DSR: fraction of sulfur removal

f_p: pore factor

G: production rate of turbulence kinetic energy

g: gravity force vector

h: heat transfer coefficient by convection

h_g: enthalpy of gas mixture

k: turbulence kinetic energy of fluid flow

k_d: mass transfer coefficient

k_r: rate constant of interface chemical reaction

k_{ov}:   overall rate coefficient

M: molecular weight

m_p: mass of a particle

$m_{i,j}$:   mass flow rate of particles of size i entering from entry port j

N_s: overall rate of sulfur removal, mole/sec

P: gas pressure

q_{cr}:   heat flux due to oxidation of particles

q_{g-w}, q_{p-w}: radiative heat flux between gas and wall and particles and wall

r: radial distance of furnace; radius of the unreacted core of a particle

r_o: particle radius

Re: Reynolds number

ROS: rate of sulfur removal, g/sec

Sc: Schmidt number

Sh: Sherwood number

S_p^Φ: source term ($\Phi = U_x$: radial momentum; U_r: radial momentum; m: mass; X_i: fractional concentration of species i; h: enthalpy of gas mixture)

t: time

T_g, T_p, T_w: gas, particle and wall temperature

U_x, U_r: axial and radial mean gas velocities

$\underline{U}$:   mean gas velocity vector

$\vec{v}$:   particle velocity vector
V_k: control volume
x: axial distance of furnace
X_i: gas volumetric fraction of i species
$\alpha_{g,w}$:   effective absorptivity of gas
τ: transport coefficient in eq.(22)
ε: emissivity of particles
ε_s: effective emissivity of particles;
ε_g: effective emissivity of gas
ε: dissipation rate of turbulence kinetic energy
μ: viscosity
ρ: gas and particle density
σ: Stefan-Boltzman constant
$\sigma_k, \sigma_\varepsilon, \sigma_{Xi}, \sigma_h$: Prandtl number for k, ε, X_i, and h_g

SUBSCRIPTS

c_r: chemical reaction
g: gas
i: species or particle size
j: entry port
k: control volume
o: original
p: particle
r: radial direction
x: axial direction

REFERENCES

1. J. Asteljoki and M. Kyto, "Alternatives for Direct Blister Copper Production", (Paper presented at the 1985 Annual Meeting of AIME, New York, Feb. 1985).

2. J. Asteljoki, L.K. Bailey, D.B. George and D.W. Rodolff, "Flash Converting-Continuous Converting of Copper Mattes", J. of Metals, (May 1985), 20-23.

3. H. Tsukada, Z. Asaki, T. Tanabe, and Y. Kondo, "Oxidation of Mixed Copper-Iron Sulfide", Metall. Trans. B, 12(Sept. 1981), 603-609.

4. V.V.V.N.S. Ramakrishna Rao and K.P. Abraham, "Kinetics of Oxidation of Copper Sulfide", Metall. Trans. B, 2(Sept. 1972) 2463-2470.

5. A. Moriyama, J. Yagi, and I. Muchi, "Rate Controlling Steps of Reduction Process of Iron-Oxide Pellet", J. Japan Inst. Metals, 29(1965), 582.

6. T.A. Henderson, "The Oxidation Rate of Lump Copper-Iron Sulfides", Trans. Instn. Min. Metall., 67(1957-1958), 437-462 (Bull. Instn. Min. Metall., London, no.619, June 1958).

7. L. Wu, Y.H. Kim and N.J. Themelis, "Rate of Gas-Solid Oxidation of Single Copper Matte Particles", (Paper presented at the AIME-TMS Annual Meeting, Feb. 1987).

8. Y.H. Kim and N.J. Themelis, "Transport Phenomena in the Roasting of Metal Sulfides", (To be Published at Can. Matall. Quarterly).

9. N.D.H. Munroe, "Simulation of Flash Smelting Phenomena in a Laboratory Reactor", (D.Eng. Sc. Thesis, Columbia Univ., New York, 1987).

10. P.C. Chaubal and H.Y. Sohn, "Intrinsic Kinetics of the Oxidation of Chalcopyrite Particles under Isothermal and Nonisothermal Conditions", Metall. Trans. B, 17B(March 1986), 51-60.

11. A. Lenchev, Rudy Met. Niezlaz, 21-9(1976) 334-337.

12. F.R.A. Jorgensen and E.R. Segnit, "Copper Flash Smelting Simulation Experiments", Proc. Australas Inst. Min. Metall., 261(1977), 39-46.

13. F.R.A. Jorgensen, "On Maximum Temperatures Attained during Single Particle Combustion of Pyrite", Trans. Instn Min. Metall., Sect. C: Min. Proc. & Extr. Metall., 90(1981), C10-C16.

14. F.R.A. Jorgensen, "Single Particle Combustion of Chalcopyrite", Proc. Australas Inst. Min. Metall., 288(1983), 37-46.

15. S. Ruottu, "The Description of a Mathematical Model for the Flash Melting of Cu-Concentrates", Combustion and Flame, 34(1979), 1-11.

16. Y.H. Kim, "Studies of the Rate Phenomena in Particulate Flash Reaction Systems: Oxidation of Metal Sulfides", (D.Eng. Sc. Thesis, Columbia Univ., New York, 1987).

17. Y.H. Kim and N.J. Themelis, "Effect of Phase Transformation and Particle Fragmentation on the Flash Reaction of Complex Metal Sulfides", Proceedings of R. Schuhmann International Symposium, ed. Gaskell et al, (TMS-AIME, Warrendale, Penn. 1986), 349-371.

18. B.E. Launder and D.B. Spalding, "The Numerical Computation of Turbulent Flows", Computer Methods in Applied Mechanics and Engineering (North-Holland Publishing Company, 1974) 269-289.

19. B.E. Launder and D.B. Spalding, Mathematical Models for Turbulence (London, Academic Press, 1972).

20. S.V. Patankar, Numerical Heat Transfer and Fluid Flow (Hemisphere Publishing Cor., 1980).

21. C.T. Crowe, M.P. Sharma and D.E. Stock, "The Particle-Source-In Cell (PSI-CELL) Model for Gas-Droplet Flows", J. of Fluids Engn., Trans. of ASME, June 1977, 325-332.

22. H.H. Kellogg and N.J. Themelis, "Principles in Sulfide Smelting", Advances in Sulfide Smelting, ed. Sohn, George and Zunkel, (TMS-AIME, 1983), 1-30.

23. I.W. Smith and A. Watts, "Heat Exchange between a Particle and its Surroundings: A Theoretical Study", (Invest. Report, CSIRO Div., Mineral Chem., no.75, 1968), 17.

24. J.J. Wormeck, "Modeling Multidimensional Systems", Pulverized-Coal Combustion and Gasification, ed. L.D. Smoot and D.T. Pratt, (Plenum Press, 1979), 263-294.

25. P.C. Chaubal and H.Y. Sohn, "The Combustion of Chalcopyrite Particles under Flash Smelting Conditions", Gas-Solid Reactions in Pyrometallurgy, ed. D.G.C. Robertson and H.Y. Sohn (The Center for Pyrometallurgy, Univ. of Missouri-Rolla, Rolla, Missouri, 1986), 17-38.

26. J. Szekely and N.J. Themelis, Rate Phenomena in Process Metallurgy (Wiley New York, 1971), 614-637.

27. V.G. Levich, V.S. Krylov and V.P. Vorotilin, "Toward the Theory of Extraction from a Falling Drop", Dokl. Akad. Nauk, SSSR, 160(6)(1965), 1358-1360.

28. V.G. Levich, V.S. Krylov and V.P. Vorotilin, "Toward the Theory of Unsteady-State Diffusion from a Moving Droplet", Dokl. Akad. Nauk, SSSR, 161(3)(1965), 648-652.

COMPUTER SIMULATION OF OXYGEN FLASH SMELTING

K. Parameswaran

ASARCO Inc.
New York, NY   10038

W.D. Marczeski and S. Jones

ASARCO Inc.
Hayden, AZ   85235

Abstract

This paper discusses the development of a mathematical model of Oxygen Flash Smelting at Asarco's Hayden plant and its application in production planning: matte grade control, fluxing, and oxygen requirements. The model makes simultaneous heat and material balance calculations to arrive at the autogenous matte grade, given tonnages and assays of feed materials. Potential uses of the model in monitoring furnace performance, such as judging condition of the furnace refractories from the heat loss value, are discussed.

Introduction:

Asarco's Hayden Plant is a custom copper smelter which receives copper concentrates and precipitates from various shippers as well as its own mines and by-products from other Asarco plants. The concentrates and precipitates are blended with dust, by products and flux to produce a furnace mix.

In November 1983, the Hayden Plant brought on line an INCO design oxygen flash furnace. The INCO furnace uses sulfur and iron in the concentrates as fuel, and tonnage oxygen (96%) as the oxidant. The exothermic chemical reaction between oxygen and sulfur provides the heat for the furnace operation. The INCO technology replaced the traditional roaster/reverb operation where natural gas was a primary source of fuel. Aspects of the retrofit have been discussed by W. D. Marczeski and T. L. Aldrich.[1]

In early operations the plant experienced 1) problems in the scheduling of converters, 2) difficulties in achieving and maintaining proper silica concentrations in flash furnace slag, and 3) a rapid increase in matte and slag shell inventory.

Converter scheduling was adversely affected by significant short-term variability in matte grade (44 - 69 % Cu on a daily basis) due to fluctuations in charge compositions, and to a lesser extent, to changes in smelting rate.

Apparent causes for the difficulties experienced in achieving and maintaining proper silica concentrations in flash furnace slag were related to an inability to accurately predict the matte grade and associated flux requirements due to changes in charge make-up. The failure of trim flux feeders to deliver furnace flux at the desired rate added to the problems.

The rapid buildup in matte and slag shell inventory was related to abnormally high rates of shell generation coupled with a limited capability for direct smelting of reverts while converting high-grade mattes, i.e., exceeding 55% Cu.

It was evident that a desirable operating strategy would require the control of matte grade around 50% Cu (contrast with the design matte grade of 56% Cu.) The need arose to develop the capability of predicting matte grade and fluxing requirements, with changes in charge composition.

At first, simplified equations were derived to predict matte grade based on concentrate analysis, the percentage of concentrates on the charge and an oxygen to concentrate ratio. These equations were used in mix calculations. The development of a mathematical model of Oxygen Flash Smelting, involving simultaneous heat and material balances, was initiated in June 1984.

The preliminary version of the model, developed at ASARCO's Central Research Department, was incorporated at Hayden in late September 1984. Program refinements were made including the facility to incorporate petroleum coke as part of the smelter charge and a more accurate calculation of the converter slag weight. The final version of the model was implemented at Hayden in December of 1984. Since then, the program has been modified to make it more user friendly and is used routinely in making mix calculations.

This paper describes the mathematical model and discusses how it is being used at the Hayden Plant.

Oxygen Flash Smelting Simulation

1. Model Description

The Hayden flash furnace computer model makes simultaneous heat and material balance calculations to arrive at the autogenous matte grade, given tonnages and assays of feed materials. The flux and oxygen requirements and the tonnage and assay of products, i.e., matte, discard slag, and off-gas, are also predicted. The program code consists of the following subprograms:

1. MIX: To create/edit mix assay files and to run the flash furnace simulation program.

2. HSIM: Current version of the flash furnace simulation program.

The development model at the Central Research Department was written in FORTRAN VII-D and implemented on a PERKIN-ELMER 3230 minicomputer. At the Hayden Plant, the development code as rewritten in FORTRAN IV-X and implemented on an HP 1000 minicomputer.

2. 'MIX' Subprogram

To initiate the program, the 'MIX' subprogram is called which returns a menu (with 11 options) and retrieves a master file containing assays of smelter feed material, i.e., blended concentrates, flash furnace flux, recycled flash furnace dust, recycled converter dust, molten converter slag, slag concentrate, lead smelter speiss and matte and miscellaneous residues. The 'MIX' program updates the assay master file from the previous mix into the new file being created. The input stream to be modified is selected by specifying a menu option and the assays for the stream are entered. One of the menu options runs the simulation program.

3. 'HSIM' Subprogram

Figure 1 (page 5) represents a flow chart of the computer program "HSIM."

After the simulation program is initiated, weights (tons/day) and volumes (scfm) of input streams are specified, following a computer prompt. Also entered are: coke on charge as percent of blended concentrates and coke assay; percent of converter slag recycled; trial matte grade; % Fe in slag; SiO_2/Fe in flash furnace slag; percent Fe in flash furnace slag; percent oxygen in tonnage oxygen, matte grade increment, heat balance criterion and an index that specifies whether the program is to be run in an iterative mode to converge to the autogenous matte grade or in a non-iterative mode to produce a specified matte grade.

The composition of streams are stored in two matrices -- A (elemental assay) and B (molecular constituent assay). The first step in the program is to convert elemental assays of streams to molecular constituent assays. As an example, for blended concentrates stream, it is assumed that all copper is present as copper sulfide (Cu_2S). All iron not as magnetite is present as iron sulfide (FeS). Lead and zinc are present as lead and zinc sulfides. Arsenic, antimony, bismuth and cadmium are present in elemental form. Similar conversion rules are specified for all other input streams. Other preliminary calculations involve conversion of volumes (scfm) of infiltrated air to tonnages.

The next section of the program deals with material balance calculations. These calculations attempt to develop a balance based on producing matte of specified or trial matte grade and a discard slag of specified composition (% Fe, SiO_2/Fe ratio). First, the weight of flash furnace dust generated is calculated based on a dusting rate (tons dust/ton

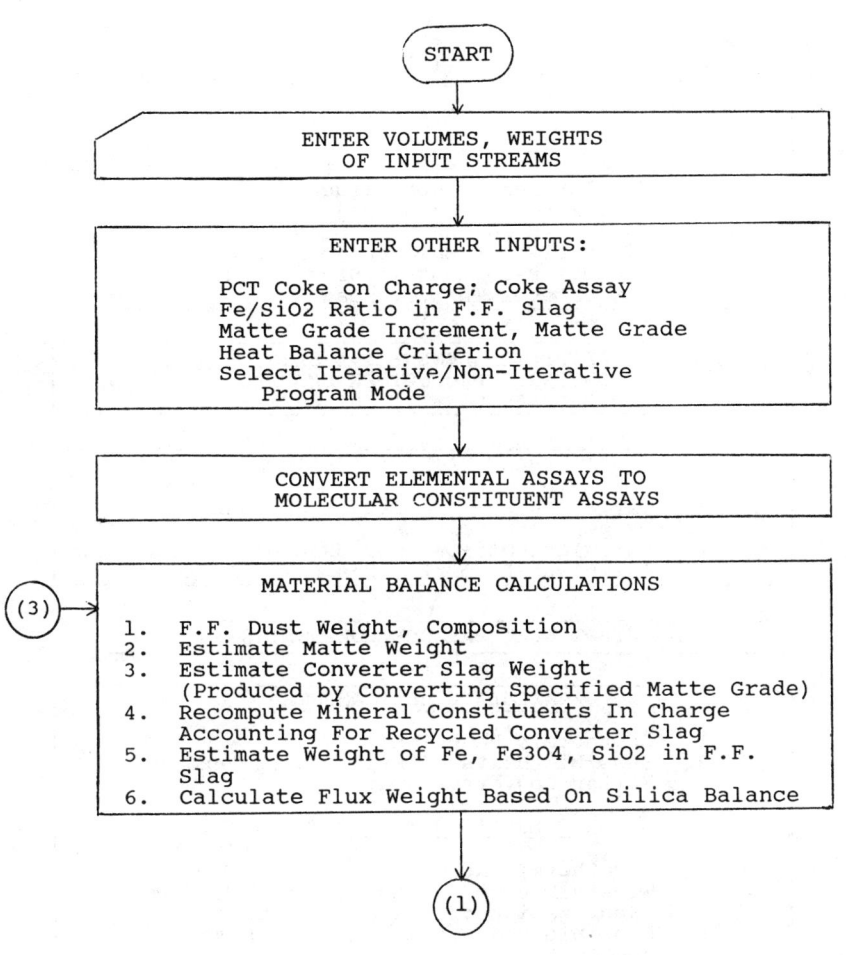

START

ENTER VOLUMES, WEIGHTS
OF INPUT STREAMS

ENTER OTHER INPUTS:

PCT Coke on Charge; Coke Assay
Fe/SiO2 Ratio in F.F. Slag
Matte Grade Increment, Matte Grade
Heat Balance Criterion
Select Iterative/Non-Iterative
    Program Mode

CONVERT ELEMENTAL ASSAYS TO
MOLECULAR CONSTITUENT ASSAYS

MATERIAL BALANCE CALCULATIONS

(3)

1. F.F. Dust Weight, Composition
2. Estimate Matte Weight
3. Estimate Converter Slag Weight
   (Produced by Converting Specified Matte Grade)
4. Recompute Mineral Constituents In Charge
   Accounting For Recycled Converter Slag
5. Estimate Weight of Fe, Fe3O4, SiO2 in F.F.
   Slag
6. Calculate Flux Weight Based On Silica Balance

(1)

Figure 1: Flow Chart of Computer
           Program HS1M

```
                                  (1)
                                   │
                                   ▼
┌─────────────────────────────────────────────────────────────────┐
│                   REFINE MATERIAL BALANCE                         │
│                                                                   │
│   1.   Recompute Matte Weight to Account For Cu                   │
│        In Flux, Minor Constituents In Matte                       │
│   2.   Calculate Weight of Major Constituents                     │
│        In F.F. Slag and F.F. Slag Weight                          │
│   3.   Minor Element Balance To Determine                         │
│        Amounts Reporting To Slag; Adjustments                     │
│        For S Level In F.F. Slag                                   │
└─────────────────────────────────────────────────────────────────┘
                                   │
                                   ▼
┌─────────────────────────────────────────────────────────────────┐
│              CALCULATE ELEMENTAL ASSAYS OF                        │
│              CONDENSED PRODUCT PHASES FROM                        │
│              MOLECULAR CONSTITUENT ASSAYS                         │
└─────────────────────────────────────────────────────────────────┘
                                   │
                                   ▼
┌─────────────────────────────────────────────────────────────────┐
│                GAS PHASE MATERIAL BALANCE                         │
│                                                                   │
│   1.   Oxygen Balance To Determine Tonnage                        │
│        Oxygen Requirements                                        │
│   2.   Calculate Off-Gas Composition                             │
│   3.   Calculate Elemental Gas Phase Assay                        │
└─────────────────────────────────────────────────────────────────┘
                                   │
                                   ▼
┌─────────────────────────────────────────────────────────────────┐
│              START HEAT BALANCE COMPUTATIONS                      │
└─────────────────────────────────────────────────────────────────┘
                                   │
                                   ▼
┌─────────────────────────────────────────────────────────────────┐
│                   CALCULATE HEAT INPUTS                           │
│                                                                   │
│   1.   Exothermic Reactions                                      │
│   2.   Sensible Heats of Dried Charge and                        │
│        Molten Converter Slag                                     │
│                                                                   │
│                   CALULATE HEAT OUTPUTS                           │
│                                                                   │
│   1.   Endothermic Reactions:                                    │
│        Mineral, Dust Decomposition,                              │
│        Vaporization of Charge Moisture                           │
│   2.   Sensible Heat of F.F. Matte, Slag                         │
│        and Dust                                                   │
│   3.   Furnace Heat Loss                                         │
│                                                                   │
│                   COMPUTE HEAT IMBALANCE                          │
└─────────────────────────────────────────────────────────────────┘
                                   │
                                   ▼
                                  (2)
```

Figure 1:   Flow Chart of Computer
            Program HS1M

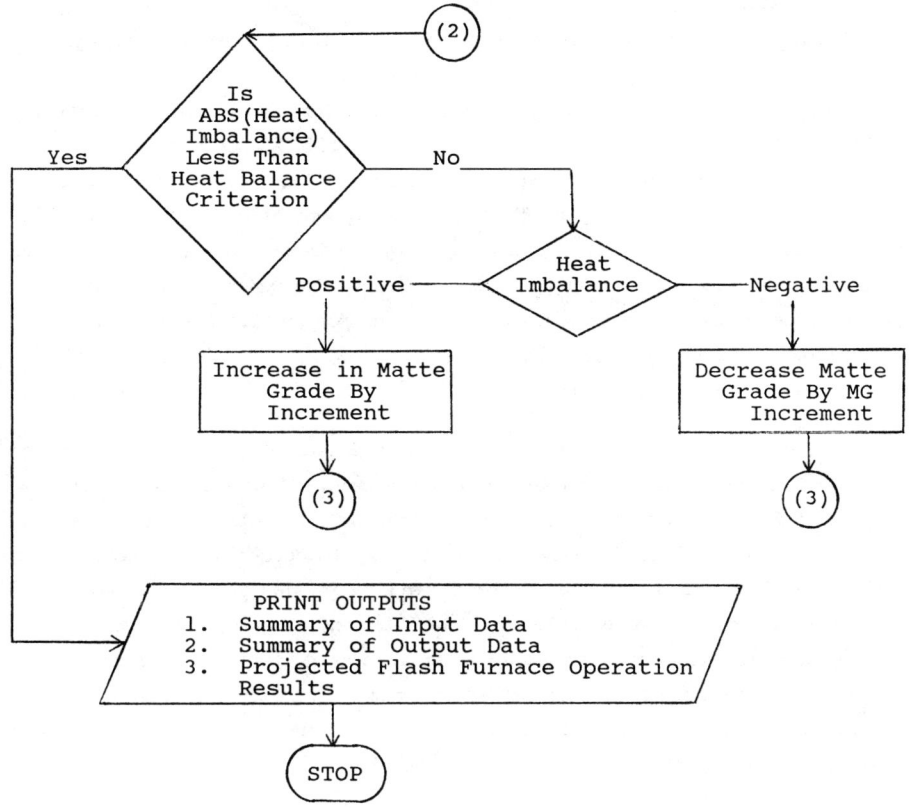

Figure 1: Flow Chart of Computer
Program HS1M

blended concentrates). Allowance is made for the burning of labile sulfur in the dust. The adjusted flash furnace dust weight and composition are calculated. Also, percent Cu_2S, FeO, Fe_3O_4, Fe total in matte are calculated from equations relating these matte constituents to matte grade. Next, the tons per day of mineral constituents in charge (blended concentrates, recycled flash furnace and converter dust, slag concentrate, lead smelter matte and speiss, miscellaneous residues and coke) are calculated. Matte weight is calculated based on tons copper in charge and the matte grade. The weight of converter slag generated by converting this matte is determined by an iterative calculation involving the amount of iron in the matte and iron assay in converter slag. The quantity of mineral constituents in charge is adjusted to account for converter slag being recycled. The flash furnace matte weight is recomputed and quantity of major matte constituents are calculated.

The next set of calculations involved a determination of the weights of flash furnace discard slag and flux requirements. The magnetite to total iron in flash furnace slag is related to the SiO2/Fe ratio in the slag. An iron balance (iron sulfide and oxides) is made to determine the amount of iron in slag. From this, magnetite and silica in slag are calculated. A silica balance is used to calculate the flux requirements, based on the effective silica (silica available after accounting for iron in flux.) The flux requirements are categorized as bed flux requirements associated with cold charge materials and trim flux requirements associated with the recycle of molten converter slag.

Further refinement of calculations involves recomputation of matte weight to account for copper from flux and the major constituents in matte. Minor constitutents in matte are computed based on distribution ratios (defined as percentage of element in charge reporting to matte.) Also, tonnages of major slag constituents are determined from a balance between charge and products other than flash furnace slag, i.e., the excess reports to slag. The weight of flash furnace discard slag is related to the sum of the major slag constituents. Adjustments involve determination of copper in flash furnace slag based on a copper distribution coefficient between matte and slag (i.e., $(\%Cu)_{slag}/(\%Cu)_{matte}$.) This slag assay is achieved by transferring an equivalent amount of matte to slag. A minor element balance, i.e., element input minus that reporting in matte and flash furnace dust determines the amounts of these elements in slag. An adjustment is also made for the specified sulfur content of flash furnace slag. Excess sulfur (over that present in matte) is assumed to be present as FeS; finally

866

an equivalent amount of FeO is reduced from slag. The condensed phase material balance being completed, A and B matrices for condensed phase products are calculated.

From the condensed phase material balance, the amount of FeS oxidized to FeO and Fe_3O_4 is calculated. The amount of pyritic sulfur consumed by sulfate decomposition reactions is calculated and adjustments are made for sulfur deficiency in lead smelter matte and speiss, and the burning of labile sulfur in dust.

Next, the oxygen requirements for various reactions are calculated. These reactions are listed in Table I. The oxygen supplied by air infiltration and oxygen in charge is deducted to arrive at tonnage oxygen requirements, taking into account the oxygen content of tonnage oxygen. Finally, the off-gas composition and amounts are calculated based on these reactions and an allowance for excess oxygen. The off-gas constituents include SO_2, N_2, O_2, CO_2 and H_2O (vapor.) The A and B matrices for the flash furnace off-gas are calculated, completing the material balance calculations.

Next, the heat balance calculations are made. On the input side are the exothermic heats of reactions (reactions 1-6,9; Table I) and the sensible heats of dried charge materials and molten converter slag. The output items include endothermic requirements for mineral and dust decomposition, vaporization of moisture in charge, sensible heats of flash furnace matte, slag and dust and furnace heat losses. The sensible heats of reactants and products and heats of reactions were selected from a number of sources (2,3,4,5.) These values are contained in data statements in the computer program. A difference between heat input and output items is calculated. If the absolute value of this difference is less than the heat balance criterion, the iteration is completed and the current matte grade is the autogenous matte grade. If not, the iteration will search for the proper matte grade. Once the iteration is completed, the program prints: 1) Summary of input data, 2) Summary of output data, and 3) Projected flash furnace operations results.

4. Illustrative Use of model in Mix Calculations

Smelter feed material are bedded in one of four beds. Each bed is identified by a mix number, e.g., Mix 94-1 (Bed 1.) As a bed is readied for smelting, a mix calculation worksheet is prepared on which are noted the

Table I
Chemical Reactions

1. $FeS + 1.5\ O_2$ $=$ $FeO + SO_2$

2. $3\ FeS + 5.0\ O_2$ $=$ $Fe_3O_4 + 3\ SO_2$

3. $S\ (labile) + O_2$ $=$ SO_2

4. $ZnS + 1.5\ O_2$ $=$ $ZnO + SO_2$

5. $PbS + 1.5\ O_2$ $=$ $PbO + SO_2$

6. $2\ FeO + SiO_2$ $=$ $2\ FeO.SiO_2$

7. $FeSO_4 + S_2$ $=$ $FeS + 2\ SO_2$

8. $2\ CuSO_4 + 3/2\ S_2$ $=$ $Cu_2S + 4\ SO_2$

9. $C + O_2$ $=$ CO_2

10. $2\ CuFeS_2$ $=$ $Cu_2S + 2\ FeS + 1/2\ S_2$

tonnages of feed materials (including flux) on the bed.

Assays of feed materials (Cu, Fe, S, CaO, SiO_2, Al_2O_3) are based on prior assay data on these materials. Also, as a bed is being prepared a composite sample (one gm per ton material) of feed materials, fluxed and unfluxed, is made and assayed.

Based on tonnages (corresponding to a specified daily smelting rate) and assays of feed materials (excluding flux,) the flash furnace simulation model is run to obtain predictions of the autogenous matte grade, flux and tonnage oxygen requirements and flash furnace discard slag composition. If the matte grade is too high, the simulation program is rerun with different amounts of coke on charge. Similarly, the impact of the addition of specific concentrates or other charge constituents can be determined. Once an acceptable matte grade and discard slag composition are obtained, correction is made to the bed as increased flux, concentrates or coke additions to get the bed make-up in line with the simulated feed.

As smelting of the mix progresses, grab samples of dried charge are taken daily. The composite feed, daily grab sample assays and feed assays based on historic data provide a means of reconciliation. Finally, shift composite assays are used to calculate daily matte and slag assays.

5. Model Maintenance

The following factors can impact the predictions made by the flash furnace simulation model: (1) actual smelting rate is different from the assumed smelting rate used in the mix calculations, (2) actual assays are different from previous month assays, (3) inaccuracies in assays or sampling and (4) estimated furnace heat loss. It is, therefore, desirable to analyze matte grade predictions with actual smelting results in an effort to track the reliability of the model and obtain indications of changes in furnace heat loss. Examples of the results of such analyses are shown in Figures 2 and 3.

Auxiliary Applications

The MIX program has value in flash furnace operations in addition to mix calculations. It has been used to estimate furnace heat loss values and tonnage effects on matte grade.

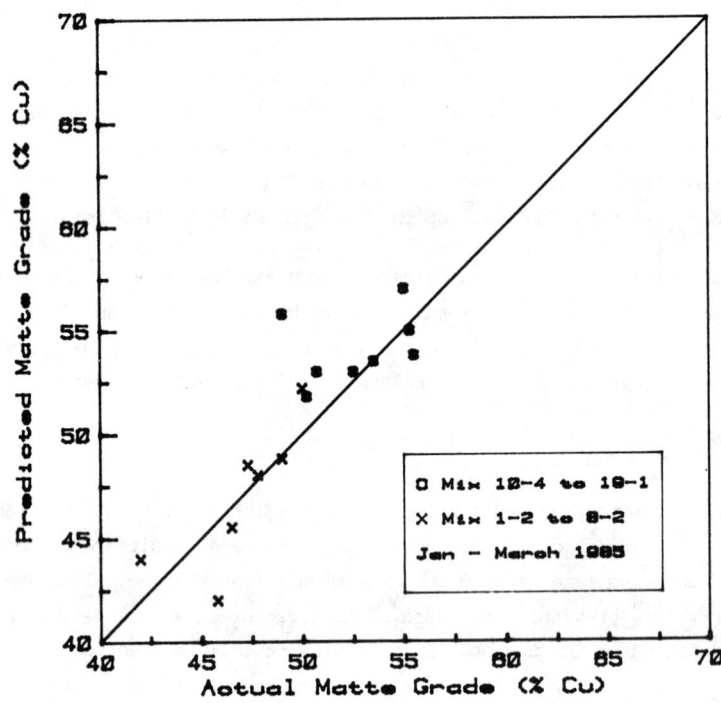

Figure 2: Comparison of matte grade predictions with actual matte grade

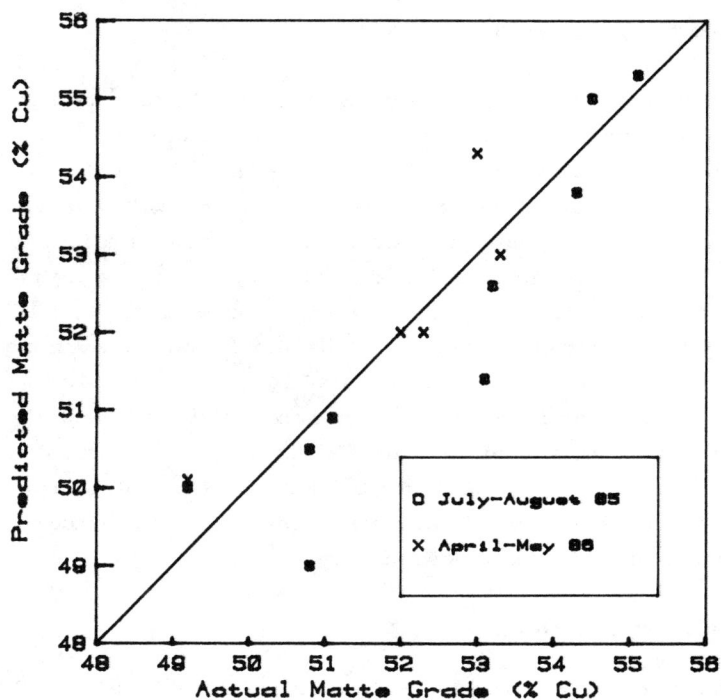

Figure 3: Comparison of matte grade predictions with actual matte grade

1.) Furnace Heat Loss Determination

As the life of the furnace increases, refractory wear causes an
increase in heat loss. Periodic checks of actual matte grades versus
predicted matte grades could indicate changes in the furnace heat loss. As
matte grade values diverge, the program can be used through successive
iterations to determine a new heat loss value.

In June 1985, cooling of the arch was initiated to protect the furnace
refractories. This necessitated a change in the furnace heat loss value
from 375 to 500 million Btu/day.

More recently in May 1987, the heat loss value of the newly rebuilt flash
furnace was estimated. During the rebuild, the furnace cooling system was
modified thus changing the heat loss value. The furnace cooling system was
installed in the majority of the furnace side and end walls, uptake and a
portion of the arch. The system was developed to provide additional cooling
of the furnace refractory. Based on previous operating data the plant
technical staff estimated the new heat loss value would be 600 million
Btu/day rather than the normal 375 million Btu/day value used in the past
for a rebuilt furnace without furnace cooling. After six weeks of
operation, matte grade data was collected and compared to the predicted
values. It was determined, by using the mix program as described above,
that the actual heat loss value for the new furnace was 705 million
Btu/day. Using this value brought the actual matte grades produced in range
with the predicted values. Refer to Figures 4 and 5.

2.) Matte Grade Control

The mix program has been used to generate graphs (similar to Figure 6)
showing the effect of smelting rate on the autogenous matte grade. As can
be seen increasing the smelting rate decreases the matte grade. Prior to
the mix program these determinations required many hours of hard
calculations.

In addition, the mix program has also been used to determine the
amount of coke required in the furnace feed to achieve a specified matte
grade. Using the program, a graph was developed comparing percent coke to
concentrates versus the reduction in matte grade from the "zero coke"

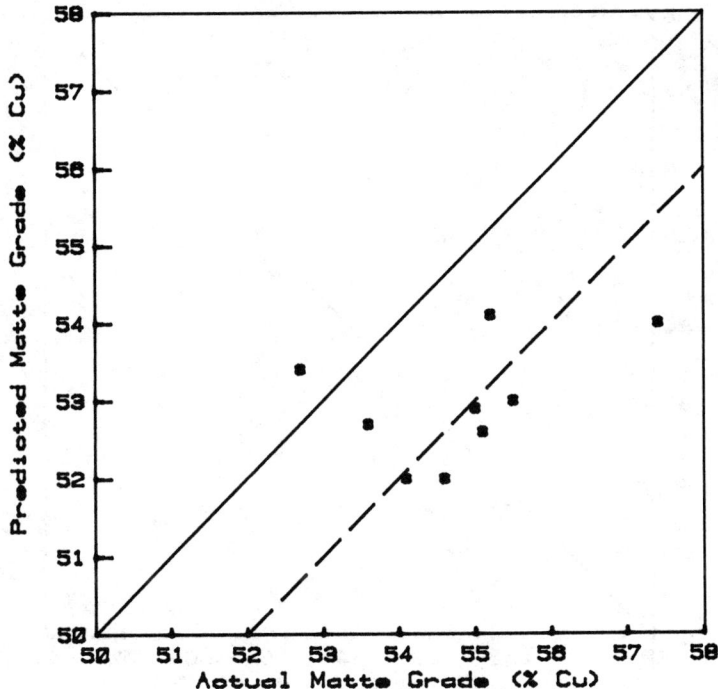

Figure 4: Comparison of predicted matte grade with
actual matte grade after most recent
furnace rebuild (before heat loss adjustment--
furnace heat loss: 600 MMBTU/day; Smelting
rate: 2200 TPD)

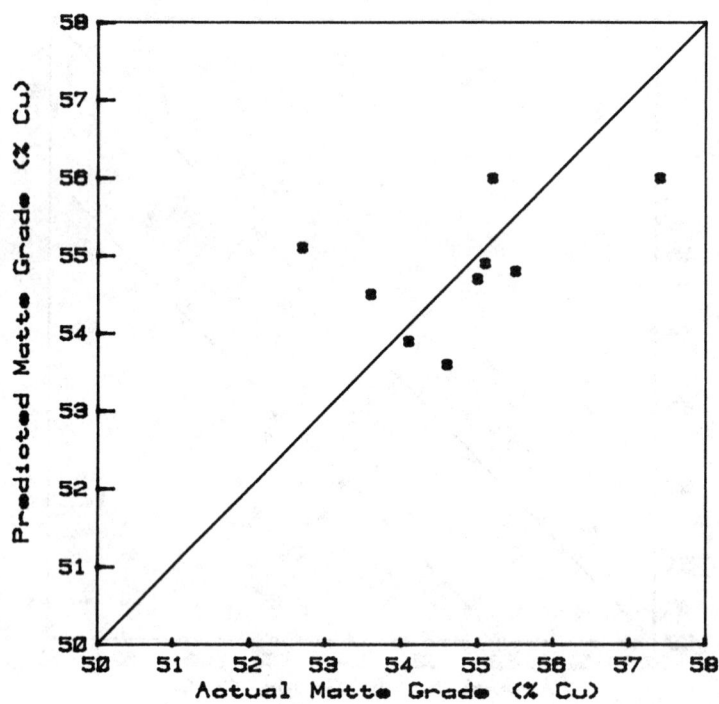

Figure 5:   Comparison of predicted matte grade with
actual matte grade after most recent furnace
rebuild (after heat loss adjustment--furnace
heat loss: 705 MMBTU/day; Smelting rate: 2200 TDP)

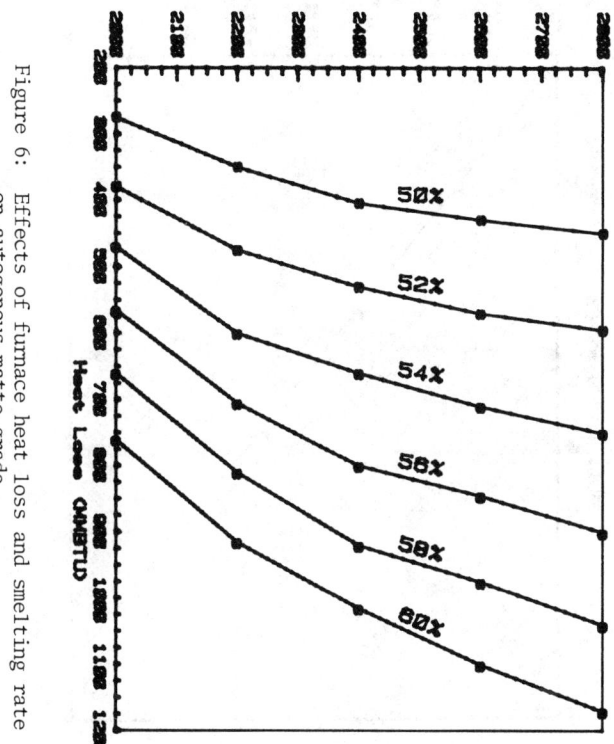

Figure 6: Effects of furnace heat loss and smelting rate on autogenous matte grade

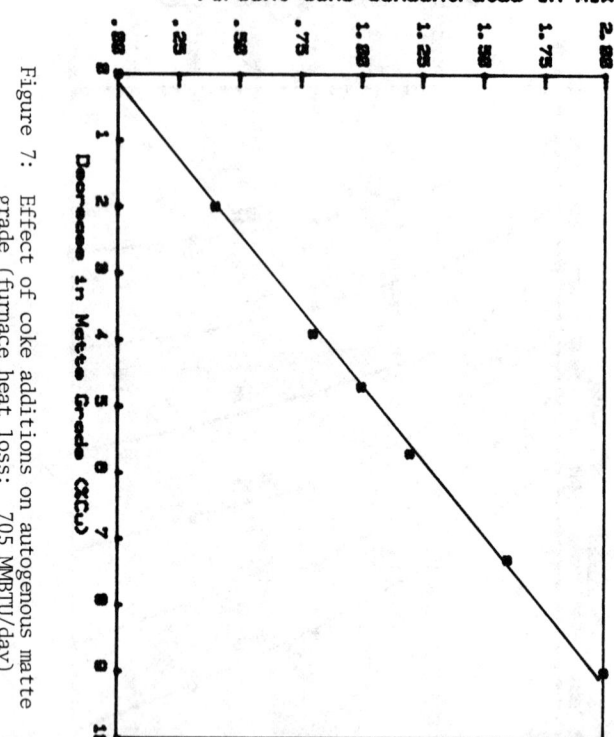

Figure 7: Effect of coke additions on autogenous matte grade (furnace heat loss: 705 MMBTU/day)

required to correct a furnace feed mix.

Model Improvements

1. Impurity Distribution in Flash Smelting

The oxygen flash furnace simulation model has been modified to determine the distribution of minor elements (Pb, Zn, As, Sb, Bi and Cd) between matte, slag and dust. Improvements include: 1) accounting of impurity elements between matte and slag through distribution coefficients (% I in matte/% I in slag) related to matte grade and 2) the quantity of impurity in dust is determined by means of a coefficient (% of total input, reporting to dust.) These improvements have been incorporated into a revised version of the model.

Model validation has not been conducted since impurity assays of the charge are not generally available. Such validation would be necessary before this revised version can be used.

2. Mix Format Modifications

Since December 1984, the model has been modified to make it more user friendly. The mix assays which were recorded by the assay laboratory and subsequently entered into the MIX program by the furnace metallurgist, are now entered into the mix program directly at the lab. This has reduced the time required to make the mix calculations. The MIX master assay file was modified to update the most recent assays entered into the model, when a new mix file is initiated.

The menu portion of the program was simplified to only three options rather than the initial eleven. All the different mix component assay options were combined into a single entry option on a spreadsheet format. The simulation program was altered to remember the previous entries into the mix file. This modification allows the metallurgist to selectively make changes to the file without reentering all the simulation data.

Summary

The mathematical model developed by ASARCO's Central Research Department has been in use at the Hayden plant since December 1984. The model has been used to determine the autogenous matte grade, furnace slag composition, oxygen and flux requirements for smelting a specified flash furnace charge. The model has also been useful in estimating the furnace heat loss (as a measure of furnace refractory wear) and controlling matte grade for the oxygen flash furnace.

References

1. W. D. Marczeski and T. L. Adlrich, "Retrofitting Hayden Plant to Flash Smelting", 1986 Annual AIME Meeting, New Orleans, Louisiana

2. A. Butts, Metallurgical Problems, 2nd ed., (New York and London, McGraw-Hill Book Company, Inc., 1948)

3. C. E. Wicks and F. E. Block, "Thermodynamic Properties of 65 Elements--Their Oxides, Halides, Carbides and Nitrides," (Bulletin 605, Bureau of Mines, U.S. Government Printing Office, Washington, D.C. 1963)

4. T. Rosenquist, Principles of Extractive Metallurgy, (New York, McGraw-Hill, 1974)

5. Richard A. Robie and David R. Waldbaum, "Thermodynamic Properties of Minerals and Related Substances at 298.15°K (25.0°C) and One Atmosphere (1.013 Bars) Pressure and at Higher Temperatures," (Washington, D.C. Geological Surrey Bulletin 1259, U.S. Government Printing Office, 1968)

REACTOR TECHNOLOGY

Session Chairman
Marc Cross, Thames Polytechnic

This session employs various simulation techniques in assessing chemical reactions in materials processing systems. Equilibrium analysis of vapor-phase epitaxy predicts the composition of quaternary phases in double heterostructures used for laser diodes. Other papers consider geometries of CVD deposition and the rate of heat transfer in plasma deposition of ultrafine metal and ceramic powders. A model of a multistage counter current flow reactor can be applied to glass melting or ore roasting processes. Finally, local equilibrium information is used to study oxygen pressure profiles in a flash smelting vessel.

SIMULATION OF OXYGEN PRESSURE PROFILE ALONG THE REACTION SHAFT

N.Kemori, Y.Ojima, Y.Mori and M.Yasukawa

Pyrometallurgical Research Centre, Niihama Research Laboratories
Sumitomo Metal Mining Co., Ltd. Saijo, Ehime 793
Japan

Abstract

A reaction mechanism between copper concentrate, flux and oxygen gas was deduced from the oxygen pressure profile along the reaction shaft of a copper flash smelting furnace at the Toyo smelter. According to the reaction mechanism, the concentrate particles must become larger in size falling through the reaction shaft, which has been proved by independent experiments in a pilot scale flash furnace. In this study the oxygen pressure profile along the reaction shaft is simulated on the basis of the reaction mechanism and local equilibrium calculations.

Introduction

Flash smelting systems developed by Outokumpu and International Nickel Co. are the major technologies for copper smelting. Especially, Outokumpu–type flash smelting is most widely used at copper smelters around the world. About 25 Outokumpu flash furnaces including 5 furnaces in Japan are in operation nowadays. In an Outokumpu flash furnace, copper concentrates with the average particle size of 50μ m are fed into a reaction shaft together with flux, supplemental fuel and reaction air to become oxidized to an appropriate extent within the retention time of about 1 second. Investigations on the combustion behavior of copper concentrates are therefore important for improving the existing furnaces and developing more intensive suspension-type reactors.

Apart from laboratory experiments on the combustion or oxidation of copper concentrates (1–10), plant investigations using a pilot– or commercial–scale flash furnace are limited. Ruottu (11) and Themelis et al. (12) have studied the rate phenomena taking place in the shaft based on mathematical models. Kemori et al. have measured oxygen pressures of the molten concentrate particles falling through the shaft (13,14) and also investigated the applicability of equilibrium calculations to the overall reaction in the shaft (15). Kimura et al. (16) have collected water-quenched particles from the shaft and studied the variation of their size along the shaft.

Reaction mechanism between copper concentrate, flux and oxygen gas, which would be useful for metallurgical engineers if it were well known, has not been cleared enough for a commercial flash furnace. However two reaction models have been proposed by Kim and Themelis (17) and Kemori et al. (14), which are completely different with respect to the interaction of concentrate particles. After examining the validity of the two models based on available information, oxygen pressure profiles along the shaft are simulated in the present study on the basis of the reaction model of Kemori et al. (14) as well as the assumption of local equilibrium.

Reaction Mechanism in the Shaft

Pyrometallurgical copper production is based mainly on oxidation of copper concentrates. If the oxygen pressure of the concentrate particles can be measured along the reaction path in an actual furnace, some important information concerning the reactions between the particles and oxygen gas will be obtained. Moreover a reaction model may also be deduced from the results.

According to Kemori et al. (13,18), the oxygen pressures of the matte and slag are not only equal in the settler of a copper flash smelting furnace, but does not change along the settler. The oxygen pressure of copper concentrate particles is therefore expected to drastically change in the vertical direction of the reaction shaft. Figs.1 and 2 (14) show the variation of the oxygen pressure of the molten concentrate particles along the shaft. Locations of the measurements are depicted in Fig.3 (14).

Although, as was expected, there is a drastic oxygen pressure change in the shaft from the roof to the bath, the oxygen pressure does not change very much from hole A to the bath. Since a three–phase coexistence of gas, slag and matte is only possible in such a small oxygen pressure change,i.e. Cu_2O (s) or Fe_3O_4 (s) is unstable, the extent of oxidation of the particles can not be evaluated by comparison with oxygen pressures fixed by particular reactions such as

$$2Cu\ (l)\ +\ \frac{1}{2}\ O_2\ (g)\ =\ Cu_2O\ (s)\quad or\quad 3FeO\ (l)\ +\ \frac{1}{2}\ O_2\ (g)\ =\ Fe_3O_4\ (s).$$

Furthermore oxygen pressure is dependent on temperature. Consequently it is convenient to introduce the concept of normalized oxygen pressure (13,14,18) for investigating how the oxidation reactions proceed along the shaft. All the data shown in Figs.1 and 2 are oxygen pressures normalized with respect to the reaction $FeO\ (l)\ +\ 1/4\ O_2\ (g)\ =\ FeO_{1.5}\ (l)$ at 1523K. Hereafter oxygen pressure means normalized oxygen pressure or oxygen pressure at 1523K throughout this paper.

Figs.1 and 2 show that the oxygen pressure of the molten concentrate particles decreases along the shaft for the operation with recycling of dust. Taking into account that the oxygen pressure in the burner cone does not depend on recycling of dust (see Fig.2), it must also decrease in the upper zone of the shaft for the operation without recycling of dust, as is shown by a broken line in Fig.2, although there is a little increase in oxygen pressure below hole A (see Fig.1). Such oxygen pressure profiles are considered to be queer in view of the fact that the main reaction in the shaft is oxidation of copper concentrates. This is probably because copper concentrates burn very quickly in the shaft. Jorgensen (3) have studied single—particle combustion of chalcopyrite concentrate in oxygen—nitrogen atmospheres and reported that the reaction times required for oxidizing the concentrate with the $\leq 53 \mu$ m size to the matte grades (MG) of 60 and 80% in air at 700°C are approximately 60 and 100ms, respectively. Assuming that the average retention time of the particles is 1 sec. in the actual shaft and that they fall through the shaft with the average speed of 8m/s, the reaction time of 100ms corresponds to

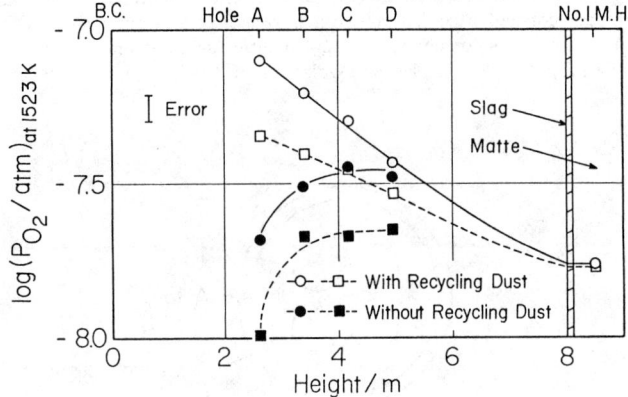

Figure 1 — Measured oxygen pressure profile along the shaft (14).

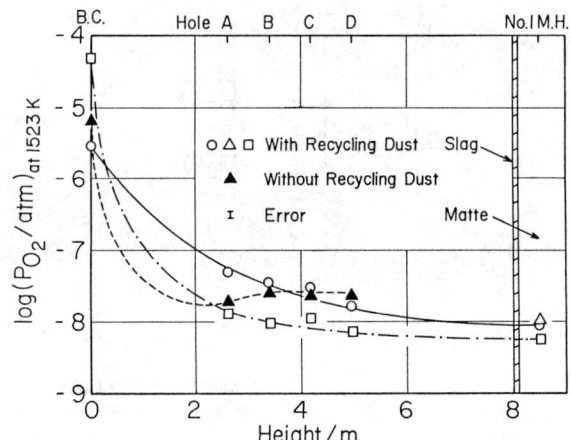

Figure 2 — Measured oxygen pressure profile along the shaft (14).

885

the position of 0.8m below the shaft roof which is situated at one-third of the distance between the roof and hole A. Kemori et al. (14) have measured SO_2 and O_2 contents of the gas in the shaft and confirmed that most of the oxygen contained in the reaction air is consumed in the upper zone of the shaft.

Based on the above results and consideration, a two-particle reaction model (14) was deduced for the operation without recycling of dust, which is reproduced in Fig.4.

Firstly, most of the oxygen in the reaction air is consumed by supplemental fuel and some easily combustible concentrate particles which become excessively oxidized. The excessively oxidized particles are then reduced by contact with the remainder of the concentrate particles while they fall through the shaft. This reduction reactoin is followed by the slag formation reaction, and all these reactions are completed at about 3m below the shaft roof. In this model supplemental fuel means oil or pulverized coal. Apart from oil, pulverized coal is not considered to always burn faster than copper concentrates. In fact, Jorgensen (19) found that coal with the $< 74 \mu$m size and volatiles of 24.3% required higher gas preheat temperature for ignition in air by about 150°C than nickel concentrates. However there has been nor evidence that copper concentrates burn faster than pulverized coal. Measurements on CO_2 and SO_2 contents of the gas along the shaft (14) showed that the combustion rate of pulverized coal was a little faster than that of copper concentrates. Therefore it is reasonable in our case to consider that the oxygen in the reaction air is consumed by supplemental fuel and some easily combustible concentrates. Incidentally the average size and volatiles of the coal (14) were 55μ m and 39%, respectively.

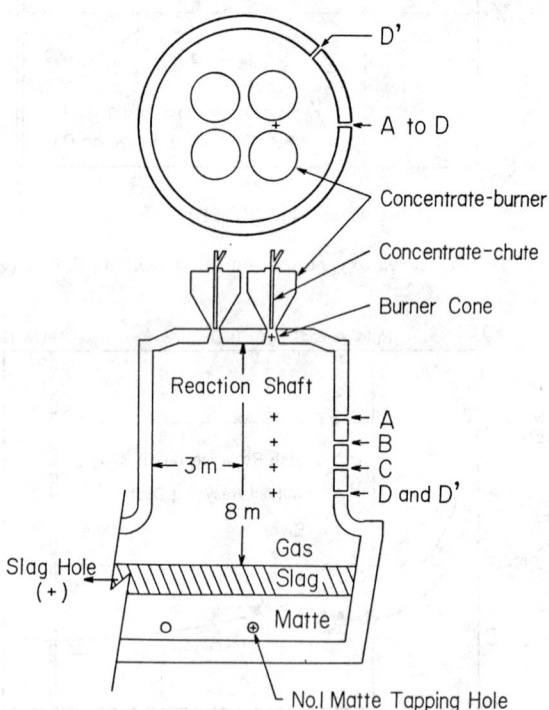

Figure 3 – Schematic diagram of flash smelting furnace showing location of emf measurements (+) (14).

According to the above two-particle model, the concentrate particles must become larger in size falling through the shaft because the reduction of the excessively oxidized molten particles is assumed to be done by contact with the less oxidized solid particles. This conclusion is completely different from a particle fragmentation model proposed by Kim and Themelis (17), but the validity of the two-particle model has been proved by Kimura et al.(16). Kimura et al. collected water-quenched particles along the shaft of a pilot scale flash furnace shown in Fig.5 and found them to grow as depicted in Fig.6.

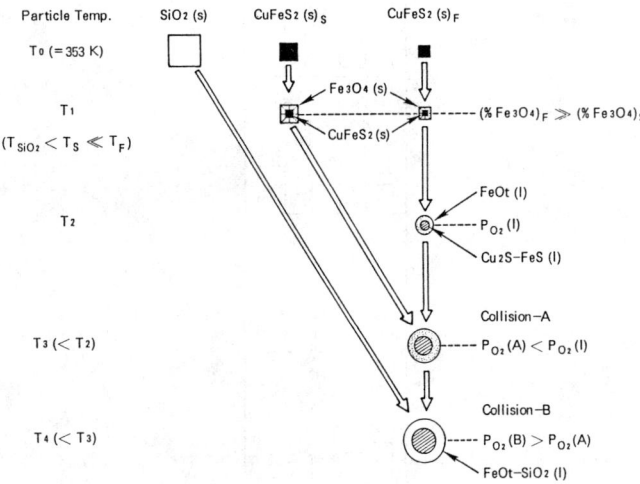

Figure 4 – Two-particle reaction model for the operation without recycling of dust (14); F = fast reacting and S = slow reacting.

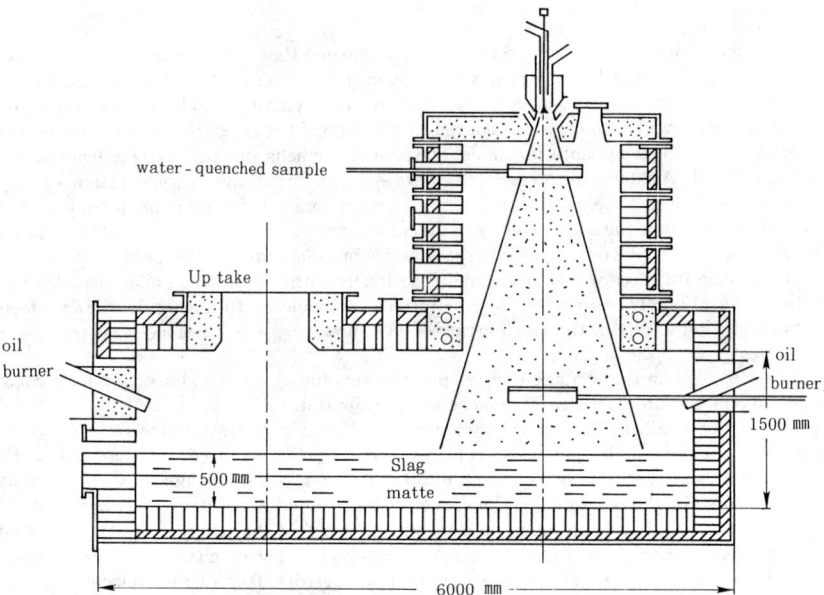

Figure 5 – Schematic diagram of pilot flash furnace (16).

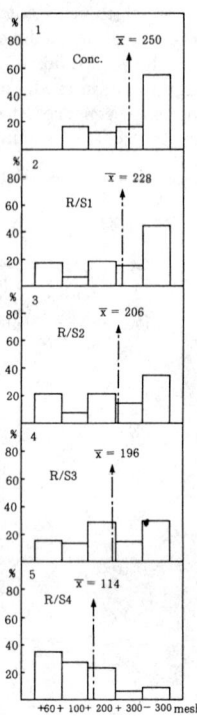

Figure 6 – Change of the size distribution of concentrate particles along the shaft (16).

Their microscopic observations, Photos 1 to 3, showed that some amounts of chalcopyrite with angular shape existed even in the water–quenched sample collected at the outlet of the shaft. These findings are not incosistent with the two–particle model at all. Because the chalcopyrite with angular shape is considered to correspond to the less oxidized solid concentrate particles. The particle fragmentation model of Kim and Themelis may be based on their inaccurate sampling method. Although Jorgensen (3) have also reported the fragmentation of copper concentrate particles, his experimental conditions are quite different from actual operating conditions, for example the space density of the concentrate is much more dilute in his experiment. If the fragmentation of copper concentrates were predominant in the shaft, operations of Outokumpt flash furnaces would be disturbed seriously with waste heat boiler troubles due to high dust generation. Therefore the fragmentation is considered to have only minor effects in an actual reaction shaft. In the two-particle model, however, it has not been cleared yet how the particles grow in the shaft.

In order to explain the strange oxygen pressure profile along the shaft, one may consider that all the concentrate particles become equally oxidized in the upper zone of the shaft and an oxygen pressure gradient occurs in each particle. However the structure of a concentrate burner may prevent all the particles from becoming equally oxidized. As depicted in Fig.7, copper concentrates with flux are charged into the reaction shaft through a concentrate chute being surrounded by the reaction air. Consequently the inner concentrates tend to be less oxidized than the outer concentrates although an appropriate speed of the reaction air at the throat of the concentrate burner and the use of a dispersion cone give better mixing of the concentrates with the reaction air. In an extreme case of the poor performance of the concentrate burner, heaps consisting of unreacted concentrates are sometimes formed in the settler under the shaft. Therefore the two-particle model is also reasonable in this respect.

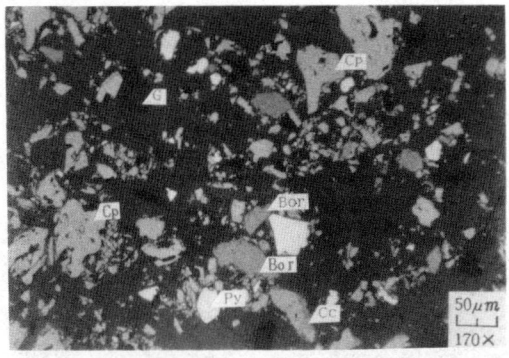

Photo 1—Original copper concentrates(16).

They consist of many angular particles which are composed mainly of chalcopyrite.

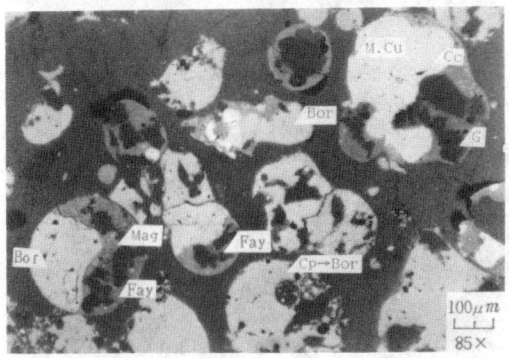

Photo 2—Water—quenched concentrate particles with the size of +100 mesh(16).

Chalcopyrite is changed to chalcocite and bornite. The particles, of which the shape is round, have matte and slag as well as even metallic copper fractions.

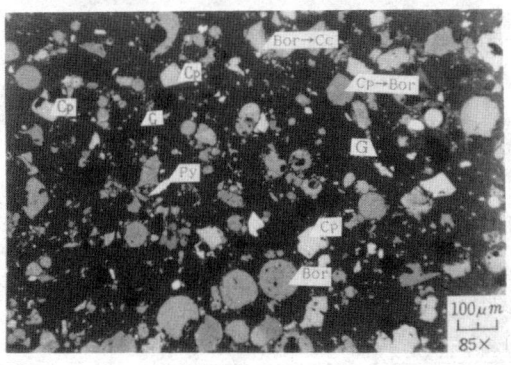

Photo 3—Water—quenched concentrate particles with the size of −100 mesh(16).

Most of the particles are composed of chalcopyrite and have angular shapes.

Cp	: Chalcopyrite	Fay	: Fayalite
Bor	: Bornite	Mag	: Magnetite
Py	: Pyrite	M.Cu	: Metallic copper
Cc	: Chalcocite	G	: Gangue

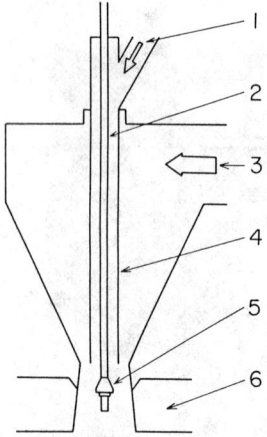

Figure 7 — Schematic configuration of concentrate burner;
(1) furnace charge and pulverized coal, (2) oil burner,
(3) reaction air, (4) concentrate chute,
(5) dispersion cone for furnace charge, and (6) reaction shaft roof.

Equilibrium Calculation

Assuming that the oxidized molten concentrates which consist of matte and slag are in equilibrium with the surrounding gas at any point along the shaft, oxygen pressures of the concentrates were calculated to quantitatively check the validity of the two-particle model based on the actual oxygen pressure profile. The overall reaction in the shaft is known to be fairly well described by equilibrium calculations (15).

Thermodynamic System

Six elements (Cu,S,Fe,O,Si and N) were taken into consideration for the present equilibrium calculations. Judging from the oxygen pressures measured in the shaft except those in the burner cone (13,14), oxygen pressures of a thermodynamic equilibrium system consisting of gas, slag and matte phases were studied. Components in each phase are listed in Table I.

Table I. Components in Gas, Slag and Matte Phases

Phase (state)	Component
Gas (g)	S_2 , SO_2 , O_2 , N_2
Slag (l)	FeO, Fe_3O_4 , SiO_2
Matte (l)	Cu_2S, FeS, FeO, Fe_3O_4

890

Thermodynamic Data

Necessary thermodynamic data for standard Gibbs free energies of formation and activity coefficients are shown in Tables Ⅱ and Ⅲ. Since the activity coefficient of Fe_3O_4 (s) in the matte phase, which was evaluated by Goto (24), gave considerably higher oxygen contents of the matte than the values measured by Kameda and Yazawa (25) and Eilliott et al. (26,27), it was multiplied by 1.5 in this study. The activity coefficients of FeO (l) and $FeO_{1.333}$ (l) in the slag phase were quoted from the work of Björkman and Eriksson (23) based on the experimental results obtained by Timucin and Morris (28) and the activities given by Korakas (29).

Table Ⅱ. Standard Gibbs Free Energies of Formation

Reaction	$\triangle G°$ (cal / mol)	Reference
$\frac{1}{2} S_2 (g) + O_2 (g) = SO_2 (g)$	$-86620 + 17.31$ T	Kubaschewski and Alcock (20)
$2 Cu (l) + \frac{1}{2} S_2 (g) = Cu_2S (l)$	$-35160 + 10.05$ T	Nagamori and Mackey (21)
$Fe (s) + \frac{1}{2} S_2 (g) = FeS (l)$	$-24250 + 5.50$ T	Nagamori and Mackey (21)
$Fe (s) + \frac{1}{2} O_2 (g) = FeO (l)$	$-54890 + 10.55$ T	Nagamori and Mackey (21)
$3 Fe (s) + 2 O_2 (g) = Fe_3O_4 (s)$	$-260870 + 72.24$ T	Nagamori and Mackey (21)
$3 Fe (s) + 2 O_2 (g) = Fe_3O_4 (l)$	$-229100 + 55.18$ T	Yazawa (22)

Table Ⅲ. Activity Coefficients

Phase	Component	Activity Coefficients	Reference
Slag	FeO (l)	$\exp [(1/T) \{2465.9 \cdot N_{FeO_{1.333}} \cdot (1 - N_{FeO}) - 3109.0 \cdot N_{SiO_2} \cdot (1 - N_{FeO}) - 395.57 \cdot N_{FeO_{1.333}} \cdot N_{SiO_2} \}]$	Björkman and Eriksson (23)
	$FeO_{1.333}$ (l)	$1.0429 / (N_{FeO_{1.333}})^{\frac{1}{2}}$	Björkman and Eriksson (23)
Matte	Cu_2S (l)	1.0	Goto (24)
	FeS (l)	$\exp [(1458/T) \ln (0.54 + 1.4 N_{FeS} \log N_{FeS} + 0.52 N_{FeS})]$	Goto (24)
	FeO (l)	$\exp [(1573/T) \{5.10 + 6.20 \ln N_{Cu_2S} + 6.41 (\ln N_{Cu_2S})^2 + 2.80 (\ln N_{Cu_2S})^3 \}]$	Goto (24)
	Fe_3O_4 (s)	$\exp [(1.5(1573/T) \{4.96 + 9.90 \ln N_{Cu_2S} + 7.43 (\ln N_{Cu_2S})^2 + 2.55 (\ln N_{Cu_2S})^3 \}]]$	Goto (24)

Furnace Charge

Chalcopyrite ($CuFeS_2$) with silica (SiO_2) was treated as copper concentrate in the present calculations. Here silica was regarded as gangue. The composition of the concentrate is as follows:

Component	Cu	S	Fe	SiO_2	$CuFeS_2$	SiO_2
wt %	32.55	32.84	28.61	6.00	94	6

Pure silica was added as flux to get an appropriate slag composition.

Calculation Conditions

All equilibrium calculations were made at 1523K. Oxygen enrichment of 35% and produced matte grade of 60% were chosen in the light of the actual usual operation. Although heat balance was not incorporated in the present equilibrium calculations, the above calculation conditions nearly satisfied the heat balance in the whole shaft.

In the actual furnace the ratio of iron to silica content of the slag is controlled at about 1.15. According to Vartiainen (30), however, such an actual slag would not be a complete liquid solution precipitating solid silica if it were regarded as the $Fe-O-SiO_2$ system (see Fig.8). Consequently the (wt% Fe)/(wt% SiO_2) ratio of 1.63, with which the slag can exist in a liquid state over the most wide oxygen pressure range at 1523K, was adopted for the equilibrium calculations. If cations in MgO, CaO and Al_2O_3 contained in the actual slag are regarded as iron ions, the (wt% Fe)/(wt% SiO_2) ratio becomes about 1.5. The value of 1.63 is therefore considered to be reasonable.

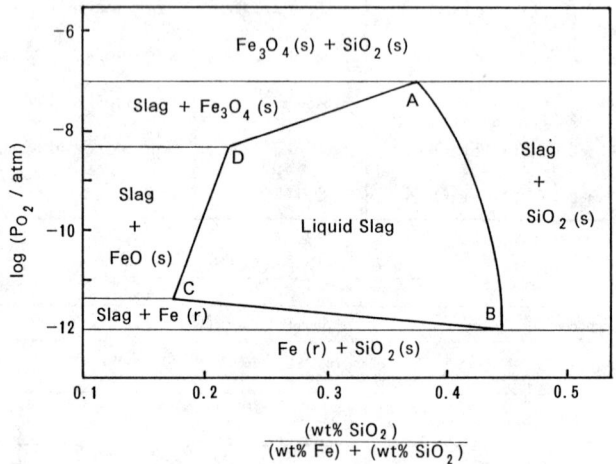

Figure 8 – $Fe-O-SiO_2$ diagram at 1523K showing composition and oxygen pressure (30);
[log (PO_2 / atm), (wt% Fe) / (wt% SiO_2)] = A (−7.0, 1.63), B (−12.0, 1.24),
C (−11.4, 4.64) and D (−8.3, 3.54).

Results and Discussion

Terminal Oxidation State

The amounts of reaction air and flux, which are necessary to produce a matte with MG = 60% and a slag with (wt% Fe)/(wt% SiO_2) = 1.63, are listed together with the resultant composition of each phase in Table Ⅳ. The calculated Fe_3O_4 content of the slag is considerably higher than usual contents of the actual slag. This is probably because (wt% Fe)/(wt% SiO_2) for the calculation was set to be higer than an actual value of 1.15. To quantitatively clear this point, the results for (wt% Fe)/(wt% SiO_2) = 1.15 are also included as case 2 in the table. Although this calculation should yield slag saturated with solid silica, the precipitation of silica is not taken into consideration by assuming hypothetical liquid slag without solid silica for simplicity. Since the activity of Fe_3O_4 in the slag is increased with increasing (wt% Fe)/ (wt% SiO_2), the oxygen content of the matte is rather high for (wt% Fe)/(wt% SiO_2) = 1.63

892

Table IV. Calculated Results

Case	$\frac{(\% Fe)}{(\% SiO_2)}$ in Slag	Air Volume of 35% O_2 $\frac{N m^3}{Conc.t}$	Silica Addition $\frac{kg}{Conc.t}$	Composition									
				Gas (vol %)			Matte (wt %)				Slag (wt %)		
				SO_2	S_2	O_2	Cu	S	Fe	O	Fe	Fe_3O_4	SiO_2
1	1.63	546.6	61	28.4	$10^{-2.30}$	$10^{-8.04}$	60.0	21.8	16.4	1.8	52.0	16.5	31.9
2	1.15	519.0	112	28.5	$10^{-2.10}$	$10^{-8.14}$	60.0	22.8	16.2	1.1	46.0	10.3	40.0
3	2.25	546.6	0	28.2	$10^{-2.40}$	$10^{-8.00}$	51.7	19.4	24.0	4.9	56.9	22.4	25.3

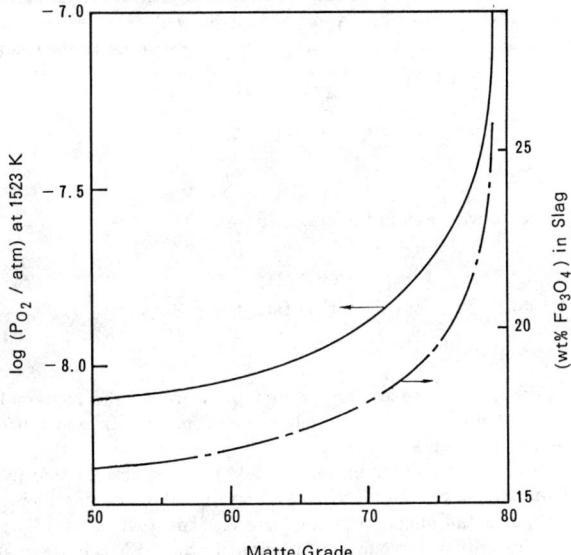

Figure 9 – Relations between oxygen pressure, magnetite content of slag, and matte grade.

although it has been made to be 1.1% for (wt% Fe)/(wt% SiO_2) = 1.15 by increasing the reported activity coefficient of Fe_3O_4 in the matte (24) by 1.5 times. As a result higher value of (wt% Fe)/(wt% SiO_2) needs larger amount of reaction air per 1 ton of copper concentrate, which results in higher oxygen pressure.

Oxygen pressures measured in the settler (13,14), which are considered to correspond to the terminal oxidation state of copper concentrates, usually range from $10^{-7.7}$ to $10^{-8.3}$ atm at 1523K. The calculated oxygen pressures are in this range, and the present equilibrium calculations can be used to evaluate the terminal oxidation state.

In order to study the application limit of the present calculations, several calculations were performed over the relatively wide MG range of 50 to 80% for (wt% Fe)/(wt% SiO_2) = 1.63, the results of which are shown in Fig.9. As mentioned before, the highest oxygen pressure which can be encountered in homogeneous liquid slags of the Fe–O–SiO_2 system at 1523K is $10^{-7.0}$ atm (see Fig.8), the Fe_3O_4 content of the slag for which is 25% (21) or 23% (31). On the other hand, thermodynamic data used in this study give the highest oxygen pressure of $10^{-6.76}$ atm for which the slag composition is 39.6 wt% FeO, 28.8 wt% Fe_3O_4 and 31.6 wt% SiO_2. It can therefore be seen in Fig.9 that the application of the present calculations should be limited up to MG = 79% corresponding to the highest oxygen pressure of $10^{-7.0}$ atm.

Influence of Silica Addition

The two−particle model says that an increase in oxygen pressure near hole A (see Fig.1) is caused by contact between molten concentrate and solid silica. To confirm this point, equilibrium calculation was performed so that 1 ton of the copper concentrate is smelted with the amount of the reaction air calculated for case 1 under conditions of no silica addition. The results are also given as case 3 in Table Ⅳ. The comparison of cases 1 and 3 reveals that silica addition does not increase the oxygen pressure of molten concentrates but decrease it. This is understood as follows. The participation of added silica in the smelting reaction results in an increase in activity coefficient of Fe_3O_4 in the slag as well as a decrease in that of FeO. As a result, Fe_3O_4 in the slag decomposes into FeO and O_2 according to the following reaction,

$$Fe_3O_4 \text{ (slag)} = 3 \text{ FeO(slag)} + \frac{1}{2} O_2 \text{ (g)}. \qquad (1)$$

Liberated oxygen gas reacts with FeS in the matte as

$$FeS \text{ (matte)} + \frac{1}{2} O_2 \text{ (g)} = \text{ FeO (slag)} + \frac{1}{2} S_2 \text{ (g)}. \qquad (2)$$

After all, liberated sulfur gas decreases the oxygen pressure of the system because SO_2 pressure is almost constant not depending on silica addition. The oxygen buffer capacity of this system is given by FeS in the matte phase.

Fig.10 shows the influence of silica addition on oxygen pressures of two systems, one of which consists of the gas and slag phases for case 3 and the other the gas phase without both S_2 and SO_2 components and the slag phase for case 3. The two systems were chosen to investigate the respective extents of oxygen buffer capacity due to FeS in the matte phase and S_2 in the gas phase. It is found in the figure that if the slag phase is only in equilibrium with the gas phase but not with the matte phase, the oxygen pressure of the slag and gas phases is a little increased by silica addition. Because the oxygen gas liberated by reaction (1) reacts with sulfur gas as

$$\frac{1}{2} S_2 \text{ (g)} + O_2 \text{ (g)} = SO_2 \text{ (g)}. \qquad (3)$$

The participation of all the added silica in the smelting reaction results in an increase of 0.07 in common logarithm of oxygen pressure, while the increases in that of measured oxygen pressure were 0.13 to 0.24. However it seems not to be realistic to assume that only the matte phase is not equilibriated with the slag and gas phases. It can also be seen in Fig.10 that silica addition significantly increases the oxygen pressure of the slag in equilibrium with only a gas phas consisting of N_2 and O_2. Because the gas phase does not have any oxygen buffer.

Judging from the above results, it is not considered that the contact between molten concentrate and solid silica causes an increase in oxygen pressure. There may be another cause or some inaccuracy in the experimental results of Fig.1.

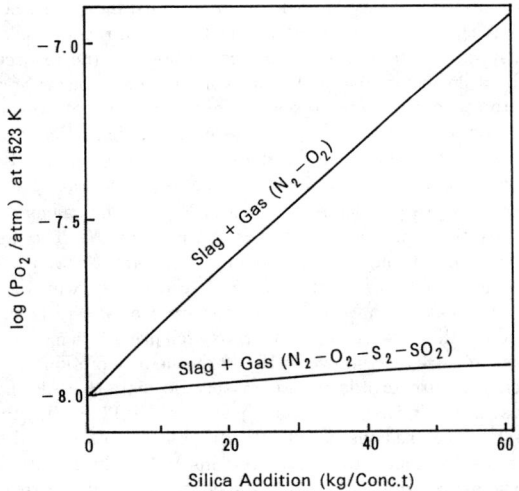

Figure 10 − Influence of silica addition on oxygen pressure.

Initial Oxidation State

Oxygen pressures measured in the burner cone (14) , which correspond to the initial oxidation state of copper concentrates, have revealed that copper concentrates are likely to be oxidized to the extent that metallic copper appears. Oxygen pressures corresponding to the initial oxidation state can therefore be evaluated by the following reaction,

$$Cu_2S \text{ (l)} + O_2 \text{ (g)} = 2Cu \text{ (l)} + SO_2 \text{ (g)}, \quad \triangle G° = -51460 + 7.26T \text{ (cal/mol)}. \tag{4}$$

Under conditions of α_{Cu_2S} (l) = $\alpha_{Cu(l)}$ = 1 and P_{SO_2} = 0.1 to 0.3 atm, the above reaction gives the oxygen pressures of $10^{-6.8}$ to $10^{-6.3}$ atm at 1523K. Those oxygen pressures are higher than the highest one of the Fe−O−SiO₂ slag, and stable oxide phases are solid magnetite and silica. In the burner cone, therefore, the reaction of FeO (l) + 1/4 O₂ (g) = FeO₁.₅ (l) should not be used for normalization of oxygen pressure. The use of this reaction for normalization seems to lead to much higher normalized oxygen pressures as shown in Fig. 2.

Fig.11 shows relations between oxygen pressure, the weight ratio of Cu (l) to total copper in Cu₂S (l) and Cu (l), and the amount of copper concentrate which is smelted with the amount of the reactoin air calculated for case 1. The amount of Cu (l) is largely increased with decreasing the amount of copper concentrate, whereas the oxygen pressure is a little increased.

Simulation of Oxygen Pressure Profile

Based on the two-particle reaction model, oxygen pressure profiles along the shaft were calculated as follows. Part of copper concentrates fed into the shaft become excessively oxidized with the whole reaction air required for producing a matte with MG = 60% to the extent that 1% of total copper appears as metallic copper. Then the excessively oxidized concentrates come

895

in contact with the remainder of the concentrates so that the amount of the remainder decreases proportionally to $(1-\text{specific distance})^n$ where specific distance is defined as a ratio of any distance from the shaft roof to a distance at which the overall reaction is completed and where n is 1/2, 1, or 2. The specific distances of 0 and 1 correspond to the respective positions of the shaft roof and of about 3m below the shaft roof for the operation without recycling of dust. For the operation with recycling of dust,the specific distance of 1 corresponds to the position of the slag surface,i.e. 8m below the shaft roof. Since the participation of added silica has little effect on oxygen pressures of the system consisting of gas, slag and matte phases, the added silica was regarded as being attached to the concentrates just like gangue.

Calculated oxygen pressure profiles are shown in Fig.12. The calculation conditions at the specific distance = 1 are the same as those for case 1 in Table Ⅳ . The unreacted concentrate ratio and oxygen pressure at the specific distance = 0 are 23.2%, i.e. the amount of the remainder = 232kg/conc. t, and $10^{-6.36}$ atm at 1523K respectively, which are shown as point A in Fig.11. Since oxygen pressures have not been measured between the specific distances of 0 and 1 for the operation without recycling of dust, oxygen pressures measured for the operation with recycling of dust (14) are also shown in Fig.12 on the assumption that recycled dust does not affect an oxygen pressure profile based on specific distance. All the measured oxygen pressures except those in the burner cone are adjusted in Fig12 so that the oxygen pressures measured at No.1 matte hole are equal to $10^{-8.04}$ atm. On the other hand, those in the burner cone are re-normalized with respect to reaction (4) instead of the equilibrium reaction between Fe^{2+} and Fe^{3+} . The measured oxygen pressures seem to be simulated relatively well by curve I with n=1/2, and the strange oxygen pressure profile has been found to be quantitatively explainable by the two-particle model.

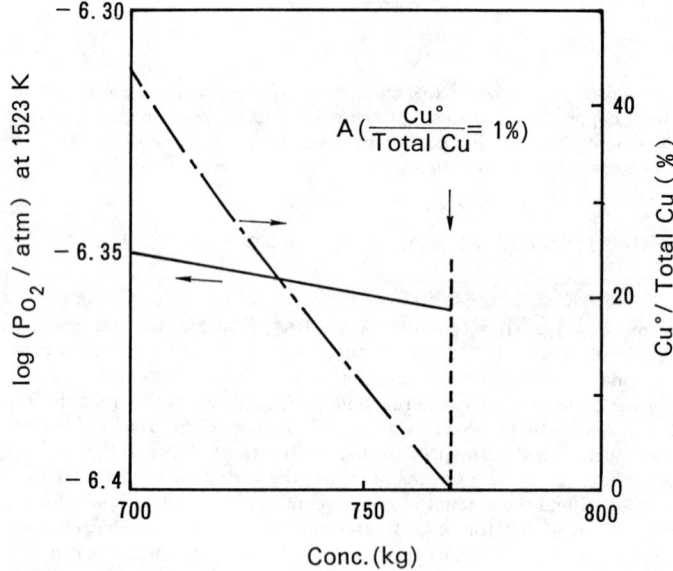

Figure 11 − Variations of oxygen pressure and the weight ratio of Cu (l)
to total copper in Cu₂S (l) and Cu (l) with the amount
of concentrate.

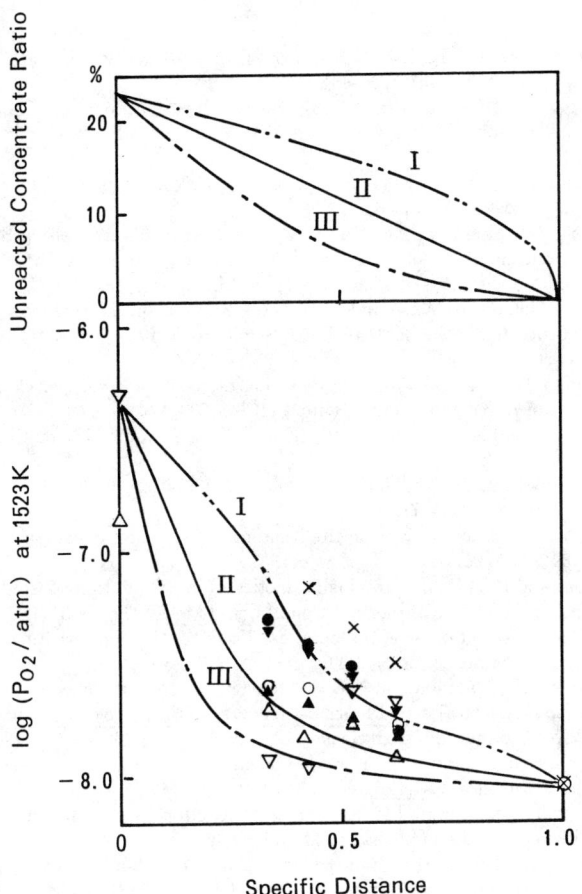

Figure 12 – Calculated oxygen pressure profiles against specific distance;
In the cases of I, II and III, unreacted concentrate ratios decrease
proportionally to $(1-\text{specific distance})^n$ where $n = 1/2$, 1 and 2, respectively.

Conclusion

A two-particle reaction model proposed by Kemori et al. (14) has been discussed on the basis of available information in the present study, and then used to simulate oxygen pressure profiles along the shaft. The model can explain quantitatively the decreasing profiles but not an increase in oxygen pressure, which takes place for the operation without recycling of dust, under conditions of the three-phase coexistence.

REFERENCES

1. F.R.A. Jorgensen and E.R. Segnit, "Copper Flash Smelting Simulation," Proc. Australas. Inst. Min. Metall.,No.261(1977), 34-46.
2. F.R.A. Jorgensen, "Combustion of Chalcopyrite, Pyrite, Galena and Sphalerite under Simulated Suspension Smelting Conditions," Australia Japan Extractive Metall. Symp., (1980), 41-51.
3. F.R.A. Jorgensen, "Single - Particle Combustion of Chalcopyrite," Proc. Australas. Inst. Min. Metall., No.288(1983), 37-46.
4. R.G. Henley, H.C. Hsiano and F.R.A. Jorgensen, "Suspension Smelting Studies on Mount Isa Copper Concentrate," Advances In Sulfide Smelting, vol.1, eds. H.Y. Sohn, D.B. George and A.D. Zunkel (The Metallurgical Society of AIME, 1983), 81-98.
5. T. Kumano, S. Ishida, K. Wase and N. Asano, "Fundamental Studies on the Oxidation of Copper Concentrates in the Shaft of Flash Smelting Furnace," J. Min. Metall. Inst. Japan, 96(1980), 559-564.
6. K. Otsuka and T. Soma, "Suspension Roasting of Pyrite and Copper Concentrate," Australia Japan Extractive Metall. Symp., (1980), 233-243.
7. H. Tsukada, Z. Asaki, T. Tanabe and Y. Kondo," Oxidation of Mixed Copper-Iron Sulfide," Metall. Trans. B, 12B (1981), 603-609.
8. Z. Asaki, A. Ueguchi, T. Tanabe and Y. Kondo, "Oxidation of Cu_2S Pellet," Trans. Japan Inst. Metals, 27(1986), 361-371.
9. I. Yusufoglu and M. Lemperle, "On the Kinetics of Oxidizing Roasting of Cuprous Sulfide," Erzmetal, 39(1986), 287-292.
10. P.C. Chaubal and H.Y. Sohn, "Intrinsic Kinetics of the Oxidation of Chalcopyrite Particles under Isothermal and Nonisothermal Conditions," Metall. Trans. B, 17B(1986), 51-60.
11. S. Ruottu,"The Description of a Mathematical Model for the Flash Melting of Cu-Concentrates," Combustion and Flame, 34(1979), 1-11.
12. N.J. Themelis, J.K. Mäkinen and N.D.H. Munroe, "Rate Phemomena in the Outokumpu Flash Smelting Reaction Shaft," Physical Chemistry of Extractive Metallurgy,eds.V. Kurdryk and Y.K. Rao (The Metallurgical Society of AIME, 1985), 289-309.
13. N. Kemori, Y. Shibata and M. Tomono, "Measurements of Oxygen Pressure in a Copper Flash Smelting Furnace," Metall. Trans. B, 17B(1986), 111-117.
14. N. Kemori, W.T. Denholm and H. Kurokawa, "Reaction Mechanism in a Copper Flash Smelting Furnace," to be published in Metall. Trans. B.
15. N. Kemori, T. Kimura, Y. Mori and S. Goto, "An Application of Goto's Model to a Copper Flash Smelting Furnace," be in press in " Pyrometallurgy' 87" of IMM.
16. T. Kimura, Y. Ojima, Y. Mori and Y. Ishii, "Reaction Mechanism in a Flash Smelting Reaction Shaft," The Reinhardt Schuhmann International Symposium, eds. D.R. Gaskell, J. P. Hager, J.E. Hoffmann and P.J. Mackey (The Metallurgical Society of AIME, 1986), 403-418.
17. Y.H. Kim and N.J. Themelis, "Effect of Phase Transformation and Particle Fragmentation on the Flash Reaction of Complex Metal Sulfides," ibid. 349-369.
18. N. Kemori, H. Kurokawa and Z. Kozuka, "Applications of Oxygen Probes to a Copper Flash Smelting Furnace," J. Min. Metall. Inst. Japan, 102(1986), 4i-47.
19. F.R.A. Jorgensen, "Ignition Testing of Materials from the Kalgoorlie Nickel Smelter," TMS Technical Paper, A86-21.
20. O. Kubaschewski and C.B. Alcock, Metallurgical Thermochemistry, 5th Edition (Pergamon International Library, 1979), 382.
21. M. Nagamori and P.J. Mackey, "Thermodynamics of Copper Matte Converting: Part 1, Fundamentals of the Noranda Process," Metall. Trans. B, 9B(1978), 255-265.
22. A. Yazawa et al., eds., Hitetsu Kinzoku Seiren (Japan Inst. Metals, 1980), 317.
23. B. Björkman and G. Eriksson, "Quantitative Equilibrium Calculations on Conventional Copper Smelting and Converting," Can. Metall. Quart., 21(1982), 329-337.
24. S. Goto, "Equilibrium Calculations between Matte, Slag and Gaseous Phases in Copper Smelting," Copper Metallurgy - Practice and Theory (Inst. Min. Met., 1974), 23-34.
25. M. Kameda and A. Yazawa, "Effects of Partial Pressures of Oxygen and Sulfur on the

Oxygen Content in Copper Matte," In Studies in Metallurgy (Tohoku University, 1969), 159-166.

26. A. Luraschi and J.F. Elliott, "Thermodynamic Behaviour of Oxygen and Sulphur in Copper-Iron - Sulphur - Oxygen Mattes at 1200 °C," Trans.- Inst. Min. Metall., Sect. C, 89(1980), C 14-C25.

27. D.L. Kaiser and J.F. Elliott, "Solubility of Oxygen and Sulfur in Copper - Iron Mattes," Metall. Trans. B, 17B(1986), 147-157.

28. M. Timucin and A.E. Morris, Metall. Trans.,1(1970), 3193.

29. N. Korakas, "Magnetite Formation during Copper Matte Converting," Trans. Inst. Min. Metall., 72(1962-1963), 35-53.

30. A. Vartiainen, "Schematic Presentation of Copper Losses within the Homogeneous Field of Fe−O−SiO$_2$ Slags," Scand. J. Metallurgy, 11(1982), 239-242.

31. R.P. Goel, H.H. Kellogg and J. Larrain, "Mathematical Description of the Thermodynamic Properties of the Systems Fe−O and Fe−O−SiO$_2$," Metall. Trans. B, 11B(1980), 107-117.

MATHEMATICAL MODELLING OF AN EQUILIBRIUM

COUNTER-CURRENT MULTISTAGE REACTOR SYSTEM

Harry E. Flynn*, Arthur E. Morris**, and Daniel Carter***

*Harry Flynn, formerly a Graduate Student at the
University of Missouri-Rolla, is now Project Engineer,
Kerr-McGee Corporation, Oklahoma City, Oklahoma 73125

**Arthur Morris is Professor of Metallurgical Engineering,
University of Missouri-Rolla, Rolla, Missouri 65401

***Dan Carter is a candidate for the Degree of BS in Computer Science,
University of Missouri-Rolla

Abstract

The chemical equilibrium program Solgasmix has been modified to simulate
a multistage reactor system consisting of up to nine stages. The program
(called STGSOL) simulates reactor systems in which the gas and condensed
phases move counter-currently. The temperature of each stage may be specified independently, material may be added at each stage, and non-ideal activity coefficients can be used for solution phases. STGSOL is configured for
use on an IBM-PC, and contains a menu-style data entry system, a text editor,
and a thermodynamic data base. Applications of the program to pyrite roasting, iron ore reduction, and nuclear waste processing are illustrated.

This research was supported by the U.S. Department of Energy through
the Waste Solidification Technology Branch; E.I. Du Pont de Nemours
& Company; Savannah River Laboratory, Aiken, S.C.; and the Department of the Interior's Mineral Institutes program administered by
the Bureau of Mines through the Generic Mineral Technology Center
for Pyrometallurgy under allotment grant number G1125129.

Introduction

The calculation of chemical equilibrium in complex systems involves solving several simultaneous non-linear equations. In recent years, a number of computer programs using various numerical techniques have been developed to perform these calculations in single-stage closed systems. A review of these techniques has been presented in an earlier paper in this volume (1). In particular, the program Solgasmix (2) has been widely used in the calculation of chemical equilibria, and it was chosen as the basis for developing two programs for modelling industrial reactor systems. The first of the programs is called ITSOL (1), and models open systems such as batch reactors by iteratively adding reactants and removing the gas phase in small increments. ITSOL is a single-stage multi-step reactor model.

The second of the programs is called STGSOL, and models steady-state multi-stage countercurrent reactor systems. In view of the proliferation of powerful microcomputers, both ITSOL and STGSOL have been configured to operate on IMB-PC or compatible computers.

STGSOL consists of four separate subroutines, each of which is menu-driven and has extensive text-editing capabilities. These are as follows:

1) FRED, a free energy database management system capable of creating multiple databases, each consisting of up to 500 species. FRED calculates the coefficients for free energy of formation equations of the type:

$$\Delta G^o_f = a + bT + cT\ln T \tag{1}$$

FRED stores the coefficients by compound name and formula, and reconstructs a table of free energy of formation data at 100° intervals.

2) SOLDAT, a reactor information data-entry program used to specify the temperature, pressure, and amounts of species to be added to each reactor stage. SOLDAT also searches FRED to select the appropriate free energy of formation data for each species thought to be present in each stage. The program can accommodate up to nine stages, with up to 14 elements, 50 species, and 19 phases. Up to four condensed-phase solution phases may be accommodated, and activity coefficients may be used by writing them into a special subroutine.

3) STGCALC, a linked group of Solgasmix free-energy minimization programs to calculate the equilibrium conditions in each stage. When the equilibrium condition is calculated for a given stage, the gas phase is passed upwards to the preceding stage, and the solid phase is passed downwards to the following stage. This process is repeated for all stages of the reactor system until a user-specified predetermined level of overall mass balance convergence is attained.

4) STGFILE, a storage location and file management system for reconfiguring the results from the last iteration of STGCALC. The user may browse through these results and select appropriate files to be printed, or the results may be stored under a selected filename.

Particular attention was given during the development of STGSOL to anticipating the need for the user to move between various subroutines and change or modify data. Wherever possible, an explanation of the various menu-screen option has been provided, and the most logical default values assigned to certain choices. A general overview of the subroutines making up STGSOL is shown in Figure 1.

902

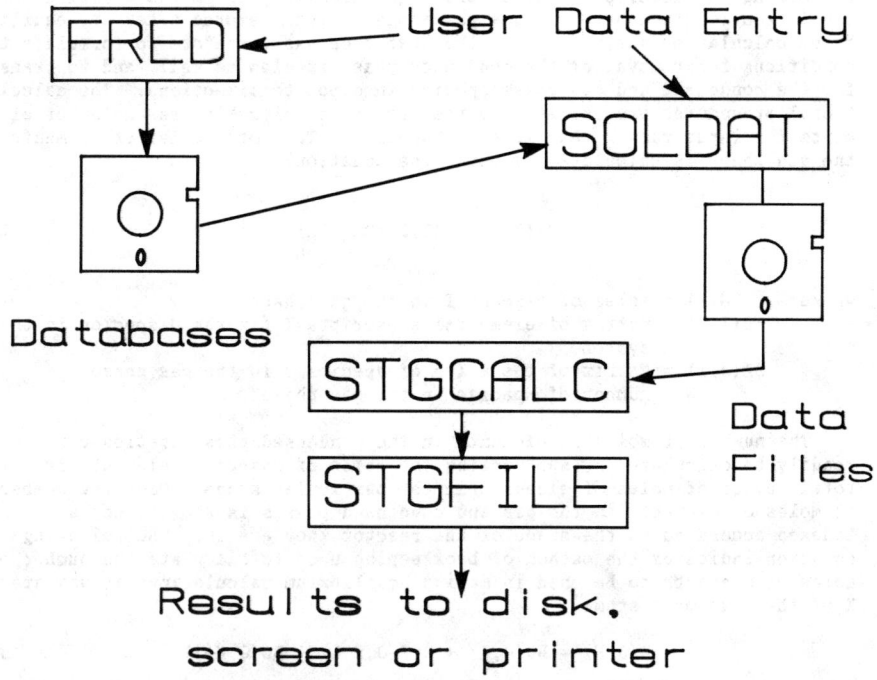

Figure 1 - Relationship between subroutines of the reactor
simulation program STGSOL.

Description of STGSOL

The features that a generic multi-stage counter-current reactor model
should have are as follows:

1) The ability of the equilibrium algorithm to converge with reasonable speed
   for a wide variety of problems (i.e., a "robust" program).

2) The ability to use activity coefficient equations for solution phases.

3) The ability to add material to any and all stages, to operate each stage
   at a different temperature and pressure, or to operate at constant volume.

4) The ability to independently specify which species and phases are to be
   considered in each stage.

5) The ability to transfer material from a particular stage to the one above
   and below it.

The version of Solgasmix used in STGSOL (3) already had features 1 and 2,
and was modified to produce a reactor system program which has all of the
above features. The program modifications that were made to incorporate
features 3 and 4 were easily accomplished. The main difficulty was to
incorporate a way to transfer species from stage to stage.

During the development of ITSOL, the Solgasmix program was modified to simulate a reactor where the gas phase species was removed after the equilibrium calculation was made (1). The next step was therefore to formulate the conditions for removal of the condensed-phase species as well, and to transfer the condensed and gas phase species in opposite directions. The calculational procedures was assisted by the fact that Solgasmix uses moles of elements for input rather than moles of species. The total moles of elements in the gas phase is calculated by using the equation:

$$BG(I) = \sum_{J=1}^{N} TM(I,J)*BA(J,1) \qquad (2)$$

where: BG(I) = moles of element I in the gas phase
         TM(I,J) = matrix of elemental subscripts I for the J species in the system
         BA(J,1) = Matrix of the moles of species J in the gas phase
              N = number of species in the gas phase.

The number of moles of elements in the condensed-phase species can readily be calculated by subtracting the moles of gas-phase elements from the total number of moles of elements in the particular stage. Once the number of moles of elements in the gas and condensed phases is known, they are indexed according to the stage of the reactor they are in. The following equation indicates the method of bookkeeping used to calculate the number of moles of elements to be used in making equilibrium calculations in any stage K of the reactor system:

$$B(J,K) = B\emptyset(J,K) + BCP(J,K-1) + BG(J,K+1) \qquad (3)$$

where   B(J,K) = number of moles of element J to be used in the Kth reactor stage
         B∅(J,K) = number of moles of element J input into the Kth reactor stage from external sources
       BCP(J,K-1) = number of moles of element J input into the Kth reactor stage from the condensed phases of the K-1 stage (next stage above)
       BG(J,K+1) = Number of moles of element J input into the Kth reactor stage from the gas phase of the K+1 stage (next stage below)

For the first cycle through the reactor BG(J,K+1) is set equal to zero.

The calculational part of the program (STGCALC) operates by sequentially calculating the equilibrium condition for each stage, then passing the elements in the gas phase to the stage above, and the elements in the condensed phase to the stage below. After the first iteration through all of the stages, a check is made on the overall mass balance for each element in each stage. After a second iteration through all stages, another check on the overall mass balance is made, and the difference between the two iterations is compared. Convergence is attained, and the program stopped, when the difference in the overall reactor mass balance for two successive iterations is less than a previously user-specified amount.

Aids to Convergence

Owing to the nature of Solgasmix, convergence can be slow if a large number of condensed phase species are stable, an even larger number is selected for consideration, and the starting estimate is "far" from the equilibrium condition. When a series of Solgasmix algorithms are linked and

904

material is transferred between them, the time for overall convergence for a multi-element multi-phase system may be several hours (on an IBM PC-XT with 8087 math co-processor chip). Two features have been incorporated in STGSOL to minimize the time required to attain overall convergence.

The first feature is the use of the equilibrium composition of a stage from the previous cycle as the initial estimate for equilibrium during the next cycle. For many systems this is a very useful feature, since the difference between a stage in one cycle and the next becomes less and less as the reactor approaches overall convergence. This does not improve the performance of the first cycle of course, but calculation times for subsequent cycles may be decreased by a factor of 50.

The second feature is the calculation and printout of the activities of all of the condensed-phase species being considered in each stage. This allows the user to assess the relative stabilities of each of the non-stable species, and to eliminate from consideration (for future runs) all species having an equilibrium activity less than about 10^{-2} (or any other low number). The fewer condensed-phase species considered by the algorithm, the faster the convergence. The activity of the condensed-phase species is calculated by:

$$AC(I) = \exp \left[-\Delta G_i^\circ / RT + \Sigma A(I,J) * PI(J) \right] \tag{4}$$

where $AC(I)$ = activity of specie I
  G_i° = free energy of formation of specie I
  R = gas constant
  T = temperature, K
  $A(I,J)$ = subscript of element J for specie I in the element coefficient matrix
  $PI(J)$ = Lagrange multiplier for element J as calculated by Solgasmix

Persons familiar with the Solgasmix algorithm used for STGSOL will remember that a gas phase must be present in any system being treated (3). For this reason, a small amount of inert gas like nitrogen should be added to each stage.

Examples of STGSOL Use

Three examples will be given to illustrate the applicability of STGSOL to determine the overall and stagewise mass flow and gas concentration in a countercurrent flow reactor system. The first example is roasting of pyrite (two stages), the second example is reduction of impure hematite (four stages) and the third example is glassification of nuclear waste (nine stages).

Pyrite Roasting

Pyrite may be roasted under controlled conditions to yield all of its contained sulfur in the elemental form. This is illustrated in the following example, in which pyrite is introduced into the top stage of a two-stage reactor, and a mixture of SO_2 and O_2 are introduced into the bottom stage. As mentioned earlier, a trace of nitrogen is introduced into both stages of the reactor to assist in the convergence. A sketch of the reactor system is shown in Figure 2.

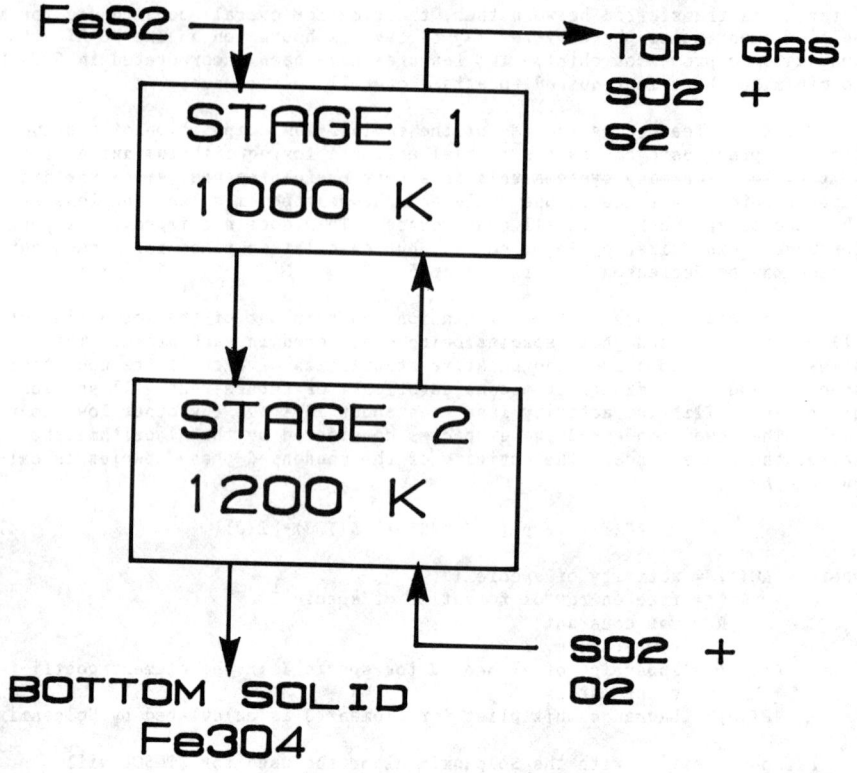

Figure 2 - Two-stage countercurrent reactor arrangement for the production of
elemental sulfur from pyrite. Trace nitrogen omitted from figure.

In the top stage (stage 1), pyrite is thermally decomposed to pyrrhotite
of composition approximately $FeS_{1.14}$, thus evolving 0.43 moles of labile
diatomic sulfur gas (S_2). In the bottom stage (stage 2), pyrrhotite is
oxidized to magnetite by a mixture of SO_2 and O_2 chosen in such a way that
the gaseous oxidation product contains all of the pyrrhotite sulfur as S_2.
The thermodynamics of pyrite/pyrrhotite oxidation have been discussed by
Rosenqvist (4) among others. A series of runs of varying ratio of SO_2/O_2
were made with STGSOL until the observed moles of SO_2 leaving in the top gas
were the same as the moles of SO_2 entering stage 2. In each case, convergence was accomplished in less than 3 minutes using a PC-AT with math co-
processor chip. A summary of results is shown in Table I, which omits the
trace of nitrogen used as a convergence aid.

Iron Ore Reduction

The reduction of hematite iron ore may produce intermediate phases such
as magnetite and wustite, before forming metallic iron. If impurities are
present, they may form other intermediate reaction products as well. In this
example, a mixture of hematite and silica is reduced in four stages to pro-

duce metallic iron (containing dissolved carbon), and silica. Phases considered are hematite, magnetite, wustite (of varying composition), fayalite, iron-carbon solid solution, silica, and carbon. A sketch of the reactor system is shown in Figure 3.

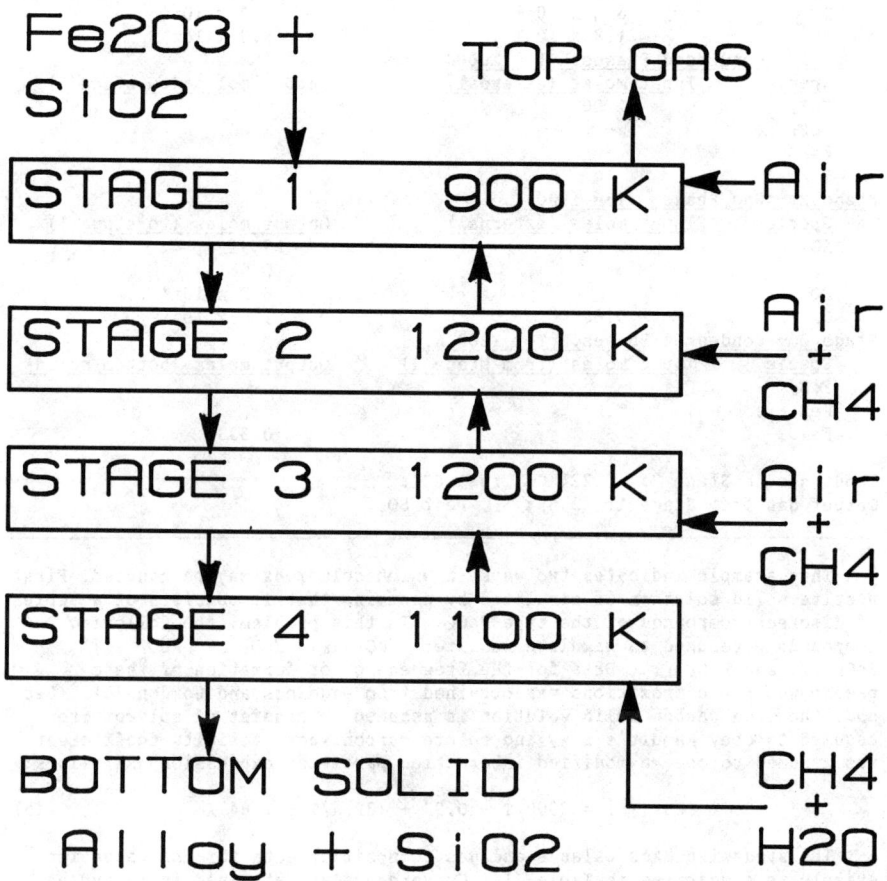

Figure 3 - Four-stage countercurrent reactor arrangement for the reduction of impure hematite ore. Trace nitrogen omitted from figure.

In stage 1, the reducing gas is burned with air to preheat the hematite. In stages 2 and 3, partial reduction occurs; a small amount of air and methane are added to simulate the addition of thermal energy and to aid convergence. In stage 4, complete reduction to an iron-carbon alloy takes place by the addition of a steam-methane mixture.

Stage 1 - Gas Phase. T = 1000 K.

Specie	Input moles (from stage 2)	Output moles (to top gas)
SO_2	29.91	29.91
S_2	0.56	1.00
SO	6.2×10^{-3}	9.0×10^{-4}
O_2	1.8×10^{-10}	9.3×10^{-14}

Stage 1 - Condensed Phases. T = 1000K.

Specie	Input moles (external)	Output moles (to stage 2)
FeS_2	1.00	-
$FeS_{1.14}$	-	1.00
Fe_3O_4	-	-

Stage 2 - Gas Phase. T = 1200 K.

Specie	Input moles (external)	Output moles (to stage 1)
SO_2	29.91	29.91
S_2	-	0.56
SO	-	6.2×10^{-3}
O_2	0.68	1.8×10^{-10}

Stage 2 - Condensed Phases. T = 1200 K.

Specie	Input moles (from stage 1)	Output moles (bottom solid)
FeS_2	-	-
$FeS_{1.14}$	1.0	-
Fe_3O_4	-	0.333

Input gas to Stage 2: 2.22% O_2, rest SO_2.
Output gas from Stage 1: 3.69% S_2, rest SO_2.

This example indicates two ways in which solutions may be handled. First, wustite solid solution is simulated by assuming that it consists of a series of discreet compounds of the type FeO_x. In this problem, the following compounds were used to simulate wustite: $FeO_{1.053}$, $FeO_{1.07}$, $FeO_{1.09}$, $FeO_{1.11}$, and $FeO_{1.13}$. Data for the free energy of formation of these pseudo-wustite compositions was obtained from Giddings and Gordon (5). Second, the iron carbon solid solution is assumed to consist of solvent iron assumed to obey Raoult's Law, and solute carbon whose activity coefficient was assumed to obey a modified interaction parameter expression as follows:

$$\log (\gamma_C) = 2300/T - 0.92 + (3254/T + 1.64)X_C \tag{5}$$

The stagewise mass balance and gas composition data for the reduction example is summarized in Table II. Convergence was attained in 13 cycles, and required about 12 minutes on a PC-AT with math co-processor chip. Convergence could have been speeded up by eliminating from consideration more of the condensed-phase species that are clearly unstable under the particular conditions.

Slurry-Feed Melter for Nuclear Waste Processing

Slurry-feed melters are being developed to convert radioactive waste to borosilicate glass for permanent disposal (6). Controlled burning of the organic materials in the waste is necessary to prevent accumulation of combustible gases or soot. The oxygen potential must also be controlled; if too high, glass foaming may occur, and if too low, a matte phase may form, which lowers melter life.

Table II. Stagewise Mass Balance for Iron Oxide Reduction Example

Stage 1 - Gas Phase. T = 900 K.

Specie	Input moles: (from stg 2)	(external)	Output moles (to top gas)
O_2	1.23×10^{-15}	2.10	0.53
N_2	0.16	7.90	8.06
CO	0.90	-	-
CO_2	0.70	-	1.60
H_2	2.44	-	-
H_2O	2.60	-	5.04
CH_4	4.5×10^{-5}	-	-

Stage 1 - Condensed Phases. T = 900 K.

Specie	Input moles (external)	Output moles (to stg 2)
Fe_2O_3	1.0	1.0
SiO_2	0.2	0.2
Fe_3O_4	-	-
Fe_2SiO_4	-	-

Stage 2 - Gas Phase. T = 1200 K.

Specie	Input moles: (from stg 3)	(external)	Output moles (to stg 1)
O_2	4.4×10^{-16}	0.021	1.23×10^{-15}
N_2	0.084	0.079	0.16
CO	1.08	-	0.90
CO_2	0.51	-	0.70
H_2	3.05	-	2.44
H_2O	1.97	-	2.60
CH_4	1.4×10^{-4}	0.01	4.5×10^{-5}

Stage 2 - Condensed Phases. T = 1200 K.

Specie	Input moles (from stg 1)	Output moles (to stg 3)
Fe_2O_3	1.00	-
Fe_3O_4	-	-
$FeO_{1.053}$	-	-
$FeO_{1.07}$	-	-
$FeO_{1.09}$	-	-
$FeO_{1.11}$	-	-
$FeO_{1.13}$	-	1.60
SiO_2	0.20	0.20
Fe_2SiO_3	-	-

Stage 3 - Gas Phase. T = 1200 K.

Specie	Input moles: (from stg 4)	(external)	Output moles (to stg 2)
O_2	2.0×10^{-19}	0.021	4.4×10^{-16}
N_2	0.005	0.079	0.084
CO	1.38	-	1.08
CO_2	1.83	-	0.51
H_2	4.38	-	3.05
H_2O	0.59	-	1.97
CH_4	0.02	0.01	1.4×10^{-4}

Stage 3 - Condensed Phases. T = 1200 K.

Specie	Input moles (from stg 2)	Output moles (to stg 4)
Fe_2O_3	-	-
Fe_3O_4	-	-
$FeO_{1.053}$	-	0.13
$FeO_{1.07}$	-	-
$FeO_{1.09}$	-	-
$FeO_{1.11}$	-	-
$FeO_{1.13}$	1.60	-
SiO_2	-	-
Fe_2SiO_4	0.20	0.20
Fe-C Alloy	-	1.47; ($X_C = 6.6 \times 10^{-4}$)

Table II - continued

Stage 4 - Gas Phase.   T = 1100 K.

Specie	Input moles (external)	Output moles (to stg 3)
O_2	–	2.0×10^{-19}
N_2	0.005	0.005
CO	–	1.38
CO_2	–	1.83
H_2	–	4.38
H_2O	1.80	0.59
CH_4	1.60	0.02

Stage 4 - Condensed Phases.   T = 1100 K.

Specie	Input moles (to stg 3)	Output moles (bottom solid)
Fe_2O_3	–	–
Fe_3O_4	–	–
$FeO_{1.053}$	0.13	–
$FeO_{1.07}$	–	–
$FeO_{1.09}$	–	–
$FeO_{1.11}$	–	–
$FeO_{1.13}$	–	–
SiO_2	–	0.20
Fe_2SiO_3	0.20	–
Fe-C alloy	1.47 ($X_c = 6.6 \times 10^{-4}$)	2.02 ($X_c = 0.0088$)

A sketch of the reactor is shown in Figure 4. The slurry enters the melter near the top. As the slurry descends, water is boiled off, hydroxides, carbonates, and sulfates decompose, and glass is formed in the bottom stage. A typical feed composition is shown in Table III.

Owing to the length of this problem, a detailed stagewise mass balance will not be presented here. However, the composition of the top gas and the composition of the bottom melt are shown in Table IV.

Table III - Major constituents (dry basis) of a typical nuclear waste slurry fed into electrically-heated melter. Feed is about 50 wt.% solids; liquid phase is dilute formic acid solution.

Specie	Fe_2O_3	SiO_2	Na_2O	Cr_2O_3	NiO	S
moles	0.58	7.0	1.53	0.004	0.27	0.02

Table IV - Top gas and bottom phase composition leaving melter (major species).

Gas Phase	N_2	H_2O	CO	CO_2	H_2			
moles	10.1	61.3	0.02	0.086	0.57			

Liquid Phase	FeO	Fe_2O_3	Fe_3O_4	SiO_2	Na_2O	NiO	Cr_2O_3	SiO_2 (solid)
moles	0.27	0.23	0.14	2.05	1.53	0.27	0.004	4.95

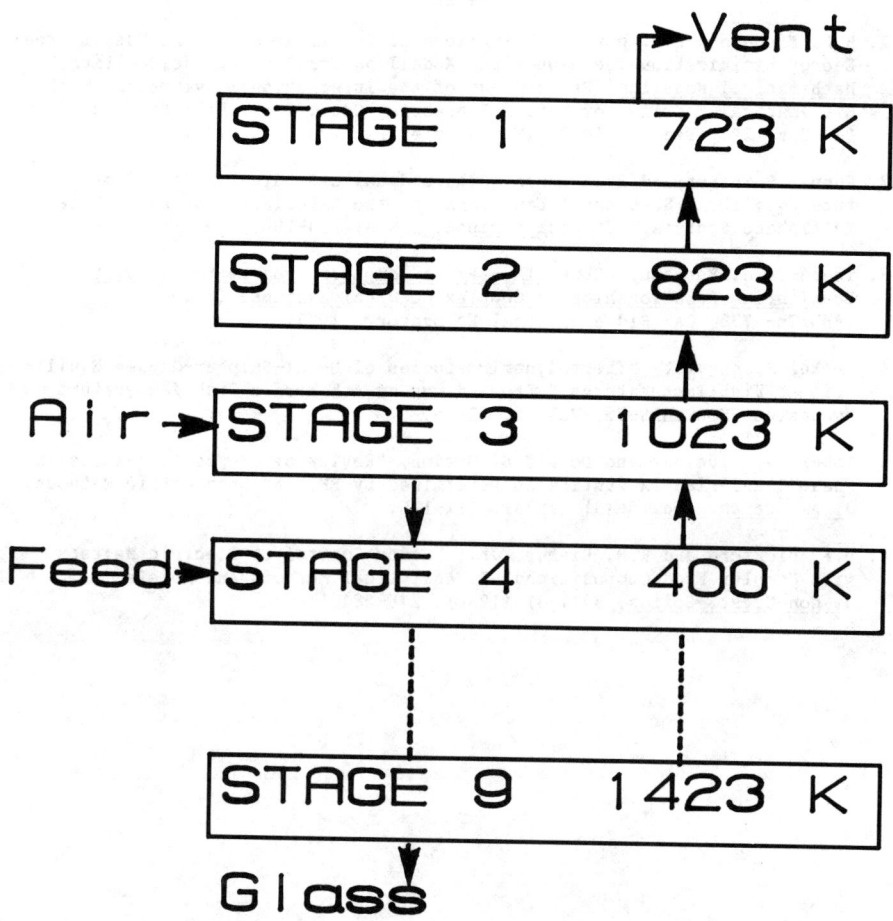

Figure 4 - Nine-stage countercurrent reactor simulation of an
electrically-heated slurry-fed waste melting furnace.

Conclusions

STGSOL is a convenient and easy-to-use program for modelling steady-state
multi-stage counter-current reactor systems where chemical equilibrium can be
assumed. The program is available in compiled Fortran 77, and is configured
to run on an IBM PC-XT or better, with a math co-processor chip and a color
graphics card. Reactors consisting of up to 9 stages may be modelled. The
program consists of subroutines to create and manage a thermodynamic database
(FRED), set up the reactor operating characteristics and search the database
(SOLDAT), calculate the equilibrium stage composition and move material between stages (STGCALC), and manage the output file (STGFILE).

References

1. H.E. Flynn and A.E. Morris, "Experimental Verification of the Use of Free-Energy Minimization Techniques for Modelling Complex Sulfide Smelting," Mathematical Modeling, Proceedings of the International Symposium on the Mathematical Modeling of Metals Processing Operations, Palm Springs, CA, Nov. 29 - Dec. 2, 1987, (Warrendale, PA: TMS-AIME, 1987).

2. Gunnar Eriksson and Erik Rosen, "Thermodynamic Studies of High Temperature Equilibria 8. General Equations for the Calculation of Equilibria in Multiphase Systems," Chemica Scripta, (1973) 193-194.

3. Theodore M. Besmann, SOLGASMIX - PV, a Computer Program to Calculate Equilibrium Relationships in Complex Chemical Systems, (Report ORNL/TM-5775, Oak Ridge National Laboratory, 1977).

4. Terkel Rosenqvist, "Thermodynamic Studies of Metal-Sulphur-Oxygen Equilibria at High Temperatures," Proceedings of 4th Nordic High Temperature Symposium -NORTEMPS-75, Vol. 1, 25-45.

5. Robert A. Giddings and Donald S. Gordon, "Review of Oxygen Activities and Phase Boundaries in Wustite as Determined by EMF and Gravimetric Methods," J. Am. Ceram. Soc. 56(3) (1973), 111-116.

6. D.F. Bickford and R.B. Diemer, Jr., "Redox Control of Electric Melters with Complex Feed Compositions. I. Analytical Methods and Models," J. Non-Cryst. Solids, 84(1-3) (1986), 276-284.

HEAT-TRANSFER RATES BETWEEN A CONFINED NON-TRANSFERRED ARC

PLASMA-JET AND A PLANE ISOTHERMAL SURFACE

F. K. Ojebuoboh, M. J. Cusick, and G. P. Martins

Department of Metallurgical Engineering
Colorado School of Mines
Golden, Colorado   80401

Abstract

The rate of heat-transfer between a plasma-jet impinging onto a plane
surface (quench plate) within an enclosure has been investigated. The
relevance of this research is associated with the subsequent rapid
quenching of reaction products produced within a plasma jet. Specifically,
the results can be interpreted within the context of producing ultrafine
(<100nm) metal or ceramic powders. Fluid-flow and temperature fields were
predicted through the solution of the flow equations for mass, momentum and
energy. The turbulent-flow behavior was described using the two equation
k-ε model. The customized digital-computer code--TEACH--was used to
produce the computer process-simulator. The heat-transfer rate to the
quench plate was computed using values of the local wall-shear-stress,
based on Reynolds analogy. The predicted values of the enthalpy-transfer
rates from the plasma were assessed from results using a custom-designed
calorimeter with geometry corresponding to the system simulated. These
computer-simulated results overpredicted the heat-transfer rates by
approximately a factor of two.

913

Introduction

In contrast to plasma spraying and coating, thermal–plasma processes for material synthesis require controlled atmospheres and, therefore, the plasma jet must be confined within a physical enclosure. In this paper, the rate of heat transfer from a confined jet to a plane, isothermal surface is reported for the case in which the jet is produced from a non–transferred–arc plasma. Figure 1 is an illustration of the confined plasma and the impingement system used in this investigation. Downstream and normal to the plasma jet is a plane, diabatic surface onto which the jet impinges.

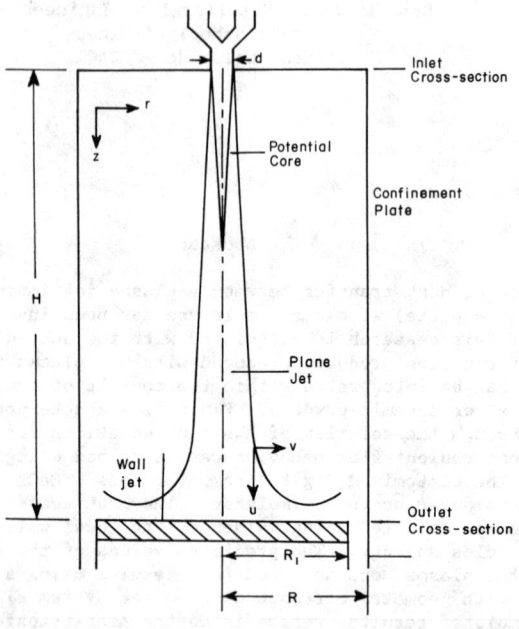

Figure 1.   Schematic representation of the confined–jet system used for the computer model.

Following accepted practice in jet–flow phenomenology (1), a stagnation flow, a wall–jet flow and the plane–jet flow can be identified. This is the flow pattern envisaged for a confined plasma–jet which is produced from a non–transferred arc, and subsequently impinges onto a plane surface. Plasma reactors of this configuration may be useful in thermal plasma processing for materials production (2,3). The production of metal and ceramic powders in a thermal plasma requires rapid cooling. In plasma–melted and rapidly–solidified (PMRS) processing, powders may be quenched in an inert or "compatible" fluid medium or onto a chilled substrate (2). In plasma chemical–synthesis (4,5), the resulting gas stream produced by reaction of a metal or ceramic precursor within a plasma jet may also be quenched into a fluid medium or by impingement onto a cooled surface.

Impingement onto a surface appears to provide superior rapid-cooling in these processes. Reported in this paper is the solution of the flow equations for mass, momentum and thermal energy used to predict the fluid-flow and temperature fields of an argon plasma in a reactor of cylindrical geometry and which contains a cooled impingement plate. The results provide the basis for predicting the heat-transfer rates to a plane surface, such as the quench plate in a plasma reactor.

Mathematical-Model Development

The problem is formulated by considering the case of axisymmetric jet-expansion into a cylindrical plenum, and the principal parameters have been defined in Figure 1. In addition, the flow in the expansion section is turbulent and the quench plate is isothermal. These details are commensurate with observed test conditions in which the plasma source is a torch of the non-transferred-arc configuration and the quench plate is a water-cooled, copper plate with a plane surface (6).

The velocity and temperature flow-fields within the confinement can be described by momentum and energy conservation-equations which incorporate an acceptable turbulent-flow description (7).

Conservation Equations

The equations for two-dimensional flow can be written in cylindrical coordinates in terms of the primitive variables, u, v, and T. The steady-state equations for the plasma-jet which is symmetrical about the flow axis may then be written as follows:

a. continuity

$$\frac{\partial}{\partial z} (\rho u) + \frac{\partial}{\partial r} (\rho r v) = 0 \tag{1}$$

b. axial momentum conservation

$$\frac{1}{r} \frac{\partial}{\partial r} (\rho r v u) + \frac{\partial}{\partial z} (\rho u^2) = - \frac{\partial P}{\partial z} + \frac{1}{r} \frac{\partial}{\partial r} \{r \mu_e (\frac{\partial u}{\partial r} + \frac{\partial v}{\partial z})\} +$$

$$2 \frac{\partial}{\partial z} (\mu_e \frac{\partial u}{\partial z}) + \rho g \tag{2}$$

c. radial momentum conservation

$$\frac{1}{r} \frac{\partial}{\partial r} (\rho r v^2) + \frac{\partial}{\partial z} (\rho u v) = - \frac{\partial P}{\partial r} + \frac{\partial}{\partial z} \{\mu_e (\frac{\partial u}{\partial r} + \frac{\partial v}{\partial z})\} +$$

$$2 \frac{\partial}{\partial r} (\mu_e r \frac{\partial v}{\partial r}) - 2 \frac{v \mu_e}{r^2} \tag{3}$$

d. energy conservation

$$\frac{1}{r} \frac{\partial}{\partial r} (\rho r v T) + \frac{\partial}{\partial z} (\rho u T) = \frac{1}{r} \frac{\partial}{\partial r} (\frac{\mu_e}{\sigma_T} r \frac{\partial T}{\partial r}) + \frac{\partial}{\partial z} (\frac{\mu_e}{\sigma_T} \frac{\partial T}{\partial z}) \tag{4}$$

915

Turbulent-Flow Equations

The turbulence of the flow is described by the two-equation, k-ε viscosity model of Launder and Spalding (7). In this model, the turbulent viscosity is given by:

$$\mu_t = \frac{C_\mu \rho k^2}{\epsilon} \tag{5}$$

and

$$\mu_e = \mu_t + \mu_\ell \tag{6}$$

k and ε are respectively the turbulent kinetic-energy and the turbulent kinetic energy dissipation rate. Equations similar in form to the energy-conservation equation are obtained which provide local values of these quantities. These equations in turn describe the local values of the length and time scales of the turbulence. These equations are:

e. turbulent kinetic-energy

$$\frac{1}{r}\frac{\partial}{\partial r}(\rho r v k) + \frac{\partial}{\partial z}(\rho u k) = \frac{\partial}{\partial z}\left(\frac{\mu_e}{\sigma_k} \cdot \frac{\partial k}{\partial z}\right) +$$

$$\frac{1}{r}\frac{\partial}{\partial r}\left(\frac{\mu_e}{\sigma_k} r \frac{\partial k}{\partial r}\right) + (G - \rho\epsilon) \tag{7}$$

f. turbulent-kinetic-energy dissipation-rate

$$\frac{1}{r}\frac{\partial}{\partial r}(\rho r v \epsilon) + \frac{\partial}{\partial z}(\rho u \epsilon) = \frac{\partial}{\partial z}\left(\frac{\mu_e}{\sigma_e} \cdot \frac{\partial \epsilon}{\partial z}\right) +$$

$$\frac{1}{r}\frac{\partial}{\partial r}\left(\frac{\mu_e}{\sigma_e} r \frac{\partial \epsilon}{\partial r}\right) + (C_1 \frac{\epsilon}{k} G - C_2 \frac{\epsilon^2}{k} \rho) \tag{8}$$

The term G is the generation of turbulence and it is given by:

$$G = 2\mu_t\left\{\left(\frac{\partial u}{\partial z}\right)^2 + \left(\frac{\partial v}{\partial r}\right)^2 + \left(\frac{v}{r}\right)^2 + \frac{1}{2}\left(\frac{\partial u}{\partial r} + \frac{\partial v}{\partial z}\right)^2\right\} \tag{9}$$

Gosman et al. (8) have shown that the similarity amongst the partial differential equations permits a standard recasting of all the flow equations in terms of the primitive variables, the exchange coefficient and the source terms. This has been done for this system of equations in Table I. It should be noted that the applicability of the equations have been restricted to the conditions in which viscous heat generation and radiant heat loss are negligible. Numerical values, suggested in reference (7), for the turbulence model constants appearing in Equations 5 through 8 are presented in Table II.

TABLE I. General Form of Recasted Differential Equations.

$$\frac{1}{r} \left[\frac{\partial}{\partial r} (\rho r v \Phi) + \frac{\partial}{\partial z} (\rho r u \Phi) - \frac{\partial}{\partial r} (r \Gamma_\Phi \frac{\partial \Phi}{\partial r}) - \frac{\partial}{\partial z} (r \Gamma_\Phi \frac{\partial \Phi}{\partial z}) \right] = S_\Phi$$

Primitive Variable Φ	Exchange Coefficient Γ_Φ	Source Term S_Φ
1	0	0
u	μ_e	$-\frac{\partial P}{\partial z} + \frac{\partial}{\partial z} (\mu_e \frac{\partial u}{\partial z}) + \frac{1}{r} \frac{\partial}{\partial r} (r \mu_e \frac{\partial v}{\partial z}) + \rho g$
v	μ_e	$-\frac{\partial P}{\partial r} + \frac{\partial}{\partial z} (\mu_e \frac{\partial u}{\partial r}) + \frac{1}{r} \frac{\partial}{\partial r} (\mu_e r \frac{\partial v}{\partial r})$ $-\frac{2v}{r^2} \mu_e$
T	μ_e / σ_T	0
k	μ_e / σ_k	$G - \rho \varepsilon$
ε	$\mu_e / \sigma_\varepsilon$	$C_1 (\frac{\varepsilon}{k}) G - C_2 (\frac{\varepsilon^2}{k}) \rho$

Boundary Conditions

The following boundary conditions were used for the numerical solution of the set of equations of Table I;

i) plasma nozzle exit (inlet cross-section); z = o:

$$v = o \ , \ u = u_{in} \ , \ T = T_{in} \qquad\qquad [10]$$

ii) axis of symmetry; r = o:

$$\frac{\partial u}{\partial r} = o \ , \ \frac{\partial T}{\partial r} = o \qquad\qquad [11]$$

iii) confinement wall; r = R:

$$v = o \ , \ u = o \ , \ T = T_{wall} \qquad\qquad [12]$$

917

iv) plane surface (quench plate); z = H:

$$v = o \; , \; u = o \; , \; T = T_{plate} \tag{13}$$

v) reactor exit; z = H and $R_1 \leq r \leq R$:

Here, the relevant variable is allowed to "drift" to a stable configuration. For the velocity, this is achieved by a mass-balance closure in the calculation domain. A similar treatment is not apparent for temperature; however, Patankar (9) has suggested the "local one-way behavior" at the outflow boundary.

TABLE II. Model Constants Used in Computer Simulation

k-ε:	C_1	1.44
	C_2	1.92
	C_μ	0.09
	σ_k	1.0
	σ_ϵ	1.3
wall function:	κ	0.4
	E	9.0
heat transfer:	σ_T	0.7

Near-Wall Conditions

"Wall functions", recommended in Launder and Spalding (7), are used to describe the near-wall conditions. They account for the rapidly changing velocity gradients near walls; in addition, they are also used to calculate the values of heat flux (a scalar quantity) at the wall and quench plate. The expression for the local shear-stress is:

$$\frac{u_n}{(\tau/\rho)_w} C_\mu^{1/4} k_n^{1/2} = \frac{1}{\kappa} \ln \left[E \left(C_\mu^{1/4} k_n^{1/2} \right) (z_w - z_n) \right] \tag{14}$$

The following expression suggested by Serag-Eldin and Spalding (10) was used for the heat flux:

$$q = C_{pn} \cdot \frac{\tau_w}{u_n} (T_n - T_w) \tag{15}$$

918

Notice that the local shear stress at the wall as well as conditions at the near-wall node, n, are required to determine the flux to the wall, w. The values of κ and E selected for the numerical computations are listed in Table II.

Enthalpy flowrates were subsequently calculated on the basis of integrating q over the surface area of the plane surface.

Computational Details

The digital computer code, TEACH, was used to produce the computer process-simulator. TEACH (Teaching Elliptic Axisymmetric flow Characteristics Heurestically) uses a finite difference scheme for computing turbulent flows, and the version used here is for the solution of two-dimensional Elliptical Flows (TEACH-2EF) (11). The 19x19 numerical grid used for this simulation is shown in Figure 2. As can be seen, the grid configuration has non-uniform spacing which provides for more computation nodes in regions with rapidly changing flow properties.

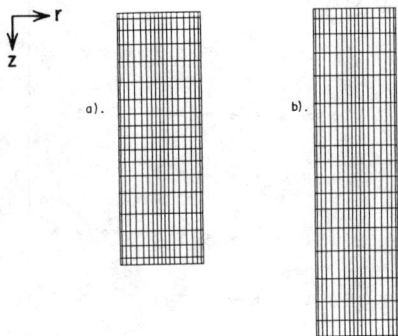

Figure 2. Grid configuration employed for the computation;
(a) H = 0.076m and (b) H = 0.1m.

The computational scheme typically utilized about 120 seconds of CPU time on the Colorado School of Mines DEC-10 computer. Convergence was achieved when the sums of normalized residuals of the primitive variables are below prescribed values and the residual of the overall-energy-balance-closure did not exceed 10%. The computational strategy was found to be most efficient if non-isothermal perturbations, due to temperature dependent properties, were effected after the flow-field had attained some stability. The temperature dependence of the physical properties of the plasma were obtained from several sources (12,13,14) and fitted to polynominals in the temperature range of interest.

Predicted Results

The primary purpose of the computer simulator was to predict the heat-transfer rates from the plasma to the plate onto which it impinges. However, this prediction requires that the fluid-flow field and temperature

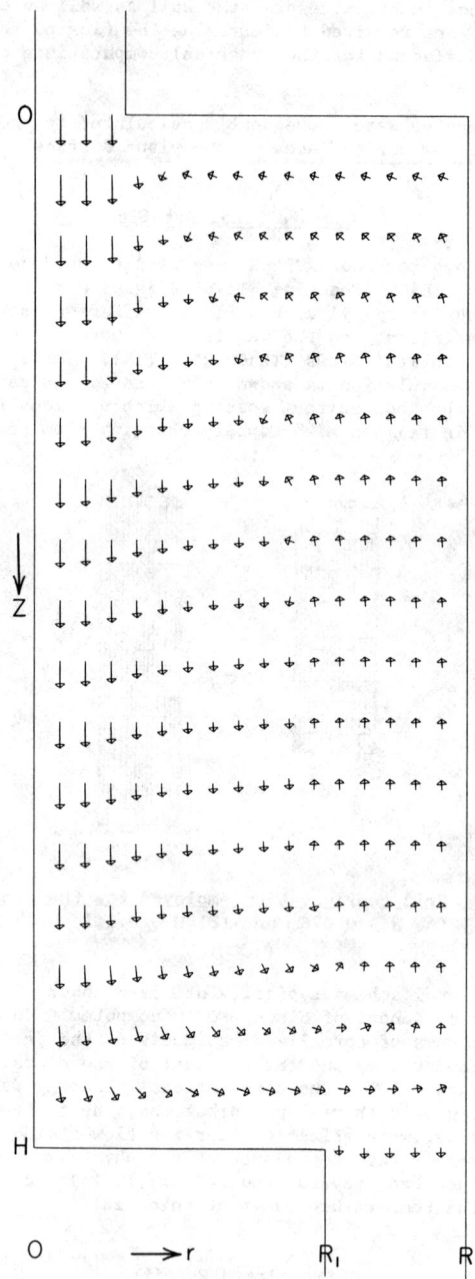

Figure 3. Velocity vector plot of predicted flow field
(100m·s⁻¹ inlet velocity).

profile of the plasma gas just prior to impingement be obtained. A
secondary objective was, therefore, to observe the characteristic features
of the plasma–jet flow-field within the enclosure. This information is
important in regard to the mixing behavior and temperature distribution
within the reactor. A selection of the model–predictions for the
conditions indicated in Table III is now presented. The operating
parameters for the plasma, 100 m·s^{-1} and 10,000K corresponds to a
working-gas flowrate (standard conditions) of approximately 14 m(m)^3·min^{-1}
(14 ℓ·min^{-1} [SLM]) at a specific enthalpy of 350 W/SLM. Thus the nozzle
(net) plasma–energy is of the order of 5 kW.

A velocity-vector plot, Figure 3, summarizes the velocity of the
plasma/gas between the entrance plane and the impingement surface. Each
vector describes the resultant axial and radial velocities at that point,
and it can be seen that the magnitude of the velocity rapidly diminishes in
the radial direction. Also, the direction of the vectors (arrow head)
indicates that a significant recirculating–region develops away from the
axis of symmetry. Although the size of this region is relatively large,
the recirculating fluid–volume is not proportionately large. The velocity
at, and close to, the symmetry axis remains relatively high until the fluid
is in close proximity to the impingement surface. This behavior of the
flow in the plane-jet region is better quantified in Figure 4 where the
axial velocity is shown as a function of the distance from the inlet
plane. The modification of the flow by the impingement surface, is clearly
confined to the near–plate region.

The same behavior is observed for the effect of the impingement
surface on the temperature profile. In Figure 5, isotherms display the
predicted temperature–profiles in the enclosure, respectively 0.076m and
0.1m long. The rapid decrease in temperature of the gas in the proximity
of the surface is evident from the compression of the contours in the

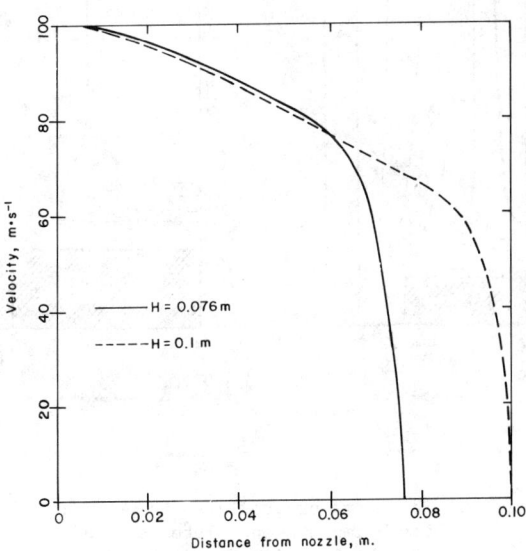

Figure 4. Axial velocity along the symmetry axis.

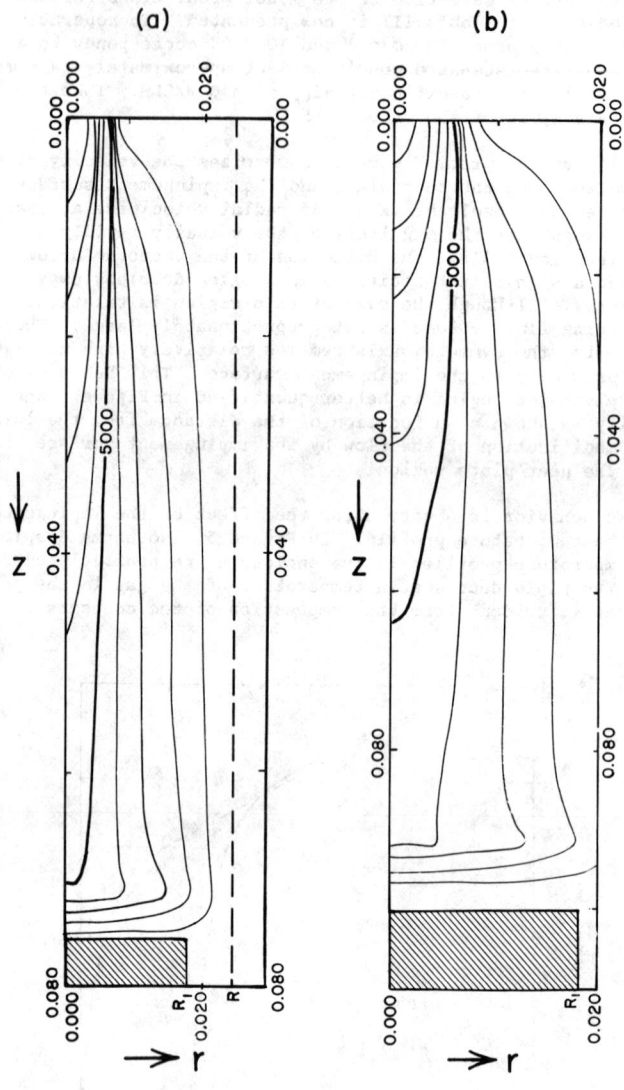

Figure 5. Isotherms of predicted temperatures for a
torch/impingement plate separation of (a) 0.076m
and (b) 0.1m (contour interval = 1000K).

TABLE III - Operating Conditions and Geometric Quantities

Inlet temperature	10,000K
Inlet velocity	100 m·s^{-1}
Nozzle diameter (I.D.)	10mm
(Torch) gas-flow-rate	14.2 SLM argon
Plate diameter	47.8mm
Plate separation from inlet	76 and 100mm
Reactor diameter	50.8mm

near-plate region. From Figure 5(b), it can be seen that the steepest change in temperature in this near-plate region occurs in the last 5% of the distance from the torch exit to the plane surface. This region is exaggerated in Figure 6, where a temperature gradient of 7.00×10^2 K·mm^{-1} is attained in the last centimeter prior to surface impingement.

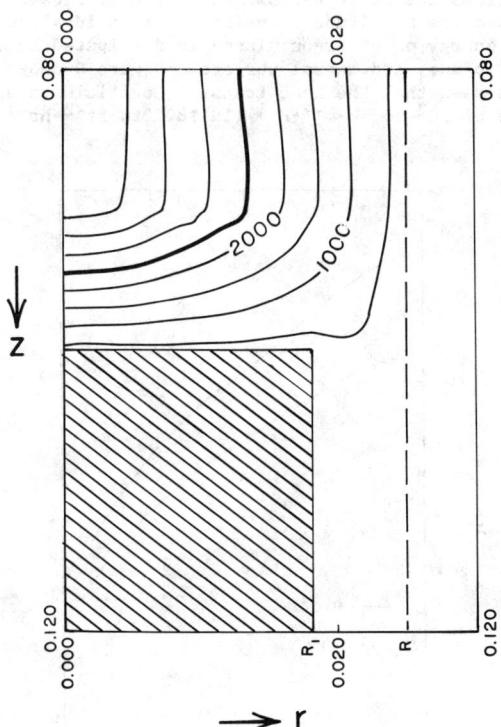

Figure 6.   Isotherms of predicted plasma temperatures in the vicinity of the impingement plate (contour interval = 500K).

In summary, the important inference to be made from these results of the fluid-flow and temperature fields is that changes in the confined plasma-flow due to impingement on a plane surface are restricted to the near-plate region.

It is recalled from Equation 9 that the local heat-fluxes to the surface are computed on the basis the plasma-gas temperature at the last grid point before the surface (3mm away) and for a prescribed surface temperature. A quench-plate temperature of 353K (80°C) was used for all computations. However, the predicted heat fluxes are insensitive to changes in the quench-plate temperature that would be envisaged for a water-cooled (thin) copper plate. The reason is evident from Figure 7 where the temperatures at the last grid points (near-plate nodes) are shown. The temperatures, especially in the stagnation-flow region, are much higher than the quench-plate temperatures. In Figure 8, the distribution of the heat fluxes to the plane surface is summarized as a function of position on the surface. As would be expected, the highest heat-fluxes (>1 $MJ \cdot m^{-2} \cdot s^{-1}$) are reported in the stagnation flow region where the plasma-jet impingement is normal to the surface.

The distribution of the local heat-transfer coefficients predicted from the model follows the pattern shown for the heat fluxes; however, it is important to note the magnitude of values in this idealized system in which the thermal energy of an argon plasma is dissipated by impingement onto a stationary, plane, isothermal surface. Figure 9 shows these results, and it is seen that the heat-transfer coefficients are predicted to be in the range of 10^2-10^3 $W \cdot m^{-2} \cdot K^{-1}$ (18-180 $Btu \cdot ft^{-2} \cdot hr^{-1} \cdot {}^{\circ}F^{-1}$).

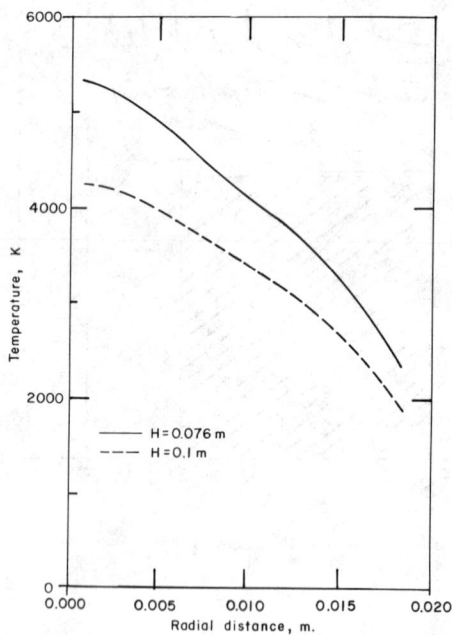

Figure 7. Fluid temperature at the near-plate nodes.

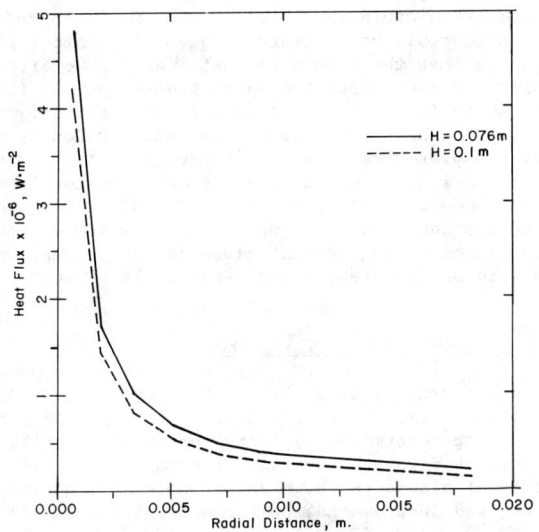

Figure 8. Distribution of the heat flux to the impingement plate.

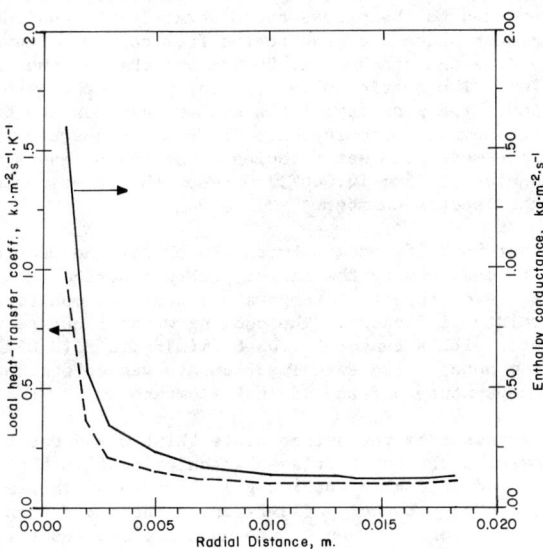

Figure 9. Plasma/Impingement plate heat transfer properties (H = 0.1m).

925

These values are in the range obtained for gas-jets impinging onto a solid surface (15) which are typically 10-50 Btu·ft^{-2}·hr^{-1}·°F^{-1}. In fact, the heat-transfer rates attainable in these jet systems is an order of magnitude larger than could be achieved by quenching into a fluid (gas) medium. It is noted that the predicted heat-transfer coefficients also compare favorably with those reported by El-Kaddah, et al. (16) for a free argon-plasma jet which impinges onto a target. The enthalpy conductance (17) (local heat-transfer coefficient divided by the local heat-capacity of the plasma gas) is also presented. It separates the heat-transfer effects induced by the local shear-stresses from those due to the temperature-dependent fluid-properties. The highest values are also attained close to the center of the quench plate, which is consistent with the largest shear-stress being present close to the stagnation point where the dissipation-rate of turbulent kinetic-energy is highest.

Calorimetry

Calculations of enthalpy flowrates in this system provides a means for assessing the computer simulation, since the enthalpy transferred in a physical system can be determined by calorimetry. The system shown in Figure 10 was fabricated for the enthalpy-flowrate measurements. It consists of a thermal-plasma torch, a water-cooled cylindrical chamber, an impingement plate, and instrumentation for measuring water flowrates and temperatures.

The plasma torch is a Metco Inc., Type 3MB, plasma flame-spray gun with a high-intensity, non-transferred, constricted arc configuration. The torch was used to produce the thermal plasma of argon, and its gross power-input was in the range of 4 to 7 kW. This corresponds to operating currents of 160 to 280A at 25V, and working-gas flowrates were in the range of 10 SLM (0.3g·s^{-1}) to 30 SLM (0.9g·s^{-1}). Typically, 30% of the gross power-input reported to the plasma-torch nozzle. The confinement chamber and the impingement plate were fabricated from copper. The diameter of the copper-tube used as the chamber was 50.8mm and that of the impingement plate was 47.7mm. The nozzle to impingement plate separation could be adjusted to 76mm. The geometry of the system was thus the same as those selected for the computer simulation. The reactor-chamber walls and the impingement plate were both water-cooled. The cooled gases were exhausted through the annulus (1.55mm [0.060"]) between the water-cooled thimble and the walls of the reactor chamber.

The temperature difference between the outlet and the inlet water streams could be measured by thermistor probes inserted in the inlet and outlet streams. A differential temperature could be monitored via a Jenco Model 747 thermistor indicator. The cooling-water flowrate was measured by a Dwyer flowmeter with a range of 0.08-8.4m(m)^3·min^{-1} (0.02-2.2 GPM -- gallons per minute). The enthalpy flowrate was determined from the differential temperature and the coolant flowrate.

The heat transfer to the quench-plate thimble and reactor-chamber walls were summed to obtain the plasma enthalpy available at the nozzle. The tests were conducted at a working-gas flowrate of 14 m(m)^3·min^{-1} (14 SLM) (0.42 g·s^{-1}). Enthalpy flowrates to the impingement plate were measured for several values of the nozzle plasma-energy and these are shown in Figure 11. The separation between the nozzle and the impingement plate was 76mm. It is seen that the enthalpy flowrate to the impingement plate increases with increasing plasma energy available at the nozzle (in the range tested of 1-2 kW). More significantly, only 40-30% of this available energy is transferred to the impingement plate. It should be noted that the specific enthalpy reporting to the torch nozzle in these tests ranged between 70-140 W/SLM. Furthermore, this is equivalent to a mean plasma-torch gas-temperature of 5000-8000K and a nozzle velocity of 50-80m·s^{-1}.

The behavior predicted by the simulator for a specific enthalpy of 350 W/SLM is also shown for comparison. It should be mentioned that only four points (0, 1.8, 2.8, and 3.2 kW) were used to draw the curve. It can be seen that the results produced by the simulator predict heat-transfer

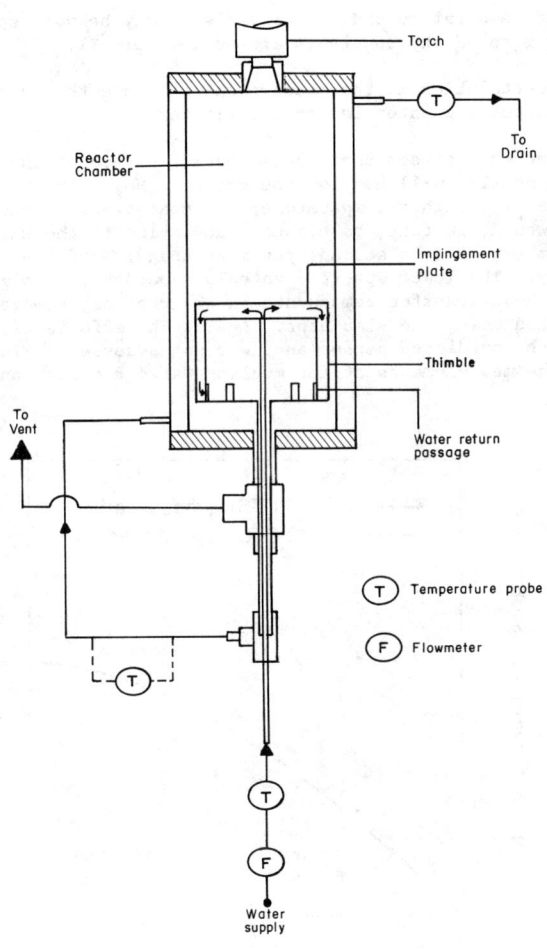

Figure 10. Schematic of the physical system used to measure enthalpy flowrates.

rates to the quench–plate which are significantly larger than measured. Thus, a larger fraction of the enthalpy input to the reactor reports to the reactor walls.

There are several reasons which might be cited to explain this discrepancy:

i)   The non–isothermal radial temperature–distribution at the plasma–torch nozzle and radiation from the plasma–jet to the walls of the reactor.

ii) The lower specific enthalpy of the plasma gas used in the calorimetric measurements.

iii) The temperature and velocity fields may be very sensitive to one or more of the constants listed in Table II.

iv) Uncertainties in the measurement of the enthalpy reporting to the cooling–water in the calorimeter.

It is not clear what effect the non–isothermal nature of the plasma–jet, as it leaves the nozzle, will have on the system. Most likely (because the center line is at a higher temperature) the heat–transfer rate to the quench plate would, in fact, be higher. Radiation to the walls of the reactor is not expected to account for more than 10% of the input–enthalpy. The lower specific enthalpy could have a significant effect on the heat–transfer components in the reactor; however, this was not investigated using the simulator. Again, the effects of the model constants on the predicted performance was not assessed. Finally, an error analysis of the measurements of the cooling–water flowrate and temperature

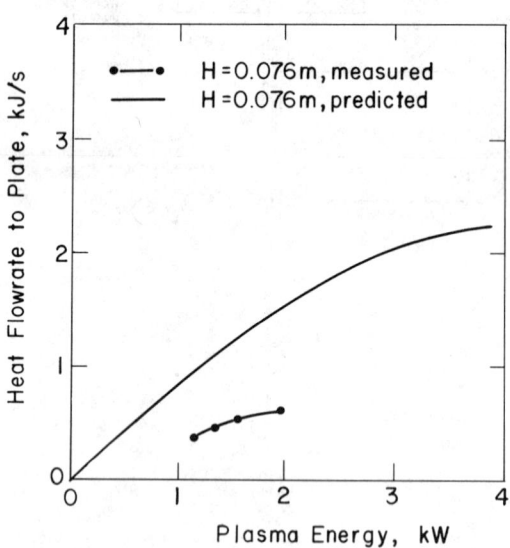

Figure 11. Heat flow rate to the impingement plates.

rise indicated that the uncertainties in these measurement would produce a cumulative effect of less than 15%.

Conclusion

The final section in this paper examines the results of the work conducted as they relate to materials synthesis. Conclusions regarding the correlation of the heat-transfer rates predicted by the model with the calorimetric measurements are then presented.

Relevance to Materials Synthesis

It is useful to discuss the results from the computer process-simulator within the contexts of plasma chemical synthesis and the PMRS process. In both processes, a confined high-velocity, high-temperature plasma stream is quenched with the objective of obtaining a metal or ceramic powder; however, direct measurements of the heat transfer during material deposition pose difficulties. Modeling provides the potential of quantifying the heat-transfer rates, and thus, cooling rates during material condensation or solidification. In their examination of the impinging jet flow as it applies to plasma spraying, Szekely et al. (16) suggest that information pertaining to the plasma behavior is critical to the characterization of the deposition process. As in that work, this mathematical model provides information pertaining to the behavior of the plasma within a confinement as a basis for describing the heat-transfer to a plane isothermal surface.

Examples of the benefit of the process-simulator are apparent in the interpretation of the predicted recirculatory flow and the heat-transfer rates. The recirculating flow-field generated in the enclosure (and presented in Figure 3) shows that the recirculating flow has velocities typically less than 10 percent of the inlet velocity. For example, $7.90 \text{ m} \cdot \text{s}^{-1}$ is the high value when an inlet velocity of $100 \text{ m} \cdot \text{s}^{-1}$ is specified. However, the location of the vortex center indicates that the recirculatory flow occupies a large portion of the reactor. In addition to the inherent reduction of heat-transfer to the quench plate, this recirculatory flow can also lead to material loss. The inclusion of a secondary gas-flow introduced at the top plane of the reactor is an obvious remedy to this problem.

The process-simulator has also predicted heat-transfer rates from the plasma to a quench surface, and the values indicate that this mode of heat-transfer is superior to quenching into a compatible fluid medium. In particular, the predicted temperatures of the reactor exit-gas are only 25-50K above the quench-plate temperature. This condition and the high temperature gradient in the impingement surface region indicate that impingement quenching on a chilled substrate can lead to the high supersaturation necessary for vapor-solid transformation and, thus, ultrafine metal- or ceramic-powders. However, in plasma chemical-synthesis where competing or reverse reactions are typically present during quenching, it should be stressed that it is desirable to consider also the constraints posed on the deposition process by the kinetics of such reactions. In plasma-melted and rapidly-solidified powder-processing, it can be concluded that cooling of the plasma stream by impingement onto a quench plate would provide for the requisite cooling-rates.

929

Conclusions Regarding Prediction of Heat-Transfer Rates

The shortcomings of the predictive capabilities of the simulator have already been discussed. It is clear that a critical assessment of the method of calculating the heat flux to the impingement plate would be well justified. Further work is planned so that the discrepancies which have been uncovered may be resolved. In particular:

a) Simulation at higher gross-power-inputs, in line with the optimum operating conditions for the Metco torch (40 kW, 40 SLM Argon) will be conducted.

b) Analysis of the uncertainties in the calorimetric measurements will be performed so that this source of error may be minimized.

The results of this upcoming work, it is hoped, will be published in the near future.

Acknowledgments

The authors are grateful to the Standard Oil Company of Ohio (SOHIO) for their financial support of this research. Our appreciation is extended to Professor J. W. Evans, University of California at Berkeley for supplying us with the initial form of the TEACH-2EF computer code. One of the authors (FKO) is indebted to SOHIO and the W. J. Kroll Institute for Extractive Metallurgy at CSM who provided funding of a research assistantship during his graduate studies. Finally, we wish to thank Ms. Nina Eads for typing the drafts and final copy of this manuscript.

Nomenclature

c_p	Heat capacity per unit fluid mass ($L^2 \cdot t^2 \cdot T^{-1}$)
c_μ, c_1, c_2	Coefficients in turbulence model ($M^\circ \cdot L^\circ \cdot t^\circ \cdot T^\circ$)
d	Nozzle internal diameter (L)
E	Wall-function constant ($M^\circ \cdot L^\circ \cdot t^\circ \cdot T^\circ$)
G	Turbulence generation term ($M \cdot L^{-1} \cdot t^3$)
H	Nozzle to impingement plate height (L)
k	Turbulent kinetic energy per unit fluid mass ($L^2 \cdot t^{-2}$)
P	Local pressure ($M \cdot L^{-1} \cdot t^{-2}$)
q	Heat flux to the wall ($M \cdot t^{-3}$)
r	Radial coordinate (L)
R	Confinement chamber radius (L)
R_1	Impingement plate radius (L)
S_Φ	Source Term (recasted equations)
T	Local jet temperature (T)
T_n	Local near-wall temperature (T)
T_w	Local wall temperature (T)
u	Local axial jet-velocity ($L \cdot t^{-1}$)
u_n	Local near-wall axial velocity ($L \cdot t^{-1}$)

v	Local radial jet-velocity $(L \cdot t^{-1})$
z	Axial coordinate (L)
Γ_Φ	Exchange coefficient (recasted equations)
ϵ	Turbulent dissipation rate per unit fluid mass $(L^2 \cdot t^{-3})$
κ	von Karman's constant $(M^\circ \cdot L^\circ \cdot t^\circ \cdot T^\circ)$
μ_e	Effective viscosity $(\mu_\ell + \mu_t)$ $(M \cdot L^{-1} \cdot t^{-1})$
μ_ℓ	Laminar viscosity of the jet $(M \cdot L^{-1} \cdot t^{-1})$
μ_t	Turbulent viscosity of the jet $(M \cdot L^{-1} \cdot t^{-1})$
ν	Kinematic viscosity (Momentum diffusivity) $(L^2 \cdot t^{-1})$
ρ	Fluid jet density $(M \cdot L^{-3})$
$\sigma_k, \sigma_T, \sigma_\epsilon$	Prandtl numbers $(M^\circ \cdot L^\circ \cdot t^\circ \cdot T^\circ)$
τ_w	Local wall shear stress $(M \cdot L^{-1} \cdot t^{-2})$
Φ	Primitive variable (recasted equations)

References

1. N. Rajaratnam, Turbulent Jets (New York, NY: Elsevier Scientific Publishing Company, 1976).

2. R. F. Cheyney, "Plasma-Melted and Rapidly Solidified Powders", Plasma Processing and Synthesis of Materials, ed. J. Szekely and D. Apelian, (New York, NY: North-Holland, 1984) 163-172.

3. K. Akashi, "Progress in Thermal Plasma Deposition of Alloys and Ceramic Fine Particles", Pure and Applied Chem., 57 (9) (1985) 1197-1206.

4. G. J. Vogt, et al., "Novel RF-Plasma System for the Synthesis of Ultrafine Ultrapure SiC and Si_3N_4", Plasma Processing and Synthesis of Materials, ed. J. Szekely and D. Apelian (New York, NY: North-Holland, 1984) 283-290.

5. K. Akashi, et al., "The Characteristics of Titanium Micro Crystals Formed From A Mixture of Titanium Tetrachloride and Hydrogen in and Argon Plasma Jet", Comm.-Zene. Symp. Int. de Chimie des Plasmas, ed. P. Fauchais, (1977) S.4.6.

6. M. Boulos, "Modeling of Plasma Processes", Plasma Processing and Synthesis of Materials, ed. J. Szekely and D. Apelian, (New York, NY: North-Holland, 1984) 53-60.

7. B. E. Launder and D. B. Spalding, "The Numerical Computation of Turbulent Flows", Computer Methods in Applied Mechanics and Engineering, 3, (1974) 269-289.

8. A. D. Gosman et al., "The Calculation of Two-Dimensional Turbulent Recirculating Flows", Turbulent Shear Flow - I, (New York, NY: Springer-Verlag, 1979) 237-255.

9. S. V. Patankar, Numerical Heat Transfer and Fluid Flow (New York, NY: Hemisphere Publishing, 1980).

10. M A. Serag-Eldin and D. B. Spalding, "Computations of Three-Dimensional Gas-Turbine Combustion Chamber Flows", Trans. of the ASME, 101, (1979) 326-336.

11. B. E. Launder, "The TEACH-2E Computer Program – General Structure", (Report CTF-14, U. Calif. Davis, Mech. Eng. Dept., 1977).

12. W. F. Ahtye, "A Critical Evaluation of Methods for Calculating Transport Coefficients of Partially and Fully Ionized Gases" (Report NASA TN D-2611, 1965).

13. A. B. Campbell, Plasma Physics and Magneto Fluid-Mechanics (New York, NY: McGraw-Hill, 1963).

14. C. F. Liu, "Numerical Analysis of High Intensity Arcs" (Ph.D. thesis, University of Minnesota, 1977).

15. S. T. Han and T. Seely, "Heat Transfer from a Solid Surface Under Air Impingement", Tappi, 48(12) (1965), 705-708.

16. N. El-Kaddah, J. McKelliget, and J. Szekely, "Heat Transfer and Fluid Flow in Plasma Spraying", Met. Trans., 15B, (1984), 59-70.

17. W. M. Kays and M. E. Crawford, Convective Heat and Mass Transfer (New York, NY: McGraw-Hill, 2nd Edition, 1980).

THE MATHEMATICAL MODELLING OF TEMPERATURE VELOCITY FIELDS

AND DEPOSITION RATES IN CVD SYSTEMS WITH A ROTATING SUBSTRATE

A.H. Dilawari* and J. Szekely**

Department of Materials Science and Engineering
Massachusetts Institute of Technology
Cambridge, MA   02139, U.S.A.

*Visiting Associate Professor, on leave from Institute of Chemical Engineering and Technology, University of the Punjab, Lahore - 20, Pakistan

**Professor of Materials Engineering, M.I.T.

Abstract

A mathematical formulation has been developed to represent the fluid flow, heat transfer and mass transfer phenomena in an axi-symmetric CVD unit involving a rotating substrate. In the formulation allowance is made for the three velocity components (axial, radial and azimuthal) and for natural convection, caused by the presence of the heated substrate. The computed results are in essential agreement with order-of-magnitude estimates which suggest that substrate rotation is unlikely to be important, except at high rpm levels; natural convection tends to be of importance in determining the flow field and the conical substrate construction is helpful in balancing the depletion of the gaseous reactant, by providing acceleration and, hence, an enhancement in the local mass transfer rate. Overall, it has been found that the various design parameters, such as the system geometry, the heating arrangements and the gas flow rate all have a marked influence on the uniformity of the deposit; thus, machine computation appears to be an essential tool in evolving an optimal design.

933

Introduction

The widespread application of chemical vapor deposition for the production of thin films for coatings and semi-conductor applications has generated extensive research in this field. In recent times it has been realized that in many instances, the understanding of the transport phenomena associated with CVD operations is critical, and may provide the key to the solution of problems associated with scale-up and optimal reactor design. This stimulated considerable research into the modelling of these systems (1-13).

The current state of modelling of CVD systems may be summarized by stating that while a useful start has been made in identifying the key factors that govern heat flow, fluid flow and mass transfer in these systems, our understanding of these phenomena is far from complete. Recent work on the modelling of three-dimensional flow phenomena in a rectangular CVD reactor has clearly shown the need to address details of the reactor design, such as the role played by the inlet configuration and the flow patterns in the near wall regions. Indeed, up to the present, while the modelling effort has been helpful in producing a general understanding, it has not yet evolved to the stage of becoming a concrete design tool.

The purpose of the work to be presented in the following paper is to develop a formulation and computed results to describe fluid flow, heat flow and mass transfer in a complex, cylindrical CVD system employing a rotating susceptor typically used for producing gallium arsenide wafers. Such systems have received very little attention up to the present, and pose a particular challenge to the engineer, because of the complex interplay of factors associated with forced and natural convection, the substrate rotation and the system geometry.

From a practical standpoint, we would seek a reactor design which allows for a spatially uniform deposition rate on the wafers, permits the maximization of the production rate, and at the same time provides the means for the efficient purging of the system when the production of layered structures is desired.

The very large number of parameters needed to characterize such a system would render a purely experimental approach to reactor optimization very lengthy and impractical. For this reason mathematical modelling can be a very effective tool for accomplishing this purpose.

In the following, we shall present a formulation, some asymptotic considerations, and computed results in order to provide a general insight into the behavior of these systems. At the same time, we shall seek to address certain specific design issues, such as the effect of the system geometry, susceptor rotation, heating arrangements and gas flow rates on the system performance.

Formulation

Fig. 1 depicts a schematic sketch of a typical CVD system. It is seen to consist of a vertical cylindrical chamber with a truncated conical susceptor containing multiple wafers onto which the material is deposited.

The key issues that have to be addressed are to define the criteria for achieving deposits of the desired structure, composition and uniformity as affected by the input conditions and suggest conditions under which the system can be purged efficiently.

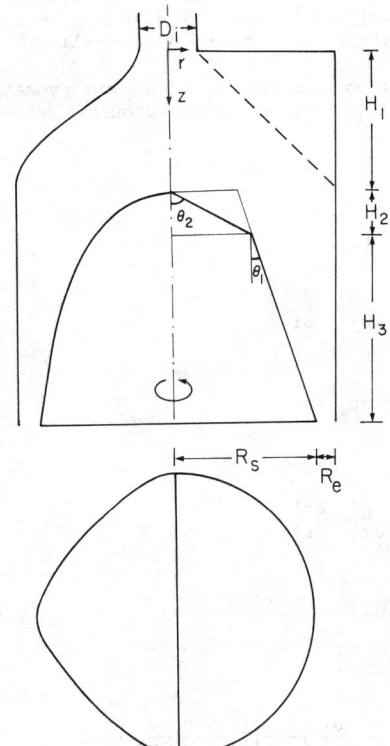

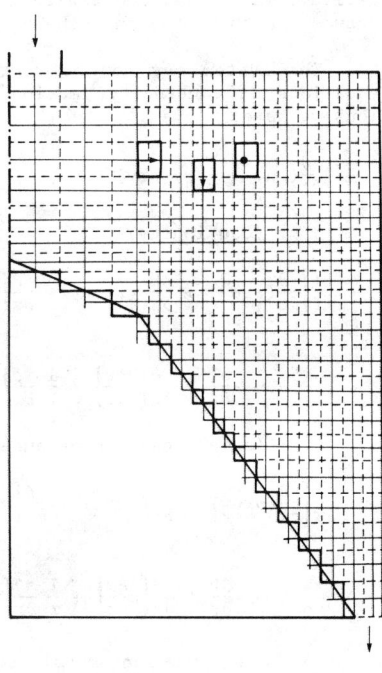

FIG. 1. A schematic sketch of a
typical CVD system. The left-hand
side shows the real system, while
the right-hand side shows the
idealization adopted.

FIG. 2. A schematic layout of the
finite difference grid.

The quantitative representation of the process requires the statement
of the

-- fluid flow equations;
-- thermal energy balance equation; and
-- the convective mass transfer equation.

The flow is assumed laminar, as the Reynolds number, based on the
maximum velocity, viscosity and density of the arsene/hydrogen mixture at
ambient temperature and diameter of the reactor chamber, is much lower than
the critical value for the flow to the turbulent. Flow non-uniformities that
may result from buoyancy-driven flow and the non-uniform transport coefficients due to non-uniform temperature within the system are considered.

An arsene/hydrogen mixture is used and the arsene is assumed to decompose at the susceptor surface if maintained at $1000^{\circ}K$, depositing arsenic.
Therefore, in the formulation of the mass transport equation, a binary
system of arsene and hydrogen is considered.

935

This is an oversimplification, ignoring the effect of tri-methyl gallium, which will be relaxed in a subsequent phase of the work. We shall assume, furthermore, that the overall rate is mass transfer-controlled.

The system may now be represented by writing down the axi-symmetric equations of continuity, motion, enthalpy balance, and component balance. Thus we have:

$$\frac{\partial}{\partial z}(\rho u) + \frac{1}{r}\frac{\partial}{\partial r}(\rho r v) = 0 \tag{1}$$

(continuity)

$$\frac{1}{r}\left[\frac{\partial}{\partial z}(\rho r u^2) + \frac{\partial}{\partial r}(\rho r v u) - \frac{\partial}{\partial z}\left(r\mu\frac{\partial u}{\partial z}\right) - \frac{\partial}{\partial r}\left(r\mu\frac{\partial u}{\partial r}\right)\right]$$

$$= -\frac{\partial P}{\partial z} + \frac{\partial}{\partial z}\left(\mu\frac{\partial u}{\partial z}\right) + \frac{1}{r}\frac{\partial}{\partial r}\left(r\mu\frac{\partial v}{\partial z}\right) + \rho g_z \tag{2}$$

(axial momentum balance)

$$\frac{1}{r}\left[\frac{\partial}{\partial z}(\rho r u v) + \frac{\partial}{\partial r}(\rho r v^2) - \frac{\partial}{\partial z}\left(r\mu\frac{\partial v}{\partial z}\right) - \frac{\partial}{\partial r}\left(r\mu\frac{\partial u}{\partial r}\right)\right]$$

$$= -\frac{\partial P}{\partial r} + \frac{\partial}{\partial z}\left(\mu\frac{\partial u}{\partial r}\right) + \frac{1}{r}\frac{\partial}{\partial r}\left(r\mu\frac{\partial v}{\partial r}\right) + \frac{\rho w^2}{r} - \frac{2\mu v}{r^2} \tag{3}$$

(radial momentum balance)

$$\frac{1}{r}\left[\frac{\partial}{\partial z}(\rho u r^2 w) + \frac{\partial}{\partial r}(\rho v r^2 w) - \frac{\partial}{\partial z}\left(r\mu\frac{\partial r w}{\partial z}\right) - \frac{\partial}{\partial z}\left(r\mu\frac{\partial r w}{\partial r}\right)\right] = -\frac{2}{r}\frac{\partial}{\partial r}(\rho r w) \tag{4}$$

(azimuthal momentum balance--needed because of the substrate rotation)

$$\frac{1}{r}\left[\frac{\partial}{\partial z}(\rho r h) + \frac{\partial}{\partial r}(\rho r h) - \frac{\partial}{\partial z}\left(r\, K/C_p\,\frac{\partial h}{\partial z}\right) - \frac{\partial}{\partial r}\left(r\, K/C_p\,\frac{\partial h}{\partial r}\right)\right] = 0 \tag{5}$$

(enthalpy balance)

Here, C_p and K are the heat capacity and the thermal conductivity of the gas mixture, respectively.

A mass balance on arsene is given as:

$$\frac{1}{r}\left[\frac{\partial}{\partial z}(\rho r m) + \frac{\partial}{\partial r}(\rho r m) - \frac{\partial}{\partial z}\left(r\rho D\frac{\partial m}{\partial z}\right) - \frac{\partial}{\partial r}\left(r\rho D\frac{\partial m}{\partial r}\right)\right] = 0 \tag{6}$$

where D is the mass diffusivity. Properties of the gases involved were taken from tabulated data (14), and the other input parameters used are given in Table 1.

The above equations alone do not specify the problem; additional appropriate information regarding boundary conditions is required.

Boundary Conditions

The transport equations are elliptic, and thus require the statement of conditions on all boundaries of the solution domain shown in Fig. 1. At the inlet, the reactant mixture (arsene/hydrogen) was assumed to have flat axial velocity and temperature profiles with a known mass fraction of arsene. The radial and azimuthal velocity components were assigned zero value, each.

At all the solid boundaries (reactor chamber walls and the susceptor surface), all the three components of velocity were zero except for the rotating susceptor, where an appropriate value was assigned to the azimuthal component. The surface temperature of the susceptor was generally taken as $1000^\circ K$; however, calculations were also made for a top surface temperature of $300^\circ K$. The chamber wall temperature was assigned a value of $300^\circ K$.

The susceptor surface, if at $1000^\circ K$, was considered reactive, and the mass fraction of arsene was assigned zero value at this surface. For all other solid surfaces, a zero flux condition was specified.

At the axis of symmetry the radial velocity v and the radial gradients of all the remaining flow properties were set to zero. The procedure was clearly invalid for the azimuthal velocity w. However, it is not w but rw which is the property governed by Eq (4), and therefore a zero radial gradient for rw was valid.

Across the exit plane, the axial pressure gradient was assumed to be zero. Also, the axial gradients of each of the other flow properties were given a zero value. This assumption may not be truly valid because of the inclined surface of the heated susceptor. The flow develops both hydrodynamically and thermally throughout the chamber length, and therefore the conditions of fully developed flow specified at the exit plane may influence the predicted results. However, the error introduced is not expected to be large.

Solution Procedure

The principles of the solution procedure employed are described in the basic "2/E/FIX" computer program (15), which provided a starting point from which the present computer code has been developed. The schematic layout of the finite difference grid and the stair-step approximation approach adopted to simulate the susceptor surface is shown in Fig. 2. The axial and the radial velocities were stored at the locations displaced from the main grid point (.) where P, rw, m, h and other auxiliary variables were stored. The differential equation for each dependent variable was integrated over the respective control volume, and the following algebraic equation for the general variable, ϕ, was obtained for each grid in the calculation domain:

$$\phi_p = A_n\phi_n + A_s\phi_s + A_w\phi_w + A_e\phi_e + S_\phi \tag{7}$$

The A's in Eq (7) consist of convective and diffusive fluxes through the control faces; e.g., A_n contains the convective and diffusive fluxes of ϕ through the north face of the control volume. Special care was needed to calculate these coefficients for the grid points adjacent to the susceptor surfaces, because of the stair-step approximation employed to simulate the inclined face.

The finite difference equations and boundary conditions constitute a system of coupled simultaneous algebraic equations. The method of solution adopted followed the "SIMPLE" procedure (16). Using this technique, the algebraic equations such as Eq (7) were solved many times, the coefficients and the source terms being updated prior to each iteration.

The computational mesh used for the calculations consisted of 35 x 27 non-uniformly distributed nodes. To reduce the computational labor, the domain of calculation was restricted to the volume through which flow had taken place. A typical run required a cpu time of less than one hour on a MicroVAX II, giving an error of less than 0.5%.

The streamline plots showing the recirculation zones and stagnation points were represented in terms of a dimensionless stream function, defined as:

$$\psi^* = \int_0^r \rho u r \, dr \ / \ \int_0^{R_o} \rho u r \, dr \tag{8}$$

The model contained digitized tables for the viscosity, the thermal conductivity, and the heat capacity at constant pressure. The viscosity of the binary mixture was estimated using Wilkes's technique, while the thermal conductivity was obtained using Brokaw's method (17). The density of the mixture was calculated using:

$$\rho = \frac{PM}{RT} \tag{9}$$

$$\text{and} \quad M = \sum_i x_i M_i \tag{10}$$

where x_i and M_i are the mole fraction and the molecular weight of species i.

The temperature of the mixture required during computation was calculated using the relation

$$h = \sum_i m_i \int_{T_o}^T C_{P_i} \, dT \tag{11}$$

where T_o is the reference temperature, and C_{P_i} is the specific heat at constant pressure of species i.

Computed Results

In presenting the computed results, we must bear in mind that although the prime practical concern is with factors that govern the uniformity of the deposit, the detailed knowledge of the velocity and the temperature fields is essential in gaining insight into the behavior of the system. For this reason, we shall start by presenting the velocity and the corresponding temperature and concentration profiles for various operating conditions, and will conclude the section with the results concerned with the deposition rates.

Before presenting the actual computed results, it is important to note that certain qualitative aspects of the system behavior may be predicted on the basis of "fluid mechanics common sense," that is, the consideration of fluid flow fundamentals. Some key points that can be made in this regard are the following:

938

(1) When a fluid is introduced into a container through an inlet nozzle which has a much smaller diameter than that of the vessel, recirculating loops will result in the vicinity of the inlet region. Such behavior may be avoided, or at least minimized, by altering these diameter ratios, or by appropriately changing the vessel shape in the inlet region.

(2) The existence of a heated vertical or near-vertical heated surface will give rise to thermal natural convection; when the resultant velocities are comparable to those produced by the forced flow, instabilities or further recirculating loops may be established.

(3) If the top surface of the susceptor is flat, the local deposition rate may become highly non-uniform in the vicinity of the stagnation point. For this reason a contoured or dome-shaped top would have to be preferred.

(4) Rotation of the substrate will modify the velocity field within the system, but only when the local azimuthal velocities are of comparable magnitude to those caused by forced and natural convection.

(5) Last but not least, since the Prandtl and Schmidt numbers are of the order of unity for gaseous systems, the velocity field will have a marked effect on the temperature and the concentration fields, and hence on the spatial variability of the deposition rate.

These considerations may be developed a little further by considering the following:

For the typical operating conditions employed in the particular system considered, the nozzle diameters ranged from 35-60 mm, with volumetric gas flow rates in the range of 2×10^{-5} to 7×10^{-5} kg/s. This gave inlet Reynolds numbers in the range of 60-80, clearly signifying laminar flow.

The characteristic gas velocity at the inlet ranged from 0.14-0.67 m/s. The bulk velocities were significantly less; however, in the exit region we had velocities of the order of 1 m/s.

The azimuthal velocities at the substrate surface ranged from 0.5-1 m/s at 1000 rpm and were, of course, proportionately less at the lower rates of revolution.

The characteristic velocities due to thermal natural convection for an idealized parallel plates system may be estimated from the following relationship (18):

$$u_m = \frac{\bar{\rho} \, \bar{\beta} \, g \, b^2 \, \Delta T}{31 \, \bar{\mu}} \tag{12}$$

in which $\bar{\rho}$, $\bar{\mu}$ and $\bar{\beta}$ are the density, the viscosity and the coefficient of volume expansion of the gas mixture, evaluated at the mean temperature. ΔT is the difference between the wall temperatures, and b is the half-spacing between the plates; Equ (12) gives a value of 0.25 m/s for the operating conditions under study.

From these data we may readily draw the following general conclusions for the operating parameters considered:

(a) Substrate rotation will be unimportant for rpm values of less than about 500.

(b) The velocities due to thermal natural convection may be comparable to those due to forced flow.

(c) Due to the relatively small inlet nozzle diameter and the comparability of the forced and the buoyant flow components, complex recirculating flow patterns may exist that can be very sensitive to relatively small changes in the system geometry or operating parameters.

However, these general considerations can provide only qualitative guidelines; in order to characterize a system quantitatively, it is necessary to perform actual numerical calculations.

Fig. 3 shows the behavior of a truncated cone susceptor with a non-heated top surface; here (a), (b) and (c) depict the streamline pattern, the temperature fields and the arsene concentration fields, respectively, for zero susceptor rotation. The two recirculating loops are readily apparent on the streamline plot. The larger of these is a direct consequence of the disparity between the inlet nozzle and the column diameter, as noted earlier. The second, smaller recirculating loop is due to thermal natural convection. The temperature and the concentration fields are markedly non-parallel in the vicinity of the "leading edge," i.e. top corner, which is a direct consequence of the fact that only the side walls were being heated in this instance. At first sight, one might expect the temperature and the concentration profiles to be nearly identical, but the disparity is due to the fact that the gaseous reactant is being consumed at the heated surface.

Fig. 4 shows the behavior of the identical system, but now with a heated top surface. Inspection of the streamline pattern shows a stronger recirculating flow field which is now augmented by natural convection. This enhanced circulation is readily apparent in the distortion of the temperature field. The concentration field is, of course, markedly different from that seen in Fig. 3, because in the present case the reactant is being consumed at the top surface also.

Fig. 5 shows the behavior of a system with a non-flat top surface which is an approximation to the dome-shaped top susceptor that is frequently employed in practice. The velocity vector plot shows evidence of two recirculating loops and of the expected acceleration of the fluid as the area available for flow is narrowed near the exit. The streamline pattern shows the two recirculating loops and the temperature and the concentration plots are relatively similar.

Fig. 6 depicts the same system, but with a significantly larger inlet nozzle. The very marked recirculation is readily apparent, as is the drastic alteration in the overall flow pattern. The computed temperature and concentration fields show evidence of this markedly altered velocity field.

In examining the effect of substrate rotation, calculations have been carried out for several substrate rotation rates, including 60, 300, 600, 1,200 and 1,500 rpm. Only values in excess of about 1,000 rpm showed a significant difference from the zero substrate rotation case, which is in good agreement with the order of magnitude estimates given earlier.

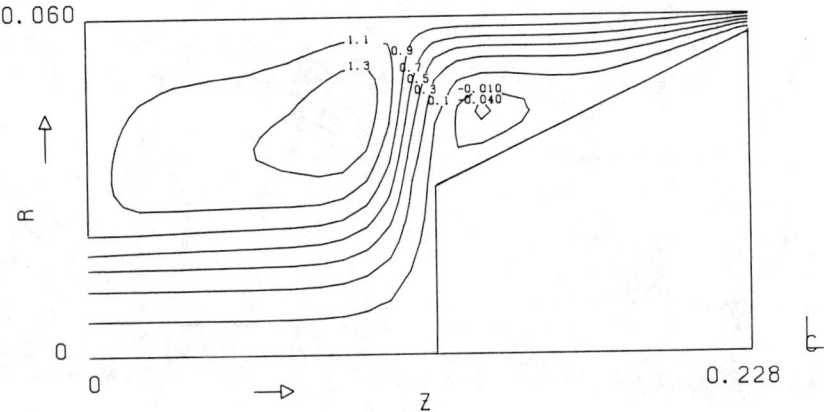

FIG. 3(a). Predicted streamline pattern for a truncated cone susceptor with a non-heated top surface.

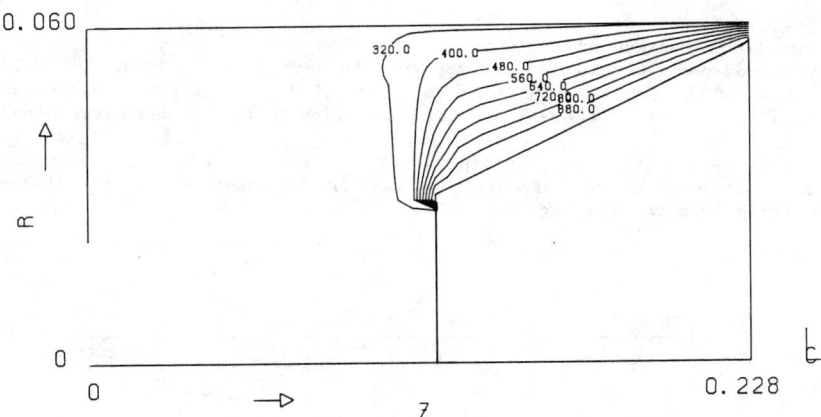

FIG. 3(b). Predicted temperature fields for a truncated cone susceptor with a non-heated top surface.

Fig. 7 shows the behavior of the system with a susceptor rotation rate of 1,500 rpm. It is seen that the structure of the flow field has changed, with the rotating substrate now acting like a pump, eliminating the effect of the natural convection-driven recirculating loop in the lower part of the vessel.

Let us now consider the spatial variability of the deposition rates. On all these plots, we show the deposition rates as a function of the distance along the inclined wall of the susceptor. Fig. 8 represents the behavior of a system with a flat top which is not heated. The solid and the

941

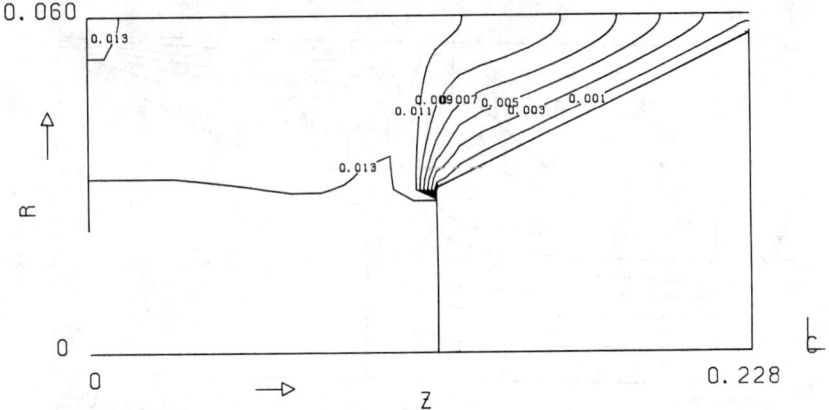

FIG. 3(c). Predicted arsene concentration fields for a truncated cone susceptor with a non-heated top surface.

broken lines correspond to a high and a low gas velocity, respectively. Very marked non-uniformities are apparent in both cases. The higher velocity case will give a more uniform deposition rate further downstream, because the effect of thermal natural convection is being suppressed here.

Fig. 9 shows the corresponding plots, but now for a heated top surface; here, relatively uniform deposition rates may be observed at some distance downstream from the leading edge.

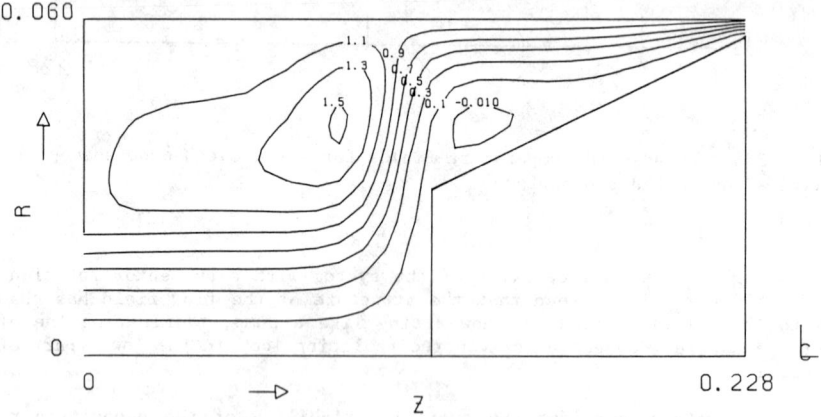

FIG. 4(a). Predicted streamline pattern for a truncated cone susceptor with a heated top surface.

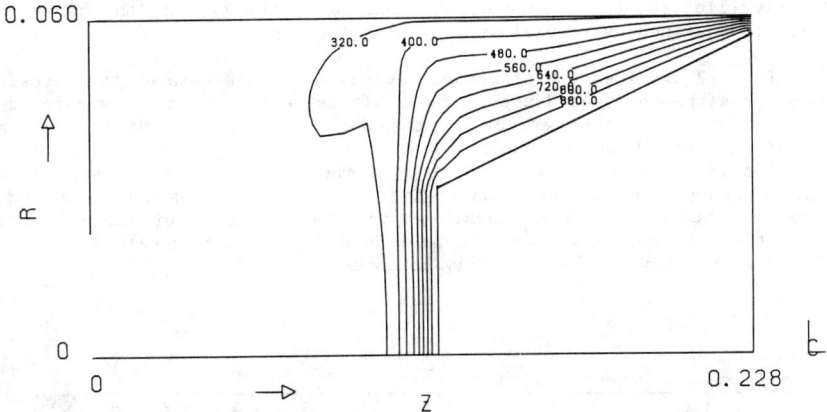

FIG. 4(b). Predicted temperature fields for a truncated cone susceptor with a heated top surface.

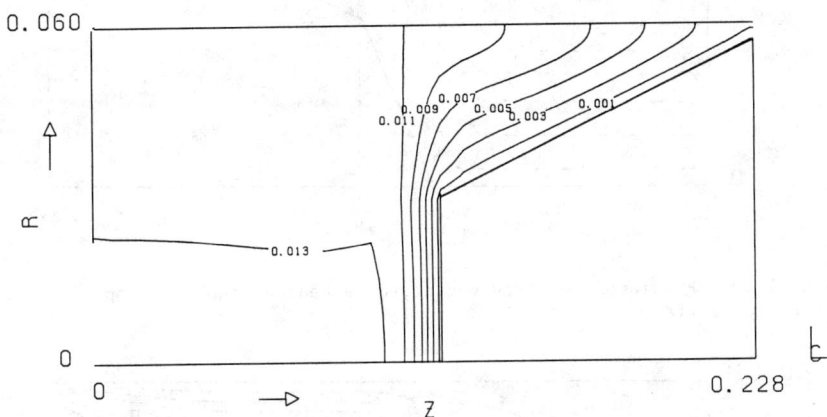

FIG. 4(c). Predicted arsene concentration fields for a truncated cone susceptor with a heated top surface.

Fig. 10 displays the effect of changing the temperature of the susceptor flat top, while maintaining the same mass input rate. The expected higher deposition rate for cold tip is readily apparent.

Fig. 11 shows the effect of changing the inlet diameter for an identical mass input, for a shaped, heated susceptor top surface. It is seen that the two curves are similar, but that the deposition rate is somewhat lower for the smaller nozzle; at first sight, this might seem unreasonable to have a lower deposition rate for the higher inlet velocity, but on closer examination this is consistent with the following physical reasoning: at the higher inlet velocity, there will be a higher deposition rate in the stagnation region, and hence the reactant will be depleted more rapidly, resulting in a lower deposition rate at the inclined substrate surface. This point clearly illustrates the very complex nature of the system, the behavior of which is governed by several interrelated factors.

Fig. 12 examines the effect of the gas mass flow rate on the deposition rate for a shaped, heated top surface. It is seen that the flow rate of the gas has a very marked effect on the uniformity of the deposition rate. Indeed, it would appear that for a given geometry there will exist a specific gas flow rate, which will give the most uniform deposition rate. This finding seems to be consistent with physical reasoning, because in order to obtain a uniform deposition rate, the effects of thermal natural convection, entrainment near the inlet and forced flow acceleration toward the exit--all have to be carefully balanced.

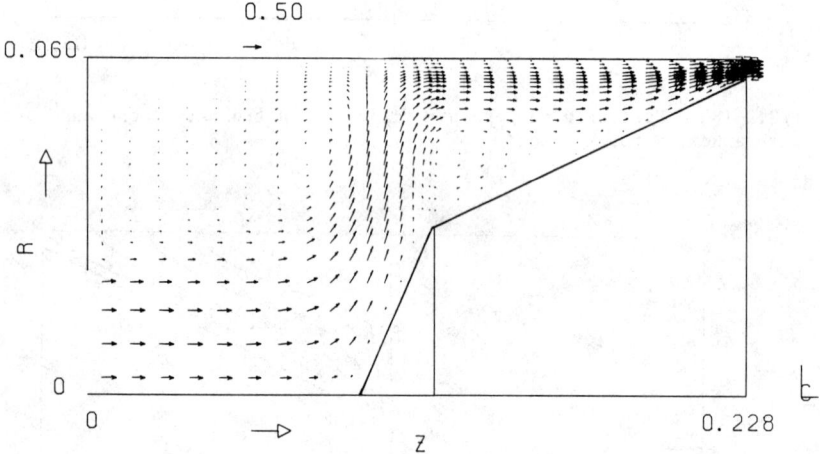

FIG. 5(a). Predicted velocity vector for a heated, non-flat top susceptor surface.

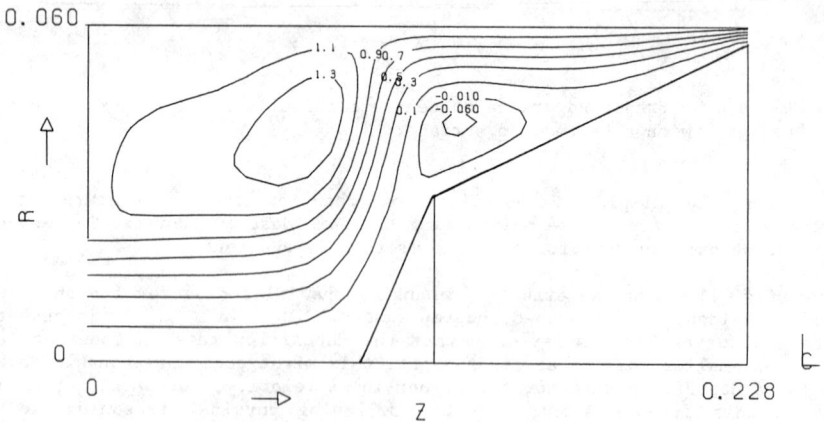

FIG. 5(b). Predicted streamline pattern for a heated, non-flat top susceptor surface.

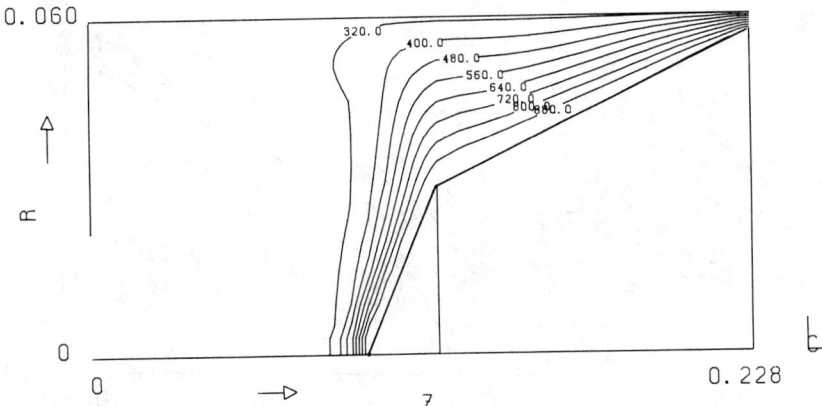

FIG. 5(c). Predicted temperature fields for a heated, non-flat top susceptor surface.

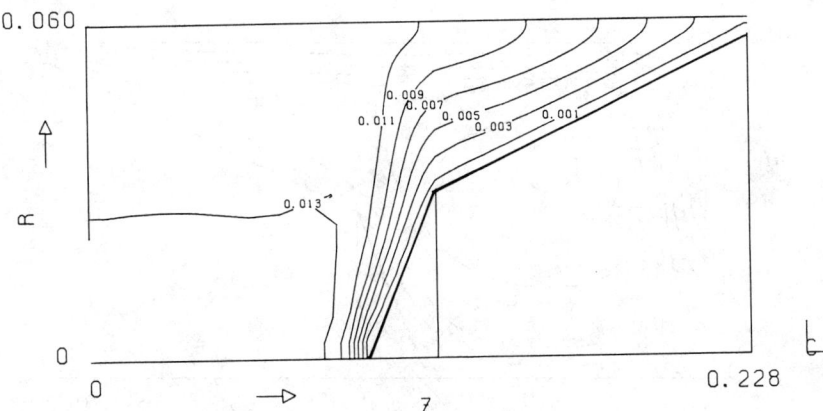

FIG. 5(d). Predicted arsene concentration for a heated, non-flat top susceptor surface.

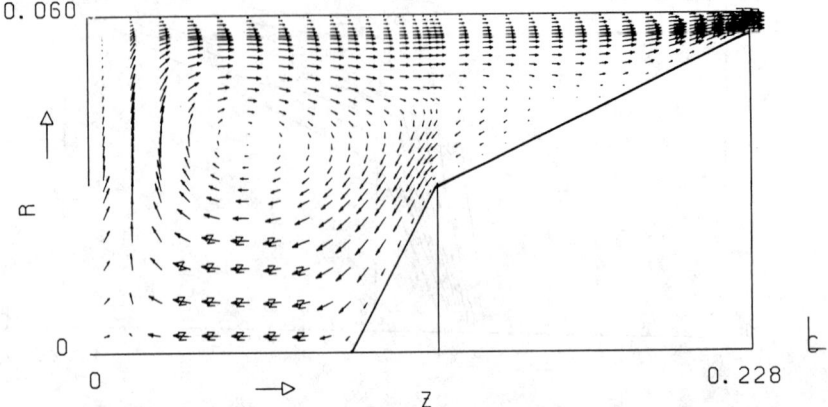

FIG. 6(a). Predicted velocity vector plot for a heated, non-flat top susceptor surface.

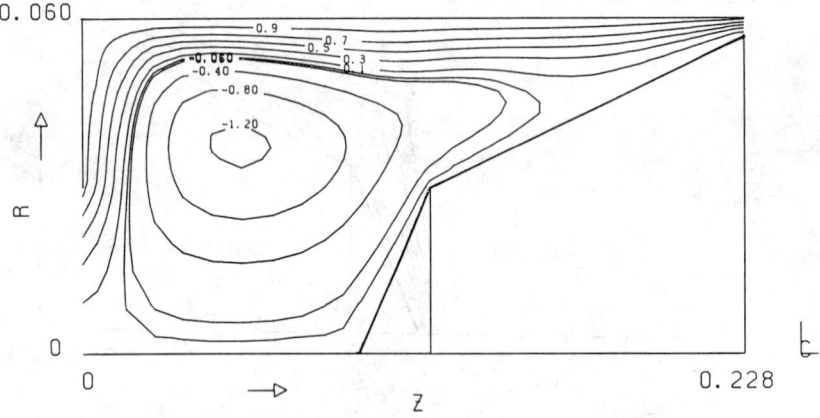

FIG. 6(b). Predicted streamline pattern for a heated, non-flat top susceptor surface.

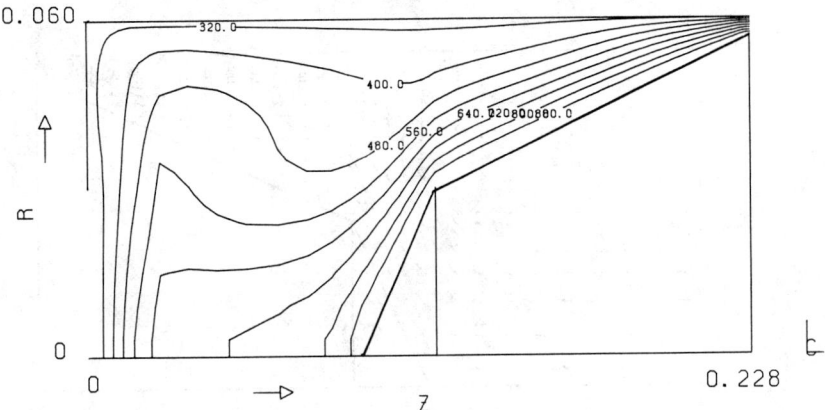

FIG. 6(c). Predicted temperature fields for a heated, non-flat top
susceptor surface.

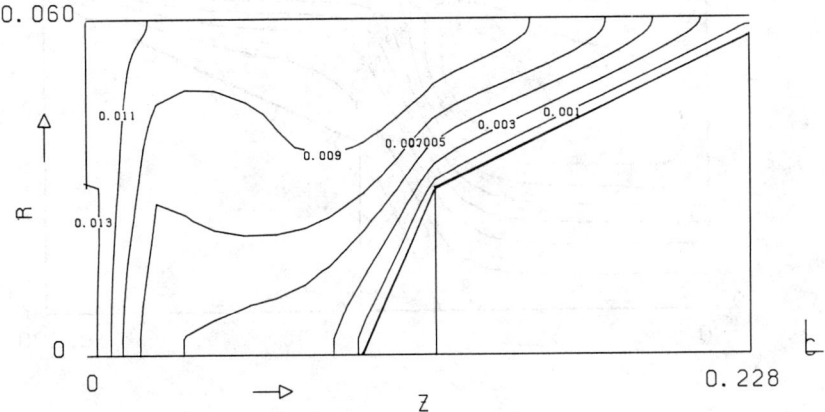

FIG. 6(d). Predicted arsene concentration for a heated, non-flat top
susceptor surface.

0.50

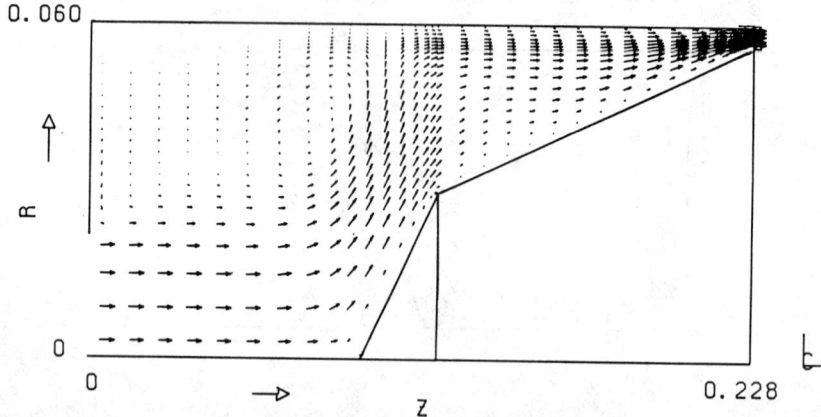

FIG. 7(a). Predicted velocity vector plots for a heated, non-flat top susceptor surface, with a rotation speed of 1500 rpm.

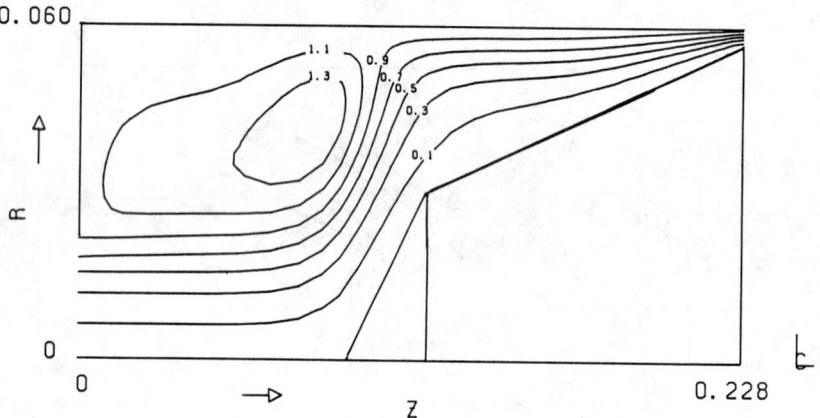

FIG. 7(b). Predicted streamline pattern for a heated, non-flat top susceptor surface, with a rotation speed of 1500 rpm.

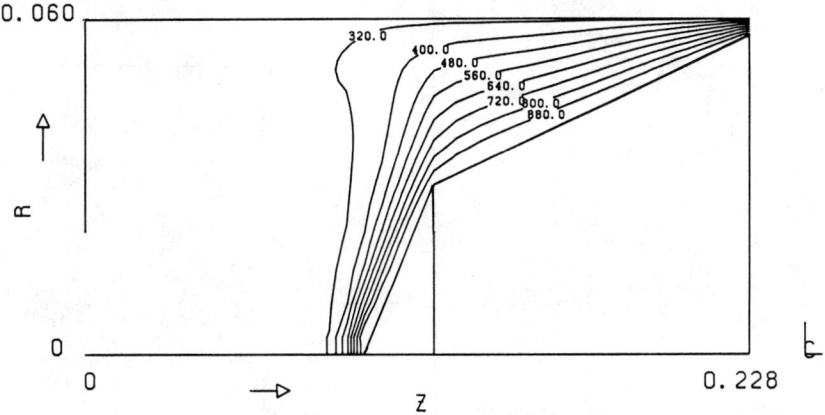

FIG. 7(c). Predicted temperature fields for a heated, non-flat top susceptor surface, with a rotation speed of 1500 rpm.

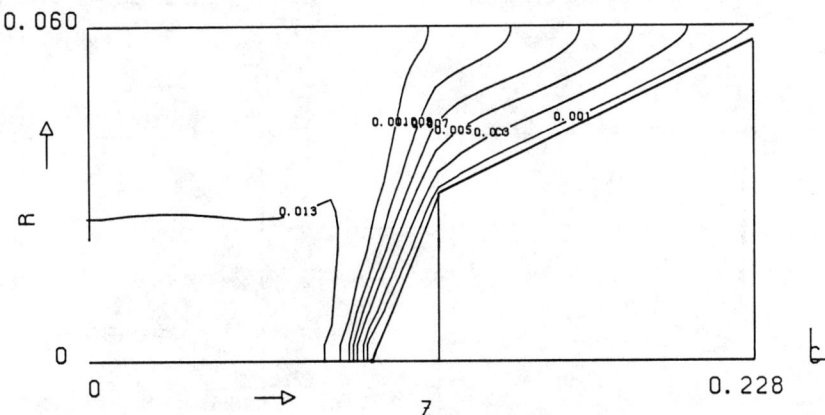

FIG. 7(d). Predicted arsene concentration for a heated, non-flat top susceptor surface, with a rotation speed of 1500 rpm.

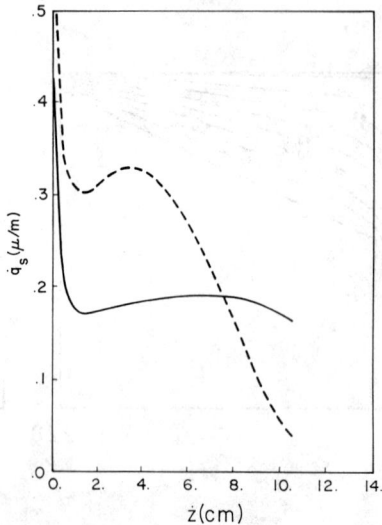

FIG. 8. Effect of gas velocity on deposition rate along the inclined surface of an unheated flat-top susceptor.

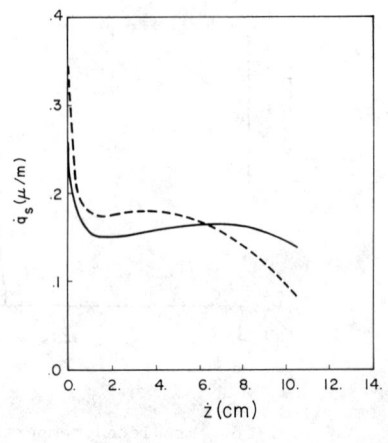

FIG. 9. Effect of gas velocity on deposition rate along the inclined surface of a susceptor with heated flat top.

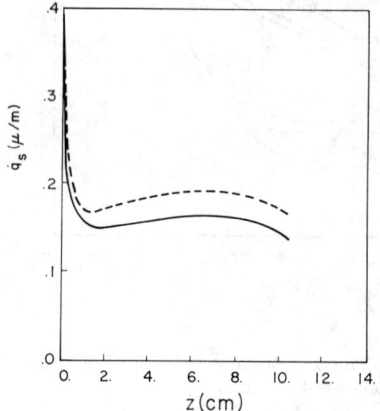

FIG. 10. The effect of temperature of the flat-top susceptor on the deposition rate.

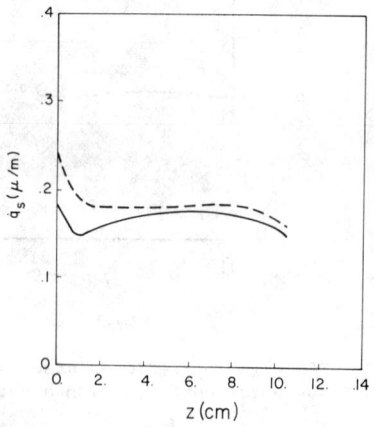

FIG. 11. The effect of the inlet nozzle diameter on the deposition rate for a shaped, heated susceptor top surface.

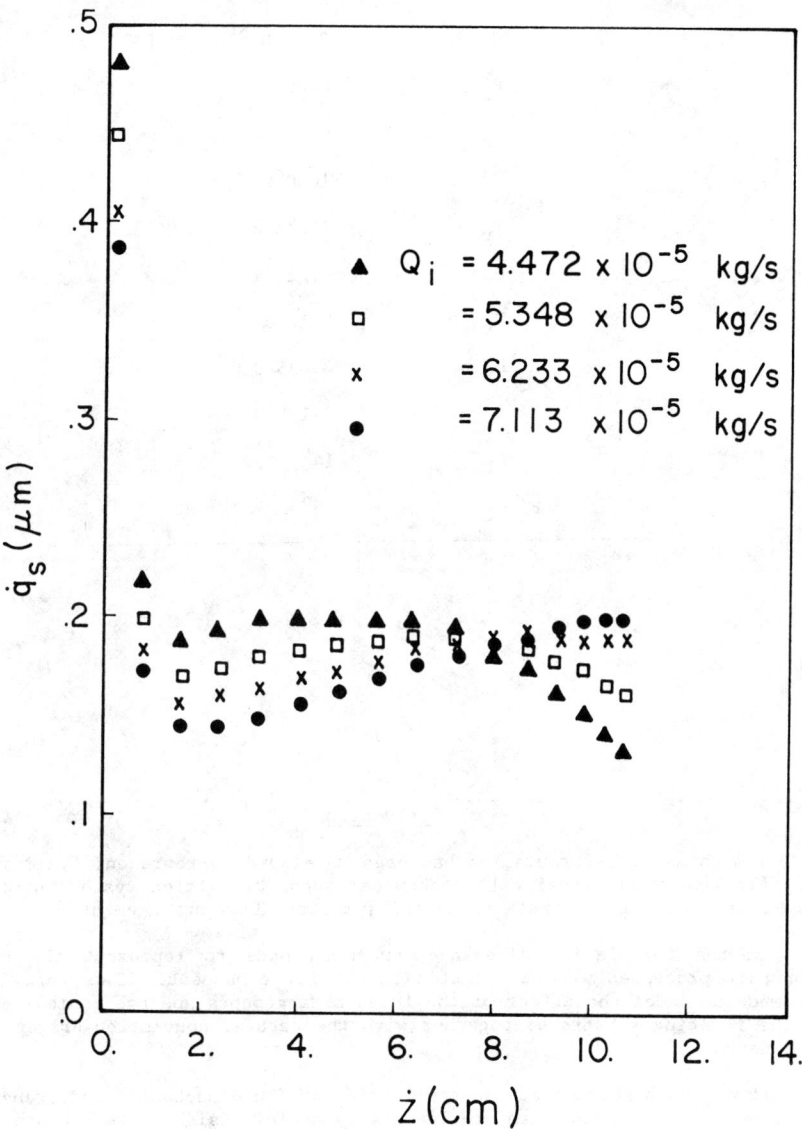

FIG. 12. The effect of the gas mass flow rate on the deposition for a shaped, heated top susceptor surface.

Table 1: Input Parameters

m	0.01-0.04 AsH_3
G	2×10^{-5} - 7×10^{-5} kg/s
R_e	0.003 m
R_s	0.057 m
T_s	$1000^\circ K$
T_t	$300^\circ K$ or $1000^\circ K$
H_r	0.228 m
D_r	0.12 m
D_i	0.035-0.06 m
ω_o	0-1,500 rpm
θ_1	14°
θ_2	45°

Discussion

A mathematical formulation has been developed to represent fluid flow, heat flow and mass transfer in a chemical vapor deposition system involving a rotating conical substrate in an axi-symmetric flow arrangement.

In the formulation allowance has been made to represent the usual transport processes with a rotational velocity component. Thus, provision was made to model the effect of the inlet arrangements and the conical shape of the rotating substrate, together with the natural convection-driven flow components.

It has been shown that the application of fluid mechanics and transport phenomena "common sense" can provide a very useful insight into the behavior of these systems. Thus, without performing any calculations, it was possible to specify the range of rpm values for the susceptor which are likely to play an important role in affecting the overall system behavior. By the same token, it was also possible to show that for the conditions considered, thermal natural convection and the forced flow components play a comparable role in affecting the flow pattern. This fact, together with the well-known behavior of fluid systems exhibiting strong recirculating flow patterns when a small-diameter fluid stream is introduced into a larger vessel, indicates extreme sensitivity of the performance of the system, even to relatively small changes in process conditions.

It is also a well-established fact that in "straight-through" flow systems, one would naturally expect strong spatial non-uniformities in the deposition rate because of boundary layer effects and the gradual depletion of the reactant. Furthermore, recirculating loops in the vicinity of the susceptor would be undesirable, because these too would cause non-uniform deposition rates.

It is reasonable to suggest that by the careful balancing of these factors, and particularly by gradually accelerating the fluid as it passes along the susceptor surface, it should be possible to obtain fairly uniform deposition rates over a major part of the domain of interest.

The specific conditions which would correspond to optimal operation will, of course, depend on the particular system and may be defined only through machine computation.

The computed results that have been presented seem to support these contentions. The principal findings that may be deduced from the machine computation may be summarized as follows:

(1) For the conditions considered, substrate rotation has a negligible effect below about 500 rpm. Above this threshold level, the rotating substrate will start acting like a centrifugal pump. Such behavior may be quite helpful in suppressing the effect of thermal natural convection and, in extreme cases, also suppressing the vortex that would have formed otherwise near the inlet nozzle. Such an arrangement may be useful, also, if abrupt changes of wafer composition are desired through the alternating use of different gas streams.

(2) For most of the conditions considered, two recirculating loops were found: one near the inlet, caused by entrainment of the incoming stream; and the other further downstream, due to thermal natural convection. The specific role of these recirculating flows in affecting the deposition rate was found to depend quite critically on the other system parameters.

(3) The specific examination of the deposition rates found that these were affected by virtually all the operating parameters, particularly by the relative size of the inlet nozzle, the gas velocity, the shape of the top susceptor surface, and whether this top surface was being heated.

(4) Due to the complex relationships between these individual process parameters, it is difficult to establish unequivocal specific guidelines regarding the design and operation of these systems. There is reason to believe that many practical systems may be quite close to optimum as a result of extensive trial and error procedures; however, this very time-consuming process would have to be started all over again, when changes in operating conditions are necessary. In this regard, the computational techniques discussed in this paper should be of considerable help.

(5) It is felt that through modelling efforts of this type, we should be approaching the CAD-CAM stage of process design for such systems.

C_p	Heat capacity at a constant pressure
D	Mass diffusivity
D_i	Inlet diameter of the reactor nozzle
D_r	Diameter of reaction chamber
G	Mass flow rate of arsene/hydrogen mixture
g_z	z-component of gravitational acceleration
H_r	Height of reaction chamber
K	Thermal conductivity
m	Mass fraction of arsene
p	Fluid pressure
$\dot{q}_s$	Deposition rate of arsenic
r	Radial coordinate
R_e	Dimension of annular exit
R_s	Radius of susceptor base
T	Temperature
T_s	Temperature of susceptor surface
T_t	Temperature of top susceptor surface
u	Axial velocity component
v	Radial velocity component
w	Azimuthal velocity component
z	Axial coordinate
$\dot{z}$	Distance along the inclined surface of the susceptor
μ	Viscosity of mixture
ρ	Density of mixture
θ_1, θ_2	Angles of susceptor surfaces
ω_o	Rotation speed of susceptor

Acknowledgement

The authors wish to thank the U.S Department of Energy for partial support of this investigation under Grant #DE-FG02-85ER-13331; the United States Education Foundation in Pakistan; and the Council for International Exchange of Scholars in Washington, DC.

References

1. E. Fujii, H. Nakamura, K. Haruna and Y. Yoga, "A quantitative calculation of the growth rate of epitaxial silicon from $SiCl_4$ in a barrel reactor", J. Electrochem. Soc. 119(8), 1106-13 (1972).

2. R. Takahashi, Y. Koga and K. Sugawara, "Gas flow pattern and mass transfer analysis in a horizontal flow reactor for chemical vapor deposition," J. Electrochem. Soc. 119(10), 1406-1412 (1972).

3. C.W. Manke and L.F. Donaghey, "Analysis of transport processes in vertical cylinder epitaxy reactors," J. Electrochem. Soc. 124(4), 561-569 (1977).

4. G. Wahl, "Hydrodynamic description of CVD processes," Thin Solid Films 40, 13-26 (1977).

5. J. Juza and J. Cermak, "Phenomenological model of the CVD epitaxial reactor," J. Electrochem. Soc. 129(7), 1627-1634 (1982).

6. G.H. Westphal, D.W. Shaw and R.A. Hartzell, "A flow channel reactor for GaAs vapor phase epitaxy," J. Crystal Growth 56, 324-331 (1982).

7. K.F. Jensen, "Modelling of chemical vapor deposition reactors," Proc. of the Ninth Intl. Conf. on Chemical Vapor Deposition, Electrochem. Soc. (McD. Robinson, C.H.J. van den Brekel, G.W. Cullen and J.M. Blocher, Jr., eds.), p. 3. Pennington, NJ: Electrochem. Soc. (1984).

8. G. Wahl, "Theoretical description of CVD processes," Proc. of the Ninth Intl. Conf. on Chemical Vapor Deposition, Electrochem. Soc. (McD. Robinson, C.H.J. van den Brekel, G.W. Cullen and J.M. Blocher, Jr., eds.), p. 60. Pennington, NJ: Electrochem. Soc. (1984).

9. Y. Kusumoto, T. Hayashi and S. Komiya, "Numerical analysis of the transport phenomena in MOCVD process," Jap. J. Appl. Phys. 24(5), 620-625 (1985).

10. S. Rhee and J. Szekely, "The analysis of CVD operations in the presence and absence of plasma enhancement," Proc. of the Third Conf. on Modelling of Casting and Welding Processes (S. Kou and R. Mehrabian, eds.), pp. 51-78. Warrendale, PA: TMS-AIME (1986).

11. Y. Sahai, paper presented at 116th Annual Meeting, AIME, Denver, CO (1987).

12. K.F. Jensen, "Micro-reaction engineering: Application of reaction engineering to processing of electronic and photonic materials," Chem. Eng. Sci. 42(5), 923-958 (1987).

13. S. Rhee, J. Szekely and O.J. Ilegbusi, "On three-dimensional transport phenomena in CVD processes," J. Electrochem. Soc., in press (1987).

14. R.A. Svehla, "Estimated viscosities and thermal conductivities of gases at high temperature," NASA TR 132 (1962).

15. W.M. Pun and D.B. Spalding, "A general computer program for two-dimensional elliptic flows," Rept. No. HTS/76/2, Imperial College of Science and Technology, London (1976).

16. S.V. Patankar and D.B. Spalding, "A calculation procedure for heat, mass and momentum transfer in three-dimensional parabolic flows," Int. J. Heat and Mass Transfer 16, 1787-1806 (1972).

17. R.C. Reid, J.M. Prausnitz and T.K. Sherwood, The Properties of Gases and Liquids, New York: McGraw-Hill (1977).

18. B.R. Bird, W.E. Stewart and E.N. Lightfoot, Transport Phenomena, New York: John Wiley and Sons (1960).

ELECTRONIC MATERIALS PROCESSING

MODELING OF GROWTH OF $Ga_xIn_{1-x}As_yP_{1-y}$ BY VPE

Y. K. Rao and H. G. Han

Department of Materials Science and Engineering
University of Washington
Seattle, Washington 98195

Abstract

The growth of $InP-Ga_xIn_{1-x}As_yP_{1-y}$ heterostructure by vapor phase
epitaxy is analyzed by the virtual equilibrium model. The quaternary
$Ga_xIn_{1-x}As_yP_{1-y}$ crystal is treated as a quasichemical solution of components
GaAs, GaP, InAs, and InP. The gas phase composed of GaCl, InCl, As_4, P_4,
HCl, H_2, and derived species is assumed to attain a state of virtual (or
near) equilibrium with the solid solution as it flows past the substrate.
The equilibrium compositions of the gas mixture and the solid phase are
computed at five different temperatures in the range 923-1023 K. The rates
of deposition of the species composing the solid solution are also determined. The respective influences of deposition-temperature, reactant gas
input partial pressures, and total pressure have been studied using the
model. The composition-transient that develops in the epitaxial layers of
the crystal once the reactive gas inputs to the reactor are terminated is
examined in detail.

Introduction

The $Ga_xIn_{1-x}As_yP_{1-y}$ quaternary solid solutions are receiving
considerable attention because of their potential applications in optoelectronic devices and fiber optical communications systems. The energy
band gap-lattice parameter relationships for the various III-V coumpounds
and their solutions are shown in Figure 1 (1). By carefully selecting the
solid solution composition, the quaternary compound can be lattice-matched
with InP over a large range of energy band gaps, between 0.7 and 1.7eV. The

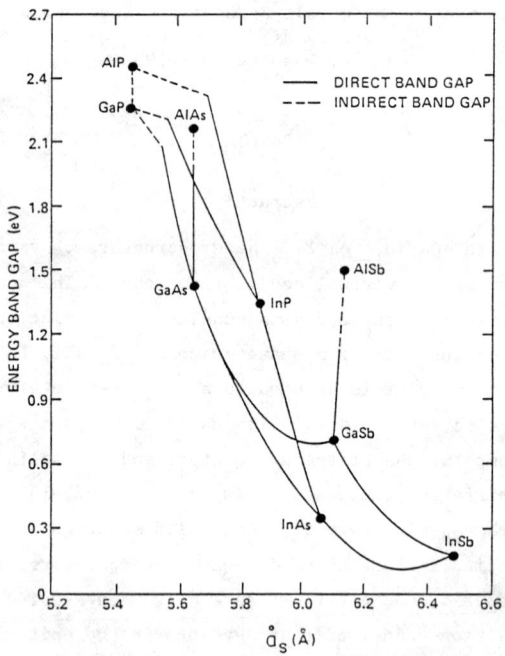

Figure 1 - Energy band gap versus lattice
parameter for III-V semiconductors (1).

devices fabricated from the $Ga_xIn_{1-x}As_yP_{1-y}$ - InP heterostructures include
lasers, light-emitting diodes (LED's) and detectors. These crystals can be
grown on indium phosphide substrates by the liquid-phase epitaxial (LPE)
method (2). Epitaxial layers (about 8μm thick) of the quaternary
$Ga_xIn_{1-x}As_yP_{1-y}$ have been grown from the melt on both (001)InP and (111)InP
substrates. The lattice parameter mismatch for the $Ga_{0.1}In_{0.9}As_{0.26}P_{0.74}$
layer on the InP substrate (a = 5.8687 Å at 20°C) is reportedly small, less
than 0.05% (2).

A disadvantage of the LPE is the dependence of the solid solution composition on the thickness of the epitaxial layer. With increasing thickness, there is loss of phosphorus, the composition changes, and the lattice parameter shifts toward that for InAs; and there is a corresponding decrease in the energy band gap. The heterojunction structures $Ga_xIn_{1-x}As_yP_{1-y}$ - InP have also been grown by vapor phase epitaxy (VPE) involving chemical vapor deposition of solid species from a suitable gas mixture (3,4,5,6). The CVD technique is more attractive than the LPE method from the viewpoint of the mass production of optoelectronic devices. Two different types of CVD systems have been used to produce quaternary III-V solid solutions: organometallic vapor phase epitaxy (OMVPE) and inorganic Ga - In - As - P - H - Cl vapor phase epitaxy. In the OMVPE, the GaInAsP crystal is deposited on an inductively-heated substrate from a gas mixture composed of metal alkyls--triethyl gallium, $Ga(C_2H_5)_3$, and triethylindium, $In(C_2H_5)_3$--arsine(AsH_3), phosphine(PH_3), and hydrogen (3).

$$Ga(C_2H_5)_3 \ + \ AsH_3(\text{or } PH_3) \ \longrightarrow \ GaAs(\text{or } GaP)(s) \ + \ 3 \ C_2H_6$$

$$In(C_2H_5)_3 \ + \ AsH_3(\text{or } PH_3) \ \longrightarrow \ InAs(\text{or } InP)(s) \ + \ 3 \ C_2H_6$$

Carbon deposition can occur by the decomposition of hydrocarbon products:

$$C_2H_6 \ \longrightarrow \ 2 \ C(s) \ + \ 3 \ H_2$$

The III-V crystals produced by the OMVPE technique tend to contain high levels of carbon impurity.

The vapor phase epitaxy of $Ga_xIn_{1-x}As_yP_{1-y}$ alloys on GaAs substrates by the hydride method has been reported by Sugiyama et al. (4) High-purity films can be grown by this inorganic chemical vapor deposition technique in which GaCl and InCl species produced by passing H_2 - HCl mixtures over Ga(ℓ) and In(ℓ), respectively, are mixed with AsH_3, PH_3, and H_2 in the mixing-zone of a flow-reactor at about 900°C; and the deposition of the quaternary solid solution from the gas phase takes place downstream on a GaAs substrate held at a temperature of 600-750°C (4). The growth of quaternaries of InGaAsP lattice-matched to InP substrates by a similar method has been described by Hyder et al. (5). Modified VPE reactors equipped with dual-growth chambers (6) and waiting chamber (7) have been used for the growth of InP/InGaAsP/InP heterostructures to ensure that the compositional-change from the quaternary to InP is rapid.

The thermodynamics of the quaternary $Ga_xIn_{1-x}As_yP_{1-y}$ solid solutions have been investigated by Nagai (8) and Onabe (9). The former (8) developed an equilibrium model for the quaternary/gas phase system. The gas species considered include GaCl, InCl, HCl, H_2, As_4, and P_4. The solid phase was treated as a homogeneous solid solution composed of GaP(s), GaAs(s), InAs(s), and InP(s); and for the activity coefficients of these species, the formulae derived by Onabe (9), on the basis of quasichemical (pair) approximation, were employed (8) to compute the composition of the gas phase in equilibrium with a solid solution of specified composition. The solid solution compositions investigated ranged from $Ga_{0.30}In_{0.70}As_{0.85}P_{0.15}$ to $Ga_{0.79}In_{0.21}As_{0.40}P_{0.60}$, at a substrate temperature of 650°C (8).

In the present work, a virtual equilibrium model is developed for the deposition of quaternary $Ga_xIn_{1-x}As_yP_{1-y}$ on an indium phosphide substrate. From the knowledge of the input gas compositions and flow rates, the model can predict the composition of the solid solution depositing, as well as the rate of its deposition. The respective effects of gas composition, deposition-zone temperature, and total pressure are investigated. The variation in the composition of the quaternary alloy in response to a change in the input gas composition is examined in some detail.

Thermodynamics of Quaternary
Solid Solutions

In the Ga-In-As-P system, strong interactions between Group III and Group V elements lead to the formation of associated species GaP, GaAs, InAs, and InP. By treating the $Ga_xIn_{1-x}As_yP_{1-y}$ alloy as a strictly regular solution composed of these four species, expressions can be deduced for the activity coefficients. Jordan (10,11,12) has made noteworthy contributions to the theory of associated regular solutions. Panish and Ilegems (13,14) have applied the regular associated solution model to analyze phase equilibria in III-V ternary systems. The chemical species model developed by Kellogg and associates (15,16) is more sophisticated and can be applied to systems with non-symmetrical interactions.

The activity coefficient γ_i of the ith component in a regular quaternary is given by (11):

$$RT \ln\gamma_i = \sum_{\substack{i=1 \\ i \neq j}}^{4} \alpha_{ij}x_j^2 + \sum_{\substack{k=1 \\ i \neq k}}^{4}\sum_{\substack{j=1 \\ i \neq j \; k<j}}^{4} x_k x_j (\alpha_{ij} + \alpha_{ik} - \alpha_{kj}) \qquad (1)$$

where x_j is the mole fraction of the jth species and α_{ij}'s are interaction parameters, one for each i-j pair. In the foregoing, $\alpha_{ii} = 0$ and $\alpha_{ij} = \alpha_{ji}$. The activity coefficients for the species i = 1,2,3, and 4, found using equation (1), satisfy the Gibbs-Duhem equation. It is readily seen that altogether six independent interaction parameters are required to describe

960

the composition-dependence of the activity coefficients for the four
species.

Onabe (9) developed a quasichemical model of the quaternary
$Ga_xIn_{1-x}As_yP_{1-y}$ alloy; only four interaction parameters are required to
express the activity coefficients of the species GaP, GaAs, InAs, and InP.

$$RT \ln\gamma_{GaAs} = \Omega_A(1-x)(1-y)(1-2x) + \Omega_B(1-x)[(1-x)y + x(1-y)]$$
$$+ \Omega_C(1-x)(1-y)(1-2y) + \Omega_D(1-y)[(1-x)y + (1-y)x] \qquad (2.a)$$

$$RT \ln\gamma_{GaP} = \Omega_A(1-x)[(1-x)(1-y) + xy] + \Omega_B(1-x)y(1-2x)$$
$$+ \Omega_C(1-x)y(2y-1) + \Omega_D y[xy + (1-x)(1-y)] \qquad (2.b)$$

$$RT \ln\gamma_{InAs} = \Omega_A x(1-y)(2x-1) + \Omega_B x[xy + (1-x)(1-y)]$$
$$+ \Omega_C(1-y)[(1-x)(1-y) + xy] + \Omega_D x(1-y)(1-2y) \qquad (2.c)$$

$$RT \ln\gamma_{InP} = \Omega_A x[(1-x)y + x(1-y)] + \Omega_B xy(2x-1)$$
$$+ \Omega_C y[(1-x)y + x(1-y)] + \Omega_D xy(2y-1) \qquad (2.d)$$

It will be noted that Ω_A, Ω_B, Ω_C, and Ω_D are the interaction parameters for
the pairs InP-GaP, InAs-GaAs, InAs-InP, and GaP-GaAs, respectively. The
composition parameters x and y are expressed by the following:

$$x = (n_{GaAs} + n_{GaP})/n_s = x_{GaAs} + x_{GaP} \qquad (3.a)$$

$$y = (n_{GaAs} + n_{InAs})/n_s = x_{GaAs} + x_{InAs} \qquad (3.b)$$

where n_i is the number of moles of the ith species in the quaternary
$Ga_xIn_{1-x}As_yP_{1-y}$ solid alloy. Note that

$$n_s = n_{GaAs} + n_{GaP} + n_{InAs} + n_{InP}$$

Furthermore, x_i is the mole fraction of the ith species in the solid
solution.

The activity coefficients defined by equations (2.a)-(2.d) must conform
to the Gibbs-Duhem equation. For the quaternary solid solution,

$$RT(x_{GaAs}d\ln\gamma_{GaAs} + x_{GaP}d\ln\gamma_{GaP} + x_{InAs}d\ln\gamma_{InAs}$$
$$+ x_{InP}d\ln\gamma_{InP}) = 0 \qquad (4)$$

Differentiation of equations (2.a)-(2.d) followed by substitution into
equation (4) yields,

$$\phi(\Omega_i, x_i, x, y)dx + \theta(\Omega_i, x_i, x, y)dy = 0 \qquad (5)$$

where the coefficients ϕ and θ of the differentials 'dx' and 'dy' are
functions of Ω_A, Ω_B, Ω_C, Ω_D, x_{GaAs}, x_{GaP}, x_{InAs}, x_{InP}, x, and y. Since x

and y can vary independently, equation (5) is viable only if the coefficients of 'dx' and 'dy' are zero. Hence

$$\phi(\Omega_i, x_i, x, y) = 0 \tag{6.a}$$

$$\theta(\Omega_i, x_i, x, y) = 0 \tag{6.b}$$

Writing out the expression (6.a) in its full complexity, we have,

$$[\Omega_A(1-y)(4x-3) + \Omega_B(4xy+1-2x-3y) + \Omega_C(1-y)(2y-1) + \Omega_D(1-y)(1-2y)]x_{GaAs}$$

$$+ [\Omega_A(2x+3y-4xy-2) + \Omega_B y(4x-3) + \Omega_C y(1-2y) + \Omega_D y(2y-1)]x_{GaP}$$

$$+ [\Omega_A(1-y)(4x-1) + \Omega_B(4xy-2x-y+1) + \Omega_C(1-y)(2y-1) + \Omega_D(1-y)(1-2y)]x_{InAs}$$

$$+ [\Omega_A(2x+y-4xy) + \Omega_B y(4x-1) + \Omega_C y(1-2y) + \Omega_D y(2y-1)]x_{InP} = 0 \tag{7}$$

This equation must hold for all compositions including the limiting values. The particular condition that is of most interest is $x_{InP} \to 1.0$ because the quaternary $Ga_x In_{1-x} As_y P_{1-y}$ solid solution is grown on an indium phosphide substrate. Note that $x_{GaAs} \to 0.0$, $x_{GaP} \to 0.0$, and $x_{InAs} \to 0.0$. Moreover, $(1-x) \to 1.0$, $(1-y) \to 1.0$; hence $y \to x$. Applying these limits to equation (7), we obtain,

$$\Omega_A(2x+x-4x^2) + \Omega_B x(4x-1) + \Omega_C x(1-2x) + \Omega_D x(2x-1) = 0$$

On cancelling 'x' throughout, the equation reduces to

$$\Omega_A(3-4x) + \Omega_B(4x-1) + \Omega_C(1-2x) + \Omega_D(2x-1) = 0$$

Taking the limit of $x \to 0.0$, we find

$$3\Omega_A - \Omega_B + \Omega_C - \Omega_D = 0$$

or $\quad \Omega_D = 3\Omega_A - \Omega_B + \Omega_C \tag{8}$

Similarly, starting with equation (6.b), we find,

$$[\Omega_A(1-x)(2x-1) + \Omega_B(1-x)(1-2x) + \Omega_C(1-x)(4y-3) + \Omega_D(4xy+1-3x-2y)]x_{GaAs}$$

$$+ [\Omega_A(1-x)(2x-1) + \Omega_B(1-x)(1-2x) + \Omega_C(1-x)(4y-1) + \Omega_D y(2x-1)]x_{GaP}$$

$$+ [\Omega_A x(1-2x) + \Omega_B x(2x-1) + \Omega_C(3x+2y-4xy-2) + \Omega_D x(4y-3)]x_{InAs}$$

$$+ [\Omega_A x(1-2x) + \Omega_B x(2x-1) + \Omega_C(x+2y-4xy) + \Omega_D x(4y-1)]x_{InP} = 0 \tag{9}$$

Following the same procedure as above, we can show that,

$$\Omega_D = \Omega_A - \Omega_B + 3\Omega_C \tag{10}$$

From equations (8) and (10), it follows that,

$$\Omega_A = \Omega_C = 0.25(\Omega_B + \Omega_D) \tag{11}$$

For the case of $Ga_x In_{1-x} As_y P_{1-y}$ solid alloy grown on a gallium arsenide substrate, the limit $x_{GaAs} \to 1$ is considered for equations (7) and (9).

Because of the particular choice of the activity coefficient expressions made here, the resulting interrelationships between the four interaction parameters are different from those given above.

Direct measurements of the interaction paramaters Ω_A, Ω_B, Ω_C, and Ω_D for the quaternary $Ga_x In_{1-x} As_y P_{1-y}$ are lacking. The oft-quoted values of Panish and Ilegems (13) were derived from the phase diagrams for the pseudobinaries InP-GaP, InAs-GaAs, InAs-InP, and GaP-GaAs. These are:

Ω_A : 3500 cal/mole (14644 J/mole) for InP-GaP

Ω_B : 3000 cal/mole (12552 J/mole) for InAs-GaAs

Ω_C : 400 cal/mole (1674 J/mole) for InAs-InP

Ω_D : 400 cal/mole (1674 J/mole) for GaP-GaAs

Stringfellow (17) reported a semi-empirical method for estimating the values of Ω_i for psuedobinaries in the III-V systems; the interaction parameter is supposed to vary directly with the second power of the lattice mismatch (Δa_m) and inversely with the $4\frac{1}{2}$-power of the lattice parameter $(\bar{a})$ for the ternary III-V solid solution. In the present work, the interaction parameters were chosen in conformity with the Gibbs-Duhem equation for the $Ga_x In_{1-x} As_y P_{1-y}$ phase growing on an indium phosphide substrate.

Ω_B = 12,552 J/mole   and Ω_D = 4,184 J/mole

Ω_A = Ω_C = 4,184 J/mole

The particular case when the quaternary is an ideal solution (Ω_A = Ω_B = Ω_C = Ω_D = 0) is not considered here.

Ga - In - As - P - H - Cl Chemical Vapor Deposition System

The deposition of quaternary $Ga_x In_{1-x} As_y P_{1-y}$ crystal from gas phase at temperatures in the range 650-750°C is examined in some detail by the methods of equilibrium thermodynamics. The deposition occurs on an InP(s) substrate supported in a flow stream composed of chlorides of gallium and indium, hydrogen, hydrogen chloride, arsenic, and phosphorus species. The input gas mixture is composed of GaCl, InCl, As_4, P_4, and H_2. As this gas stream reaches the deposition zone, several new species are formed due to chemical reactions. In the present system, there are fourteen species in the gas phase and four species in the solid solution:

GaCl, $GaCl_3$, InCl, $InCl_3$, HCl, H_2, Cl_2, As_4, As_2, P_4, P_2, AsH_3,

PH_3, and $AsCl_3$: gas phase.

GaP - InAs - GaAs - InP : solid solution.

In order to determine the number of independent reactions in the system, an atom coefficient matrix is constructed by representing each species as a vector and then reducing the matrix to an echelon form to find its rank ($c*$). For the 18 x 6 matrix that represents the 18 species present in the heterogeneous system at equilibrium, the rank $c*$ is six. By virtue of the Gibbs stoichiometric rule (18),

$$0 < r_m \leqslant (N - c*) \tag{12}$$

the maximum number (r_m) of independent equilibria is the total number of species (N) less the rank ($c*$) of the atom matrix. It is readily seen that $r_m = 18 - 6 = 12$. The following set of reactions is chosen to represent the equilibrium system. These reactions are linearly-independent and each species appears at least once in this set.

$$H_2 + Cl_2 = 2\ HCl \tag{A.1}$$

$$GaCl + \tfrac{1}{4}\ As_4 + \tfrac{1}{2}\ H_2 = GaAs(s) + HCl \tag{A.2}$$

$$GaCl + \tfrac{1}{4}\ P_4 + \tfrac{1}{2}\ H_2 = GaP(s) + HCl \tag{A.3}$$

$$InCl + \tfrac{1}{4}\ As_4 + \tfrac{1}{2}\ H_2 = InAs(s) + HCl \tag{A.4}$$

$$InCl + \tfrac{1}{4}\ P_4 + \tfrac{1}{2}\ H_2 = InP(s) + HCl \tag{A.5}$$

$$\tfrac{1}{4}\ As_4 = \tfrac{1}{2}\ As_2 \tag{A.6}$$

$$\tfrac{1}{4}\ P_4 = \tfrac{1}{2}\ P_2 \tag{A.7}$$

$$AsH_3 = \tfrac{1}{4}\ As_4 + 1\tfrac{1}{2}\ H_2 \tag{A.8}$$

$$PH_3 = \tfrac{1}{4}\ P_4 + 1\tfrac{1}{2}\ H_2 \tag{A.9}$$

$$GaCl_3 + H_2 = GaCl + 2\ HCl \tag{A.10}$$

$$InCl_3 + H_2 = InCl + 2\ HCl \tag{A.11}$$

$$AsCl_3 = \tfrac{1}{4}\ As_4 + 1\tfrac{1}{2}\ Cl_2 \tag{A.12}$$

The minimum number of degrees of freedom, f, the system has are found by applying the phase rule (18,19,20):

$$f = N - r_m - p + 2 = 18 - 12 - 2 + 2 = 6.$$

These six degrees of freedom are satisfied by specifying (1) temperature (T), (2) total pressure (P_t), (3) P°_{GaCl}, (4) P°_{InCl}, (5) $P^\circ_{As_4}$, and (6) $P^\circ_{P_4}$ where P°_i is the partial pressure of the ith species in the input gas stream to the mixing zone of the CVD reactor. It follows that

$$P^\circ_{H_2} = P_t - P^\circ_{GaCl} - P^\circ_{InCl} - P^\circ_{As_4} - P^\circ_{P_4} \tag{13}$$

If F_o (in ml/min at STP) denotes the volumetric flow rate of the input gas stream, the molar input rate Q_o is given by

$$Q_o = [F_o \cdot P_t / (82.06)(273)], \text{ mole/min} \qquad (14)$$

The input rates (in gram-atoms/min) of elements gallium, indium, arsenic, phosphorus, and hydrogen can be calculated from the molar flow rate and the input partial pressures of species containing the particular element. Thus

$$Q_{Ga}^\circ = Q_o(P_{GaCl}^\circ / P_t) \qquad (15.a)$$

$$Q_{In}^\circ = Q_o(P_{InCl}^\circ / P_t) \qquad (15.b)$$

$$Q_{As}^\circ = 4 \, Q_o(P_{As_4}^\circ / P_t) \qquad (15.c)$$

$$Q_P^\circ = 4 \, Q_o(P_{P_4}^\circ / P_t) \qquad (15.d)$$

$$Q_H^\circ = 2 \, Q_o(P_{H_2}^\circ / P_t) \qquad (15.e)$$

Since no deposition of chlorine or hydrogen occurs, conservation of these in the gas-phase is valid. In other words, the Cl/H atom ratio of the gas mixture does not change during equilibration:

$$Cl/H = RCLH = [(P_{GaCl}^\circ + P_{InCl}^\circ)/2 \, P_{H_2}^\circ] \qquad (16)$$

The standard free energy changes for the reactions (A.1)-(A.12) are drawn from published sources (19,21,22):

$$\Delta G_1^\circ = -188,196 - 12.804 \text{ T}, \text{ J}$$

$$\Delta G_2^\circ = -129,457 + 117.997 \text{ T}, \text{ J}$$

$$\Delta G_3^\circ = -156,795 + 122.951 \text{ T}, \text{ J}$$

$$\Delta G_4^\circ = -106,165 + 114.767 \text{ T}, \text{ J}$$

$$\Delta G_5^\circ = - 99,342 + 124.043 \text{ T}, \text{ J}$$

$$\Delta G_6^\circ = 55,574 - 35.893 \text{ T}, \text{ J}$$

$$\Delta G_7^\circ = 54,288 - 34.748 \text{ T}, \text{ J}$$

$$\Delta G_8^\circ = -118,984 - 86.090 \text{ T}, \text{ J}$$

$$\Delta G_9^\circ = 17,259 - 73.450 \text{ T}, \text{ J}$$

$$\Delta G_{10}^\circ = 165,904 - 127.930 \text{ T}, \text{ J}$$

$$\Delta G_{11}^\circ = 99,646 - 105.868 \text{ T}, \text{ J}$$

$$\Delta G_{12}^\circ = 296,210 - 82.297 \text{ T}, \text{ J}$$

The equilibrium constant $K(J,I)$ for the J-th reaction at temperature $T(I)$ is computed using

$$K(J,I) = \exp[-\Delta G^\circ(J,I)/RT(I)]$$

Phase Compositions
at Equilibrium

The flowing gas stream attains a state of "virtual equilibrium" with the solid phase $Ga_xIn_{1-x}As_yP_{1-y}$ in the deposition zone. The total pressure in the reactor is uniform throughout and is held constant during the process. The compositions of the gas mixture and the solid solution are computed by means of the iterative equilibrium constant method (IECM) described by the senior author in a recent work (19).

To facilitate the calculation, and additional state variable--the volumetric flow rate, F_o--must be specified. In here,

$$F_o = 1000 \text{ ml(STP)/min}$$

Furthermore, let Q^S_{GaAs}, Q^S_{GaP}, Q^S_{InAs}, and Q^S_{InP} in mole/min represent the respective amounts of GaAs, GaP, InAs, and InP that deposit from the gas stream onto the solid crystal per unit time. At the start of the calculation, reasonable guesses are made for Q^S_{GaAs}, Q^S_{GaP}, and Q^S_{InAs}; it is assumed that the gallium (or indium) content in the input gas stream is evenly divided between the arsenide and phosphide components of the solid phase. In addition, first estimates are also made for the equilibrium partial pressures of H_2, HCl, and As_4. Using these and the equilibrium constant expressions for reactions (A.1)-(A.12), the following are calculated:

(i) activity coefficients γ_1, γ_2, γ_3, and γ_4 for the components GaAs, GaP, InAs, and InP, respectively, in the $Ga_xIn_{1-x}As_yP_{1-y}$ solid solution;

(ii) activities a_{GaAs}, a_{GaP}, a_{InAs}, and a_{InP};

(iii) partial pressures of Cl_2, GaCl, P_4, InCl, As_2, P_2, AsH_3, PH_3, $GaCl_3$, $InCl_3$, and $AsCl_3$.

At the completion of the first iteration, as enumerated above, the new total pressure, the new Cl/H atom ratio, and the new volumetric flow rate of the gas mixture are as follows. For the J-th iteration,

$$P_t(J) = P_{H_2} + P_{HCl} + P_{As_4} + P_{Cl_2} + P_{GaCl} + \cdots + P_{AsCl_3}$$

$$\frac{1}{RCLH(J)} = \frac{2\,P_{H_2} + P_{HCl} + 3(P_{PH_3} + P_{AsH_3})}{2\,P_{Cl_2} + P_{HCl} + P_{GaCl} + P_{InCl} + 3(P_{GaCl_3} + P_{InCl_3} + P_{AsCl_3})}$$

By hydrogen atom balance in the gas stream, it can be shown that, the new volumetric flow rate is

$$F(J) = \frac{2 \, F_o \cdot P^\circ_{H_2}}{2 \, P_{H_2} + P_{HCl} + 3(P_{PH_3} + P_{AsH_3})}$$

The amounts (in gram-atoms/min) of elements Ga, In, As, and P present in the vapor species are found from the flow rate and the partial pressures.

$$Q^v_{Ga} = F(J)(P_{GaCl} + P_{GaCl_3})/(82.06)(273)$$

$$Q^v_{In} = F(J)(P_{InCl} + P_{InCl_3})/(82.06)(273)$$

$$Q^v_{As} = F(J)(4 \, P_{As_4} + 2 \, P_{As_2} + P_{AsH_3} + P_{AsCl_3})/(82.06)(273)$$

$$Q^v_{P} = F(J)(4 \, P_{P_4} + 2 \, P_{P_2} + P_{PH_3})/(82.06)(273)$$

Taking into consideration the amounts of Ga, In, As, and P that have deposited from the gas phase, we can write, for the J-th iteration,

$$Q_{Ga}(J) = Q^v_{Ga} + Q^s_{GaAs} + Q^s_{GaP}$$

$$Q_{In}(J) = Q^v_{In} + Q^s_{InAs} + Q^s_{InP}$$

$$Q_{As}(J) = Q^v_{As} + Q^s_{GaAs} + Q^s_{InAs}$$

$$Q_{P}(J) = Q^v_{P} + Q^s_{GaP} + Q^s_{InP}$$

These quantities are compared to the input values Q°_{Ga}, Q°_{In}, Q°_{As}, and Q°_{P} respectively, given by equations (15.a)-(15.d). The convergence limits are defined as

$$\left| \frac{Q_i(J) - Q^\circ_i}{Q^\circ_i} \right| \leq 10^{-4} \tag{17}$$

where i = Ga, In, As, and P. Similar convergence criteria were also applied to the total pressure and the Cl/H atom ratio. Normally, several iterations are required to obtain convergence. The six values that were guessed for the quantities Q^s_{GaAs}, Q^s_{GaP}, Q^s_{InAs}, P_{H_2}, P_{HCl}, and P_{As_4} at the start of the calculation are now modified as follows:

$$Q^{s,new}_{GaAs} = Q^{s,old}_{GaAs} \, [Q^\circ_{Ga}/Q_{Ga}(J)]^{0.25}[Q^\circ_{As}/Q_{As}(J)]^{0.2} \tag{18.a}$$

$$Q^{s,new}_{GaP} = Q^{s,old}_{GaP} \, [Q^\circ_{Ga}/Q_{Ga}(J)]^{0.2}[Q^\circ_{P}/Q_{P}(J)]^{0.2} \tag{18.b}$$

$$Q^{s,new}_{InAs} = Q^{s,old}_{InAs} \, [Q^\circ_{In}/Q_{In}(J)]^{0.25}[Q^\circ_{As}/Q_{As}(J)]^{0.2} \tag{18.c}$$

$$P^{new}_{H_2} = P^{old}_{H_2} \, [RCLH(J)/RCLH]^{0.2}[PT/PT(J)]^{0.4} \tag{18.d}$$

$$P_{HCl}^{new} = P_{HCl}^{old} [RCLH/RCLH(J)]^{0.3} [PT/PT(J)]^{0.5} \qquad (18.e)$$

$$P_{As_4}^{new} = P_{As_4}^{old} [Q_{As}^{\circ}/Q_{As}(J)]^{0.4} \qquad (18.f)$$

Using these modified values, the entire calculation is repeated. This is continued until convergence is achieved. The computer program developed here, and given in the appendix, has been applied to a number of input gas compositions, deposition-zone temperatures, and total pressures. The results of these computations are now considered.

Gas Phase Composition

The equilibrium gas composition in the system was determined at 1 atm total pressure for five different temperatures--923, 948, 973, 998, and 1023 K. The input gas mixture consisted of H_2, GaCl, InCl, As_4, and P_4 at the following partial pressures:

P_{GaCl}° : 0.644×10^{-3} atm ; P_{InCl}° : 0.268×10^{-2} atm

$P_{As_4}^{\circ}$: 0.102×10^{-2} atm ; $P_{P_4}^{\circ}$: 0.109×10^{-3} atm

This input gas composition was chosen so that a comparison can be made with the earlier work of Nagai (8). Figures 2 and 3 present the results in the form of semi-logarithmic plots. It is well to note that hydrogen constituted greater than 99% of the gas mixture. As shown in Figure 2, the species InCl, As_4, HCl, As_2, GaCl, and PH_3 occur in significant proportions with P_2 and P_4 becoming more important at higher temperatures. The less abundant species--$GaCl_3$, $InCl_3$, and AsH_3--are represented in Figure 3. Also shown here are the least important of the fourteen species, namely $AsCl_3$ and Cl_2.

Similar calculations were performed for twenty-two different sets of input gas mixture compositions. In each case, the equilibrium partial pressures were computed.

Composition of Solid Solution

The $Ga_x In_{1-x} As_y P_{1-y}$ crystal is best viewed as consisting of species GaAs, GaP, InAs, and InP. The phase-composition can be computed from the steady deposition-rates (in mole/min) Q_{GaAs}^{s}, Q_{GaP}^{s}, Q_{InAs}^{s}, and Q_{InP}^{s} of the respective species. It is readily seen that

$$x_{GaAs} = Q_{GaAs}^{s}/Q_{tot}^{s} \quad ; \quad x_{GaP} = Q_{GaP}^{s}/Q_{tot}^{s} \qquad (19.a,b)$$

$$x_{InAs} = Q_{InAs}^{s}/Q_{tot}^{s} \quad ; \quad x_{InP} = Q_{InP}^{s}/Q_{tot}^{s} \qquad (19.c,d)$$

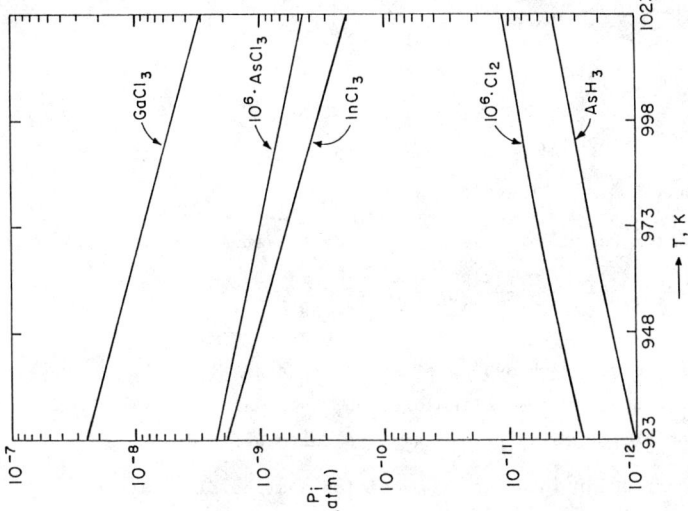

Figure 3 - Partial pressures of $GaCl_3$, $InCl_3$, AsH_3, $AsCl_3$, and Cl_2 in gas phase at equilibrium with $Ga_xIn_{1-x}As_yP_{1-y}$ solid solution.

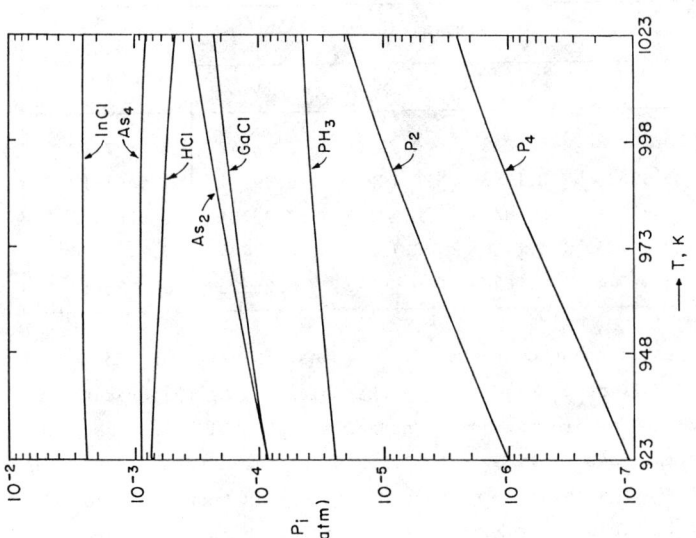

Figure 2 - Partial pressures of several species in gas phase at equilibrium with $Ga_xIn_{1-x}As_yP_{1-y}$ solid solution.

where x_i is the mole-fraction of the ith species and

$$Q_{tot}^s = Q_{GaAs}^s + Q_{GaP}^s + Q_{InAs}^s + Q_{InP}^s \qquad (20)$$

Furthermore,

$$x = x_{GaAs} + x_{GaP} \qquad (21.a)$$

$$y = x_{GaAs} + x_{InAs} \qquad (21.b)$$

$$u = x_{GaAs} \; ; \; v = x_{InAs} \qquad (21.c,d)$$

The solid solution can also be represented by the stoichiometric formula:

$$(GaAs)_u (GaP)_{x-u} (InAs)_v (InP)_{1-x-v}$$

This type of representation has the advantage of providing the complete composition of the solid phase. The use of 'x' and 'y', although it gives a more concise formula for the solid solution, is clearly not recommended if the composition of the solid phase is to be fully defined.

For the particular input gas composition and the flow rate (F_o) specified earlier, the values, at equilibrium, of x, u, v, and y were computed at five different temperatures. These results are summarized in Table I.

Table I. Solid Phase Composition and Deposition Rate (mole/min)

T(K)	Composition				Q_{tot}^s mole/min
	x	u	v	y	
923	0.74378	0.20381	0.25160	0.45541	3.3487×10^{-5}
948	0.80326	0.19883	0.19219	0.39101	2.9441×10^{-5}
973	0.84511	0.18674	0.15027	0.33701	2.6267×10^{-5}
998	0.87465	0.17361	0.12063	0.29424	2.3487×10^{-5}
1023	0.89544	0.16349	0.09977	0.26326	2.0785×10^{-5}

At 923 K, the solid phase contains 20.381% GaAs, 53.997% GaP, 25.16% InAs, and 0.462% InP, all percentages are by the mole; as the temperature is raised to 1023 K, the proportion of the gallium phosphide increases whereas those of GaAs and InAs decline. It will also be noted that in the temperature range investigated, the InP-content of this quaternary solid solution is rather small. This is probably due to the low phosphorus-content of the input gas-mixture.

The results at 923 K can be compared to those obtained by Nagai (8). The solid solution composition $Ga_{0.30}In_{0.70}As_{0.85}P_{0.15}$ reported by this author (8) differs significantly from the composition given above. This may be due to the fact that Nagai (8) used a different set of values for the interaction parameters ($\Omega_A = \Omega_B = 20{,}920$ J/mole and $\Omega_C = \Omega_D = 8{,}368$ J/mole).

Deposition Rate

The rate of deposition of the quaternary crystal $Ga_xIn_{1-x}As_yP_{1-y}$ from the gas phase was computed assuming virtual equilibrium between the solid substrate and the flowing gas stream containing species GaCl, InCl, As_4, P_4, H_2, HCl, and so forth. The deposition rate, Q_{tot}^s, is a function of input gas composition, deposition-zone temperature, total pressure, and the volumetric flow rate of the gas mixture. For the same input gas-phase as considered earlier and $F_o = 1000$ ml(STP)/min, the values of Q_{tot}^s were computed at different substrate temperatures and 1 atm total pressure and are listed in Table I. The rate decreases with increasing temperature, in conformity with the thermodynamic requirements for an exothermic process.

The room-temperature lattice constants for $Ga_xIn_{1-x}As_yP_{1-y}$ layers can be calculated by means of the Vegard's law (4):

$$\overset{\circ}{a}_s = \overset{\circ}{a}_{GaAs}xy + \overset{\circ}{a}_{InAs}(1-x)y + \overset{\circ}{a}_{GaP}x(1-y) + \overset{\circ}{a}_{InP}(1-x)(1-y) \qquad (22)$$

where the lattice constants for the individual species at 300 K are:

$$\overset{\circ}{a}_{GaAs} : 5.653 \text{ Å} \quad ; \quad \overset{\circ}{a}_{InAs} : 6.058 \text{ Å}$$

$$\overset{\circ}{a}_{GaP} : 5.451 \text{ Å} \quad ; \quad \overset{\circ}{a}_{InP} : 5.869 \text{ Å}$$

For the quaternary crystal that deposits at 973 K (Table I), the lattice constant, $\overset{\circ}{a}_s = 5.5831$ Å and the molecular weight is 122.49. This gives a theoretical density of 4.675 gm/cm^3 for the crystal. Furthermore,

$$\text{rate} = (2.6267\times10^{-5})(122.49)(60) = 0.193 \text{ gm/hr}$$

When the deposition takes place on a substrate 1.27 cm x 1.27 cm in size,

$$\text{rate of growth} = \frac{(0.193)(10^4)}{(4.675)(1.6129)} = 256 \text{ μm/hr}$$

This may be compared with the observed growth rate of 4 μm/hr reported by Hyder et al. (5) for the growth of quaternary from a $GaCl-InCl-AsH_3-PH_3-H_2$ phase at 973 K. Due to differences in the input gas compositions, and on account of the virtual equilibrium assumption, it is not surprising that the experimental growth rate is much smaller than the rate predicted. Because the number of rate measurements reported in the literature are only few, it is difficult to draw any definitive conclusions with respect to the applicability of virtual equilibrium model.

971

Effect of Input Gas Composition

The partial pressures P^o_{GaCl}, P^o_{InCl}, $P^o_{As_4}$, and $P^o_{P_4}$ in the input gas mixture can be varied independently and the resultant changes in the phase-compositions and deposition rates can be computed using the present model. Beginning with the input gas mixture listed earlier, the partial pressure P^o_{GaCl} was decreased in steps and the values of the composition-parameters x, u, v, and y were determined in each case for a deposition zone temperature of 923 K and 1 atm total pressure. These results are summarized in Table II.

Table II. Effect of P^o_{GaCl} on the Composition and Deposition Rate of $Ga_xIn_{1-x}As_yP_{1-y}$

P^o_{InCl} : 2.68×10^{-3} atm ; $P^o_{As_4}$: 1.02×10^{-3} atm ; $P^o_{P_4}$: 1.09×10^{-4} atm					
P^o_{GaCl} (atm)	Composition				Q^s_{tot} mole/min
	x	u	v	y	
6.44×10^{-4}	0.74378	0.20381	0.25160	0.45541	3.3487×10^{-5}
3.22×10^{-4}	0.54521	0.05566	0.42864	0.48430	2.4701×10^{-5}
1.07×10^{-4}	0.20124	0.02215	0.75635	0.77849	2.1298×10^{-5}
6.44×10^{-5}	0.12173	0.01460	0.83576	0.85036	2.0464×10^{-5}
1.61×10^{-5}	0.03036	0.00411	0.92848	0.93258	1.9149×10^{-5}

With decreasing P^o_{GaCl}, it is seen that InAs becomes the predominant component in the quaternary solid solution and the deposition rate decreases. Similar behavior was noted at 973 K.

Furthermore, the effects of varying P^o_{InCl}, $P^o_{As_4}$, and $P^o_{P_4}$ on the composition and deposition rate of the solid solution were investigated at 923 K and 973 K for 1 atm total pressure. A 20-fold increase in InCl-input partial pressure produces a solid phase that is InAs for the most part, at both 923 K and 973 K. On increasing the As_4-input partial pressure to 0.306 atm, a 300-fold change, the resulting quaternary phase, at 923 K, contains 17.594% GaAs, 23.603% GaP, 58.261% InAs, and 0.542% InP, showing a rise in the InAs-content and a parallel decline in the proportion of GaP. By decreasing the P_4-input partial pressure ten-fold, at 923 K, the mole fraction of GaAs is increased from 0.20381 to 0.62454 whereas that of GaP decreases from 0.53997 to 0.06497; a similar behavior is noted at 973 K. In general, the deposition-rates are less sensitive to variations in the input gas composition.

Effect of the Cut-off of Reactant-input on
Solid-phase Composition and Deposition

In the conventional flow-reactor used for the VPE of $Ga_xIn_{1-x}As_yP_{1-y}$, the gas-composition changes only slowly even if the reactant gas inputs can be turned on and off quickly. At 1000 ml(STP)/min flow rate, the mean residence time in a typical tubular reactor (2.68 cm inside diameter) is 10 sec at 923 K and 1 atm total pressure. Due to back-mixing (and diffusion), the concentrations of the reactant species GaCl, InCl, As_4 and P_4 in the hydrogen flow stream decrease slowly thus requiring much longer than the 10 sec sweep-time to reach negligible levels. The growth of the solid solution continues nevertheless, with the result that a composition-transient develops in the surface layers of the $Ga_xIn_{1-x}As_yP_{1-y}$ crystal. This phenomenon can be investigated by the "virtual equilibrium" model. The reactant partial pressures are assumed to decay exponentially once their inputs are turned off. For the input gas composition previously examined, following the cut-off of the four reactant species, with hydrogen flow held steady at 1000 ml(STP)/min, we can write, for the pre-equilibrium partial pressures,

$$P^\circ_{GaCl} = 6.44 \times 10^{-4} \exp(-t/10) \tag{23.a}$$

$$P^\circ_{InCl} = 2.68 \times 10^{-3} \exp(-t/10) \tag{23.b}$$

$$P^\circ_{As_4} = 1.02 \times 10^{-3} \exp(-t/10) \tag{23.c}$$

$$P^\circ_{P_4} = 1.09 \times 10^{-4} \exp(-t/10) \tag{23.d}$$

where t (in sec) denotes the time elapsed after the reactant gas flow has been discontinued.

The equilibrium compositions of the gas and solid phases as well as the rate of deposition of the quaternary crystal can be calculated for any specified value of t. The temperature and total pressure, for the most part, were set at 923 K and 1 atm, respectively. For t, fifteen different values were assigned, starting with t = 1 sec and concluding with t = 60 sec. The results of these calculations are now considered in some detail.

Solid Solution Composition

The composition of the $Ga_xIn_{1-x}As_yP_{1-y}$ crystal before switching off the gas supply is 20.381% GaAs, 53.997% GaP, 25.16% InAs, and 0.462% InP. The changes in the mole fractions of the components GaAs, GaP, InAs, and InP of the solid phase, with elapse of time, are depicted in Figure 4; its

973

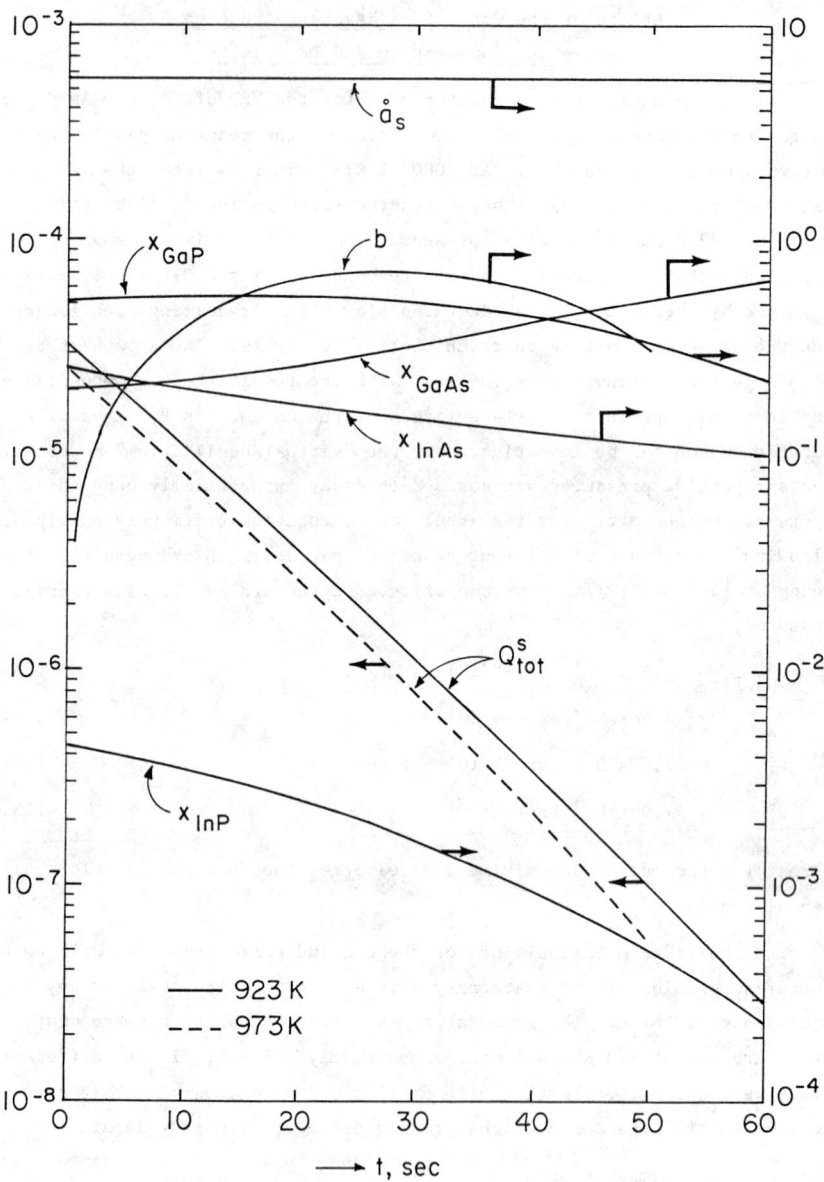

Figure 4 - Composition-transient that develops in the epitaxial layers of $Ga_xIn_{1-x}As_yP_{1-y}$ following the termination of the inputs of reactant gases GaCl, InCl, As_4, and P_4 to the flow reactor.

lattice parameter decreases from 5.6486 Å (at t = 0) to a value of 5.6083 Å
(at t = 30 sec) and then increases slowly to 5.6484 Å (at t = 60 sec). In
addition, the percentage lattice mismatch expressed relative to the lattice
parameter of the crystal grown prior to the onset of the transient, can be
calculated as follows:

$$b = (100 - 17.7 \, \overset{\circ}{a}_s)$$ (24)

The variation of b with time is also shown in Figure 4; it reaches a
maximum of about 0.7% at t = 30 sec.

As shown in Figure 4, the growth rate Q_{tot}^s (in mole/min) of the
quaternary solid solution decreases with time. For instance, at 923 K, it
declines from an initial value of 2.9994×10^{-5} mole/min at t = 1 sec to
2.9271×10^{-8} mole/min at t = 60 sec, following the termination of the
reactant gas inputs. A similar behavior is noted for a deposition-
temperature of 973 K (Figure 4).

Composition-transient

To gain further insight into the nature of solid solution depositing
under transient conditions, the average properties of the epitaxial layer
grown over a 10 sec interval are determined. The mole fractions of the four
components, the amount (Σ_s) of the quaternary crystal, and the thickness (L)
of the layer, at 923 K, are summarized in Table III.

Table III. Composition and Thickness of Solid Solution under Transient
Conditions

Δt, sec	x_{GaAs}	x_{GaP}	x_{InAs}	x_{InP}	Σ_s, mole	L, μm*
0	0.20381	0.53997	0.25160	0.00462	–	–
0–10	0.21245	0.55036	0.23298	0.00421	3.39414×10^{-5}	9.16
10–20	0.23993	0.56230	0.19458	0.00319	1.15033×10^{-5}	3.08
20–30	0.29662	0.54002	0.16116	0.00220	0.38210×10^{-5}	1.02
30–40	0.35687	0.50029	0.14141	0.00143	0.11909×10^{-5}	0.32
40–50	0.46056	0.41667	0.12278	0.00000	0.03600×10^{-5}	0.10
50–60	0.56473	0.31885	0.11642	0.00000	0.01151×10^{-5}	0.03

* L is the thickness of the epitaxial layer on a surface 1 cm^2 in total
area.

It is seen that the crystal becomes progressively richer in GaAs and
poorer in GaP and InAs components; the uppermost layers are practically

ternary because of the absence of InP. The portion of the solid solution
with variable-composition induced by the gas-composition transient during a
60 sec interval, at 923 K, amounts to 6.63 mg (13.7 μm thick). The average
composition is: 23.090% GaAs, 54.964% GaP, 21.574% InAs, and 0.372% InP; the
lattice parameter and the molecular weight are 5.6317 Å and 130.22,
respectively.

VPE of $Ga_xIn_{1-x}As_yP_{1-y}$ at Low Pressures

The effect of the total pressure in the growth reactor on the
composition and deposition-rate of the quaternary crystal was investigated
by means of the virtual equilibrium model. At 923 K, four different total
pressures were employed for the purposes of the present calculation: 1.0,
0.1, 10^{-2}, and 10^{-8} atm. The input reactant gas partial pressures were set
according to the following relations:

$$P°_{GaCl} = 6.44 \times 10^{-4}\ P \quad ; \quad P°_{InCl} = 2.68 \times 10^{-3}\ P \qquad (25.a,b)$$

$$P°_{As_4} = 1.02 \times 10^{-3}\ P \quad ; \quad P°_{P_4} = 1.09 \times 10^{-4}\ P \qquad (25.c,d)$$

where P(in atm) is the total pressure. The results are presented in
Table IV.

Table IV. Effect of Total Pressure on Solid Composition and Deposition
Rate at 923 K

P	Solid-phase Composition				Q^s_{tot}	(dL/dt)*
(atm)	x_{GaAs}	x_{GaP}	x_{InAs}	x_{InP}	mole/min	μm/min
1.0	0.20381	0.53997	0.25160	0.00462	3.3487×10^{-5}	9.09
0.1	0.12238	0.81470	0.06014	0.00278	2.1178×10^{-6}	0.54
10^{-2}	0.08392	0.89432	0.02027	0.00149	1.0308×10^{-7}	2.56×10^{-2}
10^{-8}	0.09275	0.89189	0.01440	0.00096	9.5639×10^{-18}	2.37×10^{-12}

* Calculated assuming a total surface area of 1 cm^2.

The solid-phase compositions listed in Table IV show that with
decreasing pressure, the GaP-component becomes the most predominant. The
deposition rate (Q_{tot}) is very sensitive to the total pressure. The growth
rate of the epitaxial layer was computed at different total pressures using
the data on solid compositions and deposition rates. The rate decreases
from 1515 Å/sec at 1 atm to 4.3 Å/sec at 10^{-2} atm. For the lowest pressure

$(10^{-8}$ atm) employed, the growth rate is exceedingly small, 4×10^{-10} Å/sec; only negligibly small fractions of the reactant species supplied are utilized in the growth process.

An infinitesimally small growth rate, obtainable at very low pressures, is inherently advantageous in producing double heterostructures with clean, sharp, interfaces because of the absence of composition-transients.

Low-pressure growth conditions in combination with the use of GaCl, InCl, As_4, and P_4 are potentially attractive in the molecular beam epitaxy of quaternary $Ga_xIn_{1-x}As_yP_{1-y}$ alloys; the elements As_4 and P_4 can be replaced by the hydrides AsH_3 and PH_3 without sustaining any intolerable loss in the flexibility of the operation.

Conclusions

The compositions of gas and solid phases in the Ga - In - As - P - H - Cl chemical vapor deposition system were determined by the virtual equilibrium model. The rate of deposition of the quaternary $Ga_xIn_{1-x}As_yP_{1-y}$ solution was ascertained. The influences of deposition-zone temperature, input gas partial pressures, and total pressure on the phase-compositions and growth rate were investigated. The composition-transient in the epitaxially grown $Ga_xIn_{1-x}As_yP_{1-y}$ crystal that occurs upon terminating the reactant gas inputs was examined in detail.

Acknowledgments

The financial support of this work by the Washington Technology Center, University of Washington, Seattle, Washington, is gratefully acknowledged. Thanks are also due to the Department of Energy, Washington, D.C., and the Boeing Aerospace Company, Seattle, Washington.

References

1. G. B. Stringfellow, "OMVPE Growth of $Al_xGa_{1-x}As$," Journal of Crystal Growth, 53(1981), 42-52.

2. Y. Henry, M. Moulin, and A. Laugier, "Thickness Dependence of Lattice Parameter and Band GaP in $Ga_xIn_{1-x}As_{1-y}P_y$/InP Heterostructures," Journal of Crystal Growth, 51(1981), 387-393.

3. J. P. Duchemin, J. P. Hirtz, M. Razeghi, M. Bonnet, and S. D. Hersee, "GaInAs and GaInAsP Materials Grown by Low Pressure MOCVD for Microwave and Optoelectronic Applications," Journal of Crystal Growth, 55(1981), 64-73.

977

4. K. Sugiyama, H. Kojma, H. Enda, and M. Shibata, "Vapor Phase Epitaxial Growth and Characterization of $Ga_{1-y}In_yAs_{1-x}P_x$ Quaternary Alloys," *Japanese Journal of Applied Physics*, 16(12)(1977), 2197-2203.

5. S. B. Hyder, R. R. Saxena, and C. C. Hooper, "Vapor-Phase Epitaxial Growth of Quaternary $In_{1-x}Ga_xAs_yP_{1-y}$ in the 0.75-1.35 ev Band-GaP Range," *Applied Physics Letters*, 34(1979), 584-586.

6. T. Mizutani, M. Yoshida, A. Usui, H. Watanabe, T. Yuasa, and I. Hayashi, "Vapor Phase Growth of InGaAsP/InP DH Structures by the Dual-Growth-Chamber Method," *Japanese Journal of Applied Physics*, 19(2)(1980), L113-L116.

7. G. H. Olsen, C. J. Nuese, and M. Ettenberg," Low-threshold 1.25-μm Vapor-grown InGaAsP cw Lasers," *Applied Physics Letters*, 34(1979), 262-264.

8. H. Nagai, "Thermodynamic Analysis of $Ga_xIn_{1-x}As_yP_{1-y}$ CVD : Ga - In - As - P - H - Cl system," *Journal of Crystal Growth*, 48(1980), 359-362.

9. K. Onabe, "Thermodynamics of Type $A_{1-x}B_xC_{1-y}D_y$ III-V Quaternary Solid Solutions," *Journal of Physics and Chemistry of Solids*, 43(1982), 1071-1086.

10. A. S. Jordan, "A Theory of Regular Associated Solutions Applied to the Liquidus Curves of the Zn-Te and Cd-Te Systems," *Metallurgical Transactions*, 1(1970), 239-249.

11. A. S. Jordan, "Activity Coefficients for a Regular Multicomponent Solution," *Journal of the Electrochemical Society*, 119(1972), 123-124.

12. A. S. Jordan and M. Ilegems, "Solid-Liquid Equilibria for Quaternary Solid Solutions Involving Compound Semiconductors in the Regular Solution Approximation," *Journal of Physics and Chemistry of Solids*, 36(1975), 329-342.

13. M. B. Panish and M. Ilegems, "Phase Equilibria in Ternary III-V Systems," *Progress in Solid State Chemistry, vol. 7*, ed. H. Reiss and J. O. McCaldin (Elmsford, New York: Pergamon Press, 1972), 39-83.

14. M. Ilegems and M. B. Panish, "Phase Equilibria in III-V Quaternary Systems-Application to Al - Ga - P - As," *Journal of Physics and Chemistry of Solids*, 35(1974), 409-420.

15. H. H. Kellogg, "Thermochemical Modeling of Molten Sulfides," *Physical Chemistry in Metallurgy*, Proceedings of the Darken Conference, ed.

978

R. M. Fisher, R. A. Oriani, and E. T. Turkdogan (Monroeville, Pennsylvania: U.S. Steel Research Laboratory, 1976), 49-68.

16. R. P. Goel, H. H. Kellogg, and J. Larrain, "Mathematical Description of the Thermodynamic Properties of the Systems Fe-O and Fe-O-SiO$_2$," Metallurgical Transactions B, 11B(1980), 107-117.

17. G. B. Stringfellow, "Calculation of Regular Solution Interaction Parameters in Semiconductor Solid Solutions," Journal of Physics and Chemistry of Solids, 34(1973), 1749-1750.

18. J. W. Gibbs, "On the Equilibrium of Heterogeneous Substances," The Collected Works of J. Willard Gibbs, Vol. I, Thermodynamics (New Haven, Connecticut: Yale University Press, 1957), 55-349.

19. Y. K. Rao, Stoichiometry and Thermodynamics of Metallurgical Processes (Cambridge: Cambridge University Press, 1985), 957 pp.

20. Y. K. Rao, "Extended Form of the Gibbs Phase Rule," Chemical engineering Education, 19(1)(1985), 40-43; 46-49.

21. D. T. J. Hurle and J. B. Mullin, "Thermodynamics of Gas-Phase Equilibria: The Ga:As:Cl:H System," Crystal Growth, ed. H. S. Peiser (Oxford, U.K.: Pergamon Press, 1967), 241-248.

22. L. B. Pankratz, Thermodynamic Properties of Halides, Bulletin 674, Bureau of Mines (Washington, D. C.: U.S. Department of the Interior, 1984), 25-29.

```
       PROGRAM GASINP (INPUT,OUTPUT,TAPE5=INPUT,TAPE6=OUTPUT)

C   CALCULATION OF EQUILIBRIUM COMPOSITION OF EPITAXIALLY
C   GROWN III-V COMPOUND SEMI-CONDUCTOR.
C   X/(1-X)=GA/IN ATOM RATIO
C   Y/(1-Y)=AS/P ATOM RATIO
C   CALCULATION OF VIRTUAL EQUILIBRIUM RATE OF DEPOSITION OF SOLID.
C   SEMI-CONDUCTOR IS A REGULAR SOLUTION COMPOSED OF
C   SPECIES GAAS,INAS,GAP, AND INP
C   RANK OF ECHELON MATRIX=6, NUMBER OF SPECIES=18
C   NUMBER OF INDEPENDENT REACTIONS=18-6=12, AS FOLLOWS

C    (1) H2 + CL2 = 2HCL
C    (2) GACL + 1/4AS4 + 1/2H2 = GAAS(S) + HCL
C    (3) GACL + 1/4P4 + 1/2H2 = GAP(S) + HCL
C    (4) INCL + 1/4AS4 + 1/2H2 = INAS(S) + HCL
C    (5) INCL + 1/4P4 + 1/2H2 = INP(S) + HCL
C    (6) 1/4AS4 = 1/2AS2
C    (7) 1/4P4 = 1/2P2
C    (8) ASH3 = 1/4AS4 + 3/2H2
C    (9) PH3= 1/4P4 +3/2H2
C    (10) GACL3 + H2 = GACL + 2HCL
C    (11) INCL3 + H2 = INCL + 2HCL
C    (12) ASCL3 = 1/4AS4 + 3/2CL2

C   THE QUANTITIES QSGAAS,QSGAP,QSINAS, AND QSINP ARE
C   INTRODUCED TO FACILITATE CALCULATION. THESE REPRESENT THOSE
C   AMOUNTS (IN MOL/MIN) OF GAAS,GAP,INAS, AND INP THAT DEPOSIT
C   FROM THE GAS MIXTURE.
C   DEGREES OF FREEDOM= (18-(18-6))-2+2=6
C   TEMP,TOTAL PRESSURE,INPUT PARTIAL PRESSURES POGACL,POINCL,POAS4,
C   AND POP4 ARE SPECIFIED. THE ATOM RATIO CL/H,AMOUNTS (IN MOL/MIN)
C   Q0GA,Q0IN,Q0AS, AND Q0P OF GA,IN,AS, AND P ARE CALCULATED.
C   INTERACTION ENERGY(OMEGA) VALUES FOR THE PAIRS GAP(S)-GAAS(S),
C   INP(S)-INAS(S), AND GAAS(S)-INAS(S) ARE SPECIFIED.
C   F0:TOTAL INPUT FLOW RATE IN ML/MIN (STP)

         DIMENSION T(10),G(15,10),AK(15,10)
         COMMON T,AK

         READ(5,*) PT,TOL,NT,(T(I),I=1,NT),F0
         WRITE(6,90)
         DO 60 LOOP=1,22
         READ(5,*) POGACL,POINCL,POAS4,POP4

C   Q0 IS THE MOLAR INPUT RATE(MOL/MIN) OF GAS MIXTURE

         Q0=F0*PT/(82.06*273.0)

C   POGACL,POINCL,POAS4, AND POP4 ARE INPUT PARTIAL PRESSURES(IN ATM)
C   OF GACL,INCL,AS4, AND P4.
C   RCLH=CL/H ATOM RATIO
C   DETERMINE THE ATOM RATIO AND AMOUNTS OF FIVE ELEMENTS FROM
C   THE INPUT PARTIAL PRESSURES AND FLOW RATE.

         POH2=PT-POGACL-POINCL-POAS4-POP4
         Q0GA=Q0*POGACL/PT
         Q0IN=Q0*POINCL/PT
         Q0AS=4.0*Q0*POAS4/PT
         Q0P=4.0*Q0*POP4/PT
         Q0H=2.0*Q0*POH2/PT
         RCLH=(POGACL+POINCL)/(2.0*POH2)
```

```
      DO 20 I=1,NT
      G(1,I)=-188196-12.804*T(I)
      G(2,I)=-129457+117.997*T(I)
      G(3,I)=-156795+122.951*T(I)
      G(4,I)=-106165+114.767*T(I)
      G(5,I)=-99342+124.043*T(I)
      G(6,I)=55574-35.893*T(I)
      G(7,I)=54288-34.748*T(I)
      G(8,I)=-118984-86.090*T(I)
      G(9,I)=17259-73.45*T(I)
      G(10,I)=165904-127.930*T(I)
      G(11,I)=99646-105.868*T(I)
      G(12,I)=296210-82.297*T(I)
20    CONTINUE
      RT=8.3143
      DO 30 J=1,12
      DO 30 I=1,NT
30    AK(J,I)=EXP(-G(J,I)/(RT*T(I)))

C  SIX(6) EQUILIBRIUM VALUES ARE GUESSED AT KOUNT=0

      DO 50 L=1,NT

      KOUNT=0
      QSGAAS=0.5*Q0*POGACL/PT
      QSGAP=0.5*Q0*POGACL/PT
      QSINAS=0.5*Q0*POINCL/PT
      PH2=0.9*PT
      PHCL=0.01*PT
      PAS4=0.01*PT

C  GAMMA1: ACTIVITY COEFFICIENT OF GAAS
C  GAMMA2: ACTIVITY COEFFICIENT OF GAP
C  GAMMA3: ACTIVITY COEFFICIENT OF INAS
C  GAMMA4: ACTIVITY COEFFICIENT OF INP
C  TO DETERMINE QSINP AND ACTIVITY COEFFICIENTS
C  CALL SUBROUTINE REGSOL

35    CALL REGSOL(RT,L,QSGAAS,QSGAP,QSINAS,QSINP,GAMMA1,
     1            GAMMA2,GAMMA3,GAMMA4,SIG)
      IF(SIG.EQ.0) GO TO 40
      QSTOT=QSGAAS+QSGAP+QSINAS+QSINP
      AGAAS=GAMMA1*(QSGAAS/QSTOT)
      AGAP=GAMMA2*(QSGAP/QSTOT)
      AINAS=GAMMA3*(QSINAS/QSTOT)

      AINP=GAMMA4*(QSINP/QSTOT)
      PCL2=PHCL**2.0/(AK(1,L)*PH2)
      PGACL=AGAAS*PHCL/(AK(2,L)*PAS4**0.25*PH2**0.5)
      PP4=(AGAP*PHCL/(AK(3,L)*PGACL*PH2**0.5))**4.0
      PINCL=AINAS*PHCL/(AK(4,L)*PAS4**0.25*PH2**0.5)
      PAS2=(AK(6,L)*PAS4**0.25)**2.0
      PP2=(AK(7,L)*PP4**0.25)**2.0
      PASH3=(PAS4**0.25*PH2**1.5)/AK(8,L)
      PPH3=(PP4**0.25*PH2**1.5)/AK(9,L)
      PGACL3=PGACL*PHCL**2.0/(AK(10,L)*PH2)
      PINCL3=PINCL*PHCL**2.0/(AK(11,L)*PH2)
      PASCL3=(PAS4**0.25*PCL2**1.5)/AK(12,L)
      PTJ=PH2+PCL2+PHCL+PP4+PP2+PPH3+PGACL+PGACL3+PAS4+PAS2+PASH3
     1   +PINCL+PINCL3+PASCL3
      SUMH=2.0*PH2+PHCL+3.0*(PPH3+PASH3)
      SUMCL=2.0*PCL2+PHCL+PGACL+PINCL+3.0*(PGACL3+PINCL3+PASCL3)

C  VOLUMETRIC FLOW RATE THROUGH REACTOR CHANGES DUE TO THE
C  FORMATION OF NEW SPECIES. FJ IS NEW FLOW RATE IN ML/MIN
C  (STP) AT EQUILIBRIUM. AMOUNTS (IN MOL/MIN) QVGA, QVIN,
C  QVAS, AND QVP OF ELEMENTS GA, IN, AS, AND P IN THE
C  GAS PHASE AT EQUILIBRIUM ARE CALCULATED FROM THE
C  PARTIAL PRESSURES OF RESPECTIVE SPECIES.
```

```
            FJ= (F0*2.0*P0H2)/SUMH
            QVGA= (FJ/(82.06*273.0))* (PGACL+PGACL3)
            QVIN= (FJ/(82.06*273.0))* (PINCL+PINCL3)
            QVAS= (FJ/(82.06*273.0))* (4.0*PAS4+2.0*PAS2+PASH3+PASCL3)
            QVP= (FJ/(82.06*273.0))* (4.0*PP4+2.0*PP2+PPH3)
            QJGA=QVGA+QSGAAS+QSGAP
            QJIN=QVIN+QSINAS+QSINP
            QJAS=QVAS+QSGAAS+QSINAS
            QJP=QVP+QSGAP+QSINP
            RCLHJ=SUMCL/SUMH

            DIF1=ABS((PTJ-PT)/PT)
            DIF2=ABS((RCLHJ-RCLH)/RCLH)
            DIF3=ABS((QJGA-Q0GA)/Q0GA)
            DIF4=ABS((QJIN-Q0IN)/Q0IN)
            DIF5=ABS((QJAS-Q0AS)/Q0AS)
            DIF6=ABS((QJP-Q0P)/Q0P)
            DIFMAX=AMAX1(DIF1,DIF2,DIF3,DIF4,DIF5,DIF6)
            IF(DIFMAX.LE.TOL) GO TO 45
            KOUNT=KOUNT+1
            IF(KOUNT.GT.1000) GO TO 40

C     THE GUESSED VALUES FOR THE SIX QUANTITIES ARE MODIFIED USING THE
C     FOLLOWING CONVERGENCE FORMULAS

            QSGAAS=QSGAAS*(Q0GA/QJGA)**0.25*(Q0AS/QJAS)**0.2
            QSGAP=QSGAP*(Q0GA/QJGA)**0.2*(Q0P/QJP)**0.2
            QSINAS=QSINAS*(Q0IN/QJIN)**0.25*(Q0AS/QJAS)**0.2
            PH2=PH2*(RCLHJ/RCLH)**0.2*(PT/PTJ)**0.4
            PHCL=PHCL*(RCLH/RCLHJ)**0.3*(PT/PTJ)**0.5

            PAS4=PAS4*(Q0AS/QJAS)**0.4
            GO TO 35

40          WRITE(6,97)T(L),P0GACL,P0INCL,P0AS4,P0P4
            GO TO 50

C     QSTOT IS THE RATE OF DEPOSITION OF SOLID IN GRAM-MOLES PER MIN.
C     WHEN GAS PHASE ATTAINS VIRTUAL EQUILIBRIUM WITH THE SUBSTRATE.

45          X=(QSGAAS+QSGAP)/QSTOT
            U=QSGAAS/QSTOT
            Y=(QSGAAS+QSINAS)/QSTOT
            V=QSINAS/QSTOT
            WRITE(6,92) KOUNT,T(L),P0GACL,P0INCL,P0AS4,P0P4
            WRITE(6,94) PH2,PCL2,PHCL,PP4,PP2,PPH3,PGACL,
         1              PGACL3,PAS4,PAS2,PASH3,PINCL,PINCL3,PASCL3
            WRITE(6,96) Q0GA,Q0IN,Q0AS,Q0P,RCLH,AGAAS,AGAP,
         1              AINAS,AINP,QSGAAS,QSGAP,QSINAS,QSINP,X,U,Y,V
            WRITE(6,99) QSTOT
            WRITE(6,100) PT,TOL
50          CONTINUE
60          CONTINUE

90          FORMAT(1H1,15X,"EQUIL DATA FOR GA-IN-AS-P-H-CL SYSTEM"/)
92          FORMAT(//1X,"KOUNT",5X,"TEMP",6X,"P0GACL",6X,"P0INCL",7X,
         1          "P0AS4",7X,"P0P4"//1X,I5,2X,F7.0,2X,4E12.5/)
94          FORMAT(5X,"PH2",6X,"PCL2",6X,"PHCL",7X,"PP4",7X,"PP2",8X,
         1          "PPH3",6X,"PGACL"//1X,7E10.4//
         2          4X,"PGACL3",5X,"PAS4",6X,"PAS2",6X,"PASH3",5X,
         3          "PINCL",4X,"PINCL3",4X,"PASCL3"//1X,7E10.4/)
96          FORMAT(/6X,"Q0GA",8X,"Q0IN",8X,"Q0AS",9X,"Q0P",8X,
         1"RCLH"//1X,5E12.5//17X,"AGAAS",8X,"AGAP",7X,
         2"AINAS",8X,"AINP"//13X,4E12.5//17X,"QSGAAS",7X,"QSGAP",6X,
         3"QSINAS",7X,"QSINP"//13X,4E12.5//19X,"X",11X,"U",11X,
         4"Y",11X,"V"//13X,4E12.5/)
97          FORMAT(//5X,F7.0,4F8.4,4X,"SORRY! NO SOLUTION. TRY AGAIN!")
99          FORMAT(1X,"VIRTUAL MAX. GROWTH RATE=",E12.5,1X,"G-MOL/MIN"/)
100         FORMAT(/1X,"TOTAL PRESSURE=",E12.5,1X,"ATM",12X,"TOL=",E12.5//)

            STOP
            END
```

982

```fortran
      SUBROUTINE REGSOL(RT,L,QSGAAS,QSGAP,QSINAS,QSINP,
     1                  GAMMA1,GAMMA2,GAMMA3,GAMMA4,SIG)

      DIMENSION T(10),AK(15,10)
      COMMON T,AK

C  INTERACTION PARAMETERS(IN JOULES/MOL) ARE DESIGNATED AS OMA,OMB,
C  OMC, AND OMD FOR INP-GAP,INAS-GAAS,INAS-INP, AND GAP-GAAS
C  RESPECTIVELY.

      OMA=4184.0
      OMB=12552.0
      OMC=4184.0
      OMD=4184.0

      KOUNT=0
      QSINP=0.9*QSINAS
10    QSTOT=QSGAAS+QSGAP+QSINAS+QSINP
      X=(QSGAAS+QSGAP)/QSTOT
      Y=(QSGAAS+QSINAS)/QSTOT
      EPA=OMA*(1.0-X)*(1.0-Y)*(1.0-2.0*X)+
     1    OMB*(1.0-X)*((1.0-X)*Y+X*(1.0-Y))+
     1    OMC*(1.0-X)*(1.0-Y)*(1.0-2.0*Y)+
     1    OMD*(1.0-Y)*((1.0-X)*Y+X*(1.0-Y))
      GAMMA1=EXP(EPA/(RT*T(L)))
      EPB=OMA*(1.0-X)*((1.0-X)*(1.0-Y)+X*Y)+
     1    OMB*(1.0-X)*Y*(1.0-2.0*X)+
     1    OMC*(1.0-X)*Y*(2.0*Y-1.0)+
     1    OMD*Y*(X*Y+(1.0-X)*(1.0-Y))
      GAMMA2=EXP(EPB/(RT*T(L)))
      EPC=OMA*X*(1.0-Y)*(2.0*X-1.0)+
     1    OMB*X*(X*Y+(1.0-X)*(1.0-Y))+
     1    OMC*(1.0-Y)*((1.0-X)*(1.0-Y)+X*Y)+
     1    OMD*X*(1.0-Y)*(1.0-2.0*Y)
      GAMMA3=EXP(EPC/(RT*T(L)))
      EPD=OMA*X*((1.0-X)*Y+X*(1.0-Y))+
     1    OMB*X*Y*(2.0*X-1.0)+
     1    OMC*Y*((1.0-X)*Y+X*(1.0-Y))+
     1    OMD*X*Y*(2.0*Y-1.0)
      GAMMA4=EXP(EPD/(RT*T(L)))

C  CONSIDER THE EXCHANGE REACTION GIVEN BELOW:
C  GAP(S) + INAS(S) = GAAS(S) + INP(S)

      QSINPN=AK(2,L)*AK(5,L)*QSGAP*QSINAS*GAMMA2*GAMMA3/
     1       (AK(3,L)*AK(4,L)*QSGAAS*GAMMA1*GAMMA4)
      IF(ABS((QSINP-QSINPN)/QSINP).LE.1.0E-04) GO TO 12
      KOUNT=KOUNT+1
      IF(KOUNT.GT.1000) GO TO 14
      QSINP=QSINPN
      GO TO 10
12    SIG=1
      RETURN
14    SIG=0
      RETURN
      END
```

983

CASTING

Session Chairman
Achiles Vassilicos, US Steel

This session combines heat mass transfer principles to model a number of different processes parameters in continuous casting and electromagnetic casting, ESR and VAR. Basic solidification is modelled using a finite diference code to demonstrate phenomenalogical behavior. The occurence and nature of casting defects are forecast using both qualitative and mathematical formulations.

MODELLING SOLIDIFICATION PROCESSES

Vaughan Voller

Mineral Resources Research Center
University of Minneapolis
Minnesota USA

Abstract

Solidification processes involve complex heat and mass transfer phenomena and their modelling often requires state of the art numerical techniques. A sound modelling approach is to use available numerical code. In this paper the applicability of the PHOENICS code for the analysis of metallurgical solidification is investigated. This is a finite difference code for the general analysis of heat and mass transfer systems. In particular a methodology for coupling the modelling process with the numerical code will be examined. To demonstrate the methodology various solidification phenomena, e.g. latent heat evolution, convection in melt, flow in the mushy region, etc., will be modelled.

Metallurgical solidification processes involve complex heat and mass transfer phenomena. As a result their modelling often requires the computational solution of coupled equations of heat, mass and momentum, which are collectively referred to as the Navier-Stokes equations. When written in the standard form

transient term + convection term = diffusive term + source term　　　(1)

a number of computer codes are available which will solve coupled Navier-Stokes equations, e.g., SOLA (1,2) and PHOENICS (3). In terms of mathematics equation 1 can be written as (4)

$$\frac{\partial}{\partial t} (\rho\phi) + \text{div} (\rho\underline{u}\phi) = \text{div} (\Gamma \text{ grad } \phi) + S \qquad (2)$$

where ρ is density, $\underline{u}$ is velocity, Γ is a diffusion coefficient, ϕ, is the dependent transported variable and S is the source term.

A sound approach in modelling heat and mass transfer processes is to make use of available numerical codes to solve the resulting Navier-Stokes equations. In using this approach an obvious question to ask is:

In modelling a particular process how should the source term "S" be defined and what are appropriate boundary and initial conditions?

The answer to this question will be the major theme of the current paper. The aim will be to outline the modelling process by which selection of boundary conditions and definitions of source terms are made in specific applications of Navier-Stokes numerical codes. This process will be fully illustrated on modelling a number of metallurgical solidification systems. In each of these models emphasis will be placed on the choice of appropriate source terms.

The Modelling Process

Before specific examples of the numerical modelling of metallurgical solidification phenomena are examined it will be worthwhile to define, within the context of this paper, what is meant by the term the "modelling process". The key elements in a modelling study which will involve the numerical solution of the Navier-Stokes equation is in the identification and definition of,

1) the boundary conditions, in which the interaction of the system, and its environment are modelled, and

2) the source terms, in which phenomena occurring within the system are accounted for.

At this point a parallel between the analytical solution of differential equations and the numerical modelling of processes can be made. The analytical solution of a differential equation involves two parts. In the first part a general solution is derived, that is a function that will satisfy the given equation. This general solution will have a number of free parameters. On specifying the free parameters by matching boundary and initial conditions, a particular solution, which will represent the behavior of a given system, can be obtained. In modelling heat and mass transfer systems available code can be regarded as a general solution, all-be-it not an explicit one, of the standard form of the Navier-Stokes equations. The modelling of a particular system then involves the selection of appropriate boundary conditions and source terms which will force the code to produce numerical solutions applicable to the process to be modelled. The "modelling cycle" in using this approach is illustrated in Figure 1. The initial step, and driving force, is the definition of the source terms and boundary conditions. The next step involves the numerical representation of these terms. Note that, at this stage it is important that the numerical representation is consistent with the main numerical code and that any iterative procedures, required for non-linear terms, are carefully designed. After this step the Navier-Stokes solver can be implemented and numerical solutions generated. At this point a modeler will compare results with experiment, the actual process, or seat-of-the-pants intuition and either accept the results as having reached the required standard or begin the cycle again, making changes in the definitions of the source terms or boundary conditions, in an attempt to improve the numerical predictions.

	Experiments
SUITABILITY	Process
	Intuition
RESULTS	
NUMERICAL SOLUTION	
REPRESENTATION	
	Definition of Source Terms
DRIVER	
	Definition of Boundary Conditions

Figure 1 The Modelling Cycle.

Source Terms

It is recognised that in many systems modelling and representation of the boundary conditions requires careful work. For example continuous casting which can involve multiple regimes of heat transfer along the boundaries(5). In general, however, the mechanisms involved in boundary conditions are well known and the procedure required for modelling and numerical representation are well defined. In practice it is in the definition of the source terms that a modeler has the greatest flexibility. The following points are made on the role of the source terms. The major role is to account for phenomena that occur within the system. In computer solutions of the Navier-Stokes equations, however, additional source terms are often used as part of the numerical modelling strategy. In particular source terms can be used as,

1.) a "dumping ground" in which non-linear effects and features that do not match with the standard form, given in equation 1, can be deposited,

2.) a means of a coupling dependent variables, and

3.) a numerically convenient strategy for representing boundary conditions.

The bottom line is that the definition of the source terms, to represent internal phenomena or as part of the numerical solution strategy, are the driving force in the modelling of heat and mass transfer systems. To fully establish this point the remainder of this paper will concentrate on the development of appropriate source terms in order that Navier-Stokes code can be used to model metallurgical solidification process and phenomena.

Modelling Of Solidification

A Test Geometry

A number of phase change phenomena will be studied with the aim of demonstrating how solidification source terms can be defined. So as not to lose sight of this objective thermal and physical data which has no impact on the nature of the solidification source terms is considered to be constant. In addition a common and simple test geometry of a two dimensional square cavity will be used, this geometry is illustrated in Figure 2. In the simplest case the solidification can be described as follows. Initially the cavity is filled with super heated liquid (of pure material or multi-component) temperature $T = T_i$. At time t=0 the temperature of the left hand wall is lowered and fixed at temperature $T_0 < T_i$ so that a solid layer builds up and advances into the liquid. The manner in which this layer advances into the liquid will depend on the phenomena under study and boundary conditions used. These considerations will be noted at the appropriate place in the text. Standard values of the thermal and physical data used, for each of the three models to be studied are given in Table I.

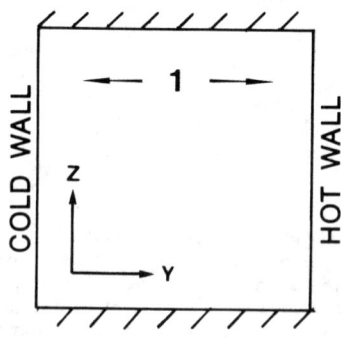

Figure 2 The Test Problem Geometry.

Table I. Test Problem Data

In All Models

Density ρ = 1; Specific Heat c =1; Viscosity μ = 1;

Conductivity K =0.001; Gravity g = 1000; Latent Heat L = 5;

Initial Temperature T_i = 0.5; Coefficient of Expansion β =0.01;

Half Mushy Range ϵ = 0.1.

In Heat Conduction and Convection Models

Hot Wall = 0.5; Cold Wall = -0.5

In Solute Transport Model

Hot wall is insulated; Cold Wall = -2.0; partition ratio k = 0.2;

Initial Concentration C_0 = 0.1.

Model 1 : Heat Conduction only

The simplest system to consider in the test geometry of Figure 2. is the case of solidification under conduction only. The major problem to deal with is the modelling of the latent heat evolution. A standard mean of dealing with this problem is to use an enthalpy formulation to describe the heat transfer i.e.,

$$\frac{\partial \rho H}{\partial t} - \text{div} (K \text{ grad } T) = 0 \qquad (3)$$

where K is the conductivity and T is temperature. The enthalpy H is defined to be the sum of sensible and latent heats,

$$H = h + \Delta H \qquad (4)$$

where the sensible heat h = $\int$ cdT, or cT if specific heat c is constant, and ΔH = f(T) (a function of temperature). On recognizing that the latent heat is associated with the local liquid fraction a general form for f(T) can be written,

$$f(T) = \begin{cases} L & T \geq T_\ell \\ L(1-f_S) & T_\ell > T \geq T_S \\ 0 & T < T_S \end{cases} \qquad (5)$$

where $f_S(T)$ is the local solid fraction, L is the latent heat of the phase change, T_ℓ is the liquidus temperature at which solid formation commences and T_S is the temperature at which full solidification is achieved. Note that the temperatures T_ℓ and T_S essentially define a so called solid +liquid

991

mushy region and the task of fully defining the nature of the latent heat evolution is that of identifying the form of the local solid fraction temperature relationship in this region. In the metallurgical literature there are a number of possible relationships for $f_S(T)$ depending on the assumptions made for the nature of solute redistribution, see Flemings(6) for examples.

In using equations 4 and 5 in an implicit numerical solution of equation 3 systems of non-linear equations result. A numerically convenient means of dealing with this, which is also physically appealing, is to rewrite equation 3 as

$$\frac{\partial \rho h}{\partial t} - \text{div } (\alpha \text{ grad } h) + S_h = 0 \qquad (6)$$

where $S_h = [\partial \rho \Delta H / \partial t]$ is a source term. In this fashion the non-linear terms are "dumped" in the source term and more efficient numerical solutions can be obtained(7). In metallurgical applications to uni-directional solidification of binary alloys Clyne(8), using equation 5 writes the source term as $\Delta H \, \partial f_S / \partial T$ with the term $\partial f_S / \partial T$ explicitly expressed via one of the local solute redistribution equations. Results from numerical implementations compare very favorably with experiments and are reported in references (8) and (9).

Model 2 : Convection in Melt

Although the case of pure conduction has metallurgical significance, as demonstrated by the work of Clyne, numerical solution of the governing equations does not involve state-of-the-art techniques. For example the Clyne model(8) outlined above has been successfully solved using a heat balance integral approach(10,11). This situation changes when convection effects have to be considered in the melt. In such a case the fluid flow considerations require that a numerical solution of the full Navier-Stokes equations is generated.

In considering convection effects, in addition to the problem of dealing with the latent heat evolution, the nature of the fluid velocity in the vicinity of the phase change needs to be accounted for. An obvious way of dealing with these problems is to use a numerical representation which will specifically account for the solid, mushy and liquid regions. A drawback is the requirement of a deforming numerical mesh. Such approaches are not consistent with available Navier-Stokes codes which, in general, are based on fixed meshes. Latent heat and velocity conditions can be accounted for in a fixed mesh environment, however, on supplying appropriate definitions to the source terms. This will be the approach adopted in the current work.

The Navier-Stokes equations are derived for the case of natural convection under the assumption Newtonian, incompressible, laminar flow. To achieve this derivation it will be helpful to consider the physical state of the system at time t >0. In a multi-component system, Figure 3, three regions will exist, a full solid region, a solid+liquid porous mushy region, and a full liquid region. Fluid flow will occur both within the full liquid region and the porous mushy region. To fully account for the convection effects it is necessary to write down two-phase equations for flow in the Mushy region. These equations are:

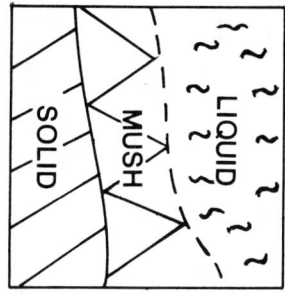

Figure 3 State Of Cavity at Time t

Conservation of Mass :-

$$\frac{\partial \rho}{\partial t} + \text{div} \ (\rho_L f_L \underline{u}) = 0 \tag{7}$$

where $\underline{u} = (v,w)$ is the fluid velocity, $\rho = \rho_S f_S + \rho_L f_L$ is the local averaged density and f_L and f_S are the local solid and liquid fractions respectively.

Conservation of Momentum :-

$$\frac{\partial}{\partial t}(\rho_L f_L v) + \text{div}(\rho_L f_L v \underline{u}) = \text{div}(f_L \mu \ \text{grad} \ v) - f_L \frac{\partial P}{\partial y} + S_y \tag{8a}$$

$$\frac{\partial}{\partial t}(\rho_L f_L w) + \text{div}(\rho_L f_L w \underline{u}) = \text{div}(f_L \mu \ \text{grad} \ w) - f_L \frac{\partial P}{\partial z} + S_z + S_b \tag{8b}$$

where P is pressure, μ is the liquid viscosity, and S_y, S_z, and S_b are source terms.

Conservation of Heat :-

$$\frac{\partial \rho h}{\partial t} + \text{div}(\rho_L f_L \underline{u} h) = \text{div}(\alpha \ \text{grad} \ h) - S_h \tag{9}$$

which is similar to equation 6 but with convection terms.

From a numerical point of view, although possible, a two-phase treatment is difficult to achieve. A more convenient, one-phase model can be derived from equations 7 - 9, on careful definition of velocity and source terms. The true nature of the mushy region is that of a distinct solid phase (dendrites) coexisting with liquid, as illustrated in Figure 4a.

The central element in the proposed one phase model is that the material within a small representative volume is considered to be a homogeneous mixture (i.e. a "mushy fluid"), see Figure 4b, with density ρ. The state of this mushy fluid will depend on the local solid fraction f_S. If $f_S = 0$ the mushy fluid behaves as the liquid phase. When $f_S = 1$ the mushy fluid behaves as the solid phase. At intermediate values the mushy fluid behaves as a homogeneous fluid with velocity $0 \leq \underline{u}^* \leq \underline{u}$. On noting that the

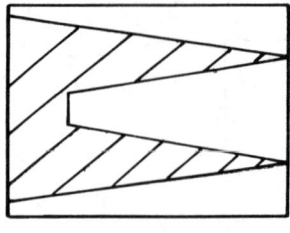 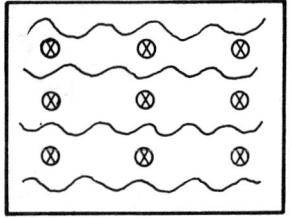

(a) Distinct Phases   **(b) Mushy Fluid**

Figure 4. Illustration of the two-phase and one phase models.

mushy region behaves as a porous medium, of porosity $\lambda = f_L$, the relationship between the actual liquid velocity and the mushy fluid velocity is given by,

$$\rho \underline{u}^* = \rho_L f_L \underline{u} \tag{10}$$

Note that, in the full liquid $f_L=1$ and $\rho_L=\rho$, hence $\underline{u}^* = \underline{u}$. With this definition for the velocity the governing equations of heat and mass transfer in the mushy fluid can be rederived.

Conservation of Mass :- Substitution of equation 8 into equation 1 will give,

$$\frac{\partial \rho}{\partial t} = -\text{div } (\rho \underline{u}^*) \tag{11}$$

Conservation of Momentum :- Following Voller & Prakash(12) the conservation of momentum equations can be written as

$$\frac{\partial}{\partial t}(\rho v^*) + \text{div}(\rho v^* \underline{u}^*) = \text{div}(\mu \text{ grad } v^*) \quad - \quad \frac{\partial P}{\partial y} + S_y \tag{12a}$$

$$\frac{\partial}{\partial t}(\rho w^*) + \text{div}(\rho w^* \underline{u}^*) = \text{div}(\mu \text{ grad } w^*) \quad - \quad \frac{\partial P}{\partial z} + S_z + S_b \tag{12b}$$

Noted that, in the full liquid region ($f_L=1$, $\rho=\rho_L$) these equations will match the two-phase momentum equations, equation 8. In the full solid and mushy regions, however, a match is not obtained, i.e., direct substitution of $\underline{u}^*$ in equations 12 will not result in the two-phase momentum equations. This discrepancy will be accounted for via the definition of the source terms S_y and S_z, a step which is fully described below.

Conservation of Heat :-

$$\frac{\partial \rho h}{\partial t} + \text{div}(\rho \underline{u}^* h) = \text{div}(\alpha \text{ grad } h) - S \tag{13}$$

Equations 11 - 13, which are consistent with the standard Navier-Stokes form, equation 1, will be used as the governing equations. Before solution, however, the various source terms have to be defined.

The S_y and S_z source terms are used to modify the momentum equations in the mushy region. This region can be regarded as a porous medium with porosity $\lambda=f_L$. The velocity in porous media is described by the superficial

994

velocity (i.e., the ensemble average velocity) which is related to the actual fluid velocity via

$$\underline{u}_a = f_L \underline{u} \tag{14}$$

Flow is then governed by a Darcy law, i.e.,

$$\underline{u}_a = -(K/\mu) \text{ grad } P \tag{15a}$$

or in terms of the mushy fluid velocity,

$$\frac{\rho}{\rho_L} \underline{u}^* = -(K/\mu) \text{ grad } P \tag{15b}$$

where K, the permeability, is a function of the porosity λ ($=f_L$) and morphology of the mushy region. As the porosity decreases the permeability and the superficial velocity also decrease, down to a limiting value of zero when the mush becomes full solid. In a numerical model this behavior can be accounted for by defining

$$S_y = -\rho A/\rho_L \ v \quad \text{and} \quad S_z = -\rho A/\rho_L \ w \tag{16}$$

where A increases from zero to a large value as the local solid fraction f_S increases from its liquid value of 0 to its solid value of 1. The effect of these sources is as follows. In the liquid region the sources take a zero value and the momentum equations are in terms of the actual fluid velocities. In the mushy region the value of A increases such that the value of the sources begin to dominate the transient, convective, and diffusive terms and the momentum equation approximates the Darcy law. As the local solid fraction approaches 1 the sources dominate all other terms in the momentum equation and force the predicted superficial velocities to values close to zero. Note that, due to the domination and suppression of the various terms in the momentum equations in the mush and solid, the discrepancies in these terms, noted above, has little influence.

In order to obtain a suitable function for 'A' one can appeal to physics. A well known equation derived form the Darcy law, which has been proposed as a general model for flow in metallurgical mushy regions(13) is the Carman-Koseny equation,

$$\text{grad } P = -C (1 - \lambda)^2/\lambda^3 \ \underline{u}_a \tag{17}$$

This equation suggests the following form for the function A

$$A = -C (1- \lambda)^2/(\lambda^3 + q) \tag{18}$$

The value of C will depend on the morphology of the mushy region and is regarded as a free parameter. The constant q is introduced to avoid divisions by zero.

The S_b source term ,in the w momentum equation, is a buoyancy term used to induce natural convection in the cavity. Assuming the Boussinesq treatment to be valid, i.e., density is constant in all terms except a gravity source term, the buoyancy source term is given by

$$S_b = \rho g \beta (h - h_{ref})/c \tag{19}$$

where β is a thermal expansion coefficient and h_{ref} is a reference value of the sensible heat.

The form of the enthalpy source term S_h takes a role similar to the conduction only source term in equation 6 , but with the addition of a convection term, i.e.,

$$S_h = \frac{\partial \rho \Delta H}{\partial t} + \text{div}(\rho \underline{u}^* \Delta H) \tag{20}$$

As noted, in the development of the conduction only model, the form of ΔH will depend on the solid fraction temperature relationship. There are a number of relationships proposed in the literature, for simplicity in the current work a linear relationship is chosen,

$$f_S(T) = \begin{cases} 0 & T \geq \epsilon \\ (\epsilon - T)/2\epsilon & \epsilon > T \geq -\epsilon \\ 1 & T < -\epsilon \end{cases} \tag{21}$$

where the temperature has been scaled such that $T=\epsilon$ and $T=-\epsilon$ are the liquidus and solidus temperatures respectively.

With the source define as above the governing equations 11 - 13 can be numerically solved without regard to the position of the solidus or liquidus fronts in the mushy region. The PHOENICS(3) code has been used to achieve this step. This code uses a fixed domain approach and is similar in many respects to the methodology outlined by Patankar(4). The code is easily able to handle the user definition of source terms and associated numerical iteration schemes. As a sample of the results obtained. Figure (5a), taken from reference (12) shows the position of the solidus ($-\epsilon$) and liquidus (ϵ) isotherms as the steady state is reached, for the data in Table I. Of particular interest is the effect of the flow in the mushy region, especially the marked effect on the shape of the liquidus isotherm. This flow will be a function of the porosity and morphology of this region. The porosity has been defined via the Carman-Kozeny equation but the morphology is a free parameter. In Figure(5a), also taken from reference (12), the morphology was modelled on setting $C = 1.6 \times 10^3$ (with q= 0.001) in equation 18. To fully specify the effect of the morphology of the mushy region a detailed study, including experiments, will be required. A crude accounting of the effect of the morphology can be seen on reference (5b) which shows the same result as Figure (5a) but with the morphology defined by $C = 1.6 \times 10^5$. In these results the effect of the reduction of flow in the mushy region is clearly observed A more detailed discussion of the effects of convection in solidification may be found in references (12) and (14).

Model 3 : Solute Transport

From a metallurgical viewpoint the model presented above is incomplete an important phenomena that will occur alongside the solidification is solute redistribution ,i.e., macrosegregation. For this case an additional equation expressing conservation of solute mass fraction needs to be solved along with the equations of heat and mass transfer. In terms of a two-phase model the conservation of solute is,

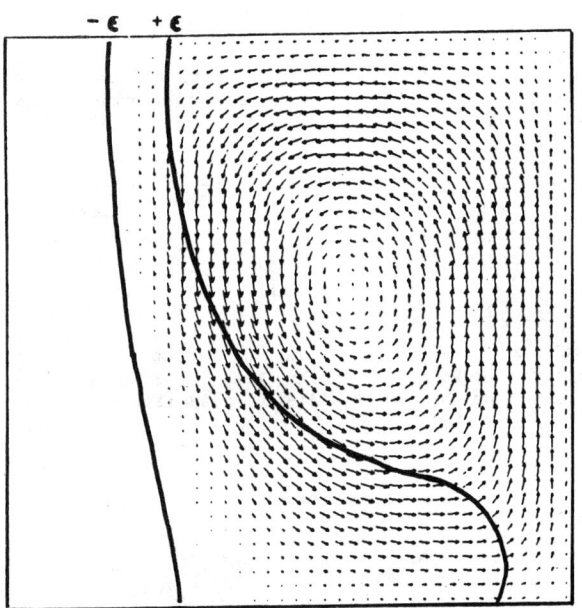

Figure 5a. Position of Mushy Region at t=2000 (C=1.6x10^3)

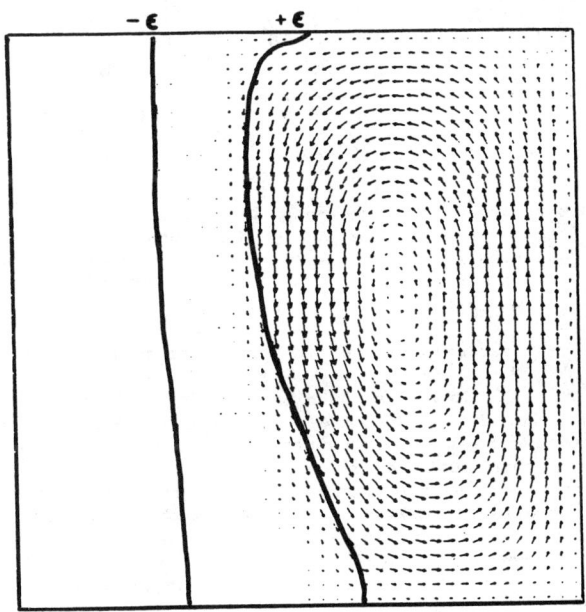

Figure 5b. Position of Mushy Region at t=2000 (C=160x10^3)

$$\frac{\partial}{\partial t}(\rho C) + \text{div}(C_L \rho_L f_L \underline{u}) = \text{div}[\ D_L f_L\ \text{grad}\ C_L + D_S f_S\ \text{grad}\ C_S] \qquad (22)$$

where C is the concentration of solute, D the diffusivity of solute, and ρC $= \rho_L f_L C_L + \rho_S f_S C_S$. On using equation 10, the solute conservation can be written in terms of the mushy velocity to give,

$$\frac{\partial}{\partial t}(\rho C) + \text{div}(\rho \underline{u}^* C_L) = \text{div}[\ D_L f_L\ \text{grad}\ C_L + D_S f_S\ \text{grad}\ C_S] \qquad (23)$$

If thermodynamic equilibrium is maintained at the solid / liquid interface a concentration discontinuity exists,

$$C_S = k\ C_L$$

where k is the equilibrium partition coefficient. In the mushy fluid model it is assumed that this relationship will hold in each computational control volume, where C_S and C_L are identified as locally averaged solid and liquid compositions. On introducing a new variable

$$V = C_L = C_S/k,$$

defining an apparent diffusivity

$$D^* = (D_L f_L + k D_S f_S),$$

and noting that

$$\rho C = \rho V - (1-k) f_S \rho_S V$$

the following one phase equation can be obtained from equation 23,

$$\frac{\partial}{\partial t}(\rho V) + \text{div}(\rho \underline{u}^* V) = \text{div}(D^*\ \text{grad}\ V) + R \qquad (24)$$

where the source term, R, is defined as

$$R = \frac{\partial}{\partial t}[\ (1-k) f_S \rho_S V\ ] \qquad (25)$$

In order to account for the fact that density variations will arise as a result of both temperature and compositional differences the momentum source S_b, resulting from the Boussinesq approximation, is modified to

$$S_b = \rho g \beta (h-h_{ref})/c + \rho g \epsilon (V-V_{ref}) \qquad (26)$$

ϵ is the thermal and solutal coefficients of expansion of the fluid.

Equations 10-13 and 24 along with the definition of source terms, i.e., equations 16, 20, 25 and 26, are the governing equations for the case of solidification with solute transport. In a numerical solution via Navier-Stokes code , care must be taken to couple the solution of the heat and solute equations (i.e., the h and V fields). In a strict sense the coupling should be controlled by the equilibrium phase change diagram. In the present work, however, interest is in the definition and implementation of source terms and the approach adopted is the one proposed by Voller(15). Coupling between the solute and enthalpy fields is achieved on assuming that the liquidus and solidus temperature are fixed and that the solid fraction is a linear function of temperature across the mushy region, viz., $f_L = (T+\epsilon)/2\epsilon$.

In obtaining the solutions of solidification with solute transport practical interest will rest in the prediction of segregation (i.e., solute redistribution). Segregation can be revealed on plotting the concentration field, given by,

$$C = V (k + (1-k)f_L) \qquad (27)$$

on full solidification. Figure 6 shows concentration contours on full solidification of the test cavity, using the data in Table 1, in the case of a nominal solute concentration of $C_0 = 0.1$. In these results areas of positive segregation ($C>C_0$) and negative segregation ($C<C_0$) are clearly recognizable which indicates the potential of the proposed model to predict segregation patterns generated by convective flows.

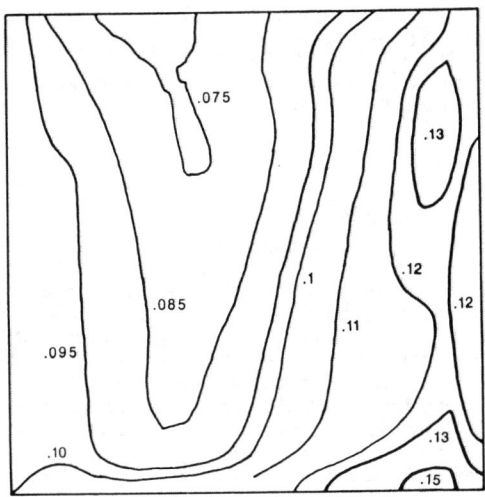

Figure 6. Concentration contours on full solidification

Summary

Many equations have been proposed above and in order to see "the wood for the trees" it will be worth while to summarize. In describing the mass and heat transfer aspects of solidification phenomena the aim is to express the governing equations in the standard form

$$\frac{\partial}{\partial t} (\rho\phi) + \text{div} (\rho\underline{u}\phi) = \text{div} (\Gamma \text{ grad } \phi) + S$$

In order to achieve this various definitions of the velocity have to be furnished and appropriate source terms defined. In the three models of phase change presented above the governing equations and sources take the following forms.

Conservation of Heat :- The dependent variable is enthalpy, defined as $h=\int cdT$. The source term defined in terms of a latent heat function ΔH is

$$S_h = \frac{\partial \rho \Delta H}{\partial t} + \text{div}(\rho \underline{u}^* \Delta H)$$

In the case of a conduction only phase change or an isothermal phase change the second term on the right will zero.

<u>Conservation of Momentum</u> :- The dependent variable is the mushy velocity $\underline{u}^*$, defined in equation 10. The source terms are; the Darcy source terms which account for the flow in the mushy region, given by

$$S_y = -\rho A/\rho_L \, v \qquad \text{and} \qquad S_z = -\rho A/\rho_L \, w$$

where the function A can be defined as

$$A = -C \, (1- \lambda)^2/(\lambda^3 + q),$$

and the Boussinesq source term which accounts for natural convection,

$$S_b = \rho g \beta (h-h_{ref})/c +\rho g \epsilon (V-V_{ref})$$

<u>Conservation of Solute</u> :- The dependent variable is V, related to the concentration C via equation 26. The source term takes the form

$$R = \frac{\partial}{\partial t} [\, (1-k)f_s \rho_s V \,]$$

Conclusions

A central theme in this paper is the viewpoint that modelling of heat and mass transfer processes and phenomena via available Navier-Stokes code requires the correct specification of source terms. This statement is somewhat lacking because it implies that such a course of action is straightforward. In practice defining appropriate source terms is far from simple. To be effective it requires a number of skills the primary of which is a sound physical understanding of the process to be modelled coupled with an ability to translate that understanding into numerical code which is consistent with general Navier-Stokes solution procedures.

In order to demonstrate the nature of the task of defining source terms a number of solidification models (i.e., governing equations + associated source terms) have been derived and investigated. As these models stand the source terms have the correct form, i.e., they force the system to behave in the appropriate manner. There is however, a fair degree of freedom in choice of free parameters, e.g., the morphology of the mushy region. If the solidification techniques presented here are to be applied in the analysis of existing systems then the source terms need to take better account of the physics. Such a step can be achieved with the input of experience metallurgists and the use of well designed experiments.

Future work in the development of numerical models for the analysis of phase change phenomena will concentrate on the modification of the source terms so that specific systems can be studied. Possible systems include: macrosegregation with full coupling between the heat and solute flow, die casting (in particular the effect of residual velocities from the filling stage) and the use of freezelineings in smelting and plasma processes.

References

1. C. W. Hirt, B. D. Nichols and N. C. Romero, "SOLA- A Numerical Solution Algorithm for Transient Fluid Flows" (Los Alamos Scientific Laboratory Report LA-5852, 1975).

2. B. D. Nichols, C. W. Hirt and R. S. Hotchkiss, "SOLA-VOF: A Solution Algorithm for Transient Fluid Flow with Multiple Free Boundaries" (Los Alamos Scientific Laboratory Report LA-8355, 1980).

3. D. B Spalding, "An Introduction to PHOENICS" (CHAM Technical Report TR/68, 1981).

4. S. V. Patankar, Numerical Heat Transfer and Fluid Flow (Washington: Hemisphere Publishing Corporation, 1980).

5. D. C. Weckman and P. Niessen, "A Numerical Simulation of the D.C. Continuous Casting Process Including Nucleate Boiling Heat Transfer" Metallurgical Transactions B, 13B (1982) 593-602.

6. M. C. Flemings, Solidification Processing (New York: McGraw-Hill, 1974).

7. V. R. Voller, "Implicit Finite-difference Solutions of the Enthalpy Formulation of Stefan Problems" IMA Journal of Numerical Analysis, 5 (1985) 201-214.

8. T. W. Clyne, "Numerical Modelling of Directional Solidification of Metallic Alloys", Metals Science, 16 (1982) 441-450.

9. T. W. Clyne, "The Use of Heat Flow Modelling to Explore Solidification Phenomena", Metallurgical Transactions B, 13B (1982) 471-477.

10. V. R. Voller, "A Heat Balance Integral Method for Estimating Practical Solidification Parameters", IMA Journal of Applied Mathematics, 35 (1985) 223-232.

11. V. R. Voller, "A Heat Balance Integral Method Based on an Enthalpy Formulation", International Journal of Heat and Mass Transfer, 30 (1987) 607-610.

12. V. R. Voller and C. Prakash, "A Fixed Grid Numerical Modelling Methodology for Convection-Diffusion Mushy Region Phase-Change Problems", International Journal of Heat and Mass Transfer, (in press).

13. D. R. Poirier, "Permeability for Flow of Interdendritic Liquid in Columnar-Dendritic Alloys", Metallurgical Transactions B, 18B (1987), 245-256.

14. V. R. Voller, M. Cross and N. C. Markatos, "An Enthalpy Method for Convection/Diffusion Phase Change", International Journal of Numerical Methods in Engineering, 24 (1987), 271-284.

15. V. R. Voller, A Numerical Method for Analysis of Solidification in Heat and Mass Transfer Systems", Numerical Methods in Thermal Problems V (in press).

MATHEMATICAL MODELING OF ELECTROMAGNETIC CASTING

J. Sakane and J. W. Evans

Department of Materials Science and Mineral Engineering
University of California
Berkeley, CA 94720

Abstract

A mathematical model for the electromagnetic casting of aluminum or other metals is described. This model computes the magnetic induction, current densities, electromagnetic forces, meniscus shape, velocities and turbulence within a caster as a function of parameters that are under control of the designer or operator. Examples of calculated results are presented such as the effect of inductor design and frequency on meniscus shape and the effect of the electromagnetic screen on the meniscus and metal stirring.

The electromagnetic casting of aluminum has grown to the point of being an established technology; the amount of aluminum cast in this way is probably well in excess of a million tons per year. The origin of this technology is the USSR, e.g. the patent of Getselev (1) and a sketch of a Russian caster appears on the left of Fig. 1. A key feature of this Russian design is the "screen" that is interposed part way between the inductor and the molten metal. This screen, typically of stainless steel, is intended to modify the electromagnetic fields so as to control the meniscus shape and stirring within the molten metal. The screen therefore absorbs a significant amount of electrical energy and can present other problems. An alternative design, appearing in Kaiser patents (e.g. (2)), attempts to achieve the same effect by shaping the inductor and thereby the screen is eliminated, as seen on the right of Fig. 1.

Henceforth for brevity these two alternatives will be distinguished by the words "Russian type" and "Kaiser type". However it is noted that this technology is proprietary; consequently the precise designs of casters are unknown to the authors and no representation is made that the results presented below are exactly those to be expected for a commercial caster.

The deformation of the meniscus and the electromagnetically driven flow occurring in electromagnetic casters are important aspects of caster performance. The former must be sufficient that the column of molten metal is adequately supported and does not contact the inductor or screen (if present). The latter is generally deterimental since oxidation, gas absorption and entrainment of inclusions are all promoted by a vigorous electromagnetic stirring of the metal near the meniscus.

Electromagnetic Casters

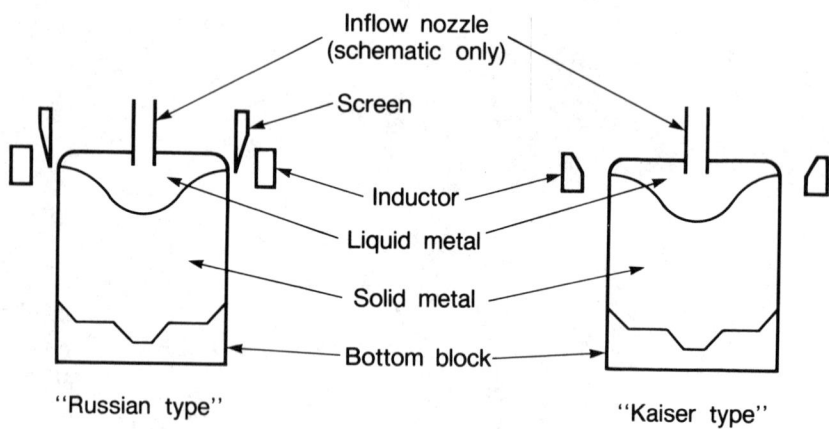

Inflow nozzle
(schematic only)

Screen

Inductor

Liquid metal

Solid metal

Bottom block

"Russian type"

"Kaiser type"

XBL 875-7720

Figure 1 - Sketches of the two types of electromagnetic caster in use within the aluminum industry

Two significant previous investigations of electromagnetic casting have been those of Lavers (3,4) and of Vives and Ricou (5). Lavers performed analytical calculations for a one dimensional model and a two dimensional semi-infinite model, as well as numerical calculations for a two dimensional finite model. Calculated results included electromagnetic field strengths, forces and pressures, as well as melt flow patterns (calculated ignoring the inflow of liquid at the nozzle). The effect of varying parameters such as the inductor geometry and frequency, and the screen geometry were calculated.

Vives and Ricou (5) measured magnetic fields within and around the metal pool of an actual caster. Current densities were also measured in the metal pool, as were velocities. Velocity measurements were carried out using an electromagnetic velocity probe. Measurements were performed under only two sets of conditions (probably because of the difficulty of performing measurements on actual casters) but were supplemented with measurements on a cold physical model of the caster. In this model the molten aluminum was simulated by mercury contained within a solid 340 mm diameter stainless steel cylinder with its axis vertical and its top surface machined to the shape of the solidification front. A significant feature of the model was that the inductor was driven at 50 Hz, compared to the 2000-3000 Hz of an actual caster; the skin depth was therefore a disproportionate fraction of the ingot diameter. Nevertheless, this investigation of Vives and Ricou is invaluable in providing experimental data with which mathematical models may be compared. The development of such a model will now be described.

Mathematical Development

The model described here is an axisymmetric one with z being the axial (vertical) co-ordinate and r the radial co-ordinate. The calculation of the parameters of interest (meniscus shape and the velocity field) starts with the solution of the electromagnetic field equations, specifically Maxwell's equations and Ohm's law for a moving conductor. The magnetic Reynolds number for the flow in question is low which implies that the $V \times B$ term in the latter equation can be neglected. In the absence of ferromagnetic materials, it is then possible (e.g. 6) to rewrite the field equations in the form

$$ J = - \frac{j\omega\sigma\mu}{4\pi} \int_{vol} \frac{J'}{|r'|} \, dV' \qquad (1) $$

Where J is the current density at the point of interest
ω is the angular frequency of the current in the coil
σ is the electrical conductivity at the point
μ is the magnetic permeability

and the integral is carried out over all adjacent regions where significant induced or imposed currents flow, r' being the vector connecting a point where the current density is J' to the point of interest. In this approach the current density is treated as a phasor, that is as a complex variable, only the real part of which has physical significance.

1005

In the present context, currents flow only in the tangential direction and equation (1) permits the induced current at any point in the metal to be calculated as a function of the current at all other points in the metal, coil and screen (if present). This is achieved by using a Gaussian integration formula for the integral in equation (1). Details are contained in a full length paper by the present authors (7) and the result is

$$ J_{ij} = -j\omega \sum_{k=1}^{N} \sum_{l=1}^{M} C_{ijkl} J_{kl} \qquad \text{(metal)} $$

$$ - j\omega \sum_{k=1}^{NS} \sum_{l=1}^{MS} C_{sijkl} J_{skl} \qquad \text{(shield)} $$

$$ - j\omega \sum_{k=1}^{NC} \sum_{l=1}^{MC} C_{cijkl} J_{ckl} \qquad \text{(inductor)} \qquad (2) $$

In this equation, the metal is represented by an M by N mesh (see Fig. 2), the screen by an MS by NS mesh and the coil by a mesh that is NC by MC. Equation (2) then gives the current at point i,j in the first of these meshes in terms of the currents at all other mesh points and the constants C_{ikjl} etc. Those constants are made up of products of geometric factors (akin to mutual inductances), weight factors of the numerical integration scheme and the conductivity at point i,j; details can be found in a previous paper (6). In the calculations that follow the current density in the inductor J_{ckl} is assumed equal to the inductor current divided by cross-sectional area; this ignores the skin effect in the inductor but preliminary calculations (where J_{ckl} was non-uniformly distributed) showed that the consequences of this approximation were insignificant. A similar set of equations can be written for J_{skl} and the two sets can then be merged to form one set of linear simultaneous equations containing only the unknown current densities. This set can then be solved by matrix inversion. Next the magnetic induction at any point is obtained from the equation

$$ \nabla \times \mathbf{J} = - j\omega\sigma\mathbf{B} \qquad (3) $$

This is achieved by numerically differentiating the calculated current density in the metal. Next the electromagnetic force distribution through the metal can be calculated from the cross product of the current density and the magnetic induction

$$ F_r = \frac{1}{2} \, \text{Re} \, \{ \, J_\theta B_z \, \} \qquad (4) $$

and

$$ F_\theta = - \frac{1}{2} \, \text{Re} \, \{ \, J_\theta B_r \, \} \qquad (5) $$

1006

where subscripts $_{r,\ \theta}$ and $_z$ indicate the radial, tangential and axial components, and Re indicates the real part of the quantity in parentheses. The electromagnetic forces calculated in this way are values averaged over a cycle of the power supply.

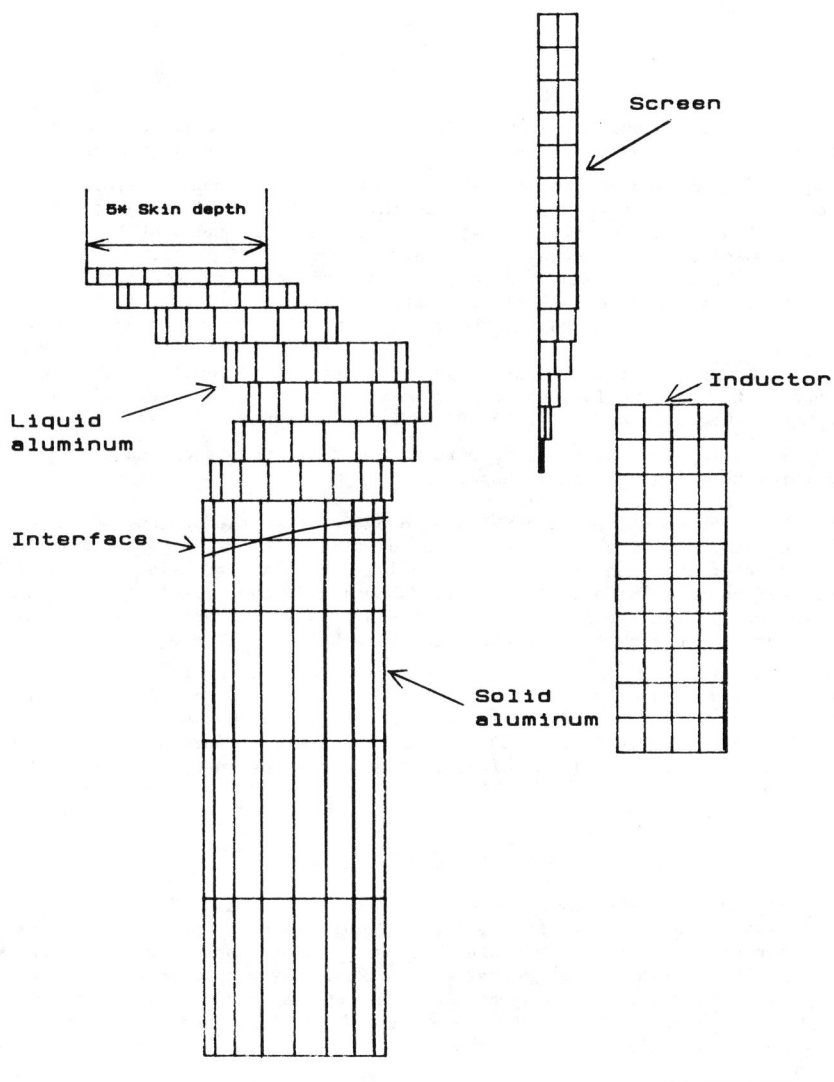

XBL 876-2883

Figure 2 - Diagram of the mesh used for calculation of the meniscus shape. The axis of symmetry is to the left of the figure.

1007

The electromagnetic forces can be broken into a rotational part, which drives the circulation of the metal, and an irrotational part which supports the column of liquid metal above the solidification front. The latter can be written (4) as the gradient of a "magnetic pressure", P_m, given by

$$P_m = \frac{B^2}{2\mu} \qquad (6)$$

A necessary condition for the stability of the meniscus is that along the meniscus this magnetic pressure equals the static pressure produced by the head of molten metal[1]. This permits a numerical calculation of the shape of the meniscus once the magnetic field is known. That calculation is an iterative one starting with a first guess of the shape of the meniscus. Referring to Figure 2, the magnetic pressure is calculated at all points on the meniscus for this first guess and compared to the static pressure (calculated for each position by multiplying the vertical distance from the top of the meniscus by the product of density and gravitational force). If the former exceeds the latter the horizontal row of elements is moved inward a small amount proportional to the distance; if the opposite, the row is moved outward. All rows having been adjusted the calculations (of J, B and P_m) are repeated and so on until the magnetic and static pressure match at all points on the meniscus, yielding the final meniscus shape.

Finally, velocities are calculated by solving the turbulent fluid flow equations. These consist of the equation of continuity, the momentum equations (containing an effective viscosity and the electromagnetic body forces calculated by (4) and (5) above) and two auxiliary partial differential equations that yield the effective viscosity. These last two equations embody the k-ϵ approach to turbulence. The equations, and their numerical solution, have been described many times in the literature (7,8,9), and they will not be repeated here. The TEACH computer program, modified to include electromagnetic forces, was employed in the fluid flow calculations. All calculations were performed on an IBM AT computer, or a compatible machine, in some cases augmented by a Definicon board.

Results and Discussion

Fig. 3 shows the computed meniscus shape for a caster of the Kaiser type (with no screen to modify the magnetic field). The meniscus shape has been computed for coils of three cross sections (shown in the key), all of 205 mm radius with center planes 35 mm below the top of the melt. Physical properties are those of aluminum and the liquid-solid interface shape is that reported by Vives and Ricou (5). It is seen that the meniscus shape is affected by the shape of the coil but not to a great extent.

---

[1] At the velocities encountered in casters the dynamic pressures and surface tension are small.

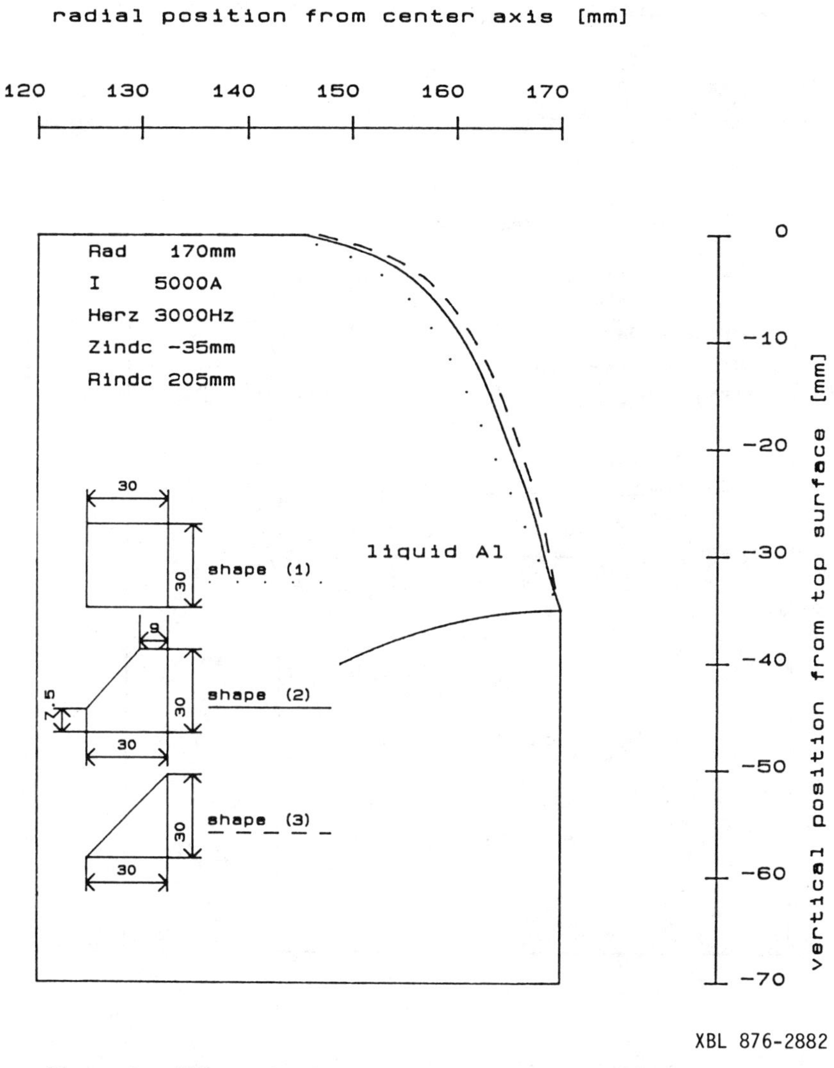

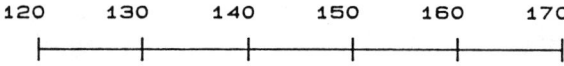

radial position from center axis [mm]
120
130
140
150
160
170
Rad 170mm
I 5000A
Herz 3000Hz
Zindc -35mm
Rindc 205mm
30
30
shape (1)
9
7.5
30
shape (2)
30
30
shape (3)
30
30
liquid Al
0
-10
-20
-30
-40
-50
-60
-70
vertical position from top surface [mm]
XBL 876-2882

Figure 3 - Effect of inductor shape on computed meniscus shape for a caster of the Kaiser type.

However, the vertical position of the coil appears to be an important variable as shown in Fig. 4. The meniscus profile calculated for the coil 50 mm below the top of the melt would be an unstable one; solidification would proceed with the outward growth of the ingot. Conversely an ingot with a meniscus steeply inclined inward would result in a reduction of ingot diameter as solidification proceeded. It appears then that a desirable meniscus shape is one that is nearly vertical at the melt-solid interface; presumably surface tension would allow some departure from this criterion before changes in ingot size result. These calculations were performed for a coil of shape 2 (Fig. 3) but similar results were obtained for one of shape 3.

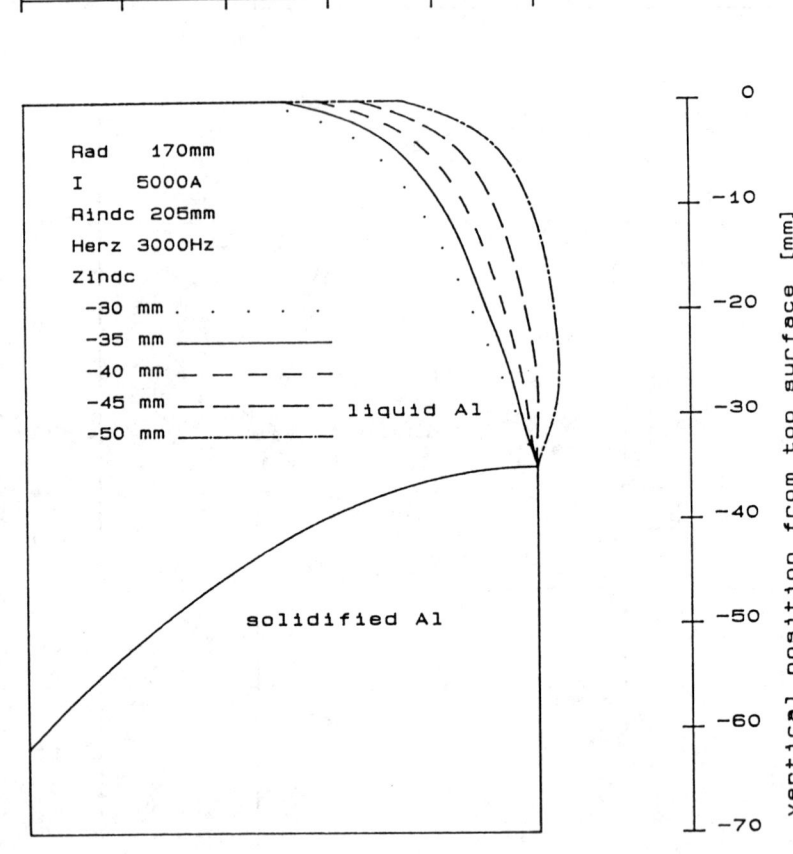

radial position from center axis [mm]

120   130   140   150   160   170

XBL 876-2880

Figure 4 - Effect of vertical position of inductor, shape (2), on the meniscus for a Kaiser type caster. Z_{indc} is the vertical distance from the top of the meniscus to the center of the inductor.

Fig. 5 shows the computed meniscus shape for a caster of the Russian type (with a screen) for various frequencies. The meniscus is seen to be less well supported at higher frequencies. This result is a consequence of the greater effect of the screen in diminishing the magnetic field and induced current in the melt at higher frequencies. In contrast, the meniscus shape was found to be insensitive to frequency for casters of the Kaiser type.

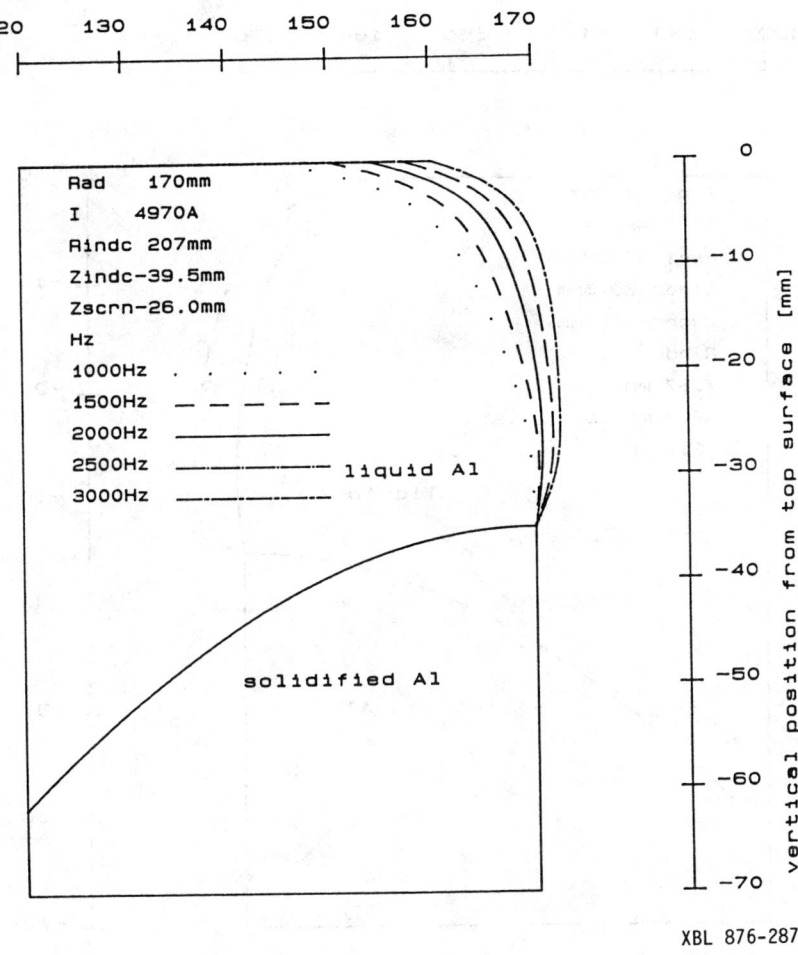

radial position from center axis [mm]

120   130   140   150   160   170

Rad   170mm
I   4970A
Rindc 207mm
Zindc-39.5mm
Zscrn-26.0mm
Hz
1000Hz . . . . .
1500Hz — — — — —
2000Hz ————————
2500Hz ——·——·—— liquid Al
3000Hz —·—·—·—·—

solidified Al

vertical position from top surface [mm]

0
-10
-20
-30
-40
-50
-60
-70

XBL 876-2879

Figure 5 - Effect of frequency on meniscus shape for a caster of the
Russian type. Z_{scrn} is the vertical distance from the top of the
meniscus to the bottom of the screen.

The sensitivity of the meniscus shape to coil radius in a caster of
the Russian type is revealed in Fig. 6. The calculations were carried
out at fixed screen geometry and may reflect the greater effect of the
screen in diminishing the magnetic field and induced current in the melt
as the meniscus approaches the screen. This sensitivity is not found in
the calculated results for the caster of the Kaiser design; variations in
coil radius of the same size as those in Fig. 6 shift the meniscus by
less than 3 mm.

radial position from center axis [mm]

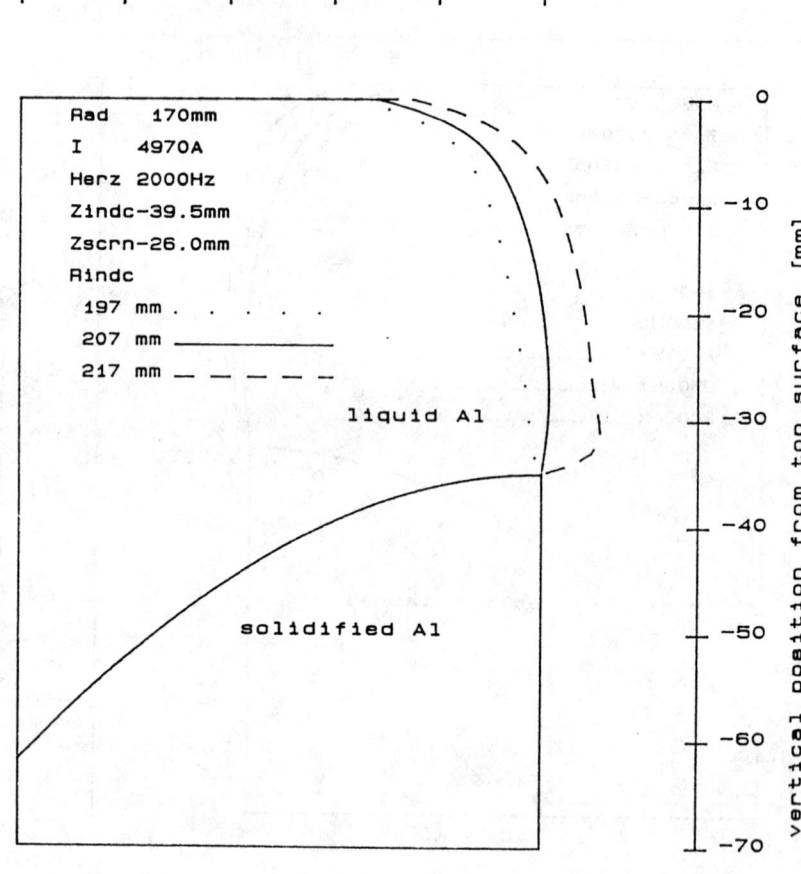

```
120      130      140      150      160      170
|----+--------+--------+--------+--------+----|
```

Rad   170mm
I   4970A
Herz 2000Hz
Zindc-39.5mm
Zscrn-26.0mm
Rindc
  197 mm . . . . .
  207 mm _____
  217 mm _ _ _ _ _ _

liquid Al

solidified Al

vertical position from top surface [mm]

0
-10
-20
-30
-40
-50
-60
-70

XBL 876-2878

Figure 6 - Effect of inductor radius (R_{indc}) on meniscus shape for a Russian type caster.

As anticipated the meniscus can be shifted inward by increasing the coil current, eliminating outward bowing and stabilizing the meniscus. Typically an increase in current of 20% was calculated to move the meniscus inward by 5 mm at mid height for the Russian type caster.

For the screen used in these calculations, (tapered and of moderate thickness and electrical conductivity) the vertical position of the screen had an effect, but not a large one, on the meniscus shape. Doubling or halving the screen electrical conductivity or thickness had a major effect, shifting the meniscus by 4-7 mm, the shift being inward for reduced conductivity or thickness.

Fig. 7 shows the computed velocities for a caster of the Russian type operated at a casting speed of 50 mm/min. Allowance is made in the calculations for the curvature of the meniscus and the inflow of metal. The lengths of the arrows indicate the magnitude of the velocity at a position indcated by the tail of each arrow. These calculated values should be compared with the measured values reported by Vives and Ricou (5) and presented in the lower half of the figure. In most of the melt the agreement is satisfactory, the exception being in the vicinity of the solidification front where natural convection (not incorporated in the model) appears to be significant. The strong recirculation due to the electromagnetic forces in the outer region of the melt and the relatively quiescent central core should be noted. The strong horizontal jet emerging from the nozzle would be a function of the nozzle design.

Fig. 8 shows the computed velocities for a caster of the Kaiser type operated at virtually the same conditions. The velocities in this caster are significantly higher than those of Fig. 7 indicating that the screen of the Russian type caster may offer an advantage if control of the metal velocity is important.

For additional results the reader is referred to a recent paper (7).

Concluding Remarks

This paper has presented a brief description of a mathematical model for a cylindrical electromagnetic caster. The model computes electromagnetic parameters (such as current density), the meniscus shape, velocities and other parameters not presented here (such as turbulence intensity).

A selection of computed results is presented. Computed velocities for a real caster match the experimental measurements of other investigators. The two caster types commonly encountered in aluminum casting show different sensitivities to such design parameters as coil radius and frequency but similar behaviour in other respects (e.g. a dependence of meniscus shape on vertical position of the coil). Both types of caster displayed a rapid circulation of metal at the outer edges of the melt, this velocity being diminished by interposing an electromagnetic screen. The results presented have been for the casting of aluminum but the model is equally applicable to the casting of other metals and the predictions should be applicable (in a qualitative sense) to casters which are not axisymmetric. This paper is an interim one describing some results from a continuing investigation, by both physical and mathematical model, of electromagnetic casting.

Acknowledgment

Support of this research by the National Science Foundation under grant no. MSM-8520783 is gratefully acknowledged, as is support granted by Nippon Steel Corp. to one author (J.S.) permitting him to spend a year at Berkeley.

1013

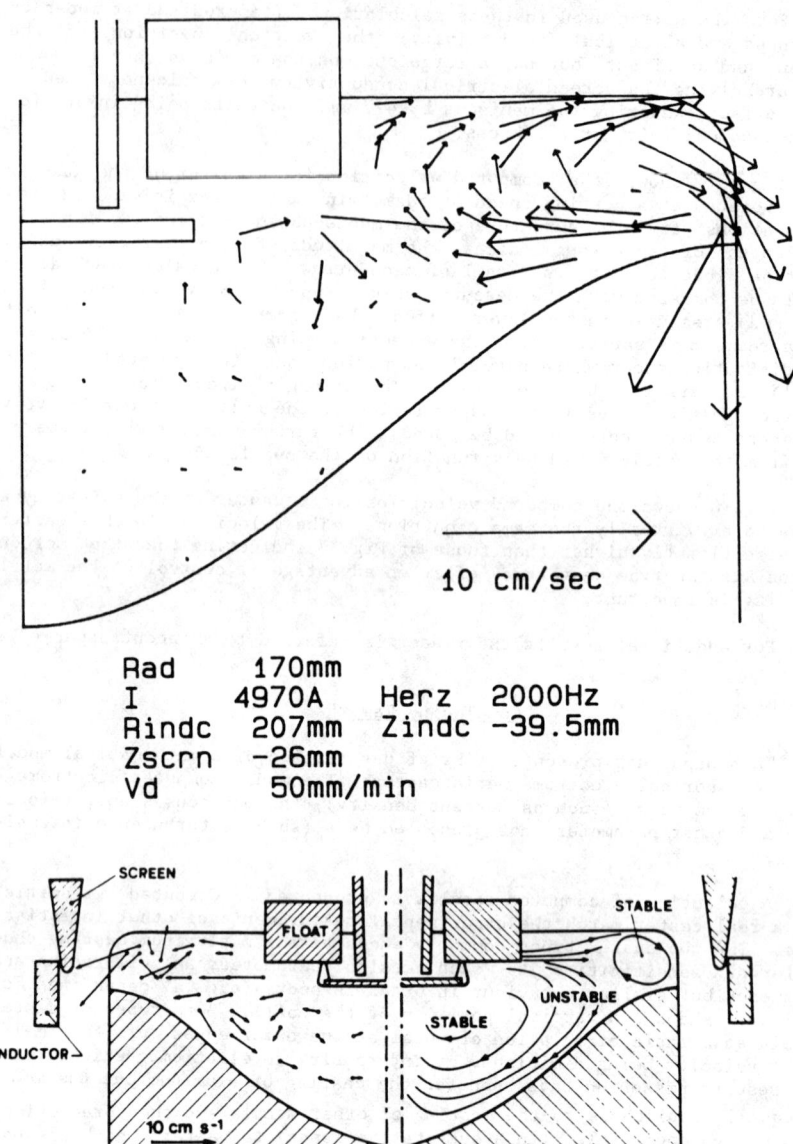

Figure 7 - A comparison of the measured velocities of Vives and Ricou (5) and computed velocities for this caster.

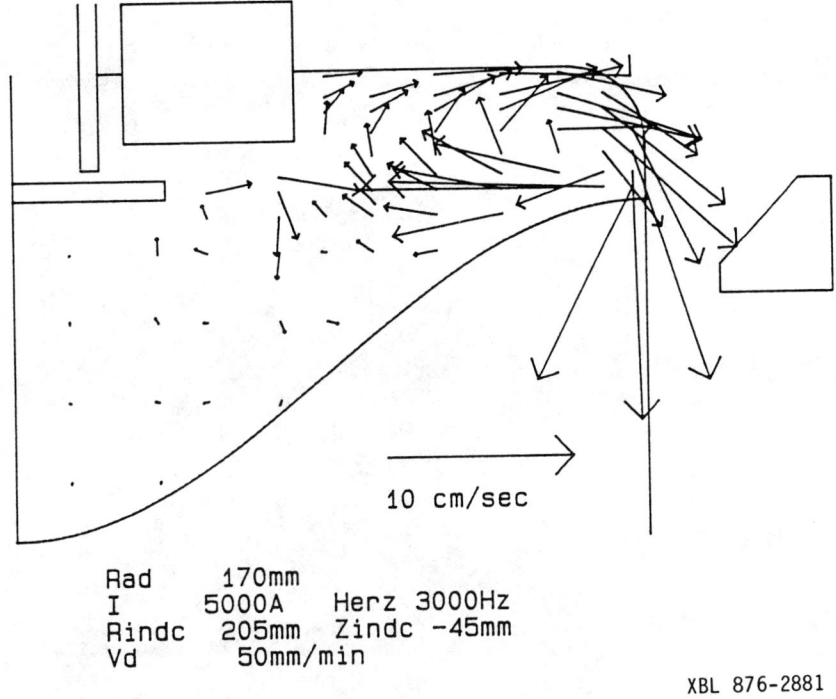

10 cm/sec

Rad	170mm	
I	5000A	Herz 3000Hz
Rindc	205mm	Zindc −45mm
Vd	50mm/min	

XBL 876-2881

Figure 8 - Computed velocities for a caster of the Kaiser type.

References

1. Z. N. Getselev, U.S. Patent 3,605,865 (Sept. 20, 1971), "Continuous Casting Apparatus with Electromagnetic Screen".
2. D. G. Goodrich, U.S. Patent 4,351,384, (Sept. 28, 1982), "Coolant Control in Electromagnetic Casting".
3. J. D. Lavers, "An Analysis of an Electromagnetic Mold for the Continuous Casting of Nonferrous Metals" (IEEE Trans. Industry Applications, 1981, IA-17), 427.
4. J. D. Lavers, "Force and Stirring Patterns Produced by an Electromagnetic Field" (S.F., IEEE-IAS Conf. Proc., Oct. 4-7, 1982), 954.
5. C. Vives and R. Ricou, "Experimental Study of Continuous Electromagnetic Casting of Aluminum Alloys", Met. Trans., 16B (1985) 377-405.
6. E. D. Tarapore and J. W. Evans, "Fluid Velocities in Induction Melting Furnaces: Part I. Theory and Laboratory Experiments", Met. Trans., 7B (1976) 343-351.
7. J. Sakane and J. W. Evans, submitted to Met. Trans.
8. W. Rodi, "Examples of Turbulence Models for Incompressible Flows", AIAA J., 20 (1982) 872-879.
9. N. El-Kaddah and J. Szekely, "The Turbulent Recirculating Flow Field in a Coreless Induction Furnace, a Comparison of Theoretical Predictions with Measurements", J. Fluid Mech., 133 (1983) 37-46.

A Knowledge Based System Approach to Predict

Longitudinal Surface Cracking in Continuous Casting

J. Suni and H. Henein

Dept. of Metallurgical Engineering and Materials Science
Carnegie-Mellon University
Pittsburgh, PA 15213 U.S.A.

Abstract

Defect prediction, using the knowledge based system approach, is being developed for the case of longitudinal surface cracks. Knowledge about longitudinal cracking is in the form of databases, mathematical models and qualitative information. Much of this comes from the technical literature on related topics, but is supplemented by plant experience, as well. Knowledge based systems provide the facility for handling such types of knowledge, with control particulars left up to the system builder. The current implementation is forward chaining and uses the OPS5 inference engine. The system involves the use of all three types of knowledge, but more mathematical models are being sought.

Introduction

Hot charging has made defect awareness and detection in the continuous casting of steel a more important problem than in the past. If slabs coming off the caster are inspected, thermal units will be lost, and if slabs are passed, unchecked, the risk is higher that finished product quality will be unacceptable to the customer and higher costs will be incurred. It would seem clear that the best approach to the situation is an accurate predictive tool that provides for the confidence that only high quality slabs are passed to the hot rolling mill.

The current practice consists of 'setting the controls', according to established practice for the particular grade (or application) being cast, i.e. a table lookup of recommended operating regions for different grades. This has a number of drawbacks. First of all, the acceptance of established procedure may be too conservative under certain conditions. For example, there may be heats which can be cast faster, or sprayed differently than the recommended amounts, in order to hold more heat for the rolling operation, or to increase productivity. Secondly, there are a number of 'response' variables which can significantly affect various defect formations, which are not immediately selectable by the operator, such as variations in casting speed or mold level. Thirdly, table lookups only account for one variable at a time. In addition, tables provide no guidelines or predictions for dispositioning of slabs.

This paper will present some of the basic types of information relating to longitudinal cracks that may be used in a knowledge based system. This modelling approach will be defined and contrasted with the more traditional algorithmic approaches. In addition, the framework developed for predicting longitudinal surface cracking will be outlined.

Information on Longitudinal Surface Cracking

The difficulty is that, while certain problems in casting, such as shell thickness determination (mathematically), are relatively straightforward and dependable, reliable defect prediction methods are not readily available. This is a consequence of the uncertainty with which defects form. In the çase of longitudinal surface cracking, it is obvious from the literature[1,2] that the problem is related to irregularity in the mold cooling, both transverse and longitudinal. Irregular heat fluxes will give rise to irregularities in the solid shell thickness and, hence, in the solid shell strength, as well. In this situation, it is easy to envision a thicker portion of the shell pulling away from the wall, while a neighboring thinner area is pushed back against the wall by ferrostatic pressure, leading to a crack, or worse, a breakout. The key issues, then, are thermal and mechanical ones. How large is the difference in neighboring shell thicknesses? How much transverse stress is the shell experiencing, and how does a particular steel respond to these stresses?

Methods are described in the literature which attempt to make predictions about cracking, based on finite element stress/strain analysis[3,4]. These have primarily been with reference to internal cracks, i.e. further down the caster, but have even been attempted for the case of longitudinal cracking[5]. Such approaches cannot account for the large amount of uncertainty involved, or for various operating features, such as mold level fluctuations. In addition, important qualitative considerations cannot be integrated.

The primary sources of knowledge for defect formation, in general, are databases, mathematical models and qualitative considerations. Database information is of two types, relating to properties or phenomenology. Property databases simply consist of tables or plots of various mold, strand and powder properties showing, for example, dependence on temperature. Phenomenological databases cover all of the information in the literature (and elsewhere) which speaks to either the defect, itself, or some important aspect of the defect. For example, peritectic steels (carbon in the range 0.08 to 0.16 %) are expected to exhibit longitudinal surface cracking greater than other grades,[6-9] as shown in Figure 1. This is related to the shrinking attendent to the peritectic transformation[10,11], as well as microsegregation effects[12]. Similar results can be found for a host of variables :

- Chemistry

- Operating variables (speed, pouring temperature, spray practice, etc.)

- Mold powder characteristics (viscosity, softening temperature, etc.)

- strand dimensions

- 'Response' variables (variations in speed, mold level, etc.)

Other chemistry effects include the Mn/S ratio, which is directly related to ductility considerations. In this case, 'database' refers to the knowledge which has been compiled on the high temperature strength and ductility of steels, as a function of composition[13,14]. Casting speed (v), and mold powder viscosity (η) have been shown to have the same kind of effects[15,16] (see Figure 2), since both will be involved in the even flow of molten flux down the side of the strand. It has even been suggested that a parameter of interest would be $\eta^* v$[17]. strand dimension effects have mainly been in the form of width to thickness ratios for slabs, higher ratios leading to more cracking[18]. 'Response variable' effects have been shown to be strong, particularly with mold level variations[19,20].

These variables are obviously important, but what about quantities such as the steepest temperature gradient at the midface (at various positions in the mold and just below), the temperature rebound below the mold, etc.? These quantities are of considerable importance, and should be integrated as well. This will lead to greater specificity of the independent variable space. For example, the (average) casting speed will greatly affect the temperature distribution, but will have separate effects, as well. These kinds of information are afforded by the vast work done on modelling in continuous casting. Heat flow modelling has been done on the strand[21-23], in the mold[24], and in the flux layer[25], as well. This is usually done in the following form (in the strand) :

$$\frac{\partial}{\partial x} \left(k \frac{\partial T}{\partial x} \right) + \frac{\partial}{\partial y} \left(k \frac{\partial T}{\partial y} \right) = \rho C_p \frac{\partial T}{\partial t} \tag{1}$$

In addition, modelling work has been done in the area of stress and strain determination[3,26,27]. This is a difficult undertaking in the mold, where mesh sizes must be extremely fine for finite element strain analysis (on solid shells only 1 or 2 cm thick), but has been done with reasonable results[27,28].

The final source of knowledge about cracking is in the form of qualitative considerations. These are often in the form of heuristics, or rules of thumb, which don't necessarily have any basis in theory, but are found to be true, nonetheless. For example, tube changes have been suggested to have a negative effect on cracking[29]. Also, it has been suggested that longitudinal cracks often happen in bunches[30].

Knowledge Based Systems

BACKGROUND

The problem of defect prediction is being approached in the present study, in two different ways, for the particular defect of longitudinal surface cracking. One approach is to take a statistical approach to partitioning the independent variable space, and assessing probabilities for cracking in each sub-space, using the package, 'Entropy Minimax'[31,32]. This is a data specific method which not only attempts to partition the variables, based on the data, but includes facilities for inserting models such as heat flow, as well. This is primarily a numerical process, however, and is not necessarily based on an understanding of what causes longitudinal surface cracking. The other approach, which is the focus of this paper, is to use the knowledge based system implementation.

Knowledge based systems can take a number of forms, depending on the architecture and

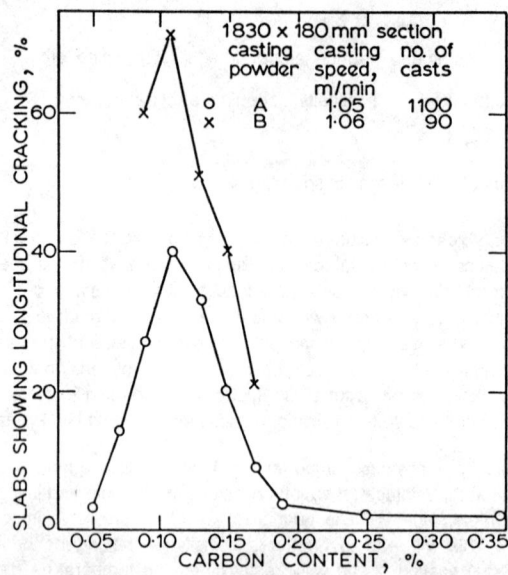

Figure 1.   Effect of carbon on longitudinal cracking
              (after Gray et al[9])

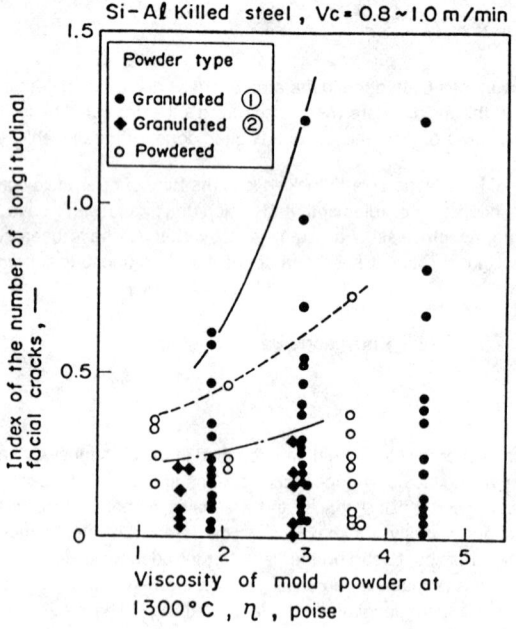

Figure 2.   Effect of flux viscosity on longitudinal cracking
              (after Suzuki et al[2])

control structure used, but are characterized, in general, by a separation of the domain knowledge (i.e. information about cracking) from the controlling mechanism of the system[33]. This is contrasted with a typical algorithmic application (e.g. in Fortran), where a majority of the coding is concerned with controlling the execution of the program. In addition, knowledge based systems are characterized by an attempt to reason through some process. For example, cracking could be thought of as depending on strain levels, which could be thought of as depending on stress levels, which could be thought of as depending on temperature gradients, strand/mold wall friction, etc., and so on.

Reasoning, as used here, can be in three different directions. Using the terms of logical formalism, these are deduction, induction and abduction. These are characterized by which of the general system components, p (input data), $p \Rightarrow q$ (rules or information) and q (results), is being sought, i.e.

- Deduction: p, $(p \Rightarrow q)$ --->   q

- Induction: p,   q   ---> $(p \Rightarrow q)$

- Abduction: $(p \Rightarrow q)$,   q   --->   p

Deduction is the most straightforward and, accordingly, the most common in knowledge based systems. Deduction is central to defect prediction problems. Induction is the process of making large statements about the process, based on experience, i.e. learning. In the context of a computer environment, this is a difficult thing to do. Obviously, induction, in the human environment, must be done before any deduction can be done. Abduction is of intermediate difficulty, and may be concerned with problems such as determining what things caused cracking in the past.

The particular knowledge based system implementation that is put into place is largely determined by the architecture and control structure that are used. Architecture, here, is used to refer to software particulars. In the present study, a package known as OPS5[34] is being used, within the framework of a larger development tool, Knowledge Craft[TM]. OPS5 is a production system, consisting, entirely, of IF/THEN rules, which are cyclically checked for candidates whose conditions are all true, as shown in Figure 3. Control structure is used here to refer to the particulars of how one goes about doing the particular tasks involved in the system. The first question in determining control structure is the direction that reasoning will proceed, i.e. forward versus backward chaining. In forward chaining, or data driven systems, reasoning proceeds from the input data to the conclusions which are desired. In backward chaining, or goal driven systems, reasoning proceeds from the conclusion of interest to the data, in order to test the conclusion's validity. Further control structure issues involve the depth of rules or information to be used.

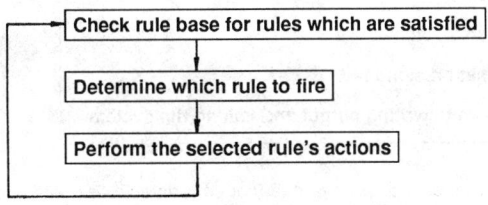

Figure 3. Production System Cycle

FORMULATION

The problem of determining whether longitudinal surface cracking will occur is being structured as shown in Figure 4, below. In this forward chaining approach, a small number of rules are written, called control rules, which insure that this sequence is stepped through each time a prediction about cracking is made. Making an inference about cracking amounts to literally stepping through what is expected to be the progression of the crack's formation. Each task in the sequence given below contains a number of rules used to make the necessary determination. For example, the first task contains rules whose actions are simply to say how smooth the heat flux is around the shell perimeter, the second contains rules to estimate the transverse regularity of the solid shell, and so on.

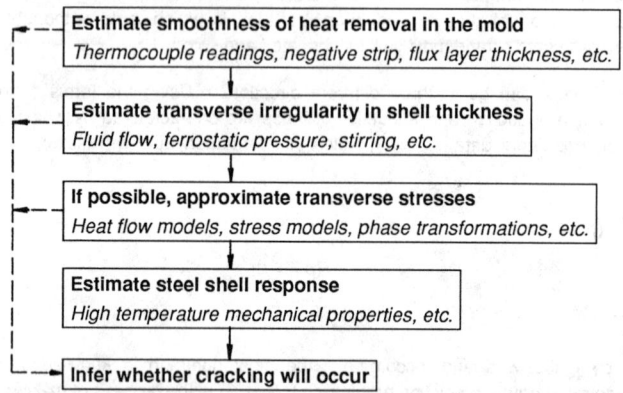

Estimate smoothness of heat removal in the mold
Thermocouple readings, negative strip, flux layer thickness, etc.

Estimate transverse irregularity in shell thickness
Fluid flow, ferrostatic pressure, stirring, etc.

If possible, approximate transverse stresses
Heat flow models, stress models, phase transformations, etc.

Estimate steel shell response
High temperature mechanical properties, etc.

Infer whether cracking will occur

Figure 4. Structure of longitudinal surface crack prediction problem

By performing a number of tasks, in series, uncertainty is being propagated down through the problem. How this uncertainty is handled is entirely up to the system builder. Much has been written on fuzzy systems[35], uncertainty factors[36], prior odds, and so on, but there is no accepted means for keeping track of uncertainty. Shell irregularity can be assigned a number, for example, from 0 to 1, or from -5 to 5. Similarly, irregularity can be assigned a qualitative tag, such as 'slight' or 'high'.

An important feature of the current structure is that 'routes' out of the problem are available at any level. For example, rules in the first task (smoothness of heat flux) can deal with the case where there are thermocouple readings in the mold. Under certain conditions of these readings a direct inference about cracking can be made. That is, some rules in this task (and in each task) have the following form :

IF the current goal is to check for irregular heat flux

AND condition 1

AND condition 2 , etc.

THEN infer that cracking is ...

AND step down to writing output and halting the system

It's clear from the formulation of the above rule that considerable flexibility is afforded by the production system architecture. Right hand sides of rules (the actions, or THEN parts) can do whatever the software and hardware configurations will allow. This includes writing output, interacting with the user, and, importantly, running external mathematical codes, written in a suitable programming language, such as Fortran. This last capability is of extreme importance, because it allows the integration of accepted modeling techniques, such as heat flow, stress and strain analysis, etc. into the problem. The insertion of models (with our configuration) is trivial, and once a model has been called, the results can themselves be conditions for further rules to act on. In addition, long-winded models may not always be necessary. The rules can be written so that the models are only run under certain conditions, and only down (along the caster) to the point of interest. Furthermore, the rules can be written so that, for particularly uncertain conditions, analyses can be run which arrive at predictions about cracking from purely mathematical criteria.

It is easy to see that the knowledge base can be, at the very least, a locust for knowledge about continuous casting, in the form of databases, models and qualitative information. In addition, the flexibility inherent in knowledge based systems affords straightforward user interfaces which can include such things as explanations of why a particular thing is being sought, references for where a particular piece of information came from, caster diagrams, showing where interest is being focused, and plots of model results.

Implementation

In the current approach to knowledge based system prediction of longitudinal surface cracking, a forward chaining OPS implementation is being used, with the control structure given above. The hardware being used is a MicroVax II, with 16 Megabytes of memory and 220 Megabytes of disk storage. The software being used is Vax Lisp (version 2.1), Vax Fortran, and Knowledge Craft^{TM}, a system development tool containing the OPS inference engine. Currently, callouts are done only for heat flow computation[37], but stress/strain models are being sought for integration, as well. The dependent variable, longitudinal surface cracking, is being considered as a ternary output variable. That is, three output values are possible, cracking, uncertain, and no cracking. This is hoped to mirror the actual choice made in dispositioning, where a strand may face scarfing (cracking), or inspection (uncertain), or may be passed to the rolling mill (no cracking). The only possible further refinement is the division of 'uncertain' to include two grades of uncertainty, one being more likely to show cracking, in order to represent the decision between hot and cold inspection. This may be a non-trivial distinction, since cold inspection can be of considerable cost.

Uncertainties are handled, implicitly, by assigning qualitative tags to various parameters. For example, ductility is assumed to be either 'Good' or 'Poor'. Shell irregularity is assumed to be 'Little', 'Slight', 'Moderate' or 'High', and so on. Cracking is based on these qualitative tags, and is checked for at various points down the mold. Points below the sprays are not considered, and only midface temperature gradients are considered. Paths out of the system are provided at each task level. When wide enough variations in mold thermocouple readings are logged, then direct statements about cracking are made. Also, when stresses are being considered, and the heat flow model is called, gradients severe enough, along with other conditions (chemistry, etc.) cause direct statements to be made, as well.

Information for the various tasks to be performed has been taken from the literature available on longitudinal surface cracking and related topics. At this point, no attempt is made to reconcile conflicting information. That is, only one point of view has been inserted for a particular piece of information. For example, if an estimate is required for the (average) thickness of molten flux between the strand and mold wall, a relationship is used from a particular source on mold powders[15]. However, it is envisioned that other approximations for such quantities will be introduced shortly, with provisions for weighting or blending these different approaches.

Summary

The literature clearly indicates that longitudinal surface cracking is closely related to the interaction of thermal and mechanical features of continuous casting. The effects of carbon, manganese, sulfur, phosphorous, casting rate, flux viscosity, mold level fluctuation and strand dimensions can all be considered in this light. In order to reason through crack formation, then, these interactions need to be accounted for. This is done with an emphasis on the problem of irregularity in mold cooling. This irregularity is assumed to have an important influence in determining the variations in solid shell thickness and, in turn, solid shell strength.

Reasoning through longitudinal crack formation in this manner can be done in a knowledge based system environment. Knowledge based systems provide not only the framework for reasoning through the process, but for collecting the primary sources of information on cracking, namely databases (experimental data, property information, etc.), mathematical models and qualitative considerations, as well. In addition, the flexibility of this approach allows for providing a number of alternative paths to the conclusion of interest. That is, the main path down through the problem, including the running of models, can be circumvented under appropriate conditions. Such conditions include, for example, wide variations in mold thermocouple readings.

The current implementation of longitudinal surface crack prediction deals in a ternary output variable, with possible values of cracking, uncertain and no cracking. Inferences are based on estimates of derived quantities (shell irregularity, temperaure gradients, strains, etc.), input quantities (chemistry, operating variables, etc.) and observations (recency of tube changes, etc.).

Acknowledgements

The authors would like to acknowledge USX Corp. for financial support.

Disclaimer

References

1. T. Saeki, S. Ooguchi, S. Mizoguchi, T. Yamamoto, H. Misumi and A. Tsuneoka. **Effect of Irregularity in Solidified Shell Thickness on Longitudinal Surface Cracks in CC Slabs**. Tetsu-To-Hagane, 1982, Vol. 68, pg. 1773.

2. M. Suzuki, S. Miyahara, Y. Miyashita and Y. Miyawaki. **Relationship Between Solidification Phenomena in the Continuous Casting Mold and Surface Defects of Slabs**. Published by TMS/AIME, Warrendale, PA, Paper # A83 - 9, 18 pgs.

3. A. Grill, J.K. Brimacombe and F. Weinberg. **Mathematical Analysis of Stresses in Continuous Casting of Steel**. Ironmaking and Steelmaking, 1976, Vol. 3, pg. 38.

4. B. Barber, B.A. Lewis and B.M. Leckenby. **Finite-Element Analysis of strand Deformation and Strain Distribution in Solidifying Shell During Continuous Slab Casting**. Ironmaking and Steelmaking, 12, 1985, pg. 171.

5. T. Matsumiya, T. Saeki, J. Tanaka and T. Ariyoshi. **Mathematical Model Analysis on the Formation Mechanism of Longitudinal Surface Cracks in Continuously Cast Slabs**. Tetsu-To-Hagane, 1982, Vol. 68, pg. 1782.

6. W.R. Irving and A. Perkins. **Basic Parameters Affecting the Quality of Continuously Cast Slabs**. Ironmaking and Steelmaking, 1977, No. 5, pg. 292.

7. Y. Miyashita, M. Suzuki, K. Taguchi, S. Uchida, H. Sato and M. Yamamura. **Improvement of Surface Quality of Continuously Cast Slab**. Nippon Kokan Technical Report Overseas, No. 36, 1982, pg. 55.

8. W.R. Brown and A.T. Kwong. **Control of Slab Quality at Stelco's Lake Erie Works**. Steelmaking Proceedings, ISS, Volume 65, 1982, pg. 313.

9. R.J. Gray, A. Perkins and B. Walker. **Quality of Continuously Cast Slabs**. Solidification and Casting of Metals, Sheffield Proceedings, 1979, pg. 300.

10. S.N. Singh and K.E. Blazek. **Heat Transfer, and Skin Formation in a Continuous Casting Mold as a Function of Steel Carbon Content.** Journal of Metals, October 1974, pg. 17.

11. A. Grill and J.K. Brimacombe. **Influence of Carbon Content on Rate of Heat Extraction in the Mould of a Continuous Casting Machine.** Ironmaking and Steelmaking, 1976, No. 2, pg. 76.

12. M. Wolf and W. Kurz. **The Effect of Carbon Content on Solidification of Steel in the Continuous Casting Mold.** Met. Trans. B, Vol. 12B, March 1981, pg. 85.

13. C.J. Adams. **Hot Ductility and Strength of strand Cast Steels up to Their Melting Points.** Open Hearth Proceedings, Vol. 54, 1971, pg. 290.

14. W.T. Lankford, Jr. **Some Considerations of Strength and Ductility in the Continuous Casting Process.** Met. Trans., Volume 3, June 1972, pg. 1331.

15. R. Gray and H. Marston. **The Influence of Mould Fluxes on Casting Operations and Surface Quality.** Steelmaking Proceedings. ISS, Vol. 62, 1979, pg. 93.

16. T. Emi, H. Nakato, Y. Iida, K. Emoto, R. Tachibana, T. Imai and Hajime Bada. **Influence of Physical and Chemical Properties of Mold Powders on the Solidification and Occurance of Surface Defects on strand Cast Slabs.** Continuous Casting, ISS-AIME, Volume 1, pg. 135.

17. T. Nakano, T. Kishi, K. Koyama, T. Komai and S. Naitoh. **Mold Powder Technology for Continuous Casting of Aluminum Killed Steel.** Transactions ISIJ, 24, 1984, pg. 950.

18. Charles R. Taylor. **Continuous Casting Update.** Met. Trans. B, Volume 6B, September 1975, pg. 359.

19. A. Delhalle, J.F. Mariotton, J.P. Birat, J. Foussal, M. Larrecq and G. Tourscher. **New Developments in Quality and Process Monitoring on Solmer's Slab Caster.** Steelmaking Proceedings, ISS, Volume 67, 1984, pg. 21.

20. H. Ohzu, H. Ohmori and J. Yamazaki. **Problems With Automatic Liquid Level Control System For a Slab Caster.** Steelmaking Proceedings. Vol. 64, 1981, pg. 36.

21. E.A. Mizikar. **Mathematical Heat Transfer Model for Solidification of Continuously Cast Steel Slabs.** Continuous Casting, ISS-AIME, Volume 2, pg. 9.

22. I. Saucedo, J. Beech and G.J. Davis. **Heat Transfer and Solidification Kinetics in Meniscus Zone During Casting.** Metals Technology, July 1982, Vol. 9, pg. 282.

23. J.K. Brimacombe. **Design of Continuous Casting Machines Based on a State-of-the-Art Review.** Continuous Casting, ISS-AIME, Volume 2, pg. 17.

24. I.V. Samarasekera and J.K. Brimacomge. **The Continuous Casting Mould.** International Metals Review, 23, 1978, pg. 284.

25. A. Delhalle, M. Larrecq, J.F. Marioton and P.V. Riboud. **Slag Melting and Behavior at Meniscus Level in a Continuous Casting Mold.** Mold Powders for Continuous Casting and Bottom Pour Teeming, ISS-AIME, 1987, pg. 15.

26. K. Miyazawa and K. Schwerdtfeger. **Computation of Bulging of Continuously Cast Slabs with Simple Bending Theory.** Ironmaking and Steelmaking, 1979, No. 2, pg. 68.

27. J.E. Kelly, K.P. Michalek, B.G. Thomas and J.A. Dantzig. **Initial Development of Thermal and Stress Fields in Continuously Cast Steel Billets.** Met. Trans. A, (in print).

28. K. Sorimachi and J.K. Brimacombe. **Improvements in Mathematical Modelling of Stresses in Continuous Casting of Steel.** Ironmaking and Steelmaking, 4, 1977, pg. 240.

29. J.H. Gallenstein. USX Corp. Private communication, 1987.

30. A.W. Cramb. Carnegie-Mellon University. Private communication, 1987.

31. R. Christensen. **Entropy Minimax Multivariate Statistical Modeling - 1 : Theory**. Int. Journal of General Systems, 1985, Vol. 11, pg. 231.

32. R. Christensen. **Entropy Minimax Multivariate Statistical Modeling - 2 : Applications**. Int. Journal of General Systems, 1986, Vol. 1, pg. 227.

33. P. Harmon and D. King. Expert Systems. John Wiley & Sons, New York, 1985.

34. L. Brownston, R. Farrel, E. Kant and N. Martin. Programming Expert Systems in OPS5. Addison-Wesley Inc., Reading, Massachusetts, 1985.

35. C.V. Negoita. Expert Systems and Fuzzy Systems. Benjamin/Cummings Publishing Co., Menlo Park, California, 1985.

36. R. Forsyth. **Fuzzy Reasoning Systems**. Expert Systems, Chapman and Hall, London, 1984, pg. 51.

37. B. Lally. **Optimization and Mathematical Modelling of Continuous Steel Casting Processes**. PhD thesis, Carnegie-Mellon University, 1987.

BEHAVIOUR OF LIQUID POOLS IN ESR AND VAR PROCESSES

A. Jardy and D. Ablitzer

Laboratoire de Science et Génie des Matériaux Métalliques, U.A. 159
Ecole des Mines, Parc de Saurupt, 54042 NANCY CEDEX - FRANCE

ABSTRACT

A mathematical model has been developed in order to simulate the behaviour of the liquid metal in the electroslag remelting (ESR) and vacuum arc remelting (VAR) processes. The steady state MAXWELL, NAVIER-STOKES and FOURIER equations are solved in an axi-symmetrical geometry. Turbulent motions and thermal transfers in the liquid pool are thus calculated by the model which takes into account the real shape of the curved liquid/solid interface. The model has been used to simulate electroslag or vacuum remelting of a fictional alloy at a pilot scale. It shows the influence of natural convection forces related to the thermal gradients, and of electromagnetic forces derived from the interaction between the electric current and induced magnetic field. Particularly, by simulating different flow modes of the electric current in VAR operation, the model allows prediction of this parameter's effect on the behaviour of the liquid pool.

1. INTRODUCTION - SUBJECT PRESENTATION

The work presented here is part of a vast study on the analysis and modelling of consumable electrode remelting processes. The object of our work is the simulation of the behaviour of liquids in ESR (electroslag remelting) and VAR (vacuum arc remelting) processes.

A schematic representation of these processes is given figure 1.

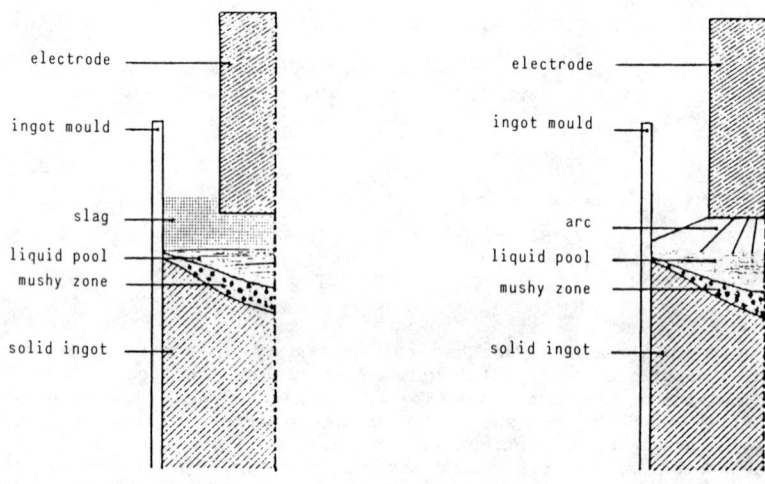

| 1a : ESR process | 1b : VAR process |

Figure 1 - Schematic representation of the ESR and VAR processes.

A consumable electrode of the desired grade is remelted and cast in an ingot mould cooled by a counter-current water circuit. The origin of the energy necessary for the melting is of an electrical nature. In the case of the VAR process, an arc is created between the electrode and the liquid pool of the secondary ingot. In electroslag remelting, the current flow in the slag heats it by Joule effect, inducing melting of the immerged electrode. In both processes, the remelted alloy cools and solidifies upon contact with the mould walls. The rise of these processes is due essentially to the metallurgical qualities of remelted ingots. Those are directly related to the solidification process of the ingot. The control of parameters characterizing this process (velocity of the liquid metal, intensity of thermal transfers), can allow reducing formation of segregated phases and solidification defects. Mathematical modelling permits an estimation of these parameters from operating data and thermophysical properties of materials.

A major part of the models of remelting processes available in literature are thermal models in transient state, which simulate thoroughly the thermal behaviour of ingots during remelting. Particularly, we mention here the modellings developed by JEANFILS et al.(1), BALLANTYNE and MITCHELL (2,3) and the present authors (4,5). Although these approaches, once validated by confrontation of results with industrial experiments, allow good prediction of the shape and depth of liquid pools, as well as local solidification times and cooling rates of the solid ingot, they remain dependent of some adjustable parameters. In particular, turbulent convective movements in the liquid pool are taken into account by introducing an "effective thermal conductivity".

The models based on numerical solution of heat and momentum transfer equations in turbulent regime are a more fundamental approach. Whereas many studies of this type have beared on the slag behaviour in ESR (6-8), the liquid pool is more difficult to treat. This is why its geometry was first extremely simplified (9). More recently, BERTRAM and ZANNER (10) for VAR, and CHOUDHARY and SZEKELY (11) for ESR, have fixed their attention to the solution of coupled equations in a liquid pool of real geometry.

The association of both these approaches (4,12) has allowed relation of the operating parameters of ESR to the local solidification times in a nickel-base superalloy ingot. The determination of the liquid metal behaviour, in ESR as well as in VAR, forms the subject of the work presented here.

2. DESCRIPTION OF THE MATHEMATICAL MODEL

The model is based on numerical solution of the MAXWELL, NAVIER-STOKES and FOURIER equations in the liquid pool of a VAR ingot or in the ingot's liquid pool and the slag in ESR.

2.1. Assumptions

- The regime is steady and the geometry axi-symmetrical. This assumption allows treatment of a bidimensional problem in cylindrical coordinates r and z.
- The position of the liquid/solid interface is fixed in advance. The determination of this position can be realized either experimentally (marking of the liquid pool), or by using a thermal model well validated, such as the one proposed by FALK et al. (5). The presence of a mushy zone is not taken into account by the model.

2.2. Maxwell equations - Boundary conditions

The model solves the classic electromagnetic equations which, considering the assumptions, can be written :

$$\sigma \, \mu_o \, \frac{\partial H_\theta}{\partial t} = \frac{\partial}{\partial r} \left[\frac{1}{r} \frac{\partial}{\partial r} (r \, H_\theta) \right] + \frac{\partial^2 H_\theta}{\partial z^2} \qquad (1)$$

where H_θ is the orthoradial magnetic field intensity

σ the electric conductivity of the material

μ_o its magnetic permeability

The term $\partial H_\theta / \partial t$ permits consideration of effects related to the frequency of the alternating current generally used in electroslag remelting.

In the case of ESR , the boundary conditions used to solve equation (1) do not create any theoretical difficulties since the whole installation

1029

is conducting and the lateral walls covered by an insulating layer of solidified slag. These conditions simply express the general symmetry, the tangential electric field continuity at the interfaces, and the Ampere's law. On the other hand, in the case of vacuum remelting, certain difficulties arise, related to the lack of knowledge on the modes of current flow, as shown schematically in figure 2.

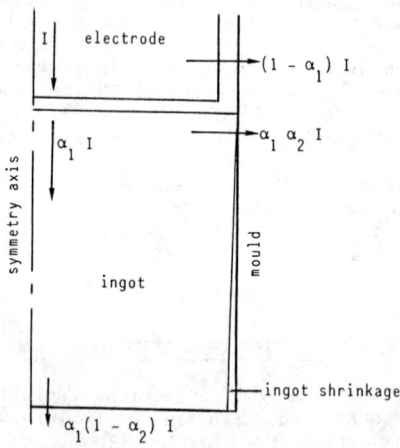

Figure 2 - Schematic representation of the flow modes
of the electric current in VAR.

Although the ingot shrinks because of its cooling, a contact zone ingot/mould remains in the top part and a fraction α_2, more or less important, of the electric current is liable to follow this path. Furthermore, because of electrode/mould short-circuits, only a fraction α_1 of the total current reaches the ingot. The values of the parameters α_1 and α_2 are difficult to evaluate, theoretically as well as experimentally : BERTRAM and ZANNER (10) propose values of $\alpha_1 = 0.55$ and $\alpha_2 = 1$, without inferring however to any universalness.

The components of the current density vector $\vec{J} = \vec{\nabla} \times \vec{H}$, Lorenz forces $\vec{F} = \mu_0 \vec{J} \times \vec{H}$ and power consumed by Joule effect $Q = \frac{1}{\sigma} \vec{J}.\vec{J}$ are then drawn from the magnetic field intensity map.

2.3. Navier-Stokes equations

Turbulent motions are represented by momentum transfer equations :

$$\vec{\nabla}(\rho \vec{V} \vec{V}) = -\vec{\nabla} P + \vec{\nabla}.[(\mu + \mu_t) \vec{\nabla} \vec{V}] + \vec{F} \qquad (2)$$

where P is the pressure
$\vec{V}$ the velocity vector
ρ the fluid density
μ its dynamic viscosity
μ_t the turbulent viscosity (ref. § 2.4)

The vector of volumetrical forces $\vec{F}$ is constituted by the sum of gravity and Lorenz forces, $\vec{F} = \mu_0 \vec{J} \times \vec{H} + \rho \vec{g}$, where $\vec{g}$ is the gravity vector. In this expression, ρ is liable to change with the temperature T following a linear relation $\rho = \rho_o (1 - \beta (T - T_o))$ where β is the thermal expansion coefficient of the fluid.

The boundary conditions of the Navier-Stokes equations express the general symmetry, the no-slip condition at liquid/solid interfaces and cancellation of the tangential shear stress on free surfaces. At the slag/metal interface, the model writes the continuity of the radial velocity and momentum flux.

2.4. Turbulence model

In equation (2), μ_t is the turbulent viscosity of the fluid, calculated by the model using the coupled model k-ε, originally developed by LAUNDER and SPALDING (13), following which :

$$\mu_t = 0.09 \ \rho \ \frac{k^2}{\varepsilon} \qquad (3)$$

k, turbulent kinetic energy by mass unit, and ε, dissipation rate of this energy, follow classic convective-diffusive equations. The model constants, which intervene in these equations, take their usual values, proposed by LAUNDER and SPALDING, without adjustment.

Calculation of k and ε near the walls needs use of wall-functions based on the logarithmic law of velocity in a turbulent boundary layer (14).

2.5. Fourier equation

The heat transfer is obtained from the equation :

$$\rho \ C_p \ \vec{V} \ . \ \vec{\nabla} T = \vec{\nabla} . \ [\ (\lambda + \lambda_t) \ \vec{\nabla} \ T \ ] + Q \qquad (4)$$

where C_p is the specific heat of the fluid
λ its thermal conductivity
Q   the power dissipated by Joule effect

The turbulent conductivity λ_t is deduced from the turbulent viscosity μ_t by the relation $C_p \mu_t / \lambda_t = 1$, proposed as a first approximation by DAVIES (15).

The boundary conditions used to solve equation (4) are generally Dirichlet-type conditions (fixed temperature) because of phase changes (liquid/solid metal or liquid/solid slag) which occur on most boundaries. In electroslag remelting, the model writes the continuity of temperature and thermal flux at the slag/metal interface. In the case of VAR, the boundary condition at the top of the ingot is a Dirichlet condition $T(r) = T_\ell + \Delta T(r)$, where T_ℓ is the alloy liquidus temperature and $\Delta T(r)$ an overheat whose value is estimated from measurements available in literature (10,16).

The transfer coefficients in turbulent boundary layers and laminar boundary under-layers are calculated from the Prandtl-Taylor analogy (14).

2.6. Numerical solution

First of all, the model solves the magnetic field transfer equation, uncoupled of other transfers, by a control-volume method. This classic method is based on the use of a mesh, regular or irregular, and on the calculation of transfer conductances between neighbouring nodes. The partial differential equation is considered as a representation of a

1031

differential balance whose integral form on a control-volume around a mesh node is written algebraically.

The Navier-Stokes equations are expressed by secondary variables, vorticity ω and stream function Ψ, following the technique described by GOSMAN et al. (17). The control-volume method is then used to solve a system of 5 coupled equations (ω, ψ, k, ε, T). Convective terms are represented using an upwind difference scheme (17,18). Classic rectangular meshes do not allow a fine treatment of curved boundaries. Therefore, the method retained for taking into account the curvature of the solidification front consists in representing the curved boundary by a series of straight segments whose extremities are occupied by some nodes of the mesh. The control-volumes around these points then have a trapezoidal shape, as shown in figure 3

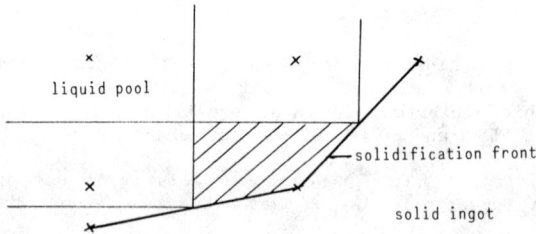

Figure 3 - Example of a mesh near the curved boundary.

The whole of the algebraic equations thus obtained is solved by an iterative method of the Gauss-Seidel type. An important under-relaxation is necessary to prevent divergence of ω, k and ε tranport equations.

The mesh is refined up to the point where further refinements will not induce a change in results. Typically, an (r, z) mesh of 31 x 34 nodes is used to cover the liquid pool and a mesh of 31 x 11 nodes for the slag.

2.7. Necessary data and informations given by the model

Subsequently in this paper, we will present simulations of ESR and VAR operations on a pilot scale, in an ingot mould of a diameter $\emptyset$ = 0.20 m. The thermophysical properties of the fictional liquid alloy considered are assembled in table I.

liquidus temperature	1300 °C
electric conductivity	10^6 $\Omega^{-1}.m^{-1}$
density	5×10^3 kg.m^{-3}
coefficient of thermal expansion	10^{-4} K^{-1}
dynamic viscosity	7.5×10^{-3} kg.m^{-1}.s^{-1}
thermal conductivity	65 W.m^{-1}.K^{-1}
specific heat	400 J.kg^{-1}.K^{-1}

Table I - Thermophysical properties of the liquid metal used for simulations.

The model allows obtention of the following results :

- calculation at any point of the ESR installation of the current density, electromagnetic force and heat dissipation by Joule effect. This calculation is restrained to the ingot in the case of vacuum remelting.

- calculation at any point of the liquids of the velocity and turbulent thermal viscosity and conductivity.

- calculation of the thermal map of fluids and the melting rate of the electrode in ESR.

3. RESULTS AND DISCUSSION - VACUUM ARC REMELTING

Some complementary data must be introduced in the model :

- the current at the generator's clamps is 5×10^3 A and the height of the ingot/mould contact zone h = 0.045 m. These values correspond to remeltings actually realized in a pilot furnace (5).

- the depth and shape of the liquid pool correspond to an experimental marking realized in the same conditions.

- the filling rate of the ingot mould is 5×10^{-4} m.s^{-1}, equal to the one calculated by the model (ref. § 4) in the case of electroslag remelting, thus allowing comparison of the two processes.

- concerning α_1, two assumptions have been simulated (ref. § 2.2). In the first case, only half of the total electric current reaches the ingot ($\alpha_1 = 0.5$). In the second case, there is no electrode/ingot mould short-circuit ($\alpha_1 = 1$).

- the value of the parameter α_2 (ref. § 2.2) is 1. Indeed, we have supposed, lacking further information, that the totality of the electric current reaching the ingot top flows directly into ingot mould by lateral contact.

3.1. Current density and electromagnetic forces

Figure 4 represents the current density vectors in the upper part of the ingot. As imposed by the boundary conditions, the model predicts that the totality of the electric current brought by the arc at the top of the ingot flows towards the zone of lateral ingot/mould contact. Therefore, the current lines strongly diverge and this divergence induces a curl of electromagnetic forces aiming to the creation of a counter-clockwise movement in the liquid pool (ref. § 3.2). According to the equation $\vec{F} = \mu_0 \vec{J} \times \vec{H}$ and since $\vec{H} = \vec{\nabla} \times \vec{J}$, the intensity of the electromagnetic forces is proportional to J^2 and, consequently, to $\alpha_1^2 \, \alpha_2^2 \, I^2$. Therefore the Lorenz forces are 4 times more important in the case $\alpha_1 = 1$ (b) than in the case $\alpha_1 = 0.5$ (a).

3.2. Motion in the liquid pool

The motion in the liquid pool is presented figure 5. The movement as a whole is strongly influenced by the natural convection forces. The liquid metal, cooler near the solidification front, goes downwards in this region and upwards in a zone comparatively hot, along the symmetry axis.

1033

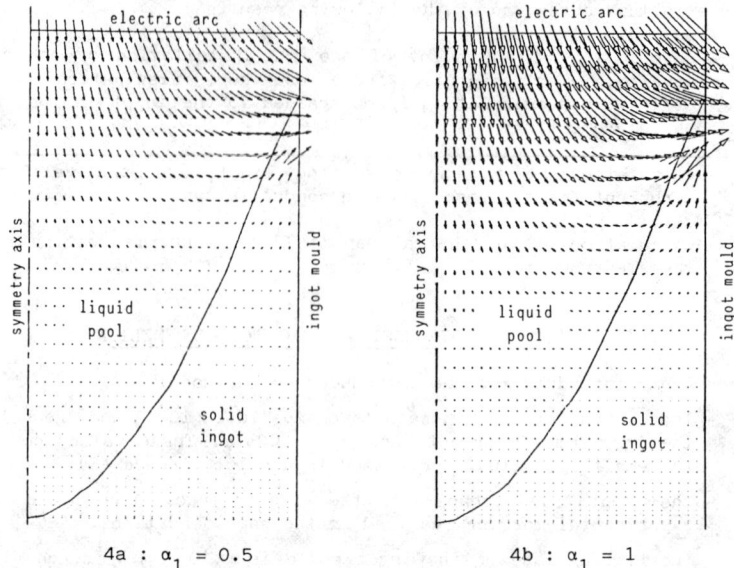

Figure 4 - Computed current density vectors (VAR).
($\downarrow$ 10^5 A.m^{-2})

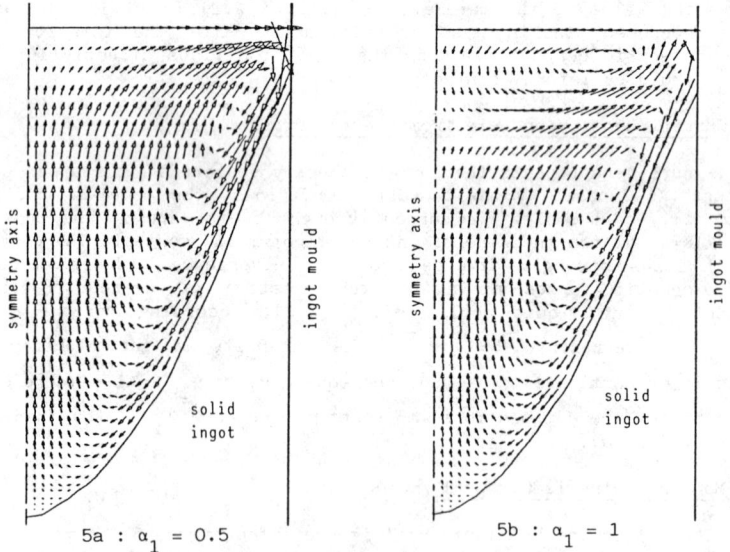

Figure 5 - Computed velocity vectors in the liquid pool of the VAR ingot.
($\downarrow$ 5 x 10^{-3} m.s^{-1})

When α_1 = 0.5, the Lorenz forces created by the divergence of electric current lines offer nearly no opposition to this movement, except at the very top of the ingot, near the symmetry axis, where the velocity cancels, owing to the competition between both volumetrical forces. On the other hand, when α_1 = 1, that is to say when all the electric current effectively flows through the ingot top, the multiplication of the electromagnetic forces by a factor 4 accounts for the creation of a second vortex, in the opposite direction to the recirculation loop due to natural convection.

In both cases, the movement at the liquid/solid interface is downwards. However, when α_1 = 0.5, the maximum velocity reached (0.013 m.s^{-1}) is much higher than the velocity reached when α_1 = 1 (0.008 m.s^{-1}), where the electromagnetic forces slow down the convection movement.

3.3. Turbulence and thermal transfers

Motion in the liquid pool, calculated by the model, is slightly turbulent, the ratio μ_t/μ reaching values comprised between 2 and 10 in the major part of the liquid pool.

Owing to the important thermal conductivity of the liquid metal, thermal transfer by turbulence has only a negligible part, and λ_t/λ is inferior to 0.5 in the bulk. In fact, the model shows that the thermal transfer prevailing is convective transfer. The attempt to take into account this transfer by convection in purely thermal models gives a high "effective thermal conductivity", generally comprised between 2 and 10 times the molecular conductivity. However, an approach which does not consider explicitly motions in the liquid pool will represent quite imperfectly the thermal phenomena in the heart of the pool. This conclusion, consistent with results of CHOUDHARY and SZEKELY (11) for electroslag remelting, is corroborated by examination of the thermal map of the liquid pool, presented figure 6.

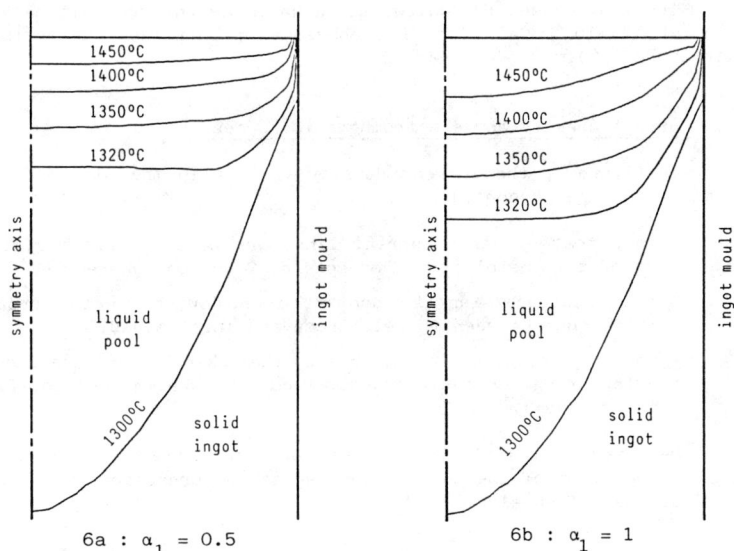

6a : α_1 = 0.5   6b : α_1 = 1

Figure 6 - Computed isotherms in the liquid pool of the VAR ingot.

We see, especially when $\alpha_1 = 0.5$ and when the Lorenz forces do not disturb too much natural convection motion, a phenomenon of horizontal thermal stratification, characteristic of a situation of natural convection where the liquid is cooled by the bottom. An important part of the liquid pool has an overheat lower than 20 degrees, which will have consequences on the solidification process and the final quality of remelted ingots. This low overheat zone is reduced when $\alpha_1 = 1$.

4. RESULTS AND DISCUSSION - ELECTROSLAG REMELTING

In order to compare results obtained in ESR and in VAR, we have supposed that, for a same melting rate, the liquid pools are identical. This artificial assumption, which does not correspond to experimental reality, is discussed further (ref. § 4.2)

For simulation of electroslag remelting, the model needs some data, given table II.

effective alternating current	5×10^3 A
current frequency	50 Hz
height of slag	0.05 m
diameter of the electrode	0.18 m

Table II - Data used by the model for ESR simulation.

The slag chosen for calculations has a weight composition 70 % CaF_2, 15 % Al_2O_3, 15 % CaO. Its thermophysical properties are estimated from literature (19).

4.1. Current density and electromagnetic forces

In figure 7, the current density vectors in the slag and upper part of the ingot are presented.

We see that electric current lines are parallel and vertical in the major part of the installation, except for 2 regions where they diverge :

- the electrode/slag/air contact zone, owing to the cross-section difference between the electrode and ingot mould.
- the top of the ingot, where the skin effect related to the alternating current frequency tends to make the current flow on the lateral wall.

Let us recall that the zone of ingot/mould lateral contact does not allow direct flow of the current because of the presence of an insulating layer of solidified slag.

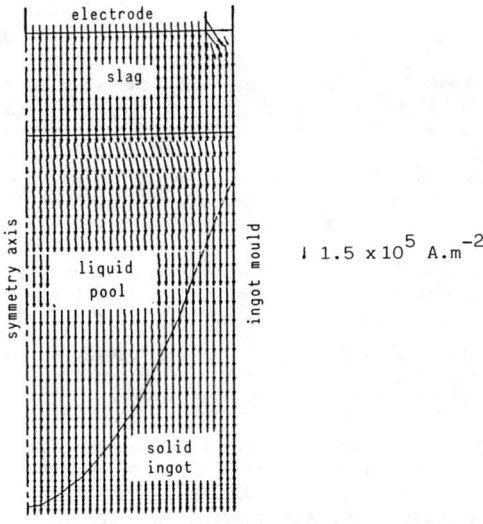

Figure 7 - Computed current density vectors (ESR).

4.2. Turbulent motion and thermal transfers in the liquid pool

We will not consider in this paper the behaviour of the slag, which agrees well with knowledge already acquired (8) : natural convection is responsible for the slag movement, result expected owing to the high value (0.81) of the fill ratio . The liquid pool motion and the isotherms in the pool are the subject of figures 8 and 9.

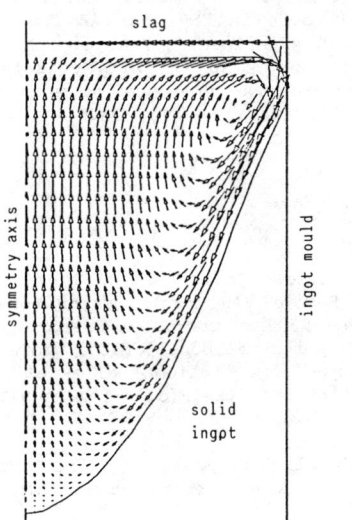

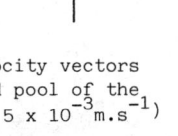

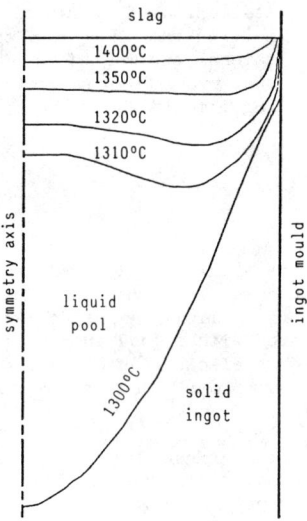

Figure 8 - Computed velocity vectors in the liquid pool of the ESR ingot.($\downarrow$ 5 x 10^{-3}m.s^{-1})

Figure 9 - Computed isotherms in the liquid pool of the ESR ingot.

We see in figure 8 that the movement of the liquid pool is a recirculating movement with one vortex, entirely due to natural convection forces. Indeed, viscous dragging by the slag is extremely low and, at this scale, the electric current lines divergence created by the skin effect is not sufficient to have a visible effect. In these conditions, the behaviour of the liquid metal in electroslag remelting is not without recalling an extrapolation of VAR behaviour with a value $\alpha_1 \alpha_2 I = 0$. The general movement, which is not at all slowed down by the electromagnetic forces, induces a curvature of the isotherms (figure 9).

The results are extremely dependent on the value taken by β, thermal expansion coefficient of the liquid metal. As a matter of fact, opposite conclusions are liable to be drawn (12) from model results if β is lowered by an order of magnitude(10^{-5} K^{-1}), totally unrealistic value for a liquid metal.

For a same melting current, melting rate and shape of the liquid pool, the whole of the above results appear to show that the behaviour of liquid metal in ESR and VAR differ only slightly. As a matter of fact, for a same melting rate, the electromagnetic behaviour and pool depth can be very different depending on the process, partly invalidating a hasty conclusion.

. Concerning the electromagnetic behaviour, we shall note that, depending on the value of $\alpha_1 \alpha_2 I$ in VAR, the velocities along the solidification front can be twice less important than in the case of ESR (figures 5a, 5b and 8).

. The depth and shape of the liquid pool directly influence the hydrodynamic and thermal behaviour of the liquid metal. Now, the assumption of identical liquid pools in ESR and VAR for a given melting rate does not agree with experimental reality, the better lateral cooling of an ESR ingot (owing to the high emissivity of the solidified slag layer and to the presence of air in the ingot/mould space) leading to shallower liquid pools.

5. CONCLUSION

The mathematical model developed allows to understand the behaviour of liquid metal in the liquid pool of ESR or VAR ingots during remelting, using as a data the effective shape of this pool. This shape can be calculated with a good approximation by a thermal model of the whole ingot. Applied to electroslag or vacuum remelting of a fictional alloy at a pilot scale, the model has allowed obtention of the following qualitative results :

- The prevailing movement in the liquid pool is a recirculating movement due to natural convection, going down along the solidification front.

- Depending on the total current and the electric flow mode in VAR, a vortex of an electromagnetic source can superimpose this movement in the upper part of the liquid pool.

- The shape of the isotherms, which can not be predicted by a purely thermal model using the notion of effective thermal conductivity, shows a marked horizontal stratification. A major part of the liquid pool has a very low overheat.

In an other stage of the study, the model will consider the shapes of the liquid pools corresponding effectively to the total electric powers and to the melting rates used during the remeltings. Those simulations will allow quantitative comparison of both ESR and VAR processes, at a pilot scale as well as industrial scale. This study will permit better understanding of the solidification process and optimization of the driving of the processes.

ACKNOWLEDGMENTS

The authors wish to thank the Centre National de la Recherche Scientifique and the Agence Française pour la Maîtrise de l'Energie for partial financial support of this study.

REFERENCES

1. C.L. Jeanfils, J.H. Chen and H.J. Klein, "Temperature Distribution in an Electroslag Remelted Ingot During Transient Conditions" (Paper presented at the 6th International Conference on Vacuum Metallurgy, San Diego, California, April 1979).

2. A.S. Ballantyne and A. Mitchell, " Modelling of Ingot Thermal Fields in Consumable Electrode Remelting Processes", Ironmaking and Steelmaking, 1977, n° 4 : 222-239.

3. A.S. Ballantyne, "Heat Flow in Consumable Electrode Remelted Ingots" (Ph. D. thesis, University of British Columbia, 1978).

4. A. Jardy, "Modélisation des Procédés de Refusion à Electrode Consommable. Application à l'Inconel 718" (Dr-Ing. thesis, Institut National Polytechnique de Lorraine, 1984).

5. L.Falk, A. Jardy, D. Ablitzer and P. Paillere, "Thermal Modelling of Vacuum Arc Remelting of Titanium or Zirconium Alloys" (Paper presented at T.M.S. Extractive and Process Metallurgy Fall Meeting, Palm Springs, California, November-December 1987).

6. J. Kreyenberg and K. Schwerdtfeger, "Stirring Velocities and Temperature Field in the Slag During Electroslag Remelting", Archiv für das Eisenhüttenwesen, 50 (1979) 1-6.

7. M. Choudhary and J. Szekely, "Modelling of Fluid Flow and Heat Transfer in Industrial-Scale ESR System", Ironmaking and Steelmaking, 1981, n° 5 : 225-232.

8. A. Jardy, D. Ablitzer and J.F. Wadier, "Modelling of Movements and Heat Transfer in ESR Slags" (Paper presented at the 8th International Conference on Vacuum Metallurgy, Linz, Austria, September-October 1985), 1152-1173.

9. A.H. Dilawari and J. Szekely, "Heat Transfer and Fluid Flow Phenomena in Electroslag Refining", Metallurgical Transactions, 9B (1978) 77-87.

10. L.A. Bertram and F. Zanner, "Interaction Between Computational Modeling and Experiments for Vacuum Consumable Arc Remelting" (Paper presented at T.M.S. Conference on Modeling of Casting and Welding Processes, Rindge, New Hampshire, August 1980), 333-349.

11. M. Choudhary and J. Szekely, "A Comprehensive Representation of Heat and Fluid Flow Phenomena in ESR Systems", Transactions of the Iron and Steel Society, 1983, n° 3 : 67-75.

12. A. Jardy, D. Ablitzer and J.F. Wadier, " Mathematical Modelling of Coupled Fluid Flow and Heat Transfer Phenomena During Electroslag Remelting of Superalloys" (Paper presented at the European Materials Research Society Conference, Strasbourg, France, June 1986), 285-294.

13. B.E. Launder and D.B. Spalding, "The Numerical Computation of Turbulent Flows", Computer Methods in Applied Mechanics and Engineering, 3 (1974) 269-289.

14. H. Schlichting, Boundary Layer Theory (New York, NY : Mc Graw-Hill, 1960).

15. J.T. Davies, Turbulence Phenomena, (London and New York : Academic Press, 1972).

16. H. Ichihashi et al., "Heat Transfer During Melting of Titanium Alloy in Vacuum Arc Furnace by Consumable Electrode", Tetsu-to-Hagane, 72 (6) (1986), 45-52.

17. A.D. Gosman et al., Heat and Mass Transfer in Recirculating Flows (London and New York : Academic Press, 1969).

18. S.V. Patankar, Numerical Heat Transfer and Fluid Flow (New York, NY : Hemisphere Publishing Corporation, 1980).

19. K.C. Mills and B.J. Keene, "Physicochemical Properties of Molten CaF_2-Based Slags", International Metals Review, 1981, n° 1 : 21-69.

THERMAL MODELLING OF VACUUM ARC REMELTING OF TITANIUM OR ZIRCONIUM ALLOYS

L. Falk*, A. Jardy*, D. Ablitzer* and P. Paillère**

* Laboratoire de Science et Génie des Matériaux Métalliques, U.A. 159
  Ecole des Mines, Parc de Saurupt, 54042 NANCY CEDEX - FRANCE
** Compagnie Européenne du Zirconium CEZUS - Centre de Recherches d'Ugine
  73400 UGINE - FRANCE

ABSTRACT

A thermal model in transient state of the vacuum arc remelting process has been developed at the Ecole des Mines de Nancy, and applied to the case of titanium or zirconium alloys. The model has been validated by comparison of its results with macrographic observations of a Zircalloy 4 experimental ingot and of an unalloyed zirconium industrial ingot. The model worked out supplies informations on the thermal state and local solidification conditions of the ingot, as well, it allows determination of the thermal balance of the process. The influence of some operating parameters, such as melting and cooling rates of the ingot has been investigated.

1. INTRODUCTION

The increasing need for metallic materials of excellent metallurgical quality asks for an optimal working of elaboration processes.

In the VAR (Vacuum Arc Remelting) process where numerous complex physical phenomena occur, mathematical modelling is a performing mean in the realization of this optimization. The principle of the VAR process is recalled schematically in figure 1. A consumable electrode of the required grade is remelted and cast in a mould. The heat necessary for the melting of the electrode is supplied by an electrical arc between it and the liquid pool of the secondary ingot. The remelted alloy cools and solidifies upon contact with the mould walls. In the elaboration cycle of titanium or zirconium alloys, many remeltings are successively done, the first electrode being made of welded and compacted sponges.

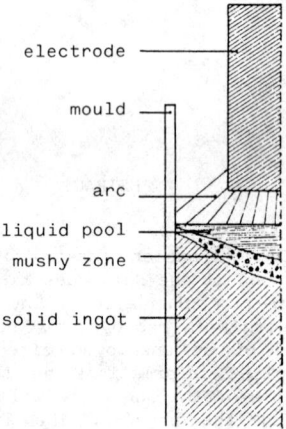

electrode

mould

arc
liquid pool
mushy zone

solid ingot

Figure 1 - Schematic representation of the VAR process.

Works on numerical modelling of the VAR process have concerned mainly the thermal aspect (1-4), except for simulation studies of the liquid metal's movement, in particular by BERTRAM and ZANNER (5) and the present authors (6). Particularly, the model of BALLANTYNE and MITCHELL (2,3), well validated by experience, allows transient-regime calculation of the thermal map and local conditions of cooling and solidification of remelted ingots. The present study is founded on the approach followed by that model.

2. PRESENTATION OF THE MATHEMATICAL MODEL

Using appropriate boundary conditions, this model is based on numerical solution of the heat transfer equation in the ingot.

2.1. Assumptions

. The assumption of an axi-symmetrical geometry reduces the problem to two space dimensions, r and z.

. The ingot growth during melting is simulated by adding a certain amount of liquid metal at the top of the ingot, at time intervals corresponding to the mould's filling rate.

. Because of the important electrical conductivity of the metallic alloy, the heat production by Joule effect in the ingot is neglected.

2.2. Fourier equation

Considering these assumptions, this equation is written :

$$\frac{\partial T}{\partial t} = \alpha_{eff} \left(\frac{\partial^2 T}{\partial r^2} + \frac{1}{r} \frac{\partial T}{\partial r} + \frac{\partial^2 T}{\partial z^2} \right)$$

where T   is the temperature

r, z, t   the space and time coordinates

α_{eff} the effective thermal diffusivity

ρ being the alloy's density, the effective thermal diffusivity is calculated using the relation :

$$\alpha_{eff} = \frac{F \lambda}{\rho C_p^*}$$

The model thus takes into account the turbulent motions in the liquid pool by multiplying the thermal conductivity λ by a factor F, worth one in the solid ingot and the mushy zone. In a first approach, the value of F in the liquid pool is an adjustable parameter of the model. Subsequently, a finer representation of hydrodynamic and thermal phenomena in the liquid pool will allow compensation of this drawback.

Furthermore, the dissipation of latent heat during the solidification and $\beta \rightarrow \alpha$ transition is taken into account by the equivalent specific heat (C_p^*) method, which consists in increasing the specific heat by a fraction of the latent heat during the phase change.

2.3. Initial and boundary conditions

2.3.1. Top of the ingot.   Heat transfer at the top of the ingot during remelting is due to the arc radiation and to the enthalpic contribution of the liquid metal's drops.

Not considering the behaviour of the electrode and the arc, the model simply assumes that the temperature profile at the top of the ingot is not time-dependent :

$$T = T(r)$$

This is in good agreement with optical pyrometry experiments realized in the pilot furnace of the CEZUS company.

The arc cut off, the upper surface cools by radiation on the mould's cold walls considered at a temperature T_{mould}. The boundary condition then expresses the continuity of thermal fluxes :

$$\phi = \frac{(T^4 - T_{mould}^4)}{\frac{1}{\varepsilon} + \frac{1}{\varepsilon_{mould}} - 1}$$

where ε is the alloy emissivity

ε_{mould} the emissivity of the mould's walls

σ the Stefan-Boltzmann constant.

2.3.2. <u>Bottom of the ingot.</u>   Transfer between the ingot and the cooling water is represented by a coefficient h_{bot} :

$$\phi = h_{bot} \ (T- T_w)$$

where T_w is the temperature of the cooling water

2.3.3. <u>Lateral surface.</u>   The phenomenon of ingot shrinkage during cooling induces a spacing between the mould and the ingot. This phenomenon which controls lateral transfer can be considered in a satisfactory manner by introducing the idea of a contraction temperature T*, slightly inferior to the solidus temperature of the alloy.

. For T > T*, $\phi = h_{lat} \ (T - T_w)$

where h_{lat} is the transfer coefficient by contact. Numerical values of h_{lat} and h_{bot} can be estimated from literature (7).

. For T < T*,

$$\phi = \frac{\sigma \ (T^4 - T^4_{mould})}{\dfrac{1}{\varepsilon} + \dfrac{1}{\varepsilon_{mould}} - 1}$$

This condition expresses the fact that the ingot and the crucible radiate upon each other.

2.3.4. <u>Initial condition.</u>   The calculation is initiated by considering a temperature field in an ingot a few centimeters high :

$$T = T_i \ (r,z)$$

2.4. <u>Numerical solution</u>

The heat equation is solved in the time interval between 2 additions of liquid metal at the top of the ingot by a control-volume finite difference method. The mesh being determined on the ingot, the model calculates the transfer conductances between neighbouring nodes. In order to avoid the problems of numerical stability related to explicit solution schemes, the technique retained is the alternating direction implicit (A.D.I.) scheme. It combines an unconditional numerical stability and a quick computer treatment based on the tri-diagonal matrix algorithm (T.D.M.A.) (8-10). The program (approximately 1000 FORTRAN instructions) runs on a mini-computer DPS 6/950 BULL. The size of the meshes used is variable. Typically, the ingot is covered by a grid of 80 x 20 nodes. The time step used during the computation is about 1 second.

2.5. Informations given by the model

2.5.1. <u>Thermal maps, liquid pool and pipe</u>. The model allows obtention of the ingot's thermal map at any moment of the remelting and of the cooling. The liquid pool is delimited by the liquidus isotherm. This does not create any particular problem in the case of the slightly charged titanium or zirconium alloys we are studying, for which the solidification interval does not exceed about ten degrees centigrade. The evolution of the shape and depth of the liquid pool during melting can thus be calculated. Also, the model can predict the position of the residual pipe, created by the density difference between the liquid alloy and the solid ingot, by calculating the position of the last point solidified. In view of validation, these results of the model can be compared with experimental observations(cf. § 3).

2.5.2. <u>Total thermal balances</u>. A special effort has been made concerning computation of total thermal balances, their consistency verifying the model's internal validity.Figure 2 illustrates the energetic balance occuring in the process :

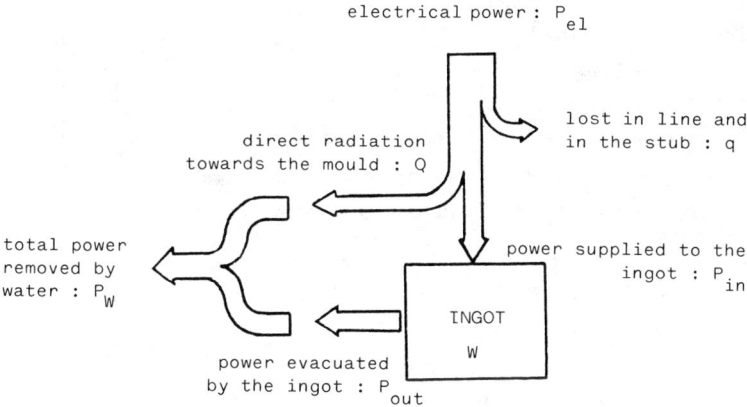

Figure 2 - Diagram of the energetic balance in the VAR process

The model allows access to some unmeasurable quantities :

. the power really provided to the ingot, P_{in}

. the power proper to the ingot's cooling, P_{out}

. the ingot's enthalpy, W.

Furthermore, it is possible to measure experimentally the following quantities :

. the total electric power consumed, P_{el}

. the thermal power removed by the cooling circuit, P_W

The comparison of these quantities then enables an estimation of the powers lost :

. in line and in the stub, q

. by direct radiation from the arc towards the mould's wall, Q

Those quantities are otherwise difficult to evaluate.

2.5.3. Phase changes. Moreover, the model calculates the local solidification time (time taken in one point to fall from the liquidus temperature to the end of solidification temperature) and the local $\beta \rightarrow \alpha$ transition time (time taken in one point to fall from the starting temperature of the transition to its ending temperature).

In short, the informations supplied by the model are :

During the melting :

. Position and shape of the liquid pool at any time

. Thermal balances :

   − Power brought to the ingot

   − Cooling power

   − Ingot enthalpy

. Thermal maps

. Phase changes :

   − Maps of the local solidification time

   − Maps of the local $\beta \rightarrow \alpha$ transition time

After cutting off the arc :

. Position of the pipe and thermal maps of the ingot

. Thermal balances :

   − Cooling power of the ingot

   − Ingot enthalpy

. Phase changes:

   − Maps of the local solidification time

   − Maps of the local $\beta \rightarrow \alpha$ transition time

3. PILOT SCALE VALIDATION OF THE MODEL

An experimental melting was realized in the pilot furnace of the CEZUS company in order to test the accuracy of the model's results.

The instrumentation being more complete than in the case of an industrial furnace, complementary informations are obtained on the physical phenomena occurring during the process.

3.1. Experimental procedure

A validating casting has been done, using a Zircalloy 4 electrode in which a tracer (Fe) has been introduced in some points.

During melting, these metal additions allow, at different moments, to mark the shape and position of the liquid pool. After cutting the remelted ingot, the comparison between the liquid pools thus visualized and the pools calculated, allows model validation. During remelting, the electrical power consumed, as well as the thermal power removed by the mould's cooling water circuit, are recorded. Those measurements are compared with results of thermal balances calculated by the model.

The experimental remelting for the validation was realized in a pilot furnace with the following characteristics :

. electrode diameter	0.16 m
. mould diameter	0.20 m
. total time of melting	0.51 h
. melting rate	258 kg/h
. mean melting current	5000 A
. filling rate of the mould	1.25 m/h

The melting was driven with a very short arc.

For the numerical simulation, we have furthermore used the following additional data :

- . The temperature profile at the top of the ingot is considered uniform with an overheat of 200°C in accordance with experiments realized by ICHIHASHI et al. (11).
- . The coefficient F is considered constant during melting. Its value is 4, order of magnitude currently admitted (3,4).

The ingot's macrography is given in figure 3a. We can see that, all along the melting, the liquid pool's volume is increasing. It shows that the quasi-steady-state regime is not reached. This confirms that only the implement of a transient-regime model will allow a satisfactory approach of reality.

3.2. Results calculated by the model

3.2.1. Melting

* Liquid pool

Figure 3b represents the evolution of the liquid pool's position and shape, calculated by the model, at times corresponding to the additions of the tracer. We can see that the results obtained by the model and experimentally show a very good agreement.

* Thermal balances

We see in figure 4 that the ingot's proper cooling power P_{out}, calculated by the model, increases during time whereas the power brought P_{in} remains constant ; therefore the ingot's enthalpy increases more and more slowly during the remelting and we tend towards a steady state, where $P_{in} \simeq P_{out}$. Also, we note that the power Q lost by direct radiation from the arc towards the mould's walls represents around 10 % of the total power consumed ; this value rather low can be related to the fact that the melting was realized in short arc.

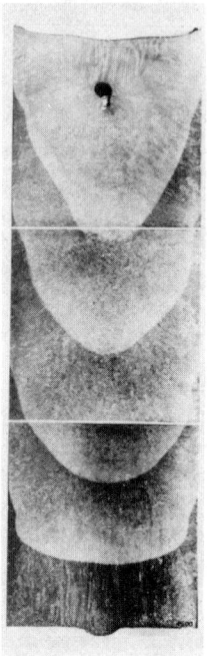 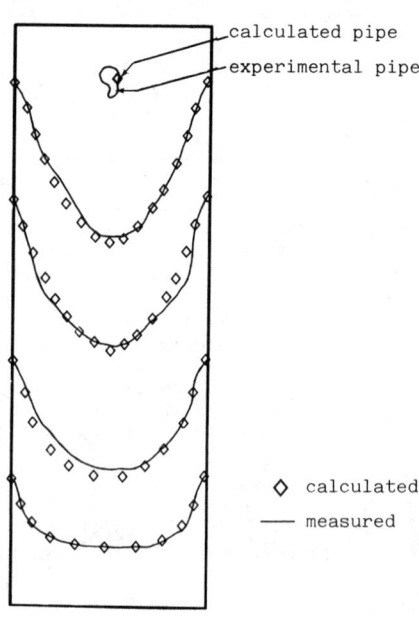

3a. Macrography of the ingot.   3b. Evolution of the liquid pool
    Marking of the liquid pool.   and residual pipe's position.
                                     Comparison between computed
                                     and measured results.

Figure 3 - Experimental remelting of Zircalloy 4 in a pilot furnace.

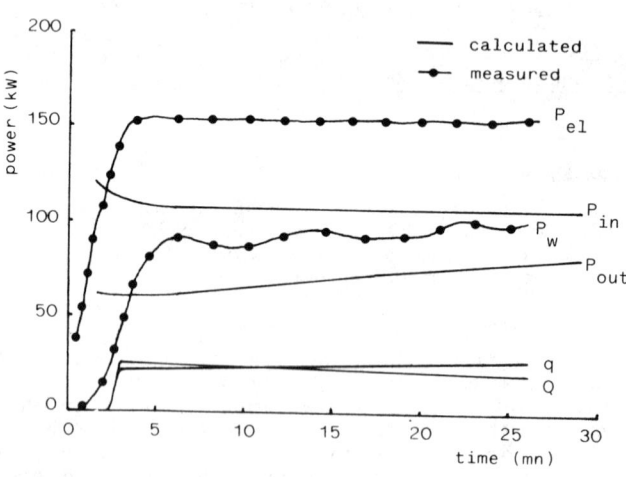

Figure 4 - Energetic exchanges during Zy4 melting in a pilot furnace.

1048

3.2.2. Cooling

* Liquid pool and pipe

After cutting off the arc, the liquid pool solidifies, leaving a pipe of which the volume and depth are characteristic of the pool's size. The model allows localization of this pipe by calculating the position of the last liquidus isotherm before complete solidification of the ingot.

* Thermal balances

After cutting off the arc, we must have $P_W = P_{out}$. Indeed, we see in figure 5 that the experimental cooling power is very near the calculated one.

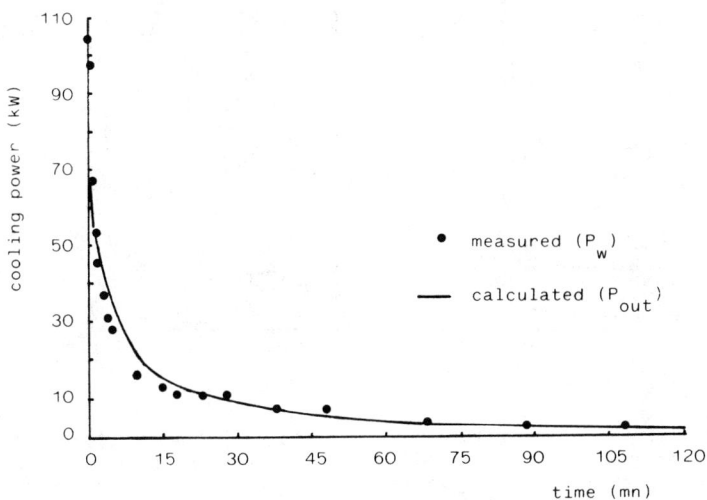

Figure 5 - Cooling power of the ingot after cutting off the arc.

4. VALIDATION AND APPLICATION OF THE MODEL ON AN INDUSTRIAL SCALE

In order to simulate remeltings on an industrial scale, it is important to validate the model for such conditions of remelting. To this end, an industrial casting of an unalloyed zirconium ingot of 0.6 m diameter has been realized.

To allow confrontation of results calculated by the model with reality, additions of a metallic tracer, introduced at different points of the electrode, permit marking of the shape and position of the liquid pool in the ingot. After cutting the ingot, comparison of these marks with liquid pools calculated validates the model. Unfortunately, on the industrial furnace, we do not have the equipment necessary to measure the power P_W and, in the end, confirm the calculation of the energetic balance. The evolution, determined experimentally, of the position and shape of the liquid pool during melting is given figure 6a. The considerable depth and

1049

volume of the liquid pool compared to the size of the ingot enable us to think that liquid motions are important and extremely turbulent. Therefore, the value of the model's coefficient F has been taken equal to 8, and constant during the melting. The results of the model are compared to the experimental observation in figure 6b :

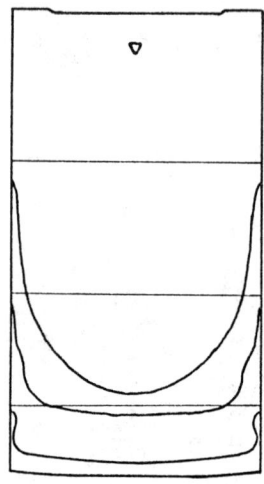
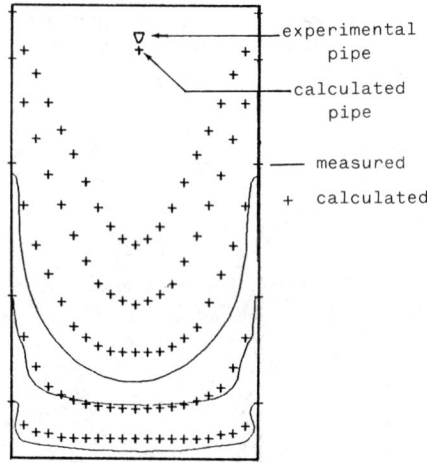

6a. Experimental determination :
    .of the evolution of the shape
    and position of the liquid pool
    .of the residual pipe's position.

6b. Evolution of the liquid pool
    and position of the residual
    pipe. Comparison between mea-
    sured and calculated results.

Figure 6 - Industrial remelting of unalloyed zirconium

We can see that the first liquid pools calculated correspond with experimental pools whereas the concordance for the third pool is less satisfactory. This can be explained by the fact that the model assumes F constant all along melting, an over-simple assumption. Indeed, with the increase of the liquid pool's volume, motions are more reliable to develop themselves and thus to increase the effective conductivity of the liquid.

On the same figure was also represented the calculated position of the liquid pool at the end of melting. The model locates the residual pipe by calculating the position of the last liquid point in the ingot. This position is in good agreement with that of the pipe observed.

5. STUDY OF THE EFFECT OF TWO OPERATING PARAMETERS : MELTING RATE AND COOLING RATE

We have showed that the development of a thermal model of the VAR process in transient regime allows good simulation of real remeltings. This model is therefore quite adequate to understand the effects of operating parameters having a direct influence on the thermal state of the ingot, particularly the melting rate and lateral cooling rate of the ingot during melting.

5.1. Effect of the melting rate

Many remeltings of a titanium alloy type TA6V have been simulated for different melting rates. The characteristics of those remeltings are the following :

- . electrode diameter   0.08 m
- . mould diameter   0.10 m
- . melting rates comprised between 30 and 100 kg/h

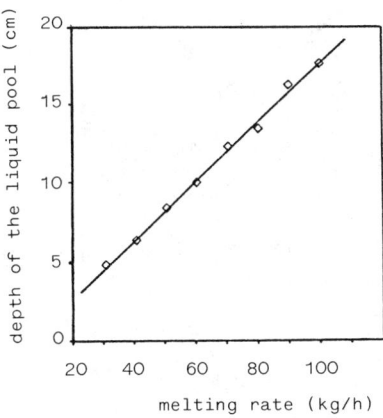
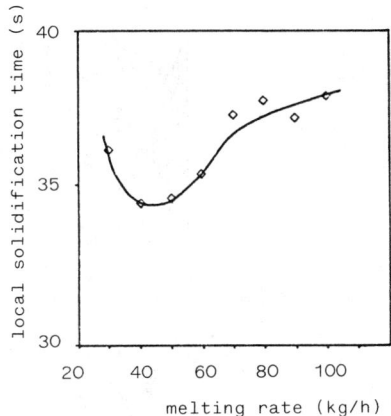

7a. Effect of the melting rate on the depth of the liquid pool.

7b. Effect of the melting rate on the mean local solidification time.

Figure 7 - Remelting of TA6V on a pilot scale.

Figure 7a represents the effect of the melting rate on the depth of the liquid pool : it shall be noted that the pool's depth, in the field of variation studied, increases linearly with the electrode's melting rate.

Also, it has an influence on the local solidification time, as shown in figure 7b. We see that the local solidification time exhibits a minimum for a melting rate of about 45 kg/h. This effect, already displayed by BALLANTYNE and MITCHELL (2) is due to the existence of two contradictory phenomena when we increase the melting rate. To understand this effect, let us imagine that a steady-state regime is almost reached, the advancing rate

of the isotherms, V, is then directly proportional to the melting rate. On
the other hand, an increase of the latter induces an increase of the liquid
pool's depth (figure 7a) which implies, if the temperature profile at the
top of the ingot does not vary, a decrease of the thermal gradient G in the
liquid pool. Because of these two effects, the solidification rate G x V
shows a maximum and the local time of solidification, which is inversely
proportional to it, a minimum.

5.2. Effect of the lateral cooling rate during melting

In vacuum arc remelting, the ingot cooling occurs mainly by the
lateral surface where the existence of a shrinkage zone restrains thermal
exchanges to radiation transfer. The latter quickly becomes a limiting step
in the cooling process of the ingot. To accelerate exchanges with the
mould, it is possible to inject in the course of remelting a gas under
slight pressure in the annular space. Therefore simulations of both
remeltings of TA6V, one under vacuum, the other with argon injection, have
been realized with a melting rate of 60 kg/h. The positions of the
isotherms calculated at the end of melting are given in figures 8a and 8b.
Table I resumes the main results obtained.

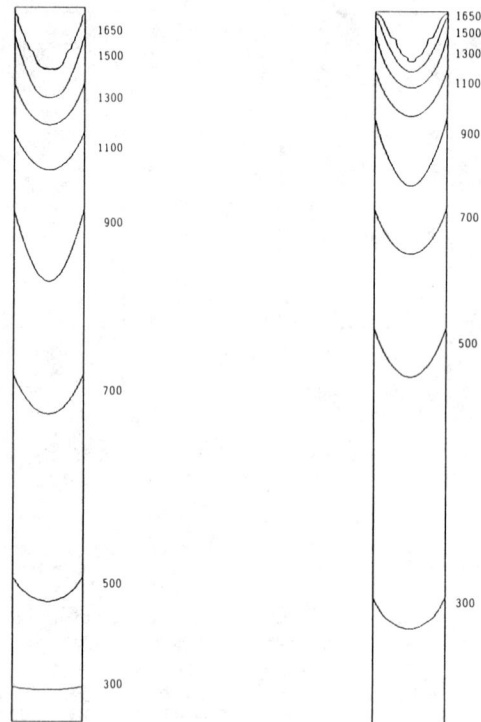

8a. Isotherms calculated at the   8b. Isotherms calculated at the end
    end of vacuum melting.   of melting with argon injection.

Figure 8 - Remelting of TA6V on a pilot scale

	Vacuum remelting	Argon remelting
depth of the pool at the end of melting	10.1 cm	6.6 cm
mean local solidi- fication time	36 s	23 s

Table I : Influence of argon on the depth of liquid pool and the mean local solidification time during remelting of TA6V on a pilot scale

We note that argon addition, thanks to an appreciable improvement of thermal transfers between the ingot and the mould, allows to decrease of about a third the depth of the liquid pool and the mean local solidi- fication time.

These studies of the effect of operating parameters on the thermal state of the ingot constitute two typical examples of the interest that a modelling study can represent in view of the process optimization.

6. CONCLUSION

Based on transient-regime solution of the heat transfer equation, a mathematical model of the Vacuum Arc Remelting process has been developed at the Ecole des Mines de Nancy and applied in the case of remeltings of titanium or zirconium alloys.

The model has been validated by comparison of its results with macrographic observations of a small ingot of Zircalloy 4 remelted in a pilot furnace of the CEZUS company on one hand, and of an industrial ingot of non alloyed zirconium on the other. The results, in terms of shape and depth of the liquid pool in the course of time, as well as of position of the residual pipe, are very satisfactory.

The model allows access to some informations otherwise difficult to evaluate such as local solidification and transition times, as well as the importance of the energetic losses at different stages of the process.

It allows to test easily by numerical simulations the influence of the variation of some operating parameters. As an example, the study of the influence of melting and cooling rates on the thermal behaviour of small ingots of TA6V remelted in a pilot furnace has been done. Thus, predictive use of the model is quite susceptible to help to the optimization of the process.

REFERENCES

1. W.B. Eisen and A. Campagna, "Computer Simulation of Consumable Melted Slabs", Metallurgical Transactions, 1 (1970) 849-856.

2. A.S. Ballantyne and A. Mitchell, "Modelling of Ingot Thermal Fields in Consumable Electrode Remelting Processes", Ironmaking and Steelmaking, 1977, n° 4 : 222-239.

3. A.S. Ballantyne, "Heat Flow in Consumable Electrode Remelted Ingots" (Ph. D. thesis, University of British Columbia, 1978).

4. A. Jardy, "Modélisation des Procédés de Refusion à Electrode Consommable. Application à l'Inconel 718" (Dr-Ing. thesis, Institut National Polytechnique de Lorraine, 1984).

5. L.A. Bertram and F. Zanner, "Interaction Between Computational Modeling and Experiments for Vacuum Consumable Arc Remelting" (Paper presented at T.M.S. Conference on Modeling of Casting and Welding Processes, Rindge, New Hampshire, August 1980), 333-349.

6. A. Jardy and D. Ablitzer, "Behaviour of Liquid Pools in ESR and VAR Processes" (Paper presented at T.M.S. Extractive and Process Metallurgy Fall Meeting, Palm Springs, California, November-December 1987).

7. P.G. Clites and R.A. Beall, "A Study of Heat Transfer to Water-Cooled Copper Crucibles During Vacuum Arc Melting" (Report of Investigations 7035, Bureau of Mines, 1967).

8. B. Carnahan, M.A. Luther, and J.O. Wilkes, Applied Numerical Methods (New York, NY : John Wiley and sons, 1969).

9. D.R. Croft and D.G. Lilley, Heat Transfer Calculations Using Finite Difference Equations (London : Applied Science Publishers, 1977).

10. S.V. Patankar, Numerical Heat Transfer and Fluid Flow (New York, NY : Hemisphere Publishing Corporation, 1980).

11. H. Ichihashi et al., "Heat Transfer During Melting of Titanium Alloy in Vacuum Arc Furnace by Consumable Electrode", Tetsu-to-Hagane, 72 (6)(1986), 45-52.

Prediction of Optimal Operating Parameters

for Continuous Casting of Billets

B. Lally[1], H. Henein[1] and L. Biegler[2]

[1]Department of Metallurgical Engineering and Materials Science
[2]Department of Chemical Engineering
Carnegie Mellon University
Pittsburgh, Pennsylvania

Abstract

Process optimization determines process parameters that maximize or minimize (optimize) some aspect of a process (the objective function), while ensuring that the process operates within established limits. In this work a mathematical model that simulates heat flow and solidification in a continuously cast steel strand is coupled with mathematical optimization techniques to predict optimal process parameters for several aspects of the continuous casting process. The optimizations are constrained so that representative process constraints are enforced. A description of the model, the optimization method, and the means of coupling are presented. The formulation of objective functions and constraints for continuous casting of billets and the predictions resulting from optimizing these formulations is also discussed.

Introduction

The continuous casting process currently accounts for more than 50% of total world crude steel production.[1] Many applications require steel of a quality level only obtainable through continuous casting. The productivity of a continuous casting operation and the quality of the resulting product are largely dependent on the casting parameters used during the casting process.[2] The operating parameters for the continuous casting process need to be chosen so that a predetermined balance between productivity, product quality and operating costs is optimized. The selection of optimal operating parameters becomes even more important as the use of direct charging of hot strands to rolling operations becomes more prevalent. The temperature distribution and total heat content in the strand must be closely controlled in order to roll high quality products.

The problem of selecting continuous casting process parameters that optimize some function of the caster operating state falls within the framework of problems known as constrained optimization problems. We desire to optimize (maximize or minimize) an objective function (a function used to determine if one operating state is more or less desirable than another), while ensuring that constraints that represent physical limits on the process are obeyed. The casting process is represented by a mathematical model, for reasons of cost and convenience and to allow the optimization process to proceed to the optimal point by paths that may include infeasible states (operating states where one or more of the process constraints are violated). In this work, only heat flow aspects of the continuous casting process are considered, although extensions to stress/strain and other aspects can certainly be made within the framework presented here. All of the objective and constraint functions are therefore stated in terms of temperature fields and thermal behavior.

The relationships between the objective function, constraint values and process variables are available only through the use of a numerical heat transfer simulation for a continuous caster. These relationships are nonlinear, hence the optimization problem is a Non-Linear Program (NLP). An optimization technique known as Successive Quadratic Programming[3, 4] (SQP) has been used in this study to solve the constrained NLP problems. This technique was chosen because it typically requires fewer function evaluations than other NLP methods[5] and function evaluations (model simulations) have been found to be quite expensive in terms of both real time and computer time.

The SQP algorithm can be derived from a Newton-Raphson approach applied to the optimality conditions for the nonlinear programming problem. Numerous applications of this method have been made to chemical process optimization problems (see the paper by Biegler[6] for a review) and currently it is the algorithm of choice for solving moderately sized optimization problems based on computationally intensive models. In addition, SQP has excellent constraint handling features and requires only function and gradient information from the process model. Based on the implementation of Biegler and Cuthrell[7] the optimization algorithm is relatively straightforward to apply to general purpose optimization problems with *smooth* objective and constraint functions.

Previous attempts at applying nonlinear constrained optimization techniques to continuous casting processes have been few. In the work by Larrecq, et.al.[2], a detailed list of process operation and product quality constraints is presented. A gradient method is used to minimize a cost function that represents violated constraints, at constant casting speed, and then the casting speed is manipulated manually until a maximum casting speed is found. Holappa, et.al., have used a similar method.[8] Neither group has allowed the casting rate to be a variable in the optimization process even though the casting rate has an extremely important effect on the temperature distribution and metallurgical structure of the cast product.

The optimization system is comprised mainly of two parts, the model and the optimizer. These are shown schematically in Figure 1. The model is further subdivided into a part that calculates the temperature field in a continuous caster and a part that uses the resulting temperature field to calculate

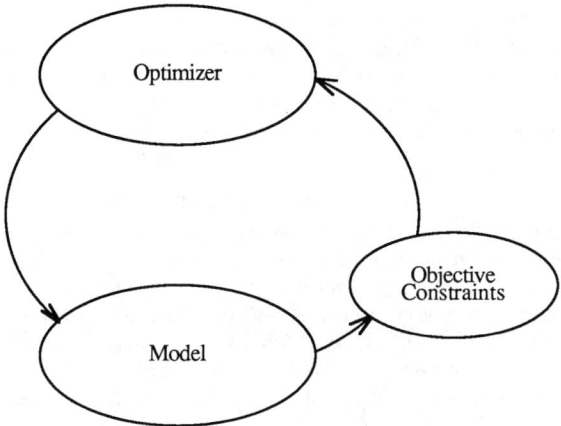

Figure 1: Schematic Representation of Optimization System

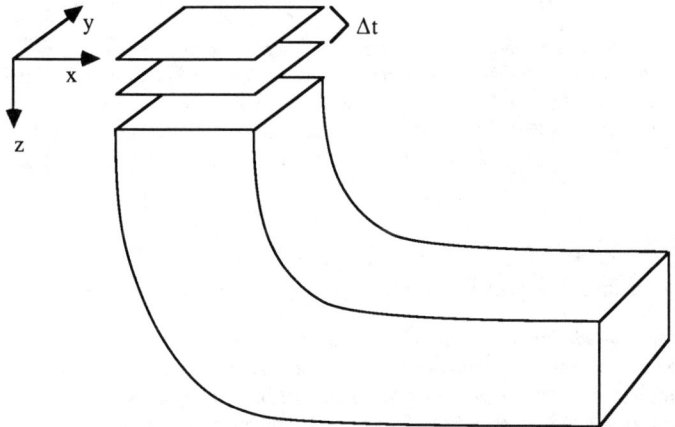

Figure 2: Schematic Representation of Slice Modelling Technique

the values of the objective and constraint functions. The optimizer is "primed" with initial values for all of the process parameters, and uses the model to calculate the objective function and the values of the constraints. New values for the process parameters are calculated by the optimizer on the basis of the objective and constraint values returned by the model, and a new set of objective and constraints are calculated. This is done iteratively, until a set of equations that describe the optimum point are satisfied. This modular approach, where the model is separated from the optimizer, allows us to easily change models to reflect other phenomena that are considered important to the problem at hand.

Description and Verification of Models

A slice technique, similar to that used by Brimacombe[9], Mizikar[10] and Perkins and Irving[11] was used to model the temperature field in the continuously cast strand. Heat flow in a two dimensional, transverse slice moving with the strand was considered. Heat flow by conduction in the direction of strand movement is small compared to the heat flow caused by bulk motion of the strand in this direction, and can be safely ignored.[12] The slice is shown schematically in Figure 2. By calculating the time dependent temperature field in the transverse slice at sufficient positions during the withdrawal of the strand, a three dimensional, steady state temperature field can be calculated for the entire caster.

The two dimensional, transient heat flow equation solved in this work is shown in equation (1).

$$\frac{\partial}{\partial x} k(T) \frac{\partial T}{\partial x} + \frac{\partial}{\partial y} k(T) \frac{\partial T}{\partial y} = \rho(T) C_p(T) \frac{\partial T}{\partial t} \tag{1}$$

The position along the caster z is related to time t through the casting rate r, as shown in equation (2).

$$z = rt \tag{2}$$

Boundary conditions and constraints are normally stated in terms of position, while the heat flow equations are most easily formulated in terms of time. The thermal conductivity k, heat capacity C_p and density ρ are allowed to be unrestricted functions of temperature. Convection in the liquid pool is modelled by using an artificially high value for the thermal conductivity, nominally 5 times normal. The effect of convection in the two phase region is modelled as a quadratic function of the fraction liquid as shown in equation (3).

$$k_{eff} = k_{T_S} + (k_{T_L} - k_{T_S}) f_L^2 \tag{3}$$

The heat of fusion is accounted for by letting C_p be a strongly varying function of temperature.

To facilitate solution of the nonlinear equations resulting from the discretization of equation (1), the Kirchoff transformation was used.[13] This transformation is shown in equation (4). Use of this transformation removes the dependence of the thermal conductivity on temperature from the left side of the equation, where it appeared within a gradient operator, and puts all temperature dependencies in one term on the right side of the equation, outside of any differential operators. The transformed heat flow equation actually solved is shown in equation (5).

$$\theta = \int_{T_0}^{T} \frac{k(T')}{k(T_0)} dT' \tag{4}$$

$$\frac{\partial^2 \theta}{\partial x^2} + \frac{\partial^2 \theta}{\partial y^2} = C(\theta) \frac{\partial \theta}{\partial t} \tag{5}$$

The initial condition used to solve the heat flow equation is given by equation (6). This condition sets the temperature at the beginning of the simulation to the pouring temperature.

$$T(x,y,0) = T_{pour} \tag{6}$$

1058

In this work, symmetry across the centerlines of the cast piece was assumed, hence equation (5) was solved for only one quarter of the cast section. The boundary conditions applied along the centerlines of the strand are given by equations (7).

$$-k\frac{\partial T}{\partial x}=0 \qquad @\,x=0$$
$$-k\frac{\partial T}{\partial y}=0 \qquad @\,y=0 \qquad (7)$$

The boundary conditions in the mold have been modelled using several empirical equations found in the literature, and a proprietary relationship developed by Inland Steel using an instrumented mold. Inland Steel is a member of the Center for Iron and Steel Research at Carnegie Mellon University. Since the examples in this study are based on an Inland Steel caster, the mold boundary conditions used were the Inland Steel conditions. They are of the form shown in equations (8), where the heat flux Q was from an experimentally determined table.

$$-k\frac{\partial T}{\partial x}=Q_x(z) \qquad @\,x=X$$
$$-k\frac{\partial T}{\partial y}=Q_y(z) \qquad @\,y=Y \qquad (8)$$

The spray zone boundary conditions are shown in equations (9), and are in terms of heat transfer coefficients. No attempt has been made in this work to relate heat transfer coefficients to water flow rates, nozzle type or spray chamber design. These correlations have previously been considered by Mizikar[14] and Muller.[15] Outside of the mold and spray chambers, radiant cooling has been assumed, with boundary conditions given by equations (10).

$$-k\frac{\partial T}{\partial x}=h_x(T-T_0) \qquad @\,x=X$$
$$-k\frac{\partial T}{\partial y}=h_y(T-T_0) \qquad @\,y=Y \qquad (9)$$

$$-k\frac{\partial T}{\partial x}=\sigma\varepsilon(T^4-T_0^4) \qquad @\,x=X$$
$$-k\frac{\partial T}{\partial y}=\sigma\varepsilon(T^4-T_0^4) \qquad @\,y=Y \qquad (10)$$

The equations have been solved using an alternate direction implicit finite difference scheme, with iterations at each time step to recalculate and average the rapidly changing thermal properties in the vicinity of the solidus and liquidus temperatures. We have found that iteration coupled with the Kirchoff transformation has allowed us to take very large steps in time and reduced the computer time required for accurate solution of these problems significantly.

The model has been verified against a model belonging to Inland Steel that is known to closely represent one of their billet casters. The result of this comparison is shown in Figures 3 and 4. The model has also been used to calculate the surface temperatures in a cast slab using the data from the paper by Larrecq, et.al.[2] In all cases the agreement has been quite satisfactory.

Optimization Method

The general statement of a constrained optimization problem is given by:

Minimize $f(p)$ $\qquad\qquad$ (11)

subject to
$$h_i(p)=0 \qquad i=1,2,\ldots I$$
$$g_j(p)\geq0 \qquad j=1,2,\ldots J$$

where p is an n dimensional vector of variables that represent the process operating parameters, $h_i(p)$

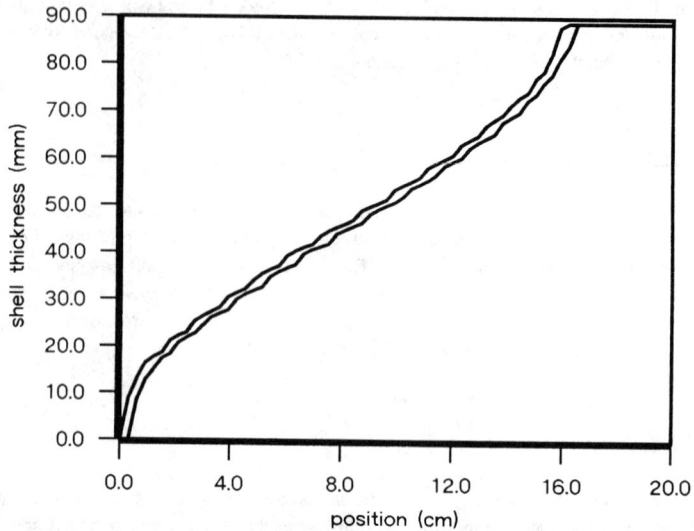

Figure 3: Verification of Model Results - Shell Thickness

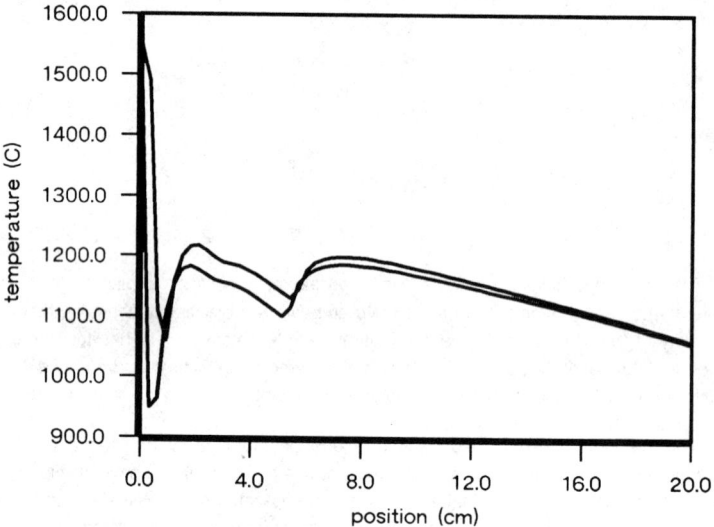

Figure 4: Verification of Model Results - Surface Temperature

are I equality constraints, $g_j(p)$ are J inequality constraints and $f(p)$ is the objective function. The inequality constraints are active if $g_j(p)=0$ at the point p; they are inactive if $g_j(p)>0$. The following conditions are satisfied at the optimum point of a nonlinear, constrained function and are known as the Kuhn-Tucker conditions:

$$\nabla L(p,u,v) = \nabla f(p) - \sum_{j=1}^{J} u_j \nabla g_j(p) - \sum_{i=1}^{I} v_i \nabla h_i(p) = 0 \tag{12}$$

$$
\begin{aligned}
g_j(p) &\geq 0 & j &= 1,2,\ldots J \\
h_i(p) &= 0 & i &= 1,2,\ldots I \\
u_j g_j(p) &= 0 & j &= 1,2,\ldots J \\
u_j &\geq 0 & j &= 1,2,\ldots J
\end{aligned}
$$

L is the Lagrangian function, the coefficients v_i are Lagrange multipliers that are applied to the equality constraints, and the coefficients u_j are similar multipliers that apply to the inequality constraints. u_j is positive if constraint j can be active, otherwise it is zero.

The Kuhn-Tucker conditions for optimality are solved using a Successive Quadratic Programming technique. SQP approximates the Lagrangian function, L, in equation (12) at a trial point with a quadratic polynomial. The constraints at the trial point are linearly approximated. The solution to the approximate problem, with linear constraints, is easily found, using a pivoting strategy, for example. This solution is used as a new trial point for another iteration. The method stops when the Kuhn-Tucker conditions are within a specified tolerance.

Gradients of the objective and constraint functions with respect to the process variables are needed for the solution of the Kuhn-Tucker conditions. These gradients have been obtained by perturbation of the independent variables.

Formulation of Optimization Problems

Four optimization problems have been solved in this work. The problems differ in the objective functions used and the limiting value for one of the constraints. The first problem investigates maximum withdrawal rates. Knowledge of the maximum withdrawal rate is needed to maximize caster throughput. Caster scheduling for sequences of uninterrupted casts requires knowledge of both the maximum and minimum casting rates possible and is discussed by Lally, Biegler and Henein.[16] The second and third problem both determine minimum possible casting rates, using different maximum values for the reheat constraints. The fourth problem addresses maximizing the internal heat content of the cast strand in preparation for direct charging of the strand to a rolling mill. The variables in the optimization formulations have been restricted to the withdrawal rate p_1, and settings for the heat transfer coefficients used in the secondary cooling system p_{2-n}. The objective and constraint functions have been formulated in terms of these variables and the 3 dimensional temperature field, $T(x,y,z)$, predicted from these variables.

Objective Functions

The formulation of the objective function for the maximum casting rate is given in equation (13). The standard form of a NLP is stated so as to minimize the objective function, hence to find a maximum, we minimize the negative of the objective function. No difficulties are introduced by making the objective function a simple linear function of one of the variables.

$$f(p) = -p_1 \tag{13}$$

To find the minimum casting rate, the following objective function is used in the optimization problem:

$$f(p) = p_1 \tag{14}$$

The formulation of the objective for the maximum heat content problem is only slightly more complicated, and is given by the integral in equation (15). This integral is evaluated over the cross sectional area of the strand at the strand cut off point and approximates the enthalpy of the strand at this point.

$$f(p) = -\int_A T(x,y,Z_{cutoff})\, dx\, dy \tag{15}$$

The same types of constraints were used in all the problems. The constraints were chosen to represent both strand quality and the mechanical limitations of the machine. The constraints involved bounds on the casting variables, and limits on shell thickness at the mold exit, metallurgical length, surface temperature and surface reheating. Each type of constraint is formulated separately in the following paragraphs.

Bounds

The simplest types of constraints to enforce are simple upper and lower bounds on the casting variables. Mechanical considerations such as maximum and minimum motor speeds, water availability and water pump capacity give rise to upper and lower bounds on the casting rate, and the heat transfer coefficients for each spray zone. The formulation of these constraints is shown in equation (16).

$$p_i^L \leq p_i \leq p_i^U \qquad\qquad i = 1, 2, \ldots n \tag{16}$$

Shell Thickness

The shell thickness at the end of the mold is required to be greater than some fixed distance d_{min}^{shell}. This requirement is used to prevent breakout conditions to be present in the optimal solutions. This constraint is calculated by first calculating the shell thicknesses at the end of the mold along both transverse centerlines and constraining them to be greater than a fixed value, as in equation (17).

$$d_{shell}^{min} \leq \left(\min X - x \text{ s.t. } T(x,0,Z_{mold}) < T_S \right)$$
$$d_{shell}^{min} \leq \left(\min Y - y \text{ s.t. } T(0,y,Z_{mold}) < T_S \right) \tag{17}$$

Metallurgical Length

In each case discussed, the point of final solidification of the casting is required to be before the unbending point of the curved strand. While this may not be a requirement for all casting operations, it is a good example of the type of positional constraints that can be applied. It is stated in equation (18) in a manner similar to the shell thickness constraint.

$$z_{unbend} \geq \left(\min z \text{ s.t. } T(0,0,z) < T_S \right) \tag{18}$$

Surface Reheating

When the strand passes from a cooling zone with a high heat transfer rate to one with a lesser heat transfer rate, the surface temperature of the strand increases. This is caused by a relaxation of the large thermal gradients created during the high heat transfer period and subsequent accumulation of enthalpy in the surface of the casting. This reheating effect must be limited, as it causes thermally induced stresses that can result in cracking. Several authors have suggested how much reheating can be tolerated.[2, 9]

Originally, the amount of reheat was defined as the greatest difference between the maximum surface temperature after the mold exit, and the minimum surface temperature that occurred prior to the maximum temperature location. This definition of reheat led to a nondifferentiable function that was extremely ill behaved and worked very poorly within the optimization framework. The difficulty occurs when the maximum and/or minimum surface temperatures abruptly change their locations from one cooling zone to another. Switching of locations leads to discontinuities in the reheat constraint gradients that are used to predict the locations of new trial points for the optimization procedure. Here, if the gradient information is valid for only a limited range (because of the discontinuities), the extrapolated predictions will be inaccurate and the optimization algorithm fails.

An alternative formulation of the reheat constraint was developed which removed these gradient discontinuities from the problem. This treatment entailed writing several reheat constraints, in the following manner. The temperature at the end of the mold and each zone end was recorded. Call these temperatures T_i^{min}. The highest temperature found in each zone was also recorded. Call these temperatures T_j^{max}. There are n_z+1 of these temperatures, where n_z is the number of spray cooling zones. The quantities shown in equation (19) were calculated, and all were required to be less than the maximum reheat allowed. By requiring all to be less than the maximum allowed, it is obvious that the greatest one will also be less than the maximum.

$$T_{reheat}^{max} \geq T_j^{max} - T_i^{min} \qquad i=1,2,\dots n_z+1, \; j=i,i+1,\dots n_z+1 \qquad (19)$$

Surface Temperature

The surface temperature is required to be less than a given value at all points of the simulation after the mold exit. This constraint is used to ensure that the solid shell has sufficient strength to contain the molten steel in the center. A separate constraint is used for each cooling zone i in order to make the problem less ill behaved, as is done for the reheat constraints. The constraints are calculated by finding the greatest temperature in each zone outside of the mold and requiring each to be less than a fixed maximum, equation (20).

$$T_{surf}^{max} \geq \max T(0,Y,z) \qquad i=1,2,\dots n_z$$
$$T_{surf}^{max} \geq \max T(X,y,z) \qquad i=1,2,\dots n_z \qquad (20)$$

Unbending Temperature

The surface temperature at the unbending point is also important, as a surface temperature within the ductility trough can cause cracking during straightening. The surface temperature at the unbending point is therefore required to be greater than the temperature that marks the onset of the ductility trough. This constraint is stated in equation (21).

$$T_{unbend}^{min} \geq T(X,0,Z)_{unbend}$$
$$T_{unbend}^{min} \geq T(0,Y,Z)_{unbend} \qquad (21)$$

Problem Solutions and Discussion

The geometry of the caster that was simulated in this work is summarized in Table I. The simulated caster is based on a billet caster in operation at Inland Steel. It casts 17.75cm x 17.75cm billets, using a 61cm mold. There are 4 spray cooling zones, with independently controlled water sprays. The process parameters chosen as optimization variables were the casting rate, p_1, and four heat transfer coefficients, p_2–p_5, that represent the effect of the cooling water sprays on the solidifying strand in each of the four spray zones. The casting rate is specified in units of m/s, and the heat transfer coefficients in units of kJ/m^2/sec/°C. The thermal physical properties of the steel were chosen to approximate a 1010 carbon steel and are shown in Table II.

The first problem that has been solved involved determining the maximum casting rate that the caster could be operated at without violating any of the chosen constraints. The objective and

Table I: Summary of Caster Geometry

section size	17.75 x 17.75 cm
mold length	61 cm
spray zone lengths	
zone 1	9 cm
zone 2	38 cm
zone 3	183 cm
zone 4	244 cm
unbending point	20.0 m

Table II: Summary of Steel Thermal Physical Properties

solidus temperature	1477°C
liquidus temperature	1522°C
heat capacity	0.682 kJ/kg°C
heat of fusion	272 kJ/kg
thermal conductivity	
solid	0.0366 kW/m
liquid	0.2622 kW/m (includes convection)
density	
solid	7400 kg/m^3
liquid	7700 kg/m^3
emissivity	0.6

Table III: Summary of Optimization Results

	Initial	Rate Problems Optimum			Enthalpy Problem Initial	Optimum
problem number ->		1	2	3		4
objective	0.0300m/s	0.0326m/s	0.0252m/s	0.0201m/s	1147°C	1196°C
p_1 m/s	0.0300	0.0326	0.0252	0.0201	0.0300	0.0300
p_2 kJ/m^2/s/°C	0.900	0.903	0.836	0.642	0.900	0.838
p_3 kJ/m^2/s/°C	0.600	0.603	0.504	0.563	0.600	0.414
p_4 kJ/m^2/s/°C	0.400	0.417	0.504	0.454	0.400	0.295
p_5 kJ/m^2/s/°C	0.350	0.325	0.504	0.310	0.350	0.189
iterations	-	4	28	31	-	10

constraint functions used have been developed in previous sections. A concise statement of the maximum rate problem is given in equation (22).

$$
\begin{aligned}
\max \quad & -p_1 \\
\text{s.t.} \quad & 0.01 \le p_1 \le 0.15 \\
& 0.0 \le p_2 \le 2.0 \\
& 0 \le p_3 \le 2.0 \\
& 0 \le p_4 \le 2.0 \\
& 0 \le p_5 \le 2.0 \\
& d^x_{shell} \ge 1cm \\
& d^y_{shell} \ge 1cm \\
& z_{met} \le 20.0m \\
& T^{reheat}_{i,j} \le 175^\circ C \\
& T^{Ux}_i \le 1200^\circ C \\
& T^{Uy}_i \le 1200^\circ C \\
& T^x_{unbend} \ge 900^\circ C \\
& T^y_{unbend} \ge 900^\circ C
\end{aligned}
\tag{22}
$$

The second and third problems involved finding the minimum rate at which the caster could be operated. The formulation of the second problem is the same as the maximum rate problem except equation (14) is used as the objective function. In the third problem, the value of the maximum allowed reheat constraint has been relaxed slightly, from 175°C to 200°C. The fourth problem (the "maximum enthalpy" problem) is somewhat different. Here p_1 was fixed at 0.03 m/s and equation (15) substituted for the objective function. Reheat was limited to 175°C. The starting points and solutions to these problems are summarized in Table III, and are discussed separately in the sections which follow.

Maximum Casting Rate

Initially the casting rate was set to 0.03 m/s. Representative values for the heat transfer coefficients were also chosen. The solution of the maximum casting rate problem predicts that this rate can be raised to 0.0326 m/s without violating any of the casting constraints. The solution yields values of the operating parameters that will result in a rate increase of 8.7%. In this case it was found that the binding, or limiting, constraint was the shell thickness constraint at the mold exit - any further increases in casting rate would result in shell thicknesses that were less than the minimum. The operating parameters p_{2-5} (the heat transfer coefficients) are not involved in calculating the shell thickness at the mold exit, hence they are not uniquely determined. They could be further optimized along the lines of the maximum enthalpy problem, if this were desired.

Minimum Casting Rate

The second and third problems calculated minimum feasible casting rates under two sets of reheat conditions. In the second problem the maximum reheat was limited to 175°C while in the third problem this constraint was relaxed to 200°C. The minimum casting rate problems were started from the same initial point. Again, the results are summarized in Table III. The second problem predicts a minimum casting rate of 0.025 m/s. The binding constraints in this case are reheat constraints. At certain points along the strand surface, the temperature has increased by the maximum amount allowed. This result provides the motivation for relaxing the maximum reheat value in the third problem. Resolving the problem with the relaxed constraint yields a minimum casting rate of 0.020 m/s. The binding constraint in this result is also a reheat constraint, but it occurs in a different cooling zone than the previous result. It is common (and intuitive) for problems with relaxed constraints to move further in the direction of the optimum than problems with tighter constraints.

Maximum Enthalpy

During the solution of the maximum enthalpy problem, the casting rate was held fixed at 0.03 m/s. This was done to demonstrate the effect that the secondary cooling system has on the heat content of the strand. The average temperature in the strand increased from 1147 °C at the starting point, to 1196 °C at the optimal point. This is an increase in average temperature of approximately 50°C. The increase occurs without violating any of the preset casting constraints, and means that strands can easily be produced with greater heat content. This extra heat is heat that will not have to be supplied in a reheating furnace if the strand is scheduled for hot charging to the rolling mill. The binding constraint in this case is the limit on the metallurgical length. If the secondary cooling is reduced further in order to increase the average temperature, then the strand will not be fully solidified at the unbending point.

Performance

The calculations were all performed on typical engineering workstation class computing hardware[*]. This small, affordable, computing hardware was chosen to demonstrate that this type of approach is feasible in a process control/design scenario. The optimization problems required between 4 and 31 SQP iterations to reach the optimal points. The first problem required the least iterations, while the third problem required the most. This translates into CPU time requirements of between 62 and 610 minutes respectively. The number of iterations required is a function of the shape and smoothness of the objective and constraint functions, as well as the tolerance to which the resulting equations must be satisfied. Each iteration required 6 model simulations (5 in the case of the maximum enthalpy problem). The number of simulations for each iteration is a result of calculating the necessary gradient information by perturbations. For each iteration a base point calculation and a perturbation of each variable must be performed. Hence, it is essential that efficient models be used to solve such optimization problems. The continuous casting model described requires approximately 2.5 CPU minutes to execute.

Figures 5 and 6 show the progress of the optimization procedure as a function of SQP iterations for the maximum enthalpy problem. Results from the other problems are similar. In these figures, the constraints have been normalized, and positive constraint values are allowable, while negative values are not. The optimizer initially makes fast progress by taking large steps, and overshoots the constraint limitations. In subsequent iterations the violated constraints are satisfied, and the optimizer fine tunes the solution.

Sensitivity

An additional result of the optimization algorithm is calculation of the shadow prices associated with the constraints. The shadow prices are measures of the sensitivity of the constraints to small changes in the operating parameters. In effect, they are the derivatives of the constraint functions with respect to the process parameters evaluated at the optimal point. They can be used to detect which constraints are most sensitive to changes in the operating parameters, and how sensitive they are. This information is useful in determining how large a safety factor should be used when calculating values for the constraints. For example, if the shell thickness constraint is extremely sensitive to changes in the casting rate, it might be desirable to set the minimum shell thickness rather conservatively to prevent small fluctuations in withdrawal speed from causing a breakout.

---

[*]DEC MicroVAX-II, MicroVMS v4.3

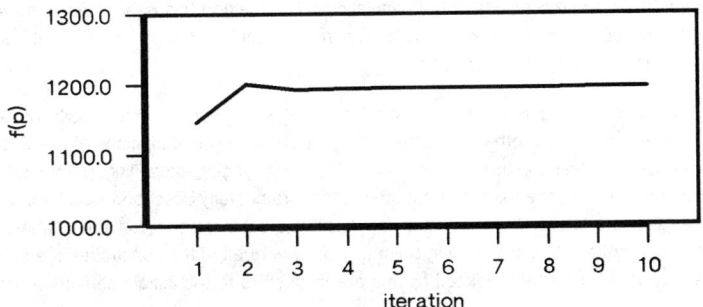

Figure 5: Objective Function Value as a Function of SQP Iterations, Problem 4

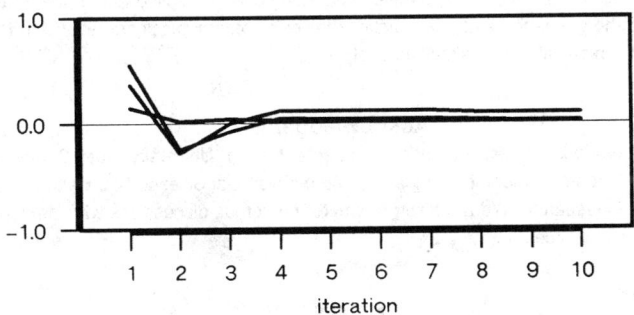

Figure 6: Representative Constraint Function Values as a Function of SQP Iterations, Problem 4

Summary

A method for determining optimal process variables from specified objective functions, subject to constraints on process operation, has been developed. The method uses a mathematical heat flow model for process simulation. Objective and constraint functions are determined from a combination of the process variables and the results of the simulation. The method has been used to solve several example problems concerning the continuous casting of steel billets in an efficient manner on engineering workstation computing hardware.

The developed method is extremely modular and flexible, hence it can easily be applied to variations of these problems, or other problems, simply by changing the definitions of the objective and constraint functions. Other casters can be simulated by changing modelling parameters. If the process to be optimized is dependent on phenomena that are not fully described by a heat flow model, other models (such as stress/strain models) can easily be substituted within this framework. The method does not require a feasible starting point (a point where all of the constraints are satisfied) for the optimization, hence the final predicted optima are insensitive to the initial estimate of the process paramters.

This work is currently being extended to solve problems involving other casters, including slab casters. This will further demonstrate the flexibility and usefulness of the approach. Extensions to other casting operations, with different constraint sets, are also being developed. The method can be used for solving design problems by using an objective that describes the desired design criteria, and adding caster design variables to the optimization problem. It may also be possible to use this type of optimization technique as part of a real time control system for continuous casting and other plant operations. This type of application will require accurate, high speed models to perform the simulations, as well as alternate algorithms to determine the required gradient information. Solution of the real time control problem will likely benefit from the higher performance computing machinery that is constantly coming available. Also, parallel processing can be applied to the perturbation approach for determining the gradients, as the simulations for each parameter perturbation are independent of each other and can be calculated simultaneously.

Acknowledgements

The authors wish to acknowledge Carnegie Mellon University, the Center for Iron and Steelmaking Research, its member companies and the National Science Foundation (grant 84-21112) for support of this research. We are also grateful for numerous discussions with Ismael Saucedo and Ken Blazek, of Inland Steel.

References

1. Continuously Cast Steel, 1977-1986, Iron and Steelmaker. Data originally from the International Iron and Steel Institute

2. M. Larrecq, J. P. Birat, C. Saguez and J. Henry, "Optimization of Casting and Cooling Conditions on Steel Continuous Casters - Implementation of Optimal Strategies on Slab and Bloom Casters", *Application of Mathematical and Physical Models in the Iron and Steel Industry*, AIME, March, 1982, pp. 273.

3. S. P. Han, "A Globally Convergent Method for Nonlinear Programming", *Journal of Optimization Theory and Applications*, July 1977, pp. 297.

4. M. J. D. Powell, "A Fast Algorithm for Nonlinearly Constrained Optimization Problems", *Dundee Conference on Numerical Analysis*, 1977.

5. W. Hock and K. Schittkowski, "Comparative Performance Evaluation for Nonlinear Programming Codes on Hand-Selected and Real Life Test Problems", *Computing*, 1983, pp. 335.

6.   L. T. Biegler, "On the Simultaneous Solution and Optimization of Large-Scale Engineering Systems", *Proceedings of CEF '87*, 1987, pp. 15.

7.   L. T. Biegler and J. E. Cuthrell, "Improved Infeasible Path Optimization of Sequential Modular Simulators - II: The Optimization Algorithm", *Computing and Chemical Engineering*, Vol. 9, No. 3, September 1985, pp. 257.

8.   L. Holappa, E. Laitinen, S. Louhenkilpi and P. Neittaanmaki, "Optimization of the Secondary Cooling in the Continuous Casting of Steel Billets", *Proceedings of the 24th Annual Conference of Metallurgists*, Vancouver, British Columbia, August, 1985.

9.   J. K. Brimacombe, "Design of Continuous Casting Machines Based on a Heat Flow Analysis: State-of-the-Art Review", *Canadian Metallurgical Quarterly*, April 1976, pp. 163.

10.   E. A. Mizikar, "Mathematical Heat Transfer Model for Solidification of Continuously Cast Steel Slabs", *Transactions of the Metallurgical Society of AIME*, November 1967, pp. 1747.

11.   A. Perkins and W. R. Irving, "Two-dimensional Heat Transfer Model for Continuous Casting", *2nd Process Technology Conference*, 1981, pp. 187.

12.   A. W. D. Hills, "Simplified Theoretical Treatment for the Transfer of Heat in Continuous-Casting Machine Molds", *Journal of The Iron and Steel Institute*, January 1965, pp. 18.

13.   H. S. Carslaw and J. C. Jaeger, *Conduction of Heat in Solids*, University Press, Oxford, 1947.

14.   E. A. Mizikar, "Spray Cooling Investigation for Continuous Casting of Billets and Blooms", *Iron and Steel Engineer*, June 1970, pp. 53.

15.   H. Muller and R. Jeschar, "Investigation of the Heat Transfer in a Simulated Secondary Cooling Zone in the Continuous Casting Process", *Arch Eisenhuttenwes.*, 1973, pp. 589.

16.   B. Lally, L. Biegler and H. Henein, "A Model for Sequencing a Continuous Casting Operation to Minimize Costs", *Iron and Steelmaker*, October 1987, accepted for publication.

REHEAT AND DEFORMATION

Session Chairman
Owen Richmond, Alcoa Laboratories

This session reviews several slab and bloom reheating algorithms to optimize furnace control, billet stacking and three dimensional temperature profiles for steel mill operations. A mathematical model of a rotary piercing process is used to streamline process design and to control operation parameters without costly trial and error iteration. A finite element analysis used to simulate hot isostatic pressing of beryllium powder incorporated a constitutive model and sensitivity analysis to modify part design to promote homogeneous mechanical behavior in process and in service.

SIMULATION AND CONTROL OF CAST BLOOM HEATING

IN SOAKING PITS

Yaz F. Bilimoria

Inland Steel Research Laboratories
3001 E. Columbus Drive
East Chicago, IN 46312

The cast blooms produced at Inland Steel Company's new No.3 Combination Caster are reheated in the soaking pits prior to being rolled at the Plant 2 Mills complex. A two-dimensional bloom cooling and heating model capable of simulating the temperature distribution in the bloom from the time it leaves the caster until it reaches the first rollstand after reheating, was developed and verified. The effects of stacking arrangement and heating practice were studied. The best combinations of stacking arrangement and heating practice were identified for a range of production situations. This paper describes the results of the analysis of the experimental data, the transfer of the simulation model to the soaking pit computer system, and the routine use of the on-line model as a production tool.

Introduction

Inland Steel Company's No.2 Basic Oxygen Furnace/Continuous Caster (No. 2 BOF/CC) began producing cast blooms in February 1986. The major proportion of these blooms are shipped to the Plant 2 Shape Products Mills for reheating and rolling. The cast blooms are heated in the soaking pits at the Plant 2 Shape Products Mills and rolled into billets. In response to a need to develop improved control of bloom heating in the soaking pits, a two-dimensional heat transfer bloom cooling and heating model was developed to simulate the thermal behavior of blooms from the time they leave the caster until their delivery to the first rollstand at the No. 2 Blooming Mill. The model was verified experimentally for its ability to accurately predict the temperature distribution in a bloom at any time during its processing path from the caster to the blooming mill. After verification, the model was tested for its ability to predict the ready-to-roll time based on the predicted temperature distribution, pre-specified desired finishing temperatures (generally based on metallurgical considerations) and rolling mill load capacity. The model was then installed in a soaking pit computer system for on-line use as a production tool. This paper presents the theoretical basis for the model, its verification, in-plant proof-of-concept evaluation, and its current status as a routine production tool.

Model Theory and Development

Mathematical Basis of Model

The bloom cooling and heating model is a two-dimensional variable thermal properties heat transfer simulation of the thermal behavior of cast blooms from the time they leave the caster upender in railroad stake cars until they are charged into the soaking pits, heated and delivered to the rollstand at No. 2 Blooming Mill. The model calculates the temperature distribution in any transverse cross-section of a bloom at any given point in time between the upender and the rollstand as a function of initial bloom temperature distribution, bloom cross-sectional dimensions, extent of the thermal insulation of the blooms during their transport from the caster to the soaking pits, the stacking arrangement of the blooms in the pit and the heating practice used to heat the blooms. Much of the discussion of the theory covered in this section is based on work done earlier by Veslocki (1) on a billet caster heat transfer model.

The model was derived by considering the heat transfer that occurs within an imaginary slice of material in a bloom as shown in Figure 1. Heat transfer through this slice can be expressed by a generalized transient heat transfer equation of the form:

$$\rho\, C_p \left\{ \frac{\partial T}{\partial t} \right\} - \left[\frac{\partial}{\partial X} \left\{ K \frac{\partial T}{\partial X} \right\} + \frac{\partial}{\partial Y} \left\{ K \frac{\partial T}{\partial Y} \right\} + \frac{\partial}{\partial Z} \left\{ K \frac{\partial T}{\partial Z} \right\} \right] = 0 \qquad (1)$$

where the terms are defined in Appendix A.

In the cooling and heating of blooms, the heat transfer in the cross-sectional direction (the X-Y plane) is much greater than in the axial direction (the Z direction), and Equation (1) reduces to:

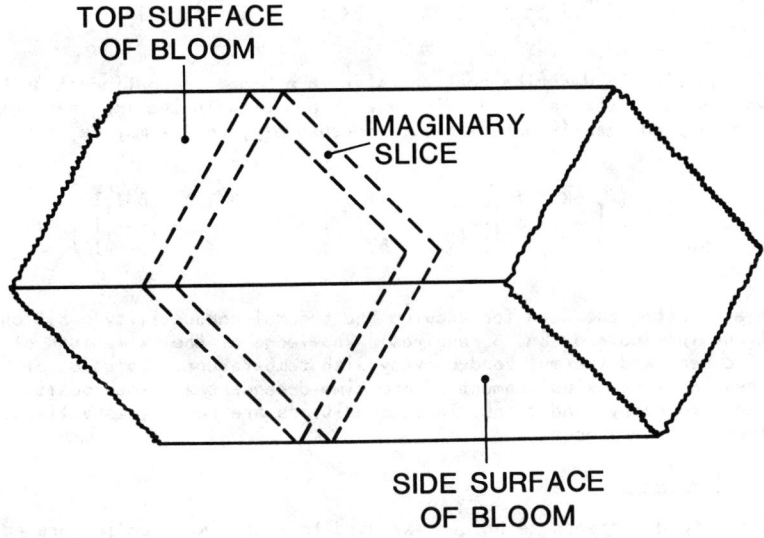

Figure 1 - Imaginary slice of material
in bloom.

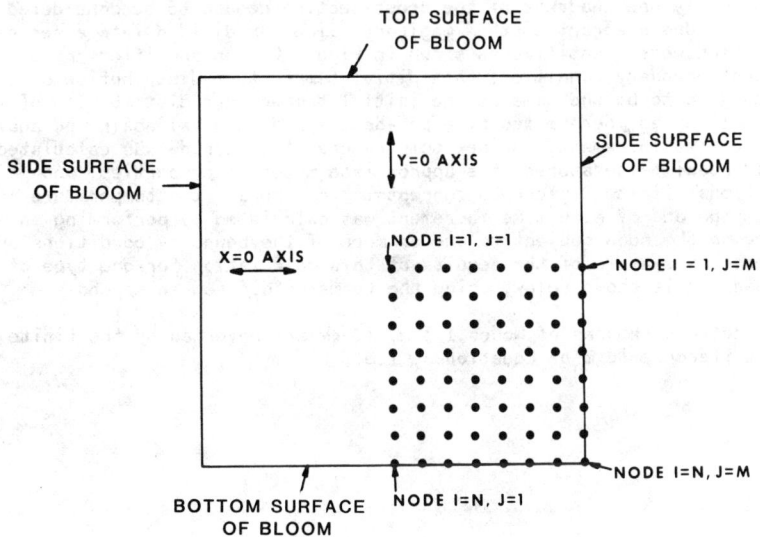

Figure 2 - Coordinate system used to
identify nodes.

$$\rho \, C_p \left\{ \frac{\partial T}{\partial t} \right\} - \left[\frac{\partial}{\partial X} \left\{ K \frac{\partial T}{\partial X} \right\} + \frac{\partial}{\partial Y} \left\{ K \frac{\partial T}{\partial Y} \right\} \right] = 0 \qquad (2)$$

The ability to describe heat transfer in a bloom is contingent on knowing the variation of thermal conductivity with position in the imaginary slice, a relationship not easily established. Fortunately, Eq. 2 may be re-written as:

$$\rho \, C_p \left\{ \frac{\partial T}{\partial t} \right\} - \left\{ \left\{ \frac{\partial K}{\partial T} \right\} \left[\left(\frac{\partial T}{\partial X} \right)^2 + \left(\frac{\partial T}{\partial Y} \right)^2 \right] + K \left[\left(\frac{\partial^2 T}{\partial X^2} \right) + \left(\frac{\partial^2 T}{\partial Y^2} \right) \right] \right\} = 0 \qquad (3)$$

which eliminates the need for knowing the thermal conductivity-position relationship; instead, Eq. 3 requires a knowledge of the variations of density, specific heat and thermal conductivity with temperature. Solution of Eq. 3 also requires the establishment of one time-dependent and four position-dependent boundary conditions. These conditions are readily established from a knowledge of the process.

Numerical Solution

A finite difference technique was used to approximate an integrated solution for the differential equation for conduction (eq. 3 above). Variable material heat capacity and thermal conductivity relationships taken from a paper by Kung (2), and a fixed material density of 470 pounds per cubic foot were used in this numerical solution. By setting up the boundary conditions to stipulate symmetrical heat transfer about the centerlines of the bloom cross-section, only one quadrant of the cross-section needed to be considered in the solution. One quadrant of the imaginary slice was divided into a series of nodes which were identified as shown in Figure 2. As specified by the time-dependent boundary condition, the initial temperature distribution of the nodes was assigned to be the same as the initial temperature distribution of the bloom. Time was incremented by a pre-determined interval again and again, and for each time increment, the new temperature of each node was calculated using the old nodal temperatures, the appropriate material properties, and conventional finite difference concepts. In general, the temperature of each node at the end of each time increment was calculated by performing an energy balance on the node subject to one or more of the boundary conditions specified earlier. An example of the results of this calculation for one type of node in the quadrant is shown below, using the terms as defined in Appendix A:

The temperatures of Nodes $1<I<N$, $1<J<M$ are governed by the finite difference form of Equation 3, i.e.,

$$T'(I,J) = \left[1 - \left(\frac{576}{3600} \right) \frac{(\Delta t)K}{\rho C_p (\Delta N)^2} \right] T(I,J) +$$

$$\left(\frac{144}{3600} \right) \left(\frac{\Delta t}{\rho C_p (\Delta N)^2} \right) \left\{ \frac{b}{4} \left[[T(I+1,J) + T(I-1,J)]^2 \right. \right.$$

$$\left. + [T(I,J+1) - T(I,J-1)]^2 \right] + K [T(I+1,J) + T(I-1,J)$$

$$+ T(I,J+1) + T(I,J-1)] \right\} \tag{4}$$

Similar finite difference equations may be written for the other types of nodes in the quadrant.

The heat flux terms, q_x and q_y, are the surface heat fluxes across the outer surfaces of those nodes that lie on the bloom surface. Since heat transfer during the cooling and heating of blooms is of interest primarily for those conditions wherein the surface temperatures are high (usually greater than 1200°F), radiation is the predominant mode of heat transfer. q_x may therefore be written as:

$$q_x = p \left\{ \sigma \epsilon (T_s^4 - T_a^4) \right\} \tag{5}$$

during cooling, and

$$q_x = F \left\{ \sigma \epsilon (T_s^4 - T_a^4) \right\} \tag{6}$$

during heating,

where the terms are defined in Appendix A. q_y can be written in a similar fashion. The attenuating factor, p, used only during cooling, scales down the heat flux from the surface. This factor is incorporated to accomodate the insulating effect of transporting the hot blooms from the caster to the soaking pits in insulated stake cars. The view factor, F, used only during heating, accomodates the shading effect of other blooms on the bloom cross-section under consideration. The emissivity of a cast bloom was set at 0.80, the average emissivity for oxidized iron (1).

The equations describing the heat transfer by conduction and radiation, the thermal properties and the boundary conditions were coded in FORTRAN 77 language, initially on a Digital Equipment DEC-10 computer system. The coded version of the model was ready for testing and verification by mid-1984.

Model Verification

After completion of the initial debugging and testing, the model was ready for verification under mill operating conditions. The verification procedure used for bloom cooling was different from that used for bloom heating. The main reason for this difference was as follows.

During cooling on the track, the blooms are loaded in stake cars in a fairly standard configuration; during heating in the pits, however, the blooms may be stacked in several different configurations in two fundamentally different types of pits (i.e., either top-fired or bottom-fired). The cooling process was therefore judged to be quite predictable. Literature data available on the cooling of ingots was used to verify the accuracy of the model during cooling. Ingot cooling data was used since no reliable data was found for bloom cooling. The assumption of predictability was not felt to be valid for the heating operation. Many variations in the heating process are possible, and all such variations were accomodated using a single tunable parameter. This parameter was the **view factor**[1] . The view factor, F_{ij}, is defined as the fraction of the radiation which leaves surface i in all directions and is intercepted by surface j (3).

Cooling Verification

The ability of the model to simulate the cooling of rectangular cross-sectional pieces of steel was verified using published experimental data from the literature. An example of the results of a comparison of the model-predicted cooling behavior of ingots with their measured behavior (4) is shown in Figure 3.

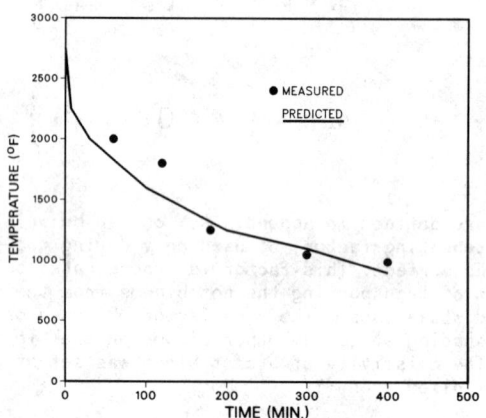

Figure 3 - Comparison of the predicted steel surface temperature with the corresponding measured temperature during cooling for a 48 X 48" ingot (4).

----

1. Terminology specific to the subject of this paper is highlighted in boldface.

On the basis of surface temperature measurements of blooms arriving at the soaking pits, the attenuation factor, p, was set at 0.80 for the transport of hot blooms using stake cars.

As noted earlier, one of the boundary conditions requires the specification of the bloom temperature at time zero. For cold blooms (defined as those cast more than 24 hours before being charged into a soaking pit), the initial temperature of the blooms is set equal to the ambient temperature. For hot blooms, the initial temperature distribution used in the simulation is the temperature distribution at the caster upender (provided by the caster computer).

<u>Heating Verification</u>

The temperature distribution in the blooms after being charged into a hot pit is determined using the model in the heating mode. Verification of this portion was done by running a set of several experiments that involved the heating of blooms instrumented with thermocouples under a range of simulated operating conditions. The temperatures predicted by the model were matched as closely as possible with the measured temperatures recorded during each experiment by adjusting the tunable parameter, namely, the view factor. Various combinations of the following operating conditions were incorporated into the set of experiments:

1. Type of pit

    i. Top-fired

    ii. Bottom-fired

2. Initial thermal condition of the blooms

    i. Cold (ambient temperature)

    ii. Hot (typical temperature distribution expected in hot blooms arriving from the caster)

3. Bloom stacking arrangement in the pit

    i. Scissors arrangement (shown in Figure 4)

    ii. Lincoln-log arrangement (shown in Figure 5)

    iii. Leaning arrangement (shown in Figure 6)

The basic experimental procedure for the verification of the model in the heating mode consisted of:

    i. drilling holes to various depths in several of the blooms to be heated in a soaking pit, loading all the blooms in a cold pit using a specific stacking arrangement,

    ii. installing the thermocouples in the blooms with the leadwires exiting the pit through one of the pit walls,

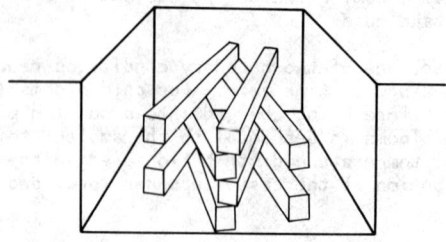

Figure 4 - Scissors stacking arrangement for 8 blooms in a top-fired pit.

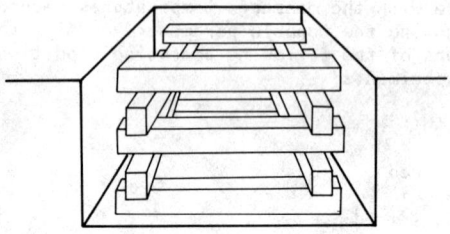

Figure 5 - Lincoln-log stacking arrangement for 10 blooms in a bottom-fired pit.

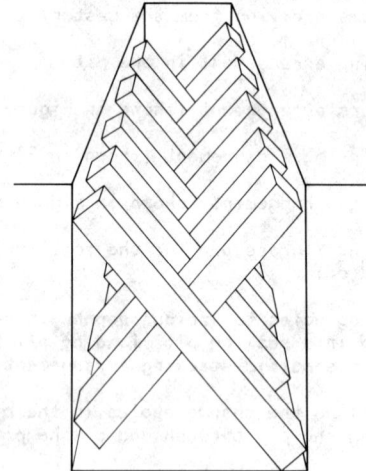

Figure 6 - Leaning stacking arrangement for 10 blooms in a top-fired pit.

   iii.   heating the pit and the blooms according to a practice
              typically employed to heat the particular steel grade used in
              the experiment, and

   iv.   monitoring and recording the bloom temperatures, the pit wall
              temperatures and other pit process variables.

The bloom temperatures and pit process variables data was collected on
magnetic tape cassettes for subsequent analysis on the Research computer. The
best match between the model-predicted temperatures and the corresponding
measured temperatures was established using the view factor in the model as the
tunable parameter. Since the measured temperatures in the pit at a given time
vary significantly from one location to another, different view factors are
obtained for each thermocouple location. Typically, bloom locations closest
to, and in direct view of, the flame are the hottest (represented by high view
factors, often close to 1.0); bloom locations heavily shaded by other blooms
are the coldest. (represented by low view factors, around 0.1). For blooms in
a scissors arrangement (shown in Figure 4), the highest view factors are
obtained in the top-most layer at the bloom surfaces closest to the flame; in
this configuration, the lowest view factors are recorded in the bottom layer.

The bloom cross-section experiencing the lowest temperatures is of special
significance since its thermal condition determines whether the entire pit is
ready to be drawn and rolled. The procedure developed for determining the
ready-to-roll time for a charge is discussed in the next section.

The tunable parameters for several key bloom section locations were
determined using the heating verification procedure outlined above. Figure 7
shows an example of the match obtained between the model-predicted and the
measured temperatures for a key bloom cross-section location in the scissors
stacking arrangement (Figure 4).

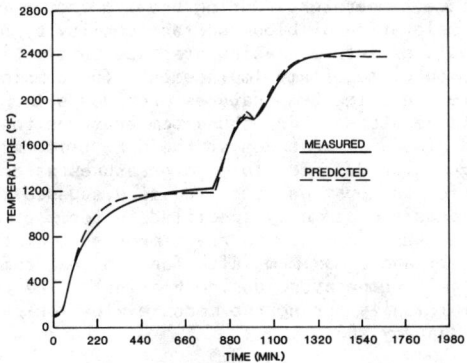

Figure 7 - Example of a comparison of
the model-predicted temperature with
the corresponding measured temperature
during heating (Scissors stacking.)

Once the tunable parameter is known for each combination of soaking pit type, bloom stacking arrangement and bloom location in the pit, the model can be used to calculate the steel temperature distribution at the cross-section under consideration. An example of the calculated temperature distributions for an arbitrary combination of process variables is shown in Table I.

Calculation of the Ready-to-Roll Time

The view factors determined during the verification process allow accurate calculation of bloom temperatures during heating for virtually any location in the pit. By linking the bloom temperatures at draw time (when the blooms are discharged from the pit for rolling) with the heat transfer processes occurring during the delivery of the blooms to the rollstand and during the subsequent rolling operation, it is theoretically possible to calculate the temperatures in the rolled section at the end of the rolling operation (commonly called the **finishing temperatures** which, strictly speaking, are the surface temperatures of the rolled product). Although calculation of the bloom temperatures during transport to the rollstand is easily done by using the model in the cooling mode, calculation of the steel temperatures during rolling is much more complex and was judged to be beyond the scope of the current project. Instead, an approach consisting of correlating the calculated bloom temperatures at the rollstand with the measured finishing temperatures, was adopted. The desired finishing temperatures, determined from metallurgical and mill load capacity considerations, served as the starting point in this approach.

Each of the above requirements usually translates into a minimum permissible finishing temperature for a particular steel grade. The need for energy-efficient heating combined with the need for high productivity translates into a requirement that the residence time in the pit be as short as possible. This requirement usually implies a maximum permissible finishing temperature. Another consideration that places additional constraints on the desired bloom surface temperatures during heating is the importance of preventing the deterioration of bloom surface quality by over-heating. This consideration translates into a heating practice that minimizes the prolonged exposure of blooms to direct flame impingement. The combination of minimum and maximum permissible finishing temperatures provides a **target window** for the actual finishing temperatures. For these temperatures to fall within the desired range, the bloom temperatures at the draw time must fall within a corresponding range. The desired bloom temperatures at draw are termed the **rollability criteria**, and must be specified as discussed above for each steel grade. These criteria are commonly specified in terms of a minimum value for the **cold-spot temperature** (defined as the temperature of the coldest region of any bloom in the pit) and a maximum value for the **soak temperature** (the steady-state maximum pit wall temperature during heating). The rollability criteria were determined empirically during the model evaluation period which ran from July 1985 through January 1986.

The **ready-to-roll** time is defined as the residence time of the blooms in the pit required to allow the calculated cold-spot temperature to meet the rollability criteria using a specific heating practice. The procedure for determining the ready-to-roll time consists of running the heating model for the proper combination of process parameters until the cold-spot temperatures satisfy the rollability criteria.

In-plant Proof-of-Concept

Extensive model evaluation experiments were conducted prior to the start-up of the caster. The evaluation experiments were designed to demonstrate the ability of the model to accurately predict bloom temperatures and ready-to-roll times under realistic plant conditions.

The evaluation experiments served two additional needs:

1. Provide a training ground for the operators to learn how to charge, heat, draw and roll blooms.

2. Identify equipment and operating problems unique to processing blooms through the soaking pits.

The evaluation procedure consisted of the following steps:

1. Conventionally poured ingots of various steel grades were rolled to one of the two bloom sizes expected to be produced at the bloom caster (cross-sectional dimensions of either 15" X 20" or 15" X 24"). These blooms were termed **breakdown blooms** and were used in all the evaluation runs made before caster start-up.

2. The breakdown blooms were charged into the soaking pits using one of the three stacking arrangements identified earlier.

3. The blooms were heated and rolled[2] according to pre-specified practices designed to simulate practices to be used to process cast blooms.

4. Pit operating variables during heating were recorded on magnetic tape cassettes for subsequent analysis; rolling mill data such as under-voltage trips, finishing temperatures, etc. were recorded manually.

5. Surface defect ratings on individual billets rolled from the blooms during subsequent processing were recorded.

6. The soaking pit data obtained from each run was used to prepare a computerized mass and energy balance for the duration of the charge. The purpose of developing the balance was to identify the sources and the uses of energy supplied to the pit during the heating operation.

In the course of the evaluation experiments, various combinations of pit process variables, identified in the Heating Verification section, were used to process a total of 30 breakdown bloom charges.

The evaluation experiments were very valuable in proving the feasibility of using the model to predict the consistent and efficient heating of blooms. As the evaluation process progressed, it became obvious that, despite the difficulties involved in running controlled experiments in a mill environment,

----

2. Bloom handling problems, problems with crane tong bits, pit and mill operating problems frequently prevented the operators from accurately following instructions provided to them.

major benefits were attainable with correct use of the model. Correct model use implies the meeting of two major requirements. First, the **heating practice** specified at the start of the experiment in terms of the stacking arrangement, the initial pit wall temperature, ramp rate and soak temperature must be followed as closely as possible during the heating process. Second, the blooms must be drawn and rolled within a few minutes of the model-predicted ready-to-roll time. The mass and energy balances prepared for each charge conclusively demonstrated that large fuel penalties are incurred by attempting to heat the charge faster than a certain rate (depending on pit condition and charge characteristics) and by holding the charge in the pit past the model-predicted ready-to-roll time.

At the conclusion of the model evaluation program, a heater training course was organized and conducted. The major goals of the training course were to educate the heaters on the rudiments of the heating model and to familiarize them with the procedures for its use.

On-line Installation of the Model

The on-line heating model became operational at the soaking pits DEC 11/73 computer system in September 1986. A brief description of the three major CRT screens used to interrogate the model is provided below:

1. TEMP: This screen provides the heater with the temperature distribution within the bloom using the most recent execution of the model for any specified pit.

2. PRAC: This screen displays either the downloaded practice (currently active practice) or the pending practice (a feature that allows a heater to design a practice off-line for future use) for a specified pit.

3. MODL: This screen allows the heater to determine the ready-to-roll time for a downloaded or a pending practice. It also allows the heater to design a practice that will result in a projected ready-to-roll time specified by the heater.

Operator inputs needed to run the model are provided to the computer system via CRT terminals located in the Pit Office and in the Computer Console Room. The routine procedure established to provide these inputs consists of the pit recording clerk entering charge information into the computer, thereby allowing the heater to select a **controlling piece**[3] , which in turn, initiates the execution of the model. Once the execution of the model has been completed, the heater may use the TEMP screen to review bloom temperatures as updated by the model, usually about every 20 minutes. In addition, the heater has the capability of using the MODL screen to review the most recent projection of the ready-to-roll time, or the PRAC screen to review or change the heating practice. The procedure also triggers an automatic re-run of the model to

----

3. The controlling piece is the specific bloom in the charge whose characteristics (such as size, steel grade, etc.) are used as inputs to the on-line model. The controlling piece is the bloom in the charge that is expected to require the longest ready-to-roll time.

generate an updated ready-to-roll time shortly after the pit wall temperature reaches the soak temperature.

Routine Use of the On-line Model

The ultimate objective of the bloom heating project is to provide the heaters with a tool which can be used routinely and easily to accurately predict the steel temperatures and the ready-to-roll times for bloom charges heated in soaking pits. To this end, the extent to which the model is used is an important measure of the success of the project. To provide a database for determining the extent of model usage, soaking pit charge histories have been transmitted back to the Research computer via magnetic tape since September 1986. Several parameters related to the extent of model usage are excerpted from these charge histories and plotted for analysis. One parameter, termed the **usage parameter**, providing an aggregate measure of the extent to which the model is used, is plotted in Figure 8. The plot shows that, although the initial level of model usage was only between 40 and 50%, the level in recent weeks has been up between 70 and 80%. It should be noted that the parameter plotted in Figure 8 represents the frequency with which the model run was deliberately initiated by the heater. For each run initiated by the heater, a model-predicted ready-to-roll time was calculated and displayed on the MODL screen.

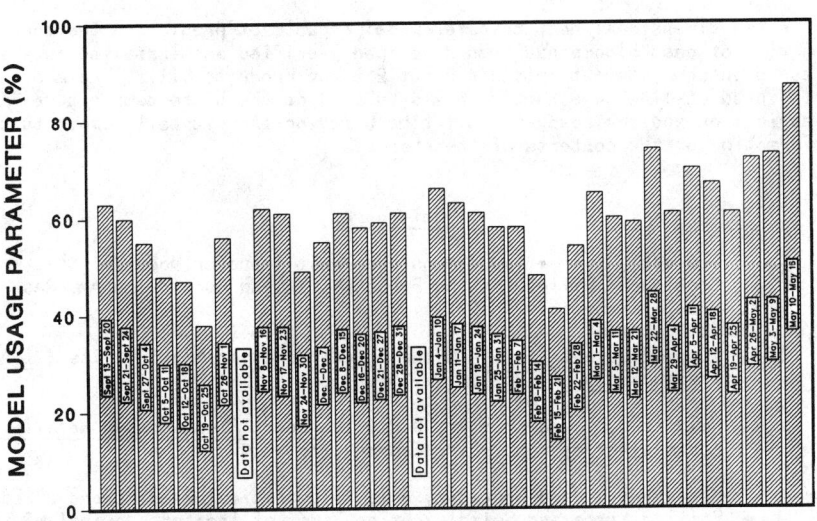

Figure 8 - Chronological plot of the on-line model usage parameter from September 1986 through May 1987.

The usage parameter plotted in Figure 8 provides no indication of whether the charge was actually drawn and rolled at, or close to, the model-predicted ready-to-roll time. The actual draw time is greatly affected, not only by when the charge is ready to be rolled, but also by many other factors such as charging delays (when the cranes are being used to charge or draw other pits), mill delays (scheduled and unscheduled interruptions in the availability of the rolling mill) and drawing delays (breakdowns in the pit cranes or transfer buggy), etc., which are usually beyond the control of the heater. If one of these delays occurs, the heater has no choice but to leave the charge in the pit past the projected ready-to-roll time.

Future Activities

The ultimate objective of any future activity in this area will be directed towards raising the usage parameter closer to 100%. In addition, an environment should be created to encourage the operators to draw the pit and roll the blooms within a short time of the model-predicted ready-to-roll time. The key to drawing the pits punctually lies in the improvement programs that are currently being put in place at the Plant 2 Shape Products Mills, namely, the standardization of practices, the preventive maintenance program and the improvements in scheduling both within the facility and in concert with other upstream and downstream facilities.

Summary

A two-dimensional heat transfer model capable of predicting the thermal behavior of cast blooms has been developed, verified and installed in a soaking pits computer system at Inland's Plant 2 Shape Products Mills. The model is being used on-line on a routine basis to predict the bloom temperature distribution and the ready-to-roll time based on the pit wall temperature and information on the contents of the charge.

References

1. Veslocki, T.A., "Two-dimensional Heat Transfer Model of the Continuous Billet Casting Process", Inland Steel Company, Research Report, February 22, 1979.

2. Kung, E.Y., et al., "A Mathematical Model of Soaking Pits," ISA Transactions, 6(1967), 162-166.

3. Geiger, G.H. and Poirier, D.R., Transport Phenomena in Metallurgy, Addison-Wesley Publishing Company,Inc., 1973.

4. Massey, I.D. and Sheridan, A.T., "Theoretical Predictions of Earliest Rolling Times and Solidification Times of Ingots", Journal of the Iron and Steel Institute, May, 1971, 391-327.

APPENDIX A

Nomenclature

b	=	rate of change of thermal conductivity with temperature $(Btu/hr\text{-}ft\text{-}°R^2)$, i.e. $(\partial K/\partial T)$
C_p	=	heat capacity of steel $(Btu/lbm\text{-}°R)$
ϵ	=	emissivity
F	=	view factor used to adjust rate of heat transfer during bloom heating in the soaking pit
I	=	node number designation in the X-direction
J	=	node number designation in the Y-direction
K	=	thermal conductivity of steel $(Btu/hr\text{-}ft\text{-}°R)$
M	=	a surface node on the broad face of a bloom
N	=	a surface node on the narrow face of a bloom
p	=	attenuating factor used to adjust the rate of bloom cooling in the stake car
q	=	surface heat flux $(Btu/hr\text{-}ft^2)$
q_x	=	surface heat flux on the broad face of a bloom $(Btu/hr\text{-}ft^2)$
q_y	=	surface heat flux on the narrow face of a bloom $(Btu/hr\text{-}ft^2)$
ρ	=	density of steel (lbm/ft^3)
σ	=	Stefan-Boltzmann constant $(Btu/hr\text{-}ft^2\text{-}°R^4)$
T	=	temperature of a node $(°R)$
$T(aa,bb)$	=	temperature of node (aa,bb) at time t $(°R)$
$T'(aa,bb)$	=	temperature of node (aa,bb) at time $t + \Delta t$ $(°R)$
T_a	=	ambient temperature $(°R)$
T_s	=	bloom surface temperature $(°R)$
t	=	elapsed time (sec)
Δt	=	time increment (sec)
x	=	distance coordinate in the bloom thickness direction (ft)
y	=	distance coordinate in the bloom width direction (ft)
z	=	distance coordinate in the bloom length direction (ft)
ΔX	=	distance increment in the X-direction (in)
ΔY	=	distance increment in the Y-direction (in)
ΔN	=	size of node (in)

TABLE I. SAMPLE INPUT AND OUTPUT FOR BLOOM COOLING AND HEATING MODEL

Inputs for Simulation Run

Bloom Cross-section : 15 X 20 inches
Bloom Initial Temperature : 1400°F
(at caster upender)
Stacking Arrangement : Scissors
(Fig.4)
Node Size for Simulation : 1 inch
Time on Track : 30 minutes
Ready-to-roll Criteria : 2275 °F {minimum
at center node (1,1)}

Heating Practice :
0 min. : 1800°F
360 min. : 2380°F
RTR Time : 2380°F

Output Cold Spot Temperatures

center
1400 .. 1400 .. 1400
: .. : .. :
1400 ..1400.. 1400
: .. : .. :
1400 ..1400.. 1400
corner
Time = 0 min.
(at upender)

center
1389 .. 1365 .. 1212
: .. : .. :
1356 .. 1301 .. 1166
: .. : .. :
1221 .. 1197 .. 1093
corner
Time on track = 30 min.
Time in pit = 0 min.
(at Finish Charge)

center
1842 .. 1871 .. 1957
: .. : .. :
1867 .. 1894 .. 1976
: .. : .. :
1939 ..1962.. 2033
corner
Time = 360 min. in pit
(at Start Soak)

center
2275 .. 2286 .. 2300
: .. : .. :
2293 .. 2296 .. 2306
: .. : .. :
2299 .. 2301 .. 2319
corner
Time in pit = 555 min.
(as Ready-to-Roll)

COMPUTER SIMULATION OF THE SLAB REHEATING FURNACE

Zongyu Li*, P.V. Barr**, J.K. Brimacombe**

*Department of Metallurgical Engineering and Materials Science
Carnegie-Mellon University
Pittsburgh, PA. 15213-3890 U.S.A.

**The Centre for Metallurgical Process Engineering
The University of British Columbia
Vancouver, B.C. V6T 1W5 Canada

ABSTRACT

The development of a mathematical model for predicting steady-state heat transfer within the slab reheating is described and preliminary model predictions presented. The model is fully three dimensional, both in the furnace gas and slab, and accounts for the presence of the skidrail structure and the furnace side walls. In addition to longitudinal temperature and heat-flux profiles, contour plots of slab temperature in the transverse plane of the furnace are predicted. At its present stage of development, the model requires knowledge of the temperature distribution in the furnace gas although future extensions will eliminate this restriction.

BACKGROUND

Despite the significant economic advantage, the hot strip mill has yet to achieve one hundred percent direct hot rolling of steel slabs. Owing primarily to the presence of surface defects, which currently require cooling of the slabs for detection and removal, and insufficient temperature following continuous casting, the reheating furnace remains essential to the hot working of steel. Since it may require decades to overcome these problems, the incentive to optimize existing furnace performance is considerable.

A typical five-zone push-type reheating furnace is shown in Fig. 1. The slabs, which are supported on fixed, watercooled skidrails, are generally charged either singly or in tandem, with the length dimension transverse to the furnace longitudinal direction. Charging of new slabs results in a continuous slab movement through the furnace with a velocity of ~ 15 m/hr. Energy for slab heating is supplied by roof and tangential burners, firing oil, natural gas or coke-oven gas fuels at about 10% excess air. Overall furnace dimensions are in the order of 35 m x 12 m x 6 m. Slab residence times are typically 2-3 h to obtain mean slab temperatures ~ 1250°C at furnace exit.

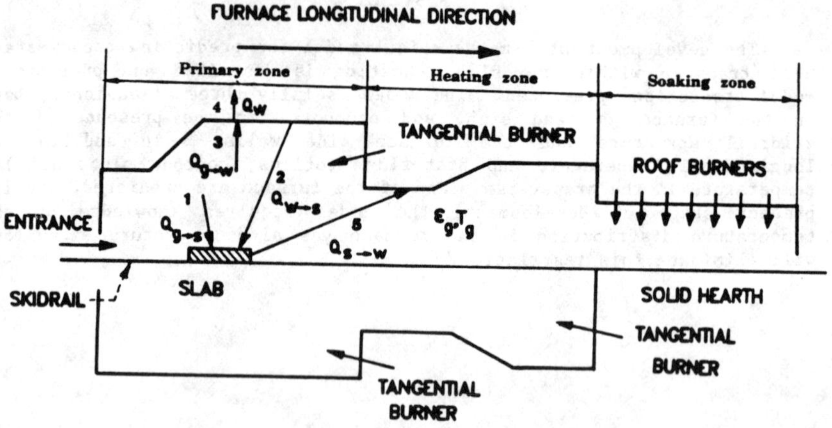

Figure 1.   Five-zone push type reheating furnace showing the heat-transfer paths.

Heat transfer within the furnace chamber occurs as radiation and convection to the slab surface and conduction within the slabs. As shown by Fig. 2, the presence of the skidrail structure may considerably distort heat transfer to the slab surfaces, both as a result of radiative shielding of the slab bottom surface and conduction of energy across the slab/skidrail contact area. The combined effect is the undesirable formation of skidmarks on the slab surface due to localized temperature depression. Strip width and thickness variations have been[1,2] closely linked to skidmark severity.

Prior to the past decade the main factors in reheating furnace design[3] were the maximization of furnace capacity, in terms of slab throughput, and minimization of capital investment. In an era of low energy costs, the development of multi-zone furnaces was forced by the

1090

demand for increased capacity. Since operating experience indicated a linkage between furnace throughput and skidmark severity, increased capacity, in turn, required the inclusion of longer hearths to soak out the otherwise accompanying increases in slab temperature nonuniformity. The reheating furnace is a significant component of the strip mill energy requirement, yet only 30-40%[14] of the energy supplied to the furnace is utilized for slab heating. Losses to the skidrail coolant may account[14] for 10-30% of the furnace losses. The stringent product quality requirements and increased energy costs associated with modern steelmaking provide an impetus for improving reheating furnace practice and design.

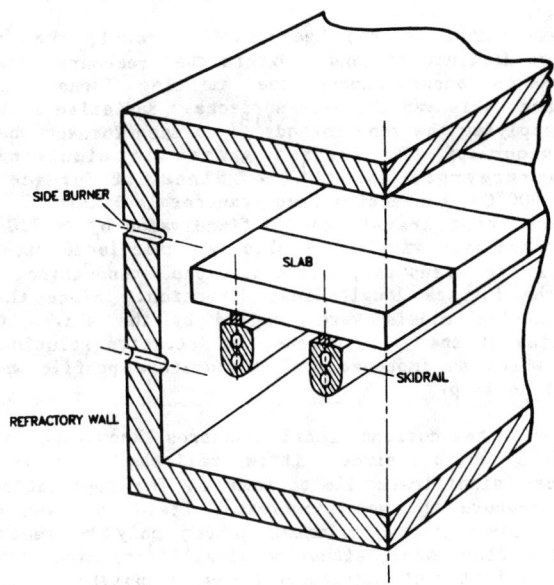

Figure 2.   The skidrail support structure for the slabs.

The development of mathematical models for the reheating furnace has been paced by the rate of improvement in the digital computer. Sophisticated but computationally complex models[4-8] have produced detailed predictions for furnace performance but simpler, less realistic, models are still required for furnace control applications[2,3,9-12]. Of the detailed furnace models, early efforts such as by Fitzgerald and Sheridan[4] or Colin[5] were basically slab conduction solutions for specified slab surface temperature or heat flux. The recent model of Veslocki and Smith[6] is capable of predicting two-dimensional temperature distributions in the slab, furnace gas and refractory wall, for both transient and steady-state conditions, and has been verified in plant trials. However, the model omits both the skidrail structure and the furnace side walls. The work of Fontana et al.[7] extends to three-dimensions in the slab, and includes the skidmark effect, but the furnace gas was assumed to be well mixed in the transverse furnace direction. Stracke and Kohne[8] derived a three-dimensional radiation model based on the zone method of Hottel and Cohen[18] but omitted the skidrail structure. Studies aimed specifically at elucidating the skidmark phenomenon[13-15] have identified radiative shading as the primary cause, although conflicting suggestions[15] for reducing severity have occasionally been proposed.

SCOPE OF THE MODELLING EFFORT

Although models for the reheating furnace have been developed, only scanty details of the modelling procedure and model predictions have appeared in the open literature. To alleviate this situation a reheating furnace modelling project was initiated at UBC and the preliminary results of this effort are the subject of this presentation. Although the furnace chosen for the study was similar to the Stelco Lake Erie Works unit in regards to geometry and burner placement, a simplified skidrail system was assumed for the model, particularly in the soaking zone where no allowance was made for the use of hot riders, exaporative cooling or skidrail offset. Therefore model predictions will not be rigorously applicable to the LEW furnace.

In the development of the heat-transfer model, the furnace was considered as two distinct regions. Within the freeboard, radiative and convective exchange occurs among the emitting furnace gases, the refractory furnace walls and the slab surfaces. Radiative exchanges were calculated by applying the zone method[18] to each furnace chamber. The emissive behaviour of the furnace gases was simulated using a clear-plus-two-gray-gas model[18]. Since, at furnace operating temperatures (> 900°C), convective heat transfer is likely[14] to account for < 5% of total heat transferred, a fixed value h_c = 7.8 $W/m^2 °K$ was employed. Heat transfer within the slab was calculated with a finite-difference model for transient, two-dimensional conduction, neglecting conduction in the furnace longitudinal direction. Since the slab and chamber heat-transfer models were coupled by the shared temperature boundary condition at the slab surface, an iterative solution technique was devised in which an improved slab temperature profile was obtained from each iteration loop.

Operation of the current model requires knowledge of the gas temperature field which, since little reliable data is currently available, places significant limitations on its application. It is impractical to measure the gas temperature field for even one set of operating conditions and this limitation can only be removed by the development of a flow model, either physical[17] or mathematical, which will allow the prediction of gas temperatures by applying energy balances to each gas zone. Pending this extension of the work, the current model was applied to investigate:

(i) The effects of varying furnace operating parameters, such as longitudinal and transverse gas temperature distributions, steel slab grade and slab push rate, etc., on slab exit temperatures and skidmark formation.

(ii) The mechanism of skidmark formation and identification of measures to eliminate or reduce it.

(iii) The effects of various heating strategies for hot-charged slabs so as to minimize fuel consumption and improve slab temperature uniformity.

THE FURNACE CHAMBER HEAT-TRANSFER MODEL

Within the furnace freeboard the primary source of radiant energy was assumed to be emission from the bipolar gases, predominantly CO_2 and H_2O, resulting from the combustion process. Since the LEW furnace fires natural and coke-oven gases through turbulent, premix burners, flame luminosity due to particulate formation during the combustion process is unlikely to be significant and was not considered. The rigorous modelling of radiative emission/absorption from gases is rendered difficult by the strongly banded nature of these emissions. The emissivity of the H_2O/CO_2 mixture present within the furnace chambers was instead simulated using a weighted summation of one clear plus two gray gases[18] for which

$$\varepsilon_g(T_g, pL) = \sum_{i=0}^{2} e_{g,n}[1 - \exp(-K_n pL)] \tag{1}$$

$$\alpha_g(T_g, T_s, pL) = \sum_{i=0}^{2} a_{g,n}[1 - \exp(-K_n pL)] \tag{2}$$

As detailed in ref.[19], the equivalent gray gas extinction coefficients, K_n, and the emissivity weighting coefficients, $e_{g,n}$, were obtained by curve fitting of the empirical emissivity data for the 2:1 mixture of $H_2O:CO_2$ present in the LEW furnace. The absorptivity weighting coefficients, $a_{g,n}$, were derived from the relationship between gas absorptivity and emissivity

$$\alpha_g(T_s, T_g, pL) = (\frac{T_g}{T_s})^{0.65} [\varepsilon_g(T_s, pL \frac{T_s}{T_g})] \tag{3}$$

For subsequent use in the heat-transfer model, the emissivity weighting coefficients were fitted to the exponential forms

$$e_{g,n}(T_g) = A_n \exp(-B_n T_g) \tag{4}$$

while a linear form was chosen for the absorptivity coefficients

$$a_{g,n}(T_g, T_s) = C_{n,T_g} T_s + D_{n,T_g} \tag{5}$$

The resulting values are summarized in Tables I and II. The clear-plus-two-gray-gas model was capable of simulating the actual gas emissivity with an accuracy of better than 5% (Fig. 3).

Table I. Coefficient Terms for Calculation of Gas Emissivity Using Eq.(4)

Gray Gas Component	A_n	B_n
0	0.3601	4.06×10^{-4}
1	0.4338	-2.79×10^{-4}
2	0.5001	-1.349×10^{-4}

Table II. **Coefficient Terms for Calculation of Gas Absorptivity Using Eq.(5)**

Gas Temp.	$a_{g,0}$		$a_{g,1}$		$a_{g,2}$	
°C	C_0	D_0	C_1	D_1	C_2	D_2
900	0.00038	0.1813	−0.00026	0.5618	−0.00012	0.2569
1000	0.00039	0.1446	−0.00028	0.6080	−0.00012	0.2474
1100	0.00041	0.1122	−0.00029	0.6493	−0.00011	0.2384
1200	0.00042	0.0834	−0.00031	0.6867	−0.00011	0.2299
1300	0.00043	0.0573	−0.00032	0.7207	−0.00010	0.2299
1400	0.00043	0.0334	−0.00033	0.7521	−0.00010	0.2145

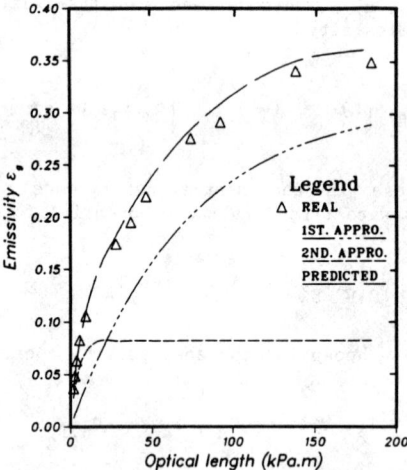

Figure 3. Accuracy of the gas emissivity model.

Having evaluated the coefficient terms required to calculate the emissivity/absorptivity of the freeboard gas mixture, radiative exchange within the furnace chambers was calculated by the zone method. Since, compared to the total chamber surface, the area available for radiative exchange between the furnace chambers is small, these openings behave (approximately) as black surfaces. The calculation of radiative exchange was considerably simplified by considering the furnace freeboard as individual chambers, with each enclosure being completed by the radiatively black transitional surface (Fig. 4). In the application of the zone method, the refractory and slab surfaces forming each chamber were subdivided into between 36 and 88 (depending on the chamber size) isothermal, uniform property, surface zones, while the emitting gas within the chamber was broken up into 12 to 32 gas volume zones (again depending on chamber size). The chamber zoning is summarized in Table III while the exchange geometry is shown by Fig. 5. For the calculation of radiative exchanges, the gas zone temperatures were assigned and the surface zone temperatures obtained by, applying energy balances at each surface. The entire furnace was assumed to be at steady-state.

Table III. Summary of Furnace Chamber Zoning

Chamber	Gas Zones	Surface Zones	Total
1	32	88	120
2	24	60	84
3	32	88	120
4	12	36	48
5	16	44	60

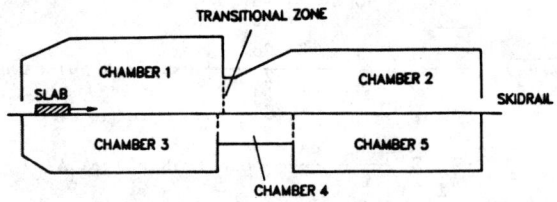

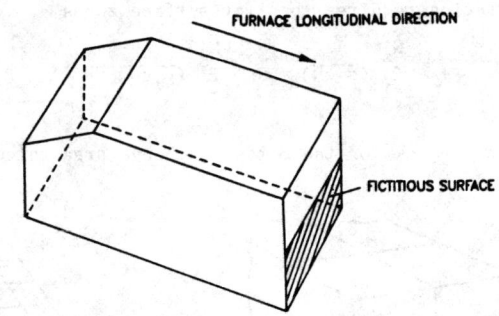

Figure 4. Division of the reheating furnace into chambers and showing the transitional surfaces.

The net rate of direct radiative heat transfer to surface zone A_j from another surface zone A_i (for energy within the nth emission band of the chamber gas) is given by

$$(Q_{A_i \to A_j})_{n,dir} = (\overline{s_i s_j})_n E_{i,n} \tag{6}$$

and the rate of heat transfer from a volume zone V_i by

$$(Q_{V_i \to A_j})_{n,dir} = (\overline{g_i s_j})_n E_{gi,n} \tag{7}$$

The direct exchange areas $(\overline{s_i s_j})_n$ and $(\overline{g_i s_j})_n$, which are functions of chamber geometry and gas partial pressure, are formulated as surface and volume integrals and usually require numerical evaluation. The required quadruple and quintuple integrations are, however, very expensive and the direct exchange areas were instead evaluated with sufficient accuracy, relative to other approximations inherent in the modelling process (eg. radiative properties of the chamber surfaces), by the approximations.

$$(\overline{s_i s_j})_n = \frac{1}{\pi} \int_{A_i} \int_{A_j} \frac{\cos \theta_i \cos \theta_j}{r_{ij}^2} \exp(-K_n pL) \, dA_j dA_i \qquad (8\text{-}a)$$

$$\approx \frac{1}{\pi} \sum_{A_i} \sum_{A_j} \frac{\cos \theta_i \cos \theta_j}{r_{ij}^2} \exp(-K_n pL) \, \Delta A_j \Delta A_i \qquad (8\text{-}b)$$

$$(\overline{g_i s_j})_n = \frac{1}{\pi} \int_{V_i} \int_{A_j} \frac{4 K_n \cos \theta_j}{r_{ij}^2} \exp(-K_n pL) \, dA_j dV_i \qquad (9\text{-}a)$$

$$\approx \frac{1}{\pi} \sum_{V_i} \sum_{A_j} \frac{4 K_n \cos \theta_j}{r_{ij}^2} \exp(-K_n pL) \, \Delta A_j \Delta V_i \qquad (9\text{-}b)$$

Energy conservation requires that for surface zones

$$\sum_{i=1}^{N_s} (\overline{s_i s_j})_n + \sum_{i=1}^{N_g} (\overline{g_i s_j}) = A_j \qquad (10)$$

which provided a check for the direct exchange area calculations.

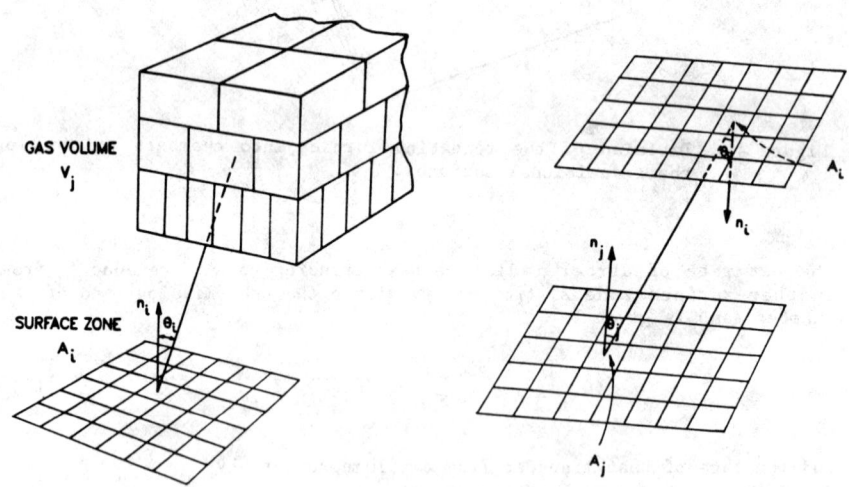

Figure 5. Geometry for gas to surface to surface radiative heat transfer.

The direct exchange expressions, Eqs. (6) and (7), do not account for radiative heat transfer involving reflection from intermediate surfaces. Expressions for the net rate of radiative heat transfer, including all reflection effects, between surface zone A_j and another surface zone A_i or volume zone V_i are

$$(Q_{A_i \rightarrow A_j})_n = (\overline{S_i S_j})_n E_{i,n} \qquad (11)$$

$$(Q_{V_i \rightarrow A_j})_n = (\overline{G_i S_i})_n E_{gi,n} \qquad (12)$$

The total exchange areas $(\overline{S_i S_j})_n$ and $(\overline{G_i S_i})_n$ were obtained from the direct exchange areas, the resulting sets of simultaneous linear equations being solved by standard matrix methods. Like the direct exchange areas, total exchange areas are functions of chamber geometry and gas partial pressure and were checked by comparing the calculated values to the requirement that

$$\sum_{i=1}^{N_s} (\overline{S_i S_j})_n + \sum_{i=1}^{N_g} (\overline{G_i S_j})_n = \varepsilon_j A_j \qquad (13)$$

Although the total exchange areas include all reflection effects, they do not account for the variation with surface temperature of the amount of energy emitted by that surface, Eq.(3), which falls within the gas absorption bands. The actual net rates of radiative heat transfer to the surface zone A_j, from another surface zone A_i or volume zone V_i, are

$$Q_{A_i \rightarrow A_j} = \overrightarrow{S_i S_j} E_i \qquad (14)$$

$$Q_{V_i \rightarrow A_j} = \overrightarrow{G_i S_j} E_{g,i} \qquad (15)$$

with the arrow indicating the direction of the radiative flux. The directed exchange areas were evaluated for the clear-plus-two-gray-gas model according to

$$\overrightarrow{S_i S_j} = \sum_{n=0}^{2} a_{n,T_i} (\overline{S_i S_j})_n \qquad (16)$$

$$\overrightarrow{G_i S_j} = \sum_{n=0}^{2} e_{n,T_i} (\overline{G_i S_j})_n \qquad (17)$$

Since the furnace was assumed to be at steady-state, energy conservation requires that, for each surface zone A_j

$$\sum_{i=1}^{N_s} \overrightarrow{S_i S_j} E_i \; + \; \sum_{i=1}^{N_g} \overrightarrow{G_i S_j} E_{g,i} \; + \; h_c A_i (T_{g,j} - T_j)$$

$$- \; \varepsilon_j A_j E_j \; = \; Q_{net,j} \tag{18}$$

which, when applied to each surface zone in the chamber, results in a set of N_s simultaneous nonlinear equations in the N_s unknown surface temperatures. These were solved by a quasi-Newton method.

THE SLAB CONDUCTION MODEL

Heat transfer within the steel slab was calculated using the two-dimensional transient conduction equation

$$\rho c_p \left(\frac{\partial T}{\partial t} \right) \; = \; \frac{\partial}{\partial x} \left(k \, \frac{\partial T}{\partial x} \right) \; + \; \frac{\partial}{\partial y} \left(k \, \frac{\partial T}{\partial y} \right) \tag{19}$$

across the furnace width (x) and slab thickness (y) directions. Slab conduction in the longitudinal furnace direction (z) was judged to be negligible. The slabs were assumed to be isothermal when charged. Three boundary conditions were specified at the slab surfaces:

(i) Exposed top and bottom slab surfaces (except in the region of skidrails). At the slab surface the net rate of heat transfer from the furnace chamber is balanced by conduction into the slab. The applicable boundary condition expression for an area A_j on the slab surface is

$$kA_i \left(\frac{\partial T}{\partial Y} \right)_{surface} \; = \; Q_{net,j} \tag{20}$$

and $Q_{net,j}$ is evaluated from Eq.(18). The (relatively) small convective component of the slab surface flux was calculated with $h_c = 7.8$ W/m^2°K, obtained from Glinkov[20].

(ii) The slab/skidrail contact region. In these areas radiation and convection from the furnace environment is partially blocked by the skid structure, the effect of which is compounded by conduction from the slab at the skidrail/slab contact. The resulting localized depression of slab temperature provides the mechanism for skidmark formation. The calculation of heat transfer to the slab surface was simplified by introducing the fictitious surface AB (Fig. 6) into the analysis. Since all radiation from the chamber gas and refractory must cross AB, the small sub-volume ABEG could be treated as an enclosure. The shadowing effect of the skidrail structure was found[22] to substantially reduce radiative heat transfer to the slab surface immediately adjacent to the skids. Conduction between the slab and skidrails was assumed to involve a contact resistance. Experimental values of contact conductance as a function of contact temperature are given by Howells et al.[13] and were used in the model. Their data indicate a rapid increase in thermal conductance from 1.4 kW/m^{2}C° at a mean contact temperature of 120°C to about 4.4 kW/m^{2}C° at 200°C and then falling to almost constant value of ~ 0.2 kW/m^{2}C° above 500°C. As in the work of Ford et al.[14] convective heat transfer to the coolant was calculated from the Dittus-Boelter result for developed pipe flow

$$Nu = 0.23 \, Re^{0.8} Pr^{0.4} \tag{21}$$

(iii) Slab edges adjacent to the furnace refractory wall. Slab thickness is typically < 6% of the slab length (recall that slabs are charged with the length dimension oriented transverse to the furnace and heat transfer to the edge surfaces will not represent a large fraction of the total input to the slab. However, the proximity of the slab edges to the furnace wall means that radiative exchange with the refractory, rather than with the furnace gas, may dominate and result in significant aberrations in slab temperature adjacent to the exposed edge. By assuming that the edges interact radiatively with the refractory side wall the boundary condition at the slab edge was determined to be

$$k\left(\frac{\partial T}{\partial x}\right)_{surface} = 0.665 \, \sigma \, (T_{ref}^4 - T^4) + h_c(T_g - T) \tag{22}$$

where the factor 0.665 accounts for both the surface emissivity and the view factor between the slab edge and the furnace wall.

Subject to the imposed initial and boundary conditions, the slab conduction expression (Eq.19) was solved using explicit finite-difference methods. A rectilinear nodal system was employed as shown in Fig. 7.

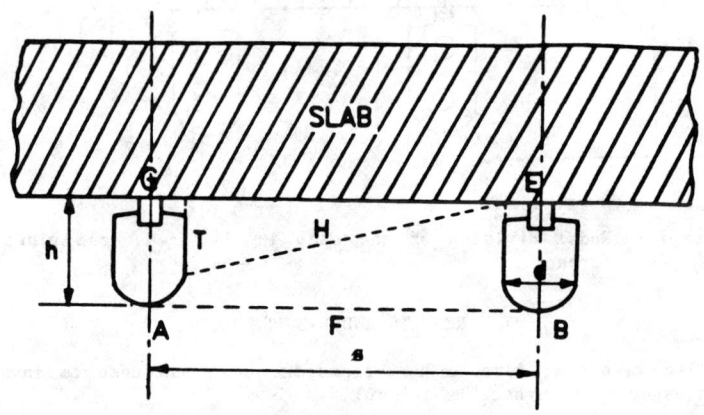

Figure 6.   Contact region between the skidrail and the slab (detail).

SOLUTION PROCEDURE

Since radiative exchange within the furnace chambers and conduction within the slab are coupled by the common boundary condition at the slab surface, the following iterative solution procedure was devised:

(i) Evaluate the direct exchange areas, Eqs.(8) and (9), for each gray gas component and each furnace chamber. These are functions of geometry only and are calculated only once.

(ii) Evaluate the total exchange areas, Eqs.(11) and (12), for each gray gas component and furnace chamber. As with the direct exchange areas, total exchange areas are temperature independent.

(iii) Assume temperature profiles for the chamber refractory surfaces.

(iv) Calculate the directed exchange areas, Eqs.(16) and (17), for the current temperature profiles. Directed exchange areas are temperature dependent and must be recalculated for each iteration.

(v) Solve the refractory energy balance expressions, Eq.(18), for the refractory surface temperature distributions in each chamber. Compare with previous values. If the absolute value of maximum deviation exceeds $10°K$ repeat (iv) → (v).

(vi) Calculate radiative heat-transfer coefficients for the slab surfaces and initiate the slab conduction model. At this stage a new set of slab surface temperature profiles are generated. Compare with previous values and repeat (iv) → (vi) if required.

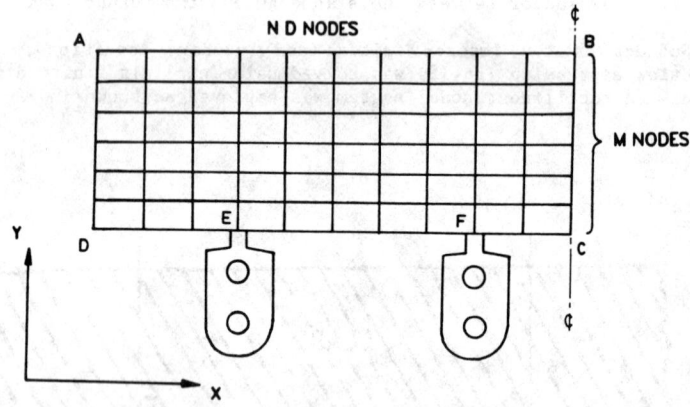

Figure 7.   Nodal division of the slab in the transverse plane of the furnace.

RESULTS AND DISCUSSION

The reheating furnace heat-transfer model was used to investigate three aspects of furnace behaviour:

(i) To predict the effects of different operational parameters (slab size, gas temperature distribution, push rate and steel grade) on the temperature distribution in the slab and refractory walls.

(ii) To calculate the nonuniform heat-flux and temperature distributions around the slab/skidrail contact region and to identify possible measures to alleviate the skidmark effect.

(iii) To develop improved heating strategies for the hot charging of slabs.

Since the emphasis in this paper is on the development of the model, the presentation of results is necessarily limited. A more complete discussion of model predictions has been submitted for publication[21].

The effect of varying each of the identified furnace operating parameters was established by holding the remaining parameters constant in the model. The standard slab selected for the calculation was medium-carbon (plate-grade) (0.23%) steel and 4.35 m wide by 0.24 m thick. Since they have similar thermal properties, most common carbon steels are expected to exhibit similar heating behaviour.

SLAB TEMPERATURE AND REFRACTORY WALL TEMPERATURE RESPONSE. Based on measurements by Stelco and previously reported data[7-11], approximate longitudinal furnace gas temperature profiles were assigned as shown in Fig. 8. Transverse gas temperatures were assumed to be uniform. Under these conditions, the slabs are heated from room temperature to a predicted exit (drop-off) temperature of 1180-1200°C, a value well above the austenizing temperature for the steel. As expected, the slab centreline temperature is lower than the surface temperature until the slab reaches the soaking zone, with the maximum temperature difference (230°C) occurring at the centre of the heating zone (about 20 m into the furnace). The lag in centreline temperature is due to the relatively low conductivity of the steel. Figure 9 shows temperature contours at three different axial furnace positions (25.2 m, 27.5 m, 32.0 m). Although at the entrance to the soaking zone (25.25 m), the slab surface temperature is significantly higher than at the centreline, the influence of the skids is relatively minor. As the slab progresses through the soaking zone the localized slab temperature depression adjacent to the skids becomes more apparent. At 27.5 m into the furnace , the temperature of the skidmarks is about 25°C lower than the unaffected surface, a value which increases to 50°C at the exit. As the slabs move through the soaking zone, the centreline temperature quickly approaches the surface temperature and, whereas in the heating sections the major heat sink in the slab is conduction into the centre, in the soaking zone the slab/skidrail contact region becomes the main heat sink and the skidmark effect grows until the slab is discharged.

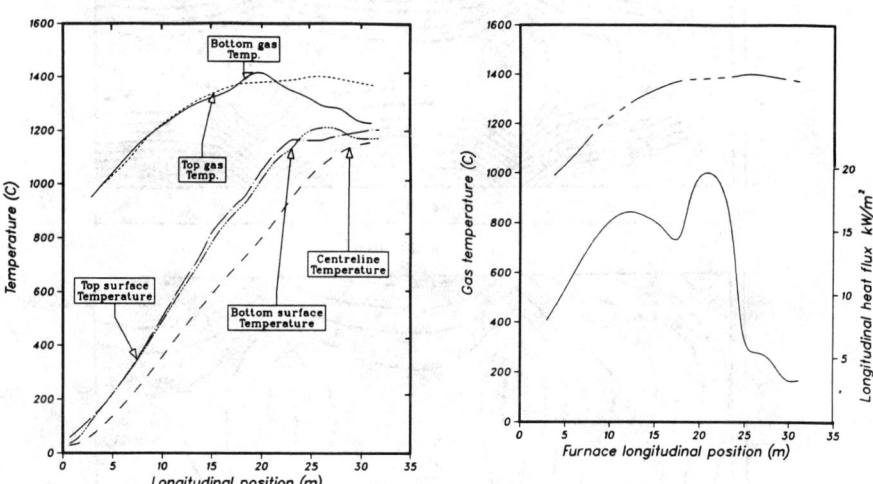

Figure 8. Predicted longitudinal slab temperature profiles for specified gas temperature profiles.

Figure 10. Variation of predicted heat flux to the slab top surface with longitudinal furnace.

Model predictions for the net surface heat flux (Fig. 10) indicate that the maximum slab heating rate occurs around the midpoint of the heating zone, about 20 m into the furnace. In the soaking zone, the radiative heat flux is much reduced, partly due to the locally high slab temperature. These results are in agreement with other studies[11,23]. Of more significance, however, are the transverse heat-flux distributions shown in Fig. 11, since these are believed to exert a significant impact on the rolling process[24]. The transverse heat-flux distribution is nonuniform even for the case of uniform transverse gas temperature. Near the furnace entrance (10.67 m), the heat-flux profile is concave due to the influence of the refractory side wall. However, as the slab approaches the furnace discharge (28 m into the furnace), the predicted heat-flux profiles are convex, with the maximum heat flux occuring in the middle of the furnace. This reversal results from the slab end temperatures becoming greater than those in the central part of the furnace so that the net radiative heat received is proportionally

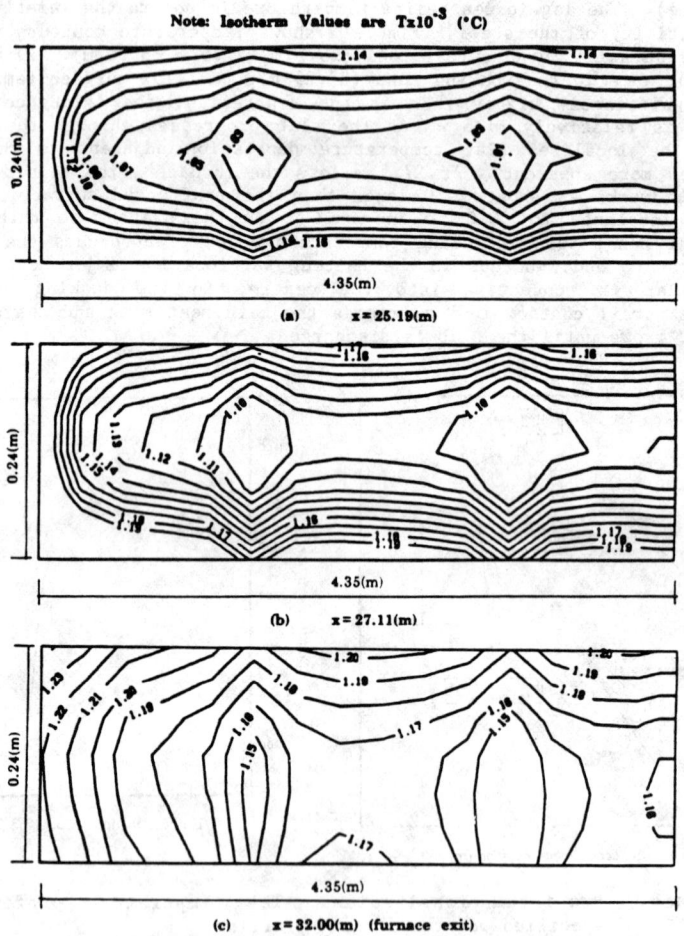

Note: Isotherm Values are $T \times 10^{-3}$ (°C)

(a)   x = 25.19(m)

(b)   x = 27.11(m)

(c)   x = 32.00(m) (furnace exit)

Figure 9.   Predicted slab temperature contours in the transverse plane of the furnace at three longitudinal positions.

decreased. Model predictions for transverse heat-flux profiles over the bottom surface of the slab (Fig. 12) clearly illustrate the severe shadowing effect of the skidrail structure. Conduction heat loss to the skidrail was found to be two orders of magnitude less than the reduction in radiative heat transfer to the slab. These results confirm[13-15] that the dominant factor in the formation of skidmarks is the radiative shielding by the skid structure. The effect of nonuniform transverse gas temperature on heat flux to the top slab surface is shown in Fig. 13. The temperature variation of 100°C imposed on the transverse gas temperature (upper curve in the figure) can be seen to significantly increase the nonuniformity of the surface heat flux, compared to the uniform gas temperature condition.

Varying the rate of slab throughput (push rate) is a common plant practice. Figure 14 shows a comparison of the centreline temperature for two push rates: 0.0034 m/s (200T/hr.) and 0.004 m/s (235T/hr.); the remaining parameters were held constant. The predicted centreline temperature (at furnace exit) is 80°C lower for the higher push rate. The temperature contours (Fig. 15) in the discharged slabs indicate that higher production rates exacerbate the problem of skidmark formation.

Note: Isotherm Values are $Tx10^{-3}$ (°C)

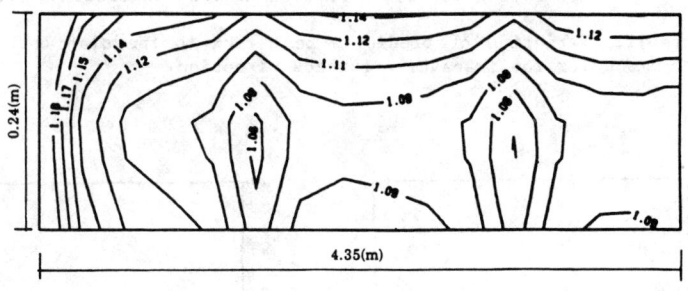

(a)   Push rate: 0.004 m/sec.

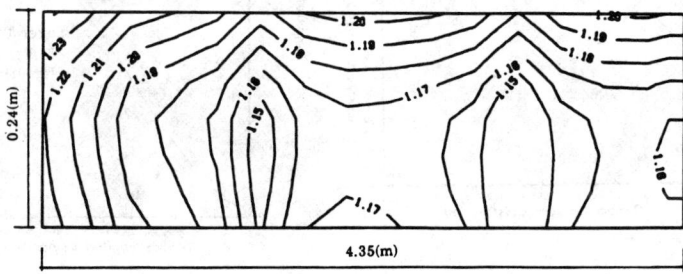

(b)   Push rate: 0.0034 m/sec.

Figure 15.   Predicted slab temperature contours at furnace exit for two push rates.

1103

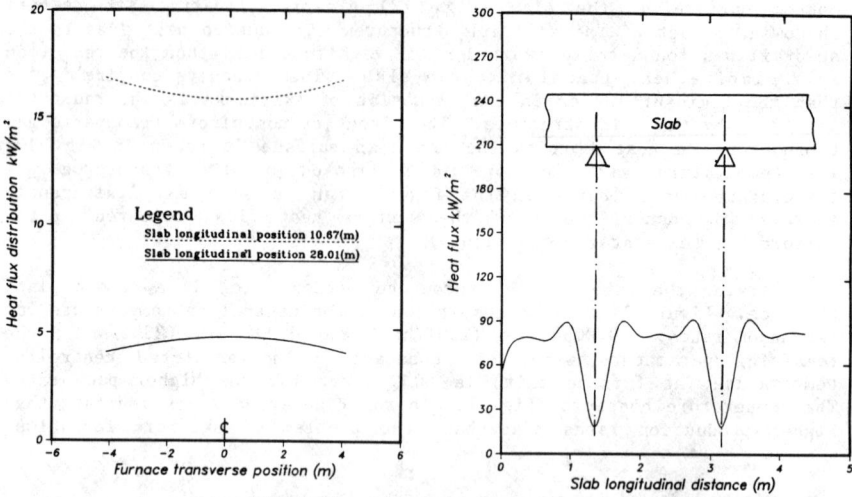

Figure 11. Variation of predicted heat flux to slab top surface for the transverse furnace direction at two longitudinal positions.

Figure 12. Variation of predicted heat flux to the slab bottom surface for the transverse furnace direction.

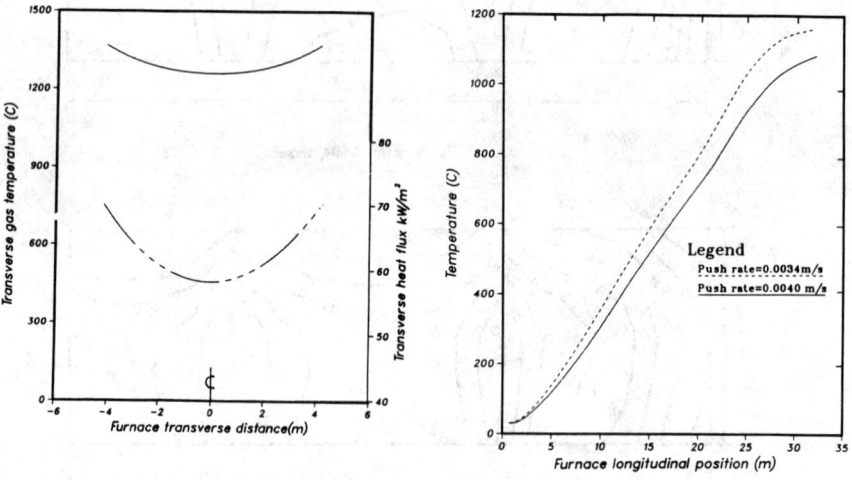

Figure 13. Variation of predicted heat flux to slab top surface for the transverse furnace direction with non-uniform transverse gas temperature distribution.

Figure 14. Predicted longitudinal variation of slab centreline temperature profiles at two slab push rates.

SUMMARY

A computer model of the reheat furnace has been developed to study the problems of furnace design and operation. Although the temperature distribution of the furnace gas, for which little data is currently available, must be specified, the model can predict, in detail, furnace performance for the specified gas temperature distribution. For example, the effects of different longitudinal or tranverse gas temperature profiles, slab thicknesses, push rates or steel grades on the temperature distribution within the slab at the furnace exit can be calculated readily. Since the presence of the skidrail structure and the furnace sidewalls are accounted for, skidmark severity and temperature anomolies at the slab edges are included in the results.

For a fixed axial gas temperature profile the effect of increasing slab throughput was to decrease the mean slab temperature at the furnace exit and to increase skidmark severity. Contrary to what might be anticipated, skidmark severity was found to increase through the soaking zone of the furnace. Radiative shadowing of the slab surface by the skidrail structure was identified as the dominant factor in the formation of skidmarks. Increasing the slab push rate was found to aggravate the skidmark problem. Due to the radiative influence of the furnace wall refractory, the distribution of heat flux to the slab surface, tranverse to the furnace axis, was found to be markedly nonuniform, even with uniform tranverse gas temperature. The maximum calculated heat flux was found to occur at the edges of the slabs early in their heating process but near the end of the cycle to occur at the furnace midplane. Transverse gradients of gas temperature significantly altered the slab surface heat flux and, except in the regions of contact with the skids, could provide the means of controlling the transverse variation of slab surface temperature. A more detailed presentation of model predictions has been prepared for publication[21].

ACKNOWLEDGEMENTS

The authors are most grateful to the Natural Sciences and Engineering Research Council of Canada and Stelco Inc. for their financial support of the modelling work and to Stelco for providing plant data and practical guidance on furnace operation.

NOMENCLATURE

A_i	Area of ith surface zone (m^2)
c_p	Specific heat at constant pressure (kJ/kg °K)
e_n	Emissivity weighting coefficient for the nth gray gas component
E_i	Black body emissive power of ith surface zone (W/m^2)
E_{gi}	Black body emissive power of the ith gas volume zone (W/m^2)
$(g_j s_i)_n$	Direct exchange area for jth gas volume zone and ith surface zone for nth gray gas component (m^2)
$(\overline{G_j S_i})_n$	Total exchange area for jth gas volume zone and ith surface zone for nth gray gas component (m^2)

$(\overrightarrow{G_j S_i})_n$ Directed exchange area for jth gas volum zone and ith surface zone for nth gray gas component (m^2)

h_c Convective heat-transfer coefficient ($W/m^2 °K$)

k Thermal conductivity ($W/m°K$)

K_n Extinction coefficient for nth gray gas component (m^{-1})

L Path length of a beam through a gas (m)

N_g, N_s Total number of gas and surface zones respectively

P Partial pressure of emitting/absorbing gases (atm)

Q Heat transfer rate (kW)

$(\overline{S_j S_i})_n$ Direct exchange area between ith and jth surface zones

$(\overline{S_j S_i})_n$ Total exchange area between ith and jth surface zones for nth gray gas component (m^2)

$(\overrightarrow{S_i S_j})$ Directed exchange area between ith and jth surface zones for nth gray gas component (m^2)

t Time (sec)

T Temperature (°K)

V_j Volume of jth gas zone (m^3)

Subscripts

g Gas

i ith surface or volume zone

n nth gray gas component

s Solid surface

Greek

α Absorptivity

ϵ Emissivity

ρ Density (kg/m^3)

Dimensionless

Nu Nusselt number for heat transfer to skidpipe coolant

Re Reynolds number for skidpipe coolant flow

Pr Prandtl number for skidpipe coolant

REFERENCES

1. H. Kay, "Design of Furnaces for Reheating", Metals Technology, (1975), 450-462.

2. J.L. Roth, H. Sierpinski, J. Chabanier and J.M. Germe, "Computer Control of Slab Furnaces Based on Physical Models", Iron and Steel Engineer, 63(8) (1986), 41-47.

3. F. Hollander, "Reheating Processes and Modifications to Rolling Mill Operations for Energy Savings", Iron and Steel Engineer, 60(6) (1983), 55-62.

4. F. Fitzgerald and A.T. Sheridan, "Heating of a Slab in a Furnace", J. of Iron and Steel Inst., 208, (1970), 18-28.

5. R. Collin, "A Flexible Mathematical Model for the Simulation of Reheating Furnace Performance", Proc. of Conf. on Reheating for Hot Working, (ISI Publications, 1968), 230-235.

6. T.A. Veslocki and C.C. Smith, "Applications of a Dynamic Mathematical Model of a Slab Reheating Furnace", Proc. of 3rd Process Technology Conference of AIME, Pittsburgh, (1982), 134-141.

7. P. Fontana, A. Boggiano and A. Furinghetti, "A Two Dimensional Mathematical Model of Slab Reheating Furnaces: Theory and Practical Utilization", Proc. of 3rd Process Technology Conference of AIME, Pittsburgh, (1982), 142-150.

8. H. Stracke and H. Kohne, "A Mathematical Model for Radiation Exchange of Fuel-Fired Furnace Chambers and its Application to Calculate Temperature Fields in Pusher Type Furnaces", Stahle u. Eisen, 100 (1980), 887-894.

9. F. Hollander and R.L. Huisman, "Computer Controlled Reheat Furnaces Optimize Hot Strip Mill Performance", AISE Yearbook, (1982), 427-440.

10. R.J. Schurko, C. Weinstein, M.K. Hanne and D.J. Pellechia, "Computer Control of Reheat Furnaces: A Comparison of Strategies and Applications", Iron and Steel Engineer, 64(5) (1987), 37-42.

11. Y. Misaka, R. Takahashi, A. Shinjo, Y. Nariai and M. Kooriki, "Computer Control of a Reheat Furnace at Kashima Steel Works Hot Strip Mill", Iron and Steel Engineer, 59(5) (1982), 51-55.

12. P.G. Fontana, A. Boggiano, A. Furinghetti and B. Pastorino, "Optimizing Heat Distribution in Continuous Furnaces Using a Dynamic Mathematical Model", BTF Special Issue, (1985), 9-14.

13. R.I.L. Howells, J. Ward, S.D. Probert, "Thermal Conductances of Contacts at High Temperatures", J. of Iron and Steel Inst., 212 (1973), 193-196.

14. R. Ford, N.V. Suryanarayana and J.H. Johnson, "Heat Transfer Model for Solid/Slab/Water-Cooled Skid Pipe in Reheat Furnace", *Iron and Steelmaking*, (3) (1980), 140-146.

15. L.A. Weaver and W.F. Barraclough, "Application of Hot Alloy Skids Imroves Quality", *Iron and Steel Engineer*, 63(10) (1986), 25-28.

16. R.L. Howells, S.D. Probert and J. Ward, "Influence of Skid Design on Skid-Mark Formation", *J. of Iron and Steel Inst.*, 210 (1972), 10-21.

17. S. Matsunaga and B. Hiraoka, "Simulation Experiment on Gas Flow Patterns in a Reheating Furnace", *Trans. ISIJ*, 12 (1972), 72-78.

18. H.C. Hottel and E.S. Cohen, "Radient Heat Exchange in a Gas Filled Enclosure: Allowance for Nonuniformity of Gas Temperature", *A.I.Ch.E. Journal*, 4(1) (1958), 3-14.

19. H.C. Hottel and A. Sarofim, *Radiative Transfer* (New York, N.Y.: McGraw-Hill, 1967).

20. M.A. Glinkov, "General Theory of Furnaces", *J. of Iron and Steel Inst.*, (1968), 584-594.

21. L. Zongyu, J.K. Brimacombe and P.V. Barr, "Results from a Computer Model of the Slab Reheating Furnace", submitted for Publication, *Canadian Metallurgical Quarterly*.

22. Z. Li, "Computing Simulation of the Slab Reheating Furnace", (M.Sc. Thesis, UBC, 1987).

23. F. Hollander and R.L. Huisman, "Computer Controlled Reheating Furnaces Optimize Hot Strip Mill Performance", *AISE Yearbook*, (1972), 427-439.

24. S. Kozo, Y. Hoshine, O. Takamori, A. Kawabata, K. Sannomiya, K. Goto and T. Matsukawa, "Development of Open Radiant Tube Type Reheating Furnace", (Technical Report), *Trans. ISIJ*, 25 (1985), 972-976.

MATHEMATICAL MODELING OF THE TWO-ROLL

ROTARY PIERCING OPERATION

Emin Erman[*]

Research Department
Bethlehem Steel Corporation
Bethlehem, PA 18016

Abstract

Over the past decade, the demand for oil country tubular goods (OCTG)
forced exploration of the new technologies, new offshore drilling sites and
the working wells previously considered unprofitable due to the extremely
severe working conditions. As a result, the demand for higher quality Oil
Country Tubular Goods (OCTG) has been increasing for closer dimensional
tolerances, greater corrosive resistance, better surface quality and collapse
strength, etc. In addition to these demands, basic requirements from a
successful modern seamless mill can be classified as: a) High product yield,
b) High Production and Operating efficiency, c) High dimensional and surface
quality, d) Low energy consumption, e) The most economical feedstock, f)
The simplest process for production. To meet this demands seamless pipe mills
have been active in developing new and improved hot mill practices which are
compatible with the known processes. The current hot mill practices are
largely based on experience and trial and error techniques that provide
practical knowledge and insight about the operations. In Piercing process,
the mill trials are quite time consuming and thus costly since there are
substantial number of the variables. Therefore, to provide design criteria
for the seamless mills, as well as better understanding of the operations
mathematical modeling should be applied. A major objective of mathematical
modelling for the piercing process is to provide necessary information for
proper design and control of this process and to reduce the number of
expensive trials necessary to setup an optimum piercing design.

Formerly at U. S. Steel Research Center, now an Engineer at the Homer Research
Laboratory of the Bethlehem Steel Corporation, Bethlehem, PA.

Introduction to Piercing

Piercing, which represents the first stage in the production of seamless tubes, can be done one of the three most used machines, namely;

a) The piercing Press
b) The rotary Piercing mill (known as mannesmann piercer)
c) The press piercing mill (PPM).

Because of its more common use, this study was based on the two roll rotary piercing mill, which is today's most important piercing unit. This type of mill, as seen in Figure 1, consists of two barrel-shaped rolls with their axes parallel to the vertical plane and angular to the horizontal plane. Both rolls turn in the same direction, thus driving the work piece in the opposite direction. A piercing plug (also called point) held in a required position between the rolls by a mandrel. The rolls were set obliquely to each other and were inclined at equal angles to the mill axis. The billet is drawn into the deformation zone by frictional tractive forces, one component of this force of the roll is in the longitudinal direction and this pulls the billet forward. The tangential component of this force is responsible for its rotation. As the billet travels along the roll, it meets the plug, this time the metal immediately starts to flow along around the point to form a tube. As the billet is pierced between the plug and the roll, it tends to elongate in a transverse direction to the workpiece and this produces a larger inside diameter than the point diameter. Because of this action, which separates the pierced material from the point while the material is tight of the point on the two sides, the point cools more than would be the case if the hot metal was right on the plug, so that it increases the point life [1]. The first piercer does not provide the reduction of the wall thickness and elongation. So that the second piercer or in some mills, mandrel type of mill is introduced for a further stage (often called elongater).

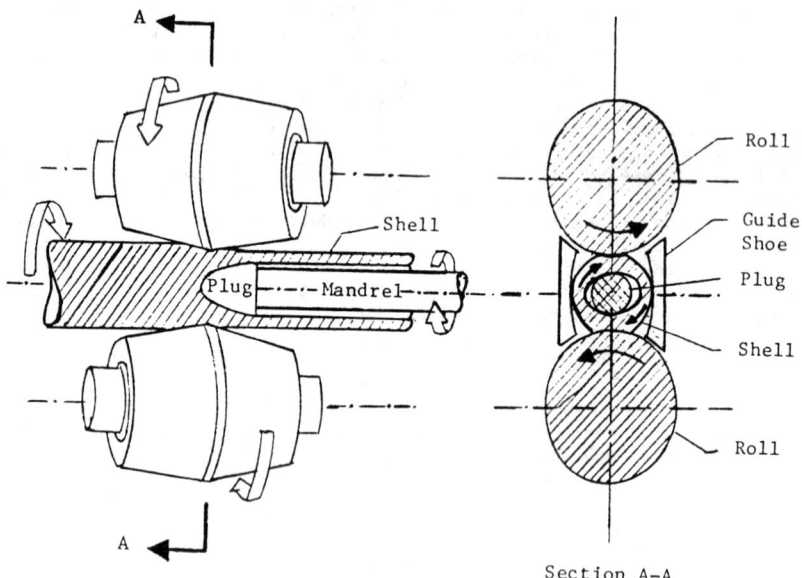

Figure 1 - Illustrations of cross-sections of a Rotary-Piercing Mill employing a round billet.

Energy Considerations In Piercing Process

In piercing process, specific (actual) work is the sum of the homogeneous work, internal shear work and friction work. The homogeneous work represents the minimum work per unit volume of material necessary to transform a solid billet into a pierced hollow without any changes that do not contribute directly to the final shape. The internal shear work (redundant work) is introduced through internal deformations which occur, but are not essential, to the required change of shape of the workpiece [2]. The frictional work, on the other hand, arises from friction at the interface between the deforming metal and the tool faces that constrain the metal. If a particular setup gives a high degree of redundancy, this means, it will be less efficient than setups with less redundancy. It is possible to reduce the amount of inhomogeneous deformation by changing the processing as well as geometric variables. That is, the actual work is related to the point/pass design because of the work which is done on the design of the pass. Thus, before assessing and devising possible means of improving the efficiency of the mills, it is just necessary to derive mathematical expressions for the energy involved in the deformation of the solid billet. These expressions, in turn, will be very helpful to maximize the efficiency of the rolling mills. This is the most desirable engineering concept, not only reduce energy consumption but also to increase product quality and productivity. It is especially true for rotary piercing which incurs large amounts of redundant work that can be dependent on the mill setup variables. The effects of these mill set up variables on the piercer mill performance can be obtained by conducting rolling trials. However, because of the substantial number of variables, such trials are quite time consuming and thus costly. In this study, the aim was to derive mathematical expressions that will describe the piercing process sufficiently to permit analyses of the piercing mill variables without having to resort to the aforementioned trials.

Deformation Modes In Piercing Process

In piercing process, certain deformations occur which are essential to the change in shape from a solid billet to a hollow shell. In addition to these homogeneous deformations, other deformations are found to occur which are not essential for conversion of solid billet into hallow shell. These unnecessary deformations are called redundant deformations. Figures 2 and 3 illustrate the homogeneous and redundant deformations in rotary piercing, respectively. As shown in Figure 2, three homogeneous (or essential) strains can be distinguished; namely, radial (ε_r), longitudinal (ε_L), and circumferential (ε_θ).

The redundant deformation strains occurring in the rotary piercing process; namely, longitudinal redundant strain, circumferential redundant strain and twist, as illustrated in Figure 3. Longitudinal redundant strain is an axial shear strain through the wall caused by forward longitudinal force exerted by the roll on the outer surface of the billet and a backward longitudinal force exerted by the plug on the ID surface as illustrated in Figure 3a. Circumferential shear strain, on the other hand, caused by tangential shear stresses through the wall of the shell resulting from the roll-shell traction forces and the oppositely directed resistance at the piercer-point surface which is illustrated in Figure 3b. A twist occurs because of the differential tangential velocity of the billet along the axis of the piercer pass (Figure 3c). The practical significance of the redundant deformation varies from process to process [3]. Although the hot-working processes are not particularly affected by the changes in yield stresses and the associated phenomena, the presence of large redundant deformations generally, and in rotary operations particularly, reduces the quality of the final product, limits the range of alloys that can be processed in a given operation and imposes limitations on the magnitude of the possible deformation. The predisposition to fracturing in the regions of inhomogeneous shear within the body of the material is generally strong, and the presence of inclusions, metallic or otherwise aggravates the situation even further.

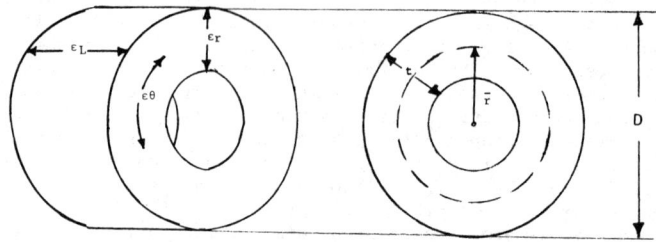

Figure 2 - Illustrations of homogeneous deformation strains in rotary piercing process.

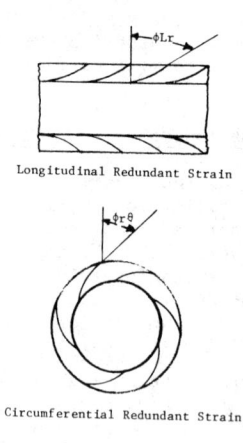

Longitudinal Redundant Strain

Circumferential Redundant Strain

Twist

Figure 3 - Illustrations of redundant deformation strains in rotary piercing process.

Since the shearing of an element during its passage through a forming pass depends for its pattern on the profile of the surface of contact between the material and the tools, it can be logically concluded that the design of the tools is of importance. In fact correctly designed tools, by allowing the material to follow the optimum flow path as closely as possible, will reduce or even possibly eliminate the incidence and magnitude of the redundant deformation. Thus, need for avoiding redundancy is very real from the point of view of the efficient and satisfactory operation of an industrial metalforming process. The solution lies in rational and logical tool design, and therefore by implication in the choice of the correct profile of the working zone of a pass.

First Piercing Operation

Parallel Pass: In the rotary-piercing process (Figure 1), a solid cylindrical billet is pierced and elongated, usually between two rolls and a piercing plug with two guide shoes to contain the workpiece in the pass. The simplest case would be that for a parallel pass; that is, the diameter of the pierced shell is the same as that of the round billet. As shown in Figure 2, three homogeneous (or essential) strains can be distinguished; namely, radial, longitudinal and circumferential. As a first assumption, the diameter, D_b, of the billet and shell remain constant throughout the piercer pass. This assumption greatly simplifies the analysis of the problem. Hence,

$$D = \text{constant} \tag{1}$$

Although this assumption ignores the change in shape of the periphery of the workpiece during the piercing process (going from round to somewhat elliptical and back to round), the degree of this deformation is quite small relative to that in the overall piercing process. The other assumptions for the derivation of the mathematical expression are;
a) the workpiece is subjected to an ideal plastic behavior, that is, the yield stress is constant throughout the deformation zone.
b) no redundant work and
c) no friction occurs between workpiece and the tools.
The radial-strain increment is defined as the relative incremental change in the wall thickness (independent of radius).

$$d\varepsilon_r = \frac{dt}{t} \tag{2}$$

Similarly, the longitudinal strain increment is defined as the relative incremental change in the cross-sectional area, A,

$$d_{\varepsilon L} = \frac{dA}{A} \tag{3}$$

and the tangential strain increment as the relative incremental change in the mean radius, $\bar{r}$.

$$d_{\varepsilon_\theta} = \frac{d\bar{r}}{\bar{r}} \tag{4}$$

The effective (equivalent) strain increment, which represents the total plastic-deformation increment, is expressed in terms of these three normal strains by the equation

$$d_{\bar{\epsilon}} = \frac{\sqrt{2}}{3} \left[(d_{\epsilon L} - d_{\epsilon_\theta})^2 + (d_{\epsilon_\theta} - d_{\epsilon r})^2 + (d_{\epsilon r} - d_{\epsilon L})^2 \right]^{\frac{1}{2}} \tag{5}$$

Expressing equations (3) and (4) in terms of diameter and wall thickness gives

$$d_{\epsilon L} = - \frac{(D-2t)}{(D-t)} \frac{dt}{t} \tag{6}$$

$$d_{\epsilon_\theta} = - \frac{t}{(D-t)} \frac{dt}{t} \tag{7}$$

Substituting equations 2, 6, and 7 into equation (5) yields the following equation for the effective strain increment.

$$d_{\bar{\epsilon}} = \frac{2}{\sqrt{3}} \frac{[3t^2 - 3Dt + D^2]^{\frac{1}{2}}}{(D-t)^2} \frac{dt}{t} \tag{8}$$

Integration of this equation for the total effective strain requires a positive result over the complete strain path. Hence, if dt/t is negative, the negative root must be taken. The requirement can be most simply satisfied by expressing the total effective strain as the absolute value of the integration. Integrating equation (8) between the limits of D/2 (billet radius) and t (the shell wall) gives the following expression for the total effective strain.

$$\bar{\epsilon} = \left| \ln \frac{-D^{\sqrt{3}}}{[2\sqrt{Q} + \sqrt{3}(2t-D)\sqrt{3}]} \cdot \frac{t}{t-D} \frac{2\sqrt{Q} + 3t - D}{2\sqrt{Q} - 3t + 2D} \right| \tag{9}$$

where $Q = 3t^2 - 3Dt + D^2$

Finally, the specific work (w) of homogeneous deformation can be calculated from the following expression,

$$W_H = \bar{\sigma} \cdot \bar{\epsilon} \tag{10}$$

Reducing and Expanding Passes: The work calculated by expression (10) represents the minimum work per unit volume of material necessary under the assumed conditions to achieve the desired change from a billet to a pierced hollow in a parallel pass. Although it does not include such factors as redundant (shear) work and losses resulting from slippage and friction, it can serve as a reference for estimating the efficiency of operations for which data are available. But in real practice, two type of piercing mill pass contours are employed: a) expanding, and b) reducing passes, rather then parallel pass. As shown in Figure 1, the diameter of the leading tube can be controlled by manipulating the top guide shoe position, the piercer point position (lead) and the gorge. It should also be remembered that in piercing, deformations are irreversible and the work done depends on the particular strain path followed by the material passing through the piercing zone, that is, the accuracy of the effective strain, depends on how good the assumed deformation path is. If the assumed deformation path is accurate, the energy of the deformation would also be accurate. For this reason, it is necessary to develop more realistic mathematical models that recognize aforementioned variables such as roll contour, piercer-point profile, point location and strain path. Derivation (9) provides a very good base for more comprehensive models that accounts the geometry of the deformation and the strain path. A piercer pass profile is illustrated in Figure 4, which shows the shape of the roll represented by a linear equation of $Y = bx+c$ and the shape of the point is represented by a parabolic equation of $Y^2 = x/a$. The area, in where a solid billet undergoes severe deformation to become a hollow shell, is the area represented by the differential difference between these equations.

Once again, the radial-strain increment is defined as the relative incremental change in the wall thickness

$$d\varepsilon_r = \frac{dt}{t} = \int_0^L \frac{1}{t} \cdot \frac{dt}{dx} \, dx \qquad (10)$$

By calculating the wall thickness in terms of geometric parameters and differentiating it with respect to x, the radial strain increment takes the form of

$$d\varepsilon_r = \int_0^{L(x)} \frac{(2\alpha p - 1)}{2\,(\alpha px - x + c)} \, dx \qquad (10)$$

And similarly, the longitudinal strain increment can be derived as

$$d\varepsilon_L = \int_0^{L(x)} \frac{2\,(\alpha^2 p^2 + a\alpha c) - 1}{\alpha p^2 x + (2a\alpha c - 1)x + c^2} \, dx \qquad (11)$$

and finally the circumferential strain increment is

$$d\varepsilon_\Theta = \int_0^{L(x)} \frac{(2\alpha p + 1)}{2\,(\alpha px + x + 2cp)} \, dx \qquad (12)$$

where, $p = \sqrt{ax}$

These expressions gives precise realistic simulation incorporating with geometric and processing parameters. They remove the black box on the deformation zone as it has been the case in parallel pass. Table I shows the computed values of the three homogeneous strains and the value of resultant effective strain for a particular expanding pass.

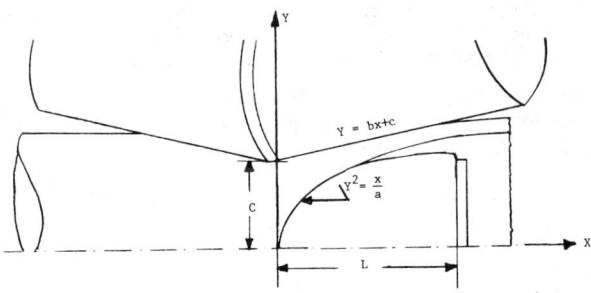

Figure 4 - Illustration of a piercer pass profile with parabolic shape of piercer point.

1115

TABLE I. Computed True Strains and Effective Strain for a
Parabolic type of Piercer Point
(parabolic shape constant, a=0.90)

		Distance From Point Nose, X (inches)	ε_r	ε_L	ε_θ	$\bar{\varepsilon}$
Gorge (inches)	5.4	1.0	−0.47	0.12	0.34	0.48
		2.5	−0.86	0.31	0.50	0.77
Roll	0.044	3.0	−0.99	0.39	0.55	0.98
Exit Angle (rad.)						
Lead (inches)	0.0	4.0	−1.26	0.56	0.62	1.24
		5.0	−1.57	0.76	0.67	1.52

If the piercing is made with a cone type of piercer point deformation
occurs in a field represented by two linear equations as illustrated in Figure
5. In this case, the radial strain increment is represented by

$$d\varepsilon_r = \int_o^L \frac{(\alpha-a)}{(\alpha-a)\,x+c}\,dx \tag{13}$$

Longitudinal strain increment is given by

$$d\varepsilon_L = \int_o^L \frac{2(a^2-\alpha^2)\,x-2\alpha c}{(\alpha^2-a^2)\,x^2+2\alpha cx+c^2}\,dx \tag{14}$$

and finally the circumferential strain increment is represented by

$$d\varepsilon_\theta = \int_o^L \frac{(a+\alpha)}{(a+\alpha)\,x+c}\,dx \tag{15}$$

Table II shows the values of the homogeneous strains and effective
strains for a cone type of piercer point. As noticed summation of the
computed strains are equal to zero. This time, because of the simplicity of
the linear equations, that do not contain some of the parameters as the
parabolic equations, give very accurate precise results.

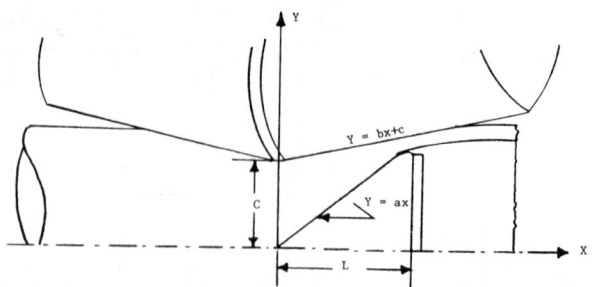

Figure 5 - Illustration of a piercer pass profile
with conical shape of piercer point.

TABLE II. Homogeneous Strains and Effective Strain
for a Cone Type of Piercer Point

		Distance From Point Nose, X (inches)	ε_r	ε_L	ε_θ	$\bar{\varepsilon}_{eff}$
Gorge (inches)	5.4	1"	-0.17	0.17	0.0	0.19
Roll Exit Angle (Rad.)	0.044	2"	-0.38	0.32	0.06	0.41
		3"	-0.64	0.45	0.19	0.65
Lead (inches)	0.0	5"	-1.55	0.66	0.89	1.55

Figure 6 and 7 illustrate the variation of the incremental homogeneous strains and the effective strains along the strain path for the parabolic and the cone type of piercer points respectively. As seen in either cases the radial homogenous strain dominates the entire deformation and there is a big competition between circumferential and longitudinal deformations. Circumferential strain dominates, over longitudinal strain almost until the end of working region in where the longitudinal strain takes over. These relations are particularly significant for design engineers to reduce the amount of undesirable deformations. These expressions can serve as a reference for estimating the efficiency of the operations and making comparisons between different passes to obtain an optimum performance. Also, knowing the strains throughout the deformation path will be of greatest significance for the state of stress in the pierced metal, deformation load, energy requirements and for the occurrence of defects under known rolling conditions. Thus, the effect of deformation strains can also be evaluated for the quality of the hollow billets.

The Rate Of Deformation

In metal-working processes theoretical analyses are being made to derive expressions for loads, torques, power requirements etc., in terms of processing and geometrical variables, and as well as the yield strength of the material. Unlike cold-forming operations, hot-working operations gives more complicated situations to determine the yield strength characteristics of the materials. In hot-working, the temperature and strain rate (rate of deformation) have a pronounced affect on the yield strength of the material. The yield strength of the material increases with strain rate, also the temperature of the workpiece is increased because of adiabatic heating. In piercing operations, since the deformation rate is dependent on the geometry of the deformation zone, it governs the final quality of the finished product. The working length of the piercer point in this case determines the rate of reduction from solid to tube, that is, the conversion from solid billet to hollow shell is controlled by the length of the piercer point. The shorter the point length means the quicker the conversion from billet to hollow shell, so that it provides conditions either higher or lower deformation rates. That means the presence of the point substantially influences the stress and strain state of the pierced materials at the head of the piercer point [4]. The magnitude of these stresses imposed by the piercer points against material flow greatly depends on the point shape which controls the deformation rate. That is, the final quality of the finished product is strongly dependent on the magnitude of deformation rate. Since it is a very important parameter in determining

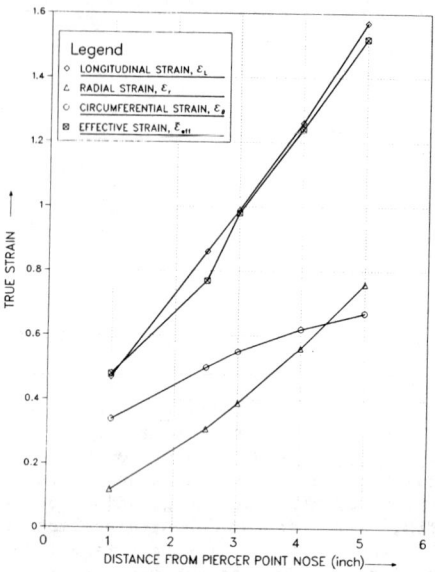

Figure 6 - Illustration of the incremental homogeneous strain
distributions in a piercing pass with parabolic shape of piercer
point.

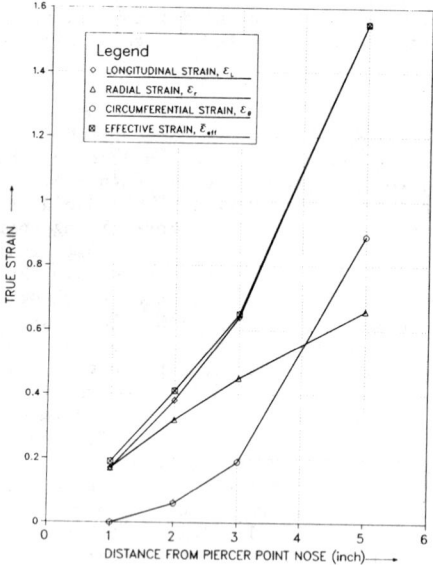

Figure 7 - Illustration of the incremental homogeneous strain
distributions in a piercing pass with conical shape of piercer
point.

the final quality, for design purposes, it was necessary to express this important piercing parameter in terms of point geometry and other processing variables, Figure 8. For the first piercing operation, a mathematical expression which describes the strain rate in terms of processing and geometrical parameters was developed and given by:

$$\dot{\varepsilon} = \frac{4\pi\eta R_g NSin\phi\, tg^\psi}{D_b + D_s - 2D_g} \cdot \bar{\varepsilon} \tag{19}$$

where $\bar{\varepsilon} = \ln\left(\frac{2t}{D_b}\right)$

This expression was also derived in the following way which is related to piercer point working length,

$$\dot{\varepsilon} = \frac{4\pi\eta\, R_g N\, \phi A_e}{L_w(A_b + A_e)} \cdot \bar{\varepsilon} \tag{20}$$

Feed Efficiency

In piercing, one of the most important relationships which influences the piercing performance is the relationship between shell delivery speed of the pierced material and the axial component of the peripheral speed of the rolls [5], Figure 9. This is expressed as follows:

$$\eta = \frac{v_o}{u_o} \tag{21}$$

By using the maximum diameter of the rolls, and determining the maximum circumferential speed of the roll from its feed angle, it is possible to obtain a very important piercing parameter that not only controls the production efficiency, but also controls the effective working volume called, feed efficiency. In piercing the feed efficiency controls the production rate and effective working volume which undergoes a relatively large amount of cross rolling in the feed region with a potential danger of incurring fractures in the center. It varies with the tooling surface condition of the rolls, billet material, piercing temperature, strain rate, roll speed and more importantly piercing set up. By its definition, a higher feed efficiency means a higher production rate through the piercer while a small feed efficiency describes a small advance per revolution of a planar section of the billet through the feed region where an extensive cross rolling takes place. The feed efficiency can be obtained in the following way:

$$\eta = \frac{v_o}{uo}\, 100\ (\%) \tag{22}$$

$$\text{where} \quad uo = u \cdot \sin^\alpha \tag{23}$$

$$\text{and} \quad u = {}^\pi D_g n \tag{24}$$

and finally feed efficiency is given by,

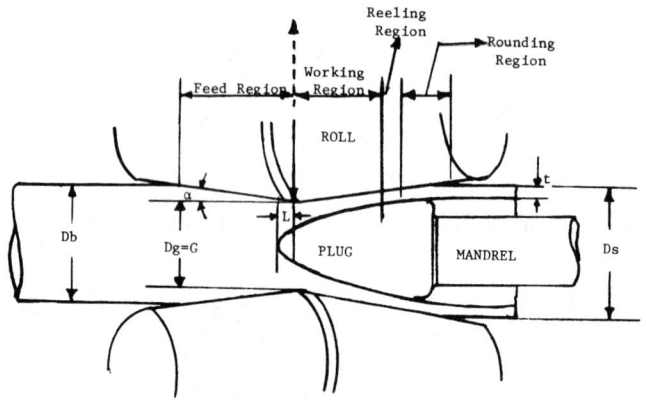

Figure 8 – Illustration of the various piercing parameters
in a rotary-piercing mill.

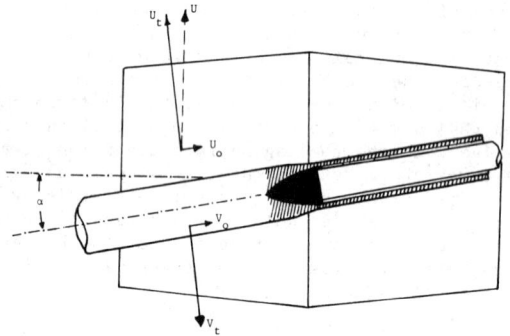

Figure 9 – Illustration of various velocity components
of the piercing rolls and the pierced material.

$$\eta = \frac{v_o}{\pi D_g n \sin \alpha} \cdot 100 (\%) \tag{25}$$

<u>Stich-Zahl</u>

In piercing in regard to surface quality, the number of revolutions of the billet (called stich-zahl) in the feed region, where cross rolling takes place, is another important factor to control the quality of the pierced shell [6]. It is important because the work done on round in the feed region is related to the numbers of the turns from roll contact to the nose of the point. A higher total number of the revolutions creates a greater propensity for billet-center heating and fracturing which can adversely affect ID surface quality. In this study mathematical expression which represents the number of revolutions of the pierced billet in the feed region was developed as follows:

$$N = \frac{D_b \cdot L_f}{4A_e \cdot \eta \cdot \alpha} \tag{26}$$

The relationship between stich-zahl and feed efficiency as a function of a particular set-up and point shape is illustrated in Figure 10 which shows an increase in feed efficiency as stich zahl decreases. The effect of feed efficiency on the horsepower is illustrated in Figure 11 which clearly shows a substantial increase in horsepower when feed efficiency increases [6].

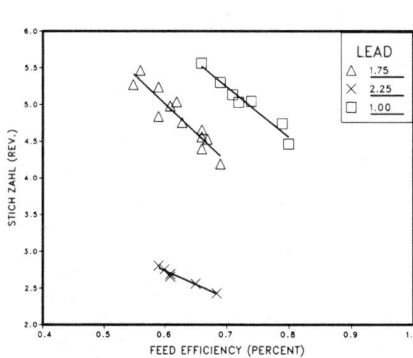

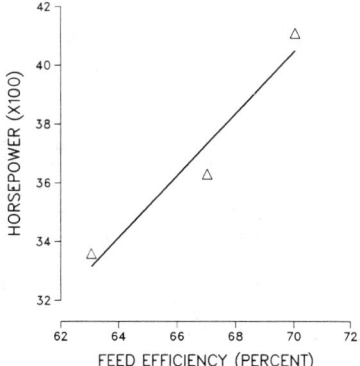

Figure 10 - Illustration of the relationship between stich-zahl and feed efficiency.

Figure 11 - Illustration of the relationship between feed efficiency and horsepower.

Shell Wall Thickness

In piercing, the dimensional quality of the shells are directly related to the accurate concentricity of the shell wall thickness. That is, the lower the eccentricity in wall thickness the better the dimensional quality. A study of the various piercer setups at a seamless pipe mill, that would yield a quality related unified practice, must necessarily include product-in and product-out dimensions. This requires mathematical expressions relating round size and pierced shell dimensions to piercer setup parameters since there is an absolute relationship between setup variables and wall thickness. In a piercing mill, wall thickness can be calculated from measurements of the length and outside diameter of the pierced shell and the billet size assuming constant volume deformation by equating the shell volume to the hot-billet metal volume. The latter volume is determined using the measured cold length, a nominal cold diameter, and allowances for thermal expansion and losses in density, as well as furnace scaling. A mathematical expression in terms of these conditions was developed and given by:

$$W = 0.5 \ [D_s - \sqrt{(D_s^2 - D_o^2) \ \frac{L_b}{L_s}} \ ] \qquad (27)$$

The wall thickness in terms of setup variables such as gorge, lead, working length of the point, point diameter, and roll-outlet face angle was developed and can also be used to compute the wall thickness, is given by:

$$W = 0.5 \ (G - D_p) + tg^\alpha (L_w + L_R - L) \qquad (28)$$

These expressions, however, do not include some variables, which adversely influence the final wall thickness, are: 1) the displacements of the roll surfaces and the piercer point during loading (elastic deformation of the rolls), 2) the errors in the dimensional measurements, 3) unpredictable billet-density loss (~2%) and, 4) thermal expansion and contraction in cooling. Thus, small insignificant differences between actual measurements and computed values would be expected.

These expressions have been used to develop piercer set up and quality control algorithms which control the entire process within the proposed boundary conditions. The program is called "control strategy program" that provides the most suitable piercing conditions, such as plug diameter, gorge, lead and distance between guide shoes by determining stich zahl, shell wall thickness, horsepower, specific work and feed efficiency. The control strategy is also capable for the adjustment of the controlling parameters during piercing in an attempt to maintain errors in the shell wall and in the stich zahl within predetermined amounts. It consists of the combination of nine allowable deviations (based on roll outlet face angle, lead and gorge) and the suggested adjustments in setup for errors in shell wall and stich zahl.

Some of the computed results describing the relationship between set-up and performance variables are illustrated in Figures 12 through 14 [6]. The relationship between horsepower and set-up values for the range of variation encountered in the experiments with a particular point shape is shown in Figure 12. As seen, it shows a sharp decrease when the lead and gorge increase. Also, the relationship between feed efficiency versus lead and gorge and stich zahl versus lead and gorge for the range of variations are shown in Figures 13 and 14 respectively. They show a decrease in feed efficiency and stich zahl as lead and gorge increase.

1122

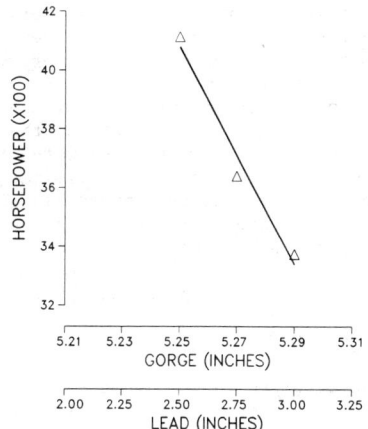

Figure 12 - Illustration of the
relationship between horsepower
and set-up variables.

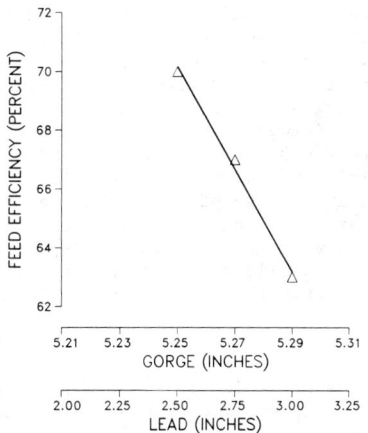

Figure 13 - Illustration of the
relationship between feed efficiency
and set-up variables.

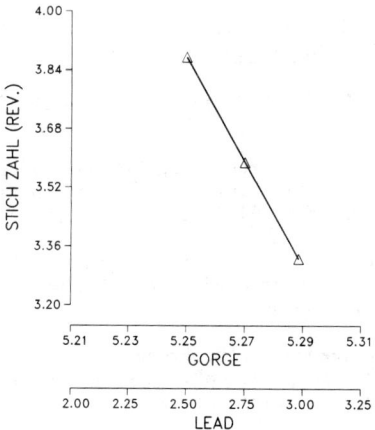

Figure 14 - Illustration of the relationship between stich-zahl
and set-up variables.

Conclusion

The work calculated by the expressions and models represent the minimum work per unit volume of material necessary under the assumed conditions to achieve the desired change from a solid billet to a pierced hollow in different passes. Even though, they do not include such factors as redundant (shear) work and losses resulting from slippage and friction, it can serve as a reference for estimating the efficiency of the operations. Also, the derivations provide a base for the other metalforming operations to derive more reliable and more realistic mathematical expressions in terms of geometrical and processing variables. To provide design criteria for the first piercer mill, as well as better understanding of its operation, the expressions for work of deformation in various passes, deformation rate, feed efficiency, stich-zahl and wall thickness were derived under very limited simplified assumptions. These expressions provide necessary information for proper design and control of the piercing process and to reduce the number of expensive trials. The validity of these equations have been verified experimentally and have been found sufficient use in the mill design and more importantly they are now very significant part of the computerized piercing mills.

Acknowledgement

The author would like to express his appreciation to the Management of the Technical Center of the USX (U S Steel) Corp. for permission to publish this paper.

REFERENCES

1. The Making, Shaping, and Treating of Steel, United States Steel (USS), Edited by W. T. Lankford, Jr. et al., Tenth Edition 1985, Pittsburgh, PA.

2. T. Z. Blazynski and Cole I. M., "An Analysis of Redundant Deformations in Rotary Piercing", Proc. Inst. Mech. Engrs., 1963-64, 178, p. 867.

3. T. Z. Blazynski, "Metal Forming, Tool Profiles and Flow", published by Halsted Press, 1976.

4. E. Erman, "The Effect of Processing Parameters on the Propensity for Central Fracturing in Piercing", Journal of Applied Metalworking, Vol. 4, No. 4, January 1987, pp. 331-341.

5. M. Snasel, "Kinetics of the Rotary Piercing Process, Practical Measurements, and Calculations of Coefficients of Slip", Hutn. Listy., 1977, 32, (2), pp. 103-106.

6. E. Erman, "The Influence of the Processing Parameters on the Performance of the Two-Roll Piercing Operation", accepted to be published in Journal of Mechanical Working Technology, November 6, 1986.

$\bar{\sigma}$ = Flow Strength of the Material at the Piercing Temperature, Strain, and Strain Rate

$\bar{\epsilon}$ = Effective Strain Increment

a = Piercer Point (parabolic) Shape Constant

c = One Half of the Shortest Distance Between Rolls (inch)

x = Axial Piercing Path (inch)

L_f = Length of Feed Region (inch)

A_e = Exit Shell Area (inch^2)

A_b = Billet Area (inch^2)

r = Billet Radius (inch)

t = Wall Thickness (inch)

η = Feed Efficiency (Assumed 0.75-0.85)

R_g = Roll Radius at Gorge (inch)

N = Number of Successive Turns of Billet in the Feed Region

ϕ = Feed Angle (rad.)

α = Roll Inclined Angle (Rad.)

D_b = Billet Diameter (inch)

D_s = Shell Diameter (inch)

D_g = Diameter of Billet at Gorge (inch)

L_b = Billet Length (inch)

L_s = Shell Length (inch)

G = Gorge (inch)

D_p = Point Diameter (inch

L_w = Point Working Length (inch)

L_R = Point Reeling Length (inch)

L = Lead (inch)

v_o = Axial Component of Roll Velocity (inch/sec)

U_o = Axial Component of the Circumferential Velocity of Rolls (inch/sec)

n = Roll Speed (rps)

u = Circumfential Velocity of the Rolls (inch/sec)

FINITE ELEMENT ANALYSIS OF

HOT ISOSTATIC PRESSING OF

BERYLLIUM POWDER INTO NEAR NET SHAPES

Tom P. Courtney
Jan L. Krankota

Rockwell International
Rocky Flats Plant
Golden, Colorado

ABSTRACT

Hot Isostatic Pressing (HIPing) of beryllium powder into hear net
shapes is being investigated at Rocky Flats. Powder is produced from a
blend of recycled scrap and high-purity, electrorefined flake. One
shape of interest is a hemispherical shell which is HIPed from powder
in the annulus of a can made from two hemispherical steel shells of
different radii welded at the brim. Initial tests resulted in the
outer wall buckling during HIPing. Finite Element Analysis
(MARC/MENTAT code) was used to construct 2-D axisymmetric models and
perform simulations of the process. This large deformation, non-
linear analysis included a constitutive model for the compaction of
the powder. Sensitivity studies were made to show the effect of
changes in process variables and shape geometries (i.e., wall thickness,
shell radius) on the deformation behavior of the powder in the
annulus. Results were used to modify can design to produce acceptable
parts for finish machining. The analysis confirmed the need to
thicken the outer wall to prevent buckling.

INTRODUCTION

Hot Isostatic Pressing (HIPing) of beryllium powder into near net part shapes is being investigated to reduce the amount of powder being processed and to minimize machining. Near net shapes are HIPed directly from either loose-packed powder or from preforms that have been cold isostatically pressed (CIPed) from powder. HIPing of loose-packed powder directly into near net shapes is the process modeled in this paper.

The hemican used for direct HIP has unusual features which can interfere with shape control. The can is made from two thin-walled, hemispherically-shaped steel shells of different sizes joined together at a common brim. HIPing of powder-filled cans produces large, nonlinear canwall deformations, mainly due to the geometrical effects characteristic of thin shell structures. These deformations often lead to buckling which seriously interfers with dimensional control of the part.

An advanced K3 version of the MARC/MENTAT Finite Element Analysis (FEA) code was used to analyze and model the HIPing of 304L stainless steel cans with 1:1 and 4:1 wall thickness ratios. This code incorporates new Rebelo contact algorithms (Reference 1) for metal forming simulation using deformable and rigid dies, and also combines necessary material property data, part geometry, and boundary condition nonlinearities to produce realistic simulations of metal working processes. User subroutines further simulate the highly nonlinear response of the metal powder both to temperature change and to the compacting isostatic pressure. In addition they provide, through an extension of classical creep and plasticity theories included in the code, a means of accounting for the large volume decrease of the part during HIPing.

LABORATORY EXPERIMENTAL PROCEDURE

Initially, small hemicans made from 1/16-inch thick stainless steel sheet pressed into hemispherically shaped shells by rubber-pad forming (Marforming) were tested so they could be HIPed in a laboratory MiniHIPer. The outer shell was nearly 3-1/2 inches in diameter, and each shell had a flat brim so they could be nested together and single-bead welded on the brim edge to form a can with a 1/4-inch annulus (Figure 1).

Large hemicans were then made from two hemispherically-shaped shells each Marformed from 1/16-inch thick stainless steel sheet. The outer shell was 9 inches in diameter, while the inner shell was 8 inches in diameter so that a 1/2-inch annulus was formed when similarly welded (Figure 2).

Laboratory HIPing experience and FEA showed better shape control was achieved with a graphite ring die and with hemicans having thicker outer shells. Hemicans were thus tested with 1/8-inch thick outer shells and 1/32-inch and 1/16-inch thick inner shells.

Beryllium powder was loaded into the annular gap through a 1/4-inch diameter fill tube located on the pole of the outer shell of the hemican. The powder (SP-200E, minus 325 mesh, less than 44 micron diameter) was vibrated to 55 percent theoretical

1128

density on a sieve shaker. Adsorbed gas contaminants were removed by heating the hemicans under a vacuum of 10^{-5} torr for more than 8 hours at 650° C. A graphite ring die was used to support the large hemicans since an initial test showed a .3-inch out-of-round distortion at the waist.

Small hemicans were HIPed in the laboratory MiniHIPer, and large hemicans were HIPed at a vendor's site and at Los Alamos National Laboratory (LANL). The cans were first heated for 105 minutes to avoid straining cold welds during the subsequent 15-minute pressurization cycle. Then, a HIP cycle of four hours at 1000° C and 15,000 psi was used. After HIP, the cans were similarly depressurized before cooling to room temperature to complete the cycle.

LABORATORY EXPERIMENTAL RESULTS

Small hemicans exhibited very little distortion during HIPing (Figure 3). The brim rotated down and away from the pole of the hemican as was predicted by FEA, and no buckling of the 1/16-inch thick shells occurred. The beryllium powder was compacted to 100 percent theoretical density with properties similar to beryllium made by conventional vacuum hot pressing (VHP).

Buckling of the outer 1/16-inch thick shell occurred in large hemicans (Figure 4). The hemicans waist distorted without the use of a graphite ring die. FEA showed buckling is eliminated by thickening the outer shell, and a graphite ring die minimizes out-of-round distortion. Subsequent graphite die supported hemicans with thicker (1/8-inch) outer shells did not buckle nor distort during HIPing (Figure 5), though the brim did rotate slightly up as predicted. The properties of the beryllium powder HIPed in large hemicans are also equivalent to those of conventional VHP beryllium.

FINITE ELEMENT ANALYTICAL PROCEDURE

PREVIOUS POWDER MATERIAL SIMULATION MODEL

FEA simulation of beryllium hemishell HIPing required a constitutive law describing the time-dependent compaction behavior of the powder at HIP temperatures/pressures. The previous equation in use at Rocky Flats was derived from density-pressure relations measured on direct HIPed powder (Reference 2) and on CIP/HIPed powder (Reference 3). However, this equation does not account for time-dependent compaction,

Strain = Pressure (K) where K = 1/8900 (1/psi) for T = 1000° C

The density change is related to the strain which also relates instantaneous compaction to final density upon pressurization,

$$\text{Strain} = 1 - \frac{\%TD_o}{\%TD_f}$$

where $\quad \%TD_o$ = the initial density (55-60% for direct HIP and 78-80% for CIP/HIP)

$\%TD_f$ = the final density after HIP

CURRENT MATERIAL SIMULATION MODEL

A recent empirical constitutive model derived by Carroll (Reference 4) now accounts for time-dependent effects during constant temperature/isostatic pressurization of powdered metal. The model is empirical in that it requires test data to determine the values of three adjustable material parameters: **mod, tau,** and **M** which account for instantaneous response, creep response and equilibrium response of the powder compact. The equation relates density increase (%TD) with time,

$$\text{Density} = 1 - (1 - D_{oo}) \left\{ 1 - \left[1 - \left(\frac{1 - D_{\infty}}{1 - Do} \right)^M \right] \left[\exp \left(\frac{PM}{mod} \right) \right] \left[\exp \left(-\frac{Mtime}{tau} \right) \right] \right\}^{-1/M} \quad (1)$$

where
D_{oo} = final relative density
D_o^{oo} = initial relative density prior to application of pressure
time = elapsed time during HIP
P = applied isostatic pressure (psi)
mod = instantaneous densification modulus upon application of pressure (1/psi)
tau = time constant similar to inverse of densification rate (1/min) or material constant
M = nondimensional exponent or creep compaction constant

Partial compaction, occurring as soon as pressure is applied, is accounted for by the instantaneous densification modulus. Next, the slowly decreasing rate of compaction (or densification) due to the powder particles deforming by creep mechanisms is accounted for by the time constant and nondimensional exponent. The final density attainable at the given pressure and temperature is accounted for by D_{oo}, which must be less than 1.0 (0.99 is used.)

Carroll has shown that this equation fits empirical metal powder data quite well, and that laboratory densification behavior can be represented by curves generated from this equation when appropriate values of the adjustable parameters are selected. Typical curves to be generated from the equation with ranges of parameters expected for beryllium powder are shown (Figure 6). Laboratory measurements of densification behavior in progress will refine these parameters.

MODEL BUILDING

MENTAT was used to construct the two-dimensional, axisymmetric half section element mesh with fill tube, which represents the powder-filled hemishell can. The number of elements (MARC element type 10, i.e., an axisymmetric ring with a quadrilateral cross-section) used in the models varied from 1153 to 1366, while the number of nodes varied from 1547 to 1759. The original model employed can walls of equal thickness (Figure 7), while subsequent models contained thicker outer walls and thinner inner walls (Figure 8). Since the analysis was originally performed to confirm the design of the seal weld joining the brims of each wall of the can, more elements were placed in the weld region to accurately represent displacements of the brims. The hemican was placed upside down in a graphite ring die, simulated in the code by a rigid die. A small gap was introduced part-way down between the can top and the powder to allow metal thermal expansion to occur before the contact algorithm comes fully into effect during the compression phase. Setup for this problem is available upon request.

MARC FEA/DESIGN PROCESS SIMULATION

A general purpose FEA code such as MARC allows either a visco-plastic formulation or an elastic-plastic formulation which includes compressibility. A creep analysis is often subsequently performed after the rate independent elastic-plastic analysis. The present approach reduces code complexity; first, by not requiring a specially written code with resultant loss of application flexibility and second, by bypassing the normally-required procedure of elastic-plastic loading (which is time-independent), followed by a steady state creep increment. The latter step is done by embedding, via subroutine, the Carroll model constitutive law in the swelling term. Since this swelling term is called at each increment when the updating of the time step takes place, access to subroutine VSWELL is automatically obtained. The Carroll model, which contains its own creep parameter, is therefore implemented during the elastic-plastic analysis phase without the requirement for a subsequent creep phase.

In general, the analysis employs a finite plasticity/large displacement/updated Lagrange formulation in time. A full Newton-Raphson iterative procedure, a mean normal method utilized in the plasticity equations, a forcing of non-positive definite systems, and a displacement convergent test (.1 maximum allowable change in displacement iteration divided by displacement increment) are used. Since the elastic-plastic stress analysis is time-independent, it is necessary that the strain increments be small enough to obtain the required accuracy in the integration of the rate equations of plasticity.

Thermal and pressure loading stages are separated to simulate the present forming process. A formally coupled, adiabatic analysis is not used since the temperature-dependent process is isothermal, and coupled creep is not presently available. Volume strain is constant for the steel in each element so this metal becomes approximately incompressible (required for structures operating in the fully plastic range).

MATERIAL SIMULATION

An isotropic constitutive law is used only for the stainless steel. Initial low strength room temperature properties are assigned for the beryllium powder; Young's modulus (5,000,000 psi), yield stress (4000 psi), and density (82 percent of theoretical). These properties remain in effect until the Carroll Model is introduced 105 minutes into the loading cycle. More current temperature-dependent beryllium material properties are then used for tracking during the remaining pressurization/creep/depressurization/cooldown phases.

Expansion/shrinking of the can and powder due to absorption/loss of heat requires temperature-dependent material properties which can be automatically tracked by the code. The work hardened

temperature-dependent yield stress is obtained through subroutine YIEL (Table 1) for the stainless steel and the solid
beryllium. The initial yield and failure stresses for these
metals at the current temperature are obtained from handbook
values. Initial parameters for a strain hardening power equation are then approximated using values for ductile materials
(Reference 5). These parameters are improved by an iterative
scheme using the percent elongation at the failure stress.
Finally, the work hardened yield stress at current plastic
strain is obtained from the power law, and a check is made to
insure the value is between the yield and failure stresses.

CARROLL POWDER MODEL IMPLEMENTATION

The Carroll model is presently only applicable at constant
temperature/pressure. It is used in a 10-step linear pressurization loading cycle which approximates the HIP process.
Subroutine VSWELL (Table 2), called at each integration point
where constitutive calculations are being performed by the
program, normally allows the user to define long term dilational
creep and swelling. A code change mechanism permits inclusion
of the Carroll model into the swelling term (SWELL). Since it
can be shown swelling ($\Delta v/v$) is equivalent to the negative of
shrinkage ($\Delta d/d$) (Table 3-1; proof) the difference in relative
densities (Equation 1) can be taken at two time increments to
compute the shrinkage term $\dot{E}$ (SWELL) (Table 3-2). Total
shrinkage accumulation is automatically performed so plotting of
this term can be accomplished during post processing.

CONTACT ALGORITHM

The contact history between these metals strongly depends on
temperature-dependent material properties, geometry, friction,
and the process pressure/temperature loading. Gap openings/-
closings between the metals are automatically provided by the
direct contact algorithm since both metals are declared as dies
(Table 4 - contact closure distance is set at .005-inch). This
tolerance was changed to .007-inch in order to speed up computations once pressurization began. During this compaction phase,
the hemicans begin to diffusion bond to the powder walls so the
contact distance requirement is no longer critical. Diffusion
bonding of the beryllium and steel elements is completed during
cooldown, turning the structure into a continuum.

FRICTION

Temperature/velocity dependent metal friction factors are
determined in subroutine UFRIC (Table 5) where a threshhold
value of .1-inch per minute sliding velocity separates kinematic
from static friction values for the different metal surfaces.
The two metals are initially in contact except on the top side
near the brim where an open space is simulated. Besides
simulating what occurs normally, this gap prevents the contact
algorithm constraints and friction considerations from being
fully satisfied when thermal expansion of the metals needs to
occur.

1132

DIE MOVEMENT

Subroutine DIEPOS (Table 6) permits changing the velocities of the dies in time, so removal of the outer cans after the HIPing process can be simulated. This permits observation of springback and any remaining residual stress effects after cooldown. This can only be obtained accurately if an elastic-plastic analysis is performed.

PRESSURE LOADING

When elements contact the graphite ring, distributed pressures at the integration points must be set to zero. When contact is broken, these distributed pressures must be reset to the current autoclave pressure. Subroutine FORCEM (Table 7) contains a list of possible contact elements with their associated surface nodes. Surface corner nodes of each element are tested against the graphite ring nodes, actively in contact with the dies, to determine whether local contact is made. The integration point distributed pressures are then adjusted accordingly.

FEA RESULTS

Final deformed shapes (Figures 9A and 9B) compared favorably with laboratory experiments. The brim of the 1:1 hemican rotated downward during the thermal loading cycle as shown in Figure 10A, using an exaggerated 1/2-inch scale to show the deformation mechanism. The maximum thermal strain needed to be restricted to 20%-50% of the yield strain so eight temperature changes of 220.25° F. were required to reach the HIP temperature. It is evident the graphite die ring (by restraining lateral motion) and the fill tube structure (by creating a sharp corner) have significant effects on deformation.

At high temperature, lower yield strengths permit plastic straining at the corner of the steel tube and at the beryllium near the brim. In addition, the brim closes up causing a dimple on the top which permits it to curve downward. The outermost elements in the beryllium at the brim tend to go inside out due to their low strength and the movement of the can walls into the open gap. Full strength properties were assigned to the right-most three powder elements to permit a solution. It is evident the outer wall must be made thicker in order to provide enough strength to act as support for these focused deformations.

At 5,000 psi pressure, the outer wall of the hemican buckled through at the pole. The 4:1 hemican (Figure 10B) displayed only plastic straining in the beryllium powder in the fill tube, which bulges slightly during the same thermal loading cycle. A portion of the residual stress state is not due simply to pressure/thermal gradients, but rather to nonuniform volumetric changes which occur in the early stages of compaction.

The outer wall of the 1:1 steel can began to compress vertically and bulged out as it wrapped around the radius on top of the graphite die. Friction shearing between the powder and the outer can was evident in the flat area just above the first point of contact with the graphite die. At the end of the thermal cycle the inner can moved away from the powder in the pole region down to 30° from the brim. Upon pressurization the inner can compressed the powder more in the region of the pole causing a ripple in the inner and outer cans at 45° from the brim. The wall of the can in contact with the graphite die flattened out. During the thermal expansion phase, rotation of the brim of the hemican, due to contact with the ring, caused the inner hemican to separate from the powder. The same type of behavior occurred in the 4:1 hemican except the force exerted inward by the inner wall toward the graphite die caused slightly more shearing in this region.

A contact problem encountered at the pole required a workaround solution. Two beryllium nodes in contact with two steel nodes on the axisymmetric axis passed through each other since the contact algorithm cannot distinguish between horizontal and vertical boundary conditions. The problem was corrected by modifying the Y coordinates of these beryllium nodes so they were no longer on the axisymmetric axis. This slight indent does not affect the solution.

Another limitation occurred in a simple checkout problem run prior to the production simulation. Due to the imbalance in pressure loading between the inner and outer cans caused by the difference in surface areas, the resisting friction of the graphite ring was exceeded. Since no boundary conditions were set in the Y-direction to offset this rigid body motion, the solution began to oscillate. For the large scale models, oscillation presented no problem since during the thermal loading phase, local buckling of the outer walls of the hemicans at the top of the ring provided enough support.

The shrinkage mechanism operated correctly since the beryllium elements decreased in volume. The modification of the distributed pressure loading was modeled correctly since these pressures were zeroed out when the elements contacted the graphite ring.

CONCLUSION

The prediction of HIPing response of complex shaped cans is difficult intuitively. FEA minimizes the costly trial and error testing program normally used to prove this HIP can design. By predicting deformation patterns in the compacting powder and strains in the seal welds, a double-walled can design was achieved for hot isostatically pressing (HIP) beryllium powder into hemishell shapes. The improvement in can design by using a thicker outer wall to avoid buckling and the use of a graphite die restraining ring showed good correlation with a limited number of laboratory tests.

An FEA model was constructed combining large scale meshes with recent micromechanistic models of particulate compaction during the constant temperature and pressure phase of the HIP cycle. Future simulation refinements will include (1) extension of the Carroll model to the heat up as well as the pressurization HIP phases, (2) coupling of these phases during the entire process using the new K3 AUTO THERM CREEP option, (3) addressing diffusion bonding of beryllium to the can walls, (4) inclusion of a dense random pore model in the elements in the major shear zone to observe pore closure, and (5) prediction of ductility/-hardness/grain size from flow stress using state variables (data has not yet been accumulated for beryllium).

REFERENCES

1. Rebelo, N.M.R.S., "Implementation of forging analysis procedures in the MARC program-Phase III Task 5", United Technologies Corp., Pratt and Whitney Manufacturing Division, East Hartford, CT; Report TR8703, prepared by: MARC Analysis Research Corp., 260 Sheridan Ave., Suite 280, Palo Alto, CA 94306, U.S.A., June 1987,

2. Pinto, N. P., "Comminution and Consolidation", Chapter 2 in Beryllium Science & Technology, D. Floyd and J.N. Lowe, Plenum Press, NY, 1979.

3. Roberts, D., "The warm isopressing of beryllium powder", Paper 24 in UCRL-89338, Vol. 2 (Paper presented at US/UK Information Exchange at Rocky Flats Plant, CO, June 1983), April 1984.

4. Carroll, M. M. "An empirical model for hot isostatic pressing of metal powders", Metallurgical Transactions A; v.17A, pp. 1977-1984, November 1986.

5. Beitscher, S., "An easy way to estimate strain hardening exponents", Metal Process, pp. 35-36, August 1985.

Figure 1. Small double-walled hemican for HIPing beryllium powder

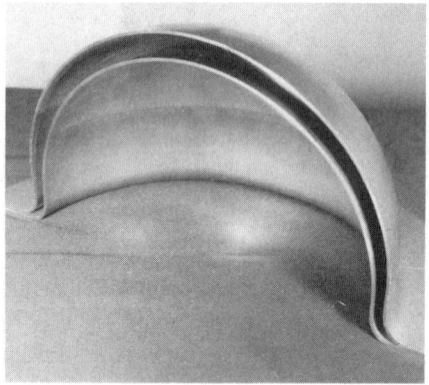

Figure 2. Large double-walled hemican for HIPing beryllium powder

Figure 3. Small double-walled hemican after HIPing of beryllium powder into hemishell

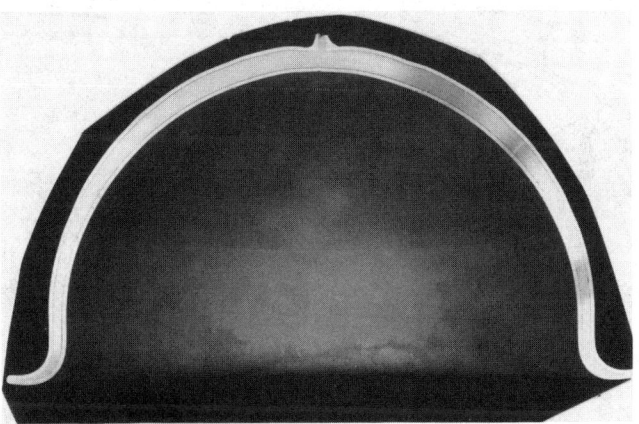

Figure 4. Large double-walled hemican after HIPing of beryllium powder into hemishell; note slight buckling of outer wall

Figure 5. Large double-walled hemican with thicker outer wall after HIPing of beryllium powder into hemishell; note absence of buckling of outer wall

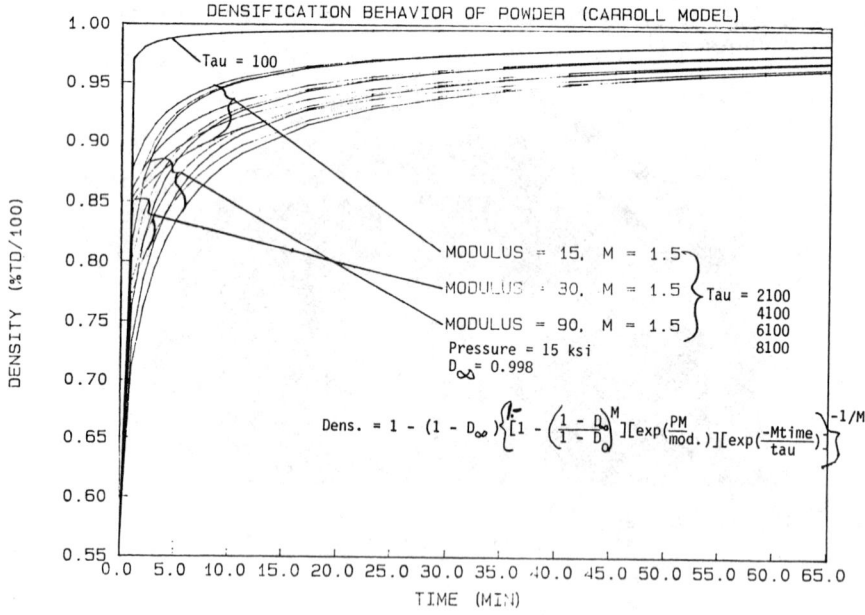

Figure 6. Expected densification behavior of beryllium powder under HIP conditions as modelled by Carroll (ref. 1)

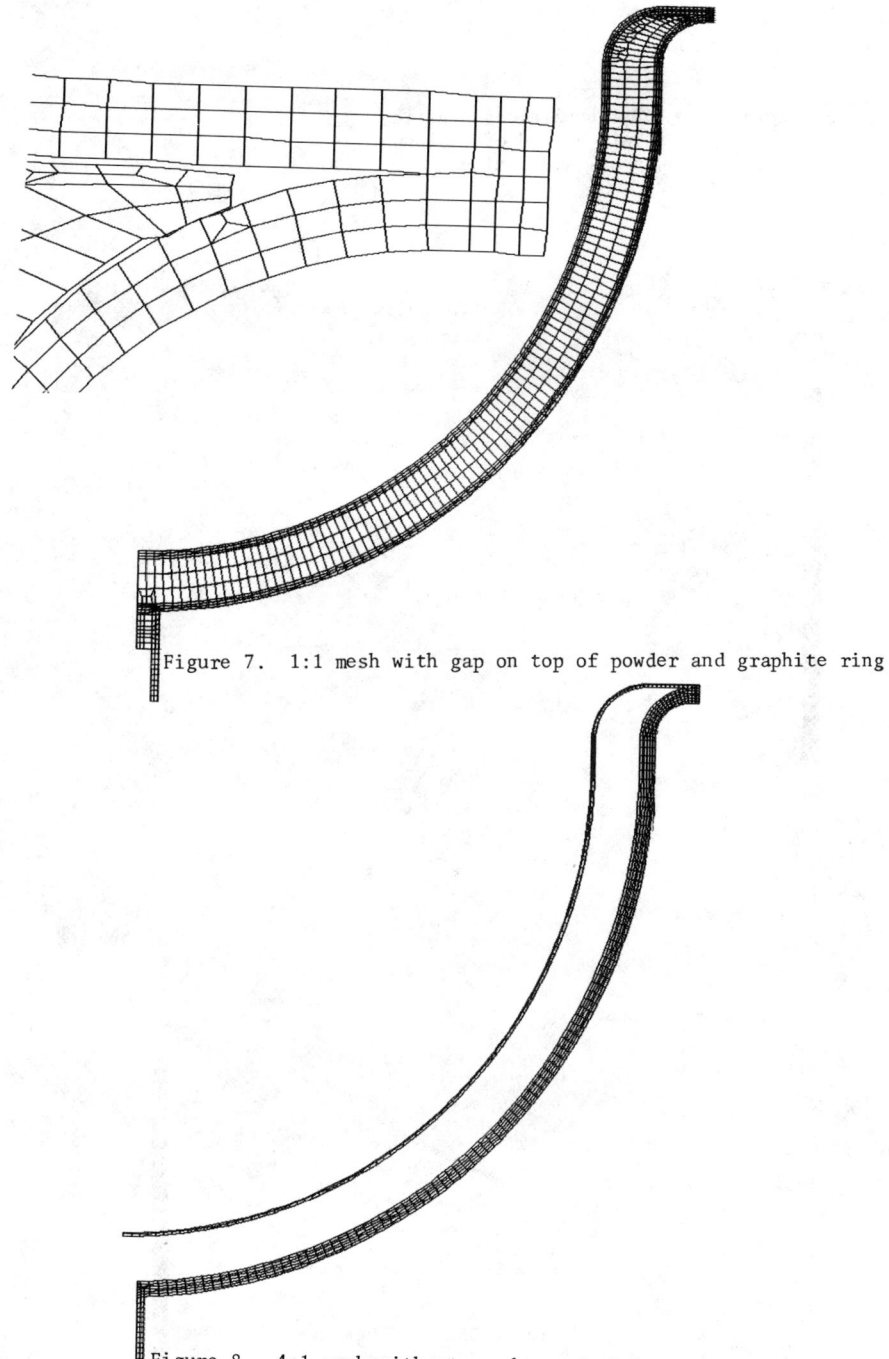

Figure 7. 1:1 mesh with gap on top of powder and graphite ring

Figure 8. 4:1 mesh without powder and with graphite ring

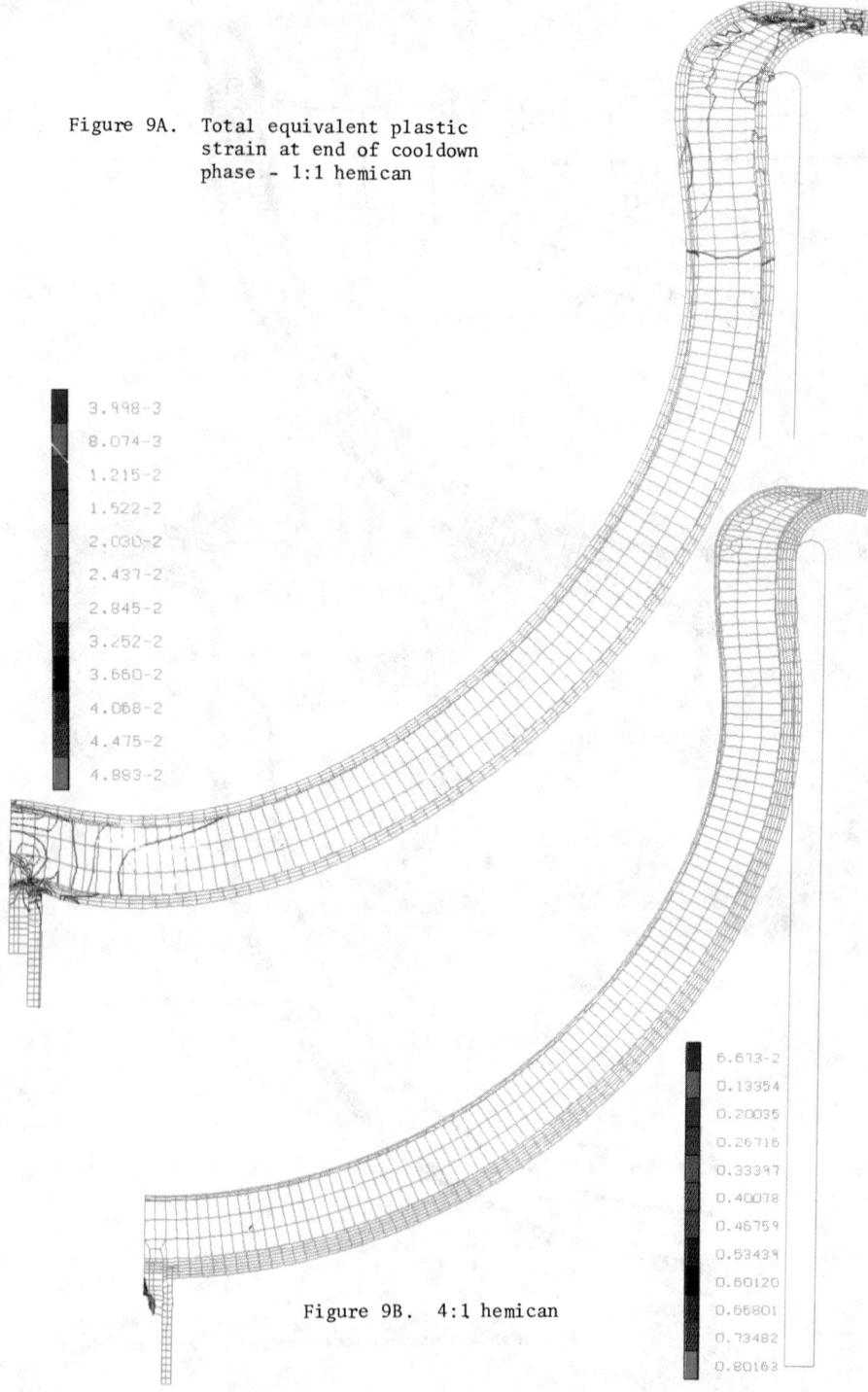

Figure 9A. Total equivalent plastic
strain at end of cooldown
phase - 1:1 hemican

3.998-3
8.074-3
1.215-2
1.522-2
2.030-2
2.437-2
2.845-2
3.252-2
3.660-2
4.068-2
4.475-2
4.883-2

Figure 9B. 4:1 hemican

6.673-2
0.13354
0.20035
0.26716
0.33397
0.40078
0.46759
0.53439
0.60120
0.66801
0.73482
0.80163

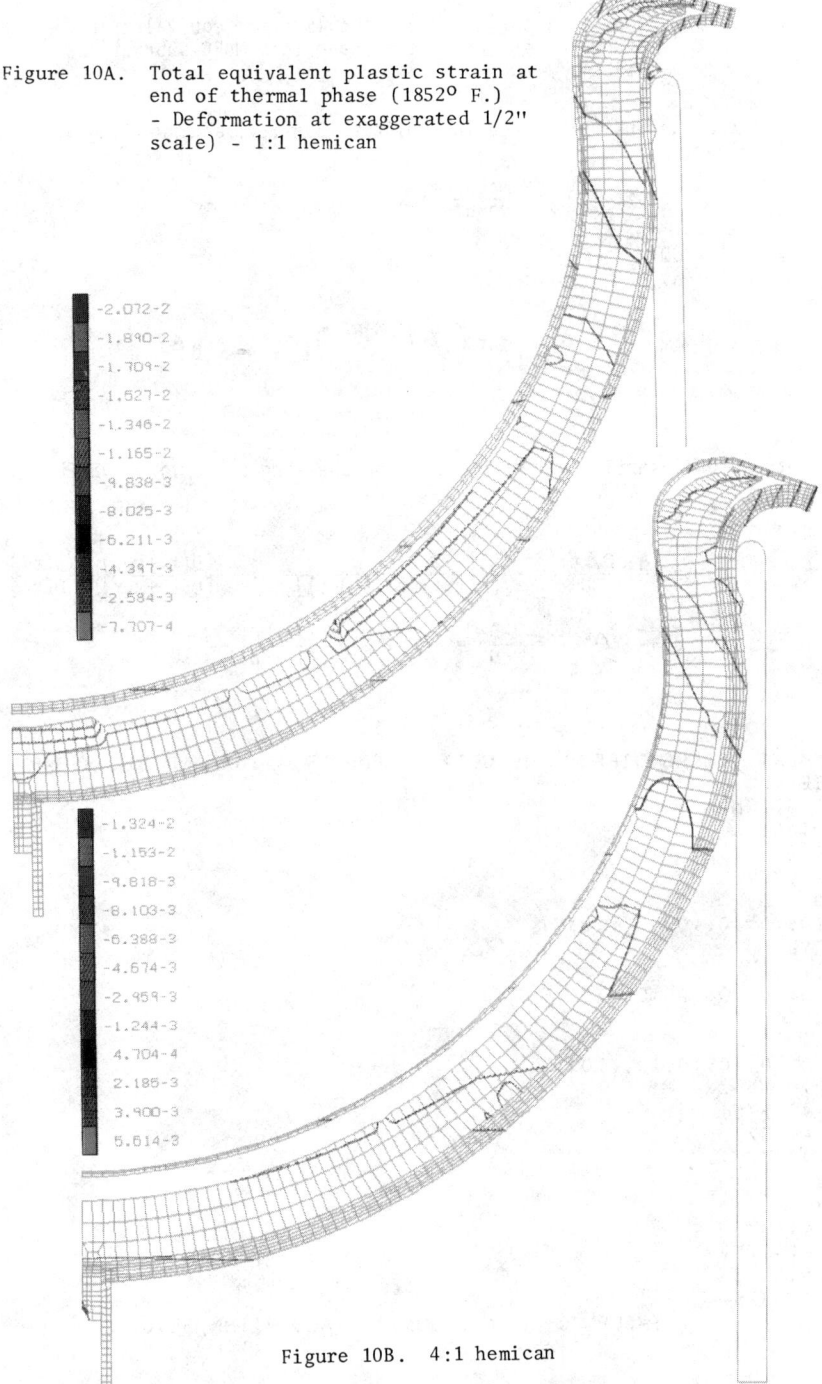

Figure 10A. Total equivalent plastic strain at
end of thermal phase (1852° F.)
- Deformation at exaggerated 1/2"
scale) - 1:1 hemican

-2.072-2
-1.890-2
-1.709-2
-1.527-2
-1.346-2
-1.165-2
-9.838-3
-8.025-3
-6.211-3
-4.397-3
-2.584-3
-7.707-4

-1.324-2
-1.153-2
-9.818-3
-8.103-3
-6.388-3
-4.674-3
-2.959-3
-1.244-3
4.704-4
2.185-3
3.900-3
5.614-3

Figure 10B. 4:1 hemican

<TABLE **3**> Proof that shrinkage and swelling are equivalent and method for inserting shrinkage into MARC subroutine VSWELL

1) Prove that shrinkage (increasing relative density) and swelling are equivalent.

$$\text{Density (D)} = \frac{\text{mass (m)}}{\text{volume (V)}}$$

$$\frac{dD}{dV} = -\frac{m}{V^2}$$

$$\Delta D \cong \frac{dD}{dV}\Delta V = -\frac{m}{V^2}\Delta V = -\frac{m}{V}\frac{\Delta V}{V} = -D\frac{\Delta V}{V}$$

$$-\frac{\Delta D}{D} \cong \frac{\Delta V}{V}$$

2) Method for inserting shrinkage into the swelling subroutine, VSWELL.

$$\overset{\circ}{D} = \frac{dD}{dt} \cong \frac{D(t+\Delta t) - D(t)}{\Delta t} \qquad \text{where } t = \text{time}$$

$$\Delta e_v = \frac{\overset{\circ}{D}\Delta t}{D} = \frac{[D(t+\Delta t) - D(t)]}{\frac{1}{2}[D(t+\Delta t) + D(t)]} = \frac{2[D(t+\Delta t) - D(t)]}{[D(t+\Delta t) + D(t)]}$$

$$\overset{\circ}{e}_v \cong \frac{\Delta e_v}{\Delta t} = \frac{-\overset{\circ}{D}}{D} \cong \frac{-2[D(t+\Delta t) - D(t)]}{\Delta t[D(t+\Delta t) + D(t)]}$$

```
COMMENT, CALLS DIEPOS AND UFRIC - FOR FRICTION AND DIE SEPAR
UDIE
UFRICTION
CONTACT
3,700,900,2,
.1,.005,
1,0,
0.,0.,0.,0.,0.,0.,,015,
BERYL
2,0,
0.,0.,0.,0.,0.,0.,,01,
SS304
3,2,
0.,0.,0.,0.,0.,0.,,01,
2,2,
.5,4.6375,
.5,4.4375,
.5,4.5375,
.1,
1,4,
.5,4.4375,
6.,4.4375,
6.,4.6375,
.5,4.6375,
```

<TABLE 4>   MARC CONTACT ALGORITHM SETUP

```
          SUBROUTINE VSWELL(SWELL,SIG,TEMP,N,NN,CPTIM,TIMINC,MAT,
         +DTEMP)
C         INCLUDE M.M. CARROLL EXPERIMENTAL MODEL (BERKELEY NOV 1986
C         VOL 17A METALLURGICAL TRANSACTIONS A) TO DESCRIBE HOT
C         ISOSTATIC TIME DEPENDENT PRESSURIZATION AND DENSIFICATION
C         OF BERYLLIUM METAL POWDER AT CONSTANT TEMPERATURE.
C         INCORPORATES:   1. INSTANTANEOUS RESPONSE
C                         2. EQUILIBRIUM RESPONSE
C                         3. CREEP RESPONSE
C         BASIC IMPLEM INVOLVES TAKING DER OF D (EQ 40) TO FORM:
C             DELTA E SUB V = D DOT DELTA T/D
C                           = 2*(D(T+DEL T)-D(T)/(D(T+DEL T)+D(T))
C         THEN: E DOT SUB V = -D DOT/D
C                           = -2*(D(T+DEL T)-D(T))/DEL T/(D(T+DEL T)+D(T))
C         USER DEF INCR OF DILATIONAL SHRINKAGE/CREEP TO BE CALLED
C         AT EACH INTEGRATION PT WHERE CONSTITUTIVE CALCULATIONS ARE
C         BEING PERFORMED. CALLED BY AUTO CREEP, CREEP ETC.
          COMMON/FAR/IDUM1,INC,IDUM2(15),M
          DIMENSION SIG(3),TEMP(1),DTEMP(1)
C         CPTIM        = TOTAL CREEP TIME
C         DTEMP(1)     = TEMPERATURE INCREMENT
C         DTEMP(2),(3) = INCREMENTS OF ADDITIONAL STATE VARIABLES
C         MAT          = MATERIAL ID NUMBER FROM PROPERTY OPTION
C         N/M          = ELEMENT NUMBER
C         NN           = INTEGRATION POINT NUMBER
C         SIG(1)       = UNIAXIAL EQUIVALENT OF J2 STRESS
C         SIG(2)       = HYDROSTATIC STRESS
C         SIG(3)       = CURRENT TOTAL SWELLING STRAIN (ACCUM FROM
C                        THIS SUBR) = UNIAXIAL COMPONENT = SUM DV/3V
C         SWELL        = USER DEF INCR OF VOL SWELLING D/DT(DV/V)
C         TEMP(1)      = TEMPERATURE
C         TEMP(2),(3)  = ADDITIONAL STATE VARIABLES READ (CREDE)
C         TIMINC       = CURRENT TIME INCREMENT
C         INPUT LOADS:   P = APPLIED PRESSURE (15000.)
C                        T = APPLIED TEMPERATURE (1832.)
C         EXPERIMENTAL EXPONENTIAL MODEL PARAMS DESCRIBE BERYLLIUM'S
C         TRANSITION FROM NON-LINEAR TO LINEAR SHRINKAGE:
C         TAU = TIME CONSTANT (2500. MIN)
C         M   = NON-DIMENSIONAL EXPONENT (2.)
C         DBO = VALUE OF B AFTER PRESSURE RISE
C             = PREVIOUS ITERATION REL DENSITY VALUE FOR CALC SWELL
C         DINF= FINAL(24 HOUR) MEASURED VALUE OF D AT PRESSURE P
          DATA TAU,XM,DBO,DINF/2500.,2.,.6,.99/
          SWELL=0.
C         SEE IF TIME IS ZERO - NO SHRINKAGE HAS OCCURED
C         RETURN IF NOT BERYLLIUM OR THREE TOE ELEMENTS
C         IF(M.GE.390.OR.M.LE.3)RETURN
          IF(TIMINC.LT.105..OR.TIMINC.GT.360.)             GO TO 1
C         COMPUTE CURRENT RELATIVE DENSITY
          DBOO=DBO-SIG(3)
C         K PARAMETER
          XK=1.-((1.-DINF)/(1.-DBOO))**XM
C         D = CURRENT REL DENSITY = VOLUMETRIC STRAIN CHANGE(DEL EV)
          D=1.-(1.-DINF)/(1.-XK*EXP(-XM*TIMINC/TAU))**(1./XM)
C         CALCULATE INCREMENT OF VOLUMETRIC SWELLING (EV DOT=DV/V)
C         SHRINKING IS NEGATIVE
          SWELL=-2.*(D-DBOO)/(D+DBOO)/TIMINC/100.
        1 RETURN
          END
            <TABLE 2>    CARROLL CONSTITUTIVE MODEL SUBROUTINE
```

1143

```
      SUBROUTINE UFRIC (MIBODY,X,FN,VREL,TEMP,YIEL,FRIC)
C     DETERMINE TEMPERATURE DEPENDENT FRICTION FACTOR
C     BELOW THRESHOLD STICKING VELOCITY - USE STATIC FRICTION.
C                               ABOVE - USE KINETIC FRICTION.
      COMMON /FAR/ IDUM1,INC,IDUM2(15),M
      DIMENSION X(2),MIBODY(3)
C     FN        = NORMAL PRESSURE BEING APPLIED AT THAT POINT
C     FRIC      = FRICTION COEFFICIENT OR FRICTION FACTOR
C     INC       = INCREMENT NUMBER
C     MIBODY(1) = ELEMENT NUMBER
C     MIBODY(2) = SIDE NUMBER
C     MIBODY(3) = SURFACE INTEGRATION POINT NUMBER
C     TEMP      = TEMPERATURE OF CONTACT POINT
C     VREL      = RELATIVE SLIDING VELOCITY AT CONTACT POINT
C     X         = UPDATED COORDS OF CONTACT POINT WHERE FRICTION
C                 IS BEING CALCULATED
C     YIEL      = WORKPIECE MATERIAL FLOW STRESS AT CONTACT PT
      A=(TEMP-69.)*1.2125E-5
      B=(TEMP-1472.)*3.161E-6
C     IF(M.GT.389)GO TO 1
C     BERYLLIUM
      IF(TEMP.LT.1472.)                   FRIC=.015+A
      IF(TEMP.GE.1472.)                   FRIC=.03-B
      IF(VREL.LT..1.AND.TEMP.LT.1472.) FRIC=.15+A
      IF(VREL.LT..1.AND.TEMP.GE.1472.) FRIC=.15-B
      GO TO 2
C     STAINLESS STEEL
    1 IF(TEMP.LT.1472.)                   FRIC=.01+A
      IF(TEMP.GE.1472.)                   FRIC=.02-B
      IF(VREL.LT..1.AND.TEMP.LT.1472.) FRIC=.1+A
      IF(VREL.LT..1.AND.TEMP.GE.1472.) FRIC=.1-B
    2 RETURN
      END
            <TABLE 5>   FRICTION SUBROUTINE




      SUBROUTINE DIEPOS (X,F,V,TIME,NDIE)
C     MOVE DIE TO SEPERATE PART FOR SPRINGBACK DETERMINATION
      DIMENSION    X(3),V(3),F(2)
      COMMON /FAR/ IDUM(17),M
C     F    = LOAD COMPONENTS FX FY
C     M    = ELEMENT NUMBER
C     NDIE = DIE SURFACE NUMBER
C     TIME = CURRENT TIME
C     V    = VELOCITY VX,VY  ANGULAR ROTATION VROT
C     X    = POS OF CENTER OF ROT X,Y AND CURRENT ANGLE ROTATED
      V(1)=ABS(0.)
      V(2)=ABS(0.)
C     MOVE DIES APART AFTER LOADING 8 HOURS TO OBTAIN RES STRESS
      IF(NDIE.EQ.2.AND.TIME.GT.480.)V(1)=-3.
      IF(NDIE.EQ.3.AND.TIME.GT.480.)V(1)=3.
      IF(NDIE.EQ.4.AND.TIME.GT.480.)V(2)=3.
      RETURN
      END
            <TABLE 6>   DIE MOTION SUBROUTINE
```

```fortran
      FUNCTION YIEL (N,YIELD,NEL,IFIRST,DT,EPLAS,ERATE)
C  TEMP DEF WORK HARDENED YIELD STRESS (YIEL), USING DUCTILE
C  POWER LAW FOR: BERYLLIUM AT TOE (3 ELS); FORMED POWDER
C  (AFTER 360 MIN); AND SS CANS.          SIGY=A*EXP**M
C  CFTIM  - TIME AT BEGINNING OF INCREMENT
C  DT     - CURRENT TOTAL TEMPERATURE
C  EPLAS  - CURRENT TOTAL EQUIV PLASTIC STRAIN
C  ERATE  - CURRENT EQUIV PLASTIC STRAIN RATE
C  IFIRST   1 FOR INITIAL YIELD - 2 10TH CYCLE ORNL STRESS
C  INC    - INCREMENT NUMBER
C  M      - ELEMENT NUMBER
C  N      - MATERIAL NUMBER
C  NEL    - NUMBER ELEMENTS IN MESH
C  TIMINC - CURRENT TIME STEP (DELTA T)
C  YIELD  - YIELD STRESS ENTERED ON PROPERTY CARD
C  YIEL   - CURRENT YIELD STRESS (USER DEFINED)
      COMMON /PAR /IDUM1,INC,IDUM2(15),M
      COMMON /CREEPS/ IDUM3,CPTIM,IDUM4(90),TIMINC
      DIMENSION TBE(9),YBE(9),UBE(9),TSS(10),YSS(10),
     .         USS(10),XM(6),UT(6),ELOM(8)
C
C  TEMP(F)  YIELD STRESS(PSI)  ULT STR(PSI)  ELONG (%/100)
C  SOLID BERYLLIUM/STEEL YIELD/ULT STRESS AS FN OF TEMP
      DATA TBE/80.,300.,500.,700.,900.,1100.,1300.,1500.,1832./,
     .  YBE/48000.,44400.,40400.,38400.,34100.,31900.,30000.,25800./,
     .  35000.,18000.,8500./, ISS/70.,400.,600.,800.,1000.,1800.,
     .  1400.,1600.,1800.,2000./, YSS/30000.,24000.,20000.,18000.,
     .  16000.,13000.,12000.,10000.,7800.,5800./, USS/75000.,
     .  71000.,68000.,63000.,58000.,43000.,27000.,15000.,8000./,
     .  6000./, NELB,NELS/9,10/, UT/3.5,2.75,2,15,1.6,1.30,1.17/,
     .  XM/.31,.285,.215,.14,.095,.065./, ELOM(8)/
C
      YIEL=YIELD
      IF(INC.EQ.0)RETURN
      TIM=CPTIM+TIMINC
C  USE REGULAR MATL PROP IF TOE ELEMENTS OR BERYLLIUM POWDER
C  AFTER 360 MIN  USE VSWELL CARROLL EQ FOR POWDER BERYLLIUM
      IF(M.LE.3.OR.M.GT.3.AND.M.LT.390.AND.TIM.GE.360.)GO TO 4
      IF(M.GT.3.AND.M.LT.390)RETURN
C  STAINLESS STEEL
      DO 1 I=1,NELS
      EL=ELOM(1)
      IF(DT.LT.TSS(I).AND.I.EQ.1)                        GO TO 2
    1 IF(DT.LE.TSS(I+1))                                 GO TO 3
      Y=YSS(NELS)
      U=USS(NELS)
                                                         GO TO 8
    2 Y=YSS(1)
      U=USS(1)
C
    3 Y=YSS(I)+(YSS(I+1)-YSS(I))*(DT-TSS(I))/(TSS(I+1)-TSS(I))
      U=USS(I)+(USS(I+1)-USS(I))*(DT-TSS(I))/(TSS(I+1)-TSS(I))
                                                         GO TO 8
C
    4 BERYLLIUM
      EL=ELOM(2)
      IF(DT.LT.TBE(I).AND.I.EQ.1)                        GO TO 7
    5 IF(DT.LE.TBE(I+1))                                 GO TO 6
      Y=YBE(NELB)
      U=UBE(NELB)
    6 Y=YBE(I)+(YBE(I+1)-YBE(I))*(DT-TBE(I))/(TBE(I+1)-TBE(I)) GO TO 8
      U=UBE(I)+(UBE(I+1)-UBE(I))*(DT-TBE(I))/(TBE(I+1)-TBE(I))
                                                         GO TO 8
    7 Y=YBE(1)
      U=UBE(1)
C  STRAIN HARDENING POWER EQ USING A,M VALUES
    8 R=U-Y
      DO 9 I=2,6
      IF(R.LT.UT(I))                                     GO TO 10
      IF(R.GT.UT(6))                                     GO TO 11
    9 IF(R.LT.UT(I))XM1=XM(I-1)+(R-UT(I-1))/(UT(I)-UT(I-1))*(XM(I)
     .  -XM(I-1))/(UT(I)-UT(1))
   10 XM1=XM(1)+(R-UT(1))*(XM(2)-XM(1)*XM5)/(UT(2)-UT(1))
   11 XM1=XM(6)+(R-UT(6))*(XM(6)-XM5)/(UT(6)-UT(5))
C  SET INITIAL VALUES FOR ITERATION
   12 A1=A1*EL**XM1
      Y1=A1*EL**XM1
      XM2=XM1-.05
      A2=Y*500.**XM2
      Y2=A2*EL**XM2
C  ITERATE TO GET CORRECT A,M FITTING Y AND U
      DO 15 I=1,50
      IF(U.LE.Y2)                                        GO TO 13
C  SHIFT POSITIONS OF Y1 AND Y2
      SAV=XM1
      XM1=XM2
      XM2=SAV
      SAV=A1
      A1=A2
      A2=SAV
      SAV=Y1
      Y1=Y2
      Y2=SAV
   13 IF(U.LT.Y2.AND.U.GT.Y1)XM3=XM2-(Y2-U)*(XM2-XM1)/(Y2-Y1)
      IF(U.LT.Y2)XM3=XM3-XM1-(Y1-Y2)*(XM2-XM1)/(Y2-Y1)
      IF(U.LT.Y1)XM3=XM1-(Y1-U)*(XM2-XM1)/(Y2-Y1)
      A3=Y*500.**XM3
      Y3=A3*EL**XM3
C  SHIFT OLD VALUES
      IF(Y3.GT.Y2.OR.Y3.LE.U)                            GO TO 14
      XM2=XM3
      A2=Y*500.**XM2
      Y2=A2*EL**XM2
C  COMPUTE NEW VALUES
   14 XM2=XM3
      A2=A3
      Y2=Y3
      IF(ABS(Y3-U).LE.1)                                 GO TO 15
   15 CONTINUE
      IF(XM2.GT.0..AND.XM2.LT.1.E10)YIEL=A2*EPLAS**XM2
      IF(YIEL.LT.Y.OR.XM2.LT.0..OR.XM2.GT.1.E10)YIEL=Y
      IF(YIEL.GT.U)YIEL=U
      RETURN
      END
```

<TABLE 1> TEMPERATURE DEPENDENT YIELD STRESS SUBROUTINE

```
      SUBROUTINE FORCEM (P,X1,X2,NN,N)
C     ELS IN CONTACT WITH RING DIE MUST NOT HAVE AUTOCLAVE
C     PRESSURE APPLIED - INTERFERS WITH CONTACT BOUNDARY CONDS
C     TOM COURTNEY RI RF JUN 87
      DIMENSION N(2),X1(1),X2(1)
      COMMON /ARRAYS  IDUM(35),INPNUM
      COMMON /DEVELF  IUDUM1(7),JOPTIT
      COMMON /SPACE   INTS(1)
      COMMON /FORM    IDUM2(6),NBCN,IDUM7(16),INBCT
      COMMON /ARRAY4  IDUM3(12),INF,ITOUCH
      COMMON /CDC     IDUM4(18),NCYCLE
      COMMON /FAR     IDUM5,INC
      COMMON /LASS    IDUM6(7),NNN
C     ARRAY TO SAVE SUMMED PRESSURES IN CASE PART MOVES OFF DIE
C     AND PRESSURE IS REINSTATED
      COMMON /MINE/ PX,PPREV(33)
C     INC    = INCREMENT NUMBER
C     INF    = POINTER TO CONTACT NODES IN DATA BASE
C     INTS   - MARC DATA BASE
C     ITOUCH - POINTER TO DIE NOS CORR TO NODES IN CONTACT IN DB
C     N(1)   = ELEMENT NUMBER
C     N(2)   = TYPE OF LOAD PARAMETER
C     NBCN   = NUMBER OF BOUNDARY NODES IN CONTACT
C     NCYCLE - NUMBER OF CYCLES
C     NNN    = INTEGRATION POINT NUMBER
C     P      - LOAD ON ELEMENT
C     X1     = FIRST COORD OF INTEGRATION POINT
C     X2     - SECOND COORD OF INTEGRATION POINT
      DIMENSION IA(33),NX(34)
C     ARRAY OF ELEMENTS (IA) AND NODES (NX) POSSIBLY
C     IN CONTACT WITH GRAPHITE RING DIE (3)
      DATA IA/1009,1013,1024,1029,1030,1032,1037,1039,1038,1036,
     +1035,1033,1034,1040,1042,1044,1050,1052,1054,1060,1063,
     +1069,1074,1081,1085,1091,1097,1103,1109,1111,1121,1127,
     +1133/,NX/1364,1368,1380,1382,1385,1391,1396,1400,1399,
     +1398,1395,1393,1392,1394,1401,1402,1406,1413,1415,1419,
     +1426,1430,1437,1445,1454,1458,1466,1474,1482,1489,1496,
     +1503,1510,1517/,NEL/33/,PX/0./,PPREV/33*0./
C     SAVE CURR PRESSURE INCR (PX) FOR FIRST EL OF POSS CONTACT
      IF(N(1).EQ.1009.AND.NNN.EQ.1.AND.NCYCLE.EQ.0)PX=PX+P
      DO 1 IX=1,NEL
    1 IF(N(1).EQ.IA(IX))                                 GO TO 2
                                                         GO TO 6
C     FIND TWO NODE NUMBERS FOR ELEMENT SIDE IN CONTACT
    2 NO1=NX(IX)
      NO2=NX(IX+1)
C     BODY OF INTEREST IN CONTACT WITH RING IS DIE 2 (STEEL)
      MZ=2
C     FIND LIST OF CONTACT NODES AND COMPARE WITH TWO EL NODES
C     NO OF NODES ON GRAPHITE RING BOUNDARY CONTACTING DIE 2
      NBCT=INTS(INBCT+MZ-1)
      DO 3 I=1,NBCT
      I1=INF+(MZ-1)*NBCN+I-1
C     INTERNAL NODE NO (J) ON BOUNDARY OF DIE 2
      J=INTS(I1)
C     CONVERT INTERNAL NODE NUMBER INTO EXTERNAL NODE NO (LEXT)
      LA2=INPNUM+J-1
      IF(JOPTIT.NE.0) LEXT=IGETSH(INTS(LA2),1)
C     SEE IF NODE IN CONTACT WITH DIE (NDIE=0 MEANS NO CONTACT)
      I2=ITOUCH+(MZ-1)*NBCN+I-1
C     FIND DIE NUMBER
      NDIE=INTS(I2)
C     SEE IF RING DIE (NDIE=3) OR NOT IN CONTACT (NDIE=0)
      IF(NDIE.NE.3)GO TO 3
C     COMPARE WITH TWO NODE NUMBERS TO SEE IF IN CONTACT
      IF(LEXT.NE.NO1.AND.LEXT.NE.NO2)P=PX
C     NOT IN CONTACT - RESTORE PRESSURE INCREMENT
      IF(LEXT.NE.NO1.AND.LEXT.NE.NO2)                    GO TO 3
C     IN CONTACT - SET PRESSURE TO NEGATIVE PREVIOUS TOTAL
C     TO ZERO OUT DISTRIBUTED PRESSURE ON INTEGRATION POINTS
      P=-PPREV(IX)
                                                         GO TO 6
    3 CONTINUE
C     SAVE ACCUM PRESSURES (PPREV) FOR ELS OF POSSIBLE CONTACT
C     AT FIRST INTEGRATION POINT AND BASE CYCLE CALL
    6 IF(NNN.EQ.1.AND.NCYCLE.EQ.0) PPREV(IX)=PPREV(IX)+P
      IF(N(1).EQ.1039.AND.INC.LE.1)WRITE(6,5)INC,N(1),P
                                                         RETURN
    5 FORMAT(' FORCEM',2I5,E16.5)
      END
```

TABLE 7. PRESSURE ZEROING SUBROUTINE

ECONOMICS

Session Chairman
Noel Jarrett, Alcoa Technical Center

This session applies cost optimizing procedures to process simulation models for the entire steel melting and casting sequences. One presentation specifically evaluates low-cost charging and another employes linear programming to the melting-ladle refining-concast system.

ARC/AOD OPTIMIZATION MODEL

D.J. McMahon
J.W. Tommaney
V.P. Ardito

Process R&D Department
Technical Center
Allegheny Ludlum Corporation
Brackenridge, PA 15014

ABSTRACT

To gain insight into a coordinated approach for raw material management, an ARC/AOD Model has been developed as a planning tool for strategic evaluation of raw material design and pricing policies. The ARC/AOD Model combines the least cost ARC initial charge model and the least cost AOD blow and refining model enabling a least cost optimization over both the Arc Furnace and the AOD vessel. The Model minimizes the raw material costs, as well as the AOD gasses, fluxes, operating and refractory costs for a particular stainless steel chemical specification. The Model calculates optimal ARC hot metal chemistries and transfer weight, breakeven prices, and charge mix variance calculations.

INTRODUCTION

A brief description of stainless steel melting is needed to understand the role of mathematics in melting. <u>Figure 1</u> illustrates the melting steps in producing stainless steel. At ALC, raw materials in various forms are purchased for melt down in a spare electric arc furnace. The melted hot metal is then transferred to a secondary refining vessel where the hot metal is refined, primarily, where oxygen is injected to react with carbon. Each of these furnaces have two distinct stages in which raw materials can be charged. This is seen in <u>Figure 2</u>:

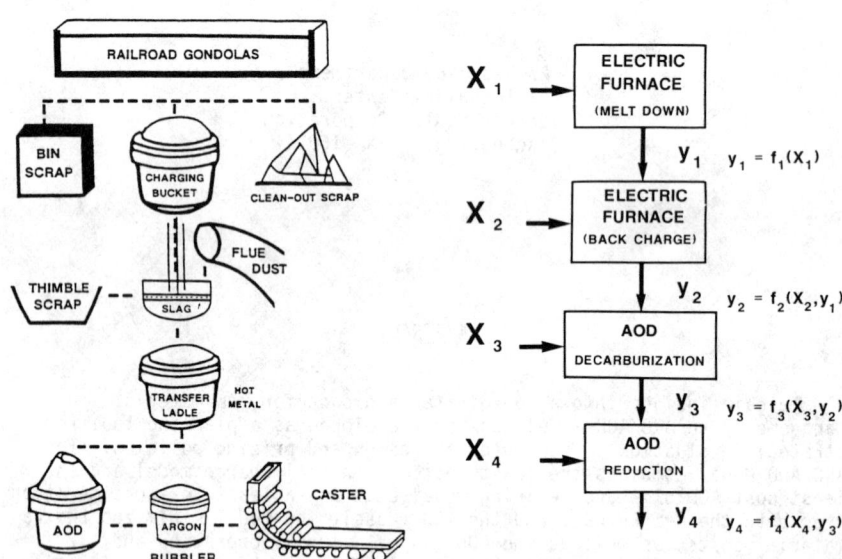

FIGURE 1: SCHEMATIC OF AN ARC/AOD MELT SHOP

FIGURE 2: 4-STAGE MELT DIAGRAM

The ARC meltdown is the initial melting of the raw materials. The ARC back charge tries to adjust the hot metal chemistry based on "surprises" from the initial meltdown. The argon/oxygen decarburization (AOD) removes carbon from the hot metal, and also adds raw materials for continued chemical adjustments. The AOD reduction stage primarily reduces chromium from the slag on top of the hot metal back into the hot metal and also adds final chemical adjustments to meet customer requirements. Thus, in what stage the raw materials are added becomes a critical cost-related decision.

The raw materials charged in the electric arc furnace (ARC) may have different yields than if they were charged in the refining vessel, an Argon-Oxygen Decarburizing vessel (AOD). Some raw materials can be added in all four stages, but others should only be added in the AOD because of the profound yield and metallic oxidation considerations. Thus, the yield considerations, both measurable and subjective, are needed as part of the mathematical formulation. The objective is, obviously, to minimize the costs of the raw materials and still meet the quality objectives of our customers.

A mathematical model is essential to understand all of the many raw material blending situations which can arise. This is essential when one considers that about 60% of the overall costs of producing stainless steels are in the raw materials. The raw materials are commodities with prices varying daily. Raw material management is essential in order to minimize the costs of ALC's stainless steel products and still maintain product quality. Management does coordinate the corporate strategic plans for raw materials concerning melt shop requirements, raw material inventory utilization, raw material purchasing strategies and the conformance to customer requirements. To help this effort, the mathematical structure for the raw material utilization can be formulated.

ARC/AOD OPTIMIZATION MODEL

The four-staged ARC/AOD optimization model can now be formulated. It would classically be solved using dynamic programming techniques. However, because some of the stages can be formulated as a linear program, linear programming techniques were selected with the nonlinear equations checked for feasibility in an iterative manner. The purpose of this paper is not to present the details of the mathematical equations, but to present the concepts of the approach, the computer support systems and several examples of how the ARC/AOD Model is used at Allegheny Ludlum Corporation.

Figure 3 shows the schematic of the melt shop control model.

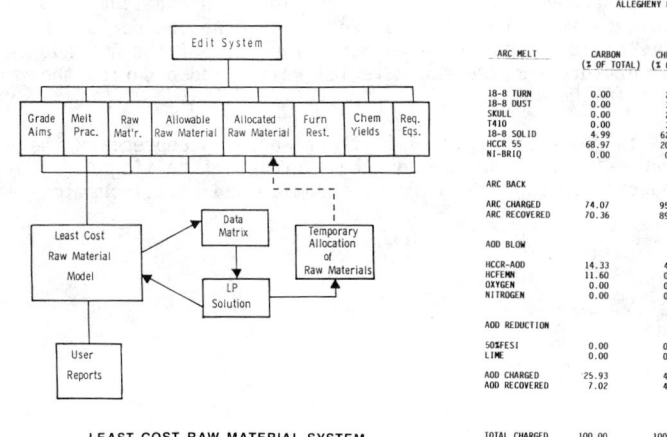

ALLEGHENY LUDLUM CORPORATION

ARC MELT	CARBON (% OF TOTAL)	CHROMIUM (% OF TOTAL)	NICKEL (% OF TOTAL)	% WEIGHT	% COST
18-8 TURN	0.00	2.85	3.83	4.32	3.72
18-8 DUST	0.00	3.99	3.94	5.18	0.03
SKULL	0.00	2.33	2.30	3.02	0.50
T410	0.00	3.92	0.16	6.48	1.63
18-8 SOLID	4.99	62.18	71.54	66.41	65.70
HCCR 55	68.97	20.42	0.00	7.13	10.39
NI-BRIQ	0.00	0.00	18.23	1.44	12.23
ARC BACK					
ARC CHARGED	74.07	95.70	100.00	93.98	94.20
ARC RECOVERED	70.36	89.00	100.00	88.78	94.20
AOD BLOW					
HCCR-AOD	14.33	4.30	0.00	1.47	1.05
HCFEMN	11.60	0.00	0.00	1.29	0.73
OXYGEN	0.00	0.00	0.00		0.84
NITROGEN	0.00	0.00	0.00		0.22
AOD REDUCTION					
50%FESI	0.00	0.00	0.00	3.26	2.35
LIME	0.00	0.00	0.00	0.00	0.61
AOD CHARGED	25.93	4.30	0.00	6.02	5.80
AOD RECOVERED	7.02	4.30	0.00	4.13	5.80
TOTAL CHARGED	100.00	100.00	100.00	100.00	100.00
TOTAL RECOVERED	7.02	93.30	100.00	92.83	100.00

LEAST COST RAW MATERIAL SYSTEM

FIGURE 3: SCHEMATIC OF MELT SHOP
CONTROL MODEL

FIGURE 4: ARC/AOD OUTPUT REPORT

1151

The ARC/AOD Model has equations for the following:

. Raw Material Chemical Balance
. Overall Material Balance
. Allowable raw materials per stage and per stainless type
. Chemical restrictions per melt furnace
. AOD blow and reduction practices
. Scrap practice definitions
. Quality restrictions based on chemistry
. Utilization restrictions based on raw material form
. Regression equation for the Metallic Oxidation Factor

The basic formulation of the ARC/AOD Model is as follows:

Minimize the cost of raw materials:

$$\text{Minimize} \sum C_i \; RM_i \tag{1}$$

subject to:

$$\sum (YRM_i) \times (YGRM_i) \times (RM_i) = (\text{Tap Weight}) \tag{2}$$

where YRMi is the sum of the elemental yields for a raw material
YGRMi is the gross yield lost during charging in furnace
RMi is the ith initial charged raw material weight

$$\sum (EL_{ij}) \times (YEL_j)(YGRM_i) \times (RM_i) < (ELAIM_j) \times (\text{Tap Weight}) + (\text{Control} \tag{3}$$
$$- \text{Limit})$$

where ELij is the chemical value for the jth element in the ith raw
    material.
YELj is the yield of the jth element
ELAIMj is the aim element requirement for this stainless steel

A typical problem has about 70 equations and 80 raw materials. So
it really is beyond the scope of this paper to decribe each equation.
The non-linearity derives from the use of a regression equation per grade
which predicts the metallic oxidation factor as a function of the ARC hot
metal weight and temperature and the raw material weight added during the
decarburization phase in the AOD.

A typical output is found in <u>Figure 4</u> so the reader can appreciate the
complexity of the raw material flow through the 4 stages of melting
stainless steel production. Only, the carbon, chromium and nickel chemical
elements are shown.

EXAMPLES USING THE MODEL

The ARC/AOD Model can be used for critical decisions in the following examples:

A.   Sensitivity Analyses on Scrap Charges

B.   Breakeven Prices on raw materials not in the optimum solution

C.   Usage within stainless meltshop operations.

D.   Quality implications of Raw Material Optimization Models.

A.   Sensitivity Analyses on Scrap Charges

The least cost, raw material solution changes as the availability of scrap type materials change. Scrap definitions primarily relate to forms of raw materials that have been refined at one time during their life and usually have a lower price than the ferro alloys. Figure 5 shows how the optimum levels for chromium and nickel in the ARC furnace change as the scrap charge is increased from 65 to 88 percent. Also seen in Figure 5, the scrap charge has an effect on the ARC hot metal carbon, the Metallic Oxidation Factor, the arc hot metal weight and the AOD refining times.

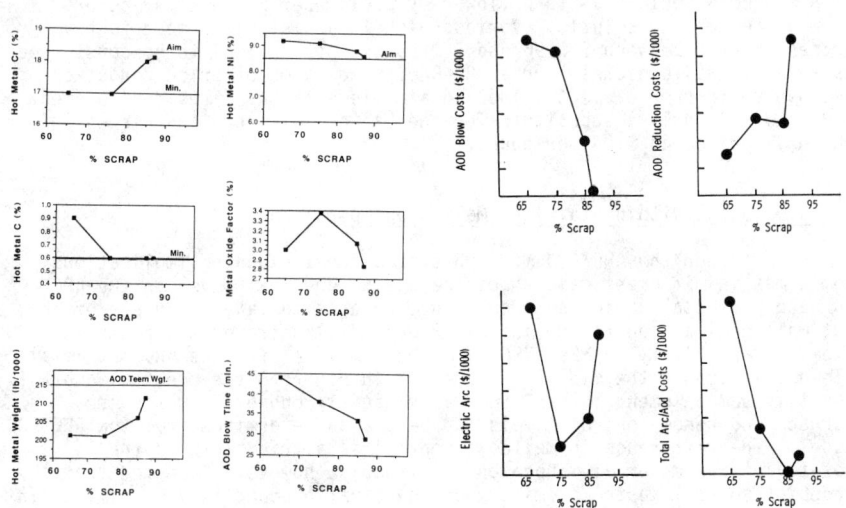

FIGURE 5:   ARC CHEMISTRY AIMS AS A
FUNCTION OF SCRAP CHARGE

FIGURE 6:   ARC/AOD COSTS SEPARATED
INTO STAGES

Figure 6 separates the costs in each stage as a function of scrap charge practice. If scrap availability is no problem, then the raw material conditions existing at the time of this study was made, indicates that the optimum scrap charge practice is about 85%. It is of interest to note that the scrap practice that minimizes the ARC costs, e.g., 75%, actually maximizes the AOD Decarburization costs. The benefits of a 4-staged dynamic program are obvious, even though linear programming techniques using a heuristic approach to resolve the non-linear equations were used.

B. Breakeven Prices On Raw Materials Not In Optimum Solution

In the example found in Figure 4, several raw materials were not selected in the least cost solution. These raw materials are summarized to illustrate the breakeven concepts:

RAW MATERIAL	PRICE	BREAKEVEN PRICE
BUSH	X1	(X1 - .0147)
25 NIP	X2	(X2 - .0196)
NI-BRIQ	X3	(X3 - .0211)
NI-BRIQ	X4	(X4 - .0259)
CU-BNDL	X5	(X5 - .8538)

BUSH was not selected because there were lower cost sources for iron and BUSH would have to be lower by $.0147 per pound alloy before it would enter the solution. Various nickel raw materials need to have prices lowered between $.0196 and $.0259 per pound alloy to be considered as an alternative nickel source. CU-BNDL needs a major price reduction because the copper element is not an aim element, but a residual element and, thus, is only a substitute for the balance element, in this case iron which is priced at $.04 per pound.

C. Usage Within Stainless Melt Shop Operations

Even though the ARC/AOD Model has tremendous planning implications, sometimes in different melt shops the actual chemical level in the ARC Furnace hot metal varies so much to not be able to take advantage of the optimal chemical control decisions. Figure 7 illustrated the chemical element variability in MELT SHOP 1. The aim level for the ARC hot metal element is set in the midpoint. However, in Figure 8 the chemical control for this same element in MELT SHOP 2 is greatly enhanced in the Arc Furnace and hence control levels can be set as determined from the ARC/AOD Model. The difference in meltdown control is attributable to the lot-to-lot raw material management. It should, however, be noted that the endpoint control in the chemistry of the final product is very similar in both Melt Shops.

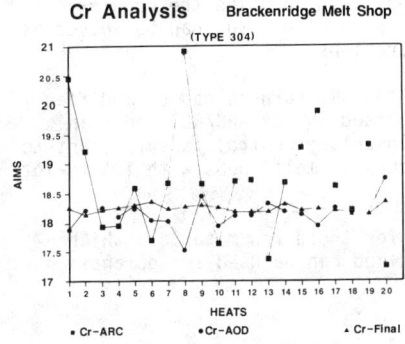

Cr Analysis — Brackenridge Melt Shop (TYPE 304)

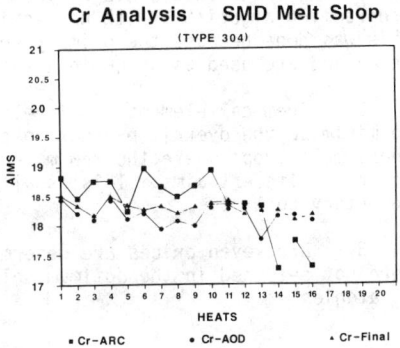

Cr Analysis — SMD Melt Shop (TYPE 304)

FIGURE 7: CHROMIUM CHARTS FOR
MELT SHOP 1

FIGURE 8: CHROMIUM CHARTS FOR
MELT SHOP 2

D.   Quality Implications of Raw Material Optimization Models

The ARC/AOD Optimization Model is only one raw material blending program which is used at Allegheny Ludlum to produce quality stainless steel. Through a combined effort of operating management and mathematics, the chemical uniformity of ALC's stainless steels have been greatly improved. Figure 9 illustrates the 13 year effort to control the chemical uniformity of one critical element comprising our stainless steels.

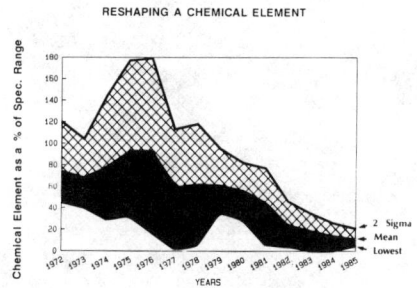

RESHAPING A CHEMICAL ELEMENT

FIGURE 9:   RESHAPING A CHEMICAL ELEMENT
LONG-TERM PROGRESS

CONCLUSIONS

1. The raw material allocation over the electric ARC and AOD vessels can be expressed in a 4-stage dynamic programming model. However, linear programming algorithms with heuristics were used to solve the problem. This was done because the primary equations in each stage can be solved as an LP and are used as an LP in actual operations.

2. Chemical element aim levels for the ARC Furnace can be modified to minimize the overall production costs based on the ARC/AOD Model in those melt shops where the raw material inventory control is sufficient to control different aims. This usually occurs in melt shops with lot-to-lot inventory control.

3. Breakeven prices are determined for those raw materials which were not selected in the optimum solution and can be used for purchasing strategies.

RECOMMENDATIONS

1. An integrated raw material management system is needed to take full advantage of the ARC/AOD Optimization Model.

A Model for Sequencing a Continuous

Casting Operation to Minimize Costs

B. Lally[1], L. Biegler[2] and H. Henein[1]

[1]Department of Metallurgical Engineering and Materials Science
[2]Department of Chemical Engineering
Carnegie Mellon University
Pittsburgh, Pennsylvania

Abstract

A model of a meltshop/ladle station/continuous casting operation is used to minimize projected costs of sequence casting. The model is stated as a Mixed Integer Linear Program (MILP) where the integer variables take the form of binary decision variables that represent the existence of breaks in the casting sequence. Operating costs are approximated linearly for the meltshop, for ladle handling and for the caster. The sum of these costs is minimized to determine a least cost casting schedule, subject to operational constraints that ensure feasible operation. The formulation of the sequence casting model is explained, and results that demonstrate how operating costs are affected by the meltshop/ladle/casting operation are presented. The sensitivity of the solutions to changes in the meltshop/ladle/caster parameters is also examined.

*This paper previously appeared in the October '87 issue of <u>Iron and Steelmaker</u>, published by the Iron and Steel Society.

Introduction

In today's competitive steel market it is essential that steelmakers produce high quality steel products at the lowest possible cost. One way to reduce costs is to make the most effective use of available facilities. The manner in which facilities are best used is not always intuitively obvious. In this paper, cost effective ways to manage the use of three sequential batch operations are examined and applied to a meltshop/ladle station/caster to minimize costs for sequence casting.

In broader terms, an optimized sequence allows us to handle interactions between the caster and the meltshop. We use the term *sequence* to denote how breaks are scheduled between casting of heats that are of predetermined sizes and in a predetermined order. We are not attempting to solve the complete optimal *scheduling* problem in which the ordering of heats is considered. The focus of this work is to assess scenarios where the caster and the meltshop must be synchronized in order to maximize productivity. It is emphasized that productivity is not just throughput, yield, or utilization, but a combination of all three.[1] The synchronization required of an electric arc furnace/continuous casting operation is discussed by Pearce.[2] High power electric arc furnaces make the synchronization of caster and meltshop even more critical. The synchronization issue involves making decisions about breaking the casting sequence between heats, when (and for how long) to hold a heat if the caster is not ready and how fast to run the caster. Slowing of the casting operation has the immediate effect of reducing caster throughput, but may be used in order to eliminate breaks in the caster sequence, and actually increase the overall throughput of the casting operation.[3] Jackson shows the importance to production of the number of casts in a sequence.[4] However, current operations rely only on a foreman to make informed decisions to coordinate the furnace with the caster.[5]

The objective of this work is to develop a systematic procedure for caster sequencing at as low a cost as possible, subject to process operating constraints that ensure feasible operation of the castshop. The cost of casting is approximated as a linear function of the independent variables; the constraints for the problem are a set of equations that relate times in the meltshop schedule to times in the caster schedule. These constraints are a set of linear equalities and inequalities. In addition, several of the sequencing decisions are of the yes/no type - should the casting sequence be broken at a given point? These decisions require the inclusion of binary variables in the optimization problem formulation. Since the objective function and the constraints are all linear, and since some of the variables are allowed only binary values, mixed integer linear programming (MILP) methods[6] are used to solve the problem.

The interactions between the meltshop and the caster can have an important effect on the caster operating parameters. There is normally some degree of freedom in the choice of casting rate for the continuous casting of various steels. This freedom can be used to help synchronize the casting process with the meltshop schedule. If this approach is used, it becomes important that the maximum *and minimum* casting rates be known. These maximum and minimum rates can be determined from operating experience, or by applying nonlinear optimization techniques to models of the continuous casting process.[7, 8, 9]

In this work, a mathematical model that represents the interactions of caster, ladle and meltshop schedules is used to determine the lowest cost casting operation possible for a given sequence of heats. The costs incurred by the casting shop are minimized subject to the operational constraints of the meltshop/castshop system. The result of this minimization is a schedule for the meltshop, for ladle operations and for the caster. This model can be applied to any three sequential processes; for example, a continuous caster, slab holding furnace and a reheat furnace for hot charging applications.

Problem Development

A sample schedule for a meltshop/continuous casting operation is shown in Figure 1. The first heat, which took time S_1 to prepare, becomes available at time T_1 (point A in Figure 1), is transferred and held in a ladle for time H_1, and is then cast (point B). Casting this first heat requires time C_1. As soon as this first heat is placed in a ladle, the meltshop begins to make another heat. The making of this next heat requires time S_2. This heat is transferred and held in a ladle for time H_2, and is cast during time interval C_2. After the casting interval C_2, the sequence is broken for a period P_2 (point C), while the caster waits for the third heat to become available.

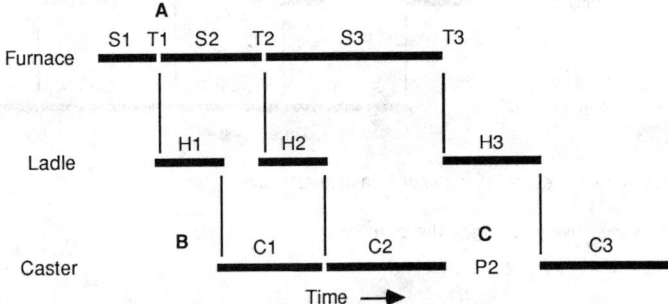

Figure 1: Example of a schedule for a meltshop/continuous casting operation.

Initially, the problem is formulated as an attempt to derive schedules where the caster can be operated continuously. The number of heats that are to be considered is given by n. The position of a specific heat in the sequence of heats produced is denoted by the subscript i. Each heat i becomes available at time T_i. The interval of time between two successive heats (the preparation time, S_i) is given by

$$S_i = T_i - T_{i-1} \qquad\qquad i = 1, 2, \ldots n \qquad\qquad (1)$$

The casting times are denoted by C_i, and have upper and lower bounds given by C_i^U and C_i^L respectively. These bounds are functions of the casting equipment, casting parameters, tonnage and grade of steel being cast. If all the intervals S_i obey the inequalities

$$C_i^L \le S_{i+1} \le C_i^U \qquad\qquad i = 1, 2, \ldots n \qquad\qquad (2)$$

then the scheduling problem is trivial. n heats can be consecutively cast by setting

$$C_{i-1} = S_i \qquad\qquad i = 1, 2, \ldots n \qquad\qquad (3)$$

Thus the arrival rate of heats to the caster and the service rate of the caster can be adjusted to be equal. If the arrival rate is greater than the service rate, the caster is operated as close to C_i^L as possible. This situation clearly implies effective ladle handling facilities.

The problem becomes more interesting if, for at least one heat, the inequalities in equation (2) are violated (i.e., the arrival rate is less than the service rate), or if the ladle handling facility is a bottleneck in the system (e.g., it can only handle one ladle at a time). Consider the situation shown in Figure 2. In this case, $C_1 = C_1^U$, $C_2 = C_2^U$, and $C_3 = C_3^L$, are all forced to occur by assuming the conditions $S_2 > C_1^U$, $S_3 > C_2^U$, and $S_4 < C_3^L$. The start of the casting of heat 1 must be delayed by a time H_1, the beginning of cast 2 by a time H_2 and the beginning of cast 4 by a time H_4. It is implicitly assumed that the casting of a heat can be delayed by holding a heat in a heated ladle, at a heated ladle treatment station, or in an unheated ladle that is insulated well enough so that the minimum and maximum

1159

casting times are not affected (i.e., the pouring temperature remains approximately constant).[*] This assumption seems reasonable, as ladles are currently used in industry as buffers between the meltshop and the caster.[5]

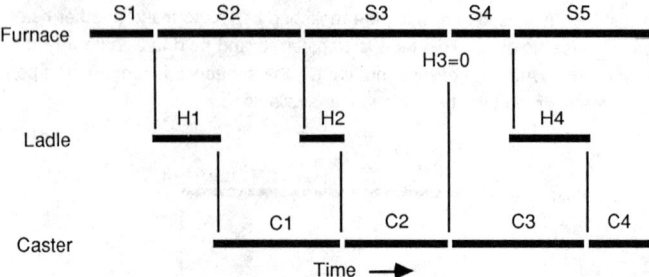

Figure 2: A more interesting example of a casting schedule.

There is incentive to minimize the quantity

$$H_1 + C_1 + H_2 + C_2 + H_3 + C_3 + \ldots H_n + C_n \tag{4}$$

as these time intervals, the holding times (H_i), and casting times (C_i), incur costs. Here the cost is actually written as:

$$\text{cost} = \sum_{i=1}^{n} w_i^H H_i + \sum_{i=1}^{n} w_i^C C_i \tag{5}$$

where w_i^H is a coefficient that represents the cost of holding heat i and w_i^C represents the cost of casting heat i. The superscript H (or C) is used to associate a cost coefficient with a time period of type H (or C).

Examining the schedule shown in Figure 2 shows that in order for the casts to occur as a single sequence, equalities of the following form must hold.

$$H_1 + C_1 = S_2 + H_2$$
$$H_1 + C_1 + C_2 = S_2 + S_3 + H_3 \tag{6}$$
$$H_1 + C_1 + C_2 + C_3 = S_2 + S_3 + S_4 + H_4$$

etc.

A more general statement of these requirements is shown in equation (7).

$$H_1 + \sum_{j=1}^{i} C_j - \sum_{j=2}^{i+1} S_j - H_{i+1} = 0 \qquad\qquad i = 1, 2, \ldots n \tag{7}$$

The holding times are also required to be non-negative,

$$H_i \geq 0 \qquad\qquad i = 1, 2, \ldots n \tag{8}$$

and the scheduling problem therefore consists of equations (5), (7) and (8). The objective function is linear (the w_i are known or assumed), as are all of the constraints. Thus the problem is stated as the following linear program (LP):

---

minimize $\qquad \sum_{i=1}^{n} w_i^H H_i + \sum_{i=n}^{n} w_i^C C_i$ $\qquad\qquad\qquad\qquad\qquad\qquad$ (9)

such that $\qquad C_i^L \le C_i \le C_i^U$ $\qquad\qquad\qquad\qquad\quad i=1,2,\dots n$

$\qquad\qquad\quad H_1 + \sum_{j=1}^{i} C_j - \sum_{j=2}^{i+1} S_j - H_{i+1} = 0 \qquad\quad i=1,2,\dots n$

$\qquad\qquad\quad H_i \ge 0 \qquad\qquad\qquad\qquad\qquad\qquad i=1,2,\dots n$

This problem can be solved very efficiently using standard linear programming packages.[10] In this formulation, neither meltshop rescheduling nor breaks in the casting sequence are allowed to occur. Furthermore, no restrictions have yet been placed on the maximum time a ladle can be held, or the number of ladles that can be handled simultaneously.

To introduce meltshop scheduling, it is assumed that there is a minimum time required to prepare a heat. This time is a lower bound for the interval S_i, and is referred to as S_i^L. The conditions

$\qquad S_i^L \le S_i$ $\qquad\qquad\qquad\qquad\qquad\quad i=2,3,\dots n$ $\qquad\qquad\qquad$ (10)

and associated costs

$\qquad \sum_{i=2}^{n} w_i^S S_i$ $\qquad\qquad\qquad\qquad\qquad\qquad\qquad\qquad\qquad\qquad$ (11)

are added to the formulation to represent the cost of operating the meltshop. The problem can now be stated as:

minimize $\qquad \sum_{i=1}^{n} w_i^H H_i + \sum_{i=1}^{n} w_i^C C_i + \sum_{i=2}^{n} w_i^S S_i$ $\qquad\qquad\qquad\qquad$ (12)

such that $\qquad C_i^L \le C_i \le C_i^U$ $\qquad\qquad\qquad\qquad\quad i=1,2,\dots n$

$\qquad\qquad\quad H_1 + \sum_{j=1}^{i} C_j - \sum_{j=2}^{i+1} S_j - H_{i+1} = 0 \qquad\quad i=1,2,\dots n$

$\qquad\qquad\quad H_i \ge 0 \qquad\qquad\qquad\qquad\qquad\qquad i=1,2,\dots n$

$\qquad\qquad\quad S_i^L \le S_i \qquad\qquad\qquad\qquad\qquad\quad i=2,3,\dots n$

This is also an LP, and can be solved in the same way.

The heat preparation time S_i can be divided into two separate times E_i and D_i, where E_i represents the actual heat preparation time and D_i represents meltshop downtime. This is shown in equation (13).

$\qquad S_i = E_i + D_i$ $\qquad\qquad\qquad\qquad\qquad\quad i=1,2,\dots n$ $\qquad\qquad\qquad$ (13)

The objective function then includes the following cost terms:

$\qquad \sum_{i=2}^{n} w_i^E E_i + \sum_{i=2}^{n} w_i^D D_i$ $\qquad\qquad\qquad\qquad\qquad\qquad\qquad\qquad$ (14)

and equation (7) is written as

$\qquad H_1 + \sum_{j=1}^{i} C_j - \sum_{j=2}^{i+1} E_j - \sum_{j=2}^{i+1} D_j - H_{i+1} = 0 \qquad i=1,2,\dots n$ $\qquad$ (15)

In cases where the downtime cost w_i^D is less than the production cost w_i^E, this will always force $E_i = E_i^L$ as less costly times D_i will be used to take up the slack in equation (13). The cost term $\sum w_i^E E_i$ will be a constant, and the effect of the original cost term $\sum w_i^S S_i$ is expressed by changes in the new cost term $\sum w_i^D D_i$. In these cases, no generality is lost by using equation (7). Cases where the downtime cost w_i^D is greater than the production cost w_i^E are rare and have not been considered further.

In order to introduce breaks in the casting sequence, the binary variables y_i are added to the formulation. The set of binary variables y_i, one for each heat, represent the decision to break the sequence after heat i ($y_i=1$) or not to break the sequence after heat i ($y_i=0$). The problem is now a Mixed Integer Linear Program (MILP). P_i is used to denote the length of time for which the casting sequence is broken after heat i. This is the situation shown in Figure 1. Instead of equation (7), the following equalities are now required:

$$H_1+C_1+P_1=S_2+H_2$$
$$H_1+C_1+C_2+P_1+P_2=S_2+S_3+H_3 \qquad (16)$$
$$H_1+C_1+C_2+C_3+P_1+P_2+P_3=S_2+S_3+S_4+H_4$$

$$etc.$$

or,

$$H_1+\sum_{j=1}^{i}C_j+\sum_{j=1}^{i}P_j-\sum_{j=2}^{i+1}S_j-H_{i+1}=0 \qquad i=1,2,\ldots n \qquad (17)$$

If $y_i=1$ (i.e., there is a break after heat i) the break time, P_i, is required to be greater than some minimum time P_i^L. The time P_i is essentially unbounded from above. If $y_i=0$ (no break), then $P_i=0$ is required. These requirements can be expressed as

$$P_i^L y_i \le P_i \le P_i^U y_i \qquad i=1,2,\ldots n \qquad (18)$$

where P_i^U is a very large, but finite number. As to the cost of a break, two types of terms are added to the cost expression. There is a fixed cost associated with the occurrence of a break of minimum length. In addition, there is a cost invoked for any additional time involved in a break beyond the minimum. This cost represents the lack of utilization of the caster. These costs are written as

$$\sum_{i=1}^{n}w_i^y y_i + \sum_{i=1}^{n}w_i^P(P_i-P_i^L) \qquad (19)$$

Also, it is unreasonable to require only that the holding times H_i be non-negative. These times must be greater than some minimum time H_i^L that represents the time required for a heat to be physically moved from the meltshop to the caster. These times must also be bounded by a maximum time H_i^U that represents the longest period of time that a heat can be held in a ladle. These requirements are stated in equation (20).

$$H_i^L \le H_i \le H_i^U \qquad i=1,2,\ldots n \qquad (20)$$

Combining problem (12) and equations (17) to (20), results in the following MILP for n heats.

minimize
$$\sum_{i=1}^{n}w_i^H H_i + \sum_{i=1}^{n}w_i^C C_i + \sum_{i=2}^{n}w_i^S S_i + \sum_{i=1}^{n}w_i^y y_i + \sum_{i=1}^{n}w_i^P(P_i-P_i^L) \qquad (21)$$

such that
$$C_i^L \le C_i \le C_i^U \qquad\qquad i=1,2,\ldots n$$
$$S_i^L \le S_i \qquad\qquad i=2,3,\ldots n$$
$$H_1+\sum_{j=1}^{i}C_j+\sum_{j=1}^{i}P_j-\sum_{j=2}^{i+1}S_j-H_{i+1}=0 \qquad i=1,2,\ldots n$$
$$H_i^L \le H_i \le H_i^U \qquad\qquad i=1,2,\ldots n$$
$$P_i^L y_i \le P_i \le P_i^U y_i \qquad\qquad i=1,2,\ldots n$$
$$y_i \in \{0,1\} \qquad\qquad i=1,2,\ldots n$$

It is reassuring to note that by eliminating breaks from this problem (requiring $y_i = 0$, $i = 1, 2, \ldots n$) the problem reduces to problem (12). By further requiring $S_i = S_i^L$, $i = 2, 3, \ldots n$, the problem reduces to the original LP, problem (9).

As an example of how additional operating criteria can be included in the problem, the use of scrap in the preparation step is considered. Part of the preparation time is used to remove carbon from the steel. The lower bound on preparation time is affected by the carbon content of the initial furnace charge. High carbon charges take longer to decarburize. This relationship has been linearly approximated. Carbon content in the initial charge can be controlled by controlling the quantity of scrap charged. The relation between scrap charged and carbon content has also been approximated as linear. For conditions when this approximation is not valid, piecewise linear relationships could be used. In this situation the constraint given by equation (10) is replaced by the combined linear relationship between percent scrap (x_i) and the lower bound on preparation time shown in equation (22), where k_1 and k_2 are constants.

$$S_i^L = k_1 x_i + k_2 \tag{22}$$

In addition, the upper and lower bounds on the percent scrap (x_i^U and x_i^L respectively) charged in a given heat given by equation (23) are enforced, as well as the upper limit on the total percentage of scrap (**X**) used during the n heat sequence given by equation (24). This scrap calculation has been included in all of the cases in which casting time was not fixed (cases 5 through 10).

$$x_i^L \leq x_i \leq x_i^U \qquad\qquad i = 1, 2, \ldots n \tag{23}$$

$$\frac{1}{n} \sum_{i=1}^{n} x_i \leq \mathbf{X} \tag{24}$$

Problem Solution

The model has been formulated as an MILP, using a linear objective function and linear equality and inequality constraints. The variables that represent existence of breaks in the casting sequence are further constrained to only have the binary values 0 or 1. The LP part of the problem is normally solved by the dual simplex method.[6] The addition of the binary variables makes the optimization problem significantly more difficult, as the solution of the LP may result in non-integral values for the integer variables. In situations where the values of the integer variables are large, the non-integral results can often be rounded or truncated to integral values. In the present case, where the values of the integer variables can only be 0 or 1, rounding or truncation can easily lead to non-optimal solutions. An alternative is to remove the integer variables from the formulation and solve an LP subproblem for each possible combination of the integer variables, choosing the solution with the greatest objective function as the optimal solution to the MILP problem. This technique is feasible only for very small quantities of integer variables, as it requires the solution of 2^m LP subproblems for a problem with m integer variables. For $m = 10$, this is over 1000 subproblems!

Instead of this brute force, complete enumeration approach, a branch and bound procedure is used. The branch and bound procedure is an efficient alternative to enumerating all of the possible solutions to the MILP problem. Assume for this discussion that it is desired to *maximize* the objective function in the MILP problem. The procedure begins by considering the problem as a binary tree. The root node of the tree is the relaxed LP, and the descendant nodes successively add integrality constraints to the integer variables in the MILP. The relaxed LP is solved using the dual simplex method. This provides an upper bound to the problem (the objective can only stay the same or decrease as more constraints are added to the problem). The integer variables are sequentially set equal to 0 or 1 until all of the variables are integer. This produces a lower bound on the optimization problem. Now, other combinations of 0s and 1s are tried, one at a time, to determine if other branches in the tree can possibly have better solutions than the one just found. By comparing the results at each newly calculated node in the tree to the upper and lower bounds, and by updating the bounds when better solutions are found, large portions of the tree can often be eliminated from further consideration in the solution of the MILP problem. The algorithms used to select which variable is next to become integral, and which of the values 0 or 1 it takes on determine the method in use. The commercially available code LINDO[10] used in this study uses a depth first branch and bound method, in which the tree is investigated by making all of the binary variables integral as soon as possible and then relaxing the resulting integral solution. An alternate approach is the breadth first search, where it is determined for each variable if it is better to make it a 0 or a 1 before proceeding to the next variable.

Results and Discussion

The model for optimizing sequence casting developed in the previous section has been applied to the casting operation of Allegheny-Ludlum Corporation, a member of the Center for Iron and Steelmaking Research at Carnegie Mellon University. The relative costs (the w_i) for this casting operation are summarized in Table I and the cases that were tested are summarized in Table II. All calculations were performed for sequences of 10 and 20 casts ($n=10$ and $n=20$). Only the results for $n=10$ are discussed here, as in all the cases the results for $n=20$ are nearly identical. The reference case (case 0) used for the cost calculations is based on current sequencing schedules in use at Allegheny-Ludlum Corporation. Realistic upper and lower bounds for processing times were also provided by Allegheny Ludlum. These bounds are summarized in Table III. All processing times are measured in minutes.

Table I: Relative Costs for Time Intervals

Reason for Cost	Symbol	Relative Cost	
furnace conversion time	w^S	1.0	($/minute)
ladle hold time	w^H	0.25	($/minute)
casting time	w^C	0.55	($/minute)
break in sequence	w^y	40.0	($)
break time	w^P	0.0	($/minute)

The base case (0) consisted of setting the heat preparation times, the casting times and the placements of breaks in the sequence equal to what was practiced in the shop. The meltshop schedule consisted of alternating 90 and 78 minute preparation times. The caster schedule consisted of casting periods of 56 and 55 minutes, followed by a break. This sequence was continued for 10 heats, and is shown pictorially in Figure 3. Note that in order to cast two heats back to back (doubles), ladle holding strategies alternate between long and short holding times.

Case 1 is similar to the base case, except that the position of breaks in the sequence and values for ladle hold times (H_i) were determined by minimizing the total cost. The preparation times and the casting times were held fixed at the same values as in case 0. The resulting schedule is shown pictorially in Figure 4. This rearrangement of the casting sequence resulted in a reduction of the cost function for the casting operation of 3.8%. Relative costs and percent reductions for the various cases are tabulated in Table II. Here the model predicts that three casts can be sequence cast back to back (triples) by effectively using the ladle handling station as a buffer. In order to cast the second set of triples, two ladles must be handled simultaneously. This situation may not be possible in some shops due to a lack ladle and/or crane availability.

Case 2 is the same situation as case 1, except that now additional constraints have been added

1164

Table II: Summary of Cases Tested by the Sequence Model (10 casts in each case)

Case	Casting Times	Furnace Times	Casting Breaks	Ladle Overlaps	Total Relative Cost	% Reduction in Total Cost	Maximum Sequence Casts
0	fixed	fixed	fixed	yes	1479	-	2
1	fixed	fixed	free	yes	1423	3.8	3
2	fixed	fixed	free	no	1438	2.8	3
3	free[1]	fixed	free	yes	1384	6.4	4
4	free[1]	fixed	free	no	1385	6.4	continuous
5	free[1]	free	free	yes	1202	18.7	continuous
6	free[1]	free	free	no	1203	18.7	continuous
7	free[2]	free	free	yes	1209	18.3	6
8	free[2]	free	free	no	1212	18.1	8
9	free[3]	free	free	yes	1223	17.3	4
10	free[3]	free	free	no	1228	17.0	4

[1] $C_i^U = 85$ min
[2] $C_i^U = 65$ min
[3] $C_i^U = 55$ min

Table III: Bounds on Processing Variables

Variable	Lower Bound	Upper Bound
C_i (min)	45	85
% scrap	40	90
total % scrap	-	65
H_i (min)	30	100
S_i (min)	-	120

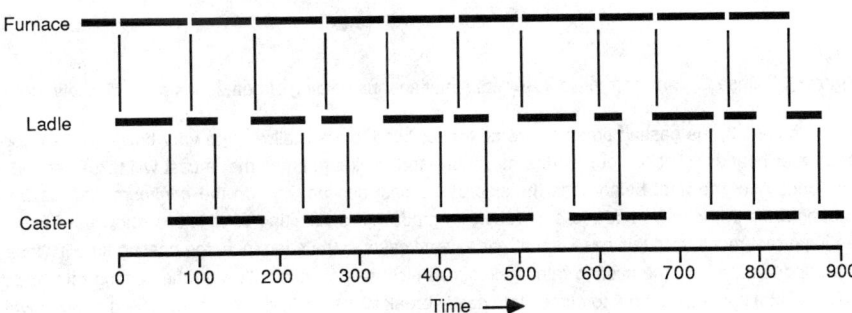

Figure 3: Case 0. Sequencing practice used as the reference case.

to the problem to prevent the simultaneous holding of heats in ladles. These constraints are written as

$$H_i \leq S_{i+1} \qquad i = 1, 2, \ldots n \qquad (25)$$

As seen in Table II, the cost of case 2 is greater than the cost of case 1. This is to be expected as the additional constraints further restrict the sequencing problem. A cost improvement of 2.8% over case 0 now occurs. The resulting schedule is shown in Figure 5. Note that the ladle holding periods do not

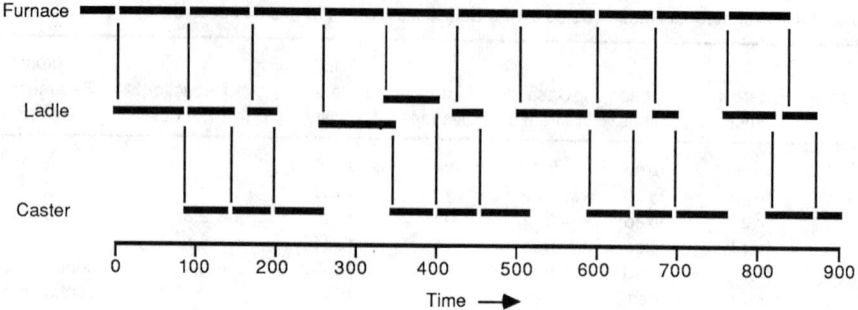

Figure 4: Case 1. Position and occurrence of breaks and ladle hold times are variable.

overlap and that the casting sequence begins with a triple and reduces to doubles. Triples could be made to be the steady state solution by altering the furnace schedule to break for approximately five minutes every third heat, with a slightly reduced productivity and slightly higher cost. On the other hand, if the sequence of breaks in case 1 was specified and ladle overlapping not allowed, with no breaks in the furnace schedule, the problem was found to be over constrained and had no feasible solution. Thus, in order to continue to cast triples, a reduction in productivity must occur by breaking the furnace shop every third heat.

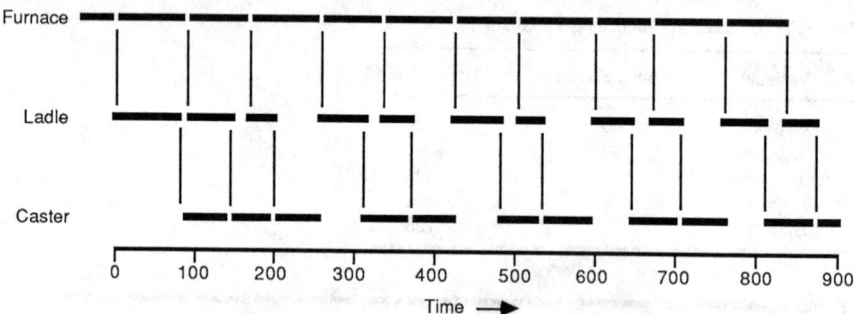

Figure 5: Case 2. Similar to case 1, except simultaneous holding of heats has been disallowed.

In case 3, the casting times C_i are no longer fixed, but are allowed to vary between their upper and lower bounds. These bounds (the maximum and minimum times that a cast will take) depend on the tonnage of the heat being cast, the size of the cast section, and on the *minimum and maximum casting rates* that can be used to cast a certain grade. The resulting schedule is shown in Figure 6. An important result from this case is that, for several casts in the sequence, the casting time is close to the upper bound, i.e., the casting rate is almost as slow as possible. Slowing the casting rate has the effect of stretching a cast out to eliminate a costly break in the casting sequence. The degree to which this stretching occurs depends on the relative cost of casting compared to the cost of interrupting the casting sequence. Note that the casting sequence starts with triples and then adjusts to quadruple sequences.

Case 4 is similar to case 3, but here ladle overlaps are disallowed. This case is shown in Figure 7. The cost is slightly higher than in case 3 because of the additional constraints, although the caster can be run continuously without breaks. This is an interesting case, as the schedule adjustments needed to make casting continuous cost slightly more than the breaks found in case 3. In case 4 the

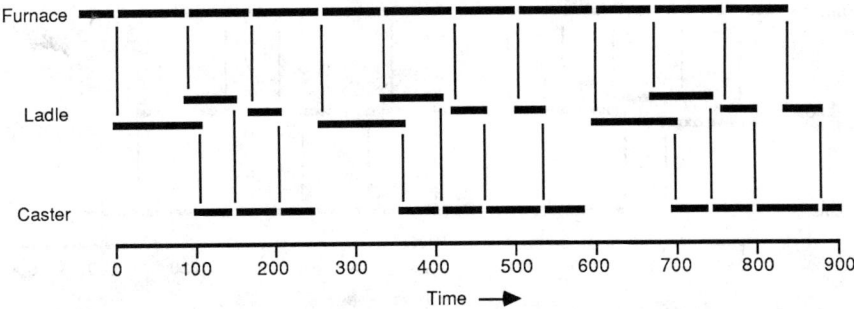

Figure 6: Optimal schedule for case 3. Casting times may vary and ladle overlaps are allowed.

casts are longer and the ladle holds shorter than in case 3. Since ladle holds are less expensive than casting, case 3 is slightly less expensive than case 4, even with the inclusion of break costs.

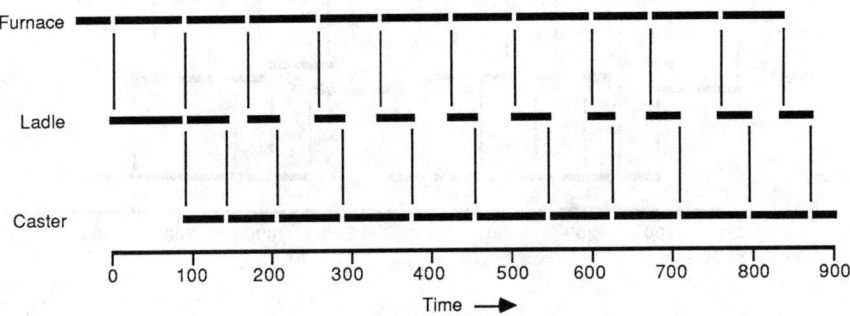

Figure 7: Optimal schedule for case 4. Casting times may vary and ladle overlaps are not allowed. Casting is continuous, although at a slightly higher cost than case 3.

In case 5, the heat preparation time S_i was also allowed to vary between the upper and lower bounds with ladle overlaps. The result is shown in Figure 8. In this case, it was found that casting could be truly continuous, with no breaks in the sequence. A reduction in cost of 18.7% was found. It is emphasized that the bounds on all of the variables were considered to be attainable. Case 6 is similar to case 5, but with no ladle overlaps. The cast sequence is again continuous, at slightly higher cost than case 5 because of the additional constraints.

In cases 7 and 9 (again, 8 and 10 are similar, but with no ladle overlaps), the maximum casting times (determined in part by the minimum casting rate) were reduced from 85 minutes to 65 and 55 minutes respectively to investigate the sensitivity of the schedule to the minimum casting rate. This tightening of the bounds resulted in cost increases compared to case 5, as expected. Breaks in the sequence were also forced to occur. The resulting schedules are shown in Figures 9 and 10. In general, the more tightly constrained sequencing problems result in higher costs and more breaks in the casting sequence.

An additional result from the solution of the MILP problem is an automatic sensitivity analysis with fixed integer variables (breaks). The sensitivity analysis gives information about how the solution of the MILP problem will change with changes in the constant coefficients and variable bounds. This information can be used to determine if a given schedule is optimal for outside fluctuations for a given

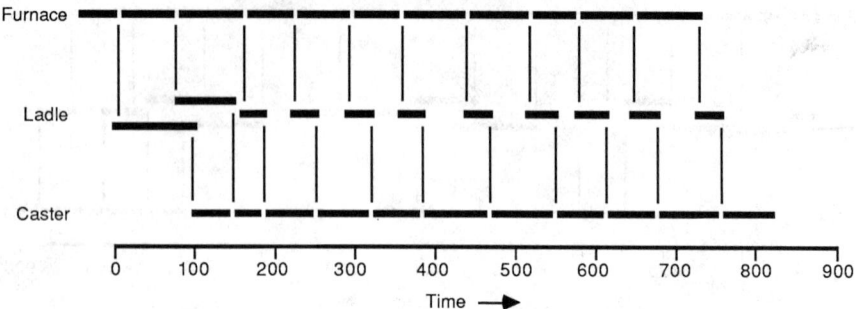

Figure 8: Optimal schedule for case 5. Heat preparation times may vary and ladle overlaps are allowed.

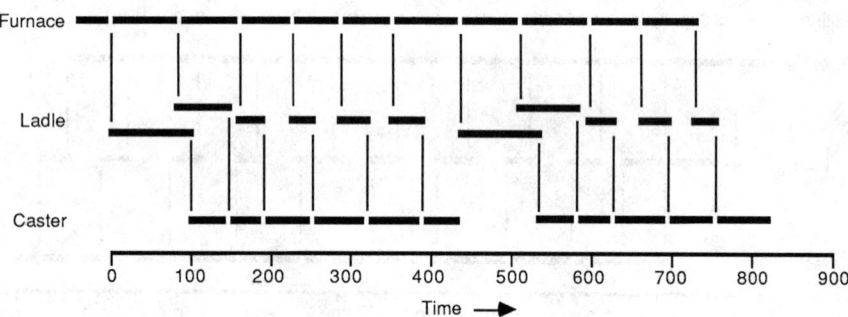

Figure 9: Optimal schedule for case 7. Maximum casting time reduced from 85 minutes to 65 minutes, with ladle overlaps.

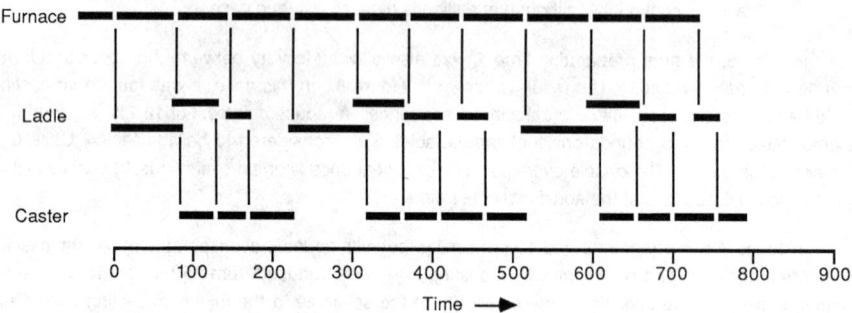

Figure 10: Optimal schedule for case 9. Maximum casting time further reduced to 55 minutes, with ladle overlaps.

set of breaks (i.e., is there a big enough safety margin to allow small changes in the schedule times), and to pinpoint critical times that must be met in the resulting schedules. This will be discussed further in the following section.

The statement and solution of the scheduling problem as a Mixed Integer Linear Program is very important to the efficient solution of the problem. While methods exist to solve optimization problems involving nonlinear objective functions and constraints, they are much more time consuming than the methods used to solve linear problems.[6] Solution of the problems discussed in this paper (for $n = 10$) took from 64 to 205 CPU seconds.[*] Since new solutions are available fairly quickly, on modest computing hardware, new schedules can be calculated to accommodate unforeseen changes in plant operation online, *while the casting operation continues.*

Sensitivity of Optimal Schedule to Input Parameters

In many situations the information for creating the MILP scheduling problem is subject to change, or not well known. For example, relative costs for casting, ladle holding and interrupting the cast might be of limited accuracy and are certainly subject to change. Similarly, bounds on holding, casting and furnace times, as well as scrap content may be variable. One way of dealing with these issues is to resolve the schedule for different parameter values. Because of the structure of the MILP problem, however, a sensitivity analysis of these parameters is easily generated as a by-product of the solution procedure. This ensures the optimality of a given schedule for a fixed break pattern even under these variable conditions.

The LINDO[10] package performs the sensitivity analysis automatically for two sets of parameters. First, costs are linearly related to shadow prices which determine which constraints and variable bounds are active. Here the range of cost values which retain the optimal active set is generated by linear extrapolation. For these ranges of cost coefficients the optimal schedule will not change, although the objective function will. LINDO also generates ranges for the variable bounds under which the optimal active set is retained. For these ranges, variables at the upper or lower bounds will have to adjust to the bounds, but the casting schedule will remain qualitatively the same. The sensitivity analysis therefore gives ranges for maintaining optimality of the schedule for one-at-a-time perturbations of costs or variable bounds.

As an example, consider the casting run of case 5 where furnace, casting and holding times were optimized and no breaks occurred in the schedule. Figures 11 and 12 show graphs of the normalized ranges for a selected set of costs and variable bounds. Here the cost of casting time for the first cast (C1 in Figure 11) may be increased indefinitely or decreased by as much as 44% before forcing a change in this optimal schedule. For comparison, in order to maintain the optimal schedule the range on the lower bound for casting time allows a decrease of 16.7%, and no increase at all (CL in Figure 12). Consequently, if the casting lower bound needs to be increased, the MILP model must be resolved to get an optimal schedule.

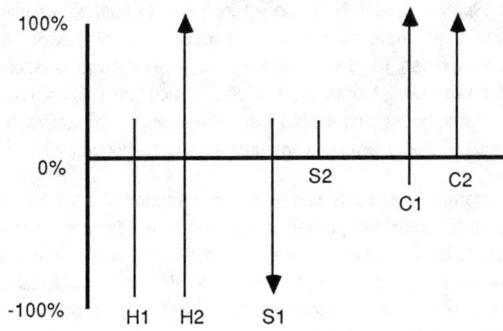

Figure 11: Normalized sensitivity of optimal solution to cost coefficients.

---

[*]All computation was performed on a DEC MicroVAX-II running MicroVMS v4.1m.

1169

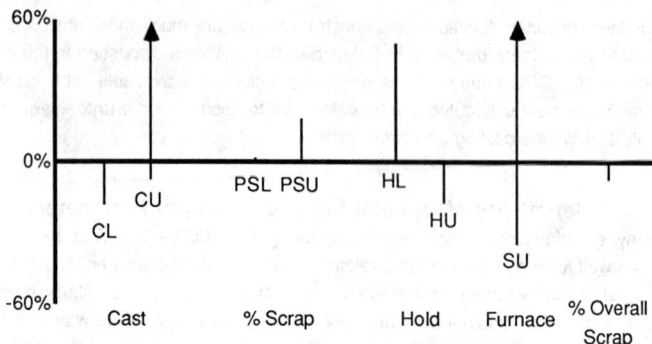

Figure 12: Normalized sensitivity of optimal solution to variable bounds.

While the above examples refer to one-at-a-time changes in cost or variable bound parameters, one can generalize this analysis for *simultaneous* changes. For example, if these parameters change by a known amount, one can still maintain the optimality of an existing schedule as long as the 100% rule

$$\sum_j \frac{\delta p_j}{\Delta p_j} < 1 \tag{26}$$

is satisfied for either bound or cost parameter variations. Here δp_j is the parameter variation in the operation of the caster and Δp_j is the range calculated from the sensitivity analysis. For example, a simultaneous decrease in the upper bounds of holding and furnace times of 10% allows the schedule to be the same, with adjustment of furnace and holding times variables to their upper bounds. This is easily seen in equation (27).

$$\frac{10\%}{15\%} + \frac{10\%}{33.3\%} = 0.967 < 1 \tag{27}$$

Conclusions

The method developed here is a useful tool for investigating relative costs associated with sequence casting. Costs and adjustable parameters can easily be modified to represent many different casting shops and combinations of unit operations. Several of the solutions to the trial cases show time periods where the casting machine has been slowed down to almost the slowest casting rate allowed. This situation results because an increase in casting time, even with an increase in casting cost, is more than offset by the decreased costs involved with not breaking the casting sequence. This strategy can therefore be effective even though it is counter-intuitive and often conflicts with current practice. Moreover, the short calculation times required to solve the MILP problem makes this method of scheduling caster operations attractive for real time sequencing scenarios, where plant conditions are likely to change on short notice.

The linear programming analysis can be easily extended to situations involving differing grades and tonnages of heats by using the proper (constant) scaling factors. The bounds on the variables would then be different for each heat. Differing bounds for each heat are allowed in the current formulation, but were not used in the calculations. Extending the work to allow the order of production of heats of different grades and tonnages to be varied as part of the solution would be a very interesting and useful exercise as would be extensions of the method to include other parts of the metal production operation (hot rolling and hot charging for example).

Acknowledgements

The authors wish to acknowledge the support the Center for Iron and Steelmaking Research, its member companies and the National Science Foundation (grant 84-21112) for support of this research. We are also grateful for numerous discussions with Dr. Don McMahon, Manager, Processing Section, Research Center, Allegheny-Ludlum Corporation, whose suggestions were particularly helpful.

References

1. H. Pielet, "Process Productivity in Continuous Casting", *Continuous Casting of Steel - Second Process Technology Conference*, 1981, pp. 246.

2. J. Pearce, "Development Trends in EAF Steelmaking", *Journal of Metals*, March 1986, pp. 38.

3. D. McMahon, Allegheny-Ludlum Corporation, personal communication, 1986.

4. C. R. Jackson, Design, Start-up and Operation of Continuous Casters, Lecture Notes.

5. J. Saillour and M. Mangin, "Slab Casting and Secondary Steelmaking at Sollac", *Iron and Steelmaker*, Vol. 12, No. 9, September 1985, pp. 22.

6. G. V. Reklaitis, A. Ravindran and K. M. Ragsdell, *Engineering Optimization,* Wiley-Interscience, New York, 1983.

7. B. Lally, H. Henein and L. Biegler, Carnegie Mellon University, unpublished research.

8. L. Holappa, E. Laitinen, S. Louhenkilpi and P. Neittaanmaki, "Optimization of the Secondary Cooling in the Continuous Casting of Steel Billets", Proceedings of the 24th Annual Conference of Metallurgists, Vancouver, British Columbia, August, 1985

9. M. Larrecq, J. P. Birat, C. Saguez and J. Henry, "Optimization of Casting and Cooling Conditions on Steel Continuous Casters - Implementation of Optimal Strategies on Slab and Bloom Casters", *Application of Mathematical and Physical Models in the Iron and Steel Industry*, AIME, March 1982, pp. 273.

10. L. E. Schrage, *User's Manual for Lindo,* 1983.

Nomenclature

n	number of heats
i	heat position in sequence
C_i	casting time
C_i^L	lower bound on casting time
C_i^U	upper bound on casting time
H_i	holding time
H_i^L	lower bound on holding time
H_i^U	upper bound on holding time
T_i	delivery time of heat
S_i	interval between delivery times
S_i^L	lower bound on heat preparation time
S_i^U	upper bound on heat preparation time
E_i	actual heat preparation time
E_i^L	lower bound on actual heat preparation time
D_i	meltshop inactivity time
w_i	cost weighting factors
y_i	binary decision variables (breaks) $y_i \in \{0,1\}$

P_i	length of breaks
P_i^L	lower bound on break time
P_i^U	upper bound on break time
x_i	percent scrap in heat i
x_i^L	lower bound on percent scrap in heat i
x_i^U	upper bound on percent scrap in heat i
X	upper limit on total percent scrap in n heats
k_1	constant
k_2	constant

MAXIMIZING PRODUCTIVITY IN A STAINLESS STEEL MELT SHOP

D.J. McMahon
J.D. Nauman
N.Y. Toker
V.P. Ardito

Process R&D Department
Technical Center
Allegheny Ludlum Corporation
Brackenridge, PA 15014

ABSTRACT

The purpose of this paper is to review the mathematical models used to produce stainless steel in Allegheny Ludlum Corporation's Stainless Melt Shop. The Melt Shop consists of four Electric Arc Furnaces, an AOD, an Argon Bubbling Station and a Continuous Slab Caster. The models described are:

1. The Arc Furnace Meltdown Model
2. The 2-stage AOD Refining Model
3. The Kinetic Decarburization Model
4. The Final Additions Model
5. The Argon Bubbling Model
6. The Secondary Cooling Model
7. The Caster Cut Model
8. Actual and Predictive Status Screens

These operational control models along with capital investments and management involvement have been instrumental in doubling the tonnage output of our stainless steel melt shop over the last 10 years.

INTRODUCTION

A brief description of stainless steel manufacturing is needed. Figure 1 illustrates the melting steps in producing stainless steel. At ALC, raw materials in various forms are purchased for melt down in an electric arc furnace. The melted hot metal is then transferred to an AOD vessel where the hot metal is refined, primarily, where oxygen is injected to react with carbon. The stainless steel chemistry is adjusted to meet customer requirements in the AOD. The ladle containing refined steel is transferred to Argon-Bubbling-Station in order to improve melt cleanliness, and temperature homogenity as well as to meet temperature requirements. The hot metal then is taken to the continuous slab caster, cast and cut into slabs having lengths based on eventual coil weight requirements.

The objective of the melt shop is to produce stainless steel slabs having quality meeting the customer's requirements at a minimum cost. The procedures needed to accomplish this need to clearly define the mathematical relationships of each of the melting operations.

The purpose of this paper is to review the mathematical models used to produce stainless steel in Allegheny Ludlum Corporation's Stainless Melt Shop. The Melt Shop consists of four Electric Arc Furnaces, an AOD, an Argon Bubbling Station and a Continuous slab caster. The models described are:

1. The Arc Furnace Meltdown Model
2. The 2-stage AOD Refining Model
3. The Kinetic Decarburization Model
4. The Final Additions Model

5. The Argon Bubbling Model
6. The Secondary Cooling Model
7. The Caster Cut Model
8. Actual and Predictive Status Screens

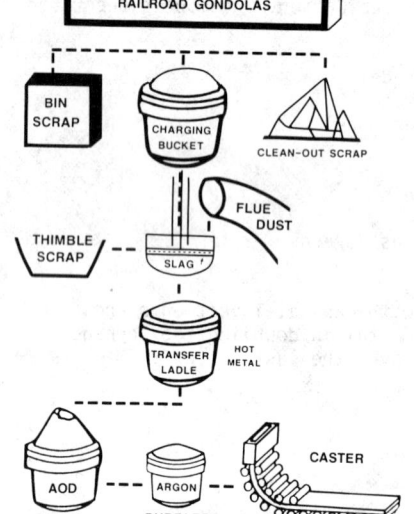

FIGURE 1: SCHEMATIC OF MELT SHOP

ALLEGHENY LUDLUM CORPORATION

	ARC MELT	CARBON (% OF TOTAL)	CHROMIUM (% OF TOTAL)	NICKEL (% OF TOTAL)	% WEIGHT	% COST
a.						
	18-8 TURN	0.00	2.85	3.83	4.32	3.72
	18-8 DUST	0.00	3.99	3.94	5.18	0.03
	SKULL	0.00	2.33	2.30	3.02	0.50
	T410	0.00	3.92	0.16	6.48	1.63
	18-8 SOLID	4.99	62.18	71.54	66.41	65.70
	HCCR 55	68.97	20.42	0.00	7.13	10.39
	NI-BRIQ	0.00	0.00	18.23	1.44	12.23
	ARC BACK					
	ARC CHARGED	74.07	95.70	100.00	93.98	94.20
	ARC RECOVERED	70.36	89.00	100.00	88.78	94.20
b.	AOD BLOW					
	HCCR-AOD	14.33	4.30	0.00	1.47	1.05
	HCFEMN	11.60	0.00	0.00	1.29	0.73
	OXYGEN	0.00	0.00	0.00		0.84
	NITROGEN	0.00	0.00	0.00		0.22
	AOD REDUCTION					
	50%FESI	0.00	0.00	0.00	3.26	2.35
	LIME	0.00	0.00	0.00	0.00	0.61
	AOD CHARGED	25.93	4.30	0.00	6.02	5.80
	AOD RECOVERED	7.02	4.30	0.00	4.13	5.80
	TOTAL CHARGED	100.00	100.00	100.00	100.00	100.00
	TOTAL RECOVERED	7.02	93.30	100.00	92.83	100.00

FIGURE 2: ARC/AOD OUTPUT REPORT

1. ARC FURNACE RAW MATERIAL MELT MODEL

Raw materials are melted in an electric arc furnace with the objective of minimizing the cost of the raw materials to meet a quality requirement. The model primarily is a set of material balance equations. The Model includes the following type restrictions:

. Material Chemical Balance
. Overall Material Balance
. Allowable raw materials per stage and per stainless type
. Chemical restrictions per melt furnace
. Scrap practice definitions
. Quality restrictions based on chemistry
. Utilization restrictions based on raw material form

The basic formulation of the ARC Model is as follows:

Minimize the cost of raw materials:

$$\text{Minimize} \sum C_i \, RM_i \qquad (1)$$

$$\sum (YRM_i) \times (YGRM_i) \times (RM_i) = (\text{Tap Weight}) \qquad (2)$$

where YRM_i is the sum of the elemental yields for a raw material
$YGRM_i$ is the gross yield lost during charging in furnace
RM_i is the initial charged raw material weight

$$\sum (EL_{ij}) \times (YEL_j)(YGRM_i) \times (RM_i) = (ELAIM_j) \times (\text{Tap Weight}) \qquad (3)$$

where EL_{ij} is the chemical value for the jth element in the ith raw material.
YEL_j is the yield of the jth element
$ELAIM_j$ is the aim element requirement for this stainless steel

A typical problem has about 30 equations and 40 raw materials. So it really is beyond the scope of this paper to decribe each equation.

A typical output is found in Figure 2A. The model has significantly lowered the raw material costs.

2. 2-STAGE AOD REFINING MODEL

This Model adds, to the previous ARC Model, equations which describe the decarburization and reduction phases of the refining process. The decarburization phase describes how oxygen, nitrogen and argon gasses are injected into the hot metal to remove carbon from the metal. Raw materials are also added to converge to the customer's chemistry requirements. Because of the affinity of other chemical elements, e.g., chromium, to react with oxygen, there needs to be a reduction phase to reduce the chromium from the slag back into the hot metal. The metals oxidized can be summarized in a metallic oxidation factor which is dependent on the weight and temperature of the ARC hot metal and the amount of raw materials added in the AOD. This generates nonlinearities to the set of linear equations. Because of the two-stages, dynamic programming techniques could be used. However, linear programming techniques are used, with the non-linear equations being satisfied in a heuristic manner.

Figure 2B also includes typical output for this Model. The Model has reduced AOD refining times significantly.

3. KINETIC DECARBURIZATION MODEL

The "AOD-BLOW-Model" is designed to predict the variation of process variables, namely temperature, weights and chemistries of alloy and slag throughout the decarburization phase of AOD refining cycle. Based on these predictions, for any given initial conditions (charge alloy chemistry, weight and temperature),the model recommends sequence of gas ratios (and their switch points) necessary for meeting end-point temperature and carbon aims while maintaining a trajectory that is optimized with respect to AOD blow time and reduction-mix requirements. The model can accept actual temperature and/or, carbon data and make future predictions based on updated information.

Model assumptions as well as a few of the representative equations are summarized below:

(1) O_2 component of the tuyere gas mixture is assumed to react instantaneously, generating a mixed oxide of composition $(Fe_{1-x-y}Mn_xCr_y)$ Cr_2O_4. Activities of mixed oxide components $(FeCr_2O_4, MnCr_2O_4, Cr_3O_4)$ are calculated from thermodynamic considerations utilizing time dependent alloy chemistry.

(2) Decarburization is assumed to take place through reduction of oxides at alloy/bubble interface and rate is controlled by mass transport within the alloy. Rate of transport of N_2 to and from bubbles is described by an expression which includes effects of mass transport and chemical reaction. At any given bulk alloy composition and temperature, CO, N_2 and Ar contents of ascending bubbles are calculated as a function of bath height and instantaneous decarburization and nitrogen absorption (or desorption) rates are calculated.

Example equations are:

$$\frac{dn_{CO}}{dx} = 5.91 \ 10^{-3} \ D_C^{\frac{1}{2}} \ V_B^{0.4} \ (\%C - Bp_{CO}) \tag{4}$$

$$\frac{dn_{N_2}}{dx} = 2.5 \ 10^{-3} \ D_N^{\frac{1}{2}} \ V_B^{0.4} \left\{ \%N + \frac{k_N^D}{2k_N^R} - \left[(\frac{k_N^D}{2k_N^R})^2 + kp_{N_2} \ f_N^{-1} + \frac{k_N^D}{k_N^R} \%N \right]^{\frac{1}{2}} \right\} \tag{5}$$

where $D_N, D_C = f(T)$: diffusivities of nitrogen and carbon in liquid alloy

$V_B = f(n_{N_2}, n_{CO}, n_{Ar}, x, T, p)$: volume of a bubble

$\%C, \%N$ = bulk composition of carbon and nitrogen in alloy

$k_N^D = f(D_N, V_B)$: mass transport coefficient of nitrogen

$k_N^R = f(T, \text{alloy chemistry})$: reaction rate constant for nitrogen

$p_{CO}, p_{N_2} = f(n_{N_2}, n_{CO}, n_{Ar}, x, p)$: partial pressure of carbon monoxide and nitrogen in a bubble

n_i = moles of species i in a bubble: i = N₂, CO, Ar

$B = f(T, \text{alloy chemistry, oxide chemistry})$: thermodynamic parameter

x = melt height above tuyeres

p = atmospheric pressure

$k = f(T, \text{alloy chemistry})$: thermodynamic parameter

T = alloy temperature

f_N = activity coefficient of nitrogen

Simultaneous solution of above equations using the Runga-Kutta methods will yield

$$n_i = f(\text{alloy chemistry}, X, T) \tag{6}$$

(3) Differential mass (for all alloying elements including oxygen, nitrogen and carbon) and heat balance equations are solved simultaneously (R-K methods) in order to determine variation of alloy and slag chemistries and weights and temperature with time.

Heat loss terms are calculated as a function of alloy temperature and time using a separate sub-routine, which involved solution of partial differential equations controlling coupled radiation-conduction heat transfer problem using finite-difference methods.

Since there is no substantial difference (except geometry) between the treatments used in calculating heat losses during AOD refining and ladle argon bubbling operations, example equations used for calculating heat loss terms are described in the section dealing with Argon-Bubble-Model.

The success of the model in providing excellent guidance to the melters has led to the achievement of the goal of one vessel turn-down/heat (2 minutes) during the entire decarburization cycle. Figure 3 is an example of this AOD Blow Model.

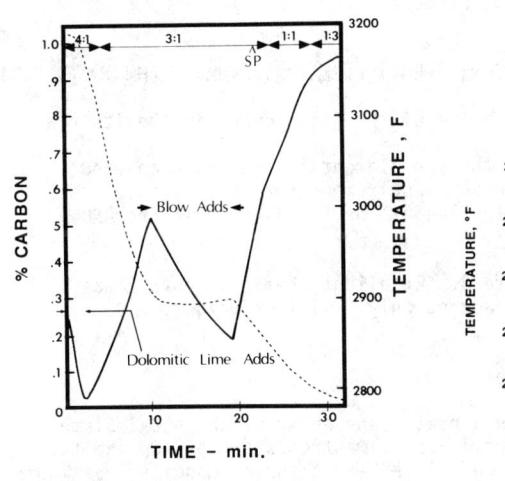

AOD Refining of Heat 889642
grade 413

TIME – min.

—— Temp.

------ Carbon

FIGURE 3: SCHEMATIC OF AOD BLOW MODEL

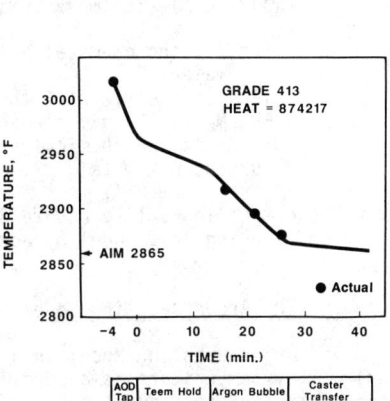

FIGURE 4: SCHEMATIC OF ARGON
BUBBLING MODEL

4. The Final Additions Model

The Final Additions Model is exactly like the ARC Model except that additional restrictions have been made on the weight of raw materials permitted to be added. At this point of the processing, it is desirable to spend as little time making these corrective additions as possible. This phase of the melting is the last chance to correct for "surprises" from the previous processing steps. It is the most critical and expensive part of the chemistry control. Thus, the least amount of raw materials should be added, but they should be added in such a manner that an integer number of full length slabs will be eventually cut.

The general linear programming problem is modified as follows:

Minimize the cost of raw materials:

$$\text{Minimize} \sum C_i \, RM_i \tag{7}$$

$$\sum (YRM_i) \times (RM_i) = (\text{Tap Weight}) \tag{8}$$

where YRMi is the sum of the elemental yields in the stage charged
RMi is the initial charged raw material weight

$$\text{Tap Weight} = \text{Existing Hot Metal Weight+Corrective Weight} \tag{9}$$
$$= HM + \sum(YRM_i) \times (RM_i)$$
known + unknown

Substituting EQ 9 into EQ 3, we get:

$$\sum [(EL_{ij}) \times (YEL_j) - (ELAIM_j) \times (YRM_i)](RM_i) = [(ELAIM_j) - ELHM_j)] \times (HM) \tag{10}$$

where ELij is the chemical value of the jth element in the ith raw material.
YELj is the yield of the jth element in the stage charged
ELAIMj is the aim element requirement for this stainless steel
ELHMj is the chemical value of the jth element in the ARC furnace hot metal

The output looks like Figure 2. Excellent chemical central has been achieved by using this model by adding only a minimum weight.

5. The Argon Bubbling Model

The Argon-Bubbling Model is a heat transfer model which calculates ladle alloy temperature as a function of time from start of AOD tap till commencement of continuous casting. This heat transfer model solves ladle refractory thermal profiles and uses information on tapping parameters - ladle thermal history before tap, alloy weight, type and quantity of ladle additions, time differential between AOD tap and start of casting, required argon bubble time and temperature after bubbling - in order to recommend AOD tap temperature necessary for meeting time/temper- ature constraints. After AOD tap, the model is updated with actual and/or expected time/temperature data before and during argon bubbling and recommends argon bubbling strategy for meeting time/temperature constraints imposed by the caster.

Use of this model, by providing smooth transition between the AOD and the caster, improves productivity of the melt-shop. A few example equations are:

(1) Heat transfer equations controlling ladle cooling in the absence of liquid metal:

$$(\rho C_p)_i^R \frac{\partial T_i}{\partial t} = \frac{\partial}{\partial x} (k_i^R \frac{\partial T_i}{\partial x}) \qquad\qquad i = 1, n \quad (11)$$

$$k_i^R \frac{\partial T_i}{\partial t} = q_i^R \quad \text{at hot face of brick work} \qquad\qquad (12)$$

$$\sum (\frac{\sigma_{ki}}{\varepsilon_i} - F_{ki} \frac{1-\varepsilon_i}{\varepsilon_i}) q_i^R = \sum (\sigma_{ki} - F_{ki}) \sigma(T_i^R)^4 \qquad \text{for } k = 1 \text{ to } n \ (13)$$

where σ = Stephan-Boltzman Constant

(ρ, C_p^R, k_i^R) = density, heat capacity, thermal conductivity of refractory segment i of ladle brick work

T_i^R = temperature of refractory segment i of brick work

q_i^R = radiation heat-flux at the surface of refractory segment i

ε_i, F_{ki} = emissivity and view factors for segment i

$\sigma_{ki} = \begin{matrix} 0 & k \neq i \\ 1 & k = i \end{matrix}$ x : distance into refractory

Where equation 11 deals with conduction heat transfer, equation 13 deals with radiation heat-transfer. These two equations are coupled through the boundary condition expressed in equation 12. After partitioning ladle walls into a number of segments, thermal profile (time-temperature) for each segment is determined using finite-difference methods.

(2) Heat-transfer equations after AOD tap:

Temperature stratification within the alloy is assumed negligible, thus giving

$$W_A \, c_P^A \, \frac{\partial T_A}{\partial t} = \sum_{i=1}^{m} A_i^R \, k_i^R \, \left(\frac{\partial T_i}{\partial x}\right)_{x=0} + A_A^R \, q_A^R + \dot{Q}_G + \Sigma \dot{Q}_J \, (dm_J/dt) \tag{14}$$

where   m = # of refractory segments below alloy top surface

A_i^R = area of segment i

A_A^R = area of alloy exposed to radiation losses

W_A, c_P^A = weight and heat capacity of alloy

q_A^R = rate of radiation heat losses from top of alloy surface

$\dot{Q}_G$ = rate of heat loss to argon gas

$\dot{Q}_J$ = heat consumed per unit weight of ladle addition type j

dm_J/dt = rate of dissolution of addition type j

Solution of equation 14, together with equation 11 and equation 13 with i=m+1 to n+1 with (n+1)th surface being the exposed surface of the alloy, yields thermal profiles for all the refractory surfaces and alloy.

Figure 4 gives an example of the Argon Bubbling Model. In sequence casting the concepts of this model have reduced bubbling time by 5 minutes for the second heat of the sequence.

6. The Secondary Cooling Model

The objective of the slab casting model in the melt shop was to provide recommendations for the design of operating practices as well as recommendations during the actual casting operations. The model was constructed to simulate the effect of various casting speeds and water sprays for steel grades and widths of interest. From these simulations, operating practices were developed to improve the productivity and the control of quality during casting. The model was also used on the shop operating computer to provide an estimation of the thickness of the solidified shell and a recommendation of optimum casting speed and water spray settings.

The slab casting model consisted of a solution to the two dimensional heat flow equation with a change of phase. The partial differential equation in two dimensions is:

$$k \frac{\partial T}{\partial x} + k \frac{\partial T}{\partial y} = cp \frac{\partial T}{\partial t} \qquad (15)$$

where T is the temperature at some place in the slab, k is the thermal conductivity at that temperature, c is the specific heat, p is the density of the metal, x is the distance in the direction of the broad width of the slab, y is the distance in the direction of the narrow face of the slab, and t is the time. The specific heat was generalized to include the effect of temperature and the effect of the heat on solidification.

To solve the equation for a rectangular slab, a two dimensional matrix was defined and the equation 15 was replaced with its finite difference form. Only the right half of the slab was solved, assuming that the left half is a mirror image. The coarsest possible element spacing was chosen to minimize the number of calculations that would be required to give a solution. A strongly implicit numerical method was selected to solve the array of finite difference equations to avoid instability at large time steps. The temperature distribution was then calculated at each time step throughout the length of the caster.

The boundary conditions at the broad and narrow faces of the cast slab were defined for the mold and each spray zone in the caster. The general condition at a boundary was described by the heat balance:

$$q(contact) + q(rad) + q(spray) = cp \frac{dT}{dt} + k \frac{dT}{dx} + k \frac{dT}{dy} \qquad (16)$$

where q(contact) is the heat flux to the mold through the mold powder, q(rad) is the heat flux from the surface when no rolls or mold was in contact with the slab, and q(spray) is the heat flux for the water sprays.

The heat flux in the mold varied with the position in the mold, depending on the casting speed, the flow of water in the mold, and the thermal contraction during solidification. The contraction of the corners away from the mold was simulated to allow accurate estimation of the shell thickness and the thermal stress and strain during cooling.

The heat flux from the water sprays was modeled with a transfer coefficient developed from operating data. The coefficient was proportional to the water flow rate of the sprays and inverse of the temperature of the slab surface. Thus, the coefficient varied from zone to zone in the spray chamber, as well as across the width and down the length of the slab in any zone. The transfer coefficient was found to decrease significantly in zones that had a single spray in the center, where the flow of water per slab surface decreased dramatically away from the center of the slab.

Throughout the caster the strain caused by the variations in cooling was calculated. The strain was converted into stress and strain rates to estimate the potential for thermal cracking on or below the surface of the slab. The stresses created by variation in cooling, as the slab moves from the mold, to the high intensity spray zones, and then to the low intensity spray zones, were high enough to cause plastic flow, so it was the strain rate that best indicated the potential for cracking.

In the real-time operating environment the caster model used both feed-forward and feed-back data from the caster to estimate the shell thickness, the potential for cracking, recommended casting speed, and recommended water sprays. The grade specifications, width, operating limits, tundish metal temperature, and the time before the delivery of the next heat were fed forward from the shop computer before the heat was cast. After casting was initiated, the heat extraction in the mold, the surface temperature of the slab in the spray chamber and straightener, and the actual water flow rates in each of the spray zones was monitored. From this set of data, the shell thickness, potential for cracking, recommended casting speed, and recommended water sprays was continuously estimated and displayed to the operator. The operator made the final decision on how to use the estimation to maximize productivity, while maintaining safety and steel quality.

Figure 5 shows the comparison between the casting model predictions and the actual data measured on the caster.

Figure 6 shows the estimation of the shell thickness and the tensile stress on the shell during a normal cast.

The use of the model has led to increases in the casting rates between 3 and 7% depending on chemical compositions.

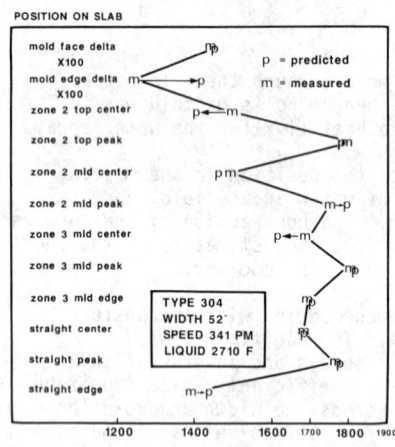

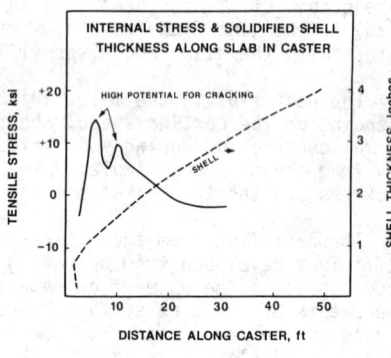

FIGURE 5: SCHEMATIC OF SECONDARY COOLING MODEL

FIGURE 6: ESTIMATION OF SHELL THICKNESS AND TENSILE STRESS

7. The Caster Cut Model

Figure 7 shows the cutting problems which occur when sections of the stainless steel continuous slab must be cut to satisfy a customer's order based on width and weight of the coil to be shipped. The random variable is the weight of the hot metal. The decision is which slabs should be lengthened or shortened and still maximize the material yield with the fewest number of slabs being generated. This problem can be solved using mixed-integer programming techniques and, at one time, was used in production. However, certain cutting rules could be established and cutting tables are now generated for each stainless steel batch.

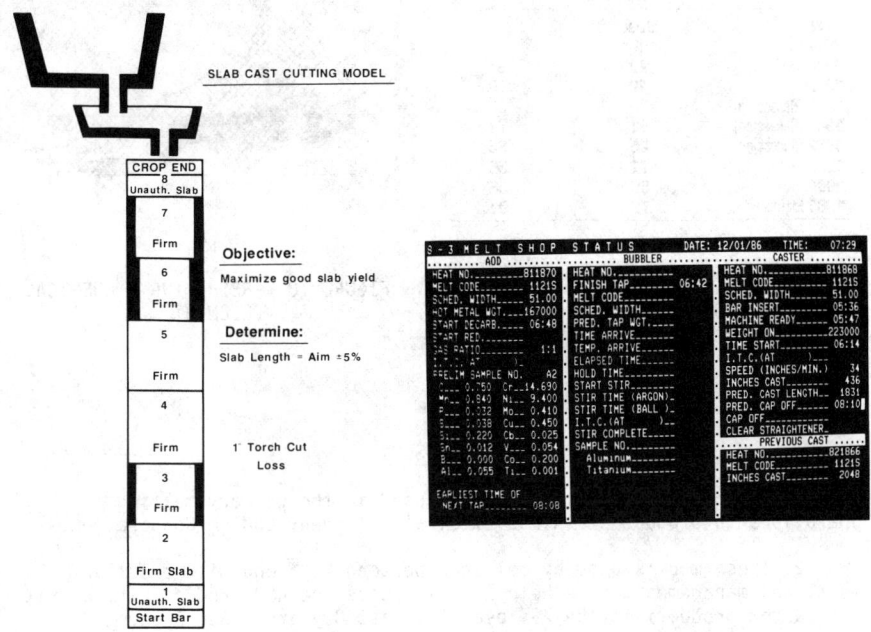

FIGURE 7: SLAB CUTTING MODEL   FIGURE 8: STATUS SCREEN EXAMPLE

8. Actual and Predictive Status Screens

In order to reduce communication time within the 8-3 melt shop, a display of critical event data was created. See Figure 8. The display consolidates current melt shop information occurring at the AOD, Bubbler, and Caster into a single video display of which copies can be located anywhere a need is shown. The display updates every 20 seconds, providing operators with the current and predicted critical events that are needed in order to synchronize the operations of the various shop areas. In this way, by reducing communication time the operators will have more time to devote to the melting process, thereby resulting in increased productivity and improvements in product quality.

Productivity

Year	Total Output Cumulative Improvement — %	Concast — %
1976	Base	
1977	9	
1978	22	31
1979	42	52
1980 (Recession)	20	66
1981 (Recession)	31	71
1982 (Recession)	25	86
1983	71	80
1984	65	86
1985 Estimated	79	91

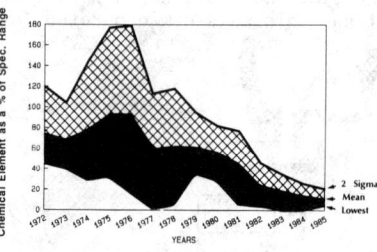

RESHAPING A CHEMICAL ELEMENT

FIGURE 9: STAINLESS STEEL PRODUCTIVITY   FIGURE 10: RESHAPING A CHEMICAL
            COMPARISON   ELEMENT

CONCLUSIONS

1. Seven mathematical models are used in the primary melting operations to produce stainless steel at Allegheny Ludlum Corporation.

2. These models used by meltshop personnel, along with excellent melt shop management, have helped to increase the AOD productivity by 91% and Caster productivity by 79% over the last 8 years. See Figure 9.

3. The quality of the stainless steels as measured by the uniformity of the chemical composition has improved by 90%. See Figure 10.

4. The use of active and predictive status screens to display the sensor data and math model calculations has improved the communications among meltshop personnel and has laid the groundwork for the Intelligent Processing of Materials.

RECOMMENDATIONS

The mathematical control models need continual modifications to integrate new computer efficiencies and the growing interfaces between the process-related objectives and the business-related objectives.

Expert system concepts need to be integrated into the control strategies of Allegheny Ludlum's stainless steel melt shops.

DYNAMIC SIMULATION STUDIES OF OPERATING PARAMETERS

IN GAS DISPERSION PROCESSES OF LIQUID STEEL

Saffet Turkan and Klaus W. Lange

Institute of Ferrous Metallurgy
Technical University of Aachen, Aachen, Germany

Abstract

With the progress in computer technology the method of mathematical modelling and dynamic simulation of continuous processes is developing itself to a standard tool of process analysis. This technique can be used effectively for the quantitative analysis of the interrelated multiphase phenomena occurring in gas-liquid dispersion processes of steel metallurgy.
In this work a full-scale simulation of a vacuum degassing process of 125 tons of steel in a ladle by simultaneous argon injection is carried out.
A series of simulation studies is made to reveal the effect of variations in operative parameters on the hydrodynamic behaviour of the system.
Varied parameters of the process are the ambient pressure (P_a); the aspect ratio of the steel bath (H/D); the flow rate of argon gas ($\dot{V}_g$). Investigated system values are the mixing power input (P_{mt}), the axial velocity (u_p) of liquid in the plume zone, the mixing time (t_m), the specific interfacial area (a) and the hold-up (ϕ_p) in the plume zone.

Introduction

In the last few years the techniques of gas injection into liquid metals have made considerable progress in high temperature metallurgical processes. The main advantages of gas injection, such as large interfacial area and intensive stirring with subsequent increase in refining rates have been well recognized and intensively utilized in ferrous and nonferrous metallurgical processes.

However, since the multiphase interrelated processes in a gas-liquid metal dispersion system are too complex for an analytical treatment and too expensive for explicit experimental investigations, previous studies concerned with the behaviour of the gas phase were limited to cases where the liquid phase was stagnant or correlations were extrapolated from cold temperature model studies. Furthermore, the effects of heat and mass transfer phenomena on the hydrodynamics of a dispersion system were usually neglected.

Although characteristic quantities of a gas dispersion system such as hold-up, interfacial area, induced circulation rate of liquid and stirring power are strongly interrelated with quantities being characteristic of the gas bubbles such as number, size, breakup, frequency of formation, residence time and relative velocity little is known about the dynamics of the relations between all these quantities. Especially the contribution of purgeable gases to these quantities is rather unknown.

On the basis of dynamic simulation method a series of studies has been made already by the authors (1-11) to improve fundamental knowledge of the behavior of the gas phase in dispersion systems. In previous dynamic simulation studies the formation, detachment, ascent and breakup of gas bubbles in stagnant and cocurrent flowing liquids and the mass transfer into the rising bubbles are evaluated. Degassing effect and the stirring work done by the dispersed gas phase and thus the induced circulation rate of the liquid are determined and verified. Results of the dynamic simulation studies yield a better understanding of complex interrelated phenomena and a better aimed engineering of gas dispersion techniques in metallurgical processes.

Objective of this study is to investigate the dynamics of the gas injection and vacuum degassing operations in a 125 tons steel bath with on-line simulations on a main-frame computer Cyber 175 to reveal the interrelations between the subprocesses and to evaluate the effect of variations of operative parameters on the hydrodynamic behaviour of the dispersion system.

Mathematical modelling and simulation techniques

Parallel to the progress in high speed and high capacity computer technology the method of dynamic simulation of physical systems and complex dynamic processes is developing itself to a standard analysis tool of operations research and system engineering in various fields of technology. The method is based on the simplified and idealized mathematical formulation of all main subsystems involved in the process individually with help of basic physical laws expressed by differential or algebraic equations, which should be tested and verified individually. To evaluate the whole process the submodels are then joined together on a time-event axis and solved simultaneously by application of numerical methods.

Some of the main advantages of this method are:
- quantitative analysis of continuous multiphase systems or complex dynamic processes where an analytical solution is not available,
- besides macro scale studies of a process also individual micro scale investigations on interrelated subsystems are possible,
- general behavior of a process can be effectively investigated through systematical variations of constructive and operating parameters, if necessary even beyond the experimental limits,
- economical and very fast investigation is provided.
Moreover, simulation often is the only recourse in design, debottlenecking or scheduling studies in which the effects of variations in operating parameters must be taken into account.

Two powerful simulation languages MIMIC (Continuous System Simulation) (12) and ACSL (Advanced Continuous Simulation Language) (13) are available at the Computer Center of the Technical University of Aachen for the simulation of continuous systems.Besides a large number of simulation oriented operators which simplify the representation of the system behavior, ACSL and MIMIC provide numerical solution of 120 or more integrations simultaneously, whereas the relative error of each integration step is less then 10^{-8}.
In this and the prior papers MIMIC was used as the simulation language for the on-line simulation runs.

The general concept of the gas-liquid dispersion system simulation used is shown in Figure 1.

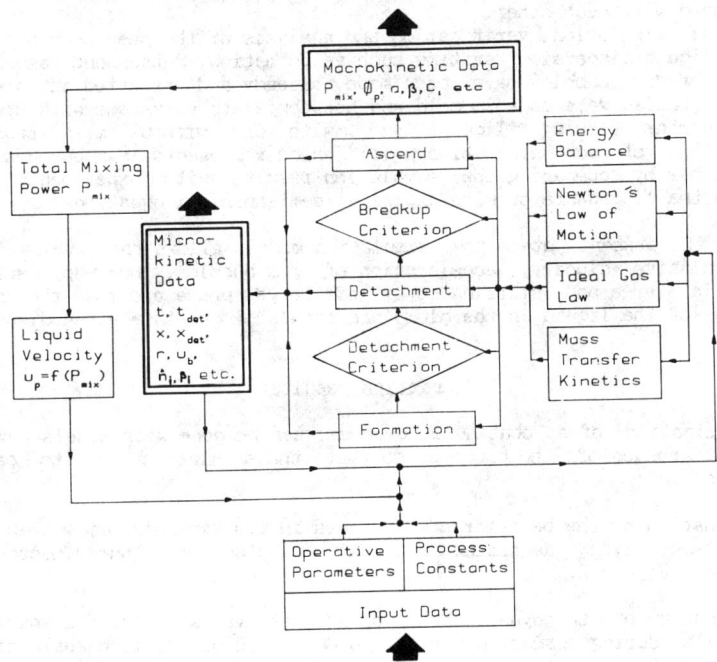

Figure 1: Simplified representation of the method used to simulate the dynamic behavior of the gas-liquid steel dispersion system.

The setup of the dispersion process is defined by "operative parameters" such as the initial temperature and flow rate of gas ; height, diameter and temperature of bath ; position, number and diameter of the orifice and the ambient pressure.

Physical parameters of the system, i.e. density and viscosity of the phases,surface tension,initial concentration and the constants of diffusion and mass action law of purgeable elements in the liquid are defined by "process constants".

First the start-up period and then the run period of the process is simulated by consecutive simulation runs. The expression "simulation run" represents here the duration of simulation which a gas bubble requires from the beginning of formation till it reaches the surface of the liquid bath.

Characteristic quantities of a dispersion process should be evaluated in steady state condition of the system. Although very important for the simulation, the start-up period of a process is not suited for comparison purposes when the intention is to investigate the response of the whole process to parameter variations.

The steady state condition of a process is obtained during the simulation with so called " feedback-loop technique" (14), which is described as follows. From main quantities (state variables) of the system such as induced liquid velocity, virtual density, hold-up in the plume and the work done by the gas phase, integral mean values are sampled at the end of each simulation run. These values are then used in the subsequent simulation run as new state conditions, and new integral mean values are calculated. This procedure is then repeated by consecutive simulation runs until the difference between the input and output values are stable and differ less than 10 %.

By this way, state variables obtained by various operational conditions can be compared with each other.

Successful simulation, verification and analysis of the phenomena occurring in gas-liquid dispersion systems such as formation, detachment, ascent and breakup of gas bubbles have been made already and reported in previous papers. Studies were made in cold and hot temperature systems with stagnant or cocurrent flowing liquids and with or without mass transfer. Verification of each simulation model which represents a subprocess was done either by comparing the simulation results with experimental data given in the literature or with empirical/semiempirical equations.

In the present state the simulation model allows the evaluation of size, relative velocity, acceleration of gas bubbles, hold-up, specific interfacial area and stirring power of the gas phase and thus the induced flowrate of the liquid in the plume and annular zones in a given dispersion system.

Simulation results

Examination of a continuous process can be done with models covering different ranges of the process so that the simulation results can be classified in :

- examination of the behavior of a process in its time-history within large time steps, i.e. degassing process (minutes or hours),macro-scale simulation, Figure 2.

- examination of the dynamic behavior of state variables within small time steps, i.e. during formation period (0.001 - 1.0 sec),micro-scale simulation, Figure 3.

- examination of the changes in state variables in different systems or by different operating conditions, Figures 4 - 8.

Figure 2 illustrates the simulation of a ladle degassing experiment*
where hydrogen was removed by simultaneous argon dispersion and pressure
decreasing.

Operating parameters were given as: weight of the steel 125 tons,
diameter of the vessel 3.0 m and the argon flow rate 25 Nm3/h through a
porous plug at the center of the vessel bottom. The ambient pressure was
reduced in three steps to 0.0013 bar during the first 6 minutes. The
initial concentration of hydrogen was 9 ppm. It was assumed that no
additional CO2 and N2 evolution had to be considered in the bath (aluminium
killed steel) during the treatment. Furthermore it was assumed that the
porous plug would contain 100 active channels with an orifice diameter of
0.25 cm.
The variation of hydrogen concentration in the melt during degassing and
stirring operation is shown in Figure 2 (plot a).

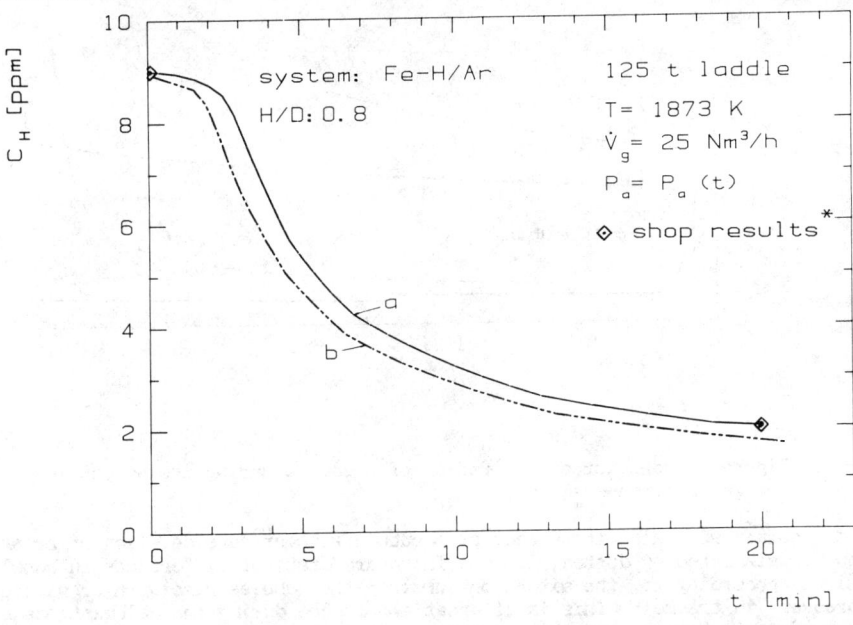

Figure 2: Simulation of the vacuum degassing process of 125 tons of
gas stirred steel.

The simulation of the 20 minutes degassing operation requires ca. 26 CPU
seconds on a Cyber 175. Consistency of the simulation result with shop data
is very good.
Another simulation with 500 active channels (plot b) reveals that in this
particular system a finer dispersion is effective mostly during the first
half period of degassing, when the hydrogen concentration is relatively
high. After 10 minutes the difference between hydrogen concentrations is
less then 0.3 ppm.
Furthermore, the slope of the plots indicates that after 20 minutes no
major changes in hydrogen concentration are to be expected, which justifies
the usual degassing period of 20 to 30 minutes applied in practice.

---

* Vacuum degassing experiment of Kloeckner Stahlforschung GmbH, private
communication, 1987.

In Figure 3, as an example of a micro-scale simulation the behavior of the size of a bubble is illustrated during its ascent in the gas stirred steel bath at two different ambient pressures in the above mentioned system.

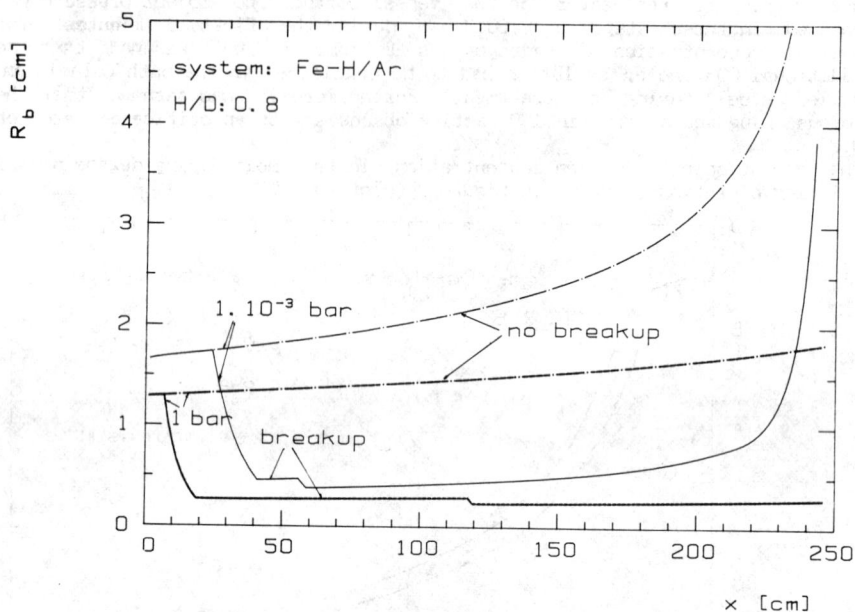

Figure 3: Behavior of the radius of a bubble during its ascent in the gas stirred steel bath.

For comparison, also the ascent of a bubble without consideration of break-up is simulated (dotted lines).There are distinct differences in bubble sizes according to the situation whether the bubbles rise with or without breakup in the melt. This is of great importance during the calculations of the dynamic behavior of the gas phase. Interesting is, e.g.,the more than 8 fold increase in bubble radius without breaking-up, short below the surface under vacuum, although breakup was allowed. Possible reason is the high velocity of the surrounding liquid, which causes a very fast transport of the bubble out of the melt. In this particular case,at 0.0013 bar ambient pressure, the calculated axial velocity of liquid in the plume zone was approximately 2 m/s,(s. Fig. 5b).

Simulation results for the size of a bubble rising in the gas stirred steel bath at atmospheric conditions and with consideration of mass transfer and breakup show good agreement with equations given in the literature and solved for the system considered: Calderbank (15) 0.1 - 0.5 cm, Sano & Mori (16) 0.2 - 0.4 cm.

In the following a series of simulation studies is made to investigate the effect of various operational conditions on the hydrodynamic behavior of the system considered.

Figures 4-8 illustrate the effect of operating parameters such as the ambient pressure (P_a), the aspect ratio (H/D) of the liquid in the vessel and the flow rate of inert gas ($\dot{V}_g$) on system values like stirring power input (P_{mt}), axial velocity of liquid (u_p) in the plume zone, mixing time (t_m), specific interfacial area (a) and hold-up (ϕ_p) in the plume zone.

Gas flow rate is varied from 0.0024 to 0.024 Nm^3/min.ton (Shop data 0.0033 Nm^3/min.ton).

The height-to-diameter ratio of the bath is varied from 0.82 to 1.4, while the liquid volume is kept constant. Furthermore, the initial hydrogen concentration in the melt is 9 ppm throughout the simulations.

Specific power input P_{mt}:

Figure 4a and 4b show the dependency of specific power input P_{mt} [W/ton] on the operating parameters.

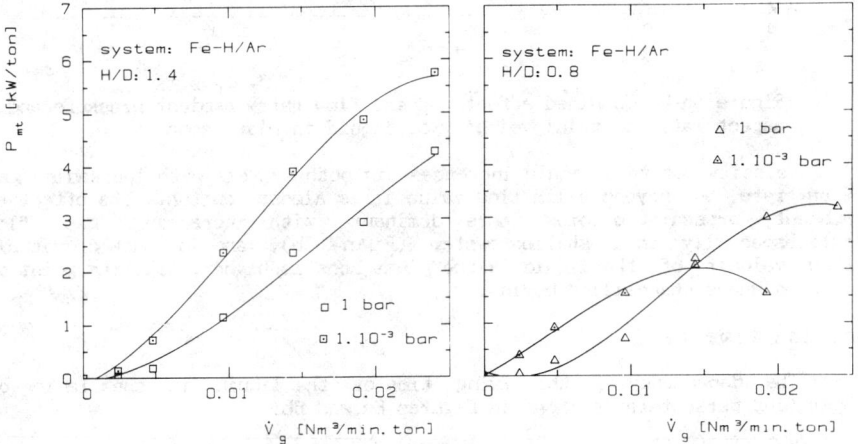

Figure 4a-b: Combined effects of gas flow rate, ambient pressure and aspect ratio of the melt on specific power input of the gas phase.

With increasing gas flow rate and aspect ratio the stirring power input increases. Decreasing the pressure leads also to better results but only when sufficient liquid height is provided. In the system with smaller height-to-diameter ratio (Figure 4b) and under vacuum, beyond a critical gas flow rate the power input begins to decrease. This is probably due to the "channeling effect", which occurs more easily at high flow rates and low pressures in shallow liquids.

Velocity of liquid u_p in plume zone:

The axial velocity of the liquid in the plume zone induced by the injected gas phase is given in Figures 5a and 5b.

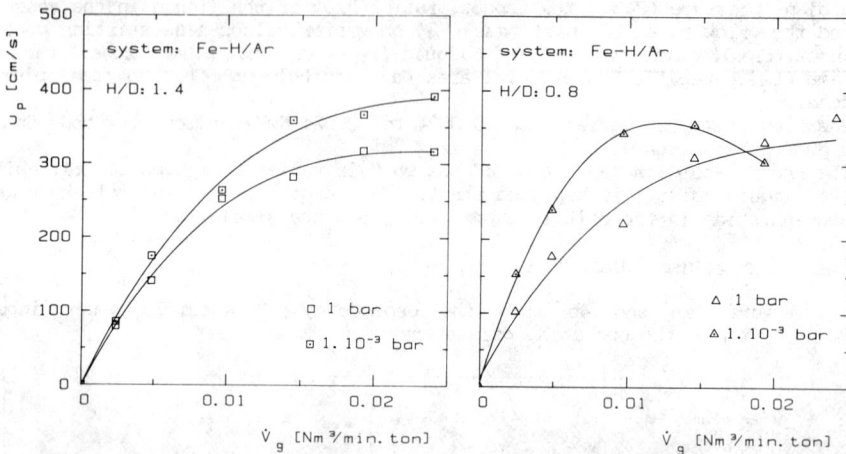

Figure 5a-b: Combined effects of gas flow rate, ambient pressure and aspect ratio on axial velocity of liquid in plume zone.

Velocity of the liquid increases in both systems with increasing gas flow rate, but beyond a limiting value it is almost constant. The effect of ambient pressure becomes more dominant with increasing gas flow rate. Especially in a shallow system (Figure 5b), and in case of vacuum, flow velocity of the liquid in the plume zone is higher, till the point is reached where channelling begins.

Mixing time t_m:

The dependency of the mixing time of the liquid in the ladle on operating parameters is shown in Figures 6a and 6b.

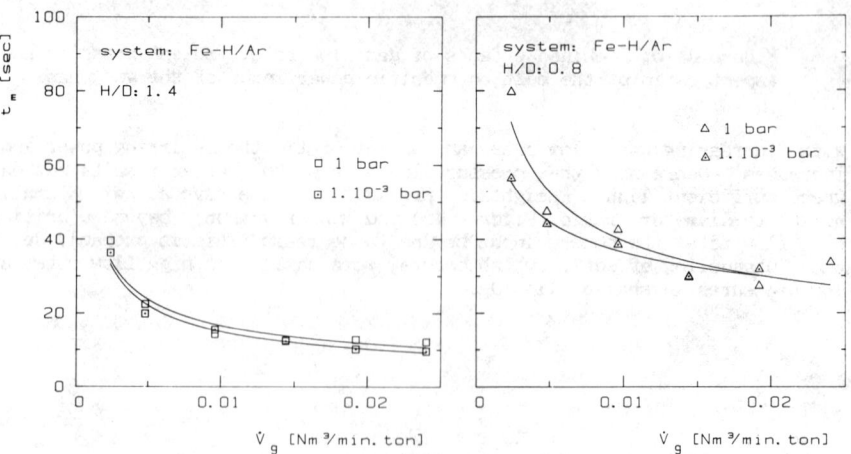

Figure 6a-b: Combined effects of gas flow rate, ambient pressure and aspect ratio on the mixing time of steel in ladle.

Mixing time is calculated from the (mean) induced recirculation rates of the liquid in annular zone and on the basis of the 5% criterion (7). Reducing the pressure has no major effect on mixing time. On the other hand the aspect ratio of the bath seems to have an effect but the aspect ratio does effect the power input anyway (s. Figs. 4a-4b). Simulation results indicate that for a given system, mixing time reaches a limiting value which is then independent of gas flow rate or power input. The simulation results are in good agreement with correlations given in the literature (17,18). But using these literature data, which correlate t_m with the specific power input, it is necessary to use the relation between the gas flow rate and the specific power input according to Fig.4.

Specific interfacial area a:

In this simulation model the specific interfacial area a [cm^2/cm^3] is calculated with mean integral values of a bubble surface and the number of all bubbles existing in the bath at steady state conditions, where at the same time also bubble breakup and mass transfer into the bubble is considered.

Simulation results in Figures 7a and 7b illustrate that aspect ratio has a considerable effect, which explains the fact that in practice bubble columns are used for mass transfer operations.

A decrease of ambient pressure is also highly effective to increase the interfacial area within the bath,provided the aspect ratio is large enough.

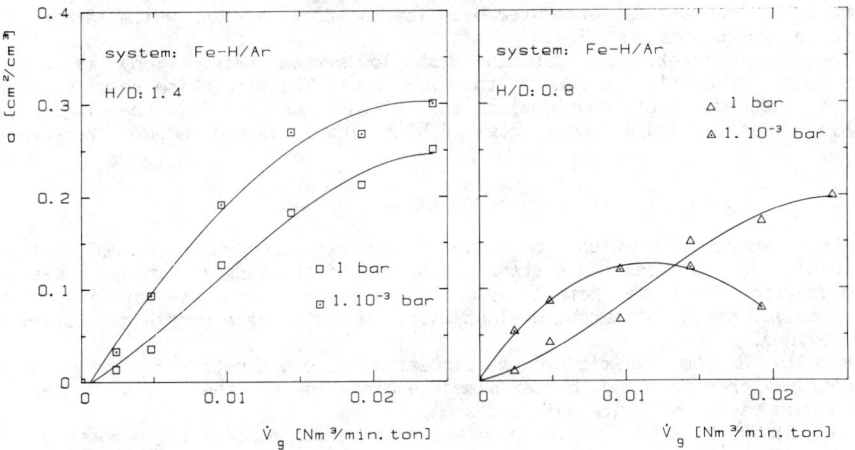

Figure 7a-b: Combined effects of gas flow rate,ambient pressure and aspect ratio on specific interfacial area in bath.

Hold-up ϕ_p in the plume zone:

The volume fraction of gas (hold-up) in the plume zone is evaluated with help of integral mean values of the volume of a bubble and the number of bubbles at steady state conditions of each investigated system.

Hold-up values are used in the simulation model to calculate the virtual density of liquid in the plume zone,which in turn is used to calculate the hydrodynamics of the system.

The behavior of hold-up, which is similar to the other state variables considered, is illustrated in Figures 8a and 8b.

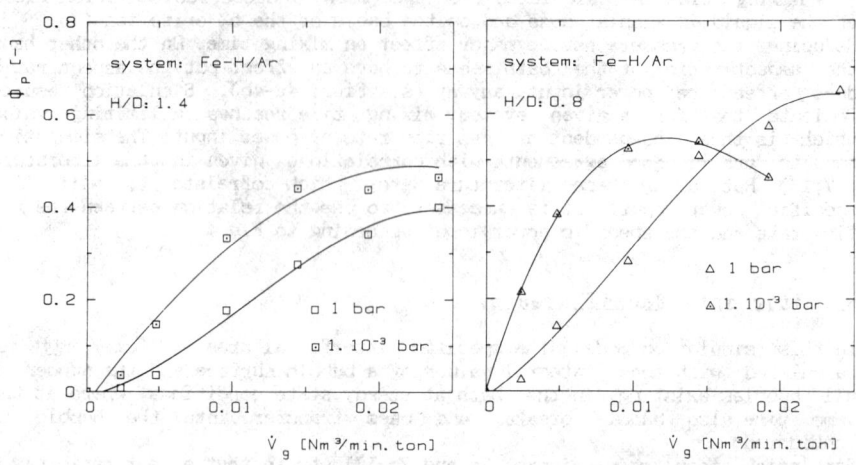

Figure 8a-b: Combined effects of gas flow rate, ambient pressure and aspect ratio on hold-up in plume zone.

Volume fraction of gas in the plume zone increases with increasing gas flow rate and decreasing pressure, but levels off to a value which is specific for a system configuration.

The volume fraction of gas in a shallow system (Figure 8b) is higher because of smaller volume of the plume zone. The simulation results reveal the necessity of considering the virtual density of the liquid in hydrodynamic calculations in case of high flow rates and reduced pressures.

Conclusions

It is shown that with the aid of dynamic simulation method complex multiphase interrelated systems such as gas stirring and vacuum degassing operations of ladle metallurgy can be analysed, interdependencies can be revealed and quantitative evaluation of characteristic system quantities is provided.

Results of the dynamic process simulations yield a better understanding of complex phenomena and better aimed engineering of the gas dispersion techniques in metallurgical processes.

In this work a full-scale simulation of a vacuum degassing process of 125 tons of steel in ladle by simultaneous argon injection is carried out.

Hereby was of particular interest to reveal the combined effects of operating parameters such as the gas flow rate, the ambient pressure and the aspect ratio of the bath on the hydrodynamic behavior of the system.

In general the specific power input, the axial velocity of the liquid and the gas hold-up in the plume zone and the interfacial area increase with gas flow rate, aspect ratio and decreasing ambient pressure. The mixing time decreases with increasing gas flow rate, aspect ratio and decreasing pressure.

Acknowledgments

The authors thank the German Research Foundation (DFG) for a research grant to carry out this work.

References

1. S.Turkan and K.W. Lange: Arch.Eisenhüttenwes.,Vol.55(1984), No.11, pp.507-513

2. S.Turkan and K.W. Lange: Arch. Eisenhüttenwes.,Vol.55(1984), No.12, pp.567-572.

3. S.Turkan and K.W. Lange: Steel research, Vol.56(1985), No.2,pp. 83-92.

4. S.Turkan and K.W. Lange: Steel research, Vol.56(1985), No.4,pp. 199-210.

5. S.Turkan and K.W. Lange: Steel research, Vol.56(1985), No.5,pp. 247-253.

6. S.Turkan and K.W. Lange: Steel research, Vol.57(1986), No.2,pp. 59-67.

7. S.Turkan and K.W. Lange: Steel research, Vol.57(1986), No.10,pp.495-502.

8. S.Turkan and K.W. Lange: Conf.Proc. 5th Intern. Iron & Steel Congress, April 6-9, 1986, Washington, D.C.,USA.
   Process Technology Proc., Vol. 6(1986) pp.115-122.

9. S.Turkan and K.W. Lange: Steel research, Vol.58(1987), No.4,pp.157-161.

10. S.Turkan and K.W. Lange: "Investigation on the Kinetics of Breakup of Gas Bubbles in Liquid Metals with Digital Simulation Method", Steel research, Vol.58(1987), No.9.

11. S.Turkan and K.W. Lange: "Computer Simulation on the Hydrodynamics of Gas Injection Process in Liquid Steel" Steel research, Vol.58(1987), No.10.

12. MIMIC-A Digital Simulation Language Reference Manual. Control Data Corporation, Minneapolis, Minnesota, Publ. No.44610400,Rev. D.(1970).

13. ACSL-Advanced Continuous Simulation Language User/Reference Manual. Mitchell & Gauthier Associates, Inc., Concord, Mass., 2nd ed. (1975).

14. D.M. Himmelblau and K.B. Bischoff: Process Analysis and Simulation, John Wiley & Sons,Inc., New York 1986.

15. P.H. Calderbank: Brit. Chem. Eng. 1(1956) pp. 206-209.

16. M. Sano and K. Mori: Trans. ISIJ, Vol. 20(1980) pp. 668-674.

17. M. Sano and K. Mori: Tr__ . ISIJ, Vol. 23(1983) pp. 169-175.

18. A. Murthy and J. Szekely: Met. Trans. AIME, Vol. 17B(1986),pp.487-490.

AUTHOR INDEX